Sources in the Development of Mathematics

The discovery of infinite products by Wallis and infinite series by Newton marked the beginning of the modern mathematical era. The use of series allowed Newton to find the area under a curve defined by any algebraic equation, an achievement completely beyond the earlier methods of Torricelli, Fermat, and Pascal. The work of Newton and his contemporaries, including Leibniz and the Bernoullis, was concentrated in mathematical analysis and physics. Euler's prodigious mathematical accomplishments dramatically extended the scope of series and products to algebra, combinatorics, and number theory. Series and products proved pivotal in the work of Gauss, Abel, and Jacobi in elliptic functions; in Boole and Lagrange's operator calculus; and in Cayley, Sylvester, and Hilbert's invariant theory. Series and products still play a critical role in the mathematics of today. Consider the conjectures of Langlands, including that of Shimura-Taniyama, leading to Wiles's proof of Fermat's last theorem.

Drawing on the original work of mathematicians from Europe, Asia, and America, Ranjan Roy discusses many facets of the discovery and use of infinite series and products. He gives context and motivation for these discoveries, including original notation and diagrams when practical. He presents multiple derivations for many important theorems and formulas and provides interesting exercises, supplementing the results of each chapter.

Roy deals with numerous results, theorems, and methods used by students, mathematicians, engineers, and physicists. Moreover, since he presents original mathematical insights often omitted from textbooks, his work may be very helpful to mathematics teachers and researchers.

RANJAN ROY is the Ralph C. Huffer Professor of Mathematics and Astronomy at Beloit College. Roy has published papers and reviews in differential equations, fluid mechanics, Kleinian groups, and the development of mathematics. He co-authored *Special Functions* (2001) with George Andrews and Richard Askey, and authored chapters in the *NIST Handbook of Mathematical Functions* (2010). He has received the Allendoerfer prize, the Wisconsin MAA teaching award, and the MAA Haimo award for distinguished mathematics teaching.

Cover image by NFN Kalyan; Cover design by David Levy.

Sources in the Development of Mathematics

Infinite Series and Products from the Fifteenth to the Twenty-first Century

RANJAN ROY

Beloit College

CAMBRIDGE UNIVERSITY PRESS
Cambridge, New York, Melbourne, Madrid, Cape Town,
Singapore, São Paulo, Delhi, Tokyo, Mexico City

Cambridge University Press
32 Avenue of the Americas, New York, NY 10013-2473, USA

www.cambridge.org
Information on this title: www.cambridge.org/9780521114707

© Ranjan Roy 2011

This publication is in copyright. Subject to statutory exception
and to the provisions of relevant collective licensing agreements,
no reproduction of any part may take place without the written
permission of Cambridge University Press.

First published 2011

Printed in the United States of America

A catalog record for this publication is available from the British Library.

ISBN 978-0-521-11470-7 Hardback

Cambridge University Press has no responsibility for the persistence or accuracy of URLs for external or third-party Internet Web sites referred to in this publication and does not guarantee that any content on such Web sites is, or will remain, accurate or appropriate.

Contents

Preface		*page* xvii
1	Power Series in Fifteenth-Century Kerala	1
	1.1 Preliminary Remarks	1
	1.2 Transformation of Series	4
	1.3 Jyesthadeva on Sums of Powers	5
	1.4 Arctangent Series in the *Yuktibhasa*	7
	1.5 Derivation of the Sine Series in the *Yuktibhasa*	8
	1.6 Continued Fractions	10
	1.7 Exercises	12
	1.8 Notes on the Literature	14
2	Sums of Powers of Integers	16
	2.1 Preliminary Remarks	16
	2.2 Johann Faulhaber and Sums of Powers	19
	2.3 Jakob Bernoulli's Polynomials	20
	2.4 Proof of Bernoulli's Formula	24
	2.5 Exercises	25
	2.6 Notes on the Literature	26
3	Infinite Product of Wallis	28
	3.1 Preliminary Remarks	28
	3.2 Wallis's Infinite Product for π	32
	3.3 Brouncker and Infinite Continued Fractions	33
	3.4 Stieltjes: Probability Integral	36
	3.5 Euler: Series and Continued Fractions	38
	3.6 Euler: Products and Continued Fractions	40
	3.7 Euler: Continued Fractions and Integrals	43
	3.8 Sylvester: A Difference Equation and Euler's Continued Fraction	45
	3.9 Euler: Riccati's Equation and Continued Fractions	46
	3.10 Exercises	48
	3.11 Notes on the Literature	50

4		The Binomial Theorem	51
	4.1	Preliminary Remarks	51
	4.2	Landen's Derivation of the Binomial Theorem	57
	4.3	Euler's Proof for Rational Indices	58
	4.4	Cauchy: Proof of the Binomial Theorem for Real Exponents	60
	4.5	Abel's Theorem on Continuity	62
	4.6	Harkness and Morley's Proof of the Binomial Theorem	66
	4.7	Exercises	67
	4.8	Notes on the Literature	69
5		The Rectification of Curves	71
	5.1	Preliminary Remarks	71
	5.2	Descartes's Method of Finding the Normal	73
	5.3	Hudde's Rule for a Double Root	74
	5.4	Van Heuraet's Letter on Rectification	75
	5.5	Newton's Rectification of a Curve	76
	5.6	Leibniz's Derivation of the Arc Length	77
	5.7	Exercises	78
	5.8	Notes on the Literature	79
6		Inequalities	81
	6.1	Preliminary Remarks	81
	6.2	Harriot's Proof of the Arithmetic and Geometric Means Inequality	87
	6.3	Maclaurin's Inequalities	88
	6.4	Jensen's Inequality	89
	6.5	Reisz's Proof of Minkowski's Inequality	90
	6.6	Exercises	91
	6.7	Notes on the Literature	96
7		Geometric Calculus	97
	7.1	Preliminary Remarks	97
	7.2	Pascal's Evaluation of $\int \sin x \, dx$	100
	7.3	Gregory's Evaluation of a Beta Integral	101
	7.4	Gregory's Evaluation of $\int \sec \theta \, d\theta$	104
	7.5	Barrow's Evaluation of $\int \sec \theta \, d\theta$	106
	7.6	Barrow and the Integral $\int \sqrt{x^2 + a^2} \, dx$	108
	7.7	Barrow's Proof of $\frac{d}{d\theta} \tan \theta = \sec^2 \theta$	110
	7.8	Barrow's Product Rule for Derivatives	111
	7.9	Barrow's Fundamental Theorem of Calculus	114
	7.10	Exercises	114
	7.11	Notes on the Literature	118
8		The Calculus of Newton and Leibniz	120
	8.1	Preliminary Remarks	120
	8.2	Newton's 1671 Calculus Text	123
	8.3	Leibniz: Differential Calculus	126

	8.4	Leibniz on the Catenary	129
	8.5	Johann Bernoulli on the Catenary	131
	8.6	Johann Bernoulli: The Brachistochrone	132
	8.7	Newton's Solution to the Brachistochrone	133
	8.8	Newton on the Radius of Curvature	135
	8.9	Johann Bernoulli on the Radius of Curvature	136
	8.10	Exercises	137
	8.11	Notes on the Literature	138
9	De Analysi per Aequationes Infinitas		140
	9.1	Preliminary Remarks	140
	9.2	Algebra of Infinite Series	142
	9.3	Newton's Polygon	145
	9.4	Newton on Differential Equations	146
	9.5	Newton's Earliest Work on Series	147
	9.6	De Moivre on Newton's Formula for $\sin n\theta$	149
	9.7	Stirling's Proof of Newton's Formula	150
	9.8	Zolotarev: Lagrange Inversion with Remainder	152
	9.9	Exercises	153
	9.10	Notes on the Literature	156
10	Finite Differences: Interpolation and Quadrature		158
	10.1	Preliminary Remarks	158
	10.2	Newton: Divided Difference Interpolation	163
	10.3	Gregory–Newton Interpolation Formula	165
	10.4	Waring, Lagrange: Interpolation Formula	165
	10.5	Cauchy, Jacobi: Lagrange Interpolation Formula	166
	10.6	Newton on Approximate Quadrature	168
	10.7	Hermite: Approximate Integration	170
	10.8	Chebyshev on Numerical Integration	172
	10.9	Exercises	173
	10.10	Notes on the Literature	175
11	Series Transformation by Finite Differences		176
	11.1	Preliminary Remarks	176
	11.2	Newton's Transformation	181
	11.3	Montmort's Transformation	182
	11.4	Euler's Transformation Formula	184
	11.5	Stirling's Transformation Formulas	187
	11.6	Nicole's Examples of Sums	190
	11.7	Stirling Numbers	191
	11.8	Lagrange's Proof of Wilson's Theorem	194
	11.9	Taylor's Summation by Parts	195
	11.10	Exercises	196
	11.11	Notes on the Literature	199

12	The Taylor Series	200
	12.1 Preliminary Remarks	200
	12.2 Gregory's Discovery of the Taylor Series	206
	12.3 Newton: An Iterated Integral as a Single Integral	209
	12.4 Bernoulli and Leibniz: A Form of the Taylor Series	210
	12.5 Taylor and Euler on the Taylor Series	211
	12.6 Lacroix on d'Alembert's Derivation of the Remainder	212
	12.7 Lagrange's Derivation of the Remainder Term	213
	12.8 Laplace's Derivation of the Remainder Term	215
	12.9 Cauchy on Taylor's Formula and l'Hôpital's Rule	216
	12.10 Cauchy: The Intermediate Value Theorem	218
	12.11 Exercises	219
	12.12 Notes on the Literature	220
13	Integration of Rational Functions	222
	13.1 Preliminary Remarks	222
	13.2 Newton's 1666 Basic Integrals	228
	13.3 Newton's Factorization of $x^n \pm 1$	230
	13.4 Cotes and de Moivre's Factorizations	231
	13.5 Euler: Integration of Rational Functions	233
	13.6 Euler's Generalization of His Earlier Work	234
	13.7 Hermite's Rational Part Algorithm	237
	13.8 Johann Bernoulli: Integration of $\sqrt{ax^2+bx+c}$	238
	13.9 Exercises	239
	13.10 Notes on the Literature	243
14	Difference Equations	245
	14.1 Preliminary Remarks	245
	14.2 De Moivre on Recurrent Series	247
	14.3 Stirling's Method of Ultimate Relations	250
	14.4 Daniel Bernoulli on Difference Equations	252
	14.5 Lagrange: Nonhomogeneous Equations	254
	14.6 Laplace: Nonhomogeneous Equations	257
	14.7 Exercises	258
	14.8 Notes on the Literature	259
15	Differential Equations	260
	15.1 Preliminary Remarks	260
	15.2 Leibniz: Equations and Series	268
	15.3 Newton on Separation of Variables	270
	15.4 Johann Bernoulli's Solution of a First-Order Equation	271
	15.5 Euler on General Linear Equations with Constant Coefficients	272
	15.6 Euler: Nonhomogeneous Equations	274
	15.7 Lagrange's Use of the Adjoint	276
	15.8 Jakob Bernoulli and Riccati's Equation	278
	15.9 Riccati's Equation	278

	15.10	Singular Solutions	279
	15.11	Mukhopadhyay on Monge's Equation	283
	15.12	Exercises	285
	15.13	Notes on the Literature	287

16 Series and Products for Elementary Functions — 289
- 16.1 Preliminary Remarks — 289
- 16.2 Euler: Series for Elementary Functions — 292
- 16.3 Euler: Products for Trigonometric Functions — 293
- 16.4 Euler's Finite Product for $\sin nx$ — 294
- 16.5 Cauchy's Derivation of the Product Formulas — 295
- 16.6 Euler and Niklaus I Bernoulli: Partial Fractions Expansions of Trigonometric Functions — 298
- 16.7 Euler: Dilogarithm — 301
- 16.8 Landen's Evaluation of $\zeta(2)$ — 302
- 16.9 Spence: Two-Variable Dilogarithm Formula — 304
- 16.10 Exercises — 306
- 16.11 Notes on the Literature — 310

17 Solution of Equations by Radicals — 311
- 17.1 Preliminary Remarks — 311
- 17.2 Viète's Trigonometric Solution of the Cubic — 316
- 17.3 Descartes's Solution of the Quartic — 318
- 17.4 Euler's Solution of a Quartic — 319
- 17.5 Gauss: Cyclotomy, Lagrange Resolvents, and Gauss Sums — 320
- 17.6 Kronecker: Irreducibility of the Cyclotomic Polynomial — 324
- 17.7 Exercises — 325
- 17.8 Notes on the Literature — 325

18 Symmetric Functions — 326
- 18.1 Preliminary Remarks — 326
- 18.2 Euler's Proofs of Newton's Rule — 331
- 18.3 Maclaurin's Proof of Newton's Rule — 332
- 18.4 Waring's Power Sum Formula — 334
- 18.5 Gauss's Fundamental Theorem of Symmetric Functions — 334
- 18.6 Cauchy: Fundamental Theorem of Symmetric Functions — 335
- 18.7 Cauchy: Elementary Symmetric Functions as Rational Functions of Odd Power Sums — 336
- 18.8 Laguerre and Pólya on Symmetric Functions — 337
- 18.9 MacMahon's Generalization of Waring's Formula — 340
- 18.10 Exercises — 343
- 18.11 Notes on the Literature — 344

19 Calculus of Several Variables — 346
- 19.1 Preliminary Remarks — 346
- 19.2 Homogeneous Functions — 351
- 19.3 Cauchy: Taylor Series in Several Variables — 352

	19.4	Clairaut: Exact Differentials and Line Integrals	354
	19.5	Euler: Double Integrals	356
	19.6	Lagrange's Change of Variables Formula	358
	19.7	Green's Integral Identities	359
	19.8	Riemann's Proof of Green's Formula	361
	19.9	Stokes's Theorem	362
	19.10	Exercises	365
	19.11	Notes on the Literature	365
20	Algebraic Analysis: The Calculus of Operations		367
	20.1	Preliminary Remarks	367
	20.2	Lagrange's Extension of the Euler–Maclaurin Formula	375
	20.3	Français's Method of Solving Differential Equations	379
	20.4	Herschel: Calculus of Finite Differences	380
	20.5	Murphy's Theory of Analytical Operations	382
	20.6	Duncan Gregory's Operational Calculus	384
	20.7	Boole's Operational Calculus	387
	20.8	Jacobi and the Symbolic Method	390
	20.9	Cartier: Gregory's Proof of Leibniz's Rule	392
	20.10	Hamilton's Algebra of Complex Numbers and Quaternions	393
	20.11	Exercises	397
	20.12	Notes on the Literature	398
21	Fourier Series		400
	21.1	Preliminary Remarks	400
	21.2	Euler: Trigonometric Expansion of a Function	406
	21.3	Lagrange on the Longitudinal Motion of the Loaded Elastic String	407
	21.4	Euler on Fourier Series	410
	21.5	Fourier: Linear Equations in Infinitely Many Unknowns	412
	21.6	Dirichlet's Proof of Fourier's Theorem	417
	21.7	Dirichlet: On the Evaluation of Gauss Sums	421
	21.8	Exercises	424
	21.9	Notes on the Literature	425
22	Trigonometric Series after 1830		427
	22.1	Preliminary Remarks	427
	22.2	The Riemann Integral	429
	22.3	Smith: Revision of Riemann and Discovery of the Cantor Set	431
	22.4	Riemann's Theorems on Trigonometric Series	432
	22.5	The Riemann–Lebesgue Lemma	436
	22.6	Schwarz's Lemma on Generalized Derivatives	436
	22.7	Cantor's Uniqueness Theorem	437
	22.8	Exercises	439
	22.9	Notes on the Literature	443

23	The Gamma Function	444
	23.1 Preliminary Remarks	444
	23.2 Stirling: $\Gamma(1/2)$ by Newton–Bessel Interpolation	450
	23.3 Euler's Evaluation of the Beta Integral	453
	23.4 Gauss's Theory of the Gamma Function	457
	23.5 Poisson, Jacobi, and Dirichlet: Beta Integrals	460
	23.6 Bohr, Mollerup, and Artin on the Gamma Function	462
	23.7 Kummer's Fourier Series for $\ln \Gamma(x)$	465
	23.8 Exercises	467
	23.9 Notes on the Literature	474
24	The Asymptotic Series for $\ln \Gamma(x)$	476
	24.1 Preliminary Remarks	476
	24.2 De Moivre's Asymptotic Series	481
	24.3 Stirling's Asymptotic Series	483
	24.4 Binet's Integrals for $\ln \Gamma(x)$	486
	24.5 Cauchy's Proof of the Asymptotic Character of de Moivre's Series	488
	24.6 Exercises	489
	24.7 Notes on the Literature	493
25	The Euler–Maclaurin Summation Formula	494
	25.1 Preliminary Remarks	494
	25.2 Euler on the Euler–Maclaurin Formula	499
	25.3 Maclaurin's Derivation of the Euler–Maclaurin Formula	501
	25.4 Poisson's Remainder Term	503
	25.5 Jacobi's Remainder Term	505
	25.6 Euler on the Fourier Expansions of Bernoulli Polynomials	507
	25.7 Abel's Derivation of the Plana–Abel Formula	508
	25.8 Exercises	509
	25.9 Notes on the Literature	513
26	L-Series	515
	26.1 Preliminary Remarks	515
	26.2 Euler's First Evaluation of $\sum 1/n^{2k}$	521
	26.3 Euler: Bernoulli Numbers and $\sum 1/n^{2k}$	522
	26.4 Euler's Evaluation of Some L-Series Values by Partial Fractions	524
	26.5 Euler's Evaluation of $\sum 1/n^2$ by Integration	525
	26.6 N. Bernoulli's Evaluation of $\sum 1/(2n+1)^2$	527
	26.7 Euler and Goldbach: Double Zeta Values	528
	26.8 Dirichlet's Summation of $L(1,\chi)$	532
	26.9 Eisenstein's Proof of the Functional Equation	535
	26.10 Riemann's Derivations of the Functional Equation	536
	26.11 Euler's Product for $\sum 1/n^s$	539
	26.12 Dirichlet Characters	540
	26.13 Exercises	542
	26.14 Notes on the Literature	545

27		The Hypergeometric Series	547
	27.1	Preliminary Remarks	547
	27.2	Euler's Derivation of the Hypergeometric Equation	555
	27.3	Pfaff's Derivation of the $_3F_2$ Identity	556
	27.4	Gauss's Contiguous Relations and Summation Formula	557
	27.5	Gauss's Proof of the Convergence of $F(a,b,c,x)$ for $c-a-b>0$	559
	27.6	Gauss's Continued Fraction	560
	27.7	Gauss: Transformations of Hypergeometric Functions	561
	27.8	Kummer's 1836 Paper on Hypergeometric Series	564
	27.9	Jacobi's Solution by Definite Integrals	565
	27.10	Riemann's Theory of Hypergeometric Functions	567
	27.11	Exercises	569
	27.12	Notes on the Literature	572
28		Orthogonal Polynomials	574
	28.1	Preliminary Remarks	574
	28.2	Legendre's Proof of the Orthogonality of His Polynomials	578
	28.3	Gauss on Numerical Integration	579
	28.4	Jacobi's Commentary on Gauss	582
	28.5	Murphy and Ivory: The Rodrigues Formula	583
	28.6	Liouville's Proof of the Rodrigues Formula	585
	28.7	The Jacobi Polynomials	587
	28.8	Chebyshev: Discrete Orthogonal Polynomials	590
	28.9	Chebyshev and Orthogonal Matrices	594
	28.10	Chebyshev's Discrete Legendre and Jacobi Polynomials	594
	28.11	Exercises	596
	28.12	Notes on the Literature	597
29		q-Series	599
	29.1	Preliminary Remarks	599
	29.2	Jakob Bernoulli's Theta Series	605
	29.3	Euler's q-series Identities	605
	29.4	Euler's Pentagonal Number Theorem	606
	29.5	Gauss: Triangular and Square Numbers Theorem	608
	29.6	Gauss Polynomials and Gauss Sums	611
	29.7	Gauss's q-Binomial Theorem and the Triple Product Identity	615
	29.8	Jacobi: Triple Product Identity	617
	29.9	Eisenstein: q-Binomial Theorem	618
	29.10	Jacobi's q-Series Identity	619
	29.11	Cauchy and Ramanujan: The Extension of the Triple Product	621
	29.12	Rodrigues and MacMahon: Combinatorics	622
	29.13	Exercises	623
	29.14	Notes on the Literature	625

30	Partitions	627
	30.1 Preliminary Remarks	627
	30.2 Sylvester on Partitions	638
	30.3 Cayley: Sylvester's Formula	642
	30.4 Ramanujan: Rogers–Ramanujan Identities	644
	30.5 Ramanujan's Congruence Properties of Partitions	646
	30.6 Exercises	649
	30.7 Notes on the Literature	651
31	q-Series and q-Orthogonal Polynomials	653
	31.1 Preliminary Remarks	653
	31.2 Heine's Transformation	661
	31.3 Rogers: Threefold Symmetry	662
	31.4 Rogers: Rogers–Ramanujan Identities	665
	31.5 Rogers: Third Memoir	670
	31.6 Rogers–Szegő Polynomials	671
	31.7 Feldheim and Lanzewizky: Orthogonality of q-Ultraspherical Polynomials	673
	31.8 Exercises	677
	31.9 Notes on the Literature	679
32	Primes in Arithmetic Progressions	680
	32.1 Preliminary Remarks	680
	32.2 Euler: Sum of Prime Reciprocals	682
	32.3 Dirichlet: Infinitude of Primes in an Arithmetic Progression	683
	32.4 Class Number and $L_\chi(1)$	686
	32.5 De la Vallée Poussin's Complex Analytic Proof of $L_\chi(1) \neq 0$	688
	32.6 Gelfond and Linnik: Proof of $L_\chi(1) \neq 0$	689
	32.7 Monsky's Proof That $L_\chi(1) \neq 0$	691
	32.8 Exercises	692
	32.9 Notes on the Literature	694
33	Distribution of Primes: Early Results	695
	33.1 Preliminary Remarks	695
	33.2 Chebyshev on Legendre's Formula	701
	33.3 Chebyshev's Proof of Bertrand's Conjecture	705
	33.4 De Polignac's Evaluation of $\sum_{p \leq x} \frac{\ln p}{p}$	710
	33.5 Mertens's Evaluation of $\prod_{p \leq x} \left(1 - \frac{1}{p}\right)^{-1}$	710
	33.6 Riemann's Formula for $\pi(x)$	714
	33.7 Exercises	717
	33.8 Notes on the Literature	719
34	Invariant Theory: Cayley and Sylvester	720
	34.1 Preliminary Remarks	720
	34.2 Boole's Derivation of an Invariant	729

	34.3	Differential Operators of Cayley and Sylvester	733
	34.4	Cayley's Generating Function for the Number of Invariants	736
	34.5	Sylvester's Fundamental Theorem of Invariant Theory	740
	34.6	Hilbert's Finite Basis Theorem	743
	34.7	Hilbert's Nullstellensatz	746
	34.8	Exercises	746
	34.9	Notes on the Literature	747
35	Summability	749
	35.1	Preliminary Remarks	749
	35.2	Fejér: Summability of Fourier Series	760
	35.3	Karamata's Proof of the Hardy–Littlewood Theorem	763
	35.4	Wiener's Proof of Littlewood's Theorem	764
	35.5	Hardy and Littlewood: The Prime Number Theorem	766
	35.6	Wiener's Proof of the PNT	768
	35.7	Kac's Proof of Wiener's Theorem	771
	35.8	Gelfand: Normed Rings	772
	35.9	Exercises	775
	35.10	Notes on the Literature	777
36	Elliptic Functions: Eighteenth Century	778
	36.1	Preliminary Remarks	778
	36.2	Fagnano Divides the Lemniscate	786
	36.3	Euler: Addition Formula	790
	36.4	Cayley on Landen's Transformation	791
	36.5	Lagrange, Gauss, Ivory on the agM	794
	36.6	Remarks on Gauss and Elliptic Functions	800
	36.7	Exercises	811
	36.8	Notes on the Literature	813
37	Elliptic Functions: Nineteenth Century	816
	37.1	Preliminary Remarks	816
	37.2	Abel: Elliptic Functions	821
	37.3	Abel: Infinite Products	823
	37.4	Abel: Division of Elliptic Functions and Algebraic Equations	826
	37.5	Abel: Division of the Lemniscate	830
	37.6	Jacobi's Elliptic Functions	832
	37.7	Jacobi: Cubic and Quintic Transformations	834
	37.8	Jacobi's Transcendental Theory of Transformations	839
	37.9	Jacobi: Infinite Products for Elliptic Functions	844
	37.10	Jacobi: Sums of Squares	847
	37.11	Cauchy: Theta Transformations and Gauss Sums	849
	37.12	Eisenstein: Reciprocity Laws	852
	37.13	Liouville's Theory of Elliptic Functions	858
	37.14	Exercises	863
	37.15	Notes on the Literature	865

38		Irrational and Transcendental Numbers	867
	38.1	Preliminary Remarks	867
	38.2	Liouville Numbers	878
	38.3	Hermite's Proof of the Transcendence of e	880
	38.4	Hilbert's Proof of the Transcendence of e	884
	38.5	Exercises	885
	38.6	Notes on the Literature	886
39		Value Distribution Theory	887
	39.1	Preliminary Remarks	887
	39.2	Jacobi on Jensen's Formula	892
	39.3	Jensen's Proof	894
	39.4	Bäcklund Proof of Jensen's Formula	895
	39.5	R. Nevanlinna's Proof of the Poisson–Jensen Formula	896
	39.6	Nevanlinna's First Fundamental Theorem	898
	39.7	Nevanlinna's Factorization of a Meromorphic Function	901
	39.8	Picard's Theorem	902
	39.9	Borel's Theorem	902
	39.10	Nevanlinna's Second Fundamental Theorem	903
	39.11	Exercises	905
	39.12	Notes on the Literature	906
40		Univalent Functions	907
	40.1	Preliminary Remarks	907
	40.2	Gronwall: Area Inequalities	914
	40.3	Bieberbach's Conjecture	916
	40.4	Littlewood: $|a_n| \leq en$	917
	40.5	Littlewood and Paley on Odd Univalent Functions	918
	40.6	Karl Löwner and the Parametric Method	920
	40.7	De Branges: Proof of Bieberbach's Conjecture	923
	40.8	Exercises	927
	40.9	Notes on the Literature	928
41		Finite Fields	929
	41.1	Preliminary Remarks	929
	41.2	Euler's Proof of Fermat's Little Theorem	932
	41.3	Gauss's Proof that $\mathbb{Z}_p^\times$ Is Cyclic	932
	41.4	Gauss on Irreducible Polynomials Modulo a Prime	933
	41.5	Galois on Finite Fields	936
	41.6	Dedekind's Formula	939
	41.7	Exercises	940
	41.8	Notes on the Literature	941
References			943
Index			959

Preface

But this is something very important; one can render our youthful students no greater service than to give them suitable guidance, so that the advances in science become known to them through a study of the sources. – Weierstrass to Casorati, December 21, 1868

The development of infinite series and products marked the beginning of the modern mathematical era. In his *Arithmetica Infinitorum* of 1656, Wallis made groundbreaking discoveries in the use of such products and continued fractions. This work had a tremendous catalytic effect on the young Newton, leading him to the discovery of the binomial theorem for noninteger exponents. Newton explained in his *De Methodis* that the central pillar of his work in algebra and calculus was the powerful new method of infinite series. In letters written in 1670, James Gregory presented his discovery of several infinite series, most probably by means of finite difference interpolation formulas. Illustrating the very significant connections between series and finite difference methods, in the 1670s Newton made use of such methods to transform slowly convergent or even divergent series into rapidly convergent series, though he did not publish his results. Illustrating the importance of this approach, Montmort and Euler soon used new arguments to rediscover it. Newton further wrote in the *De Methodis* that he conceived of infinite series as analogues of infinite decimals, so that the four arithmetical operations and root extraction could be carried over to apply to variables. Thus, he applied infinite series to discover the inverse function and implicit function theorems. Newton concentrated largely on analysis and mathematical physics; Euler's prodigious intellect broadened Newton's conception to apply infinite series and products to number theory, algebra, and combinatorics; this legacy continues unabated even today.

Infinite series have numerous manifestations, including power series, trigonometric series, q-series, and Dirichlet series. Their scope and power are evident in their pivotal role in many areas of mathematics, including algebra, analysis, combinatorics, and number theory. As such, infinite series and products provide access to many mathematical questions and insights. For example, Maclaurin, Euler, and MacMahon studied symmetric functions using infinite series; Euler, Dirichlet, Chebyshev, and Riemann employed products and series to get deep insight into the distribution of primes. Gauss employed q-series to prove the law of quadratic reciprocity and Jacobi applied the triple product identity, also discovered by Gauss, to determine the number of representations of integers as sums of squares. Moreover, the correspondence between Daniel Bernoulli and Goldbach in the 1720s introduced the problem of determining whether a given series of rational numbers was irrational or transcendental. The 1843 publication of their letters prompted Liouville to lay the foundations of the theory of transcendental numbers.

The detailed table of contents at the beginning of this book may prove even more useful than the index in locating particular topics or questions. The preliminary remarks in each chapter provide some background on the origins and motivations of the ideas discussed in the subsequent, more detailed, and substantial sections of the chapter. The exercises following these sections offer references so that the reader may perhaps consult the original sources with a specific focus in mind. Most works cited in the notes at the end of each chapter should be readily accessible, especially since the number of books and papers online is increasing steadily.

Mathematics teachers and students may discover that the old sources, such as Simpson's books on algebra and calculus, Euler's *Introductio*, or the correspondence of Euler and Goldbach and the Bernoullis, are fruitful resources for calculus projects or undergraduate or graduate seminar topics. Since early mathematicians often omitted to mention the conditions under which their results would hold, analysis students could find it very instructive to work out the range of validity of those results. For example, Landen's formula for the dilogarithm, while very insightful and significant, is incorrect for a range of values, even where the series converge. At an advanced level, important research has arisen out of a study of old works. Indeed, by studying Descartes and Newton, Laguerre revived a subject others had abandoned for two hundred years and did his excellent work in numerical solutions of algebraic equations and extensions of the rule of signs. Again, André Weil recounted in his 1972 Ritt lectures on number theory that he arrived at the Weil conjectures through a study of Gauss's two papers on biquadratic residues.

It is edifying and a lot of fun to read the noteworthy works of long ago; this is common practice in literature and is equally appropriate and beneficial in mathematics. For example, a calculus student might enjoy and learn from Cotes's 1714 paper on logarithms or Maria Agnesi's 1748 treatment of the same topic in her work on analysis. At a more advanced level, Euler gave not just one or two but at least eight derivations of his famous formula $\sum 1/n^2 = \pi^2/6$. Reading these may serve to enlighten us on the variety of approaches to the perennial problem of summing series, though most of these approaches are not mentioned in textbooks. Students of literature routinely learn from and enjoy reading the words of, say, Austen, Hawthorne, Turgenev, or Shakespeare. We may likewise deepen our understanding and enjoyment of mathematics by reading and rereading the original works of mathematicians such as Barrow, Laplace, Chebyshev, or Newton. It might prove rewarding if mathematicians and students of mathematics were to make such reading a regular practice. In the introduction to his *Development of Mathematics in the 19th Century*, Felix Klein wrote, "Thus, it is impossible to grasp even *one* mathematical concept without having assimilated all the concepts which led up to its creation, and their connections."

Wherever practical, I have tried to present a mathematician's own notational methods. Seeing an argument in its original form is often instructive and can give us insight into its motivations and underlying rationale. Because of the numerous notations for logarithms, for simplicity I have denoted the logarithm of a real value by the familiar ln; in the case of complex or non-e-based logarithms, I have used log.

I am indebted to many persons who helped me in writing this book. I would first like to thank my wife, Gretchen Roy, for her invaluable assistance in editing and preparing the manuscript. I am grateful to Kalyan for his beautiful cover art. I thank my colleagues: Paul Campbell for his expert and generous assistance with the indexes and typesetting and Bruce Atwood for so cheerfully and accurately preparing the figures as they now appear. I am grateful to Ashish Thapa for his skillful typesetting and figure construction. Many thanks to Doreen Dalman, who typeset the majority of the book and did valuable troubleshooting. I am obliged to Paul Campbell and David Heesen for their meticulous work on the bibliography. I benefited from the input of those who read preliminary drafts of some chapters: Richard Askey, George Andrews, Lonnie Fairchild, Atar Mittal, Yu Shun, and Phil Straffin. I was fortunate to receive assistance from very capable librarians: Cindy Cooley and Chris Nelson at Beloit College, Travis Warwick at the Kleene Mathematics Library at the University of Wisconsin, the efficient librarians at the University Library in Cambridge, and the kind librarians at St. Andrews University Library. A. W. F. Edwards, of Gonville and Caius College, also gave me helpful guidance. I am grateful for financial and other assistance from Beloit College; thanks to John Burris, Lynn Franken, and Ann Davies for their encouragement and support. Heartfelt gratitude goes to Maitreyi Lagunas, Margaret Carey, Mihir Banerjee, Sahib Ram Mandan, and Ramendra Bose. Finally, I am deeply indebted to my parents for their intellectual, emotional, and practical support of my efforts to become a mathematician. I dedicate this book to their memory.

1

Power Series in Fifteenth-Century Kerala

1.1 Preliminary Remarks

The Indian astronomer and mathematician Madhava (c. 1340–c. 1425) discovered infinite power series about two and a half centuries before Newton rediscovered them in the 1660s. Madhava's work may have been motivated by his studies in astronomy, since he concentrated mainly on the trigonometric functions. There appears to be no connection between Madhava's school and that of Newton and other European mathematicians. In spite of this, the Keralese and European mathematicians shared some similar methods and results. Both were fascinated with transformation of series, though here they used very different methods.

The mathematician-astronomers of medieval Kerala lived, worked, and taught in large family compounds called illams. Madhava, believed to have been the founder of the school, worked in the Bakulavihara illam in the town of Sangamagrama, a few miles north of Cochin. He was an Emprantiri Brahmin, then considered socially inferior to the dominant Namputiri (or Nambudri) Brahmin. This position does not appear to have curtailed his teaching activities; his most distinguished pupil was Paramesvara, a Namputiri Brahmin. No mathematical works of Madhava have been found, though three of his short treatises on astronomy are extant. The most important of these describes how to accurately determine the position of the moon at any time of the day. Other surviving mathematical works of the Kerala school attribute many very significant results to Madhava. Although his algebraic notation was almost primitive, Madhava's mathematical skill allowed him to carry out highly original and difficult research.

Paramesvara (c.1380–c.1460), Madhava's pupil, was from Asvattagram, about thirty-five miles northeast of Madhava's home town. He belonged to the Vatasreni illam, a famous center for astronomy and mathematics. He made a series of observations of the eclipses of the sun and the moon between 1395 and 1432 and composed several astronomical texts, the last of which was written in the 1450s, near the end of his life. Sankara Variyar attributed to Paramesvara a formula for the radius of a circle in terms of the sides of an inscribed quadrilateral. Paramesvara's son, Damodara, was the teacher of Jyesthadeva (c. 1500–c. 1570) whose works survive and give us all the surviving proofs of this school. Damodara was also the teacher of Nilakantha (c. 1450–c. 1550) who composed the famous treatise called the *Tantrasangraha* (c. 1500), a digest of the mathematical and astronomical knowledge of his time. His works allow us determine his approximate dates since in his *Aryabhatyabhasya*, Nilakantha refers to his observation of solar eclipses in 1467 and 1501. Nilakantha made several efforts to establish new parameters for the mean motions of the planets and vigorously defended the necessity of continually correcting astronomical parameters on the basis of observation. Sankara Variyar (c. 1500–1560) was his student.

The surviving texts containing results on infinite series are Nilakantha's *Tantrasangraha*, a commentary on it by Sankara Variyar called *Yuktidipika*, the *Yuktibhasa* by Jyesthadeva and the *Kriyakramakari*, started by Variyar and completed by his student Mahisamangalam Narayana. All these works are in Sanskrit except the *Yuktibhasa*, written in Malayalam, the language of Kerala. These works provide a summary of major results on series discovered by these original mathematicians of the indistinct past:

A. Series expansions for arctangent, sine, and cosine:

1. $\theta = \tan\theta - \frac{\tan^3\theta}{3} + \frac{\tan^5\theta}{5} - \cdots$,
2. $\sin\theta = \theta - \frac{\theta^3}{3!} + \frac{\theta^5}{5!} - \cdots$,
3. $\cos\theta = 1 - \frac{\theta^2}{2!} + \frac{\theta^4}{4!} - \cdots$,
4. $\sin^2\theta = \theta^2 - \frac{\theta^4}{(2^2-2/2)} + \frac{\theta^6}{(2^2-2/2)(3^2-3/2)} - \frac{\theta^8}{(2^2-2/2)(3^2-3/2)(4^2-4/2)} + \cdots$.

In the proofs of these formulas, the range of θ for the first series was $0 \leq \theta \leq \pi/4$ and for the second and third was $0 \leq \theta \leq \pi/2$. Although the series for sine and cosine converge for all real values, the concept of periodicity of the trigonometric functions was discovered much later.

B. Series for π:

1. $\frac{\pi}{4} \approx 1 - \frac{1}{3} + \frac{1}{5} - \cdots \mp \frac{1}{n} \pm f_i(n+1)$, $i = 1, 2, 3$, where

$$f_1(n) = 1/(2n), \quad f_2(n) = n/(2(n^2+1)),$$

and

$$f_3(n) = (n^2+4)/(2n(n^2+5));$$

2. $\frac{\pi}{4} = \frac{3}{4} + \frac{1}{3^3-3} - \frac{1}{5^3-5} + \frac{1}{7^3-7} - \cdots$;
3. $\frac{\pi}{4} = \frac{4}{1^5+4\cdot 1} - \frac{4}{3^5+4\cdot 3} + \frac{4}{5^5+4\cdot 5} - \cdots$;
4. $\frac{\pi}{2\sqrt{3}} = 1 - \frac{1}{3\cdot 3} + \frac{1}{5\cdot 3^2} - \frac{1}{7\cdot 3^3} + \cdots$;
5. $\frac{\pi}{6} = \frac{1}{2} + \frac{1}{(2\cdot 2^2-1)^2-2^2} + \frac{1}{(2\cdot 4^2-1)^2-4^2} + \frac{1}{(2\cdot 6^2-1)^2-6^2} + \cdots$;
6. $\frac{\pi-2}{4} \approx \frac{1}{2^2-1} - \frac{1}{4^2-1} + \frac{1}{6^2-1} - \cdots \mp \frac{1}{n^2-1} \pm \frac{1}{2((n+1)^2+2)}$.

These results were stated in verse form. Thus, the series for sine was described:

The arc is to be repeatedly multiplied by the square of itself and is to be divided [in order] by the square of each even number increased by itself and multiplied by the square of the radius. The arc and the terms obtained by these repeated operations are to be placed in sequence in a column, and any last term is to be subtracted from the next above, the remainder from the term then next above, and so on, to obtain the jya (sine) of the arc.

So if r is the radius and s the arc, then the successive terms of the repeated operations mentioned in the description are given by

$$s \cdot \frac{s^2}{(2^2+2)r^2}, \quad s \cdot \frac{s^2}{(2^2+2)r^2} \cdot \frac{s^2}{(4^2+4)r^2}, \dots$$

and the equation is

$$y = s - s \cdot \frac{s^2}{(2^2+2)r^2} + s \cdot \frac{s^2}{(2^2+2)r^2} \cdot \frac{s^2}{(4^2+4)r^2} - \dots$$

where $y = r\sin(s/r)$. Nilakantha's *Aryabhatiyabhasya* attributes the sine series to Madhava. The *Kriyakramakari* attributes to Madhava the first two cases of B.1, the arctangent series, and series B.4; note that B.4 can be derived from the arctangent by taking $\theta = \pi/6$. The extant manuscripts do not appear to attribute the other series to a particular person. The *Yuktidipika* gives series B.6, including the remainder; it is possible that this series is due to Sankara Variyar, the author of the work. We can safely conclude that the power series for arctangent, sine, and cosine were obtained by Madhava; he is, thus, the first person to express the trigonometric functions as series. In the 1660s, Newton rediscovered the sine and cosine series; in 1671, James Gregory rediscovered the series for arctangent.

The series for $\sin^2\theta$ follows directly from the series for $\cos\theta$ by an application of the double angle formula, $\sin^2\theta = \frac{1}{2}(1-\cos 2\theta)$. The series for $\pi/4$ (B.1) has several points of interest. When $n \to \infty$, it is simply the series discovered by Leibniz in 1673. However, this series is not useful for computational purposes because it converges extremely slowly. To make it more effective in this respect, the Madhava school added a rational approximation for the remainder after n terms. They did not explain how they arrived at the three expressions $f_i(n)$ in B.1. However, if we set

$$\frac{\pi}{4} = 1 - \frac{1}{3} + \frac{1}{5} - \dots \mp \frac{1}{n} \pm f(n), \tag{1.1}$$

then the remainder $f(n)$ has the continued fraction expansion

$$f(n) = \frac{1}{2} \cdot \frac{1}{n+} \frac{1^2}{n+} \frac{2^2}{n+} \frac{3^2}{n+} \dots, \tag{1.2}$$

when $f(n)$ is assumed to satisfy the functional relation

$$f(n+1) + f(n-1) = \frac{1}{n}. \tag{1.3}$$

The first three convergents of this continued fraction are

$$\frac{1}{2n} = f_1(n), \quad \frac{n}{2(n^2+1)} = f_2(n), \quad \text{and} \quad \frac{1}{2}\frac{n^2+4}{n(n^2+5)} = f_3(n). \tag{1.4}$$

Although this continued fraction is not mentioned in any extant works of the Kerala school, their approximants indicate that they must have known it, at least implicitly. In fact, continued fractions appear in much earlier Indian works. The *Lilavati* of Bhaskara (c. 1150) used continued fractions to solve first-order Diophantine equations and Variyar's *Kriyakramakari* was a commentary on Bhaskara's book.

The approximation in equation B.6 is similar to that in B.1 and gives further evidence that the Kerala mathematicians saw a connection between series and continued fractions. If we write

$$\frac{\pi-2}{4} = \frac{1}{2^2-1} - \frac{1}{4^2-1} + \frac{1}{6^2-1} - \cdots \pm \frac{1}{n^2-1} \pm g(n+1), \text{ then} \quad (1.5)$$

$$g(n) = \frac{1}{2n} \cdot \frac{1}{n+} \frac{1 \cdot 2}{n+} \frac{2 \cdot 3}{n+} \frac{3 \cdot 4}{n+} \cdots, \text{ and} \quad (1.6)$$

$$g_1(n) = \frac{1}{2n}, \quad g_2(n) = \frac{1}{2(n^2+2)}. \quad (1.7)$$

Newton, who was very interested in the numerical aspects of series, also found the $f_1(n) = 1/(2n)$ approximation when he saw Leibniz's series. He wrote in a letter of 1676 to Henry Oldenburg:

> By the series of Leibniz also if half the term in the last place be added and some other like device be employed, the computation can be carried to many figures.

Though the accomplishments of Madhava and his followers are quite impressive, the members of the school do not appear to have had any interaction with people outside of the very small region where they lived and worked. By the end of the sixteenth century, the school ceased to produce any further original works. Thus, there appears to be no continuity between the ideas of the Kerala scholars and those outside India or even from other parts of India.

1.2 Transformation of Series

The series in equations B.2 and B.3 are transformations of

$$\sum_{k=1}^{\infty} \frac{(-1)^{k-1}}{k}$$

by means of the rational approximations for the remainder. To understand this transformation in modern notation, observe:

$$\frac{\pi}{4} = (1 - f_1(2)) - \left(\frac{1}{3} - f_1(2) - f_1(4)\right) + \left(\frac{1}{5} - f_1(4) - f_1(6)\right) - \cdots. \quad (1.8)$$

The $(n+1)$th term in this series is

$$\frac{1}{2n+1} - f_1(2n) - f_1(2n+2) = \frac{1}{2n+1} - \frac{1}{4n} - \frac{1}{4(n+1)} = \frac{-1}{(2n+1)^3 - (2n+1)}. \quad (1.9)$$

Thus, we arrive at equation B.2. Equation B.3 is similarly obtained:

$$\frac{\pi}{4} = (1 - f_2(2)) - \left(\frac{1}{3} - f_2(2) - f_2(4)\right) + \left(\frac{1}{5} - f_2(4) - f_2(6)\right) - \cdots, \quad (1.10)$$

and here the $(n+1)$th term is

$$\frac{1}{2n+1} - \frac{n}{(2n)^2+1} - \frac{n+1}{(2n+2)^2+1} = \frac{4}{(2n+1)^5 + 4(2n+1)}. \quad (1.11)$$

Clearly, the nth partial sums of these two transformed series can be written as

$$s_i(n) = 1 - \frac{1}{3} + \frac{1}{5} - \frac{1}{7} + \cdots \mp \frac{1}{2n-1} \pm f_i(2n), \quad i = 1, 2. \quad (1.12)$$

Since series (1.8) and (1.10) are alternating, and the absolute values of the terms are decreasing, it follows that

$$\frac{1}{(2n+1)^3 - (2n+1)} - \frac{1}{(2n+3)^3 - (2n+3)} < \left|\frac{\pi}{4} - s_1(n)\right|$$

$$< \frac{1}{(2n+1)^3 - (2n+1)^3}. \text{ Also,} \quad (1.13)$$

$$\frac{4}{(2n+1)^5 + 4(2n+1)} - \frac{4}{(2n+3)^5 + 4(2n+3)} < \left|\frac{\pi}{4} - s_2(n)\right|$$

$$< \frac{4}{(2n+1)^5 + 4(2n+1)}. \quad (1.14)$$

Thus, taking fifty terms of $1 - \frac{1}{3} + \frac{1}{5} - \cdots$ and using the approximation $f_2(n)$, the last inequality shows that the error in the value of π becomes less than 4×10^{-10}. The Leibniz series with fifty terms is normally accurate in computing π up to only one decimal place; by contrast, the Keralese method of rational approximation of the remainder produces numerically useful results.

1.3 Jyesthadeva on Sums of Powers

The Sanskrit texts of the Kerala school with few exceptions contain merely the statements of results without derivations. It is therefore extremely fortunate that Jyesthadeva's Malayalam text *Yuktibhasa*, containing the methods for obtaining the formulas, has survived. Sankara Variyar's *Yuktidipika* is a modified Sanskrit version of the *Yuktibhasa*. It seems that the *Yuktibhasa* was the text used by Jyesthadeva's students at his illam. From this, one may surmise that Variyar, a student of Nilakantha, also studied with Jyesthadeva whose illam was very close to that of Nilakantha.

A basic result used by the Kerala school in the derivation of their series is that

$$\lim_{n \to \infty} \frac{1}{n^{p+1}} \sum_{k=1}^{n} k^p = \frac{1}{p+1}. \quad (1.15)$$

This relation has a long history; sums of powers of integers have been used in the study of area and volume problems at least since Archimedes. Archimedes summed

$$S_n^{(p)} = \sum_{k=1}^{n} k^p$$

for $p=1$ and $p=2$. For $p=2$, he proved the more general result: If $A_1, A_2, \ldots, A_n$ are n lines (we may take them to be numbers) forming an ascending arithmetical progression in which the common difference is equal to A_1 (the least term), then

$$(n+1)A_n^2 + A_1(A_1 + A_2 + \cdots + A_n) = 3(A_1^2 + A_2^2 + \cdots + A_n^2). \tag{1.16}$$

This implies that

$$3(1^2 + 2^2 + \cdots + n^2) = n^2(n+1) + (1 + 2 + 3 + \cdots + n). \tag{1.17}$$

Archimedes used this formula in his work on spirals and in computing the volume of revolution of a segment of a parabola about its axis. The celebrated Arab mathematician al-Haytham (c.965–1039), known also as Alhazen, generalized Archimedes's formula to find the volume of revolution of segment of a parabola about its base. The calculation involved sums of cubes and fourth powers of integers. Al-Haytham proved his generalization by means of a diagram; it can be expressed in modern notation by

$$(n)S_n^{(p)} = S_n^{(p)} + S_1^{(p-1)} + S_2^{(p-1)} + \cdots + S_{n-1}^{(p-1)}. \tag{1.18}$$

It is interesting that the statement of Jyesthadeva's first lemma leading to the proof of (1.15) is a restatement of al-Haytham's formula; Jyesthadeva's result was stated:

> Whenever we wish to obtain the sum (sankalitam) of any given powers [say the pth powers of natural numbers, up to an assigned limit n], we multiply the sankalitam of the next lower powers [that is, $(p-1)$th powers, up to the given limit n] by the limit [n]. The result will contain the required sankalitam and also the sankalitam of all the sankalitams of all lower powers up to various limits.

Jyesthadeva's next lemma stated:

> Multiply the lower [power] sankalitam [up to the limit of n] by the limit [n]. Subtract from this product the result of dividing the product by an integer one more than the given power [p]. The result will be [asymptotically equal to] the desired sankalitam.

Thus

$$nS_n^{(p-1)}\left(1 - \frac{1}{p+1}\right) \sim S_n^{(p)} \text{ as } n \to \infty. \tag{1.19}$$

Jyesthadeva proved this result inductively, but he did not perform the induction completely. It is easy to see that (1.19) is equivalent to (1.15) and thus Jyesthadeva assumed that

$$S_n^{(p-1)} \sim n^p/p,$$

which is certainly true for $p = 1$. From this it can be deduced that

$$S_1^{(p-1)} + S_2^{(p-1)} + \cdots + S_n^{(p-1)} \sim \frac{1^p + 2^p + \cdots + n^p}{p} = \frac{S_n^{(p)}}{p} \quad \text{as } n \to \infty.$$

Jyesthadeva asserted this but verified it only for $p = 2$ and 3. But once we fill in the gap by proving this for all p, equation (1.18) implies that

$$(n+1)S_n^{(p-1)} \sim S_n^{(p)} + \frac{S_n^{(p)}}{p} \quad \text{as } n \to \infty.$$

Hence by the inductive hypothesis it follows that

$$S_n^{(p)} \sim \frac{n^{p+1}}{p+1} \quad \text{as } n \to \infty.$$

This was Jyesthadeva's argument for (1.15).

1.4 Arctangent Series in the *Yuktibhasa*

The following derivation of the arctangent series, attributed to Madhava, boils down to the integration of $1/(1+x^2)$, as do the methods of Gregory and Leibniz.

In Figure 1.1, AC is a quarter circle of radius one with center O; $OABC$ is a square. The side AB is divided into n equal parts of length δ so that $n\delta = 1$ and $P_{k-1}P_k = \delta$. EF and $P_{k-1}D$ are perpendicular to OP_k. Now, the triangles OEF and $OP_{k-1}D$ are similar, implying that

$$\frac{EF}{OE} = \frac{P_{k-1}D}{OP_{k-1}} \quad \text{or} \quad EF = \frac{P_{k-1}D}{OP_{k-1}}.$$

The similarity of the triangles $P_{k-1}P_kD$ and OAP_k gives

$$\frac{P_{k-1}P_k}{OP_k} = \frac{P_{k-1}D}{OA} \quad \text{or} \quad P_{k-1}D = \frac{P_{k-1}P_k}{OP_k}.$$

Thus,

$$EF = \frac{P_{k-1}P_k}{OP_{k-1}OP_k} \simeq \frac{P_{k-1}P_k}{OP_k^2} = \frac{P_{k-1}P_k}{1 + AP_k^2} = \frac{\delta}{1 + k^2\delta^2}.$$

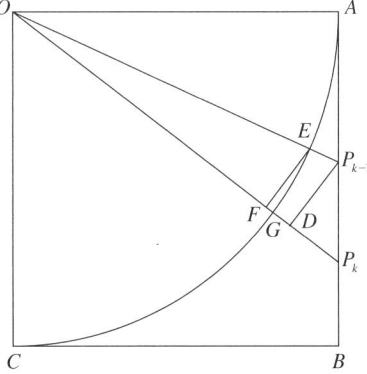

Figure 1.1. Rectifying a circle by the arctangent series.

Now
$$\text{arc } EG \simeq EF \simeq \frac{\delta}{1+k^2\delta^2},$$
and if we write $AP_k = x = \tan\theta$, where $\theta = A\widehat{O}P_k$, then
$$\arctan x = \lim_{k\to\infty} \sum_{j=1}^{k} \frac{\delta}{1+j^2\delta^2}. \tag{1.20}$$

To compute this limit, Jyesthadeva expanded $\frac{1}{1+j^2\delta^2}$ as a geometric series. He derived the series by an iterative procedure:
$$\frac{1}{1+x} = 1 - x\left(\frac{1}{1+x}\right) = 1 - x\left(1 - x\left(\frac{1}{1+x}\right)\right).$$

Thus, (1.20) is converted to
$$\arctan x = \lim_{k\to\infty}\left(\delta\sum_{j=1}^{k}1 - \delta^3\sum_{j=1}^{k}j^2 + \delta^5\sum_{j=1}^{k}j^4 - \cdots\right)$$
$$= \lim_{k\to\infty}\left(\frac{x}{k}\sum_{j=1}^{k}1 - \frac{x^3}{k^3}\sum_{j=1}^{k}j^2 + \frac{x^5}{k^5}\sum_{j=1}^{k}j^4 - \cdots\right)$$
$$= x - \frac{x^3}{3} + \frac{x^5}{5} - \cdots.$$

The last step follows from (1.15). Note that this is the Madhava–Gregory series for $\arctan x$ and the series for $\pi/4$ follows by taking $x=1$.

1.5 Derivation of the Sine Series in the *Yuktibhasa*

Once again, Madhava's derivation of the sine series has similarities with Leibniz's derivation of the cosine series. In Figure 1.2, suppose that $A\widehat{O}P = \theta$, $OP = R$, P is the midpoint of the arc $P_{-1}P_1$, and PQ is perpendicular to OA, where O is the origin of the coordinate system. Let $P = (x, y)$, $P_1 = (x_1, y_1)$, and $P_{-1} = (x_{-1}, y_{-1})$. From the similarity of the triangles $P_{-1}Q_1P_1$ and OPQ, we have
$$\frac{P_{-1}P_1}{OP} = \frac{x_{-1} - x_1}{y} = \frac{y_1 - y_{-1}}{x}. \tag{1.21}$$

For a small arc $P_{-1}P = \Delta\theta/2$, identified by Jyesthadeva with the line segment $P_{-1}P$, we can write (1.21) as
$$\cos\left(\theta + \frac{\Delta\theta}{2}\right) - \cos\left(\theta - \frac{\Delta\theta}{2}\right) = -\sin\theta\,\Delta\theta \text{ and} \tag{1.22}$$
$$\sin\left(\theta + \frac{\Delta\theta}{2}\right) - \sin\left(\theta - \frac{\Delta\theta}{2}\right) = \cos\theta\,\Delta\theta. \tag{1.23}$$

1.5 Derivation of the Sine Series in the Yuktibhasa

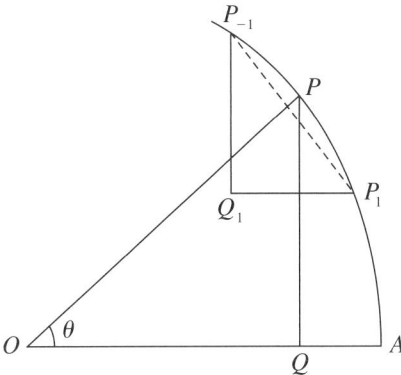

Figure 1.2. Derivation of the sine series.

In fact, Bhaskara earlier stated this last relation and proved it in the same way; he applied it to the discussion of the instantaneous motion of planets. Interestingly, in the 1650s, Pascal used a very similar argument to show that $\int \cos\theta\, d\theta = \sin\theta$ and $\int \sin\theta\, d\theta = -\cos\theta$.

From (1.22) and (1.23) Jyesthadeva derived the result, given in modern notation:

$$\sin\theta - \theta = -\int_0^\theta \int_0^t \sin u\, du\, dt. \tag{1.24}$$

We also note that Leibniz found the series for cosine using a similar method of repeated integration. In Jyesthadeva, the integrals are replaced by sums and double integrals by sums of sums. The series is then obtained by using successive polynomial approximations for $\sin\theta$. For example, when the first approximation $\sin u \approx u$ is used in the right-hand side of (1.24), the result is

$$\sin\theta - \theta \sim -\frac{\theta^3}{3!} \quad\text{or}\quad \sin\theta \sim \theta - \frac{\theta^3}{3!}.$$

When this approximation is employed in (1.24), we obtain

$$\sin\theta - \theta \sim -\frac{\theta^3}{3!} + \frac{\theta^5}{5!}.$$

Briefly, Jyesthadeva arrived at the sums approximating (1.24) by first dividing AP into n equal parts using division points $P_1, P_2, \ldots, P_{n-1}$. Denote the midpoint of the arc $P_{k-1}P_k$ as $P_{k-1/2}$. Then by (1.21)

$$x_{k+1/2} - x_{k-1/2} = -\frac{\Delta\theta}{2R} y_k, \quad k = 1, 2, \ldots, n-1. \tag{1.25}$$

We also have

$$(y_{k+1} - y_k) - (y_k - y_{k-1}) = \frac{\Delta\theta}{2R}\left(x_{k+1/2} - x_{k-1/2}\right), \quad k = 1, \ldots, n-1 \text{ or} \tag{1.26}$$

$$y_{k+1} - 2y_k + y_{k-1} = -\left(\frac{\Delta\theta}{2R}\right)^2 y_k, \quad k = 1, 2, \ldots, n-1. \tag{1.27}$$

Now start with $k = n-1$ and multiply the equations by $1, 2, \ldots, n-1$ respectively and sum up the resulting equations. We then have

$$\begin{aligned} y_n - ny_1 &= -\left(\frac{\Delta\theta}{2R}\right)^2 (y_{n-1} + 2y_{n-2} + \cdots (n-1)y_1) \\ &= -\left(\frac{\Delta\theta}{2R}\right)^2 (y_1 + (y_1 + y_2) + \cdots + (y_1 + y_2 + \cdots y_{n-1})). \end{aligned} \quad (1.28)$$

This is the result corresponding to (1.24). To obtain the successive polynomial approximations, Jyesthadeva had to work with sums of powers of integers; in order to deal with these sums, he applied the same lemma (1.15) he had used for the arctangent series.

1.6 Continued Fractions

The noted twelfth-century Indian mathematician Bhaskara, who lived and worked in the area now known as Karnataka, used continued fractions in his c. 1150 *Lilavati*. The Kerala school was certainly familiar with Bhaskara's work, since they commented on it. It is therefore possible that they were aware of the specific continued fractions (1.2) and (1.6) for the error terms, even though they mentioned only the first few convergents of these fractions. They did not indicate how they obtained these convergents. Some historians have suggested that Madhava may have found the approximations for the error term, without knowing the continued fractions, by comparing the first few partial sums of the series with a known rational approximation of π. Others speculate that Madhava may have used a method of Wallis.

Whether or not Madhava knew it, Wallis's technique can be used to derive the continued fractions of which the Kerala school gave the convergents; this may be of interest. Start with the functional equation (1.4) for $f(n)$,

$$f(n+1) + f(n-1) = \frac{1}{n}. \quad (1.29)$$

It is obvious that a first approximation for $f(n)$ is given by $f(n) \approx \frac{1}{2n}$. As a first step toward the continued fraction for $f(n)$, set

$$f(n) = \frac{1}{2r_n^{(0)}} \quad \text{and} \quad r_n^{(0)} = n + \frac{1}{r_n^{(1)}}. \quad (1.30)$$

It follows from (1.29) that $r_n^{(0)}$ satisfies

$$\left(2r_{n+1}^{(0)} - n\right)\left(2r_{n-1}^{(0)} - n\right) = n^2. \quad (1.31)$$

From (1.30)

$$2r_{n+1}^{(0)} - n = n + 2 + \frac{2}{r_{n+1}^{(1)}},$$

1.6 Continued Fractions

and a similar relation holds for $r_{n-1}^{(0)}$. When these values are substituted in (1.31), some calculation gives us

$$\left(2r_{n+1}^{(1)} - (n-2)\right)\left(2r_{n-1}^{(1)} - (n+2)\right) = n^2. \tag{1.32}$$

Once again, $r_n^{(1)} \approx n$. So assume $r_n^{(1)} = n + 1/s_n^{(2)}$ and substitute in (1.32) to get, after simplification,

$$16 s_{n-1}^{(2)} s_{n-1}^{(2)} - 2(n+4) s_{n+1}^{(2)} - 2(n-4) s_{n-1}^{(2)} - 4 = 0.$$

To obtain an equation such as (1.31) or (1.32), multiply this last equation by 4, set

$$s_n^{(2)} = r_n^{(2)}/2^2,$$

and add n^2 to both sides to get

$$\left(2r_{n+1}^{(2)} - (n-4)\right)\left(2r_{n-1}^{(2)} - (n+4)\right) = n^2. \tag{1.33}$$

We then have $r_n^{(1)} = n + 2^2/r_n^{(2)}$. A similar calculation shows that

$$r_n^{(2)} = n + 3^2/r_n^{(3)}$$

satisfies the equation

$$\left(2r_{n+1}^{(3)} - (n-6)\right)\left(2r_{n-1}^{(3)} - (n+6)\right) = n^2. \tag{1.34}$$

Inductively, it can be shown that if

$$r_n^{(k-1)} = n + k^2/r_n^{(k)}, \text{ and}$$

$$\left(2r_{n+1}^{(k-1)} - (n-2(k-1))\right)\left(2r_{n-1}^{(k-1)} - (n+2(k-1))\right) = n^2, \text{ then}$$

$$\left(2r_{n+1}^{(k)} - (n-2k)\right)\left(2r_{n-1}^{(k)} - (n+2k)\right) = n^2.$$

It follows that $f(n)$ has the continued fraction expansion (1.2). In a similar way, we may obtain the continued fraction (1.6) for $g(n)$ if we start with the functional relation

$$g(n-1) + g(n+1) = \frac{1}{n^2 - 1}.$$

It may be instructive to consider another method for finding the continued fractions of the Kerala school, also a method for obtaining the successive convergents. There is certainly no clear indication that this method was discovered before Gauss did so in his work on approximate quadrature, published in 1813. Start with a series of the form

$$f(n) = \frac{a_1}{n} + \frac{a_2}{n^2} + \frac{a_3}{n^3} + \cdots. \tag{1.35}$$

Note that it is always possible to associate a continued fraction with (1.35) by applying successive division. Write (1.35) as

$$f(n) = \frac{1}{n/a_1 - a_2/a_1^2 + t_1(n)}. \tag{1.36}$$

From this we can see that

$$t_1(n) = \frac{b_1}{n} + \frac{b_2}{n^2} + \frac{b_3}{n^3} + \cdots,$$

a series of the same kind as (1.35). So the process can be continued, and the result is a continued fraction for $f(n)$. To find the numbers $a_1, a_2, a_3, a_4, \ldots$, substitute (1.35) in (1.29). The first four values are

$$a_1 = 1/2, \ a_2 = 0, \ a_3 = -1/2, \ a_4 = 0.$$

From these values, we obtain the second convergent of the continued fraction for $f(n)$ by applying the process described in (1.36):

$$\frac{1}{2n + \frac{2}{n}} = \frac{n}{2(n^2 + 1)}.$$

By also using $a_5 = 5/2$ and $a_6 = 0$, we obtain the third convergent: $(n^2+4)/(2n(n^2+5))$. One problem that arises in the computation of a_1, a_2, a_3, etc., is finding the series expansions of $1/(n+1)^2$, $1/(n-1)^2$, $1/(n+1)^3$, etc. Although this may appear to require knowledge of the binomial theorem for negative integer powers, observe that the series may be obtained by repeatedly multiplying the geometric series by itself. In our chapter on the binomial theorem, we see that Newton verified the correctness of his binomial theorem by multiplying series.

1.7 Exercises

1. Prove that if C is the circumference and D the diameter of a circle, then

$$C = 3D + 6D\left(\frac{1}{(2 \cdot 2^2 - 1)^2 - 2^2} + \frac{1}{(2 \cdot 4^2 - 1)^2 - 4^2} + \frac{1}{(2 \cdot 6^2 - 1)^2 - 6^2} + \cdots\right).$$

This result (equivalent to B.5) is easily derived from series B.2; it is contained in the *Karanapaddhati*, by an unknown author from the Putumana illam in Sivapur, Kerala. The result is described: "Six times the diameter is divided separately by the square of twice the squares of even integers minus one, diminished by the squares of the even integers themselves. The sum of the resulting quotients increased by thrice the diameter is the circumference." See Bag (1966). Also see Srinivasienger (1967), p. 149.

2. Compute

$$1 - \frac{1}{3} + \frac{1}{5} - \cdots + \frac{1}{149} - f_3(150)$$

where f_3 is defined in (1.4). This gives π correct to eleven decimal places. In one of his astronomical works, Madhava gave a value of π: "For a circle of diameter 9×10^{11} units, the circumference is 2,827,433,38,233 units." This gives the approximate value of π as 3.14159265359, correct to eleven decimal places. The *Sadratnamala* by Sankara Verman of unknown date gives π to seventeen decimal places. See Parameswaran (1983), p. 194, and Srinivasiengar (1967).

3. Prove al-Haytham's formula (1.18).
4. This exercise outlines the proof of Paramesvara's formula for the radius of the circle circumscribing a cyclic quadrilateral, as given in the *Kriyakramakari*. First, prove that the product of the flank sides of any triangle divided by the diameter of its circumscribed circle is equal to the altitude of the triangle. This result follows from a rule given by Brahmagupta (c. 628) in an astronomical work, the *Brahmasphutasiddhanta*.

   Next, prove that the area of the cyclic quadrilateral is given by

   $$A = \sqrt{s(s-a)(s-b)(s-c)} \text{ where } s = (a+b+c+d)/2$$

   and a, b, c, d are the lengths of the sides of the quadrilateral. This was also stated by Brahmagupta. The *Yuktibhasa* contains a complete proof. See also Kichenassamy (2010), who convincingly argues that Brahmagupta had a proof and reconstructs it from indications in *Brahmasphutasiddhanta*.

   Then, let $ABCD'$ be the quadrilateral obtained from $ABCD$ by interchanging the sides AD and CD, so that $AD' = CD = c$ and $CD' = AD = d$. Show that if x, y, z denote the three diagonals AC, BD, BD', respectively, then

   $$yz = ab + cd, zx = bc + da, xy = ca + bd.$$

   This is, of course, Ptolemy's theorem. Ptolemy's formula is equivalent to the addition formula for the sine function; his *Almagest*, containing this relation, is heavily indebted to the *Chords in a Circle* of Hipparchus. Bhaskara defined the three diagonals in his *Lilavati*. See Boyer and Merzbach (1991) and Maor (1998), pp. 87–94. Finally, prove that the radius of the circle circumscribing the cyclic quadrilateral is

   $$r = \sqrt{\frac{(ad+bc)(ac+bd)(ab+cd)}{(b+c+d-a)(c+d+a-b)(d+a+b-c)(a+b+d-d)}}.$$

   This is Paramesvara's formula, sometimes attributed to S. A. J. L'Huillier who published it in 1782. See Gupta (1977).
5. Use Archimedes's formula (1.17) to show that

   $$\sum_{k=1}^{n} k^2 = \frac{1}{3}n^3 + \frac{1}{2}n^2 + \frac{1}{6}n.$$

6. Use al-Haytham's formula (1.19) to obtain

$$\sum_{k=1}^{n} k^3 = \frac{1}{4}n^4 + \frac{1}{2}n^3 + \frac{1}{4}n^2 \text{ and}$$

$$\sum_{k=1}^{n} k^4 = \frac{1}{5}n^5 + \frac{1}{2}n^4 + \frac{1}{3}n^3 - \frac{1}{30}n.$$

7. Prove that with the notation as in (1.18),

$$\frac{S_n^{(p-1)}}{n^p} \sim \frac{1}{p} \text{ implies that } \frac{S_1^{(p-1)} + S_2^{(p-1)} + \cdots + S_n^{(p-1)}}{1^p + 2^p + \cdots + n^p} \sim \frac{1}{p} \text{ as } n \to \infty.$$

This proves the missing step in the *Yuktibhasa*.

1.8 Notes on the Literature

The Weierstrass quotation at the beginning of the preface is a translation by the author of a passage in a letter to Casorati; see Neuenschwander (1978b), p. 73. It seems that the work of Madhava and his followers on series became known outside India only when a British civil servant and Indologist, Charles M. Whish, wrote a paper on the subject, posthumously published in the *Transactions of the Royal Asiatic Society of Great Britain and Ireland* in 1835. This journal was founded by British Indologists in the early 1830s, though Sir William Jones had first conceived the idea about fifty years earlier. Unfortunately, Whish's paper had little impact. Interest in the Kerala school was renewed in the twentieth century by the efforts of C. Rajagopal and his associates, who published several papers on the topic. See Rajagopal (1949), Rajagopal and Aiyar (1951), Rajagopal and Venkataraman (1949), Rajagopal and Rangachari (1977), and Rajagopal and Rangachari (1986). The translation of the verse enunciating the series for sine is taken from Rajagopal and Rangachari (1977), p. 96. The translations of verses in the *Yuktibhasa* concerning sums of powers are contained in Rajagopal and Aiyar (1951), p. 70. Newton's letter is quoted from Newton (1959–60), p. 140. We discussed two methods for deriving continued fractions for the remainder term; for further details, see Srinivasiengar (1967), pp. 149–151 and Rajagopal and Rangachari (1977). The latter paper makes use of Whiteside's (1961) reconstruction of Wallis's incomplete derivation of Brouncker's continued fraction for π.

The *Yuktibhasa* of Jyesthadeva and the *Tantrasangraha* of Nilakantha have recently been published with commentaries in English by Sarma (1977) and (2008). Sarma (2008) was posthumously published with additional notes by Ramasubramanian, Srinivasa, and Sriram. This two-volume translation with extensive and informative commentary contains both the mathematical and astronomical portions; the original Malayalam text extends to 300 pages. Sarma (1972) also discusses the Kerala school, but from the astronomical point of view. Biographical information on the members of the Kerala school, as well as numerous other ancient and medieval Indian astronomers and mathematicians, can be found in David Pingree's five-volume work (1970–1994).

1.8 Notes on the Literature

Readers who wish to read more on the Indian work on series, but with modern notation, may consult Roy (1990), Katz (1995), and Bressoud (2002). These papers are conveniently available in Anderson, Katz, and Wilson (2004). Also see the papers by Parameswaran (1983) on Madhava, Bag (1966) on the *Karanapaddhati*, Gupta (1977) on Paramesvara's rule for radius of the cyclic quadrilateral, and Sarma and Hariharan (1991) on the *Yuktibhasa*. Plofker (2009) presents a scholarly, detailed, and readable discussion of Kerala mathematics, with several excerpts on π translated from Sankara Variyar's *Kriyakramakari*. She also presents the derivation of the sine series with translations from the *Yuktidipika* and describes Takao Hayashi's suggested reconstruction of Madhava's remainder term results. In order to derive the continued fraction, Hayashi and his collaborators have compared the values of partial sums of Madhava's series for π with the then-known rational approximations for π. Van Brummelen (2009), on the history of trigonometry, discusses the contributions of the Kerala school and relates them to the astronomical work of medieval India. In the context of the development of astronomy, Van Brummelen (2009) presents the *Yuktibhasa* derivation of the sine series. This accessible presentation is very helpful, since the mathematics of the Kerala school was largely motivated by an interest in astronomy.

2

Sums of Powers of Integers

2.1 Preliminary Remarks

Archimedes, and most probably the ancient Babylonian mathematicians, discovered the formulas for the sum of the first n integers and for the sum of the squares of these integers. Archimedes applied these formulas to area and volume problems. Moreover, the great mathematician and physicist al-Haytham extended Archimedes's results to cubes and fourth powers and gave a recursive procedure for finding each successive power. Recall that Madhava and his successors employed this method in the course of their work on power series for trigonometric functions. When evaluating integrals, they were focused on the asymptotic value of the sums of powers:

$$S_n^{(p)} \sim \frac{n^{p+1}}{p+1} \quad \text{as} \quad n \to \infty, \tag{2.1}$$

where sum $S_n^{(p)} = 1^p + 2^p + \cdots + n^p$.

The work of Archimedes and al-Haytham showed that $S_n^{(p)}$ could be expressed as a polynomial in n; the asymptotic value simply yields the term with the highest power. Because Madhava and his school were primarily interested in integration, and thus in the highest power, they failed to note the full significance of the polynomial itself.

In the seventeenth century, Fermat was very interested in asymptotic values, since he too wished to evaluate $\int_0^a x^p \, dx$. While the Indians followed the geometric approach of al-Haytham, Fermat arrived at his results through figurate numbers.

By contrast, in the early seventeenth century, Johann Faulhaber (1580–1635) initiated an approach to the topic of sums of powers, taking an algebraic and number theoretic point of view. Thus, he gave the expression of sums of powers as a polynomial, of which the asymptotic value was just the first term. Faulhaber's approach was also motivated by his fascination with figurate numbers.

The figurate numbers have been studied since ancient times. The one-dimensional figurate numbers are merely the consecutive positive integers $1, 2, 3, \ldots, n$. The two-dimensional figurate numbers are the triangular numbers, where the nth triangular

2.1 Preliminary Remarks

number is the sum of the first n consecutive numbers:

$$1, 1+2=3, 1+2+3=6, 1+2+3+4=10, \ldots, n(n+1)/2.$$

The three-dimensional figurate numbers are the pyramidal or tetrahedral numbers such that the nth pyramidal number is the sum of the first n triangular numbers:

$$1, 4, 10, 20, \ldots, n(n+1)(n+2)/6, \ldots.$$

There exists an ancient Egyptian papyrus (c. 300 BC) containing the formula for the nth triangular and tetrahedral numbers. These formulas in modern notation can be written as

$$\sum_{k=1}^{n} \binom{k}{1} = \binom{n+1}{2}, \tag{2.2}$$

$$\sum_{k=1}^{n} \binom{k+1}{2} = \binom{n+2}{3}. \tag{2.3}$$

When written this way, it is clear that the figurate numbers are related to the number of combinations of k things chosen from m different things, for appropriate m and k. It appears that the connection between figurate numbers and combinations was recognized by Narayana Pandita whose *Ganita Kaumudi* of 1356 makes this explicit.

Narayana Pandita also algebraically extended the figurate numbers by taking sums of sums of sequences. So the sequence after the tetrahedral numbers would be

$$1, 1+4=5, 1+4+10=15, 1+4+10+20=34, \ldots.$$

Some of the earlier mathematicians may have refrained from doing this because they did not conceive of dimensionality beyond three as meaningful. In effect, Narayana had the formula

$$\sum_{k=1}^{n} \binom{k+p-1}{p} = \binom{n+p}{p+1}, \quad p=1, 2, 3, \ldots. \tag{2.4}$$

Fermat rediscovered (2.4) around 1635, though he apparently never wrote down a proof. In the margin of his copy of Diophantus's *Book on Polygonal Numbers*, he wrote that he had discovered this proposition and called it beautiful and wonderful. He also noted that the margin was too small to contain his proof, though we may surmise it to have been an inductive proof. It was the common practice of mathematicians up to the nineteenth century to work out a number of special cases as evidence for the correctness of the general result. But in his work *Plane Loci* Fermat proved a proposition by induction and included the crucial n to $n+1$ step; one may surmise that Fermat accomplished this around 1630. It is possible that Fermat had learned of the need to supply this step from F. Maurolico's *Arithmeticorum Libri Duo*, written in 1557 and published in 1575. Maurolico proved by complete induction that the sum of the first n odd integers was n^2. Also, in 1654 Pascal gave a lucid exposition of mathematical induction in his treatise on the arithmetical triangle; he too was familiar with Maurolico's work.

In fact, Fermat used (2.4) to determine the asymptotic values of sums of powers. To see this, observe that since

$$\binom{n+p-1}{p} = \frac{n(n+1)(n+2)\cdots(n+p-1)}{p!}$$
$$= \frac{n^p}{p!} + \text{terms of lower order},$$

we can write (2.4) as

$$\sum_{k=1}^{n}\left(\frac{k^p}{p!} + \text{lower order terms}\right) = \frac{n^{p+1}}{(p+1)!} + \text{lower order terms}.$$

This implies (2.1).

The work of the English mathematicians Thomas Harriot and Henry Briggs on problems related to interpolation shows that they also understood formula (2.4). The German algebraist and arithmetician Johann Faulhaber also independently discovered (2.4), but his motivation was an interest in numbers and in particular the figurate numbers. However, his results do not seem to have influenced Fermat, Harriot, or Briggs.

Faulhaber (1580–1635) was born in Ulm, Germany, and learned the weaving trade from his father. His love of computation led him to study mathematics. His knowledge of Latin was not very good, so in the course of his studies, he laboriously translated several mathematical texts, ancient and modern, into German. He founded a school for engineers in the early 1600s and wrote treatises on arithmetical questions. Faulhaber gave an algorithm for expressing $S_n^{(p)}$ as a polynomial in n; though he worked with Bernoulli numbers, he failed to note their significance. It was not the practice in Faulhaber's time to give proofs of algorithms. Two centuries later, in a paper on the Euler–Maclaurin formula, Jacobi provided proofs of some of Faulhaber's formulas.

Around 1700, Jakob Bernoulli gave a simple method for computing the polynomial in n for $S_n^{(p)}$. Bernoulli numbers, a sequence of rational numbers, play a significant role in the determination of this polynomial. Bernoulli's interest in the summation of finite and infinite series was connected with his study of probability theory. Jakob Bernoulli (1654–1705) was the eldest in an illustrious scientific and mathematical family, including his brother Johann, nephews Niklaus I, Niklaus II, Daniel, and Johann II. In 1676, Bernoulli received a degree in theology from the University of Basel, intending to go into the ministry. He then traveled in Europe, coming into contact with the Dutch mathematician Hudde and members of the Royal Society. These experiences aroused his interest in science and mathematics. In the 1680s, he taught himself mathematics by reading short treatments by Leibniz on differentiation and integration; he then taught this subject to his younger brother Johann. One of the first mathematicians to grasp Leibniz's calculus, Jakob Bernoulli proceeded to apply it to fundamental problems in mechanics and to differential equations. The study of Huygens's treatise on games of chance led Bernoulli to a study of probability theory, on which he wrote the first known full-length text, *Ars Conjectandi*. From 1687 until his death, Bernoulli happily served as professor of mathematics at Basel, in spite of a salary more meager than he would have received as a clergyman. This professorship was occupied by a member of the Bernoulli family for one hundred years.

Although he spent many years on the problems contained in his probability treatise, he never completed it. It appears that he wished to include several problems arising out of "civil, moral, and economic matters," i.e., applications to practical situations. For example, even a year before his death, he repeated his earlier request to Leibniz for a hard-to-find copy of Jan de Witt's work on annuities and life expectancy. *Ars Conjectandi* was posthumously published in 1713 by Jakob Bernoulli's son Niklaus with a foreword by his nephew Niklaus I. Publication was delayed when Jakob's immediate family, fearing academic dishonesty, refused to hand over the manuscript to Johann or to Niklaus I. The first part of the book consisted in Huygens's treatise with extensive annotations.

2.2 Johann Faulhaber and Sums of Powers

Faulhaber published his *Academia Algebrae* in 1631. In spite of the Latin title, the text is in German and in it he discussed the sums of powers of integers. He wished to study the polynomial in n determined by $\sum_{k=1}^{n} k^p$. He explicitly wrote down the polynomial expressions for $p = 1, 2, \ldots 17$ and an encryption at the end of the book indicates that he had carried out the computation up to $p = 25$. He also expressed these polynomials in powers of the triangular numbers $N = n(n+1)/2$ with rational coefficients and gave an algorithm to determine these coefficients. For example, Faulhaber presented the formulas

$$\sum_{k=1}^{n} k^{15} = \tfrac{1}{12}(192N^8 - 768N^7 + 1792N^6 - 2816N^5 + 2872N^4 - 1680N^3 + 420N^2),$$

$$\sum_{k=1}^{n} k^{17} = \tfrac{1}{45}\left(1280N^9 - 6720N^8 + 21120N^7 - 46880N^6 + 72912N^5 - 74220N^4 \right.$$
$$\left. + 43404N^3 - 10851N^2\right).$$

More generally he showed that

$$\sum_{k=1}^{n} k^{2m} = \left(\sum_{k=1}^{n} k^2\right)\left(b_1 + b_2 N + b_3 N + \cdots + b_m N^{m-1}\right), \tag{2.5}$$

$$\sum_{k=1}^{n} k^{2m+1} = N^2\left(c_1 + c_2 N + c_3 N^2 + \cdots + c_m N^{m-1}\right). \tag{2.6}$$

Faulhaber gave a procedure for finding b_i from c_i, amounting to

$$3(j+1)c_j = 2(2m+1)b_j, \quad j = 1, 2, \ldots m. \tag{2.7}$$

He also gave a method for obtaining the c_i themselves.

Like Narayana Pandita, Faulhaber also considered sums of sums. To understand Faulhaber's insight here, consider

$$\sum^2 n^m \equiv \sum_{j=1}^n \sum_{k=1}^j k^m \quad \text{and} \quad \sum^3 n^m \equiv \sum_{j=1}^n \sum_{h=1}^j \sum_{k=1}^h k^m$$

and so on. Faulhaber saw that the r-fold sum $\sum^r n^{2m}$ could be written as a polynomial in $n(n+r)$ times $\sum^r n^2$, while $\sum^r n^{2m+1}$ would be such a polynomial times $\sum^r n$. He explicitly wrote down the 17th degree polynomial equal to $\sum^{11} n^6$. In fact, this expression was verified in 1993 by Donald Knuth.

In 1834, Jacobi gave a slightly different formulation of the Faulhaber results connected with (2.5) and (2.6). He stated that

$$\frac{1}{2m+1} \frac{d}{dn} \left(\sum_{k=1}^n k^{2m+1} \right) = \sum_{k=1}^n k^{2m}. \tag{2.8}$$

Moreover, he observed that for $u = n(n+1)$ and

$$\sum_{k=1}^n k^{2m+1} = \frac{1}{2m+2} \left(u^{m+1} - d_1 u^m + d_2 u^{m-1} - \cdots (-1)^{m-1} d_{m-1} u^2 \right), \tag{2.9}$$

the relation between the coefficients $d_1, d_2, \ldots, d_{m-1}$ and the coefficients $e_1, e_2, \ldots, e_{m-2}$ in

$$\sum_{k=1}^n k^{2m-1} = \frac{1}{2m} \left(u^m - e_1 u^{m-1} + e_2 u^{m-2} - \cdots + (-1)^m e_{m-2} u^2 \right) \tag{2.10}$$

would be given by

$$(2m+1)(2m+2)e_k$$
$$= (2m-2k+1)(2m-2k+2)d_k - (m-k+1)(m-k+2)d_{k-1}, \tag{2.11}$$

where

$$k = 1, 2, \ldots, m-1, \quad d_0 = 1, \quad e_{m-1} = 0.$$

Jacobi's result shows that if the sum of the first n integers, each taken to an odd power, is given as a polynomial in $n(n+1)$, then this same sum, with each integer taken to the next odd power, can be determined as a polynomial in $n(n+1)$. Moreover, the polynomial expression for a sum of integers taken to an even power can be obtained by differentiating the polynomial for the sum of the next odd powers.

2.3 Jakob Bernoulli's Polynomials

The second part of Bernoulli's great probability treatise contains results on permutations and combinations. He rigorously worked out the connection between binomial

2.3 Jakob Bernoulli's Polynomials

coefficients and figurate numbers. He thought that he was the first to do this, but Pascal anticipated him in 1654. Bernoulli also rediscovered the formula (2.4) and applied it to the problem of finding the sums of powers of integers. Herein he made his enduring discovery of the role played by the sequence of rational numbers now named after him. Bernoulli found a pattern in the coefficients of the polynomials for $S_n^{(p)}$ that had been missed by so outstanding an arithmetician as Faulhaber.

Bernoulli began by explicitly expressing $S_n^{(p)}$ for $p = 1, 2, \ldots, 10$ as polynomials in n:

Sums of Powers

$$\int n = \frac{1}{2}nn + \frac{1}{2}n.$$

$$\int n^2 = \frac{1}{3}n^3 + \frac{1}{2}nn + \frac{1}{6}n.$$

$$\int n^3 = \frac{1}{4}n^4 + \frac{1}{2}n^3 + \frac{1}{4}nn.$$

$$\int n^4 = \frac{1}{5}n^5 + \frac{1}{2}n^4 + \frac{1}{3}n^3 * - \frac{1}{30}n.$$

$$\int n^5 = \frac{1}{6}n^6 + \frac{1}{2}n^5 + \frac{5}{12}n^4 * - \frac{1}{12}nn.$$

$$\int n^6 = \frac{1}{7}n^7 + \frac{1}{2}n^6 + \frac{1}{2}n^5 * - \frac{1}{6}n^3 + \frac{1}{42}n.$$

$$\int n^7 = \frac{1}{8}n^8 + \frac{1}{2}n^7 + \frac{7}{12}n^6 * - \frac{7}{24}n^4 * \frac{1}{12}nn.$$

$$\int n^8 = \frac{1}{9}n^9 + \frac{1}{2}n^8 + \frac{2}{3}n^7 * - \frac{7}{15}n^5 * + \frac{2}{9}n^3 * - \frac{1}{30}n.$$

$$\int n^9 = \frac{1}{10}n^{10} + \frac{1}{2}n^9 + \frac{3}{4}n^8 * - \frac{7}{10}n^6 * + \frac{1}{2}n^4 * - \frac{1}{12}nn.$$

$$\int n^{10} = \frac{1}{11}n^{11} + \frac{1}{2}n^{10} + \frac{5}{6}n^9 * - 1n^7 * 1n^5 * - \frac{1}{2}n^3 * + \frac{5}{66}n.$$

A. W. F. Edwards has noted that the last term in the polynomial for $\int n^9$ should be $-\frac{3}{20}n^2$, rather than $-\frac{1}{12}n^2$.

Bernoulli went on:

> Any one who carefully observed the symmetry properties of this table will easily be able to continue it. If we let c denote an arbitrary exponent, we have
>
> $$\int n^c = \frac{1}{c+1}n^{c+1} + \frac{1}{2}n^c + \frac{c}{2}An^{c-1} + \frac{c(c-1)(c-2)}{2 \cdot 3 \cdot 4}Bn^{c-3}$$
> $$+ \frac{c(c-1)(c-2)(c-3)(c-4)}{2 \cdot 3 \cdot 4 \cdot 5 \cdot 6}Cn^{c-5}$$
> $$+ \frac{c(c-1)(c-2)(c-3)(c-4)(c-5)(c-6)}{2 \cdot 3 \cdot 4 \cdot 5 \cdot 6 \cdot 7 \cdot 8}Dn^{c-7} + \cdots, \quad (2.12)$$
>
> the exponents of n decreasing by 2 until n or nn is reached. The capitals A, B, C, D, etc. denote, in order, the last terms in the expressions of $\int nn$, $\int n^4$, $\int n^6$, $\int n^8$ etc. namely
>
> $$A = 1/6, \ B = -1/30, \ C = 1/42, \ D = -1/30.$$

But these coefficients are so established that each of the coefficients along with the others of its order adds up to one. Thus $D = -1/30$, since

$$\frac{1}{9} + \frac{1}{2} + \frac{2}{3} - \frac{7}{15} + \frac{2}{9} + D = 1.$$

Using this table, it took me less than a quarter of an hour to compute the tenth powers of the first 1000 integers; the result is

$$91, 409, 924, 241, 424, 243, 424, 241, 924, 242, 500.$$

This example shows the uselessness of the book *Arithmetica Infinitorum* by Ismael Bullialdus, which is entirely devoted to a tremendously large computation of the sums of the six first powers – less than what I have accomplished in a single page.

Bernoulli left it to the reader to use "the symmetry properties of this table" to figure out how he obtained his general formula for the sums of powers. A little earlier in his book, Bernoulli presented a "table of combinations or figurate numbers" and analyzed it columnwise. Apply this idea to Bernoulli's table on sums of powers. The progression in the first column is easy to understand. Now in the second column, factor out $1/2$ to obtain the progression $1, 1, 1, \ldots$; these form the first column of Bernoulli's table of figurate numbers. Next factor out the first number in the third column, $1/6$, to obtain the sequence $1, 3/2, 2, 5/2, 3, \ldots$, and this turns out to be $1/2$ of the sequence $2, 3, 4, 5, 6, \ldots$ appearing in the second column of the figurate numbers table. The fourth column of the sums of powers table consists of only zeros but factor out $-1/30$ from the fifth column to obtain the progression $1, 5/2, 5, 35/4, 14, \ldots$. This last sequence is equal to $1/4$ of the fourth column in the figurate numbers table. These observations clarify Bernoulli's comment on the "symmetry properties of the table." Note that the Bernoulli numbers are formed by the sequence of coefficients of n in the polynomial expansions of the various sums of powers. Today, however, we take the first Bernoulli number to be $-1/2$ rather than $1/2$ so that the signs alternate.

In modern terminology, Bernoulli's formula (2.12) can be written as

$$\sum_{k=1}^{n-1} k^c = \frac{1}{c+1}\left(B_0 n^{c+1} + \binom{c+1}{1} B_1 n^c + \binom{c+1}{2} \right.$$
$$\left. \times B_2 n^{c-1} + \binom{c+1}{3} B_3 n^{c-2} + \cdots + \binom{c+1}{c} B_c n \right), \qquad (2.13)$$

where

$$B_0 = 1, \ B_1 = -\frac{1}{2}, \ B_{2k+1} = 0 \text{ for } k = 1, 2, \ldots,$$

$$B_2 = \frac{1}{6}, \ B_4 = -\frac{1}{30}, \ B_6 = \frac{1}{42}, \ B_8 = -\frac{1}{30}, \text{ etc.}$$

Bernoulli noted that these numbers could be calculated using

$$\frac{1}{c+1}\left(B_0 + \binom{c+1}{1} B_1 + \binom{c+1}{2} B_2 + \cdots + \binom{c+1}{c} B_c \right) = 1. \qquad (2.14)$$

2.3 Jakob Bernoulli's Polynomials

This equation follows when $n = 1$ in (2.13). Faulhaber also noted this fact, useful in successively finding the values of the Bernoulli numbers. Thus, if B_2, B_4 and B_6 are known, take $c = 8$ in (2.14) to get

$$\frac{1}{9}\left(1 + \binom{9}{1}\frac{1}{2} + \binom{9}{2}\frac{1}{6} + \binom{9}{4}\left(-\frac{1}{30}\right) + \binom{9}{6}\frac{1}{42} + 9B_8\right) = 1,$$

or

$$\frac{1}{9} + \frac{1}{2} + \frac{2}{3} - \frac{7}{15} + \frac{2}{9} + B_8 = 1$$

or

$$B_8 = -\frac{1}{30}.$$

Although Bernoulli left us no proof of his formula, he may well have had one; in fact, fairly straightforward inductive proofs exist. We note that the name Bernoulli numbers was given in 1730 by de Moivre, when he used (2.12) in his work on the asymptotic behavior of $n!$ for large n. The notation B_k was introduced in the nineteenth century by a number of mathematicians, though it often stood for B_{2k}.

The Bernoulli polynomials are essentially those polynomials given by Faulhaber and Bernoulli for the sums of powers; in the 1840s, Joseph Raabe formally defined them:

$$B_c(x) = x^c + \binom{c}{1} B_1 x^{c-1} + \binom{c}{2} B_2 x^{c-2} + \cdots + \binom{c}{c} B_c. \tag{2.15}$$

From (2.13), it follows that the sums of powers can be explicitly obtained from the Bernoulli polynomials by the equation

$$S_{n-1}^{(c)} = \frac{1}{c+1}\left(B_{c+1}(n) - B_{c+1}\right), \quad c = 1, 2, 3, \ldots. \tag{2.16}$$

In the early 1730s, Euler found a generating function for the Bernoulli numbers, apparently unaware that Bernoulli had already defined these numbers in a different way. His generating function is given by

$$\frac{t}{e^t - 1} = \sum_{k=0}^{\infty} \frac{B_k}{k!} t^k.$$

This exponential generating function arose naturally and automatically in his work on the Euler–Maclaurin summation formula. In fact, the function he expanded was

$$\frac{e^t}{e^t - 1} = 1 + \frac{1}{e^t - 1}.$$

In his 1834 paper mentioned earlier, Jacobi gave the generating function for the Bernoulli polynomials:

$$\frac{te^{xt}}{e^t - 1} = \sum_{k=0}^{\infty} \frac{B_k(x)}{k!} t^k.$$

Once again, this discovery took place in the context of the Euler–Maclaurin summation formula, as Jacobi was deriving an expression for the remainder term in the Euler–Maclaurin series. Since the odd Bernoulli numbers B_{2k+1} vanish for $k \geq 1$, Jacobi worked only with the even Bernoulli polynomials.

2.4 Proof of Bernoulli's Formula

In the first part of his 1755 differential calculus book *Institutiones Calculi Differentialis*, Euler suggested deriving Bernoulli's formula, (2.12), by means of finite differences, but he did not provide any details. In the second part of his book, Euler derived (2.12) using the Euler–Maclaurin summation formula. Sylvestre F. Lacroix (1765–1843), in the third volume of his important text on calculus, summarized Euler's ideas on finite differences and then indicated how they could be worked into a proof. Lacroix investigated partial differential equations under the tutelage of Gaspard Monge but did not pursue mathematical research. Rather, at the urging of Condorcet, he decided that his broad knowledge of eighteenth-century mathematics should be put to use in the writing of elementary and advanced mathematics textbooks. These books were widely popular, going into numerous editions and translations. I here summarize Lacroix's treatment of Bernoulli's formula.

Using Lacroix's notation, let $\sum x^m$ denote the sum $\sum_{k=1}^{x-1} k^m \equiv S(x)$. Note that, unlike Euler and Bernoulli, Lacroix took the sum up to $x - 1$, rather than x. With this in mind, he had

$$S(x+1) - S(x) = x^m.$$

Now Lacroix did not use subscripts such as A_k, but we use the modern notation: Assume

$$\sum x^m = \sum_{k=0}^{m+1} A_k x^{m+1-k}.$$

Then

$$x^m = \sum_{k=0}^{m+1} A_k \left((x+1)^{m+1-k} - x^{m+1-k} \right)$$

$$= \sum_{k=0}^{m+1} A_k \left(\binom{m+1-k}{1} x^{m-k} + \binom{m+1-k}{2} x^{m-k-1} + \cdots + \binom{m+1-k}{m-k} x + 1 \right).$$

Equate the powers of x to get

$$F A_0 = \frac{1}{m+1}; \ A_1 = -A_0 \frac{(m+1)}{2} = -\frac{1}{2}; \ A_2 = -A_0 \frac{(m+1)m}{2 \cdot 3} - A_1 \frac{m}{2} = \frac{1}{6} \cdot \frac{m}{2};$$

$$A_3 = -A_0 \frac{(m+1)(m(m-1))}{2 \cdot 3 \cdot 4} - A_1 \frac{m(m-1)}{2 \cdot 3} - A_2 \frac{(m-1)}{2} = 0; \ \text{etc.}$$

Lacroix wrote that from these equations one could successively obtain the values of the coefficients A_k, and he explicitly gave the values of $A_k, k = 0, 1, \ldots, 20$. Although he did not work out the general case, it is easy to do: If we take the coefficient of x^{m-k}, then we obtain

$$A_0 \frac{(m+1)m(m-1)\cdots(m-k+1)}{(k+1)!}$$
$$+ A_1 \frac{m(m-1)\cdots(m-k+1)}{k!} + \cdots + A_k(m-k+1) = 0. \qquad (2.17)$$

We must show that

$$A_s = C_s m(m-1)\cdots(m-s+2) \quad \text{for} \quad s = 2, 3, 4, \ldots,$$

where C_s is some rational number independent of m. Now the result is true for $m = 2$. Assume that $A_3, A_4, \ldots, A_{k-1}$ are of the required form. Then each term except the last in (2.17) is of the form $m(m-1)\cdots(m-k+1)$ multiplied by a rational number independent of m. Thus, by equation (2.17) the result follows. We may therefore write

$$A_s = \binom{m}{s-1} B_s/s,$$

where B_s can be seen to be the sth Bernoulli number. This completes Lacroix's demonstration of Bernoulli's formula.

2.5 Exercises

1. Prove Narayana's formula (2.4).
2. Prove the inequality of Roberval

$$S_{n-1}^{(p)} < \frac{n^{p+1}}{p+1} < S_n^{(p)}.$$

   Roberval mentioned this inequality in a letter to Torricelli as a means of computing $\int_0^1 x^p \, dx$. Hofmann (1990), vol. 2, pp. 232 and 471 give references as well as connections with the work of Fermat and Archimedes.
3. Prove Pascal's formula

$$\binom{m+1}{1} d \sum_{k=0}^{n-1}(a+kd)^m + \binom{m+1}{2} d^2 \sum_{k=0}^{n-1}(a+kd)^{m-1} + \cdots$$
$$+ \binom{m+1}{m} d^m \sum_{k=0}^{n-1}(a+kd)$$
$$= (a+nd)^{m+1} - (a^{m+1} + nd^{m+1}). \qquad (2.18)$$

   By taking $a = 0$ and $d = 1$, Pascal obtained an explicit recursion formula for the sum of the mth powers. See Boyer (1943).
4. Prove Bernoulli's formula (2.12) or (2.13) by induction.

5. Use Bernoulli's formula (2.12) to prove Jacobi's statement that

$$\frac{d}{dn}\left(\sum_{k=1}^{n} k^{2m+1}\right) = (2m+1)\sum_{k=1}^{n} k^{2m} \text{ for } m = 1, 2, 3, \ldots.$$

6. Prove Jacobi's recurrence relations (2.11) for the coefficients of the Faulhaber polynomials.
7. Prove that the Faulhaber coefficients $e_0, e_1, \ldots, e_{m-2}$ in (2.10) satisfy Knuth's relations

$$e_0 = 1, \quad \sum_{j=0}^{k}\binom{m-j}{2k+1-2j}e_j = 0. \tag{2.19}$$

Expand $u^k = n^k(n+1)^k$ in powers of n and then use the fact that $B_3 = 0$, $B_5 = 0$, $B_7 = 0$, etc. See Knuth (2003), pp. 61–84.

8. Use (2.16) and (2.10) to express $e_1, e_2, \ldots e_{m-2}$ in terms of Bernoulli numbers. See Knuth (2003), pp. 61–84.

2.6 Notes on the Literature

Faulhaber's *Academia Algebrae* is now a very rare book. According to Knuth (2003), p. 83, "An extensive search of printed indexes and electronic indexes indicates that no copies have ever been recorded to exist in America, in the British Library, or the Bibliothèque Nationale." A copy is available at the Cambridge University Library, and Knuth placed an annotated photocopy at the Mathematical Sciences Library at Stanford University. Knuth (2003), pp. 61–84, has given an excellent summary of the results in Faulhaber's book, adding some details on the methods Faulhaber may have used. Ivo Schneider (1983) also offers a discussion of the methods of Faulhaber. As we have mentioned, Jacobi gave sophisticated proofs of some of these results without referring to Faulhaber; this paper, *De Usu Legitimo Formulae Summatoriae Maclaurinianae*, can be found on pp. 64–75 of vol. VI of Jacobi's collected papers. However, A. W. F. Edwards discovered that the copy of Faulhaber's book in the Cambridge University Library had once belonged to Jacobi. Moreover, before Jacobi, it may well have been in the possession of J. F. Pfaff. Luckily, Edwards brought Faulhaber's previously neglected work to the attention of Knuth and the mathematical community. In the past twenty years, two biographies of Faulhaber have appeared, by Schneider and by Kurt Hawlitschek.

Jakob Bernoulli's important *Ars Conjectandi* is now available in an English translation by Edith D. Sylla; see Bernoulli (2006). The quotation from Bernoulli on sums of powers is from Bernoulli (2006), pp. 215–216. The translator's helpful 126-page introduction puts the book in historical context. Schneider's nice article in Grattan-Guinness (2005) provides a summary of the contents and explains its importance in the development of probability theory.

2.6 Notes on the Literature

The material quoted from Lacroix can be found on pp. 82–85 of the third volume of Lacroix (1819). Many old mathematical works have been difficult to find, although the internet is changing this very rapidly. However, anyone wishing easy access to eighteenth century analysis can do no better than to read the three volumes of Lacroix on calculus. The first two discuss differential and integral calculus, while the third deals with the calculus of finite differences. Lacroix performed a great service to mathematics, not only by giving a clear exposition of the received mathematics of his day but also by including a well-organized, section-by-section list of all the papers and books from which he gleaned his expertise. Grattan–Guinness (2005) also includes J. C. Domingues's informative article on the first edition of Lacroix's calculus work. Illustrating the popularity of Lacroix, Domingues mentions that in 1812 Lacroix was translated into Portugese in Brazil.

To gain further background in the topic of sums of powers and in the mathematics contributing to it, one may consult A. W. F. Edwards (2002), Mahoney (1994), and (on Narayana) Bressoud (2002). One may also look at André Weil's review, contained in the third volume of his collected papers, of the first edition of Mahoney. This review, although overly critical of Mahoney, contains a highly insightful and concise summation of Fermat's mathematical work.

3

Infinite Product of Wallis

3.1 Preliminary Remarks

In 1655, John Wallis produced the following very important infinite product:

$$\frac{4}{\pi} = \frac{3}{2} \cdot \frac{3}{4} \cdot \frac{5}{4} \cdot \frac{5}{6} \cdots. \tag{3.1}$$

This result appeared in his *Arithmetica Infinitorum*, published in 1656; in 1593 François Viète gave the only earlier example of an infinite product, a calculation of the value of π by inscribing regular polygons in a circle. The passage of 350 years has not diminished the beauty and significance of Wallis's result, the culmination of a series of remarkable mathematical insights and audacious guesses; his book exercised great influence on the early mathematical work of Newton and Euler.

John Wallis (1616–1703) apparently received little mathematical training at school or at Emmanuel College, Cambridge. He taught himself elementary arithmetic from textbooks belonging to his younger brother, who was going into a trade. It was only during the English Civil War (1642–1648) that Wallis's mathematical inclinations began to be evident as he decoded letters for Parliament. The code operated by replacing letters with numerical values. Wallis gained a feeling for numerical relationships through this experience, and he applied it to his mathematical researches for the *Arithmetica Infinitorum*. In fact, the manner in which he presented and analyzed the mathematical data in his book is reminiscent of the way in which he decoded messages.

It was probably around 1646 that Wallis began delving more deeply into mathematics, by studying the famous *Clavis Mathematicae* by William Oughtred (1574–1660), inventor of the slide rule. First published in 1631 and composed for the instruction of the son of the Earl of Arundel, this book was widely studied and exerted a tremendous influence on seventeenth-century English mathematics. A second edition in English and then in Latin appeared in 1647 and 1648. The second edition was among the first mathematical texts studied by Newton as a student in 1664. In the 1690s Newton recommended that the book be reprinted for the new generation of students of mathematics. The *Clavis* introduced Wallis to algebraic notation and to the method of applying algebra to geometric problems in the manner developed by Viète in the 1590s.

3.1 Preliminary Remarks

In 1649, Wallis was appointed to the Savilian Chair of Geometry at Oxford. The valuable service Wallis had provided to the winning side in the Civil War helped him attain this post. The Savilian Chair, endowed by Sir Henry Savile in 1619 to promote development of mathematics in England, was the second endowed mathematics chair in England; the first was founded in 1597 at Gresham College, London. With the rapid advancement of mathematics after 1550, it had become clear that university instruction in mathematics was essential, especially since this subject was proving useful in navigation and military matters. In fact, Italy and France had already established a number of mathematics professorships.

At the time of his appointment, Wallis knew little more than the contents of the *Clavis*. But the professorship gave him access to the Savile Library with its fine collection of mathematics books. Wallis was most influenced by Frans van Schooten's 1649 Latin translation of René Descartes's *La Géométrie* and Evangelista Torricelli's *Opera Geometrica* of 1644. Although Oughtred and Viète had employed algebra in the study of geometry, Descartes took the process to a higher level by reducing the study of curves to algebraic equations by means of coordinate systems. At around the same time, Pierre Fermat (1607–1666) also made this major step, but his expositions on this and other topics were unfortunately published only posthumously. Wallis's first book, *De Sectionibus Conicis*, written in 1652 and published in 1656, was clearly inspired by Descartes. He obtained properties of conic sections algebraically, making extensive use of the symbolic algebra developed by Harriot and Descartes. Wallis defined the parabola, hyperbola, and ellipse by means of algebraic equations; he remarked that "It is no more necessary that a parabola is the section of a cone by a plane parallel to a side than that a circle is a section of a cone by a plane parallel to the base, or that a triangle is a section through the vertex."

Wallis learned of Bonaventura Cavalieri's method of indivisibles from Torricelli; Wallis regarded his own *Arithmetica Infinitorum* as a continuation of Cavalieri, an accurate assessment. Wallis spent a fair amount of his book computing the area under $y = x^m$ when m was a positive integer, using an arithmetical approach, as contrasted with Torricelli's geometrical method. Wallis then extended the result to the case $m = 1/n$ where n was a positive integer, by observing that the curve $y = x^{1/n}$ was identical to $x = y^n$ when seen from the y-axis. Now when the area under $y = x^n$ on the interval $(0, 1)$ was added to the area under $x = y^n$, taken on the same interval on the y-axis, the result was a square of area 1. But since Wallis had already found the area under $y = x^n$ to be $1/(n+1)$, the area under $y = x^{1/n}$ turned out to be

$$1 - \frac{1}{n+1} = \frac{n}{n+1} = \frac{1}{(1/n)+1}. \tag{3.2}$$

Wallis then jumped to the conclusion that the area under $y = x^{m/n}$ over the unit interval, where m and n were positive integers, was

$$\frac{1}{(m/n)+1}. \tag{3.3}$$

In the *Arithmetica*, Wallis's aim was to obtain the arithmetical quadrature of the circle. In modern term this means that he wished to evaluate the integral $\int_0^1 (1-x^2)^{1/2} dx$

using numerical calculations. Since (3.3) gave the value of $\int_0^1 x^{m/n} dx$, Wallis's plan of attack was to compute $\int_0^1 (1 - x^{1/p})^q dx$ for positive integer values of p and q and then interpolate the values of the integral for fractional p and q. The area of the quarter circle was obtained when $p = q = 1/2$. To compute $\int_0^1 (1 - x^{1/p})^q dx$ for integer q, Wallis expanded the integrand and integrated term by term. For example, for $q = 3$ one has (in modern notation)

$$\int_0^1 (1 - x^{1/p})^3 dx = \int_0^1 (1 - 3x^{1/p} + 3x^{2/p} - x^{3/p}) dx = 1 - \frac{3}{\frac{1}{p}+1} + \frac{3}{\frac{2}{p}+1} - \frac{1}{\frac{3}{p}+1}.$$

In proposition 131, he tabulated thirty-six values of these integrals (or areas) for $1 \leq p, q \leq 6$, of which we present the reciprocals:

$$\begin{array}{cccccc} 2 & 3 & 4 & 5 & 6 & 7 \\ 3 & 6 & 10 & 15 & 21 & 28 \\ 4 & 10 & 20 & 35 & 56 & 84 \\ 5 & 15 & 35 & 70 & 126 & 210 \\ 6 & 21 & 56 & 126 & 252 & 462 \\ 7 & 28 & 84 & 210 & 462 & 924 \end{array}.$$

Here the rows are given by p and the columns by q. Wallis observed that these were figurate numbers. For example, the second row/column consisted of triangular numbers, the third row/column of pyramidal numbers, and so on. It was already known (though Wallis may have rediscovered this) that these numbers could be expressed as ratios of two products. Thus the numbers in the pth row were given by

$$\frac{(q+1)(q+2)\cdots(q+p)}{p!}.$$

Therefore, if

$$w(p, q) = \frac{1}{\int_0^1 (1 - x^{1/p})^q dx},$$

then Wallis had

$$w(p, q) = \frac{(q+1)(q+2)\cdots(q+p)}{p!}. \tag{3.4}$$

Wallis then assumed that the formula continued to hold when q was a half integer. Of course, p could not be taken to be a half integer since neither the denominator nor the numerator would have meaning in that case. However, for $p = 1/2$ the integral would be $\int_0^1 (1 - x^2)^q dx$; this could be easily computed when q was an integer. So Wallis had a row corresponding to $p = 1/2$, and he got the values of $w(\frac{1}{2}, q)$ for $q = 0, 1, 2, 3, \ldots$ as

$$1, \frac{3}{2} = \frac{\frac{1}{2}+1}{1!}, \frac{15}{8} = \frac{(\frac{1}{2}+1)(\frac{1}{2}+2)}{2!}, \frac{105}{48} = \frac{(\frac{1}{2}+1)(\frac{1}{2}+2)(\frac{1}{2}+3)}{3!}, \ldots.$$

3.1 Preliminary Remarks

To find $w(\frac{1}{2}, q)$ when q was a half integer, he observed that (3.4) implied (in our notation)

$$w(p, q+1) = w(p, q) \frac{p+q+1}{q+1}. \tag{3.5}$$

From this relation, he could get the value of $w(\frac{1}{2}, \frac{1}{2}+n)$, for integer n, in terms of $w(\frac{1}{2}, \frac{1}{2})$. So if A denoted $w(\frac{1}{2}, \frac{1}{2}) = 4/\pi$, then the row corresponding to $p = 1/2$ and $q = -\frac{1}{2}, 0, \frac{1}{2}, 1, \frac{3}{2}, 2, \frac{5}{2}, \cdots$ would be

$$\frac{1}{2}A, \ 1, \ A, \ \frac{3}{2}, \ \frac{4}{3}A, \ \frac{3 \times 5}{2 \times 4}, \ \frac{4 \times 6}{3 \times 5}A, \ \frac{3 \times 5 \times 7}{2 \times 4 \times 6}, \cdots. \tag{3.6}$$

Wallis understood the rule for forming the subsequence of the first, third, fifth, ... terms and of the subsequence of the second, fourth, sixth, ... terms, but he was initially unable to see the law for constructing the full sequence. Wallis's research was stalled at this stage in the spring of 1652. He consulted a number of his mathematical friends at Oxford including Christopher Wren, the famous architect, but none could help him. Three years later, he informed Oughtred of the progress he had made and where he was still stymied, ending his letter with the request, "wherein if you can do me the favour to help me out; it will be a very great satisfaction to me, and (if I do not delude myself) of more use than at the first view it may seem to be." Apparently, Oughtred could provide no assistance and eventually in the spring of 1655, Wallis requested help from Brouncker, who sent back an infinite continued fraction to solve the problem. It is likely that Brouncker's solution inspired Wallis to discover his own very different one, though some speculate that Wallis made his discovery independently.

William Brouncker (c. 1620–1684) may have studied at Oxford around 1636, though he told his friend John Aubrey that he was "of no university." However, Brouncker was very proficient in languages as well as mathematics. He did all his surviving mathematical work in association with Wallis, with the exception of his series for ln 2. In addition to the continued fraction for π, he wrote a short piece on the rectification of the semicubical parabola $y = x^{\frac{3}{2}}$, probably after seeing William Neil's work. He also gave a method for solving Fermat's problem of finding integer solutions of $x^2 - Ny^2 = 1$ for a given positive integer N. This solution can also be described in terms of continued fractions, but when Wallis wrote up Brouncker's method, he did not use that form. A letter of 1669 from Collins to James Gregory suggests that Brouncker found the series for $(1-x^2)^{\frac{1}{2}}$ independently of Newton. Indeed, Charles II chose Brouncker as the inaugural President of the Royal Society, a post he held from 1662 to 1677. The Society's *Philosophical Transactions* was founded during his tenure; his proof of the formula

$$\ln 2 = \frac{1}{1 \cdot 2} + \frac{1}{3 \cdot 4} + \frac{1}{5 \cdot 6} + \cdots \tag{3.7}$$

appeared in the April 1668 issue.

Brouncker provided no explanation of how he obtained his very intriguing result on the continued fraction for π and in his book, Wallis presented only a sketch of Brouncker's argument. In the course of this discussion, Wallis included a short account of a few fundamental results on continued fractions, including the recurrence relations satisfied by the numerators and denominators of the successive convergents of a continued fraction. Brouncker's result, as well as Wallis's exposition of it, suggests connections between continued fractions and series, products, integrals, and rational approximations. It is surprising to note that, although Huygens and Cotes gave isolated results, no mathematician before Euler made a systematic study of continued fractions. Wallis's book had a tremendous impact on Euler who, at the age of 22, used it as his starting point for his theory of gamma and beta functions. At about the same time, Euler began his investigations into continued fractions, as indicated by a 1731 letter from Euler to his friend Goldbach. He explained how he had applied continued fractions to solve a Riccati equation. Shortly after that, he began researching the relation between continued fractions and infinite series, infinite products, and integrals. It is a remarkable fact that when Euler chanced upon a mathematical avenue or by-path, such as those suggested by Wallis, he explored it with vigor and almost always found numerous results of interest and value.

3.2 Wallis's Infinite Product for π

Wallis, we may recall, was searching for a rule capable of describing (3.6) in some form. He eventually arrived at the deep insight that the sequence of the reciprocals was logarithmically convex. This allowed him to express the first term of the sequence as an infinite product. To reach his insight, Wallis first denoted the numbers in the sequence (3.6) by the letters α, a, β, b, γ, c, δ, d etc. He observed that the ratios

$$\frac{\beta}{\alpha} = \frac{2}{1}, \frac{b}{a} = \frac{3}{2}, \frac{\gamma}{\beta} = \frac{4}{3}, \frac{c}{b} = \frac{5}{4}, \frac{\delta}{\gamma} = \frac{6}{5}, \frac{d}{c} = \frac{7}{6}, \text{ etc.}$$

were decreasing. He then assumed the same for the ratios a/α, β/a, b/β, γ/b, etc. This meant that $a^2 > \alpha\beta$, $\beta^2 > ab$, $b^2 > \beta\gamma$, and so on. So, if we denote three consecutive members of (3.6) by a_{n-1}, a_n, a_{n+1}, we must have

$$a_n^2 > a_{n-1}a_{n+1}, \ a_0 = \frac{1}{2}A. \tag{3.8}$$

Since this indicates logarithmic concavity, the reciprocals must be logarithmically convex. Wallis wrote down the first few of these inequalities explicitly. Thus, $c^2 > \gamma\delta$ and $\delta^2 > cd$ gave him

$$A < \frac{3 \times 3 \times 5 \times 5}{2 \times 4 \times 4 \times 6}\sqrt{\frac{6}{5}},$$

$$A > \frac{3 \times 3 \times 5 \times 5}{2 \times 4 \times 4 \times 6}\sqrt{\frac{7}{6}}.$$

In general, we can write these inequalities as

$$\frac{3 \times 3 \times \cdots \times 2n-1 \times 2n-1}{2 \times 4 \times \cdots \times 2n-2 \times 2n} \sqrt{\frac{2n+1}{2n}} < A$$
$$< \frac{3 \times 3 \times \cdots \times 2n-1 \times 2n-1}{2 \times 4 \times \cdots \times 2n-2 \times 2n} \sqrt{\frac{2n}{2n-1}}.$$

Studying the pattern evident in only the first few cases of these two inequalities, Wallis concluded that

$$A = \frac{4}{\pi} = \frac{3 \times 3 \times 5 \times 5 \times 7 \times 7 \times \cdots}{2 \times 4 \times 4 \times 6 \times 6 \times 8 \times \cdots}. \tag{3.9}$$

Newton studied Wallis as a student in the winter of 1664–65 and made notes in a notebook now held by the University Library, Cambridge. Here Newton observed that Wallis's proof of (3.9) could be simplified. He noted that the sequence (3.6) was increasing, though he did not explain why. Observe, however, that the terms of the sequence are the reciprocals of the integrals

$$\int_0^1 (1-x^2)^m dx, \ m = -\frac{1}{2}, 0, \frac{1}{2}, 1, \ldots.$$

The integrand decreases as m increases and hence so does the integral. Therefore, we see that

$$\frac{3 \times 5 \times \cdots \times 2n-1}{2 \times 4 \times \cdots \times 2n-2} < \frac{4 \times 6 \times \cdots \times 2n}{3 \times 5 \times \cdots \times 2n-1} A < \frac{3 \times 5 \times \cdots \times 2n-1 \times 2n+1}{2 \times 4 \times \cdots \times 2n-2 \times 2n}.$$

And these two inequalities together imply (3.9). Newton's argument certainly shortened the proof of Wallis. But Wallis's use of (3.8) gave a deep insight into the connection between interpolation of factorials and logarithmic convexity. Note that the inequality (3.8) implies the logarithmic convexity of the sequence $1/a_n$. This was fully understood only in the 1920s, when Bohr and Mollerup showed that logarithmic convexity was one of the defining properties of the gamma function, by which the factorial is interpolated. Thus, as Bourbaki also commented, Wallis's methods are very similar to those used today in the theory of the gamma function. It is possible that by 1890 the Dutch mathematician T. J. Stieltjes had also gained an understanding of the significance of logarithmic convexity as it related to the gamma function.

3.3 Brouncker and Infinite Continued Fractions

Indian mathematicians between 700 and 1500 discussed finite continued fractions. We have noted that it is possible that the Kerala school also had a conception of infinite continued fractions. It appears, however, that the first explicit discussions of infinite continued fractions appeared in the works of two professors of mathematics at the University of Bologna: Rafael Bombelli (1526–1572) and Pietro Antonio Cataldi (1548–1626). In 1572, Bombelli described a method for computing $\sqrt{13}$, amounting to the continued fraction expansion

$$\sqrt{13} = 3 + \frac{4}{6+} \frac{4}{6+} \cdots.$$

Though Cataldi's work appeared later than that of Bombelli, he may fairly be regarded as the first discoverer of infinite continued fractions. He explained how to expand the square root of a number in terms of fractions in such a way as to clearly show that an infinite continued fraction must result. Moreover, he introduced a modern notation for continued fractions, also used by Wallis. Cataldi also gave the recurrence relations satisfied by the successive convergents of the continued fraction representation of a quadratic irrational. Finally, he showed that the convergents were successively larger and smaller than the continued fraction and that they converged to it.

Brouncker was the first British mathematician to work with continued fractions. He applied them to present an ingenious solution to Wallis's longstanding problem of finding the law of formation of the sequence (3.6). He stated that the continued fraction

$$\phi(n) = n + \frac{1^2}{2n+} \frac{3^2}{2n+} \frac{5^2}{2n+} \cdots, \quad n = 0, 1, 2, 3, \ldots \qquad (3.10)$$

had the two properties:

$$\phi(n-1)\phi(n+1) = n^2, \quad n = 0, 1, 2, \ldots \qquad (3.11)$$

$$\text{and} \quad \phi(1) = 4/\pi \equiv A. \qquad (3.12)$$

It follows from these properties that the mth term of the sequence (3.6), starting at 1 rather than at $A/2$, is given by

$$\frac{A}{2} \cdot \frac{2}{\phi(1)} \cdot \frac{4}{\phi(3)} \cdots \frac{2m}{\phi(2m-1)}, \quad m = 1, 2, 3, \ldots. \qquad (3.13)$$

If we take the empty product in (3.13) to be 1, then for $m = -1$, we also get the term $A/2$ in (3.6).

Wallis was able to prove (3.12) from his formula (3.9) combined with (3.11). We note briefly that by (3.11),

$$\phi(1) = \frac{2^2}{\phi(3)} = \frac{2^2}{4^2}\phi(5) = \frac{2^2}{4^2} \cdot \frac{6^2}{\phi(7)} = \frac{2^2 \cdot 6^2 \cdots (4m-2)^2}{4^2 \cdot 8^2 \cdots (4m)^2}\phi(4m+1)$$

$$= \frac{1}{2} \cdot \frac{3^2 \cdot 5^2 \cdots (2m-1)^2}{2 \cdot 4^2 \cdots (2m-1)^2 2m} \cdot \frac{\phi(4m+1)}{2m}. \qquad (3.14)$$

Now by (3.10), $n < \phi(n) < n+1$, and, therefore,

$$\frac{4m+1}{2m} < \frac{\phi(4m+1)}{2m} < \frac{4m+2}{2m}.$$

If we let $m \to \infty$ in (3.14), then these inequalities and Wallis's formula imply that $\phi(1) = 4/\pi$. Wallis did not give a complete proof of (3.11), but one may reconstruct his thought from the arguments he gave. He wrote that Brouncker had noticed that the

3.3 Brouncker and Infinite Continued Fractions

product of two consecutive odd or even numbers was one less than a square, since $(n-1)(n+1) = n^2 - 1$. He then asked by what fraction the factors should be increased so that one obtained n^2 rather than $n^2 - 1$. We may say that he looked for a function $\phi(n)$ such that

$$\phi(n-1)\phi(n+1) = n^2.$$

Since $\phi(n) = n$ gives $n^2 - 1$, we take

$$\phi(n) = n + \frac{\alpha_1}{\phi_1(n)}, \quad (3.15)$$

where α_1 is a constant to be determined. If we substitute this in (3.11), we get

$$-\phi_1(n-1)\phi_1(n+1) + \alpha_1(n+1)\phi_1(n+1) + \alpha_1(n-1)\phi_1(n-1) + \alpha_1^2 = 0. \quad (3.16)$$

The symmetry of (3.11) is preserved if we take $\alpha_1 = 1$ for then (3.16) can be written as

$$(\phi_1(n-1) - (n+1))(\phi_1(n+1) - (n-1)) = n^2. \quad (3.17)$$

Now let

$$\phi_1(n) = 2n + \frac{\alpha_2}{\phi_2(n)}, \quad (3.18)$$

so that (3.17) simplifies to

$$-9\phi_2(n-1)\phi_2(n+1) + \alpha_2(n+3)\phi_2(n+1) + \alpha_2(n-3)\phi_2(n-1) + \alpha_2^2 = 0. \quad (3.19)$$

If we take $\alpha_2 = 3^2$, then we get an equation similar to (3.17):

$$(\phi_2(n-1) - (n+3))(\phi_2(n+1) - (n-3)) = n^2.$$

So set

$$\phi_2(n) = 2n + \frac{\alpha_3}{\phi_3(n)},$$

and it turns out that $\alpha_3 = 5^2$. One can continue in this way to get the continued fraction expansion (3.10).

Wallis's contribution to the theory of continued fractions was to develop the recurrence relations for the convergents of a general continued fraction. Take a continued fraction

$$C = b_0 + \frac{a_1}{b_1+} \frac{a_2}{b_2+} \cdots, \quad (3.20)$$

and set the nth convergent (or approximant) of the continued fraction to be

$$C \equiv \frac{P_n}{Q_n} \equiv b_0 + \frac{a_1}{b_1+} \frac{a_2}{b_2+} \cdots \frac{a_n}{b_n}, \quad n = 1, 2, 3, \ldots \quad (3.21)$$

with $P_0 = b_0$, $P_{-1} = 1$, $Q_0 = 1$, and $Q_{-1} = 0$. Then Wallis's recurrence relations for the numerators and denominators P_n, Q_n of the convergents can be written as

$$P_n = b_n P_{n-1} + a_n P_{n-2}, \tag{3.22}$$

and $\quad Q_n = b_n Q_{n-1} + a_n Q_{n-2}. \tag{3.23}$

Wallis wrote the continued fraction (3.20) with $b_0 = 0$ as

$$\cfrac{a}{\alpha \cfrac{b}{\beta \cfrac{c}{\gamma \cfrac{d}{\sigma \cfrac{e}{\epsilon}}}}} \text{ etc.}$$

and gave the first four convergents. He stated the rules (3.22) and (3.23) in words and showed how it worked by an example. He remarked that these results allowed one to compute the convergents by starting at the beginning of the fraction rather than from the end. The twelfth-century Indian mathematician, Bhaskara, in his *Lilaviti*, also gave the rules (3.22) and (3.23). Since he considered continued fractions of only rational numbers, the value of a_k was 1.

3.4 Stieltjes: Probability Integral

It appears from Stieltjes's papers and from his letters to Hermite that he was quite familiar with Wallis's work. In fact, in 1890, more than two centuries after Wallis, Stieltjes evaluated the probability integral $\int_0^\infty e^{-u^2} du$ by Wallis's method of defining a logarithmically convex sequence whose first term was the integral to be evaluated. Stieltjes's insightful use of Wallis's method showed that he perceived the depth of Wallis's approach. Stieltjes defined the sequence

$$I_n = \int_0^\infty u^n e^{-u^2} du, \quad n = 0, 1, 2, \ldots \tag{3.24}$$

and observed that integration by parts implied

$$I_n = \frac{n-1}{2} I_{n-2}. \tag{3.25}$$

Stieltjes applied (3.25) repeatedly to obtain the formulas

$$I_{2k} = \frac{1 \cdot 3 \cdot 5 \cdots (2k-1)}{2^k} I_0 \tag{3.26}$$

$$I_{2k+1} = \frac{1 \cdot 2 \cdot 3 \cdots k}{2}. \tag{3.27}$$

He obtained convexity by observing that for an arbitrary real number x,

$$I_{n+1} + 2x I_n + x^2 I_{n-1} = \int_0^\infty u^{n-1} (u+x)^2 e^{-u^2} du > 0.$$

3.4 Stieltjes: Probability Integral

Note that this is equivalent to

$$(xI_{n-1} + I_n)^2 > I_n^2 - I_{n-1}I_{n+1},$$

so that Stieltjes could conclude that

$$I_n^2 < I_{n-1}I_{n+1}. \tag{3.28}$$

To see this, simply take $x = -I_n/I_{n-1}$. From (3.25) and (3.28), Stieltjes found

$$I_n^2 < \frac{n}{2}I_{n-1}^2. \tag{3.29}$$

Inequalities (3.28) and (3.29) produced the two inequalities

$$I_{2k}^2 > \frac{2}{2k+1}I_{2k+1}^2, \quad I_{2k}^2 < I_{2k-1}I_{2k+1}.$$

Therefore by (3.27), Stieltjes had

$$I_{2k}^2 > \frac{(1 \cdot 2 \cdot 3 \cdots k)^2}{4k+2}, \quad I_{2k}^2 < \frac{(1 \cdot 2 \cdot 3 \cdots k)^2}{4k},$$

$$\text{or} \quad I_{2k}^2 = \frac{(1 \cdot 2 \cdot 3 \cdots k)^2}{4k+2}(1+\epsilon), \quad 0 < \epsilon < \frac{1}{2k}.$$

At this point, Stieltjes used (3.26) to conclude that

$$2I_0^2 = \frac{(2 \cdot 4 \cdot 6 \cdots 2k)^2}{(1 \cdot 3 \cdot 5 \cdots (2k-1))^2(2k+1)}(1+\epsilon).$$

$$2I_o^2 = \pi/2, \quad I_0 = \sqrt{\pi}/2.$$

Note that Wallis's sequence (3.6) can be written as the reciprocal of

$$I_n = \int_0^1 (1-u^2)^{n/2}\,du,$$

where we start at $n = -1$. Wallis obtained the infinite product for I_1 by assuming convexity, and since I_1 was the area of the quarter circle, he had the infinite product for π. Note that convexity follows immediately by Stieltjes's method.

We observe that the gamma integral

$$I_n = \int_0^\infty u^{x+n-1}e^{-u}\,du \quad \text{where } 0 < x < 1,$$

can be similarly treated. As we shall see in the chapter on the gamma function, in 1922 Bohr and Mollerup were the first to use logarithmic convexity to explicitly work out the derivation of this product representation from the gamma integral. The details of their argument were different from those of Stieltjes. However, it is clear that we can give credit to Stieltjes for providing a rigorous exposition of Wallis's method.

We may also raise the interesting question of whether Wallis was very close to being able to evaluate the probability integral. In 1652, Wallis found that

$$\int_0^1 (1-x^2)^n \, dx = \frac{2 \cdot 4 \cdot 6 \cdots 2n}{3 \cdot 5 \cdot 7 \cdots (2n+1)}.$$

If we set $x = y/\sqrt{n}$, then we have

$$\int_0^{\sqrt{n}} \left(1 - \frac{y^2}{n}\right)^n dy = \frac{2 \cdot 4 \cdot 6 \cdots 2n}{3 \cdot 5 \cdot 7 \cdots (2n+1)} \sqrt{n}.$$

The probability integral follows by letting $n \to \infty$ and applying Wallis's formula. This argument was given by Euler, although he expressed it as the integral

$$\int_0^1 (\ln 1/x)^{-1/2} \, dx.$$

It should be remembered, however, that it was not until the 1690s that Leibniz and Johann Bernoulli developed the calculus of the exponential functions. Moreover, it took another forty years for Euler to find the formula

$$\lim_{n \to \infty} (1 - y^2/n)^n = e^{-y^2},$$

required in the above evaluation. So even though Wallis appears to have been only a change of variables and a limit away from discovering the probability integral, it was a very long step and took mathematicians almost a century to complete it. It is important to keep these things in mind when reading older works.

3.5 Euler: Series and Continued Fractions

Wallis's discussion of Brouncker's continued fractions convinced Euler of their importance in analysis. Quite early in his career, he found a connection with the Riccati equation and saw the necessity of relating continued fractions with series, products, and definite integrals. In this way, Euler succeeded in fleshing out the methods of which Wallis and Brouncker had only given key examples.

Euler presented his general theorems on the conversion of series to continued fractions in such a way that the nth partial sum of the series and the nth convergent of the continued fraction were identical. Euler's first paper on this topic, of 1737, treated this topic somewhat briefly but the second one, of 1739, was more detailed. It explicitly stated the formulas for obtaining the corresponding series starting with a given continued fraction and, conversely, for obtaining the continued fraction from the given series. We present Euler's result for converting a fraction to a series, and the converse, in his own notation, except that he described some steps in words instead of symbolically.

The continued fraction

$$a + \cfrac{\alpha}{b + \cfrac{\beta}{c + \cfrac{\gamma}{d + \cfrac{\sigma}{e + \text{etc.}}}}}$$

is equal to the series

$$a + \frac{\alpha}{1 \cdot b} - \frac{\alpha\beta}{b(bc+\beta)} + \frac{\alpha\beta\gamma}{(bc+\beta)(bcd+\beta d+\gamma b)}$$
$$- \frac{\alpha\beta\gamma\sigma}{(bcd+\beta d+\gamma b)(bcde+\cdots)} + \text{etc.}$$

This is equivalent to

$$b_0 + \frac{a_1}{b_1+} \frac{a_2}{b_2+} \cdots = b_0 + \frac{a_1}{Q_1} - \frac{a_1 a_2}{Q_1 Q_2} + \frac{a_1 a_2 a_3}{Q_2 Q_3} - \frac{a_1 a_2 a_3 a_4}{Q_3 Q_4} + \cdots \qquad (3.30)$$

where Q_n is the denominator of the nth convergent P_n/Q_n of the continued fraction. To prove (3.30), multiply equation (3.22) by Q_{n-1} and equation (3.23) by P_{n-1} and subtract to obtain

$$P_n Q_{n-1} - P_{n-1} Q_n = -a_n(P_{n-1} Q_{n-2} - P_{n-2} Q_{n-1}).$$

Iterate this formula n times and divide the final result by $Q_n Q_{n-1}$ to arrive at

$$\frac{P_n}{Q_n} - \frac{P_{n-1}}{Q_{n-1}} = (-1)^{n-1} \frac{a_1 a_2 \cdots a_n}{Q_{n-1} Q_n}. \qquad (3.31)$$

Observe that because of cancellation

$$\frac{P_n}{Q_n} = \left(\frac{P_n}{Q_n} - \frac{P_{n-1}}{Q_{n-1}}\right) + \left(\frac{P_{n-1}}{Q_{n-1}} - \frac{P_{n-2}}{Q_{n-2}}\right) + \cdots + \left(\frac{P_1}{Q1} - \frac{P_0}{Q_0}\right) + \frac{P_0}{Q_0}.$$

The series on the right is the nth partial sum of the series in (3.30). The result follows by letting $n \to \infty$.

Next suppose we start with the finite series

$$c_1 - c_2 + c_3 - c_4 + \cdots + (-1)^{n-1} c_n. \qquad (3.32)$$

To obtain the corresponding continued fraction, compare this with (3.30) to set

$$c_1 = a_1, \; c_2 = a_1 a_2, \; c_3 = a_1 a_2 a_3, \ldots \quad \text{and} \quad Q_k = 1 \text{ for } k \geq 0.$$

$$\text{Thus,} \quad a_1 = c_1, \; a_2 = c_2/c_1, \; a_3 = c_3/c_2, \ldots,$$

and we have the numerators in the continued fraction. To find the denominators b_k, recall that $Q_k = b_k Q_{k-1} + a_k Q_{k-2}$. Since $Q_k = 1$ for $k \geq 0$ and $a_{-1} = 0$, we have

$$b_k = 1 - a_k, \quad k \geq 2 \text{ and } b_1 = 1. \quad \text{Hence,}$$

$$\sum_{k=1}^{n}(-1)^{k-1} c_k = \frac{c_1}{1+} \frac{c_2/c_1}{1 - c_2/c_1 +} \frac{c_3/c_2}{1 - c_3/c_2 +} \cdots \frac{c_n/c_{n-1}}{1-} \frac{c_n}{c_{n-1}}. \qquad (3.33)$$

The result for the infinite series follows by letting $n \to \infty$. Euler illustrated this with two examples:
$$\ln 2 = 1 - \frac{1}{2} + \frac{1}{3} - \frac{1}{4} + \cdots \quad \text{and}$$
$$\frac{\pi}{4} = 1 - \frac{1}{3} + \frac{1}{5} - \frac{1}{7} + \cdots.$$

Euler used equation (3.33) to convert these series to the continued fractions:

$$\ln 2 = \frac{1}{1+} \frac{1/2}{1-1/2+} \frac{2/3}{1-2/3+} \frac{3/4}{1-3/4+} \cdots = \frac{1}{1+} \frac{2^2}{1+} \frac{3^2}{1+} \cdots, \qquad (3.34)$$

$$\frac{\pi}{4} = \frac{1}{1+} \frac{1/3}{1-1/3+} \frac{3/5}{1-3/5+} \frac{5/7}{1-5/7+} \cdots = \frac{1}{1+} \frac{1}{2+} \frac{3^2}{2+} \frac{5^2}{2+} \cdots. \qquad (3.35)$$

Thus, he derived Brouncker's continued fraction for π. He then noted that the two series were equal to the integrals

$$\int_0^1 \frac{dx}{1+x} \quad \text{and} \quad \int_0^1 \frac{dx}{1+x^2}.$$

More generally, he observed that

$$\int_0^1 \frac{x^{n-1}}{1+x^m} dx = \int_0^1 x^{n-1}(1 - x^m + x^{2m} + \cdots) dx$$
$$= \frac{1}{n} - \frac{1}{m+n} + \frac{1}{2m+n} - \frac{1}{3m+n} + \cdots$$
$$= \frac{1}{n+} \frac{n^2}{m+} \frac{(m+n)^2}{m+} \frac{(2m+n)^2}{m+} \frac{(3m+n)^2}{m+} \cdots. \qquad (3.36)$$

The last step follows from (3.33).

In equation (3.35), Euler derived Brouncker's continued fraction from the Madhava–Leibniz series for $\pi/4$. In spite of repeated attempts, he was unable to obtain this continued fraction from Wallis's product as Brouncker had apparently done. Euler deeply regretted that Brouncker never wrote up his derivation, allowing it to sink into oblivion. This failure, however, did not prevent Euler from discovering several methods of converting infinite products, including the one found by Wallis, into some striking continued fractions.

3.6 Euler: Products and Continued Fractions

In his 1739 paper, "De Fractionibus Continuis, Observationes," Euler investigated the relation of certain infinite products with continued fractions. For example, he considered the infinite product

$$\frac{p(p+2q+r)(p+2r)(p+2q+3r)\cdots}{(p+2q)(p+r)(p+2q+2r)(p+3r)\cdots}. \qquad (3.37)$$

3.6 Euler: Products and Continued Fractions

This and other similar products had earlier appeared prominently in Euler's work on the gamma and beta functions, to be discussed in a later chapter. In the previously-mentioned paper, Euler noted that the product was a ratio of beta integrals:

$$\int_0^1 y^{p+2q-1}(1-y^{2r})^{-1/2}\,dy \Big/ \int_0^1 y^{p-1}(1-y^{2r})^{-1/2}\,dy. \tag{3.38}$$

In an analysis quite similar to one carried out by Wallis, he associated a sequence $A_0, A_1, A_2, \ldots$ (he wrote $A, B, C, \ldots$) with the product. The sequence was defined by the relations

$$A_0 A_1 = p/(p+2q), \quad A_2 A_3 = (p+2r)/(p+2q+2r),$$

$$A_4 A_5 = (p+4r)/(p+2q+4r), \ldots. \tag{3.39}$$

He added the requirement that

$$A_1 A_2 = (p+r)/(p+2q+r), \quad A_3 A_4 = (p+3r)/(p+2q+3r),$$

$$A_5 A_6 = (p+5r)/(p+2q+5r), \ldots. \tag{3.40}$$

Euler desired a continued fraction representation for the infinite product A_0. To eliminate the denominators, Euler defined another sequence $a_0, a_1, a_2, \ldots$ by the relations

$$A_0 = \frac{a_0}{p+2q-r}, \quad A_1 = \frac{a_1}{p+2q}, \quad A_2 = \frac{a_2}{p+2q+r}, \quad A_3 = \frac{a_3}{p+2q+2r}, \ldots, \tag{3.41}$$

so that $a_0 a_1 = (p+2q-r)p, \quad a_1 a_2 = (p+2q)(p+r),$

$$a_2 a_3 = (p+2q+r)(p+2r), \cdots. \tag{3.42}$$

He then set

$$a_0 = m - r + \frac{1}{\alpha_1}, \quad a_1 = m + \frac{1}{\alpha_2},$$

$$a_2 = m + r + \frac{1}{\alpha_3}, \quad a_4 = m + 2r + \frac{1}{\alpha_4}, \cdots$$

to obtain a continued fraction for a_0 or A_0 dependent on m. He then chose special values of m such as $p-r$, $p+q$, $p+2q$ to obtain several interesting continued fractions for A_0.

To simplify the equations satisfied by $\alpha_1, \alpha_2, \alpha_3, \ldots$, Euler set

$$P = p(p+2q-r) - m(m-r) \quad \text{and} \quad Q = 2r(p+q-m).$$

Then he had the relations

$$P\alpha_1\alpha_2 - (m-r)\alpha_1 = m\alpha_2 + 1,$$

$$(P+Q)\alpha_2\alpha_3 - m\alpha_2 = (m+r)\alpha_3 + 1,$$

$$(P+2Q)\alpha_3\alpha_4 - (m+r)\alpha_3 = (m+2r)\alpha_4 + 1,$$

$$(P+3Q)\alpha_4\alpha_5 - (m+2r)\alpha_4 = (m+3r)\alpha_5 + 1,$$

and so on. From these equations, he deduced

$$\alpha_1 = \frac{m\alpha_2 + 1}{P\alpha_2 - (m-r)} = \frac{m}{P} + \frac{p(p+2q-r) : P^2}{-(m-r) : P + \alpha_2} \left(\text{where } x:y = \frac{x}{y}\right),$$

$$\alpha_2 = \frac{(m+r)\alpha_3 + 1}{(P+Q)\alpha_3 - m} = \frac{m+r}{P+Q} + \frac{(p+r)(p+2q) : (P+Q)^2}{-m : (P+Q) + \alpha_3},$$

$$\alpha_3 = \frac{(m+2r)\alpha_4 + 1}{(P+2Q)\alpha_4 - (m+r)} = \frac{m+2r}{P+2Q} + \frac{(p+2r)(p+2q+r) : (P+2Q)^2}{-(m+r) : (P+2Q) + \alpha_4}, \text{ etc.}$$

To write the resulting continued fraction for α_1 in simpler form, Euler set

$$R = p^2 + 2pq - mp - mq + qr \quad \text{and} \quad S = pr + qr - mr$$

and obtained the continued fraction for α_1 as

$$\alpha_1 = \frac{m}{P} + \frac{p(p+2q-r) : P^2}{2rR : P(P+Q)+} \frac{(p+r)(p+2q) : (P+Q)^2}{2r(R+S) : (P+Q)(P+2Q)+} \quad (3.43)$$

$$\frac{(p+2r)(p+2q+r) : (P+2Q)^2}{2r(R+2S) : (P+2Q)(P+3Q)+} \cdots. \quad (3.44)$$

He could therefore write down the continued fraction for a_0 after transforming the denominators in the fractions of α_1:

$$a_0 = m - r + \frac{P}{m+} \frac{p(p+2q-r)(P+Q)}{2rR+}$$
$$+ \frac{(p+r)(p+2q)P(P+2Q)}{2r(R+S)+} \frac{(p+2r)(p+2q+r)(P+Q)(P+3Q)}{2r(R+2S)+} \cdots. \quad (3.45)$$

To derive the continued fractions related to Wallis's product, Euler took $p = 2q = r = 1$. In this case the ratio of the beta integrals was

$$A_0 = a_0 = \frac{\int_0^1 \frac{y\,dy}{\sqrt{1-y^2}}}{\int_0^1 \frac{dy}{\sqrt{1-y^2}}} = \frac{2}{\pi};$$

the values of P, Q, R, S were, respectively, $1 + m - m^2$, $3 - 2m$, $(5 - 3m)/2$, $(3 - 2m)/2$. Thus,

$$a_0 = m - 1 + \frac{1+m-m^2}{m+} \frac{1^2(4-m-m^2)}{5-3m+}$$
$$\frac{2^2(1+m-m^2)(7-3m-m^2)}{8-5m+} \frac{3^2(4-m-m^2)(10-5m-m^2)}{11-7m+} \cdots. \quad (3.46)$$

As special cases he presented the continued fractions

$$\frac{\pi}{2} = \frac{1}{1-}\frac{1}{2+}\frac{1\times 2}{1+}\frac{2\times 3}{1+}\frac{3\times 4}{1+}\cdots = 1+\frac{1}{1+}\frac{1\times 2}{1+}\frac{2\times 3}{1+}\frac{3\times 4}{1+}\cdots \quad (3.47)$$

and $$\frac{\pi}{2} = \frac{1}{1:2+}\frac{1:4}{3:2+}\frac{1^2}{2+}\frac{2^2}{2+}\frac{3^2}{2+}\cdots = 2 - \frac{1}{2+}\frac{1^2}{2+}\frac{2^2}{2+}\frac{3^2}{2+}\cdots. \quad (3.48)$$

Thus, the continued fractions produced by Euler's method for the product for π were different from the continued fraction found by Brouncker. This makes the question of Brouncker's method quite intriguing.

3.7 Euler: Continued Fractions and Integrals

In the later part of his 1739 paper, Euler gave another method for drawing a relationship between definite integrals and continued fractions. His approach was to find functions P and R such that a relation

$$(a+n\alpha)\int_0^1 PR^n\,dx = (b+n\beta)\int_0^1 PR^{n+1}\,dx + (c+n\gamma)\int_0^1 PR^{n+2}\,dx \quad (3.49)$$

would hold for some a, b, c, α, β and γ. Then

$$\frac{\int_0^1 PR^n\,dx}{\int_0^1 PR^{n+1}\,dx} = \frac{b+n\beta}{a+n\alpha} + \frac{c+n\gamma}{a+n\alpha}\cdot\frac{1}{\int_0^1 PR^{n+1}\,dx/\int_0^1 PR^{n+2}\,dx}. \quad (3.50)$$

Thus, Euler obtained the continued fraction

$$\frac{\int_0^1 P\,dx}{\int_0^1 PR\,dx} = \frac{b}{a} + \frac{c:a}{(b+\beta):(a+\alpha)+}\frac{(c+\gamma):(a+\alpha)}{(b+2\beta):(a+2\alpha)+}\frac{(c+2\gamma):(a+2\alpha)}{(b+3\beta):(a+3\alpha)+}\cdots, \quad (3.51)$$

and for the reciprocal with the fractions rationalized he had

$$\frac{\int_0^1 PR\,dx}{\int_0^1 P\,dx} = \frac{a}{b+}\frac{(a+\alpha)c}{b+\beta+}\frac{(a+2\alpha)(c+\gamma)}{b+2\beta+}\frac{(a+3\alpha)(c+2\gamma)}{b+3\beta+}\cdots. \quad (3.52)$$

Euler's method for finding P and R was to consider the relation

$$(a+n\alpha)\int PR^n\,dx + R^{n+1}S = (b+n\beta)\int PR^{n+1}\,dx + (c+n\gamma)\int PR^{n+2}\,dx \quad (3.53)$$

such that $R^{n+1}S$ vanished at $x = 0$ and $x = 1$. Euler took the derivative of this equation and divided by R^n to get

$$(a+n\alpha)P\,dx + R\,dS + (n+1)S\,dR = (b+n\beta)PR\,dx + (c+n\gamma)PR^2\,dx. \quad (3.54)$$

He then argued that since this relation was true for all n, it implied the two relations

$$aP\,dx + R\,dS + S\,dR = bPR\,dx + cPR^2\,dx, \tag{3.55}$$

$$\alpha P\,dx + S\,dR = \beta PR\,dx + \gamma PR^2\,dx. \tag{3.56}$$

Euler solved these equations for $P\,dx$ to get

$$P\,dx = \frac{R\,dS + S\,dR}{bR + cR^2 - a} = \frac{S\,dR}{\beta R + \gamma R^2 - \alpha}, \quad \text{whence} \tag{3.57}$$

$$\frac{dS}{S} = \frac{(b-\beta)R\,dR + (c-\gamma)R^2\,dR - (a-\alpha)\,dR}{\beta R^2 + \gamma R^3 - \alpha R}$$
$$= \frac{(a-\alpha)\,dR}{\alpha R} + \frac{(\alpha b - \beta a)\,dR + (\alpha c - \gamma a)R\,dR}{\alpha(\beta R + \gamma R^2 - \alpha)}. \tag{3.58}$$

To illustrate this method, Euler took the continued fraction

$$r + \frac{fh}{r+} \frac{(f+r)(h+r)}{r+} \frac{(f+2r)(h+2r)}{r+} \cdots, \tag{3.59}$$

and compared it with the general formula

$$\frac{a\int_0^1 P\,dx}{\int_0^1 PR\,dx} = b + \frac{(a+\alpha)c}{b+\beta+} \frac{(a+2\alpha)(c+\gamma)}{b+2\beta+} \frac{(a+3\alpha)(c+2\gamma)}{b+3\beta+} \cdots. \tag{3.60}$$

In this manner, he obtained $b = r$, $\beta = 0$, $\alpha = r$, $\gamma = r$, $a = f - r$, $c = h$. He substituted these values in the differential equation (3.58) for S to obtain

$$\frac{dS}{S} = \frac{rR\,dR + (h-r)R^2\,dR - (f-2r)\,dR}{rR^3 - rR}$$
$$= \frac{(f-2r)\,dR}{rR} + \frac{r\,dR + (h-f+r)R\,dR}{r(R^2-1)}. \tag{3.61}$$

Then after integration

$$\ln S = \frac{f-2r}{r}\ln R + \frac{h-f}{2r}\ln(R+1) + \frac{h-f+2r}{2r}\ln(R-1) + \ln C$$

$$\text{or} \quad S = CR^{(f-2r)/r}(R^2-1)^{(h-f)/2r}(R-1). \tag{3.62}$$

Thus, $R^{n+1}S = CR^{(f+(n-1)r)/r}(R^2-1)^{(h-f)/2r}(R-1)$

$$\text{and} \quad P\,dx = \frac{CR^{(f-2r)/r}(R^2-1)^{(h-f)/2r}\,dR}{r(R+1)}. \tag{3.63}$$

Recall that Euler required that $R^{n+1}S$ vanish for $x = 0$ and $x = 1$. This would happen when $R = x^r$ and $0 < f - 2r < h$ and in that case

$$P\,dx = C_1 \frac{x^{f-r-1}(1-x^{2r})^{(h-f)/2r}}{1+x^r}\,dx, \tag{3.64}$$

for a constant C_1. Thus, Euler was able to express the continued fraction (3.59) as a ratio of two definite integrals. In a similar way, he also considered the more general continued fraction

$$s + \frac{ch}{s+} \frac{(c+r)(h+r)}{s+} \frac{(c+2r)(h+2r)}{s+} \cdots. \tag{3.65}$$

In this case $b = s$ in (3.58), and the values of $a, c, \alpha, \beta, \gamma$ remained the same as in the previous example.

3.8 Sylvester: A Difference Equation and Euler's Continued Fraction

In 1869, J. J. Sylvester rediscovered Euler's formula (3.47) while investigating a difference equation arising out of the successive involutes to a circle. The successive convergents of the continued fraction in Euler's formula produced the partial products of Wallis's infinite product. This infinite product was not useful for deriving approximations of π, but Sylvester showed that its continued fraction could be modified to yield good approximations. The continued fraction representation often provides better approximations than other representations, as was noted by Euler in his first paper on the topic. In his work on successive involutes, Sylvester was led to study the difference equation

$$v_{n+1} - v_{n-1} = \frac{1}{n} v_n. \tag{3.66}$$

He found two sequences as particular solutions of this equation:

$$\beta_1 = 1, \; \beta_{2n} = \beta_{2n+1} = \frac{2 \cdot 4 \cdot 6 \cdots 2n}{1 \cdot 3 \cdot 5 \cdots 2n-1}, \; n = 1, 2, 3, \ldots$$

$$\alpha_{2n-1} = \alpha_{2n} = \frac{3 \cdot 5 \cdot 7 \cdots 2n-1}{2 \cdot 4 \cdot 6 \cdots 2n-2}, \quad n = 1, 2, 3, \ldots.$$

From Wallis's formula for π, Sylvester concluded that

$$\frac{\pi}{2} = \lim_{n \to \infty} \frac{\beta_n}{\alpha_n}.$$

From equations (3.22), (3.23), and (3.66), we see that β_n/α_n is the nth convergent of a continued fraction

$$b_0 + \frac{a_1}{b_1+} \frac{a_2}{b_2+} \frac{a_3}{b_3+} \cdots,$$

where $a_n = 1$ and $b_n = 1/n$ for $n \geq 1$. Thus, Sylvester could write that

$$\frac{\pi}{2} = 1 + \frac{1}{1+} \frac{1}{2^{-1}+} \frac{1}{3^{-1}+} \cdots = 1 + \frac{1}{1+} \frac{2}{1+} \frac{6}{1+} \frac{12}{1+} \frac{20}{1+} \cdots. \tag{3.67}$$

To give a sense of Sylvester's inimitable style, we quote the sentence immediately following the last continued fraction, from Sylvester's paper: "This is obviously the simplest form of continued fraction for π that can be given, and yet, strange to say, has not, I believe, before been observed. Truly wonders never cease!" Though Sylvester's result was of course not new, his method was original, and he also explained that this continued fraction could be used to improve the approximation obtained from Wallis's product or, equivalently, from the Madhava–Leibniz series.

Note that if u_n is the remainder after n terms of the fraction, then

$$u_n = \frac{n(n+1)}{1+} \frac{(n+1)(n+2)}{1+} \cdots \qquad (3.68)$$

$$\text{and } u_n = \frac{n^2+n}{1+u_{n+1}}. \qquad (3.69)$$

This shows that u_n is unbounded as $n \to \infty$, and hence $u_n u_{n+1} \approx n^2 + n$ and $u_n \approx n$ for large n. Thus, for large n we may write, following Sylvester,

$$\frac{\pi}{2} = 1 + \frac{1}{1+} \frac{2}{1+} \frac{6}{1+} \cdots \frac{n(n-1)}{1+n}. \qquad (3.70)$$

This correction by n at the end of the formula improves the nth approximant obtained from the continued fraction. Thus, Sylvester noted that the convergents for $n = 4$ and $n = 5$ were

$$\frac{64}{45} = 1.4222 \quad \text{and} \quad \frac{384}{225} = 1.7056,$$

while the corrected values given by (3.70) were

$$\frac{128}{81} = 1.5802 \quad \text{and} \quad \frac{352}{225} = 1.5644.$$

For comparison, note that the actual value of $\pi/2$ to four decimal places is 1.5708; thus, the continued fraction has an advantage over the Wallis product.

3.9 Euler: Riccati's Equation and Continued Fractions

Euler found continued fractions for e, its square and cube roots, and other related numbers. In his first paper on the topic, "De Fractionibus Continuis Dissertatio" written in 1737 and published in 1744, he explained that he had initially found these expansions by studying the patterns in the continued fractions for the rational approximations of these numbers. It was only later that he attempted to prove the results. In the process, he discovered a connection with the Riccati equation and he employed this to establish his formulas. It is interesting that Euler gave the main theorem of this paper in a 1731 letter to Goldbach. For e he had the expansion

$$e = 2 + \frac{1}{1+} \frac{1}{2+} \frac{1}{1+} \frac{1}{1+} \frac{1}{4+} \frac{1}{1+} \frac{1}{1+} \frac{1}{6+} \cdots. \qquad (3.71)$$

This was obtained by taking the approximation $e = 2.71828182845904$ and applying the division algorithm. Cotes had earlier given this expansion by applying the same

3.9 Euler: Riccati's Equation and Continued Fractions

procedure. Similarly, Euler took $\sqrt{e} = 1.6487212707$ and found

$$\sqrt{e} = 1 + \cfrac{1}{1+} \cfrac{1}{1+} \cfrac{1}{1+} \cfrac{1}{5+} \cfrac{1}{1+} \cfrac{1}{1+} \cfrac{1}{9+} \cfrac{1}{1+} \cfrac{1}{1+} \cfrac{1}{13+} \cdots. \qquad (3.72)$$

Then again

$$\frac{e^{\frac{1}{3}} - 1}{2} = 0.1978062125 = \cfrac{1}{5+} \cfrac{1}{18+} \cfrac{1}{30+} \cfrac{1}{42+} \cfrac{1}{54+} \cdots \qquad (3.73)$$

and $\quad \dfrac{e+1}{e-1} = 2 + \cfrac{1}{6+} \cfrac{1}{10+} \cfrac{1}{14+} \cfrac{1}{18+} \cfrac{1}{22+} \cfrac{1}{26+} \cdots. \qquad (3.74)$

He observed that in (3.71) and (3.72), the arithmetic progressions of the denominators $2, 4, 6, \ldots$ and $1, 5, 9, 13, \ldots$ were interrupted by consecutive 1's, whereas in (3.73) and (3.74) they were not. He showed how to convert the interrupted progressions into non-interrupted progressions. When he applied this procedure to (3.71) and (3.72), he got

$$e = 2 + \cfrac{1}{1+} \cfrac{2}{5+} \cfrac{1}{10+} \cfrac{1}{14+} \cfrac{1}{18+} \cfrac{1}{22+} \cfrac{1}{26+} \cdots \qquad (3.75)$$

and $\quad \sqrt{e} = 1 + \cfrac{2}{3+} \cfrac{1}{12+} \cfrac{1}{20+} \cfrac{1}{28+} \cdots. \qquad (3.76)$

Euler then noted that he had not really proved any of these expansions and that it was only probable that the arithmetic progressions continued in the manner indicated. He wrote that after some exertion he had found a rigorous though peculiar proof that related the problem to differential equations. He stated without proof the theorem that if

$$q = \cfrac{1}{p+} \cfrac{1}{3/p+} \cfrac{1}{5/p+} \cdots \cfrac{1}{(2n-1)/p+} \cfrac{1}{1/x^{2n/(2n+1)}y}, \qquad (3.77)$$

where $p = (2n+1)x^{1/(2n+1)}$, then y satisfied the differential equation

$$dy + y^2 dx = x^{-4n/(2n+1)} dx. \qquad (3.78)$$

Euler's expression for q also contained a parameter a but this can be taken to be equal to 1 without loss of generality.

It is possible to give an inductive proof of this theorem and it is very likely that Euler had discovered that argument. Note that when $n = 0$ and when $n = 1$, (3.77) takes the form

$$q = y \quad \text{and} \quad q = \cfrac{1}{p+} \cfrac{1}{1/x^{2/3}y}. \qquad (3.79)$$

The corresponding differential equations would be

$$dy + y^2 dx = dx \qquad (3.80)$$

$$dy + y^2 dx = x^{-4/3} dx. \qquad (3.81)$$

In this way, the solution of (3.81) required the solution of (3.80). More generally, the solution of the Riccati equation (3.78) depended on that of (3.80). However, Euler easily solved (3.80) by observing that it was equivalent to

$$\frac{dy}{1-y^2} = dx \text{ or } \frac{1}{2}\ln\frac{1+y}{1-y} = x.$$

Since $x = p$ and $y = q$ for $n = 0$, Euler wrote the solution as

$$q = \frac{e^{2p}+1}{e^{2p}-1}, \tag{3.82}$$

or, in modern terms, $q = \coth p$. Euler observed that when n was an infinite number in (3.77), then

$$q = \frac{1}{p+} \frac{1}{3/p+} \frac{1}{5/p+} \frac{1}{7/p+} \cdots. \tag{3.83}$$

This result is now called Lambert's continued fraction, although Euler found it earlier. Now, since $e^{2p} = 1 + \frac{2}{q-1}$, Euler saw that

$$e^{2p} = 1 + \frac{2}{(1-p)/p+} \frac{1}{3/p+} \frac{1}{5/p+} \frac{1}{7/p+} \cdots,$$

$$\text{or} \quad e^{1/s} = 1 + \frac{2}{2s-1+} \frac{1}{6s+} \frac{1}{10s+} \frac{1}{14s+} \cdots. \tag{3.84}$$

He then noted that (3.84) would in fact produce all those continued fractions he had obtained experimentally by using rational approximations.

3.10 Exercises

1. Evaluate (3.65) by Euler's method for the continued fraction (3.59) and find the values of S and P corresponding to equations (3.62) and (3.63). See Eu. I-14, pp. 339–340.

2. Prove that

$$s + \frac{1}{s+} \frac{4}{s+} \frac{9}{s+} \frac{16}{s+} \cdots = \left(2\int_0^1 \frac{x^s dx}{1+x^2}\right)^{-1}.$$

See Eu. I-14, pp. 292–297.

3. Prove that

$$\frac{\int_0^\infty \sinh^{\beta-1} u \cosh^{-\alpha} u e^{-xu}\, du}{(\beta-1)\int_0^\infty \sinh^{\beta-2} u \cosh^{1-\alpha} u e^{-xu}\, du}$$

$$= \frac{1}{x+} \frac{\alpha\beta}{x+} \frac{(\alpha+1)(\beta+1)}{x+} \frac{(\alpha+2)(\beta+2)}{x+} \cdots.$$

See Stieltjes (1993), vol. II, p. 391.

4. From Stieltjes's formula in the previous exercise, deduce that

$$\frac{\Gamma\left(x-\tfrac{1}{2}a+\tfrac{1}{4}\right)\Gamma\left(x+\tfrac{1}{2}a+\tfrac{3}{4}\right)}{\Gamma\left(x+\tfrac{1}{2}a+\tfrac{1}{4}\right)\Gamma\left(x-\tfrac{1}{2}a+\tfrac{3}{4}\right)} = 1 + \frac{2a}{4x-a+} \frac{1^2-a^2}{4x+} \frac{2^2-a^2}{4x+} \frac{3^2-a^2}{4x+} \cdots,$$

$$\frac{\Gamma\left(x-\tfrac{1}{2}a+\tfrac{1}{4}\right)\Gamma\left(x+\tfrac{1}{2}a+\tfrac{1}{4}\right)}{\Gamma\left(x-\tfrac{1}{2}a+\tfrac{3}{4}\right)\Gamma\left(x+\tfrac{1}{2}a+\tfrac{3}{4}\right)} = \frac{4}{4x+} \frac{1^2-4a^2}{8x+} \frac{3^2-4a^2}{8x+} \frac{5^2-4a^2}{8x+} \frac{7^2-4a^2}{8x+} \cdots.$$

Also show that if

$$\frac{1 \cdot 3 \cdot 5 \cdots (2n-1)}{2 \cdot 4 \cdot 6 \cdots 2n} = \frac{1}{\sqrt{(\pi(n+\epsilon))}},$$

then $\phi(n) = 1 + \dfrac{2}{8n-1+} \dfrac{1 \cdot 3}{8n+} \dfrac{3 \cdot 5}{8n+} \dfrac{5 \cdot 7}{8n+} \dfrac{7 \cdot 9}{8n+} \cdots.$

See Stieltjes (1993), vol. II, pp. 396–398.

5. Show that

$$a + \frac{1}{m+} \frac{1}{n+} \frac{1}{b+} \frac{1}{m+} \frac{1}{n+} \frac{1}{c+} \frac{1}{m+} \frac{1}{n+} \frac{1}{d+} \cdots$$

$$= \frac{1}{mn+1}\left((mn+1)a + n + \frac{1}{(mn+1)b+m+n+} \frac{1}{(mn+1)c+m+n+} \cdots\right).$$

See Euler (1985), p. 313, and Eu. I-14, p. 205.

6. Show that

$$a + \frac{1}{m+} \frac{1}{n+} \frac{1}{p+} \frac{1}{q+} \frac{1}{b+} \frac{1}{m+} \cdots$$

$$= \frac{1}{p}\left(P + npq + n + q + \frac{1}{Pb+Q+} \frac{1}{Pc+Q+} \frac{1}{Pd+q+} \cdots\right),$$

where $P = mnpq + mn + mq + pq$ and $Q = mnp + npq + m + n + p + q$. See Euler (1985), p. 318, and Eu. I-14, p. 208.

7. Show that

$$\sqrt{2} = 1 + \frac{1}{2+} \frac{1}{2+} \frac{1}{2+} \cdots$$

$$\sqrt{3} = 1 + \frac{1}{1+} \frac{1}{2+} \frac{1}{1+} \frac{1}{2+} \cdots.$$

See Euler (1985), pp. 307–308 and Eu. I-14, p. 200.

8. Show that if $x = a + \frac{1}{b+} \frac{1}{b+} \cdots$, then $x = a - \frac{b}{2} + \sqrt{1 + \frac{b^2}{4}}$. See Euler (1985), p. 308, and Eu. I-14, p. 201.

9. Show that if $x = a + \frac{1}{a_1+} \frac{1}{a_2+} \cdots \frac{1}{a_n+} \frac{1}{a_1+} \cdots$, that is, if x is a periodic continued fraction, then x satisfies a quadratic equation. See Eu. I-14, p. 203. Euler stated the result in words, as opposed to symbolically.

10. Show that

$$\tan \frac{\pi x}{4} = \frac{x}{1+} \frac{1-x^2}{2+} \frac{3-x^2}{2+} \frac{5-x^2}{2+} \cdots.$$

See Stieltjes's March 4, 1891 letter to Hermite in Baillaud and Bourget (1905), p. 157.

3.11 Notes on the Literature

The introduction to Stedall's excellent English translation of Wallis's 1656 *Arithmetica Infinitorum*, Wallis (2004), discusses the evolution of Wallis's ideas and the influence of his book on his contemporaries and mathematical heirs. The two quotes attributed to Wallis were taken from Stedall's introduction. The article by Stedall in Grattan-Guinness (2005) may also be helpful, especially for its insight into how Wallis's work influenced Newton. The fruitful collaboration of Wallis and Brouncker is also the subject of two interesting notes by Stedall (2000). Wallis's letter of Feb. 28, 1655, can be found in Rigaud (1841), pp. 85–86.

To read Newton's simplification of Wallis's proof of the infinite product for π, see Newton (1967–1981), vol. I, p. 103, giving Newton's annotations of Wallis. Newton first summarized Wallis and then gave his own modifications: "Thus Wallis doth it, but it may bee [sic] done thus." In his Cambridge thesis, Whiteside (1961) reconstructed Wallis's attempt to recreate the continued fraction formula communicated to him without proof by Brouncker. This thesis is an informative and perceptive resource on seventeenth-century mathematics. Bourbaki (1994), p. 187 presents an enlightening commentary on Wallis's interpolation results. Brezinski (1991) is a very useful book; it contains excerpts from original works accompanied by interesting historical commentary.

The two papers by Euler, "De Fractionibus Continuis, Dissertatio," and "De Fractionibus Continuis, Observationes," can be found in Eu. I-14, pp. 187–216 and 291–349. Surprisingly, a translation into English of the "Dissertatio" appeared in the applied mathematics journal *Mathematical Systems Theory* (1985); the editors requested this translation, since they thought Euler's discussion of Riccati's equation could be useful to their readers. Khrushchev (2008) contains an English translation of Euler's "Observationes." Khrushchev gives a systematic and well-organized summary of the work on continued fractions by Wallis, Brouncker, Huygens, the Bernoullis, Euler, Lagrange, Gauss, Chebyshev, Stieltjes, and others. Khrushchev illustrates the process by which the ideas of earlier researchers in continued fractions have evolved into important modern theories, such as that of orthogonal polynomials. Euler's 1731 letter to Goldbach can be found in Fuss (1968), pp. 57–59.

The results of Stieltjes discussed in this chapter appear on pp. 267–268 of Stieltjes (1993), vol. II. And Sylvester's paper on Euler's continued fraction can be seen on pp. 691–693 of vol. II of Sylvester (1973). Papers leading up to this, on the successive involutes to a circle, are also contained in the same volume. One of these papers takes the naturalist Thomas Huxley to task for claiming that "mathematics is that study which knows nothing of observation, nothing of experiment, nothing of induction, nothing of causation." Sylvester gives a list of illustrious mathematicians who would have to disagree with Huxley and explains that his paper on the successive involutes to a circle in fact "owes its origin" to a practical problem of self-reversing military gun carriages.

4

The Binomial Theorem

4.1 Preliminary Remarks

The discovery of the binomial theorem for general exponents exerted a tremendous impact on the development of analysis, especially the theory of power series. It also led to an understanding that an exponential function was defined by the property $f(a+b) = f(a)f(b)$. The binomial theorem was pivotal not only in the initial discovery of series for other important functions but also in the eventual consolidation of the foundations of analysis as a whole. The development of the theorem is particularly fascinating because it was independently found by both Newton and Gregory; because of the various approaches to its proof, including one by Euler; and because the validation of these proofs elicited the efforts of the best mathematicians of the nineteenth century.

Islamic mathematicians were the original discoverers of the binomial theorem for positive integral exponents, although they did not have the notation to write the expansion for arbitrary integers. But they knew how to find the coefficients for any given integral exponent. The two important rules for binomial coefficients appear in the work of al-Kashi of around 1427, and it is likely that earlier Islamic mathematicians such as al-Tusi, Omar Khayyam, and al-Karji were also aware of them: Let $C_{n,k}$ denote the coefficient of x^k in the expansion of $(1+x)^n$. Then

$$C_{n,k} = C_{n-1,k} + C_{n-1,k-1} \text{ and } C_{n,k} = \frac{n(n-1)\cdots(n-k+1)}{k!}.$$

The first formula, the additive rule for binomial coefficients, leads to the expansion of $(1+x)^n$, by means of the expansion of $(1+x)^{n-1}$; the second formula, the multiplicative rule, immediately yields the expansion of $(1+x)^n$. Henry Briggs (1561–1630) appears to be the first European to explicitly state both formulas, though Cardano may have known the results around 1570. In 1654, Pascal gave a proof by complete induction of the second formula.

Newton discovered the general binomial theorem in the winter of 1664–65, while he was still a student at Cambridge. He was motivated by this discovery to develop his "method of infinite series" and apply it to several important problems. Indeed, the binomial theorem played a basic role in his approach to such topics as algebraic equations in two variables and differential equations. James Gregory independently found this theorem between 1668 and 1670, and it formed an important part of his original work on infinite series.

Newton discussed particular cases of his theorem in two papers written in 1669 and 1671. However, the first explicit statement of the general theorem for rational exponents appeared in a June 13, 1676 letter from Newton to Oldenburg. This letter was a response to an inquiry from Leibniz, who had learned of Newton's series for $\arcsin x$ and $\sin x$ from the Danish mathematician Georg Mohr. Newton's letter also introduced his new notation for exponents, as he explained:

> These are the foundation of these reductions: but extractions of roots are much shortened by this theorem,
>
> $$(P+PQ)^{m/n} =$$
>
> $$P^{m/n} + \frac{m}{n}AQ + \frac{m-n}{2n}BQ + \frac{m-2n}{3n}CQ + \frac{m-3n}{4n}DQ + \text{etc.}$$
>
> where $P + PQ$ signifies the quantity whose root or even any power, or the root of a power, is to be found: P signifies the first term of that quantity, Q the remaining terms divided by the first, and m/n the numerical index of the power of $P + PQ$, whether that power is integral or (so to speak) fractional, whether positive or negative. For as analysts, instead of $aa, aaa,$, etc., are accustomed to write a^2, a^3, etc., so instead of $\sqrt{a}, \sqrt{a^3}, \sqrt{c}: a^5$, etc. I write $a^{\frac{1}{2}}, a^{\frac{3}{2}}, a^{\frac{5}{3}}$, and instead of $1/a, 1/aa, 1/a^3$, I write a^{-1}, a^{-2}, a^{-3}. And so for
>
> $$\frac{aa}{\sqrt{c}:(a^3+bbx)}$$
>
> I write $aa(a^3+bbx)^{-\frac{1}{3}}$, and for $\dfrac{aab}{\sqrt{c}:(a^3+bbx)(a^3+bbx)}$ I write $aab(a^3+bbx)^{-\frac{2}{3}}\cdots$.

In Newton's formula, A denotes the first term, B the second term, and so on, such notation being common at that time. Note also that $\sqrt{c}:x$ stands for the cube root of x.

Intrigued by Newton's groundbreaking work, Leibniz responded with some of his own discoveries on series and requested details about the origin and derivation of Newton's results, especially the binomial theorem. Newton wrote a lengthy reply amounting to nineteen printed pages in his October 24, 1676 letter, again through Oldenburg. Newton explained that in 1664–65, he was inspired by Wallis's *Arithmetica Infinitorum* to consider the integral $\int_0^x (1-t^2)^{1/2}\,dt$ and to expand the integrand. He looked at the absolute values of the coefficients of the polynomials

$$(1-x^2)^0 = 1,\ (1-x^2)^1 = 1-x^2,\ (1-x^2)^2 = 1-2x^2+x^4,$$
$$(1-x^2)^3 = 1-3x^2+3x^4-x^6,\ (1-x^2)^4 = 1-4x^2+6x^4-4x^6+x^8,\ldots$$

4.1 Preliminary Remarks

and asked how the (absolute) values of the first two coefficients of any of these polynomials could produce the remaining coefficients.

> I found that on putting m for the second figure [coefficient], the rest could be produced by a continual multiplication of the terms of this series,
>
> $$\frac{m-0}{1} \times \frac{m-1}{2} \times \frac{m-2}{3} \times \frac{m-3}{4} \times \frac{m-4}{5}, \text{ etc.}$$

For example, let $m = 4$, and $4 \times \frac{1}{2}(m-1)$, that is 6 will be the third term, and $6 \times \frac{1}{3}(m-2)$, that is 4 the fourth, and $4 \times \frac{1}{4}(m-3)$, that is 1 the fifth, and $1 \times \frac{1}{5}(m-4)$, that is 0 is the sixth, at which term in this case the series stops. According, ..., for the circle, ..., I put $m = 1/2$ and the terms arising were

$$\frac{1}{2} \times \frac{\frac{1}{2}-1}{2} \text{ or } -\frac{1}{8}, \quad -\frac{1}{8} \times \frac{\frac{1}{2}-2}{3} \text{ or } +\frac{1}{16}, \quad \frac{1}{16} \times \frac{\frac{1}{2}-3}{4} \text{ or } -\frac{5}{128},$$

and so to infinity.

Thus, Newton learned how to generate the binomial series when the exponent was any number m and, by taking $m = 1/2$, he obtained the expansion

$$(1-x^2)^{1/2} = 1 - \frac{1}{2}x^2 - \frac{1}{8}x^4 - \frac{1}{16}x^6 - \frac{5}{128}x^8 - \cdots$$

from which he derived the value of the integral $\int_0^x (1-t^2)^{1/2} \, dt$ as an infinite series.

It is curious that Newton was unaware of the work of Briggs, Pascal, and others on the multiplicative formula for binomial coefficients. It seems that the mathematical texts Newton studied as a student did not contain the multiplicative formula. In fact, Wallis wrote in 1685 that he had not known this formula when he wrote his *Artithmetica Infinitorum*. This is surprising because this work included the multiplicative expression for figurate numbers, intimately connected with binomial coefficients. In any case, Wallis's book was apparently sufficiently suggestive for Newton to make his discovery about $C_{n,k}$ for integral n and then extend it to fractional n by following Wallis once again. Newton attempted to verify his theorem by the interpolation methods he had learned from Wallis but he soon found more satisfactory techniques, described in his October 24, 1676 letter.

> For in order to test these processes, I multiplied
>
> $$1 - \frac{1}{2}x^2 - \frac{1}{8}x^4 - \frac{1}{16}x^6, \text{ etc.} \qquad (4.1)$$
>
> into itself; and it became $1 - x^2$, the remaining terms vanishing by the continuation of the series to infinity. And even so $1 - \frac{1}{3}x^2 - \frac{1}{9}x^4 - \frac{5}{81}x^6$, etc. multiplied twice into itself also produced $1 - x^2$. And as this was not only sure proof of these conclusions so too it guided me to try whether, conversely, these series, which it thus affirmed to be roots of the quantity $1 - x^2$, might not be extracted out of it in an arithmetical manner. And the matter turned out well. This was the form of

the working in square roots.

$$1 - x^2(1 - \frac{1}{2}x^2 - \frac{1}{8}x^4 - \frac{1}{16}x^6, \text{ etc.}$$

$$\begin{array}{r} 1 \\ \hline 0 - x^2 \\ -x^2 + \frac{1}{4}x^4 \\ \hline -\frac{1}{4}x^4 \\ -\frac{1}{4}x^4 + \frac{1}{8}x^6 + \frac{1}{64}x^8 \\ \hline 0 \quad -\frac{1}{8}x^6 - \frac{1}{64}x^8. \end{array}$$

After getting this clear I have quite given up the interpolation of series, and have made use of these operations only, as giving more natural foundations.

Newton realized that all the algebraic operations could be applied to infinite series and that series could be viewed as the algebraic analogs of infinite decimals. Just as the latter appear when division and root extraction are performed on integers, infinite series result when these operations are performed on polynomials. In the preceding quote, Newton explained that when he applied the square root algorithm, the result was the series for $(1 - x^2)^{1/2}$. For division, Newton gave the example of the geometric series

$$\frac{1}{d+e} = \frac{1}{d} - \frac{e}{d^2} + \frac{e^2}{d^3} - \frac{e^3}{d^4} + \text{ etc.} \tag{4.2}$$

In searching for the proof of the binomial theorem, Newton looked no further than a few cases, and he verified these by multiplication. We shall see that this method is the basis for one proof of the binomial theorem, due to Euler.

James Gregory first revealed his discovery of the binomial theorem in a letter to his longtime correspondent, John Collins. First, on March 24, 1670, Collins wrote to Gregory, mentioning some mysterious work done by Newton: "Mr Newtone of Cambridge sent the following series for finding the Area of a Zone of a Circle to Mr. Dary, to compare with the said Dary's approaches, putting R the radius and B the parallell [sic] distance of a Chord from the Diameter the Area of the space or Zone betweene [sic] them is $= 2RB - \frac{B^3}{3R} - \frac{B^5}{20R^3} - \frac{B^7}{56R^5} - \frac{5B^9}{576R^7}$." This area is given by the integral $2\int_0^B \sqrt{R^2 - x^2}\, dx$. We note that Newton obtained the series by expanding the integrand as a binomial series and then doing term-by-term integration.

Gregory then formulated the binomial theorem in a November 30, 1670 letter to Collins, stated as the solution of a problem: Use the numbers $b, b+d$ and the values of their logarithms, e and $e + c$, respectively, to find the number whose logarithm is $e + a$. Gregory wrote that the desired number was given by the series

$$b + \frac{a}{c}d + \frac{a(a-c)}{c \cdot 2c}\frac{d^2}{b} + \frac{a(a-c)(a-2c)}{c \cdot 2c \cdot 3c}\frac{d^3}{b^2} + \text{ etc.} \tag{4.3}$$

4.1 Preliminary Remarks

Since

$$\ln(b(1+d/b)^{a/c}) = \ln b + (a/c)(\ln(b+d) - \ln b) \quad (4.4)$$
$$= e + (a/c)(e+c-e) = e+a,$$

we see that the series is the binomial expansion of $b\left(a + \frac{d}{b}\right)^{a/c}$. In this letter, Gregory also stated his general interpolation formula; the manner in which he stated this formula and the binomial theorem suggests that the latter was derived from the former.

In spite of the fact that he had already found the binomial theorem, it was not until December 1670 that Gregory could perceive the origin of Newton's series. Gregory explained in a letter of December 19 to Collins, that he had derived numerous series for the circle and had mistakenly expected Newton's to be a corollary of at least one of them. He added, "I admire much my own dulness [sic], that in such a considerable time, I had not taken notice" that Newton's series followed from a binomial expansion. Note that the interpolation formula can be written as

$$f(x) = f(0) + x\Delta f(0) + \frac{x(x-1)}{2!}\Delta^2 f(0) + \frac{x(x-1)(x-2)}{3!}\Delta^3 f(0) + \cdots.$$

To derive the binomial theorem, take

$$f(x) = b\left(1 + \frac{d}{b}\right)^x, \text{ so that } \Delta f(x) = f(x+1) - f(x) = b\left(1 + \frac{d}{b}\right)^x \frac{d}{b},$$

$$\Delta^2 f(x) = \Delta f(x+1) - \Delta f(x) = b\left(1 + \frac{d}{b}\right)^x \frac{d^2}{b^2}, \quad \Delta^3 f(x) = b\left(1 + \frac{d}{b}\right)^x \frac{d^3}{b^3}, \text{etc.}$$

Thus,

$$\Delta f(0) = d, \quad \Delta^2 f(0) = d^2/b, \quad \Delta^3 f(x) = d^3/b^2 \text{ etc.},$$

and we get Gregory's series (4.3) by taking $x = a/c$ in the interpolation formula.

Interestingly, Gregory's derivation is logically more sound than Newton's original argument by a Wallis interpolation. Yet the two derivations both involve interpolation with respect to the exponent. In spite of their highly imaginative and useful work in this area, both Newton and Gregory failed to give well-founded derivations of the binomial series. Eighteenth-century mathematicians made very interesting attempts to fill this gap, but it took until the nineteenth century to find a completely rigorous derivation.

In the eighteenth century, it was generally known that the binomial expansion for $f(x) = (1+x)^\alpha$ could be obtained as the series solution of the equation

$$(1+x)\frac{dy}{dx} = \alpha y. \quad (4.5)$$

Of course, whether a series could be differentiated term by term and whether a differential equation had a series solution were not then seen as problems. The point that bothered the English mathematician John Landen (1719–1790) was that the proof used derivatives (fluxions) to obtain a result in algebra. In his 1758 *Discourse Concerning the Residual Analysis*, Landen applied the algebraic identity

$$\frac{x^{\frac{m}{n}} - v^{\frac{m}{n}}}{x - v} = x^{\frac{m}{n}-1} \times \frac{1 + q + q^2 + q^3 + \cdots + q^{m-1}}{1 + q^{\frac{m}{n}} + q^{\frac{2m}{n}} + q^{\frac{3m}{n}} + \cdots + q^{(n-1)\frac{m}{n}}}, \qquad (4.6)$$

where $q = v/x$ and m and n were integers, to avoid differentiation.

Euler took a different approach, presenting a proof using Newton's idea of multiplication of series. He showed that if

$$f(m) = 1 + \frac{m}{1}x + \frac{m}{1} \cdot \frac{m-1}{2} x^2 + \text{etc., then}$$

$$f(m+n) = f(m) \cdot f(n). \qquad (4.7)$$

His proof consisted in demonstrating that the coefficients of x^k on both sides of equation (4.7) were the same. This was sufficient to derive the binomial theorem for rational exponents, except that he did not address convergence questions, particularly in the case of the product of two series. We must note that seventeenth- and eighteenth-century mathematicians had more or less clear ideas of convergence of series, but only occasionally did they apply these ideas to the series arising in their work. As examples of rigor, Grégoire St. Vincent (1584–1667) gave an entirely rigorous treatment of the geometric series in his *Opus Geometricum* of 1647. Twenty years later, Wallis discussed the logarithmic series with a careful analysis of the remainder term, obtainable from the remainder in a geometric series. Then, in the 1680s, Leibniz gave a thorough account of alternating series with decreasing terms. Some gems from the eighteenth century include Stirling's 1717 criterion for convergence based on second differences of the terms of a series (though it required an amendment) and Maclaurin's statement and proof of the integral test in his 1742 *Treatise of Fluxions*. Moreover, d'Alembert made some comments on the convergence of series from which the ratio test can be developed. We mention that Gauss greatly extended this ratio test in his famous work on hypergeometric series.

To extend Euler's proof to all real exponents, it was necessary to give a precise definition of continuity. Bernard Bolzano (1781–1848) and A. L. Cauchy (1789–1856) independently accomplished this. Bolzano was a professor of theology at Prague; his main interests were in philosophy and mathematics. He defined continuity in an 1817 paper on the intermediate value theorem: A function fx varies according to the law of continuity for all values of x inside or outside certain limits if the difference $f(x+w) - f(x)$ can be made smaller than any given quantity, provided w can be taken as small as we please. Bolzano's definition leaves little to be desired.

Cauchy emphasized rigor in analysis from the very beginning of his teaching career at the École Polytechnique. His published lectures from 1821 and 1823 discussed the concepts of limits, continuity, and convergence. His 1821 lectures *Analyse algébrique* gave his definition of continuity, not quite as good as Bolzano's: The function $f(x)$ will be a continuous function of the variable x between two assigned bounds if, for each value of x between these bounds, the numerical value of the differences $f(x+\alpha) - f(x)$ decrease indefinitely with α.

Cauchy derived the continuity of the series $f(m)$ in (4.7) from the erroneous result that if every term of an infinite series is continuous and the series is convergent, then the series is continuous. In fact, in his 1826 paper on the binomial theorem, Abel noted that Cauchy's theorem on the continuity of a series admits of exception. For example,

$$\sin\phi - \frac{1}{2}\sin 2\phi + \frac{1}{3}\sin 3\phi - \cdots \qquad (4.8)$$

was discontinuous for every value $(2m+1)\pi$ of ϕ, where m is a whole number. Abel then proceeded to state and prove his famous continuity theorem for power series, using the method of summation by parts. This method had been known for over a century, but Abel was the first to apply it to problems of convergence of series. Dirichlet profited from Abel's paper and used these ideas very effectively in his study of L-series less than a decade later. Interestingly, Abel gleaned ideas of mathematical rigor from Cauchy's lectures, obtained from his friend Crelle's library; Abel's paper appeared in Crelle's newly founded journal.

The concept of uniform convergence was implied in Abel's continuity theorem, but its explicit formulation came later. First, C. Gudermann observed in an 1838 paper on modular functions, published in Crelle's journal, that he had obtained a certain series having the same convergence rate for all values of the variable. A year later, K. Weierstrass was the only student in Gudermann's course on modular functions. Weierstrass introduced the term uniform convergence, understood its importance, and gave its definition in an 1841 paper "Zur Theorie der Potenzreihen", submitted as part of his examination for teaching certification. Gudermann declared, "The candidate hereby enters by birthright into the ranks of discoverers crowned with glory." Unfortunately, the paper was not published until 1894. During the winter of 1859–60, Weierstrass lectured at the University of Berlin on the foundations of analysis, but it took some time before his ideas spread to other European countries and to America. In a letter of 1881 to his former student Hermann A. Schwarz, Weierstrass observed that people in France were finally grasping the importance of the idea of uniform convergence. Finally, in the last two decades of the nineteenth century, textbooks containing Weierstrassian ideas appeared in several languages. In 1848, the British mathematical physicist G. G. Stokes (1819–1903) and Dirichlet's student P. Seidel (1821–1896), also wrote on concepts related to uniform convergence, though their papers did not have much influence.

Interestingly, Cauchy wrote a paper in 1853 acknowledging his mistake on continuity, noting that it was easy to rectify. He then proceeded to work with uniform convergence without naming the concept, so it is not clear whether he fully realized the wider significance of the idea.

4.2 Landen's Derivation of the Binomial Theorem

In 1758, John Landen presented the standard eighteenth-century derivation of the binomial theorem, in which one assumes the series expansion

$$(1+x)^{m/n} = 1 + ax + bx^2 + cx^3 + dx^4 + \text{etc.} \qquad (4.9)$$

Now take the derivative of each side to get

$$(m/n)(1+x)^{\frac{m}{n}-1} = a + 2bx + 3cx^2 + 4dx^3 + \text{etc.} \tag{4.10}$$

Multiply the last equation by $1+x$ to see that

$$(m/n)(1 + ax + bx^2 + cx^3 + \text{etc.}) = (1+x)(a + 2bx + 3cx^2 + 4dx^3 + \text{etc.}).$$

Equate coefficients to obtain

$$a = \frac{m}{n}, \quad 2b + a = \frac{m}{n}a, \quad 3c + 2b = \frac{m}{n}b, \quad 4d + 3c = \frac{m}{n}c, \ldots \text{ so that}$$

$$b = \left(\frac{m}{n}\left(\frac{m}{n}-1\right)\right)/2!, \quad c = \left(\frac{m}{n}\left(\frac{m}{n}-1\right)\left(\frac{m}{n}-2\right)\right)/3!$$

$$d = \left(\frac{m}{n}\left(\frac{m}{n}-1\right)\left(\frac{m}{n}-2\right)\left(\frac{m}{n}-3\right)\right)/4!, \ldots,$$

proving the binomial theorem. Landen also gave an alternative method, avoiding differentiation, by starting with (4.9) and applying (4.6) to get

$$\frac{(1+x)^{\frac{m}{n}} - (1+y)^{\frac{m}{n}}}{x-y} = (1+x)^{\frac{m}{n}-1} \times \frac{1 + \frac{1+y}{1+x} + \cdots + \left(\frac{1+y}{1+x}\right)^{m-1}}{1 + \left(\frac{1+y}{1+x}\right)^{\frac{m}{n}} + \cdots + \left(\frac{1+y}{1+x}\right)^{(n-1)\frac{m}{n}}}$$

$$= a + b(x+y) + c(x^2 + xy + y^2) + d(x^3 + x^2y + xy^2 + y^3) + \text{etc.}$$

He then observed that the last equation is an algebraic identity true for all values of y and so that he could take $y = x$ to obtain (4.10).

Almost seven decades later, Abel objected to differentiation in this context, not because he perceived the binomial series as algebraic but, as he wrote from Berlin to his friend and former teacher Holmboe, he thought it impermissible to apply operations on infinite series as if they were finite. He noted that it had not been proved that the derivative of an infinite series could be obtained by taking the derivative of each term, and that there were numerous counterexamples. For example, he observed, the sum of the series (4.8) was $\phi/2$ in the interval $-\pi < \phi < \pi$. Taking derivatives gave

$$\frac{1}{2} = \cos\phi - \cos 2\phi + \cos 3\phi - \text{etc.},$$

a clearly false result, because the series was divergent. In 1841, Weierstrass finally addressed Abel's concerns when he developed the theorems for differentiation and integration of series.

4.3 Euler's Proof for Rational Indices

When he presented his 1774 proof of the binomial theorem, Euler explained that to avoid circularity, he wished to give a demonstration not using differentiation, since he had used the binomial theorem in his differential calculus book of 1755 to find the

4.3 Euler's Proof for Rational Indices

derivative of x^n. Moreover, in 1763, the German scientist Franz Aepinus published an inductive proof for positive integral exponents in the Petersburg Academy journal. Euler thought that the argument, while ingenious, was quite obscure. Euler started by observing that since $(a+b)^n = a^n \left(1+\frac{b}{a}\right)^n$, it was sufficient to obtain the expansion of $(1+x)^n$. He set

$$[m] = 1 + \frac{m}{1}x + \frac{m}{1} \cdot \frac{m-1}{2}x^2 + \text{etc.} \tag{4.11}$$

with the aim of proving that $[m] = (1+x)^m$ when m was a fraction. Note that he already knew that the result was true when m was a positive integer. The important step in his proof was to show that $[m] \cdot [n] = [m+n]$. He took

$$[n] = 1 + \frac{n}{1}x + \frac{n}{1} \cdot \frac{n-1}{2}x^2 + \text{etc. so that}$$

$$[m] \cdot [n] = 1 + \frac{m}{1}x + \frac{m}{1} \cdot \frac{m-1}{2}x^2 + \text{etc.}$$
$$+ \frac{n}{1}x + \frac{m}{1} \cdot \frac{n}{1}x^2 + \text{etc.}$$
$$+ \frac{n}{1} \cdot \frac{n-1}{2}x^2 + \text{etc.}$$

Thus, the product had the form $1 + Ax + Bx^2 + Cx^3 + \text{etc.}$, where

$$A = m+n, \quad B = \frac{mm}{2} + mn + \frac{nn}{2} = \frac{m+n}{1} \cdot \frac{m+n+1}{2}.$$

Euler then observed that it was very laborious to compute C, D, E, etc., by this method. To see in modern terms what this involved, note that the coefficient of x^k in the product is

$$\binom{m}{k}\binom{n}{0} + \binom{m}{k-1}\binom{n}{1} + \cdots + \binom{m}{0}\binom{n}{k}. \tag{4.12}$$

Since m and n are not necessarily positive integers, we define

$$\binom{x}{j} = \frac{x(x-1)\cdots(x-j+1)}{j!}.$$

Euler had to prove that the sum (4.12) reduced to

$$\binom{m+n}{k} = \frac{(m+n)(m+n-1)\cdots(m+n-k+1)}{k!}. \tag{4.13}$$

So, to compute the coefficients more simply, Euler noted that when m and n were positive integers, then by the known result $[m] = (1+x)^m$ and $[n] = (1+x)^n$,

$$[m] \cdot [n] = (1+x)^m \cdot (1+x)^n = (1+x)^{m+n} = [m+n]. \tag{4.14}$$

Hence, the sum (4.12) reduced to (4.13) when m and n were positive integers. From this, he concluded that (4.12) and (4.13) were equal for all real m and n. This depended on the facts that both (4.12) and (4.13) were polynomials in m and n and that they were equal for infinitely many values of m and n. Since

$$[m] \cdot [n] = [m+n] \qquad (4.15)$$

was true for all real m and n, Euler could deduce the binomial theorem for rational exponents. He supposed $m = p/q$ where p and q were positive integers. Then by (4.14)

$$(1+x)^p = [p] = \left[\frac{p}{q} + \frac{p}{q} + \cdots + \frac{p}{q}\right] = \left[\frac{p}{q}\right] \cdot \left[\frac{p}{q}\right] \cdots \left[\frac{p}{q}\right] = \left[\frac{p}{q}\right]^q, \text{ or}$$

$$\left[\frac{p}{q}\right] = (1+x)^{p/q}, \qquad (4.16)$$

proving the theorem for positive rational exponents. Euler extended this result to negative rational exponents by noting that

$$[m][-m] = [m-m] = [0] = 1 \quad \text{and, therefore,}$$

$$\left[\frac{p}{q}\right]\left[-\frac{p}{q}\right] = 1.$$

By (4.16), this meant that $(1+x)^{p/q}\left[-\frac{p}{q}\right] = 1$ and thus $\left[-\frac{p}{q}\right] = (1+x)^{-p/q}$. Euler did not discuss convergence questions. He certainly knew that the series for $[m]$ converged when $|x| < 1$, but he had not given thought to the more subtle questions related to the convergence of the products of infinite series. Cauchy, Abel, and Dirichlet eventually addressed such issues.

4.4 Cauchy: Proof of the Binomial Theorem for Real Exponents

In his lectures at the École Polytechnique, Cauchy attempted to put the work of his predecessors on a more solid foundation, although his students did not appreciate his efforts and apparently complained about it. In order to make Euler's work of the previous section more rigorous, Cauchy had to define a continuous function, and he also needed to work out the definitions of convergence, absolute convergence, and the product of infinite series.

Cauchy defined the convergence of a series

$$\mu_0 + \mu_1 + \mu_2 + \cdots \text{ and let } s_n = \mu_0 + \mu_1 + \mu_2 + \cdots + \mu_n.$$

Cauchy stated that if s_n approached a fixed limit as n increased indefinitely, then the series would converge; otherwise, it diverged. The limit of s_n was said to be s if for any small number ϵ, the limit of s_n fell between $s - \epsilon$ and $s + \epsilon$ for large enough n.

4.4 Cauchy: Proof of the Binomial Theorem for Real Exponents

He stated and proved the ratio test for convergence of a series and deduced that the binomial series for $|x| < 1$ converged. He also defined what is known as the Cauchy product of two series

$$u_0 + u_1 + u_2 + \cdots \quad \text{and} \quad v_0 + v_1 + v_2 + \cdots \quad \text{as}$$

$$u_0 v_0 + (u_0 v_1 + u_1 v_0) + (u_0 v_2 + u_1 v_1 + u_2 v_0) + \cdots$$

$$+ (u_0 v_n + u_1 v_{n-1} + \cdots + u_n v_0) + \cdots.$$

He then proved that if the two series were absolutely convergent and converged to s and s', then the product series converged to $s'' = ss'$. For the case in which all u and v were positive, Cauchy observed that

$$s_{m+1} s'_{m+1} < s''_n < s_n s'_n,$$

where $m = \frac{n-1}{2}$ for all n odd and $m = \frac{n-2}{2}$ for n even. The two inequalities implied the theorem for series with positive terms. When the terms $u_0, u_1, u_2, \ldots$ and $v_0, v_1, v_2 \ldots$ were positive as well as negative, Cauchy first observed that

$$s_n s'_n - s''_n = u_{n-1} v_{n-1} + (u_{n-1} v_{n-1} + u_{n-2} v_{n-2}) + \cdots + (u_{n-1} v_1 + \cdots u_1 v_{n-1}).$$

He denoted the absolute values of the u and v by $P_0, P_1, P_2, \ldots$ and $P'_0, P'_1, P'_2, \ldots$ respectively and remarked that, from the result on the convergence of the product of series with all positive terms,

$$P_{n-1} P'_{n-1} + (P_{n-1} P'_{n-2} + P_{n-2} P'_{n-1}) + \cdots + (P_{n-1} P'_1 + \cdots + P_1 P'_{n-1})$$

tended to zero as $n \to \infty$. Since this expression bounded $|s_n s'_n - s''_n|$, it followed that $s_n s'_n - s''_n \to 0$ as $n \to \infty$, proving the result.

With these theorems in hand, Cauchy could close the gaps in Euler's proof of the binomial theorem. The binomial series $[m]$ and $[n]$ in (4.11) were absolutely convergent for $|x| < 1$. Hence by Euler's argument and Cauchy's result on products of series, $[m] \cdot [n] = [m+n]$. Finally, if $[m]$ was a continuous function of m, and if for integer p and q, $[p/q] = (1+x)^{p/q}$, then $[m] = (1+x)^m$ for all real exponents m. Recall that Cauchy's proof of the continuity of $[m]$ was inadequate, and the gap was filled by Abel.

Cauchy gave another proof of the binomial theorem as a corollary of Taylor's or Maclaurin's theorem. It was well known in the eighteenth century that the binomial theorem could be formally derived from Maclaurin's theorem, but Cauchy was the first to understand how an analysis of the remainder term could be applied to obtain a rigorous proof of the binomial theorem for real exponents. Recall that Cauchy gave the remainder in two different forms, as discussed in our chapter on Taylor series:

$$f(x) = f(0) + \frac{x}{1} f'(0) + \frac{x^2}{1 \cdot 2} f''(0) + \cdots + \frac{x^{n-1}}{1 \cdot 2 \cdots (n-1)} f^{n-1}(0) + R_n,$$

where

$$R_n = \frac{x^n}{1 \cdot 2 \cdots n} f^{(n)}(\theta x), \quad 0 < \theta < 1, \text{ or}$$

$$R_n = \frac{x^n}{1 \cdot 2 \cdots (n-1)} (1-\theta_1)^{n-1} f^{(n)}(\theta_1 x), \quad 0 < \theta_1 < 1.$$

When $f(x) = (1+x)^\mu$, he had

$$f^{(k)}(x) = \mu(\mu-1) \cdots (\mu-k+1)(1+x)^{\mu-k} \text{ and hence}$$

$$\frac{f^{(k)}(0)}{k!} = \frac{\mu(\mu-1) \cdots (\mu-k+1)}{k!},$$

so the binomial series was obtained. To determine the values of x for which the series equaled $(1+x)^\mu$, it was necessary to find the values of x for which $R_n \to 0$ as $n \to \infty$. Taking m large enough that $|\mu/m| < 1$, Cauchy noted that

$$\mu_{n-1} = \frac{\mu(\mu-1) \cdots (\mu-n+1)}{1 \cdot 2 \cdot 3 \cdots n} x^{n-1}$$

$$= \frac{\mu(\mu-1) \cdots (\mu-m+1)}{1 \cdot 2 \cdot 3 \cdots m} x^{m-1} \cdot \left(1 - \frac{\mu+1}{m+1}\right) \cdots \left(1 - \frac{\mu+1}{n}\right)(-x)^{n-m},$$

and hence for $|x| < 1$, $\mu_{n-1} \to 0$ as $n \to \infty$. Now the first form of the remainder would be

$$R_n = \frac{\mu(\mu-1) \cdots (\mu-n+1)}{1 \cdot 2 \cdot 3 \cdots n} x^n (1+\theta x)^{\mu-n}$$

$$= \mu_{n-1} \cdot x(1+\theta x)^\mu \left(\frac{1}{1+\theta x}\right)^n.$$

The factor $\left(\frac{1}{1+\theta x}\right)^n$ was bounded only for positive x, and he could deduce that $R_n \to 0$ as $n \to \infty$ for $0 \leq x < 1$. Cauchy needed the second form of the remainder to be able to deduce the binomial theorem for $|x| < 1$. Using the second form, Cauchy had

$$R_n = \mu_{n-1} \cdot x(1+\theta_1 x)^{\mu-1} \left(\frac{1-\theta_1}{1+\theta_1 x}\right)^{n-1}.$$

Clearly, $\left(\frac{1-\theta_1}{1+\theta_1 x}\right)^n$ was bounded for $|x| < 1$ and $R_n \to 0$ as $n \to \infty$ for $|x| < 1$.

4.5 Abel's Theorem on Continuity

Abel's continuity theorem was a response to Cauchy's 1821 result requiring uniform convergence, not developed until Weierstrass. Implicitly accounting for uniform convergence in his result, Abel proved a theorem yielding the binomial theorem for real

4.5 Abel's Theorem on Continuity

exponents. Though Abel found a mistake in Cauchy's work, he acknowledged his indebtedness to Cauchy and wrote in his paper that every analyst who loved rigor in mathematics should study Cauchy's *Cours d'analyse*; we note that this work is also called *Analyse algébrique*. Abel's basic result on power series is now called Abel's continuity theorem; in modern terms, it states that if an infinite series $\sum a_n$ converges, then the series $\sum a_n x^n$ converges uniformly for $0 \leq x \leq 1$ and also tends to $\sum a_n$ as x tends to 1^-. In 1897, Alfred Tauber (1866–1942) proved a conditional converse, leading to the extensive Tauberian theory, developed and named by Hardy and Littlewood.

In a January 1826 letter to Holmboe, Abel discussed his continuity theorem:

> Let $a_0 + a_1 + a_2 + a_3 + a_4 +$ etc. be any infinite Series and thus You know that a very useful Manner of adding up this Series is to seek the sum of the following: $a_0 + a_1 x + a_2 x^2 + a_3 x^3 + \cdots$ and then later, put $x = 1$ in the Results. This is correct; but it seems to me that one cannot accept it without Proof.

Abel applied his theorem to show that if A and B were convergent infinite series and their Cauchy product C was convergent, then $AB = C$. Abel's continuity theorem was based on a lemma using summation by parts: If $t_0, t_1, \ldots, t_m, \ldots$ denoted a sequence of arbitrary quantities, and if the quantity $p_m = t_0 + t_1 + \cdots + t_m$ was less than a definite quantity δ, then

$$r = \epsilon_0 t_0 + \epsilon_1 t_1 + \cdots + \epsilon_m t_m < \delta \epsilon_0, \tag{4.17}$$

where $\epsilon_0, \epsilon_1, \epsilon_2, \ldots$ were positive decreasing quantities. To prove this result, Abel noted that

$$r = \epsilon_0 p_0 + \epsilon_1 (p_1 - p_0) + \epsilon_2 (p_2 - p_1) + \cdots + \epsilon_m (p_m - p_{m-1})$$
$$= p_0 (\epsilon_0 - \epsilon_1) + p_1 (\epsilon_1 - \epsilon_2) + \cdots + p_{m-1} (\epsilon_{m-1} - \epsilon_m) + p_m \epsilon_m$$
$$< \delta (\epsilon_0 - \epsilon_1 + \epsilon_1 - \epsilon_2 + \cdots + \epsilon_{m-1} - \epsilon_m + \epsilon_m) = \delta \epsilon_0.$$

The last step in this proof was valid because $\epsilon_0 - \epsilon_1, \epsilon_2 - \epsilon_1, \ldots$ were positive. Next, for the continuity theorem, Abel wrote that if the series

$$f(\alpha) = v_0 + v_1 \alpha + v_2 \alpha^2 + \cdots + v_m \alpha^m + \cdots$$

converged for $\alpha = \delta$, it would also converge for every smaller value of α; likewise, $f(\alpha - \beta)$, for continually decreasing values of β, would come arbitrarily close to the limit $f(\alpha)$, with α equal to or smaller than δ. To prove this, Abel let

$$v_0 + v_1 \alpha + \cdots + v_{m-1} \alpha^{m-1} = \phi(\alpha),$$
$$v_m \alpha^m + v_{m+1} \alpha^{m+1} + \cdots = \psi(\alpha). \text{ Then}$$

$$\psi(\alpha) = \left(\frac{\alpha}{\delta}\right)^m \cdot v_m \delta^m + \left(\frac{\alpha}{\delta}\right)^{m+1} + \cdots < \left(\frac{\alpha}{\delta}\right)^m p,$$

where p was the maximum of

$$v_m\delta^m, \; v_m\delta^m + v_{m+1}\delta^{m+1}, \; v_m\delta^m + v_{m+1}\delta^{m+1} + v_{m+2}\delta^{m+2}, \ldots.$$

Note that the inequality followed from his lemma proving (4.17). Then for $0 \leq \alpha \leq \delta$, m could be chosen large enough so that $\psi(\alpha) = w$. We observe that Abel used the symbol w to denote an arbitrarily small quantity. Next,

$$f(\alpha) = \phi(\alpha) + \psi(\alpha), \text{ and hence}$$
$$f(\alpha) - f(\alpha - \beta) = \phi(\alpha) - \phi(\alpha - \beta) + w.$$

Since $\phi(\alpha)$ was a polynomial, β could be taken small enough that $\phi(\alpha) - \phi(\alpha - \beta) = w$ and hence $f(\alpha) - f(\alpha - \beta) = w$, proving the theorem.

To address the defect in Cauchy's proof of the binomial theorem, Abel stated and proved the theorem: Let $v_0 + v_1\delta + v_2\delta^2 + \cdots$ be a convergent series, in which $v_0, v_1, v_2, \ldots$ are continuous functions of a variable quantity x between $x = a$ and $x = b$; then the series $f(x) = v_0 + v_1\alpha + v_2\alpha^2 + \cdots$, where $\alpha < \delta$, will be convergent and a continuous function of x between the same limits. As in the proof of the previous theorem, Abel set

$$v_0 + v_1\alpha + \cdots + v_{m-1}\alpha^{m-1} = \psi(x) \text{ and } v_m\alpha^m + v_{m+1}\alpha^{m+1} + \cdots = \phi(x).$$

Then

$$\psi(x) = \left(\frac{\alpha}{\delta}\right)^m v_m\delta^m + \left(\frac{\alpha}{\delta}\right)^{m+1} v_{m+1}\delta^{m+1} + \left(\frac{\alpha}{\delta}\right)^{m+2} v_{m+2}\delta^{m+2} + \cdots.$$

By the summation by parts lemma, if $\theta(x)$ denoted the largest of the quantities

$$v_m\delta^m, \; v_m\delta^m + v_{m+1}\delta^{m+1}, \; v_m\delta^m + v_{m+1}\delta^{m+1} + v_{m+2}\delta^{m+2}, \ldots,$$

then $\psi(x) < \left(\frac{\alpha}{\delta}\right)^m \theta(x)$. Thus, for m large enough, $\psi(x) = w$ and $f(x) = \phi(x) + w$, where w was less than any assignable quantity. Similarly,

$$f(x) - f(x - \beta) = \phi(x) - \phi(x - \beta) + w.$$

Since $\phi(x)$ was a finite sum of continuous functions, it was also continuous and hence $\phi(x) - \phi(x - \beta) = w$. Therefore, $f(x) - f(x - \beta) = w$, which meant that $f(x)$ was continuous. It was here that Abel pointed out in a footnote that Cauchy's theorem on an infinite sum of continuous functions had some exceptions. But Abel succeeded in filling the gap, so that the proof of the binomial theorem for real exponents was complete.

4.5 Abel's Theorem on Continuity

Abel went on to prove the binomial theorem for a complex variable x and complex exponent. He finally stated his result as

$$1 + \frac{m+ni}{1}(a+bi) + \frac{(m+ni)(m-1+ni)}{1\cdot 2}(a+bi)^2$$
$$+ \frac{(m+ni)(m-1+ni)(m-2+ni)}{1\cdot 2\cdot 3}(a+bi)^3 + \cdots$$
$$+ \frac{(m+ni)(m-1+ni)\cdots(m-\mu+1+ni)}{1\cdot 2\cdot 3\cdots \mu}(a+bi)^\mu + \cdots$$
$$= [\cos(m\arctan\frac{b}{1+a} + \frac{1}{2}n\ln[(1+a)^2+b^2])$$
$$+ i\sin(m\arctan\frac{b}{1+a} + \frac{1}{2}n\ln[(1+a)^2+b^2])]$$
$$\times [(1+a)^2+b^2]^{\frac{m}{2}} e^{-n\arctan\frac{b}{1+a}}.$$

Note that Abel wrote $\sqrt{-1}$ for i and log for ln; the right-hand side was the principal value of $(1+a+bi)^{m+ni}$.

Liouville found Abel's proof of the continuity theorem difficult to understand and asked Dirichlet to explain it. Dirichlet presented a proof on the spot; Liouville then used it in his lectures at the Collège de France. After Dirichlet's death, Liouville published the proof in honor of his friend. He stated and proved the theorem:

If the series $a_0 + a_1 + a_2 + \cdots$ converges to A, then

$$\lim_{p\to 1^-} \sum_{n=0}^\infty a_n p^n = A.$$

Let $\delta_n = a_0 + a_1 + \cdots + a_n$ and $0 < p < 1$. Then

$$s = a_0 + a_1 p + a_2 p^2 + \cdots + a_n p^n + \cdots$$
$$= \delta_0 + (\delta_1 - \delta_0)p + (\delta_2 - \delta_1)p^2 + \cdots + (\delta_n - \delta_{n-1})p^n + \cdots$$
$$= (1-p)(\delta_0 + \delta_1 p + \delta_2 p^2 + \cdots + \delta_n p^n + \cdots).$$

The last equation is true because the first $n+1$ terms of the two series differ by $\delta_n p^{n+1}$, and this tends to zero as $n \to \infty$. Next, break up the last series into two parts:

$$S(p) = (1-p)(\delta_0 + \delta_1 p + \cdots + \delta_{n-1}p^{n-1}) + (1-p)(\delta_n p^n + \delta_{n+1}p^{n+1} + \cdots).$$

Let P_n be a number between the maximum and minimum values of the sequence

$$\delta_n, \delta_{n+1}, \delta_{n+2}, \ldots$$

such that the second series is equal to

$$(1-p)P_n(p^n + p^{n+1} + p^{n+2} + \cdots) = P_n p^n.$$

Clearly, $P_n \to A$ as $n \to \infty$. So if we let $p \to 1$ and then let $n \to \infty$, the finite series tends to zero and the other series tends to A and the theorem is proved.

4.6 Harkness and Morley's Proof of the Binomial Theorem

Weierstrass promulgated his fundamental ideas through his teaching. Thus, it was left to others to write up and disseminate these ideas. For example, in 1898, J. Harkness of Bryn Mawr and F. Morley of Haverford College, wrote *Introduction to the Theory of Analytic Functions*. They explained: "we recognized that readers approaching the subject for the first time could not fail to be hampered by the non-existence in English of any text-book giving a consecutive and elementary account of the fundamental concepts and processes employed in the theory of functions." In his delightful article, "A Mathematical Education," the great English analyst J. E. Littlewood mentioned that his study of Harkness and Morley's book was one of the bright spots in his education up to the time he took his Tripos examination in 1905.

Harkness and Morley's proof of the binomial theorem is different from the other proofs presented here, and it considers the general case where the variable and exponent are both complex numbers. The proof depends on a theorem attributed to Weierstrass: Let u_q, $q = 0, 1, 2, \ldots$ be series in powers of x:

$$u_q = a_{q0} + a_{q1}x + a_{q2}x^2 + \cdots + a_{qn}x^n + \cdots.$$

Given that the separate series u_q and the collective series $\sum u_q$ converge within the circle (R) and that the series $\sum u_q$ converges uniformly along every circle (R_1) where $R_1 < R$, then within the circle (R) we have

$$\sum_{q=0}^{\infty} u_q = \sum_{n=0}^{\infty} a_n x^n,$$

where a_n is the sum of the coefficients of x^n in the series of us. Now consider the function $(1-a)^{-x} = \exp(-x \log(1-a))$, where a and x are complex with $|a| < 1$ and where log takes its principal value. Then

$$(1-a)^{-x} = 1 + u + u^2/2! + u^3/3! + \cdots \quad \text{where}$$

$$u = -x \log(1-a) = xa + xa^2/2 + xa^3/3 + \cdots.$$

It is clear that the series in u is absolutely and uniformly convergent in every circle $|u| \leq R$, while the series for $u^n/n!$ in powers of a is absolutely and uniformly convergent in $|a| \leq 1 - \delta$ for every $\delta > 0$. Therefore, by Weierstrass's theorem

$$\sum_{n=0}^{\infty} u^n/n! = \sum_{n=0}^{\infty} x_n a^n/n!$$

for $|a| < 1$. It remains to find x_n. Note that only $u, u^2, \ldots, u^n$ contribute to the expression. So

$$x_n = x^n + \cdots + (n-1)!x$$

is a polynomial of degree n in x. When $x = 0, -1, -2, \ldots, -n+1$, we have $(1-a)^{-x} = (1-a)^m$ where $m = 0, 1, 2, \ldots, n-1$. For these values of m, the coefficient of a^n in $(1-a)^m$ is zero. Hence, $x_n = 0$ for $x = 0, -1, -2, \ldots, -n+1$, and we have

$$x_n = x(x+1)(x+2)\cdots(x+n-1) \quad \text{and thus}$$

$$(1-a)^{-x} = \sum_{n=0}^{\infty} \frac{x(x+1)\cdots(x+n-1)}{n!} a^n.$$

4.7 Exercises

1. Following Newton, apply the procedure for finding the square root of a number to the algebraic expression $1 - x^2$ and show that you get the series

$$1 - \frac{1}{2}x^2 - \frac{1}{8}x^4 - \frac{1}{16}x^6 - \cdots.$$

2. Apply the Gregory–Newton difference formula to the function $f(\alpha) = (1+x)^\alpha$ and show that you get the binomial series.
3. Prove that the Cauchy product of the series $\sum_{n=0}^{\infty} \frac{(-1)^n}{\sqrt{n+1}}$ with itself diverges. Cauchy gave this example in his *Analyse algébrique*, chapter 6.
4. Cauchy stated the ratio test in chapter 6 of his *Analyse algébrique*: If for n increasing and positive, the ratio u_{n+1}/u_n converges to a fixed limit k, then the series $u_0 + u_1 + u_2 + \cdots$ converges when $k < 1$ and diverges when $k > 1$. Prove this theorem.
5. Cauchy's *Analyse algébrique*, chapter 6, also gave the condensation test: A series of positive and decreasing terms $u_0 + u_1 + u_2 + \cdots$ converges if and only if $u_0 + 2u_1 + 4u_3 + 8u_7 + \cdots$ converges. Prove this theorem, and use it to prove that $\sum_{n=1}^{\infty} \frac{1}{n^\mu}$ converges for $\mu > 1$ and diverges for $\mu \leq 1$. Cauchy gave this application. Earlier, the fourteenth-century French theologian and scientific thinker N. Oresme used the condensation test in the case $\mu = 1$.
6. Prove Abel's theorem on products, that if $c_n = a_0 b_n + a_1 b_{n-1} + \cdots + a_n b_0$, and $A = \sum_{n=0}^{\infty} a_n$, $B = \sum_{n=0}^{\infty} b_n$, $C = \sum_{n=0}^{\infty} c_n$ are all convergent, then $AB = C$. See Abel (1965), vol. 1, p. 226.
7. Prove F. Mertens's extension of Cauchy's product theorem: If A is absolutely convergent and B conditionally convergent, then $AB = C$. See Mertens (1875).
8. Prove that if A and B are convergent, and $|a_n| \leq k/n$, $|b_n| \leq k/n$ for all n, then C is convergent. This theorem is due to G. H. Hardy; see Hardy (1966–1979), vol. 6, p. 414–416.
9. Follow Abel in proving that $\sum_{k=2}^{\infty} \frac{1}{k \ln k}$ diverges. Use $\ln(1+x) < x$ to show that that $\ln \ln(1+n) < \ln \ln n + \frac{1}{n \ln n}$. Conclude that $\ln \ln(1+n) < \ln \ln 2 + \sum_{k=1}^{n} \frac{1}{k \ln k}$. See Abel (1965), vol. 1, pp. 399–400.
10. Show that if $A_n = a_1 + a_2 + \cdots + a_n$, $|A_n|$ is bounded for all n, and $\sum_{k+1}^{n} |b_{k+1} - b_k|$ is bounded for all n, and $b_n \to 0$ as $n \to \infty$, then $\sum_{n=1}^{\infty} a_n b_n$ converges.

11. (a) Suppose ϕ is a real valued continuous function and $\phi(x+y) = \phi(x) + \phi(y)$. Show that $\phi(x) = ax$ for some constant a.
    (b) Suppose ϕ is a real valued continuous function and $\phi(x+y) = \phi(x)\phi(y)$. Show that $\phi(x) = A^x$ for some positive constant A.
    (c) Suppose ϕ is a real valued continuous function and for $x > 0$, $y > 0$, $\phi(xy) = \phi(x) + \phi(y)$. Show that $\phi(x) = a \ln x$.
    (d) Suppose ϕ is a real valued continuous function and for $x > 0$, $y > 0$, $\phi(xy) = \phi(x)\phi(y)$. Show that $\phi(x) = x^a$ for $x > 0$.
    (e) Suppose ϕ is a real valued continuous function and $\phi(y+x) + \phi(y-x) = 2\phi(x)\phi(y)$. Show that if $0 \le \phi(x) \le 1$ and ϕ is not constant, then $\phi(x) = \cos(ax)$ where a is a constant. If $\phi(x) \ge 1$, then there exists a positive constant A such that $\phi(x) = A^x$. The solutions of these five problems take up all of chapter 5 of Cauchy's *Analyse algébrique* of 1821.

12. Let

$$\phi(x) = 1 + \frac{x}{1} + \frac{x^2}{1 \cdot 2} + \frac{x^3}{1 \cdot 2 \cdot 3} + \cdots.$$

Show that $\phi(x+y) = \phi(x) \cdot \phi(y)$ and hence that $\phi(x) = (\phi(1))^x = e^x$. See Cauchy (1989), pp. 168–169.

13. Prove the binomial theorem: Let

$$f_\alpha(x) = \sum_{k=0}^{\infty} \frac{\alpha(\alpha+1)\cdots(\alpha+k-1)}{k!} x^k.$$

Show that $\frac{d}{dx} f_\alpha(x) = \alpha f_{\alpha+1}(x)$ and $f_{\alpha+1}(x) - f_\alpha(x) = x f_{\alpha+1}(x)$. Deduce that $f_\alpha(x) = (1-x)^{-\alpha}$. Gauss did not explicitly give this proof, but the first two steps are very special cases of results in his paper on hypergeometric functions.

14. With the development of set theory and the language of sets by Cantor, Dedekind and others in the period 1870–1900, mathematicians could ask whether the equation $\phi(x+y) = \phi(x) + \phi(y)$ had solutions other than $\phi(x) = ax$, a a constant, and, if so, whether a condition weaker than continuity would imply $\phi(x) = ax$. Hilbert's student Georg Hamel (1877–1954) used Zermelo's result, that the set of real numbers can be well-ordered, to obtain a basis B for the vector space of real numbers over the field of rational numbers. Thus, B has the property that for every real number x, $x = r_1\alpha_1 + r_2\alpha_2 + \cdots + r_n\alpha_n$ for some n, rationals $r_1, \ldots, r_n$ and basis elements $\alpha_1, \ldots, \alpha_n$. Use a Hamel basis to show that $\phi(x+y) = \phi(x) + \phi(y)$ has solutions different from $\phi(x) = ax$, where a is a constant. See G. Hamel (1905).

15. Prove that if ϕ is measurable and satisfies $\phi(x+y) = \phi(x) + \phi(y)$, then $\phi(x) = ax$. This was proved by M. Fréchet in 1913. For a simple proof, see Kac (1979), pp. 64–65.

4.8 Notes on the Literature

The two letters from Newton to Oldenburg, intended for Leibniz, on the binomial theorem are in Newton's correspondence. See Newton (1959–60), pp. 20–47 and 110–161). The original letters are in Latin, but English translations with further explanatory notes are also given. These letters give an excellent summary of Newton's work on series and related topics during the period 1664–1676. In 1712, Newton used them as documentary evidence in his bitter priority dispute with Leibniz. The early work of Newton leading to the discovery of the binomial theorem is printed in vol. 1 of his *Mathematical Papers*. Pages 104–110 and 122–134 show how Newton used Wallis interpolation to first discover the binomial theorem and then to attempt to establish it. He later abandoned this method of proof.

James Gregory's letter of November 23, 1670, on the binomial theorem, appears in Turnbull (1939), pp. 118–137. This and other letters of Gregory to Collins, now contained in Turnbull (1939), were used to establish Newton's absolute priority in the invention of the calculus. In that context, it was argued that Gregory was led to his work on series after he saw Newton's series on the area of the zone of a circle. Consequently, Gregory's work was relegated to a secondary position, although his discoveries were independent. However, the publication of the Gregory memorial volume has helped to establish Gregory's reputation as one of the greatest mathematicians of the seventeenth century. For Euler's proof of the binomial theorem, see Eu. I-15, pp. 207–216.

Cauchy started teaching analysis at the École Polytechnique in 1817. He divided his course into two parts, the first dealing with infinite series of real and complex variables and the second with differential and integral calculus. Following eighteenth century usage, he called the first part algebraic analysis. These lectures were published in 1821 with the title *Analyse algébrique*. Bradley and Sandifer (2009) present an English translation with useful notes. This was the first textbook dealing fairly rigorously with the basic concepts of infinite series: limits, convergence, and continuity. For Cauchy's derivation of the binomial theorem by use of Taylor's theorem with remainder, see chapters 8 and 9 of Cauchy (1829).

Abel's paper on the binomial theorem appeared in the first issue of A. L. Crelle's *Journal für die reine und angewandte Mathematik* in 1826. It is the oldest mathematical journal still being published today. A large majority of Abel's papers were published in this journal, helping this journal quickly secure a high standing in the mathematics community. There are two editions of Abel's collected papers. Abel's college teacher B. M. Holmboe (1795–1850) published the first in 1837. Unfortunately, Holmboe could not include Abel's great paper, "Mémoire pour une propriété générale d'une classe très-étendue des fonctions transcendantes," presented by Abel to the French Academy in 1826. This manuscript was lost and found several times before it was finally published in Paris in 1841. The manuscript was again lost, possibly stolen by G. Libri; in 1952 a portion was recovered in Florence by the Norwegian mathematician V. Brun. In 2000, Andrea Del Cantina discovered the remaining parts, with the exception of four pages. Abel's "Mémoire" was included in the second and larger edition of his collected work, edited by L. Sylow and S. Lie, published in 1881. Abel (1965) is a reprint of this edition; the first volume contains Abel's paper on the binomial theorem; see pp. 219–250. Abel's letter to Holmboe is quoted from Stubhaug (2000), p. 345. For Dirichlet's paper on Abel's theorem, see Dirichlet (1862), a paper published by Liouville.

Landen (1758), pp. 5–7 gives his remarks on the binomial theorem. His 43-page booklet was a contribution to the mathematical tendency of that time, to employ algebra to avoid infinitesimals and fluxions. Also part of this tradition, Hutton (1812) discusses the binomial theorem, expressing appreciation for Landen's proof. Hutton's three-volume work is entertaining reading, with articles on building bridges, experiments with gunpowder, histories of trigonometric and logarithmic tables, and a long history of algebra. See p. 134 of Harkness and Morley (1898) for a statement of Weierstrass's theorem; p. 169 for the proof of the binomial theorem. For Gudermann's praise of Weierstrass, see Klein (1979), p. 263.

5

The Rectification of Curves

5.1 Preliminary Remarks

Up until the seventeenth century, geometry was pursued along the Greek model. Thus, second-order algebraic curves were studied as conic sections, though higher-order curves were also considered. Algebraic relationships among geometric quantities were considered, but algebraic equations were not used to describe geometric objects. In the course of his attempts during the late 1620s to recreate the lost work of Apollonius, it occurred to Fermat that geometry could be studied analytically by expressing curves in terms of algebraic equations. Now conic sections are defined by second-degree equations in two variables, but this new perspective expanded geometry to include curves of any degree. Fermat's work in algebraic geometry was not published in his lifetime, so its influence was not great. But during the 1620s, René Descartes (1596–1650) developed his conception of algebraic geometry and his seminal work, *La Géométrie*, was published in 1637. The variety of new curves thus made possible, combined with the development of the differential method, spurred the efforts to discover a general method for determining the length of an arc. In the late 1650s, Hendrik van Heuraet (1634–c. 1660) and William Neil(e) (1637–1670) gave a solution to this problem by reducing it to the problem of finding the area under a related curve. In this and other areas, Descartes's new approach to geometry served as a guiding backdrop.

Descartes's early training in mathematics included the study of the classical texts of the fourth-century Greek mathematician Pappus, the *Arithmetica* of Diophantus, and the contemporary algebra of Peter Roth and Christoph Clavius. Descartes's meeting with Johann Faulhaber in the winter 1619–20 also contributed to his understanding of algebra. However, from the very beginning, Descartes was determined to develop and follow his own methods, and this eventually led him to a symbolic algebra whose notation was very similar to the one we now use.

In the first part of his *Géométrie*, Descartes explained how geometric curves could be reduced to polynomial equations in two variables. He did not consistently use what are now called the Cartesian orthogonal axes but chose the angle of his axes to suit the problem. On the subject of the rectification of curves, Descartes wrote, "geometry should not include lines that are like strings, in that they are sometimes straight and sometimes curved, since the ratios between straight and curved lines are not known, and I believe cannot be discovered by human minds." But in a November 1638 letter to Mersenne, Descartes nevertheless discussed the rectification of the logarithmic spiral, a curve he defined as making a constant angle with the radius vector at each point. To understand this apparent contradiction in his thinking, note that Descartes made a distinction between geometric and mechanical curves. Geometric curves were defined by algebraic equations; mechanical curves would today be termed transcendental. So when Descartes referred to unrectifiable curves, he meant the geometric ones. He maintained that the study of geometry should be restricted to algebraic curves, for which algebraic methods should be used. Thus, for constructing tangents and normals to curves, he used algebraic methods, as opposed to the infinitesimal methods of Fermat. Descartes considered Fermat's methods inappropriate for geometric (for us, algebraic) curves. And, since the length of an arc could not be found by algebraic methods, he stated that the length of geometric curves could not be obtained. He allowed, however, that the lengths of mechanical (transcendental) curves could be determined by infinitesimal methods. In fact, in 1638 Descartes succeeded in rectifying the equiangular (or logarithmic) spiral, and he was therefore one of the earliest mathematicians to find the length of a noncircular arc. Soon after this, Torricelli also rectified this spiral. We note that in approximately 1594, Harriot worked with this spiral in connection with his researches related to navigation, but he did not publish his results.

Frans van Schooten (1615–1660) played an important role in the solution of the rectification problem. A Dutch mathematician of considerable ability, he was certainly one of the great teachers of mathematics. He attracted a number of talented young students to mathematics, even when their primary interests lay in other disciplines. Van Schooten studied at the University of Leiden where his father, Frans van Schooten the Elder, was a professor of mathematics. The younger van Schooten received a thorough grounding in the Dutch mathematical tradition. In 1635, he met Descartes, and by the summer of 1637 he had seen his *Géométrie*, though he did not immediately understand it. In 1646 he inherited his father's professorship and in 1647, van Schooten published a Latin translation of Descartes's book with his own commentary. This translation made Descartes's ideas accessible to many more mathematicians and simultaneously helped build van Schooten's reputation. A second edition of this translation appeared in 1659. The work was about a thousand pages long, ten times longer than *Géométrie*; it included several significant contributions by van Schooten's students: Christiaan Huygens, Jan Hudde, Jan de Witt, and Hendrik van Heuraet.

Van Heuraet entered the University of Leiden as a medical student. He was inspired to study the rectification problem by Huygens's 1657 discovery that the arc length of a parabola could be measured by the quadrature of an equilateral hyperbola. In modern terms, this means that the arc length of $y = x^2$ can be computed by the integral $\int (1 + 4x^2)^{1/2} dx$. Sometime in 1658, van Heuraet solved the general problem; he communicated his work to van Schooten in a letter dated January 13, 1659. In the course of applying his method to the semicubical parabola $y^2 = ax^3$, he used a rule of Hudde concerning multiple roots of polynomials.

Jan Hudde (1628–1704) studied law at Leiden around 1648 and later served as burgomaster of Amsterdam for 21 years. He stated his rule in the article "De Maximis et

Minimis," communicated to van Schooten in a letter of February 26, 1658. His rule provided a method for determining maxima and minima of functions and a simplification of Descartes's method for finding the normal to an algebraic curve. In unpublished work from 1654, Hudde used the logarithmic series in the form $x + x^3/3 + x^5/5 + \cdots$. In his manuscript of 1693, David Gregory attested to this work of Hudde. Note that N. Mercator published this series in 1668 and Newton's unpublished results on this topic date from 1665.

Around the same time as van Heuraet, the English mathematician William Neil gave a method for rectifying the semicubical parabola; this method could also be generalized to other curves. Wallis included Neil's work in his *Tractatus de Cycloide* of 1659. The methods of Neil and van Heuraet were lacking in rigor, but Pierre Fermat very soon filled the gap. He showed in his *Comparatio Curvarum Linearum* of 1660 that a monotonically increasing curve will have a length. James Gregory, apparently independently of Fermat, also found a rigorous proof of this fact, technically better than Fermat's, using the same basic idea. Gregory's proof appeared in his *Geometriae Pars Universalis* of 1668. The inspiration behind both Fermat and Gregory was Christopher Wren's rectification of the cycloid in 1659.

5.2 Descartes's Method of Finding the Normal

Descartes asserted that properties of curves depended on their angles of intersection with other lines. He defined the angle between two curves as the angle between the normals to the curves at the point of intersection. Before giving the method for finding the normal, he wrote:

> This is my reason for believing that I shall have given here a sufficient introduction to the study of curves when I have given a general method of drawing a straight line making right angles with the curve at an arbitrarily chosen point upon it. And I dare say that this is not only the most useful and most general problem in geometry that I know, but even that I have desired to know.

Summarizing Descartes's method, let CA in Figure 5.1 be an algebraic curve. Note that in the original book, Descartes interchanged x and y. But we will suppose CP is normal to the curve at C and let $AM = x$ and $CM = y$. The problem is to find $v = AP$. Descartes took the x-axis such that the center P of the required circle fell on the axis.

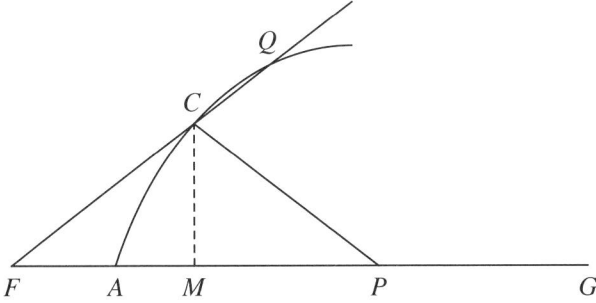

Figure 5.1. Descartes's construction of a normal.

The equation of the circle was $(x-v)^2+y^2=s^2$, where $CP=s$ was the radius. He used this equation to eliminate y from the equation of the curve, obtaining an expression in x with a double root when the circle was tangent to the curve. He could then write this expression such that it had a factor $(x-e)^2$. Descartes then found the required result by equating the powers of x. He explained his method by applying it to some examples such as the ellipse $\frac{r}{q}x^2-rx+y^2=0$, in which case he used the equation of the circle to eliminate y and obtain the equation

$$x^2+\frac{qr-2qv}{q-r}x+\frac{q(v^2-s^2)}{q-r}=0.$$

He set the left-hand side as $(x-e)^2$ to find the necessary result:

$$-2e=\frac{q(r-2v)}{q-r}, \quad \text{or} \quad v=\frac{r}{2}+\frac{q-r}{q}e=\frac{r}{2}+\frac{q-r}{q}x.$$

Claude Raubel's 1730 commentary on Descartes's *Géométrie* gave the example of the parabola $y^2=rx$. In this case,

$$x^2+(r-2v)y+v^2-s^2=0$$

must have a double root. The resulting equations, when the left-hand side is set equal to $(x-e)^2$, are

$$r-2v=-2e, \quad v^2-s^2=e^2.$$

So $v=\frac{r}{2}+e$, and this implies that $v=\frac{r}{2}+x$, since $x=e$.

It is easy to see that Descartes's method would become cumbersome for curves of higher degree. After equating coefficients, one would end up with a large number of equations. It is for this reason that Hudde searched for a simpler approach.

5.3 Hudde's Rule for a Double Root

Hudde gave a rule to determine conditions for a polynomial to have a double root:

> If in an equation two roots are equal and this is multiplied by an arbitrary arithmetical progression, naturally the first term of the equation by the first term of the progression, the second term of the equation by the second term of the progression and so on: I say that the product will be an equation in which one of the afore-mentioned roots will be found.

Hudde gave a proof for a fifth-degree polynomial, and it works in general. In modern notation, suppose b is a double root of $f(x)=0$. Then

$$f(x)=(x-b)^2(c_0+c_1x+c_2x^2+\cdots+c_{n-2}x^{n-2})$$
$$=\sum_{k=0}^{n-2}c_k(x^{k+2}-2bx^{k+1}+b^2x^k).$$

If this equation is multiplied term-wise by an arithmetic progression $p+qk$ where p and q are integers and $k = 0, 1, \ldots, n-2$, then we arrive at the polynomial

$$g(x) = \sum_{k=0}^{n-2} c_k((p+q(k+2))x^{k+2} - 2b((p+q(k+1))x^{k+1} + b^2(p+qk)x^k))$$

$$= \sum_{k=0}^{n-2} c_k(p+qk)x^k(x^2 - 2bx + b^2) + \sum_{k=0}^{n-2} c_k 2q(x^{k+2} - bx^{k+1})$$

$$= (x-b)^2 \sum_{k=0}^{n-2}(p+qk)x^k + 2q(x-b)\sum_{k=0}^{n-2} c_k x^{k+1}.$$

Clearly, $x = b$ is a root of $g(x)$. For a modern proof of Hudde's rule, observe that $g(x) = pf(x) + qxf'(x)$, where $f'(x)$ is the derivative of $f(x)$. By writing $f(x) = (x-a)^2 h(x)$, we get $f'(x) = 2(x-a)h(x) + (x-a)^2 h'(x)$. Thus, if $x = a$ is a double root of $f(x)$, then $x = a$ is also a root of $f'(x)$ and conversely.

Hudde's rule greatly reduced the computation required for Descartes's method, especially if the arithmetic progression was chosen judiciously. Van Schooten's book gave several examples of the application of Hudde's rule, and van Heuraet used it in his work on rectification.

5.4 Van Heuraet's Letter on Rectification

We present van Heuraet's 1659 rectification method largely in his own terms with some modification. Concerning Figure 5.2, van Heuraet wrote, in the classical style, "CM is to CQ as Σ to MI," where Σ denoted a fixed line segment. We describe this relationship as $CQ/CM = MI$, eliminating the reference to Σ. Unlike van Heuraet, we now view Σ as a number and set it equal to 1. Van Heuraet set out to find the length of the curve ACE. CQ was normal to the curve, and CN was the tangent at C. The point I was determined so that $CQ/CM = MI$, where MI was perpendicular to AQ. The locus of the point I then determined the curve GIL, and the area under this curve rectified ACE. Once again, recall van Heuraet's perspective: He wrote that the area under the curve GIL equaled the area of the rectangle with one side as Σ and the other side equal to the length of the curve ACE. To prove this, he observed that the similarity of the triangles STX and CMQ gave

$$\frac{ST}{SX} = \frac{CQ}{CM} = MI, \quad \left(=\frac{MI}{\Sigma}\right), \quad \text{or} \quad ST = MI \cdot SX.$$

Thus, the length of ST was the area of the rectangle of base SX and height MI. The lengths of the tangents taken at successive points along AE approximated the length of the curve ACE; when the number of points was increased to infinity, the length of the curve equaled the area under GIL.

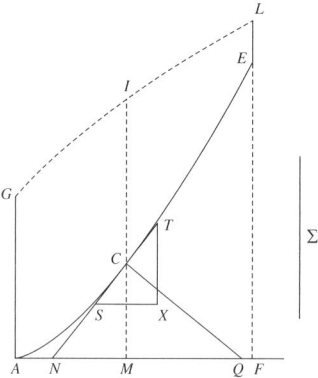

Figure 5.2. Van Heuraet's diagram.

Van Heuraet then explained how the result might be applied to the semicubical parabola $y^2 = x^3/a$, where $AM = x$ and $MC = y$. He let $AQ = s$, $CQ = v$. Then

$$s^2 - 2sx + x^2 + \frac{x^3}{a} = v^2. \tag{5.1}$$

Following Descartes, van Heuraet noted that there were two equal roots of the equation implied by the simultaneous equations $y^2 = \frac{x^3}{a}$ and $(s-x)^2 + y^2 = v^2$. So he multiplied equation (5.1), according to Hudde's method, by 0, 1, 2, 3, 0 to get

$$-2sx + 2x^2 + \frac{3x^3}{a} = 0 \quad \text{or} \quad s = x + \frac{3x^2}{2a}.$$

Thus, $\quad MI = \frac{CQ}{CM} = \frac{v}{y} = \left(1 + \frac{9}{4a}x\right)^{1/2},$

and the area under the curve GIL could be expressed as

$$\frac{8a}{27}\left(1 + \frac{9}{4a}x\right)^{3/2} - \frac{8a}{27}.$$

Van Heuraet then pointed out that the lengths of the curves defined by $y^4 = x^{5/a}$, $y^6 = x^{7/a}$, $y^8 = x^{9/a}$, and so on to infinity, could be found in a similar way. However, in the case of a parabola $y = x^2/a$, one had to compute the area under the hyperbola $y = \sqrt{4x^2 + a^2}$. He concluded, "From this exactly we learn that the length of the parabolic curve cannot be found unless at the same time the quadrature of the hyperbola is found and vice versa."

5.5 Newton's Rectification of a Curve

Isaac Newton carefully studied van Schooten's book, so he understood van Heuraet's method. Newton worked out a simpler method of rectification, based on his conception

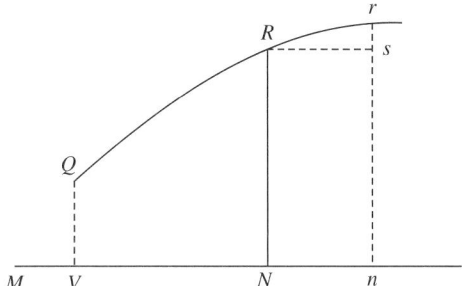

Figure 5.3. Newton's rectification of a curve.

of a curve as a dynamic entity, or as a moving point. In his 1671 treatise, *Of the Method of Fluxions and Infinite Series*, Newton treated arc length by the approach he developed in his 1666 tract on calculus (or fluxions). Referring to Figure 5.3, he explained his derivation in the text presented by the 1737 editor, modified to include Newton's later "dot" notation:

> The Fluxion of the Length is discovered by putting it equal to the square root of the sum of the squares of the Fluxion of the Absciss and of the Ordinate. For let RN be the perpendicular Ordinate moving upon the Absciss MN. And let QR be the proposed Curve, at which RN is terminated. Then calling $MN = s$, $NR = t$, and $QR = v$, and their Fluxions $\dot{s}$, $\dot{t}$, and $\dot{v}$, respectively; conceive the line NR to move into the place nr infinitely near the former, and letting fall Rs perpendicularly to nr; then Rs, sr and Rr, will be contemporaneous moments of the lines MN, NR, and QR, by the accession of which they become Mn, nr, and Qr; but as these are to each other as the Fluxions of the same lines, and because of the Rectangle Rsr, it will be $\sqrt{\overline{Rs}^2 + \overline{sr}^2} = Rr$, or $\sqrt{\dot{s}^2 + \dot{t}^2} = \dot{v}$.

Later in the treatise, he added that one may take $\dot{s} = 1$. This gives exactly the formula we have in textbooks now. It is interesting that some of his examples still appear in modern textbooks. For example, Newton considered the equation $y = \frac{z^3}{aa} + \frac{aa}{12z}$, with a a constant. Taking $\dot{z} = 1$, he had $\dot{y} = \frac{3zz}{aa} - \frac{aa}{12zz}$ and $\sqrt{1 + \dot{y}^2} = \frac{3zz}{aa} + \frac{aa}{12zz}$. Thus, the arc length was given by $\frac{z^3}{aa} - \frac{aa}{12z}$. Newton also went on to find the constant of integration.

5.6 Leibniz's Derivation of the Arc Length

Leibniz derived a formula for arc length in 1673, during his earliest mathematical researches. At that time he had not yet understood the significance of the derivative. However, under the influence of Pascal, he used the characteristic triangle.

The vertical line in the Figure 5.4 is of length a; t is the length of the tangent between the x-axis and the vertical line. From the two similar triangles, Leibniz had

$$a\,ds = t\,dy, \quad \text{or} \quad a\int ds = \int t\,dy.$$

The Rectification of Curves

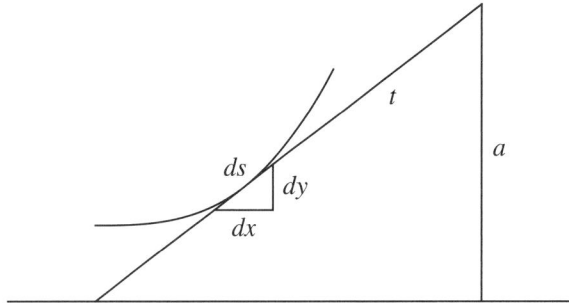

Figure 5.4. Leibniz on arclength.

Leibniz left the result in this form, since he had reduced the arc length to an area or quadrature problem. It is easy to see that

$$\frac{t}{a}dy = \sqrt{1 + \left(\frac{dx}{dy}\right)^2}\,dy.$$

In a manuscript of 1677, he merely noted that

$$ds = \sqrt{(dx)^2 + (dy)^2} = \sqrt{1 + \left(\frac{dy}{dx}\right)^2}\,dx,$$

and thus $\quad s = \int ds = \int \sqrt{1 + \left(\frac{dy}{dx}\right)^2}\,dx.$

5.7 Exercises

1. Find the lengths, between two arbitrary points, of curves defined by the equations: $y = 2(a^2 + z^2)^{3/2}/3a^2$; $ay^2 = z^3$; $ay^4 = z^5$; $ay^6 = z^7$; $ay^{2n} = z^{2n+1}$; $ay^{2n-1} = z^{2n}$; $y = (a^2 + bz^2)^{1/2}$. See Newton (1964–1967), vol. 1, pp. 181–187. Note that in $y = (a^2 + bz^2)^{1/2}$, Newton expressed the arc length as an infinite series.
2. Find the length of any part of the equiangular or logarithmic spiral. Note that this spiral in polar coordinates is defined by $r = ae^{\theta \cot \alpha}$ where a and α are constants.
3. Find the length of any arc of the Archimedean spiral defined by $r = a\theta$. Stone (1730) worked out the examples in exercises 2 and 3. Edmund Stone (1700–1768) was the son of the gardener of the Duke of Argyll. He taught himself mathematics, Latin, and French and translated mathematical works from these languages into English. Elected to the Royal Society in 1725, Stone was financially supported by the Duke whose death in 1743 left him destitute. See Pierpoint (1997).

4. Take a triangle ABC whose sides $AB = x$, $AC = a$, $BC = x^2/a$, and the perpendicular (altitude) $CD = a^2/x$ form a geometric progression. Show that

$$x = a\sqrt{1/2 + \sqrt{5}/2}.$$

See Newton (1964–1967), p. 63. This exercise is taken from Newton's book on algebra, illustrating that algebra could be used to solve geometric problems. Viète had earlier used algebra in this way. Note, however, that this use of algebra is different from the algebraic geometry of Descartes and Fermat, in which algebraic equations are used from the outset to define curves.

5. In a triangle ABC, let $AC = a$, $BC = b$, $AB = x$. Let $CD = c$ bisect the angle at C. Show that $x = (a+b)\sqrt{(ab-cc)/ab}$. See Simpson (1800), p. 261. This is an example of algebra being used in the service of geometrically defined problems.

6. Suppose ABC is a triangle such that the length of the bisectors of the angles B and C are equal. Use the result of the previous exercise to prove that the triangle is isosceles. In 1840, this theorem, now known as the Lehmus-Steiner theorem or the internal bisectors problem, was suggested as a problem by D. C. Lehmus (1780–1863) to the great Swiss geometer, Jacob Steiner (1796–1867). As a high school student during the 1930s, A. K. Mustafy rediscovered Simpson's result and applied it to solve this problem. A. S. Mittal has pointed out to me that if the trisectors or n-sectors are equal, the triangle must be isosceles; the reader may wish to prove this also.

5.8 Notes on the Literature

Descartes (1954), translated by Smith and Latham, gives both an English translation and the original French of Descartes's book on geometry, in which the discussion on the normal to a curve appears on pp. 95–112. The quotation on rectification is on p. 91; Descartes's remark on his construction of the normal is on p. 95. A detailed account of van Heuraet's mathematical work is given in van Maanen (1984). A translation into Dutch and English of van Heuraet's Latin letter can be found in Grootendorst and van Maanen (1982); see p. 107 for the Hudde quotation on double roots. The reader may also enjoy reading Bissell (1987) for a well-written summary of the 1650–1660 work of the Dutch mathematicians. Hudde's method is also discussed in these three articles. Newton's rectification procedure is taken from Newton (1964–67), vol. I, pp. 173–174. This is a 1737 anonymous English translation of Newton's 1671 text on calculus. Whiteside preferred this translation to the 1736 version by John Colson, who was later Lucasian professor at Cambridge.

Child (1920) has English translations of excerpts from some Leibniz manuscripts written during 1673–1680. These indicate the progression in Leibniz's thought as he developed the calculus. Scriba (1964) presents a careful discussion of the evolution of Leibniz's ideas; Hofmann (1974) contains an interesting chapter on rectification. Edmund Stone's 1730 book had two parts: a translation of G. l'Hôpital's 1696 differential calculus text and a treatment of the integral calculus. Out of regard for Newton,

Stone changed l'Hôpital's differentials to fluxions, even though in 1715 Newton warned against this identification, since the fluxion or velocity was finite, whereas the differential was an infinitesimal. Gucciardini's (1989) discussion of other eighteenth-century British calculus textbooks provides much interesting information on those books and their authors. For example, Gucciardini suggests that Bishop (George) Berkeley may have used Stone's presentation as mathematical background for his 1734 work, containing his philosophical objection to the concept of the infinitesimal. Pierpoint (1997) gives Stone's dates as 1700–1768, as opposed to others who give 1695 as his date of birth.

6

Inequalities

6.1 Preliminary Remarks

In his 1928 presidential address to the London Mathematical Society, G. H. Hardy observed, "A thorough mastery of elementary inequalities is to-day one of the first necessary qualifications for research in the theory of functions." He also recalled, "I think that it was Harald Bohr who remarked to me that that 'all analysts spend half their time hunting through the literature for inequalities which they want to use and cannot prove.'" It is surprising, however, that the history of one of our most basic inequalities, the arithmetic and geometric means inequality (AMGM), is tied up with the theory of algebraic equations. Inequalities connected with the symmetric functions of the roots of an equation were used to determine the number of that equation's positive and negative roots. In 1665–66, Newton laid down the foundation in this area when, in order to determine the bounds on the number of positive and negative roots of equations, he stated a far-reaching generalization of Descartes's rule of signs.

The arithmetic and geometric means inequality states that if there are n nonnegative numbers $a_1, a_2, \ldots, a_n$ and there are n positive numbers $q_1, q_2, \ldots, q_n$, such that $\sum_{i=1}^{n} q_i = 1$, then

$$a_1^{q_1} a_2^{q_2} \cdots a_n^{q_n} \leq \sum_{i=1}^{n} q_i a_i, \qquad (6.1)$$

where equality holds only when all the a_i are equal. The theory of equations has an interesting connection with the case for which $q_i = 1/n$:

$$(a_1 a_2 \cdots a_n)^{1/n} \leq \sum_{i=1}^{n} a_i / n. \qquad (6.2)$$

Note that when $n = 2$, the AMGM is simply another form of

$$(a_1 - a_2)^2 \geq 0; \qquad (6.3)$$

this case can probably be attributed to Euclid. The nature of the relationship between the AMGM and algebraic equations is clear from (6.2). To see this, suppose that

$a_1, a_2, \ldots, a_n$ are the roots of

$$x^n - A_1 x^{n-1} + A_2 x^{n-2} + \cdots + (-1)^n A_n = 0.$$

Then (6.2) is identical with the inequality $A_n^{1/n} \leq A_1$.

The three-dimensional case of the AMGM was first stated and proved by Thomas Harriot (c. 1560–1621) to analyze the roots of a cubic. Before this, François Viète (1540–1603) gave the condition under which a cubic could have distinct positive roots; Harriot then noted that Viète's condition was insufficient and that one also required $A_3^{1/3} < A_1$. Much of Harriot's algebraic work arose from his attempts to improve on both the notation and the results of the algebraist Viète. One of Harriot's innovations was to make algebra completely symbolic. Much of his work dates from about 1594 to the early 1600s, but his book on algebra *Artis Analyticae Praxis* was published in 1631, ten years after his death. And even this book omitted significant portions of Harriot's original text and, in places, changed the order of presentation so that the text lost its clarity. Harriot set up a new, convenient notation for inequality relations, and it is still in use, although Harriot's inequality symbol was very huge. William Oughtred, whose work was done later, independently introduced a different notation for inequality. One can see this cumbersome notation in the early manuscripts of Newton, but it soon fell into disuse. Harriot, with his effective notation, showed that one could carry out algebraic operations without using explanatory sentences or words and he demonstrated the superiority of his notation by rewriting Viète's expressions. The Harriot scholar J. Stedall points out some key examples of this: Where Viète wrote: If to $\frac{A \text{ plane}}{B}$ there should be added $\frac{Z \text{ squared}}{G}$, the sum will be G times A plane + $\frac{B \text{ times } Z \text{ squared}}{B \text{ times } G}$; Harriot wrote: $\frac{ac}{b} + \frac{zz}{g} = \frac{acg+bzz}{bg}$. Viète, under the influence of the Greek mathematicians, wrote A plane, meaning that A was a two-dimensional object; Harriot instead had ac. Then again, Viète described his example of antithesis:

> *A squared* minus *D plane* is supposed equal to *G squared* minus *B* times *A*. I say that *A squared* plus *B* times *A* is equal to *G squared* plus *D plane* and that by this transposition and under opposite signs of conjuction the equation is not changed.

Harriot's streamlined notation gives us:

> Suppose $aa - dc = gg - ba$. I say that $aa + ba = gg + dc$ by antithesis.

Descartes made similar advances in notation and in some places went beyond Harriot. For example, Descartes wrote a^3 or a^4, in place of aaa or $aaaa$, retaining aa for a^2. Moreover, Descartes also stated a rule giving an upper bound for the number of positive (negative) roots of an equation. Its extension by Newton contributed further to the discovery of some important inequalities. In his 1637 *La Géométrie* Descartes stated his rule: "We can determine also the number of true [positive] and false [negative] roots that any equation can have as follows: An equation can have as many true roots as it contains changes of sign from + to − or from − to +; and as many false roots as the number of times two + signs or two − signs are found in succession." Thus, in Descartes's example, $x^4 - 4x^3 - 19x^2 + 106x - 120 = 0$, the term $+x^4$ followed by the term $-4x^3$, then $-19x^2$ followed by $106x$ and, finally, $106x$ followed by -120 net a total of three changes of sign. According to Descartes's recipe, there can therefore be three positive roots. In fact, the roots are 2, 3, 4, and -5. The negative root is indicated by one repeated sign: $-4x^3$ followed by $-19x^2$. A rudimentary form of this rule of signs can be found in the earlier work of Faulhaber and Roth. Indeed, some historians have suggested with some justification that Faulhaber and Descartes may have collaborated in analyzing the work of Roth.

Since Descartes's rule gave an upper bound for the number of positive roots and for the number of negative roots, it also determined a lower bound for the number of complex roots. Newton gave an extension of this rule, yielding a more accurate lower bound for many cases. This extension also directly connected this problem with certain inequalities satisfied by the coefficients of the given polynomial. Such inequalities included the AMGM. Newton's rule involved the consideration of another sequence of polynomials, quadratic in the coefficients of the original polynomial. Newton stated this rule, called by Sylvester Newton's incomplete rule, in his *Arithmetica Universalis*, in the section "Of the nature of the roots of an equaiton."

> But you may know almost by this rule how many roots are impossible.
>
> *Make a series of fractions, whose denominators are numbers in this progression 1, 2, 3, 4, 5, &c. going on to the number which shall be the same as that of the dimensions of the equation; and the numerators the same series of numbers in a contrary order. Divide each of the latter fractions by each of the former. Place the fractions that come out over the middle terms of the equation. And under any of the middle terms, if its square, multiplied into the fraction standing over its head, is greater than the rectangle of the terms on both sides, place the sign +; but if it be less, the sign −. But under the first and last term place the sign +. And there will be as many impossible roots as there are changes in the series of the under-written signs from + to −, and − to +.*

He made the following remarks for the case in which two or more successive terms of the polynomial were zero:

> Where two or more terms are wanting together, under the first of the deficient terms you must write the sign −, under the second sign +, under the third the sign −, and so on, always varying the signs, except that under the last of such deficient terms you must always place +, when the terms next on both sides the deficient terms have contrary signs. As in the equations
>
> $$x^5 + ax^4 * * * + a^5 = 0, \quad \text{and} \quad x^5 + ax^4 * * * - a^5 = 0;$$
> $$+ \quad + \quad - + - \quad + \qquad\qquad + \quad + \quad - + + \quad +$$
>
> the first whereof has four, and the latter two impossible roots. Thus also the equation,
>
> $$x^7 - 2x^6 + 3x^5 - 2x^4 + x^3 * * -3 = 0$$
> $$+ \quad - \quad + \quad - \quad + - \quad +$$
>
> has six impossible roots.

To understand Newton's incomplete rule, we use modern notation. Let the polynomial be
$$a_0 x^n + a_1 x^{n-1} + a_2 x^{n-2} + \cdots + a_{n-1} x + a_n.$$
In the sequence of fractions $\frac{n}{1}, \frac{n-1}{2}, \frac{n-2}{3}, \ldots, \ldots, \frac{1}{n}$, divide the second term by the first, the third by the second and so on to get $\frac{n-1}{2n}, \frac{2(n-2)}{3(n-1)}, \frac{3(n-3)}{4(n-2)}, \ldots$. So $\frac{n-1}{2n}$ is placed over a_1, $\frac{2(n-2)}{3(n-1)}$ over a_2 and so on. If $\frac{(n-1)}{2n} a_1^2 > a_0 a_2$, then a + sign is placed under a_1 and a − sign if the inequality is reversed. Similarly, if $\frac{2(n-2)}{3(n-1)} a_2^2 > a_1 a_3$, then place a + sign

under a_2 and so on. These inequalities take a simpler form if we follow J. J. Sylvester's notation from his 1865 paper in which Newton's rule was proved for the first time, almost two hundred years after it was discovered. Write the polynomial as

$$f(x) = p_0 x^n + n p_1 x^{n-1} + \frac{1}{2} n(n-1) p_2 x^{n-2} + \cdots + n p_{n-1} x + p_n. \tag{6.4}$$

The inequalities become $p_1^2 > p_0 p_2$ or $p_1^2 - p_0 p_2 > 0$, $p_2^2 - p_1 p_3 > 0$, and so on. Thus, Newton's sequence of signs is obtained from the sequence of numbers

$$A_0 = p_0^2, \ A_1 = p_1^2 - p_0 p_2, \ A_2 = p_2^2 - p_1 p_3, \ldots, \ A_{n-1} = p_{n-1}^2 - p_{n-2} p_n, \ A_n = p_n^2.$$

A_0 and A_n are always positive, while a plus sign is written under a_k if

$$A_k = p_k^2 - p_{k-1} p_{k+1} > 0$$

and a minus sign if

$$A_k = p_k^2 - p_{k-1} p_{k+1} < 0.$$

Newton gave several examples of his method, including the polynomial equation $x^4 - 6xx - 3x - 2 = 0$. Here $a_1 = 0$ and the signs of A_1, A_1, A_2, A_3, A_4 came out to be $+ + + - +$. So, Newton wrote that there were [at least] two "impossible roots."

In a 1728 paper appearing in the *Philosophical Transactions*, the Scottish mathematician George Campbell published an incomplete proof of the incomplete rule (Newton's rule for complex roots); his efforts were sufficient to obtain the AMGM. A little later, in 1729, Colin Maclaurin published a paper in the same journal proving similar results. A priority dispute arose in this context, although it is generally recognized that Maclaurin's work was independent. Both Campbell and Maclaurin use the idea that the derivative of a polynomial with only real roots also had only real roots. Note that, in fact, this follows from Rolle's theorem, published in 1692, though neither Campbell nor Maclaurin referred to Rolle. Campbell wrote that the derivative result was well-known to algebraists "and is easily made evident by the method of the maxima and minima." He began his paper by stating the condition under which a quadratic would have complex roots. He then showed that a general polynomial would have complex roots if, after repeated differentiation, it produced a quadratic with complex roots. Extremely little is known of Campbell's life; he was elected to the Royal Society on the strength of his paper in the *Transactions*.

In his 1729 paper, Maclaurin stated and proved, among other results, that

$$p_1 \geq p_2^{1/2} \geq p_3^{1/3} \geq \cdots \geq p_n^{1/n}, \tag{6.5}$$

where the p_k were all positive and defined by (6.4) with $p_0 = 1$. Most of Maclaurin's work in algebra arose out of his efforts to prove Newton's unproven statements and he presented them in his *Treatise of Algebra*, unfortunately published only posthumously in 1748.

Later, especially in the nineteenth century, the arithmetic and geometric means and related inequalities became objects of study on their own merits; they were then stated and proved independent of their use in analyzing the roots of algebraic equations. In the

6.1 Preliminary Remarks

1820s, Cauchy gave an inductive proof of AMGM in his lectures at the École Polytechnique. He started with $n = 2$ and then proved it for all powers of 2. He then obtained the result for all positive integers by a proof containing an interesting trick. In 1906, Jensen discovered that Cauchy's method could be generalized to convex functions, a fruitful concept he discovered and named, though it is implicitly contained in the work of Otto Hölder. To understand Jensen's motivation, observe that the two-dimensional case of (6.1) could be written as

$$e^{x_1} + e^{x_2} \geq 2e^{(x_1+x_2)/2}.$$

This led Jensen to define a convex function on an interval $[a, b]$ as a continuous function satisfying

$$\phi(x_1) + \phi(x_2) \geq 2\phi\left(\frac{x_1 + x_2}{2}\right) \tag{6.6}$$

for all pairs of numbers x_1, x_2 in $[a, b]$. Cauchy's proof could be applied in this situation without any change and Jensen was able to prove that for any n numbers $x_1, x_2, \ldots, x_n$ in $[a, b]$,

$$\phi\left(\frac{x_1 + x_2 + \cdots + x_n}{n}\right) \geq \frac{\phi(x_1) + \phi(x_2) + \cdots + \phi(x_n)}{n}. \tag{6.7}$$

Jensen did not require continuity up to this point, but he needed it for a generalization of (6.1).

Johan Jensen (1859–1925) was a largely self-taught Danish mathematician. He studied in an engineering college where he took courses in mathematics and physics. To support himself, he took a job in a Copenhagen telephone company in 1881. His energy and intelligence soon got him a high position in the company where he remained for the rest of his life. The rapid early development of telephone technology in Denmark was mainly due to Jensen. His spare time, however, was devoted to the study of mathematics; the function theorist Weierstrass, also self-taught to a great extent, was his hero. Jensen himself made a significant contribution to the theory of complex analytic functions, laying the foundation for Nevanlinna's theory of meromorphic functions of the 1920s. Jensen wrote his generalization of (6.1) in the form

$$\phi\left(\frac{\sum a_\mu x_\mu}{a}\right) \leq \frac{\sum a_\mu \phi(x_\mu)}{a}, \tag{6.8}$$

where $a = \sum a_\mu$ and $a_\mu > 0$. He took $\phi(x) = x^p, p > 1, x > 0$ to obtain the important inequality named after Hölder, one form of which states that for $p, b_\mu,$ and c_μ all positive, if $\frac{1}{p} + \frac{1}{q} = 1$, then

$$\sum b_\mu c_\mu \leq \left(\sum b_\mu^p\right)^{1/p} \left(\sum c_\mu^q\right)^{1/q}. \tag{6.9}$$

Interestingly, in 1888, L. J. Rogers was the first to state and prove the inequality (6.9) named after Hölder; he first proved (6.1) and then derived several corollaries, including the Hölder inequality. A year later, Hölder gave the generalization (6.8), except that he took $\phi(x)$ to be differentiable, with $\phi'(x) \geq 0$. It is not difficult to prove that such functions are convex in Jensen's sense. Hölder noted that his work was based on that of Rogers, and Jensen also credited Rogers.

The case $p = q = 2$ of (6.9) is called the Cauchy–Schwarz inequality. Cauchy derived it from an identity with the form, in three dimensions,

$$(ax + by + cz)^2 + (ay - bx)^2 + (az - cx)^2 + (bz - cy)^2$$
$$= (a^2 + b^2 + c^2)(x^2 + y^2 + z^2). \qquad (6.10)$$

It is clear that the identity implies

$$\sum ax \leq \left(\sum a^2\right)^{1/2} \left(\sum x^2\right)^{1/2} \qquad (6.11)$$

and that equality would hold if $ay = bx, az = cx$, and $bz = cy$, that is, $\frac{a}{x} = \frac{b}{y} = \frac{c}{z}$. It is evident that identity (6.10) can be extended to any number of variables. Eighteenth-century mathematicians such as Euler and Lagrange applied this and other identities involving sums of squares to physics problems, number theory and other areas.

In 1885, Hermann Schwarz (1843–1921) gave the integral analog of the inequality with which his name is now associated. But this analog was actually presented as early as 1859 by Viktor Bunyakovski (1804–1859), a Russian mathematician with an interest in probability theory who had studied with Cauchy in Paris. Though Bunyakovski made no claim to this result, it is sometimes called the Cauchy–Schwarz-Bunyakovski inequality. Bunyakovski was very familiar with Laplace's work in probability theory, a subject in which he did his best work and for which he worked out a Russian terminology, introducing many terms which have became standard in that language.

One of the earliest applications of the infinite form of the Cauchy–Schwarz and Hölder inequalities was in functional analysis, dealing with infinite series and integrals. For example, in a pioneering paper of 1906, David Hilbert defined l^2 spaces consisting of sequences of complex numbers $\{a_n\}$ such that the sum of the squares of absolute values converged. The infinite form of the Cauchy–Schwarz inequality may be employed to show that an inner product can be defined on l^2. In a paper of 1910, the Hungarian mathematician Frigyes Riesz (1880–1956) generalized Hilbert's work. Dieudonné called this paper "second only in importance for the development of Functional Analysis to Hilbert's 1906 paper." Riesz kept well abreast of the work of Hilbert, Erhard Schmidt, Ernst Hellinger, Otto Toeplitz, Ernst Fischer, Henri Lebesgue, Jacques Hadamard, and Maurice Frèchet. With such inspiration, Riesz was able to define and develop the theory of l^p and L^p spaces. By using Minkowski's inequality, he proved that these were vector spaces; employing the Hölder inequality, he showed that l^q and L^q were duals of l^p and L^p, where $q = (p-1)/p$ and $p > 1$. In a proof very different from Minkowski's proof related to the geometry of numbers, Riesz demonstrated that Minkowski's inequality could be obtained from Hölder's. Thus, inequalities originating in the study of algebraic equations eventually led to inequalities now fundamental to analysis.

6.2 Harriot's Proof of the Arithmetic and Geometric Means Inequality

Harriot proved the AMGM only in the cases of two and three dimensions, but his motivation, notation, and mode of presentation are worthy of note. Harriot began by proving the inequality for dimension 2:

> Lemma I Suppose $b, a, \frac{aa}{b}$ are in continued proportion and suppose $b > a$. I say that $b + \frac{aa}{b} > 2a$ that is $bb + aa > 2ab$ so $bb - ba > ba - aa$ that is
>
> $$\begin{array}{c|c} b-a \\ \hline b \end{array} > \begin{array}{c|c} b-a \\ \hline a \end{array} \qquad (6.12)$$
>
> so $b > a$ and this is so. Therefore the lemma is true.

Note that the expression on the left was Harriot's notation for $(b-a)b$. Harriot used this lemma to analyze the different forms taken by a cubic with one positive root. He proved the three-dimensional case in connection with a result of Viète. In his *De Numerosa Potestatum Resolutione*, Viète discussed a condition for a cubic to have three distinct roots: "A cubic affected negatively by a quadratic term and positively by a linear term is ambiguous [has distinct roots] when three times the square of one-third the linear coefficient [of the square term] is greater than the plane coefficient [of the first power]." Viète's example was $x^3 - 6x^2 + 11x = 6$. Here $3\left(\frac{6}{3}\right)^2 > 11$ and the roots were 1, 2, 3. Harriot commented that Viète's statement required an amendment; in order to get three positive roots, he required that "the cube of a third of the coefficient of the square term is greater than the given constant." This would yield the three-dimensional case of the inequality. Harriot went on to give an example, showing why Viète's condition was inadequate. He noted that $aaa - 6aa + 11a = 12$ had only one positive root (namely, 4) even though Viète's condition was satisfied. In a similar way, he amended Viète's remarks for the case of equal roots.

Harriot stated and proved additional lemmas, of which we give two; he gave the comment, "But what need is there for verbose precepts, when with the formulae from our reduction, it is possible to show all the roots directly, not only for these cases, but for any other case you like. However, if a demonstration of these precepts is required, we adjoin the three following lemmas."

$$3, \begin{array}{c|c} \frac{b+c+d}{3} \\ \hline \frac{b+c+d}{3} \end{array} > bc + cd + bd \quad \text{and} \quad \begin{array}{c|c} \frac{b+c+d}{3} \\ \hline \frac{b+c+d}{3} \\ \hline \frac{b+c+d}{3} \end{array} > bcd.$$

These inequalities are particular cases of (6.5) and can be written as

$$3((b+c+d)/3)^2 > bc + cd + bd \; ; \; ((b+c+d)/3)^3 > bcd.$$

As we noted previously, Descartes made advances over Harriot in terms of notation, though he continued to write aa instead of a^2; in fact, this practice continued well into the nineteenth century as one may see in the work of Gauss, Riemann, and others. The notation for the fractional or irrational power was introduced by Newton in his earliest mathematical work. Surprisingly, Descartes claimed that he had not seen Harriot's book, published six years before his own 1637 work. He even claimed not to have read Viète.

6.3 Maclaurin's Inequalities

Maclaurin's novel proof of the arithmetic and geometric means inequality is worth studying, though it used an unproved assumption on the existence of a maximum. The proof consists of two steps, lemmas V and VI, contained in his 1729 paper on algebraic equations.

> Lemma V Let the given line AB be divided anywhere in P and the rectangles of the parts AP and PB will be a maximum when the parts are equal.

In algebraic symbols, Maclaurin's lemma would be stated: If $AB = a$ and $AP = x$, then $x(a - x)$ is maximized when $x = a/2$ for x in the interval $0 \leq x \leq a$. Maclaurin wrote that this followed from Euclid's *Elements*. He then stated and proved the following generalization:

> Lemma VI If the line AB is divided into any numbers of parts AC, CD, DE, EB, the product of all those parts multiplied into one another will be a maximum when the parts are equal among themselves.
>
> $\overline{ACDEeB}$
>
> For let the point D be where you will, it is manifest that if DB be bisected in E, the product $AC \times CD \times DE \times EB$ will be greater than $AC \times CD \times De \times eB$, because $DE \times EB$ is greater than $De \times eB$; and for the same reason CE must be bisected in C and D; and consequently all the parts AC, CD, DE, EB must be equal among themselves, that their product may be a maximum.

In other words, Maclaurin argued that if $\alpha_1, \alpha_2, \ldots, \alpha_n$ are positive quantities not all equal to each other and their sum $\sum \alpha_i = A$ is a constant, then there exist $\alpha'_1, \alpha'_2, \ldots, \alpha'_n$ with $\sum \alpha'_i = A$ and $\alpha'_1 \alpha'_2 \cdots \alpha'_n > \alpha_1 \alpha_2 \cdots \alpha_n$. Thus, if a maximum value of the product exists, then it must occur when all the α are equal. Maclaurin assumed that such a maximum must exist; proving this would boil down to showing that the continuous function of $n - 1$ variables

$$\alpha_1 \alpha_2 \cdots \alpha_{n-1} (A - \alpha_1 - \alpha_2 - \cdots - \alpha_{n-1})$$

has a maximum in the closed domain $\alpha_1 \geq 0, \alpha_2 \geq 0, \ldots, \alpha_{n-1} \geq 0, \alpha_1 + \alpha_2 + \cdots + \alpha_{n-1} \leq A$. It was common for eighteenth-century mathematicians to assume the existence of such a maximum. Lagrange did this extensively in his derivation of the Taylor theorem with remainder. The inequality for the arithmetic and geometric means follows from these lemmas. We see that if all values of α_i are equal, then $\alpha_i = A/n$ and we can conclude that

$$\alpha_1 \alpha_2 \cdots \alpha_n \leq \left(\frac{A}{n}\right)^n = \left(\frac{\alpha_1 + \alpha_2 + \cdots + \alpha_n}{n}\right)^n.$$

Moreover, equality holds if and only if all the α_i are identical.

6.4 Jensen's Inequality

Jensen proved (6.7) by following Cauchy's proof of (6.2) in detail. From the definition of convexity (6.6), he deduced that

$$\phi(x_1) + \phi(x_2) + \phi(x_3) + \phi(x_4) \geq 2\phi\left(\frac{x_1+x_2}{2}\right) + 2\phi\left(\frac{x_1+x_2}{2}\right)$$
$$\geq 4\phi\left(\frac{x_1+x_2+x_3+x_4}{4}\right).$$

He showed by an inductive argument that

$$\sum_{v=1}^{2^m} \phi(x_v) \geq 2^m \phi\left(2^{-m} \sum_{v=1}^{2^m} x_v\right).$$

This proved the inequality for the case in which the number of xs was a power of two. To prove the theorem for any number of xs, Jensen, still following Cauchy, applied Cauchy's ingenious idea: For any positive integer n, choose m so that $2^m > n$ and set

$$x_{n+1} = x_{n+2} = \cdots = x_{2^m} = \frac{x_1 + x_2 + \cdots + x_n}{n}.$$

Then $\displaystyle\sum_{v=1}^{n} \phi(x_v) + (2^m - n)\phi\left(\frac{1}{n}\sum_{v=1}^{n} x_v\right) \geq 2^m \phi\left(\frac{1}{n}\sum_{v=1}^{n} x_v\right)$

or $\displaystyle\phi\left(\frac{1}{n}\sum_{v=1}^{n} x_v\right) \leq \frac{1}{n}\sum_{v=1}^{n} \phi(x_v).$

Jensen then used the continuity of ϕ to get the more general inequality (6.8). He supposed $a_1, a_3, \ldots, a_m$ to be m positive numbers with sum a, as in (6.8). He chose sequences of positive integers $n_1, n_2, \ldots, n_m$ with $n_1 + n_2 + \cdots + n_m = n$ such that

$$\lim_{n\to\infty} \frac{n_1}{n} = \frac{a_1}{a}, \quad \lim_{n\to\infty} \frac{n_2}{n} = \frac{a_2}{a}, \quad \cdots, \quad \lim_{n\to\infty} \frac{n_{m-1}}{n} = \frac{a_{m-1}}{a}.$$

Consequently, he could write

$$\lim_{n\to\infty} \frac{n_m}{n} = \frac{a_m}{a}.$$

Now (6.7) implied that

$$\phi\left(\frac{n_1 x_1 + n_2 x_2 + \cdots + n_m x_m}{n}\right) \leq \frac{n_1}{n}\phi(x_1) + \frac{n_2}{n}\phi(x_2) + \cdots + \frac{n_m}{n}\phi(x_m);$$

from this Jensen got (6.8) by letting $n \to \infty$ and using the continuity of ϕ. Jensen also gave an integral analog of this inequality. He supposed that $a(x)$ and $f(x)$ were integrable on $(0, 1)$ and $a(x)$ was positive; $\phi(x)$ was assumed to be convex and continuous in the interval (g_0, g_1), where g_0 and g_1 were, respectively, the inferior and superior limits of $f(x)$ in $(0, 1)$. Then he had

$$\phi\left(\frac{\sum_{v=1}^{n} a(v/n) f(v/n) 1/n}{\sum_{v=1}^{n} a(v/n) 1/n}\right) \leq \frac{\sum_{v=1}^{n} a(v/n) \phi(f(v/n)) 1/n}{\sum_{v=1}^{n} a(v/n) 1/n}.$$

By letting $n \to \infty$, he found

$$\phi\left(\frac{\int_0^1 a(x)f(x)\,dx}{\int_0^1 a(x)\,dx}\right) \le \frac{\int_0^1 a(x)\phi(f(x))\,dx}{\int_0^1 a(x)\,dx}.$$

6.5 Reisz's Proof of Minkowski's Inequality

Riesz's derivation of Hölder's and Minkowski's inequalities were contained in his letter to Leonida Tonelli of February 5, 1928. Although Riesz had worked out these ideas almost two decades earlier and had presented them in papers, his object in this letter was to present proofs of the inequalities without any mention of the applications. These proofs are essentially the same as our standard derivations of all these inequalities. Stating that he did this work around 1910, Riesz started with

$$A^\alpha B^{1-\alpha} \le \alpha A + (1-\alpha)B, \quad 0 < \alpha < 1, \quad A \ge 0, \quad B \ge 0.$$

This followed immediately from the convexity of the exponential function, but Riesz gave a simpler proof. After this proof, he supposed $f(x)$ and $g(x)$ were nonnegative functions defined on a measurable set E such that

$$\int_E f^p\,dx = \int_E g^{\frac{p}{p-1}}\,dx = 1, \quad p > 1.$$

He then took $A = f^p$, $B = g^{\frac{p}{p-1}}$, $\alpha = \frac{1}{p}$ to get

$$fg \le \frac{1}{p}f^p + \frac{p-1}{p}g^{p/p-1}; \text{ thus, } \int_E fg\,dx \le \frac{1}{p} + \frac{p-1}{p} = 1.$$

For general f and g, he replaced f and g by $|f|/\left|\int_E |f|^p\,dx\right|^{1/p}$ and $|g|/\left|\int_E |g|^{\frac{p}{p-1}}\,dx\right|^{\frac{p-1}{p}}$, respectively, to obtain

$$\int_E |fg|\,dx \le \left|\int_E |f|^p\,dx\right|^{1/p} \left|\int_E |g|^{\frac{p}{p-1}}\,dx\right|^{\frac{p-1}{p}}.$$

He next cleverly observed that

$$\int_E (f+g)^p\,dx = \int_E f(f+g)^{p-1}\,dx + \int_E g(f+g)^{p-1}\,dx.$$

With $f \ge 0$ and $g \ge 0$, he had

$$\int_E (f+g)^p\,dx \le \left(\int_E f^p\,dx\right)^{1/p} \left(\int_E (f+g)^p\,dx\right)^{\frac{p-1}{p}}$$
$$+ \left(\int_E g^p\,dx\right)^{1/p} \left(\int_E (f+g)^p\,dx\right)^{\frac{p-1}{p}}.$$

Dividing across by $\left(\int_E (f+g)^p\, dx\right)^{\frac{p-1}{p}}$, he could obtain

$$\left(\int_E (f+g)^p\, dx\right)^{1/p} \le \left(\int_E f^p\, dx\right)^{1/p} + \left(\int_E g^p\, dx\right)^{1/p},$$

and this was Minkowski's inequality, stated and used by Minkowski for sums in geometry of numbers.

6.6 Exercises

1. Let $s_n = u_0 + u_1 + \cdots + u_n$, where the terms u_i are positive.

   (a) Show that $\ln \frac{s_n}{s_{n-1}} < \frac{u_n}{s_{n-1}}$.

   (b) Deduce that
   $$\frac{u_1}{s_0} + \frac{u_2}{s_1} + \cdots + \frac{u_n}{s_{n-1}} > \ln s_n - \ln s_0.$$

   (c) Prove that if $\sum_{n=1}^\infty u_n$ is divergent, then $\sum_{n=1}^\infty \frac{u_n}{s_n^\alpha}$ is divergent for $\alpha \le 1$.

   (d) Show that when $\alpha > 0$,
   $$s_{n-1}^{-\alpha} - s_n^{-\alpha} = (s_n - u_n)^{-\alpha} - s_n^{-\alpha} > s_n^{-\alpha} + \alpha s_n^{-\alpha-1} u_n - s_n^{-\alpha} = \alpha \cdot \frac{u_n}{s_n^{1+\alpha}}.$$

   (e) Deduce that if $\sum_{n=1}^\infty u_n$ is divergent, then $\sum_{n=1}^\infty (u_n / s_n^{1+\alpha})$ is convergent for $\alpha > 0$. See Abel (1965), vol. 2, pp.197–98.

2. Suppose $p > 1$ and $a_i > 0$. Suppose that the series $L = \sum_{i=1}^\infty a_i x_i$ converges for every system of positive numbers x_i ($i = 1, 2, \ldots$) such that $\sum_{i=1}^\infty x_i^p = 1$. Use Abel's result in exercise 1 to prove that $\sum_{i=1}^\infty a_i^{p/(p-1)}$ is convergent and that $L \le \left(\sum_{i=1}^\infty a_i^{p/(p-1)}\right)^{(p-1)/p}$. See Landau (1907).

3. Prove that if h is measurable and $\int_a^b |f(x) h(x)|\, dx$ exists for all functions $f \in L^p(a,b)$, then $h \in L^{p/(p-1)}(a,b)$. See Riesz (1960), vol. 1, pp. 449–451.

4. Show that
$$\sum_{m=1}^\infty \sum_{n=1}^\infty \frac{a_m b_n}{m+n} < 2\pi \left(\sum_{m=1}^\infty a_m^2\right)^{1/2} \left(\sum_{n=1}^\infty b_n^2\right)^{1/2}.$$

   Hilbert presented this result in his lectures on integral equations. It was first published in 1908 in Hermann Weyl's doctoral dissertation. I. Schur proved that the constant 2π could be replaced by π. See Steele (2004).

5. Where $p_1, p_2, \ldots, p_n$ are real, let
$$f(x,y) = (x + \alpha_1 y)(x + \alpha_2 y) \cdots (x + \alpha_n y)$$
$$= x^n + n p_1 x^{n-1} y + \binom{n}{2} p_2 x^{n-2} y^2 + \cdots + \binom{n}{n} p_n y^n.$$

(a) Derive the quadratic polynomial obtained by first taking the rth derivative of $f(x, y)$ with respect to y and then the $(n-r-2)$th derivative with respect to x of $f_y^{(r)}(x, y)$.

(b) Use the quadratic polynomial to show that if all $\alpha_1, \alpha_2, \ldots, \alpha_n$ are real, then $p_{r+1}^2 \geq p_r p_{r+2}$. This is in effect the argument George Campbell gave to show that if $p_{r+1}^2 < p_r p_{r+2}$ for some r, then $f(x, 1)$ has at least two complex roots. In fact, he did not use the variable y; instead, he applied the lemma he stated and proved:

> Whatever be the number of impossible roots in the equation
> $$x^n - Bx^{n-1} + Cx^{n-2} - Dx^{n-3} + \cdots \pm dx^3 \mp cx^2 \pm bx \mp A = 0,$$
> there are just as many in the equation
> $$Ax^n - bx^{n-1} + cx^{n-2} - dx^{n-3} + \cdots \pm Dx^3 \mp Cx^2 \pm Bx \mp 1 = 0.$$
> For the roots of the last equation are the reciprocals of those of the first as is evident from common algebra.

This lemma is also contained in Newton's *Arithmetica Universalis*. Newton explained that the equation for the reciprocals of the roots of $f(x)$ was given by $x^n f(1/x) = 0$.

6. Suppose that $\alpha_1, \alpha_2, \ldots, \alpha_n$ in exercise 5 are positive. Show that
$$p_2(p_1 p_3)^2 (p_2 p_4)^3 \cdots (p_{k-1} p_{k+1})^k < p_1^2 p_2^4 p_3^6 \cdots p_k^{2k}.$$

Deduce Maclaurin's inequality (6.5) that $p_{k+1}^{1/(k+1)} < p_k^{1/k}$. See Hardy, Littlewood, and Pólya (1967).

7. Fourier's proof of Descartes's rule of signs: Suppose that the coefficients of the given polynomial have the following signs:

$$+ + - + - - - + + - + -.$$

Multiply this polynomial by $x - p$ where p is positive. The result is

$$\begin{array}{c}+ + - + - - - + + - + - \\ \underline{- - + - + + + - - + - +} \\ + \pm - + - \mp \mp + \pm - + - +\end{array}.$$

The ambiguous sign $\pm$ appears whenever there are two terms with different signs to be added. Show that in general the ambiguous sign appears whenever $+$ follows $+$ or $-$ follows $-$. Next show that the number of sign variations is not diminished by choosing either of the ambiguous signs. Also prove that there is always one variation added at the end, whether or not the original polynomial ends with a variation, as in our example. Show by induction that these facts, taken together, demonstrate Descartes's rule. Descartes indicated no proof for his rule. In 1728, J. A. von Segner gave a proof and in 1741 the French Jesuit priest J. de Gua de Malves gave a similar proof, apparently independently. We remark that de Gua also wrote a short history of algebra in which he emphasized

French contributions to algebra at the expense of the English, in order to counter Wallis's 1685 history, emphasizing the opposite. Fourier presented the method described in this exercise in his lectures at the École Polytechnique, soon after its inauguration in November 1794. In 1789, Fourier communicated a paper on the theory of equations to the Académie des Sciences in Paris but due to the outbreak of the French Revolution the paper was lost. In the late 1790s, Fourier's interests turned to problems of heat conduction; it was not until around 1820 that he returned to the theory of equations. His book on equations was published posthumously in 1831.

8. Gauss's proof of Descartes's rule: With his extraordinary mathematical insight, Gauss saw the essence of Fourier's argument and presented it in a general form. He supposed

$$x^{n+1} + A_1 x^n + A_2 x^{n-1} + \cdots + A_{n+1} = (x-p)(x^n + a_1 x^{n-1} + a_3 x^{n-2} + \cdots + a_n),$$

and that the sign changes occurred at $a_{k_1}, a_{k_2}, \ldots, a_{k_s}$. Show that $A_{k_j} = a_{k_j} - p a_{k_{j-1}}$ and that this in turn implies that the signs of A_{k_j} and a_{k_j} are identical for $j = 1, 2, \ldots, s$. Deduce also that there is an odd number of sign changes between $A_{k_{i-1}}$ and A_{k_i}. Conclude, by induction, that the number of sign changes is an upper bound for the number of positive roots and that the two differ by an even number. Gauss published this result in 1828 in the newly founded *Crelle's Journal*. Note that Gauss did not use subscripts; we use them for convenience. See Gauss (1863–1927), vol. 3, pp. 67–70.

9. Fourier's extension of Descartes's rule gives an upper bound on the number of real roots of a polynomial $f(x)$ of degree n in an interval (a,b). Suppose r is the number of real roots in (a,b), m is the number of sign changes in the sequence

$$f(x), f'(x), f''(x), \ldots, f^{(n)}(x)$$

when $x = a$, and k is the number of sign changes when $x = b$. Prove that then $(m-k) - r = 2p$, where p is a nonnegative integer. Descartes's rule follows when $a = 0$ and $b = \infty$. In his 1831 book, Fourier gave a very leisurely account of this theorem with numerous examples.

10. Ferdinand François Budan's (1761–1840) extension of Descartes's rule: With the notation as in the previous exercise, suppose that m is the number of sign changes in coefficients of powers of x in $f(x+a)$, and that k is the corresponding number in $f(x+b)$. Then, $r \leq m - k$. Prove this theorem and also prove that it follows from Fourier's theorem. Budan was born in Haiti and was a physician by training. In 1807, he wrote a pamphlet on his theorem; then in 1811 he presented a paper to the Paris Academy. Lagrange and Legendre recommended it be published, but the Academy's journal was not printed until 1827, partly due to political problems. With the appearance of Fourier's papers in 1818 and 1820, Budan felt compelled to republish his pamphlet with the paper as an appendix. In response, Fourier pointed out that he had lectured on this theorem in the 1790s, as some of his students were willing to testify. Some of Fourier's lecture notes from this period have survived; they contain a discussion of algebraic equations,

in particular Descartes's rule, but they do not discuss Fourier's more general theorem. See the monograph, Budan (1822).

11. Let $f_0(x) = f(x)$ and $f_1(x) = f'(x)$. Apply the Euclidean algorithm to f_0 and f_1, but take the negatives of the remainders. Thus,

$$f_0(x) = q_1(x)f_1(x) - f_2(x),$$
$$f_1(x) = q_1(x)f_2(x) - f_3(x),$$
$$\cdots\cdots\cdots\cdots\cdots\cdots\cdots\cdots$$
$$f_{m-2}(x) = q_{m-1}(x)f_{m-1}(x) - f_m(x).$$

Consider the sequence $f_0(x), f_1(x), \ldots, f_m(x)$. Prove that the difference between the number of changes of sign in the sequence when $x = a$ is substituted and the number when $x = b$ is substituted gives the actual number of real roots in the interval (a, b). Charles Sturm (1803–1855) published this theorem in 1829. Sturm was a great friend of Liouville; they jointly founded the spectral theory of second order differential equations. He also worked as an assistant to Fourier, who helped him in various ways. See Sturm (1829).

12. Let $F(x) = Ax^p + \cdots + Mx^r + Nx^s + \cdots + Rx^u$, and let the powers of x run in increasing (or decreasing) order. Let m be the number of variations of signs of the coefficients and let α be an arbitrary real number. Prove that the number of positive roots of $xF'(x) - \alpha F(x) = 0$ is one less than the number of positive roots of $F(x) = 0$. Prove also that if α lies between r and s, then the number of sign variations in the coefficients of $xF' - \alpha F$ is the same as the number of sign variations in the sequence $A, \ldots, M, -N, \ldots, -R$; in other words, $m - 1$. From this, deduce Descartes's rule and prove that the equation

$$x^3 - x^2 + x^{1/3} + x^{1/7} - 1 = 0$$

has at most three positive roots and no negative roots. These results were given by Laguerre in 1883. See Laguerre (1972), vol. 1, pp. 1–3.

13. Prove de Gua's observation that when $2m$ successive terms of an equation have 0 as coefficient, the equation has $2m$ complex roots; if $2m + 1$ successive terms are 0, the equation has $2m + 2$, or $2m$ complex roots, depending on whether the two terms, between which the missing terms occur, have like or unlike signs. See Burnside and Panton (1960), vol. 1, chapter 10.

14. In his book on the theory of equations, Robert Murphy took $f(x) = x^3 - 6x^2 + 8x + 40$ to illustrate Sturm's theorem in exercise 5. Carry out the details. See Murphy (1839), p. 25.

15. Suppose $f(x)$ is a polynomial of degree n. Prove Newton's rule that if $f(a), f'(a), \ldots, f^{(n)}(a)$ are all positive, then all the real roots of $f(x) = 0$ are less than a. Newton gave this rule in his *Arithmetica Universalis* in the section, "Of the Limits of Equations."

16. Following Fourier, let $f(x) = x^5 - 3x^4 - 24x^3 + 95x^2 - 46x - 101$. Consider the sequence $f^V(x), f^{IV}(x), \ldots, f'(x), f(x)$ and find the number of sign variations when $x = -10, x = -1, x = 0, x = 1$, and $x = 10$. What does your analysis show

about the real roots of $f(x)$? Now apply Sturm's method to this polynomial. The tediousness of this computation explains why one might wish to rely on Fourier's procedure.

17. Let
$$f_0(x) = A_0 x^m + A_1 x^{m-1} + A_2 x^{m-2} + \cdots + A_{m-1} x + A_m.$$
Set $f_m(x) = A_0$, and $f_i(x) = x f_{i+1}(x) + A_{m-i}$, $i = m-1, m-2, \ldots, 0$. Prove that the number of variations of sign in $f_m(a), f_{m-1}(a), \ldots, f_0(a)$, $a > 0$, is an upper bound for the number of roots of $f_0(x)$ greater than a; show that the two numbers differ by an even number. This result is due to Laguerre. See Laguerre (1972), vol. 1, p. 73.

18. After his examples of the incomplete rule, Newton moved on to state what has become known as Newton's complete rule for complex roots. In 1865, J. J. Sylvester offered a description of this rule:

Let $fx = 0$ be an algebraical equation of degree n. Suppose
$$fx = a_0 x^n + n a_1 x^{n-1} + \frac{1}{2}(n-1) a_2 x^{n-2} + \cdots + n a_{n-1} x + a_n;$$
$a_0, a_1, a_2, \ldots, a_n$ may be termed the simple elements of fx. Suppose
$$A_0 = a_0^2, \ A_1 = a_1^2 - a_0 a_2, \ A_2 = a_2^2 - a_1 a_3, \ldots A_{n-1} = a_{n-1}^2 - a_{n-2} a_n, \ A_n = a_n^2;$$
$A_0, A_1, A_2, \ldots, A_n$ may be termed the quadratic elements of fx. a_r, a_{r+1} is a succession of simple elements, and A_r, A_{r+1} of quadratic elements.

$$\left.\begin{array}{c} a_r \\ A_r \end{array}\right\} \text{ is an } \textit{associated} \text{ couple of elements;}$$

$$\left.\begin{array}{cc} a_r & a_{r+1} \\ A_r & A_{r+1} \end{array}\right\} \text{ is an associated couple of } \textit{successions}.$$

A succession may contain a permanence or a variation of signs, and will be termed for brevity a permanence or variation, as the case may be. Each succession in an associated couple may be respectively a *permanence* or a *variation*. Thus an associated couple may consist of two permanences or two variations, or a superior permanence and inferior variation, or an inferior permanence and superior variation; these may be denoted respectively by the symbols pP, vV, pV, vP, and termed *double* permanences, *double* variations, permanence variations, variation permanences. The meaning of the simple symbols p, v, P, V speaks for itself.

Newton's rule in its complete form may be stated as follows–On writing the complete series of quadratic under the complete series of simple elements of fx in their natural order, the number of double permanences in the associated series, or pair of progressions so formed, is a superior limit to the number of negative roots, and the number of variation permanences in the same is a superior limit to the number of positive roots in fx. Thus the number of negative roots = or $< \sum pP \ldots$, positive roots = or $< \sum vP$. This is the Complete Rule as given in other terms by Newton. The rule for negative roots is deducible from that for positive, by changing x into $-x$. As a corollary, the total number of real roots = or $< \sum pP + \sum vP$, that is = or $< \sum P$. Hence, the number of imaginary roots

$$= \text{or} > n - \sum P, \text{ that is } = \text{or} > \sum V.$$

> This is Newton's incomplete rule, or *first part* of complete rule, the rule as stated by every author whom the lecturer has consulted except Newton himself.

Read Sylvester's proof of this rule. Though Newton did not write down a proof, Sylvester wrote in another paper of the same year, "On my mind the internal evidence is now forcible that Newton was in possession of a proof of this theorem (a point which he has left in doubt and which has often been called into question), and that, by singular good fortune, whilst I have been enabled to unriddle the secret which has baffled the efforts of mathematicians to discover during the last two centuries, I have struck into the very path which Newton himself followed to arrive at his conclusions." See Sylvester (1973), vol. 2, pp. 494 and 498–513. See also Acosta (2003).

6.7 Notes on the Literature

Newton's *Arithmetica Universalis*, written in 1683, contains his account of the undergraduate algebra course he taught at Cambridge in the 1670s. This was partly based on Newton's extensive notes on N. Mercator's Latin translation of Gerard Kinckhuysen's 1661 algebra text in Dutch. The later parts of the *Arithmetica* present Newton's own researches in algebra, carried out in the 1660s. This work was first published in 1707, in Latin; Newton was reluctant to have it published, perhaps because the first portion depended much on Kinckhuysen. An English translation appeared in 1720, motivating Newton to make a few changes and corrections and publish a new Latin version in 1721. In 1722, the English translation was republished with the same minor changes. Whiteside published the 1722 version in vol. 2 of Newton (1964–67); for the Newton quotations in this chapter, see pp. 103–105. The quote comparing Harriot's and Viète's notations is in Stedall (2003), pp. 8–11. This book presents Harriot's original text on algebra for the first time, although in English. The 1631 book published as Harriot's algebra was in fact a mutilated and somewhat confused version. Stedall's introduction explains this unfortunate occurrence. For Harriot's results presented in this chapter, see p. 195 and pp. 233–34. For other results, see G. Campbell (1728), Maclaurin (1729), Jensen (1906), Riesz (1960), vol. 1, pp. 519–21. Note II of Cauchy's *Analyse algébrique* gives the results on inequalities and their proofs, presented by Cauchy to his students in the early 1820s. For Viète's rule for a cubic with distinct roots, see Viète (1983), p. 360. A good source for references to early work on inequalities is Hardy, Littlewood, and Pólya (1967), though they omit Campbell. See Grattan–Guinness (1972) for an interesting historical account of Fourier's work on algebraic equations and Fourier series. For Dieudonné's remark on Riesz's paper, see p. 124 of Dieudonné (1981). This work is an excellent history of functional analysis and covers the period 1900–1950, from Hilbert and Riesz to Grothendieck. For functional analysis after 1950, see the comprehensive history of Pietsch (2007).

7

Geometric Calculus

7.1 Preliminary Remarks

During the decade 1660–1670, the discoveries of the previous quarter century on the mathematics of infinitesimals were systematized, unified, and extended. Those earlier discoveries included the integration of $y = x^{m/n}$, $y = \sin x$ or $\cos x$; the connection of the area under the hyperbola with the logarithm; the reduction of the problem of finding arc length to that of quadrature; the method for finding the tangent to a curve; and the procedure for determining the maximum or minimum point on a curve. Before 1660, the interdependence between problems on construction of tangents and problems concerning areas under curves had been evident only in special cases, but during this decade, Isaac Barrow (1630–1677), James Gregory (1638–1675), and Isaac Newton (1642–1727) independently discovered the fundamental theorem of calculus. Gregory and Barrow stated this result as a theorem in geometry, whereas Newton, deeply influenced by Descartes's algebraic approach, gave it in a form recognizable even today. Later on, Newton adopted the geometric perspective of his *Principia*. It is interesting to see that evaluations of the trigonometric functions took a very simple form when performed geometrically. In the geometric calculus, one got direct visual contact with the elementary functions and their properties, whereas the abstract approach, while more general and widely applicable, gave less insight into its underpinnings.

Barrow entered St. Peter's College, Cambridge in 1643, and later Trinity College. He had wide intellectual interests, including anatomy, botany, mathematics, astronomy, and divinity. He was ordained in 1659. In 1660, Barrow was appointed professor of Greek at Cambridge where his inaugural lectures were on Aristotle's *Rhetoric*. Two years later, he became professor of mathematics in Gresham College and in 1664, became the first Lucasian Professor of Mathematics. The Lucasian Professorship was the first endowed chair of mathematics in Cambridge, as the importance of mathematics was then beginning to be recognized in England.

Barrow resigned from this professorship in favor of Newton in 1669 so he could devote himself entirely to divinity. Barrow's lectures on geometry were printed in 1670 and two decades later Jakob Bernoulli as well as l'Hôpital saw that they contained the elements of calculus in geometric form. Indeed, Barrow stated the sum, product, and quotient rules for derivatives; gave derivatives and integrals of several specific functions; and presented the fundamental theorem of calculus. Following Descartes, Barrow described his results on tangents in terms of the subtangent segment. Thus, his theorem on the derivative of the product $w = yz$, when converted to analytic form, appears as

$$\frac{1}{w}\frac{dw}{dx} = \frac{1}{y}\frac{dy}{dx} + \frac{1}{z}\frac{dz}{dx}.$$

Barrow found derivatives for $\tan x$, $\sqrt{x^2 - a^2}$, $\sqrt{a^2 - x^2}$, and x^n for n rational. He thought that his demonstration of the result for x^n was the first rigorous proof. In his twelfth and final lecture, Barrow also gave interesting and simple geometric arguments for the evaluations of $\int \sec\theta\, d\theta$, $\int \tan\theta\, d\theta$, $\int a^x\, dx$, and $\int dx/\sqrt{x^2 + a^2}$. Apparently, he had not intended publishing these results, since they dealt with particular curves and were not general theorems, but he included them "to please a friend who thinks them worth the trouble." From Barrow's correspondence, we can determine that this friend was John Collins.

When Barrow became professor of mathematics in 1664, Newton was a student at Cambridge and may well have attended some of Barrow's lectures. Barrow's 1664–66 lectures on the fundamentals of mathematics, published in 1683 under the title *Lectiones Mathematicae*, were philosophical in approach with little connection with the calculus. Newton's researches of this period, inspired by the writings of Descartes, Wallis, and van Schooten, took a strongly analytic form. This approach stood in contrast to Barrow's geometric calculus lectures, published later on. Later in life, Newton wrote, concerning his October 1666 derivation of the fundamental theorem of calculus, that he might have been indebted to Barrow for his dynamic view of a curve as a moving point. In the 1640s, Torricelli and Gilles de Roberval (1602–1675) had applied dynamical or kinematic methods to problems on spirals. And in the 1650s, Christopher Wren and others had also used this method to study the cycloid. Since this method had been employed so widely, Newton may have been later uncertain about where he learned it. In addition, note that in the 1630s, Roberval succeeded by geometric methods in integrating $\sin x$ and $\sin^2 x$ over the interval $(0, 2\pi)$. Later, Blaise Pascal used the same method to determine the indefinite integral of these functions.

Thus, it is possible that Barrow, or perhaps others, influenced Newton to adopt the geometrical approach employed in the *Principia*. Until Whiteside, it had generally been assumed that Newton initially derived his results in analytic form, and afterwards transcribed them into geometric form. But Newton actually developed the ideas of his *Principia* at every stage in geometric terms. Later, Euler spent several years converting the proofs of the propositions in the *Principia* into analytic form. Because calculus is now taught and understood completely in an analytic context, it is very often assumed that the analytic form is simpler or easier to understand. However, Newton himself thought otherwise; to him, the geometric method was more rigorous, intuitively clearer, more directly related to the work of the ancients, and avoided concepts or symbols lacking in referential content. In a letter of June 13, 1676, to Leibniz via Oldenburg, Newton discussed infinite series and quadrature, but added, "For I write rather shortly because these theories long ago began to be distasteful to me, to such an extent that I have refrained from them for nearly five years." Thus, although Newton first discovered his calculus in analytic form, he soon found the geometric approach of Barrow and others more congenial. Here we may be instructed by the opinion of a modern scholar of the *Principia*, the Nobel Prize winning physicist, S. Chandrasekhar: "I first constructed [analytic] proofs [of the *Principia* propositions] for myself. Then I compared my proofs with those [geometric ones] of Newton. The experience was a sobering one. Each time, I was left in sheer wonder at the elegance, the careful arrangement, the imperial style, the incredible originality, and above all the astonishing lightness of Newton's proofs; and each time I felt like a schoolboy admonished by his master."

James Gregory published his work somewhat before Barrow. His *Geometriae Pars Universalis* appeared at the end of his 1668 visit to Italy, where he studied with Stephano degli Angeli (1623–1700), a student of Cavalieri. After his return to Scotland, Gregory studied the recently published *Logarithmotechnia* of Nicholas Mercator (Kaufmann) (1620–1687). This inspired him to compose a short work titled *Exercitationes Geometricae*. These two books contain all of Gregory's publications on the mathematics of infinitesimals. He continued to make interesting discoveries, especially in differential calculus and infinite series, but died before he could publish these results.

In his *Pars Universalis*, Gregory proved general theorems on tangents, arc length, and area. He considered and defined convex monotonic curves and gave a rigorous proof in the style of Archimedes that the arc length of such a curve was given by an area under a suitable curve. Van Heuraet had already stated this result, but Gregory added the necessary rigor. In the course of the proof, he also demonstrated the fundamental theorem of calculus in geometric form. It is important to make clear that Gregory did not think in terms of the processes of integration and differentiation and their inverse relationship. Rather, he conceived and stated his results by means of tangents and areas. Yet he discovered several remarkable theorems, including one in which he proved that if $f(x)$ and $g(x)$ were functions such that $f'(x) = g'(x)$ and $f(0) = g(0)$, then $f(x) = g(x)$.

The most important results of Gregory's *Exercitationes* were his evaluations of $\int \sec\theta\, d\theta$ and $\int \tan\theta\, d\theta$ in terms of the logarithm. The first integral was well known for its usefulness in navigation as Gerhard Mercator's projection, but its exact value was unproved until the appearance of Gregory's booklet. He computed the integral by transforming it into a double integral and then changing the order of integration. Naturally, he did all this in wholly geometric terms.

Barrow studied Gregory before he published the *Lectiones Geometricae* and some of his important results such as the tangent and area duality can also be found in Gregory. However, it is probable that Barrow made his discoveries independently, including his simpler proofs of some special results in Gregory, such as the evaluations of the two integrals, $\int \sec\theta\, d\theta$ and $\int \tan\theta\, d\theta$. It is generally accepted that Barrow shunned any form of plagiarism. In connection with another result of Gregory, Barrow gave some extensions, remarking "I do not like to *put my sickle into another man's harvest,* but it is permissible to interweave amongst these propositions one or two little observations…which *have obtruded themselves upon my notice whilst I have been working at something else.*" On the other hand, Gregory was motivated to go more deeply into the differential calculus by a study of Barrow's *Lectiones*. For example, on the blank space in the letter from Collins dated November 1, 1670, he noted several results on derivatives, including the formula, expressed in modern notation: $\sqrt{af}/(\sqrt{af})' = 2ce/(c+e)$, where $c = a/a'$, $e = f/f'$. The accent denotes differentiation with respect to a variable, say x.

It is difficult to determine exactly what these three very great mathematicians, Barrow, Gregory, and Newton, owed to each other, but the evidence strongly suggests that they developed their most important ideas independently.

7.2 Pascal's Evaluation of $\int \sin x \, dx$

We have noted that Gregory and Barrow derived the integral of the secant in the late 1660s. A decade earlier, in his 1659 paper "Traité des sinus du quart de cercle," Blaise Pascal (1623–1662) computed the simpler indefinite integral of the sine function; Pascal based his work on earlier results of Roberval. In his 1659 paper, Pascal also stated the change of variables formula

$$\int \sin^n \theta \, d\theta = -\int \sin^{(n-1)} \theta \, d \cos \theta;$$

in another work, he presented particular cases of the integration by parts formula. The basic idea behind Pascal's work is contained in the lemma illustrated by Figure 7.1: The rectangle formed by the sine DI and the tangent EE is equal to the rectangle formed by a portion of the base (that is RR) and the radius AB.

Note that for Pascal, as for all writers up to the end of the seventeenth century, the sine of an angle was not a ratio, but a line. To obtain the modern sine, first defined by Euler, one divides Pascal's sine, DI, by the radius $AD = AB$. Pascal's lemma follows immediately from the similarity of the $\triangle$s ADI and EEK. In modern notation, we may set $D\hat{A}C = \phi$ to write the result as

$$\sin \phi \, d\phi = -dx = -d \cos \phi, \text{ since } EE = AD \, d\phi.$$

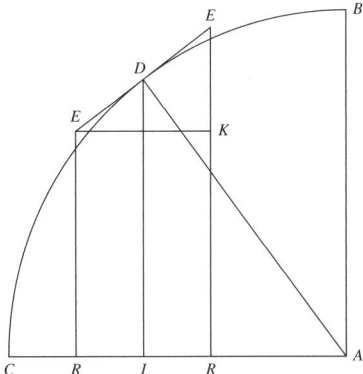

Figure 7.1. Pascal's integration of sine.

Pascal then stated the following propositions: *Proposition* I. The sum of the sines of any arc of a quadrant is equal to the portion of the base between the extreme sines, multiplied by the radius. *Proposition* II. The sum of the squares of those sines is equal to the sum of the ordinates of the quadrant that lie between the extreme sines, multiplied by the radius. *Proposition* III. The sum of the cubes of the same sines is equal to the sum of the squares of the same ordinates between the extreme sines, multiplied by the radius. *Proposition* IV. The sum of the fourth powers of the sines is equal to the sum of the cubes of the same ordinates between the extreme sines, multiplied by the radius. And so on to infinity. Pascal's proof of Proposition I consisted in summing all the equations $\sin\phi\,d\phi = -d\cos\phi$ to get

$$\int_{\phi_0}^{\phi_1} \sin\phi\,d\phi = \cos\phi_0 - \cos\phi_1.$$

Pascal used no algebraic symbolism; he appears to have ignored the notational innovations of Harriot and Descartes. He gave his work on the binomial theorem descriptively; this can prove challenging to the modern reader. But his geometrical argument for Proposition I is fairly straightforward and we give the translation of the main part of the proof:

> Indeed, let us draw at all the points D the tangents DE, each of which intersects its neighbor at the points E; if we drop the perpendiculars ER it is clear that each sine DI multiplied by the tangent EE is equal to each distance RR multiplied by the radius AB. Therefore, all the quadrilaterals formed by the sines DI and their tangents EE (which are all equal to each other) are equal to all the quadrilaterals formed by all portions RR with the radius AB; that is (since one of the tangents EE multiplies each of the sines, and since the radius AB multiplies each of the distances), the sum of the sines, DI, each of them multiplied by one of the tangents EE, is equal to the sum of the distances RR, each multiplied by AB. But each tangent EE is equal to each one of the equal arcs DD. Therefore the sum of the sines multiplied by one of the equal small arcs is equal to the distance AO multiplied by the radius.

7.3 Gregory's Evaluation of a Beta Integral

Gregory gave a beautiful and ingenious geometric argument to prove the integral formula

$$\int_0^{\pi/2} \cos^{2n}\theta\,d\theta = \frac{3\cdot 5\cdot 7\cdots 2n-1}{4\cdot 6\cdot 8\cdots 2n}\frac{\pi}{4}. \tag{7.1}$$

His derivation was effectively based on the integration by parts formula

$$\int \cos^{n+1}\theta\,d\theta = \frac{1}{n+1}\cos^n\theta\sin\theta + \frac{n}{n+1}\int\cos^{n-1}\theta\,d\theta \tag{7.2}$$

from which he obtained the recursion relation

$$\int_0^{\pi/2}\cos^{n+1}\theta\,d\theta = \frac{n}{n+1}\int_0^{\pi/2}\cos^{n-1}\theta\,d\theta. \tag{7.3}$$

It is easy to see that (7.3) implies (7.1). Thus, Gregory provided the first rigorous proof of Wallis's formula for π.

Wallis used, but did not prove, a formula equivalent to (7.1) in his derivation of the infinite product for π; thus, a proof of this formula was clearly in order. Note that Wallis's integral $\int_0^1 (1-x^2)^{n/2} dx$ is obtained when $x = \sin\theta$ is substituted in $\int_0^{\pi/2} \cos^{n+1}\theta \, d\theta$.

Pietro Mengoli (1626–1686) apparently discovered Wallis's formula at about this same time, though he published it in the early 1670s. Gregory's motivation for investigating formulas (7.1), (7.2), and (7.3) was to give a rigorous proof of Mengoli's result, of which Gregory became aware through a May 28, 1673 letter from Collins. Collins wrote that Mengoli had recently published a book in which he discussed Wallis's formula

$$\frac{2 \cdot 4 \cdot 4 \cdots 2n \cdot 2n}{3 \cdot 3 \cdot 5 \cdots (2n-1)(2n+1)} < \frac{\pi}{4} < \frac{2 \cdot 4 \cdot 4 \cdots 2n \cdot (2n+2)}{3 \cdot 3 \cdot 5 \cdots (2n+1) \cdot (2n+1)}.$$

This follows easily from the obvious inequalities

$$\int_0^{\pi/2} \cos^{2n+2}\theta \, d\theta < \int_0^{\pi/2} \cos^{2n+1}\theta \, d\theta < \int_0^{\pi/2} \cos^{2n}\theta \, d\theta$$

and the recursion formula (7.3).

Gregory wrote up his proof on the back of an April 1674 letter from Collins, where it lay undiscovered until Turnbull examined Gregory's papers: "I first examined the documents at St Andrews in 1932, when it was discovered that Gregory, the original recipient of the letters, had used their blank spaces for recording his own mathematical thoughts. As a result of careful scrutiny it has been established that Gregory made several remarkable and unsuspected discoveries, particularly in the calculus and the theory of numbers, which he never published. He was, for example, employing Taylor and Maclaurin expansions more than forty years in advance of anyone else."

The proof depends on a change of variables formula, stated in geometric form as proposition 11 of Gregory's *Geometriae Pars Universalis* of 1668. In fact, this work also contains a result equivalent to the $n=2$ case of (7.2). Gregory saw that his method could be generalized, as illustrated in Figure 7.2.

In Figure 7.2, let HDK be a quadrant of a circle with center K and radius r. Denote angle $O\widehat{K}D$ by θ and let $EI = r\cos^n\theta$ and $LI = r\cos^{n-2}\theta$. Thus, the curves HED and HLD may be taken as the graphs of $r\cos^n\theta$ and $r\cos^{n-2}\theta$, respectively, as θ varies from 0 to $\pi/2$. Let NC, the tangent to HED at E, meet the vertical line KD at C and let BC be parallel to HK and meet the vertical line IO at B. Finally, let the curve BD be the locus of B obtained by constructing a tangent at each point of the curve ED.

On the back of the letter from Collins, James Gregory wrote in Latin:

On p. 113 *Geom Pars Univers.*, let $DK = r$, $OI = e$, $EI = e^n/r^m$, $m = n-1$, $p = n-2$, $q = n-3$, $\frac{EB}{n} = \frac{r^2 e^p - e^n}{r^m}$; whence EO exceeds EB/n by an excess $e - e^p/r^q$. For let $ODE = S$, $EDF = T$; and put $LI = e^p/r^q$, and $LOD = V$; then S exceeds T/n by an excess V, or ODE exceeds DEF/n by an excess LOD; that is $LDE = DEF/n$.

7.3 Gregory's Evaluation of a Beta Integral

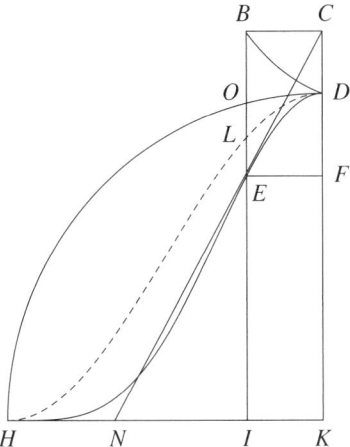

Figure 7.2. Gregory's method for integration by parts of $\cos^n \theta$.

He gave a brief outline of a proof; we expand upon that: The curve HED is defined by $IE = y = e^n/r^{n-1}$ and HLD by e^{n-2}/r^{n-3}. Let $IK = a$ so that $a^2 + e^2 = r^2$. Then the subtangent $NI = y/\frac{dy}{da}$, or

$$NI = \frac{yr^{n-1}}{ne^{n-1}de/da} = \frac{e^2}{na}.$$

The similar $\triangle$s NIE and CEF give

$$CF = EI \cdot \frac{EF}{NI} = \frac{e^n}{r^{n-1}} \cdot \frac{na^2}{e^2} = \frac{ne^{n-2}a^2}{r^{n-1}}.$$

Thus, $BE = CF = nLE$, since $LE = IL - IE = \dfrac{e^n}{r^{n-1}} - \dfrac{e^{n-2}}{r^{n-3}} = \dfrac{e^{n-2}a^2}{r^{n-1}}$.

Because the relation $BE = nLE$ holds at every point along the curve ED, we can conclude that the area of the curvilinear region BDE is n times the area of the region LDE. Now observe that $a\frac{de}{da} = BE$, and as a varies, the quantity $a\frac{de}{da}$ gives the vertical distance between the curve BD and ED. Gregory then proved that area $BDE =$ area DEF by applying proposition 11 of his *Pars Universalis*. This is a change of variables result, easily understandable in the Leibniz notation:

$$\text{area } DEF = \int a\, de = \int a \frac{de}{da} da = \text{area } BDE.$$

We now have

$$\text{area } ODE - \text{area } LDE = \text{area } LOD,$$

or $\text{area } ODE - \dfrac{1}{n} \text{area } DEF = \text{area } LOD.$

The last equation is equivalent to the integration by parts formula given at the beginning of this section. Note that if we denote $O\widehat{K}D$ by θ and let $r = 1$, then

$$\text{area } ODE = \int (\cos\theta - \cos^n\theta)\, d\sin\theta = \int (\cos^2\theta - \cos^{n+1}\theta)\, d\theta,$$

$$\text{area } DEF = \int \cos^{n+1}\theta\, d\theta - \cos^n\theta\sin\theta,$$

$$\text{area } LOD = \int (\cos^2\theta - \cos^{n-1}\theta)\, d\theta,$$

and the integration (7.2) formula follows.

Gregory did not write down the general formula (7.1), but he explicitly wrote down the values of the integral for $n = 3, 5, 7$, and 9 and then noted that the process could be continued indefinitely. Unfortunately, Gregory died before he could publish these remarkable results, which were rediscovered and generalized by Euler in the 1730s. From these theorems and his penetrating reasoning, we get a sense of Gregory's power as a mathematician and of the extent to which, in those early days, he had developed the methods of calculus.

7.4 Gregory's Evaluation of $\int \sec\theta\, d\theta$

We have already noted the connection of $\int_0^x \sec\theta\, d\theta$ with Mercator's projection, a significant tool in navigation. The great Portugese mathematician, Pedro Nuñez (1502–1578), or Nunes, after whom the nonius is named, defined the navigational problem. He saw that loxodromes, curves intersecting meridians at a constant angle, were distinct from Great Circles. He asked how loxodromes could be represented as straight lines on a map such that a compass direction could be set. Gerhard Mercator (1512–1594) employed his projection to answer this question, publishing his *Great World Map* in 1569. In the 1590s, Harriot effectively evaluated the secant integral by a stereographic projection of a loxodrome from the South Pole into a logarithmic spiral, though he did not publish his results. In 1599, Edward Wright, a Cambridge professor of mathematics, published an important book called *Certaine Errors in Navigation Corrected*, containing tables of numerical values of this integral, computed by means of the continued addition of the secants of 1', 2', 3', etc. Henry Bond analyzed the pattern of these tables and observed that the values could be closely approximated by $\ln\tan\left(\frac{\pi}{4} + \frac{x}{2}\right)$. He published this observation in the 1645 edition of Richard Norwood's *Epitome of Navigation*. Bond's observation quickly became well known, and a theoretical proof was greatly desired. Collins mentioned this need in his 1659 book on navigation, and in 1666 Nicholas Mercator (Kaufmann) offered a sum of money for a demonstration. Collins brought these facts to the notice of his friend James Gregory who provided a very difficult proof in 1668. Barrow soon presented a simpler proof, performing integration by partial fractions for the first time in any published work. Some notes of Newton on plane and spherical trigonometry dating from 1665–66 also contain some useful remarks on this problem.

7.4 Gregory's Evaluation of $\int \sec\theta\, d\theta$

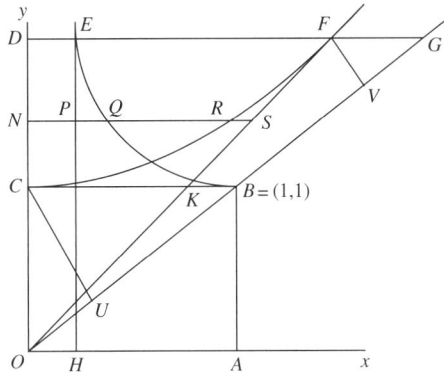

Figure 7.3. Gregory's integration of $\sec\theta$.

Gregory's evaluation of the secant integral was wholly geometric. For convenience, we follow Turnbull to describe Gregory's method in analytic terms.

The curve BQE in Figure 7.3 is parametrically given by $(1-\sin\theta, \sec\theta)$ and the curve CRF is the rectangular hyperbola $y^2 - x^2 = 1$; E is the point $(1-\sin\alpha, \sec\alpha)$, where α is fixed. The straight line OG, asymptotic to the hyperbola, is $y=x$ and the straight line OKF is easily seen to be $y = x\csc\alpha$.

Gregory converted the integral $\int_0^\alpha \sec\theta\, d\theta$ to a double integral. His first step was to do a change of variables:

$$\int_0^\alpha \sec\theta\, d\theta = \int_0^\alpha \sec^2\theta\, d(1-\sin\theta) = \int_0^{1-\sin\alpha} z\, dx,$$

where $x = 1-\sin\theta$ and $z = \sec^2\theta$. Now, in brief, with $y = \sec\theta$, we may write $dz = 2y\, dy$ and $2\int_0^\theta \sec t\, d(\sec t) = \sec^2\theta - 1$; it follows that

$$\int_0^\alpha \sec\theta\, d\theta = 2 \int\int y\, dy\, dx,$$

where the double integral is evaluated over the region $EBAH$. Changing the order of integration gives

$$2\int\int y\, dx\, dy = 2\int (x' - x'')y\, dy,$$

where x' is the value of x on the boundary $ABQE$ and $x'' = 1 - \sin\alpha$ is the value of x on HE. Gregory then divided the last integral into two parts: the first below the line CB, where $x' = 1$ and the second above the line CB, where $x' = 1 - \sin\theta$. When $x' = 1$, we have $x' - x'' = \sin\alpha$ and the value of the double integral for the first part (below CB) is equal to $\sin\alpha$. Since the x coordinate of K from $y = x\csc\alpha$ is $\sin\alpha$, it follows that twice the area of triangle OCK is equal to the first part of the double integral.

For the second part, note that $x' - x'' = \sin\alpha - \sin\theta$, so that the integrand $y(x' - x'')$ is equal to RS for $ON = y$. Thus, the double integral over the region above CB is double the area $CRFK$. Therefore, the double integral, or the sum of the two parts, equals twice the area OCF, that is, twice the area between the hyperbolic segment

$y^2 - x^2 = 1$ and the straight line $y = x \csc \alpha$. It is not difficult to show that this area $= \ln(\sec \alpha + \tan \alpha)$.

7.5 Barrow's Evaluation of $\int \sec \theta \, d\theta$

In Figure 7.4, AB is a quadrant of a circle with center C. MT is the tangent to the arc AB at M, meeting the vertical line CA at T. The line TX is parallel to BC and meets the vertical line PM at X. The curve XA is obtained by applying this process to every point on the arc MA. The hyperbola LEO is defined by the equation $LP = \frac{BC^2}{BP}$. In particular, $CE = \frac{BC^2}{BC} = BC$ and $OQ = \frac{BC^2}{BQ}$.

Note that BQ was understood simply as a line segment and not as an axis in the modern sense; likewise for ET. Thus, the curve LEO did not lie below the x-axis, since there was no negative portion of an axis. The concept of negative and positive portions of axes having a fixed origin did not arise until more than a century after Barrow's work. The purpose of a curve was to express a relation between variables (line segments), and the purpose of the axes was to signify the values of those line segments. Each line segment was thus simply a length with no negative value attached; the equation of the hyperbola in the diagram would be $y = 1/x$ and not $y = -1/x$.

Barrow stated his theorem:

> Let CQ be taken equal to CP, and QO be drawn parallel to CE, meeting the hyperbola LEO in O; then the hyperbolic space $PLOQ$ multiplied by the radius CB (or the cylinder on the base $PLOQ$ of height CB) is double the sum of the squares on the straight lines ... PX, belonging to the arc AM, and applied to the straight line CB.

In alternative notation, Barrow's sum of squares would be written as $\int_P^C y^2 \, dx$, where y is the ordinate of the curve XA and $x = CP$. If we let $\widehat{ACM} = \theta$ and set the radius

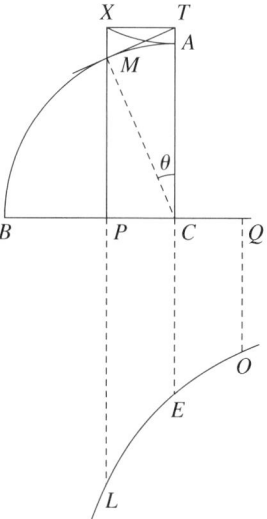

Figure 7.4. Barrow's integral of secant as an area under a hyperbola.

7.5 Barrow's Evaluation of $\int \sec\theta \, d\theta$

$CB = 1$, then $y = PX = CT = \sec\theta$ and $x = CP = \sin\theta$ and

$$\int y^2 \, dx = \int \sec^2\theta \, d(\sin\theta) = \int \sec\theta \, d\theta.$$

Let us now follow the steps of Barrow's argument. Observe that from the values of LP and OQ,

$$\frac{PL}{QO} = \frac{BQ}{BP} = \frac{BC+CP}{BC-CP}.$$

Adding one to the extreme sides of the above equation, we obtain

$$\frac{PL+QO}{QO} = \frac{2BC}{BC-CP}.$$

Then again

$$\frac{QO}{BC} = \frac{BC}{BQ} = \frac{BC}{BC+CP},$$

and thus

$$\frac{PL+QO}{BC} = \frac{PL+QO}{QO} \cdot \frac{QO}{BC} = \frac{2BC^2}{BC^2 - CP^2} = \frac{2BC^2}{PM^2}.$$

Since the similarity of the triangles TMC and PMC implies $BC/PM = PX/BC$, we arrive at

$$\frac{2PX^2}{BC^2} = \frac{PL+QO}{BC} \text{ or } 2PX^2 = (PL+QO)BC.$$

This proves Barrow's contention; in modern notation, Barrow proved that

$$2\int \sec\theta \, d\theta = \ln(1+\sin\theta) - \ln(1-\sin\theta) = \ln\frac{1+\sin\theta}{1-\sin\theta}.$$

The final result follows immediately from this. Note that in Barrow's relation

$$(PL+QO)/BC = 2BC^2/(BC^2 - CP^2),$$

the left side of the equation is equivalent to $BC/BP + BC/BQ$, and this gives the partial fractions decomposition of the right side. Barrow's proof of the formula $\int \tan\theta \, d\theta = -\ln\cos\theta$ was even simpler. We present Barrow's statement of the result and the figure accompanying it (Figure 7.5); the reader may wish to fill in the details.

ABC is a quadrant of a circle with AS a tangent at A while KZZ is a hyperbola with asymptotes CA and CY. (Barrow tended to denote more than one point by the same letter, but this practice was usually not confusing.) The hyperbola is defined, as in the previous case, by the equation $y = FZ = CA^2/CF = R^2/x$, where R is the radius. Barrow then stated:

> The straight line CS is equal to FZ; thus the sum of the secants belonging to the arc AM, applied to the line AC, is equal to the hyperbolic space $AFZK$.

As a corollary he had the integral of $\tan\theta$: "the sum of the tangents to the arc AM...is equal to the hyperbolic space $AFZK$."

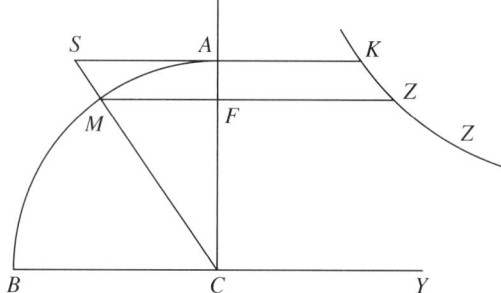

Figure 7.5. Barrow's integration of the tangent function.

7.6 Barrow and the Integral $\int \sqrt{x^2 + a^2}\, dx$

Although Barrow did not explicitly evaluate this integral, it turns out to be a corollary of his theorem related to rectification, illustrated by Figure 7.6.

> Take as you may any right-angled trapezial area (of which you have sufficient knowledge), bounded by two parallel straight lines AK, DL, a straight line AD, and any line KL whatever; to this let another such area $ADEC$ be so related that, when any straight line FH is drawn parallel to DL, cutting the lines AD, CE, KL in the points F, G, H, and some determinate straight line Z is taken, then the square on FH is equal to the squares on FG and Z. Moreover, let the curve AIB be such that, if the straight line GFI is produced to meet it, the rectangle contained by Z and FI is equal to the space $AFGC$; then the rectangle contained by Z and the curve AB is equal to the space $ADLK$. The method is just the same, even if the straight line AK is supposed to be infinite.

To understand this theorem, let $AF = x$, $IF = y = g(x)$ and the let the constant $Z = a$. Then $FG = ag'(x)$, and we have the length of the arc $AIB = \int \sqrt{1 + (g'(x))^2}\, dx$. As an example, Barrow considered the case where KL was an equilateral hyperbola with center A and axis AK, as in Figure 7.7. He concluded that in this case CGE was a straight line and AIB a parabola.

If we write the parabola as $y = g(x) = x^2/2a$, then $FG = x$ so that $y = x$ is the line AE asymptotic to the hyperbola KL given by $y = \sqrt{x^2 + a^2}$. Barrow did not go beyond this point, but he could easily have computed $\int \sqrt{x^2 + a^2}\, dx =$ area $AFHK$ if necessary. He knew from the work of Grégoire St. Vincent, Alphonse de Sarasa, and N. Mercator that since $y = x$ is asymptotic to the hyperbola $(y - x)(y + x) = a^2$, area $KHPQ = a^2(\ln AQ - \ln AP)$. Here note that KP and HQ are perpendicular to CE. Now

$$\text{area } AFHK = \text{area } AFG + \text{area } AKP + \text{area } KHQP - \text{area } GHQ$$

$$= \frac{1}{2}x^2 + \frac{1}{4}a^2 + a^2 \ln\left(\frac{1}{\sqrt{2}}(\sqrt{x^2 + a^2} - x) + \sqrt{2}x\right) - a^2 \ln\frac{a}{\sqrt{2}} - \frac{1}{4}(\sqrt{x^2 + a^2} - x)^2$$

$$= \frac{1}{2}x\sqrt{x^2 + a^2} + a^2 \ln\left(\frac{x + \sqrt{x^2 + a^2}}{a}\right).$$

7.6 Barrow and the Integral $\int \sqrt{x^2+a^2}\,dx$ 109

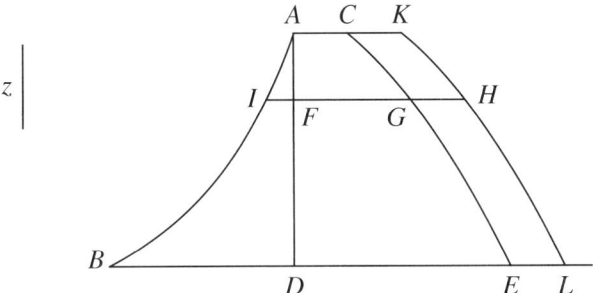

Figure 7.6. Barrow's preliminary result on rectification.

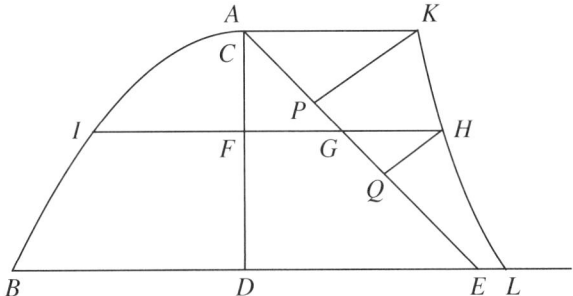

Figure 7.7. Barrow's quadrature of a rectangular hyperbola.

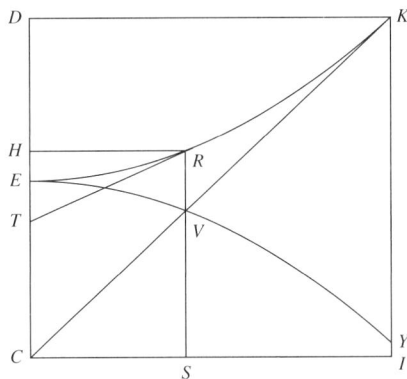

Figure 7.8. Barrow's integration of an algebraic function.

We outline the main steps of Barrow's very interesting evaluation of $\int \frac{dx}{\sqrt{a^2+x^2}}$, omitting some of his geometric reasoning; please refer to Figure 7.8.

Suppose ERK is any curve and RT is tangent to it at R. EVY is a curve such that $CT = VS$, where RS is a vertical line parallel to CD and KI. One of Barrow's general results was that the region $EDKY$ bounded by the curve EVY had twice the area of the region EDK bounded by the curve ERK. Analytically, this means that if we write ERK as $y = f(x)$, then EVY is given by $y = f(x) - xf'(x)$. Barrow's theorem then follows upon integration by parts, for it can be stated as

$$2\left(xf(x) - \int_0^x f(t)\,dt\right) = xf(x) - \int_0^x (f(t) - tf'(t))\,dt.$$

Barrow applied this result to the case in which $f(x) = \sqrt{x^2 + z^2}$ so that the curve EVY was given by $y = a^2/\sqrt{x^2 + a^2}$. We now reproduce his evaluation of $\int dx/\sqrt{x^2 + a^2}$ in his language:

> Let there be an *equilateral Hyperbola* ERK, *viz.* having equal Axes, and let KI, KD be Ordinates to these Axes (CED, CI). Also let the Curve EVY be such, that assuming the point R at pleasure in the *Hyperbola*, and drawing the right Line RVS parallel to DC, let SR, CE, SV be continual Proportionals [author's remark: This means that $SR : CE :: CE : SV$; this can be seen from our earlier remarks, since $SR = \sqrt{x^2 + a^2}$, $CE = a$, and $SV = a^2/\sqrt{x^2 + a^2}$.] Then joyning the right Line CK; the Space $CEIY$ will be the double of the *Hyperbolick Sector* KCE. For draw RT touching the *Hyperbola*, and RH parallel to CI. Then it is $CH : CE :: CE : CT$. Therefore, $CT = SV$, or $HT = RV$. Consequently the *Space* $EDKY$ is twice the *Segment* EDK. Also the Rectangle $IKDC$ is twice the Triangle CDK. Consequently the *remaining Space* $CEYI$ is twice the *remaining Sector* ECK.

We have already established the fact that the sector ECK is given by $\frac{a^2}{2}\ln(x + \sqrt{a^2 + a^2})$. Thus, Barrow's result is that

$$\int \frac{dx}{\sqrt{x^2 + a^2}} = a^2 \ln(x + \sqrt{x^2 + a^2}).$$

7.7 Barrow's Proof of $\frac{d}{d\theta}\tan\theta = \sec^2\theta$

Barrow used more algebraic notation in this proof than in any other proof in his lectures. He also neglected second-order infinitesimals in his computations and we shall do the same without comment.

In Figure 7.9, let DEB be a quadrant of a circle to which BX is a tangent. AMO is a curve such that AP is equal in length to the arc BE and $PM = BG = CB\tan\theta$.

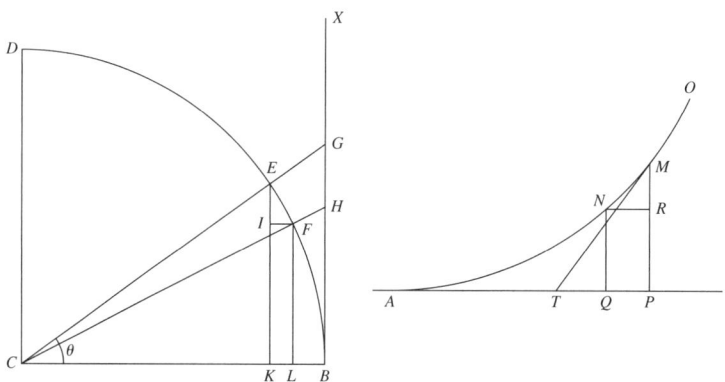

Figure 7.9. Barrow's diagrams for the derivative of $\tan\theta$.

The infinitesimal arc $EF = e = QP$. Also let EK and FL be perpendicular to CB. Denote the radius by r, $CK = f$, $KE = g$, $GB = m$ and the infinitesimal $GH = a$. The problem is to compute $\frac{a}{e} = \frac{d}{d\theta}\tan\theta$.

Since the triangle CEK is similar to the differential triangle EFI, $LK = ge/r$. Thus,

$$CL = f + \frac{ge}{r}, \text{ and } LF = \sqrt{(r^2 - f^2 - 2fge/r)} = \sqrt{(g^2 - 2fge/r)}.$$

Now from the triangles CFL and CHB, $CL : LF = CB : BH$, or

$$f + ge/r : \sqrt{(g^2 - 2fge/r)} = r : m - a.$$

Squaring this relation gives

$$f^2 + 2fge/r : g^2 - 2fge/r = r^2 : m^2 - 2ma.$$

From triangles CBG and CEK we get $fm = rg$, and hence the previous relation simplifies to $rfma = gr^2e + gm^2e$. Thus,

$$\frac{a}{e} = \frac{g(r^2 + m^2)}{rfm} = \frac{r^2 + m^2}{r^2} = \sec^2\theta.$$

7.8 Barrow's Product Rule for Derivatives

Barrow's derivation of the product rule depended on a reduction theorem stating, in analytic terms, that if (a, b) is an intersection point of the curves $y = f(x)$ and $y = g(x)$, then the curves $y = \left(1 - \frac{r}{n}\right)f(x) + \frac{r}{n}g(x)$ and $y = f(x)^{1-\frac{r}{n}}g(x)^{\frac{r}{n}}$ have a common tangent at (a, b). Barrow used this theorem to reduce problems on the derivatives of products to those on derivatives of sums. Of course, this is equivalent to taking the logarithm of the product to get a sum of logarithms. In Barrow's time, the logarithm was viewed as the means for associating an arithmetic progression with a geometric progression. For his purposes, Barrow developed a few elementary properties of these progressions and in particular, he proved the inequality, given here analytically:

$$1 + \frac{r}{n}x > (1 + x)^{r/n}, \text{ for } r < n \text{ and } x > 0$$

and, using this inequality, Barrow derived his result.

Changing Barrow's notation slightly, we give the statement and proof of the theorem upon which he based his derivation of the product rule. He denoted curves by EBE or FBF, etc. so that the symbol E or F, etc. would appear on more than one part of the curve. We shall avoid this.

In Figure 7.10, QBE and ABF are two convex curves and BR is a tangent to QBE. We summarize Barrow's description of the diagram and theorem: Let VD, TB be straight lines meeting in T, and let a straight line BD, given in position, fall across them. In our terminology, BD and TD are the coordinate axes. Let the two curves be such that if any straight line PG is drawn parallel to BD, PF is always an arithmetic

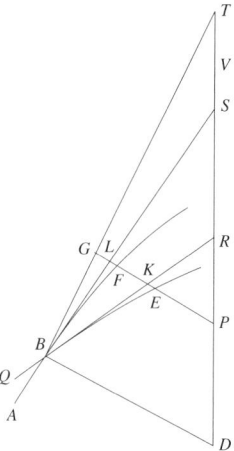

Figure 7.10. Barrow's figure for the product rule.

mean of the same order between PG and PE. This means that there are positive integers M and N, $M > N$ such that

$$M.PF = (M-N)PG + N.PE.$$

If the point S on TD is such that

$$\frac{N.TD + (M-N)RD}{M.TD} = \frac{RD}{DS}, \tag{7.4}$$

then BS touches the curve ABF. Note here that Barrow thought of a tangent to a curve as a straight line touching the curve at one point. He dealt only with convex (or concave) curves. Moreover, he consistently used ratios of magnitudes such as $RD : SD$ instead of RD/SD, although some may prefer the latter notation in a long argument.

Now in an earlier lecture, Barrow proved the lemma

$$\frac{LG.TD + KL.RD}{KG.TD} = \frac{RD}{SD}, \tag{7.5}$$

also pointing out that $EF/FG = M/N$; therefore,

$$\frac{FG.TD}{EF.TD} = \frac{N.TD}{M.ND} \text{ and } \frac{EF.RD}{EG.TD} = \frac{(M-N).RD}{M.TD}.$$

After adding the two equations, we have

$$\frac{FG.TD + EF.RD}{EG.TD} = \frac{N.TD + (M-N)RD}{M.TD} = \frac{RD}{SD}.$$

7.8 Barrow's Product Rule for Derivatives

When this is combined with the equation in the lemma (7.5), we get

$$\frac{FG.TD + EF.RD}{EG.TD} = \frac{LG.TD + KL.RD}{KG.TD}.$$

Thus, $\dfrac{EG/EF}{KG/KL} = \dfrac{FG/EF + RD/TD}{LG/KL + RD/TD}$

$$= \frac{EF/EF - RD/TD}{KL/KL - RD/TD}.$$

Therefore, $EG/EF = KG/KL$ or $EG/FG = KG/GL$. Now since $EF > KG$, we must have $FG > LG$. This implies that L falls outside the curve. Thus, SB is tangent to the curve. By use of his reduction theorem, Barrow next extended this result to curves which satisfy $PF^M = PG^{M-N}.PE^N$. He merely remarked that if PF was assumed to be a geometric mean between PG and PE (that is, the given equation held for PF) and

$$\frac{N.TD + (M-N)RD}{M.TD} = \frac{RD}{SD},$$

then BS will touch the curve ABF, proving the theorem. The product rule follows quickly after this; note that the last relation can be written as

$$\frac{N}{RD} + \frac{M-N}{TD} = \frac{M}{SD}.$$

So if the straight lines BD and TD are taken orthogonal to each other, it follows that RD, TD and SD are subtangents and

$$\frac{1}{DS} = \frac{1}{y}\frac{dy}{dx}.$$

See Figure 7.11.

Thus, we can see that Barrow found the product rule for $w^M = y^{M-N}z^N$.

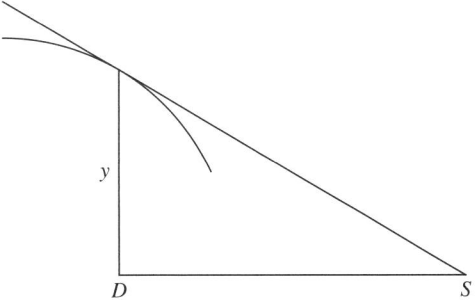

Figure 7.11. Subtangent DS in terms of the derivative.

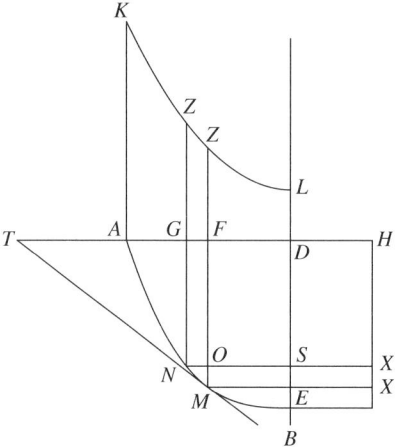

Figure 7.12. Barrow: Fundamental theorem of calculus in geometric form.

7.9 Barrow's Fundamental Theorem of Calculus

In his eleventh lecture, Barrow gave an expression of the duality between the derivative and the integral, a form of the fundamental theorem of calculus. He showed that if $R.dz/dx = y$, then $R.z = \int y\,dx$ where R was a constant. Earlier in his lectures, he gave the converse.

Following Barrow's proof very closely and using Figure 7.12, let AMB be a curve with axis AD, and let BD be perpendicular to AD. Let KZL be another curve defined as follows: When MT is tangent to the curve AMB, and MFZ is parallel to DB, cutting KZ in Z and AD in F, and when R is a line of fixed length, we then have $TF : FM = R : FZ$. The curve KZL thus defined is such that the space $ADLK$ is equal to the rectangle contained by R and DB.

Barrow's proof is very brief. He argued that if $DH = R$ and MN was an infinitesimal arc of the curve AB, then $NO : MO = TF : FM = R : fz$ and, therefore, $NO.FZ = MO.R$ and $FG.FZ = ES.EX$. "Hence," he wrote, "since the sum of such rectangles as $FG.FZ$ differs only in the least degree from the space $ADLK$, and the rectangles $ES.EX$ form the rectangle $DHIB$, the theorem is quite obvious."

7.10 Exercises

1. Barrow's lemma: BR and BS are two straight lines through B and GP is parallel to BD and intersects the two lines at K and L, respectively, as in Figure 7.13. Show that
$$\frac{LG.TD + KL.RD}{KG.TD} = \frac{RD}{SD}.$$
See Child (1916), p. 79.

2. Prove the result that if A, B, C, D, E, F are in arithmetic progression, and $A, M, N, O, P,$ and Q are in geometric progression, and the last term F is not less

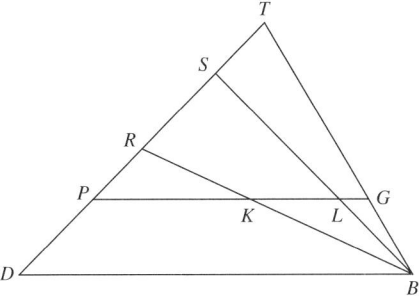

Figure 7.13. Barrow's lemma.

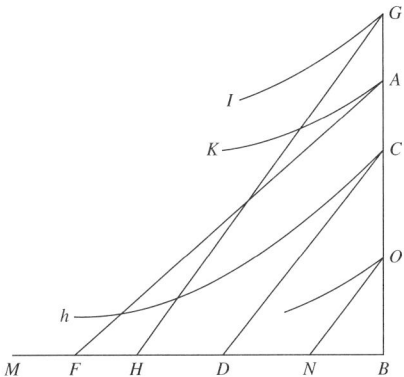

Figure 7.14. Gregory's diagram for product and quotient rules.

than the last term Q, then $B > M, C > N$ and so on. Show that this result is equivalent to the statement $1 + \alpha x > (1 + x)^\alpha$, $0 < \alpha < 1$, with α rational. See Child (1916), pp. 84–85.

3. Let Ch, AK be two curves; let MB be a straight line, and let GI be a curve such that GB is always equal to the sum of AB and CB, as in Figure 7.14.

   - Let the straight lines CD, AF touch the curves Ch, AK. Put $CB = a$, $BD = b$, $AB = c$, $FB = d$ and let $BH = \frac{adb + cdb}{da + bd}$. Prove that GH will touch the curve GI.
   - Prove that if $GB : AB :: CB : OB$, and $BH = a$, $BF = b$, $BD = c$, then
   $$BN = \frac{acb}{ab + ac - cb}.$$

   These results are equivalent to the product and quotient rules for derivatives. Gregory obtained these results after reading Barrow. See Turnbull (1939), pp. 347–349.

4. In Figure 7.15, let AB be the quadrant of a circle with center C. The hyperbola LEO is defined by $BP.LP = BC^2$. AS is tangent to the circle at A and MY is parallel to AS. The curve AY is such that $FY = AS$. The space $ACQYA$ (that

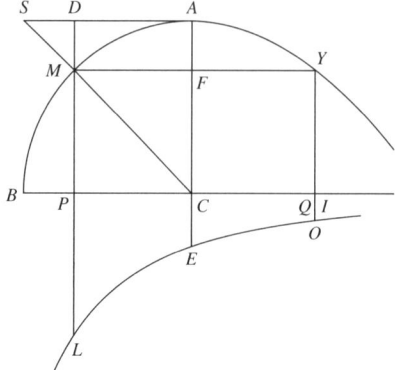

Figure 7.15. Integration by parts.

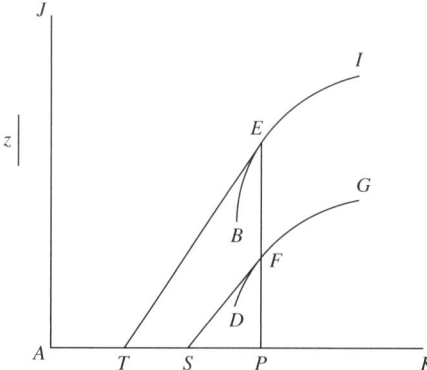

Figure 7.16. Derivative of an algebraic function.

is, the sum of the tangents belonging to the arc AM, and applied to the straight line AC, together with the rectangle $FCQY$) is one-half of the hyperbolic space $PLOQ$. Show that this result is equivalent to the integration by parts formula:

$$\int \tan\theta \, d(\cos\theta) = \tan\theta \cos\theta - \int \cos\theta \, d(\tan\theta).$$

See Child (1916), pp. 167 and 186.

5. In Figure 7.16, let BEI and DFG be two curves such that if PFE is any line parallel to the fixed line JA, then the square on PE is equal to the square on PF plus the square on a given straight line Z. Also, let the straight line TE touch BEI and let S be such that $PE^2/PF^2 = PT/PS$. Then SF will touch the curve DFG.

Show that this theorem is equivalent to the result that if $s^2 = t^2 - a^2$, then

$$ds/dt = t/\sqrt{t^2 - a^2}.$$

See Child (1916), p. 96.

7.10 Exercises

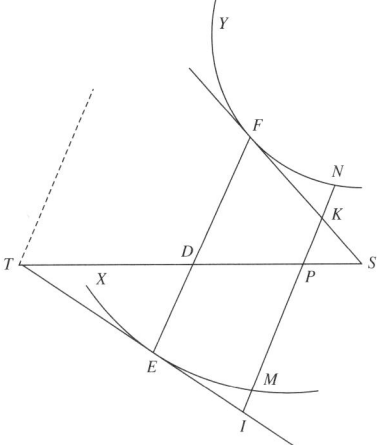

Figure 7.17. Quotient rule for derivatives.

6. In Figure 7.17, let XEM, YFN be two curves such that if any straight EDF is drawn parallel to a fixed line, the rectangle contained by DE and DF is equal to an arbitrary given area. Also the straight line ET touches the curve XEM at E. Prove that if DS is taken equal to DT, then FS will touch the curve YFN at F.

    Draw IN parallel to EF, cutting the given lines as shown and prove the following:

    - $TP/PM > TD/DE$.
    - $(TP \cdot SP)/(PM \cdot PK) > (TD \cdot SD)/(TP \cdot SP)$.
    - $TD \cdot SD > TP \cdot SP$.
    - $PM \cdot PK < PM \cdot PN$. Thus, the line FS lies outside the curve YFN and therefore is tangent to the curve.

    Show that the preceding result implies that if $u = 1/v$, then $\frac{1}{u}\frac{du}{dx} = -\frac{1}{v}\frac{dv}{dx}$. Combine this with the product rule, proved in the text, to obtain the quotient rule. See Child (1916), p. 93.

7. In Figure 7.18, ZGE is an increasing curve, that is, the ordinates VZ, PG, and DE are increasing. VIF is a line such that the rectangle contained by DF and a given length R is equal to the area $VDEZ$; and $DE/DF = R/DT$. Then TF is tangent to VIF at F. Show that this theorem is a form of the fundamental theorem of calculus. Let ZGE be $y = f(x)$ and VIF be $z = g(x)$. If $Rz = \int_0^x y\,dx$, then $y = R dz/dx$. Prove the theorem as follows: Take IL to be an infinitesimal.

    - Show that $R \times LF = LK \times DE = $ area $PDEG$.
    - Conclude that $LK < DP = LI$ and hence FK is outside the curve VIF.
    - Now take the point I on the other side of F and prove a similar result.

    See Child (1916), p. 117.

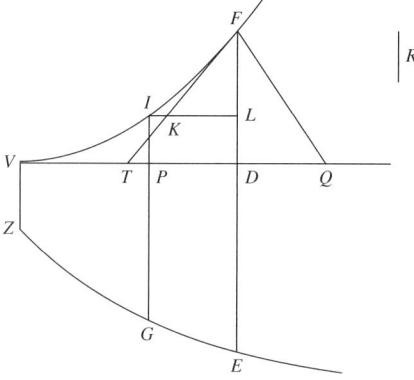

Figure 7.18. Another form of the fundamental theorem of calculus.

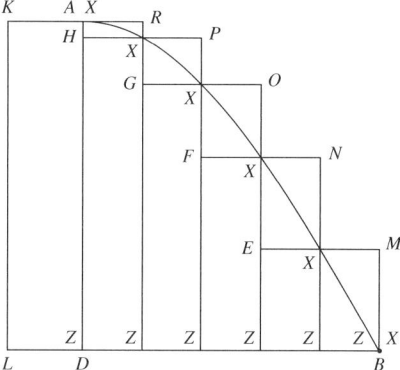

Figure 7.19. Barrow's and Newton's illustration of the integrability of a monotonic function.

8. In Figure 7.19, let AB be a decreasing curve lying over the abscissa DB; DB is divided into several parts by points denoted by Z. Observe that each base ZZ forms the base of a smaller rectangle and the base of a larger rectangle containing the smaller (e.g., $XPZZ$ and $GXZZ$). Show that the difference between the sum of the areas of the larger rectangles and the sum of the areas of the smaller rectangles yields the area of the rectangle $KADL$. Then show that this proves that a monotonic function is integrable. See Child (1920), p. 172.

7.11 Notes on the Literature

The best source for the study of Barrow's work on the calculus is his *Lectiones Geometricae*. These lectures have been translated into English twice, first in 1735 by Edmund Stone, and more recently in 1916 in abridged form by J. M. Child. Stone's edition is a straightforward translation, but Child's work contains a long introduction to put Barrow's lectures in historical perspective, mentioning where in the lectures a specific calculation can be found. Child also added notes at the end of each chapter, making it easier to understand Barrow's results. Child's purpose in translating the *Lectiones Geometricae* was to establish the thesis, stated at the beginning of his preface:

> Isaac Barrow was the first inventor of the Infinitesimal Calculus; Newton got the main idea of it from Barrow by personal communication; and Leibniz also was in some measure indebted to Barrow's work, obtaining confirmation of his own original ideas, and suggestions for their further development, from the copy of Barrow's book that he purchased in 1673.

Child is mistaken in his comments on Newton, since he did not have a chance to see all of Newton's early manuscripts; Whiteside's edition of Newton's complete papers was not then available. In spite of this, Child's book is worth reading. Child's exaggerated claims led to a decline of attention to the work of Barrow, but there has been a recent renewal of interest; see Feingold (1990) and (1993). For another treatment of Barrow and Newton, see Arnold (1990).

The James Gregory Memorial Volume, edited by Turnbull, published in 1939, contains the mathematical discoveries Gregory was unable to publish in his time. Turnbull's remark on when he examined Gregory's papers and notes is given on p. vi. This volume contains all of Gregory's mathematical and scientific correspondence. The majority of his letters are to John Collins, who was similar to Marin Mersenne in being a very interested amateur who kept up a vast correspondence with several scholars and informed them of the results of others. Apparently Gregory evaluated the integral $\int_0^{\pi/2} \cos^n \theta \, d\theta$ after receiving news from Collins that Mengoli had written a book in which he had rediscovered Wallis's formula for π. Gregory recorded his derivation on the back of a letter from Collins. To note important discoveries on the back of letters was a common practice with Gregory. Turnbull contains all such recordings and makes fascinating reading; it also has detailed summaries of Gregory's published works.

Gregory's evaluation of the beta integral is given in Turnbull (1939) pp. 378–382 and his evaluation of the integral of $\sec x$ can be found on pp. 463–464. See also Baron (1987), a thorough history of the origins of calculus, for a nice discussion of Gregory and Barrow. Our discussion of Barrow's work is based largely on Child (1916). Barrow's quote about his sickle may be found on p. 190 of Child (1916). Barrow's derivations of the special integrals are in lecture twelve and in the first appendix to that lecture. He obtained the product and quotient rules for derivatives in chapter nine; the derivative of $\tan x$ in chapter ten; the fundamental theorem of calculus in chapter eleven. See pp. 101–109, 123, 135, 160–168, and 183–186. Child's translation is abridged, so for Barrow's evaluation of $\int \sec x$, by means of partial fractions, consult p. 236 of Barrow (1735) and see pp. 238–239 for the quotation concerning the evaluation of $\int dx/\sqrt{x^2+a^2}$. The English translation of Pascal's evaluation of $\int \sin x$ is in Struik (1969), pp. 239–241. For the quote of Chandrasekhar, see Wali (1991), p. 243, where he refers to Chandrasekhar's unpublished manuscript on the *Principia*.

8

The Calculus of Newton and Leibniz

8.1 Preliminary Remarks

Newton was a student at Cambridge University from 1661 to 1665, but he does not appear to have undertaken a study of mathematics until 1663. According to de Moivre, Newton purchased an astrology book in the summer of 1663; in order to understand the trigonometry and diagrams in the book, he took up a study of Euclid. Soon after that, he read Oughtred's *Clavis* and then Descartes's *Géométrie* in van Schooten's Latin translation. By the middle of 1664, Newton became interested in astronomy; he studied the work of Galileo and made notes and observations on planetary positions. This in turn required a deeper study of mathematics and Newton's earliest mathematical notes date from the summer of 1664. On July 4, 1699, Newton wrote in his 1664–65 annotations on Wallis's work that a little before Christmas 1664 he bought van Schooten's commentaries and a Latin translation of Descartes and borrowed Wallis's *Arithmetica Infinitorum* and other works. In fact, his meditations on van Schooten and Wallis during the winter of 1664–65 resulted in his discovery of his method of infinite series and of the calculus.

Following the methods of van Schooten's commentaries, Newton devoted intense study to problems related to the construction of the subnormal, subtangent, and the radius of curvature at a point on a given curve. Newton's analyses of these problems gradually led him to discover a general differentiation procedure based on the concept of a small quantity, denoted by o, that ultimately vanished. Later in life, Newton wrote that he received a hint of this method of Fermat from the second volume of van Schooten's commentaries, although this gave only a brief summary based on P. Hérigone's 1642 outline of Fermat's method of finding the maximum or minimum of a function. Newton found the derivative, just as Fermat had, by expanding $f(x+o) = f(x) + of'(x) + O(o^2)$. Newton realized that the derivative was a powerful tool for the analyses of the subtangent, subnormal, and curvature and by the middle of 1665 he had worked out the standard algorithms for derivatives in general. Wallis's work motivated Newton to research the integration of rational and algebraic functions. Newton combined this with a study of van Heuraet's rectification of curves; in the summer of 1665, he began to understand the inverse method of tangents, that is, the connection between derivatives and integrals.

Newton left Cambridge in summer 1665 due to the plague, and returned to his home in Lincolnshire for two years. This gave him the opportunity to organize his thoughts on calculus and several other subjects. He gave up the idea of infinitesimal increments and adopted the concepts of fluents and fluxions as the new foundation for calculus. Fluents were flowing quantities; their finite instantaneous speeds were called fluxions, for which he later used the dot notation, such as $\dot{x}$, where x was the fluent. From this point of view, Newton regarded it as obvious that the fluxion of the area generated by the ordinate y along the x-axis would be y itself. In other words, the derivative of the area function was the ordinate. In the fall of 1665, Newton ran into trouble with an uncritical application of the parallelogram of forces method, but he soon realized his mistake and by the spring of 1666 he was able to apply the method to an analysis of inflection points. Note that in 1640, the French mathematician G. Roberval warned that a curve could be viewed as the result of a moving point, but that there were pitfalls to using the parallelogram of forces method to find the tangent. What was the origin of Newton's conception of a curve as a moving point? A half century later, Newton wrote that, though his memory was unclear, he might have learned of a curve as a moving point from Barrow. Another possible source was Galileo but Newton did not mention him in this connection. In any case, Newton organized his concentrated research on calculus into a short thirty-page essay without title; he later referred to it as the October 1666 tract, published only in 1967 in the first volume of Whiteside's edition of Newton's mathematical papers.

In 1671, Newton wrote up the results of his researches on calculus and infinite series as a textbook on methods of solving problems on tangents, curvature, inflection, areas, volumes, and arclength. The portions of this work on infinite series were expanded from his 1669 work *De Analysi*. Whiteside designated the 1671 book as *De Methodis Serierum et Fluxionum* because Newton once referred to it this way, but Newton's original title is unknown because the first page of the original manuscript was lost. English translations of 1736 and 37 were given the title *The Method of Fluxions and Infinite Series*. Unfortunately, Newton was unable to publish this work in the 1670s, though he made several attempts. At that time, the market for advanced mathematics texts was not good; the publisher of Barrow's lectures on geometry, for example, went bankrupt. The controversy with Leibniz, causing wasted time and effort, would have been avoided had Newton succeeded in publishing his work.

Newton's *De Methodis* dealt with fluxions analytically, but it was never actually completed; in some places he merely listed the topics for discussion. However, when he revised the text in the winter of 1671–72, Newton added a section on the geometry of fluxions, developed axiomatically; he later called this the synthetic method of fluxions. Note that in the *Principia* Newton employed his geometric approach.

As Newton was completing his researches on the calculus and infinite series, Gottfried Leibniz (1646–1716) was starting his mathematical studies. He studied law at the University of Leipzig but received his degree from the University of Altdorf, Nuremberg in 1666. At that time, he conceived the idea of reducing all reasoning to a symbolic computation, although he had not yet studied much mathematics. Leibniz's mathematical education started with his meeting with Huygens in 1672, at whose suggestion he studied Pascal and then went on to read Grégoire St. Vincent's *Opus Geometricum* and other mathematical works.

From the beginning, Leibniz searched for a general formalism, or symbolic method, capable of handling infinitesimal problems in a unified way. In a paper of 1673, Leibniz began to denote geometric quantities associated with a curve, such as the tangent, normal, subtangent and subnormal, as functions. He began to set up tables of specific curves and their associated functions in order to determine the relations among these quantities. Thus, he raised the question of determining the curve, given some property of the tangent line. In 1673, Leibniz came to the conclusion that this problem, the inverse tangent problem, was reducible to the problem of quadratures. By the end of the 1670s, Leibniz had independently worked out his differential and integral calculus. In 1684, his first paper on differentiation appeared, and in 1686 his first paper on integration was published. The notation of Leibniz, including the differential and integral signs, gave insight into the processes and operations being performed. The Bernoulli brothers were among the first to learn and exploit the calculus of Leibniz and in the 1690s, they began to make contributions to the development of calculus in tandem with Leibniz.

In the May 1690 issue of the *Acta Eruditorum*, Jakob Bernoulli proposed the problem of finding the curve assumed by a chain/string hung freely from two fixed points, named a catenary by Leibniz. Leibniz was the first to solve the problem, announcing his construction without details in the July 1691 issue of the *Acta*. Johann Bernoulli (1667–1748) soon published a solution, in which he explained that he and his brother had been surprised that this everyday problem had not attracted anyone's attention. But in his paper Leibniz wrote that the problem had been well known since Galileo had articulated it; moreover, Leibniz stated that he would refrain from publishing his solution by means of differential calculus, to give others a chance to work out a solution. Jakob had trouble with this question, since he initially thought the curve was a parabola, until Johann corrected him. Jakob, however, went on to develop a general theory of flexible strings.

In his *Mathematical Discourses Concerning Two New Sciences*, Galileo wrote that the shape of the curve formed by a chain hanging from two fixed points could be approximated by a parabola and that this approximation improved as the curvature was reduced. In his first important publication, in 1645 Christiaan Huygens (1629–1695) showed explicitly that the catenary could not be a parabola. In the 1690s, Huygens offered a geometric solution to the problem posed by Bernoulli, using classical methods of which he was a master. In their approach, Leibniz and Johann Bernoulli applied mechanical principles to determine the differential equation of the catenary, making use of the work of Pardies. The Jesuit priest Ignace-Gaston Pardies (1636–1673) published a 1673 work on theoretical mechanics, developing his original idea of tension along the string, a concept fully clarified by Jakob Bernoulli. Thus, Leibniz and Johann found the differential equation of the catenary: $dy/dx = s/a$ where s was the length of the curve. They showed that the solution of this differential equation was the integral $x = \int a\,dy/\sqrt{y^2 - a^2}$.

In his 1691 paper, Leibniz presented a geometric figure and explained that the points on the catenary could be found from an exponential curve, called by Leibniz the logarithmic line. Details of this proof can be found in his letters to Huygens and von Bodenhausen. In modern notation, the solution would be expressed as $y = \frac{a}{2}(e^{x/a} + e^{-x/a})$. Johann Bernoulli also failed to publish details but presented two geometric constructions of the catenary, one using the area under a curve related to a hyperbola and the other using the length of an arc of a parabola. In the 1690s, this kind of solution would have been acceptable, because the coordinates of any point on the catenary were then described in terms of geometric quantities related to known curves such as the hyperbola and parabola. In modern terms, the area and length can be written as the integrals

$$\int \frac{dx}{\sqrt{(x-a)^2 - a^2}} \quad \text{and} \quad \int \sqrt{\frac{2a+x}{x}}\, dx.$$

At this time, Johann was not familiar with the analytic form of the logarithm, though Leibniz soon wrote him a 1694 letter on properties of the logarithm defined by the integral $\int dx/x$. Soon after this, Bernoulli wrote a paper on the calculus of the exponential, explaining how to apply the logarithm to find the derivatives of m^n or m^{n^p}, etc., where m, n and p were varying quantities. In particular, he showed that if $y = x^x$, then

$$dy = x^x dx + x^x \ln x\, dx, \text{ or } dy : dx = y : \text{subtang}$$

where the subtangent $= 1/(1 + \ln x)$. Bernoulli also gave the area under x^x over the interval $(0,1)$ as the curious series

$$1 - \frac{1}{2^2} + \frac{1}{3^3} - \frac{1}{4^4} + \frac{1}{5^5} \&c.$$

Bernoulli gave details of his solution in his 1691–92 lectures to l'Hôpital, and published these lectures and his lectures on the integral calculus in the third volume of his collected works in 1742.

8.2 Newton's 1671 Calculus Text

The *De Methodis Serierum et Fluxionum* of Newton began by considering the general problem, called Problem 1, of determining the relation of the fluxions, given the relations to one another of two flowing quantities. As an example, Newton took

$$x^3 - ax^2 + axy - y^3 = 0. \tag{8.1}$$

His rule for finding the fluxional equation was to first write the equation in decreasing powers of x, as in (8.1), and then multiply the terms by $3\dot{x}/x, 2\dot{x}/x, \dot{x}/x$, and 0, respectively, to get

$$3\dot{x}x^2 - 2\dot{x}ax + \dot{x}ay. \tag{8.2}$$

Thus, if the term were $x^n y^m$, then it would be multiplied by $n\dot{x}/x$. He next wrote the equation in powers of y: $-y^3 + axy + (x^3 - ax^2)$ and multiplied the terms by

$-3\dot{y}/y, \dot{y}/y$ and 0 to obtain

$$-3\dot{y}y^2 + a\dot{y}x. \tag{8.3}$$

In order to obtain the equation expressing the relation between the fluxions $\dot{x}$ and $\dot{y}$ Newton added (8.2) and (8.3) and set the sum equal to zero:

$$3\dot{x}x^2 - 2a\dot{x}x + a\dot{x}y - 3\dot{y}y^2 + a\dot{y}x = 0.$$

From this it followed that

$$\dot{x} : \dot{y} = (3y^2 - ax) : (3x^2 - 2ax + ay).$$

Newton also presented more examples, involving more complex expressions such as

$$\sqrt{a^2 - x^2} \text{ and } \sqrt{ay + x^2}.$$

Explaining why his rule for finding fluxional (differential) equations worked, Newton pointed out that a fluent quantity x with speed $\dot{x}$ would change by $\dot{x}o$ during the small interval of time o. So the fluent quantity x would become $x + \dot{x}o$ at the end of that time interval. Hence, the quantities $x + \dot{x}o$ and $y + \dot{y}o$ would satisfy the same relation as x and y, and when substituted in (8.1) gave him

$$(x^3 + 3\dot{x}ox^2 + 3\dot{x}^2o^2x + \dot{x}^3o^3) - (ax^2 + 2a\dot{x}ox + a\dot{x}^2o^2)$$
$$+(axy + a\dot{x}oy + a\dot{y}ox + a\dot{x}\dot{y}o^2) - (y^3 + 3\dot{y}oy^2 + 3\dot{y}^2o^2y + \dot{y}^3o^3) = 0. \tag{8.4}$$

After subtracting (8.1) from (8.4) and dividing by o, Newton had

$$3\dot{x}x^2 + 3\dot{x}^2ox + \dot{x}^3o^2 - 2a\dot{x}x - a\dot{x}o + a\dot{x}y + a\dot{y}x$$
$$-a\dot{x}\dot{y}o - 3\dot{y}y^2 - 3\dot{y}^2oy - \dot{y}^3o^2 = 0.$$

Here Newton explained that quantities containing the factor o could be neglected, "since o is supposed to be infinitely small so that it be able to express the moments of quantities, terms which have it as a factor will be equivalent to nothing in respect of the others. I therefore cast them out and there remains

$$3\dot{x}x^2 - 2a\dot{x}x + a\dot{x}y + a\dot{y}x - 3\dot{y}y^2 = 0."$$

Note that this amounts to the result of implicit differentiation with respect to a parameter. Actually, Newton here used the letters m and n for $\dot{x}$ and $\dot{y}$, respectively. He introduced the dot notation in the early 1690s. From this he had the slope

$$\dot{y} : \dot{x} = 3x^2 - 2ax + ay : 3y^2 - ax. \tag{8.5}$$

Observe that to construct the tangent, rather than work with slope, it is better to find the point where the tangent intersects the x-axis and join this point to the point of tangency on the curve. Now if (x, y) is the point on the curve, then the length of the segment of the x-axis from $(x, 0)$ to the intersection with the tangent is given by the magnitude of $y/\frac{dy}{dx}$. In his discussion of the tangent, Newton computed this quantity to obtain $(3y^3 - axy)/(3x^2 - 2ax + ay)$.

In the section of his book on maxima and minima, Newton gave a method and then two examples and nine exercises to be solved using that method. He never completed this section of his book. To find a maximum or minimum, he explained, the derivative should be set equal to zero at an extreme value:

> When a quantity is greatest or least, at that moment its flow neither increases nor decreases; for if it increases, that proves that it was less and will at once be greater than it now is, and conversely so if it decreases. Therefore seek its fluxion by Problem 1 [above] and set it equal to nothing.

In the first application of this principle, he sought the greatest value of x in equation (8.1) by setting $\dot{x} = 0$ in the fluxional equation (8.2) to get

$$-3y^2 + ax = 0. \tag{8.6}$$

This last result should be used in the original equation to obtain the largest value of x. Newton remarked that equation (8.6) illustrated the "celebrated Rule of Hudde, that, to obtain the maxima or minima of the related quantity, the equation should lie ordered according to the dimensions of the correlate one and then multiplied by an arithmetical progression." He added that his method extended to expressions with surd quantities, whereas the earlier rules and techniques did not. As an example, he gave the problem of finding the greatest value of y in the equation

$$x^3 - ay^2 + by^3/(a+y) - x^2\sqrt{ay + x^2} = 0.$$

Newton wrote that the equation for the fluxions of x and y would come out to be

$$3\dot{x}x^2 - 2a\dot{y}y + \frac{3ab\dot{y}y^2 + 2b\dot{y}y^3}{a^2 + 2ay + y^2} - \frac{4a\dot{x}xy + 6\dot{x}x^3 + a\dot{y}x^2}{2\sqrt{ay + x^2}} = 0.$$

He then observed that by hypothesis $\dot{y} = 0$ and hence, after substituting this in the equation and dividing by $\dot{x}x$,

$$3x - (2ay + 3x^2)/\sqrt{(ay + x^2)} = 0 \text{ or } 4ay + 3x^2 = 0.$$

Newton noted that this equation should be used to eliminate x or y from the original equation; the maximum would be obtained by solving the resulting cubic.

The next section of Newton's book discussed the problem of constructing tangents to curves and he mentioned seven problems solvable by the principles he explained. For example:

1. To find the point in a curve where the tangent is parallel to the base (or any other straight line given in position) or perpendicular to it or inclined to it at any given angle.
2. To find the point where a tangent is most or least inclined to the base or to another straight line given in position – to find, in other words, the bound of contrary flexure. I have already displayed an example of this above in the conchoid.

By "the bound of contrary flexure" Newton meant the point of inflection. At this point, $d^2y/dx^2 = 0$. In the example of the conchoid of Nichomedes, defined by $yx = (b+y)\sqrt{c^2 - y^2}$, Newton actually minimized the x-intercept of the tangent given by

$$x - y\,dx/dy. \tag{8.7}$$

Note that in 1653, Huygens determined the inflection points of this conchoid by this method, using Fermat's procedure to obtain the minimum value. Newton was most likely aware of Huygens' work and wanted to show that calculus algorithms could simplify Huygens's calculation. It should be noted that Huygens's criterion that the inflection points in general could be obtained by minimizing (8.7) is false, though it is true for the conchoid. As we noted before, Newton suggested the maximization (or minimization) of the tangent slope to find the inflection point.

Newton was very interested in problems related to curvature and intended to devote several chapters to the topic, but many of these are barely outlines. However, he presented a procedure for finding radius of curvature, and we explain this later on. He also included sections on arclength and the area of surface of revolution.

8.3 Leibniz: Differential Calculus

Leibniz gave a very terse account of his differential calculus in his 1684 *Acta Eruditorum* paper, starting with the basic rules for the differentials of geometric quantities (variables). Leibniz's approach was not to find the derivative of a function. As he conceived things, geometric quantities had differentials; when the quantities stood in a certain relationship to one another, then the differentials also satisfied certain relations. To determine these relations, for a constant a and variable quantities v, x, y, etc., he stated the rules for the differentials dv, dx, dy, etc.:

$$da = 0, \quad d(ax) = a\,dx, \quad d(z - y + w + x) = dz - dy + dw + dx,$$

$$d(xv) = x\,dv + v\,dx, \quad d\frac{v}{y} = \pm\frac{v\,dy \mp y\,dv}{yy}.$$

Concerning signs of differentials, Leibniz explained that if the ordinate v increased, then dv was positive, and when v decreased, dv was negative.

In only one paragraph, Leibniz described in terms of differentials: maxima and minima, concavity or convexity, and inflection points of curves. He explained that at a maximum or minimum for an ordinate v, $dv = 0$ since v was neither increasing nor decreasing. For concavity, the difference of the differences $d\,dv$ had to be positive, and for convexity $d\,dv$ had to be negative. At a point of inflection $d\,dv = 0$. After this, Leibniz gave the rules for the differentials of powers and roots, that is

$$dx^a = ax^{a-1}\,dx, \quad \text{and} \quad d\sqrt[b]{x^a} = \frac{a}{b}\sqrt[b]{x^{a-b}}. \tag{8.8}$$

He wrote that with this differential calculus, he could solve problems dealing with tangents and with maxima and minima by a uniform technique, lacking in the earlier expositions. To demonstrate the power of his method, he found the tangent to a curve defined by a complicated algebraic relation between the variables x and y. As another application of the differential calculus, he gave the derivation of Snell's law in optics, one of the standard examples in modern textbooks.

As a final example, Leibniz considered the problem that de Beaune proposed to Descartes in 1639. Florimond de Beaune (1601–1652) was a jurist who carefully studied Descartes's book on geometry. He observed that, though Descartes had given a method for finding the tangent to a curve, he had not indicated how to obtain the curve, given a property of the tangent. One of de Beaune's problems was to find the curve for which the subtangent was the same for each point of the curve. This problem translates to the differential equation $\frac{dy}{dx} = \frac{y}{a}$, where a is a constant. It is well known that the solution is $\ln y = ax + c$ and Descartes came close to a solution. In the course of his work, he obtained, without mentioning logarithms, particular cases of the inequality, written in modern notation as

$$\frac{1}{n} + \frac{1}{n+1} + \cdots + \frac{1}{m-1} > \ln \frac{m}{n} > \frac{1}{n+1} + \frac{1}{n+2} + \cdots + \frac{1}{m}.$$

To tackle this problem, Leibniz described the differential equation by saying that y was to a as dy was to dx. He then noted that dx could be chosen arbitrarily and hence could be taken to be a constant b. Then

$$dy = \frac{b}{a} y, \quad \text{or} \quad y = \frac{a}{b} dy.$$

He observed that this implied that if the x formed an arithmetic progression, then the y formed a geometric progression. Leibniz did not explain or prove this statement, but it is easy to check that if

$$y(x) = \frac{a}{b} dy(x), \text{ then}$$

$$y(x+dx) = y(x) + dy(x) = \left(1 + \frac{b}{a}\right) y(x).$$

Again,

$$y(x+2dx) = y(x+dx) + dy(x+dx)$$

$$= y(x+dx) + \frac{b}{a} y(x+dx)$$

$$= \left(1 + \frac{b}{a}\right) y(x+dx)$$

$$= \left(1 + \frac{b}{a}\right)^2 y(x).$$

Similarly,

$$y(x+3dx) = y(x+2dx) + \frac{b}{a}dy(x+3dx)$$
$$= \left(1 + \frac{b}{a}\right)^3 y(x),$$

and in general

$$y(x+ndx) = \left(1 + \frac{b}{a}\right)^n y(x).$$

This proves Leibniz's claim, since he took dx to be a constant; thus, $x, x+dx, x+2dx, \ldots$ is an arithmetic progression, and the values of y at these points form a geometric progression. This idea illustrates the seventeenth century understanding of logarithms. In fact, Leibniz was already suggesting that the logarithm be defined by means of the integral $\int dx/x$ and later on, he did so.

Leibniz could have integrated to get $\ln y = \int dy/y = 1/a \int dx = x/a$, but in his 1684 paper, he did not use or discuss integration, though he had been aware of it for several years. Surprisingly, he gave a brief exposition of his ideas on integration in a review of John Craig's 1685 book on quadrature. In the review, Leibniz introduced the symbol $\int$ for the summation of infinitesimal quantities and gave an illustration of its power when used in conjunction with differentials. Leibniz also pointed out that the integral symbol could be used to represent transcendental quantities such as the arcsine or logarithm, in such a way that it revealed a property of the quantity.

In his 1684 paper on the differential calculus, Leibniz gave a derivation of Snell's law by applying Fermat's principle of least time.

Light traveled from point C to point E and the line QP separated an upper medium of density r from a lower medium of density h (Figure 8.1). Leibniz explained that density should be understood to be with respect to the resistance to a ray of light.

Let $QF = x$, $QP = p$, $CP = c$, and $EQ = e$. Then $FC = \sqrt{cc + pp - 2px + xx} =$ (in short) $\sqrt{l}$; $EF = \sqrt{ee + xx} =$ (in short) $\sqrt{m}$. Leibniz gave the quantity to be

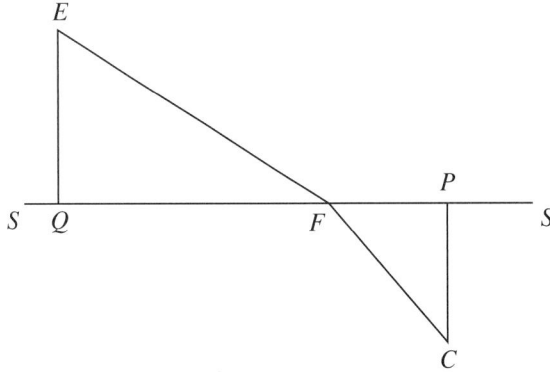

Figure 8.1. Leibniz's figure for derivation of Snell's law.

minimized when the densities were taken into account as $w = h\sqrt{l} + r\sqrt{m}$. He then argued that to minimize, set $dw = 0$, to obtain $0 = hdl : 2\sqrt{l} + rdm : 2\sqrt{m}$. Note that Leibniz specified that he would denote x/y by $x : y$. He then observed that $dl = -2(p - x)$ and $dm = 2xdx$; hence he had Snell's law: $h(p-x) : \sqrt{l} = rx : \sqrt{m}$.

8.4 Leibniz on the Catenary

Leibniz developed a theory of second- and higher-order differentials in order to apply differential calculus to geometry and mechanics. In his applications, including the catenary problem, he often took one of the variables, say y, to be such that the second-order differential ddy was 0 or, equivalently, that the first-order differential dy was a constant. This amounted to taking y to be the independent variable. To describe the curve of the catenary, Leibniz used Pardies's important mechanical principle that for any portion AC of the curve made by the string, the vertical line through the center of gravity of AC, and the tangents at A and at C intersected at one point (Figure 8.2).

Leibniz's letters to Huygens and Bodenhausen offered the following details of his derivation of the catenary. In Leibniz's figure (Figure 8.3), A is the lowest point of the catenary; CT is the tangent at a point C on the catenary; and $C\beta$, AB are perpendicular to $A\beta$, the tangent at A. We follow Leibniz's notation and argument. Let $AB = x$, $BC = y$, $T\beta = xdy : dx$ and $AT = y - xdy : dx$. Then, by Pardies's theorem, the y coordinate of the center of gravity of the arc AC of length c is $\frac{1}{c}\int y\,dc$. Thus,

$$\frac{1}{c}\int y\,dc = y - x\,dy : dx. \tag{8.9}$$

Now multiply both sides by c and differentiate to get

$$y\,dc = c\,dy + y\,dc - x\overline{dy:dx}\,dc - c\,dy - cxd,\overline{dy:dx}. \tag{8.10}$$

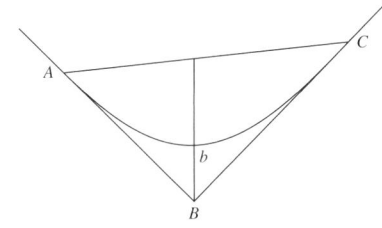

Figure 8.2. Pardies's theorem.

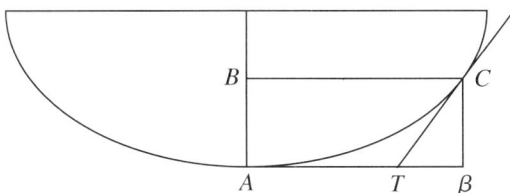

Figure 8.3. Leibniz's figure of catenary, made for Huygens.

Note that Leibniz used a comma to separate the operator d from the quantity $\overline{dy:dx}$. Upon simplification, obtain

$$dc\,\overline{dy:dx} + cd, \overline{dy:dx} = 0. \tag{8.11}$$

Suppose that y increases uniformly, so that dy is constant and $ddy = 0$. This implies by the quotient rule that

$$d, \overline{dy:dx} = -dy\,ddx : \overline{dx\,dx},$$

so that (8.11) is transformed into $dc\,dx - c\,ddx = 0$.

By differentiating $dx : c = dy : a$ we get the previous equation, indicating that this is the integral of that equation. Rewrite this integral as

$$a\,dx = c\,dy. \tag{8.12}$$

This is the differential equation of the catenary, and its differential is

$$a\,ddx = dc\,dy. \tag{8.13}$$

Following Leibniz, one may solve this equation by observing that in general, since c denotes arclength,

$$dc\,dc = dy\,dy + dx\,dx. \tag{8.14}$$

Differentiate this, using $ddy = 0$ and (8.13), to obtain

$$dc\,ddc = dy\,ddy + dx\,ddx = dx\,ddx = dx\,dc\,dy/a,$$

by (8.13). By integration (Leibniz used the term summation), we arrive at $a\,dc = (x+b)dy$, where b is a constant. Next set $z = x+b$ to rewrite, obtaining $a\,dc = z\,dy$. Combining this with $dc\,dc = dz\,dz + dy\,dy$, the result emerges as

$$aa\,dz\,dz + aa\,dy\,dy = zz\,dy\,dy. \tag{8.15}$$

Thus,

$$y = a \overline{\int dz : \sqrt{zz - aa}} \tag{8.16}$$

gives the area under the curve with ordinate $a/\sqrt{z^2 - a^2}$. This integral can be computed in terms of the logarithm. Although we today would wish to evaluate the integral, and write it as the logarithm of a specific function, mathematicians of the seventeenth century were satisfied with a result expressed in terms of areas or arclengths of known curves, so that from Leibniz's point of view, this result was sufficient to define the catenary.

We remark that the meaning of the logarithm function within the calculus was not fully understood in 1690, and so Leibniz's paper devoted some space to what he called the logarithmic line, now written as $y = a^x$. Leibniz had to explain the concept to

Huygens. In a similar context, l'Hôpital raised questions on the dimensionality of geometric objects, asking Johann Bernoulli: Since a and x were magnitudes of lines, what did it mean to have one as the power of the other? To deal with problems of this kind, Leibniz, the Bernoullis and their followers began to define their quantities in terms of formulas instead of geometric objects.

8.5 Johann Bernoulli on the Catenary

In his 1691–1692 lectures on integral calculus, Bernoulli gave details to supplement the treatment of the catenary in his 1691 paper. He first set down the mechanical principles required to obtain the fundamental equation.

In Figure 8.4, Bernoulli set $BG = x$, $GA = y$, $Ha = dx$, $HA = dx$, and $BA = s$. He then applied the laws of statics, and in effect Pardies's law, to obtain the differential

$$\frac{dx}{dy} = \frac{ks}{ka} = \frac{s}{a} \qquad (8.17)$$

for some constant a. Bernoulli's solution, like that of Leibniz, amounted to showing that (8.17) was equivalent to

$$dy = \frac{a\,dx}{\sqrt{x^2 - a^2}}. \qquad (8.18)$$

To show this, he wrote (8.17) as

$$dy = \frac{a\,dx}{s} \quad \text{or} \quad dy^2 = \frac{a^2}{s^2}dx^2.$$

Therefore,

$$ds^2 = dx^2 + dy^2 = \frac{s^2 dx^2 + a^2 dx^2}{s^2}$$

$$\text{and } ds = \frac{dx\sqrt{s^2 + a^2}}{s}, \quad \text{or} \quad dx = \frac{s\,ds}{\sqrt{s^2 + a^2}}.$$

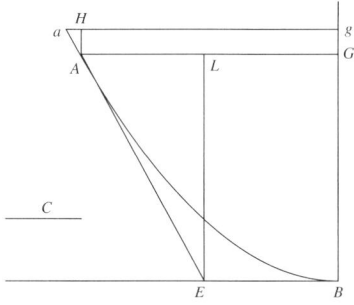

Figure 8.4. Bernoulli's diagram for a catenary.

By integration,
$$x = \sqrt{s^2 + a^2} \text{ or } s = \sqrt{x^2 - a^2}.$$

By differentiating, he got
$$ds = \frac{x\,dx}{\sqrt{x^2 - a^2}} = \sqrt{dx^2 + dy^2}.$$

Squaring this, he had
$$x^2 dy^2 - a^2 dy^2 = a^2 dx^2,$$

equivalent to (8.18).

8.6 Johann Bernoulli: The Brachistochrone

In 1696, Bernoulli conceived of and solved the brachistochrone problem: Given two points in a vertical plane but not vertically aligned, find the curve along which a point mass must fall under gravity, starting at one point and passing through the other in the shortest possible time. He argued that this mechanical problem was identical to an optical problem of the path taken by light moving from one point to another, following the curve of least time, passing through a medium whose ever-changing density is inversely proportional to the velocity of a falling body. As light passes continuously from one medium to another, the quantity $\frac{\sin \alpha}{v}$ remains constant, where α is the angle between the vertical and the direction of the path and v is the velocity.

We change Bernoulli's notation slightly; he used t for the velocity and interchanged the x and y. So in Figure 8.5, let $AC = y, CM = x, mn = dx, Cc = dy, Mm = ds$, and $\alpha = \angle nMm$. Since $\sin \alpha / v$ is a constant, we have
$$\frac{dx}{ds} = \frac{v}{a}, \text{ or } a^2 dx^2 = v^2(dx^2 + dy^2),$$

where a is a constant. Since for a falling body $v^2 = 2gy$, we get $dx = \sqrt{y/(c-y)}\,dy$, where $c = a^2/2g$. Now
$$dy\sqrt{\frac{y}{c-y}} = \frac{1}{2}\frac{c\,dy}{\sqrt{cy - y^2}} - \frac{1}{2}\frac{c\,dy - 2y\,dy}{\sqrt{cy - y^2}}.$$

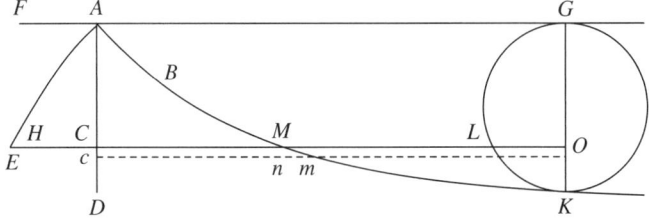

Figure 8.5. Bernoulli's diagram to derive the brachistochrone.

Integrating this, obtain $CM = \text{arc}\, GL - LO$ and, since $MO = CO - \text{arc}\, GL + LO = \text{arc}\, LK + LO$, it follows that $ML = \text{arc}\, LK$. Thus, the curve is a cycloid.

Bernoulli was particularly proud of having linked mechanics with optics. In his brachistochrone paper he wrote, "In this way I have solved at one stroke two important problems – an optical and a mechanical one – and have achieved more than I have demanded from others: I have shown that the two problems, taken from entirely separate fields of mathematics, have the same character." Bernoulli mentioned the link between geometrical optics and mechanics more than once in his works, but this concept was not developed until the 1820s when the Irish mathematician William Rowan Hamilton independently worked out the same idea.

8.7 Newton's Solution to the Brachistochrone

In 1696, although he had already solved the problem, Johann Bernoulli made a public challenge of the brachistochrone problem, perhaps directed at Newton. At that time, Newton was serving in London as warden of the mint, having given up mathematics. However, upon receiving the problem after a full day's work, he set upon it immediately and reportedly solved it within twelve hours. Whiteside commented that, although this was a marvelous feat, Newton was then out of practice, and thus he took hours instead of minutes for this problem. We note that in 1685, Newton had addressed a mathematically similar problem, of the solid of revolution of least resistance in a uniform fluid; his solution was included in the *Principia*. In 1697, Newton published a very short note with an accompanying diagram in the *Philosophical Transactions*, stating that the solution to Bernoulli's problem was a cycloid; then in 1700, he wrote up the details, apparently for the purpose of explaining the solution to David Gregory, nephew of James Gregory. In his brief note in the *Transactions*, Newton gave Figure 8.6 and stated:

> From the given point A draw the unbounded straight line $APCZ$ parallel to the horizontal and upon this same line describe both any cycloid AQP whatever, meeting the straight line AB (drawn and, if need be, extended) in the point Q, and then another cycloid ADC whose base and height [as $AC : AP$] shall be to the previous one's base and height respectively as AB to AQ. This most recent cycloid will then pass through the point B and be the curve in which a heavy body shall, under the force of its own weight, most swiftly reach the point B from the point A. As was to be found.

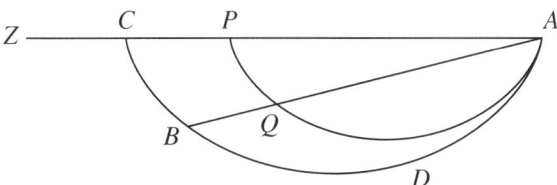

Figure 8.6. Newton's solution to the brachistochrone problem.

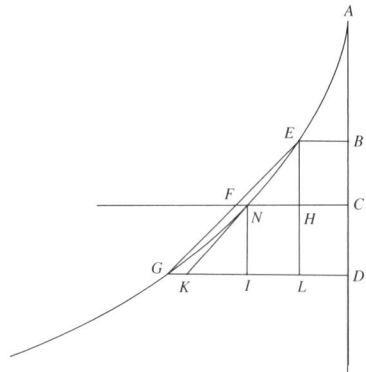

Figure 8.7. Newton's solution for David Gregory.

We summarize Newton's solution for David Gregory, based upon his diagram, Figure 8.7, and Whiteside's commentary. Let $AB = x$, $BC = o = CD$, $BE = y(= y(x))$. By Taylor's expansion

$$CN = y(x+o) = y + \dot{y}o + \frac{1}{2}\ddot{y}o^2,$$
$$DG = y(x+2o) = y + 2\dot{y}o + 2\ddot{y}o^2.$$

From this it follows that

$$HN = IK = \dot{y}o + \frac{1}{2}\ddot{y}o^2,\ IG = \dot{y}o + \frac{3}{2}\ddot{y}o^2,\ \text{and}\ LG = 2\dot{y}o + 2\ddot{y}o^2.$$

Define p and q by $FN = q$ and $GL = 2p$.

The time taken to travel from E to G is to be minimized as q varies. The expression for time is given by

$$\frac{\sqrt{o^2 + (p-q)^2}}{\sqrt{x}} + \frac{\sqrt{o^2 + (p+q)^2}}{\sqrt{o+x}} = R + S,$$

where

$$R^2 = \frac{o^2 + (p-q)^2}{x} \ \text{and}\ S^2 = \frac{o^2 + (p+q)^2}{o+x}.$$

Taking the derivative with respect to q,

$$2R\dot{R} = \frac{-2p\dot{q} + 2q\dot{q}}{x} \ \text{and}\ 2S\dot{S} = \frac{2p\dot{q} + 2q\dot{q}}{x+o}.$$

So the condition for minimum time is that

$$\frac{-p\dot{q} + q\dot{q}}{Rx} + \frac{p\dot{q} + q\dot{q}}{S(x+o)} = 0,$$

or

$$\frac{(p-q)\sqrt{x}}{\sqrt{(p-q)^2+o^2}} = \frac{(p+q)\sqrt{x+o}}{\sqrt{(p+q)^2+o^2}}.$$

This condition implies that $p\sqrt{x}/\sqrt{p^2+o^2}$ must be a constant and since $o/p = \dot{x}/\dot{y}$, we have

$$\sqrt{x}/\sqrt{1+(\dot{x}/\dot{y})^2} = \text{constant} \quad \text{or} \quad 1 + \left(\frac{dx}{dy}\right)^2 = x/c.$$

Thus, $dy = \sqrt{\frac{x}{c-x}}\, dx$, and we have the differential equation of a cycloid.

8.8 Newton on the Radius of Curvature

From the time he studied van Schooten's book on Descartes's *Géometrie*, Newton was interested in the problem of finding the radius of curvature at a point on the curve. In 1664–1665, he grappled with this question, and after several attempts he found a solution. In the 1737 anonymous translation of his treatise on fluxions and series, he wrote:

> There are few Problems concerning Curves more elegant than This, or that give a greater insight into their nature. In order to its resolution, I must premise the following general considerations....
> 2. If a Circle touches any Curve on its concave side in a given point, and its magnitude be such that no other Tangent Circle can be interscribed in the Angle of contact nearer that point, that Circle will be the same Curvature as the Curve is of in that point of contact. For that circle which comes between the curve and another Circle at the point of contact, varies less from the Curve and makes a nearer approach to its Curvature, than that other Circle does; and therefore that Circle approaches nearest to its Curvature, between which and the Curve no other Circle can intervene.
> 3. Therefore the Center of Curvature at any point of a curve, is the Center of a Circle equally curved, and thus the Radius or Semi-diameter of Curvature is part of the perpendicular which is terminated at that Center.

After some discussion of properties of the center of curvature, he described one method for finding the radius of curvature by constructing normals at two infinitely close points, D and d. The intersection of the normals gave the center C of the circle of curvature and therefore CD was the radius of curvature. Referring to Figure 8.8, he explained how to find CD.

> At any point D of the Curve AD, let DT be a Tangent, DC a Perpendicular, and C the Center of Curvature, as before. And let AB be the Absciss, to which let DB be applied at right angles, which DC meets in P. Draw DG parallel to AB, and CG perpendicular to it, in which take Cg of any given magnitude, and draw $g\delta$ perpendicular to it, which meets DC in δ. Then it will be $Cg : g\delta :: (TB : BD ::)$ as the Fluxion of the Absciss to the Fluxion of the Ordinate. Likewise imagine the point D to move in the Curve an infinitely little distance Dd, and drawing de perpendicular to DG, and Cd perpendicular to the Curve, let Cd meet DG in F, and δg in f. Then will De be the *momentum* of the Absciss, de the *momentum* of the Ordinate, and δf the contemporaneous *momentum* of the RightLine $g\delta$. Therefore $DF = De + \frac{de \times de}{De}$. Having therefore the ratios of these *momenta*, or which is the same thing, of their generating Fluxions, you will

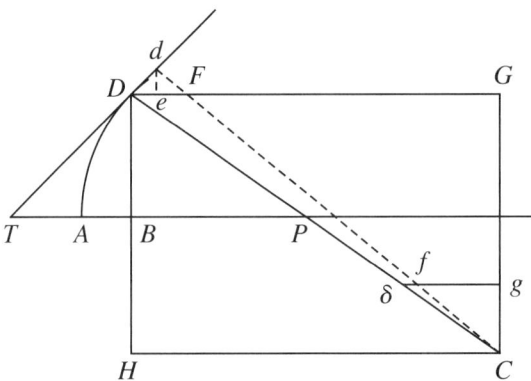

Figure 8.8. Newton's derivation of the radius of curvature.

have the ratio of GC to the given line Cg, which is the same as that of DF to δf. And thence the point C will be determined.

Therefore let $AB = x$, $BD = y$, $Cg = 1$, and $g\delta = z$. Then it will be $1 : z :: \dot{x} : \dot{y}$, or $z = \frac{\dot{y}}{\dot{x}}$. Now let the *momentum* δf of z be $\dot{z} \times o$, (that is the product of the velocity and of an infinitely small quantity o,) therefore the *momentum* $De = \dot{x} \times o$, $de = \dot{y} \times o$, and thence $DF = \dot{x}o + \frac{\dot{y}\dot{y}o}{\dot{x}}$. Therefore it is

$$Cg(1) : CG :: (\delta f : DF ::) \dot{z}o : \dot{x}o + \frac{\dot{y}\dot{y}o}{\dot{x}}, \text{ that is,}$$

$CG = \frac{\dot{x}\dot{x}+\dot{y}\dot{y}}{\dot{x}\dot{z}}$. And whereas we are at liberty to ascribe whatever velocity we please to the Fluxion of the Absciss, to which as to an equable Fluxion the rest may be referred, make $\dot{x} = 1$, and then $\dot{y} = z$, and $CG = \frac{1+zz}{\dot{z}}$; whence $GD = \frac{z+z^3}{\dot{z}}$; and $DC = \frac{\overline{1+zz}\sqrt{1+zz}}{\dot{z}}$.

8.9 Johann Bernoulli on the Radius of Curvature

In 1691, Guillaume l'Hôpital (1661–1704) met Johann Bernoulli, who informed him that he had found a formula for the radius of curvature. A keen student of mathematics, l'Hôpital was fascinated, and requested Bernoulli give him a course of lectures. In 1691–92, Bernoulli delivered these lectures, in which was necessarily included an elaboration of the calculus. L'Hôpital proceeded to write his famous differential calculus textbook, popular for a century. Bernoulli included his integral calculus lectures as *Lectiones Mathematicae* in vol. 3 of his *Opera Omnia*; he mentions l'Hôpital in the subtitle of the lectures. The derivation of the formula for the radius of curvature is contained in Lecture 16 of this work.

In Figure 8.9, the lines OD and BD are radii normal to the curve, with O and B infinitely close, so that BD is the radius of curvature. Bernoulli let $AE = x$, $EB = y$ with $BF = dx$, $FO = dy$. Then he could write

$$FC = \frac{dy^2}{dx} [= \frac{dy}{dx} dy] \quad \text{and therefore} \quad BC = \frac{dx^2 + dy^2}{dx}.$$

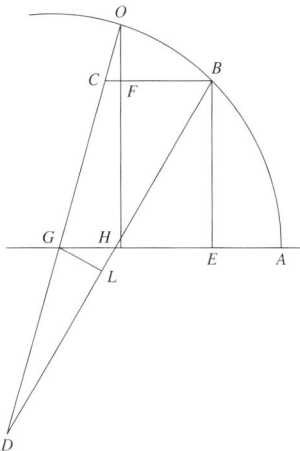

Figure 8.9. Bernoulli's figure for the radius of curvature.

Now $BF : FO = BE : EH$, so that

$$EH = y\frac{dy}{dx}, \quad BH = \frac{y\sqrt{dx^2+dy^2}}{dx}, \quad \text{and} \quad AH = x + y\frac{dy}{dx},$$

and, taking $d^2x = 0$, the differential of AH could be written as

$$HG = dx + \frac{dy^2 + y d^2 y}{dx}.$$

Then because $BC : HG = BD : HD$, he had $(BC - HG) : BC = BH : BD$ and Bernoulli obtained the formula for the radius of curvature:

$$BD = \frac{(dx^2+dy^2)\sqrt{dx^2+dy^2}}{-dx\,d^2y}.$$

8.10 Exercises

1. Let b be a root of $y^3 + a^2 y - 2a^3 = 0$. Show that if $y^3 + a^2 y + axy - 2a^3 - x^3 = 0$ and $a^2 + 3b^2 = c^2$, then

$$y = b - \frac{abx}{c^2} + \frac{a^4 bx^2}{c^6} + \frac{x^3}{c^2} + \frac{a^3 b^3 x^3}{c^8} - \frac{a^5 bx^3}{c^8} + \frac{a^5 bx^3}{c^{10}} + \cdots.$$

2. Suppose $y^3 + y^2 + y - x^3 = 0$ where x is known to be large. Show that

$$y = x - \frac{1}{3} - \frac{2}{9x} + \frac{7}{81x^2} + \frac{5}{81x^3} \quad \text{etc.}$$

See Newton (1964–1967), vol. 1, pp. 46–47 for the above two exercises.

3. Show that if $\dot{y} = 3xy^{2/3} + y$, then $y^{1/3} = \frac{1}{2}x^2 + \frac{1}{15}x^3 + \frac{1}{216}x^4$ etc. See Newton (1964–1967), vol. 1, p. 63; see p. 66 for the next exercise.
4. In a given triangle, find the dimensions of the greatest inscribed rectangle.
5. Show that in the parabola $ax = yy$, the point at which the radius of curvature is of length f is given by $x = -\frac{1}{4}a + \sqrt[3]{\frac{1}{16}af^2}$.
6. Find the locus of the center of curvature of the parabola $x^3 = a^2y$ and of the hyperbola (of the second kind) $xy^2 = a^3$. See Newton (1964–1967), vol. 1, p. 87 for this exercise and p. 85 for the previous exercise. Note that Newton called a polynomial equation $y = p(x)$ a parabola. Similarly, he called $y = a/\sqrt{x}$ a hyperbola of the second kind.
7. Find the asymptotes of the curve $y^3 - x^3 = axy$. See Stone (1730), p. 19.
8. Take a point E on the line segment AB. Find E such that the product of the square of AE times EB is the greatest. See Stone (1730), p. 58. Recall that this part of Stone's book was a translation of l'Hôpital's differential calculus book.
9. Find the volume of a parabolical conoid generated by the rotation of the parabola $y^m = x$ about its axis. See Stone (1730), p. 121 of the appendix.
10. Show that

$$\int x^3 \ln^3 x\, dx = \frac{1}{4}x^4 \ln^3 x - \frac{3}{4^2}x^4 \ln^2 x + \frac{3 \cdot 2}{4^3}x^4 \ln x - \frac{3 \cdot 2}{4^4}x^4.$$

More generally, find $\int x^m \ln^m x\, dx$. From this result, deduce that

$$\int_0^1 x^x\, dx = 1 - \frac{1}{2^2} + \frac{1}{3^3} - \frac{1}{4^4} + \frac{1}{5^5} - \frac{1}{6^6} + \text{etc.}$$

See Joh. Bernoulli (1968), vol. 3, pp. 380–381.

8.11 Notes on the Literature

Even though Newton was unable to publish his 1671 calculus treatise, the text was published several times, starting in the 1730s, in both Latin and English. However, Whiteside found that the translations were not completely adequate. Consequently, in vol. 3 of Newton (1967–1981), Whiteside presented his own translation accompanied by Newton's Latin text. The discussion of equation (8.4) and the quotation following it is from p. 81; the quotation on maxima and minima on p. 117; on tangents, p. 149. However, we quote Newton's derivation of the radius of curvature from the 1737 anonymous translation, in order to give the flavor of the text. For a discussion of Newton's brachistochrone work along with Whiteside's commentary, see pp. 86–90 of vol. 8 of Newton (1967–1981). Apparently, Newton wrote these notes on the brachistochrone for David Gregory, to help him understand Nicolas Fatio de Duillier's complicated and peculiar 1699 paper, tangentially dealing with this topic. Newton's brief description of his solution, given in 1697, appears on p. 75 of vol. 8 and the accompanying diagram is on p. 72.

For Descartes's solution of de Beaune's problem, see the discussion in Hofmann (1990), vol. 2, pp. 279–84. An English translation of Leibniz's differential calculus paper is in Struik (1969), pp. 272–280. A reprint of the original 1684 paper in Latin can be found in Leibniz (1971), vol. 3. The mathematical details of Leibniz's work on the catenary were given in his letters to Huygens and to von Bodenhausen. See vol. 7 of Leibniz (1971), pp. 370–372 for the letter to von Bodenhausen. Truesdell quotes the complete derivation of the equation of the catenary from Leibniz's letter to Huygens, but the letter is also included in Huygens's collected works, referenced by Truesdell (1960). Truesdell presents an interesting commentary on the work of Pardies, Leibniz, Huygens, and Johann Bernoulli relating to the catenary. See especially pp. 64–75.

We have not dealt with Leibniz's higher differentials in any detail. An interesting account appears in Bos (1974). Euler showed that the complicated theory of higher differentials could be avoided by using dependent and independent variables. An English translation of Bernoulli's brachistochrone paper appears in Struik (1969), pp. 392–396. A reprint of Bernoulli's original work is in Joh. Bernoulli (1968), pp. 187–193. Simmons (1992) gives an entertaining account in modern terminology. Bernoulli's derivation of the radius of curvature appears in his collected works, Joh. Bernoulli (1968), vol. 3, pp. 488–505. Find his paper on the calculus of the exponential function on pp. 179–187. The reader may enjoy seeing Knoebel, Laubenbacher, Lodder, and Pengelley (2007) for their discussion of Newton's derivation of the radius of curvature. They review this work of Newton in the context of the general notion of curvature and include a discussion of the ideas of Huygens, Euler, Gauss, and Riemann on the topic. Bourbaki (1994) contains a deep but concise summary of the development of the calculus; see pp. 166–198.

9

De Analysi per Aequationes Infinitas

9.1 Preliminary Remarks

Newton's groundbreaking paper, revealing the power of infinite series to resolve intractable problems in algebra and calculus, was probably written in the summer of 1669. Before Newton, the only infinite series to be studied, besides the infinite geometric series, was the logarithmic series, by J. Hudde and N. Mercator. Inspired by Wallis's work on the area of a quadrant of a circle, Newton, in the winter of 1664–65, considered the more general problem of finding the area under $y = \sqrt{1-t^2}$ on the interval $(0, x)$, for $x \leq 1$. This question led Newton to make the extraordinary inquiry into the value of $(1-t^2)^{1/2}$ in powers of t; Newton thus discovered the binomial theorem, first for exponent $1/2$ and soon for all rational exponents. He very quickly perceived the tremendous significance of this result, and more generally, the importance and usefulness of infinite series to analysis.

In this paper Newton resolved the general problem, at least in principle, of finding the area under a curve defined explicitly or implicitly. He showed that by means of infinite series the problem could be reduced to that of integrating $x^{m/n}$, where m and n were integers. If the equation was given explicitly as $y = f(x)$ with $f(x)$ a rational or algebraic function, then $f(x)$ could be expanded as an infinite series by the binomial theorem. The area under the curve could then be obtained after term-by-term integration. Among the examples he gave were the curves $y = 1/(1+x^2)$, $y = (2x^{1/2} - x^{3/2})/(1+x^{1/2} - 3x)$ and $y = \sqrt{(1+ax^2)}/\sqrt{(1-bx^2)}$. He wrote that the quadrature of the last example yielded the length of an elliptic arc.

The problem of integrating even these elementary functions would have been too difficult for the mathematicians before Newton, but his work on the integration of implicitly defined functions took algebra and analysis to a new level. In 1664, Newton learned from the books of Viète and Oughtred how to solve algebraic equations $f(x) = 0$ by the method of successive approximation. One chose an approximate solution, and on that basis, one derived successively better ones. The Islamic mathematician Jamshid al-Kashi (1380–1429) had used a primitive form of this method to solve cubic equations and to compute roots of numbers, that is, to solve equations $x^p - N = 0$. With the concept of infinite series in hand, and the technical skill to work with it, Newton showed how to solve the equation $f(x, y) = 0$ in the form $y = g(x)$, where $g(x)$ was an infinite series. He obtained higher and higher powers of x by successive approximation. Newton then found the area under a curve defined implicitly as $f(x, y) = 0$ by integrating $g(x)$ term by term. Newton gave this method in the *De Analysi*, but he realized that one did not always obtain the solution to $y = g(x)$ as a power series. In a longer treatise on calculus and infinite series of 1671, Newton gave examples where the solution around $x = 0$ was of the form $y = x^\alpha g(x)$, with $g(0) \neq 0$ and α a fraction. He realized that for only certain values of α could $g(0)$ be determined; for those values, the functions $y = x^\alpha g(x)$ were solutions of $f(x, y) = 0$ in the neighborhood of $x = 0$. He devised a method now called Newton's polygon to determine the allowable values of α. This method has important applications in algebraic geometry and analysis. Newton extended his method for solving $f(x, y) = 0$ to obtain the inverses of functions defined by infinite series. He knew that his formula, mentioned earlier, for the area of a sector of a circle was equivalent to the series for arcsine. By inversion he found the series for sine and from that the series for cosine. We have seen that Madhava earlier obtained the series for these functions by a different method.

Newton uncharacteristically wrote up his results on infinite series because by 1668 others were beginning to make similar discoveries. In letters to James Gregory in 1669, John Collins reported that N. Mercator had found series for sine and for the segment of a circle, and that Brouncker could expand the square root as an infinite series. In 1668, Mercator also published a book, *Logarithmotechnia* in which he expanded $1/(1 + x)$ as a series and integrated term by term to obtain his series for $\ln(1 + x)$; he applied his result to the computation of logarithmic values. In the spring of 1665, Newton had done exactly the same thing; after Mercator's publication, he realized that he would lose credit for his discoveries unless he made them known.

Newton submitted his paper to Isaac Barrow, then Lucasian Professor of Mathematics at Cambridge, who mentioned it to Collins in a letter of July 20, 1669 with the words:

> A friend of mine here, that hath a very excellent genius to those things, brought me the other day some papers, wherein he hath sett downe methods of calculating the dimensions of magnitudes like that of Mr Mercator concerning the hyperbola, but very generall; as also of resolving aequations; which I suppose will please you; and I shall send you them by the next.

He wrote Collins again on August 20, 1669:

> I am glad my friends paper giveth you so much satisfaction. his name is Mr Newton; a fellow of our College, & very young (being but the second yeest [youngest] Master of Arts) but of an extraordinary genius & proficiency in these things. you may impart the papers if you please to my Ld Brounker [sic].

Collins made a complete copy of Newton's paper and communicated some of its results to his correspondents in Britain, France, and Italy. He and Barrow urged Newton to publish his paper as an appendix to Barrow's optical lectures but Newton resisted, perhaps because he had a much larger work in mind, finally written in 1671. This long but incomplete tract is referred to as *De Methodis Serierum et Fluxionum*, though its original title or whether it even had one is unclear, since the first page of the original manuscript is lost and mathematicians of Newton's own time referred to it by various titles. In the *De Methodis*, Newton showed how to find the derivative by implicit differentiation of the equation for the curve $f(x, y) = 0$. He applied this to problems on tangents, normals, and radii of curvature. Conversely, given a fluxional (differential) equation, he explained how it could be solved, particularly with infinite series. The equations he worked with here were algebraic differential equations.

It is interesting to note that Newton wrote up his results on series only when he realized that others were working on similar problems; this exercise gave him the opportunity to rethink his ideas and improve upon them. This happened to him several times. For example, in the spring of 1684, David Gregory, nephew of James, published a fifty-page tract *Exercitatio Geometrica de Dimensione Figurarum*, discussing his uncle's results on infinite series related to the binomial theorem. He also promised to write a sequel with more results. This immediately spurred Newton to compose the "Matheseos Universalis Specimina," in the first part of which he gave a brief history of his work on series and the results on this topic he had communicated to Collins and to Leibniz. He then went on to develop some new ideas on finite differences and series. The paper was not completed and in fact ended in the middle of a sentence. Very soon after this, he reorganized his ideas and presented them in a paper called "De Computo Serierum." Here he left out the history but further clarified the new mathematical idea on series and differences, framed as the transformation formula now often called Euler's transformation. Unfortunately, Newton never published these papers. Similarly, in 1691 he wrote and rewrote the tract *De Quadratura Curvarum*, containing the first explicit statement of Taylor's theorem; he published only a portion of this work some years later.

9.2 Algebra of Infinite Series

Newton pointed out in his *De Analysi* that just as infinite decimals were needed to divide by numbers, extract roots of numbers, and solve equations with numerical coefficients, infinite series were needed to divide by polynomials, extract roots of algebraic expressions, and solve equations with algebraic coefficients. To illustrate division, Newton considered the equation $y = a^2/(b+x)$ and showed that the process led to the series

$$y = \frac{a^2}{b} - \frac{a^2 x}{b^2} + \frac{a^2 x^2}{b^3} - \frac{a^2 x^3}{b^4} + \cdots. \tag{9.1}$$

From this he concluded that the area under the curve $y = a^2/(b+x)$ could be expressed as

$$\frac{a^2 x}{b} - \frac{a^2 x^2}{2b^2} + \frac{a^2 x^3}{3b^3} - \frac{a^2 x^4}{4b^4} + \cdots, \tag{9.2}$$

if the required area was taken over the interval $(0, x)$. If, however, the required area under $y = 1/(1 + x^2)$ was over (x, ∞), then with $a = b = 1$, and x replaced by x^2, he started with the series

$$y = \frac{1}{x^2 + 1} = x^{-2} - x^{-4} + x^{-6} - x^{-8} \quad \text{etc.} \tag{9.3}$$

9.2 Algebra of Infinite Series

The area was then given by

$$-x^{-1} + \frac{1}{3}x^{-3} - \frac{1}{5}x^{-5} + \frac{1}{7}x^{-7} \quad \text{etc.} \tag{9.4}$$

Newton noted that x should be small in (9.2), but should be large in (9.4), though he did not specify how small or how large. At the end of the paper, he made some remarks on convergence. He observed that if $x = 1/2$, then x would be half of all of $x + x^2 + x^3 + x^4$ etc. and x^2 half of all of $x^2 + x^3 + x^4 + x^5$ etc. So if $x < 1/2$, then x would be more than half of all of $x + x^2 + x^3$ etc. and x^2 more than half of all of $x^2 + x^3 + x^4$ etc. He then extended the argument to the case x/b where b was a constant.

In his second example, Newton applied the algorithm for finding square roots of numbers to $\sqrt{(a^2 + x^2)}$, obtaining the infinite series

$$a + \frac{x^2}{2a} - \frac{x^4}{8a^3} + \frac{x^6}{16a^5} - \frac{5x^8}{128a^7} + \frac{7x^{10}}{256a^9} - \frac{21x^{12}}{1024a^{11}} \quad \text{etc.} \tag{9.5}$$

A little later in the paper, Newton explained his method of successive approximations to solve polynomial equations $f(x, y) = 0$. To illustrate the method, he first took an equation with constant coefficients:

$$y^3 - 2y - 5 = 0. \tag{9.6}$$

An approximate solution would be 2, so he set $y = 2 + p$ to transform (9.6) to

$$p^3 + 6p^2 + 10p - 1 = 0. \tag{9.7}$$

He argued that since p was small, the terms $p^3 + 6p^2$ could be neglected, though he noted that a better approximation would be obtained if only p^3 were neglected. Thus, he had $p = 0.1$, and he substituted $p = 0.1 + q$ in (9.7) to obtain

$$q^3 + 6.3q^2 + 11.23q + 0.061 = 0.$$

Newton linearized this equation to $11.23q + 0.061 = 0$, solved for q to get $q = -0.0054$, set $q = -0.0054 + r$, and wrote that one could continue in this manner. In a similar way, he resolved the equation

$$y^3 + a^2y - 2a^3 + axy - x^3 = 0 \tag{9.8}$$

for small values of x. He set $x = 0$ to obtain

$$y^3 + a^2y - 2a^3 = 0 \tag{9.9}$$

so that $y = a$ was a solution; he set $y = a + p$ in (9.8) and took the linear part of the equation to get $p = -\frac{1}{4}x$. In this manner, he had the series

$$y = a - \frac{1}{4}x + \frac{x^2}{64a} + \frac{131x^3}{512a^2} + \frac{509x^4}{16384a^3} \quad \text{etc.} \tag{9.10}$$

He then used the example

$$y^3 + axy + x^2y - a^3 - 2x^3 = 0 \qquad (9.11)$$

to illustrate how to obtain a solution for large values of x. Here he started with the highest power terms in the equation (9.10) to get

$$y^3 + x^2y - 2x^3 = 0.$$

Since $y = x$ was a solution of this, he set $y = x + p$ in (9.11) and proceeded as before.

Finally, Newton showed that similar methods could be applied when the equation had an infinite number of terms. The problem of interest was to solve $y = f(x)$, for x in terms of y, where $f(x)$ was an infinite series. This gave him a series for the inverse function and in the *De Analysi*, he applied it to the cases where $f(x) = -\ln(1-x) = x + \frac{1}{2}x^2 + \frac{1}{3}x^3 + \cdots$ and where $f(x) = \arcsin x = x + \frac{1}{6}x^3 + \frac{3}{40}x^5 + \frac{5}{112}x^7 + \cdots$. Thus, he found the series for the exponential and sine functions.

Observe that Newton's method of successive approximations is actually equivalent to the method of undetermined coefficients, learned by Newton through his careful study of Descartes. One assumes that if z denotes the series, say for $-\ln(1-x)$, then

$$x = a_0 + a_1 z + a_2 z^2 + \cdots,$$

and the values of $a_0, a_1, a_2, \ldots$ are obtained by substituting back in the series and equating the coefficients of the powers of z on both sides of the equation. Newton must have understood this because in his October 1676 letter to Oldenburg he wrote:

Let the equation for the area of an hyperbola be proposed

$$z = x + \frac{1}{2}x^2 + \frac{1}{3}x^3 + \frac{1}{4}x^4 + \frac{1}{5}x^5, \quad \text{etc.}$$

and its terms being multiplied into themselves, there results

$$z^2 = x^2 + x^3 + \frac{11}{12}x^4 + \frac{5}{6}x^5, \quad \text{etc.,}$$

$$z^3 = x^3 + \frac{3}{2}x^4 + \frac{7}{4}x^5, \quad \text{etc.,}$$

$$z^4 = x^4 + 2x^5, \quad \text{etc.,}$$

$$z^5 = x^5, \quad \text{etc.}$$

Now subtract $\frac{1}{2}z^2$ from z, and there remains

$$z - \frac{1}{2}z^2 = x - \frac{1}{6}x^3 - \frac{5}{24}x^4 - \frac{13}{60}x^5, \quad \text{etc.}$$

To this I add $\frac{1}{6}z^3$, and it becomes

$$z - \frac{1}{2}z^2 + \frac{1}{6}z^3 = x + \frac{1}{24}x^4 + \frac{3}{40}x^5, \quad \text{etc.}$$

I subtract $\frac{1}{24}z^4$ and there remains

$$z - \frac{1}{2}z^2 + \frac{1}{6}z^3 - \frac{1}{24}z^4 = x - \frac{1}{120}x^5, \quad \text{etc.}$$

I add $\frac{1}{120}z^5$ and it becomes

$$z - \frac{1}{2}z^2 + \frac{1}{6}z^3 - \frac{1}{24}z^4 + \frac{1}{120}z^5 = x$$

as nearly as possible; or

$$x = z - \frac{1}{2}z^2 + \frac{1}{6}z^3 - \frac{1}{24}z^4 + \frac{1}{120}z^5, \quad \text{etc.}$$

He then went on to state two general theorems:
Let $z = ay + by^2 + cy^3 + dy^4 + ey^5 +$ etc. Then conversely will

$$y = \frac{z}{a} - \frac{b}{a^3}z^2 + \frac{2b^2 - ac}{a^5}z^3 + \frac{5abc - 5b^3 - a^2d}{a^7}z^4$$
$$+ \frac{3a^2c^2 - 21ab^2c + 6a^2bd + 14b^4 - a^3e}{a^9}z^5 + \quad \text{etc.} \quad (9.12)$$

Let $z = ay + by^3 + cy^5 + dy^7 + ey^9 +$ etc. Then conversely will

$$y = \frac{z}{a} - \frac{b}{a^4}z^3 + \frac{3b^2 - ac}{a^7}z^5 + \frac{8abc - a^2d - 12b^3}{a^{10}}z^7$$
$$+ \frac{55b^4 - 55ab^2c + 10a^2bd + 5a^2c^2 - a^3e}{a^{13}}z^9 + \quad \text{etc.} \quad (9.13)$$

Newton observed that if he took $a = 1, b = \frac{1}{6}, c = \frac{3}{40}, d = \frac{5}{112}$, etc., in the second series, then the series for $\sin z$ would follow. We note that actually Newton wrote the series for $r \sin z$, where r was the radius of the circle, in powers of z/r because z was the length of an arc of the circle. Euler later eliminated the role of the radius and defined the trigonometric functions as we do now.

9.3 Newton's Polygon

In his *De Methodis*, Newton explained his method of solving $f(x, y) = 0$ by means of the Newton polygon, where the solution took the form $y = x^\alpha y_1$, with α rational and y_1 a power series in x. To find the possible values of α, he plotted the points (b, a) for each term $cx^a y^b$ in $f(x, y)$, such that b ran along the horizontal axis and a along the vertical. He then took the lower portion of the convex hull of these points, consisting of the straight line(s) joining the vertical to the horizontal axis. Although it does not enclose an area, this lower portion is called the Newton polygon and the slope(s) of these line(s) gave him the values of α. For example, if m were the slope of a line in the polygon, then one value of α would be given by $-1/m$. These values of α permitted the evaluation of a nonzero value of $y_1(0)$. Note here that the lines joining the other pairs of points (b, a) could allow for zero values of $y_1(0)$.

Newton considered the example

$$y^6 - 5xy^5 + \frac{x^3}{a}y^4 - 7a^2x^2y^2 + 6a^3x^3 + b^2x^4 = 0,$$

where he had the points (6,0), (5,1), (4,3), (2,2), (0,3), and (0,4). The line joining (0,3), (2,2), and (6,0) formed the polygon and gave Newton the terms

$$y^6 - 7a^2x^2y^2 + 6a^3x^3;$$

setting these equal to zero, he obtained the lowest-order term in the expansion of y as a series in x. The slope of the line in the polygon was $-1/2$ so in the case $y = cx^{1/2}$, the terms had the same power in x. Newton could then set $y = v\sqrt{ax}$ to reduce the equation to $v^6 - 7v^2 + 6 = 0$. He obtained $v = \pm 1, \pm\sqrt{2}, \pm\sqrt{-3}$ but rejected the complex roots. Thus, he had four possible initial values of y: $\pm\sqrt{ax}, \pm\sqrt{2ax}$. He wrote that all four expressions were acceptable initial values for y; by successive approximations, he went on to find more terms.

Newton's solutions of $f(x, y) = 0$ are the first known examples of the implicit function theorem. Significantly, though Newton did not give an existence proof, he presented an algorithm for deriving the solution. S. S. Abhyankar pointed out that this algorithm is applicable to existence proofs in analysis and can also produce the formal solutions required in algebraic geometry.

9.4 Newton on Differential Equations

In his 1670–71 treatise *De Methodis Serierum et Fluxionum*, Newton discussed how to find the derivative from the equation $f(x, y) = 0$ and, conversely, how to find the relation between x and y, given a first-order differential equation $f(x, y, \dot{x}, \dot{y}) = 0$. Note that in the 1690s, Newton began to use the dot notation to indicate a fluxion, or derivative. In his earliest work, including his work in the 1670s, he employed the letters p, q, or m, n.

To illustrate how series could be used to solve differential equations, Newton considered several examples, including

$$\dot{y}/\dot{x} = 1 + y/(a-x),$$
$$\dot{y}^2 = \dot{x}\dot{y} + \dot{x}^2x^2, \quad \text{and}$$
$$\dot{y}^3 + ax\dot{x}^2\dot{y} + a^2\dot{x}^2\dot{y} - \dot{x}^3x^3 - 2\dot{x}^3a^3 = 0.$$

He rewrote the first equation as

$$\dot{y}/\dot{x} = 1 + y/a + xy/a^2 + x^2y/a^3 + x^3y/a^4 \quad \text{etc.}$$

and then showed how to obtain particular solutions of this equation by assuming a series solution. He rewrote the second equation as a quadratic in $\dot{y}/\dot{x}$ to get

$$\dot{y}^2/\dot{x}^2 = \dot{y}/\dot{x} + x^2$$

and solved the quadratic algebraically to obtain

$$\dot{y}/\dot{x} = \frac{1}{2} \pm \sqrt{\frac{1}{4} + x^2}.$$

After expanding $(1/4+x^2)^{1/2}$ by the binomial theorem, Newton integrated the resulting infinite series term by term. He apparently did not observe that $(1/4+x^2)^{1/2}$ could be integrated directly in terms of the logarithm. As we have seen, Barrow gave the integral of $(a^2+x^2)^{1/2}$ in his *Lectiones Geometricae* of 1670 and Newton knew this work quite well. However, unlike Leibniz, Newton may not have been particularly interested in closed-form solutions. Newton changed the third equation into a cubic in $\dot{y}/\dot{x}$:

$$(\dot{y}/\dot{x})^3 + ax(\dot{y}/\dot{x}) + a^2(\dot{y}/\dot{x}) - x^3 - 2a^3 = 0,$$

the same cubic as in (9.8). So from (9.10) he saw that

$$\dot{y}/\dot{x} = a - \frac{x}{4} + \frac{x^2}{64a} + \frac{131x^3}{512a^2} \quad \text{etc.}$$

and hence

$$y = ax - \frac{x^2}{8} + \frac{x^3}{192a} + \frac{131x^4}{2048a^2} \quad \text{etc.}$$

9.5 Newton's Earliest Work on Series

In some of the earliest material recorded in his mathematical notebooks, Newton raised the problem of finding x, given $\sin x$, observing that the problem was equivalent to finding the area of a segment of a circle. Newton did this work, inspired by Wallis's book, in winter 1664–65.

In Figure 9.1, let aec be a quarter of the circle of radius one with center p and let $pq = x$. If the angle $ape = \theta$, then $x = \sin \theta$ and the area of the sector ape would

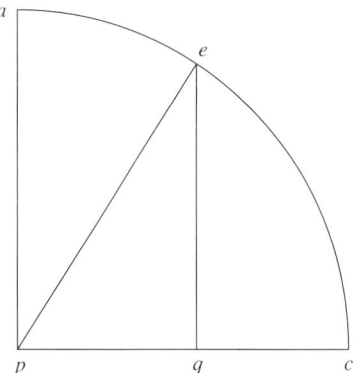

Figure 9.1. Newton's figure for derivation of the arcsine series.

be $\frac{1}{2}\theta = \frac{1}{2}\arcsin x$. Newton's problem was to find an expression for the area given by $\frac{1}{2}\arcsin x$, when x was known. He knew, from a study of Wallis's *Arithmetica Infinitorum*, that the area $aeqp$ was equal to the area under the curve $y = \sqrt{1-t^2}$ over the interval $[0, x]$. The area of the triangle peq was $\frac{1}{2}x\sqrt{1-x^2}$. So his formula in modern notation would be given as

$$\text{area of sector } aep = \frac{1}{2}\arcsin x = \int_0^x \sqrt{1-t^2}\,dt - \frac{1}{2}x\sqrt{1-x^2}. \tag{9.14}$$

In the course of this work, he discovered the binomial theorem. This gave him the result

$$(1-x^2)^{1/2} = 1 - \frac{x^2}{2} - \frac{x^4}{8} - \frac{x^6}{16} - \frac{5x^8}{128} - \frac{7x^{10}}{256} - \frac{21x^{12}}{1024} \text{ etc.}$$

He substituted this series in the integral and integrated term by term to get the solution of his problem:

$$\arcsin x = x + \frac{x^2}{6} + \frac{3x^5}{40} + \frac{5x^7}{112} + \frac{35x^9}{1152} \text{ etc.} \tag{9.15}$$

In the later *De Analysi*, Newton found the series for $\arcsin x$ by determining the arclength of the arc of a circle. In this case, he had to integrate $1/\sqrt{1-t^2}$. If this were combined with (9.14), the result would be

$$\int_0^x \frac{dt}{\sqrt{1-t^2}} = \arcsin x.$$

Thus, Newton was aware of the integral for arcsine as well as the formula

$$\int_0^x \sqrt{1-t^2}\,dt = \frac{1}{2}x\sqrt{1-x^2} + \frac{1}{2}\int_0^x \frac{dt}{\sqrt{1-t^2}}.$$

Modern textbooks usually derive the last formula by means of integration by parts. After 1666, Newton was effectively aware of substitution and integration by parts, but to obtain a result more simply, he often gave geometric arguments, similar to the preceding one; note, however, that he derived the series for arcsine in 1664–65. Newton discovered another interesting formula from which the series for arcsine can be easily derived; the first mention of it occurs in his June 13, 1676 letter to Oldenburg, in response to Leibniz:

> If an arc is be taken in a given ratio to another arc, let d be the diameter, x the chord of the given arc, and the required arc be to that given arc as $n : 1$. Then the chord of the arc required will be
>
> $$nx + \frac{1-n^2}{2\times 3d^2}x^2 A + \frac{9-n^2}{4\times 5d^2}x^2 B + \frac{25-n^2}{6\times 7d^2}x^2 C + \frac{36-n^2}{8\times 9d^2}x^2 D + \frac{49-n^2}{10\times 11d^2}x^2 E + \text{ etc.}$$
>
> Here note that when n is an odd number the series is no longer infinite and becomes the same as that which results by common algebra for multiplying the given angle by that number n.

As Newton explained in his letter, A stood for the first term, B for the second term, C for the third term, etc. Observe that this formula is equivalent to

$$\sin n\theta = n\sin\theta - \frac{n(n^2-1)}{3!}\sin^3\theta + \frac{n(n^2-1)(n^2-3^2)}{5!}\sin^5\theta \\ - \frac{n(n^2-1)(n^2-3^2)(n^2-5^2)}{7!}\sin^7\theta + \cdots. \quad (9.16)$$

Note that the series for arcsine is obtained by dividing (9.16) by n and letting n tend to zero. The corresponding cosine series is given by

$$\cos n\theta = 1 - \frac{n^2}{2!}\sin^2\theta + \frac{n^2(n^2-2^2)}{4!}\sin^4\theta - \frac{n^2(n^2-2^2)(n^2-4^2)}{6!}\sin^6\theta + \cdots. \quad (9.17)$$

This series does not appear in the extant papers of Newton, although one may safely assume he must have known this result. Note that if we subtract 1 from both sides and then divide the equation by θ^2, then we obtain the series for $\arcsin^2 x$ when θ tends to zero.

9.6 De Moivre on Newton's Formula for $\sin n\theta$

In 1698, de Moivre gave a derivation of Newton's series for $\sin n\theta$ in "A Method of Extracting the Root of an Infinite Equation" published in the *Philosophical Transactions*. Now de Moivre had already seen the method of undetermined coefficients used in this context, as presented in the letter of Newton for Leibniz. Note that British mathematicians of the 1690s were aware of these letters, since Wallis had published portions of them in his 1685 book on algebra and had presented more complete accounts in the 1690s. In his paper, de Moivre considered the situation in which the series on the left side of the equation was in terms of a variable different from that on the right; one variable had to be determined in terms of the other. He stated the main result at the very beginning of his paper:

If $az + bzz + cz^3 + dz^4 + ez^5 + fz^6 +$ etc. $= gy + hyy + iy^3 + ky^4 + ly^5 + my^6 +$ etc., then z will be

$$= \frac{g}{a}y + \frac{h-bAA}{a}y^2 \\ + \frac{i-2bAB-cA^3}{a}y^3 \\ + \frac{k-bBB-2bAC-3cAAB-dA^4}{a}y^4 \\ + \frac{l-2bBC-2bAD-3cABB-3cAAC-4dA^3B-eA^5}{a}y^5 \\ + \text{etc.}$$

Note that de Moivre also included the coefficient of y^6. Each capital letter denoted the coefficient of the preceding term. Thus, $A = \frac{g}{a}$ and $B = \frac{h-bAA}{a}$, and so on. His proof first assumed that z had a series expansion $Ay + Byy + Cy^3 + Dy^4 +$ etc., then substituted this for each z on the left side of the initial equation, and then equated coefficients of powers of y to get A, B, C, D. To apply de Moivre's formula to get Newton's formula, recall that the latter involved the expansion of $z = \sin n\theta$ in powers of $y = \sin\theta$. Clearly, we can write

$$\arcsin z = n \arcsin y.$$

When the arcsines are replaced by their power series expansions, we have

$$z + \frac{z^3}{6} + \frac{3z^5}{40} + \frac{5z^7}{112} + \cdots = ny + \frac{ny^3}{6} + \frac{3ny^5}{40} + \frac{5ny^7}{112} + \cdots.$$

De Moivre applied his general result to this special equation to obtain

$$A = n, \ B = 0, \ C = -\frac{n(n^2-1)}{6}, \ D = 0, \ E = \frac{n(n^2-1)(n^2-3^2)}{5!}, \ldots,$$

thereby completing a proof of Newton's formula.

9.7 Stirling's Proof of Newton's Formula

By studying Stirling's unpublished notebooks, Ian Tweddle discovered that Stirling gave yet another proof of Newton's formula, by means of differential equations. This work was probably done before 1730. Stirling took variables $y = r\sin\theta$ and $v = r\sin n\theta$, where n was any positive number. He used geometric considerations to define these variables and these required that $0 \leq n\theta \leq \pi/2$, but his proof is actually valid for $0 \leq |\theta| \leq \pi/2$ with n any real number. Since $\theta = \arcsin(y/r)$, $\dot\theta = r\dot y/\sqrt{r^2 - y^2}$ and similarly $n\dot\theta = r\dot v/\sqrt{r^2 - v^2}$, Stirling obtained the fluxional equation

$$\frac{n\dot y}{\sqrt{rr - yy}} = \frac{\dot v}{\sqrt{rr - vv}}; \tag{9.18}$$

after squaring, he got

$$\frac{n^2\dot y^2}{rr - yy} = \frac{\dot v^2}{rr - vv}.$$

He cross multiplied to obtain

$$n^2 r^2 \dot y^2 - n^2 \dot y^2 v^2 = \dot v^2 r^2 - y^2 \dot v^2.$$

We note that until the middle of the nineteenth century, mathematicians sometimes wrote xx and sometimes x^2. He took the fluxion (derivative) of this equation, assuming that $\dot y$ was uniform. This meant that $\ddot y = 0$. Thus, he had the equation

$$-n^2 \dot y^2 v\dot v = \dot v \ddot v r^2 - \dot v \ddot v y^2 - y\dot y \dot v^2, \text{ or}$$
$$-n^2 \dot y^2 v = \ddot v r^2 - \ddot v y^2 - y\dot y \dot v.$$

He next set $\dot{y} = 1$, without loss of generality, to obtain

$$\ddot{v}(r^2 - y^2) - y\dot{v} + n^2 v = 0. \tag{9.19}$$

Assuming a series solution, Stirling set

$$v = Ay + By^3 + Cy^5 + Dy^7 \text{ etc.} \tag{9.20}$$

After substituting (9.20) in (9.19), Stirling found the coefficients, completing the derivation:

$$B = \frac{1-n^2}{2 \cdot 3r^2} A, \quad C = \frac{9-n^2}{4 \cdot 5r^2} B, \quad D = \frac{25-n^2}{6 \cdot 7r^2} C, \text{ etc.}$$

In his 1730 *Methodus Differentialis*, proposition 15, Stirling briefly explained why he used a series of the form (9.20) to solve the differential equation. He set $v = Ay^m$ and substituted this expression in the differential equation to get

$$(m^2 - m) Ay^{m-2} + (n^2 - m^2) Ay^m = 0. \tag{9.21}$$

To obtain the lowest power of y in the series solution, he then set $m^2 - m = 0$ to obtain $m = 0$ or $m = 1$. Thus, the lowest power of y was either 0 or 1. By (9.21), the powers had to increase by two, so that either v was given by the series (9.20), or else

$$v = A + By^2 + Cy^4 + Dy^6 + \cdots. \tag{9.22}$$

By using (9.22) in a similar way, he obtained the formula (9.17) for $\cos n\theta$.

Newton did not state the cosine formula, but he must have known it from his 1664 study of Viète's booklet on angular sections, written in 1591 but published in 1615 with proofs supplied by Alexander Anderson, an uncle of the great Scottish mathematician James Gregory. In this paper, Viète expressed in geometric terms the formulas for $\cos n\theta$ and $\sin n\theta$ in powers of $\cos\theta$, with n an integer. He explicitly pointed out the appearance of the figurate numbers as coefficients of these polynomials. As a student, Newton made annotations on this work of Viète, though they do not indicate that he knew his formula (9.16) at that time. The manner in which he wrote the coefficients of the powers of $\sin\theta$ in his letter for Leibniz suggests that he found the result after his discovery of the binomial theorem. It is even likely that he found the formula while reviewing his old notes before writing his first letter for Leibniz in June 1676.

The methods employed by de Moivre and Stirling to prove Newton's formulas were familiar to Newton in 1676. In fact, it is very likely that Newton had already found a proof. It is possible that he first came upon the formula for $\sin n\theta$ by interpolation, but he wrote in his letter for Leibniz that he had discarded interpolation as a method of proof. Since Newton was very cautious, he must have had an alternative derivation when he communicated it to Leibniz, though he gave no hint of what it was.

In 1812, Gauss applied (9.16) to produce an unusual proof of Euler's gamma function formula $\Gamma(x)\Gamma(1-x) = \pi/\sin\pi x$. He also briefly mentioned that he could prove

(9.16) using transformations of hypergeometric functions. Yet another proof was given by Cauchy in his École Polytechnique lectures of 1821. He also observed that the series for $\arcsin x$ could be derived by equating the coefficient of n on both sides of (9.16). Similarly, by equating the coefficient of n^2 in (9.17), Cauchy obtained

$$\frac{1}{2}(\sin^{-1}x)^2 = \sum_{n=0}^{\infty} \frac{2^{2n}(n!)^2}{(2n+2)!} x^{2n+2}, \qquad (9.23)$$

a result he attributed to J. de Stainville who published it in 1815. In 1738, the particular case of the above series where $x = 1$ was discovered by Johann Bernoulli who communicated it to his former student Euler. Euler responded by using differential equations to prove the more general formula (9.23). As we shall see later, Bernoulli's method can be modified slightly to prove the general case. Even before Bernoulli and Euler, Takebe Katahiro published these two series in his 1722 *Yenri Tetsujutsu*.

9.8 Zolotarev: Lagrange Inversion with Remainder

Newton's statement of his two theorems on the inversion of series suggests that he got them by using the method of undetermined coefficients, though his related work employs successive approximation. In 1769, Lagrange published a more interesting result now referred to as the Lagrange inversion formula. This work was done in connection with an application to celestial mechanics. Lagrange's formula stated that if $z = a + x\phi(z)$, then

$$F(z) = F(a) + x\phi(a)F'(a) + \frac{x^2}{1\cdot 2}\frac{d}{da}(\phi^2(a)F'(a)) + \frac{x^3}{1\cdot 2\cdot 3}\frac{d^2}{da^2}(\phi^3(a)F'(a)) + \cdots.$$

In support of this formula, Lagrange gave a complicated argument using divergent series. In 1861, A. Popoff was the first to determine the remainder term for the Lagrange series, and in 1876 Zolotarev gave a simple proof of the Lagrange series with remainder:

$$F(z) = F(a) + \sum_{k=1}^{n} \frac{x^k}{k!}\frac{d^{k-1}}{da^{k-1}}(\phi^k(a)F'(a)) + \frac{1}{n!}\frac{d^n}{da^n}\left(\int_a^z (x\phi(u)+a-u)^n F'(u)\,du\right).$$

He proved this formula by setting

$$S_n = \int_a^z (x\phi(u)+a-u)^n F'(u)\,du$$

and observing that differentiation with respect to a immediately yielded

$$\frac{dS_n}{da} = nS_{n-1} - x^n \phi^n(a) F'(a).$$

By setting $n = 1, 2, \ldots, n$, he arrived at the n relations

$$S_0 = x\phi(a)F'(a) + \frac{dS_1}{da},$$

$$2S_1 = x^2\phi^2(a)F'(a) + \frac{dS_2}{da},$$

$$\ldots$$

$$nS_{n-1} = x^n\phi^n(a)F'(a) + \frac{dS_n}{da}.$$

Noting that

$$S_0 = \int_a^z F'(u)\,du = F(z) - F(a),$$

and by substituting the nth equation into the $(n-1)$th equation and continuing the process, he obtained the required formula.

9.9 Exercises

1. Show that Newton's series in (9.1) and (9.3) can be obtained by repeated division.
2. Apply the method of finding square roots to the polynomial $a^2 + x^2$ to obtain Newton's series (9.5).
3. Carry out Newton's procedure for successive approximation of a solution of (9.8) to obtain the series (9.10).
4. Verify Newton's two theorems on the reversion of series, given in equations (9.12) and (9.13).
5. Show that

$$\frac{\pi^3}{48} = \frac{1}{3} \cdot \frac{1}{2} + \frac{1}{5} \cdot \frac{1 \cdot 3}{2 \cdot 4}\left(1 + \frac{1}{3^2}\right) + \frac{1}{7} \cdot \frac{1 \cdot 3 \cdot 5}{2 \cdot 4 \cdot 6}\left(1 + \frac{1}{3^2} + \frac{1}{5^2}\right) + \cdots;$$

$$\cos\frac{\pi x}{3} = 1 - \frac{x^2}{2!} + \frac{x^2(x^2-1^2)}{4!} - \frac{x^2(x^2-1^2)(x^2-2^2)}{6!} + \cdots;$$

$$\sin\frac{\pi x}{3} = \frac{\sqrt{3}}{2}\left(x - \frac{x(x^2-1^2)}{3!} + \frac{x(x^2-1^2)(x^2-2^2)}{5!} - \cdots\right).$$

See Schellbach (1854); and Glaisher, (1878) vol. 7, pp. 76–77.

6. Prove that for any real number n and non-negative integer s,

$$\binom{\frac{n}{2}}{s} + \binom{n}{2}\binom{\frac{n-2}{2}}{s-1} + \binom{n}{4}\binom{\frac{n-4}{2}}{s-2} + \cdots$$

$$= \frac{n^2(n^2-2^2)(n^2-4^2)\cdots(n^2-(2s-2)^2)}{(2s)!};$$

$$\binom{n}{1}\binom{\frac{n-1}{2}}{s} + \binom{n}{3}\binom{\frac{n-3}{2}}{s-1} + \binom{n}{5}\binom{\frac{n-5}{2}}{s-2} + \cdots$$

$$= \frac{n(n^2-1^2)(n^2-3^2)\cdots(n^2-(2s-1)^2)}{(2s+1)!}.$$

Cauchy used these identities without proof in his *Analyse algébrique* to prove Newton's formulas (9.16) and (9.17). A proof depends on the Chu–Vandermonde identity; see chapter 27 on hypergeometric series.

7. Show that for any real number n and $|\theta| \leq \pi/4$

$$\cos n\theta = \cos^n \theta - \frac{n(n-1)}{1 \cdot 2} \cos^{n-2}\theta \sin^2\theta$$
$$+ \frac{n(n-1)(n-2)(n-3)}{1 \cdot 2 \cdot 3 \cdot 4} \cos^{n-4}\theta \sin^4\theta - \cdots,$$

$$\sin n\theta = \frac{n}{1} \cos^{n-1}\theta \sin\theta - \frac{n(n-1)(n-2)}{1 \cdot 2 \cdot 3} \cos^{n-3}\theta \sin^3\theta + \cdots.$$

Viète came close to stating these formulas. Cauchy pointed out in his *Analyse algébrique* that $|\theta| \leq \pi/4$ was necessary to expand $(\cos\theta + i\sin\theta)^n$ by the binomial theorem when n was not a positive integer.

8. Replace $\cos^k\theta$ by $(1 - \sin^2\theta)^{k/2}$ in exercise 7 and expand by the binomial theorem. Then use (9.5) to deduce Newton's formulas (9.16) and (9.17) for $|\theta| \leq \pi/4$.

9. Prove that if $z = g(a + x\phi(z))$, then

$$f(z) = f(g(a)) + \sum_{k-1}^{n} \frac{x^k}{k!} \frac{d^{k-1}}{da^{k-1}}\left(\phi^k(g(a)) \frac{d}{da} f(g(a))\right) + \frac{1}{n!} \frac{d^n}{da^n} I_n(a),$$

where $I_n(a) = \int_a^{g^{-1}} (x\phi(g(t)) + a - t)^n f'(g(t))\,dt.$

See Edwards (1954b), vol. I, pp. 373–74. An equivalent result was published by Emory McClintock (1881), pp. 96–97. McClintock (1840–1916), who served as president of the American Mathematical Society and was instrumental in the founding of the *Bulletin* and the *Transactions*, was an actuary by profession. See Johnson (2007) for historical remarks on the Lagrange series.

10. Show that if $g(0) = 0$, then

$$g^{-1}(x) = x\left(\frac{x}{g(x)}\right)_{x=0} + \frac{x^2}{2!}\left(\frac{d}{dx}\left(\frac{x}{g(x)}\right)^2\right)_{x=0} + \frac{x^3}{3!}\left(\frac{d^2}{dx^2}\left(\frac{x}{g(x)}\right)^3\right)_{x=0} + \cdots.$$

See Edwards (1954a), p. 459.

11. Show that Newton's differential equation $\dot{y}/\dot{x} = 1 + y/(a-x)$ can be written in the form

$$(a - x + y)(y' - 1) - yy' = 0,$$

where $y' = \dot{y}/\dot{x}$. Show that this can be directly integrated. This observation is due to Whiteside; see Newton (1967–81), vol. III, p. 101.

12. Show that Newton's second differential equation $\dot{y}^2/\dot{x}^2 = \dot{y}/\dot{x} + x^2$ can be integrated in closed form in terms of the logarithmic function.

13. Show that
$$\frac{\pi^2}{8} = 1 + \frac{1}{6} + \frac{1 \cdot 4}{6 \cdot 15} + \frac{1 \cdot 4 \cdot 9}{6 \cdot 15 \cdot 28} + \cdots,$$
$$\frac{\pi^2}{4} = 1 + \frac{2}{6} + \frac{2 \cdot 8}{6 \cdot 15} + \frac{2 \cdot 8 \cdot 18}{6 \cdot 15 \cdot 28} + \cdots.$$

This was proved by Tanzan Shokei in his 1728 *Yenri Hakki*.

14. Show that
$$\frac{\pi^2}{9} = 1 + \frac{1^2}{3 \cdot 4} + \frac{1^2 \cdot 2^2}{3 \cdot 4 \cdot 5 \cdot 6} + \frac{1^2 \cdot 2^2 \cdot 3^2}{3 \cdot 4 \cdot 5 \cdot 6 \cdot 7 \cdot 8} + \cdots,$$
$$\frac{\pi^2}{3} = 1 + \frac{1^2}{4 \cdot 6} + \frac{1^2 \cdot 3^2}{4 \cdot 6 \cdot 8 \cdot 10} + \frac{1^2 \cdot 3^2 \cdot 5^2}{4 \cdot 6 \cdot 8 \cdot 10 \cdot 12 \cdot 14} + \cdots.$$

These were presented by Matsunaga Ryohitsu in his *Hoyen Sankyo* of 1738.

15. Prove that
$$\frac{\pi}{4} = 1 - \frac{1}{2 \cdot 3} - \frac{1}{8 \cdot 5} - \frac{3}{48 \cdot 7} - \frac{15}{384 \cdot 9} - \frac{105}{3840 \cdot 11} - \cdots.$$

This was presented by Hasegawa Ko in his *Kyuseki Tsuko* of 1844. Prove also that
$$2\pi = 1 - \frac{1}{2^2} - \frac{3}{8^2} - \frac{3 \cdot 15}{48^2} - \frac{15 \cdot 105}{384^2} - \cdots.$$

This was given in an anonymous manuscript, *Sampo Yenri Hyo*, discussed in Mikami (1974). For exercises 14–16, see Mikami (1974), pp. 213–215.

16. Define
$$f(m) = 1 + m(i \sin \phi) + \frac{m^2}{2!}(i \sin \phi)^2 + \frac{m(m^2 - 1^2)}{3!}(i \sin \phi)^2$$
$$+ \frac{m^2(m^2 - 2^2)}{4!}(i \sin \phi)^4 + \cdots$$

where m is real and $i = \sqrt{-1}$.

(a) Show that
$$f(m_1) f(m_2) = f(m_1 + m_2).$$

(b) Use the method of exercise 8 to prove that Newton's formulas holds for $|\theta| \leq \pi/2$ when n is a positive integer. Deduce that $f(m) = \cos m\phi + i \sin m\phi$ when m is a positive integer.

(c) Show that $f(p/q) = \cos\frac{p\phi}{q} + i\sin\frac{p\phi}{q}$ when p and q are integers.

(d) Show that $f(m)$ is continuous and deduce Newton's formulas for $|\theta| \leq \pi/2$.

See Hobson (1957a), pp. 273–277.

9.10 Notes on the Literature

The *De Analysi* was first published by William Jones in 1711, over forty years after it was written. An English translation was published in 1745. This translation was reprinted in Newton (1964–1967). Whiteside's English translation is contained in Newton (1967–1981), vol. II. Newton wrote this paper so that he should not completely lose his priority in the discovery of the methods of infinite series; he circulated the manuscript privately to several people who were interested in the topic, but he did not want to publish it. For the purpose of publication, he wrote a much longer tract, *De Methodis Serierum et Fluxionum*, in 1671. Due to the difficulties of finding a publisher and other concerns, Newton did not complete the work or publish it. In 1736 John Colson published an English translation, soon retranslated by Castillione into Latin; in 1799 Samuel Horsley published the original Latin version. Whiteside remarked that a comparison of these two translations provides "an instructive check on the clarity and fluency of Newton's Latin style."

Newton used the material in the *De Methodis* to construct his two letters to Oldenburg for Leibniz in 1676. The second letter was quite long, and in it Newton gave a fairly complete account of his work on infinite series. In 1712, he included these letters in the *Commercium Epistolicum*, produced by a Royal Society committee headed by Newton to establish conclusively that he was the 'first inventor' of the calculus. The letters have been republished with English translations in the second volume of Newton's *Correspondence*, Newton (1959–1960).

The quotations from Barrow's letters to Collins can be found in Newton (1959–1960), vol. I, p. 13–15. And vol. II, p. 146, has Newton's derivation of the series for the exponential function. For Newton's solutions by infinite series of certain algebraic differential equations, see Newton (1967–1981), vol. III, pp. 89–101. Newton originally derived the series for arcsine while studying Wallis's book; see Newton (1967–1981), vol. I, pp. 104–111. Newton's statement of his series for $\sin n\theta$ is given in his first letter for Leibniz; see Newton (1959–1960), vol. II, p. 36.

It is remarkable that Newton's mathematical works were published in their entirety only 250 years after his death. Early attempts to accomplish this task were abandoned because his papers were in a state of disarray and were stored in several different locations. It was even assumed that all of Newton's significant results were already published. Thus, before Whiteside's monumental work, the world was unaware of a number of the results of Newton discussed in this book: transformation of series by finite differences, the first clear statement of Taylor's formula, and the expression of an iterated integral as a single integral. D. T. Whiteside (1932–2008) studied French and Latin at Bristol University; he was self-taught in mathematics. As a graduate student at Cambridge, he became deeply interested in the history of mathematics; his doctoral thesis on seventeenth-century mathematics became a classic. In the course of his studies, Whiteside asked to see the papers of Newton, still piled in boxes, and soon resolved to sort and edit them. Cambridge University Press, the world's oldest continually operating press, chartered by Henry VIII, published the eight handsome volumes between 1967 and 1982 from Whiteside's handwritten manuscript and hand-drawn diagrams, with facing pages giving the English and Latin. Whiteside's commentary and notes are extensive and invaluable. Whiteside executed this prodigious task in twenty years and single-handedly, with the excellent assistance of his thesis advisor Michael Hoskin and Adolf Prag, a teacher at Westminster School.

See Tweddle (1988), pp. 67–68, for Stirling's proof of Newton's series. Lagrange's first proof of his inversion formula was republished in Lagrange (1867–1892), vol. 3, p. 25. See Zolotarev (1876) for the remainder term in Lagrange's series. For some interesting historical remarks on Lagrange inversion series, see Johnson (2007); Johnson also fills out the details of Lagrange's proof, sketched in the *Théorie des fonctions analytiques*.

For a discussion of Takebe's work, see Mikami (1974). The formula for $(\arcsin x)^2$ was also discovered by Ming An-tu who was Manchurian by birth. It appeared in his *Ko-yuan Mi-lu Chieh-fa* of 1774, some years after his death. Ming had not completed the work before he died, and his son Hsin finished it. Ming An-tu's work on infinite series was inspired by the three infinite series communicated to Chinese mathematicians by the French Jesuit Pierre Jartoux in 1702. These were Newton's series for sine, cosine and arcsine.

Pierre Jartoux (1670–1720) was a French Jesuit missionary who entered China in 1701. He is said to have communicated either three or nine series for trigonometric functions to Chinese mathematicians. There is some doubt as to how much information he brought from Europe and how much the Chinese and Japanese mathematicians independently discovered. There is no doubt that he communicated the series for sine, cosine, and arcsine. But there is some question about the other six formulas, one of which is Takebe's series for $(\arcsin x)^2$. Though Jartoux's original notes are lost, Smith and Mikami (1914) suggested that the series for $(\arcsin x)^2$ was also introduced by Jartoux, who had been in correspondence with Leibniz. This appears to be unlikely. Jartoux was not a mathematician, and his correspondence with Leibniz was on an astronomical topic. If Jartoux knew the series for $(\arcsin x)^2$, he would have informed Leibniz, and perhaps others, because this would have been a new discovery. In fact, in 1737, when Euler and Bernoulli rediscovered this result and its particular case dealing with π^2, they regarded their formulas as original. And these mathematicians were very well aware of the works of all European mathematicians at that time. We may conclude that Takebe was the first to find the series for $(\arcsin x)^2$ and the corresponding series for π^2, while Ming's discovery was independent, though inspired by a knowledge of the series communicated by Jartoux.

10

Finite Differences: Interpolation and Quadrature

10.1 Preliminary Remarks

The method of interpolation for the construction of tables of trigonometric functions has been used for over two thousand years. On this method, one may tabulate the values of a function $f(x)$ constructed from first principles (definitions) for $x = a$ and $x = a + h$, where h is small, and then interpolate the values between a and $a + h$, without further computation from first principles. For sufficiently small h, one may approximate the function $f(x)$ by a linear function on the interval $[a, a + h]$. This means that, in order to interpolate the values of the function in this interval, one may use the approximation $f(a + \lambda h) \approx f(a) + \lambda(f(a + h) - f(a))$, $0 \leq \lambda \leq 1$. In his *Almagest* of around 150 AD, Ptolemy applied linear interpolation to construct a table of lengths of chords of a circle as a function of the corresponding arcs. These are the oldest trigonometric tables in existence, though Hipparchus may well have constructed similar tables almost three centuries earlier. In Ptolemy's table, the length of the chord was given as $2R \sin \theta$, where R was the radius and 2θ was the angle subtended by the arc. Later mathematicians in India, on the other hand, tabulated the half chord; when divided by the radius, this gives our sine. In his 628 work, *Dhyanagrahopadesadhyaya*, the Indian mathematician and astronomer Brahmagupta used a second order approximation equivalent to the second order Newton–Stirling interpolation formula. In addition to the sine, the Indian mathematicians tabulated the cosine (multiplied by the radius). Later on, Islamic mathematicians, including Al Biruni (973–1048), expanded the tables to include the tangent and cotangent functions; in fact, in their hands, the study of plane and spherical trigonometry was elevated to a mathematical discipline.

By the seventeenth century, the requirements of navigation and astronomy demanded finer tables of trigonometric and related functions; this led to the invention of the logarithm and better interpolation methods. Motivated by the needs of navigation, in 1611 or a little earlier, Thomas Harriot wrote a remarkable treatise, *De Numeris Triangularibus et inde de Progressionibus Arithmeticis: Magisteria Magna*, considering finite differences of third and higher order. He gave the fifth-order interpolation formula, expressed in modern notation as

$$f(x) = f(0) + \binom{x}{1}\Delta f(0) + \binom{x}{2}\Delta^2 f(0) + \cdots + \binom{x}{5}\Delta^5 f(0), \text{ where} \quad (10.1)$$

$$\binom{x}{k} = \frac{x(x-1)(x-2)\cdots(x-k+1)}{k!} \quad (10.2)$$

were the binomial coefficients and $\Delta f(0) = f(1) - f(0)$, $\Delta^2 f(0) = \Delta(\Delta f(0)) = \Delta f(1) - \Delta f(0) = f(2) - f(1) - (f(1) - f(0))$, etc. In Harriot's work, x took rational values and he used his formula to interpolate between unit values of the argument. He understood the values of (10.2) in terms of figurate numbers, instead of binomial coefficients, when x was an integer.

Unfortunately, Harriot did not publish his work; some of his methods were rediscovered soon afterwards by Henry Briggs (1561–1631). Briggs was the first professor of mathematics at Gresham College, London, as well as the first Savilian Professor at Oxford. In his *Arithmetica Logarithmica* of 1624, Briggs mentioned that the nth-order differences of the nth powers of integers were constants. This work contained tables of logarithms obtained by second-order interpolation, that is, taking the first three terms on the right side of Harriot's formula (10.1). Observe that if the second differences are approximately identical, then the third and higher differences are approximately zero and can be neglected. More generally, if the nth differences are approximately constant, then $f(x)$ can be approximated by the polynomial of degree n obtained by extending Harriot's formula (10.1) to nth differences.

Briggs also wrote *Trigonometria Britannica*, a book of trigonometric tables with a very long introduction giving details of his methods. Briggs's friend, Henry Gellibrand, had this work published in 1633, after Briggs's death. Unfortunately, the many users of these trigonometric tables did not bother to read the more important introduction in which Briggs gave some very interesting results, including the binomial theorem for exponent $1/2$. However, the Scottish mathematician James Gregory studied Briggs's introduction and thereby learned interpolation methods. Thus, also making use of advances in algebraic notation, and employing N. Mercator's discovery of infinite series, Gregory obtained interpolation formulas containing up to an infinite number of terms. In an important letter to Collins, dated November 23, 1670, he communicated his formula, given below in my translation, describing it as "both more easie and universal than either Briggs or Mercator's, and also performed without tables."

> I remember you did once desire of me my method of finding the proportional parts in tables, which is this: In figure 8 of my exercises [*Exercitationes Geometricae*], on the straight line AI consider any segment $A\alpha$, to which there is a perpendicular $\alpha\gamma$, such that γ lies on the curve ABH, the rest remaining the same; let there be an infinite series [sequence] $\frac{a}{c}$, $\frac{a-c}{2c}$, $\frac{a-2c}{3c}$, $\frac{a-3c}{4c}$, etc., and let the product of the first two terms of this series be $\frac{b}{c}$, of the first three terms $\frac{k}{c}$, of the first four terms $\frac{l}{c}$, of the first five terms $\frac{m}{c}$, etc., to infinity; the straight line $\alpha\gamma = \frac{ad}{c} + \frac{bf}{c} + \frac{kh}{c} + \frac{li}{c} +$ etc. to infinity.

Gregory defined d, f, h, i, etc., as the successive differences of the ordinates, at equal intervals c. He took $f(0) = 0$, so that $d = f(c) - f(0) = f(c)$, $f = f(2c) - 2f(c)$, etc. After inserting the values of $a, b, k, l, \ldots$, Gregory's formula can be written as

$$f(a) = \frac{a}{c}\Delta f(0) + \frac{a(a-c)}{2c^2}\Delta^2 f(0) + \frac{a(a-c)(a-2c)}{6c^3}\Delta^3 f(0)$$
$$+ \frac{a(a-c)(a-2c)(a-3c)}{24c^4}\Delta^4 f(0) + \cdots. \quad (10.3)$$

This result is now known as the Gregory–Newton forward difference formula, but it may also be called the Harriot–Briggs formula.

Newton's interest in finite differences and interpolation appears have been a response to an appeal from one John Smith for help with the construction of tables of square, cube, and fourth roots of numbers. Collins broadcast this appeal; he wrote in a letter of November 23, 1674, to Gregory, "We have one Mr. Smith here taking pains to afford us tables of the square and cube roots of all numbers from unit to 10000, which will much facilitate Cardan's rules." Smith was an accountant and compiler of tables whom Newton had helped five years earlier with the making of tables for the areas of segments of circles. Newton again assisted Smith, writing to him on May 8, 1675, giving details for the construction of tables of roots. Newton explained to Smith that he should tabulate the roots of every hundredth number n. From these, he should construct the roots of every tenth number $n \pm 10, n \pm 20, \ldots$ with a constant third difference and thence the roots of $n \pm 1, n \pm 2, \ldots$ with a constant second difference. Newton also cautioned that all computations should be done to the tenth or eleventh decimal place so as to obtain a table accurate to eight places. Newton's ideas on finite difference interpolation developed quite rapidly after this. A year later, on October 24, 1676, he set forth some of his insights in a draft of his second letter for Leibniz through Henry Oldenburg. Newton later eliminated this portion of the letter because he saw a copy of a February 1673 letter from Leibniz to Oldenburg, showing that Leibniz had independently found the Harriot–Briggs formula. Newton perhaps assumed from this that Leibniz had made progress parallel to his own in the study of finite differences, though this was not the case. However, this assumption spurred Newton to write down his ideas systematically in the manuscript "Regula Differentiarum," unpublished until 1970; this contained all the important ideas presented in Newton's *Methodus Differentialis* of 1711. In particular, Newton gave an exposition of interpolation by central differences and derived the Newton–Stirling and Newton–Bessel formulas. In 1708, Roger Cotes (1682–1716) found the latter independently; it might be more appropriate to call this the Newton–Cotes formula, since Bessel employed it only in his numerical work. In the course of addressing several practical problems, Newton also considered the more general problem of interpolating a set of points whose abscissas were not necessarily equidistant, leading to his theory of divided differences.

Newton was the single most significant contributor to the theory of finite differences; although many formulas in this subject are attributed jointly to Newton and some other mathematician, they are actually all due originally to Newton, with the exception of the Gregory–Newton formula. The secondary mathematicians usually made use of these formulas in their numerical work. As early as 1730, Stirling pointed this out in his *Methodus Differentialis*: "After Newton several celebrated geometers have dealt with the description of the curve of parabolic type [defined by a polynomial] through any number of given points. But all their solutions are the same as those which have just been shown; indeed these differ scarcely from *Newton's* solutions." It is amusing that Stirling was subsequently honored by having his name attached to a formula he explicitly and modestly attributed to Newton.

Newton's divided difference formula, in the notation of the French mathematicians A. M. Ampère (1775–1836) and A. L. Cauchy (1789–1857), was written as

$$f(x) = f(x_1) + (x - x_1) f(x_1, x_2) + (x - x_1)(x - x_2) f(x_1, x_2, x_3) + \cdots$$
$$+ (x - x_1) \cdots (x - x_{n-1}) f(x_1, x_2, \ldots, x_n)$$
$$+ (x - x_1) \cdots (x - x_n) f(x_1, \ldots, x_n, x), \quad \text{where} \tag{10.4}$$

$$f(x_1, x_2) = \frac{f(x_1) - f(x_2)}{x_1 - x_2},$$
$$f(x_1, x_2, \ldots, x_k) = \frac{f(x_1, \ldots, x_{k-1}) - f(x_2, \ldots, x_k)}{x_1 - x_k}. \tag{10.5}$$

If we denote the last term in (10.4) by $R_n(x)$, and the remaining sum as $P_{n-1}(x)$, then $P_{n-1}(x)$ is a polynomial of degree $n-1$ equal to $f(x_i)$ for $i = 1, 2, \ldots, n$. Note that this is true because $R_n(x_i) = 0$, $i = 1, \ldots, n$. Thus, $P_{n-1}(x)$ is the interpolating polynomial for a function $f(x)$ whose values are known at $x_1, x_2, \ldots, x_n$. In the 1770s, Lagrange and Waring gave a different expression for this polynomial, more convenient for many purposes, especially for numerical integration. The Lagrange–Waring interpolating polynomial is easy to obtain, yet it is interesting to see different proofs presented in the 1820s by Cauchy and Jacobi.

James Gregory was the first mathematician to use interpolating polynomials to approximately evaluate the area under a curve. He communicated his quadrature formula to Collins in the letter containing his interpolation formula, deriving it by integrating the interpolating polynomial, just as Newton did in his *Methodus Differentialis*. Newton derived his three-eighths rule by integrating the third-degree polynomial obtained by taking the first four terms of (10.1). He explained:

> If, for instance, there be four ordinates positioned at equal intervals, let A be the sum of the first and fourth, B the sum of the second and third, and R the interval between the first and the fourth, and then the new ordinate in the midst of all will be $\frac{1}{16}(9B - A)$ and the total area between the first and fourth will be $\frac{1}{8}(A + 3B)R$.

In 1707, Cotes, unaware of Newton's then unpublished work in this area, composed a treatise on approximate quadrature. He wrote down formulas for areas when the number of ordinates was $3, 4, 5, \ldots, 11$. The coefficients became fairly large after six ordinates; for example, his formula for eight ordinates was

$$\frac{751A + 3577B + 1323C + 2989D}{17280} R,$$

where A was sum of the extreme ordinates, B the sum of the ordinates closest to the extremes, C the sum of the next ones, and D the sum of the two in the middle. Cotes's paper, published posthumously in 1722, contained no proofs of his formulas.

Meanwhile, in 1719, Stirling published a paper in the *Philosophical Transactions* on the same topic, presenting formulas for approximate areas for only the odd number of ordinates 3, 5, 7 and 9. He remarked that the approximations with an odd number of ordinates were more accurate than those with an even number. He did not prove this, though it is true. For example, it can be demonstrated that if $h = R/n$, where $4n$ is the number of ordinates, then the error will be $O(h^{n+2})$ for odd n but $O(h^{n+1})$ for even n.

The Newton–Cotes method of numerical integration was used for a century before Gauss developed a new approach, including a formula exact for any polynomial of degree $2n - 1$ or less when n interpolation points were judiciously constructed. The Newton–Cotes formulas are exact only for polynomials of degree at most $n - 1$. Gauss's procedure will be discussed in a later chapter in connection with orthogonal polynomials. A drawback of the Newton–Cotes and Gauss formulas was that the coefficients of the ordinates were unequal. The Russian mathematician P. L. Chebyshev (1821–1894) observed that when the ordinates $f(x_i)$ were experimentally obtained, they were liable to errors. Assuming that the probability of error in each of the ordinates was the same, the linear combination of the ordinates with equal coefficients had the least probable error among all the linear combinations with a given fixed sum of coefficients. Chebyshev observed that a quadrature formula with equal coefficients might often be preferable. Chebyshev studied mathematics at Moscow University from 1837–1841. He was interested in building mechanical gadgets and some of his papers deal with the mathematics involved with these. Chebyshev was of the view that his job as a mathematician was to consider practical problems and to give solutions both theoretically satisfying and practically useful. He repeatedly professed this opinion in his lectures and advocated it in several papers; his work on numerical integration may be seen as an example of this perspective.

In 1874, Chebyshev wrote a paper on quadrature with equal coefficients, considering formulas of the type

$$\int_{-1}^{1} f(x)\phi(x)\,dx \equiv k(f(x_1) + f(x_2) + \cdots + f(x_n)), \tag{10.6}$$

where $\phi(x)$ was the weight function and k was the common coefficient of the ordinates. He found a method, exact for polynomials of degree less than n, for determining the interpolation points $x_1, x_2, \ldots, x_n$ and the constant k, such that they depended upon ϕ but not on f. He worked out the details with the weights given as $\phi(x) = 1$ and $\phi(x) = 1/\sqrt{1-x^2}$. In particular, he showed that when $\phi(x) = 1$, then $k = 2/n$ and $x_1, x_2, \ldots, x_n$ were the roots of that polynomial given by the polynomial portion of the expression

$$z^n e^{-n/(2\cdot 3z^2) - n/(4\cdot 5z^4) - n/(6\cdot 7z^6) - \cdots}.$$

He computed the zeros of these polynomials for $n = 2, 3, 4, 5, 6, 7$. Interestingly, Chebyshev was inspired to do this work by Hermite's 1873 Paris lectures on the case $\phi(x) = 1/\sqrt{1-x^2}$. We observe that, even before Hermite, Brice Bronwin gave the formula for this case in a paper of 1849 in the *Philosophical Magazine*. In the chapter

10.2 Newton: Divided Difference Interpolation

Newton started his work on interpolation in the mid-1770s, but had made sufficient progress to make a brief mention of it in his October 24, 1776 letter for Leibniz. While discussing the problem of determining the area under a curve, especially when the expression for the curve led to difficult calculations of series, he wrote:

> But I make little of this because, when simple series are not manageable enough, I have another method not yet communicated by which we have access to our solution at will. Its basis is a convenient, rapid and general solution of this problem, *To draw a geometrical curve which shall pass through any number of given points*. Euclid showed how to draw a circle through three given points. A conic section also can be described through five given points, and a curve of the third degree through eight given points; (so that I have it fully in my power to describe all the curves of that order, which can be determined by eight points only.) These things are done at once geometrically with no calculation intervening. But the above problem is of the second kind, and though at first it looks unmanageable, yet the matter turn out otherwise. For it ranks among the most beautiful of all that I could wish to solve.

Interestingly, Stirling quoted just this passage at the end of proposition 18 of his book. Clearly, Newton was pleased with the result of his researches on interpolation, so he did not neglect the chance to include at least one result in the *Principia*, as Lemma V, Book III. Newton gave his method of interpolation by divided differences in the *Principia* without proof; he provided details in his very short *Methodus Differentialis*. The first proposition stated that if one started with a polynomial, then the divided differences would also be polynomials of degree one less:

> If the abscissa of a curvilinear figure be composed of any given quantity A and the indeterminate quantity x, and its ordinate consist of any number of given quantities $b, c, d, e, \ldots$ multiplied respectively into an equal number of terms of the geometric progression $x, x^2, x^3, x^4, \ldots$ and if ordinates be erected at the corresponding number of given points in the abscissa: I assert that the first differences of the ordinates are divisible by the intervals between them, and that the differences of the differences so divided are divisible by the intervals between every second ordinate, and so on indefinitely.

We describe Newton's method in the Ampére–Cauchy notation: If $f(x)$ is a polynomial, then the first divided difference

$$f(x_1, x_2) = \frac{f(x_1) - f(x_2)}{x_1 - x_2} \tag{10.7}$$

is also a polynomial, as is the second divided difference

$$f(x_1, x_2, x_3) = \frac{f(x_1, x_2) - f(x_2, x_3)}{x_1 - x_3}, \tag{10.8}$$

and, in general, the so is the nth divided difference, defined inductively by

$$f(x_1, x_2, \ldots, x_n, x_{n+1}) = \frac{f(x_1, x_2, \ldots, x_n) - f(x_2, x_3, \ldots, x_{n+1})}{x_1 - x_{n+1}}. \tag{10.9}$$

Newton explicitly worked out all the divided differences for a fourth-degree polynomial. In the second proposition, he explained how the original polynomial or function could be constructed from the divided differences:

> With the same suppositions, and taking the number of terms $b, c, d, e, \ldots$ to be finite, I assert that the last of the quotients will be equal to the last of the terms $b, c, d, e, \ldots$, and that the remaining terms $b, c, d, e, \ldots$ will be yielded by means of the remaining quotients; and that once these are determined there will be given the curve of parabolic kind which shall pass through the end-points of all the ordinates.

If this procedure is applied in general, we obtain Newton's divided difference formula (10.4). In the case of fourth differences we have

$$f(x, x_1) = \frac{f(x)}{x - x_1} - \frac{f(x_1)}{x - x_1},$$

$$f(x, x_1, x_2) = \frac{f(x, x_1)}{x - x_2} - \frac{f(x_1, x_2)}{x - x_2},$$

$$f(x, x_1, x_2, x_3) = \frac{f(x, x_1, x_2)}{x - x_3} - \frac{f(x_1, x_2, x_3)}{x - x_3},$$

$$f(x, x_1, x_2, x_3, x_4) = \frac{f(x, x_1, x_2, x_3)}{x - x_4} - \frac{f(x_1, x_2, x_3, x_4)}{x - x_4}.$$

Thus, in each step, the values from the previous equation are substituted for the terms on the right-hand side and the resulting equation is multiplied by $(x - x_1)(x - x_2)(x - x_3)(x - x_4)$, yielding

$$\begin{aligned} f(x) = &f(x_1) + (x - x_1)f(x_1, x_2) + (x - x_1)(x - x_2)f(x_1, x_2, x_3) \\ &+ (x - x_1)(x - x_2)(x - x_3)f(x_1, x_2, x_3, x_4) \\ &+ (x - x_1)(x - x_2)(x - x_3)(x - x_4)f(x, x_1, x_2, x_3, x_4). \end{aligned} \tag{10.10}$$

In the third proposition, Newton derived his central difference formulas for the case where the points were equidistant. When the number of interpolating points was odd, he presented what is now known as the Newton–Stirling formula and, for the even case, the so-called Newton-Bessel formula. He did not write down details of the derivation, but it is most likely that he obtained it from his general divided difference formula, employed in modern textbooks.

In the case of an odd number of points, Newton let k denote the middle ordinate; l denote the average of the two middle first differences of the ordinates, the number of first differences being even, since the number of ordinates was odd; m denote the middle second difference; n the average of the two middle third differences, etc. Then

$$f(x) = k + x \cdot l + \frac{x^2}{1 \cdot 2} \cdot m + \frac{x(x^2 - 1^2)}{1 \cdot 2 \cdot 3} \cdot n + \frac{x^2(x^2 - 1^2)}{1 \cdot 2 \cdot 3 \cdot 4} \cdot o + \text{etc.} \tag{10.11}$$

Newton's formula for an even number of ordinates is discussed in the chapter on the gamma function. Stirling made very effective use of this method, computing the values of the gamma function at a few half-integral values.

10.3 Gregory–Newton Interpolation Formula

The Gregory–Newton formula (10.3) is important not only in numerical analysis but also in the study of sequences whose nth differences, for some n, are constant. These sequences are now studied as a part of combinatorial analysis, but in the seventeenth and early eighteenth centuries they arose in elementary number theory and in probability theory. It is therefore interesting to consider the methods by which mathematicians of that period proved this formula. Unfortunately, Gregory did not leave us a proof. It is possible that he had the simple inductive argument given by Stirling in proposition 19 of his *Methodus Differentialis*. Stirling assumed that there existed some unknown coefficients, $A, B, C, D, \ldots$ such that

$$f(z) = A + Bz + C\frac{z(z-1)}{1 \cdot 2} + D\frac{z(z-1)(z-2)}{1 \cdot 2 \cdot 3} + \cdots.$$

Clearly $A = f(0)$. Moreover,

$$\Delta f(z) = f(z+1) - f(z)$$
$$= B\Delta z + C\Delta \frac{z(z-1)}{1 \cdot 2} + D\Delta \frac{z(z-1)(z-2)}{1 \cdot 2 \cdot 3} + \cdots.$$

Observing that for $n = 2, 3, 4, \ldots$

$$\Delta \frac{z(z-1)\cdots(z-n+1)}{1 \cdot 2 \cdots n} = \frac{z(z-1)\cdots(z-n+2)}{1 \cdot 2 \cdots (n-1)}, \tag{10.12}$$

and that $\Delta z = (z+1) - z = 1$, he obtained $B = \Delta f(0)$. Continuing this process, he got $C = \Delta^2 f(0), D = \Delta^3 f(0), \ldots$, completing the proof. Note that Gregory's version of the formula, given by (10.3), would be obtained by taking $A = 0$ and $z = a/c$.

10.4 Waring, Lagrange: Interpolation Formula

Edward Waring and Joseph Lagrange independently but nearly simultaneously took up the interpolation problem of finding the polynomial of degree $n-1$, taking prescribed values at n given points: $y_1, y_2, \ldots, y_n$ at $x_1, x_2, \ldots, x_n$. Of course, this result may readily be derived by writing the Newton divided differences in symmetric form, but Lagrange and Waring gave the solution in a convenient and useful form. In fact, Waring remarked in his 1779 paper on the topic that he could state and prove the result without any "recourse to finding the successive differences." We state Waring's theorem in modern notation: Let y be a polynomial of degree $n-1$ and let the values of y at

$x_1, x_2, \ldots, x_n$ be given by $y_1, y_2, \ldots, y_n$. Then

$$y = \frac{(x-x_2)(x-x_3)\cdots(x-x_n)}{(x_1-x_2)(x_1-x_3)\cdots(x_1-x_n)}y_1 + \frac{(x-x_1)(x-x_3)\cdots(x-x_n)}{(x_2-x_1)(x_2-x_3)\cdots(x_2-x_n)}y_2$$
$$+ \cdots + \frac{(x-x_1)(x-x_2)\cdots(x-x_{n-1})}{(x_n-x_1)(x_n-x_2)\cdots(x_n-x_{n-1})}y_n. \tag{10.13}$$

Waring's proof consisted in the observation that, when $x = x_1$, the first term on the right was y_1, while, because of the factor $x - x_1$, all the other terms were zero. Continuing this argument, taking successive values of x, Waring completed his proof. Lagrange proved this result in a similar manner at about the same time, but did not immediately publish it; it appeared in his 1797 *Fonctions analytiques*.

10.5 Cauchy, Jacobi: Lagrange Interpolation Formula

The Lagrange interpolation formula is easy to prove, as Waring's demonstration shows. It is nevertheless interesting to consider other proofs such as those of Cauchy and Jacobi. Cauchy's argument, presented in his 1821 *Analyse algébrique* in the chapter on symmetric and alternating functions, was based on an interesting evaluation of the so-called Vandermonde determinant, without using modern notation for determinants. Lagrange had used this evaluation in a different context almost fifty years earlier. Cauchy was an expert on determinants, a term he borrowed from Gauss. He wrote an important 1812 paper on this topic, in which he also proved results on permutation groups and alternating functions. In his book, Cauchy considered the system of linear equations

$$\alpha^j x + \alpha_1^j x_1 + \cdots + \alpha_{n-1}^j x_{n-1} = k_j, \tag{10.14}$$

where $j = 0, 1, \ldots, n-1$. We have used subscripts more freely than Cauchy; he set

$$f(\alpha) = (\alpha - \alpha_1)(\alpha - \alpha_2)\cdots(\alpha - \alpha_{n-1}) = \alpha^{n-1} + A_{n-2}\alpha^{n-2} + \cdots + A_1\alpha + A_0,$$

so that

$$\alpha_i^{n-1} + A_{n-2}\alpha_i^{n-2} + \cdots + A_1\alpha_i + A_0 = 0, \text{ for } i = 1, 2, \ldots, n-1.$$

Cauchy multiplied the first equation of the system (10.14) by A_0; the second, when $j = 1$, by A_1; ...; and the last, when $j = n - 1$, by 1 and then added to get

$$(A_0 + A_1\alpha + \cdots + \alpha^{n-1})x = k_0 A_0 + k_1 A_1 + \cdots + k_{n-2}A_{n-2} + k_{n-1} \text{ or }$$
$$x = \frac{k_{n-1} + A_{n-2}k_{n-2} + \cdots + A_0 k_0}{f(\alpha)}. \tag{10.15}$$

He derived the values of $x_1, x_2, \ldots, x_{n-1}$ in a similar way. Cauchy applied this result to obtain the Lagrange interpolation polynomial. He supposed $u_0, u_1, \ldots, u_{n-1}$ to be values

10.5 Cauchy, Jacobi: Lagrange Interpolation Formula

of some function at the numbers $x_0, x_1, \ldots, x_{n-1}$. It was required to find a polynomial of degree $n-1$

$$u = a_0 + a_1 x + a_2 x^2 + \cdots + a_{n-1} x^{n-1},$$

such that its values were $u_0, u_1, \ldots, u_{n-1}$ at $x_0, x_1, \ldots, x_{n-1}$, respectively. Then

$$u_j = a_0 + a_1 x_j + a_2 x_j^2 + \cdots + a_{n-1} x_j^{n-1},$$

where $j = 0, 1, \ldots, n-1$. Cauchy multiplied these n equation by unknown $X_0, X_1, \ldots, X_{n-1}$ and subtracted their sum from the equation for u to get

$$\begin{aligned}
&u - X_0 u_o - X_1 u_1 - X_2 u_2 - \cdots - X_{n-1} u_{n-1} \\
&= (1 - X_0 - X_1 - X_2 - \cdots - X_{n-1}) a_0 \\
&\quad + (x - x_0 X_0 - x_1 X_1 - \cdots - x_{n-1} X_{n-1}) a_1 \\
&\quad + (x^2 - x_0^2 X_0 - x_1^2 X_1 - \cdots - x_{n-1}^2 X_{n-1}) a_2 + \cdots \\
&\quad + (x^{n-1} - x_0^{n-1} X_0 - x_1^{n-1} X_1 - \cdots - x_{n-1}^{n-1} X_{n-1}) a_{n-1}.
\end{aligned} \quad (10.16)$$

To determine $X_0, X_1, \ldots, X_{n-1}$ so that

$$u = X_0 u_0 + X_1 u_1 + \cdots + X_{n-1} u_{n-1},$$

he set equal to zero all the coefficients of $a_0, a_1, \ldots, a_{n-1}$ on the right-hand side of (10.5). Thus, he had the system of equations

$$x_0^j X_0 + x_1^j X_1 + \cdots + x_{n-1}^j X_{n-1} = x^j,$$

with $j = 0, 1, \ldots, n-1$. He could solve these as before to get

$$X_0 = \frac{f(x)}{f(x_0)} = \frac{(x-x_1)(x-x_2)\cdots(x-x_{n-1})}{(x_0-x_1)(x_0-x_2)\cdots(x_0-x_{n-1})}, \quad (10.17)$$

$$X_1 = \frac{(x-x_0)(x-x_2)\cdots(x-x_{n-1})}{(x_1-x_0)(x_1-x_2)\cdots(x_1-x_{n-1})},$$

and so on. This gave him the Lagrange–Waring interpolation formula.

Jacobi's method employed partial fractions; he presented it in his doctoral dissertation on this topic as well as in his 1826 paper on Gauss quadrature. He let

$$g(x) = (x-x_0)(x-x_1)\cdots(x-x_{n-1})$$

and $u(x)$ be the polynomials of degree $n-1$ whose values were $u_0, \ldots, u_{n-1}$ at $x_0, \ldots, x_{n-1}$, respectively. Then by a partial fractions expansion he got

$$\frac{u(x)}{g(x)} = \frac{B_0}{x-x_0} + \frac{B_1}{x-x_1} + \cdots + \frac{B_{n-1}}{x-x_{n-1}}; \quad (10.18)$$

by setting $x = x_j$, he obtained

$$B_j = u_j / ((x_j - x_0) \cdots (x_j - x_{j-1})(x_j - x_{j+1}) \cdots (x_j - x_{n-1})). \quad (10.19)$$

Jacobi arrived at Lagrange's formula by multiplying across by $g(x)$. We note that Jacobi's dissertation also discussed the case in which some of the x_i were repeated.

10.6 Newton on Approximate Quadrature

The *Methodus Differentialis* stated the three-eighths rule for finding the approximate area under a curve when four values of the function were known; one proposition suggests that Newton most probably derived the formula by integrating the interpolating cubic for the four points. However, in October 1695, he wrote a very short manuscript, though he left it incomplete, presenting his derivation of some rules for approximate quadrature. Surprisingly, he did not obtain these rules by integrating the interpolating polynomials but by means of heuristic and somewhat geometric reasoning. Since interpolation calculations tend to become very unwieldy, perhaps Newton sought a short cut, though it is not clear what stimulated him to write this short note. Whiteside conjectured that Newton may have been working on his contemporaneous amplified lunar theory where he used some of the results.

Newton wrote his results consecutively for two, three, four, ... ordinates. In Figure 10.1, he took equally spaced points $A, B, C, D, \ldots$ on the abscissa (x-axis) and points $K, L, M, N, \ldots$ on the curve ($y = f(x)$) such that $AK, BL, CM, DN, \ldots$ were the ordinates, or y values of the corresponding points on the curve. For two points, he gave the trapezoidal rule labeled as Case 1.

If there be given two ordinates AK and BL, make the area $(AKLB) = \frac{1}{2}(AK + BL)AB$.

He next obtained Simpson's rule, published by Thomas Simpson in his *Mathematical Dissertations* of 1743; Simpson gave an interesting geometric proof. We note that since Simpson's books were quite popular, his name got attached to the rule. In 1639, Cavalieri gave particular cases of this rule to determine the volume of a symmetrical wine cask. In 1668, in his *Exercitationes Geometricae*, Gregory too presented this rule to approximate $\int_0^h \tan x \, dx$. Newton derived Simpson's and the three-eights rule as Cases 2 and 3, where the box notation denotes area:

Case 2. If there be given three AK, BL and CM, say that

$$\frac{1}{2}(AK + CM)AC = \Box(AM), \quad \text{and again, by Case 1,}$$

$$\frac{1}{2}\left(\frac{1}{2}(AK + BL) + \frac{1}{2}(BL + CM)\right)AC = \frac{1}{4}(AK + 2BL + CM)AC = \Box(AM),$$

and that the error in the former solution is to the error in the latter as AC^2 to AB^2 or 4 to 1, and hence the difference $\frac{1}{4}(AK - 2BL + CM)AC$ of the solutions is to the error in the latter as 3 to 1,

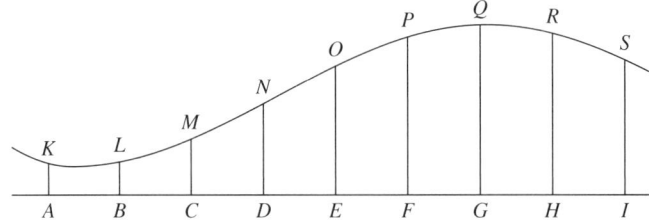

Figure 10.1. Newton's approximate quadrature.

10.6 Newton on Approximate Quadrature

and the error in the latter will be

$$\frac{1}{12}(AK - 2BL + CM)AC.$$

Take away this error and the latter solution will come to be

$$\frac{1}{6}(AK + 4BL + CM)AC = \square(AM), \quad \text{the solution required.}$$

Case 3. If there be given four ordinates AK, BL, CM and DN, say that $\frac{1}{2}(AK + DN)AD = \square(AN)$; likewise, that

$$\frac{1}{3}\left(\frac{1}{2}(AK + BL) + \frac{1}{2}(BL + CM) + \frac{1}{2}(CM + DN)\right)AD,$$

that is, $\frac{1}{6}(AK + 2BL + 2CM + DN)AD = \square(AN)$. The errors in the solutions will be as AD^2 to AB^2 or 9 to 1, and hence the difference in the errors (which is the difference $\frac{1}{6}(2AK - 2BL - 2CM + 2DN)AD$ in the solutions) will be to the error in the latter as 8 to 1. Take away this error and the latter will remain as

$$\frac{1}{8}(AK + 3BL + 3CM + DN)AD = \square(AN).$$

We observe that in these three cases and others, Newton assumed without justification that when $n+1$ equidistant ordinates were given, the corresponding ratio of the errors in using the trapezoidal rule would be $n^2 : 1$. Newton went on to consider cases with five, seven, and nine ordinates, but his results in the last two cases were not the same as the ones obtained by integrating the interpolating polynomials.

To describe Newton's proof of Simpson's formula in somewhat more analytic terms, let $[a,b]$ in Case 2 be the interval with $b = a + 2h$, and let $y = f(x)$ be the function on that interval. By the trapezoidal rule,

$$\int_a^b f(x)\,dx \approx \frac{1}{2}(f(a) + f(b))(2h) \equiv I_1.$$

If this rule is applied to each of the intervals $[a, a+h], [a+h, b]$, then

$$\int_a^b f(x)\,dx \approx \frac{1}{2}\left(\frac{1}{2}(f(a) + f(a+h)) + \frac{1}{2}(f(a+h) + f(b))\right)(2h)$$

$$= \frac{1}{4}(f(a) + 2f(a+h) + f(b))(2h) \equiv I_2.$$

Let the errors in the two formulas be e_1 and e_2, so that

$$\int_a^b f(x)\,dx = I_1 + e_1 = I_2 + e_2.$$

Newton assumed without proof that $e_1/e_2 = 4$. Hence,

$$(I_2 - I_1)/e_2 = (e_1 - e_2)/e_2 = 3 \quad \text{so that}$$

$$e_2 = \frac{1}{3}(I_2 - I_1) = -\frac{1}{12}(f(a) - 2f(a+h) + f(b))2h.$$

When this value of e_2 is added to I_2, we get Simpson's approximation

$$\frac{1}{6}(f(a)+4f(a+h)+f(b)). \tag{10.20}$$

10.7 Hermite: Approximate Integration

The formulas of Newton, Cotes, and Stirling for numerical integration were used without change for a century. In the nineteenth century, mathematicians began to present new methods, starting with Gauss, whose work in this area is discussed in our treatment of orthogonal polynomials. Charles Hermite (1822–1901) was a professor at the École Polytechnique. He gave a series of analysis lectures in 1873; these and other such lectures were published and serve as a valuable resource even today. For example, Hermite discussed an original method for the numerical evaluation of integrals of the form

$$\int_{-1}^{+1} \frac{\phi(x)\,dx}{\sqrt{1-x^2}}, \tag{10.21}$$

where $\phi(x)$ was an analytic function. He started with the nth-degree polynomial $F(x)$ defined by

$$F(x) = \cos n(\arccos x). \tag{10.22}$$

By taking the derivative, he obtained

$$F'(x) = n\sin n(\arccos x)\frac{1}{\sqrt{1-x^2}} = n\frac{\sqrt{1-F^2(x)}}{\sqrt{1-x^2}}.$$

Hence

$$\frac{1}{\sqrt{x^2-1}} = \frac{F'(x)}{nF(x)}\left(1 - 1/F^2(x)\right)^{-1/2}$$

$$= \frac{F'(x)}{nF(x)}\left(1 + \frac{1}{2}\frac{1}{F^2(x)} + \frac{1\cdot 3}{2\cdot 4}\frac{1}{F^4(x)} + \cdots\right).$$

Hermite observed that the last expression without the first term could be written in decreasing powers of x in the form

$$\frac{\lambda_0}{x^{2n+1}} + \frac{\lambda_1}{x^{2n+2}} + \frac{\lambda_2}{x^{2n+3}} + \cdots.$$

Consequently,

$$\frac{1}{\sqrt{x^2-1}} = \frac{F'(x)}{nF(x)} + \frac{\lambda_0}{x^{2n+1}} + \frac{\lambda_1}{x^{2n+2}} + \frac{\lambda_2}{x^{2n+3}} + \cdots.$$

10.7 Hermite: Approximate Integration

At this point, Hermite invoked the integral formula

$$\int_{-1}^{+1} \frac{dz}{(x-z)\sqrt{1-z^2}} = \frac{\pi}{\sqrt{x^2-1}}, \quad \text{to obtain} \qquad (10.23)$$

$$\frac{1}{\pi}\int_{-1}^{+1} \frac{dz}{(x-z)\sqrt{1-z^2}} = \frac{F'(x)}{nF(x)} + \frac{\lambda_0}{x^{2n+1}} + \frac{\lambda_1}{x^{2n+2}} + \cdots$$

$$= \frac{1}{n}\sum_{j=1}^{n} \frac{1}{x-a_j} + \frac{\lambda_0}{x^{2n+1}} + \frac{\lambda_1}{x^{2n+2}} + \cdots,$$

where $a_1, a_2, \ldots, a_n$ were the n roots of $F(x) = 0$. An application of the geometric series $1/(x-z) = \sum z^n/x^{n+1}$ gave him

$$\frac{1}{\pi}\int_{-1}^{+1}\left(1 + \frac{z}{x} + \frac{z^2}{x^2} + \cdots\right)\frac{dz}{\sqrt{1-z^2}}$$

$$= \frac{1}{n}\sum_{j=1}^{n}\left(1 + \frac{a_j}{x} + \frac{a_j^2}{x^2} + \cdots\right) + \frac{\lambda_0}{x^{2n}} + \frac{\lambda_1}{x^{2n+1}} + \cdots.$$

Equating the coefficients of $1/x^l$ on both sides yielded

$$\frac{1}{\pi}\int_{-1}^{1}\frac{z^l}{\sqrt{1-z^2}}dz = \frac{1}{n}\sum_{j=1}^{n}a_j^l \quad \text{when } l < 2n,$$

$$= \frac{1}{n}\sum_{j=1}^{n}a_j^{2n+s} + \lambda_s \quad \text{when } l = 2n+s, \; s \geq 0.$$

So, Hermite wrote $\phi(z) = k_0 + k_1 z + k_2 z^2 + \cdots + k_n z^n + \cdots$ in order to obtain the formula

$$\frac{1}{\pi}\int_{-1}^{1}\frac{\phi(z)}{\sqrt{1-z^2}}dz = \frac{1}{n}\sum_{j=1}^{n}\phi(a_j) + R, \quad \text{where} \qquad (10.24)$$

$$R = \lambda_0 k_{2n} + \lambda_1 k_{2n+1} + \cdots.$$

Hermite also noted that since the roots a_j of $F(x)=0$ were given by

$$a_j = \cos\left(\frac{2j-1}{2n}\pi\right), \; j = 1, 2, \ldots, n, \quad \text{he obtained}$$

$$\int_0^{\pi}\phi(\cos\theta)\,d\theta = \frac{\pi}{n}\sum_{j=1}^{n}\phi\left(\cos\left(\frac{2j-1}{2n}\pi\right)\right) + \pi R. \qquad (10.25)$$

Observe that the expression for the error R shows that it must be zero for any polynomial ϕ of degree less than $2n$. Hermite may have been unaware that in 1849, Brice Bronwin derived formula (10.25) by a different method, but without the error term.

10.8 Chebyshev on Numerical Integration

A nice feature of the Bronwin–Hermite formula is that it allows us to find an approximate value of the integral by simply adding the values of the function $\phi(x)$ at the zeros of $F(x)$ and then multiplying by π/n. Chebyshev's interest in applications led him to seek similar formulas for other weight functions. Thus, the purpose of Chebyshev's 1874 paper was to find a constant k, and numbers $x_1, x_2, \ldots, x_n$ such that $\int_{-1}^{1} F(x)\phi(x)\,dx$ could be approximated by $k(\phi(x_1)+\phi(x_2)+\cdots+\phi(x_n))$. Note that $\phi(x)$ was the function to be integrated with respect to the weight function $F(x)$. In general, Chebyshev required that the approximation be exact for polynomials of degree at most $n-1$, so he looked for a formula of the form

$$\int_{-1}^{+1} \frac{F(x)}{z-x}\,dx$$
$$= k(\phi(x_1)+\phi(x_2)+\cdots+\phi(x_n)) + k_1\phi^{(n+1)}(0) + k_2\phi^{(n+2)}(0) + \cdots, \qquad (10.26)$$

where $\phi^{(m)}$ denoted the mth derivative of ϕ and $k_1, k_2, \ldots$ were constants. Following Hermite, he considered the case $\phi(x) = 1/(z-x)$ to obtain

$$\int_{-1}^{+1} \frac{F(x)}{z-x}\,dx = k\left(\frac{1}{z-x_1}+\cdots+\frac{1}{z-x_n}\right) + \frac{(n+1)!k_1}{z^{n+2}} + \frac{(n+2)!k_2}{z^{n+3}} + \cdots. \qquad (10.27)$$

He set $f(z) = (z-x_1)(z-x_2)\cdots(z-x_n)$, so that after multiplying by z, the last relation became

$$z\int_{-1}^{+1} \frac{F(x)}{z-x}\,dx$$
$$= kz\frac{f'(z)}{f(z)} + \frac{1\cdot 2\cdot 3\cdots(n+1)k_1}{z^{n+1}} + \frac{1\cdot 2\cdot 3\cdots(n+2)k_2}{z^{n+2}} + \cdots. \qquad (10.28)$$

He let $z \to \infty$ to get

$$\int_{-1}^{+1} F(x)\,dx = nk, \text{ or } k = \frac{1}{n}\int_{-1}^{+1} F(x)\,dx. \qquad (10.29)$$

He thus had the value of k, and it remained for him to find the polynomial $f(z)$ whose zeros would be the numbers $x_1, x_2, \ldots, x_n$. For this purpose, he integrated equation (10.27) with respect to z to obtain

$$\int_{-1}^{1} F(x)\ln(z-x)\,dx = k\ln\frac{f(z)}{C} - \frac{n!k_1}{z^{n+1}} - \frac{(n+1)!k_2}{z^{n+2}} - \cdots,$$

where C was a constant. Hence, by exponentiation he could write

$$f(z)e^{-n!k_1/(kz^{n+1})-(n+1)!k_2/(kz^{n+2})-\cdots} = C\exp\left(\frac{1}{k}\int_{-1}^{+1} F(x)\ln(z-x)\,dx\right).$$

Chebyshev then noted that the exponential on the left differed from 1 by a series of powers of z less than z^{-n}; hence, he noted that $f(z)$ was the polynomial part of the exponential on the right-hand side. He deduced Hermite's formula by taking $F(x) = 1/\sqrt{1-x^2}$, so that

$$k = \frac{1}{n}\int_{-1}^{+1} \frac{dx}{\sqrt{1-x^2}} = \frac{\pi}{n}$$

and

$$\int_{-1}^{+1} F(x)\ln(z-x)\,dx = \int_{-1}^{1} \frac{\ln(z-x)}{\sqrt{1-x^2}}\,dx = \pi \ln\frac{z+\sqrt{z^2-1}}{2}. \tag{10.30}$$

Chebyshev could then conclude that the polynomial $f(z)$ in this case was in fact the polynomial part of

$$e^{n\ln\frac{z+\sqrt{z^2-1}}{2}} = \left(\frac{z+\sqrt{z^2-1}}{2}\right)^n,$$

and he wrote that it was equal to $\frac{1}{2^{n-1}}\cos(n\arccos z)$. He then considered the case where $F(x) = 1$, to obtain by (10.29): $k = 2/n$ and

$$\int_{-1}^{+1} \ln(z-x)\,dx = \ln\frac{(z+1)^{z+1}}{(z-1)^{z-1}} - 2.$$

Thus, Chebyshev arrived at the result he wanted:

$$\int_{-1}^{1} \phi(x)\,dx = \frac{2}{n}(\phi(x_1) + \phi(x_2) + \cdots + \phi(x_n)), \tag{10.31}$$

where $x_1, x_2, \ldots, x_n$ were the zeros of the polynomial given by the polynomial part of the expression

$$(z+1)^{n(z+1)/2}(z-1)^{-n(z-1)/2} = z^n e^{-\frac{n}{2\cdot 3z^2} - \frac{n}{4\cdot 5z^4} - \frac{n}{6\cdot 7z^6} - \cdots}.$$

He also computed the cases in which $n = 2, 3, 4, 5, 6, 7$ to get the polynomials $z^2 - 1/3$, $z^3 - 1/2z$, $z^4 - 2/3z^2 + 1/45$, $z^5 - 5/6z^3 + 7/72z$, $z^6 - z^4 + 1/5z^2 - 1/105$, $z^7 - 7/6z^5 + 119/360z^3 - 149/6480z$. He calculated the zeros of these polynomials to six decimal places. At this juncture, Chebyshev pointed out that in (10.31) the sum of the squares of the coefficients had the smallest possible value, because they were all equal; thus, his formula might sometimes even be an improvement on Gauss's quadrature formula.

10.9 Exercises

1. Let $A, B, C, E, \ldots$ be points on the x-axis and $K, L, M, N, \ldots$ corresponding points on the curve. Then the ordinates are given by $AK, BL, CM, DN, \ldots$.

Newton described the following formulas for the approximate area under the curve in the case of seven and nine ordinates:

If there be seven ordinates there will come to be

$$\frac{1}{280}(17AK + 54BL + 51CM + 36DN + 51EO + 54FP + 17GQ)AG = \Box AQ,$$

While if there be given nine there will come

$$\frac{(217AK + 1024BL + 352CM + 1024DN + 436EO + 1024FP + 352GQ + 1024HR + 217IS)AI}{5670}$$
$$= \Box AS.$$

Derive these formulas using Newton's ideas, as explained in the text and as presented by Newton in his "Of Quadrature by Ordinates." Recall that these formulas, making use of Newton's assumption on the proportionality of the errors, differ from those obtained by integrating the interpolating polynomial. See Newton (1967–1981), vol. 7, p. 695, including Whiteside's footnotes.

2. Suppose

$$u = \frac{a + bx + cx^2 + \cdots + hx^{n-1}}{\alpha + \beta x + \gamma x^2 + \cdots + \theta x^m},$$

and $u(x_k)$, $x = 0, 1, 2, \ldots, n+m-1$. Determine the values of the coefficients $a/\alpha, b/\alpha, \cdots, h/\alpha, \beta/\alpha, \cdots \theta/\alpha$. In particular, show that when $m = 1$ and $n = 2$,

$$u = \frac{u_0 u_1 \frac{x-x_2}{(x_0-x_2)(x_1-x_2)} + u_0 u_2 \frac{x-x_1}{(x_0-x_1)(x_2-x_1)} + u_1 u_2 \frac{x-x_0}{(x_1-x_0)(x_2-x_0)}}{u_0 \frac{x_0-x}{(x_0-x_1)(x_0-x_2)} + u_1 \frac{x_1-x}{(x_1-x_0)(x_1-x_2)} + u_2 \frac{x_2-x}{(x_2-x_0)(x_2-x_1)}}.$$

Cauchy discussed this interpolation by rational functions after he deduced the Waring–Lagrange formula in his lectures. See Cauchy (1989), pp. 527–529.

3. Chebyshev computed the zeros of the polynomials, $z^5 - \frac{5}{6}z^3 + \frac{7}{72}z$ and $z^6 - z^4 + \frac{1}{5}z^2 - \frac{1}{105}$ for use in (10.31). His results were

$$\pm 0.832497, \pm 0.374541, 0 \quad \text{and} \quad \pm 0.866247, \pm 0.422519, \pm 0.266635.$$

Check Chebyshev's computations.

4. Show that Chebyshev's result in (10.30) implies Hermite's formula (10.24).

5. Prove the following formulas of Stieltjes:

$$\int_{-1}^{+1} \sqrt{1-x^2} f(x)\,dx = \frac{\pi}{n+1} \sum_{k=1}^{n} \sin^2 \frac{k\pi}{n+1} f\left(\cos \frac{k\pi}{n+1}\right) + \text{corr.}$$

$$\int_{-1}^{+1} \sqrt{\frac{1-x}{1+x}} f(x)\,dx = = \frac{4\pi}{2n+1} \sum_{k=1}^{n} \sin^2 \frac{k\pi}{2n+1} f\left(\cos \frac{2k\pi}{2n+1}\right) + \text{corr.}$$

The correction is zero when $f(x)$ is a polynomial of degree $\leq 2n-1$.

$$\int_0^1 \frac{f(x)}{\sqrt{x(1-x)}} = \frac{\pi}{n} \sum_{k=1}^n f\left(\cos^2 \frac{(2k-1)\pi}{4n}\right) + \text{corr},$$

$$\int_0^1 \sqrt{x(1-x)} f(x)\,dx = \frac{\pi}{4n+4} \sum_{k=1}^n \sin^2 \frac{k\pi}{n+1} f\left(\cos^2 \frac{k\pi}{2n+2}\right) + \text{corr},$$

$$\int_0^1 \sqrt{\frac{1-x}{x}} f(x)\,dx = \frac{2\pi}{2n+1} \sum_{k=1}^n \sin^2 \frac{k\pi}{2n+1} f\left(\cos^2 \frac{k\pi}{2n+1}\right) + \text{corr}.$$

See Stieltjes (1993), vol. 1, pp. 514–515.

10.10 Notes on the Literature

For the November 23, 1674 letter of Collins, see Turnbull (1939), pp. 290–292. Stirling's comments on Newton's contribution to interpolation theory may be found on p. 122 of Tweddle (2003). Newton (1964–1967), vol. 2, pp. 168–173, provides an English translation of Newton's *Methodus*, presenting his divided difference method and its applications to various interpolation and quadrature formulas. Newton (1967–1981), vol. 7, pp. 690–99, contains his approximate quadrature method. For the quotation from Newton's letter, see Newton (1959–1960), vol. 2, p. 137. Waring (1779) presents his derivation of the Lagrange–Waring interpolation formula. Cauchy's derivation of this result appears in Note V of his *Analyse algébrique*; Jacobi's argument can be found in Jacobi (1969), vol. 6, p. 5. See Hermite (1873), pp. 452–454 for his approximate quadrature, and see Chebyshev (1899–1907), vol. 2, pp. 165–180, for his extension of Hermite.

Chapters 10 and 11 of J. L. Chabert (1999) contain interesting observations on interpolation and quadrature with excerpts from original authors. Thomas Harriot's manuscript *De Numeris Triangularibus*, containing his derivations of symbolic interpolation formulae and their applications, has now been published in Beery and Stedall (2009), almost four centuries after it was written. Beery and Stedall provide a commentary to accompany the almost completely nonverbal presentation of Harriot. They also discuss the work on interpolation of several British mathematicians in the period 1610–1670.

11

Series Transformation by Finite Differences

11.1 Preliminary Remarks

Around 1670, James Gregory found a large number of new infinite series, but his methods remain somewhat unclear. From circumstantial evidence and from the form of some of his series, it appears that he was the first mathematician to systematically make use of finite difference interpolation formulas in finding new infinite series. The work of Newton, Gregory, and Leibniz made the method of finite differences almost as important as calculus in the discovery of new infinite series. We observe that interpolation formulas usually deal with finite expressions because in practice the number of interpolating points is finite. By theoretically extending the number of points to infinity, Gregory found the binomial theorem, the Taylor series, and numerous interesting series involving trigonometric functions. Gregory most likely derived these theorems from the Gregory–Newton (or Harriot–Briggs) interpolation formula. Gregory's letter of November 23, 1670, to Collins explicitly mentions these results, and also contains some other series, not direct consequences of the Harriot–Briggs result. Instead, these other series seem to require the Newton–Gauss interpolation formula; one is compelled to conclude that Gregory must have obtained this interpolation formula, though it is not given anywhere in his surviving notes and letters. In a separate enclosure with his letter to Collins, Gregory wrote several formulas, including:

Given an arc whose sine is d, and sine of the double arc is $2d - e$, it is required to find another arc which bears to the arc whose sine is d the ratio a to c. The sine of the arc in question

$$= \frac{ad}{c} - \frac{be}{c} + \frac{ke^2}{cd} - \frac{le^3}{cd^2} + \cdots \qquad (11.1)$$

where $\quad \dfrac{b}{c} = \dfrac{a(a^2 - c^2)}{2 \cdot 3 \cdot c^3}, \quad \dfrac{k}{c} = \dfrac{a(a^2 - c^2)(a^2 - 4c^2)}{2 \cdot 3 \cdot 4 \cdot 5 \cdot c^5}$, etc.

11.1 Preliminary Remarks

In modern notation, $c = r\theta$, $d = r\sin\theta$, $2d - e = r\sin 2\theta$; hence, $e = 2r\sin\theta(1-\cos\theta)$ and the series takes the form

$$\sin\frac{a\theta}{c} = \sin\theta\left(\frac{a}{c} - \frac{a(a^2-c^2)}{3!c^3}2(1-\cos\theta)\right.$$
$$\left. + \frac{a(a^2-c^2)(a^2-4c^2)}{5!c^5}(2(1-\cos\theta))^2 - \cdots\right). \tag{11.2}$$

Gregory noted at the end of the enclosure that an infinite number of other ways of measuring circular arcs could be deduced from his calculations.

Gregory did not publish his work on series and his mathematical letters to Collins were not printed until later. Meanwhile, Newton developed his profound ideas on interpolation and finite differences starting in the mid 1670s. In the early 1680s, he applied the method of differences to infinite series and in June–July of 1684, he wrote two short treatises on the topic. He was provoked into writing up his results upon receiving a work from David Gregory, the nephew of James, *Exercitatio Geometrica de Dimensione Figurarum*. In this treatise, David Gregory discussed several aspects of infinite series, apparently without knowledge of Newton's work. Newton evidently wished to set the record straight; he first wrote "Matheseos Universalis Specimina," in which he pointed out that James Gregory and Leibniz were indebted to him in their study of series. He did not finish this treatise, but instead started a new one, called "De Computo Serierum" in which he eliminated all references to Gregory and Leibniz. The first chapter of the second treatise dealt with infinite series in a manner similar to that of his early works of 1669 and 1671. However, the second chapter employed the entirely new idea of applying finite differences to derive an important transformation of infinite series, often called Euler's transformation. In modern notation, this is given by

$$A_0 t + A_1 t^2 + A_2 t^3 + \cdots = A_0 z + \Delta A_0 z^2 + \Delta^2 A_0 z^3 + \cdots, \tag{11.3}$$

where $z = t/(1-t)$, $\Delta A_0 = A_1 - A_0$, $\Delta^2 A_0 = A_2 - 2A_1 + A_0$, etc.

Newton noted one remarkable special case of his transformation:

$$\tan^{-1} t = t - \frac{1}{3}t^3 + \frac{1}{5}t^5 - \frac{1}{7}t^7 + \cdots$$

$$= \frac{t}{1+t^2}\left(1 + \frac{2}{1\cdot 3}\frac{t^2}{1+t^2} + \frac{2\cdot 4}{1\cdot 3\cdot 5}\left(\frac{t^2}{1+t^2}\right)^2 + \frac{2\cdot 4\cdot 6}{1\cdot 3\cdot 5\cdot 7}\left(\frac{t^2}{1+t^2}\right)^3 + \cdots\right). \tag{11.4}$$

Observe that when $t = 1$ we have $t^2/(1+t^2) = 1/2$ so that while the first series converges very slowly for this value of t, the second series converges much more rapidly. Moreover when $t = \sqrt{3}$, the first series is divergent while the second is convergent. In fact, Newton wrote:

> The chief use for these transformations is to turn divergent series into convergent ones, and convergent series into ones more convergent. Series in which all terms are of the same sign cannot

diverge without simultaneously coming to be infinitely great and on that account false. These, consequently, have no need to be turned into convergent ones. Those, however, in which the terms alternate in sign and proceed regularly, are so moderated by the successive addition and subtraction of those terms as to remain true even in divergence. But in their divergent form their quantity cannot be computed and they must be turned into convergent ones by the rule introduced, while when they are sluggishly convergent the rule must be applied to make them converge more swiftly. Thus the series $y = x - \frac{1}{3}x^3 + \frac{1}{5}x^5 \cdots$, when it converges or diverges slowly enough and has been turned into this one

$$y = x - \frac{1}{3}x^3 + \frac{1}{5}x^5 - \frac{1}{7}x^7 + \frac{1}{9}x^9 - \frac{1}{11}x^{11} + \frac{1}{13}x^{13} - \frac{1}{15}x^{15}$$
$$+ x^{15}\left(\frac{1}{17}z + \frac{2}{17\cdot 19}z^2 + \frac{2\cdot 4}{17\cdot 19\cdot 21}z^3 + \frac{2\cdot 4\cdot 6}{17\cdot 19\cdot 21\cdot 23}z^4 \cdots\right),$$

will speedily enough be computed to many places of decimals. If the same series proves swiftly divergent it must be turned into the convergent

$$xy = z + \frac{2}{1\cdot 3}z^2 + \frac{2\cdot 4}{1\cdot 3\cdot 5}z^3 \cdots \tag{11.5}$$

and then by what is presented in the following chapter it can be computed. It is, however, frequently convenient to reduce the coefficients $A, B, C, \ldots$ to decimal fractions at the very start of the work.

We note that Newton's $A, B, C, \ldots$ are the $A_0, A_1, A_2 \ldots$ in (11.3) and $z = x^2/(1+x^2)$, as in (11.4). We do not know why Newton discontinued work on this treatise. Perhaps it was because Edmond Halley visited Cambridge in August 1684 and urged Newton to work on problems of planetary motion. As is well known, Newton started work on the *Principia* soon after this visit and for the next two years had time for little else.

Newton's transformation (11.4) for the arctangent series is obviously important, so it is not surprising that others rediscovered it, since Newton's paper did not appear in print until 1970. In August 1704, Jakob Bernoulli communicated the $t = 1$ case of (11.4) to Leibniz as a recent discovery of Jean Christophe Fatio de Duillier. Jakob Hermann, a student of Bernoulli, found a proof for this and sent it to Leibniz in January 1705. This proof is identical with that of Newton's when specialized to $t = 1$. Johann Bernoulli, and probably others, succeeded in deriving the general form of (11.4). Bernoulli, in fact, applied the general form in a paper of 1742 and thereby derived a remarkable series for π^2 found earlier by Takebe Katahiro by a different technique. In 1717, the French mathematician Pierre Rémond de Montmort (1678–1719) rediscovered Newton's more general transformation (11.3) with a different motivation and method of proof.

Montmort was born into a wealthy family of the French nobility and was self-taught in mathematics. An admirer of both Newton and Leibniz, he remained neutral but friendly with followers of both mathematicians during the calculus priority dispute in the early eighteenth century. He mainly worked in probability and combinatorics but also made contributions to the theory of series. His paper on series was inspired by Brook Taylor's 1715 work *Methodus Incrementorum*; consequently, Montmort's "De Seriebus Infinitis Tractatus" was published in the *Philosophical Transactions* with an appendix by Taylor, then Secretary of the Royal Society. Montmort's paper dealt with those finite as well as infinite series to which the method of differences could be applied. He first worked out the transformation of a finite power series and then obtained Newton's formula (11.3) as a corollary. He also quoted from a 1715 letter of Niklaus I Bernoulli, showing that Niklaus had found a result similar to that of Newton.

In 1717, François Nicole (1683–1758), a pupil of Montmort, also published a paper on finite differences. He too wrote that his ideas were suggested by Taylor's book of 1715. The title of his paper, "Traité du calcul des différences finies," indicated that he viewed the calculus of finite differences as a new topic in mathematics, separate from geometry, calculus, or algebra. By means of examples, he showed that the shifted factorial expression

$$(x)_n = x(x+h)(x+2h)\cdots(x+(n-1)h) \tag{11.6}$$

behaved under differencing as x^n under differentiation. Thus,

$$\Delta_h(x)_n = (x+h)_n - (x)_n = (n)h(x+h)_{n-1}. \tag{11.7}$$

Also, the difference relation

$$\frac{1}{(x)_n} - \frac{1}{(x+h)_n} = \frac{nh}{(x)_{n+1}} \tag{11.8}$$

showed that the analog of x^{-n} was $1/(x)_n$. Moreover, the inverse of a difference would be a sum. And just as the derivative of a function indicated the integral of the derived function, so also one could use the difference to find the sum. He gave an example: From

$$f(x) = x(x+1)(x+2)/3, \ f(x+1) - f(x) = (x+1)(x+2), \tag{11.9}$$

he got

$$1\cdot 2 + 2\cdot 3 + \cdots + x(x+1) = \frac{x(x+1)(x+2)}{3} + C. \tag{11.10}$$

By taking $x = 1$, he obtained the constant as zero.

In 1723, in his second memoir, Nicole discussed the problem of computing the coefficients $a_0, a_1, a_2, \ldots$ in

$$f(x+m) - f(x) = a_0 + a_1(x+h) + a_2(x+h)(x+2h)$$
$$+ \cdots + a_{k-1}(x+h)\cdots(x+(k-1)h), \tag{11.11}$$

where $f(x) = x(x+h)\cdots(x+(k-1)h)$. His method employed a long inductive process, but simpler procedures have since been found. In his second memoir and in his third memoir of 1724, Nicole solved a similar problem for the inverse factorial $1/x(x+h)\cdots(x+(k-1)h)$.

Both Montmort and Nicole mentioned Taylor as the source of their inspiration; we note that Taylor gave a systematic exposition of finite differences and derivatives with their inverses, sums, and integrals. Many of these ideas were already known but Taylor explicitly laid out some of the concepts, such as the method of summation by parts.

In a letter of November 14, 1715, Montmort also attributed to Taylor the summation formula

$$\frac{1}{b-a} = \frac{1}{b} + \frac{a}{b(b+d)} + \frac{a(a+d)}{b(b+d)(b+2d)} + \cdots. \tag{11.12}$$

There are several ways of proving this, but it is likely that Taylor proved it by the Gregory–Newton interpolation formula, since he used this to prove his famous series.

The Scottish mathematician James Stirling (1692–1770) took Nicole's work much further. Stirling's book, *Methodus Differentialis*, is sometimes called the first book on the calculus of finite differences. Like all prominent British mathematicians of the early eighteenth century, he was a disciple of Newton. His first paper, "Lineae Tertii Ordinis Newtonianae," was an account with some extensions of Newton's theory of cubic curves. His second paper, "Methodus Differentialis Newtoniana Illustrata," developed some of Newton's ideas on interpolation. He later expanded this paper into the *Methodus*. Stirling received his early education in Scotland. In 1710 he traveled to Oxford and graduated from Balliol College the same year. He stayed on at Oxford on a scholarship, but he lost his support after the first Jacobite rebellion of 1715, as his family had strong Jacobite sympathies. He then spent several years in Italy and was unable to obtain a professorship there. Athough details of his time in Italy are unknown, his second paper was communicated from Venice. After returning to Britain in 1722, he was given financial assistance by Newton, making him one of Newton's devoted friends. After teaching in a London school, in 1735 Stirling began service as manager of the Leadhills Mines in Scotland where he was very successful, looking after the welfare of the miners as well as the interests of the shareholders. In the early 1750s, he also surveyed the River Clyde in preparation for a series of navigational locks.

Stirling started his book where Nicole ended. In the introduction, he defined the Stirling numbers of the first and second kinds. These numbers appeared as coefficients when z^n was expanded in terms of $z(z-1)\cdots(z-k+1)$, and $1/z^n$ was expanded in terms of $1/z(z+1)\cdots(z+k-1)$. These expansions were required in order to apply the method of differences to functions or quantities normally expressed in terms of powers of z. Stirling constructed two small tables of these coefficients to make the transformation easy to use. In the first few propositions of his book, Stirling considered problems similar to those of Nicole, but he very quickly enlarged the scope of those methods. He applied his new method to the approximate summation of series such as $1 + 1/4 + 1/9 + 1/16 + \cdots$, whose approximate value had also been computed by Daniel Bernoulli, Goldbach and Euler in the late 1720s. It was a little later that Euler brilliantly found the exact value of the series to be $\pi^2/6$. Stirling also applied his method of differences to derive several new and interesting transformations of series. For example, observe propositions 7 and 8 of his *Methodus Differentialis* presented in modernized notation:

$$1 + \frac{z-n}{z(1-m)} + \frac{(z-n)(z-n+1)}{z(z+1)(1-m)^2} + \cdots$$
$$= \frac{m-1}{m}\left(1 + \frac{n}{zm} + \frac{n(n+1)}{z(z+1)m^2} + \cdots\right) \tag{11.13}$$

and

$$1 + \frac{(z-m)(z-n)}{z(z-n+1)} + \frac{(z-m)(z-m+1)(z-n)(z-n+1)}{z(z+1)(z-n+1)(z-n+2)} + \cdots$$
$$= \frac{z-n}{m}\left(1 + \frac{nm}{z(m+1)} + \frac{n(n+1)m(m+1)}{z(z+1)(m+1)(m+2)} + \cdots\right). \quad (11.14)$$

Note that Newton's transformation (11.4) for arctan t follows by taking $n = 1$, $z = 3/2$, and $m = 1 + 1/t^2$ in (11.13). We shall see later that these formulas are particular cases of transformations of hypergeometric series. The hypergeometric generalization of (11.13) was discovered by Pfaff in 1797, and the generalization of (11.14) was found by Kummer in 1834. Thus, the methods of hypergeometric series provide the right context with the appropriate degree of generality to study the series (11.4), (11.13), and (11.14). Moreover, Gauss extended Stirling's finite difference method to the theory of hypergeometric series and derived his well-known and important contiguous relations for hypergeometric series. Even today, contiguous relations continue to provide unexplored avenues for research.

We also note that since expressions of the form $z(z-1)\cdots(z-k+1)$ appear in finite difference interpolation formulas, Stirling numbers of the first kind also appear in those formulas. For this reason, in the early 1600s, Harriot computed these numbers. Stirling numbers also cropped up in Lagrange's 1770s proof of Wilson's theorem that $(p-1)! + 1$ was divisible by p if and only if p was prime. In fact, Lagrange's proof gave the first number theoretic discussion of Stirling numbers of the first kind.

Like Gregory, Leibniz, Taylor, and Nicole, Euler saw the intimate connections between the calculus of finite differences and the calculus of differentiation and integration. His influential 1755 book on differential calculus began with a chapter on finite differences. The second chapter on the use of differences in the summation of series discussed examples such as those in Nicole's work. In the second part of his book, Euler devoted the first chapter to Newton's transformation (11.3). He gave a proof different from Newton's and from Montmort's; this in turn led him to a further generalization of the formula. Suppose

$$S = a_0 + a_1 x + a_2 x^2 + a_3 x^3 + a_4 x^4 + \cdots.$$

Euler then got the generalization

$$A_0 a_0 + A_1 a_1 x + A_2 a_2 x^2 + A_3 a_3 x^3 + \cdots$$
$$= A_0 S + \Delta A_0 \frac{x}{1!}\frac{dS}{dx} + \Delta^2 A_0 \frac{x^2}{2!}\frac{d^2 S}{dx^2} + \Delta^3 A_0 \frac{x^3}{3!}\frac{d^3 S}{dx^3} + \cdots. \quad (11.15)$$

The Newton–Montmort formula followed by taking $a_0 = a_1 = a_2 = \cdots = 1$.

11.2 Newton's Transformation

In his 1684 "Matheseos," Newton attempted to change slowly convergent series into more rapidly convergent ones. He considered the method of taking differences of the coefficients, but it was not until a little later that he arrived at the explicit and useful transformation (11.3) contained in the "De Computo." He wrote the initial series as

$$v = At \pm Bt^2 + Ct^3 \pm Dt^4 + Et^5 \pm Ft^6 \&c. \tag{11.16}$$

Newton explained his transformation:

> Here $A, B, C, \ldots$ are to denote the coefficients of the terms whose ultimate ratio, if the series be extended infinitely, is one of equality, and 1 to t is the remaining ratio of the terms; while the sign $\pm$ is ambiguous and the converse of the sign $\mp$. Collect the first differences $b, b_2, b_3, \ldots$ of the terms $A, B, C, \ldots$; then their second ones $c, c_2, c_3, \ldots$, third ones $d, d_2, d_3, \ldots$, and any number of following ones. Collect these, however, by always taking a latter term from the previous one: B from A, C from $B, \ldots$; b_2 from b, b_3 from $b_2, \ldots$; d_2 from d, d_3 from $d_2, \ldots$, and so on. Then make $t/(1 \mp t) = z$ and when the signs are appropriately observed there will be

$$v = Az \mp bz^2 + cz^3 \mp dz^4 + ez^5 \mp fz^6 + \cdots.$$

He took the differences in reverse order of the modern convention. He had $A - B$, $B - C$, $C - D, \ldots$ for the first differences instead of $B - A$, $C - B$, $D - C, \ldots$ and similarly for the higher-order differences. The revised version of the "De Computo" did not include a proof but notes of an earlier version suggest the following iterative argument:

$$\begin{aligned} v &= z(1 \mp t)(A \pm Bt + Ct^2 \pm Dt^3 + Et^4 \pm Ft^5 \cdots) \\ &= z(A \mp (A - B)t - (B - C)t^2 \mp (C - D)t^3 - (D - E)t^4 \mp \cdots) \\ &= Az \mp z((A - B)t \pm (B - C)t^2 + (C - D)t^3 \pm (D - E)t^4 + \cdots). \end{aligned} \tag{11.17}$$

Now the last series in parentheses is of the same form as the original series except that the coefficients are the differences of the coefficients of the original. So the procedure can be repeated to give

$$\begin{aligned} v &= Az \mp z(z(A - B) \mp z((A - 2B + C)t \\ &\quad \pm (B - 2C + D)t^2 + (C - 2D + E)t^3 \pm \cdots)) \\ &= Az \mp (A - B)z^2 + (A - 2B + C)z^3 \mp \cdots. \end{aligned} \tag{11.18}$$

The final formula results from an infinite number of applications of this procedure; Newton applied this formula to the logarithmic and arctangent series. In the case of the logarithm, the transformation amounted to the equation $\ln(1 + x) = -\ln\left(1 - \frac{x}{1+x}\right)$. Newton's main purpose was to use the transformation for numerical computation and this explains why he applied the transformation after the eight term $\frac{1}{15}x^{15}$ in (11.5) rather than immediately at the outset. Note also that the first step in (11.17) was an example of the summation by parts discussed explicitly by Taylor and later used by Abel to study the convergence of series.

11.3 Montmort's Transformation

In his 1717 paper, "De Seriebus Infinitis Tractatus," Montmort started with elementary examples, but toward the end of the paper he posed the problem of summing or

11.3 Montmort's Transformation

transforming the series

$$S = \frac{a_0}{h} + \frac{a_1}{h^2} + \frac{a_2}{h^3} + \cdots. \tag{11.19}$$

He also discussed partial sums of this series, written as

$$S_0 = \frac{A_0}{h}, \quad S_1 = \frac{A_1}{h^2}, \ldots, S_p = \frac{A_p}{h^{p+1}}, \ldots, \text{ where} \tag{11.20}$$

$$A_0 = a_0, \quad A_1 = a_0 h + a_1, \quad A_2 = a_0 h^2 + a_1 h + a_2, \ldots. \tag{11.21}$$

He noted that a simple relation existed between the differences of the sequence $A_0, A_1, A_2, \ldots$ and the sequence $a_0, a_1, a_2, \ldots$. He wrote down just the first three cases: For $q = h - 1$,

$$\Delta A_0 = h a_0 + \Delta a_0,$$
$$\Delta^2 A_0 = q h a_0 + h \Delta a_0 + \Delta^2 a_0,$$
$$\Delta^3 A_0 = q^2 h a_0 + q h \Delta a_0 + h \Delta^2 a_0 + \Delta^3 a_0, \text{ etc.}$$

It is not difficult to write out the general relation from these examples. He then proved the result that if for $k \geq l$, $\Delta^k a_0 = 0$, then

$$\Delta^k A_0 = q^{k-l} \Delta^l A_0. \tag{11.22}$$

In fact, he verified this for $l = 1$ and 2, but noted that was sufficient to see that the result was true in general. Montmort then proceeded to evaluate the partial sums of (11.19) under the assumption that $\Delta^l a_0 = 0$ for some positive integer l. The basic result used here and in other examples was that for any sequence $b_0, b_1, b_2, \ldots$ and a positive integer p,

$$b_p = b_0 + \binom{p}{1} \Delta b_0 + \binom{p}{2} \Delta^2 b_0 + \cdots + \binom{p}{p} \Delta^p b_0. \tag{11.23}$$

For example, to evaluate A_p for $p > l$ and $\Delta^l a_0 = 0$, he employed (11.22) to obtain

$$A_p = \left(A_0 + \binom{p}{1} \Delta A_0 + \cdots + \binom{p}{l-1} \Delta^{l-1} A_0 \right)$$
$$+ \binom{p}{l} \Delta^l A_0 + \binom{p}{l+1} \Delta^{l+1} A_0 + \cdots + \binom{p}{p} \Delta^p A_0$$
$$= A_0 + \binom{p}{1} \Delta A_0 + \cdots + \binom{p}{l-1} \Delta^{l-1} A_0$$
$$+ \frac{\Delta^l A_0}{q^l} \left(\binom{p}{l} q^l + \binom{p}{l+1} q^{l+1} + \cdots + \binom{p}{p} q^p \right)$$
$$= A_0 + \binom{p}{1} \Delta A_0 + \cdots + \binom{p}{l-1} \Delta^{l-1} A_0$$
$$+ \frac{\Delta^l A_0}{q^l} \left(h^p - 1 - \binom{p}{1} q - \cdots - \binom{p}{l-1} q^{l-1} \right). \tag{11.24}$$

As an application of this formula, he gave the sum of the finite series

$$1 \cdot 3 + 3 \cdot 3^2 + 6 \cdot 3^3 + 10 \cdot 3^4 + 15 \cdot 3^5 + 21 \cdot 3^6.$$

Here $h = 1/3$, $q = -2/3$, $p = 5$, and $\Delta^3 a_0 = 0$. Next, as a corollary, Montmort stated without proof the transformation formula

$$\sum_{k=0}^{\infty} \frac{a_k}{h^{k+1}} = \sum_{k=0}^{\infty} \frac{\Delta^k a_0}{(h-1)^{k+1}}. \tag{11.25}$$

Of course, this is Newton's formula. Since equation (11.23) was Montmort's basic formula, we may assume that he applied it to give a formal calculation to justify (11.25). Indeed, this is easy to do:

$$\sum_{k=0}^{\infty} \frac{a_k}{h^{k+1}} = \sum_{k=0}^{\infty} \frac{1}{h^{k+1}} \sum_{j=0}^{k} \binom{k}{j} \Delta^j a_0$$

$$= \sum_{k=0}^{\infty} \Delta^k a_0 \sum_{j=0}^{\infty} \binom{k+j}{j} \frac{1}{h^{k+1+j}}$$

$$= \sum_{k=0}^{\infty} \frac{\Delta^k a_0}{(h-1)^{k+1}}.$$

Unfortunately, he did not give any interesting examples of this formula. The three cases he explicitly mentioned follow from the binomial theorem just as easily. We mention that Zhu Shijie (c. 1260–1320), also known as Chu Shi-Chieh, knew (11.23) and used it to sum finite series in his *Siyuan Yujian* of 1303.

11.4 Euler's Transformation Formula

As we mentioned earlier, Newton did not publish his transformation formula. It is not certain whether or not Euler saw Montmort's paper on this topic. In any case, Euler's approach differed from those of Newton and Montmort. Euler's proof of (11.3) applied the change of variables in the first step. We present the proof as Euler set it out. He let

$$S = ax + bx^2 + cx^3 + dx^4 + ex^5 + \&c.$$

and let $x = y/(1+y)$ and replaced the powers of x by the series

$$x = y - y^2 + y^3 - y^4 + y^5 - y^6 + \&c.$$
$$x^2 = y^2 - 2y^3 + 3y^4 - 4y^5 + 5y^6 - 6y^7 + \&c.$$
$$x^3 = y^3 - 3y^4 + 6y^5 - 10y^6 + 15y^7 - 21y^8 + \&c.$$
$$x^4 = y^4 - 4y^5 + 10y^6 - 20y^7 + 35y^8 - 56y^9 + \&c.$$

11.4 Euler's Transformation Formula

Thus,

$$S = ay - ay^2 + ay^3 - ay^4 + ay^5 \&c.$$
$$+b - 2b + 3b - 4b$$
$$+c \quad -3c + 6c$$
$$+d - 4d.$$

Note that the coefficients of the various powers of y were presented in columns. Since $y = x/(1-x)$,

$$S = a\frac{x}{1-x} + (b-a)\frac{x^2}{(1-x)^2} + (c-2b+a)\frac{x^3}{(1-x)^3} \&c,$$

yielding the transformation formula.

Note that the series for x is the geometric series, while the series for $x^2, x^3, \ldots$ can be obtained by the binomial theorem or by the differentiation of the series for x. In fact, Euler must have had differentiation in mind, since his proof of the second transformation formula (11.15) was obtained by writing the right-hand side of (11.15) as

$$\alpha S + \beta \frac{x}{1!}\frac{dS}{dx} + \gamma \frac{x^2}{2!}\frac{d^2S}{dx^2} + \delta \frac{x^3}{3!}\frac{d^3S}{dx^3} + \cdots$$

and then substituting the series for $S, dS/dx, d^2S/dx^2, \ldots$. By equating the coefficients of the various powers of x, he found $\alpha = A_0$, $\beta = A_1 - A_0$, $\gamma = A_2 - 2A_1 + A_0$, and so on.

Euler gave several examples of these formulas in his differential calculus book. At times writing xx and at other times using x^2, he considered the problem of summing the series

$$1x + 4xx + 9x^3 + 16x^4 + 25x^5 + \text{etc.} \tag{11.26}$$

The first and second differences of the coefficients 1, 4, 9, 16, 25, ... were 3, 5, 7, 9, ... and 2, 2, 2, ... Therefore, the third differences were zero, and by equation (11.3), the sum of the series (11.26) came out to be

$$\frac{x}{1-x} + \frac{3x^2}{(1-x)^2} + \frac{2x^3}{(1-x)^3} = \frac{x+xx}{(1-x)^3}. \tag{11.27}$$

To sum the finite series

$$1x + 4x^2 + 9x^3 + 16x^4 + \cdots + n^2x^n, \tag{11.28}$$

Euler subtracted $(n+1)^2 x^{n+1} + (n+2)^2 x^{n+2} + \cdots$ from the series in the previous example. He found the sum of this infinite series to be

$$\frac{x^{n+1}}{1-x}\left((n+1)^2 + (2n+3)\frac{x}{1-x} + \frac{2x^2}{(1-x)^2}\right). \tag{11.29}$$

Euler observed the general rule, already established by Montmort, that if a power series had coefficients such that $\Delta^n A_0 = 0$ for $n \geq k$, then the series would sum to a rational function.

When Euler discovered a new method of summation or a new transformation of series, he applied it to divergent as well as convergent series. He believed that divergent series could be studied and used in a meaningful way. He explained that whenever he assigned a sum to a divergent series by a given method, he arrived at the same sum by alternative methods, leading him to conclude that divergent series could be legitimately summed. Applying the differences method, he found the sums of various divergent series, including $1 - 4 + 9 - 16 + 25 - \cdots$; $\ln 2 - \ln 3 + \ln 4 - \ln 5 + \cdots$; and $1 - 2 + 6 - 24 + 120 - 720 + \cdots$. Observe that the terms in the third example are $1!, 2!, 3!, 4!, \ldots$. This was one of Euler's favorite divergent series. By taking twelve terms of the transformed series and using (11.3), he found the sum to be 0.4036524077. He must have later realized that this sum was very inaccurate, since he reconsidered it in a 1760 paper in which he took the function

$$f(x) = 1 - 1!x + 2!x^2 - 3!x^3 + \cdots$$

and proceeded to express $f(x)$ as an integral and as a continued fraction:

$$f(x) = \frac{1}{x} \int_0^x \frac{1}{t} e^{-1/t} dt \text{ and}$$

$$f(x) = \frac{1}{1+} \frac{x}{1+} \frac{x}{1+} \frac{2x}{1+} \frac{2x}{1+} \frac{3x}{1+} \frac{3x}{1+} \text{ etc.}$$

Euler then computed the value of the original series $f(1)$. To approximately evaluate the integral, he divided the interval $[0,1]$ into ten equal parts and used the trapezoidal rule. By this method, $f(1) = 0.59637255$. The continued fraction, on the other hand, gave $f(1) = 0.59637362123$, correct to eight decimal places. Even earlier, in a letter to Goldbach of August 7, 1745, Euler wrote that he had worked out the continued fraction for the divergent series $f(1)$ and found the value to be approximately 0.5963475922, adding the remark that in a small dispute with Niklaus I Bernoulli about the value of a divergent series, he himself had argued that all series such as $f(1)$ must have a definite value.

In his differential calculus book, Euler gave a few applications of his more general formula (11.15), including the derivation of the exponential generating function of a sequence whose third difference was zero. He began with the difference table in Figure (11.1).

$$\begin{array}{ccccccccccc}
2 & & 5 & & 10 & & 17 & & 26 & & 37 \\
& 3 & & 5 & & 7 & & 9 & & 11 & \\
& & 2 & & 2 & & 2 & & 2 & &
\end{array}$$

Figure 11.1. Difference Table

From this, he derived

$$2 + 5x + \frac{10x^2}{2} + \frac{17x^3}{6} + \frac{26x^4}{24} + \text{etc.}$$
$$= e^x(2 + 3x + xx) = e^x(1 + x)(2 + x).$$

More generally, he noted that when $S = e^x$, the result was

$$A_0 + A_1\frac{x}{1} + A_2\frac{x^2}{1 \cdot 2} + A_3\frac{x^3}{1 \cdot 2 \cdot 3} + A_4\frac{x^4}{1 \cdot 2 \cdot 3 \cdot 4} + \text{etc.}$$
$$= e^x\left(A_0 + \frac{x}{1}\Delta A_0 + \frac{x^2}{1 \cdot 2}\Delta^2 A_0 + \frac{x^3}{1 \cdot 2 \cdot 3}\Delta^3 A_0 + \frac{x^4}{1 \cdot 2 \cdot 3 \cdot 4}\Delta^4 A_0 + \text{etc.}\right).$$

In fact, this result is equivalent to $(1+\Delta)^n A_0 = A_n$, as is quickly verified by writing e^x as a series and multiplying the two series. Jacobi gave a very interesting application of Euler's general transformation formula (11.15) to the derivation of the Pfaff–Gauss transformation for hypergeometric functions and we discuss this in the chapter on algebraic analysis.

11.5 Stirling's Transformation Formulas

Stirling's new generalization of Newton's transformation of the arctangent series (11.4) was a particular case of a hypergeometric transformation discovered by Pfaff in 1797. Stirling stated his formula as proposition 7 of his 1730 book and his proof made remarkable use of difference equations. Beginning with a series satisfying a certain difference equation, or recurrence relation, he then showed that the transformed series satisfied the same difference equation. Adhering closely to Stirling's exposition, we state the theorem: If the successive terms T and T' of a series S satisfied $(z-n)T + (m-1)zT' = 0$, then S could be transformed to

$$S = \frac{m-1}{m}T + \frac{n}{z} \times \frac{A}{m} + \frac{n+1}{z+1} \times \frac{B}{m} + \frac{n+2}{z+2} \times \frac{C}{m} + \frac{n+3}{z+3} \times \frac{D}{m} + \text{etc.}$$

Stirling's notation made unusual use of the symbol z. In the equation $(z-n)T + (m-1)zT' = 0$, T and T' represented any two successive terms of the series S. The value of z changed by one when Stirling moved from one pair to the next. To see how this worked, take $S = T_0 + T_1 + T_2 + T_3 + \cdots$. The initial relation (between the first two terms) could then be expressed as $(z-n)T_0 + (m-1)zT_1 = 0$ and, in general, the relation between two successive terms would be

$$T_{k+1} = -\frac{z-n+k}{(z+k)(m-1)}T_k. \tag{11.30}$$

Thus, the relation between successive terms produced the series

$$S = T_0\left(1 + \frac{z-n}{z(1-m)} + \frac{(z-n)(z-n+1)}{z(z+1)(1-m)^2} + \cdots\right). \tag{11.31}$$

In Stirling's notation, $A, B, C, \ldots$ each represented the previous term so that

$$A = \frac{m-1}{m}T\left(=\frac{m-1}{m}T_0\right),\ B = \frac{n}{z}\times\frac{A}{m},\ C = \frac{n+1}{z+1}\times\frac{B}{m}, \text{etc.}$$

Hence, the transformed series could be written as

$$T_0\left(1 + \frac{n}{z}\cdot\frac{1}{m} + \frac{n(n+1)}{z(z+1)}\cdot\frac{1}{m^2} + \frac{n(n+1)(n+2)}{z(z+1)(z+2)}\cdot\frac{1}{m^3} + \cdots\right). \tag{11.32}$$

Thus, Stirling's transformation formula is equivalent to the statement that the series in (11.31) equals the series in (11.32). We use modern notation to derive Stirling's difference equation. Let

$$S_k = \sum_{n=k}^{\infty} T_n = T_k y_k$$

so that $\quad T_{k+1} y_{k+1} = S_{k+1} = S_k - T_k = T_k(y_k - 1).$

By relation (11.30), the last equation would become

$$(m-1)y_k + y_{k+1} - \frac{n}{z+k}y_{k+1} - m + 1 = 0. \tag{11.33}$$

Since Stirling wrote y and y' for any two successive y_k and y_{k+1}, he could write z instead of $z+k$ to get the recurrence relation

$$(m-1)y + y' - \frac{n}{z}y' - m + 1 = 0. \tag{11.34}$$

In proving his transformation formula, Stirling assumed that

$$y = a + \frac{b}{z} + \frac{c}{z(z+1)} + \frac{d}{z(z+1)(z+2)} + \cdots, \quad \text{so} \tag{11.35}$$

$$y' = a + \frac{b}{z+1} + \frac{c}{(z+1)(z+2)} + \frac{d}{(z+1)(z+2)(z+3)} + \cdots. \tag{11.36}$$

Before substituting these expressions for y and y' in (11.34), Stirling observed that

$$y' = a + \frac{b}{z} + \frac{c-b}{z(z+1)} + \frac{d-2c}{z(z+1)(z+2)} + \frac{e-3d}{z(z+1)(z+2)(z+3)} + \cdots. \tag{11.37}$$

We can easily see this to be true by taking the term-by-term difference of the series for y and y' in (11.35) and (11.36). Next, he substituted the expression (11.37) for y' in (11.34) but used (11.36) for the term $-\frac{n}{z}y'$ in (11.34). After these substitutions, equation (11.34) became

$$ma - m + 1 + \frac{mb - na}{z} + \frac{mc - (n+1)b}{z(z+1)} + \frac{md - (n+2)c}{z(z+1)(z+2)} + \text{etc.} = 0.$$

On setting the like terms equal to zero, he got

$$a = \frac{m-1}{m},\ b = \frac{n}{m}a,\ c = \frac{n+1}{m}b,\ d = \frac{n+2}{m}c, \ldots.$$

11.5 Stirling's Transformation Formulas

When these values were substituted back in y, Stirling got his result.

Stirling applied this transformation to the approximate summation of the series

$$\frac{1}{2}\left(1 + \frac{1}{3} + \frac{1 \cdot 2}{3 \cdot 5} + \frac{1 \cdot 2 \cdot 3}{3 \cdot 5 \cdot 7} + \frac{1 \cdot 2 \cdot 3 \cdot 4}{3 \cdot 5 \cdot 7 \cdot 9} + \cdots\right). \quad (11.38)$$

Now this is the series one gets upon taking $t = 1$ in Newton's formula (11.4), and its value is therefore $\pi/4$. It is interesting to note that, after posing the problem of approximately summing the above series, Newton had abruptly ended the second chapter of his unpublished "Matheseos" of 1684 with the word "Inveniend" (to be found). Thus, although the "Matheseos" went unpublished, Stirling took up the very problem left pending by Newton. Stirling began by adding the first twelve terms of the series and applied his transformation (11.13) to the remaining (infinite) part of the series, yielding

$$\frac{12!}{3 \cdot 5 \cdot 25}\left(1 - \frac{1}{27} + \frac{1 \cdot 3}{27 \cdot 29} - \frac{1 \cdot 3 \cdot 5}{27 \cdot 29 \cdot 31} + \cdots\right). \quad (11.39)$$

To approximate this sum, he took the first twelve terms of this series and found the value of $\pi/4$ as 0.78539816339. Since the terms are alternating and decreasing, Stirling could also have easily determined bounds on the error by using results of Leibniz dating from 1676.

Proposition 8 of Stirling's *Methodus Differentialis* was a transformation of what we now call a generalized hypergeometric series, noteworthy as an important particular case of a formula discovered by Kummer in 1836 in the course of his efforts to generalize Gauss's 1812 theory of hypergeometric series. After having seen the manner in which Stirling stated his propositions, we state Stirling's eighth proposition in a form more immediately understandable to modern readers:

$$\frac{1}{z-n} + \frac{z-m}{z(z-n+1)} + \frac{(z-m)(z-m+1)}{z(z+1)(z-n+2)} + \frac{(z-m)(z-m+1)(z-m+2)}{z(z+1)(z+2)(z-n+3)} + \cdots$$

$$= \frac{1}{m} + \frac{n}{z(m+1)} + \frac{n(n+1)}{z(z+1)(m+2)} + \frac{n(n+1)(n+2)}{z(z+1)(z+2)(m+3)} + \cdots. \quad (11.40)$$

Let S_k denote the sum of the series on the left after the first k terms have been removed:

$$S_k = \frac{(z-m)\cdots(z-m+k-1)}{z(z+1)\cdots(z+k-1)}$$

$$\times \left(\frac{1}{z-n+k} + \frac{z-m+k}{(z+k)(z-n+k+1)} + \frac{(z-m+k)(z-m+k+1)}{(z+k)(z+k+1)(z-n+k+2)} + \cdots\right).$$

Denote the sum in parentheses by y_k. It is simple to check that

$$y_k - y_{k+1} + \frac{m}{z+k}y_{k+1} - \frac{1}{z-n+k} = 0.$$

Stirling wrote this relation as

$$y - y' + \frac{m}{z}y' = \frac{1}{z-n}. \quad (11.41)$$

He then assumed that for some $a, b, c, d, \ldots$

$$y = \frac{a}{m} + \frac{b}{(m+1)z} + \frac{c}{(m+2)z(z+1)} + \frac{d}{(m+3)z(z+1)(z+2)} + \cdots.$$

He substituted this into the left side of (11.41) and after a simple calculation obtained

$$y - y' + \frac{m}{z}y' = \frac{a}{z} + \frac{b}{z(z+1)} + \frac{c}{z(z+1)(z+2)} + \frac{d}{z(z+1)(z+2)(z+3)} + \cdots. \tag{11.42}$$

Stirling applied Taylor's formula (11.12) to the right side of (11.41) to get

$$\frac{1}{z-n} = \frac{1}{z} + \frac{n}{z(z+1)} + \frac{n(n+1)}{z(z+1)(z+2)} + \frac{n(n+1)(n+2)}{z(z+1)(z+2)(z+3)} + \cdots \tag{11.43}$$

and equated the coefficients in (11.42) and (11.43) to obtain

$$a = 1, \ b = n, \ c = n(n+1), \ d = n(n+1)(n+2), \ldots.$$

This proves the transformation formula (11.40). As we mentioned, a century later Kummer obtained a more general result but he did not seem to have been aware of Stirling's work.

Newton pointed out in his second letter for Leibniz that $\sin^{-1}\frac{1}{2}$ was more convenient than $\sin^{-1} 1$ for computing π because it converged rapidly. Stirling showed that by applying his transformation to $\sin^{-1} 1$, he could cause it to converge sufficiently rapidly as to make it useful for computation. Stirling summed up the first twelve terms directly and then applied the transformation to the remainder, thereby achieving the value of π to eight decimal places.

11.6 Nicole's Examples of Sums

The method of finite differences is also useful in the summation of series, as noted by Mengoli, Leibniz, Jakob Bernoulli, and Montmort. Nicole, student of Montmort, wished to establish a new subject devoted to the calculus of finite differences. Analogous to integration in the calculus, summation of series had to be developed within this new subject. Nicole attacked this problem by summing examples of certain kinds of series and Montmort gave similar examples. The basic idea was that, given a function $g(x)$ such that $g(x+h) - g(x) = f(x)$, the sum would be $f(a) + f(a+h) + \cdots + f(a+(n-1)h) = g(a+nh) - g(a)$. When the sum was indefinite, Nicole wrote that the sum of the $f(x)$ was $g(x)$. For example, the sum of the terms $(x+2)(x+4)(x+6)$ was $x(x+2)(x+4)(x+6)/8$ because

$$\frac{(x+2)(x+4)(x+6)(x+8)}{8} - \frac{x(x+2)(x+4)(x+6)}{8} = (x+2)(x+4)(x+6).$$

Similarly, he wrote that the integral (sum) of $1/[x(x+2)(x+4)(x+6)]$ was $1/[6x(x+2)(x+4)]$. As an application of the first type of sum, Nicole considered the series

$$1 \cdot 4 \cdot 7 \cdot 10 + 4 \cdot 7 \cdot 10 \cdot 13 + 7 \cdot 10 \cdot 13 \cdot 16 + 10 \cdot 13 \cdot 16 \cdot 19 + \text{ etc.}$$

He gave the general term as $(x+3)(x+6)(x+9)(x+12)$ and its integral as $x(x+3)(x+6)(x+9)(x+12)/15$. Now to find the constant of integration, note that the starting value was $x = -2$ and the corresponding value of the integral was $(-2) \cdot 1 \cdot 4 \cdot 7 \cdot 10/15$. Thus, Nicole got the value of the series as $x(x+3)(x+6)(x+9)(x+12)/15 + (2 \cdot 1 \cdot 4 \cdot 7 \cdot 10)/15$. Similarly, he computed

$$\frac{1}{1 \cdot 3 \cdot 5 \cdot 7} + \frac{1}{3 \cdot 5 \cdot 7 \cdot 9} + \frac{1}{5 \cdot 7 \cdot 9 \cdot 11} + \frac{1}{7 \cdot 9 \cdot 11 \cdot 13}$$

$$+ \cdots + \frac{1}{x(x+2)(x+4)(x+6)} \text{ etc.}$$

by observing that

$$\frac{1}{x(x+2)(x+4)(x+6)} + \frac{1}{(x+2)(x+4)(x+6)(x+8)} + \text{etc.}$$

$$= \frac{1}{6(x+2)(x+4)(x+6)}.$$

Since the original sum started at $x = 1$, Nicole gave its value as $1/(6 \cdot 1 \cdot 3 \cdot 5) = 1/90$. As another example, Nicole then considered a slightly more difficult sum:

$$\frac{4}{1 \cdot 4 \cdot 7 \cdot 10 \cdot 13 \cdot 16} + \frac{49}{4 \cdot 7 \cdots 19} + \frac{225}{7 \cdots 22} + \text{etc.}$$

Note that the general term was $\frac{1}{36} \frac{(x+2)^2(x+3)^2}{x(x+3)\cdots(x+15)}$, where x was replaced by $x = 3$ as one moved from one term to the next. Nicole wrote the numerator as

$$\frac{1}{36}(A + Bx + Cx(x+3) + Dx(x+3)(x+6) + Ex(x+3)(x+6)(x+9))$$

and found the coefficients A to E by taking $x = 0, -3, -6,$ and -9. Nicole then expressed the general term as

$$\frac{1}{36} \left(\frac{A}{x(x+3)\cdots(x+15)} + \cdots + \frac{E}{(x+12)(x+15)} \right)$$

and wrote the integral as

$$\frac{1}{36} \left(\frac{A}{15x(x+3)\cdots(x+12)} + \cdots + \frac{E}{3(x+12)} \right). \tag{11.44}$$

He found the sum by taking $x = 1$ in the sum, or integral, as he called it.

11.7 Stirling Numbers

In the introduction to his *Methodus Differentialis*, Stirling explained that the series satisfying difference equations were best expressed by using terms of the factorial form $x(x-1)\cdots(x-m+1)$ or $1/x(x+1)\cdots(x+m-1)$ instead of x^m or $1/x^m$. He defined the Stirling numbers in order to facilitate the conversion of series expressed in powers of x into series with terms in factorial form. He gave a table for the Stirling numbers of the first kind.

1							
1	1						
2	3	1					
6	11	6	1				
24	50	35	10	1			
120	274	225	85	15	1		
720	1764	1624	735	175	21	1	
5040	13068	13132	6769	1960	322	28	1
40320	109584	118124	67284	22449	4536	546	36	1

Stirling described how to construct the table: "Multiply the terms of this progression $n, 1+n, 2+n, 3+n$, etc. repeatedly by themselves, and let the results be arranged in the following table in order of the powers of the number n, only the coefficients having been retained." Thus, to get the fourth row take $n(n+1)(n+2)(n+3) = 6n + 11n^2 + 6n^3 + n^4$ and the coefficients will be the numbers 6, 11, 6, 1 in the fourth row.

Stirling applied these numbers to the expansion of $1/z^{n+1}, n = 1, 2, 3, \ldots$ as an inverse factorial series. The numbers in the first column then appeared as numerators in the expansion of $1/z^2$; the numbers in the second column appeared in that of $1/z^3$, and so on. He wrote down three expansions explicitly:

$$\frac{1}{z^2} = \frac{1}{z(z+1)} \left(1 + \frac{1}{z+2} + \frac{2}{(z+2)(z+3)} + \frac{6}{(z+2)(z+3)(z+4)} + \cdots \right)$$

$$\frac{1}{z^3} = \frac{1}{z(z+1)(z+2)} \left(1 + \frac{3}{z+3} + \frac{11}{(z+3)(z+4)} + \cdots \right)$$

$$\frac{1}{z^4} = \frac{1}{z(z+1)(z+2)(z+3)}$$
$$\times \left(1 + \frac{6}{z+4} + \frac{35}{(z+4)(z+5)} + \frac{225}{(z+4)(z+5)(z+6)} + \&c. \right).$$

Since the inverse factorial series could be summed by the Montmort-Nicole method mentioned earlier, Stirling could apply the preceding formulas to the approximate evaluation of the series $\sum_{k=1}^{\infty} 1/k^n$. Observe that for $n = 2$,

$$\sum_{k=z}^{\infty} \frac{1}{k^2} = \sum_{k=z}^{\infty} \frac{1}{k(k+1)} + \sum_{k=z}^{\infty} \frac{1}{k(k+1)(k+2)} + \sum_{k=z}^{\infty} \frac{2}{k(k+1)(k+2)(k+3)} + \cdots$$

$$= \frac{1}{z} + \frac{1}{2z(z+1)} + \frac{2}{3z(z+1)(z+2)} + \cdots.$$

Stirling took $z = 13$ and then took thirteen terms of the last series to get 0.079957427, and when he added this quantity to $\sum_{k=1}^{12} 1/k^2$, a sum approximately equal to 1.564976638, he had the approximate value 1.644934065 for the series $1 + \frac{1}{4} + \frac{1}{9} + \frac{1}{16} +$ etc. Euler showed that this series was equivalent to $\pi^2/6$, implying that Stirling's value was correct to eight decimal places.

As defined today, the Stirling numbers of the first kind $S(m,k)$ are given by

$$x(x-1)(x-2)\cdots(x-m+1) = \sum_{k=1}^{m} S(m,k)x^k, \qquad (11.45)$$

so we can see that Stirling took his numbers as absolute values $|S(m,k)|$. Stirling's series expansion can then be expressed in modern notation as

$$\frac{1}{z^{n+1}} = \sum_{m=n}^{\infty} \frac{|S(m,n)|}{z(z+1)\cdots(z+m)}. \qquad (11.46)$$

Stirling defined numbers of the second kind $s(m,k)$ by the equation

$$x^m = \sum_{k=1}^{m} s(m,k)x(x-1)\cdots(x-k+1). \qquad (11.47)$$

The introduction to his book also contains a table of these numbers, as shown here.

1	1	1	1	1	1	1	1	1
	1	3	7	15	31	63	127	255
		1	6	25	90	301	966	3025
			1	10	65	350	1701	7770
				1	15	140	1050	6951
					1	21	266	2646
						1	28	462
							1	36
								1

Using his table, Stirling wrote down the expansion for z, z^2, z^3, z^4 and z^5. For example, the fourth column gave him the expansion for z^4:

$$z^4 = z + 7z(z-1) + 6z(z-1)(z-2) + z(z-1)(z-2)(z-3).$$

It is clearly not easy to determine the Stirling numbers of the second kind from his definition. Stirling therefore provided a generating function to find these numbers:

$$\frac{1}{(x-1)(x-2)\cdots(x-m)} = \sum_{n=m}^{\infty} \frac{s(m,n)}{x^n}. \qquad (11.48)$$

The significance of the Stirling numbers was not fully realized until the twentieth century when they became very useful in combinatorics. In his 1939 book on the calculus of finite differences, the Hungarian mathematician Charles Jordan (1871–1959) wrote, "Since Stirling's numbers are as important as Bernoulli's, or even more so, they should occupy a central position in the Calculus of Finite Differences. The demonstration of a great number of formulae is considerably shortened by using these numbers, and many new formulae are obtained by their aid; for instance, those which express differences by derivatives or vice versa."

11.8 Lagrange's Proof of Wilson's Theorem

Lagrange was the first mathematician to investigate the arithmetical properties of Stirling numbers and he did so in the process of proving Wilson's theorem. This proposition, also found by al-Haytham in the tenth or eleventh century, named after Edward Waring's best friend John Wilson (1741–1793), states that for $n > 1$, $(n-1)! + 1$ is divisible by n if and only if n is prime. The statement of Wilson's theorem was first published in Waring's *Meditationes Algebraicae* of 1770. Waring was certain of the truth of the theorem but was unable to prove it. Lagrange provided a proof, using Stirling numbers, in 1771. For this purpose, he considered the product

$$(x+1)(x+2)(x+3)(x+4)\cdots(x+n-1)$$
$$= x^{n-1} + A'x^{n-2} + A''x^{n-3} + A'''x^{n-4} + \cdots + A^{(n-1)}.$$

We can see that the coefficients $A', A'', A''', \ldots A^{(n-1)}$ are in fact absolute values of Stirling numbers of the first kind, though Lagrange did not mention Stirling. Lagrange replaced x by $x+1$ to get

$$(x+2)(x+3)(x+4)(x+5)\cdots(x+n)$$
$$= (x+1)^{n-1} + A'(x+1)^{n-2} + A''(x+1)^{n-3} + A'''(x+1)^{n-4} + \cdots + A^{(n-1)}.$$

It was then easy to see that

$$(x+n)(x^{n-1} + A'x^{n-2} + A''x^{n-3} + A'''x^{n-4} + \cdots + A^{(n-1)})$$
$$= (x+1)^n + A'(x+1)^{n-1} + A''(x+1)^{n-2} + A'''(x+1)^{n-3} + \cdots + A^{(n-1)}(x+1).$$

Expanding both sides of this equation in powers of x Lagrange obtained

$$x^n + (n+A')x^{n-1} + (nA'+A'')x^{n-2} + (nA''+A''')x^{n-3} + \cdots$$
$$= x^n + (n+A')x^{n-1} + \left(\frac{n(n-1)}{2} + (n-1)A' + A''\right)x^{n-2}$$
$$+ \left(\frac{n(n-1)(n-2)}{2\cdot 3} + \frac{(n-1)(n-2)}{2}A' + (n-2)A'' + A'''\right)x^{n-3} + \cdots.$$

He next equated the coefficients on both sides to get recurrence relations for the Stirling numbers of the first kind:

$$n + A' = n + A',$$

$$nA' + A'' = \frac{n(n-1)}{2} + (n-1)A' + A'',$$

$$nA'' + A''' = \frac{n(n-1)(n-2)}{2\cdot 3} + \frac{(n-1)(n-2)}{2}A' + (n-2)A'' + A''', \cdots,$$

or

$$A' = \frac{n(n-1)}{2},$$

$$2A'' = \frac{n(n-1)(n-2)}{2\cdot 3} + \frac{(n-1)(n-2)}{2}A',$$

$$3A''' = \frac{n(n-1)(n-2)(n-3)}{2\cdot 3\cdot 4} + \frac{(n-1)(n-2)(n-3)}{2\cdot 3}A'$$

$$+ \frac{(n-2)(n-3)}{2}A'', \cdots$$

$$(n-1)A^{(n-1)} = 1 + A' + A'' + \cdots + A^{(n-2)}.$$

Lagrange noted that if n were an odd prime p, then the first equation (in the second set of equations) showed that A' was divisible by p; the third equation showed A'' divisible by p, and so on. The last but one equation implied that p divided $A^{(n-2)}$. Next, observing that $A^{(n-1)} = (n-1)! = (p-1)!$, Lagrange perceived that the last equation implied Wilson's theorem that $A^{(n-1)} + 1 = (p-1)! + 1$ was divisible by p. As an application of this theorem, Lagrange determined the quadratic character of -1 modulo a prime p. That is, he deduced that if p were a prime of the form $4n+1$, then there had to exist an integer x such that $x^2 + 1$ was divisible by p. Note that Euler had given a remarkable proof of this result using repeated differences of the sequence 1^n, 2^n, 3^n, 4^n, ….

11.9 Taylor's Summation by Parts

The method of summation by parts is usually attributed to Abel who used it in a rigorous discussion of convergence of series. However, a century earlier, in the 1717 *Philosophical Transactions*, Taylor explicitly worked out this idea as an analog of integration by parts. Moreover, one can see in the work of Newton, Leibniz, and others that they were implicitly aware of this method.

Taylor's result is actually an indefinite summation formula in which the constants of summation are not explicitly written. Because Taylor's presentation is obscure, we present the 1819 derivation below from Lacroix in which $\sum$ and Δ were taken as inverse operations. In Lacroix's notation, P and Q were functions of an integer variable x and $P_1 = P + \Delta P$, $P_2 = P_1 + \Delta P_1$, etc. In this notation, Taylor's formula took the form

$$\sum PQ = Q\sum P - \sum(\Delta Q \sum P_1). \tag{11.49}$$

To prove this following Lacroix, first suppose that

$$\sum PQ = Q \sum P + z.$$

Apply the difference operator Δ to both sides to get

$$\Delta(\sum PQ) = \Delta(Q \sum P + z),$$

$$\text{or } PQ = (Q + \Delta Q) \sum (P + \Delta P) - Q \sum P + \Delta z$$

$$= Q \sum \Delta P + \Delta Q \sum (P + \Delta P) + \Delta z$$

$$= QP + \Delta Q \sum (P + \Delta P) + \Delta z.$$

Hence,

$$\Delta z = -\Delta Q \sum (P + \Delta P) \text{ or } z = -\sum (\Delta Q \sum (P + \Delta P)) = -\sum (\Delta Q \sum P_1),$$

and Lacroix's proof of Taylor's formula (11.49) was complete. In his book, Lacroix attributed the result to Taylor.

Just as one can perform repeated integration by parts, one may also do repeated summation by parts if necessary. Thus, Lacroix gave this formula for repeated summation by parts:

$$\sum PQ = Q \sum P - \Delta \sum{}^2 P_1 + \Delta^2 Q \sum{}^3 P_2 - \Delta^3 Q \sum{}^4 P_3 + \text{etc.} \qquad (11.50)$$

where $\sum^2, \sum^3, \ldots$ denoted double, triple, $\ldots$ summation. To derive this formula, he replaced Q by ΔQ and P by $\sum P_1$ in (11.49). He then had

$$\sum (\Delta Q \sum P_1) = \Delta Q \sum{}^2 P_1 - \sum (\Delta^2 Q \sum{}^2 P_2).$$

Combining this with (11.49), he obtained

$$\sum PQ = Q \sum P - \Delta Q \sum{}^2 P_1 + \sum (\Delta^2 Q \sum{}^2 P_2).$$

A continuation of this process would yield formula (11.50).

11.10 Exercises

1. Show that for a finite sequence of positive decreasing numbers $a_0, a_1, \ldots, a_n$

$$a_0 = \sum_{i=0}^{n-1} (a_i - a_{i+1}) + a_n.$$

The sequence can be infinite; in that case "the last number" of the sequence a_n should be replaced by the limit of the sequence. Then deduce the sum of the convergent infinite geometric series. For a reference to this 1644 result of Torricelli, see Weil (1989b).

2. Use the inequality
$$\frac{1}{a-1} + \frac{1}{a} + \frac{1}{a+1} > \frac{3}{a}$$
to prove that $1 + \frac{1}{2} + \frac{1}{3} + \frac{1}{4} + \frac{1}{5} + \cdots$
diverges. Apply Torricelli's formula in the previous problem to sum
$$\frac{1}{3} + \frac{1}{6} + \frac{1}{10} + \frac{1}{15} + \cdots.$$
See Weil (1989b) for the reference to these 1650 results of Mengoli.

3. Show that
$$\frac{1}{3} + \frac{1}{15} + \frac{1}{35} + \frac{1}{63} + \frac{1}{99} + \cdots = \frac{1}{2}.$$
Leibniz also mentioned this result in his *Historia et Origo* of 1714, written in connection with the calculus controversy, where he explained that in a 1682 article in the *Acta Eruditorum*, he had extended the inverse relationship between differences and sums to differentials and integrals. See Leibniz (1971), vol. 5, p. 122.

4. Find the sums of the reciprocals of the figurate numbers. For example, the sum of the reciprocals of the pyramidal numbers is given as
$$\frac{1}{1} + \frac{1}{4} + \frac{1}{10} + \frac{1}{20} + \frac{1}{35} + \cdots = \frac{3}{2}.$$
See Jakob Bernoulli (1993–99), vol. 4, p. 66.

5. Show that if
$$x = \frac{1}{1^m} + \frac{1}{2^m} + \frac{1}{3^m} + \frac{1}{4^m} + \cdots \text{ and } y = \frac{1}{a^m} + \frac{1}{3^m} + \frac{1}{5^m} + \cdots,$$
then $x - y = \frac{x}{2^m}$.

See Jakob Bernoulli (1993–99), vol. 4, p. 74.

6. Let m, n, and p be integers and $x = a + mn$. Show that
$$\sum_{k=0}^{m}(a+kn)(a+(k+1)n)\cdots(a+(k+p-1)n)$$
$$= \frac{x(x+n)\cdots(x+pn) - (a-n)a(a+n)\cdots(a+(p-1)n)}{(p+1)n}.$$
Deduce the values of the sums
$$1 + 2 + 3 + 4 + \cdots + x,$$
$$1 + 3 + 6 + 10 + \cdots \text{ to } x \text{ terms,}$$
$$1 \cdot 3 \cdot 5 + 3 \cdot 5 \cdot 7 + 5 \cdot 7 \cdot 9 + \cdots \text{ to } x \text{ terms.}$$

7. Sum the series

$$\frac{5}{3\cdot 5\cdot 7\cdot 9\cdot 11\cdot 13} + \frac{41}{5\cdot 7\cdot 9\cdot 11\cdot 13\cdot 15}$$
$$+ \frac{131}{7\cdot 9\cdot 11\cdot 13\cdot 15\cdot 17} + \frac{275}{9\cdots 19} + \frac{473}{11\cdots 21} + \cdots.$$

Montmort (1717) computed the sum to be $\frac{283}{80\cdot 3\cdot 5\cdot 7\cdot 9\cdot 11}$.

8. Prove Taylor's summation formula (11.12) by an application of the Harriot–Briggs, usually known as the Gregory–Newton, formula.
9. Find the values of $A, B, \ldots, E$ to obtain the sum of Nicole's series (11.44).
10. Prove Stirling's formula (11.46). Note that Stirling stated this without proof in the introduction to his *Methodus Differentialis*.
11. Prove Stirling's generating function formula (11.48) for Stirling numbers of the second kind.
12. Derive Gregory's formula (11.1) from the Newton-Gauss interpolation.
13. Use Wilson's theorem to prove that if $p = 4n+1$ is a prime, there exists an integer x such that p divides $x^2 + 1$. See Lagrange (1867–1892) vol. 3, pp. 425–438.
14. Prove Wilson's theorem by using Stirling numbers of the first kind and Fermat's little theorem that $z^{p-1} \equiv 1 \pmod{p}$ when p is prime and a is an integer not divisible by p:

    (a) Let
    $$(x-1)(x-2)\cdots(x-p+1) = x^{p-1} + A_1 x^{p-2} + A_x x^{p-3} + \cdots + A_{p-1},$$
    and $A_0 = 1 + A_{p-1}$. Observe that $x^{p-1} + A_{p-1} \equiv A_0 \pmod{p}$ for $x = 1, 2, \ldots, p-1$. Prove that
    $$A_0 + k^{p-2} A_1 + k^{p-3} A_2 + \cdots + k A_{p-2} \equiv 0 \pmod{p}$$
    for $k = 1, 2, \ldots, p-1$.

    (b) Show that the determinant of the system of $p-1$ equations in $A_0, A_1, \ldots, A_{p-2}$ has a nonzero determinant modulo p.

    (c) Deduce that $A_0 \equiv 0, A_1 \equiv 0, \ldots, A_{p-2} \equiv 0 \pmod{p}$. Sylvester published this result in 1854. See Sylvester (1973), vol. 2, p. 10.

15. Show that if a prime $p = 4n+1$, the $2n$th difference of the sequence
    $$1^{2n}, 2^{2n}, \ldots, (p-1)^{4n}$$
    is not divisible by p. Deduce that $a^{2n} - 1$ is not divisible by p for all $1 \le a \le p-1$. For this result of Euler, see his correspondence in Fuss (1968), p. 494. See also Weil (1984), p. 65, and Edwards (1977), p. 47.

11.11 Notes on the Literature

Gregory's November 23, 1670 letter contains several mathematical discoveries, and we have mentioned some in earlier chapters. The letter can be found in Turnbull (1939). Turnbull (1933) shows how, from the Newton–Bessel formula, Gregory could have derived some of his series. Newton's "De Computo" and "Matheseos" were first published in vol. 4 of Newton (1967–1981). This volume covers the period 1674–1684 of Newton's mathematical researches, including his work on finite differences and their applications to infinite series. The quotation containing equation (11.5) is from pp. 611–613 of this volume; the other quotation is on pp. 605–607; they are translations by Whiteside from Newton's original in Latin. Also note Whiteside's footnotes on these pages.

Montmort (1717) contains all his results mentioned in this chapter, including his version of the transformation formula. For Zhu Shijie's use of (11.23), see Hoe (2007), p. 401. Nicole (1717) has the examples we discuss; see Tweedie (1917–1918) for an evaluation of Nicole's researches in the calculus of finite differences. Tweddle (2003), an annotated English translation of Stirling's original *Methodus Differentialis*, gives an extensive discussion of propositions 7 and 8 of Stirling, including a numerical analysis of Stirling's examples. See the same work for the Stirling numbers. Whittaker may have been the first to notice the connection of proposition 7 with the transformation of hypergeometric series. See p. 286 of Whittaker and Watson (1927).

Euler's 1755 book on differential calculus has been reprinted in Eu. I-10. For J. D. Blanton's English translation of the first part of this book, including results on finite differences, see Euler (2000). Taylor's summation by parts is taken from pp. 91–92 of Lacroix (1819), and Taylor's formula (11.12) is mentioned in Bateman (1907). Finally, Lagrange's proof of Wilson's theorem is in Lagrange (1867–1892), vol. 3, pp. 425–438.

12

The Taylor Series

12.1 Preliminary Remarks

In 1715, Brook Taylor (1685–1731) published one of the most basic results in the theory of infinite series, now known as Taylor's formula. Taylor published his formula fifty years after Newton's seminal work on series and twenty-five years after Newton discovered, but did not publish, this same formula. In modern notation, Taylor's formula takes the form

$$f(x) = f(a) + (x-a)\frac{f'(a)}{1!} + (x-a)^2\frac{f''(a)}{2!} + \cdots. \qquad (12.1)$$

Taylor communicated this result to John Machin in a letter of July 26, 1712. We note here that in 1706, Machin calculated π to 100 digits by employing the formula

$$\frac{\pi}{4} = 4\arctan\frac{1}{5} - \arctan\frac{1}{239}.$$

In his letter to Machin, Taylor wrote

> I fell into a general method of applying Dr. Halley's Extraction of roots to all Problems, wherein the Abscissa is required, the Area being given which, for the service that it may be of calculations, (the only true use of all corrections) I cannot conceal. And it is comprehended in this Theorem. ... If α be any compound of the powers of z and given quantities whether by a finite or infinite expression rational or surd. And β be the like compound of p and the same coefficients, and $z = p + x$, and $\dot{p} = 1 = \dot{z}$. Then will
>
> $$\alpha - \beta = \frac{\dot{\beta}}{1}x + \frac{\ddot{\beta}}{1\cdot 2}x^2 + \frac{\dddot{\beta}}{1\cdot 2\cdot 3}x^3 + \frac{\ddddot{\beta}}{1\cdot 2\cdot 3\cdot 4}x^4 \quad \&c.$$
> $$= \frac{\dot{\alpha}}{1}x - \frac{\ddot{\alpha}}{1\cdot 2}x^2 + \frac{\dddot{\alpha}}{1\cdot 2\cdot 3}x^3 - \frac{\ddddot{\alpha}}{1\cdot 2\cdot 3\cdot 4}x^4 \quad \&c.$$
>
> Where $\dot{\alpha}, \ddot{\alpha}, \dddot{\alpha}$ &c. are formed in the same manner of z and the given quantities, as $\dot{\beta}, \ddot{\beta}$, &c. are formed of p. &c. So that having given α, β, and one of the abscissae z or p, the other may be found by extracting x, their difference, out of this aequation. Or you may apply this to the invention of α or β, having given z, p and x. But you will easily see the uses of this.

12.1 Preliminary Remarks

Newton discovered and extensively used infinite series in the period 1664–1670, but during that time it does not appear that he observed the connection between the derivatives of a function and the coefficients of its series expansion. This connection is the essence of the Taylor series, and it can be applied to obtain power series expansions of many functions. Newton discovered infinite series before he had investigated the concept of derivatives. Thus, he found expansions for several functions by using his binomial expansion, term-by-term integration, and reversion of series. He first indicated his awareness of the connection between derivatives and coefficients in his 1687 *Principia*, wherein he expanded $(e^2 - 2ao - o^2)^{\frac{1}{2}}$ in powers of o and interpreted the coefficients as geometric quantities directly related to the derivatives of the function. It is very possible that Newton was aware of the Taylor expansion at this time. In fact, forty years later, this *Principia* result inspired James Stirling to consider whether it could be generalized, leading him to an independent discovery of the Maclaurin series, published in 1717 in the *Philosophical Transactions*.

In 1691–92, Newton gave an explicit statement of Taylor's formula as well as the particular case now named for Maclaurin. These appear in his *De Quadratura Curvarum*, composed in the winter of 1691–92 and never fully completed; in 1704 parts of this text were published under the title *Tractatus de Quadratura Curvarum*. Unfortunately, the published portions omitted the Taylor and Maclaurin theorems. As we shall see, Gregory used this result in 1670 to construct series for numerous functions, but Newton was the first to give its clear, though unpublished, statement. In his *De Quadratura*, finally published in 1976 by Whiteside, Newton discussed the problem of solving algebraic differential equations by means of infinite series. In this context, he stated the Taylor and Maclaurin expansions and then wrote the word "Example" and left a blank space. Apparently, he intended to give a solution of an algebraic differential equation but could not hit upon a satisfactory example. According to Whiteside, Newton's worksheets from this period show that he made several attempts to solve the equation $\sqrt{1 + \dot{y}^2} \times \ddot{y} = n\dot{\ddot{y}}$ without complete success. This may explain why he did not include these results in the published work, although his corollaries three and four contain Newton's own formulation of the Taylor and Maclaurin series:

Corollary 3. Hence, indeed, if the series proves to be of this form

$$y = az + bz^2 + cz^3 + dz^4 + ez^5 + \text{etc.}$$

(where any of the terms a, b, c, d etc. can either be lacking or be negative), the fluxions of y, when z vanishes, are had by setting $\dot{y}/\dot{z} = a$, $\ddot{y}/\dot{z}^2 = 2b$, $\dddot{y}/\dot{z}^3 = 6c$, $\ddddot{y}/\dot{z}^4 = 24d$, $\dddddot{y}/\dot{z}^5 = 120e$. *Corollary* 4. And hence if in the equation to be resolved there be written $w + x$ for z, as in Case 3, and by resolving the equation there should result the series $[y =]ex + fx^2 + gx^3 + hx^4 +$ etc. the fluxions of y for any assumed magnitude of z whatever will be obtained in finite equations by setting $x = 0$ and so $z = w$. For the equations of this sort gathered by the previous Corollary will be $\dot{y}/\dot{z} = e$, $\ddot{y}/\dot{z}^2 = 2f$, $\dddot{y}/\dot{z}^3 = 6g$, $\ddddot{y}/\dot{z}^4 = 24h$ etc.

Even before Newton, the Scottish mathematician James Gregory discovered and used Maclaurin's series to obtain power series expansions of some fairly complicated functions. In a letter to John Collins dated February 15, 1671, Gregory gave the series expansions for $\arctan x$, $\tan x$, $\sec x$, $\ln \sec x$, $\ln \tan(\frac{\pi}{4} + \frac{x}{2})$, $\text{arcsec}(\sqrt{2}e^x)$,

$2\arctan(\tanh(x/2))$. Of the seven series, the first two are inverses of each other, as are the fifth and the seventh; the fourth and sixth are inverses of each other, except for a constant factor. It does not seem that Gregory applied reversion of series, a technique used by Newton, to obtain the inverses. On the back of a letter from Shaw, Gregory noted the first few derivatives of some of the seven functions. From this, we can see that Gregory derived the series for the second, third, sixth, and seventh functions by taking the derivatives; the series for the fourth and fifth using term-by-term integration of the series for the second and third functions; and the series for $\arctan x$ by integration of the series for $1/(1+x^2)$. As we shall see below, a key mistake in Gregory's calculations gives us evidence that he used the derivatives of a function to find its series.

Gregory knew that Newton had made remarkable advances in the theory of series, though he had seen only one example of Newton's work. He concluded that Newton must have known Taylor's expansion, since that could be used to find the power series expansion of any known function. Before giving his seven series expansions to Collins, Gregory wrote, "As for Mr. Newton's universal method, I imagine I have some knowledge of it, both as to geometrik and mechanick curves, however I thank you for the series [of Newton] ye sent me, and send you these following in requital."

In 1694, Johann Bernoulli published a result in the *Acta Eruditorum* equivalent to Taylor's formula, though it did not as easily produce the series expansion:

$$\int n\,dz = nz - \frac{z^2}{1\cdot 2}\frac{dn}{dz} + \frac{z^3}{1\cdot 2\cdot 3}\frac{d\,dn}{dz^2} - \frac{z^4}{1\cdot 2\cdot 3\cdot 4}\frac{d^3n}{dz^3} + \text{etc.} \qquad (12.2)$$

He used this to solve first-order differential equations by infinite series and also applied it to find the series for $\sin x$ and $\ln(a+x)$ and to solve de Beaune's problem concerning the curve whose subtangent remained a constant. Unfortunately, Bernoulli could not use this formula to derive the series for $\sin x$; he obtained only a ratio of two series for $y/\sqrt{a^2-y^2}$ where $y = a\sin x$. He commented that, though this method had this drawback, it was commendable for its universality. Bernoulli communicated his series to Leibniz in a letter of September 2, 1694, before the paper was printed. In reply, Leibniz observed that he had obtained similar results almost two decades earlier by using the method of differences of varying orders. He gave a detailed exposition of how that method would produce Bernoulli's series. We note that in 1704, Abraham de Moivre published an alternative proof of Bernoulli's series and four years later he communicated this to Bernoulli.

It seems very likely that Gregory derived the Taylor expansion from the Gregory–Newton, or rather, the Harriot-Briggs, interpolation formula:

$$f(x) = f(a) + \frac{(x-a)}{h}\Delta f(a) + \frac{(x-a)(x-a-h)}{2!h^2}\Delta^2 f(a) + \cdots,$$

where $\Delta f(a) = f(a+h) - f(a)$, $\Delta^2 f(a) = f(a+2h) - 2f(a+h) + f(a)$, ... As $h \to 0$, the number of interpolating points tends to infinity, and $\Delta f(a)/h \to f'(a)$, $\Delta^2 f(a)/h^2 \to f''(a)$, ... The resulting series is Taylor's expansion. This proof is not rigorous, but the same argument was given by de Moivre in his letter to Bernoulli and then again by Taylor in his *Methodus Incrementorum Directa et Inversa* of 1715. Leibniz too started with a formula involving finite differences to derive Bernoulli's series. On the other hand, the unpublished argument of Newton, also independently found by Stirling and Maclaurin, assumed that the function had a series expansion and then, by repeated differentiation of the equation, showed that the coefficients of the series were the derivatives of the function computed at specific values. This is called the method of undetermined coefficients.

We can see that there were three different methods by which the early researchers on the Taylor series discovered the expansion: (a) the method of taking the limit of an appropriate finite difference formula, by Gregory, Leibniz, de Moivre, and Taylor, (b) the method of undetermined coefficients, by Newton, Stirling, and Maclaurin, and (c) repeated integration by parts, or, equivalently, repeated use of the product rule, by Johann Bernoulli.

Infinite series, including power series, were used extensively in the eighteenth century for numerical calculations. On the basis of considerable experience, mathematicians usually had a good idea of the accuracy of their results even though they did not perform error analyses. It was only in the second half of the eighteenth century that a few mathematicians started considering an explicit error term. In the specific case of a binomial series, Jean d'Alembert (1717–1783) obtained bounds for the remainder of the series after the first n terms. In 1754, he also found a more general but not very useful method by which he expressed the remainder of a Taylor series using an iterative process, and when worked out, this would have resulted in an n-fold iterated integral. Surprisingly, in 1693 Newton proved a result that converted an iterated integral into a single integral. If d'Alembert had used this, he would have obtained the remainder given in many textbooks; Lagrange, de Prony, and Laplace discovered this remainder term using a different method.

In an undated manuscript determined by Whiteside to date from 1693, apparently written while he was revising *De Quadratura*, Newton worked out the nth fluent (integral) of $\dot{y}$, the fluxion of y. The formula in modern notation takes the form

$$\frac{1}{n!}\left(z^n \int_0^z \dot{y}\, dt - nz^{n-1}\int_0^z t\dot{y}\, dt + \frac{n(n-1)}{2}z^{n-2}\int_0^z t^2\dot{y}\, dt - \cdots\right)$$
$$+ a_0\frac{z^n}{n!} + a_1\frac{z^{n-1}}{(n-1)!} + a_2\frac{z^{n-2}}{(n-2)!} + \cdots + a_n.$$

The expression without the polynomial part can be simplified by the binomial theorem so that we have

$$\frac{1}{n!}\int_0^z (z-t)^n \dot{y}\, dt.$$

If we instead take the nth iterated integral of y, then this expression takes the form

$$\frac{1}{(n-1)!}\int_0^z (z-t)^{n-1} y\, dt. \tag{12.3}$$

Newton's result is but one step away from Taylor's formula with the remainder as an integral. Compare Newton's result with Cauchy's work on the equation $\frac{d^n y}{dx^n} = f(x)$, given later in this chapter. It is not clear whether Newton was aware that the Taylor series followed easily from his result, but he certainly revised his monograph in that connection. Thus, it is possible that Newton was aware of the relation between his integral and Taylor series. Interestingly, Newton included an equivalent of this result in geometric garb, but without proof, in his 1704 *Tractatus*. In 1727, Benjamin Robins published a proof in the *Philosophical Transactions*.

Newton's formula for the nth iterate of an integral appears to have escaped the notice of the continental mathematicians of the eighteenth century. Thus, it remained for Lagrange to discover an expression for the remainder term. This appeared in his *Théorie des fonctions analytiques* of 1797. Owing to his algebraic conception of the calculus, Lagrange avoided the use of integrals in this work. So, to find bounds for the remainder, he wrote down an expression for its derivative. In a later work of 1801 entitled *Leçons sur les calcul des fonctions*, he generalized the mean value theorem and consequently obtained the well-known expression of the remainder as an nth derivative, now called the Lagrange remainder. He applied this to a discussion of the maximum or minimum of a function and also to his theory of the degree of contact between two curves. Without defining area, he also used the remainder to prove that the derivative of the area was the function itself.

Though the integral form of the remainder followed immediately from his work, Lagrange never explicitly stated it. In his lectures of 1823, Cauchy wrote that in 1805 Gaspard Riche de Prony (1755–1839) used integration by parts to obtain Taylor's theorem with the integral remainder. De Prony was a noted mathematician of his time and taught at the École Polytechnique. He is now remembered as a leader in the construction of mathematical tables. To fill the need for the numerous human calculators required for this process, de Prony gave training in arithmetic to many hairdressers, left unemployed by the French Revolution. Pierre-Simon Laplace (1749–1827) included a derivation of Taylor's theorem with remainder, using integration by parts, in his famous *Théorie analytique des probabilités*, published in 1812. The third volume of Lacroix's book on calculus, of 1819, referred to Laplace but not to de Prony; Cauchy may have mentioned de Prony in order to set the record straight.

Lagrange's derivation of the remainder had significant gaps, though his outline was essentially correct. He regarded it as well known and therefore did not provide a proof that functions – by which he meant continuous functions, though he did not define continuity – had the intermediate value property as well as the maximum value property on a closed interval. It was not until about 1817 that Bolzano and Cauchy gave a precise definition of the continuity of a function and proved the intermediate value property. Bolzano's definition, similar to our modern definition, was that $f(x)$ was continuous if the difference $f(x+\omega) - f(x)$ could be made smaller than any given quantity, with ω chosen as small as desired.

Bernard Bolzano studied philosophy and mathematics at Charles University in Prague from 1796 through 1800. Although he did not particularly enjoy his mathematics courses, he studied the work of Euler and Lagrange on his own. However, Bolzano was fully converted to mathematics by the study of Eudoxus in Eulcid's *Elements*. Bolzano served as professor of theology at Prague from 1807 to 1819 but published several mathematics papers during this period, including his 1817 work on the intermediate value theorem. He based this theorem on his lemma that if a property M were true for all x less than u, but not for all x, then, among all values of u for which this was true, there existed a greatest, U. To prove this lemma, he applied a form of the Bolzano–Weierstrass theorem; lacking a complete theory of real numbers, he could not prove this last result. In the 1830s, realizing that there was a need for such a theory, Bolzano made an unsuccessful attempt to develop it. However, Bolzano's insight into real analysis was deep; he was the first mathematician to construct an example of a continuous nowhere differentiable function. In 1861, Karl Weierstrass finally made use of Cauchy's method of repeated division of the interval to prove the existence of a maximum (or minimum) value. Hilbert commented that the establishment of this last theorem created an indispensable tool for refined analytical and arithmetical investigations.

In modern calculus books, the remainder term for the Taylor series of a function $f(x)$ is used to determine the values of x for which the series represents the function. This approach is due to Cauchy; in his courses at the École Polytechnique in the 1820s, he used this method to find series for the elementary functions. Cauchy's use of the remainder term for this purpose was consistent with his pursuit of rigor; we also note that in 1822 he discovered and published the fact that all the derivatives at zero of the function $f(x) = e^{-1/x^2}$ when $x \neq 0$ and $f(0) = 0$ were equal to zero. Thus, the Taylor series at $x = 0$ was identically zero; it therefore represented the function only at $x = 0$. This example would have come as a great surprise to Lagrange who believed that all functions could be represented as series and even attempted to prove it. He built the whole theory of calculus on this basis. He defined the derivative of $f(x)$, for example, as the coefficient of h in the series expansion of $f(x+h)$. He was thereby attempting to eliminate vague concepts such as fluxions, infinitesimals, and limits in order to reduce all computations to the algebraic analysis of finite quantities. Cauchy, by contrast, rejected Lagrange's foundations for analysis but accepted with small changes some of Lagrange's proofs.

The proof of Taylor's theorem based on Rolle's theorem, now commonly given in textbooks, seems to have first been given in J. A. Serret's 1868 text on calculus; he attributed the result to Pierre Ossian Bonnet (1819–1892). In fact, Rolle proved the theorem only for polynomials. Serret did not mention Rolle explicitly in the course of his proof, but did mention him in his algebra book. Michel Rolle (1652–1719) was a paid member of the Academy of Sciences of Paris. In a small book published in 1692, Rolle established that the derivative of a polynomial $f(x)$ had a zero between two successive real zeros of $f(x)$. Since he did not initially accept the validity of calculus, Rolle worked out an algebraic procedure called the method of cascades, by which one could obtain the derivative of a polynomial. Euler, Lagrange, and Ruffini made mention of Rolle's result, but it did not occupy a central place in calculus at that time because it was seen as a theorem about polynomials. Once it was extended to all differentiable functions, its significance was greatly increased.

The modern conditions for the validity of Rolle's theorem were given in substance by Bonnet, but they were more carefully and exactly stated by the Italian mathematician, Ulisse Dini (1845–1918) in his lectures at the University of Pisa in 1871–1872. After this, mathematicians began investigating the consequences of relaxing the conditions. In the exercises, we state a 1909 result of W. H. Young and Grace C. Young, using left-hand and right-hand derivatives. Grace Chisholm (1868–1944) studied at Girton College, Cambridge and then went on to receive a doctorate in mathematics from the University of Göttingen in 1896. Her best work was done in real variables theory and she was among the very few women mathematicians of her generation with an international reputation. William Young (1863–1942) studied at Cambridge and became a mathematical coach there. He coached his future wife for the Tripos exam and took up mathematical research after their marriage in 1896. He published over 200 papers and was one of the most profound English mathematicians of the early twentieth century. It appears from a letter W. H. Young wrote to his wife that several papers published under his name alone were in fact joint efforts. In recognition of this, a volume of their selected papers was published in 2000 under both names.

Returning to Bolzano and Cauchy's proofs of the intermediate value theorem, we note that they both had gaps. Bolzano assumed the existence of a least upper bound and Cauchy's argument produced a sequence of real numbers a_n, $n = 1, 2, 3, \ldots$ such that $a_{n+1} - a_n = \frac{1}{2}(a_n - a_{n-1})$; he assumed that such a sequence must have a limit. A theory of real numbers was required to shore up these proofs. Although it seems that by the 1830s, Bolzano had begun to understand the basic problem here, it was not until the second half of the nineteenth century that mathematicians were able to construct the necessary framework. Richard Dedekind (1831–1916) was one of the first to develop it and he described his motivation in his famous paper on the theory of real numbers:

> As professor in the Polytechnique School in Zurich I found myself for the first time obliged to lecture upon the elements of the differential calculus and felt more keenly than ever before the lack of a really scientific foundation for arithmetic. In discussing the notion of the approach of a variable magnitude to a fixed limiting value, and especially in proving the theorem that every magnitude which grows continually, but not beyond all limits, must certainly approach a limiting value, I had recourse to geometric evidences. Even now such resort to geometric intuition in a first presentation of the differential calculus, I regard as exceedingly useful, from the didactic standpoint, and indeed indispensable, if one does not wish to lose too much time. But that this form of introduction into the differential calculus can make no claim to being scientific, no one will deny. For myself this feeling of dissatisfaction was so overpowering that I made the fixed resolve to keep meditating on the question till I should find a purely arithmetic and perfectly rigorous foundation for the principles of infinitesimal calculus.

Dedekind published his theory in 1872, though he had completed it by November 1858. Meanwhile, before 1872, the theories of Charles Méray, Eduard Heine, and Cantor, equivalent to Dedekind's, were published. Though he had discovered it some years before, Weierstrass presented his own independently developed theory of real numbers as part of his lectures in Berlin during the 1860s.

12.2 Gregory's Discovery of the Taylor Series

In 1671, Gregory gave power series expansions of the seven functions mentioned earlier. His notation was naturally different from the one we now use. For example, he described the series for $\tan x$ and $\ln \sec x$:

12.2 Gregory's Discovery of the Taylor Series

If radius $=r$, arcus $=a$, tangus $=t$, secans artificialis $=s$, then

$$t = a + \frac{a^3}{3r^2} + \frac{2a^5}{15r^4} + \frac{17a^7}{315r^6} + \frac{3233a^9}{181440r^8} + \quad \text{etc.,} \tag{12.4}$$

$$s = \frac{a^2}{2r} + \frac{a^4}{12r^3} + \frac{a^6}{45r^5} + \frac{17a^8}{2520r^7} + \frac{3233a^{10}}{1814400r^9} \quad \text{etc.}$$

Gregory's descriptions of the $\ln\tan(\frac{\pi}{4}+\frac{x}{2})$ and $\text{arcsec}(\sqrt{2}e^x)$ functions were slightly more complicated. In his letter to Collins, he gave no indication of how he obtained his seven series, but H. W. Turnbull determined that, except for the series for $\arctan x$, Gregory obtained them by using their derivatives. While examining Gregory's unpublished notes in the 1930s, Turnbull noticed that Gregory had written the successive derivatives of some trigonometric and logarithmic functions on the back of a January 29, 1671 letter from Gideon Shaw, an Edinburgh stationer. For example, he gave the first seven derivatives of $r\tan\theta$ with respect to θ expressed as polynomials in $\tan\theta = q/r$. He denoted the function and its derivatives by m so that he had

1st $m = q$,

2nd $m = r + \frac{q^2}{r}$,

3rd $m = 2q + \frac{2q^3}{r^2}$,

4th $m = 2r + \frac{8q^2}{r} + \frac{6q^4}{r^3}$,

5th $m = 16q + \frac{40q^3}{r^2} + \frac{24q^5}{r^4}$,

6th $m = 16r + \frac{136q^2}{r} + \frac{240q^4}{r^3} + \frac{120q^6}{r^5}$,

7th $m = 272q + \frac{987q^3}{r^2} + 1680\frac{q^5}{r^4} + 720\frac{q^7}{r^6}$,

8th $m = 272r + 3233\frac{q^2}{r} + 11361\frac{q^4}{r^3} + 13440\frac{q^6}{r^5} + 5040\frac{q^8}{r^7}$.

Note that since the derivative of $\tan\theta$ is $\sec^2\theta = 1 + q^2/r^2$, one can move from one value of m to the next by taking the derivative of the initial m with regard to q and then multiplying by $r+q^2/r$. This suggests that Gregory used a method equivalent to the chain rule; indeed, this conclusion is supported by his computational mistake in finding the seventh m from the sixth:

$$\left(272\frac{q}{r} + 960\frac{q^3}{r^3} + 720\frac{q^5}{r^5}\right)\left(r + \frac{q^2}{r}\right).$$

The coefficient of q^3/r^2 is $272 + 960 = 1232$, whereas Gregory had 987. Evidently, he had miscopied 272 from the previous step as 27 to get $27 + 960 = 987$. This in turn produced an error in the coefficient of q^2/r in the eighth m; this should be 3968 instead of Gregory's 3233. If this computation, with Gregory's mistake, is continued to the tenth m, that is, the ninth derivative of $r\tan\theta$, then the first term of the derivative would be $2 \times 3233 = 6466$, as given in his notes.

Note that the Maclaurin series for $\tan\theta$ is obtained by computing the derivatives at $\theta = 0$. According to Gregory's mistaken calculation, the coefficient of θ^9 in the series for $\tan\theta$ in the letter to Collins would be $6466/9!$. This simplifies to $3233/181440$, just as Gregory noted in (12.4). As Turnbull pointed out, the appearance of this key error in Gregory's letter fortunately allows us to see that the calculations on the back of Shaw's letter were for the purpose of constructing the series. Thus, though no explicit statement of Maclaurin's formula has been found in Gregory's papers, we may conclude that Gregory was implicitly aware of it, since he made use of it in so many instances.

In 1713, Newton, then president of the Royal Society, insisted that the society publish relevant portions of Gregory's letters to Collins in the *Commercium Epistolicum* to prove his own absolute priority in the discovery of the calculus. Recall that Gregory's letters referred to the series of Newton communicated to him by Collins. But in the published accounts, Gregory's computational error was corrected.

Gregory found the series for $\mathrm{arcsec}(\sqrt{2}e^x)$ by taking the derivatives of $r\theta$ with respect to $\ln\sec\theta$. In his notes, he wrote down the first five derivatives employed to construct the series in the letter to Collins. If we write $y = r\theta$, $x = \ln\sec\theta$, $q = r\tan\theta$, then we have

$$\frac{dy}{dx} = \frac{1}{\tan\theta}\frac{dy}{d\theta} = \frac{r^2}{q} \quad \text{and} \quad \frac{dq}{dx} = \frac{r^2}{q} + q.$$

This implies that the successive derivatives can be found by taking the derivative with respect to q and multiplying by $\frac{r^2}{q} + q$:

$$\frac{d^2y}{dx^2} = \frac{d}{dq}\left(\frac{r^2}{q}\right)\cdot\frac{dq}{dx} = -\frac{r^2}{q^2}\left(\frac{r^2}{q}+q\right), \quad \frac{d^3y}{dx^3} = \frac{r^2}{q} + \frac{4r^4}{q^3} + \frac{3r^6}{q^5}, \text{ etc.}$$

Except for the signs of the derivatives, Gregory wrote precisely these expressions in his notes. He also wrote, without signs, expressions for the next two derivatives (without signs):

$$\frac{r^2}{q} + 13\frac{r^4}{q^3} + 27\frac{r^6}{q^5} + 15\frac{r^8}{q^7}, \quad \frac{r^2}{q} + 40\frac{r^4}{q^3} + 174\frac{r^6}{q^5} + 240\frac{r^8}{q^7} + 105\frac{r^{10}}{q^9}.$$

Gregory then expanded $y = r\left(\theta - \frac{\pi}{4}\right)$ as a series in $x = \ln\frac{\sec\theta}{\sqrt{2}}$ about $x = 0$. Now when $x = 0$, then $\theta = \pi/4$ and $q = r$. Hence, he had the series, given in modern notation:

$$\theta = \frac{\pi}{4} + x - x^2 + \frac{4x^3}{3} - \frac{7x^4}{3} + \frac{14x^5}{3} - \frac{452x^6}{45} + \cdots.$$

To see how the constants in this series are produced, consider the coefficient of x^3 obtained from the third derivative with $q = r$. From the preceding expression for d^3y/dx^3, this value can be given as $r + 4r + 3r = 8r$, and since this has to be divided by $3! = 6$, we arrive at $4r/3$ or simply $4/3$ after dividing by r.

12.3 Newton: An Iterated Integral as a Single Integral

Newton wrote up his evaluation of the nth iterated integral as a single integral sometime around 1693, but did not publish it. His main idea was to repeatedly use integration by parts, combined with the fundamental theorem of calculus, to reduce a double integral to a single integral. We reproduce his derivation, though we change his notation. He used the letters $A, B, C, \ldots$ to denote areas under $\dot{y}, z\dot{y}, z^2\dot{y}, \ldots$ but we shall use these letters to denote the areas under $y, zy, z^2y, \ldots$ to obtain the result in standard form. We also employ the Fourier-Leibniz notation of a definite integral to denote area. Let $y = f(t)$ be a curve, and let $A = \int_0^z y \, dt$, $B = \int_0^z ty \, dt$, $C = \int_0^z t^2 y \, dt$, $D = \int_0^z t^3 y \, dt$, $\ldots$. Then for some constant a

$$\int_a^z y \, dt = A + g,$$

where g is a constant. The second iterated integral of y is

$$\int_a^z (A+g) \, dt = zA - \int_a^z t \frac{dA}{dt} \, dt + gz + h_1$$
$$= zA - B + gz + h,$$

where h is some constant. Integration of this expression gives

$$\frac{1}{2}z^2 A - zB - \int_a^z \left(\frac{1}{2}t^2 \frac{dA}{dt} - t \frac{dB}{dt} \right) dt + \frac{1}{2}gz^2 + hz + i_2$$
$$= \frac{1}{2}z^2 A - zB - \int_a^z \left(\frac{1}{2} \frac{dC}{dt} - \frac{dC}{dt} \right) dt + \frac{1}{2}gz^2 + hz + i_1$$
$$= \frac{1}{2}z^2 A - zB + \frac{1}{2}C + \frac{1}{2}gz^2 + hz + i.$$

This is the third iterated integral of y. The integral of its first three terms is

$$\frac{1}{2}\left(\frac{1}{3}z^3 A - z^2 B + zC\right) - \frac{1}{2}\int_a^z \left(\frac{1}{3}t^3 \frac{dA}{dt} - t^2 \frac{dB}{dt} + t \frac{dC}{dt}\right) dt + \text{constant}$$
$$= \frac{1}{6}(z^3 A - 3z^2 B + 3zC) - \frac{1}{2}\int_a^z \frac{1}{3} \frac{dD}{dt} \, dt + \text{constant}$$
$$= \frac{1}{6}(z^3 A - 3z^2 B + 3zC - D) + \text{constant}.$$

Hence the fourth iterated integral is

$$\frac{1}{6}(z^3 A - 3z^2 B + 3zC - D) + \frac{1}{6}gz^3 + \frac{1}{2}hz^2 + iz + k.$$

Newton worked out another iterate to obtain

$$\frac{1}{24}(z^4 A - 4z^3 B + 6z^2 C - 4zD + E) + \frac{1}{24}gz^4 + \frac{1}{6}hz^3 + \frac{1}{2}iz^2 + kz + l.$$

By induction, he wrote the general nth iterate, in our notation, as

$$\frac{1}{(n-1)!}\left(z^{n-1}\int_a^z y\,dt - (n-1)z^{n-2}\int_a^z ty\,dt + \frac{(n-1)(n-2)}{2}z^{n-3}\int_a^z t^2 y\,dt - \cdots\right)$$
$$+ \frac{1}{(n-1)!}gz^{n-1} + \frac{1}{(n-2)!}hz^{n-2} + \cdots. \tag{12.5}$$

Newton left the integral in this form, although it is clear that he could easily have applied the binomial theorem to obtain the integral in the form (12.3).

12.4 Bernoulli and Leibniz: A Form of the Taylor Series

Johann Bernoulli's 1794 result on series was stated in a paper and in a letter to Leibniz as

$$\text{Integr.}n\,dz = +nz - \frac{zz\,dn}{1\cdot 2\cdot dz} + \frac{z^3\,ddn}{1\cdot 2\cdot 3\cdot dz^3} - \frac{z^4\,dddn}{1\cdot 2\cdot 3\cdot 4\cdot dz^4} \quad \text{etc.} \tag{12.6}$$

Here Integr.$n\,dz$ stood for the integral of n, or $\int n\,dz$. In fact, the term "integral" was first used by the Bernoulli brothers, Jakob and Johann, who conceived of it as the antiderivative. In a letter to Johann, Leibniz once wrote that he preferred to think of the integral as a sum instead of as an antiderivative. Bernoulli's proof of this result was very simple:

$$n\,dz = +n\,dz + z\,dn - z\,dn - \frac{zz\,ddn}{1\cdot 2\cdot dz} + \frac{zz\,ddn}{1\cdot 2\cdot dz} + \frac{z^3\,dddn}{1\cdot 2\cdot 3\cdot dz^2} \quad \text{etc.}$$

He took the terms on the right in pairs to get

$$n\,dz = d(nz) - d\left(\frac{z^2}{1\cdot 2}\frac{dn}{dz}\right) + d\left(\frac{z^3}{1\cdot 2\cdot 3}\frac{ddn}{dz^2}\right) - \cdots.$$

The required result followed upon integration. This process amounts to repeated integration by parts applied to $\int n\,dz$. Bernoulli applied his formula to three questions: de Beaune's problem; determination of the series for $\ln(a+x)$; and the determination of the series for $\sin x$. He was not completely successful with the third problem and was only able to find $\sin x/\cos x$ as a ratio of two series.

In reply to Bernoulli's 1794 letter containing the preceding result, Leibniz outlined his own derivation of the formula, instructive as an illustration of Leibniz's conception of the analogy between finite and infinitesimal differences, leading to his characteristic approach to the calculus. We change Leibniz's notation slightly in the initial part of his derivation; he himself used neither subscripts nor the difference operator. Supposing the sequence $a_0, a_1, a_2, \ldots$ decreases to zero, Leibniz started with the equation

$$a_0 = -(\Delta a_0 + \Delta a_1 + \Delta a_2 + \cdots).$$

Since $\quad a_n = (1+\Delta)^n a_0 = a_0 + \frac{n}{1}\Delta a_0 + \frac{n(n-1)}{1\cdot 2}\Delta^2 a_0 + \cdots,$

Leibniz could rewrite the first equation as

$$a_0 = -\left(\Delta a_0 + \Delta a_0 + \Delta^2 a_0 + \Delta a_0 + 2\Delta^2 a_0 + \Delta^3 a_0 + \cdots\right)$$
$$= -\left(\Delta a_0 (1+1+1+\cdots) + \Delta^2 a_0 (1+2+3+\cdots)\right)$$
$$- \left(\Delta^2 a_0 (1+3+6+\cdots) + \cdots\right).$$

He then observed that a similar relation continued to hold when the differences were infinitely small and he replaced $a_0, -\Delta a_0, \Delta^2 a_0, -\Delta^3 a_0, \ldots$ by $y (= y(x)), dy, ddy, d^3y, \ldots$ respectively; moreover, by letting the infinitely small dx become 1, he set $1+1+1+\cdots$ equal to x, $1+2+3+\cdots = \int x$, $1+3+6+10+\cdots = \int\int x$, etc. Since $\int x = \frac{1}{1\cdot 2}xx$, $\int\int x = \frac{1}{1\cdot 2\cdot 3}x^3, \ldots$, Leibniz obtained

$$y = \frac{1}{1}x\frac{dy}{dx} - \frac{1}{1\cdot 2}xx\frac{ddy}{dx^2} + \frac{1}{1\cdot 2\cdot 3}x^3\frac{d^3y}{dx^3} - \frac{1}{1\cdot 2\cdot 3\cdot 4}x^4\frac{d^4y}{dx^4} \text{ etc.}$$

He then noted that Bernoulli's formula followed upon replacing y, dy, ddy, etc. by $\int y, y, dy$, etc., respectively.

12.5 Taylor and Euler on the Taylor Series

In Taylor's book of 1715, he obtained his namesake series from the well-known interpolation formula by letting the distance between the equidistant points on the axis tend to zero. We shall follow Euler's exposition of 1736, since Euler used a more convenient and easily understandable notation. Euler divided the interval from x to $x + a$ into m equal parts, each equal to dx. He let y be a function of x and then let $dy = y(x+dx) - y(x)$, $ddy = y(x+2dx) - 2y(x+dx) + y(x), \ldots$ be the first, second, $\ldots$ differences at x. He then had

$$y(x+2dx) = y + 2dy + ddy, \quad y(x+3dx) = y + 3dy + 3ddy + d^3y,$$

$$\cdots\cdots$$

$$y(x+a) = y(x+mdx) = y + mdy + \frac{m(m-1)}{1\cdot 2}ddy + \frac{m(m-1)(m-2)}{1\cdot 2\cdot 3}d^3y + \text{ etc.}$$

Next, he let m be an infinite number and dx infinitely small so that mdx was finite and equal to a. Then

$$y(x+a) = y + mdy + \frac{m^2}{1\cdot 2}ddy + \frac{m^3}{1\cdot 2\cdot 3}d^3y + \text{ etc.}$$
$$= y + a\frac{dy}{dx} + \frac{a^2}{1\cdot 2}\frac{ddy}{dx^2} + \frac{a^3}{1\cdot 2\cdot 3}\frac{d^3y}{dx^3} + \text{ etc.}$$

Gregory, de Moivre, and Taylor derived the Taylor series by means of essentially the same argument.

Euler then showed how Bernoulli's series could be derived from Taylor's formula. He set $y(0) = 0$ and $a = -x$ to get

$$0 = y - \frac{x}{1}\frac{dy}{dx} + \frac{x^2}{1\cdot 2}\frac{ddy}{dx^2} - \frac{x^3}{1\cdot 2\cdot 3}\frac{d^3y}{dx^3} + \text{etc.}$$

This implied $\quad y = \frac{x}{1}\frac{dy}{dx} - \frac{x^2}{1\cdot 2}\frac{ddy}{dx^2} + \frac{x^3}{1\cdot 2\cdot 3}\frac{d^3y}{dx^3} - \text{etc.}$

Euler then replaced y by $\int y\,dx$, as Leibniz did, and obtained Bernoulli's formula. See the exercises for the converse.

12.6 Lacroix on d'Alembert's Derivation of the Remainder

In his 1754 book *Recherches sur différents points importants du système du monde*, d'Alembert obtained the n-dimensional iterated integral for the remainder in the Taylor series. In his 1819 book, Sylvestre Lacroix (1765–1843) presented the essence of d'Alembert's proof in notation more familiar to us: Lacroix let $u' = u(x+h)$ and $u = u(x)$ and set $u' = u + P$. Then

$$\frac{du'}{dh} = \frac{dP}{dh}, \quad \text{and hence} \quad P = \int \frac{du'}{dh}\,dh.$$

Note that the derivatives of u' are partial derivatives; for now, we follow Lacroix's notation. Thus, he had

$$u' = u + \int \frac{du'}{dh}\,dh. \quad \text{Next, he let} \quad \frac{du'}{dh} = \frac{du}{dx} + Q,$$

$$\text{so that} \quad \frac{d^2u'}{dh^2} = \frac{dQ}{dh}, \quad \text{or} \quad Q = \int \frac{d^2u'}{dh^2}\,dh,$$

$$\frac{du'}{dh} = \frac{du}{dx} + \int \frac{d^2u'}{dh^2}\,dh, \quad \int \frac{du'}{dh}\,dh = \frac{du}{dx}\frac{h}{1} + \int\int \frac{d^2u'}{dh^2}\,dh^2,$$

$$u' = u + \frac{du}{dx}\frac{h}{1} + \int\int \frac{d^2u'}{dh^2}\,dh^2.$$

Setting $\quad \dfrac{d^2u'}{dh^2} = \dfrac{d^2u}{dx^2} + R, \quad$ he had

$$\frac{d^3u'}{dh^3} = \frac{dR}{dh}, \quad \text{or} \quad R = \int \frac{d^3u'}{dh^3}\,dh, \quad \frac{d^2u'}{dh^2} = \frac{d^2u}{dx^2} + \int \frac{d^3u'}{dh^3}\,dh,$$

$$u' = u + \frac{du}{dx}\frac{h}{1} + \frac{d^2u}{dx^2}\frac{h^2}{1\cdot 2} + \int\int\int \frac{d^3u'}{dh^3}\,dh^3.$$

Continuing in the same manner, he had in general

$$u' = u + \frac{du}{dx}\frac{h}{1} + \frac{d^2u}{dx^2}\frac{h^2}{1\cdot 2} + \cdots + \frac{d^{n-1}u}{dx^{n-1}}\frac{h^{n-1}}{1\cdot 2\cdots(n-1)} + \int^n \frac{d^nu'}{dh^n}\,dh^n,$$

where the n-fold multiple integral $\int^n$ was zero when $h = 0$. If Newton's formula for $\int^n$ had been used here, then the remainder would have emerged in the form

$$R_n(h) = \frac{1}{(n-1)!} \int_0^h (h-t)^{n-1} \frac{d^n}{dx^n} u(x+t) \, dt. \tag{12.7}$$

In his 1823 lectures on calculus, Cauchy showed the equivalence of the two remainders by proving

$$\frac{d^n R_n}{dh^n} = \frac{d^n}{dx^n} u(x+t).$$

He applied the fundamental theorem of calculus and Leibniz's formula for the derivative of an integral to obtain

$$\begin{aligned}\frac{dR_n}{dh} &= \frac{1}{(n-1)!} \int_0^h \frac{d}{dh}(h-t)^{n-1} \frac{d^n}{dx^n} u(x+t) \, dt \\ &= \frac{1}{(n-2)!} \int_0^h (h-t)^{n-2} \frac{d^n}{dx^n} u(x+t) \, dt.\end{aligned} \tag{12.8}$$

Cauchy derived the desired result, and in effect a proof of Newton's formula, by performing this process n times. Lacroix did not provide a proof of (12.7), merely noting that

$$\int^n H \, dh^n = \frac{1}{1 \cdot 2 \cdots (n-1)} \left(h^{n-1} \int H \, dh - \frac{n-1}{1} h^{n-2} \int Hh \, dh + \frac{(n-1)(n-2)}{1 \cdot 2} h^{n-3} \int Hh^2 \, dh - \cdots \right).$$

This result is equivalent to Newton's formula (12.5) and, as we have noted, Newton actually stated it in this form. Lacroix could have proved this inductively, using integration by parts. He did not mention Newton in this context; one may assume that he was not aware of Newton's work, since Lacroix was very meticulous in stating his sources.

12.7 Lagrange's Derivation of the Remainder Term

In his 1797 book, *Fonctions analytiques*, Joseph-Louis Lagrange (1736–1813) obtained the remainder term of the Taylor series as a single integral. He started with

$$fx = f(x - xz) + xP,$$

where $P = 0$ at $z = 0$. The derivative with respect to z of this equation gave $0 = -xf'(x - xz) + xP'$ or $P' = f'(x - xz)$. For the second-order remainder, Lagrange wrote

$$fx = f(x - xz) + xzf'(x - xz) + x^2 Q$$

and obtained $Q' = zf''(x - xz)$. Similarly,

$$fx = f(x - xz) + xzf'(x - xz) + \frac{x^2 z^2}{2} f''(x - xz) + x^3 R$$

and, after taking the derivative and simplifying, he got $R' = \frac{z^2}{2} f'''(x - xz)$. Lagrange did not write the general expression for the remainder and gave only the recursive procedure. This method gives the remainder as an integral, though Lagrange did not write it in that form, since he avoided the use of integrals in this book. It is easy to see that

$$R(z) = \frac{1}{2!} \int_0^z t^2 f'''(x - xt)\, dt = \frac{1}{2! x^3} \int_{x-xz}^x (x - u)^2 f'''(u)\, du. \qquad (12.9)$$

Had he stated it explicitly, the general form of Lagrange's formula would have been

$$fx = f(x - xz) + xz f'(x - xz) + \frac{x^2 z^2}{2} f''(x - xz) + \cdots$$
$$+ \frac{x^{n-1} z^{n-1}}{(n-1)!} f^{(n-1)}(x - xz) + x^n R_n,$$

$$\text{where} \quad R_n = \frac{1}{(n-1)!} \int_{x-xz}^x (x - u)^{n-1} f^{(n)}(u)\, du.$$

If we replace $x - xz$ by a, Lagrange's formula becomes

$$f(x) = f(a) + (x - a) f'(a) + \frac{(x - a)^2}{2!} f''(a) + \cdots$$
$$+ \frac{(x - a)^{n-1}}{(n-1)!} f^{(n-1)}(a) + \frac{1}{(n-1)!} \int_a^x (x - t)^{n-1} f^{(n)}(t)\, dt.$$

Thus, here the multiple integral remainder of d'Alembert was replaced by a single integral. However, Lagrange himself gave only the derivative of the remainder. In his 1799 lectures on the calculus, published in 1801 as *Leçons sur le calcul des fonctions*, he presented the remainder as it appears in modern texts, as an nth derivative.

Lagrange proved a lemma for the purpose of determining bounds for R_n: If $f'x$ is positive for all values of x between $x = a$ and $x = b$ with $b > a$, then $fb - fa > 0$. To prove this statement, Lagrange set $f(x + i) = fx - iP$, where P was a function of x and i, such that at $i = 0$, $P = f'(x) > 0$. So $P(x, i) > 0$ for i sufficiently small, and it followed that $f(x + i) - f(x) > 0$ for small i. Next, he divided the interval $[a, b]$ into n equal parts, each of length $(j = (b - a)/n)$, with n sufficiently large that in each subinterval $[a + kj, a + (k + 1)j]$, $k = 0, 1, \ldots, n - 1$, he had $f(a + (k + 1)j) - f(a + kj) > 0$. By adding up these inequalities, he got $fb - fa > 0$.

Lagrange's lemma was correct but his proof was obviously inadequate. For example, he assumed that the same j would work in all parts of the interval. But he went on to use the result to derive a different form of the remainder. He supposed $f'(q)$ and $f'(p)$ to be the maximum and minimum values, respectively, of $f'(x)$ in an interval. Then $g'(i) = f'(x + i) - f'(p)$ and $h'(i) = f'(q) - f'(x + i)$ were both positive. Lagrange's lemma then gave

$$g(i) = f(x + i) - f(x) - i f'(p) \geq 0, \quad h(i) = i f'(q) - f(x + i) + f(x) \geq 0.$$

These inequalities implied bounds for $f(x+i)$:

$$f(x) + if'(p) \leq f(x+i) \leq f(x) + if'(q) \text{ and thus}$$

$$f(x+i) = f(x) + if'(x+i\theta), \quad 0 < \theta < 1.$$

The last equation followed from the intermediate value theorem, implicitly assumed by Lagrange; similarly, he assumed that f' had a maximum/minimum in an interval. More generally, Lagrange showed that $f(x+i)$ lay between

$$f(x) + if'(x) + \frac{i^2}{2!}f''(x) + \cdots + \frac{i^u}{u!}f^{(u)}(p) \quad \text{and}$$

$$f(x) + if'(x) + \frac{i^2}{2!}f''(x) + \cdots + \frac{i^u}{u!}f^{(u)}(q),$$

where p and q were the values at which $f^{(u)}$ had a minimum and maximum, respectively, in the given interval. Once again, an application of the intermediate value theorem would yield Taylor's formula with the remainder as a derivative:

$$f(x+i) = f(x) + if'(x)\frac{i^2}{2!}f''(x) + \cdots + \frac{i^u}{u!}f^{(u)}(x+\theta i), \quad 0 < \theta < 1.$$

12.8 Laplace's Derivation of the Remainder Term

After being launched in his career by d'Alembert, Laplace used his tremendous command of analysis to make groundbreaking contributions in his areas of interest, celestial mechanics and probability. In his famous 1812 work, *Théorie analytique des probabilités*, Laplace used repeated integration by parts in a direct way to obtain the remainder term. He started with the observation that

$$\int dz\, \phi'(x-z) = \phi(x) - \phi(x-z), \tag{12.10}$$

when the lower limit of integration was $z = 0$. This result, the fundamental theorem of calculus, would be written in modern notation:

$$\int_0^z \phi'(x-t)\, dt = \phi(x) - \phi(x-z).$$

Using Laplace's notation, integration by parts gave

$$\int dz\, \phi'(x-z) = z\phi'(x-z) + \int z\, dz\, \phi''(x-z),$$

$$\int z\, dz\, \phi''(x-z) = \frac{1}{2}z^2 \phi''(x-z) + \int \frac{1}{2}z^2\, dz\, \phi'''(x-z) \text{ etc.}$$

Hence, in general

$$\int dz\, \phi'(x-z) = z\phi'(x-z) + \frac{z^2}{1\cdot 2}\phi''(x-z) + \cdots + \frac{z^n}{1\cdot 2\cdot 3\cdots n}\phi^{(n)}(x-z)$$
$$+ \frac{1}{1\cdot 2\cdot 3\cdots n}\int z^n dz\, \phi^{(n+1)}(x-z).$$
(12.11)

Combined with (12.10), this equation provided Taylor's theorem with remainder. Laplace then converted this remainder to the Lagrange form. Since $\int z^n dz = z^{n+1}/(n+1)$, the integral in (12.11) lies between $mz^{n+1}/(n+1)$ and $Mz^{n+1}/(n+1)$ where m and M are the smallest and largest values of $\phi^{(n+1)}(x-z)$ in the interval of integration. Hence, the value of the integral in (12.11) lies in between these values and is given by

$$\frac{z^{n+1}}{n+1}\phi^{(n+1)}(x-u),$$

where u is some value between 0 and z. Thus, the remainder term in (12.11) can be written as

$$\frac{z^{n+1}}{1\cdot 2\cdot 3\cdots (n+1)}\phi^{(n+1)}(x-u).$$

This completed Laplace's derivation of the two forms of the remainder in Taylor's theorem.

12.9 Cauchy on Taylor's Formula and l'Hôpital's Rule

In his lectures published in 1823, Cauchy took an interesting approach to Newton's n-fold integral. He started with the differential equation

$$\frac{d^n y}{dx^n} = f(x).$$

Repeated integration of this equation yielded

$$\frac{d^{n-1}y}{dx^{n-1}} = \int_{x_0}^{x} f(z)\, dz + C, \quad \frac{d^{n-2}y}{dx^{n-2}} = \int_{x_0}^{x}(x-z)f(z)\, dz + C(x-x_0) + C_1,$$

......

$$y = \int_{x_0}^{x}\frac{(x-z)^{n-1}}{1\cdot 2\cdots (n-1)} f(z)\, dz + \frac{C(x-x_0)^{n-1}}{1\cdot 2\cdots (n-1)} + \cdots + C_{n-2}(x-x_0) + C_{n-1},$$

where $C, C_1, \ldots, C_{n-1}$ were arbitrary constants. Here Cauchy used the result expressed in equation (12.8) to integrate in each step of the argument. The reader might compare this with Newton's formula (12.5). Cauchy then proceeded to obtain Taylor's theorem with remainder. He let $y = F(x)$ be a specific solution of $y^{(n)} = f(x)$ to obtain

$$C_{n-1} = F(x_0), \quad C_{n-2} = F'(x), \quad \ldots, \quad C = F^{(n-1)}(x_0),$$

and with these values, he had

$$F(x) = F(x_0) + \frac{F'(x_0)}{1!}(x-x_0) + \cdots + \frac{F^{(n-1)}(x_0)}{1 \cdot 2 \cdots (n-1)}(x-x_0)^{n-1}$$
$$+ \int_{x_0}^{x} \frac{(x-z)^{n-1} F^{(n)}(z)}{1 \cdot 2 \cdots (n-1)} dz.$$

Cauchy gave another proof modeled along the lines of Lagrange's second proof. He started with the lemma: Suppose $f(x)$ and $F(x)$ are continuously differentiable in $[x_0, x]$ with $f(x_0) = F(x_0) = 0$, and $F'(x_0) > 0$ in this interval. For x in this interval, if

$$A \leq \frac{f'(x)}{F'(x)} \leq B, \quad \text{then} \quad A \leq \frac{f(x)}{F(x)} \leq B.$$

To prove this, Cauchy noted that since $F'(x) > 0$, he had

$$f'(x) - AF'(x) \geq 0 \quad \text{and} \quad f'(x) - BF'(x) \leq 0.$$

He then applied Lagrange's lemma to the functions

$$f(x) - AF(x) \quad \text{and} \quad f(x) - BF(x)$$

to obtain the required result. Cauchy then took $x = x_0 + h$ and applied the intermediate value theorem to derive

$$\frac{f(x_0 + h)}{F(x_0 + h)} = \frac{f'(x_0 + \theta h)}{F'(x_0 + \theta h)}, \quad \text{where} \quad 0 < \theta < 1.$$

In the situation where $f(x_0)$ and $F(x_0)$ were nonzero, he replaced $f(x_0 + h)$ and $F(x_0 + h)$ by $f(x_0 + h) - f(x_0)$ and $F(x_0 + h) - F(x_0)$, respectively, to get the generalized mean value theorem:

$$\frac{f(x_0 + h) - f(x_0)}{F(x_0 + h) - F(x_0)} = \frac{f'(x_0 + \theta h)}{F'(x_0 + \theta h)}, \quad \text{where} \quad 0 < \theta < 1. \tag{12.12}$$

He next supposed $f'(x_0) = f''(x_0) = \cdots = f^{(n-1)}(x_0) = 0 = F'(x_0) = F''(x_0) = \cdots = F^{(n-1)}(x_0)$ and $F^{(n)} \neq 0$, and that all the derivatives were continuous. By an iteration of the process used to find the generalized mean value theorem, Cauchy deduced that

$$\frac{f(x_0 + h)}{F(x_0 + h)} = \frac{f^{(n)}(x_0 + \theta h)}{F^{(n)}(x_0 + \theta h)}, \quad \text{where} \quad 0 < \theta < 1. \tag{12.13}$$

He then let $h \to 0$ to deduce l'Hôpital's rule

$$\lim_{x \to x_0} \frac{f(x)}{F(x)} = \lim_{x \to x_0} \frac{f^{(n)}(x)}{F^{(n)}(x)}.$$

From this result, Cauchy derived Taylor's formula with Lagrange's remainder by taking $F(x) = (x - x_0)^n$ and replacing $f(x)$ by

$$g(x) = f(x) - f(x_0) - f'(x_0)(x - x_0) - \cdots - \frac{f^{(n-1)}(x_0)}{(n-1)!}(x - x_0)^{n-1};$$

this vanished at x_0, along with its first $n - 1$ derivatives. Then by (12.13),

$$g(x_0 + h) = \frac{h^n}{n!} g^{(n)}(x_0 + \theta h), \quad 0 < \theta < 1.$$

Since $g^{(n)}(x) = f^{(n)}(x)$, the required result followed.

Cauchy also obtained another form of the remainder by defining a function $\phi(a)$ by the equation

$$f(x) = f(a) + \frac{x-a}{1} f'(a) + \frac{(x-a)^2}{1 \cdot 2} f''(a) + \cdots + \frac{(x-a)^{n-1}}{1 \cdot 2 \cdots (n-1)} f^{(n-1)}(a) + \phi(a).$$

In (12.12), taking $F(x) = x, x_0 = a, h = x - a$, and $f = \phi$, he had

$$\phi(a) = \phi(x) + (a - x)\phi'(a + \theta(x - a)), \quad 0 < \theta < 1.$$

Since $\phi(x) = 0$ and $\phi'(a) = -\frac{(x-a)^{n-1}}{1 \cdot 2 \cdots (n-1)} f^{(n)}(a),$

he concluded

$$\phi(a) = \frac{(1-\theta)^{n-1}(x-a)^n}{1 \cdot 2 \cdots (n-1)} f^{(n)}(a + \theta(x - a)). \tag{12.14}$$

This remainder, called Cauchy's remainder, can also be obtained from the integral form of the remainder.

12.10 Cauchy: The Intermediate Value Theorem

Recall that the intermediate value theorem was regarded as intuitively or geometrically obvious by eighteenth-century mathematicians. For example, Lagrange and Laplace assumed it in their derivation of the remainder. Bolzano and Cauchy saw the need for a proof and each provided one. Cauchy stated and proved the theorem in his lectures, published in 1821: Suppose $f(x)$ is a real function of x, continuous between x_0 and X. If $f(x_0)$ and $f(X)$ have opposite signs, then the equation $f(x) = 0$ is satisfied by at least one value between x_0 and X. In his proof, Cauchy first divided $[x_0, X]$ of length $h = X - x_0$ into m parts to consider the sequence $f(x_0), f(x_0 + h/m)$, $f(x_0 + 2h/m), \ldots, f(X - h/m), f(X)$. Since $f(x_0)$ and $f(X)$ had opposite signs, he had two consecutive terms, say, $f(x_1)$ and $f(X')$ with opposite signs. Clearly

$$x_0 \leq x_1 \leq X' \leq X \quad \text{and} \quad X' - x_1 = h/m = (X - x_0)/m.$$

We remark that Cauchy's notation was slightly different in that he used $<$ for $\leq$. He repeated the preceding process for the interval $[x_1, X']$ to get $x_1 \leq x_2 \leq X'' \leq X'$ with

$X'' - x_2 = (X - x_0)/m^2$. Continuation of this procedure produced two sequences, $x_0 \leq x_1 \leq x_2 \leq \cdots$ and $X \geq X' \geq X'' \geq \cdots$, such that the differences between corresponding members of the two sequences became arbitrarily small. Thus, he had the two sequences converging to a common limit a. Now since f was continuous between $x = x_0$ and $x = X$, the two sequences $f(x_0), f(x_1), f(x_2), \ldots$ and $f(X), f(X'), f(X''), \ldots$ converged to $f(a)$. Since the signs of the numbers in the first sequence were opposite to the signs of the numbers in the second sequence, it followed by continuity that $f(a) = 0$. Since $x_0 \leq a \leq X$, Cauchy had the required result. Observe that Cauchy assumed that a sequence, now called a Cauchy sequence, must converge; this was later proved by Dedekind. Bolzano's slightly earlier proof of the intermediate value theorem had a similar deficiency, as he himself recognized in the 1830s.

12.11 Exercises

1. Following Johann Bernoulli, consider the differential equation $dy = \frac{\sqrt{a^2-y^2}}{a} dx$ for $y = \sin x$, and take $n = \frac{\sqrt{a^2-y^2}}{a}$, $dz = dx$ in Bernoulli's formula (12.6) to obtain

$$\frac{y}{\sqrt{a^2-y^2}} = \frac{x - \frac{x^3}{2\cdot 3 a^2} + \frac{x^5}{2\cdot 3\cdot 4\cdot 5 a^4} - \cdots}{a - \frac{x^2}{2a} + \frac{x^4}{2\cdot 3\cdot 4 a^3} - \cdots}.$$

Next consider the equation $dy = a\,dx/(a+x)$ and apply Bernoulli's method to obtain his series for $a \ln(\frac{a+x}{a})$:

$$y = \frac{ax}{a+x} + \frac{ax^2}{2(a+x)^2} + \frac{ax^3}{3(a+x)^3} + \frac{ax^4}{4(a+x)^4} + \cdots.$$

See Joh. Bernoulli (1968), vol. I, pp. 127–128. This paper was published in the *Acta Eruditorum* in 1694.

2. Complete de Moivre's outline of a method to obtain Bernoulli's series for $\int y\,dz$. Note that this is similar to Newton's method of successive approximation. Let the fluent of $\dot{z}y$ be $zy - q$ so that $\dot{z}y = \dot{z}y + z\dot{y} - \dot{q}$ or $\dot{q} = z\dot{y}$. Now let $\dot{y} = \dot{z}v$ so that $\dot{q} = z\dot{z}v$. Take the fluent of each side to get $q = \frac{1}{2}zzv - r$ for some r. Then $z\dot{z}v = z\dot{z}v + \frac{1}{2}zz\dot{v} - \dot{r}$, so that $\dot{r} = \frac{1}{2}zz\dot{v}$. Set $\dot{v} = z\dot{s}$ and continue as before. De Moivre gave this argument in 1704. See Feigenbaum (1985), p. 93.

3. Show that Bernoulli's series (12.6) is obtained by applying integration by parts to $\int n\,dz$ and then repeating the process infinitely often.

4. In the Bernoulli series for $\int n\,dz$, set $n = f'(z)$ to obtain

$$f(z) - f(0) = zf'(z) - \frac{1}{2!}z^2 f''(z) + \frac{1}{3!}z^3 f'''(z) - \cdots. \tag{12.15}$$

Similarly, find the series for $f'(z) - f'(0)$, $f''(z) - f''(0)$, $f'''(z) - f'''(0)$, ... and use them to eliminate $f'(z)$, $f''(z)$, $f'''(z)$, ... from the right-hand side of (12.15). Show that the result is the Maclaurin series for $f(z)$. See Whiteside's footnote in Newton (1967–1981), vol. VII, p. 19.

5. Show that all the derivatives of $f(x) = e^{-1/x^2}$ $(x \neq 0)$, $0(x = 0)$ are zero. Cauchy remarked that the two functions e^{-x^2} and $e^{-x^2} + e^{-1/x^2}$ had the same Maclaurin series. See Cauchy (1823), p. 230 and Cauchy (1829), p. 105.

6. Show that the remainder in Taylor's theorem can be expressed in the form

$$\frac{h^n}{p(n-1)!}(1-\theta)^{n-p} f^{(n)}(x+\theta h), \quad 0 < \theta < 1,$$

where p is a positive integer $\leq n$. This was first given by Schlömilch and also published a decade later by E. Roche. See Prasad (1931), p. 90, Hobson (1957b), vol. 2, p. 200, and Schlömilch (1847), p. 179.

7. Prove that if f, ϕ, and F are differentiable, then

$$\begin{vmatrix} f(x+h) & \phi(x+h) & F(x+h) \\ f(x) & \phi(x) & F(x) \\ f'(x+\theta h) & \phi'(x+\theta h) & F'(x+\theta h) \end{vmatrix} = 0,$$

for some $0 < \theta < 1$. This result and its generalization to $n+1$ functions is stated in Giuseppe Peano's *Calcolo differenzial* of 1884.

8. Let $h > 0$. Set $m(x_1, x_2) = \frac{f(x_1) - f(x_2)}{x_1 - x_2}$. Define the four derivatives of f:

$$f^+(x) = \overline{\lim_{h \to 0}} \, m(x+h, x),$$

$$f_+(x) = \underline{\lim_{h \to 0}} \, m(x+h, x),$$

$$f^-(x) = \overline{\lim_{h \to 0}} \, m(x-h, x),$$

$$f_-(x) = \underline{\lim_{h \to 0}} \, m(x-h, x).$$

Show that if $f(x)$ is continuous on $[a, b]$, then there is a point x in (a, b) such that either

$$f^+(x) \leq m(a, b) \leq f_-(x) \quad \text{or} \quad f^-(x) \geq m(a, b) \geq f_+(x).$$

The generalized mean value theorem is a corollary: If there is no distinction with respect to left and right with regard to the derivatives of $f(x)$, then there is a point x in (a, b) at which f has a derivative and its value is equal to $m(a, b)$. See Young and Young (1909).

12.12 Notes on the Literature

H. Bateman (1907) contains Taylor's letter to Machin. For Gregory's seven series and the quote from his letter, see Turnbull (1939), pp. 168–176; for Gregory's notes on these series, see pp. 349–360. Malet (1993) explains that Gregory could have obtained the Taylor rule without being in possession of a differential or equivalent technique. The correspondence of Leibniz and Bernoulli concerning (12.2) can be found in Leibniz (1971) vol. 3-1, pp. 143–157 and series (12.6) is in Bernoulli (1968), vol. I, pp. 123–128.

12.12 Notes on the Literature

Euler's 1736 paper with the derivations of the Taylor and Bernoulli series appears in Eu 1-14, pp. 108–123. This paper also contains Euler's first derivation of the Euler–Maclaurin formula.

Feigenbaum (1985) presents a thorough discussion of Taylor's book, *Methodus Incrementorum*, as well as a treatment of the work of earlier mathematicians who contributed to the Taylor series. See also Feigenbaum (1981), containing an English translation of the *Methodus*. For later work on the Taylor series, especially the remainder term, see Pringsheim (1900).

Grabiner (1981) and (1990) are interesting sources for topics related to the work on series of Lagrange and Cauchy. Grabiner shows that, although Cauchy did not accept Lagrange's ideas on the foundations of calculus, Lagrange's use of algebraic inequalities nevertheless exerted a significant influence on Cauchy. She further points out that, a half century earlier, Maclaurin made brilliant use of inequalities to prove theorems in calculus. See her article, "Was Newton's Calculus a Dead End?" in Van Brummelen and Kinyon (2005). Newton's calculations for the nth iterated integral as a single integral are given in Newton (1967–1981), vol. VII, pp. 164–169; for his statements of the Maclaurin and Taylor series, see pp. 97–99.

Cauchy's proof of the intermediate value theorem is in Note 3 of his 1821 *Analyse algébrique*. Cauchy (1823), pp. 208–213 gives the remainder in a Taylor series as an integral; pp. 360–61 contain Cauchy's form of the remainder (12.14). Cauchy (1829), pp. 69–79 also treats the remainder. These books are also conveniently found in his collected works, Cauchy (1882–1974). A look at Cauchy's 1820s lectures on calculus from a modern viewpoint is in Bressoud (2007). The quotation from Dedekind can be found in Dedekind (1963), p. 1. Laplace (1812), pp. 176–177 gives a derivation of Taylor's theorem with remainder using integration by parts. For Lacroix's exposition of d'Alembert's work on the remainder term, see Lacroix (1819), pp. 396–397.

13

Integration of Rational Functions

13.1 Preliminary Remarks

The integrals of rational functions form the simplest class of integrals; they are included in a first course in calculus. Yet some problems associated with the integration of rational function have connections with the deeper aspects of algebra and of analysis. Examples are the factorization of polynomials and the evaluation of beta integrals. These problems have challenged mathematical minds as great as Newton, Johann Bernoulli, de Moivre, Euler, Gauss, and Hermite; indeed, they have their puzzles for us even today. For example, can a rational function be integrated without factorizing the denominator of the function?

Newton was the first mathematician to explicitly define and systematically attack the problem of integrating rational and algebraic functions. Of course, mathematicians before Newton had integrated some specific rational functions, necessary for their work. The Kerala mathematicians found the series for arctangent; N. Mercator and Hudde worked out the series for the logarithm. But Newton's work was made possible by his discovery, sometime in mid-1665, of the inverse relation between the derivative and the integral. At that time, he constructed tables, extending to some pages, of functions that could be integrated because they were derivatives of functions already explicitly or implicitly defined. He extended his tables by means of substitution or, equivalently, by use of the chain rule for derivatives. He further developed the tables by an application of the product rule for derivatives, or integration by parts, in his October 1666 tract on fluxions. In this work, Newton viewed a curve dynamically: The variation of its coordinates x and y could be viewed as the motion of two bodies with velocities p and q, respectively. He posed the problem of determining y when q/p was known and noted: "Could this ever bee done all problems whatever might bee resolved. But by y^e following rules it may be very often done."

After giving the already known rules for integrating $ax^{m/n}$ when $m/n \neq -1$ and when $m/n = -1$, Newton went on to consider examples, such as the integrals of $x^2/(ax+b)$, $x^3/(a^2-x^2)$, and $c/(a+bx^2)$. He did not take more complicated rational functions, perhaps because of a lack of an understanding of partial fractions. Instead, he evaluated integrals of some algebraic functions involving square roots. One result stated: If $\frac{cx^{3n}}{x\sqrt{ax^n+bx^{2n}}} = \frac{q}{p}$. Make $\sqrt{a+bx^n} = z$. y^n [then] is

$$\frac{cx^n}{2nb}\sqrt{ax^n+bx^{2n}} - \Box\frac{3ac}{2nbb}\sqrt{\frac{zz-a}{b}} = y. \tag{13.1}$$

The square symbol was the equivalent of our integral with respect to z, representing area.

Newton was led deeper into the integration of rational functions by an August 17, 1676 letter from Leibniz, addressed to Oldenburg, but intended for all British mathematicians. In this letter, Leibniz presented his series for π,

$$\frac{\pi}{4} = 1 - \frac{1}{3} + \frac{1}{5} - \frac{1}{7} + \cdots. \tag{13.2}$$

To obtain this series, Leibniz applied transmutation, a somewhat ad hoc method of finding the area of a figure by transforming it into another figure with the same area. St. Vincent, Pascal, Gregory, and others had employed this method before Leibniz. In his reply of October 1676, Newton listed an infinitely infinite family of rational and algebraic functions, saying that he could integrate them. These included the four rational functions

$$\frac{dz^{\eta-1}}{e+fz^\eta+gz^{2\eta}}, \quad \frac{dz^{2\eta-1}}{e+fz^\eta+gz^{2\eta}} \quad \text{etc.,}$$

$$\frac{dz^{\frac{1}{2}\eta-1}}{e+fz^\eta+gz^{2\eta}}, \quad \frac{dz^{\frac{3}{2}\eta-1}}{e+fz^\eta+gz^{2\eta}} \quad \text{etc.,}$$

where d, e, f, and g were constants and in the third and fourth expressions, in case e and g had the same sign, $4eg$ had to be $\leq f^2$. Newton went on to observe that the expressions could become complicated, "so that I hardly think they can be found by the transformation of the figures, which Gregory and others have used, without some further foundation. Indeed I myself could gain nothing at all general in this subject before I withdrew from the contemplation of figures and reduced the whole matter to the simple consideration of ordinates alone."

Newton then observed that Leibniz's series would be obtained by taking $\eta = 1$ and $f = 0$ in the first function. In fact,

$$\frac{\pi}{4} = \arctan 1 = \int_0^1 \frac{dz}{1+z^2} = \int_0^1 (1-z^2+z^4+\cdots)\,dz = 1 - \frac{1}{3} + \frac{1}{5} - \frac{1}{7} + \cdots.$$

As another series, he offered

$$\frac{\pi}{2\sqrt{2}} = 1 + \frac{1}{3} - \frac{1}{5} - \frac{1}{7} + \frac{1}{9} + \frac{1}{11} - \cdots \tag{13.3}$$

and explained that it could be obtained by means of a calculation, setting $2eg = f^2$ and $\eta = 1$. He did not clarify any further, and Leibniz did not understand him since even a quarter century later Leibniz had trouble with the integral arising in this situation.

Some of Newton's unpublished notes from this period suggest that he considered integrals of the form $\int 1/(1 \pm x^m)$, since these integrals would lead to simple and interesting series. To express them in terms of standard integrals (that is, in terms of elementary functions such as the logarithm and arctangent), Newton had to consider the problem of factorizing $1 \pm x^m$ so that he could resolve the integrals into simpler ones by the use of partial fractions. We note that, in several examples in his 1670–71 treatise on fluxions and infinite series, he had broken up rational fractions into a sum of two fractions. In the 1710s, Cotes and Johann Bernoulli and, to a lesser extent, Leibniz pursued the algebraic topic of partial fractions with more intensity than did Newton. It may be noted in this context that even in 1825 Jacobi was able to make an original contribution to partial fractions in his doctoral dissertation.

Newton's method for finding the quadratic factors of $1 \pm x^m$ was to start with

$$(1 + nx + x^2)(1 - nx + px^2 - qx^3 + rx^4 - \cdots) = 1 \pm x^m \tag{13.4}$$

and then determine the pattern of the algebraic equations satisfied by n for different values of m. In this way he factored $1 - x^3$, $1 + x^4$, $1 - x^5$, $1 + x^6$, $1 \pm x^8$, $1 \pm x^{12}$, though he apparently was unable to resolve the equation for n when $m = 10$. As an example, Newton found the equation for $1 \pm x^4$ to be $n^3 - 2n = 0$, or $n = \pm\sqrt{2}$ and $n = 0$. This would yield

$$x^4 + 1 = (x^2 + \sqrt{2}x + 1)(x^2 - \sqrt{2}x + 1), \tag{13.5}$$

and, of course, $\quad x^4 - 1 = (x^2 - 1)(x^2 + 1)$,

though Newton did not bother to write this last explicitly. Note that this factorization of $x^4 + 1$ was just what he needed to derive his series for $\pi/(2\sqrt{2})$ in (13.3). He also recognized that values of n were related to cosines of appropriate angles. He was just a step away from Cotes's factorization of $x^n \pm a^n$.

Newton also considered the binomial $1 \pm x^7$ and found the equation for n to be $n^6 - 5n^4 + 6nn - 1 = 0$. Note that the solution involved cube roots; Newton did not write the values of n^2, apparently because he wanted to consider only those values expressible, at worst, by quadratic surds. One wonders whether it occurred to him to ask which values of m would lead to equations in n solvable by quadratic radicals. In 1796, Gauss resolved this problem in his theory of constructible regular polygons.

In 1702, since Newton's work remained unpublished, Johann Bernoulli and Leibniz in separate publications discussed the problem of factorizing polynomials, in connection with the integration of rational functions. In general, Leibniz and Bernoulli were of the opinion that integration of rational functions could be carried out by partial fractions, but the devil lay in the details. Leibniz factored

$$x^4 + a^4 = \left(x + a\sqrt{\sqrt{-1}}\right)\left(x - a\sqrt{\sqrt{-1}}\right)\left(x + a\sqrt{-\sqrt{-1}}\right)\left(x - a\sqrt{-\sqrt{-1}}\right). \tag{13.6}$$

13.1 Preliminary Remarks

He was puzzled by this factorization and wondered whether the integrals $\int dx/(x^4+a^4)$ and $\int dx/(x^8+a^8)$ could be expressed in terms of logarithms and inverse trigonometric functions. Bernoulli's paper also observed that the arctangent was related to the logarithm of imaginary values because

$$\int \frac{dx}{a^2+x^2} = \frac{1}{2a}\left(\int \frac{dx}{a+ix} + \int \frac{dx}{a-ix}\right). \tag{13.7}$$

Cotes made the connection between the logarithm and the trigonometric functions even more explicit with his discovery of the formula

$$\ln(\cos\theta + i\sin\theta) = i\theta. \tag{13.8}$$

Roger Cotes (1679–1716) is known for his factorization theorem, his work on approximate quadrature, and for editing the 1713 edition of the *Principia*. He studied at Cambridge and became Fellow of Trinity College in 1704 and Plumian Professor of Astronomy and Experimental Philosophy in 1705. Unfortunately, he published only one paper in his lifetime, on topics related to the logarithmic function. Robert Smith published Cotes's mathematical writings in a 1722 work titled *Harmonia Mensurarum*. Formula (13.8) was stated geometrically:

> For if some arc of a quadrant of a circle described with radius CE has sine CX, and sine of the complement of the quadrant XE, taking radius CE as modulus, the arc will be the measure of the ratio between $EX + XC\sqrt{-1}$ and CE, the measure having been multiplied by $\sqrt{-1}$.

Observe that this statement translates to $iR\ln(\cos\theta + i\sin\theta) = R\theta$, which is not quite correct, because the i should be on the other side of the equation.

Cotes's factorization theorem was stated as a property of the circle. In Figure 13.1, the circumference of a circle of radius a with center O is divided into n equal parts. A point P lying on the line OA_1 and inside the circle is joined to each division point $A_1, A_2, A_3, \ldots, A_n$. Then

$$PA_1 \cdot PA_2 \cdots PA_n = a^n - x^n, \tag{13.9}$$

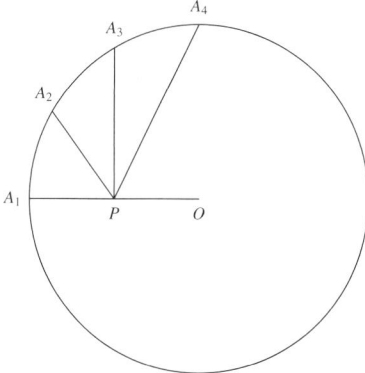

Figure 13.1. Cotes's factorization of $x^n - a^n$ as a property of the circle.

where $x = OP$. Cotes noted that if P lay outside the circle, the product equaled $x^n - a^n$, and he had a similar result for the factorization of $x^n + a^n$.

Cotes wrote to William Jones on May 5, 1716, that he had resolved by a general method the questions raised by Leibniz in his 1702 paper on the integration of rational functions. Unfortunately, Cotes died two months later, but Smith searched among his papers and unearthed the new method. Smith's note in his copy of the *Harmonia* stated: "Sir Isaac Newton, speaking of Mr. Cotes said 'if he had lived we might have known something.'"

It is very likely that the source of Cotes's inspiration was Bernoulli's paper on the integration of rational functions, pointing out the connection between the logarithm and the arctangent. In the second part of the *Logometria* on integration published posthumously in 1722, Cotes wrote that the close connection between the measure of angles and measure of ratios (logarithms) had persuaded him to propose a single notation to designate the two measures. He used the symbol

$$R \left| \frac{R+T}{S} \right. \tag{13.10}$$

to stand for $R \ln((R+T)/S)$ when R^2 was positive; when R^2 was negative, it represented $|R|\theta$, where θ was an angle such that the radius, tangent, and secant were in the ratio $R : T : S$. We should keep in mind that for Cotes, tangent and secant stood for $R \tan \theta$ and $R \sec \theta$. In his tables, he gave the single value

$$\frac{2}{e} R \left| \frac{R+T}{S} \right.$$

for the fluent of $\dot{x}/(e + fx^2)$; that is, for $\int dx/(e+fx^2)$, when $R = \sqrt{-e/f}$, $T = x$ and $S = \sqrt{(x^2 + e/f)}$. Recall that when $f < 0$, the integral is a logarithm and when $f > 0$, the integral is an arctangent. Cotes's notation distinguished the two cases by the interpretation of the symbols depending on R. This notation implies that when R is replaced by iR in the logarithm, we get the angular measure provided that S and T are replaced by $R \sec \theta$ and $R \tan \theta$. Thus, we have

$$iR \ln \frac{iR + R \tan \theta}{R \sec \theta} = R\theta + C, \text{ where } C \text{ is the constant of integration.}$$

This yields

$$\ln(\cos \phi + i \sin \phi) = i\phi,$$

when we take $C = -R\pi/2$ and $\theta - \pi/2 = \phi$. It may be of interest to note that, as de Moivre and Euler showed, this result connecting logarithms with angles also served as the basis for Cotes's factorization formula. Surprisingly, Johann Bernoulli did not make any use of his discovery of the connection between the logarithm and the arctangent (13.7). When Euler pointed out to Bernoulli that his formula implied that the logarithm of -1 was imaginary, he refused to accept it, maintaining that it had to be zero.

British mathematicians such as Newton and Cotes were ahead of the Continental European mathematicians in the matter of integration of rational functions, but by 1720 the Continental mathematicians had caught up. In 1718, Brook Taylor challenged them to integrate rational functions of the form

$$\frac{x^{m-1}}{e+fx^m+gx^{2m}}.$$

Johann Bernoulli and Jakob Hermann, a former student of Jakob Bernoulli, responded with solutions. In particular, they explained how the denominator could be factored into two trinomials of the form $a+bx^{m/2}+cx^m$.

We can describe the Newton–Cotes–Bernoulli–Leibniz algorithm for integrating a rational function $f(x)$ with real coefficients by writing

$$f(x) = P(x) + \frac{N(x)}{D(x)},$$

where P, N, D are polynomials with degree $N <$ degree D and where N and D have 1 as their greatest common divisor. Factorize $D(x)$ into linear and quadratic factors so that their coefficients are real:

$$D(x) = c \prod_{i=1}^{n}(x-a_i)^{e_i} \prod_{j=1}^{m}(x^2+b_jx+c_j)^{f_j}. \tag{13.11}$$

Then there are real numbers A_{ik}, B_{jk}, and C_{jk} such that

$$f(x) = P + \sum_{i=1}^{n}\sum_{k=1}^{e_i}\frac{A_{ik}}{(x-a_i)^k} + \sum_{j=1}^{m}\sum_{k=1}^{f_j}\frac{B_{jk}x+C_{jk}}{(x^2+b_jx+c_j)^k}. \tag{13.12}$$

From this it is evident that the result of the integration of $f(x)$ contains an algebraic part, consisting of a rational function, and a transcendental part, consisting of arctangents and logarithms.

Though Leibniz and Bernoulli had in principle solved the problem of the integration of rational functions, the practical problem of computing the constants a, b, c and A, B, C was formidable. In 1744, Euler tackled this problem in two long papers, taking up 140 pages of the Petersburg Academy Journal (or 125 pages of vol. 17 of Euler's *Opera Omnia*). In these papers he explained in detail how to compute A, B, C in (13.12) when the roots of the denominator were known. He also worked out a large number of special integrals of the form

$$\int \frac{x^m}{1 \pm x^n} dx, \tag{13.13}$$

where m and n were integers. By evaluating these integrals, Euler gained insight into several important topics. In fact, they provided him with new proofs of evaluations of zeta and L-series values; the reflection formula for the gamma function; and of the infinite product expressions for trigonometric functions. It is no wonder that Euler published several hundred pages on the integration of rational functions.

The problem of factoring a polynomial is a difficult one, so the partial fractions method has its drawbacks. A question raised in the nineteenth century was whether a part or all of the integral of a rational function could be obtained without factorizing the denominator. The Russian mathematician M. V. Ostrogradski published an algorithm in 1845 by which the rational part after integration could be obtained without factorization. In 1873, C. Hermite published a different algorithm and taught it in his courses at the École Polytechnique.

With the development of general computer algebra systems, the problem of mechanizing integration, including the integration of rational functions, has received new attention. The methods of Ostrogradski and Hermite, along with others, have been important in the development of symbolic integration. The question of obtaining the logarithmic or arctangent portion of the integral of a rational function, without factorization of the denominator, has been resolved by a host of researchers. In these symbolic integration methods, the problem of factorization is replaced by the much more accessible problems of obtaining the greatest common denominators and/or resultants of polynomials. These last procedures in turn require polynomial division and the elimination of variables. Contributors to symbolic integration are many, including M. Bronstein, R. Risch, and M. F. Singer.

13.2 Newton's 1666 Basic Integrals

In the beginning sections of his October 1666 tract on calculus, Newton tackled the problems of finding the areas under the curves $y = 1/(c+x)$ and $y = c/(a+bx^2)$, equivalent to evaluating integrals of those functions. Recall, however, that seventeenth-century mathematicians thought in terms of curves, even those defined by equations, rather than functions. The variables in an equation were regarded as quantities or magnitudes on the same footing, rather than dependent and independent variables. Newton's two integrals were the building blocks for the more general integrals of rational functions. It is interesting to read what he said about these integrals. He first noted the rule that if

$$\frac{q}{p} = ax^{\frac{m}{n}}, \quad \text{then} \quad y = \frac{na}{m+n} x^{\frac{m+n}{n}}.$$

Note that in Newton's 1690s notation q/p was written as $\dot{y}/\dot{x}$, whereas Leibniz wrote dy/dx. Newton next observed:

> Soe [so] if $\frac{a}{x} = \frac{q}{p}$. Then is $\frac{a}{0}x^0 = y$. soe y^t [that] y is infinite. But note y^t in this case x & y increase in y^e [the] same proportion y^t numbers & their logarithmes doe [do], y being like a logarithme added to an infinite number $\frac{a}{0}$. [That is, $\int_0^x \frac{a}{t} dt = a \ln x - a \ln 0 = a \ln x + \frac{a}{0}$.] But if x bee diminished by c, as if $\frac{a}{c+x} = \frac{q}{p}$, y is also diminished by y^e infinite number $\frac{a}{0}c^0$ & becomes finite like a logarithme of y^e number x. & so x being given, y may bee mechanically found by a Table of logarithmes, as shall be hereafter showne.

Here Newton was explaining how the logarithm could be obtained by an application of the power rule, since he regarded this as the fundamental rule for integration. Newton clearly saw the difficulty of division by zero when the rule was applied to $\frac{a}{x}$. His method of dealing with this stumbling block can now be seen as an attempt to define

13.2 Newton's 1666 Basic Integrals

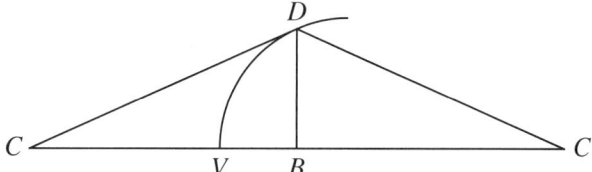

Figure 13.2. Newton's integration of a rational function.

the logarithm as a limit. We may say that Newton was describing the calculation:

$$y = \int_0^x \frac{a}{c+t}\,dt = \lim_{\epsilon \to 0} \int_0^x \frac{a}{(c+t)^{1-\epsilon}}\,dt$$
$$= \lim_{\epsilon \to 0}\left(\frac{a}{\epsilon}((c+x)^\epsilon - c^\epsilon)\right) = a\ln\frac{c+x}{c}. \quad (13.14)$$

As for the integral of $\frac{1}{1+x^2}$, we would now give the result as $\arctan x$. In Newton's time, the trigonometric quantities or functions were conceived of as line segments and their ratios constructed in relation to arcs of circles. It was therefore natural for Newton to connect the area under $y = \frac{1}{1+x^2}$ to the area of simpler or more well-known geometric objects such as conic sections. For this reason, he reduced the integral to the area of a sector of an ellipse. Consider the diagram in Newton's tract, Figure 13.2. Set $BD = v(x)$ and $CB = z(x)$, where C is the point on the right-hand side. Let $z(t) = 1/\sqrt{1+t^2}$ so that $tz(t) = \sqrt{1-z^2(t)} = v(t)/2$. Thus, the curve VD is an arc of the ellipse $(v/2)^2 + z^2 = 1$. Note that $CV = z(0) = 1$. In a one-line argument, Newton showed that $\int_0^x dt/(1+t^2)$ was equal to the area of sector CVD. To see this, observe that

$$\int_0^x \frac{1}{1+t^2}\,dt = \int_0^x z^2\,dt = xz^2(x) - \int_1^{z(x)} 2zt\,dz. \quad (13.15)$$

Since $2zt = v$, the rightmost integral represents the area under the ellipse from V to B; when the negative sign is included with the integral, the area under the ellipse from B to V is obtained. Moreover, $xz^2(x) = v(x)z(x)/2 = BD \cdot CB/2$, and hence $xz^2(x)$ represents the area of the triangle DBC, completing the proof.

Newton may also have known at this point that he could relate this area to an arc of the circle. Since $2zt = 2\sqrt{1-z^2}$, the integral on the right of (13.15) is twice the area under the circle $y = \sqrt{1-z^2}$ from $z(x)$ to 1. Recall that Newton had already related this area to the arcsine almost two years earlier when generalizing a result of Wallis, to obtain (9.14). Thus, he knew that

$$\int_{z(x)}^1 2zt\,dz = 2\int_{z(x)}^1 \sqrt{1-z^2}\,dz = 2\left(\frac{\pi}{4} - \frac{1}{2}z\sqrt{1-z^2} - \frac{1}{2}\arcsin z\right). \quad (13.16)$$

When (13.15) is combined with (13.16), we get the integral in terms of the arctangent:

$$\int_0^x \frac{1}{1+t^2}\,dt = \frac{\pi}{2} - \arcsin z(x) = \arccos\left(1/\sqrt{1+x^2}\right) = \arctan x. \quad (13.17)$$

13.3 Newton's Factorization of $x^n \pm 1$

Most probably around 1676, Newton wrote his very sketchy notes on this factorization, of which Whiteside has given a very helpful clarification. From these sources we learn that Newton's method of factoring $1 \pm x^m$ was to write

$$\left(1+nx+x^2\right)\left(1-nx+px^2-qx^3+rx^4-\cdots\pm x^{m-2}\right) \equiv 1 \pm x^m \qquad (13.18)$$

and equate the coefficients of x^i for $2 \le i \le m-1$ to 0 in order to obtain equations satisfied by $n, p, q, r, \ldots$. By eliminating $p, q, r, \ldots$, he obtained the algebraic equation satisfied by n. For example, when $m=4$, Newton's equations were $p+1-n^2=0$ and $pn-n=0$. Note that the first equation multiplied by n gives $pn+n-n^3=0$, and hence, by the second equation, $n^3 - 2n = 0$. One can then write $n = 0, \pm\sqrt{2}$. The first case gives the factorization $1-x^4 = (1-x^2)(1+x^2)$ and the second gives

$$1+x^4 = \left(1-\sqrt{2}x+x^2\right)\left(1+\sqrt{2}x+x^2\right). \qquad (13.19)$$

Recall that Newton applied this factorization to derive his series for $\pi/(2\sqrt{2})$. His cryptic remark on his method was apparently insufficient for Leibniz to decipher, so in 1702 Leibniz could obtain only (13.6), leading him to wonder if $\int dx/(a^4+x^4)$ could be expressed in terms of logarithms and arctangents. One may get a sense of the ill will existing at that time between the supporters of Newton and those of Leibniz from a remark in a 1716 letter from Roger Cotes to William Jones:

> M. Leibnitz, in the Leipsic Acts of 1702 p. 218 and 219, has very rashly undertaken to demonstrate that the fluent of $\dot{x}/(x^4+a^4)$ cannot be expressed by measures of ratios and angles; and he swaggers upon the occasion (according to his usual vanity), as having by this demonstration determined a question of the greatest moment.

Using the same method as before, $m=5$ gave Newton the equation $n^4 - 3n^2 + 1 = 0$, or $n^2 = \frac{3}{2} \pm \sqrt{\frac{5}{4}}$, or $n = \frac{1 \pm \sqrt{5}}{2}$. Thus, he got the factorization

$$1-x^5 = \left(1+\frac{1+\sqrt{5}}{2}x+x^2\right)\left(1-\frac{1+\sqrt{5}}{2}x+\frac{1+\sqrt{5}}{2}x^2-x^3\right).$$

Of course, the second factor is $(1-x)(1+\frac{1-\sqrt{5}}{2}x+x^2)$, though Newton did not write this out explicitly. Newton explicitly gave the factorization of x^6+1. Here the equations satisfied by the coefficients n, p, q, r are $q=nr$, $p=qn-r$, $n=pn-q$, and $n^2-p=1$. This implies that

$$p = n^2-1, \quad q = n^3-2n, \quad r = n^4-3n^2+1, \qquad (13.20)$$

and hence the equation satisfied by n is

$$n^5 - 4n^3 + 3n = 0.$$

This indicates the values of n to be 0, ± 1, $\pm\sqrt{3}$. When $n = 0$, we have $r = 1$, $q = 0$, $p = -1$, and the factorization

$$1 + x^6 = \left(1 + x^2\right)\left(1 - x^2 + x^4\right). \tag{13.21}$$

When $n = \sqrt{3}$, we have $r = 1$, $q = \sqrt{3}$, $p = 2$, and the factorization given by Newton was

$$1 + x^6 = \left(1 + \sqrt{3}x + x^2\right)\left(1 - \sqrt{3}x + 2x^2 - \sqrt{3}x^3 + x^4\right). \tag{13.22}$$

It follows that the second factor in (13.21) is $1 - x^2 + x^4 = (1 + \sqrt{3}x + x^2)(1 - \sqrt{3}x + x^2)$, and the second factor in (13.22) can be written as $(1 + x^2)(1 - \sqrt{3}x + x^2)$.

Newton wrote down the polynomials satisfied by n for $m = 3$ to $m = 12$ and solved the polynomials for those cases when n could be expressed in terms of quadratic surds. In the case $m = 7$, he had the equation $n^6 - 5n^4 + 6n^2 - 1 = 0$. He wrote "$n^2 =$" next to the equation and filled in no value when he realized it would involve cube roots. For $m = 10$, he did not appear to expect the solutions to be in terms of quadratic surds and wrote nothing after the equation for n. He appears to have missed the factorization, noted by Whiteside,

$$n^9 - 8n^7 + 21n^5 - 20n^3 + 5n = n\left(n^4 - 5n^2 + 5\right)(n^4 - 3n^2 + 1),$$

yielding the quadratic surds for n when $m = 10$.

Newton seems to have grasped the connection between the values of n and the cosines of $\frac{2k\pi}{m}$ or $\frac{(2k-1)\pi}{m}$. He drew a diagram of a right triangle with one angle as $22\frac{1}{2}° = \pi/8$ and noted that $2\cos(\pi/8)$ gave a value of n when $m = 8$. At this point, he was just one step away from Cotes's factorization of $a^n \pm x^n$. Moreover, the number $\cos\frac{2\pi}{m}$ is related to the length of a side of a regular polygon of m sides. Such a polygon is constructible when $\cos\frac{2\pi}{m}$ can be expressed in terms of quadratic surds. It is unlikely that Newton considered constructible polygons, but he may have wondered about conditions for n to be expressed in quadratic surds. Thus, we have an interesting connection between Newton and Gauss, although Gauss could not have been aware of it because Newton did not publish his work, since it was incomplete.

13.4 Cotes and de Moivre's Factorizations

De Moivre presented his method of factorizing the more general trinomial $x^{2n} - 2\cos n\theta\, x^n + 1$ in his *Miscellanea Analytica* of 1730. His method depended on a formula he stated without proof: Let l and x be cosines of arcs A and B, respectively, of the unit circle where A is to B as the integer n to one. Then

$$x = \frac{1}{2}\sqrt[n]{l + \sqrt{l^2 - 1}} + \frac{1}{2}\frac{1}{\sqrt[n]{l + \sqrt{l^2 - 1}}}. \tag{13.23}$$

Note that this is equivalent to the formula named after de Moivre:

$$\cos\theta = \frac{1}{2}\left((\cos n\theta + i\sin n\theta)^{1/n} + (\cos n\theta - i\sin n\theta)^{1/n}\right). \tag{13.24}$$

De Moivre had published a similar result without proof for sine in a *Philosophical Transactions* paper of 1707. In the chapter on difference equations, we present a proof by Daniel Bernoulli in which he solved a difference equation obtained from the addition formula for cosine. Here we present Euler's simple proof given in his *Introductio in Analysin Infinitorum* of 1748. Observe that by the addition formulas for sine and cosine

$$(\cos y \pm i\sin y)(\cos z \pm i\sin z) = \cos(y+z) \pm i\sin(y+z). \tag{13.25}$$

By taking $y = z$, Euler had

$$(\cos y \pm i\sin y)^2 = (\cos 2y \pm i\sin 2y).$$

When both sides were multiplied by $\cos y \pm i\sin y$, he got

$$(\cos y \pm i\sin y)^3 = \cos 3y \pm i\sin 3y.$$

Finally, it followed by induction that for a positive integer n,

$$(\cos y \pm i\sin y)^n = \cos ny \pm i\sin ny, \tag{13.26}$$

completing the proof. To obtain the factorization, de Moivre set $z = \sqrt[n]{l + \sqrt{l^2 - 1}}$ so that

$$z^n - l = \sqrt{l^2 - 1} \text{ or } z^{2n} - 2lz^n + 1 = 0, \quad \text{where } l = \cos n\theta.$$

By de Moivre's formula (13.23), $x = (z + 1/z)/2$, or $z^2 - 2zx + 1 = 0$, where $x = \cos\theta$. De Moivre's theorem was therefore equivalent to the statement that $z^{2n} - 2\cos n\theta z^n + 1 = 0$, when $z^2 - 2\cos\theta z + 1 = 0$. Thus, $z^2 - 2xz + 1$ was a factor of $z^{2n} - 2lz^n + 1$. To obtain the other $n-1$ factors, de Moivre observed that

$$(\cos A \pm i\sin A)^{1/n} = \cos\left(\frac{2k\pi \pm A}{n}\right) + i\sin\left(\frac{2k\pi \pm A}{n}\right), \quad k = 0, 1, 2, \ldots.$$

The factorization thus obtained after taking $\theta = A/n$ may be written in modern notation as

$$z^{2n} - (2\cos A)z^n + 1 = \prod_{k=0}^{n-1}\left(z^2 - 2\cos\left(\frac{2k\pi + A}{n}\right)z + 1\right). \tag{13.27}$$

We note that de Moivre used the symbol C for 2π. Cotes's factorization theorems are actually corollaries of de Moivre's (13.27). For example, let C be a circle of radius a and center O with B a point on the circumference and P a point on OB such that $OP = x$. Also let $A_1, A_2, \ldots, A_n$ be points on the circumference such that, for $k = 1, 2, \ldots, n$, the angle $BOA_k = (2k-1)\pi/n$. Then the product $A_1P \cdot A_2P \cdots A_nP$ is equal to $x^n + a^n$. This result of Cotes can be derived by taking $A = \pi$ in (13.27). But by taking $A = 0$, we obtain the result of Cotes (13.9).

13.5 Euler: Integration of Rational Functions

In his 1744 papers on the integration of rational functions, Euler assumed that the factors of the denominator of the rational function were known. He then gave explicit formulas for decomposing the rational functions to partial fractions. From these he obtained the integral. Euler then gave applications to specific integrals where he could factorize the denominators by Cotes's formula or, more generally, by de Moivre's formula. When these specific integrals are taken over the interval $(0, \infty)$, they become extremely important examples of beta integrals. In fact, Euler's results using these integrals in effect provided a new proof of his reflection formula for the gamma function. His representation of the integral as an infinite series of partial fractions then gave Euler the partial fraction expansion of the cosecant function – and hence the infinite products for the sine and other trigonometric functions. Euler worked out these connections in the 1740s. But he never really abandoned any area of study, and so he returned to this topic thirty years later with new insights, allowing him to streamline computations and make the details more transparent.

Thus, in his first paper of 1744 on integration of rational functions, Euler took the polynomial in the denominator to be

$$N(x) = (1 + px)(1 + qx)(1 + rx) \cdots,$$

where some factors could be repeated. Suppose $M(x)$ is a polynomial of degree less than that of $N(x)$. Now allow that $1 + px$ is repeated exactly n times and that

$$N(x) = (1 + px)^n A(x),$$

$$\frac{M(x)}{N(x)} = \frac{C_1}{1 + px} + \frac{C_2}{(1 + px)^2} + \cdots + \frac{C_n}{(1 + px)^n} + \frac{D(x)}{A(x)}, \qquad (13.28)$$

$$\text{and} \quad V(p) = p^{n-1} \frac{M(-1/p)}{A(-1/p)}. \qquad (13.29)$$

Euler showed that

$$C_k = \frac{p}{(n-k)!} \frac{d^{n-k}}{dp^{n-k}} (V/p^k), \; k = 1, 2, \ldots, n. \qquad (13.30)$$

In the second paper, Euler presented the formula for the case where the factors were of the form $(p + qx)^n$. He let $N(x) = (p + qx)^n S$ and

$$\frac{M(x)}{N(x)} = \frac{b_0}{(p+qx)^n} + \frac{b_1}{(p+qx)^{n-1}} + \cdots + \frac{b_{n-1}}{p+qx} + \frac{D(x)}{S(x)}.$$

He showed that

$$b_j = \frac{1}{j! q^j} \left(\frac{d^j}{dx^j} (M/S) \right)_{x=-p/q}, \; j = 0, 1, 2, \ldots, n-1. \qquad (13.31)$$

In both papers, Euler worked out some important specific examples of integrals of rational functions in which the denominator had quadratic factors. Consider his important example related to the beta integral:

$$\int \frac{x^m\, dx}{1+x^{2n}}, \quad m < 2n. \tag{13.32}$$

Euler started with a formula well known in his time:

$$\int \frac{(P+Qx)\,dx}{1+px+qxx} = \frac{Q}{2q}\ln(1+px+qxx) + \frac{2Pq-Qp}{q\sqrt{(4q-pp)}}\arctan\frac{x\sqrt{(4q-pp)}}{2+px}, \tag{13.33}$$

when $4q > p^2$. Note that Euler wrote A. tang. for arctangent. He then observed that the factors of $1+x^{2n}$ were

$$1+2x\cos\frac{k\pi}{2n}+xx, \quad k=1,3,\ldots,2n-1. \tag{13.34}$$

He referred to de Moivre's *Miscellanea Analytica* for this result, but, as we have seen, this factorization was first given by Cotes. Euler showed that the partial fractions of $x^m/(1+x^{2n})$ were of the form

$$\frac{(-1)^m}{n}\cdot\frac{\cos\frac{mk\pi}{2n}+x\cos\frac{(m+1)k\pi}{2n}}{1+2x\cos\frac{k\pi}{2n}+x^2}, \quad \text{where}\ \ k=1,3,\ldots,2n-1. \tag{13.35}$$

Hence by (13.33), the integral (13.32) was a sum of terms of the form

$$\frac{(-1)^m}{2n}\cos\frac{(m+1)k\pi}{2n}\ln\left(1+2\cos\frac{k\pi}{2n}+xx\right)$$
$$+\frac{(-1)^m}{n}\sin\frac{(m+1)k\pi}{2n}\arctan\frac{x\sin(k\pi/2n)}{1+x\cos(k\pi/2n)}. \tag{13.36}$$

Further details of these evaluations are given in the next section in connection with Euler's calculations for a slightly more general integral.

13.6 Euler's Generalization of His Earlier Work

Late in his life, Euler used his earlier method to evaluate a more general integral, with gratifying results. In the 1740s, he used Cotes's factorization; later on, by using de Moivre's factorization applied to a more general polynomial, Euler was able to obtain the Fourier series for $\cos\lambda x$ and $\sin\lambda x$. This result appeared in the paper "Investigatio Valoris Integralis," published posthumously in 1785. Using modern notation, we give a brief sketch of his evaluation of the integral

$$\int_0^\infty \frac{x^{m-1}\,dx}{1-2x^k\cos\theta+x^{2k}}.$$

By de Moivre's product for the denominator of the integrand

$$\frac{x^{m-1}}{1-2x^k\cos\theta+x^{2k}} = \sum_{s=0}^{k-1}\frac{A_s+B_s x}{1-2x\cos\left(\frac{2s\pi+\theta}{k}\right)+x^2}.$$

13.6 Euler's Generalization of His Earlier Work

Let $\omega_s = (2s\pi + \theta)/k$ and

$$\frac{A_s + B_s x}{1 - 2x \cos w_s + x^2} = \frac{f_s}{x - e^{iw_s}} + \frac{g_s}{x - e^{-iw_s}}$$

so that

$$B_s = f_s + g_s \text{ and } A_s = (f_s - g_s)i \sin w_s - (f_s + g_s) \cos w_s.$$

To find f_s, observe that

$$\frac{x^{m-1}(x - e^{iw_s})}{1 - 2x^k \cos \theta + x^{2k}} = f_s + R_s(x - e^{iw_s}),$$

where R_s consists of the remaining partial fractions. The limit as $x \to e^{iw_s}$ gives f_s and here Euler applied l'Hôpital's rule to find

$$f_s = \frac{e^{imw_s}}{2ki \sin \theta e^{i\theta}}; \quad \text{similarly,} \quad g_s = \frac{e^{-imw_s}}{-2ki \sin \theta e^{-i\theta}}.$$

Therefore, $\quad B_s = \dfrac{\sin(mw_s - \theta)}{k \sin \theta}, \quad A_s = -\dfrac{\sin((m-1)w_s - \theta)}{k \sin \theta}.$

Now

$$\int \frac{(A_s + B_s x)\,dx}{1 - 2x \cos w_s + x^2} = \int \left(\frac{B_s(x - \cos \omega_s)}{1 - 2x \cos w_s + x^2} + \frac{(A_s + B_s \cos \omega_s)}{1 - 2x \cos w_s + x^2} \right) dx$$

$$= \frac{1}{2} B_s \ln(1 - 2x \cos w_s + x^2) + \frac{A_s + B_s \cos \omega_s}{\sin w_s} \arctan \frac{x \sin w_s}{1 - x \cos w_s}.$$

The original integral is a sum of these integrals as s ranges from 0 to $k-1$. We have also to integrate over $(0, \infty)$. At $x = 0$, the logarithm is zero, as is the arctangent. For large x, the logarithm behaves as $2 \ln x$. So the logarithmic part can be written as

$$\ln x \sum_{s=0}^{k-1} B_s = \frac{\ln x}{k \sin \theta} \sum_{s=0}^{k-1} \sin\left((m-k)\frac{\theta}{k} + 2s\frac{m\pi}{k} \right)$$

$$= \frac{\ln x}{k \sin \theta} \sum_{s=0}^{k-1} \sin(2s\alpha + \zeta),$$

where $\alpha = m\pi/k$ and $\zeta = (m - k)\theta/k$. Observe that

$$2 \sin \alpha \sin(2s\alpha + \zeta) = \cos((2s - 1)\alpha - \zeta) - \cos((2s + 1)\alpha + \zeta),$$

and hence, after cancellation,

$$2 \sin \alpha \sum_{s=0}^{k-1} \sin(2s\alpha + \zeta) = \cos(\alpha - \zeta) - \cos((2k - 1)\alpha + \zeta).$$

Now note that the sum of the angles $\alpha - \zeta$ and $(2k-1)\alpha + \zeta$ is $2k\alpha = 2m\pi$ and hence $\sum B_s = 0$. Thus, we know that the integral has no logarithmic part. For large x, $\arctan(x \sin w_s/(1 - x \cos w_s))$ behaves like

$$\arctan(-\tan w_s) = \pi - w_s,$$

and by a short calculation,

$$\frac{A_s + B_s \cos \omega_s}{\sin w_s} = \frac{\cos(mw_s - \theta)}{k \sin \theta}.$$

Therefore, the sum of the arctangents at $x = \infty$ is

$$\frac{1}{k \sin \theta} \sum_{s=0}^{k-1} (\pi - w_s) \cos(mw_s - \theta).$$

Set $\frac{\pi}{k} = \beta$, $\pi - \frac{\theta}{k} = \gamma$ and let α, ζ be as before. Denote the preceding sum, without the factor $1/k \sin \theta$, by S. An application of the addition formula for the sine function followed by an easy simplification gives

$$2S \sin \alpha = \gamma \sin(\alpha - \zeta) + (\gamma - (2k-1)\beta) \sin((2k-1)\alpha + \zeta) + \beta T \qquad (13.37)$$

where $T = 2(\sin(\alpha + \zeta) + \sin(3\alpha + \zeta) + \cdots + \sin((2k-3)\alpha + \zeta))$.

The series T is summed similarly. Yet another application of the addition formula for the cosine function results in

$$T \sin \alpha = \cos \frac{\theta(k-m)}{k} - \cos \frac{2m\pi + \theta(k-m)}{k} = 2 \sin \frac{m\pi + \theta(k-m)}{k} \sin \frac{m\pi}{\alpha}$$

$$= 2 \sin(\alpha - \zeta) \sin \alpha.$$

Substitute this value of T in (13.37) to obtain

$$2S \sin \alpha = (\gamma + 2\beta) \sin(\alpha - \zeta) + (\gamma - 2(k-1)\beta) \sin((2k-1)\alpha + \zeta)$$
$$= (\gamma + 2\beta)(\sin(\alpha - \zeta) + \sin((2k-1)\alpha - \zeta)) - 2k\beta \sin((2k-1)\alpha + \zeta)$$
$$= (2\gamma + 4\beta) \sin \alpha k \cos((k-1)\alpha + \zeta) - 2k\beta \sin((2k-1)\alpha + \zeta).$$

Now $\sin \alpha k = \sin m\pi = 0$. So

$$S = -k\beta \frac{\sin((2k-1)\alpha + \zeta)}{\sin \alpha} = \pi \frac{\sin \frac{m\pi + \theta(k-m)}{k}}{\sin \frac{m\pi}{k}}.$$

Thus, the final result is

$$\int_0^\infty \frac{x^{m-1}}{1 - 2x^k \cos \theta + x^{2k}} dx = \frac{\pi \sin \frac{m(\pi - \theta) + k\theta}{k}}{k \sin \theta \sin \frac{m\pi}{k}}.$$

The special case $\theta = \pi/2$ gave Euler the value of the beta integral

$$\int_0^\infty \frac{x^{m-1}}{1+x^{2k}} dx = \frac{\pi}{2k \sin \frac{m\pi}{2k}}, \qquad (13.38)$$

and from $\theta = \pi$, he obtained

$$\int_0^\infty \frac{x^{m-1} dx}{(1+x^k)^2} = \frac{(1-\frac{m}{k})\pi}{k \sin \frac{m\pi}{k}}.$$

Observe that this integral can be obtained from the previous beta integral by integration by parts.

13.7 Hermite's Rational Part Algorithm

As a professor at the École Polytechnique, Hermite lectured on analysis. This gave him the opportunity to rethink several elementary topics. He often came up with new proofs and presentations of old material. In his lectures, published in 1873, Hermite gave a method for finding the rational part of the integral of a rational function, by employing the Euclidean algorithm. He first found the square-free factorization of the denominator $Q(x)$ of the rational function $P(x)/Q(x)$:

$$Q = Q_1 Q_2^2 \cdots Q_n^n, \qquad (13.39)$$

where $Q_1, Q_2, \ldots, Q_n$ were the relatively prime polynomials with simple roots. This decomposition could be accomplished by the Euclidean algorithm, but Hermite did not give details in his published lectures. Note that there existed polynomials $P_1, P_2, \ldots, P_n$ such that

$$\frac{P}{Q} = \frac{P_1}{Q_1} + \frac{P_2}{Q_2^2} + \cdots + \frac{P_n}{Q_n^n}. \qquad (13.40)$$

As a first step in the derivation of this relation, Hermite observed that $U = Q_1$, and that $V = Q_2^2 \cdots Q_n^n$ were relatively prime and hence by the Euclidean algorithm, there existed polynomials P_1 and $\tilde{P}_1$ such that

$$P = P_1 V + \tilde{P}_1 U$$

or $\quad \dfrac{P}{Q} = \dfrac{P_1}{Q_1} + \dfrac{\tilde{P}_1}{Q_2^2 \cdots Q_n^n}. \qquad (13.41)$

Hermite obtained the required result by a repeated application of this procedure. Since

$$\int \frac{P}{Q} = \int \frac{P_1}{Q_1} + \int \frac{P_2}{Q_2^2} + \cdots + \int \frac{P_n}{Q_n^n}, \qquad (13.42)$$

he needed a method to reduce $\int \frac{P_k}{Q_k^k}$ to $\int \frac{E}{Q_k^{k-1}}$, for some polynomial E, when $k > 1$. Since Q_k had simple roots, Q_k and its derivative Q_k' were relatively prime. Thus, there existed polynomials C and D such that

$$P_k = CQ_k + DQ'_k \quad \text{and}$$

$$\frac{P_k}{Q_k^k} = \frac{C}{Q_k^{k-1}} + \frac{DQ'_k}{Q_k^k} = \frac{C}{Q_k^{k-1}} - \frac{D}{k-1}\frac{d}{dx}\left(\frac{1}{Q_k^{k-1}}\right).$$

After integration by parts, he obtained the necessary reduction

$$\int \frac{P_k}{Q_k^k} = -\frac{D}{(k-1)Q_k^{k-1}} + \int \frac{C + D'/(k-1)}{Q_k^{k-1}}. \tag{13.43}$$

Again, by a repeated application of this algorithm, Hermite had

$$\int \frac{P}{Q} = R + \int \frac{P_1}{Q_1} + \int \frac{\tilde{P}_2}{Q_2} + \int \frac{\tilde{P}_3}{Q_3} + \cdots + \int \frac{\tilde{P}_n}{Q_n}, \tag{13.44}$$

where R was a rational function. Since $Q_1, Q_2, \ldots, Q_n$ were pairwise relatively prime and had simple roots, the integrals on the right-hand side formed the transcendental part of the original integral.

13.8 Johann Bernoulli: Integration of $\sqrt{ax^2 + bx + c}$

We have seen that Isaac Barrow geometrically evaluated the integrals $\int \sqrt{x^2 + a^2}\,dx$ and $\int dx/\sqrt{x^2 + a^2}$ and that his results could be immediately converted to analytic form. Roger Cotes included in his tables of integrals those of the form

$$\int R(x,t)\,dx \quad \text{with} \quad t = \sqrt{ax^2 + bx + c},$$

where the integrand was a rational function of x and t. Clearly, seventeenth- and eighteenth-century mathematicians knew how to handle such integrals. But Johann Bernoulli pointed out in his very first lecture on integration, contained in vol. 3 of his *Opera Omnia*, that there was another method, related to Diophantine problems. By a substitution used in the study of Diophantine equations, the integral $\int R(x,t)\,dx$ could be rationalized. At the end of his lecture, Bernoulli illustrated this idea by means of an example: His problem was to integrate $a^3 dx : x\sqrt{ax - x^2}$; his method was to rewrite the quantity within the root as a square containing x and a newly introduced variable. In this case, he had $ax - x^2 = a^2x^2 : m^2$. Thus, $x = am^2 : (m^2 + a^2)$, $dx = 2a^3 m\,dm : (m^2 + a^2)^2$ and

$$\int \frac{a^3\,dx}{x\sqrt{ax - x^2}} = \int \frac{2a^3\,dm}{m^2} = -\frac{2a^3}{m}.$$

We note that a general substitution of the form $ax^2 + bx + c = (u + \sqrt{a}x)^2$ could be used to rationalize integrals involving $\sqrt{ax^2 + bx + c}$.

13.9 Exercises

1. Prove Newton's formula (13.3) by showing

   (a) $\int_0^x \frac{1+t^2}{1+t^4} dt = \frac{1}{\sqrt{2}} \arctan \frac{x\sqrt{2}}{1-x^2}.$

   (b) $\int_0^x \frac{1+t^2}{1+t^4} dt = \frac{x}{1} + \frac{x^3}{3} - \frac{x^5}{5} - \frac{x^7}{7} + \frac{x^9}{9} + \frac{x^{11}}{11} - \cdots,$ for $0 \leq x \leq 1$.

2. In his October 24, 1676 letter to Oldenburg, Newton remarked that Leibniz's series and his own variant of it were unsuitable for the approximate evaluation of π: "For if one wished by the simple calculation of this series $1 + \frac{1}{3} - \frac{1}{5} - \frac{1}{7} + \frac{1}{9} +$ etc. to find the length of the quadrant to twenty decimal places, it would need about 5000000000 terms of the series, for the calculation of which 1000 years would be required." He recommended his series for arcsin for this purpose. He suggested another formula to evaluate π:

$$\frac{\pi}{4} = \frac{a}{1} - \frac{a^3}{3} + \frac{a^5}{5} - \frac{a^7}{7} + \text{etc.} + \frac{a^2}{1} + \frac{a^5}{3} - \frac{a^8}{5} - \frac{a^{11}}{7} + \frac{a^{14}}{9} + \frac{a^{17}}{11} - \text{etc.}$$

$$+ \frac{a^4}{1} - \frac{a^{10}}{3} + \frac{a^{16}}{5} - \frac{a^{22}}{7} + \frac{a^{28}}{9} - \text{etc.},$$

where $a = 1/2$. Prove this formula and show that it is equivalent to

$$\frac{\pi}{4} = \arctan \frac{1}{2} + \frac{1}{2} \arctan \frac{4}{7} + \frac{1}{2} \arctan \frac{1}{8}.$$

Also prove Newton's formula:

$$1 - \frac{1}{7} + \frac{1}{9} - \frac{1}{15} + \frac{1}{17} - \frac{1}{23} + \frac{1}{25} - \frac{1}{31} + \frac{1}{33} + \text{etc.} = \frac{\pi}{4}(1 + \sqrt{2}).$$

3. Derive Newton's equations for n defined by (13.18) when $m = 3, 4, 5, 6, 7, 8, 9, 10, 11$, and 12. For these values of m, Newton had, respectively,

$$nn - 1 = 0, \quad n^3 - 2n = 0, \quad n^4 - 3nn + 1 = 0, \quad n^5 - 4n^3 + 3n = 0,$$

$$n^6 - 5n^4 + 6nn - 1 = 0, \quad n^7 - 6n^5 + 10n^3 - 4n = 0, \quad n^8 - 7n^6 + 15n^2 - 10nn + 1 = 0,$$

$$n^9 - 8n^7 + 21n^5 - 20n^3 + 5n = 0, \quad n^{10} - 9n^8 + 28n^6 - 35n^4 + 15nn - 1 = 0,$$

$$n^{11} - 10n^9 + 36n^7 - 56n^5 + 35n^3 - 6n = 0.$$

Use Newton's equation for $m = 7$ to show that the cubic equation satisfied by $2\cos\frac{2\pi}{7}$ is $x^3 + x^2 - 2x - 1 = 0$.

4. Prove Newton's integration formula (13.1).

5. Show that for $e > 0$ and $f < 0$

$$\int \frac{dx}{d+fx^2} = \frac{1}{e} R \ln \frac{R+T}{S} + \frac{1}{e} R \ln \frac{1}{i},$$

where $R = \sqrt{-e/f}$, $T = x$, $S = \sqrt{x^2 + e/f}$. Then show that for $e > 0$ and $f > 0$

$$\int \frac{dx}{e + fx^2} = \frac{1}{e} R \arctan \frac{x}{R},$$

where $R = \sqrt{e/f}$. Compare these results with comments on Cotes's notation for integrals of rational functions in the preliminary remarks for this chapter.

6. Prove that

$$\int \frac{dx}{e + fx^2 + gx^4} =$$

(a) (when $4eg < f^2$ and $b^2 = e$) $\int \frac{\alpha \, dx}{b + mx^2} + \int \frac{\beta \, dx}{b + nx^2}$, where α, β, m, n

can be determined in terms of e, f, g,

(b) (when $4eg \geq f^2$ and $b^2 = e$) $\int \frac{(\alpha + \gamma x) \, dx}{b + nx + mx^2} + \int \frac{(\beta + \epsilon x) \, dx}{b - nx + nx^2}$, where

$\alpha, \beta, \gamma, \epsilon, m, n$ can be determined in terms of e, f and g.

Bernoulli published an entertaining paper containing this result in the *Acta Eruditorum* in response to a challenge from Brook Taylor. See Joh. Bernoulli (1968), vol. II, p. 409.

7. Use Hermite's reduction formula (13.43) to show that the integral

$$\int \frac{4x^9 + 21x^6 + 2x^3 - 3x^2 - 3}{(x^7 - x + 1)^2} \, dx$$

has only the rational part $-(x^3 + 3)/(x^7 - x + 1)$ and no transcendental part. See G. H. Hardy (1905), pp. 14–15.

8. Prove that

$$a^{2n-2} \int \frac{dx}{(x^2 - a^2)^n} = (-1)^{n-1} \frac{1 \cdot 3 \cdots (2n - 3)}{2 \cdot 4 \cdots (2n - 2)} \left(\int \frac{dx}{x^2 - a^2} + f_{n-1}(x) \right),$$

where

$$f_n(x) = \frac{x}{x^2 - a^2} \left(1 + \sum_{k=1}^{n-1} (-1)^k \frac{2 \cdot 4 \cdots (2k)}{3 \cdot 5 \cdots (2k + 1)} \frac{a^{2k}}{(x^2 - a^2)^k} \right).$$

See Hermite (1905–1917), vol. 3, p. 50.

9. Show that

$$\int \frac{(1 - x) \, dx}{x^4 (2x - 1)^3 (3x - 2)^2 (4x - 3)} = C - \frac{1}{36x^3} - \frac{7}{18xx} - \frac{1879}{432x} + \frac{24499}{1296} \ln x$$

$$+ \frac{8}{(2x - 1)^2} + \frac{48}{2x - 1} - 272 \ln(2x - 1) + \frac{729}{16(3x - 2)}$$

$$+ \frac{3645}{16} \ln(3x - 2) + \frac{2048}{81} \ln(4x - 3).$$

See Eu. I-17, p. 165.

10. Show that

$$\int \frac{dx}{1+x^5} = \frac{1}{5}\ln(1+x) + \frac{1}{5}\cos\frac{2\pi}{5}\ln\left(1+2x\cos\frac{2\pi}{5}+x^2\right)$$
$$+\frac{2}{5}\sin\frac{2\pi}{5}\arctan\frac{x\sin(2\pi/5)}{1+x\cos(2\pi/5)} - \frac{1}{5}\cos\frac{\pi}{5}$$
$$\times \ln\left(1-2x\cos\frac{\pi}{5}+x^2\right) + \frac{2}{5}\sin\frac{\pi}{5}\arctan\frac{x\sin(\pi/5)}{1-x\cos(\pi/5)},$$

$$\cos\frac{\pi}{5} = \frac{1+\sqrt{5}}{4}, \quad \sin\frac{\pi}{5} = \frac{\sqrt{(10-2\sqrt{5})}}{4},$$
$$\cos\frac{2\pi}{5} = \frac{-1+\sqrt{5}}{4}, \quad \sin\frac{2\pi}{5} = \frac{\sqrt{(10+2\sqrt{5})}}{4}.$$

Also show that

$$\int \frac{dx}{1+x^6} = \frac{\sqrt{3}}{12}\ln\frac{1+x\sqrt{3}+x^2}{1-x\sqrt{3}+x^2} + \frac{1}{3}\arctan x$$
$$+ \frac{1}{6}\arctan\frac{x}{2+x\sqrt{3}} + \frac{1}{6}\arctan\frac{x}{2-x\sqrt{3}}.$$

Note that the second formula implies

$$1 - \frac{1}{7} + \frac{1}{13} - \frac{1}{19} + \frac{1}{25} - \frac{1}{31} + \cdots = \frac{\pi}{6} + \frac{\sqrt{3}}{12}\ln\frac{2+\sqrt{3}}{2-\sqrt{3}}.$$

See Eu. I-17, pp. 131 and 120, respectively, for the two formulas.

11. Show that for $0 < m < 2k$,

$$\int_0^\infty \frac{x^{m-1}\,dx}{1+2x^k\cos\theta+x^{2k}} = \frac{\pi\sin\left(1-\frac{m}{k}\right)\theta}{k\sin\theta\sin\frac{m\pi}{k}}.$$

See Eu. I-18, p. 202.

12. Show that for $0 < m < k$

$$\int_0^\infty \frac{x^{m-1}\,dx}{(1+x^k)^n} = \frac{\pi}{k\sin\frac{m\pi}{k}}\prod_{s=1}^{n-1}\left(1-\frac{m}{sk}\right).$$

See Eu. I-18, p. 188.

13. (a) Show that

$$\frac{\sin\eta}{1+2x^k\cos\eta+x^{2k}} = \sin\eta - x^k\sin 2\eta + x^{2k}\sin 3\eta - x^{3k}\sin 4\eta + \cdots.$$

(b) Write the integral in the previous exercise as

$$\int_0^1 \frac{x^{m-1}+x^{2k-m-1}}{1+2x^k\cos\eta+x^{2k}}\,dx.$$

Next, apply part (a) to obtain

$$\frac{\pi \sin\left(1 - \frac{m}{k}\right)\eta}{k \sin \frac{m\pi}{k}} = \sum_{s=1}^{\infty} (-1)^{s-1} \left(\frac{1}{m + (s-1)k} + \frac{1}{(s+1)k - m} \right) \sin s\eta.$$

(c) Deduce that

$$\frac{\pi \sin(n\eta/k)}{2k^2 \sin(n\pi/k)} = \frac{\sin \eta}{k^2 - n^2} - \frac{2\sin 2\eta}{4k^2 - n^2} + \frac{3\sin 3\eta}{9k^2 - n^2} - \frac{4\sin 4\eta}{16k^2 - n^2} + \cdots,$$

$$\frac{\pi n \cos(n\eta/k)}{2k^3 \sin(n\pi/k)} = \frac{\cos \eta}{k^2 - n^2} - \frac{4\cos 2\eta}{4k^2 - n^2} + \frac{9\cos 3\eta}{9k^2 - n^2} - \frac{16\cos 4\eta}{16k^2 - n^2} + \cdots,$$

$$\frac{\pi \cos(n\eta/k)}{2nk \sin(n\pi/k)} = \frac{1}{2n^2} + \frac{\cos \eta}{k^2 - n^2} - \frac{\cos 2\eta}{4k^2 - n^2} + \frac{\cos 3\eta}{9k^2 - n^2} - \frac{\cos 4\eta}{16k^2 - n^2} + \cdots.$$

Consult Eu. I-18 pp. 202–208.

14. Apply the factorization

$$1 - x^{2n} = (1 - x^2) \prod_{k=1}^{n-1} \left(1 + 2x \cos \frac{k\pi}{n} + x^2 \right)$$

to show that

(a)

$$\int_0^x \frac{x^{m-1}}{1 - x^{2n}} dx = \frac{-1}{2n} \ln(1 - x) + \frac{(-1)^{m-1}}{2n} \ln(1 + x)$$

$$+ \frac{(-1)^{m-1}}{2n} \sum_{k=1}^{n-1} \cos \frac{km\pi}{n} \ln\left(1 + 2x\cos \frac{\pi}{n} + x^2\right)$$

$$+ \frac{(-1)^{m-1}}{n} \sum_{k=1}^{n-1} \sin \frac{km\pi}{n} \arctan \frac{x \sin(k\pi/n)}{1 + x\cos(k\pi/n)}.$$

(b)

$$\int_0^x \frac{x^{m-1} - x^{2n-m-1}}{1 - x^{2n}} dx = \frac{2(-1)^{m-1}}{n} \sum_{k=1}^{n-1} \sin \frac{km\pi}{n} \arctan \frac{x \sin(k\pi/n)}{1 + x\cos(k\pi/n)}.$$

(c)

$$\int_0^1 \frac{x^{m-1} - x^{2n-m-1}}{1 - x^{2n}} dx = \frac{\pi}{2n} \cot \frac{m\pi}{2n}.$$

See Eu. I-17, pp. 35–69.

15. Show that

$$\int_0^{\infty} \frac{dx}{(x^4 + 2ax^2 + 1)^{m+1}} = \frac{\pi}{2^{m+3/2}(a+1)^{m+1/2}} P_m(a),$$

where

$$P_m(a) = 2^{-2m} \sum_{k=0}^{m} 2^k \binom{2m-2k}{m-k} \binom{m+k}{m} (a+1)^k.$$

See Moll (2002) or Boros and Moll (2004), p. 154 for this example and for some intriguing open questions related to the integration of rational functions.

13.10 Notes on the Literature

Newton's October 1666 tract, first published in vol. 1 of Newton (1967–1981), gives in fewer than fifty pages the first sketch of an exposition of calculus. It contains six pages of tables of integrals of rational and algebraic functions. Newton increased the size of this table in later expositions. Newton's notes on the factorization of $x^n \pm 1$ can be found in Newton (1967–81), vol. 4, pp. 205–213. Note that Newton's October 1676 letter is on p. 138 of vol. 2 of his correspondence, Newton (1959–1960). Newton's two letters were first published in full in *Commercium Epistolicum D. Johannis Collins, et aliorum de analysi promota* of 1712. Newton had this work published in order to document his priority in the calculus. Primarily concerned with the claims of Leibniz, this work also gave the indirect impression that Gregory's results were probably corollaries of Newton's work; thus, Gregory did not receive much attention until Turnbull studied his unpublished notes in the 1930s.

Gowing (1983) gives a thorough discussion of the various mathematical works of Roger Cotes with a chapter on the Cotes factorization and its application by Cotes to the evaluation of integrals of rational functions. The quote by Smith can be found on pp. 141–42 of Gowing; the quote from the *Harmonia Mensurarum* is on p. 50. See Rigaud (1841), vol. I, p. 271, for the Cotes letter to Jones on Leibniz. De Moivre's more general factorization formula is given in the first two pages of de Moivre (1730b). Smith (1959), on pp. 440–454 of vol. II, gives an English translation of parts of de Moivre's works on this factorization and related topics.

The original paper of Leibniz on the integration of rational functions appeared in 1701 in the *Journal de Trevoux* and in 1702 in the *Acta Eruditorum*, the first German scientific journal, founded in 1682 by Otto Mencke (1644–1707), a Leipzig professor of moral and political philosophy. This paper was reprinted in Leibniz (1971), vol. 5, pp. 350–361. Leibniz's first published mathematical paper, on the arithmetic quadrature of the circle, giving the integral for the arctangent, appeared in the *Acta* in February, 1682.

Johann Bernoulli's 1702 paper, showing the connection between the logarithmic function and the arctangent, was published by the French Academy and can be found in vol. 1 of Johann Bernoulli (1968), pp. 393–400. His second paper on this topic appeared in the *Acta Eruditorum* in 1719. See Joh. Bernoulli (1968), vol. II, pp. 402–418. Euler wrote two to three hundred pages on the integration of rational functions. His results are available in volumes 17 and 18 of his *Opera Omnia*. See Euler's "Investigatio Valoris Integralis $\int \frac{x^{m-1} dx}{1 - 2x^k \cos\theta + x^{2k}}$ a termino $x = 0$ usque ad $x = \infty$ extensi", Eu. I-18, pp. 190–208. Hermite (1905–1917), vol. 3, pp. 40–44, contains his algorithm for finding the rational part of the integral. Our discussion makes use of the account in Hardy (1905), pp. 13–16. Hardy's book also deals with the evaluation of integrals in finite terms, significant for symbolic integration. See Bronstein (1997). Even today, mathematicians continue to find interesting results on integrals of rational functions. See Boros and Moll (2004) and Moll (2002).

14

Difference Equations

14.1 Preliminary Remarks

Difference equations occur in discrete problems, such as are encountered in probability theory, where recursion is an oft-used method. In the mid seventeenth century, probability was developing as a new discipline; Pascal and Huygens used recursion, or first-order difference equations, in working out some elementary probability problems. Later, in the early eighteenth century, Niklaus I Bernoulli, Montmort, and de Moivre made use of more general difference equations. By the 1710s, it was clear that a general method for solving linear difference equations would be of great significance in probability and in analysis. Bernoulli and Montmort corresponded on this topic, discussing their methods for solving second-order difference equations with constant coefficients. In particular, they found the general term of the Fibonacci sequence. In 1712, Bernoulli also solved a special homogeneous linear equation of general degree with constant coefficients. He accomplished this in the course of tackling the well-known Waldegrave problem, involving the probability of winning a game, given players of equal skill. Then in 1715, Montmort rediscovered and communicated to de Moivre Newton's transformation, (11.3). This revealed the connection between difference equations and the summation of infinite series. It was an easy consequence of the Newton–Montmort transformation formula that the difference equation

$$\Delta^n A_k = A_{n+k} - \binom{n}{1} A_{n+k-1} + \binom{n}{2} A_{n+k-2} - \cdots + (-1)^n A_k = 0, \tag{14.1}$$

$k = 0, 1, 2, \ldots$, implied that the series $\sum_{k=0}^{\infty} A_k x^k$ was a rational function with $(1-x)^n$ as denominator. More generally, de Moivre called a series recurrent if its coefficients satisfied the recurrence relation

$$a_0 A_{n+k} + a_1 A_{n+k-1} + \cdots + a_n A_k = 0, \tag{14.2}$$

where $a_0, a_1, \ldots, a_n$ were constants and $k = 0, 1, 2, \ldots$. De Moivre was the first to present a general theory of recurrent series. He proved that such a series could be represented by a rational function and showed how to find this function. He then applied partial fractions to obtain the general expression for A_n in terms of the roots of the denominator of the rational function. De Moivre was therefore the first mathematician to solve a general linear difference equation with constant coefficients by generating functions. He expounded this theory without proofs in the first edition of his *Doctrine of Chances*, published in 1717. He omitted the proofs, but he provided proofs in his 1730 *Miscellanea Analytica*.

In the 1720s, several mathematicians turned their attention toward recurrent series. Daniel Bernoulli (1700–1782) made some very early investigations into this topic without making much headway, being unaware of the results of de Moivre, Niklaus I Bernoulli, and Montmort. In his *Exercitationes* of 1724, he stated that there was no formula for the general term of the sequence 1, 3, 4, 7, 11, 18, ... Niklaus informed his cousin Daniel that this was false; that the general term should be $\left((1+\sqrt{5})/2\right)^n + \left((1-\sqrt{5})/2\right)^n$. Apparently, Daniel Bernoulli subsequently became familiar with the work of de Moivre and others. At the end of 1728, he wrote a paper explaining the method, still contained in our textbooks, for giving special solutions for homogeneous linear difference equations with constant coefficients, in which the form of the solution is assumed and then substituted into the equation. The values of the parameter in the assumed solution can then be determined by means of an algebraic equation. He obtained the general solution by taking an arbitrary linear combination of the special solutions.

Though the connection between differentials and finite differences had become clear by 1720, simultaneous advances in the two topics did not occur; one area seemed to make progress in alternation with the other. For example, D. Bernoulli's 1728 method of solution was not matched by a similar advance in the area of differential equations with constant coefficients until 1740. Euler, having defined the number e and the corresponding exponential function, then gave the general solution of a differential equation as a combination of special exponential functions. He used the exponential function as the form of the solution in giving a method for solving a differential equation with constant coefficients.

Then, from the 1730s through the 1750s, the theory of differential equations made great strides, partially due to the application of this subject to physics problems, such as hanging chains and vibration of strings. In fact, d'Alembert, Euler, and Clairaut initiated the study of partial differential equations in this context. However, no corresponding progress took place in the area of difference equations until 1759 when Lagrange had the inspiration of applying the progress made in differential equations to difference equations. He found a technique, analogous to d'Alembert's method for differential equations, for solving a non-homogeneous difference equation by reducing its degree by one. A repeated application of this technique reduced a general nth degree difference equation to a first-degree equation, already treated by Taylor in 1715. Similarly, Lagrange adapted his method of variation of parameters for solving differential equations to the case of difference equations. In the 1770s, Laplace published several papers, extending Lagrange's method and using other techniques to solve linear difference equations with variable coefficients. In a paper written in 1780 and published in 1782, he introduced the term "generating functions". He developed this theory by the symbolic methods introduced in 1772 by Lagrange, courtesy of Leibniz.

Laplace also applied generating functions of two variables to solve partial difference equations.

De Moivre's work on recurrent series also contained interesting, albeit implicit, results on infinite series, a topic of his earliest research and a life-long interest. In the 1737 second edition of his *Doctrine of Chances*, he solved the problem of summing the series $\sum_{n=0}^{\infty} a_{np+k} x^n$, where p and k were integers and $0 \leq k < p$, when the sum of $\sum_{n=0}^{\infty} a_n x^n$ was known. In his solution, de Moivre dealt only with recurrent series, but in 1758, Simpson published a paper tackling the general problem. Even a year earlier, Waring also gave a general solution for summing the series $\sum_{n=0}^{\infty} a_{np+k} x^n$. An expert in the area of symmetric functions, he obtained his solution by taking specific symmetric functions of roots of unity. He did not publish the paper, but communicated it to the Royal Society. Waring later wrote that he believed Simpson's proof was based on this result, since Simpson was an active member of the society.

Another of de Moivre's results on series implied that if the recurrent series $\sum a_n x^n$ and $\sum b_n x^n$ had singularities at α and β, respectively, then $\sum a_n b_n x^n$ had a singularity at $\alpha\beta$. In 1898 Hadamard extended this result to arbitrary power series, though it does not seem that he was motivated by de Moivre's theorem.

14.2 De Moivre on Recurrent Series

In his *Doctrine of Chances*, de Moivre wrote that the summation of series was required for the solution of several problems relating to chance, that is, to probabilistic problems. He then presented a list of nine propositions connected with recurrent series, series whose coefficients satisfied a linear recurrence relation. Thus, a series $\sum_{n=0}^{\infty} a_n x^n$ was a recurrent series if there were constants $\alpha_1, \alpha_2, \ldots, \alpha_k$ such that a_n satisfied the difference equation

$$a_n + \alpha_1 a_{n-1} + \alpha_2 a_{n-2} + \cdots + \alpha_k a_{n-k} = 0, \tag{14.3}$$

for $n = k, k+1, k+2, \ldots$. De Moivre started with the example of the series

$$1 + 2x + 3xx + 10x^3 + 34x^4 + 97x^5 + \cdots, \tag{14.4}$$

whose coefficients satisfied the equation

$$a_n - 3a_{n-1} + 2a_{n-1} - 5a_{n-3} = 0, \tag{14.5}$$

for $n = 3, 4, 5, \ldots$. His terminology is no longer used. For example, instead of the recurrence relation (14.5), he called $3x - 2xx + 5x^3$, or simply $3 - 2 + 5$, the scale of relation of the series. This scale of relation was used to sum the series. Thus, if we let S denote the series (14.4), then

$$-3xS = -3x - 6x^2 - 9x^3 - 30x^4 - 102x^5 - \cdots$$
$$+2x^2 S = 2x^2 + 4x^3 + 6x^4 + 20x^5 + \cdots$$
$$-5x^3 S = -5x^3 - 10x^4 - 15x^5 - \cdots.$$

Now add these three series to the original series (14.4) for S. Because of the recurrence relation (14.5) satisfied by the coefficients, or because of the scale of relation of the series, we get

$$(1 - 3x + 2x^2 - 5x^3)S = 1 - x - x^2.$$

All other terms on the right-hand side cancel and we have the sum of S:

$$S = \frac{1 - x - x^2}{1 - 3x + 2x^2 - 5x^3}. \tag{14.6}$$

De Moivre called the expression in the denominator the differential scale, since it was obtained by subtracting the scale of relation from unity.

De Moivre's purpose in summing S was to find the numerical value of the coefficient a_n of the series, or, in modern terms, to solve the difference equation (14.5). Once he had S, he could factorize the denominator and obtain the partial fractions decomposition of the rational function S. Actually, he did not discuss this algebraic process in his book. He merely noted the form of a_n when the denominator was a polynomial of degree m with roots $\alpha_1, \alpha_2, \ldots, \alpha_m$ in the cases $m = 2, 3, 4$. Moreover, he wrote the solutions for only those cases in which the roots were distinct. One can be sure that he knew how to handle the case of repeated roots, because only a knowledge of the binomial theorem for negative integral powers was required. Thus, all series $\sum_{n=0}^{\infty} a_n x^n$, whose coefficients a_n satisfy a linear difference equation with constant coefficients as in (14.3), must be rational functions. Conversely, the power series expansions of rational functions whose numerators are of degree less than the corresponding denominators are recurrent series. Euler devoted a chapter of his *Introductio in Analysin Infinitorum* to recurrent series. Since de Moivre gave very few examples, we consider two examples from Euler's exposition, illustrating the method of using generating functions to solve linear difference equations such as (14.3). In the first example, the recurrence relation was the same as the one satisfied by the Fibonacci sequence, though the initial values were different. The coefficients of the series

$$\sum_{n=0}^{\infty} a_n x^n = 1 + 3x + 4x^2 + 7x^3 + 11x^4 + 18x^5 + 29x^6 + 47x^7 + \cdots$$

satisfied the recurrence relation $a_n = a_{n-1} + a_{n-2}$ for $n \geq 2$. Note from our discussion of de Moivre's work that the above series would sum to a rational function whose denominator would be $1 - x - x^2$. In fact, the sum was

$$\frac{1 + 2x}{1 - x - x^2} = \frac{(1 + \sqrt{5})/2}{1 - \left(\frac{1+\sqrt{5}}{2}\right)x} + \frac{(1 - \sqrt{5})/2}{1 - \left(\frac{1-\sqrt{5}}{2}\right)x}.$$

Hence Euler had the solution of the difference equation:

$$a_n = \left((1 + \sqrt{5})/2\right)^{n+1} + \left((1 - \sqrt{5})/2\right)^{n+1}.$$

14.2 De Moivre on Recurrent Series

In the second example, there were repeated roots as well as complex roots. Euler explained earlier in his book precisely how to obtain the partial fractions in this situation. The difference equation would be

$$a_n - a_{n-1} - a_{n-2} + a_{n-4} + a_{n-5} - a_{n-6} = 0, \tag{14.7}$$

and the initial conditions would yield the sum of the series as

$$\frac{1}{1 - x - x^2 + x^4 + x^5 - x^6} = \frac{1}{(1-x)^3(1+x)(1+x+x^2)}$$
$$= \frac{1}{6(1-x)^3} + \frac{1}{4(1-x)^2} + \frac{17}{72(1-x)} + \frac{1}{8(1+x)} + \frac{2+x}{9(1+x+x^2)}. \tag{14.8}$$

Euler obtained the general term a_n by expanding the partial fractions using the binomial theorem. Thus, he had

$$a_n = \frac{n^2}{12} + \frac{n}{2} + \frac{47}{72} \pm \frac{1}{8} \pm \frac{4\sin\frac{(n+1)\pi}{3} - 2\sin\frac{n\pi}{3}}{9\sqrt{3}}, \tag{14.9}$$

where a positive sign was used for n even and negative sign for n odd.

In his *Doctrine of Chances*, de Moivre stated a few specific examples but did not work out details for obtaining the general term. Of his nine theorems, the first six dealt in general terms with the ideas in the above two examples from Euler. The last three propositions applied to more general series, though de Moivre worked wholly in terms of recurrent series. In the seventh proposition, on the even and odd parts of a rational function, de Moivre supposed that $\sum_{n=0}^{\infty} a_n x^n$ was a recurrent series and hence representable as a rational function. He then gave a method for representing the even series $\sum_{n=0}^{\infty} a_{2n} x^{2n}$ and the odd series $\sum_{n=0}^{\infty} a_{2n+1} x^{2n+1}$ as rational functions. In this connection, he explained that if $A(x)$ was the denominator, or differential scale, for $\sum_{n=0}^{\infty} a_n x^n$, then the common differential scale for the two series with the even and odd powers was the polynomial obtained by eliminating x from the equations $A(x) = 0$ and $x^2 = z$. More generally, de Moivre wrote that if

$$a_0 + a_1 x + a_2 x^2 + \cdots = B(x)/A(x), \tag{14.10}$$

then the m series

$$a_j x^j + a_{m+j} x^{m+j} + a_{2m+j} x^{2m+j} + \cdots, \quad j = 0, 1, \ldots, m-1, \tag{14.11}$$

had the common differential scale obtained by eliminating x from $A(x) = 0$ and $x^m = z$. We may state a more general problem: Given $f(x) = \sum_{n=0}^{\infty} a_n x^n$, express $g(x) = \sum_{k=0}^{\infty} a_{km+j} x^{km+j}$ in terms of values of $f(\alpha x)$, α a root of unity. This was the problem solved by Simpson and Waring in the late 1750s. The essence of their method was to use appropriate mth roots of unity and those roots were implicit in de Moivre's use of the equation $x^m = z$.

The eighth proposition of de Moivre explained how to find the differential scale for $\sum (a_n + b_n) x^n$ when the differential scales of $\sum a_n x^n$ and $\sum b_n x^n$ were known.

This was straightforward. In the very interesting ninth proposition, de Moivre worked out the differential scale for $\sum a_n b_n x^n$, but only for the case where the scales for $\sum a_n x^n$ and $\sum b_n x^n$ were quadratic polynomials. His result is stated as an exercise at the end of this chapter. An immediate consequence of this result is that $\sum a_n b_n x^n$ has a singularity at $\alpha\beta$ if $\sum a_n x^n$ and $\sum b_n x^n$ have singularities at α and β respectively. In 1898, Hadamard, probably unaware of this result of de Moivre, stated and proved a beautiful generalization, usually called Hadamard's multiplication of singularities theorem: If $\sum_{n=0}^{\infty} a_n z^n$ has singularities at $\alpha_1, \alpha_2, \ldots$, and $\sum_{n=0}^{\infty} b_n z^n$ at $\beta_1, \beta_2, \ldots$, then the singularities of $\sum_{n=0}^{\infty} a_n b_n z^n$ are among the points $\alpha_i \beta_j$.

14.3 Stirling's Method of Ultimate Relations

Stirling extended de Moivre's recurrent series method to sequences satisfying difference equations with nonconstant coefficients. In the preface to his 1730 *Methodus*, he wrote:

> For I was not unaware that De Moivre had introduced this property of the terms into algebra with the greatest success, as the basis for solving very difficult problems concerning recurrent series: And so I decided to find out whether it could also be extended to others, which of course I doubted since there is so great a difference between recurrent and other series. But, the practical test having been made, the matter has succeeded beyond hope, for I have found out that this discovery of De Moivre contains very general and also very simple principles not only for recurrent series but also for any others in which the relation of the terms varies according to some regular law.

In the statement of the proposition 14, Stirling explained the term ultimate relation: Let T be the zth term of a series and T' the next term; let r, s, a, b, c, d, be constants. Suppose that the relation

$$r(z^2 + az + b)T + s(z^2 + cz + d)T' = 0 \tag{14.12}$$

held between the successive terms. Then the ultimate relation of the terms was defined as

$$rT + sT' = 0. \tag{14.13}$$

Stirling used the term ultimate because he understood that z was a very large integer, so that $az + b$ and $cz + d$ could be neglected in comparison to z^2. This made it clear that (14.13) followed from (14.12). Similarly, if the equation were

$$r(z+a)T + s(z+b)T' + t(z+c)T'' = 0, \tag{14.14}$$

then the ultimate relation would be

$$rT + sT' + tT'' = 0. \tag{14.15}$$

In modern notation, if $\sum A_n$ is the series, then (14.14) takes the form

$$r(k+a)A_k + s(k+b)A_{k+1} + t(k+c)A_{k+2} = 0. \tag{14.16}$$

14.3 Stirling's Method of Ultimate Relations

Stirling stated his theorem as proposition 14:

> Every series $A + B + C + D + E + \&c.$ in which the ultimate relation of the terms is $rT + sT' + tT'' = 0$ splits into the following

$$(s+t) \times \left(\frac{A}{n} + \frac{A_2}{n^2} + \frac{A_3}{n^3} + \frac{A_4}{n^4} + \frac{A_5}{n^5} + \&c. \right)$$

$$+ t \times \left(\frac{B}{n} + \frac{B_2}{n^2} + \frac{B_3}{n^3} + \frac{B_4}{n^4} + \frac{B_5}{n^5} + \&c. \right)$$

where $n = r + s + t$ and

$A_2 = rA + sB + tC,$ $\quad A_3 = rA_2 + sB_2 + tC_2, A_4 = rA_3 + sB_3 + tC_3, \&c.$

$B_2 = rB + sC + tD,$ $\quad B_3 = rB_2 + sC_2 + tD_2, B_4 = rB_3 + sC_3 + tD_3, \&c.$

$C_2 = rC + sD + tE,$ $\quad C_3 = rC_2 + sD_2 + tE_2. C_4 = rC_3 + sD_3 + tE_3, \&c.$

$D_2 = rD + sE + tF,$ $\quad D_3 = rD_2 + sE_2 + tF_2, D_r = rD_3 + sE_3 + tF_3, \&c.$

$E_2 = rE + sF + tG,$ $\quad E_3 = rE_2 + sF_2 + tG_2, E_4 = rE_3 + sF_3 + tG_3, \&c.$

$\&c.$

This result generated some interest in its time. A reviewer of the *Methodus* wrote in 1732 that the result was very powerful and complicated. In a letter to Stirling dated June 8, 1736, Euler wrote

> But before I wrote to you, I searched all over with great eagerness for your excellent book on the method of differences, a review of which I had seen a short time before in the *Actae Lipslienses*, until I achieved my desire. Now that I have read through it diligently, I am truly astonished at the great abundance of excellent methods contained in such a small volume, by means of which you show how to sum slowly converging series with ease and how to interpolate progressions which are very difficult to deal with. But especially pleasing to me was prop. XIV of Part I in which you give a method by which series, whose law of progression is not even established, may be summed with great ease using only the relation of the last terms; certainly this method extends very widely and is of the greatest use. In fact the proof of this proposition, which you seem to have deliberately withheld, caused me enormous difficulty, until at last I succeeded with very great pleasure in deriving it from the preceding results, which is the reason why I have not yet been able to examine in detail all the subsequent propositions.

Stirling gave three examples of this theorem. The first example, similar to the second, was the summation of the series

$$1 + 4x + 9x^2 + 16x^3 + 25x^4 + 36x^5 + \text{ etc.} \tag{14.17}$$

Recall that Euler summed this series in his *Institutiones Calculi Differentialis* of 1755 by applying Newton or Montmort's transformation. Stirling was aware that the series could be summed by that method and mentioned Montmort explicitly. Stirling observed that the difference equation for the terms of the series was

$$(z^2 + 2z + 1)xT - z^2 T' = 0. \tag{14.18}$$

For example, for the third term, $z = 3$, $T = 9x^2$, and $T' = 16x^3$. The ultimate relation was $xT - T' = 0$, so that $r = x$, $s = -1$, $t = 0$, and $n = x - 1$. It followed that $A = 1$, $A_2 = -3x$, $A_3 = 2x^2$, $A_4 = 0$, and the series transformed to

$$-1\left(\frac{1}{x-1} - \frac{3x}{(x-1)^2} + \frac{2x^2}{(x-1)^3}\right) = \frac{1+x}{(1-x)^3}.$$

In the third example, he considered the series

$$1 - 6x + 27x^2 - 104x^3 + 366x^4 - 1212x^5 + 3842x^6 - 11784x^7 + \text{etc.}$$

defined by the difference equation

$$x^2(z+4)T - 2x(z+2)T' - zT'' = 0; \qquad (14.19)$$

the ultimate relation was

$$x^2 T - 2xT' - T'' = 0. \qquad (14.20)$$

Hence $r = x^2$, $s = -2x$, $t = -1$ and $n = x^2 - 2x - 1$. Stirling computed the values of the A and B as

$$A = 1, \; A_2 = -14x^2, \; A_3 = 29x^4, \; A_4 = 0,$$
$$B = -6x, \; B_2 = 44x^3, \; B_3 = -70x^5, \; B_4 = 0.$$

Thus, the sum of the series was

$$-(2x+1)\left(\frac{1}{x^2-2x-1} - \frac{14x^2}{(x^2-2x-1)^2} + \frac{29x^4}{(x^2-2x-1)^3}\right)$$
$$-\left(\frac{-6x}{x^2-2x-1} + \frac{44x^3}{(x^2-2x-1)^2} - \frac{70x^5}{(x^2-x-1)^3}\right).$$

Note that in (14.19) z takes the values $2, 3, 4, \ldots$ while in (14.18) z starts at 1. Thus, in the second series when $T = 1$, $T' = -6x$, and $T'' = 27x^2$, we take $z = 2$. Stirling's normal practice was to start at $z = 1$.

14.4 Daniel Bernoulli on Difference Equations

In 1728, while at the St. Petersburg Academy, Bernoulli presented to the academy a method for solving a difference equation in which the form of the solution was assumed; this particular approach is often given in elementary textbooks. Unlike Bernoulli, we use subscripts to write the equation

$$a_n = \alpha_1 a_{n-1} + \alpha_2 a_{n-2} + \cdots + \alpha_k a_{n-k}, \qquad (14.21)$$

with $\alpha_1, \alpha_2, \ldots, \alpha_k$ constants. Bernoulli assumed $a_x = \lambda^x$, substituted in the equation and divided by λ^{n-k} to arrive at

$$\lambda^k = \alpha_1 \lambda^{k-1} + \alpha_2 \lambda^{k-2} + \cdots + \alpha_k. \qquad (14.22)$$

14.4 Daniel Bernoulli on Difference Equations

Bernoulli stated that if $\lambda_1, \lambda_2, \ldots, \lambda_k$ were the k distinct solutions of the algebraic equation (14.22), then the general solution of (14.21) would be an arbitrary linear combination of the particular solutions λ_i^x, that is

$$a_x = A_1\lambda_1^x + A_2\lambda_2^x + \cdots + A_k\lambda_k^x. \tag{14.23}$$

However, if $\lambda_1 = \lambda_2$, then the first two terms of (14.23) would be replaced by $(A_1 + A_2 x)\lambda_1^x$. More generally, if a root λ_j was repeated m times, then that part of the solution (14.23) corresponding to λ_j would be replaced by

$$(A_j + A_{j+1}x + \cdots + A_{m+j-1}x^{m-1})\lambda_j^x.$$

Daniel Bernoulli considered examples of distinct roots and of repeated roots. He first took the Fibonacci sequence 0, 1, 1, 2, 3, 5, 8, 13, …, leading to the difference equation $a_n = a_{n-1} + a_{n-2}$ and the algebraic equation $\lambda^2 = \lambda + 1$. The solutions were $\lambda_1 = (1+\sqrt{5})/2$, $\lambda_2 = (1-\sqrt{5})/2$ so that

$$a_x = A_1\lambda_1^x + A_2\lambda_2^x.$$

To find A_1 and A_2, Bernoulli took $x = 0$ and $x = 1$ to get

$$A_1 + A_2 = 0 \text{ and } A_1\left(\frac{1+\sqrt{5}}{2}\right) + A_2\left(\frac{1-\sqrt{5}}{2}\right) = 1.$$

Solving these equations, Bernoulli found $A_1 = 1/\sqrt{5}$ and $A_2 = -1/\sqrt{5}$. Recall that Montmort and Niklaus I Bernoulli in their correspondence of 1718–1719 had already solved the problem of the general term in the Fibonacci sequence.

As an example of a difference equation leading to repeated roots, Daniel Bernoulli considered the sequence $0, 0, 0, 0, 1, 0, 15, -10, 165, -228$, etc., generated by the difference equation

$$a_n = 0a_{n-1} + 15a_{n-2} - 10a_{n-3} - 60a_{n-4} + 72a_{n-5}.$$

He found the roots of the corresponding algebraic equation to be $2, 2, 2, -3, -3$; the general term of the sequence was then

$$((1026 - 1035x + 225xx) \cdot 2^x + (224 - 80x) \cdot (-3)^x)/9000.$$

As a final example, Bernoulli set $a_n = \sin nx$ and applied the addition formula for sine to get, in modern notation,

$$a_{n+1} + a_{n-1} = \sin(n+1)x + \sin(n-1)x = 2\cos x \sin nx = 2\cos x a_n.$$

This produced the algebraic equation $\lambda^2 - 2\cos x\lambda + 1 = 0$ whose roots were given by

$$\lambda_1 = \cos x + \sqrt{\cos^2 x - 1} = \cos x + \sqrt{-1}\sin x,$$
$$\lambda_2 = \cos x - \sqrt{\cos^2 x - 1} = \cos x - \sqrt{-1}\sin x.$$

This gave him the formula for sine:

$$a_n = \sin nx = \frac{(\cos x + \sqrt{-1}\sin x)^n - (\cos x - \sqrt{-1}\sin x)^n}{2\sqrt{-1}}.$$

Note that this gives a new proof of de Moivre's formula. Bernoulli also made an interesting observation about the root largest in absolute value of the algebraic equation (14.22), noting that such a root could be obtained from the sequence satisfying the corresponding difference equation. Taking λ to be the root largest in absolute value, and writing the sequence as $a_1, a_2, a_3, \ldots, a_m, \ldots$, he observed that as m went to infinity, a_{m+1}/a_m would approach the value λ. Also, the root smallest in absolute value could be found by setting $\lambda = 1/\mu$ in (14.22). Bernoulli was quite proud of this result, writing in a February 20, 1728 letter to Goldbach that even if it were not useful, it was among the most beautiful theorems on the topic. Euler must have agreed with Bernoulli, since he devoted a whole chapter of his *Introductio* of 1748 to finding roots of algebraic equations by solving difference equations.

Illustrating that his beautiful theorem was in fact useful, Bernoulli showed how to find the approximate solution of $xx = 26$. He began by setting $x = y + 5$ to get $1 = 10y + yy$. Of the two roots of this last equation, he needed the smaller in absolute value, so he set $y = 1/z$ to obtain $z^2 = 10z + 1$. The corresponding difference equation was $a_n = 10a_{n-1} + a_{n-2}$, and Bernoulli took the two initial values of the sequence to be 0 and 1. The difference equation then gave him the sequence 0, 1, 10, 101, 1020, 10301, 104030, …. To obtain an approximate value of y, Bernoulli took the ratio of the seventh and sixth terms of the sequence, obtaining $x = \sqrt{26} = 5 + 10301/104030 = 5.09901951360$. He then computed $\sqrt{26}$ by the usual method and got 5.0990151359. Bernoulli employed this idea to find the smallest roots of Laguerre polynomials of low degree. In his work with hanging chains, the roots of these polynomials yielded the frequencies of the oscillations.

14.5 Lagrange: Nonhomogeneous Equations

In 1759, Lagrange published a method for solving a nonhomogeneous linear difference equation with constant coefficients and this method can be seen as the analog of d'Alembert's method for the corresponding differential equation. Lagrange started with a third-order equation to illustrate the technique. In brief, let the equation be $y + A\Delta y + B\Delta^2 y + C\Delta^3 y = X$; set $\Delta y = p$ and $\Delta p = q$ so that the equation can be written as $y + Ap + Bq + C\Delta q = X$. For the arbitrary constants a and b we have

$$y + (A+a)p + (B+b)q - a\Delta y - b\Delta p + C\Delta q = X. \tag{14.24}$$

Choose a and b such that

$$\Delta y + (A+a)\Delta p + (B+b)\Delta q = \Delta y + \frac{b}{a}\Delta p - \frac{C}{a}\Delta q. \tag{14.25}$$

Then $A + a = b/a$, $B + b = -C/a$. These equations imply that a satisfies the cubic $a^3 + Aa^2 + Ba + C = 0$. Moreover, by (14.25), equation (14.24) is reduced to the

14.5 Lagrange: Nonhomogeneous Equations

first-order equation

$$z - a\Delta z = X, \quad \text{where} \tag{14.26}$$

$$z = y + (A+a)p + (B+b)q. \tag{14.27}$$

The problem is now reduced to solving (14.26). Suppose it has been solved for each of the three values a_1, a_2, a_3 of a, obtained from the cubic. Let z_1, z_2 and z_3 be the corresponding values of z from (14.26). Now we have three linear equations

$$y + (A+a_1)p + (B+b_1)q = z_1, \quad y + (A+a_2)p + (B+b_2)q = z_2,$$

$$y + (A+a_3)p + (B+b_3)q = z_3,$$

and these can be solved to obtain $y = Fz_1 + Gz_2 + Hz_3$ for some constants F, G, and H. Finally, to solve the first-order equation (14.26), Lagrange considered the more general equation

$$\Delta y + My = N, \tag{14.28}$$

where M and N were functions of an integer variable x. He set $y = uz$ to get

$$u\Delta z + z\Delta u + Mzu = N. \tag{14.29}$$

He let u be such that $(\Delta u + Mu)z = 0$, or u was a solution of the homogeneous part of (14.28). Thus,

$$u(x) - u(x-1) = -M(x-1)u(x-1) \text{ or } u(x) = (1 - M(x-1))u(x-1).$$

By iteration,

$$u(x) = (1 - M(x-1))(1 - M(x-2)) \cdots (1 - M(1)).$$

For this u, equation (14.29) simplified to

$$z(x) - z(x-1) = \frac{N(x-1)}{u(x-1)}.$$

Therefore

$$z(x) = \frac{N(x-1)}{u(x-1)} + z(x-1) = \frac{N(x-1)}{u(x-1)} + \frac{N(x-2)}{u(x-2)} + \cdots + \frac{N(1)}{u(1)} + z(1).$$

Laplace later observed that (14.28) could be solved directly by iteration:

$$y(x) = y(x-1) + N(x-1) - M(x-1)y(x-1)$$
$$= (1 - M(x-1))y(x-1) + N(x-1)$$
$$= N(x-1) + (1 - M(x-1))N(x-2) + (1 - M(x-2))y(x-2) \text{ etc.}$$

As Lagrange himself pointed out, this method could obviously be generalized to a nonhomogeneous equation of any order. Of course, the question of solving the corresponding algebraic equation of arbitrary degree would be a separate problem.

Lagrange found another method of solving difference equations, using the device of the variation of parameters. Again presenting Lagrange's work in brief, suppose we have a third-order difference equation

$$y_{x+3} + P_x y_{x+2} + Q_x y_{x+1} + R_x y_x = V_x. \tag{14.30}$$

Let z_x, z'_x, z''_x be three independent solutions of the corresponding homogeneous equation

$$y_{x+3} + P_x y_{x+2} + Q_x y_{x+1} + R_x y_x = 0. \tag{14.31}$$

The general solution of this equation is $Cz_x + C'z'_x + C''z''_x$ where C, C', and C'' are constants. Now suppose C_x, C'_x and C''_x are functions of x, determined by the condition that

$$y_x = C_x z_x + C'_x z'_x + C''_x z''_x \tag{14.32}$$

is a solution of the nonhomogeneous equation. Changing x to $x+1$, we have

$$y_{x+1} = C_{x+1} z_{x+1} + C'_{x+1} z'_{x+1} + C''_{x+1} z''_{x+1}$$
$$= C_x z_{x+1} + C'_x z'_{x+1} + C''_x z''_{x+1} + \Delta C_x z_{x+1} + \Delta C'_x z'_{x+1} + \Delta C''_x z''_{x+1}.$$

Now suppose that C_x, C'_x, C''_x are such that

$$z_{x+1} \Delta C_x + z'_{x+1} \Delta C'_x + z''_{x+1} \Delta C''_x = 0. \text{ Then}$$

$$y_{x+1} = C_x z_{x+1} + C'_x z'_{x+1} + C''_x z''_{x+1}. \tag{14.33}$$

If in the equation for y_{x+2}, we again change x to $x+1$, the result is

$$y_{x+3} = C_x z_{x+3} + C'_x z'_{x+3} + C''_x z''_{x+3} + \Delta C_x z_{x+3} + \Delta C'_x z'_{x+3} + \Delta C''_x z''_{x+3}.$$

Thus,

$$y_{x+3} = C_x z_{x+3} + C'_x z'_{x+3} + C''_x z''_{x+3}. \tag{14.34}$$

We also have an equation for y_{x+1} resembling the equation for y_x. If we make a similar $x \to x+1$ change in the equation for y_{x+1}, we can require that

$$z_{x+2} \Delta C_x + z'_{x+2} \Delta C'_x + z''_{x+2} \Delta C''_x = 0. \tag{14.35}$$

Multiply equation (14.32) by R_x; multiply equation (14.33) by Q_x; multiply (14.34) by P_x. Now add the results to (14.35). From (14.30) and the fact that z_x, z'_x, z''_x satisfy (14.31), it follows that

$$z_{x+3} \Delta C_x + z'_{x+3} \Delta C'_x + z''_{x+3} \Delta C''_x = V_x.$$

Consider this last equation together with the two equations

$$z_{x+2} \Delta C_x + z'_{x+2} \Delta C'_x + z''_{x+2} \Delta C''_x = 0,$$
$$z_{x+1} \Delta C_x + z'_{x+1} \Delta C'_x + z''_{x+1} \Delta C''_x = 0.$$

Recall that we required ΔC to satisfy these last two equations. Thus, we have three linear equations yielding ΔC_x, $\Delta C'_x$, and $\Delta C''_x$. Suppose we obtain $\Delta C_x = H_x$, $\Delta C'_x = H'_x$, and $\Delta C''_x = H''_x$. These first-order equations can be solved for C_x, C'_x, and C''_x, and hence we have y_x from (14.32).

As an example of the method of variation of parameters, Lagrange considered a nonhomogeneous equation with constant coefficients. In this case, briefly, z_x will be of the form m^x for some constant m. Suppose we have a second-order equation for which $z_x = m^x$ and $z'_x = m_1^x$. Then the equations for ΔC_x and $\Delta C'_x$ are

$$m^{x+1}\Delta C_x + m_1^{x+1}\Delta C'_x = 0,$$

$$m^{x+2}\Delta C_x + m_1^{x+2}\Delta C'_x = V_x.$$

Solving for ΔC_x and $\Delta C'_x$, we have

$$\Delta C_x = \frac{V_x}{m^{x+1}(m - m_1)},$$

$$\Delta C'_x = \frac{V_x}{m_1^{x+1}(m_1 - m)}; \quad \text{therefore,}$$

$$C_x = \frac{V_x(m^x - 1)}{(m - 1)(m - m_1)m^x} + C_0 \quad \text{and} \quad C'_x = \frac{V_x(m_1^x - 1)}{(m_1 - 1)(m_1 - m)m_1^x} + C'_0.$$

14.6 Laplace: Nonhomogeneous Equations

The method of Laplace for solving a non-homogeneous equation differed from the variation of parameters of Lagrange, but was analogous to Lagrange's method for equations with constant coefficients (14.25). Suppose the equation to be

$$y_{x+n} + P_x y_{x+n-1} + Q_x y_{x+n-2} + \cdots + T_x y_{x+1} + U_x y_x = V_x.$$

Laplace assumed that there existed functions p_x and q_x such that $y_{x+1} = p_x y_x + q_x$. This implied

$$y_{x+2} = p_{x+1} y_{x+1} + q_{x+1} \cdots,$$

$$y_{x+n} = p_{x+n-1} y_{x+n-1} + q_{x+n-1}.$$

Laplace introduced functions $\alpha_1, \alpha_2, \ldots, \alpha_{n-1}$ to obtain

$$y_{x+n} = p_{x+n-1} y_{x+n-1} + q_{x+n-1}$$
$$= (p_{x+n-1} - \alpha_{n-1}) y_{x+n-1} + (\alpha_{n-1} p_{x+n-2} - \alpha_{n-2}) y_{x+n-2}$$
$$+ (\alpha_{n-2} p_{x+n-3} - \alpha_{n-3}) y_{x+n-3} + \cdots + \alpha_1 p_x y_x$$
$$+ q_{x+n-1} + \alpha_{n-1} q_{x+n-2} + \cdots + \alpha_1 q_x.$$

He then chose $\alpha_1, \alpha_2, \ldots, \alpha_{n-1}$ such that

$$P_x = p_{x+n-1} - \alpha_{n-1},$$
$$Q_x = \alpha_{n-1}p_{x+n-2} - \alpha_{n-2},$$
$$R_x = \alpha_{n-2}p_{x+n-3} - \alpha_{n-3},$$
$$\ldots$$
$$T_x = \alpha_2 p_{x+1} - \alpha_1,$$
$$U_x = \alpha_1 p_x.$$

Therefore, p_x satisfied

$$\prod_{i=0}^{n-1} p_{x+i} = P_x \prod_{i=0}^{n-2} p_{x+i} + Q_x \prod_{i=0}^{n-3} p_{x+i} + \cdots + p_x T_x + U_x;$$

q_x satisfied an equation of order $m-1$:

$$V_x = q_{x+n-1} + \alpha_{n-1} q_{x+n-2} + \cdots + \alpha_1 q_x.$$

By successive reduction, q_x could be determined, although the equation satisfied by p_x was more difficult to handle.

14.7 Exercises

1. In Proposition VII for recurrent series of the *Doctrine of Chances*, de Moivre showed that if

$$a_0 + a_1 x + a_2 x^2 + \cdots = B(x)/(1 - fx + gx^2),$$

with $B(x)$ a linear function, then the two series

$$\sum_{n=0}^{\infty} a_{2n} x^{2n} \quad \text{and} \quad \sum_{n=0}^{\infty} a_{2n+1} x^{2n+1}$$

sum to a rational function with denominator $1 - (f^2 - 2g)x^2 + g^2 x^4$. If the denominator of the original series was $1 - fx + gx^2 - hx^3$, then the denominator of the two series would be

$$1 - (f^2 - 2g)x^2 - (2fh - g^2)x^4 - h^2 x^6.$$

Work out the details by following de Moivre's method described in the text. Extend the results to the case where the original series is divided into three parts. See de Moivre (1967).

2. Simpson showed that if p, q, and r were the three cube roots of unity and $f(x) = \sum_{n=0}^{\infty} a_n x^n$, then $p + q + r = p^2 + q^2 + r^2 = 0$, $p^3 + q^3 + r^3 = 3$, and

$$(f(px) + f(qx) + f(rx))/3 = \sum_{n=0}^{\infty} a_{3n} x^{3n}.$$

He also explained how to generalize this to sum $\sum_{n=0}^{\infty} a_{mn+j} x^{mn+j}$, $j = 0$, $1, \ldots, m-1$ by using mth roots of unity. Prove Simpson's result and obtain the generalization. Compare Simpson's results with de Moivre's in exercise 1. See Simpson (1759). Thomas Simpson (1710–1761) was a self-taught mathematician who contributed to the popularization of mathematics and other intellectual pursuits during that period in England. He was an editor of the *Ladies Diary* and was one of the earliest mathematics professors at the Royal Military Academy at Woolwich. See the excellent account by Clarke (1929).

3. In Proposition IX for recurrent series of his *Doctrine*, de Moivre stated that if $\sum a_n x^n$ and $\sum b_n x^n$ have the differential scales $1 - fx + gx^2$ and $1 - mx + px^2$, respectively, then the differential scale of $\sum a_n b_n x^n$ is

$$1 - fmx + (f^2 p + m^2 g - 2gp)x^2 - fgmpx^3 + g^2 p^2 x^4.$$

Prove this result. Compare with Hadamard's theorem on the multiplication of singularities in Hadamard (1898). See de Moivre (1967).

4. Solve the recurrence relation $a_{n+5} = a_{n+4} + a_{n+1} - a_n$, with $a_0 = 1, a_1 = 2$, $a_2 = 3, a_3 = 3, a_4 = 4$ by recurrent series (generating function) as well as by letting $a_x = \lambda^x$. See Euler (1988), p. 195.

5. Use recurrent series to find the largest root of the equation $y^3 - 3y + 1 = 0$. See Euler (1988), p. 288.

6. Find the smallest root of $y^3 - 6y^2 + 9y - 1 = 0$. This value is $2(1 - \sin 70°)$. See Euler (1988), p. 290.

14.8 Notes on the Literature

For de Moivre's work on recurrent series, see de Moivre (1967), pp. 220–229. Euler's treatment of this topic can be found in Euler (1988), pp. 181–203. Daniel Bernoulli's 1728 paper on recurrent series is contained in D. Bernoulli (1982–1996), vol. 2, pp. 49–64. U. Bottazzini has discussed D. Bernoulli's early mathematical work, including recurrent series, and put it into historical perspective; for this, see D. Bernoulli (1982–1996), vol. 1, pp. 133–189. Lagrange's 1759 paper on difference equations was reprinted in Lagrange (1867–1892), vol. 1, pp. 23–36; his discussion of the method of variation of parameters can be found in vol. 4, pp. 151–60. Hald (1990) gives a fine treatment of the history of difference equations in the eighteenth century, providing references to the significant papers of Laplace. For Euler's letter to Stirling, see Tweddle (1988), pp. 141–154, especially p. 141. Stirling (2003), pp. 88–91 contains his proposition 14 and the examples.

15

Differential Equations

15.1 Preliminary Remarks

In the seventeenth century, before the development of calculus, problems reducible to differential equations began to appear in the study of general curves and in navigation. Interestingly, these differential equations were often related to the logarithm or exponential curve. For example, Harriot obtained the logarithmic spiral by projecting the loxodrome onto the equatorial plane. And in 1638, I. F. de Beaune (1601–1652) posed to Descartes the problem of finding a curve such that the subtangent at each point was a constant. Note that the problem actually leads to the simple differential equation $\frac{dy}{dx} = \frac{y}{a}$. Descartes replied with a solution involving the logarithmic function, though he did not explicitly recognize it. In 1684, Leibniz gave the first published solution by explicitly stating the problem as a differential equation.

Newton understood the significance of the differential equation as soon as he started developing calculus. In his October 1666 tract on calculus, written a year after he graduated from Cambridge, he wrote

> If two Bodys $A \& B$, by their velocitys $p \& q$ describe y^e [the] lines x and y. & an Equation bee given expressing y^e relation twixt one of y^e lines x, & y^e ratio $\frac{q}{p}$ of their motions $q \& p$; To find the other line y. Could this ever bee done all problems whatever might be resolved.

So Newton's problem to solve all problems was: Given $f\left(x, \frac{dy}{dx}\right) = 0$, find y. In a treatise prepared five years later, Newton gave a classification of first-order differential equations $\frac{dy}{dx} = f(x, y)$.

In the 1660s, Isaac Barrow and James Gregory too dealt with differential equations, arising from geometric problems. Gregory considered the question of determining a curve whose area of surface of revolution produced a given function. This translates to the differential equation

$$y^2 \left(1 + \left(\frac{dy}{dx}\right)^2\right) = f(x),$$

15.1 Preliminary Remarks

where $f(x)$ is the given function. In connection with this, Barrow gave a geometric solution for the differential equation

$$y\left(1 + \frac{dy}{dx}\right) = a,$$

by expressing the solution in terms of areas under hyperbolas. Geometrically, the problem would be to find a curve $y = f(x)$ such that the sum of the ordinate y and the subnormal yy' is a constant.

By the 1690s, it was clear to Newton, Leibniz, the two Bernoullis, and the other mathematicians with whom they corresponded that differential equations were intimately connected with curves and their properties. If they knew some property of a geometric object, such as the subtangent or curvature, then the problem of finding the curve itself usually led them to a differential equation. They had begun to recognize or get a glimpse of some general methods of solving these equations, such as separation of variables and multiplication of the equation by an integrating factor.

Newton encountered differential equations in the geometrical and astronomical problems of the *Principia*; in his *De Quadratura Curvarum* of 1691, Newton once again emphasized the importance of differential equations, or fluxional equations, discussing a number of special methods for solving them, as well as the general separation of variables method. He wrote, "Should the equation involve both fluent quantities, but can be arranged so that one side of the equation involves but a single one together with its fluxion and the other the second alone with its fluxions." The term separation of variables was first used by Johann Bernoulli in his May 9, 1694 letter to Leibniz and then in a related paper published in November 1694. Bernoulli also noted that there were important equations unable to be solved by this method, such as $aady = xxdx + yydx$. Observe that this is an equation between the differentials dx and dy; it is hence given the name "differential equation." We would now write it as $a^2 dy/dx = x^2 + y^2$, a particular case of Riccati's equation, to which we will return later.

The first person to discover the integrating factor technique seems to be the Swiss mathematician Nicolas Fatio de Duillier (1664–1753). In 1687, he mastered the elements of differential and integral calculus by his own unaided efforts. Since so little on this subject had been published, Fatio's achievement was remarkable. He exchanged several letters on calculus with Huygens, to whom in February 1691 Fatio first communicated his method of multiplying an equation by $x^\mu y^\nu$ to possibly put it into integrable form. Huygens in turn wrote Leibniz concerning Fatio's method for solving the differential equations

$$-2xy\,dx + 4x^2\,dy - y^2\,dy = 0 \quad \text{and} \quad -3a^2 y\,dx + 2xy^2\,dx - 2x^2 y\,dy + a^2 x\,dy = 0.$$

Observe that, after multiplying across by y^{-5}, the first equation becomes the differential of $-x^2 y^{-4} + \frac{1}{2} y^{-2} = c$. And the second equation, when multiplied by x^{-4}, can be integrated to yield $a^2 x^{-3} y - x^{-2} y^2 = c$. Fatio later told Newton about his method, and Newton included it in his *De Quadratura*, giving credit to Fatio. The technique, as Newton explained it, was to multiply $f_1(x, y)\dot{x} + f_2(x, y)\dot{y} = 0$ by $x^\mu y^\nu$ to get $M(x, y)\dot{x} + N(x, y)\dot{y} = 0$, where M and N were polynomials or even algebraic functions of x and y. The basic idea was to compute $\frac{\partial}{\partial y}\int M(x,y)\,dx$ and choose μ and ν so that this quantity became equal to $N(x,y)$. If this was possible, then $\int M(x,y)\,dx = c$ was the solution of the differential equation. Newton even extended this method to second, third- and higher-order differential equations. Regrettably, Newton did not include these results on differential equations in the published version of *De Quadratura*; they were rediscovered by Leibniz and the two elder Bernoullis. It should be noted that Newton's acknowledgment of Fatio's contribution was unusual; it showed the depth of their friendship at that time. Unfortunately, it appears that in 1693, this friendship was abruptly and emotionally terminated.

In the tenth of his 1691–92 lectures on integral calculus to l'Hôpital, Johann Bernoulli gave an ingenious application of integrating factors to solve the separable equation $ax\,dy - y\,dx = 0$. He multiplied the equation by y^{a-1}/x^2 to obtain $\frac{ay^{a-1}}{x}dy - \frac{y^a}{x^2}dx = 0$. Since the left-hand side was the differential of y^a/x, integration yielded $y^a/x = c$. Note that, after separation of variables in the original differential equation, one gets a logarithm on each side. But it seems that at that time Bernoulli found some difficulty in working with logarithms in an analytic setting and found the solution by a method avoiding the logarithm. It was only after an exchange of letters with Leibniz that Bernoulli understood logarithms; in fact, in 1697 he published a paper on exponentials and logarithms.

In the 1690s, Leibniz and the Bernoullis also learned to handle first-order linear differential equations. In a 1695 paper, Jakob Bernoulli raised the question of how to solve the nonlinear equation $a\,dy = yp\,dx + by^n q\,dx$ where p and q were functions of x and a, b were constants. In response, Leibniz as well as Johann Bernoulli observed that the equation could be linearized by the substitution $v = y^{1-n}$. Bernoulli found an interesting method, applicable to linear equations as well, for solving the equation by setting $y = mz$. This technique showed that in the case of linear equations, m could be chosen to be $e^{-\int p\,dx}$, the reciprocal of the integrating factor. Three decades later, in 1728, Euler wrote a paper in which he solved the linear equation by making use of an integrating factor, making due reference to his teacher Bernoulli.

The theory of linear differential equations with constant coefficients was developed much more slowly than one might expect. Recall that in 1728 Daniel Bernoulli solved linear difference equations with constant coefficients by substituting x^n in the difference equation to obtain an algebraic equation in x, whose solutions $x_1, x_2, \ldots$ determined the possible values of x. The general solution was then a linear combination $c_1 x_1^n + c_2 x_2^n + \cdots$ of the special solutions. Yet it took Euler nearly a decade to perceive that he could solve the corresponding differential equation in a similar way. For more discussion on the alternating development of difference and differential equations, see the chapter on difference equations.

The search for a general solution for a linear differential equation with constant coefficients seems to have started with Daniel Bernoulli's May 4, 1735 letter to Euler, describing his work on the transverse vibration of a hanging elastic band fixed at one end to a wall. Bernoulli wrote that he found the equation for the curve of vibration to be $n\,d^4 y = y\,dx^4$, n a constant. He requested Euler's help in solving the equation, noting that if p divided m, then the solutions of $\alpha\,d^p y = y\,dx^p$ were contained in those of $n\,d^m y = y\,dx^m$. It followed, he observed, that the logarithm satisfied both his equation

15.1 Preliminary Remarks

and $n^{1/2}ddy = ydx^2$, but that it was not general enough for his purpose. Euler too was unable to solve the equation except as an infinite series. Commenting on this, C. Truesdell remarked, "These are two great mathematicians who have just shown themselves not fully familiar with the exponential function; we must recall that this is 1735!"

In his *Principia*, Newton gave a full description of his geometric treatment of simple harmonic motion. In a 1728 paper on simple harmonic motion, Johann Bernoulli gave a more analytic treatment, solving the second-order linear differential equation $d^2y/dx^2 = -y$ by reducing it to a first-order equation; Hermann did similar work in a 1716 paper in the *Acta Eruditorum*. These are the earliest treatments of simple harmonic motion by the integration of the differential equation describing this motion.

In a May 5, 1739 letter to Johann Bernoulli, Euler wrote that he had succeeded in solving the third-order equation $a^3dy^3 = ydx^3$, where dx was assumed constant. Note that this meant that x was the independent variable. Euler gave the solution as

$$y = be^{x/a} + ce^{-x/2a} \cdot \text{Sinum Arcus } \frac{(f+x)\sqrt{3}}{2a}.$$

He gave no indication of how he found this, but he probably did not use a general method, because in his letter of September 15, 1739 to Bernoulli, he wrote that he had recently found a general method for solving in finite terms the equation

$$y + a\frac{dy}{dx} + b\frac{ddy}{dx^2} + c\frac{d^3y}{dx^3} + d\frac{d^4y}{dx^4} + e\frac{d^5y}{dx^5} + \text{etc.} = 0.$$

He expressed surprised that the solution depended on the roots of the algebraic equation

$$1 - ap + bp^2 - cp^3 + dp^4 - ep^5 + \text{etc.} = 0.$$

As an example, he explained that the solution of Daniel Bernoulli's equation $d^4y = k^4ydx^4$ was determined by the algebraic equation $1 - k^4p^4 = 0$. Thus, the solution of the differential equation emerged as

$$y = Ce^{-x/k} + De^{x/k} + E\sin(x/k) + F\cos(x/k).$$

In a letter of January 19, 1740, Euler mentioned that he could also solve

$$0 = y + ax\frac{dy}{dx} + bx^2\frac{d^2y}{dx^2} + cx^3\frac{d^3y}{dx^3} + \cdots. \tag{15.1}$$

Johann Bernoulli replied to Euler in a letter of April 16, 1740, that he too had solved (15.1). He reduced its order by multiplying it by x^p and choosing p appropriately. He remarked that he had actually done this before 1700 and also wrote that he had found a special solution similar to that found by Euler for the equation with constant coefficients. However, he was puzzled as to how the imaginary roots could lead to sines and cosines. For about a year, he discussed this with Euler. Finally, Euler pointed out that the equation $ddy + ydx^2 = 0$ had the obvious solution $y = 2\cos x$, also taking the form $y = e^{x\sqrt{-1}} + e^{-x\sqrt{-1}}$. Bernoulli ended his letter by asking whether it was possible to reduce the equation

$$yxx\,dx^2 + a\,ddy = 0 \quad \left(\text{i.e., } a\frac{d^2y}{dx^2} + x^2 y = 0\right)$$

to a first-order equation. Euler answered that by the substitution $y = e^{\int z\,dx}$, the equation reduced to

$$xx\,dx + a\,dz + azz\,dx = 0 \tag{15.2}$$

and noted that this was a particular case of the Riccati equation $dy = yy\,dx + ax^m\,dx$ on which he had already written papers.

Interestingly, in a paper on curves and their differential equations, published 46 years earlier, Bernoulli stated an equation almost identical to (15.2) and wrote that he had not solved it; he noted that, of course, separation of variables would not work. His older brother Jakob made persistent efforts to solve the equation and finally succeeded in 1702. In a letter to Leibniz dated November 15, 1702, some of which was devoted to the relation of the sum $\sum 1/n^2$ with integrals of the form $\int x^l \ln(1+x)\,dx$, he mentioned in passing that he could solve $dy = yy\,dx + xx\,dx$ by reducing it to $ddy = x^2 y\,dx^2$ and then applying separation of variables. When Leibniz asked for details, Jakob provided them in a letter of October 3, 1703. In modern notation, he defined a new function z by the equation $y = -\frac{1}{z}\frac{dz}{dx}$ to reduce

$$\frac{dy}{dx} = x^2 + y^2 \quad \text{to the form} \quad \frac{d^2z}{dx^2} + x^2 z = 0.$$

He solved this second order linear equation by an infinite series for z from which he obtained y as a quotient of two infinite series. After performing the division of one series by the other, his result was

$$y = \frac{x^3}{3} + \frac{x^7}{2 \cdot 2 \cdot 7} + \frac{2x^{11}}{3 \cdot 3 \cdot 3 \cdot 7 \cdot 11} + \frac{13x^{15}}{3 \cdot 3 \cdot 3 \cdot 3 \cdot 5 \cdot 7 \cdot 7 \cdot 11} + \cdots. \tag{15.3}$$

Now Newton would have been satisfied with this infinite series solution, whereas Leibniz and the Bernoullis had a different general outlook. They strove to find solutions in finite form, using the known elementary functions. Perhaps this may explain why mathematicians in the Leibniz-Bernoulli school took scant note of Jakob Bernoulli's new method for dealing with the Riccati equation. However, Euler himself later rediscovered this method and generalized it.

Jacopo Riccati (1676–1754) studied law at the University of Padua, but was encouraged to pursue mathematics by Stephano degli Angeli, who had earlier taught James Gregory. Riccati became interested in the equation named after him upon studying Gabriele Manfredi's treatise *De Constructione Aequationum Differentialium* in which he considered the equation

$$nxx\,dx - nyy\,dx + xx\,dy = xy\,dx,$$

15.1 Preliminary Remarks

a special case of what is now known as the generalized Riccati equation

$$\frac{dy}{dx} = P + Qy + Ry^2,$$

where P, Q, and R are functions of x. Around 1720, Riccati and others worked on the special case

$$ax^m dx + yy dx = b dy, \qquad (15.4)$$

and Riccati attempted a solution using separation of variables. Amusingly, Riccati corresponded on this topic with all the then living Bernoulli mathematicians: Niklaus I, Johann, and his two sons Niklaus II and Daniel. The latter two determined by different methods a sequence of values of m for which the equation could be solved in finite terms. Riccati published his work in 1722 with a note by D. Bernoulli, who gave the announcement of the solution of the Riccati equation (15.4) as an anagram. About a year later, without reference to his anagram, D. Bernoulli published details of his solution. Briefly, his result was that the equation could be solved in terms of the logarithmic, exponential and algebraic functions when $n = \frac{-4m}{2m \pm 1}$, that is, when n was a number in the sequence

$$0; -\frac{4}{1}, -\frac{4}{3}; -\frac{8}{3}, -\frac{8}{5}; -\frac{12}{5}, -\frac{12}{7}; -\frac{16}{7}, -\frac{16}{9}, \ldots.$$

His method was to show that the substitutions $\frac{x^{n+1}}{n+1} = u$, $y = -\frac{1}{v}$ in the equation $dy/dx = ax^n + by^2$ produced another equation of the same form, but with n changed to $-\frac{n}{n+1}$. Then again, the substitutions $x = \frac{1}{u}$, $y = -\frac{u}{b} - vu^2$ also produced another equation of the same form, but with n changed to $-n - 4$. Now when $n = 0$, the equation was the integrable $dy/dx = a + by^2$. It followed that when $n = -4$, the equation would still be integrable, and the same was true when $n = -\frac{-4}{-4+1} = -\frac{4}{3}$. The result was the sequence given by Daniel Bernoulli. Note that when $m \to \infty$, $n = \frac{-4m}{2m \pm 1} \to -2$; it turns out that when $n = -2$, the equation is still integrable. In that case, $dy/dx = a/x^2 + by^2$ and the substitution $y = v/x$ produces the separable equation

$$x \frac{dv}{dx} = a + v + bv^2.$$

Euler published many papers on the Riccati equation. In the 1730s, he elucidated its relation to continued fractions, and solved it as a ratio of two infinite series in the manner of Jakob Bernoulli. Euler's method here also demonstrated that these series would be finite for those values of m defined by Daniel Bernoulli and his brother. In the 1760s, Euler demonstrated that the generalized Riccati equation could be transformed to a linear second-order equation, and conversely. He also showed that if one particular solution, y_0, of the generalized Riccati equation were known, then, by the substitution $y = y_0 + 1/v$, Riccati's equation could be reduced to the linear equation

$$\frac{dv}{dx} + (Q + 2Ry_0)v + R = 0. \qquad (15.5)$$

On the other hand, if two solutions, y_0 and y, were known, then $w = (y - y_0)/(y - y_1)$ satisfied the simpler equation

$$\frac{1}{w}\frac{dw}{dx} = R(y_0 - y_1). \tag{15.6}$$

Interestingly, in 1841, Joseph Liouville proved the converse of D. Bernoulli's theorem on the Riccati equation. Liouville was very interested in the general problem of integration in finite terms, using the elementary and algebraic functions, a topic now seeing renewed interest in the area of symbolic integration. Liouville proved that if $dy/dx = ax^n + by^2$ could be solved in finite terms, then n had to be one of the numbers determined by Bernoulli.

In a 1753 paper, Euler solved nonhomogeneous linear equations by the technique of multiplying the equation by an appropriate function to reduce its order. Later, in 1762, Lagrange found another method for reducing the order of such an equation, leading him to the concept of an adjoint, a label apparently first used in this context by Lazarus Fuchs about a century later. Briefly, Lagrange took the differential equation to be

$$Ly + M\frac{dy}{dt} + N\frac{d^2y}{dt^2} + \cdots = T, \tag{15.7}$$

where $L, M, N, \ldots, T$ were functions of t. He then multiplied the equation by some function $z(t)$ and integrated by parts. Since

$$\int Mz\frac{dy}{dt}dt = Mzy - \int \frac{d}{dt}(Mz)y\,dt,$$

$$\int Nz\frac{d^2y}{dt^2}dt = Nz\frac{dy}{dt} - \frac{d}{dt}(Nz)y + \int \frac{d^2}{dt^2}(Nz)y\,dt,$$

and so on, the original equation was transformed to

$$y\left(Mz - \frac{d}{dt}(Nz)\right) + \frac{dy}{dt}Nz + \cdots$$

$$+ \int \left(Lz - \frac{d}{dt}(Mz) + \frac{d^2}{dt^2}(Nz) + \cdots\right) y\,dt = \int Tz\,dt. \tag{15.8}$$

Lagrange then took z to be the function satisfying

$$Lz - \frac{d}{dt}(Mz) + \frac{d^2}{dt^2}(Nz) - \cdots = 0. \tag{15.9}$$

This was the adjoint equation, and if z satisfied it, then the expression within parentheses in the integral on the left-hand side of (15.8) would vanish. The remaining equation would then be of order $n - 1$. In this way, the order of the equation (15.7) was reduced by one, and the process could be continued. When Lagrange applied this procedure to the adjoint equation (15.9) to reduce its order, he obtained the homogeneous part of the equation (15.7). Thus, he saw that the adjoint of a homogeneous equation was, in fact, that equation itself. Lagrange also discovered the general method of variation of parameters in order to obtain the solution of a nonhomogeneous equation, once the solution of the corresponding homogeneous equation was known. Lagrange did this work around 1775, but in 1739, Euler applied this same method to the special equation $\frac{d^2y}{dx^2} + ky = X$.

We have seen that in his 1743 paper, Euler introduced the concepts of general and particular solutions of linear equations. By choosing appropriate constants in the general solution, any particular solution could be obtained. Taylor in 1715 and Clairaut in 1734 found solutions for some special nonlinear equations, solutions not producible by choosing constants in the general solutions. Of one solution, Taylor remarked that it was singular, and so they were named. Euler also studied singular solutions; he found it paradoxical that they could not be obtained from the general solutions. He first encountered such a situation in the course of his study of mechanics in the 1730s. In his paper of 1754, he posed a number of geometric problems leading to singular solutions, commenting that the paradox of singular solutions was not a mere aberration of mechanics.

It appears that the French mathematician Alexis Claude Clairaut (1713–1765) was the first to give a geometric interpretation of a singular solution; this appeared in his 1734 paper on differential equations. He considered the equation

$$y = (x+1)\frac{dy}{dx} - \left(\frac{dy}{dx}\right)^2 \equiv (x+1)p - p^2.$$

Note here that the more general solution $y = xy' + f(y')$ is now called Clairaut's equation. Briefly describing his solution, we differentiate with respect to p and simplify the equation to get

$$dp(x+1-2p) = 0.$$

The first factor gives $p = c$, a constant, so that $y = c(x+1) - c^2$. The second factor gives $p = (x+1)/2$ so that $y = (x+1)^2/4$ is a solution. Now the envelope of the family of straight lines $y = c(x+1) - c^2$ is found by first eliminating c from this equation and its derivative with respect to c, given by $0 = x + 1 - 2c$. Thus, the envelope is given by $y = (x+1)^2/4$ and in this case, the singular solution $y = (x+1)^2/4$ is the envelope of the family of integral curves.

D'Alembert, Euler, and Laplace also studied singular solutions. These efforts culminated in the general theory developed in the 1770s by Lagrange. Considering equations of the form

$$f\left(x, y, \frac{dy}{dx}\right) = a_n(x, y)\left(\frac{dy}{dx}\right)^n + \cdots + a_1(x, y)\frac{dy}{dx} + a_0(x, y) = 0,$$

he gave two approaches to the study of singular solutions. In the second of these methods, he differentiated the equation with respect to x to obtain the expression for the second derivative

$$\frac{d^2y}{dx^2} = -\frac{\partial f/\partial x + (\partial f/\partial y)\,dy/dx}{\partial f/\partial y'}.$$

Lagrange asserted that at the points of the singular solutions, the numerator and the denominator both vanish. Hence, both the equations

$$\frac{\partial f}{\partial y'} = 0 \quad \text{and} \quad \frac{\partial f}{\partial x} + \frac{\partial f}{\partial y}\frac{\partial y}{\partial x} = 0$$

had to be satisfied along the singular solutions. Note that these relations are true for Clairaut's singular solution. In fact, the second equation is identically true in that case.

In 1835, Cauchy was the first to treat the question of the existence of a solution of a differential equation. Earlier mathematicians had presented methods for solving the equation on the assumption that solutions existed. Cauchy worked with a system of first-order differential equations; given in its simplest form with only one equation, his result can be stated: Suppose $f(x, y)$ is analytic in the variables x and y in the neighborhood of a point (x_0, y_0); then the equation $dy/dx = f(x, y)$ has a unique analytic solution $y(x)$ in a neighborhood of x_0 such that $y(x_0) = y_0$. Cauchy's first proof of this theorem had gaps and in the 1840s, he published papers giving a more detailed exposition of the result. The French mathematicians C. Briot and J. Bouquet worked out a clearer and more complete presentation of this method, called the method of majorant. For a brief description of this method, take $x_0 = y_0 = 0$ and suppose

$$f(x, y) = \sum a_{kl} x^l y^k \text{ for } |x| \leq A, \ |y| \leq B.$$

Also, let M be the maximum of $|f(x, y)|$ in this region. Then the function

$$F(x, y) = \frac{M}{((1 - x/A)(1 - y/B))} \text{ is such that if}$$

$$F(x, y) = \sum A_{kl} x^l y^k \text{ then } |a_{kl}| \leq |A_{kl}|.$$

The differential equation $dy/dx = F(x, y)$ can now be solved explicitly. First, it can be shown that the coefficients of the formal power series solution of $dy/dx = f(x, y)$ are bounded by the coefficients of the explicit solution of $dy/dx = F(x, y)$, and this fact may then be used to show that the formal solution is an actual solution and is unique. In the 1870s, Kovalevskaya showed that this method could be extended to a certain system of partial differential equations. The paper containing this result formed a part of her doctoral dissertation, supervised by Weierstrass.

15.2 Leibniz: Equations and Series

Leibniz's approach, as contrasted with Newton's, called for only occasional use of infinite series. Nevertheless, Leibniz discussed the connection between series and calculus; he derived series for elementary functions in several ways. In a paper of 1693, referring to Mercator and Newton, he derived the logarithmic and exponential series. As early as 1674, while working his way toward his final conception of the calculus, Leibniz discovered the series for $\exp(x)$. He started with the series

$$y = x + \frac{x^2}{2!} + \frac{x^3}{3!} + \cdots,$$

15.2 Leibniz: Equations and Series

and integrated to get

$$\int_0^x y\,dx = \frac{x^2}{2!} + \frac{x^3}{3!} + \frac{x^4}{4!} + \cdots = y - x. \tag{15.10}$$

Note that at the time Leibniz did this calculation, he was still using the "omn." notation for the integral. By taking the differential of both sides, he obtained

$$y\,dx = dy - dx \text{ or } dx = \frac{dy}{y+1}.$$

Now Leibniz knew from the work of N. Mercator that this last equation implied $x = \ln(y+1)$ so that

$$\exp(x) = 1 + y = 1 + x + \frac{x^2}{2!} + \frac{x^3}{3!} + \cdots.$$

Although Leibniz did not explicitly write the last equation, he clearly understood the relation of the series to the logarithm.

Leibniz wrote a letter to Huygens, dated September 4/14, 1694 explaining his new calculus techniques. Here note that Leibniz gave two dates because of the ten-day difference between the Julian and Gregorian calendars during that period. Huygens, at the close of an outstanding scientific career, was still eager to learn of the new advances in mathematics. He had already glimpsed the power of calculus from the work of Leibniz and the two Bernoullis on the catenary problem. Leibniz's first example for Huygens was the derivation of the infinite series for $\cos x$ from its differential equation. He started by observing that if y was the arc of a circle of radius a, and x denoted $\cos y$, then

$$y = a \int \frac{dx}{\sqrt{a^2 - x^2}}, \tag{15.11}$$

$$\text{and } dy = \frac{a\,dx}{\sqrt{a^2 - x^2}}, \tag{15.12}$$

$$\text{or } \sqrt{a^2 - x^2}\,dy = a\,dx. \tag{15.13}$$

Note here that the integral (15.11) was Leibniz's definition of arcsine or arccosine, as given in his 1686 paper. We follow Leibniz almost word for word. He set

$$v = \sqrt{a^2 - x^2} \tag{15.14}$$

$$\text{so that } v\,dy = a\,dx. \tag{15.15}$$

By differentiating this equation he found

$$v\,ddy + dv\,dy = a\,ddx. \tag{15.16}$$

Leibniz then assumed that the arcs y increased uniformly, that is, dy was a constant or ddy was zero. Recall that in our terms this meant that he was taking y as the independent variable and x and v as functions of y. Thus, he had $ddy = 0$, and equation (15.16) reduced to

$$dv\,dy = a\,ddx. \tag{15.17}$$

To eliminate v, he observed that from (15.14), $v^2 = a^2 - x^2$, and therefore $v\,dv = -x\,dx$ or

$$dv = -x\frac{dx}{v}. \tag{15.18}$$

By (15.15) and (15.18),

$$dv = -\frac{x\,dy}{a}; \tag{15.19}$$

then, by (15.17) and (15.19), he arrived at the required differential equation,

$$-x\,dy\,dy = a^2\,ddx. \tag{15.20}$$

In order to derive the series for $a\cos(y/a)$ from this equation, Leibniz set

$$x = a + by^2 + cy^4 + ey^6 + \cdots.$$

He substituted this in (15.20) and equated coefficients. After a detailed calculation, he arrived at

$$x = \frac{1}{1}a - \frac{1}{1\cdot 2\cdot a}y^2 + \frac{1}{1\cdot 2\cdot 3\cdot 4\cdot a^3}y^4 - \frac{1}{1\cdot 2\cdot 3\cdot 4\cdot 5\cdot 6\cdot a^5}y^6 + \text{etc.}$$

15.3 Newton on Separation of Variables

To get an idea of Newton's thinking and notation, we consider one simple example on separation of variables from *De Quadratura*. He began with the equation $-ax\dot{x}y^2 = a^4\dot{y} + a^3x\dot{y}$. He separated the variables and integrated to get

$$\left(\frac{aa}{a+x} - a\right)\dot{x} = \frac{a^3}{y^2}\dot{y},$$

and then $\boxed{\dfrac{aa}{a+x}} - ax = -\dfrac{a^3}{y}.$

In Newton's notation, the square box denoted an integral. Sometimes, he replaced the box by the letter Q, denoting the Latin expression for area. He had no special notation for the logarithm and merely referred to it as the area of the hyperbola with ordinate $aa/(a+x)$ and abscissa x. He then rewrote the last equation, omitting the constant term, as an infinite series

$$\frac{a^2}{y} = \frac{1}{2}xx - \frac{x^3}{3a} + \frac{x^4}{4aa} - \cdots.$$

"Or again, if c^2 is any given quantity, then is

$$\frac{a^2}{y} = c^2 + \frac{1}{2}xx - \frac{1}{3}\frac{x^3}{a} + \frac{1}{4}\frac{x^4}{aa}\cdots$$

the equation to be found." Newton illustrated Fatio's method of integrating factors, starting with

$$9x\dot{x}y - 18\dot{x}y^2 - 18xy\dot{y} + 5x^2\dot{y} = 0. \tag{15.21}$$

Applying Fatio's technique, he multiplied the equation by $x^\mu y^\nu$ to get

$$9\dot{x}x^{\mu+1}y^{\nu+1} - 18\dot{x}x^\mu y^{\nu+2} - 18x^{\mu+1}y^{\nu+1}\dot{y} + 5x^{\mu+2}y^\nu\dot{y} = 0.$$

He then integrated with respect to x the terms forming the coefficient of $\dot{x}$ to obtain

$$\frac{9}{\mu+2}x^{\mu+2}y^{\nu+1} - \frac{18}{\mu+1}x^{\mu+1}y^{\nu+2} + g.$$

The fluxion of this expression would then reproduce the terms containing $\dot{x}$, but the terms with $\dot{y}$ would be given by

$$\frac{9(\nu+1)}{\mu+2}x^{\mu+2}y^\nu\dot{y} - \frac{18(\nu+2)}{\mu+1}x^{\mu+1}y^{\nu+1}\dot{y}.$$

For this to agree with (15.21), Newton required that $\frac{9(\nu+1)}{\mu+2} = 5$ and $\frac{\nu+2}{\mu+1} = 1$; or that $\mu = 5/2$ and $\nu = 3/2$. Hence, with g a constant, the solution of the fluxional equation was

$$2x^{9/2}y^{5/2} - \frac{36}{7}x^{7/2}y^{7/2} + g = 0.$$

15.4 Johann Bernoulli's Solution of a First-Order Equation

We have seen that quite early in the study of differential equations, mathematicians noticed that separation of variables and integrating factors were applicable in many special situations. At the same time, they observed that there were simple first-order equations to which these methods could not be directly applied. Jakob Bernoulli, in the November 1695 issue of the *Acta Eruditorum*, posed the problem of solving the differential equation

$$a\,dy = yp\,dx + by^n q\,dx, \tag{15.22}$$

where p and q were functions of x and a was a constant. In 1696, Leibniz noted that this problem could be reduced to a linear equation, though he did not give details. A year later, Johann Bernoulli wrote that $y = v^{1/1-n}$ would reduce the equation to the linear equation

$$\frac{1}{1-n}a\,dv = vp\,dx + bq\,dx. \tag{15.23}$$

However, his alternative method was to take y to be a product of two new variables so that the extra variable could be appropriately chosen. Thus, he set $y = mz$ so that $dy = m\,dz + z\,dm$ and equation (15.22) would take the form

$$az\,dm + am\,dz = mzp\,dx + bm^n z^n q\,dx. \tag{15.24}$$

He then set $am\,dz = mzp\,dx$ or

$$a\,dz : z = p\,dx. \tag{15.25}$$

Hence, z could be found in terms of x. Bernoulli denoted this function by ξ. In the notation developed by Euler in the late 1720s, one would write $\xi = c^{\frac{1}{a}\int p\,dx}$ or $e^{\frac{1}{a}\int p\,dx}$. Note that Euler changed the c to e in the 1730s. Equation (15.24) was then reduced to

$$az\,dm = bm^n z^n q\,dx \quad \text{or} \quad a\xi\,dm = bm^n \xi^n q\,dx$$

$$\text{or} \quad am^{-n}\,dm = b\xi^{n-1} q\,dx. \tag{15.26}$$

After integration, he had

$$\frac{a}{-n+1} m^{-n+1} = b\int \xi^{n-1} q\,dx. \tag{15.27}$$

In 1728, Euler made use of the integrating factor suggested by Bernoulli's method to solve nonhomogeneous linear equations of the first order. The specific equation he faced was

$$dz + \frac{2z\,dt}{t-1} + \frac{dt}{tt-t} = 0.$$

He found the integrating factor by taking the exponential of the integral of the coefficient of z. So he multiplied the equation by $c^{2\int dt/(t-1)} = (t-1)^2$ and then integrated to obtain the solution

$$(t-1)^2 z + \int \frac{t-1}{t}\,dt = a.$$

In the general situation, this would require the solution of $\frac{dy}{dx} + py = q$; multiplying by $e^{\int p\,dx}$, Euler wrote

$$\frac{d}{dx}(ye^{\int p\,dx}) = qe^{\int p\,dx}$$

and hence $y = e^{-\int p\,dx} \int qe^{\int p\,dx}\,dx.$

15.5 Euler on General Linear Equations with Constant Coefficients

Euler seems to have been the first mathematician to apply linear superposition of special solutions to obtain the general solution of a linear differential equation. One may

15.5 Euler on General Linear Equations with Constant Coefficients

contrast this with the method employed by Johann Bernoulli to solve the equation for simple harmonic motion,

$$n^2 a^2 \frac{d^2 x}{dy^2} = a - x.$$

Note that we have slightly modernized Bernoulli's notation; he wrote the equation as

$$nnaaddx : dy^2 = a - x.$$

Note also that dy^2 stands for $(dy)^2$. Bernoulli multiplied the equation by dx (or dx/dy) and integrated to get

$$\frac{n^2 a^2}{2} \left(\frac{dx}{dy}\right)^2 = ax - \frac{1}{2}x^2.$$

Here observe that Bernoulli used $\int dx ddx = \frac{1}{2}(dx)^2$; next, he had

$$n \int \frac{a\, dx}{\sqrt{2ax - x^2}} = \int dy = y, \text{ or } y = na \arcsin \frac{x-a}{a} + C.$$

Bernoulli presented his solution in this form with no mention of linear superposition of particular sine and cosine solutions. In 1739, Euler considered an equation of the form

$$0 = Ay + B\frac{dy}{dx} + C\frac{d^2 y}{dx^2} + D\frac{d^3 y}{dx^3} + \cdots + N\frac{d^n y}{dx^n} \tag{15.28}$$

and observed that if $y = u$ was a solution, then so was $y = \alpha u$, with α a constant. Moreover, if n particular solutions $y = u, y = v, \ldots$ could be found, then the general solution would be $y = \alpha u + \beta v + \cdots$. To obtain these special solutions, he took $y = e^{\int p\, dx}$; Euler then wrote out the derivatives of y:

$$y = e^{\int p\, dx}$$

$$\frac{dy}{dx} = e^{\int p\, dx} p$$

$$\frac{d^2 y}{dx^2} = e^{\int p\, dx} \left(pp + \frac{dp}{dx}\right)$$

$$\frac{d^3 y}{dx^3} = e^{\int p\, dx} \left(p^3 + 3p\frac{dp}{dx} + \frac{d^2 p}{dx^2}\right)$$

$$\frac{d^4 y}{dx^4} = e^{\int p\, dx} \left(p^4 + 6pp\frac{dp}{dx} + 4p\frac{d^2 p}{dx^2} + \frac{d^3 p}{dx^3}\right).$$

When these values were substituted in the differential equation, the expression would be simplest when p was a constant, for in that case the derivatives of p would vanish; p would then satisfy the algebraic equation

$$0 = A + Bz + Czz + Dz^3 + \cdots + Nz^n.$$

If $z = s/t$ was a root of this equation, then $s - tz = 0$ satisfied the nth-degree algebraic equation, and the solution $y = \alpha e^{sx/t}$ of the differential equation

$$sy - t\frac{dy}{dx} = 0$$

also satisfied the nth-order differential equation. When there was a repeated root, so that $(s - tz)^2 = ss - 2stz + ttzz = 0$ was a factor of the nth-degree polynomial, then there would be a corresponding factor

$$ssy - 2st\frac{dy}{dx} + tt\frac{ddy}{dx^2} = 0$$

of (15.28). To solve this second-order equation, Euler set $y = e^{sx/t}u$ to find that u satisfied

$$\frac{ddu}{dx^2} = 0.$$

Thus, $u = \alpha x + \beta$ and $y = e^{sx/t}(\alpha x + \beta)$. Similarly, when there were three repeated roots, then $y = e^{sx/t}(\alpha x^2 + \beta x + \gamma)$. In the general case where a root was repeated k times, the solution would turn out to be $e^{sx/t}$ times a general polynomial of degree $k - 1$. In 1743, Euler then considered complex roots, making use of a result he had published three years earlier, that the general solution of the equation

$$\frac{d^2y}{dx^2} + ky = 0 \text{ was } A\cos\sqrt{k}x + B\sin\sqrt{k}x.$$

So he supposed that $a - bz + cz^2 = 0$ was a quadratic factor of the polynomial; this yielded $z = \frac{b \pm \sqrt{b^2 - 4ac}}{2c}$. Then he assumed $\frac{b}{2\sqrt{ac}} < 1$ and set $\cos\phi = b/(2\sqrt{ac})$. Euler wrote the quadratic factor as $a - 2z\sqrt{ac}\cos\phi + cz^2$ and the corresponding differential equation as

$$0 = ay - 2\sqrt{ac}\cos\phi\frac{dy}{dx} + c\frac{d^2y}{dx^2}.$$

The substitution $y = e^{\sqrt{ac}x\cos\phi}u$ reduced the equation to

$$c\frac{d^2u}{dx^2} + \left(ac^2\cos^2\phi - 2ac\cos^2\phi + a\right)u = 0;$$

note that this was of the required form $\frac{d^2u}{dx^2} + ku = 0$. Euler then proceeded to the case where there were repeated complex roots.

15.6 Euler: Nonhomogeneous Equations

Euler suggested a series method to solve a nonhomogeneous equation

$$X = Ay + B\frac{dy}{dx} + C\frac{ddy}{dx^2} + \cdots + \Delta\frac{d^n y}{dx^n}, \qquad (15.29)$$

where $A, B, C, \ldots$ were constants and X was any power series. This method, using power series and equating coefficients, was presented to the Petersburg Academy in 1750–51 and was published in 1753. But in this paper, Euler also discussed in detail a second method, starting with the first-order equation

$$X = Ay + B\frac{dy}{dx}; \tag{15.30}$$

he multiplied by $e^{ax} dx$ to get

$$e^{ax} X\, dx = Ae^{ax} y\, dx + Be^{ax}\, dy \tag{15.31}$$

and then assumed

$$\int e^{ax} X\, dx = Be^{ax} y. \tag{15.32}$$

Differentiation of this equation gave Euler

$$e^{ax} X\, dx = aBe^{ax} y\, dx + Be^{ax}\, dy.$$

Comparing this equation with (15.31), he concluded that $a = A/B$, and by (15.32), the solution of (15.30) emerged as

$$y = \frac{a}{A} e^{-ax} \int e^{ax} X\, dx.$$

Euler then considered a second-order equation to illustrate how the method could be generalized. This second order equation, after multiplication by $e^{ax} dx$, yielded

$$e^{ax} X\, dx = Ae^{ax} y\, dx + Be^{ax}\, dy + Ce^{ax} \frac{ddy}{dx}. \tag{15.33}$$

He assumed

$$\int e^{ax} X\, dx = e^{ax} \left(A'y + B' \frac{dy}{dx} \right) \tag{15.34}$$

in this case, so that, after differentiation, he obtained

$$e^{ax} X\, dx = e^{ax} \left(aA' y\, dx + a'\, dy + b' \frac{ddy}{dx} + aB'\, dy \right). \tag{15.35}$$

By comparing (15.33) and (15.35), he obtained

$$B' = C, \quad A' = B - aC, \quad \text{and} \quad aA' = A. \tag{15.36}$$

Hence, a came out to be a solution of the quadratic $0 = A - aB + a^2 C$. Moreover, the second-order equation was reduced to a first-order equation

$$e^{-ax} \int e^{ax} X\, dx = A'y + B' \frac{dy}{dx}, \tag{15.37}$$

where $A' = B - aC$ and $B' = C$. Euler could then solve (15.37) in the manner used to solve (15.30) and, in fact, he gave full details. For the general case (15.29), he set

$$\int e^{ax} X\, dx = e^{ax}\left(A'y + B'\frac{dy}{dx} + C'\frac{ddy}{dx^2} + \cdots + \Delta'\frac{d^{n-1}y}{dx^{n-1}}\right)$$

and showed that a had to be a solution of

$$A - Ba + Ca^2 - Da^3 + \cdots \pm \Delta a^n = 0 \text{ and that}$$

$$A' = A/a, \quad B' = (B - aA)/a^2, \ C' = (Ca^2 - Ba + A)/a^3, \quad \text{etc.}$$

Thus, (15.29) was reduced to an equation of a lower order:

$$e^{ax}\int e^{ax} X\, dx = A'y + B'\frac{dy}{dx} + \cdots + \Delta'\frac{d^{n-1}y}{dx^{n-1}}.$$

15.7 Lagrange's Use of the Adjoint

In 1761–62, Lagrange illustrated how to use the knowledge of the solutions of a homogeneous equation of second-order to solve the corresponding nonhomogeneous equation. He supposed that y_1 and y_2 were (independent) solutions of the homogeneous part of the second order equation

$$Ly + M\frac{dy}{dt} + N\frac{d^2y}{dt^2} = T. \tag{15.38}$$

By multiplying by z and applying integration by parts, the adjoint equation was obtained:

$$y\left(Mz - \frac{dNz}{dt}\right) + \frac{dy}{dt}Nz + \int\left(Lz - \frac{d(Mz)}{dt} + \frac{d^2(Nz)}{dt^2}\right)dt = \int Tz\,dt. \tag{15.39}$$

He then set

$$Lz - \frac{d(Mz)}{dt} + \frac{d^2(Nz)}{dt^2} = 0,$$

so that

$$y\left(Mz - \frac{d(Nz)}{dt}\right) + \frac{dy}{dt}Nz = \int Tz\,dt. \tag{15.40}$$

Multiplying this adjoint equation by a function y and applying the same integration by parts procedure, Lagrange arrived at

$$y\left(Mz - \frac{d(Nz)}{dt}\right) + \frac{dy}{dt}Nz - \int\left(Ly + M\frac{dy}{dt} + N\frac{d^2y}{dt^2}\right)z\,dt = \text{constant}.$$

15.7 Lagrange's Use of the Adjoint

When $y = y_1$ or $y = y_2$, the integral vanished, and he got the two relations

$$z\left[\left(M - \frac{dN}{dt}\right)y_1 + N\frac{dy_1}{dt}\right] - \frac{dz}{dt}Ny_1 = c_1 \tag{15.41}$$

$$z\left[\left(M - \frac{dN}{dt}\right)y_2 + N\frac{dy_2}{dt}\right] - \frac{dz}{dt}Ny_2 = c_2, \tag{15.42}$$

where c_1 and c_2 were constants. Lagrange solved for z to obtain

$$z = \frac{c_1 y_2 - c_2 y_1}{N\left(y_2 \frac{dy_1}{dt} - y_1 \frac{dy_2}{dt}\right)}. \tag{15.43}$$

It may be helpful to note that the denominator here is N times the Wronskian of y_1 and y_2. Lagrange then took $c_1 = 0$ and then $c_2 = 0$, denoting the corresponding values of z by z_1 and z_2, respectively. When these were substituted in (15.40), he obtained

$$y\left(Mz_1 - \frac{d(Nz_1)}{dt}\right) + \frac{dy}{dt}Nz_1 = \int Tz_1\,dt,$$

$$y\left(Mz_2 - \frac{d(Nz_2)}{dt}\right) + \frac{dy}{dt}Nz_2 = \int Tz_2\,dt.$$

He solved for y and thereby arrived at a solution for (15.38):

$$y = \frac{z_2 \int Tz_1\,dt - z_1 \int Tz_2\,dt}{N\left(z_1 \frac{dz_2}{dt} - z_2 \frac{dz_1}{dt}\right)}.$$

Lagrange then considered the case where just one solution y_1 was known so that he had only equation (15.41). Now (15.41) was a linear first order equation in z solvable by Euler's integrating factor method, so that

$$z = \frac{y_1}{N}e^{\int \frac{M}{N}dt}\left(c - c_1 \int \frac{e^{-\int \frac{M}{N}dt}}{y_1^2}dt\right).$$

Again, by taking $c_1 = 0$ and then $c = 0$, he obtained two values, z_1 and z_2, of z. Thus, Lagrange taught us that when two solutions of the homogeneous equation are known, the solution of the nonhomogeneous equation may be obtained by solving two sets of linear equations. But when only one solution is known, one must solve a first-order equation and a pair of linear equations. Lagrange pointed out that when L, M, and N were constants, then the homogeneous equation could easily be solved by Euler's method, and hence the nonhomogeneous equation could also be solved in general for this case. In the case for which L, M, and N were constants and k_1 and k_2 were solutions of $L + Mk + Nk^2 = 0$, $(k_1 \neq k_2)$, Lagrange gave the solution of (15.38) by the formula

$$y = \frac{e^{k_2 t}\int Te^{-k_2 t}dt - e^{k_1 t}\int Te^{-k_2 t}dt}{N(k_2 - k_1)}.$$

15.8 Jakob Bernoulli and Riccati's Equation

In his 1703 letter to Leibniz, Bernoulli gave the derivation for the solution of Riccati's equation, $dy = yy\,dx + xx\,dx$. Part of the problem here was to reduce the equation to a separable one. Bernoulli used an interesting substitution to accomplish this, ending up with a second-order instead of a first-order equation. In order to solve the second order equation, he had to use infinite series. He began by setting $y = -dz : z\,dx$ $\left(y = -\frac{1}{z}\frac{dz}{dx}\right)$, so that by the quotient rule for differentials, the differential equation took the form

$$dx\,dz^2 - z\,dx\,ddz : zz\,dx^2 = dy = yy\,dx + xx\,dx = dz^2 : zz\,dx + xx\,dx;$$

this simplified to

$$-z\,dx\,ddz = xxzz\,dx^3,$$

a separable equation expressible as

$$-ddz : z = xx\,dx^2. \tag{15.44}$$

Now this was the second-order equation $-\frac{1}{z}\frac{d^2z}{dx^2} = x^2$. Bernoulli observed that if you had an equation $-z^e\,ddz = x^v\,dx^2$ and you sought a solution $z = ax^m$, then by substituting this in the equation you would find that $m = (v+2) : (e+1)$. However, in (15.44), $e = -1$ and therefore no solution of the form ax^m would be possible. He then drew an analogy with the first-order equation $dz : z = x^v\,dx$, for which no algebraic solution was possible when $v = -1$. So Bernoulli concluded that no algebraic solution was possible for (15.44) and that he must take recourse in infinite series. He obtained the series solution as

$$z = 1 - \frac{x^4}{3\cdot 4} + \frac{x^8}{3\cdot 4\cdot 7\cdot 8} - \frac{x^{12}}{3\cdot 4\cdot 7\cdot 8\cdot 11\cdot 12} + \frac{x^{16}}{3\cdot 4\cdot 7\cdot 8\cdot 11\cdot 12\cdot 15\cdot 16} - \cdots.$$

Since $y = -\frac{1}{z}\frac{dz}{dx}$, he could write his solution as a ratio of two infinite series. By dividing the series for $-dz/dx$ by the series for z, he obtained the first few terms of the series for y given in (15.3).

15.9 Riccati's Equation

We have seen that in the 1730s, Euler wrote on Riccati's equation, and returned to the topic sometime around 1760, then composing an important paper on first-order differential equations, published in 1763. In that paper, he explained how to obtain the general solution of the Riccati equation if one particular solution were known. His method was to use the known solution to reduce the Riccati equation to a first-order linear differential equation. He supposed v to be a solution of the equation

$$dy + Py\,dx + Qyy\,dx + R\,dx = 0, \tag{15.45}$$

and observed that $y = v + 1/z$ reduced this equation to

$$-\frac{dz}{zz} + \frac{P\,dx}{z} + \frac{2Qv\,dx}{z} + \frac{Q\,dx}{zz} = 0 \quad \text{or}$$
$$dz - (P + 2Qv)z\,dx - Q\,dx = 0. \tag{15.46}$$

He then noted that $S = e^{-\int (P+2Qv)dx}$ was an integrating factor. Hence his solution of (15.46) was $Sz - \int QS\,dx =$ Constant. Later in the paper, Euler considered the particular case of (15.45), discussed by Riccati and the Bernoullis:

$$dy + yy\,dx = ax^m\,dx. \tag{15.47}$$

Euler could find a special solution of this equation, by use of which he could determine the general solution by the already described method. He set $a = cc$, $m = -4n$ and

$$y = cx^{-2n} + \frac{1}{z}\frac{dz}{dx} \tag{15.48}$$

so that (15.47) was converted to the linear second-order equation

$$\frac{ddz}{dx^2} + \frac{2c}{x^{2n}}\frac{dz}{dx} - \frac{2nc}{x^{2n+1}}z = 0. \tag{15.49}$$

He then solved this equation as a series:

$$z = Ax^n + Bx^{3n-1} + Cx^{5n-2} + Dx^{7n-3} + Ex^{9n-4} + \text{etc.}$$

After substituting this in (15.49), he found

$$B = \frac{-n(n-1)A}{2(2n-1)c}, \quad C = \frac{-(3n-1)(3n-2)B}{4(2n-1)c}, \quad D = \frac{-(5n-2)(5n-3)C}{6(2n-1)c}, \text{ etc.}$$

Euler did not write the general case but if we let A_k denote the coefficient of $x^{(2k+1)n-k}$, starting at $k = 0$, then the recurrence relation would be

$$A_k = \frac{-((2k-1)n - k + 1)((2k-1)n - k)}{2k(2n-1)C} A_{k-1}, \quad k = 1, 2, 3, \ldots. \tag{15.50}$$

Note that if for some k, $A_k = 0$, then $A_n = 0$ for $n = k+1, k+2, \ldots$. In this case, the series reduced to a polynomial and from (15.50) one could determine the general condition to be $n = (k-1)/(2k-1)$, or $n = k/(2k-1)$. For these values, the solution could be written in finite form.

15.10 Singular Solutions

In his 1715 book, *Methodus Incrementorum*, Brook Taylor presented some techniques for solving differential equations. In proposition VIII, he explained that solutions in finite form might be found if the equation could be suitably transformed. In describing one method of doing this, he considered the differential equation

$$4x^3 - 4x^2 = (1+z^2)^2 \dot{x}^2 \tag{15.51}$$

and found a singular solution. He set $x = v^\theta y^\gamma$, where θ and γ were parameters to be chosen appropriately later on, so that

$$\dot{x} = (\theta \dot{v} y + \gamma \dot{y} v) v^{\theta-1} y^{\gamma-1}. \tag{15.52}$$

He substituted these values of x and $\dot{x}$ in (15.51) to obtain

$$4v^{3\theta} y^{3\gamma} - 4v^{2\theta} y^{2\gamma} = (1+z^2)^2 (\theta \dot{v} y + \gamma \dot{y} v)^2 v^{2\theta-2} y^{2\gamma-2}. \tag{15.53}$$

Taylor then chose $v = 1 + z^2$ and assumed that z was flowing uniformly, that is $\dot{z} = 1$ so that $\dot{v} = 2z$. Substituting these values in (15.53), he arrived at

$$4v^\theta y^{\gamma+2} - 4y^2 = (2\theta z y + \gamma \dot{y} v)^2. \tag{15.54}$$

Taylor then took $\gamma = -2$ to eliminate y in the first term on the left to obtain

$$v^\theta - y^2 = (\theta z y - \dot{y} v)^2 \quad \text{or}$$

$$v^\theta = (\theta^2 z^2 + 1) y^2 - 2\theta z v y \dot{y} + v^2 \dot{y}^2. \tag{15.55}$$

At this point, he set $\theta = 1$ so that $\theta^2 z^2 + 1 = z^2 + 1 = v$; then dividing by v, equation (15.55) reduced to $1 = y^2 - 2zy\dot{y} + v\dot{y}\dot{y}$. Taking fluxions, he found

$$0 = 2y\dot{y} - 2y\dot{y} - 2zy\ddot{y} - 2z\dot{y}\dot{y} + \dot{v}\dot{y}\dot{y} + 2v\dot{y}\ddot{y}.$$

Since $\dot{v} = 2z$, he had $-2zy\ddot{y} + 2\dot{v}\ddot{y} = 0$. (15.56)

This implied that either $\ddot{y} = 0$ or $-2zy + 2v\dot{y} = 0$. Then the second equation gave him

$$-\dot{v}y + 2v\dot{y} = 0 \quad \text{or} \quad y^2 = v. \tag{15.57}$$

Now since he had taken $\theta = 1$ and $\gamma = -2$, he got a solution,

$$x = v^\theta y^\gamma = vy^{-2} = 1,$$

where the last relation followed from (15.57). At this point, he remarked that $x = 1$ was "a certain singular solution of the problem." For the equation $\ddot{y} = 0$, he picked the initial values in such a way as to obtain as solution

$$y = a + \sqrt{1 - a^2} z, \quad \text{and hence}$$

$$x = vy^{-2} = \frac{1+z^2}{a + \sqrt{1-a^2} z}. \tag{15.58}$$

Observe that the solution $x = 1$ is not obtained from the general (15.58) for any value of a.

15.10 Singular Solutions

Euler discussed singular solutions in a 1754 paper on paradoxes in integral calculus. He there noted the paradoxical fact was that there were differential equations easier to solve by differentiating than by integrating. He wrote that he had encountered such equations in his work on mechanics but that his purpose was to explain that there were easily stated geometric problems from which similar types of equations could arise. Euler started by presenting the problem of finding a curve such that the length of the perpendicular from a given point to any tangent to the curve was a constant. By using similar triangles, he found that the differential equation would be

$$y\,dx - x\,dy = a\sqrt{dx^2 + dy^2}, \tag{15.59}$$

where a denoted a constant. After squaring and solving for dy he obtained

$$(a^2 - x^2)\,dy + xy\,dx = a\,dx\sqrt{x^2 + y^2 - a^2}. \tag{15.60}$$

He substituted $y = u\sqrt{a^2 - x^2}$ to transform the equation into the separable equation

$$\frac{du}{\sqrt{u^2 - 1}} = \frac{a\,dx}{a^2 - x^2}. \tag{15.61}$$

After integration he obtained

$$\ln(u + \sqrt{u^2 - 1}) = \frac{1}{2}\ln\frac{n^2(a+x)}{a-x},$$

where n was a constant. He simplified this to

$$u = \frac{n}{2}\sqrt{\frac{a+x}{a-x}} + \frac{1}{2n}\sqrt{\frac{a-x}{a+x}} \quad \text{or}$$

$$y = u\sqrt{a^2 - x^2} = \frac{n}{2}(a+x) + \frac{1}{2n}(a-x). \tag{15.62}$$

Note that equation (15.61), when written as

$$du = \frac{a\sqrt{u^2 - 1}\,dx}{a^2 - x^2},$$

shows that $u \equiv 1$ is also a solution because both sides vanish. When Euler used this solution in the first equation in (15.62), he got $y = \sqrt{a^2 - x^2}$, or

$$x^2 + y^2 = a^2. \tag{15.63}$$

Thus, the solution of (15.60) turned out to be a family of straight lines (15.62) as well as the circle (15.63). In the same paper, Euler next set out to show that he could derive these solutions by differentiation. Now note that this paper was written before his book on differential calculus in which he explained how higher differentials could be completely replaced by higher differential coefficients. So he explained that he would assume that $dy = p\,dx$, with p a differential coefficient, to remove difficulties associated with further differentiation. His equation (15.59) then became

$$y = px + a\sqrt{1 + p^2}. \tag{15.64}$$

Euler then differentiated (instead of integrated) this equation to get

$$dy = p\,dx + x\,dp + \frac{ap\,dp}{\sqrt{1+p^2}}$$

$$\text{or}\quad 0 = x\,dp + \frac{ap\,dp}{\sqrt{1+p^2}} \tag{15.65}$$

$$\text{or}\quad x = -\frac{ap}{\sqrt{1+p^2}}.$$

Hence by (15.64), $y = \frac{a}{\sqrt{1+p^2}}$. By eliminating p, he obtained the solution $x^2 + y^2 = a^2$ and noted that he could also find the family of straight lines by this method. For that purpose, he observed that (15.65) also had the solution $dp = 0$. This implied $p = $ constant $= n$, and so by (15.64) he obtained $y = nx + a\sqrt{1+n^2}$, the required system of straight lines.

Euler then remarked that equation (15.59) could be modified in such a way that the new equation could be solved more easily by the second method than the first. He considered the equation

$$y\,dx - x\,dy = a(dx^3 + dy^3)^{1/3}, \tag{15.66}$$

or $y = px + a(1 + p^3)^{1/3}$. As solutions, he found a sixth-order curve

$$y^6 + 2x^3y^3 + x^6 - 2a^3y^3 + 2a^3x^3 + a^6 = 0 \tag{15.67}$$

and the family of straight lines

$$y = nx + a(1+n^3)^{1/3}. \tag{15.68}$$

Euler gave three other geometric examples leading to differential equations with singular solutions. One of them yielded

$$y\,dx - (x-b)\,dy = \sqrt{(a^2-x^2)dx^2 + a^2dy^2}, \tag{15.69}$$

an equation difficult to integrate. He solved it by differentiation to obtain the solutions: the ellipse

$$\frac{(x-b)^2}{a^2} + \frac{y^2}{a^2-b^2} = 1,$$

and the family of straight lines

$$y = -n(b-x) + \sqrt{a^2(1+n^2) - b^2}.$$

Later in the paper, Euler remarked that he found it strange that integration, which introduced arbitrary constants, did not produce the general solution, while differentiation did.

15.11 Mukhopadhyay on Monge's Equation

In his September 4/14, 1694 letter to Huygens, Leibniz noted that the differential equation for the circle $x^2 + y^2 = a^2$ could be expressed as $dy/dx = -x/y$. Now start with the equation for a general circle,

$$(x-a)^2 + (y-b)^2 = c^2,$$
$$\text{or } x^2 + y^2 = 2ax + 2by + c^2 - a^2 - b^2. \tag{15.70}$$

To obtain the differential equation, we temporarily let p, q, r denote the first three derivatives of y with respect to x; differentiation of the last equation gives $x + yp = a + bp$ and hence $1 + p^2 + yq = bq$. Therefore, $1 + p^2 = (b-y)q$ and $x - a = (b-y)p$; then, after a short calculation,

$$c = (1+p^2)^{3/2}/q. \tag{15.71}$$

In his book on differential equations, Boole remarked that since the right-hand side of (15.71) was the expression for the radius of curvature, this equation was the differential equation for a circle of radius c. To obtain the equation for a circle of arbitrary radius, take the derivative of (15.71) to get

$$3pq^2 - r(1+p^2) = 0. \tag{15.72}$$

George Salmon (1819–1904) offered an interesting geometric interpretation of this equation in his book *Higher Plane Curves*, first published in 1852, also in later editions. Salmon defined "aberrancy of a curve" where the curve was $y = f(x)$. Let P be a point on the curve and V the midpoint of a chord AB drawn parallel to the tangent at P. Let δ denote the limit of the angles made by the normal at P with the line PV as A and B tend to P. Salmon called δ the aberrancy because $\delta = 0$ for a circle. He noted that

$$\tan\delta = p - \frac{(1+p^2)r}{3q^2}. \tag{15.73}$$

Thus, the geometric meaning of (15.72) was that the aberrancy vanished at any point of any circle.

Boole also stated the differential equation of a general conic

$$ax^2 + 2nxy + by^2 + 2gx + 2fy + c = 0:$$

$$9\left(\frac{d^2y}{dx^2}\right)^2 \frac{d^5y}{dx^5} - 45\frac{d^2y}{dx^2}\frac{d^3y}{dx^3}\frac{d^4y}{dx^4} + 40\left(\frac{d^3y}{dx^3}\right)^3 = 0. \tag{15.74}$$

This differential equation was published in 1810 by the French geometer Gaspard Monge (1746–1818). Boole remarked on this equation, "But here our powers of geometrical interpretation fail, and results such as this can scarcely be otherwise useful than as a registry of integrable forms."

The Indian mathematician Asutosh Mukhopadhyay (also Mookerjee) (1864–1924) published an 1889 paper in the *Journal of the Asiatic Society of Bengal*, showing an interesting geometric interpretation of Monge's equation (15.74). Concerning this result, much appreciated by some British mathematicians, Edwards wrote in the 1892 second edition of his treatise on differential calculus:

> A remarkable interpretation which calls for notice has, however, been recently offered by Mr. A. Mukhopadhyay, who has observed that the expression for the radius of curvature of the locus of the centre of the conic of five pointic [sic] contact with any curve (called the centre of aberrancy) contains as a factor the left-hand member of Monge's equation, and this differential equation therefore expresses that the "*radius of curvature of the 'curve of aberrancy' vanishes* for any point of any curve."

Mukhopadhyay received an M.A. in mathematics from Calcutta University in 1886. He studied much on his own, as is shown by entries in his diary: "Rose at 6.15 a.m. Read Statesman [Newspaper], and Boole's Diff. Equations in the morning. Read Fourier's Heat at noon." And "At noon read from Messenger of Math. Vol. 2, Prof. Cayley's Memoir on Singular Solutions–to my mind, the simplest but the most philosophical account of the subject yet given; read from Forsyth on the same subject." Mukhopadhyay published papers on topics in differential geometry, elliptic functions and hydrodynamics. The following abstract of his 1889 paper, "On a Curve of Aberrancy," may give a sense of his mathematical work:

> The object of this note is to prove that the aberrancy curve (which is the locus of the centre of the conic of closest contact) of a plane cubic of Newton's fourth class is another plane cubic of the same class, the invariants of which are proportional to the invariants of the original cubic; it is also proved that the two cubics have only one common point of intersection, which is the point of inflection for both.

Mukhopadhyay thus gave evidence of a fine mathematical mind but his dream of spending his life in mathematical research could not be realized because there was no support for such endeavors in nineteenth century Indian universities. As one biographer wrote, "Sir Asutosh's contributions to mathematical knowledge were due to his unaided efforts while he was only a college student." Interestingly, after serving as a judge, in 1906 Mukhopadhyay became Vice-Chancellor of Calcutta University and his first order of business was to have the University "combine the functions of teaching and original investigation." He appointed Syamadas Mukhopadhyay (1866–1937) professor of mathematics and encouraged him to pursue research. S. Mukhopadhyay subsequently produced several interesting results, including the well-known four vertex theorem, published in the *Bulletin of the Calcutta Mathematical Society*, founded by Asutosh. Syamadas stated the theorem: "The minimum number of cyclic points on an oval is four." Asutosh created a physics department at the university with the appointment of the experimental physicist C. V. Raman, whom he persuaded to leave his post as an officer in the Indian Accounts Department. Raman went on to win a Nobel Prize in physics. In applied mathematics, he made two outstanding appointments: S. N. Bose, known for his statistical derivation of Planck's law leading to Bose-Einstein statistics, and M. N. Saha, who discovered the Saha ionization law in astrophysics.

15.12 Exercises

1. Show that the polynomial

$$y_k = L_k(x) = \sum_{j=0}^{k} \frac{k(k-1)\cdots(k-j+1)}{j!\,j!}(-x)^j,$$

the kth Laguerre polynomial, is a solution of the recurrence relation

$$(k+1)y_{k+1} - (2k+1-x)y_k + ky_{k-1} = 0, \quad k = 0, 1, 2, \ldots.$$

D. Bernoulli and Euler encountered this equation in their works on the discrete analog of the problem of the small oscillations of a hanging chain. They discussed the discreet and the continuous forms of the problem while they were colleagues at the Petersburg Academy in the late 1720s and early 1730s. Bernoulli submitted his results to the academy before his departure in 1733 and a year later presented his proofs. Upon seeing Bernoulli's work, Euler, who had obtained similar results, submitted his work. They took $x = a/\alpha$, where a was the distance between the weights and α was related to the angular frequency ω by $\omega^2 = g/\alpha$; g was the acceleration due to gravity. The y_k was the simultaneous displacement of the kth weight. The chain was assumed to hang from the nth weight, so $y_n = 0$. This gave $L_n(a/\alpha) = 0$ as the equation determining the frequencies. Euler discovered the polynomial solution of the difference equation; in that sense, he and D. Bernoulli were the first mathematicians to use Laguerre polynomials. Bernoulli found the smallest roots of these polynomials for some values of k, using his method of sequences; see the previous chapter for Daniel Bernoulli's work on difference equations.

2. The differential equation satisfied by the displacement y at a distance x from the point of suspension of a heavy chain was determined by D. Bernoulli to be

$$\alpha x \frac{d^2y}{dx^2} + \alpha \frac{dy}{dx} + y = 0.$$

Show that

$$y = AJ_0\left(2\sqrt{\frac{x}{\alpha}}\right) = A\sum_{j=0}^{\infty} \frac{1}{j!\,j!}\left(\frac{-x}{\alpha}\right)^j$$

is a solution of this equation. Note that Bernoulli did not use the J_0 notation but gave the corresponding series. The value of α is determined from the equation $J_0\left(2\sqrt{l/\alpha}\right) = 0$. Bernoulli stated that this equation had an infinite number of roots and gave the first value $\alpha/l = 0.691$, a good approximation. About 50 years later, Euler gave the first three roots. Bernoulli and Euler may have conjectured the existence of infinitely many roots because the solution of the difference equation in the previous exercise, with a slight change of variables, approximates the solution of the differential equation given in this exercise. We remark that this is not surprising, as the first is a discrete analog of the second. In fact, $L_k(x/k) \to J_0(2\sqrt{x})$. And since $L_k(x) = 0$ has k zeros, J_0 must have infinitely many roots; in fact, the rth root of L_k must tend to the rth root of J_0. This can be proved rigorously by the theory of analytic functions of a complex variable, though Bernoulli and Euler obviously had no knowledge of this theory.

3. In the works mentioned in exercises 1 and 2, Euler also treated the equation

$$\frac{x}{n+1}\frac{d^2y}{dx^2} + \frac{dy}{dx} + \frac{y}{\alpha} = 0.$$

Show, with Euler, that the equation has the series solution

$$y = Aq^{-n/2}I_n(2\sqrt{q}), \quad \text{where} \quad q = -\frac{(n+1)x}{\alpha}, \quad \text{and}$$

$$I_n(x) = (x/2)^n \sum_{k=0}^{\infty} \frac{(x/2)^{2k}}{k!\,\Gamma(k+n+1)}.$$

This appears to be the first occurrence of the Bessel function of arbitrary real index n. Euler also proved that for $n = -1/2$, the solution was $y = A\cos\sqrt{2x/\alpha}$. Prove this and show that this is equivalent to the result $I_{-1/2}(x) = \sqrt{\frac{2}{\pi x}}\cos x$.

4. Euler also obtained the solution of the differential equation in the previous exercise as the definite integral

$$y/A = \frac{\int_0^1 (1-t^2)^{\frac{2n-1}{2}} \cosh\left(2t\sqrt{\frac{(n+1)x}{\alpha}}\right)dt}{\int_0^1 (1-t^2)^{\frac{2n-1}{2}}dt}.$$

Prove Euler's result. It is equivalent to the Poisson integral representation

$$J_n(x) = \frac{2}{\sqrt{\pi}} \frac{(x/2)^n}{\Gamma(n+1/2)} \int_0^{\pi/2} \cos(x\sin\phi)\cos^{2n}\phi\,d\phi.$$

Prove this. According to Truesdell, this may be the earliest example of solution of a second-order differential equation by a definite integral. For this and the previous three exercises, see Truesdell (1960), pp. 154–65 and Cannon and Dostrovsky (1981), pp. 53–64. The references to the original papers of Euler and Bernoulli may be found there.

5. Solve the equation

$$(ydx - xdy)(ydx - xdy + 2bdy) = c^2(dx^2 + dy^2)$$

by Euler's method for finding singular solutions.

6. Let $f(x, y)$ be bounded and continuous on a domain G. Show that then at least one integral curve of the differential equation $dy/dx = f(x, y)$ passes through each interior point (x_0, y_0) of G. This result is due to the Italian mathematician Giuseppe Peano (1838–1932) who graduated from the University of Turin (Torino), where he heard lectures by Angelo Genocchi and Faà di Bruno. Peano developed some aspects of mathematical logic in order to bring a higher degree

of clarity to proofs of theorems in analysis. This led him to produce several counterexamples to intuitive notions in mathematics; his most famous example is that of a space-filling curve, dating from 1890. Bertrand Russell wrote that Peano's ideas on logic had a profound impact on him. See Peano (1973), pp. 51–57 for Peano's 1885 formulation and not completely rigorous proof of the theorem stated in this exercise. A more stringent proof may be found in Petrovski (1966), pp. 29–33. In 1890, Peano generalized this theorem to systems of differential equations. That paper also contained the first explicit formulation of the axiom of choice; interestingly, Peano rejected it as a possible component of the logic of mathematics. He wrote, "But as one cannot apply infinitely many times an *arbitrary* rule by which one assigns to a class *A* an individual of this class, a *determinate* rule is stated here." See Moore (1982), p. 4. Nevertheless, a logical equivalent of the axiom of choice, in the form of Zorn's lemma, has turned out to be of fundamental importance in algebra. For the origins of Zorn's lemma, see Paul Campbell (1978).

15.13 Notes on the Literature

For the quote from Newton's October 1666 tract, see Newton (1967–1981), vol. 1, p. 403, and for his method of separation of variables, see vol. 7, p. 73. Newton's example of the use of an integrating factor, presented in the text, may be found on pp. 78–81 of vol. 7. See also Whiteside's footnote concerning Fatio de Duillier on pp. 78–79 of that volume. For the correspondence of Leibniz with Johann and Jakob Bernoulli, see Leibniz (1971), vol. III, parts 1 and 2. For Leibniz's letter to Huygens (Leibniz wrote Hugens), see Leibniz (1971), vol. 2, pp. 195–196. For Leibniz's early work on series, see Scriba (1964). Jakob Bernoulli's solution of a Riccati equation can be found on pp. 74–75 of Leibniz (1971), vol. III/1. See Johann Bernoulli (1968), vol. 1, pp. 174–76 and Eu. I-22, pp. 10–12 in connection with the integrating factor. See Euler's paper, "De Integratione Aequationum Differentialium Altiorum Graduum," Eu. I-22, pp. 108–41, for his solution of linear differential equations with constant coefficients; Eu. I-22, pp. 214–36 for his paper on singular solutions; Eu. I-22, pp. 181–213 for his work on nonhomogeneous equations; Eu. I-22, pp. 334–94 for the paper "De Integratione Aequationum Differentialium," containing the material on Riccati's equation.

Taylor's work on singular solutions is in proposition eight of his 1715 *Methodus Incrementorum*. An English translation was a part of Feigenbaum's (1981) Yale doctoral dissertation. For Lagrange's 1762 paper on integral calculus, discussing the adjoint and its applications, see Lagrange (1867–1892), vol. 1, pp. 471–478. See Edwards (1954a), p. 436, for his remarks on Mukhopadhyay's equation. See Boole (1877), p. 20, for his comment on Monge's equation. For some discussion of Mukhopadhyay's role in the development of mathematics in India, see Narasimhan (1991). For the quotations from his diary, see Mukhopadhyay (1998); for biographical information, see Sen Gupta (2000). Katz (1998) and (1987) contain a discussion of how, in May 1739, Euler may have solved $a^3 d^3 y - y dx^3 = 0$. A lively history of the Riccati equation is available in

Bottazzini's article, "The Mathematical Writings from Daniel Bernoulli's Youth", contained in D. Bernoulli (1982–1996), vol. I, pp. 142–66. Ferraro (2004) discusses Euler's concept of the differential coefficient and its relation to differentials. See Truesdell (1960), p. 167, for the quote on Euler and D. Bernoulli's understanding of the exponential function. The reader may also wish to see Burn (2001) for the development of the concept of the logarithm in the second half of the seventeenth century, starting with the 1649 work of Alphonse de Sarasa. This topic is also discussed in Hofmann's article on differential equations in the seventeenth century; see Hofmann (1990), vol. 2, pp. 277–316.

16

Series and Products for Elementary Functions

16.1 Preliminary Remarks

Euler was the first mathematician to give a systematic and coherent account of the elementary functions, although earlier mathematicians had certainly paved the way. These functions are comprised of the circular or trigonometric, the logarithmic, and the exponential functions. Euler's approach was a departure from the prevalent, more geometric, point of view. On the geometric perspective, the elementary functions were defined as areas under curves, lengths of chords, or other geometric conceptions. Euler's 1748 *Introductio in Analysin Infinitorum* defined the elementary functions arithmetically and algebraically, as functions.

At that time, infinite series were regarded as a part of algebra, though they had been obtained through the use of calculus. The general binomial theorem was considered an algebraic theorem. So in his *Introductio*, Euler used the binomial theorem to produce new derivations of the series for the elementary functions. Interestingly, in this book, where he avoided using calculus, Euler gave no proof of the binomial theorem itself; perhaps he had not yet found any arguments without the use of calculus. In a paper written in the 1730s, Euler derived the binomial theorem from the Taylor series, as Stirling had done in 1717. It was only much later, in the 1770s, that Euler found an argument for the binomial theorem depending simply on the multiplication of series.

We recall that before Euler, between 1664 and 1666, Newton found the series for all the elementary functions, using a combination of geometric arguments, integration, and reversion of series. In the course of his work, he also discovered the general binomial theorem. Later on, Gregory, Leibniz, Johann Bernoulli, and others used methods of calculus to obtain infinite series for elementary functions. Even before Newton, unknown to European mathematicians, the Kerala school had derived infinite series for some trigonometric functions, also using a form of integration.

Although the series for e^x and e were already known, one of Euler's major innovations was to explicitly define the exponential function. To understand this peculiar fact, recall that Newton and N. Mercator discovered the series for $y = \ln(1+x)$ in the mid-1660s. Soon afterwards, Newton applied reversion of series to obtain x as a series in y. And note that for the eighteenth-century mathematician who took the geometric point of view, the basic object of study was not the function, but the curve. From this perspective, there was hardly any need to distinguish between the function and its inverse, since both curves would take same form, although with differing orientations. So the series for e^x was perceived as but another description of the logarithmic curve.

In a 1714 paper published in the *Philosophical Transactions*, Roger Cotes, in the spirit of Halley's earlier work of 1695, took the step of setting up an analytic definition of the logarithm. Cotes used this definition to derive the logarithmic series and then, by inversion, the series for the exponential. He proceeded to use the series for e to compute its value to thirteen decimal places. Incidentally, he also gave continued fractions for e and $1/e$ to obtain rational approximations of e. However, Cotes focused on the logarithm, rather than its inverse. To understand how the lack of a clear conception of the exponential handicapped mathematics, consider that in the early 1730s, Daniel Bernoulli was unable to fully solve the differential equation $K^4 d^4 y/dx^4 = y$. He observed that the logarithm, meaning the inverse or exponential, satisfied this equation as well as the equation $K^2 d^2 y/dx^2 = y$, but that no such logarithm was sufficiently general. Euler was also stumped by this problem, until he gave an explicit definition of the exponential and developed its properties in the mid-1730s.

To derive the series for elementary functions, Euler made considerable use of infinitely large and infinitely small numbers. This method can be made rigorous by an appropriate use of limits, as accomplished by Cauchy in the 1820s. Following Euler's style, Cauchy divided analysis into two parts, algebraic analysis and calculus. The former dealt with infinite series and products without using calculus, yet employed the ideas of limits and convergence. It is interesting to note here that Lagrange had earlier attempted to make differential calculus a part of algebraic analysis by defining the derivative of $f(x)$ as the coefficient of h in the series expansion of $f(x+h)$. Gauss, Cauchy, and their followers rejected this idea as invalid. Besides providing greater rigor, Cauchy's lectures presented original and insightful derivations of some of Euler's results.

In addition to defining elementary functions, Euler also showed that functions could be represented by infinite products and partial fractions. The latter could be obtained from products by applying logarithmic differentiation, a process he worked out in his correspondence with Niklaus I Bernoulli. In his *Introductio*, Euler presented fascinating ways of avoiding the methods of calculus. He also gave an exposition on the connection, discovered earlier by Cotes, between the trigonometric and the exponential functions. Cotes had found the relation $\log(\cos x + i \sin x) = ix$, although this equation was more useful when Euler wrote it as $\cos x + i \sin x = e^{ix}$. Of course, Cotes was unable to take this last step because he did not explicitly define the exponential e^x. Euler, on the other hand, made use of this relationship to derive important results such as the infinite products for the trigonometric functions.

At the very beginning of his career, Euler discovered the simple and useful dilogarithm function. The dilogarithm is defined by

$$\text{Li}_2(x) = -\int_0^x \frac{\ln(1-t)}{t} dt = \frac{x}{1^2} + \frac{x^2}{2^2} + \frac{x^3}{3^2} + \cdots, \qquad (16.1)$$

where the series converges for $|x| \leq 1$. Euler initially investigated this function in 1729–30 to evaluate the series $\sum_{n=1}^{\infty} 1/n^2$. He succeeded at that time only in determining its approximate value, but in the 1730s he found the exact value by the factorization of $\sin x$.

In the 1740s, the amateur English mathematician and surveyor John Landen began publishing his mathematical problems in the *Ladies Diary*. In the late 1750s, he discovered that the dilogarithm could be used to exactly evaluate $\sum_{n=1}^{\infty} 1/n^2$, provided logarithms of negative numbers were employed. Euler had already developed his theory of logarithms of complex numbers at that time but his work had not appeared in print. So Landen's determination of $\ln(-1) = \pm\sqrt{-1}\pi$ in 1758 was an important and independent discovery. And he went further, by repeated integration, to define the more general polylogarithm,

$$\text{Li}_k(x) = \frac{x}{1^k} + \frac{x^2}{2^k} + \frac{x^3}{3^k} + \cdots, \tag{16.2}$$

for $k = 1, 2, 3, \ldots$. He could then evaluate the series $\sum_{n=1}^{\infty} 1/n^{2k}$ for $k = 1, 2, 3$, etc.

The polylogarithm was further studied by the Scottish mathematician William Spence (1777–1815) who published his book, *Essay on Logarithmic Transcendents*, in 1809. He derived several interesting results on dilogarithms, including the theorem for which he is known today. As a student in the 1820s, Abel rediscovered this formula, having been inspired to study the dilogarithm by reading Legendre's three volumes on the integral calculus. This work discussed numerous results of Euler. Spence was apparently self-taught but, unlike many other British mathematicians of his time, he read Bernoulli, Euler, Lagrange, and other continental mathematicians. In the preface to his work, Spence commented on the disadvantage of British insularity:

> Our pupils are taught the science by means of its applications; and when their minds should be occupied with the contemplation of general methods and operations, they are usually employed on particular processes and results, in which no traces of the operations remain. On the Continent, Analysis is studied as an independent science. Its general principles are first inculcated; and then the pupil is led to the applications; and the effects have been, that while we have remained nearly stationary during the greater part of the last century, the most valuable improvements have been added to the science in almost every other part of Europe. The truth of this needs no illustration. Let any person who has studied Mathematics only in British authors look into works of the higher analysts of the Continent, and he will soon perceive that he has still much to learn.

Interestingly, other British mathematicians were independently arriving at this conclusion. In 1813, a few students at Cambridge University formed the Analytical Society in order to promote broadening mathematical studies to include the works of non-British mathematicians. Among the members of this new Society was John Herschel, who collected, published, and annotated the works of Spence, an example of the progress facilitated by broader mathematical horizons. In one of his extensive notes on Spence's work, Herschel presented, without fanfare, his own discovery of the Schwarzian derivative. Interestingly, in 1781 Lagrange also found this derivative, but in the context of cartography. We also note that Kummer, before he became a committed number theorist, wrote a very long 1840 paper on the dilogarithm; this paper contained a wealth of results, including the rediscovery of Spence's formula.

16.2 Euler: Series for Elementary Functions

Euler defined the exponential functions by explaining the meaning of a^z, first for z as an integer and then for a z as a rational number. He remarked that for irrational z the concept was more difficult to understand but that $a^{\sqrt{7}}$, for example, had a value between a^2 and a^3 when $a > 1$. He noted that the study of a^z for $0 < a < 1$ could be reduced to the case where $a > 1$. He then defined the logarithm: If $a^z = y$, then z is called the logarithm of y to the base a. Euler did not have a notation for the base of a logarithm and always expressed the base in words. It seems that in 1821, A. L. Crelle, founder of the famous journal and a friend of Abel, introduced the notation for the base of the logarithm, writing the base a on the upper left-hand side of the log. However, we employ the modern notation: $\log_a$.

To obtain the series for a^z, Euler observed that since $a^0 = 1$, he could write $a^w = 1 + kw$, where w was an infinitely small number. Thus, by the binomial theorem, called the universal theorem by Euler,

$$a^{jw} = (1+kw)^j = 1 + \frac{j}{1}kw + \frac{j(j-1)}{1 \cdot 2}k^2w^2 + \frac{j(j-1)(j-2)}{1 \cdot 2 \cdot 3}k^3w^3 + \cdots. \quad (16.3)$$

He took j to be infinitely large so that $jw = z$, a finite number, and equation (16.3) was transformed to

$$a^z = \left(1 + \frac{kz}{j}\right)^j = 1 + \frac{1}{1}kz + \frac{1(j-1)}{1 \cdot 2j}k^2z^2 + \frac{1(j-1)(j-2)}{1 \cdot 2j \cdot 3j}k^3z^3 + \cdots. \quad (16.4)$$

For infinitely large j, he concluded that $\frac{j-1}{2j} = \frac{1}{2}$, $\frac{(j-1)(j-2)}{2j \cdot 3j} = \frac{1}{2 \cdot 3}$ etc. and hence he had the series

$$a^z = 1 + \frac{kz}{1} + \frac{k^2z^2}{1 \cdot 2} + \frac{k^3z^3}{1 \cdot 2 \cdot 3} + \cdots. \quad (16.5)$$

He then set $z = 1$ to obtain the equation for k:

$$a = 1 + \frac{k}{1} + \frac{k^2}{1 \cdot 2} + \frac{k^3}{1 \cdot 2 \cdot 3} + \cdots. \quad (16.6)$$

He denoted by e the value of a when $k = 1$ and computed it to 23 decimal places. From (16.5) and (16.6), with $a = e$, Euler obtained these famous equations:

$$e^z = 1 + z + \frac{z^2}{1 \cdot 2} + \frac{z^3}{1 \cdot 2 \cdot 3} + \cdots, \quad (16.7)$$

$$e = 1 + 1 + \frac{1}{1 \cdot 2} + \frac{1}{1 \cdot 2 \cdot 3} + \cdots. \quad (16.8)$$

It also followed from (16.6) and (16.7) that $k = \ln a$ where $\ln$ stands for the natural logarithm. To find the series for $\log_a(1+x)$, Euler set $a^{jw} = (1+kw)^j = 1+x$, so that

$$jw = \log_a(1+x) = \frac{j}{k}\left((1+x)^{1/j} - 1\right). \quad (16.9)$$

He expanded the expression in parentheses by the binomial theorem to get

$$\log_a(1+x) = \frac{j}{k}\left(\frac{x}{j} - \frac{1(j-1)}{j \cdot 2j}x^2 + \frac{1(j-1)(2j-1)}{j \cdot 2j \cdot 3j}x^3 - \cdots\right)$$
$$= \frac{1}{k}\left(x - \frac{x^2}{2} + \frac{x^3}{3} - \cdots\right). \tag{16.10}$$

The second equation followed from the condition that j was an infinitely large number. When $k = 1$,

$$\ln(1+x) = \log_e(1+x) = x - \frac{x^2}{2} + \frac{x^3}{3} - \cdots. \tag{16.11}$$

To obtain the series for $\sin x$ and $\cos x$, Euler started with de Moivre's formulas

$$\cos nz = \frac{(\cos z + \sqrt{-1}\sin z)^n + (\cos z - \sqrt{-1}\sin z)^n}{2}, \tag{16.12}$$

$$\sin nz = \frac{(\cos z + \sqrt{-1}\sin z)^n - (\cos z - \sqrt{-1}\sin z)^n}{2\sqrt{-1}}. \tag{16.13}$$

By the binomial theorem, equation (16.12) could be written as

$$\cos nz = (\cos z)^n - \frac{n(n-1)}{1 \cdot 2}(\cos z)^{n-2}(\sin z)^2$$
$$+ \frac{n(n-1)(n-2)(n-3)}{1 \cdot 2 \cdot 3 \cdot 4}(\cos z)^{n-4}(\sin z)^4 - \cdots.$$

Euler took n infinitely large and z infinitely small, such that that $nz = x$ was finite. He then concluded that $\sin z = z = x/n$ and $\cos z = 1$, and hence

$$\cos x = 1 - \frac{x^2}{1 \cdot 2} + \frac{x^4}{1 \cdot 2 \cdot 3 \cdot 4} - \frac{x^6}{1 \cdot 2 \cdot 3 \cdot 4 \cdot 5 \cdot 6} + \cdots. \tag{16.14}$$

Similarly, Euler found the series for $\sin x$ from (16.13).

16.3 Euler: Products for Trigonometric Functions

Euler derived the infinite products for the sine and cosine functions from the Cotes factorization of $x^n \pm y^n$. Note that Cotes's formula for n odd may be expressed as

$$x^n \pm y^n = (x \pm y)\prod_{k=1}^{(n-1)/2}\left(x^2 \pm 2xy\cos\frac{2k\pi}{n} + y^2\right). \tag{16.15}$$

Euler observed that the series for the exponential, cosine and sine functions yielded the relations:

$$\cos x = \frac{e^{xi} + e^{-xi}}{2} = \frac{(1+xi/j)^j + (1-xi/j)^j}{2}, \tag{16.16}$$

$$\sin x = \frac{e^{xi} - e^{-xi}}{2i} = \frac{(1+xi/j)^j - (1-xi/j)^j}{2i}. \tag{16.17}$$

He first determined the factors of $e^x - 1 = (1 + x/j)^j - 1$ by Cotes's formula. He noted that one factor was $(1 + x/j) - 1 = x/j$ and the quadratic factors were of the form $(1 + x/j)^2 - 2(1 + x/j)\cos(2k\pi/j) + 1$. He also noted that every factor could be obtained by taking all positive even integers $2k$. Euler then set $\cos(2k\pi/j) = 1 - \frac{2k^2\pi^2}{j^2}$ by taking the first two nonzero terms of the series expansion for cosine and then simplified the quadratic factor to

$$\frac{x^2}{j^2} + \frac{4k^2}{j^2}\pi^2 + \frac{4k^2}{j^3}\pi^2 x = \frac{4k^2\pi^2}{j^2}\left(1 + \frac{x}{j} + \frac{x^2}{4k^2\pi^2}\right).$$

He observed at this point that though x/j was infinitesimal, it could not be neglected because there were $j/2$ factors, producing a nonzero term $x/2$. This remark shows us that Euler had some pretty clear ideas about the convergence of infinite products. To eliminate this difficulty, Euler then considered the factors of $e^x - e^{-x} = (1+x/j)^j - (1-x/j)^j$. In this case, he simplified the general quadratic factor to

$$1 + \frac{x^2}{k^2}\pi^2 - \frac{x^2}{j^2}.$$

The contribution of the term x^2/j^2 after multiplication of $j/2$ factors was x^2/j and he could now neglect this. So Euler determined the quadratic factors to be $1 + x^2\pi^2/k^2$ and wrote the formula

$$\frac{e^x - e^{-x}}{2} = x\left(1 + \frac{x^2}{\pi^2}\right)\left(1 + \frac{x^2}{4\pi^2}\right)\left(1 + \frac{x^2}{9\pi^2}\right)\left(1 + \frac{x^2}{16\pi^2}\right)\cdots. \quad (16.18)$$

Similarly, he got

$$\frac{e^x + e^{-x}}{2} = \left(1 + \frac{4x^2}{\pi^2}\right)\left(1 + \frac{4x^2}{9\pi^2}\right)\left(1 + \frac{4x^2}{25\pi^2}\right)\left(1 + \frac{4x^2}{49\pi^2}\right)\cdots. \quad (16.19)$$

To obtain the products for $\sin x$ and $\cos x$, he changed x to ix to find

$$\sin x = x\left(1 - \frac{x^2}{\pi^2}\right)\left(1 - \frac{x^2}{4\pi^2}\right)\left(1 - \frac{x^2}{9\pi^2}\right)\left(1 - \frac{x^2}{16\pi^2}\right)\cdots, \quad (16.20)$$

and

$$\cos x = \left(1 - \frac{4x^2}{\pi^2}\right)\left(1 - \frac{4x^2}{9\pi^2}\right)\left(1 - \frac{4x^2}{25\pi^2}\right)\left(1 - \frac{4x^2}{49\pi^2}\right)\cdots. \quad (16.21)$$

16.4 Euler's Finite Product for $\sin nx$

By repeated use of the addition formula for the sine function, Euler showed inductively that when n was an odd number $2m + 1$

$$\sin nx = n\sin x - \frac{n(n^2-1^2)}{1\cdot 2\cdot 3}\sin^3 x + \frac{n(n^2-1^2)(n^2-3^2)}{1\cdot 2\cdot 3\cdot 4\cdot 5}\sin^5 x - \cdots$$
$$+ (-1)^m \frac{n(n^2-1^2)\cdots(n^2-(2m-1)^2)}{1\cdot 2\cdot 3\cdots(2m+1)}\sin^n x.$$

Recall that Newton found this formula in his student days in the course of studying Viéte. The last term in the preceding formula can be shown to be $(-1)^m 2^{n-1}$. The right-hand side is a polynomial of degree n in $y = \sin x$ and in his *Introductio* Euler expressed this polynomial equation as

$$0 = 1 - \frac{n}{\sin nx} y + \frac{n(n^2-1)}{1 \cdot 2 \cdot 3 \cdot \sin nx} y^3 - \cdots \pm \frac{2^{n-1}}{\sin nx} y^n, \tag{16.22}$$

where a plus sign applied in the last term if $n = 4l - 1$ and a minus sign applied otherwise. He observed that the roots of this equation were $\sin x$, $\sin\left(x + \frac{2\pi}{n}\right)$, $\sin\left(x + \frac{4\pi}{n}\right)$, ..., $\sin\left(x + \frac{2(n-1)\pi}{n}\right)$. He factorized the polynomial on the right-hand side of (16.22) in the form

$$\left(1 - \frac{y}{\sin x}\right)\left(1 - \frac{y}{\sin(x + \frac{2\pi}{n})}\right) \cdots \left(1 - \frac{y}{\sin(x + \frac{2(n-1)\pi}{n})}\right)$$

and equated the coefficients of y, y^n, and y^{n-1} to obtain:

$$\frac{n}{\sin nx} = \frac{1}{\sin x} + \frac{1}{\sin(x + \frac{2\pi}{n})} + \frac{1}{\sin(x + \frac{4\pi}{n})} + \cdots + \frac{1}{\sin(x + \frac{2(n-1)\pi}{n})}, \tag{16.23}$$

$$\pm \frac{\sin nx}{2^{n-1}} = \sin x \sin\left(x + \frac{2\pi}{n}\right) \sin\left(x + \frac{4\pi}{n}\right) \cdots \sin\left(x + \frac{2(n-1)\pi}{n}\right), \tag{16.24}$$

$$0 = \sin x + \sin\left(x + \frac{2\pi}{n}\right) + \sin\left(x + \frac{4\pi}{n}\right) + \cdots + \sin\left(x + \frac{2(n-1)\pi}{n}\right). \tag{16.25}$$

Euler made use of all these significant formulas. We emphasize the second formula; it was used by Cauchy to derive the infinite product for the sine function by a method different from Euler's, without using Cotes's factorization.

16.5 Cauchy's Derivation of the Product Formulas

In his lectures at the École Polytechnique, published in 1821 under the title *Analyse algébrique*, Cauchy gave a rigorous treatment of series and products. He then applied these ideas to a discussion of elementary functions; his discourse on infinite products was presented in note IX as the last topic in the work. Suppose $u_0, u_1, u_2, \ldots$ to be real numbers with $|u_n| \leq 1$. Cauchy began with a definition of convergence. The infinite product

$$(1 + u_0)(1 + u_1)(1 + u_2) \cdots \tag{16.26}$$

was said to converge if $\lim_{n \to \infty} \left((1+u_0) \cdots (1+u_n) \right)$ existed and was different from zero. He then stated the theorem: If the series

$$u_0 + u_1 + u_2 + \cdots \quad \text{and} \tag{16.27}$$

$$u_0^2 + u_1^2 + u_2^2 + \cdots \tag{16.28}$$

are convergent, then the infinite product (16.26) converges. However, if (16.28) is convergent and (16.27) is divergent, then the infinite product diverges to zero. In his proof, Cauchy observed that for large enough n,

$$\ln(1+u_n) = u_n - \frac{1}{2}u_n^2 + \frac{1}{3}u_n^3 - \cdots = u_n - \frac{1}{2}u_n^2(1 \pm \epsilon_n),$$

with ϵ_n infinitesimally small. He concluded that

$$\begin{aligned}&\ln(1+u_n) + \ln(1+u_{n+1}) + \cdots + \ln(1+u_{n+m-1}) \\ &= u_n + u_{n+1} + \cdots + u_{n+m-1} - \frac{1}{2}(u_n^2 + u_{n+1}^2 + \cdots + u_{n+m-1}^2)(1 \pm \epsilon),\end{aligned} \tag{16.29}$$

when all the u had absolute value less than one and $1 \pm \epsilon$ was the average of $1 \pm \epsilon_n, 1 \pm \epsilon_{n+1}, \ldots$. Formula (16.29) completed the proof of the theorem, because the infinite product converged if and only if the series $\ln(1+u_0) + \ln(1+u_1) + \ln(1+u_2) + \cdots$ converged. As examples of this theorem, Cauchy noted that the product

$$(1+x^2)\left(1+\frac{x^2}{2^2}\right)\left(1+\frac{x^2}{3^2}\right)\cdots$$

converged for all x, while the product

$$(1+1)\left(1-\frac{1}{\sqrt{2}}\right)\left(1+\frac{1}{\sqrt{3}}\right)\left(1-\frac{1}{\sqrt{4}}\right)\cdots$$

diverged to zero.

Although earlier mathematicians did not explicitly state such a theorem on infinite products, it can hardly be doubted that Euler, with his enormous experience manipulating and calculating with infinite products and series, intuitively understood this result. However, Cauchy's presentation – with clearer, more precise and explicit definitions of fundamental concepts such as limits, continuity, and convergence – paved the way for future generations of mathematicians. His work led to higher standards of clarity and rigor in definitions, statements of theorems, and proofs. We present a slightly condensed form of Cauchy's derivation of the infinite product for $\sin x$, first noting that

$$\sin\left(x + \frac{2\pi}{n}\right)\sin\left(x + \frac{2(n-1)\pi}{n}\right) = \sin^2 x - \sin^2 \frac{2\pi}{n},$$

$$\sin\left(x + \frac{4\pi}{n}\right)\sin\left(x + \frac{2(n-2)\pi}{n}\right) = \sin^2 x - \sin^2 \frac{4\pi}{n}, \quad \text{etc.}$$

16.5 Cauchy's Derivation of the Product Formulas

Hence, by Euler's formula (16.24) for odd n,

$$\sin nx = 2^{n-1} \sin x \left(\sin^2 \frac{2\pi}{n} - \sin^2 x \right) \left(\sin^2 \frac{4\pi}{n} - \sin^2 x \right) \cdots \left(\sin^2 \frac{(n-1)\pi}{n} - \sin^2 x \right). \tag{16.30}$$

It is easy to see that the set of $\frac{n-1}{2}$ numbers $\sin^2 \frac{2\pi}{n}, \sin^2 \frac{4\pi}{n}, \ldots, \sin^2 \frac{(n-1)\pi}{n}$ is identical to the set of $\frac{n-1}{2}$ numbers $\sin^2 \frac{\pi}{n}, \sin^2 \frac{2\pi}{n}, \ldots, \sin^2 \frac{(n-1)\pi}{2n}$. Thus,

$$\sin nx = 2^{n-1} \sin x \left(\sin^2 \frac{\pi}{n} - \sin^2 x \right) \left(\sin^2 \frac{2\pi}{n} - \sin^2 x \right) \cdots \left(\sin^2 \frac{(n-1)\pi}{2n} - \sin^2 x \right).$$

Let $x \to 0$ in this formula to get

$$n = 2^{n-1} \sin^2 \frac{\pi}{n} \sin^2 \frac{2\pi}{n} \cdots \sin^2 \frac{(n-1)\pi}{2n}.$$

After replacing nx by x,

$$\sin x = n \sin \frac{x}{n} \left(1 - \frac{\sin^2 \frac{x}{n}}{\sin^2 \frac{\pi}{n}} \right) \left(1 - \frac{\sin^2 \frac{x}{n}}{\sin^2 \frac{2\pi}{n}} \right) \cdots \left(1 - \frac{\sin^2 \frac{x}{n}}{\sin^2 \frac{(n-1)\pi}{n}} \right).$$

When $n \to \infty$, we get

$$1 - \frac{\sin^2 x/n}{\sin^2 k\pi/n} \longrightarrow 1 - \frac{x^2}{k^2 \pi^2}.$$

Euler might have found the argument up to this point sufficient to obtain the product formula for $\sin x$. Cauchy, on the other hand, was more careful; we present an abridged version of the rest of his argument, illustrating his more rigorous attention to questions of convergence.

Let m be any fixed number less than $(n-1)/2$ and write

$$\sin x = n \sin \frac{x}{n} \left(1 - \frac{\sin^2 \frac{x}{n}}{\sin^2 \frac{\pi}{n}} \right) \cdots \left(1 - \frac{\sin^2 \frac{x}{n}}{\sin^2 \frac{m\pi}{n}} \right)$$

$$\cdot \left(1 - \frac{\sin^2 \frac{x}{n}}{\sin^2 \frac{(m+1)\pi}{n}} \right) \cdots \left(1 - \frac{\sin^2 \frac{x}{n}}{\sin^2 \frac{(n-1)\pi}{2n}} \right).$$

For large enough n,

$$n \sin \frac{x}{n} \left(1 - \frac{\sin^2 \frac{x}{n}}{\sin^2 \frac{\pi}{n}} \right) \cdots \left(1 - \frac{\sin^2 \frac{x}{n}}{\sin^2 \frac{m\pi}{n}} \right) = x \left(1 - \frac{x^2}{\pi^2} \right) \cdots \left(1 - \frac{x^2}{m^2 \pi^2} \right) (1 + \alpha),$$

where α is small. Moreover, because $\prod_1^n (1 - \alpha_i) > 1 - \sum_1^n \alpha_i$, we have

$$\left(1 - \frac{\sin^2 \frac{x}{n}}{\sin^2 \frac{(m+1)\pi}{n}} \right) \cdots \left(1 - \frac{\sin^2 \frac{x}{n}}{\sin^2 \frac{(n-1)\pi}{2n}} \right)$$

$$> 1 - \sin^2 \frac{x}{n} \left(\csc^2 (m+1) \frac{\pi}{n} + \cdots + \csc^2 \frac{(n-1)\pi}{2n} \right). \tag{16.31}$$

Since for $0 < x < \pi/2$, we have $x < 2\sin x$, it follows that

$$\csc^2 \frac{k\pi}{n} < \frac{4n^2}{k^2\pi^2}, \text{ for } k = m+1, \ldots, \frac{n-1}{2}.$$

Now because $\sin x < x$ for $x > 0$, we get

$$\sin^2 \frac{x}{n} \csc^2 \frac{k\pi}{n} < \frac{4x^2}{\pi^2 k^2}. \tag{16.32}$$

Thus, the product in (16.31) is greater than

$$1 - \frac{4x^2}{\pi^2}\left(\frac{1}{(m+1)^2} + \frac{1}{(m+2)^2} + \cdots + \frac{4}{(n-1)^2}\right) > 1 - \frac{4x^2}{\pi^2(m+1)},$$

and less than one. Therefore, it is equal to $1 - \frac{x^2}{\pi^2(m+1)}\theta$, where θ lies between 0 and 1. Thus, we may write

$$\sin x = x\left(1 - \frac{x^2}{\pi^2}\right)\cdots\left(1 - \frac{x^2}{m^2\pi^2}\right)\left(1 - \frac{4x^2}{\pi^2(m+1)}\theta\right)(1+\alpha).$$

The result follows by letting $m \to \infty$.

16.6 Euler and Niklaus I Bernoulli: Partial Fractions Expansions of Trigonometric Functions

Recall that Euler obtained the partial fractions expansions of $\csc x$ and $\cot x$ by the use of calculus methods and, later on, by other methods. The former approach depended upon the evaluation of the integrals $\int_0^\infty x^{p-1} dx/(1 \pm x^q)$ in two different ways. First, Euler used Cotes's factorization of $1 \pm x^q$ to express $1/(1 \pm x^q)$ as partial fractions; he then integrated to find

$$\int_0^\infty \frac{x^{p-1}}{1+x^q} dx = \frac{\pi}{q \sin \frac{p\pi}{q}} \text{ and} \tag{16.33}$$

$$\int_0^\infty \frac{x^{p-1}}{1-x^q} dx = \frac{\pi}{q \tan \frac{p\pi}{q}} \tag{16.34}$$

where $0 < p < q$ and p, q were integers. For Euler's derivation of (16.33), see Section 13.6. He found (16.34) in a similar way. The integral in (16.34) is a principal value, although this concept was not explicitly defined until Cauchy introduced it in the 1820s. Dedekind gave a number of proofs of (16.33), including a streamlined form of Euler's proof, and we present it in Section 23.5.

For Euler's derivation of the partial fractions expansion of $\csc x$, note that the change of variable $y = 1/x$ shows that

$$\int_1^\infty \frac{x^{p-1}}{1+x^q} dx = \int_0^1 \frac{y^{q-p-1}}{1+y^q} dy.$$

So Euler could rewrite the integral (16.33) as

$$\int_0^1 \frac{x^{p-1}+x^{q-p-1}}{1+x^q}\,dx = \int_0^1 (x^{p-1}+x^{q-p-1})(1-x^q+x^{2q}-x^{3q}+\cdots)\,dx$$

$$= \int_0^1 \left(x^{p-1}+x^{q-p-1}-x^{q+p-1}-x^{2q-p-1}+x^{2q+p-1}+x^{3q-p-1}-\cdots \right) dx$$

$$= \frac{1}{p}+\frac{1}{q-p}-\frac{1}{q+p}-\frac{1}{2q-p}+\frac{1}{2q+p}+\frac{1}{3q-p}-\cdots$$

$$= \frac{\pi}{q\sin\frac{p\pi}{q}}.$$

Thus, he had the partial fractions expansion for $\csc x$

$$\frac{\pi}{\sin\pi x} = \frac{1}{x}+\frac{1}{1-x}-\frac{1}{1+x}-\frac{1}{2-x}+\frac{1}{2+x}+\frac{1}{3-x}-\text{ etc.} \quad (16.35)$$

$$= \frac{1}{x}-\frac{2x}{x^2-1^2}+\frac{2x}{x^2-2^2}-\frac{2x}{x^2-3^2}+\text{ etc.}$$

In a similar way, he obtained the partial fractions expansion of $\cot x$:

$$\frac{\pi}{\tan\pi x} = \frac{1}{x}-\frac{1}{1-x}+\frac{1}{1+x}-\frac{1}{2-x}+\frac{1}{2+x}-\frac{1}{3-x}+\frac{1}{3+x}-\text{ etc.} \quad (16.36)$$

$$= \frac{1}{x}+\frac{2x}{x^2-1^2}+\frac{2x}{x^2-2^2}+\frac{2x}{x^2-3^2}+\text{ etc.}$$

Note that by integrating the last formula, Euler had another way to obtain the product for $\sin x$.

In a letter of January 16, 1742, Euler communicated these results to Niklaus I Bernoulli, also observing that partial fractions expansions of other functions could be found by repeated differentiation of (16.35) and (16.36). In particular, he had

$$\frac{\pi^2\cos\pi x}{(\sin\pi x)^2} = \frac{1}{x^2}-\frac{1}{(1-x)^2}-\frac{1}{(1+x)^2}+\frac{1}{(2-x)^2}+\frac{1}{(2+x)^2}-\frac{1}{(3-x)^2}-\text{ etc.} \quad (16.37)$$

and

$$\frac{\pi^2}{(\sin\pi x)^2} = \frac{1}{x^2}+\frac{1}{(1-x)^2}+\frac{1}{(1+x)^2}+\frac{1}{(2-x)^2}+\frac{1}{(2+x)^2}+\text{ etc.} \quad (16.38)$$

In his reply of July 13, 1742, Bernoulli noted that the logarithmic differentiation of Euler's product for $\sin x$ would immediately produce (16.36). To see this, we note that

$$\ln\sin\pi x = \ln(\pi x)+\sum_{n=1}^\infty \ln\left(1-\frac{x^2}{n^2}\right);$$

differentiation of this equation yields the required formula. In his subsequent letter of October 24, 1742, Bernoulli noted that (16.35) could also be obtained by logarithmic differentiation. In his letter, Bernoulli wrote the differential $d \ln x$ as differ.$\ln x$. His notation for the natural logarithm was log. We maintain our practice of writing $\ln x$, and present Bernoulli's argument as he wrote it:

$$\text{differ.} \ln \frac{\sin \pi x}{\cos \pi x} = \frac{\text{differ.} \sin \pi x}{\sin \pi x} - \frac{\text{differ.} \cos \pi x}{\cos \pi x}$$

$$= \frac{\pi\, dx \cos \pi x}{\sin \pi x} + \frac{\pi\, dx \sin \pi x}{\cos \pi x} = \frac{\pi\, dx}{\sin \pi x \cos \pi x}$$

$$= \frac{2\pi\, dx}{\sin 2\pi x} = \text{differ.} \ln \frac{\pi x(1-xx)(1-\frac{1}{4}xx)(1-\frac{1}{9}xx)\,\text{etc.}}{(1-4xx)(1-\frac{4}{9}xx)(1-\frac{4}{25}xx)\,\text{etc.}}$$

Replace $2x$ by x to get

$$\frac{\pi\, dx}{\sin \pi x} = \text{differ.} \ln \frac{\frac{1}{2}\pi x(1-\frac{1}{4}xx)(1-\frac{1}{16}xx)(1-\frac{1}{36}xx)\,\text{etc.}}{(1-xx)(1-\frac{1}{9}xx)(1-\frac{1}{25}xx)\,\text{etc.}}$$

$$= dx\left(\frac{1}{x} + \frac{1}{1-x} - \frac{1}{1+x} - \frac{1}{2-x} + \frac{1}{2+x} + \text{etc.}\right),$$

which yields (16.35) after division by dx.

In his *Introductio*, Euler gave an alternate derivation of the partial fractions expansions of the trigonometric functions, avoiding the use of integration and differentiation. He first showed by Cotes's formula that the quadratic factors of

$$e^y + e^{c-y} = \left(1 + \frac{y}{j}\right)^j + \left(1 + \frac{c-y}{j}\right)^j$$

were of the form

$$1 - \frac{4cy - 4y^2}{m^2\pi^2 + c^2},$$

with odd m, and hence

$$\frac{e^y + e^c \cdot e^{-y}}{1 + e^c} = \left(1 - \frac{4cy - y^2}{\pi^2 + c^2}\right)\left(1 - \frac{4cy - y^2}{9\pi^2 + c^2}\right)\left(1 - \frac{4cy - y^2}{25\pi^2 + c^2}\right)\text{ etc.,}$$

where the denominator on the left side was chosen so that value of both sides of the equation was 1 for $y = 0$. Euler then took $c = i\pi x$ and $y = i\pi v/2$, so that the left side reduced to $\cos(\pi v/2) + \tan(\pi x/2)\sin(\pi v/2)$. This gave him the formula

$$\cos\frac{\pi v}{2} + \tan\frac{\pi x}{2}\sin\frac{\pi v}{2} = \left(1 + \frac{v}{1-x}\right)\left(1 - \frac{v}{1+x}\right)\left(1 + \frac{v}{3-x}\right)\left(1 - \frac{v}{3+x}\right)\text{ etc.}$$

In a similar way he derived

$$\cos\frac{\pi v}{2} + \cot\frac{\pi x}{2}\sin\frac{\pi v}{2} = \left(1 + \frac{v}{x}\right)\left(1 - \frac{v}{2-x}\right)\left(1 + \frac{v}{2+x}\right)$$

$$\times \left(1 - \frac{v}{4-x}\right)\left(1 + \frac{v}{4+x}\right)\text{ etc.}$$

By equating the coefficients of v in the two equations, he had respectively

$$\frac{\pi}{2}\tan\frac{\pi x}{2} = \frac{1}{1-x} - \frac{1}{1+x} + \frac{1}{3-x} - \frac{1}{3+x} + \frac{1}{5-x} - \frac{1}{5+x} + \text{etc.}$$

and

$$\frac{\pi}{2}\cot\frac{\pi x}{2} = \frac{1}{x} - \frac{1}{2-x} + \frac{1}{2+x} - \frac{1}{4-x} + \frac{1}{4+x} \text{ etc.}$$

Since

$$\frac{\pi}{2}\left(\tan\frac{\pi x}{2} + \cot\frac{\pi x}{2}\right) = \frac{\pi}{\sin \pi x},$$

Euler also obtained the partial fractions expansion of $\pi/\sin \pi x$.

16.7 Euler: Dilogarithm

The dilogarithm function $\text{Li}_2(x)$ can be defined for $-1 \leq x \leq 1$ by the series

$$\text{Li}_2(x) = x + \frac{x^2}{2^2} + \frac{x^3}{3^2} + \cdots.$$

This series is obtained when the series for $\frac{-1}{t}\ln(1-t)$ is integrated term by term. Thus, we have

$$\text{Li}_2(x) = -\int_0^x \frac{\ln(1-t)}{t} dt = \int_0^x \left(1 + \frac{t}{2} + \frac{t^2}{3} + \cdots\right) dt.$$

This series arose in a 1729 paper of Euler, the purpose of which was to evaluate $\zeta(2) = \sum_{n=1}^{\infty} 1/n^2$. Though Euler succeeded only in finding an approximate value for $\zeta(2)$, he obtained an interesting and useful formula for the dilogarithm:

$$\text{Li}_2(x) + \text{Li}_2(1-x) = \sum_{n=1}^{\infty} \frac{1}{n^2} - \ln x \ln(1-x). \tag{16.39}$$

Euler wished to find an approximation for the sum of the right-hand side of (16.39). Note that this series converges very slowly. Euler took $x = 1/2$ and got

$$\sum_{n=1}^{\infty} \frac{1}{n^2} = \sum_{n=1}^{\infty} \frac{1}{2^{n-1}n^2} + (\ln 2)^2. \tag{16.40}$$

The series on the right-hand side in (16.40) converges much more rapidly and Euler found its approximate value to be 1.64481. He approximated $(\ln 2)^2$ as 0.480453 and obtained

$$\sum_{n=1}^{\infty} \frac{1}{n^2} \approx 1.644934.$$

Euler's proof of (16.39) was slightly complicated, partly because he proved a more general result. Therefore, we reproduce instead the brief argument given by Abel:

$$\text{Li}_2(x) + \text{Li}_2(1-x) = -\int_0^x \frac{\ln(1-t)}{t}\,dt - \int_0^{1-x} \frac{\ln(1-t)}{t}\,dt$$

$$= -\int_0^1 \frac{\ln(1-t)}{t}\,dt - \int_0^x \frac{\ln(1-t)}{t}\,dt + \int_{1-x}^1 \frac{\ln(1-t)}{t}\,dt$$

$$= \sum_{n=1}^{\infty} \frac{1}{n^2} - \int_0^x \left(\frac{\ln(1-t)}{t} - \frac{\ln t}{1-t} \right) dt$$

$$= \sum_{n=1}^{\infty} \frac{1}{n^2} - \ln x \ln(1-x).$$

The last step follows from the observation that the integrand in the last integral is the derivative of $\ln t \ln(1-t)$.

16.8 Landen's Evaluation of $\zeta(2)$

Landen used the logarithm of -1 to evaluate $\sum_{n=1}^{\infty} 1/n^2$ and more generally $\sum_{n=1}^{\infty} 1/n^{2k}$. He showed that the dilogarithm could be used for this purpose but one had to employ complex numbers. Thus, Landen here succeeded where Euler had failed. Landen started his paper of 1760, "A New Method of Computing the Sums of Certain Series," with the determination of the values of $\log(-1)$. He observed that if $x = \sin z$, then $\dot{z} = \dot{x}/\sqrt{1-x^2}$ or $\dot{z}/\sqrt{-1} = \dot{x}/\sqrt{x^2-1}$. He integrated, taking $z = 0$ where $x = 0$ to get

$$\frac{z}{\sqrt{-1}} = \log \frac{x + \sqrt{x^2-1}}{\sqrt{-1}}.$$

For $z = \pi/2$ and $x = 1$, he had $\log \sqrt{-1} = -\frac{\pi}{2\sqrt{-1}}$. Since the square root has two values, Landen concluded that

$$\log(-1) = 2\log\sqrt{-1} = \pm\pi/\sqrt{-1}. \qquad (16.41)$$

Landen's fundamental relation for the dilogarithm was

$$\text{Li}_2(x) = \frac{\pi^2}{3} + \frac{\pi}{\sqrt{-1}} \log x - \frac{1}{2}(\log x)^2 - \text{Li}_2(1/x). \qquad (16.42)$$

His proof was straightforward; note that he decided to use the minus sign in (16.41):

$$x^{-1} + \frac{x^{-2}}{2} + \frac{x^{-3}}{3} + \cdots = \log\frac{1}{1 - 1/x}$$

$$= \log x + \log \frac{1}{1-x} + \log(-1)$$

$$= -\frac{\pi}{\sqrt{-1}} + \log x + \log\frac{1}{1-x}.$$

16.8 Landen's Evaluation of $\zeta(2)$

Landen divided the equation by x and integrated to get

$$-\frac{x^{-1}}{1} - \frac{x^{-2}}{2^2} - \frac{x^{-3}}{3^2} - \cdots = -\frac{\pi}{\sqrt{-1}} \log x + \frac{1}{2}(\log x)^2 + \text{Li}_2(x) + C. \qquad (16.43)$$

He set $x = 1$ to find that

$$C = -2 \sum_{n=1}^{\infty} \frac{1}{n^2}.$$

Landen derived the value of this series by setting $x = -1$ in (16.43) to obtain

$$-C = 2 \sum_{n=1}^{\infty} \frac{1}{n^2} = 2\text{Li}_2(-1) - \frac{\pi}{\sqrt{-1}} \log(-1) + \frac{1}{2}\Big(\log(-1)\Big)^2. \qquad (16.44)$$

Since

$$2\text{Li}_2(-1) = -2\left(1 - \frac{1}{2^2} + \frac{1}{3^2} - \cdots\right) = -2\left(1 - \frac{2}{2^2}\right)\left(1 + \frac{1}{2^2} + \frac{1}{3^2} + \cdots\right)$$

$$= -\sum_{n=1}^{\infty} \frac{1}{n^2} = \frac{C}{2},$$

equation (16.44) became

$$-C = \frac{C}{2} - \frac{\pi}{\sqrt{-1}} \cdot \frac{\pi}{\sqrt{-1}} + \frac{1}{2}\left(\frac{\pi}{\sqrt{-1}}\right)^2. \qquad (16.45)$$

Landen took $\log(-1) = -\pi/\sqrt{-1}$ to derive (16.43), but he took $\log(-1) = \pi/\sqrt{-1}$ in the last equation. Although he did not explain this, he clearly wished to obtain a positive value for the series $\sum_{n=1}^{\infty} 1/n^2$. So (16.45) simplified to

$$-\frac{1}{2}C = \sum_{n=1}^{\infty} \frac{1}{n^2} = \frac{\pi^2}{6}. \qquad (16.46)$$

This completed Landen's proof of (16.42). Note that this formula holds for $x \geq 1$, as is suggested by the first step of his proof and by the fact that he chose the negative sign in the term containing $\pi/\sqrt{-1}$. Note, however, that we cannot set $x = e^{i\theta}$ in (16.42) unless we change the sign of $\pi/\sqrt{-1}$ in Landen's formula. The need for this change was understood only after the development of complex analysis. After this change of sign, the formula is equivalent to the Fourier expansion of the second Bernoulli polynomial. Similar changes in equations (16.49) through (16.51) produce the Fourier series for the third, fourth, and fifth Bernoulli polynomials. Of course, Landen did not mention these expansions; they were discovered by Daniel Bernoulli and Euler.

Landen also derived formulas for the polylogarithms, $\text{Li}_3(x)$, $\text{Li}_4(x)$, $\text{Li}_5(x)$, etc., where

$$\text{Li}_n(x) = \frac{x}{1^n} + \frac{x^2}{2^n} + \frac{x^3}{3^n} + \frac{x^4}{4^n} + \cdots. \qquad (16.47)$$

Note that
$$\mathrm{Li}_n(x) = \int_0^x \frac{\mathrm{Li}_{n-1}(t)}{t} dt. \tag{16.48}$$

Though he did not employ the factorial notation, Landen obtained the formulas

$$\mathrm{Li}_3(x) = \frac{\pi^2}{3}\log x + \frac{\pi}{2\sqrt{-1}}(\log x)^2 - \frac{1}{2\cdot 3}(\log x)^3 + \mathrm{Li}_3(1/x), \tag{16.49}$$

$$\mathrm{Li}_4(x) = 2\sum_{n=1}^{\infty}\frac{1}{n^4} + \frac{\pi^2}{6}(\log x)^2 + \frac{\pi}{2\sqrt{-1}}\frac{1}{3}(\log x)^3 - \frac{1}{4!}(\log x)^4 - \mathrm{Li}_4(1/x), \tag{16.50}$$

$$\mathrm{Li}_5(x) = \left(2\sum_{n=1}^{\infty}\frac{1}{n^4}\right)\log x + \frac{\pi^2}{18}(\log x)^3 + \frac{\pi}{2\sqrt{-1}}\frac{1}{3\cdot 4}(\log x)^4 - \frac{1}{5!}(\log x)^5 + \mathrm{Li}_5(1/x). \tag{16.51}$$

He got (16.49) by dividing equation (16.42) by x, integrating, and using (16.48). In a similar way, he derived (16.50) from (16.49) and got (16.51) from (16.50). Landen put $x = 1/\sqrt{-1}$ in (16.49) to produce, after some manipulation,

$$1 - \frac{1}{3^3} + \frac{1}{5^3} - \frac{1}{7^3} + \cdots = \frac{\pi^3}{32}. \tag{16.52}$$

On the other hand, $x = -1$ in (16.50) gave him

$$1 + \frac{1}{2^4} + \frac{1}{3^4} + \frac{1}{4^4} + \cdots = \frac{\pi^4}{90}. \tag{16.53}$$

Landen pointed out that these formulas can also be continued indefinitely, but he did not indicate a connection with Bernoulli numbers.

16.9 Spence: Two-Variable Dilogarithm Formula

In his essay on logarithmic transcendents, William Spence remarked that Euler and Bernoulli had only one variable in their formulas for the dilogarithm, whereas he himself used more unknowns. Spence perhaps did not fully appreciate Landen's work, saying that Landen "added nothing of consequence to the discoveries of Bernoulli and Euler." Spence was perhaps confused as to the identities of the Bernoullis. Also, Spence most likely saw only the republication of Landen's independent results of 1859; this appeared after the publication of Euler's work, of the early 1860s, on the polylogarithm. Spence worked with the function defined by

$$\overset{2}{L}(1+x) = \int_0^x \frac{dt}{t}\ln(1+t). \tag{16.54}$$

Note that he wrote $\ln(1+t)$ as $\overset{1}{L}(1+t)$. Spence observed that when $|x| < 1$:

$$\frac{\ln(1+t)}{t} = 1 - \frac{1}{2}t + \frac{1}{3}t^2 - \frac{1}{4}t^3 + \cdots \quad -1 < t < 1,$$

and so
$$\overset{2}{L}(1+x) = \frac{x}{1^2} - \frac{x^2}{2^2} + \frac{x^3}{3^2} - \frac{x^4}{4^2} + \cdots \quad -1 \le x \le 1.$$

He then gave a simple proof of the formula

$$\overset{2}{L}\left((1+mx)(1+nx)\right) = \overset{2}{L}(1+mx) + \overset{2}{L}(1+nx) - \overset{2}{L}\left(\frac{m+n+mnx}{m}\right)$$
$$-\overset{2}{L}\left(\frac{m+n+mnx}{n}\right) + \ln\left(\frac{m+n+mnx}{m}\right) \cdot \ln\left(\frac{n(1+mx)}{m}\right)$$
$$+ \ln\left(\frac{m+n+mnx}{n}\right) \cdot \ln\left(\frac{m(1+nx)}{n}\right) - \frac{1}{1 \cdot 2}\left(\ln\frac{m}{n}\right)^2 + 2\overset{2}{L}(2). \quad (16.55)$$

He expressed the formula (16.54) as

$$\overset{2}{L}(1+x) = \int \frac{dx}{x} \ln(1+x), \quad (16.56)$$

and worked with the integral as if it were an indefinite integral where the constant of integration was computed in the final step. With this in mind, he replaced x by $(m+n)x + mnx^2$ to get

$$\overset{2}{L}\left((1+mx)(1+nx)\right) = \int \left(\frac{dx}{x} + \frac{mndx}{m+n+mnx}\right) \ln\left((1+mx)(1+nx)\right)$$
$$= \int \frac{dx}{x} \ln(1+mx) + \int \frac{dx}{x} \ln(1+nx) + \int \frac{dx}{m+n+mnx} \ln(1+mx) \quad (16.57)$$
$$+ \int \frac{dx}{m+n+mnx} \ln(1+nx).$$

By definition, the first two integrals were $\overset{2}{L}(1+mx)$ and $\overset{2}{L}(1+nx)$, respectively. Letting z denote the sum of the last two integrals and setting $v = m+n+mnx$, he obtained

$$z = \int \frac{dv}{v} \ln\left(\frac{v-m}{n}\right) + \int \frac{dv}{v} \ln\left(\frac{v-n}{m}\right)$$
$$= \int \frac{dv}{v} \ln\left(\frac{v}{m}-1\right) + \int \frac{dv}{v} \ln\left(\frac{v}{n}-1\right)$$
$$= \ln\left(\frac{v}{m}\right) \ln\left(\frac{v}{m}-1\right) - \overset{2}{L}\left(\frac{v}{m}\right) + \ln\left(\frac{v}{n}\right) \ln\left(\frac{v}{n}-1\right) - \overset{2}{L}\left(\frac{v}{n}\right) + C.$$

The last step involved integration by parts and the value of the constant C was found by setting $x = 0$. This completed Spence's proof.

Abel rediscovered Spence's formula, with a different proof; it first appeared in his collected papers in 1839. Abel stated his formula as

$$\text{Li}_2\left(\frac{x}{1-x} \cdot \frac{y}{1-y}\right) = \text{Li}_2\left(\frac{y}{1-x}\right) + \text{Li}_2\left(\frac{x}{1-y}\right) - \text{Li}_2(y) - \text{Li}_2(x) - \ln(1-y)\ln(1-x).$$
$$(16.58)$$

Abel gave a simple and elegant proof of this formula: He let a denote a constant. Then it was easy to check that

$$\operatorname{Li}_2\left(\frac{a}{1-a}\cdot\frac{y}{1-y}\right) = -\int\left(\frac{dy}{y}+\frac{dy}{1-y}\right)\ln\frac{1-a-y}{(1-a)(1-y)}$$

$$= -\int\frac{dy}{y}\ln\left(1-\frac{y}{1-a}\right) + \int\frac{dy}{y}\ln(1-y) - \int\frac{dy}{1-y}\ln\left(1-\frac{a}{1-y}\right)$$

$$+ \int\frac{dy}{1-y}\ln(1-a)$$

$$= \operatorname{Li}_2\left(\frac{y}{1-a}\right) - \operatorname{Li}_2(y) - \ln(1-a)\ln(1-y) - \int\frac{dy}{1-y}\left(1-\frac{a}{1-y}\right).$$

The first equation could be verified by taking the derivative of both sides. To evaluate the last integral, Abel set

$$z = \frac{a}{1-y} \quad \text{or} \quad 1-y = \frac{a}{z} \quad \text{and} \quad dy = \frac{a\,dz}{z^2} \quad \text{so that}$$

$$-\int\frac{dy}{1-y}\ln\left(1-\frac{a}{1-y}\right) = -\int\frac{dz}{z}\ln(1-z) = \operatorname{Li}_2(z) + C = \operatorname{Li}_2\left(\frac{a}{1-y}\right) + C.$$

Thus, he had

$$\operatorname{Li}_2\left(\frac{a}{1-a}\cdot\frac{y}{1-y}\right) = \operatorname{Li}_2\left(\frac{y}{1-a}\right) + \operatorname{Li}_2\left(\frac{a}{1-y}\right) - \operatorname{Li}_2(y) - \ln(1-a)\ln(1-y) + C.$$

To find C, Abel let $y = 0$ to get $C = -\operatorname{Li}_2(a)$. This proved the formula after a was replaced by the variable x.

16.10 Exercises

1. Show that the series for $\cos x$ can be obtained by a repeated integration of the equation $\cos x = 1 - \int_0^x \int_0^t \cos u\,du\,dt$. This method of deriving the series for cosine is due to Leibniz. See Newton (1959–1960), vol. 2, p. 74.
2. Let $n = 2m+1$. Show that

$$\sin x - \sin\left(x+\frac{\pi}{n}\right) - \sin\left(x-\frac{\pi}{n}\right) + \sin\left(x-\frac{2\pi}{n}\right)$$

$$+ \sin\left(x+\frac{2\pi}{n}\right) - \sin\left(x-\frac{3\pi}{n}\right)$$

$$- \sin\left(x+\frac{3\pi}{n}\right) + \cdots \pm \sin\left(x+\frac{m\pi}{n}\right) \pm \sin\left(x-\frac{m\pi}{n}\right) = 0,$$

where the plus sign is used when m is even and the minus sign otherwise. See Euler (1988), p. 208

16.10 Exercises

3. With n as in the previous problem, prove Euler's formula

$$n \csc nx = \csc x - \csc\left(x + \frac{\pi}{n}\right) - \csc\left(x - \frac{\pi}{n}\right) + \csc\left(x + \frac{2\pi}{n}\right) + \csc\left(x - \frac{2\pi}{n}\right)$$

$$- \cdots \pm \csc\left(x + \frac{m\pi}{n}\right) \pm \csc\left(x - \frac{m\pi}{n}\right).$$

See Euler (1988), p. 209.

4. Prove the following formulas:

$$\cos nx = 2^{n-1} \cos\left(x + \frac{n-1}{n}\pi\right) \cos\left(x - \frac{n-1}{n}\pi\right)$$

$$\cdot \cos\left(x + \frac{n-3}{n}\pi\right) \cos\left(x - \frac{n-3}{n}\pi\right) \cdots,$$

where there are n factors;

$$n \cot nx = \cot x + \cot\left(x + \frac{\pi}{n}\right) + \cot\left(x + \frac{2\pi}{n}\right) + \cdots + \cot\left(x + \frac{n-1}{n}\pi\right).$$

Also show that the sum of the squares of the cotangents is $\frac{n^2}{(\sin x)^2} - n$. See Euler (1988), pp. 214 and 218.

5. Show that Landen's formula (16.42) can be correctly and comprehensively stated by Kummer's 1840 result:

$$\text{Li}_2(re^{i\theta}) = \text{Li}_2(r, \theta) + \frac{i}{2}[2\omega \log r + \text{Cl}_2(2\omega) + \text{Cl}_2(2\theta) - \text{Cl}_2(2\omega + 2\theta)]$$

where $\tan \omega = r \sin\theta/(1 - r\cos\theta)$, and

$$\text{Li}_2(r, \theta) = -\frac{1}{2}\int_0^r \frac{\log(1 - 2r\cos\theta + r^2)}{r} dr.$$

See Kummer (1840), pp. 74–90 and Lewin (1981), pp. 120–121.

6. Prove Kummer's formula

$$\text{Li}_2\left(\frac{x(1-y)^2}{y(1-x)^2}\right) = \text{Li}_2\left(\frac{x - xy}{x - 1}\right) + \text{Li}_2\left(\frac{1-y}{xy-y}\right) + \text{Li}_2\left(\frac{x - xy}{y - xy}\right)$$

$$+ \text{Li}_2\left(\frac{1-y}{1-x}\right) + \frac{1}{2}(\ln y)^2.$$

See Kummer (1975), vol. 2, p. 238.

7. Prove that Spence's formula (16.55) and Abel's formula (16.58) are equivalent.

8. Show that

$$\frac{\sin x}{\sin y} = \frac{x}{y}\left(\frac{\pi - x}{\pi - y}\right)\left(\frac{\pi + x}{\pi + y}\right)\left(\frac{2\pi - x}{2\pi - y}\right)\left(\frac{2\pi + x}{2\pi + y}\right)\left(\frac{3\pi - x}{3\pi - y}\right)\left(\frac{3\pi + x}{3\pi + y}\right)\cdots.$$

Derive the product for $\cos x$ by replacing x by $\frac{\pi}{2} - x$, y by $\pi/2$, and applying Wallis's formula. See Cauchy's *Analyse algébrique*, note IV.

9. Prove that

$$\text{Li}_2(x) + \text{Li}_2(y) - \text{Li}_2(xy) = \text{Li}_2\left(\frac{x(1-y)}{1-xy}\right) + \text{Li}_2\left(\frac{y(1-x)}{1-xy}\right)$$
$$+ \ln\left(\frac{1-x}{1-xy}\right) \ln\left(\frac{1-y}{1-xy}\right).$$

See L. J. Rogers (1907).

10. Prove the two inequalities for $0 < x \leq \pi/2$ used in Cauchy's proof of (16.30):

$$\sin x < x \quad \text{and} \quad \frac{x}{\sin x} < 2.$$

See note IX of Cauchy's *Analyse algébrique*.

11. Prove that for even m

$$\cos mx = \prod_{k=1}^{m/2}\left(1 - \frac{\sin^2 x}{\sin^2 \frac{(2k-1)\pi}{2m}}\right),$$

$$\sin mx = m \sin x \cos x \prod_{k=1}^{(m-2)/2}\left(1 - \frac{\sin^2 x}{\sin^2 \frac{2k\pi}{2m}}\right).$$

State and prove a similar formula for odd m. See Cauchy, *Analyse algébrique*, note IX.

12. Use the formulas in exercise 11 and (16.30) to show that for m even

$$\cos mx = 2^{\frac{m}{2}-1} \prod_{k=1}^{m/2}\left(\cos 2x - \cos \frac{(2k-1)\pi}{m}\right),$$

$$\sin mx = 2^{\frac{m}{2}-1} \sin 2x \prod_{k=2}^{m/2}\left(\cos 2x - \cos \frac{(2k-1)\pi}{m}\right).$$

State and prove similar results for m odd. See Cauchy, *Analyse algébrique*, note IX.

13. Suppose $\phi_0(x) = \phi_0(1/x)$ and

$$\phi_n(x) = \int \frac{dx}{x} \phi_{n-1}(x), \quad n = 1, 2, 3, \ldots.$$

Prove that

(a) $\phi_{2n}(x) - \phi_{2n}(1/x) = 2 \sum_{k=0}^{n-1} \phi_{2n-2k-1}(1)(\ln x)^{2k+1}$,
(b) $\phi_{2n+1}(x) - \phi_{2n+1}(1/x) = 2 \sum_{k=0}^{n} \phi_{2n-2k+1}(1)(\ln x)^{2k}$,
(c) $\phi_{2n+1}(1) - \phi_{2n+1}(-1) = \sum_{k=1}^{n} (-1)^{k-1} \frac{\pi^{2k}}{(2k)!} \phi_{2n-2k+1}(1)$.

Find $\phi_1(1)$, when $\phi_0(x) = \frac{x^p + x^{-p}}{x^m + a + x^{-m}}$. See Spence (1819), pp. 139–143.

14. Observe that $1/(1-a) = 1 + a/(1-a)$ and $1/((1-a)(1-b)) = 1 + a/(a1-a) + b/((1-a)(1-b))$. Generalize to prove the partial fractions formula

$$\frac{\alpha}{\alpha-a} \cdot \frac{\beta}{\beta-b} \cdot \frac{\gamma}{\gamma-c} \cdot \frac{\delta}{\delta-d} \cdots = 1 + \frac{\alpha}{\alpha-a} + \frac{\alpha b}{(\alpha-a)(\beta-b)}$$
$$+ \frac{\alpha\beta c}{(\alpha-a)(\beta-b)(\gamma-c)} + \cdots.$$

Note that from the product expansion for $\sin x$, we obtain

$$x\csc(\pi x/2) = \frac{3}{3-(x^2-1)} \cdot \frac{15}{15-(x^2-1)} \cdot \frac{35}{35-(x^2-1)} \cdots.$$

Use the partial fractions formula to deduce that

$$x\csc(\pi x/2) = 1 + \frac{x^2-1}{4-x^2}\left(1 + \sum_{n=1}^{\infty} \frac{(2^2-1)(4^2-1)\cdots((2n)^2-1)}{(4^2-x^2)(6^2-x^2)\cdots((2n+2)^2-x^2)}\right),$$

$$\frac{2}{\pi} = 1 - \left(\frac{1}{2}\right)^2 - \frac{1}{3}\left(\frac{1\cdot 3}{2\cdot 4}\right)^2 - \frac{1}{5}\left(\frac{1\cdot 3\cdot 5}{2\cdot 4\cdot 6}\right)^2 - \frac{1}{7}\left(\frac{1\cdot 3\cdot 5\cdot 7}{2\cdot 4\cdot 6\cdot 8}\right)^2 - \cdots.$$

In a similar way, show that

$$\sec(\pi x/2) = 1 + \frac{x^2}{1-x^2}\sum_{n=1}^{\infty} \frac{(1\cdot 3\cdots(2n-1))^2}{(3^2-x^2)\cdots((2n+1)^2-x^2)}.$$

What famous result is obtained by dividing the last equation by x^2 and letting x tend to 0? Also show that

$$\cos(\pi x/2) = 1 - x^2 - (1-x^2)\frac{x^2}{9}$$
$$- (1-x^2)(1-x^2/9)x^2/25$$
$$- (1-x^2)(1-x^2/9)(1-x^2/25)x^2/49 - \cdots,$$

$\sin(\pi x/2)$
$$= x + \frac{x(1-x^2)}{1\cdot 3} + \frac{x(1-x^2)(4-x^2)}{(1\cdot 3)^2 5} + \frac{x(1-x^2)(4-x^2)(16-x^2)}{(1\cdot 3\cdot 5)^2 7} + \cdots,$$

$\cos(\pi x/3)$
$$= 1 - \frac{x^2}{2!} + \frac{x^2(x^2-1)}{4!} - \frac{x^2(x^2-1)(x^2-4)}{6!} + \frac{x^2(x^2-1)(x^2-4)(x^2-9)}{8!} - \cdots,$$

$\sin(\pi x/3)$
$$= \frac{\sqrt{3}}{2}\left(x - \frac{x(x^2-1)}{3!} + \frac{x(x^2-1)(x^2-4)}{5!} - \frac{x(x^2-1)(x^2-4)(x^2-9)}{7!} + \cdots\right).$$

See Schellbach (1854), pp. 233–236. Karl Schellbach (1805–1892) was one of Eisenstein's teachers in secondary school at Berlin.

16.11 Notes on the Literature

Euler's derivations of the series and products for elementary functions are taken from Euler (1988), chapters 7–9 and 14. Euler published several papers on the partial fractions expansions of trigonometric functions. It is perhaps most interesting to read his correspondence on this topic with Niklaus I Bernoulli. See Eu. 4A-2, pp. 483–550. For Euler's first work on the dilogarithm, see Eu. I-14, pp. 38–41. For Abel's derivation of Euler's dilogarithm formula (16.39), see Abel (1965), vol. 2, p. 190. Landen's paper appeared in the *Philosophical Transactions*, 1760. See Spence (1819), a work containing his 1809 book and other writings, for his dilogarithm formula (16.55). Abel's proof of the same formula (16.58) is in Abel (1965), vol. 2, pp. 192–193. Lewin (1981) gives a modern treatment of the dilogarithm and polylogarithm as functions of complex variables.

17

Solution of Equations by Radicals

17.1 Preliminary Remarks

The problem of solving algebraic equations has been of interest to mathematicians for about four thousand years. Clay tablets from 1700 BC Babylon contain the essence of the quadratic formula for solutions of second-degree equations. Unfortunately, the tablets do not indicate how the Babylonians arrived at the methods they described. Greek mathematicians later considered the problem from a geometric point of view. Indian mathematicians from the second century AD made significant advances in algebra, especially in the development of the kind of notational system by which the symbolic algebra of the seventeenth century was made possible. These Indian mathematicians also described a method for solving the quadratic using factorization by completing the square. Medieval Islamic mathematicians continued this algebraic tradition with several original contributions of their own. For example, they considered the cubic equation and gave algebraic and geometric methods for solving special cubics, although a general method for solving the cubic was not found until the sixteenth century.

Artis Magnae, sive de Regulis Algebraicis by Girolamo Cardano (1501–1576) presented the first known general method for solving a cubic, as well as a method for solving a quartic. Published in 1545, this book contained the work of Scipione del Ferro, Niccolò Tartaglia, Lodovico Ferrari, and Cardano himself. As Cardano wrote in the first chapter:

> In our own days Scipione del Ferro of Bologna has solved the case of the cube and the first power equal to a constant, a very elegant and admirable accomplishment. Since this art surpasses all human subtlety and the perspicuity of moral talent and is a truly celestial gift and a very clear test of the capacity of men's minds, whoever applies himself to it will believe that there is nothing he cannot understand. In emulation of him, my friend Niccolò Tartaglia of Brescia, wanting not to be outdone, solved the same case when he got into a contest with his [Scipione's] pupil, Antonio Maria Fior, and, moved by my many entreaties, gave it to me. For I had been deceived by the words of Luca Paccioli, who denied that any more general rule could be discovered than his own. Notwithstanding the many things which I had already discovered, as is well known, I had despaired and had not attempted to look any further. Then, however, having received Tartaglia's solution and seeking for the proof of it, I came to understand that there were a great many other things that

could also be had. Pursuing this thought and with increased confidence, I discovered these others, partly by myself and partly through Lodovico Ferrari, formerly my pupil.

We note that Cardano did not use the term first power; this more modern terminology has been employed by the translator for greater clarity. Thus, del Ferro (1465–1526) was the original discoverer of the solution of the cubic in the form $x^3 + ax = b$; he made this discovery around 1515 when he was a professor of mathematics in Bologna, where his original papers are kept. Before his death, he passed on the solution to his students Annibale della Nave, who was also his son-in-law, and Antonio Maria Fior. In 1535, Fior challenged Tartaglia to a public problem-solving contest. All the problems posed by Fior amounted to solving del Ferro's cubic. Luckily for Tartaglia, he was able to find a method for solving this cubic the night before the contest, and Fior was defeated. When Cardano heard of this contest, he requested Tartaglia to give him the solution. Tartaglia refused, hoping to write his own book on the subject, but later visited Cardano in Milan and told him the solution in the form of a poem. When Cardano's book was published, Tartaglia was angry, claiming that Cardano had given a most solemn promise never to divulge the solution. Ferrari, a young servant in Cardano's house at the time of Tartaglia's visit, reported that he was present at the meeting and no such promise was made. This dispute led to a scholarly contest between Tartaglia and Ferrari, won by the latter. Ferrari's prestige was greatly enhanced, and he received several offers of important positions. Eventually he became the professor of mathematics at Bologna.

In Cardano's work, the coefficients of the equations were always positive; hence he distinguished between the equations $N = x^3 + ax$ and $N + ax = x^3$. He classified thirteen different cubics, and much of the book is taken up by a discussion of these cases. He mentioned the different cases of the quartic, without elaboration: "For as *positio* [first power] refers to a line, *quadratum* [the square] to a surface, and *cubum* [the cube] to a solid body, it would be very foolish for us to go beyond this point. Nature does not permit it." Thus, his arguments were framed in geometric language, accompanied by diagrams.

Cardano's method for solving the cubic was essentially to assume the form of the solution, $x = u + v$, and then show that u^3 and v^3 were the solutions of a quadratic. He then solved the quadratic to obtain x. Notice that this method of assuming the correct form of the solution to solve the equation may be applied to the quadratic equation as well. The method used up to Cardano's time was factorization after completing the square. The ninth-century Indian mathematician Sridhara employed this general method for solving $ax^2 + bx = c$. He multiplied the equation by $4a$ and added b^2 to obtain

$$4a^2x^2 + 4abx + b^2 = b^2 + 4ac \quad \text{or} \quad (2ax + b)^2 = \left(\pm\sqrt{b^2 + 4ac}\right)^2.$$

Interestingly, Ferrari used the method of completing the square to solve the quartic equation. In chapter 39 of his book Cardano wrote, "For example, divide 10 into three proportional parts, the product of the first and second of which is 6. This was proposed by Zuanne de Tonini da Coi, who said it would not be solved. I said it could, though I did not yet know the method. This was discovered by Ferrari." Cardano showed that the equation satisfied by x, the mean, was $60x = x^4 + 6x^2 + 36$. By adding $6x^2$ to both sides, he had $x^4 + 12x^2 + 36 = 6x^2 + 60x$. Next he added a quadratic expression $2bx^2 + b^2 + 12b$ to both sides so that the left-hand side was a square and the right-hand side became a square, provided b satisfied a certain cubic. This method is easily generalized to all quartics. Thus, for the quartic $x^4 + 2ax^2 + bx + c = 0$, write $x^4 + 2ax^2 = -bx - c$ or

$$x^4 + (2a + 2d)x^2 + (d+a)^2 = 2dx^2 - bx + (d+a)^2 - c.$$

The right-hand side is a perfect square if $8d\left((d+a)^2 - c\right) = b^2$, a cubic in d; thus, it can be solved.

In his 1637 *La Géométrie*, Descartes used a different factorization, solving the quartic in which the coefficient of x^3 was zero. He then demonstrated that a general quartic could be reduced to this case. Harriot had already performed this reduction, while Cardano applied a similar method to a cubic. Although Cardano solved the cubic by assuming the form of the solution, it was two centuries later that Euler first used this method to solve a quartic. Euler assumed $x = \sqrt{p} + \sqrt{q} + \sqrt{r}$ to be the solution of the quartic $x^4 + ax^2 + bx + c = 0$. He then solved the quartic by determining the cubic whose solutions were p, q, and r. Denoting the assumed solution as x_1, the other solutions would be

$$x_2 = \sqrt{p} - \sqrt{q} - \sqrt{r},$$
$$x_3 = -\sqrt{p} + \sqrt{q} - \sqrt{r},$$
$$x_4 = -\sqrt{p} - \sqrt{q} + \sqrt{r}.$$

At first sight, it might appear that Euler assumed an overly specific form of the solution, but a brief argument explains why it worked. Since the coefficient of x^3 in the quartic is zero, we must have $x_1 + x_2 + x_3 + x_4 = 0$ and thus

$$x_1 + x_2 = -(x_3 + x_4),$$
$$x_1 + x_3 = -(x_2 + x_4),$$
$$x_1 + x_4 = -(x_2 + x_3).$$

Euler's cubic is actually a polynomial of degree 3 in y^2. This cubic has the six solutions $x_i + x_j$, $(i > j)$, where $i, j = 1, 2, 3, 4$. Thus, $p = (x_1 + x_2)^2 = (x_3 + x_4)^2$ and so on.

The British mathematician Edward Waring appears to be the first to study the relation between the roots of a given quartic (or cubic) and the roots of the resolvent cubic (or quadratic). Later on, Lagrange and Alexandre Theophile Vandermonde (1835–1896) systematically exploited this relationship to determine whether a given algebraic equation could be solved by radicals. Waring noted in the third and final edition of his *Meditationes Algebraicae* that in 1763 he sent a copy of the first edition to Euler and in 1770 sent the second edition to d'Alembert, Bézout, Montucla, Lagrange, and Frisi. Lagrange praised Waring's researches in his long paper on algebraic equations. Nevertheless, Waring perhaps felt that his work had not been sufficiently recognized and in his third edition, pointed out several instances of his priority.

Waring's book showed that if α, β, γ were the roots of a cubic, then $Z = \alpha + w\beta + w^2\gamma$, where $w^3 = 1$, satisfied a quadratic in Z^3. Hence, $(\alpha + w\beta + w^2\gamma)^3$ was a solution of the resolvent quadratic. More generally, Waring considered the expression $x = a\sqrt[n]{p} + b\sqrt[n]{p^2} + c\sqrt[n]{p^3} + \cdots$. In the particular case $p = 1$, so that $\sqrt[n]{p}$ was an nth root of unity, he showed that when $a, b, c, \ldots$ were roots of a general nth degree polynomial, then x was the resolvent of the equation, now called the Lagrange resolvent. Waring found expressions for the roots $a, b, c, \ldots$ in terms of the resolvents obtained by taking the n different roots of unity. He also noted that Ferrari's method of solving a quartic with roots $\alpha, \beta, \gamma, \delta$ produced a (resolvent) cubic with roots $(\alpha\beta + \gamma\delta)/2$, $(\alpha\gamma + \beta\delta)/2$, and $(\alpha\delta + \beta\gamma)/2$. Waring also discussed other resolvents of the quartic, such as $(\alpha + \beta - \gamma - \delta)^2$ and $(\alpha\beta - \gamma\delta)^2$. In short, Waring had the conception of the resolvent of an equation, in particular the so-called Lagrange resolvent. He also had rudimentary ideas on the important role of symmetry (permutations) in the study of the roots of the equation. In the preface to the third edition of the *Meditationes*, he explained:

> In the first edition I gave a resolution, shown below, of a biquadratic equation $x^4 + 2px^3 = qx^2 + rx + s$, in which the second term is not removed: to each side of the equation the quantity $(p^2 + 2n)x^2 + 2pnx + n^2$ is added, giving the equation $(x^2 + px + n)^2 = (p^2 + 2n + q)x^2 + (2pn + r)x + s + n^2$. Now we require that $4(p^2 + 2n + q)(s + n^2) = (2pn + r)^2$, or equivalently $8n^3 + 4qn^2 + (8s - 4rp)n + 4qs + rp^2s - r^2 (= A) = 0$, whence $x^2 + px + n = \sqrt{p^2 + 2n + q}\,x + \sqrt{s + n^2}$. Then in the second edition I proved that $\frac{\alpha\beta + \gamma\delta}{2}$, $\frac{\alpha\gamma + \beta\delta}{2}$, $\frac{\alpha\delta + \beta\gamma}{2}$ are the three values of the quantity n, where $\alpha, \beta, \gamma, \delta$ are the roots of the given equation, and $\frac{\alpha+\beta-\gamma-\delta}{2}$, $\frac{\gamma+\delta-\alpha-\beta}{2}$, $\frac{\alpha+\gamma-\beta-\delta}{2}$, $\frac{\beta+\delta-\alpha-\gamma}{2}$, $\frac{\alpha+\delta-\beta-\gamma}{2}$, and $\frac{\beta+\gamma-\alpha-\delta}{2}$, are the six values of the quantity $\sqrt{p^2 + 2n + q}$, and finally $\frac{\alpha\beta - \gamma\delta}{2}$, $\frac{\gamma\delta - \alpha\beta}{2}$, $\frac{\alpha\gamma - \beta\delta}{2}$, ... are the six values of the quantity $\sqrt{s + n^2}$. The same result was later published by Lagrange. Using all three roots of the cubic ($A = 0$), we have a resolution with 24 values, of which 12, as I showed in the second edition, are extraneous, while the other 12 are true, i.e. they are roots of the given biquadratic, each of them occurring three times.

Because of his obscure and difficult style, Waring's work was not much read; his discoveries did not become known, and he did not receive much credit for his innovative work on symmetric functions and the theory of equations. J. J. Sylvester commented, "Written in Latin, and when the proper language of algebra was yet unformed, it is frequently a work of much labour to follow Waring's demonstrations and deductions, and to distinguish his assertions from his proofs."

Lagrange's great advance in the problem of solving algebraic equations by radicals was his 1771 discovery of a general technique underlying the various methods of solving equations of degrees two, three, and four. His essential idea was to place this problem within the framework of the theory of symmetric functions, the elements of which had been established by Newton. On this basis, Lagrange worked out Waring's sketchy ideas on the number of values a function of the roots could assume under permutations. Lagrange observed that if a polynomial in the roots assumed k different expressions when all the permutations of the roots were taken, then any elementary symmetric function of the k expressions would be symmetric, and hence a polynomial in the coefficients. This implied that the coefficients of the equation with exactly these k expressions as roots would be polynomials in the coefficients of the original equation. To understand Lagrange's idea, consider the simplest case where x_1 and x_2 are solutions

17.1 Preliminary Remarks

of the quadratic equation $x^2 + bx + c = 0$. Here the expression $x_1 - x_2$ takes the two values $x_1 - x_2$ and $x_2 - x_1$ under all the permutations. The coefficients of the equation

$$(x - (x_1 - x_2))(x + (x_1 - x_1)) = 0 \quad \text{or} \quad x^2 - (x_1 - x_2)^2 = 0$$

are polynomials in b and c. In fact, the coefficient of x is zero and $(x_1 - x_2)^2 = b^2 - 4c$. So $x_1 - x_2 = \pm\sqrt{b^2 - 4c}$, and since we also have $x_1 + x_2 = -b$,

$$x_1 = \frac{-b \pm \sqrt{b^2 - 4c}}{2}, \quad x_2 = \frac{-b \mp \sqrt{b^2 - 4c}}{2}.$$

Note that here $x_1 - x_2$ is the resolvent of the quadratic.

To solve a cubic $x^3 + ax + b = 0$, Lagrange took the resolvent to be a similar linear expression in the roots x_1, x_2, x_3: $x_1 + wx_2 + w^2 x_3$, where w was a nontrivial cube root of unity. Permutation of the roots gave six different expressions, but Lagrange observed that permutations of the roots in the cubed expression $(x_1 + wx_2 + w^2 x_3)^3$ gave only two different values. Thus,

$$\left(x - (x_1 + wx_2 + w^2 x_3)^3\right)\left(x - (x_1 + w^2 x_2 + wx_3)^3\right) = x^2 - 27b^3 x - (27)^2 a^3.$$

The solutions of this quadratic were

$$\frac{27b^3 \pm \sqrt{(27b^3)^2 + 2(27)^2 a^3}}{2} = \frac{27}{2}\left(b^3 \pm \sqrt{b^6 + 4a^3}\right).$$

Lagrange then obtained the value of the root x_1 from the relations

$$x_1 + x_2 + x_3 = 0,$$

$$x_1 + wx_2 + q^2 x_3 = \sqrt[3]{\frac{27}{2}\left(b^3 \pm \sqrt{b^6 + 4a^3}\right)},$$

$$x_1 + w^2 x_2 + wx_3 = \sqrt[3]{\frac{27}{2}\left(b^3 \mp \sqrt{b^6 + 4a^3}\right)}.$$

Adding the three equations, he got

$$3x_1 = \sqrt[3]{\frac{27}{2}\left(b^3 + \sqrt{b^6 + 4a^3}\right)} + \sqrt[3]{\frac{27}{2}\left(b^3 - \sqrt{b^6 + 4a^3}\right)}.$$

In this manner, Lagrange once again obtained Cardano's solution.

For the quartic with roots x_1, x_2, x_3, x_4, the obvious resolvent would be the linear expression $x_1 + ix_2 - x_3 - ix_4$, taking twenty-four different values, although the corresponding equation of degree twenty-four could be reduced to a cubic. However, Lagrange noticed that it would be simpler to work with $x_1 - x_2 + x_3 - x_4$. The twenty-four permutations in this case produced only six different expressions: $\pm(x_1 - x_2 + x_3 - x_4)$, $\pm(x_1 - x_2 - x_3 + x_4)$, and $\pm(x_1 + x_2 - x_3 - x_4)$. The squares of these quantities satisfied a cubic whose coefficients were polynomials in the coefficients of the quartic. Moreover, the solutions of the quartic could be obtained from the solutions of the cubic. For example, $4x_1$ would be the sum of the known quantity $(x_1 + x_2 + x_3 + x_4)$ and the roots

$$(x_1 - x_2 + x_3 - x_4), ((x_1 - x_2 - x_3 + x_4), (x_1 + x_2 - x_3 - x_4).$$

In his 1771 paper, Lagrange also developed some general results, now a part of group theory, in connection with roots of equations. Suppose $x_1, x_2, \ldots, x_n$ are the roots of an nth degree algebraic equation and $f(x_1, x_2, \ldots, x_n)$ is a polynomial in n variables. Lagrange raised and answered the question: What was the number of different values taken by $f(x_1, x_2, \ldots, x_n)$ when the roots were permuted? Lagrange first noted that there were $n!$ permutations of the n roots and some of these might leave f invariant. He then proved the impressive general result that if m were the number of permutations leaving f invariant, then m divided $n!$ and $n!/m$ would be the number of different values assumed by f. This is now known as Lagrange's theorem; in group-theoretic language, it states that the order of the subgroup divides the order of the group. As an example of this theorem, observe that $f = x_1 - x_2 + x_3 - x_4$ is unchanged by the permutation α in which x_1 and x_3 are interchanged, or by β when x_2 and x_4 are interchanged, or by $\alpha\beta$. Including the identity permutation, we thus have four permutations leaving f invariant, so $x_1 - x_2 + x_3 - x_4$ takes $24/4 = 6$ different values.

Lagrange did not develop a convenient notation for expressing a general permutation of n objects. This made it difficult for him to describe and prove his results. However, he succeeded in proving the following important and interesting theorem: If t and y are two functions (polynomials) of $x_1, x_2, \ldots, x_n$ such that every permutation leaving t unchanged also leaves y unchanged, then y is a ratio of polynomials in t and the elementary symmetric functions of $x_1, x_2, \ldots, x_n$. More generally, if y takes m different values, then y satisfies an equation of degree m whose coefficients are polynomials in t and the elementary symmetric functions of $x_1, x_2, \ldots, x_n$. A consequence of this theorem is that if t is invariant for only the identity permutation, then every y is a rational function of t and of the elementary symmetric functions of $x_1, x_2, \ldots, x_n$. We note that such a t was constructed and used by Ruffini and then by Galois, becoming known as the Galois resolvent.

Vandermonde independently discovered some of Lagrange's ideas and applied them to the equation $x^{11} - 1 = 0$. In this he anticipated aspects of Gauss's 1795–98 more general work. Gauss showed that $x^n - 1 = 0$ could be solved by radicals, though he omitted proofs of some crucial results. Abel and Galois later relied on the foundation laid by Lagrange and Gauss to construct their theories of algebraic equations.

17.2 Viète's Trigonometric Solution of the Cubic

While most mathematicians of the time concentrated on solving algebraic equations by working with the coefficients, by means of algebraic operations, including taking roots, François Viète (1540–1603) considered the possibility of solving algebraic equations by employing transcendental functions. In fact, in his *De Aequationum Recognitione et Emendatione Tractatus Duo*, published posthumously in 1615, he solved the cubic using the cosine function, by application of the triple-angle formula. In another work,

17.2 Viète's Trigonometric Solution of the Cubic

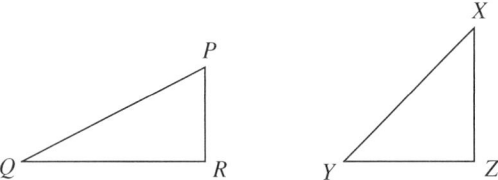

Figure 17.1. Viète's trigonometric solution of a cubic.

Ad Angularium Sectionum Analyticen Theoremata, also published in 1615, he expressed $\cos nx$ as a polynomial in $\cos x$ for several values of n, by applying algebra to trigonometry. In spite of the political turmoil of his time and difficulties in publishing his work, Viète made significant contributions to algebra and its applications to geometry and trigonometry. He introduced better algebraic notation and his 1591 *Isagoge in Artem Analyticem* had a profound influence on Harriot. In the nineteenth century, when Ruffini, Abel, and Galois showed that general fifth-degree equations could not be solved algebraically, Viète's solution of the cubic motivated a search for transcendental solutions. In particular, in 1858, Charles Hermite solved the quintic by using elliptic modular functions.

Viète's angular section analysis gave him a theorem, stated in his *Tractatus Duo*, applicable to the solution of irreducible cubics. Figure 17.1 may help in understanding the theorem.

If $$A^3 - 3B^2 A = B^2 D$$

and if, moreover, B is greater than half of D, then

$$3B^2 E - E^3 = B^2 D$$

and there are two right triangles of equal hypotenuse, B, such that the acute angle subtended by the perpendicular of the first is three times the acute angle subtended by the perpendicular of the second and twice the base of the first is D, making A twice the base of the second. The base of the second, shortened or extended by a length that, [if] raised to the square, [is] three times the perpendicular of the same, is then E.

To understand this statement in modern terms, let $A = x$ and $E = y$ so that the equations can be written as $x^3 - 3B^2 x = B^2 D$ and $3B^2 y - y^3 = B^2 D$. Note that the left side of the second equation is obtained from the first by taking $y = -x$; Viète needed these two equations because he did not use negative numbers. Thus, the two triangles may be defined by the relations:

$$P\hat{Q}R = \alpha, \ X\hat{Y}Z = 3\alpha, \ PQ = XY = B,$$
$$A = 2QR = 2B\cos\alpha, \ D = 2YZ = 2B\cos 3\alpha.$$

So $x = 2B\cos\alpha$ and $D = 2B\cos 3\alpha$. Note that we have the triple angle formula $\cos 3\alpha = 4\cos^3 \alpha - 2\cos\alpha$ or

$$B^2(2B\cos 3\alpha) = (2B\cos\alpha)^3 - 3B^2(2B\cos\alpha).$$

Thus, from $\cos 3\alpha = D/2B < 1$, we can obtain α and hence solve the cubic to get $x = 2B \cos \alpha$. Also observe that $\cos(\alpha + 2\pi/3)$ and $\cos(\alpha + 4\pi/3)$ would work equally well in the triple angle formula. Hence,

$$x = 2B\cos(\alpha + 2\pi/3) = -B\cos\alpha - B\sqrt{3}\sin\alpha,$$

$$x = 2B\cos(\alpha + 4\pi/3) = -B\cos\alpha + B\sqrt{3}\sin\alpha$$

are also solutions of the cubic. The former value is clearly negative and Viète obtained it from the second equation. To illustrate his theorem, he took the example $x^3 - 300x = 432$ and showed how all three roots were obtained. He obtained the roots 18, $9 + \sqrt{57}$, and $9 - \sqrt{57}$. He was clearly aware that a cubic in general had three roots.

17.3 Descartes's Solution of the Quartic

René Descartes discussed algebraic equations in the third book of his very influential 1637 work, *La Géométrie*. He therein stated his rule of signs for the number of positive and negative roots, his method for finding that equation whose roots (each minus a constant) were the same as the roots of a given equation, and his methods for solving equations of various degrees. His basic technique for solving equations was factorization, and he applied this method to the solution of the quartic. He first explained how to remove the cubed term: If the coefficient of x^3 were $-2a$, he set $x = z + a/2$, to obtain an equation in z where the coefficient of z^3 would be zero. To solve a quartic, Descartes first cast it in the form $x^4 \pm px^2 \pm qx \pm r = 0$ so that the coefficient of x^3 was zero. He showed that the solution of this quartic depended on the solution of the equation

$$y^6 \pm 2py^4 + (p^2 \pm 4r)y^2 - q^2 = 0,$$

a cubic in y^2. Descartes noted that if a value y^2 could be obtained from the cubic, then the quartic could be reduced to two quadratic equations

$$+x^2 - yx + \frac{1}{2}y^2 \pm \frac{1}{2}p \pm \frac{q}{2y} = 0, \quad +x^2 + yx + \frac{1}{2}y^2 \pm \frac{1}{2}p \pm \frac{q}{2y} = 0.$$

The sign of $\frac{1}{2}p$ above would be chosen to be the same as that of p in the quartic; with $+q$ in the quartic, he had $-yx + \frac{q}{2y}$ in one equation and $+yx - \frac{q}{2y}$ in the other. For $-q$, the signs were reversed. Descartes then went on to illustrate his method by solving special examples. He did not explain in detail how he obtained the cubic in y^2. For us, this appears straightforward, but a mathematician of Descartes's time might have found this challenging. Thus, suppose

$$x^4 + px^2 + qx + r = (x^2 - yx + c)(x^2 + yx + d)$$
$$= x^4 + (c + d - y^2)x^2 + (cy - dy)x + cd.$$

By equating coefficients of various powers of x, we get

$$c + d - y^2 = p, \quad y(c - d) = q, \quad cd = r.$$

Now solve for c and d in the two equations $c+d = p+y^2$, $c-d = q/y$ to obtain $c = \frac{1}{2}(y^2 + p + \frac{q}{y})$, $d = \frac{1}{2}(y^2 + p - \frac{q}{y})$. Since $cd = r$, we see that

$$(y^2+p)^2 - \frac{q^2}{y^2} = 4r \quad \text{or} \quad y^6 + 2py^4 + (p-4r)y^2 - q^2 = 0.$$

This is Descartes's cubic in y^2. By this method, Descartes solved the quartic $x^4 - 17x^2 - 20x - 6 = 0$, factoring it into $x^2 - 4x - 3 = 0$ and $x^2 + 4x + 2 = 0$.

17.4 Euler's Solution of a Quartic

In his first paper on algebraic equations, *De Formis Radicum Aequationum Cuiusque Ordinis Conjectatio*, published in 1733, Leonhard Euler solved the cubic and the quartic by the method of assuming the form of the solution. He also noted that the method seemed to break down for the fifth-degree equation. In a paper presented to the St. Petersburg Academy in 1759, published in 1764, he returned to this topic and worked with resolvents containing roots of unity. Note that in 1763 Waring sent a copy of his book to Euler; Waring believed that Euler had first learned the method of resolvents with roots of unity from that book. It is clear, however, that Euler's work was independent. We remark that, while it was not the custom in Euler's time to extensively give credit to other researchers, it is abundantly clear that Euler had no interest in taking credit for anyone else's discoveries. Euler began his solution of the quartic by letting $x = \sqrt{p} + \sqrt{q} + \sqrt{r}$ and supposing that p, q, and r were the roots of the third-degree equation $z^3 - fz^2 + gz - h = 0$. Then he could write

$$p+q+r = f, \quad pq+pr+qr = g, \quad \text{and} \quad pqr = h.$$

It followed in turn from these relations that

$$x^2 = f + 2\sqrt{pq} + 2\sqrt{pr} + 2\sqrt{qr}.$$

Moving f to the left-hand side and squaring again, he had

$$x^4 - 2fx^2 + f^2 = 4pq + 4pr + 4qr + 8\sqrt{p^2qr} + 8\sqrt{pq^2r} + 8\sqrt{pqr^2}$$
$$x^4 - 2fx^2 + f^2 - 4g = 8\sqrt{pqr}(\sqrt{p} + \sqrt{q} + \sqrt{r}) = 8\sqrt{h}x.$$

Thus, x satisfied the equation

$$x^4 - 2fx^2 - 8\sqrt{h}x + f^2 - 4g = 0.$$

Euler next supposed that the equation to be solved took the form $x^4 - ax^2 - bx - c = 0$. To solve this equation, he compared its coefficients with those in the previous equation to obtain $f = a/2$, $h = b^2/64$, $g = a^2/16 + c/4$. Once he had f, g, h, he was able to solve the cubic $z^3 - fz^2 + gz - h = 0$, whose three roots were p, q, and r. He then had one root of the quartic: $x = \sqrt{p} + \sqrt{q} + \sqrt{r}$. But Euler observed that he could actually find all four roots because there were two signs connected with each of the square roots.

In fact, there appeared to be eight different values of x, but he could choose the signs in such a way that $\sqrt{pqr} = \sqrt{h} = b/8$. So if $b > 0$, Euler gave the four roots as

$$\sqrt{p} + \sqrt{q} + \sqrt{r}, \quad \sqrt{p} - \sqrt{q} - \sqrt{r}, \quad -\sqrt{p} + \sqrt{q} - \sqrt{r}, \quad \sqrt{p} - \sqrt{q} + \sqrt{r}.$$

If $b < 0$, then the roots would be

$$\sqrt{p} + \sqrt{q} - \sqrt{r}, \quad \sqrt{p} - \sqrt{q} + \sqrt{r}, \quad -\sqrt{p} + \sqrt{q} + \sqrt{r}, \quad -\sqrt{p} - \sqrt{q} - \sqrt{r}.$$

Note that if $b = 0$, then the quartic was really a quadratic equation in x^2.

17.5 Gauss: Cyclotomy, Lagrange Resolvents, and Gauss Sums

Euler and Vandermonde showed that the equation $x^n - 1 = 0$ could be solved by radicals for positive $n \leq 11$. In his *Disquisitiones*, Gauss extended the proof to any positive integer n. He did not provide complete details of some important steps, perhaps because of constraints on the length of the book or on his time. Entries in Gauss's mathematical diary suggest that he could have filled in the gaps.

In a paper written in 1741 but published ten years later, Euler considered the equation $x^n - 1 = 0$ for values of $n \leq 10$. For prime n, his idea was to factor out $x - 1$, then reduce the degree of the other factor by dividing it by $x^{(n-1)/2}$ and setting $y = -(x + 1/x)$. For example, when $n = 5$, he had

$$x^5 - 1 = (x-1)(x^4 + x^3 + x^2 + x + 1) = 0.$$

The second factor divided by x^2 gave

$$x^2 + 1/x^2 + x + 1/x + 1 = 0.$$

Setting $y = -(x + 1/x)$ transformed this equation to $y^2 - y - 1 = 0$. After solving this quadratic, he obtained the values of x by solving two more quadratic equations:

$$x^2 + \frac{1+\sqrt{5}}{2}x + 1 = 0, \quad \text{and} \quad x^2 + \frac{1-\sqrt{5}}{2}x + 1 = 0.$$

Similarly, for $n = 7$, he had $y^3 - y^2 - 2y + 1 = 0$. He solved this cubic and then the three quadratic equations corresponding to the three solutions. Since only quadratic and cubic equations had to be solved, all the solutions could be expressed in terms of radicals. Note that if we write $x = e^{i\theta}$, then $x + 1/x = 2\cos\theta$. Thus, the solutions of the cubic are $2\cos 2\pi/7$, $2\cos 4\pi/7$, and $2\cos 6\pi/7$.

In a paper published in 1771, Vandermonde discussed the more difficult $n = 11$ case; the resulting equation for y was the fifth-degree equation

$$y^5 - y^4 - 4y^3 + 3y^2 + 3y - 1 = 0.$$

Vandermonde solved this equation by radicals; his work anticipated some important features of Gauss's work. Gauss became interested in solving the equation $x^n - 1 = 0$

17.5 Gauss: Cyclotomy, Lagrange Resolvents, and Gauss Sums

at a very early stage in his career. Up through 1795, Gauss's interests seem to have been equally divided between languages and mathematics. As he worked with this equation in late March 1796, he gained some insights on the basis of which he made up his mind to be a mathematician. With his future career clearly fixed in his mind, Gauss started keeping a mathematical diary. The first entry dated March 30, 1796, stated, "The principles upon which the division of the circle depend, and geometrical divisibility of the same into seventeen parts, etc." This was a remarkable result, of which Gauss was always proud; Gauss published a notice in the *Allgemeine Literatur Zeitung* of April, 1796:

> It is known to every beginner in geometry that various regular polygons, viz., the triangle, tetragon, pentagon, 15-gon, and those which arise by the continued doubling of the number of sides of one of them, are geometrically constructible.
>
> One was already that far in the time of Euclid, and, it seems, it has generally been said since then that the field of elementary geometry extends no farther: at least I know of no successful attempt to extend its limits on this side. ... that besides those regular polygons a number of others, e.g., the 17-gon, allow of a geometrical construction. This discovery is really only a special supplement to a theory of greater inclusiveness, not yet completed, and is to be presented to the public as soon as it has received its completion.

Gauss's work on the constructibility of regular polygons was a byproduct of his work in number theory and algebra. Gauss wrote in a letter of January 6, 1819, to his student, the physicist and mathematician Christian Gerling, that in 1795 he had thought of dividing the complex pth roots of unity, where p was an odd prime, into two groups according as the exponent of $\zeta = e^{2\pi i/p}$ was a quadratic residue or nonresidue modulo p. He considered the sums

$$\eta_0 = \sum_{j=1}^{(p-1)/2} \zeta^{j^2} \quad \text{and} \quad \eta_1 = \sum_{j=1}^{(p-1)/2} \zeta^{lj^2}, \tag{17.1}$$

where l was a nonresidue modulo p. Note that the exponents of ζ in η_0 are quadratic residues and those in η_1 are nonresidues. In 1795, Gauss was able to prove that for $p = 2m+1$,

$$(x-\eta_0)(x-\eta_1) = x^2 + x - (-1)^m (m/2), \quad \text{and also} \tag{17.2}$$
$$4(x^{p-1} + x^{p-2} + \cdots + x + 1) = Y^2 + (-1)^{m+1} p Z^2,$$

where Y and Z were polynomials with integer coefficients. These results, contained in articles 356 and 357 of his *Disquisitiones*, show that in 1795 Gauss had already perceived the connection between cyclotomy, quadratic residues, and quadratic irrationalities. However, Gauss did not fully explain how he moved from these to the construction of regular polygons. In his *Disquisitiones* and in his letter to Gerling, Gauss discussed the solutions of $x^p - 1 = 0$ for $p = 17$; this case provides an illustration of his method of constructing solutions. Gauss took 3 as the primitive root of the congruence $x^{16} \equiv 1 \pmod{17}$ and used it to group the complex roots of the equation $x^{17} - 1 = 0$. We denote a root of this equation by ζ; in particular, we take $\zeta = e^{2\pi i/17}$.

Gauss first divided the sixteen complex roots into two periods of eight terms each:

$$(8,1) = \zeta + \zeta^9 + \zeta^{13} + \zeta^{15} + \zeta^{16} + \zeta^8 + \zeta^4 + \zeta^2,$$
$$(8,3) = \zeta^3 + \zeta^{10} + \zeta^5 + \zeta^{11} + \zeta^{14} + \zeta^7 + \zeta^{12} + \zeta^6.$$

To understand the construction of these periods, consider the powers of 3 and their residues (mod 17):

$$3^0, \ 3^1, \ 3^2, \ 3^3, \ 3^4, \ 3^5, \ 3^6, \ 3^7, \ 3^8, \ 3^9, \ 3^{10}, \ 3^{11}, \ 3^{12}, \ 3^{13}, \ 3^{14}, \ 3^{15};$$
$$1, \ \ 3, \ \ 9, \ \ 10, \ \ 13, \ \ 5, \ \ 15, \ \ 11, \ \ 16, \ \ 14, \ \ 8, \ \ 7, \ \ 4, \ \ 12, \ \ 2, \ \ 6.$$

The exponents of ζ in the period (8,1) are all quadratic residues (mod 17) and consist of residues of $3^0, 3^2, 3^4, \ldots, 3^{14}$. The second set (8,3) consists of exponents three times those in (8,1). Note also that $(8,1) = 2\cos 2\pi/17 + 2\cos 4\pi/17 + 2\cos 8\pi/17 + 2\cos 16\pi/17$ and a similar result holds for (8,3).

Gauss observed that (8,1) and (8,3) were solutions of the quadratic

$$x^2 + x - 4 = 0. \tag{17.3}$$

Note that $(8,1) + (8,3) = -1$, and to verify that $(8,1) \cdot (8,3) = -4$, use $2\cos a \cos b = \cos(a+b)\cos(a-b)$. Gauss then further divided each of the two periods into two further periods of four terms each:

$$(4,1) = \zeta^{3^0} + \zeta^{3^4} + \zeta^{3^8} + \zeta^{3^{12}} = \zeta + \zeta^{13} + \zeta^{16} + \zeta^4 = 2\cos\frac{2\pi}{17} + 2\cos\frac{8\pi}{17},$$

$$(4,9) = \zeta^9 + \zeta^{15} + \zeta^8 + \zeta^2 = 2\cos\frac{4\pi}{17} + 2\cos\frac{16\pi}{17},$$

$$(4,3) = \zeta^3 + \zeta^5 + \zeta^{14} + \zeta^{12} = 2\cos\frac{6\pi}{17} + 2\cos\frac{10\pi}{17},$$

$$(4,10) = \zeta^{10} + \zeta^{11} + \zeta^7 + \zeta^6 = 2\cos\frac{14\pi}{17} + 2\cos\frac{12\pi}{7}.$$

He next observed that

$$(x - (4,1))(x - (4,9)) = x^2 - (8,1)x - 1, \tag{17.4}$$
$$(x - (4,3))(x - (4,10)) = x^2 - (8,3)x - 1. \tag{17.5}$$

Since the values of (8,1) and (8,3) could be obtained from the quadratic (17.3), the values of (4,1), (4,9), and so on could be found from the last two quadratic equations. The periods with four terms could then be used to find the periods with two terms:

$$(2,1) = \zeta + \zeta^{16} = 2\cos\frac{2\pi}{17}, \quad (2,13) = \zeta^{13} + \zeta^4 = 2\cos\frac{8\pi}{17}, \quad \text{etc.}$$

Then he had

$$(x - (2,1))(x - (2,13)) = x^2 - (4,1)x + (4,2).$$

17.5 Gauss: Cyclotomy, Lagrange Resolvents, and Gauss Sums

Since the values of (4,1) and (4,3) were known, he obtained $(2,1) = 2\cos 2\pi/17$, etc. In article 365 of the *Disquisitiones*, Gauss gave

$$\cos\frac{2\pi}{17} = -\frac{1}{16} + \frac{\sqrt{17}}{16} + \frac{\sqrt{(34-2\sqrt{17})}}{16}$$
$$+ \frac{\sqrt{(17+3\sqrt{17} - \sqrt{(34-2\sqrt{17})} - 2\sqrt{(34+2\sqrt{17})})}}{8}.$$

For a general prime $p = 2m+1$, Gauss let g denote a primitive root modulo p, so that $g, g^2, \ldots, g^{p-1}$ (mod p) produced the numbers $1, 2, \ldots, p-1$, though not in this order. In article 55 of the *Disquisitiones*, he gave a proof of the existence of a primitive root, and he used this to define, in general, e periods with f terms where $ef = p-1$. He set

$$(f, g^j) = \zeta^{g^j} + \zeta^{g^{e+j}} + \zeta^{g^{2e+j}} + \cdots + \zeta^{g^{(f-1)e+j}}, \quad 0 \le j \le e-1. \quad (17.6)$$

He proved that the product of any two periods of f terms would be a linear combination of periods of f terms. In fact, the coefficients in the combination were integers. He also showed that if f_1 and f_2 were divisors of $p-1$ and if f_2 divided f_1, then any period with f_2 terms was a root of an equation of degree f_1/f_2 whose coefficients were rational functions of a period with f_1 terms. To prove this last result, Gauss used a particular form of the Lagrange resolvent, later known as a Gauss sum. We define the Lagrange resolvent in the particular case where $f_1 = p-1$ and $f_2 = f$. Letting R be a primitive eth root of unity, the resolvents were given by

$$G_{R^i} = (f, 1) + R^i(f, g) + R^{2i}(f, g^2) + \cdots + R^{(e-1)i}(f, g^{e-1}), \quad (17.7)$$

for $1 \le i \le e-1$. This expression could also be written as

$$G_{R^i} = \zeta + R^i \zeta^g + R^{2i} \zeta^{g^2} + \cdots + R^{(p-2)i} \zeta^{g^{p-2}} \quad (17.8)$$

and in this form it would be called a Gauss sum. When $e = 2$, and $R = -1$, this is the familiar quadratic Gauss sum reducible to

$$\sum_{k=0}^{p-1} e^{2\pi i k^2/p}. \quad (17.9)$$

In article 360, Gauss sketched a proof that G_{R^i} could be obtained as the eth root of a known quantity. Moreover, he obtained the periods (f, g^i) in terms of G_{R^i}. He did not prove that the resolvents G_{R^i} were not zero, but in a posthumously published manuscript, he showed that

$$\frac{G_{R^i} G_{R^j}}{G_{R^{i+j}}} \quad (17.10)$$

was a polynomial in R with integer coefficients. He also showed that the resolvents could not be zero by proving

$$G_{R^i} G_{R^{-i}} = (-1)^i p. \quad (17.11)$$

In 1827, Jacobi independently studied quotients given by (17.10), now called Jacobi sums; he noted the analogy between Jacobi sums and the beta integral. On Jacobi's view, the Gauss sum can be seen as an analog of the gamma function, relation (17.11) as an analog of Euler's reflection formula $\Gamma(1+x)\Gamma(1-x) = \pi x/\sin \pi x$, and (17.10) as an analog of the formula expressing the beta integral in terms of gamma functions.

17.6 Kronecker: Irreducibility of the Cyclotomic Polynomial

In the course of his work on cyclotomy, Gauss demonstrated the irreducibility of the polynomial $X(x) = 1 + x + x^2 + \cdots + x^{p-1}$, where p was prime. He gave a proof of this result in article 341 of the *Disquisitiones* by considering four separate cases. Simpler proofs were found later. The 1846 proof by A. Schoenemann now appears in many textbooks, but is attributed to Eisenstein. A year earlier, in his first paper, Kronecker gave a neat proof, though it does not appear to be well known. Kronecker first proved a lemma: Suppose α is a pth root of unity, $a_0, a_1, \ldots, a_{p-1}$ are integers, and

$$f(\alpha) = a_0 + a_1\alpha + a_2\alpha^2 + \cdots + \alpha^{p-1}.$$

Then $f(\alpha)f(\alpha^2) \cdots f(\alpha^{p-1}) \equiv f(1)^{p-1} \pmod{p}$.

To prove this result, Kronecker considered the product

$$f(x)f(x^2) \cdots f(x^{p-1}) = A_0 + A_1 x + A_2 x^2 + \cdots.$$

He set $x = 1, \alpha, \alpha^2, \ldots, \alpha^{p-1}$ and added the p equations. By using the simple relation

$$A_n(1 + \alpha^n + \alpha^{2n} + \cdots + \alpha^{(p-1)n}) = 0, \quad \text{if} \quad p \nmid n,$$
$$= pA_n, \quad \text{if} \quad p \mid n,$$

he obtained

$$f(1)^{p-1} + f(\alpha)f(\alpha^2) \cdots f(\alpha^{p-1}) + f(\alpha^2)f(\alpha^4) \cdots f(\alpha^{2(p-1)}) + \cdots$$
$$= p(A_0 + A_p + A_{2p} + \cdots).$$

He then observed that for any number r between 1 and $p-1$ inclusive, the set of numbers $\alpha^r, \alpha^{2r}, \ldots, \alpha^{(p-1)r}$ was merely a rearrangement of the set of number $\alpha, \alpha^2, \ldots, \alpha^{(p-1)}$. Thus, the last relation could be rewritten as

$$f(1)^{p-1} + (p-1)f(\alpha)f(\alpha^2) \cdots f(\alpha^{p-1}) = p(A_0 + A_p + A_{2p} + \cdots).$$

This implied the lemma. To deduce the corollary, he assumed the contrary, that X could be factorized: $X = f(x)g(x)$, where $f(x)$ and $g(x)$ had to have integer coefficients by Gauss's theorem on the coefficients of the factors of a monic polynomial with integer coefficients. Setting $x = 1$, he had $X(1) = p = f(1)g(1)$. Since p was prime, he assumed that $f(1) = 1$. Since α^j was a root of $f(x)$ for some j, he had $f(\alpha)f(\alpha^2) \cdots f(\alpha^{p-1}) = 0$. This contradicted the theorem that

$$f(\alpha)f(\alpha^2) \cdots f(\alpha^{p-1}) \equiv f(1)^{p-1} = 1 \pmod{p}.$$

17.7 Exercises

1. Suppose that the monic polynomials $f(x)$ and $g(x)$ have rational coefficients, not all of which are integers. Prove that the coefficients of $f(x)g(x)$ cannot all be integers. See article 42 of Gauss's *Disquisitiones*.
2. Suppose p is prime and C is a number in a field K, but $C^{1/p}$ is not in K. Prove that $x^p - C$ is irreducible over K. See Abel (1965), vol. 2, pp. 217–43.
3. Prove that an irreducible and algebraically solvable equation of odd prime degree with rational coefficients either has all real roots or only one real root. See Kronecker (1968), vol. 4, pp. 25–37.
4. Show that the eighteen complex roots of $x^{19} - 1 = 0$ can be divided into three groups denoted by (6,1), (6,2) and (6,4). Show that the equation, whose roots are these three periods of six terms, is given by

$$x^3 + x^2 - 6x - 7 = 0.$$

Divide the period (6,1) into three periods of two terms each, with the periods denoted by (2,1), (2,7), and (2,8). Show that these periods are roots of

$$x^3 - (6,1)x^2 + ((6,1) + (6,4))x - 2 - (6,2) = 0.$$

Note that $(2,1) = 2\cos(2\pi/19)$, if we choose $\zeta = e^{2\pi i/19}$ to be the initial primitive root of unity. See Gauss (1966), article 353.

5. Solve $x^4 - 4x^3 - 19x^2 + 106x - 120 = 0$ by Descartes's method of solving a quartic and use the solution, as Descartes did, to illustrate the rule of signs. See Descartes (1954), ably translated by Smith and Latham, pp. 159–164.

17.8 Notes on the Literature

See Cardano (1993), translated by Witmer, pp. 8, 9, and 239, for the quotations from Cardano, and see Waring (1991), p. xli for the quote from Waring. For Sylvester's comment on Waring, see Sylvester (1973), vol. 2, p. 381. For Viète's trigonometric solution, see Viète (1983), pp. 174–175 and for Descartes's method of solving a quartic, see Descartes (1954), pp. 180–192. See Eu. I-6, pp. 1–19 for Euler's 1733 paper. Sandifer (2007), on pp. 106–113 comments on significant portions of that paper. For Gauss's work on cyclotomy, see section seven of his *Disquisitiones*, reprinted in Gauss (1863–1927), vol. I. There are French, German, and English translations of this work. See Gauss (1966) for the English. The excerpts of Gauss's notice of his construction of the 17-gon is from Dunnington (2004), p. 28.

See Neumann's (2007a) and (2007b) for perceptive commentary on Vandermonde and Gauss's work on the cyclotomic equation. Neumann's first article appeared in Bradley and Sandifer (2007) and his second appeared in C. Goldstein et al. (2007). These two collections provide many excellent articles on the development of mathematical ideas. For a thorough account of the theory of equations, and the cyclotomic equation in particular, see Tignol (1988). See Edwards (1984) for a more brief discussion. Dunham (1990) has a very entertaining chapter on Cardano.

18

Symmetric Functions

18.1 Preliminary Remarks

The study of symmetric functions originated in the early seventeenth century, as mathematicians took an increasingly sophisticated approach to solving algebraic equations. For example, instead of simply solving or attempting to solve an equation, attention was turned toward the relationships among roots and coefficients. Consider, for example, the cubic

$$(x-\alpha_1)(x-\alpha_2)(x-\alpha_3) = x^3 - c_1 x^2 + c_2 x - c_3.$$

When the product on the left is multiplied out and the coefficients equated, we get

$$\alpha_1 + \alpha_2 + \alpha_3 = c_1, \quad \alpha_1\alpha_2 + \alpha_1\alpha_3 + \alpha_2\alpha_3 = c_2, \quad \alpha_1\alpha_2\alpha_3 = c_3. \tag{18.1}$$

The expressions on the left-hand sides are called the elementary symmetric functions of $\alpha_1, \alpha_2, \alpha_3$. More generally, if there are n symbols $\alpha_1, \alpha_2, \ldots, \alpha_n$, then the elementary symmetric functions are

$$\sum_i \alpha_i, \quad \sum_{i<j} \alpha_i\alpha_j, \quad \sum_{i<j<k} \alpha_i\alpha_j\alpha_k, \quad \ldots, \quad \alpha_1\alpha_2\cdots\alpha_n.$$

A polynomial in $\alpha_1, \alpha_2, \ldots, \alpha_n$ is called a symmetric function if every permutation of $\alpha_1, \alpha_2, \ldots, \alpha_n$ leaves the polynomial invariant. For example, the expression $\alpha_1^2\alpha_2 + \alpha_1\alpha_2^2 + \alpha_1^2\alpha_3 + \alpha_1\alpha_3^2 + \alpha_2^2\alpha_3 + \alpha_2\alpha_3^2$ is symmetric in $\alpha_1, \alpha_2, \alpha_3$. Relation (18.1) reveals the connection between the elementary symmetric functions of the roots of a polynomial and the coefficients of that polynomial. It appears that algebraists of the early seventeenth century were aware of this relationship; as early as the 1540s, the colorful Italian mathematician Girolamo Cardano noted a particular example for the sum of the roots of a cubic. One important question in this area is whether every symmetric function of the roots can be expressed as a polynomial in the coefficients. In his 1629 book, *Invention nouvelle en l'algébre*, Albert Girard (1595–1632) showed that the first four cases of the sums of powers of the roots could be written as polynomials in the coefficients. His notation was close to that of Viéte and lacked the mature algebraic

18.1 Preliminary Remarks

symbolism of Harriot and Descartes. Girard's book defined the elementary symmetric functions:

> When several numbers are proposed, the entire sum may be called the first *faction*; the sum of all the products taken two by two may be called the second *faction*; the sum of all the products taken three by three may be called third *faction*; and always thus to the end, but the product of all numbers is the last *faction*. Now, there are as many *factions* as proposed numbers.

Soon after these definitions, Girard presented a statement of the fundamental theorem of algebra and the connection between the elementary symmetric functions and the coefficients of a polynomial:

> Every algebraic equation except the incomplete ones admits of as many solutions as the denomination of the highest quantity indicates. And the first faction of the solutions is equal to the number of the first mixed quantity, the second faction of them is equal to the number of the second mixed quantity, the third to the third, and so on, so that the last faction is equal to the closing quantity – all this according to the signs that can be noted in the alternating order.

Girard then explained his results on sums of the powers of the roots:

> It might seem to some that the *factions* would be also explicable otherwise than above. That instead of saying the sum, the products two by two, the products three by three, etc., one could say more simply, the sum, the sum of the squares, the sum of the cubes, etc., which however is not so, for when there are several solutions, the sum will be for the first mixed quantity, the sum of the products two by two for the second, etc., as has been sufficiently explicated. But it is not the case for any *factions* of the powers that someone might offer.

Example

Let:

A be the first quantity,

B the second,

C the third,

D the fourth,

etc.

Then, in every type of equation,

A

$A\,\text{sq} - B2$

$A\,\text{cub} - AB3 + C3$

$A\,\text{sq-sq} - A\,\text{sq}\,B4 + AC4 + B\,\text{sq}\,2 - D4$

will be the sum, respectively, of the

solutions

squares

cubes

square-squares

Thus, in modern notation, where $-c_1, c_2, -c_3, c_4$ are the coefficients of the polynomial and $\alpha_1, \alpha_2, \alpha_3, \alpha_4$ are the roots, Girard had

$$\sum \alpha_i = c_1, \quad \sum \alpha_i^2 = c_1^2 - 2c_2, \quad \sum \alpha_i^3 = c_1^3 - 3c_1c_2 + 3c_3,$$

$$\sum \alpha_i^4 = c_1^4 - 4c_1^2 c_2 + 4c_1 c_3 + 2c_2^2 - 4c_4.$$

Girard gave no indication of any motivation for considering the sums of powers of the roots, beyond his expressed purpose of showing the difference between these sums and the sums of the roots taken two, three, etc., at a time.

Newton rediscovered this theorem in 1665–1666. It is unlikely that he had seen it in Girard's book because Newton's knowledge of French was weak. For instance, when Collins asked Newton to comment on van Heuraet's book on optics, Newton excused himself by saying that he could not read French without continual use of a dictionary. His knowledge of algebra came from Oughtred, Viète, and Descartes, whom he read in van Schooten's Latin translation. Newton also found a recurrence relation, not in Girard, for the sums of powers of the roots. He included these and other algebraic results from 1665–66 in his lectures on algebra, given in the 1670s and early 1680s, published in 1707 as *Arithmetica Universalis*. To state Newton's rule for sums of powers of roots, we first note that he used the symbols $-p, q, -r, \ldots$ for coefficients; for example,

$$x^n - px^{n-1} + qx^{n-2} - rx^{n-3} + sx^{n-4} - tx^{n-5} + vx^{n-6} - \cdots = 0.$$

Newton stated his rule in his 1707 *Arithmetica*:

> Let us suppose now, that the known Quantities of the Terms of any Equation under their Signs changed, are p, q, r, s, t, v, &c. viz. that of the second p, that of the third q, of the fourth r, of the fifth s, and so on. And the Signs of the Terms being rightly observed, make $p = a$, $pa + 2q = b$, $pb + qa + 3r = c$, $pc + qb + ra + 4s = d$, $pd + qc + rb + sa + 5t = e$, $pe + qd + rc + sb + ta + 6v = f$, and so on *in infinitum*, observing the Series of the Progression. And a will be the Sum of the Roots, b the Sum of the Squares of each of the Roots, c the Sum of the Cubes, d the Sum of the Biquadrates, e the Sum of the Quadrato-Cubes, f the Sum of the Cubo-Cubes, and so on. As in the Equation $x^4 - x^3 - 19xx + 49x - 30 = 0$, where the known Quantity of the second Term is -1, of the third -19, of the fourth $+49$, of the fifth -30; you must take $1 = p$, $19 = q$, $-49 = r$, $30 = s$. And there will thence arise $a = (p =)1$, $b = (pa + 2q = 1 + 38 =)39$, $c = (pb + qa + 3r = 39 + 19 - 147 =) -89$, $d = (pc + qb + ra + 4s = -89 + 741 - 49 + 120 =)723$. Wherefore the Sum of the Roots will be 1, the Sum of the Squares of the Roots 39, the Sum of the Cubes -89, and the Sum of the Biquadrates 723, viz. the Roots of that Equation are 1, 2, 3, and -5, and the Sum of these $1 + 2 + 3 - 5$ is 1; the Sum of the Squares, $1 + 4 + 9 + 25$, is 39; the Sum of the Cubes, $1 + 8 + 27 - 125$, is -89; and the Sum of the Biquadrates, $1 + 16 + 81 + 625$, is 723.

If we let $\sigma_1, \sigma_2, \sigma_3, \ldots$ denote the elementary symmetric functions of the roots $\alpha_1, \alpha_2, \ldots, \alpha_n$, with the understanding that $\sigma_j = 0$ for $j > n$, and let s_k denote the sum of the kth powers of the roots, then Newton's recurrence rule in modern notation would be

$$s_k - \sigma_1 s_{k-1} + \sigma_2 s_{k-2} - \cdots + (-1)^{k-1} \sigma_{k-1} s_1 + (-1)^k k \sigma_k = 0, \qquad (18.2)$$

for $k = 1, 2, 3, \ldots$. Newton could obtain the formula for s_k by first working out the formulas for $s_1, s_2, \ldots s_{k-1}$. For example, to find s_4 take $k = 1, 2, 3, 4$ successively to get

$$s_1 - \sigma_1 = 0, \quad s_2 - \sigma_1 s_1 + 2\sigma_2 = 0, \quad s_3 - \sigma_1 s_2 + \sigma_2 s_1 - 3\sigma_3 = 0,$$
$$s_4 - \sigma_1 s_3 + \sigma_2 s_2 - \sigma_3 s_1 + 4\sigma_4 = 0. \quad \text{Thus,}$$
$$s_1 = \sigma_1, \quad s_2 = \sigma_1^2 - 2\sigma_2, \quad s_3 = \sigma_1^3 - 3\sigma_1 \sigma_2 + 3\sigma_3, \tag{18.3}$$
$$s_4 = \sigma_1^4 - 4\sigma_1^2 \sigma_2 + 4\sigma_1 \sigma_3 + 2\sigma_2^2 - 4\sigma_4. \tag{18.4}$$

In his notes dating from 1665, Newton explicitly wrote down the formulas up to s_8. He did not give a general proof of this theorem; he generally did not give proofs of his algebraic results. From his examples and remarks, however, it is clear that his insights would most probably have enabled him to give valid proofs.

Newton's notes suggest that he was led to a study of symmetric functions by a problem in elimination theory: When do two polynomials have a common root? In his notes, Newton answered this question by means of symmetric functions and illustrated using two cubics. Newton's result stated that in order for two polynomials to have a common root, their coefficients had to satisfy a certain relation. In notes judged by Whiteside to have been written in early 1665, Newton worked out the details for two cubics $x^3 + bxx + cx + d = 0$ and $fx^3 + gxx + hx + k = 0$. He assumed that r, s, t were the roots of the first polynomial; if one denoted the second polynomial by $p(x)$, then there would be a common root if $p(r)p(s)p(t) = 0$. The product produced symmetric polynomials in r, s, t, and Newton computed these in terms of b, c, d. For example, in his notes Newton denoted the expression $r^2s + rs^2 + s^2t + st^2 + t^2r + r^2t$ as "every rrs" and found its value to be $bc - 3d$. His final expression for $p(r)p(s)p(t)$ had thirty-four terms. In modern terminology, this expression is called the resultant of the two polynomials, and setting it equal to zero gives the condition for the polynomials to have a common root. It appears that Newton worked out sufficiently many examples to convince himself that any symmetric polynomial of the roots of an equation could be expressed as a polynomial in the coefficients (elementary symmetric functions of the roots) of the equation. Naturally, he gave no suggestion of a general proof. In his 1762 *Meditationes Algebraicae*, Edward Waring presented the first known proof. The result is of such fundamental importance that later researchers, including Gauss and Cauchy, searched for and found new and different proofs.

Most probably in the 1720s, Colin Maclaurin composed a treatise on algebra, from which he lectured at the University of Edinburgh. This work offered more or less complete proofs of several of Newton's algebraic theorems, including a full proof of the recursion formula for the sums of powers of the roots of an equation. In a paper presented to the Berlin Academy in 1747, Euler provided two proofs of this formula. The first, also independently obtained by John Landen, employed logarithmic differentiation of the polynomial and then expansion into an infinite series. The first part of Euler's second proof was the same as Maclaurin's, but the second part was new and was called elegant and exact by Kronecker.

After Maclaurin and Euler, the English mathematician Edward Waring (c. 1736–1798) took major steps in the study of symmetric functions. In the first chapter of his 1762 book, he gave several proofs of Newton's basic result on symmetric functions and presented an explicit formula expressing a monomial symmetric function in terms of elementary symmetric functions. Note that to construct a monomial symmetric function in n variables, we start with a monomial in those n variables and add to it all monomials obtained by permutation. Since any symmetric function is obviously a sum of monomials, Newton's result follows. Waring also explained how to express the elementary symmetric functions in terms of the sums of powers symmetric functions. In effect, he showed that the set of the monomial symmetric functions, or the sums of powers symmetric functions, and the elementary symmetric functions each formed a basis for the set of all symmetric functions. Waring was not interested in general, abstract, or existence results. For him, it was not sufficient to know, for example, that any sums of powers symmetric function could be expressed as a polynomial in the elementary functions. Even an algorithm for producing such a polynomial was not enough for Waring. He demanded a specific formula. Thus, the first theorem in his *Meditationes Algebraicae* stated:

$$s_m = m \sum (-1)^{m+k_1+k_2+\cdots+k_s} \cdot \frac{(k_1+k_2+\cdots+k_s-1)!}{k_1!k_2!\cdots k_s!} \sigma_1^{k_1} \sigma_2^{k_2} \cdots \sigma_s^{k_s},$$

where s_m was the sum of the mth powers of the roots of an equation of degree n; , σ_1, $\sigma_2, \ldots, \sigma_n$ were the elementary symmetric functions of the roots; and $k_1 + 2k_2 + \cdots + sk_s = m$. The number of terms in the sum was therefore related to the number of partitions of m. Waring's notation was not well-developed, to say the least, so he explicitly wrote several terms to convey the idea of the progression of the terms. His proof of this formula used induction, combined with Newton's recursive relation for sums of powers. This meant that he had to first guess the formula. This task was not too difficult; to see this, note that the degree of the elementary symmetric function σ_j is j and hence the summation has to be over $(k_1, k_2, \ldots, k_s)$ such that $k_1 + 2k_2 + \cdots + sk_s = m$, and the formula is easy to check for small values of m, say up to 20. Then again, the multinomial theorem suggests that it is only natural to consider, as the coefficient for $\sigma_1^{k_1}, \sigma_2^{k_2}, \ldots \sigma_s^{k_s}$, the expression $(k_1+k_2+\cdots+k_s)!/k_1!k_2!\cdots k_s!$. Moreover, the first few cases of s_m indicate that the numerator should be divided by $k_1+k_2+\cdots+k_s$. Perhaps Waring reasoned in a similar way. Also, note that the multinomial theorem was already well known in Waring's time. Leibniz had found it around 1680 but did not publish the result. Jakob, Johann, and Niklaus I Bernoulli independently rediscovered the formula, as did de Moivre. In 1713, the French probabilist Pierre Raymond de Montmort became the first to publish the result, having independently discovered it.

Waring's desire for explicit formulas often made it difficult for him to state his results in an attractive manner. Consider the problem of expressing a monomial symmetric function as a polynomial in the sums of powers symmetric functions. An inductive argument shows that this can always be done. But this result was apparently not enough for Waring. It took him two pages to describe his explicit formula. Over a century later, Percy Alexander MacMahon's (1854–1929) new concepts and notation allowed him to write Waring's formulas and his own deeper and more general results in much more

18.2 Euler's Proofs of Newton's Rule

succinct form. In 1971, Peter Doubilet showed that several classical formulas, including Waring's, could be written very briefly by employing Rota's concept of Möbius inversion on a partially ordered set. Thus, Waring's quest for explicit formulas has turned out to have mathematical relevance.

18.2 Euler's Proofs of Newton's Rule

In modern notation, Euler's proof of Newton's recursive relation (18.2) started off by setting

$$f(x) = (x - \alpha_1)(x - \alpha_2) \cdots (x - \alpha_n) = x^n - \sigma_1 x^{n-1} + \sigma_2 x^{n-2} - \cdots + (-1)^n \sigma_n.$$

Taking the logarithmic derivative, he obtained

$$\frac{f'(x)}{f(x)} = \sum_{k=1}^{n} \frac{1}{x - \alpha_k} = \sum_{k=1}^{n} \left(\frac{1}{x} + \frac{\alpha_k}{x^2} + \frac{\alpha_k^2}{x^3} + \frac{\alpha_k^3}{x^4} + \cdots \right)$$

$$= \frac{n}{x} + \frac{s_1}{x^2} + \frac{s_2}{x^3} + \frac{s_3}{x^4} + \cdots,$$

where $s_i = \sum_{k=1}^{n} \alpha_k^i$. After multiplying both sides by $f(x)$, Euler could write

$$nx^{n-1} - (n-1)\sigma_1 x^{n-2} + (n-2)\sigma_2 x^{n-3} + \cdots + (-1)^{n-1} \sigma_{n-1}$$
$$= \left(x^n - \sigma_1 x^{n-1} + \sigma_2 x^{n-2} + \cdots + (-1)^n \sigma_n \right) \left(\frac{n}{x} + \frac{s_1}{x^2} + \frac{s_2}{x^3} + \frac{s_3}{x^4} + \cdots \right). \quad (18.5)$$

Equating coefficients gave him the required result:

$$-(n-1)\sigma_1 = s_1 - n\sigma_1, \quad (n-2)\sigma_2 = s_2 - \sigma_1 s_1 + n\sigma_2,$$
$$-(n-3)\sigma_3 = s_3 - \sigma_1 s_2 + \sigma_2 s_1 - n\sigma_3,$$
$$(n-4)\sigma_4 = s_4 - \sigma_1 s_3 + s\sigma_2 s_2 - \sigma_3 s_1 + n\sigma_4, \ldots.$$

Euler gave a second proof, here presented in his own notation. He started with

$$x^n - Ax^{n-1} + Bx^{n-2} - Cx^{n-3} + Dx^{n-4} - Ex^{n-5} + \cdots \pm N = 0.$$

He assumed the roots were $\alpha, \beta, \gamma, \delta, \ldots, \nu$, substituted these values for x, and added the equations to get

$$\int \alpha^n = A \int \alpha^{n-1} - B \int \alpha^{n-2} + C \int \alpha^{n-3} - \cdots \pm M \int \alpha \mp nN.$$

Next he multiplied the original equation by x^m and again set $x = \alpha, \beta, \ldots, \nu$ and added to obtain

$$\int \alpha^{n+m} = A \int \alpha^{n+m-1} - B \int \alpha^{n+m-2} + \cdots \pm M \int \alpha^{m+1} \mp N \int \alpha^m, \quad (18.6)$$

where $\int \alpha^k = \alpha^k + \beta^k + \gamma^k + \delta^k + \cdots + \nu^k$. This proved Newton's theorem for sums of powers $\geq n$. For powers less than n, Euler illustrated his method, praised by Kronecker as elegant, taking an equation of degree five, $x^5 - Ax^4 + Bx^3 - Cx^2 + Dx - E = 0$, and the related equations of lower degree

$$x - A = 0,$$
$$x^2 - Ax + B = 0,$$
$$x^3 - Ax^2 + Bx - C = 0,$$
$$x^4 - Ax^3 + Bx^2 - Cx + D = 0.$$

He denoted generic roots of these five equations by α, p, q, r, s respectively. Thus, he had

$$\int \alpha = \int s = \int r = \int q = \int p,$$
$$\int \alpha^2 = \int s^2 = \int r^2 = \int q^2,$$
$$\int \alpha^3 = \int s^3 = \int r^3,$$
$$\int \alpha^4 = \int s^4.$$

Next, by (18.6), he completed the proof by obtaining the result

$$\int p = A, \quad \int q^2 = A \int q - 2B, \quad \int r^3 = A \int r^2 - B \int r + 3C,$$
$$\int s^4 = A \int s^3 - B \int s^2 + C \int s - 4D.$$

18.3 Maclaurin's Proof of Newton's Rule

As Euler did in his second proof, Maclaurin broke up his proof into two cases: When the power of the roots was at least n (the degree of the equation) and when it was less than n. His proof of the first case was identical to Euler's but the proof of the second case was different and instructive. It is interesting to read this in Maclaurin's notation and language, although we have condensed his ten-page argument. Maclaurin stated the lemma:

> That if A is the Coefficient of one Dimension, or the Coefficient of the second Term, in an Equation, G any other Coefficient, H the Coefficient next after it; the Difference of the Dimensions of G and A being $r - 2$: if likewise $A' \times G'$ represent the Sum of all those Terms of the Product $A \times G$ in which the Square of any Root, as a^2, or b^2, or c^2, &c. is found; then will $A' \times G' = AG - rH$.

We observe that since H is the rth term, the right-hand side of the last equation is $\sigma_1 \sigma_{r-1} - r \sigma_r$ and $A' \times G'$ consists of all terms in $\sigma_1 \sigma_{r-1}$ containing a square of a root.

Maclaurin explained this by means of an example, but his argument works in general. Thus, if we write $\sigma_1 \sigma_{r-1} = (\alpha_1 + \alpha_2 + \alpha_3 + \cdots + \alpha_n)(\alpha_1 \alpha_2 \cdots \alpha_{r-1} + \cdots)$, then a term such as $\alpha_1 \alpha_2 \cdots \alpha_r$ occurs r times: $\alpha_1 \times \alpha_2 \cdots \alpha_r, \alpha_2 \times \alpha_1 \alpha_3 \cdots \alpha_r, \ldots, \alpha_r \times \alpha_1 \alpha_2 \cdots \alpha_{r-1}$. Note that the other terms contain squares of roots. This proves the lemma.

Maclaurin then broke each coefficient $G = \sigma_{r-1}$ into two parts. The first part contained every term with a specific root (say a) in it and the second part contained every term without an a. He denoted the first part by $G^{(+a)}$ and the second part by $G^{(-a)}$. It followed that for roots $a, b, \ldots$

$$G = G^{(+a)} + G^{(-a)} \qquad H^{(+a)} = aG^{(-a)}, \qquad (18.7)$$

$$G = G^{(+b)} + G^{(-b)} \qquad H^{(+b)} = bG^{(-b)},$$

and so on. Maclaurin considered the equation

$$x^n - Ax^{n-1} + Bx^{n-2} - Cx^{n-3} + \cdots = 0,$$

with roots $a, b, c, \ldots$ and wrote it in the form

$$x^r - Ax^{r-1} + Bx^{r-2} - Cx^{r-3} \cdots + Gx - H + \frac{I}{x} - \frac{K}{x^2} + \frac{L}{x^3} \cdots + \frac{M}{x^{n-r}} = 0. \quad (18.8)$$

He observed that from relations (18.7)

$$Ga = aG^{(+a)} + aG^{(-a)}$$

$$-H = \qquad -aG^{(-a)} - H^{(-a)}$$

$$\frac{I}{a} = \qquad +H^{(-a)} + \frac{I^{(-a)}}{a}$$

$$-\frac{K}{a^2} = \qquad -\frac{I^{(-a)}}{a} - \frac{K^{(-a)}}{a^2}$$

$$\frac{L}{a^3} = \qquad -\frac{K^{(-a)}}{a^2} + \frac{L^{(-a)}}{a^3} \qquad \text{etc.}$$

Upon adding these equations he got

$$Ga - H + \frac{I}{a} - \frac{K}{a^2} + \frac{L}{a^3} - \cdots \pm \frac{M}{a^{n-r}} = aG^{(-a)}.$$

When similar relations for the other roots $b, c, d, \ldots$ were added, he summed the right-hand side to be

$$aG^{(+a)} + bG^{(+b)} + cG^{(+c)} + \cdots = A' \times G' = AG - rH.$$

Next, Maclaurin put $x = a, b, c, \ldots$ in (18.8) and added to get

$$(a^r + b^r + c^r + \cdots) - A(a^{r-1} + b^{r-1} + \cdots) + B(a^{r-2} + b^{r-2} + \cdots) + \cdots + AG - rH = 0.$$

He viewed the last A as the sum of the first powers of the roots; this proved Newton's formula.

18.4 Waring's Power Sum Formula

Waring's formula

$$s_m = m \sum (-1)^{m+k_1+k_2+\cdots+k_s} \frac{(k_1+k_2+\cdots+k_s-1)!}{k_1! k_2! \cdots k_s!} \sigma_1^{k_1} \sigma_2^{k_2} \cdots \sigma_s^{k_s},$$

where the sum is over all partitions of m, that is, all $k_1, \ldots, k_s$ such that $k_1 + 2k_2 + \cdots + sk_s = m$, is obviously true for $m = 1$. To prove the inductive step, Waring applied Newton's formula

$$s_m = \sigma_1 s_{m-1} - \sigma_2 s_{m-2} + \sigma_3 s_{m-3} + \cdots.$$

Suppose Waring's formula is true up to $m-1$. To obtain the coefficient of $\sigma_1^{k_1} \sigma_2^{k_2} \cdots \sigma_s^{k_s}$ in s_m by Newton's formula, we add up the coefficients of

$$\sigma_1^{k_1-1} \sigma_2^{k_2} \cdots \sigma_s^{k_s}, \quad \sigma_1^{k_1} \sigma_2^{k_2-1} \cdots \sigma_s^{k_s}, \quad \text{etc.},$$

in s_{m-1}, s_{m-2}, etc., respectively. By Waring's formula, these coefficients are

$$(m-1)k_1 \frac{(k_1+k_2+\cdots+k_s-2)!}{k_1! k_2! \cdots k_s!}, \quad (m-2)k_2 \frac{(k_1+k_2+\cdots+k_s-2)!}{k_1! k_2! \cdots k_s!}, \text{ etc.}$$

When added, the sum is

$$\frac{(k_1+k_2+\cdots+k_s-2)!}{k_1! k_2! \cdots k_s!} \left((m-1)k_1 + (m-2)k_2 + \cdots + (m-s)k_s \right).$$

The expression in parentheses can be written as $m(k_1 + k_2 + \cdots + k_{s-1})$, because $k_1 + 2k_2 + \cdots + sk_s = m$. This gives us the required result. Note that the signs come out correctly because of the alternating signs in Newton's result.

18.5 Gauss's Fundamental Theorem of Symmetric Functions

Gauss's elegant demonstration of this theorem has now become one of the standard proofs given in textbooks. This proof was published in an 1816 paper on an alternative proof of the fundamental theorem of algebra, since Gauss gave his first proof in his Ph.D. thesis. Although his proof of the theorem on symmetric functions is very brief even in the original, Gauss's notation does not include subscripts. We summarize using subscripts. When terms are written in lexicographic order, a term $x_1^{\alpha_1} x_2^{\alpha_2} x_3^{\alpha_3} \cdots$ is of higher-order than $x_1^{\beta_1} x_2^{\beta_2} x_3^{\beta_3} \cdots$ when the first of the nonvanishing differences $\alpha_1 - \beta_1, \alpha_2 - \beta_2, \alpha_3 - \beta_3, \ldots$ is positive. This implies that the highest order terms of the elementary symmetric functions $\sigma_1, \sigma_2, \sigma_3, \ldots$ are, respectively, $x_1, x_1 x_2, x_1 x_2 x_3, \ldots$. So the highest-order term of $\sigma_1^{\alpha_1} \sigma_2^{\alpha_2} \sigma_3^{\alpha_3} \cdots$ would be

$$x_1^{\alpha_1+\alpha_2+\alpha_3+\cdots} x_2^{\alpha_2+\alpha_3+\cdots} x_3^{\alpha_3+\cdots} \cdots.$$

Now observe that the two expressions $\sigma_1^{\alpha_1}\sigma_2^{\alpha_2}\sigma_3^{\alpha_3}\cdots$ and $\sigma_1^{\beta_1}\sigma_2^{\beta_2}\sigma_3^{\beta_3}\cdots$ have the same highest-order term if and only if

$$\alpha_1 + \alpha_2 + \alpha_3 + \cdots = \beta_1 + \beta_2 + \beta_3 + \cdots,$$
$$\alpha_2 + \alpha_3 + \cdots = \beta_2 + \beta_3 + \cdots,$$
$$\alpha_3 + \cdots = \beta_3 + \cdots.$$

Thus, they have the same highest-order term if and only if

$$\alpha_1 = \beta_1, \alpha_2 = \beta_2, \alpha_3 = \beta_3, \ldots.$$

Now let $P(x_1, x_2, \ldots, x_n)$ be a symmetric polynomial with highest-order term $x_1^{\alpha_1}x_2^{\alpha_2}x_3^{\alpha_3}\cdots$. By symmetry, $x_1^{\alpha_2}x_2^{\alpha_1}x_3^{\alpha_3}\cdots$ is also a term in P and so we have $\alpha_1 \geq \alpha_2$. Continuation of this argument gives $\alpha_1 \geq \alpha_2 \geq \alpha_3 \geq \cdots$. Observe that the expression

$$\sigma_1^{\alpha_1-\alpha_2}\sigma_2^{\alpha_2-\alpha_3}\sigma_3^{\alpha_3-\alpha_4}\cdots\sigma_n^{\alpha_n} \text{ has } x_1^{\alpha_1}x_2^{\alpha_2}x_3^{\alpha_3}\cdots$$

as the highest-order term. Therefore,

$$P(x_1, x_2, \ldots, x_n) - \sigma_1^{\alpha_1-\alpha_2}\sigma_2^{\alpha_2-\alpha_3}\sigma_3^{\alpha_3-\alpha_4}\cdots\sigma_n^{\alpha_n}$$

is a symmetric polynomial with highest term of a lower order than P. A repeated application of this process shows that $P(x_1, x_2, \ldots, x_n)$ is a polynomial in the elementary symmetric functions. This completes our summary of Gauss's proof.

18.6 Cauchy: Fundamental Theorem of Symmetric Functions

Cauchy's proof appeared in his 1829 *Exercices de mathématiques* and utilized induction on the number of variables. He wrote at some length, looking at particular cases and giving examples. As in several of his proofs of algebraic results, his argument finally depended on the solution of a system of linear equations involving a Vandermonde determinant. In his 1895 algebra book, Heinrich Weber presented a succinct proof, based on Cauchy's paper. Weber began by supposing that $\alpha_1, \alpha_2, \ldots, \alpha_n$ were the roots of the equation

$$f(x) = x^n + a_1 x^{n-1} + a_2 x^{n-2} + \cdots + a_n = 0,$$

and that $S = S(\alpha_1, \alpha_2, \ldots, \alpha_n)$ was a symmetric polynomial in $\alpha_1, \alpha_2, \ldots, \alpha_n$. In order to apply induction, he expressed S in the form

$$S = S_0 \alpha_1^\mu + S_1 \alpha_1^{\mu-1} + \cdots + S_{\mu-1}\alpha_1 + S_\mu,$$

where $S_0, S_1, \ldots, S_\mu$ were symmetric polynomials in $\alpha_2, \alpha_3, \ldots, \alpha_n$. Since S was symmetric, α_1 could be replaced by α_2, and the new coefficients would be the same polynomials $S_0, S_1, \ldots, S_\mu$, but in the variables $\alpha_1, \alpha_3, \ldots, \alpha_n$. The same argument applied to $\alpha_3, \alpha_4, \ldots, \alpha_n$. Next, from the equation

$$x^n + a_1 x^{n-1} + \cdots + a_n = (x - \alpha_1)\left(x^{n-1} + a_1' x^{n-2} + \cdots + a_{n-1}'\right),$$

Weber obtained the relations

$$a'_1 = \alpha_1 + a_1,\ a'_2 = \alpha_1^2 + a_1\alpha_1 + a_2,\ a'_3 = \alpha_1^3 + a_1\alpha_1^2 + a_2\alpha_1 + a_3, \ldots.$$

Continuing the process of induction, Weber assumed the result true for $n-1$ or fewer variables. Since $a'_1, a'_2, \ldots, a'_{n-1}$ were elementary symmetric functions of $\alpha_2, \alpha_3, \ldots, \alpha_n$, the functions $S_0, S_1, \ldots, S_\mu$ had to be polynomials in $a'_1, a'_2, \ldots, a'_{n-1}$. He substituted the value of a'_i into the expression for S to obtain

$$S = A_0\alpha_1^m + A_1\alpha_1^{m-1} + \cdots + A_{m-1}\alpha_1 + A_m,$$

where $A_0, A_1, \ldots, A_m$ were polynomials in $a_1, a_2, \ldots, a_n$. Weber then set

$$\Phi(x) = A_0 x^m + A_1 x^{m-1} + \cdots + A_{m-1}x + A_m,$$

and, by an application of the division algorithm, got $\Phi(x) = Q(x)f(x) + \Psi(x)$, where $\deg \Psi < \deg f = n$, and the coefficients of $\Psi(x)$ were polynomials in the elementary symmetric functions $a_1, a_2, \ldots, a_n$. He then supposed

$$\Psi(x) = C_0 x^{n-1} + C_1 x^{n-2} + \cdots + C_{n-2}x + C_{n-1}.$$

Clearly,

$$S = \Phi(\alpha_1) = \Psi(\alpha_1); \text{ by symmetry } S = \Psi(\alpha_2) = \Psi(\alpha_3) = \cdots = \Psi(\alpha_n).$$

So Weber had a system of equations in the unknowns $C_0, C_1, \ldots, C_{n-1}$:

$$C_0\alpha_1^{n-1} + C_1\alpha_1^{n-2} + \cdots + C_{n-2}\alpha_1 + (C_{n-1} - S) = 0,$$
$$C_0\alpha_2^{n-1} + C_2\alpha_2^{n-2} + \cdots + C_{n-2}\alpha_2 + (C_{n-1} - S) = 0,$$
$$\ldots\ldots$$
$$C_0\alpha_n^{n-1} + C_2\alpha_n^{n-2} + \cdots + C_{n-2}\alpha_n + (C_{n-1} - S) = 0.$$

Since the Vandermonde determinant was not zero, Weber could conclude that $C_0 = 0, C_1 = 0, \ldots, C_{n-2} = 0$, and $C_{n-1} = S$. Since C_{n-1} was a polynomial in $a_1, a_2, \ldots, a_n$, the result followed.

18.7 Cauchy: Elementary Symmetric Functions as Rational Functions of Odd Power Sums

In an 1837 paper on the resolution of algebraic equations, Cauchy derived the very interesting result that the elementary symmetric functions $\sigma_1, \sigma_2, \sigma_3, \ldots$ could be expressed as rational functions of the odd power sum symmetric functions $s_1, s_3, s_5, \ldots$. Laguerre rediscovered this result in 1878, and it is often named for him. We present Cauchy's argument with a slight change in notation. Let

$$f(z) = 1 - \sigma_1 z + \sigma_2 z^2 - \cdots + (-1)^m \sigma_m z^m = (1-\alpha_1 z)(1-\alpha_2 z)\cdots(1-\alpha_m z),$$

$$f(-z) = 1 + \sigma_1 z + \sigma_2 z^2 + \cdots + \sigma_m z^m = (1+\alpha_1 z)(1+\alpha_2 z)\cdots(1+\alpha_m z),$$

and $f(z)/f(-z) = e^{-2zT}$. Then

$$T = -\frac{1}{2z} \ln\left(\frac{1-\alpha_1 z}{1+\alpha_1 z} \cdot \frac{1-\alpha_2 z}{1+\alpha_2 z} \cdots \frac{1-\alpha_m z}{1+\alpha_m z}\right) = s_1 + \frac{s_3}{3} z^2 + \frac{s_5}{5} z^4 + \frac{s_7}{7} z^6 + \text{etc.}$$

and

$$\frac{\sigma_1 + \sigma_3 z^2 + \sigma_5 z^4 + \cdots}{1 + \sigma_2 z^2 + \sigma_4 z^3 + \cdots} = \frac{1}{z} \frac{e^{zT} - e^{-zT}}{e^{zT} + e^{-zT}}$$

$$= \frac{1}{6} \cdot \frac{2^2-1}{1\cdot 2} 2^2 T - \frac{1}{30} \frac{2^4-1}{1\cdot 2\cdot 3\cdot 4} 2^4 z^2 T^3 + \frac{1}{42} \frac{2^6-1}{1\cdot 2\cdot 3\cdot 4\cdot 5\cdot 6} 2^6 z^4 T^5 - \text{etc.}$$

$$= A_1 + A_2 z^2 + A_3 z^4 + A_4 z^6 + \text{etc,} \qquad (18.9)$$

where $1/6, -1/30, 1/42, \ldots$ are Bernoulli numbers and $A_1, A_2, \ldots$ are given by

$$A_1 = s_1, \quad A_2 = (s_3 - s_1^3)/3, \quad A_3 = s_5/5 - s_1^2 s_3/3 + 2s_1^5/15, \ldots . \qquad (18.10)$$

We note, with Cauchy, that in each of the polynomials $A_2, A_3, A_4 \ldots$, the sum of the coefficients is zero. After multiplying equation (18.9) by the denominator of the left-hand side and equating coefficients of the powers of z, we get

$$\sigma_1 = A_1, \quad \sigma_3 = \sigma_2 A_1 + A_2, \quad \sigma_5 = \sigma_4 A_1 + \sigma_2 A_2 + A_3, \ldots .$$

By solving these equations for $m = 4$, for example, we arrive at the required result that

$$\sigma_1 = A_1, \quad \sigma_2 = \frac{A_1 A_4 - A_2 A_3}{A_2^2 - A_1 A_3}, \quad \sigma_3 = A_1 \frac{A_1 A_4 - A_2 A_3}{A_2^2 - A_1 A_3} + A_2, \quad \sigma_4 = \frac{A_3^2 - A_2 A_4}{A_2^2 - A_1 A_3}. \qquad (18.11)$$

Note that by (18.10) $A_1, A_2, A_3, \ldots$ are polynomials in $s_1, s_3, s_5, \ldots$. Cauchy also observed that if the roots $\alpha_1, \alpha_2, \alpha_3, \ldots$ were numbers instead of literal symbols, then it could happen that the coefficients of the equation given by $-\sigma_1, \sigma_2, -\sigma_3, \ldots$ could not be determined. For example, if the roots of a fourth-degree polynomial were such that $s_1 = s_3 = s_5 = s_7 = 1$, then $A_1 = 1, A_2 = 0, A_3 = 0, A_4 = 0$ and $\sigma_1 = 1, \sigma_3 = \sigma_2$, $\sigma_4 = 0$. This implied that, although the coefficients of x^2 and x^3 were the same, they were undetermined, and the quartic reduced to $(x^2 + \sigma_2)(x^2 - x) = 0$. Cauchy went on to give a general analysis of this indeterminacy.

18.8 Laguerre and Pólya on Symmetric Functions

Edmond Laguerre (1834–1886) wrote many papers on elementary topics in algebra and analysis. It may have appeared that the study of these areas had been exhausted, but Laguerre's work contained many novel and intriguing ideas. His studies included Descartes's rule of signs, approximate roots of equations, and determining real roots of algebraic and transcendental equations. Henri Poincaré, one of the editors of the works of Laguerre, commented that Laguerre's work in algebraic equations was his most remarkable, and that Laguerre had the ability to harvest new revelations in a field from which it was thought that no more could be gleaned. For example, in an 1877 paper, he proved that the coefficients of an algebraic equation of degree n could be expressed as rational functions of the first n sums of odd-power symmetric functions. In 1837 Cauchy had anticipated this result, though Laguerre was unaware of this. But, as George Pólya showed, Laguerre's method of proof leads to interesting generalizations. To state Pólya's generalization, suppose

$$f(x) = (x-\alpha_1)(x-\alpha_2)\cdots(x-\alpha_n) = x^n + a_1 x^{n-1} + a_2 x^{n-2} + \cdots + a_n,$$
$$g(x) = (x-\beta_1)(x-\beta_2)\cdots(x-\beta_n) = x^n + b_1 x^{n-1} + b_2 x^{n-2} + \cdots + b_n,$$
$$s_k = \alpha_1^k + \alpha_2^k + \cdots + \alpha_n^k, \quad t_k = \beta_1^k + \beta_2^k + \cdots + \beta_n^k,$$
$$s_k - t_k = u_k, \text{ and } R = \prod_{i=1}^{n}\prod_{j=1}^{n}(\alpha_i - \beta_j).$$

Laguerre's theorem stated that the elementary symmetric functions

$$-a_1, a_2, -a_3, \ldots, (-1)^n a_n$$

were rational functions of $s_1, s_3, \ldots, s_{2n-1}$. Pólya's generalization was that the $2n+1$ functions

$$R, Ra_1, Ra_2, \ldots, Ra_n; \quad Rb_1, Rb_2, \ldots, Rb_n$$

were polynomials in $u_1, u_2, \ldots, u_{2n}$ and that the expression for R did not contain u_{2n}. To obtain Laguerre's result, Pólya took $\beta_i = -\alpha_i$ in his theorem. In this special case, $u_{2m-1} = 2s_{2m-1}$ and $u_{2m} = 0$ for $m = 1, 2, 3, \ldots$, so that the result followed. Following Laguerre's basic idea, Pólya computed

$$\ln \frac{g(x)}{f(x)} = \sum_{i=1}^{n} \ln \frac{x-\beta_i}{x-\alpha_i} = \sum_{i=1}^{n} \left(\frac{\alpha_i - \beta_i}{x} + \frac{\alpha_i^2 - \beta_i^2}{2x^2} + \frac{\alpha_i^3 - \beta_i^3}{3x^3} + \cdots \right)$$
$$= \frac{s_1 - t_1}{x} + \frac{s_2 - t_2}{2x^2} + \frac{s_3 - t_3}{3x^3} + \cdots$$
$$= \frac{u_1}{x} + \frac{u_2}{2x^2} + \frac{u_3}{3x^3} + \cdots.$$

He then let

$$U(x) = \frac{u_1}{x} + \frac{u_2}{2x^2} + \cdots + \frac{u_{2n}}{2nx^{2n}},$$

and, employing Laguerre's notation, he let $((x^{-m}))$ denote a series of the form

$$\frac{c_m}{x^m} + \frac{c_{m+1}}{x^{m+1}} + \frac{c_{m+2}}{x^{m+2}} + \cdots.$$

18.8 Laguerre and Pólya on Symmetric Functions

Thus, he could write

$$\ln \frac{g(x)}{f(x)} = U(x) + ((x^{-2n-1})), \text{ or } \frac{g(x)}{f(x)} = e^{U(x)} + ((x^{-2n-1})),$$

$$\text{or } f(x)e^{U(x)} = g(x) + ((x^{-n-1})).$$

He next set

$$e^{U(x)} = 1 + \frac{v_1}{x} + \frac{v_2}{x^2} + \frac{v_3}{x^3} + \cdots = 1 + U(x) + \frac{1}{2!}U^2(x) + \frac{1}{3!}U^3(x) + \cdots$$

$$= 1 + \left(\frac{u_1}{x} + \frac{u_2}{2x^2} + \cdots\right) + \frac{1}{2!}\left(\frac{u_1}{x} + \frac{u_2}{2x^2} + \cdots\right)^2 + \cdots.$$

Equating the coefficients of $1/x, 1/x^2, 1/x^3, \ldots$, he got

$$v_1 = u_1, \quad 2v_2 = u_1^2 + u_2,$$
$$6v_3 = u_1^3 + 2u_3 + 3u_1 u_2,$$
$$24v_4 = u_1^4 + 3u_2^2 + 8u_1 u_3 + 6u_1^2 u_2 + 6u_4, \quad \ldots.$$

These relations showed that v_n could be expressed as a polynomial in $u_1, u_2, \ldots, u_n$ for $n = 1, 2, 3, \ldots$. Then, by $f(x)e^{U(x)} = g(x) + ((x^{-n-1}))$, he could write

$$(x^n + a_1 x^{n-1} + a_2 x^{n-2} + \cdots + a_n)\left(1 + \frac{v_1}{x} + \frac{v_2}{x^2} + \cdots + \frac{v_n}{x^n} + \frac{v_{n+1}}{x^{n+1}} + \cdots\right)$$
$$= g(x) + ((x^{-n-1})).$$

This implied that the coefficients of $1/x, 1/x^2, \ldots, 1/x^n$ on the left-hand side all had to vanish, and he got the linear relations in $a_1, a_2, \ldots, a_n$:

$$v_1 a_n + v_2 a_{n-1} + v_3 a_{n-2} + \cdots + v_n a_1 = -v_{n+1},$$
$$v_2 a_n + v_3 a_{n-1} + v_4 a_{n-2} + \cdots + v_{n+1} a_1 = -v_{n+2},$$
$$\ldots\ldots$$
$$v_n a_n + v_{n+1} a_{n-1} + v_{n+2} a_{n-2} + \cdots + v_{2n-1} a_1 = -v_{2n}.$$

Pólya noted that the determinant of this linear system

$$V_n = \begin{vmatrix} v_1 & v_2 & \ldots & v_n \\ v_2 & v_3 & \ldots & v_{n+1} \\ \ldots & \ldots & \ldots & \ldots \\ v_n & v_{n+1} & \ldots & v_{2n-1} \end{vmatrix}$$

was a homogeneous polynomial of degree n^2 in $\alpha_1, \alpha_2, \ldots, \alpha_n; \beta_1, \beta_2, \ldots, \beta_n$ and was also symmetric in $\alpha_1, \alpha_2, \ldots, \alpha_n$ as well as in $\beta_1, \beta_2, \ldots, \beta_n$. Pólya proceeded to show that $V_n = (-1)^{n(n-1)/2} R$. In order to prove this, he supposed $\alpha_1 = \beta_1 = \lambda$. He argued that the variations of λ did not alter the values of $u_1, u_2, \ldots, u_{2n}$ or, therefore, the values $v_1, v_2, \ldots, v_{2n}$. However, some $a_1, a_2, \ldots, a_n$ would vary with λ, so $V_n = 0$ when $\alpha_1 = \beta_1$. This implied that $\alpha_1 - \beta_1$ was a factor of V_n and, by symmetry, $\alpha_i - \beta_j$, $i = 1, \ldots, n$ and $j = 1, 2, \ldots, n$, were all factors of V_n. This in turn meant that R divided V_n, and they both had the same degrees n^2 in α_i and β_j. Therefore, V_n/R was a constant, found by taking $\alpha_1 = \alpha_2 = \cdots = \alpha_n = 0, \beta_1 = \beta_2 = \cdots = \beta_n = -1$. Pólya's theorem followed from Cramer's formulas for $a_1, a_2, \ldots, a_n$. Pólya commented that the formulas for $b_1, b_2, \ldots, b_n$ could be obtained by interchanging the roles of the a and the b.

18.9 MacMahon's Generalization of Waring's Formula

One of the original insights of MacMahon's 1889 *Memoir on a New Theory of Symmetric Functions* was that Waring's formula could be obtained from (18.5), the formula used by Euler and Landen to derive Newton's recursive rule for the power sum symmetric function. MacMahon called this observation important, noting, "This fact seems to have hitherto escaped the notice of writers upon the subject." In this paper, MacMahon did not write out details of the new derivation for Waring's formula, but we present one, rewriting (18.5) as

$$\frac{\sigma_1 x - 2\sigma_2 x^2 + 3\sigma_3 x^3 - \cdots}{1 - \sigma_1 x + \sigma_2 x^2 - \sigma_3 x^3 + \cdots} = s_1 x + s_2 x^2 + s_3 x^3 + \cdots.$$

Expanding the denominator as a geometric series, we get

$$\left(\sum_{i=1}^{n}(-1)^{i-1}i\sigma_i x^i\right)\left(\sum_{l=0}^{\infty}(\sigma_1 x - \sigma_2 x^2 + \sigma_3 x^3 - \cdots)^l\right) = \sum_{m=1}^{\infty} s_m x^m.$$

On expanding the lth power term by the multinomial theorem, we obtain terms of the form

$$\frac{l!}{l_1! l_2! \cdots l_s!}(\sigma_1 x)^{l_1}(-\sigma_2 x^2)^{l_2} \cdots (\pm\sigma_s x^s)^{l_s}, \tag{18.12}$$

where $l_1 + l_2 + \cdots + l_s = l$. Since s_m, the mth power sum symmetric function, is the coefficient of x^m on the right-hand side, we need only gather together the coefficients of x^m on the left-hand side. Note that the expression (18.12) has to be multiplied by terms of the form $(-1)^{i-1}i\sigma_i x^i$ so we require that

$$l_1 + 2l_2 + \cdots + i(l_i + 1) + sl_s = m. \tag{18.13}$$

Moreover, the absolute value of the coefficient of the term after multiplication will be

$$\frac{i(l_i + 1)l!}{l_1! l_2! \cdots (l_i + 1)! l_{i+1}! \cdots l_s!}.$$

18.9 MacMahon's Generalization of Waring's Formula

It is now easy to check that by (18.13), the absolute value of the coefficient of x^m on the left can be found by summing the terms of the form

$$\frac{m(k_1+k_2+\cdots+k_s-1)!}{k_1!k_2!\cdots k_s!}\sigma_1^{k_1}\sigma_2^{k_2}\cdots\sigma_s^{k_s}, \tag{18.14}$$

$$\text{where} \quad k_1+2k_2+\cdots+sk_s=m, \tag{18.15}$$

proving Waring's formula.

In order to conveniently and clearly state his generalization of Waring, MacMahon developed a terminology. He defined the monomial symmetric function $\sum \alpha_1^{p_1}\alpha_2^{p_2}\cdots\alpha_n^{p_n}$ with the condition that the sum be over all those permutations of the indices resulting in different terms. He denoted this monomial symmetric function by $(p_1 p_2 \cdots p_n)$, where the numbers p were written in descending order. In case the p_i were repeated, he used exponents; for example, for $(p_1 p_1 p_2)$ he wrote $(p_1^2 p_2)$. Thus, MacMahon designated the elementary symmetric functions as $\sum \alpha_i = (1)$, $\sum \alpha_i \alpha_j = (1^2), \cdots, \alpha_1\alpha_2\cdots\alpha_n = (1^n)$. He called the sum $(p_1+p_2+p_3+\cdots+p_n) = w$ the weight and called p_1 the degree of the symmetric function. Note that MacMahon also referred to $(p_1 p_2 \ldots p_n)$ as a partition of w. Products of monomial symmetric functions were to be denoted, for example, as

$$\sum \alpha_1^{p_1}\alpha_2^{p_3} \sum \alpha_1^{p_2}\alpha_2^{p_4}\cdots\alpha_{n-2}^{p_{n-2}} = (p_1 p_3)(p_2 p_4 \cdots p_{n-2}).$$

MacMahon named $(p_1 p_2 \cdots p_{n-2})$ a partition of

$$p_1 + p_2 + \cdots + p_{n-2},$$

and

$$(p_1 p_3)(p_2 p_4 \cdots p_{n-2})$$

a separation of the partition, where $(p_1 p_3)$ and $(p_2 p_4 \cdots p_{n-2})$ were called the separates. He also wrote the separates in descending order according to their weights. If the successive weights of the separates were $w_1, w_2, w_3, \ldots$ then he named $(w_1 w_2 w_3 \cdots)$ the specification of the separation. MacMahon wrote the general form of a partition as

$$(p_1^{\pi_1} p_2^{\pi_2} p_3^{\pi_3} \cdots)$$

and that of a separation as

$$(J_1)^{j_1}(J_2)^{j_2}(J_3)^{j_3}\ldots, \quad \text{where} \quad J_1, J_2, J_3\ldots$$

were the distinct separates. It is interesting to note that MacMahon presented his generalization of Waring's formula, complete with proof, in his 1910 *Encyclopedia Britannica* article on algebraic forms:

Theorem.—The function symbolized by (n), viz. the sum of the nth powers of the quantities, is expressible in terms of functions which are symbolized by separations of *any* partition $(n_1^{\nu_1} n_2^{\nu_2} n_3^{\nu_3} \cdots)$ of the number n. The expression is —

$$(-)^{\nu_1+\nu_2+\nu_3+\cdots}\frac{(\nu_1+\nu_2+\nu_3+\cdots-1)!}{\nu_1!\nu_2!\nu_3!\cdots}(n)$$

$$= \sum(-)^{j_1+j_2+j_3+\cdots}\frac{(j_1+j_2+j_3+\cdots-1)!}{j_1!j_2!j_3!\cdots}(J_1)^{j_1}(J_2)^{j_2}(J_3)^{j_3}\cdots,$$

$(J_1)^{j_1}(J_2)^{j_2}(J_3)^{j_3}\cdots$ being a separation of $(n_1^{\nu_1}n_2^{\nu_2}n_3^{\nu_3}\cdots)$ and the summation being in regard to all such separations. For the particular case $(n_1^{\nu_1}n_2^{\nu_2}n_3^{\nu_3}\cdots)=(1^n)$

$$(-)^n\frac{1}{n}(n)=\sum(-)^{j_1+j_2+j_3+\cdots}\frac{(j_1+j_2+j_3+\cdots-1)!}{j_1!j_2!j_3!\cdots}(1)^{j_1}(1^2)^{j_2}(1^3)^{j_3}\cdots.$$

To establish this write —

$$1+\mu X_1+\mu^2 X_2+\mu^3 X_3+\cdots=\prod_a(1+\mu a_1 x_1+\mu^2 a_1^2 x_2+\mu^3 a_1^3 x_3+\cdots),$$

the product on the right involving a factor for each of the quantities $a_1, a_2, a_3 \ldots$, and μ being arbitrary.

Multiplying out the right-hand side and comparing coefficients

$$X_1=(1)x_1,$$
$$X_2=(2)x_2+(1^2)x_1^2,$$
$$X_3=(3)x_3+(21)x_2x_1+(1^3)x_1^3,$$
$$X_4=(4)x_4+(31)x_3x_1+(2^2)x_2^2+(21^2)x_2x_1^2+(1^4)x_1^4,$$
$$\ldots X_m=\sum(m_1^{\mu_1}m_2^{\mu_2}m_3^{\mu_3}\cdots)x_{m_1}^{\mu_1}x_{m_2}^{\mu_2}x_{m_3}^{\mu_3}\cdots,$$

the summation being for all partitions of m.

Auxiliary Theorem. —The coefficient of $x_{l_1}^{\lambda_1}x_{l_2}^{\lambda_2}x_{l_3}^{\lambda_3}\cdots$ in the product

$$\frac{X_{m_1}^{\mu_1}X_{m_2}^{\mu_2}X_{m_3}^{\mu_3}\cdots}{\mu_1!\mu_2!\mu_3!\cdots}$$

is

$$\sum\frac{(J_1)^{j_1}(J_2)^{j_2}(J_3)^{j_3}\cdots}{j_1!j_2!j_3!\cdots}$$

where

$$(J_1)^{j_1}(J_2)^{j_2}(J_3)^{j_3}\cdots\text{ is a separation of }(l_1^{\lambda_1}l_2^{\lambda_2}l_3^{\lambda_3}\cdots)$$

of specification $(m_1^{\mu_1}m_2^{\mu_2}m_3^{\mu_3}\cdots)$ and the sum is for all such separations. To establish this observe the result.

$$\frac{1}{p!}X_3^p=\sum\frac{(3)^{\pi_1}(21)^{\pi_2}(1^3)^{\pi_3}}{\pi_1!\pi_2!\pi_3!}x_3^{\pi_1}x_2^{\pi_2}x_1^{\pi_2+3\pi_3}$$

and remark that $(3)^{\pi_1}(21)^{\pi_2}(1^3)^{\pi_3}$ is a separation of $(3^{\pi_1}2^{\pi_2}1^{\pi_2+3\pi_3})$ of specification (3^p). A similar remark may be made in respect of

$$\frac{1}{\mu_1!}X_{m_1}^{\mu_1},\ \frac{1}{\mu_2!}X_{m_2}^{\mu_2},\ \frac{1}{\mu_3!}X_{m_3}^{\mu_3},\ldots$$

and therefore of the product of those expressions. Hence the theorem.

Now

$$\log(1+\mu X_1+\mu^2 X_2+\mu^3 X_3+\cdots)$$
$$=\sum_a\log(1+\mu a_1 x_1+\mu^2 a_1^2 x_2+\mu^3 a_1^3 x_3+\cdots)$$

whence, expanding by the exponential and multinomial theorems, a comparison of the coefficients of μ^n gives

$$(n) \sum (-)^{v_1+v_2+v_3\cdots-1} \frac{(v_1+v_2+v_3+\cdots-1)!}{v_1! v_2! v_3! \cdots} x_{n_1}^{v_1} x_{n_2}^{v_2} x_{n_3}^{v_3} \cdots$$

$$= \sum (-)^{v_1+v_2+v_3+\cdots-1} \frac{(v_1+v_2+v_3+\cdots-1)!}{v_1! v_2! v_3! \cdots} X_{n_1}^{v_1} X_{n_2}^{v_2} X_{n_3}^{v_3} \cdots$$

and, by the auxiliary theorem, any term $X_{m_1}^{\mu_1} X_{m_2}^{\mu_2} X_{m_3}^{\mu_3} \cdots$ on the right-hand side is such that the coefficient of $x_{n_1}^{v_1} x_{n_2}^{v_2} x_{n_3}^{v_3} \cdots$ in $\frac{1}{\mu_1! \mu_2! \mu_3! \cdots} X_{m_1}^{\mu_1} X_{m_2}^{\mu_2} X_{m_3}^{\mu_3} \cdots$ is

$$\sum \frac{(J_1)^{j_1} (J_2)^{j_2} (J_3)^{j_3} \cdots}{j_1! j_2! j_3! \cdots},$$

where, since $(m_1^{\mu_1} m_2^{\mu_2} m_3^{\mu_3} \cdots)$ is the specification of $(J_1)^{j_1} (J_2)^{j_2} (J_3)^{j_3} \cdots$, $\mu_1 + \mu_2 + \mu_3 + \cdots = j_1 + j_2 + j_3 + \cdots$. Comparison of the coefficients of $x_{n_1}^{v_1} x_{n_2}^{v_2} x_{n_3}^{v_3} \cdots$ therefore yields the result

$$(-)^{v_1+v_2+v_3+\cdots} \frac{(v_1+v_2+v_3+\cdots-1)!}{v_1! v_2! v_3! \cdots} (n)$$

$$= \sum (-)^{j_1+j_2+j_3+\cdots} \frac{(j_1+j_2+j_3+\cdots-1)!}{j_1! j_2! j_3! \cdots} (J_1)^{j_1} (J_2)^{j_2} (J_3)^{j_3} \cdots,$$

for the expression of $\sum a^n$ in terms of products of symmetric functions symbolized by separations of $(n_1^{v_1} n_2^{v_2} n_3^{v_3} \cdots)$.

18.10 Exercises

1. Express any given monomial symmetric function in terms of sums of powers symmetric functions. For Waring's result, see Waring (1991), pp. 9–11.

2. Let $p_1 + p_2 + \cdots + p_k = m$ be the weight of the monomial symmetric function $\sum \alpha_1^{p_1} \alpha_2^{p_2} \cdots \alpha_k^{p_k}$. Denote by h_m the sum of all monomial symmetric functions of weight m. Prove that

$$\frac{1}{1 - \alpha_1 x + \alpha_2 x^2 - \cdots + (-1)^n \alpha_n x^n} = 1 + h_1 x + h_2 x^2 + h_3 x^3 + \cdots.$$

Also prove that

$$h_m = \sum (-1)^{m+k_1+k_2+k_3+\cdots} \frac{(k_1+k_2+k_3+\cdots)!}{k_1! k_2! k_3! \cdots} \alpha_1^{k_1} \alpha_2^{k_2} \alpha_3^{k_3} \cdots,$$

where the sum is over all partitions of m, that is, all $k_1, k_2, k_3, \ldots$ such that $k_1 + 2k_2 + 3k_3 + \cdots = m$. Derive Faà di Bruno's formula for $m \leq n$,

$$h_m = \begin{vmatrix} \alpha_1 & \alpha_2 & \alpha_3 & \cdots & \alpha_m \\ 1 & \alpha_1 & \alpha_2 & \cdots & \alpha_{m-1} \\ 0 & 1 & \alpha_1 & \cdots & \alpha_{m-2} \\ 0 & 0 & 1 & \cdots & \alpha_{m-3} \\ \cdots & & & & \\ 0 & 0 & 0 & -1 & \alpha_1 \end{vmatrix}$$

Faà di Bruno published this result in 1882; MacMahon and Franklin gave short proofs in 1884. See MacMahon (1978), vol. I, pp. 51–52.

3. Let S_n^r denote the sum of all monomial symmetric functions related to partitions of m into r parts. Prove that

$$S_m^r - \sigma_1 S_{m-1}^r + \sigma_2 S_{m-2}^r - \cdots + (-1)^{m+1} \frac{m!}{r!(m-r)!} \sigma_m = 0.$$

Newton's formula corresponds to $r = 1$. MacMahon published this generalization of Newton's formula in an 1884 paper presenting several results on expanding monomial symmetric functions in terms of elementary symmetric functions. See MacMahon (1978), vol. I, pp. 41–42.

4. With S_m^r as in exercise 3, prove that for $k = k_1 + k_2 + k_3 + \cdots,$:

$$S_m^r = \sum (-1)^{w_2 + k + r - 1} \frac{(k-1)! w_{r+1}}{k_1! k_2! k_3!} \sigma_1^{k_1} \sigma_2^{k_2} \sigma_3^{k_3} \cdots$$

where $w_{r+1} = k_r + \binom{r+1}{1} k_{r+1} + \binom{r+2}{2} k_{r+2} + \cdots.$

See MacMahon (1978), vol. I, p. 42.

5. Show that if $a_1 + a_2 + a_3 = 0$, then

$$\left(\frac{a_1}{a_2 - a_3} - \frac{a_2}{a_1 - a_3} + \frac{a_3}{a_1 - a_2} \right) \left(\frac{a_2 - a_3}{a_1} - \frac{a_1 - a_3}{a_2} + \frac{a_1 - a_2}{a_3} \right) = 3^2.$$

Generalize to n quantities under the conditions

$$\sum_{i=1}^{n} a_i^k = 0, \quad \text{with} \quad k = 1, \ldots, n-2.$$

See MacMahon (1978), vol. I, pp. 20–22.

6. Give a combinatorial proof of the Cauchy–Laguerre theorem. See Kung (1995), p. 463.

18.11 Notes on the Literature

De Beaune, Girard, and Viéte (1986) contains English translations of algebra texts by de Beaune, Girard, and Viéte. The quotation from Girard is taken from this source. These three works together give an idea of the notation and methods used by French algebraists in the early seventeenth century. The reader may see Manders (2006) for the "cossic" algebra of Roth and Faulhaber during the same period.

For Newton's statement of his rule given in the text, see Newton (1964–67), vol. 2, pp. 107–108. Euler gave his two proofs of this rule in his 1750 paper reprinted in Eu. I-6, pp. 20–30; also see pp. 263–286. Maclaurin's proof can be found in his 1748 *Treatise of Algebra*; his lemma appears on p. 291. For proofs of the fundamental theorem for symmetric functions, see Gauss (1863–1927), vol. 3, pp. 36–38 and Weber (1895), vol. I, pp. 161–63. MacMahon's 1889 memoir is reprinted in MacMahon (1978), vol. I, pp. 109–143. See p. 130 for his remark on Waring's formula. His *Encyclopedia Britannica* article is reprinted in vol. II, pp. 620–641. The latter resource contains extensive commentaries by George Andrews relating the papers of MacMahon to more recent mathematical work.

Although earlier writers attributed Cauchy's 1837 result to Laguerre, Lascoux (2003) has pointed out Cauchy's paper. As a side point, Lascoux, among others, observes that Leibniz himself, in his communications with the Paris Academy of Science, signed his name Leibnitz.

19
Calculus of Several Variables

19.1 Preliminary Remarks

During the eighteenth and nineteenth centuries, Niklaus I Bernoulli, Clairaut, Euler, Fontaine, Lagrange, Gauss, Green, Cauchy, Ostrogradski, Jacobi, and others laid the foundations for the calculus of several variables. Most of this work was done in the context of mathematical physics. Not surprisingly, however, it is possible to see traces of this subject in the earliest work of both Newton and Leibniz. In his study of calculus conducted in May 1665, Newton expressed the curvature of $f(x, y) = 0$ in terms of homogenized first- and second-order partial derivatives of $f(x, y)$. He even worked out a notation for these partial derivatives, though he did not make use of it in any of his other work. Newton used partial derivatives as a computational device without a formal definition. Again, in his calculus book of 1670–71, Newton described the method for finding the fluxional equation of $f(x, y) = 0$, identical to the equation $f_x \dot{x} + f_y \dot{y} = 0$, where the dot notation denoted the derivative with respect to some parameter.

Gottfried Leibniz (1646–1716), whose name was written Leibnitz in most of the early literature, published two papers, in 1792 and 1794, in which he developed a procedure to find the envelope of a family of curves: a curve touching, at each of its points, a member of the given family. He showed that the methods of calculus could be extended to a family of curves by treating the parameter defining the family as a differentiable quantity. In his paper of 1794, Leibniz explained how to obtain an envelope of a family of curves using differential calculus. He noted that his method was a generalization of the technique developed by Fermat and extended by Hudde. To illustrate his method, he considered a family of circles $(x - b)^2 + y^2 = ab$ defined by the parameter b. He differentiated with respect to b and obtained $2(b - x)db = adb$ or $b = x + a/2$. By eliminating b from both equations, he found the envelope to be the parabola $y^2 = ax + a^2/4$.

At l'Hôpital's suggestion, Johann Bernoulli immediately applied these ideas to a problem originally solved by Torricelli in 1644. Torricelli had shown that the family of parabolic trajectories of cannon balls shot with the same initial velocity but different angles of elevation had a parabolic envelope called the "safety parabola," defining the range of the cannon. However, Leibniz and the Bernoullis sought new applications for

Leibniz's ideas on families of curves, including the brachistochrone and the orthogonal trajectories problems – both posed by Johann Bernoulli in 1696–97. The second problem was to find a family of curves orthogonal to a given family of curves. For example, the family of rays emanating from one fixed point would be orthogonal to the family of concentric circles with the fixed point as their common center. As a result of questions arising out of his correspondence with Johann Bernoulli, in 1697 Leibniz discovered his famous theorem on differentiation under the integral sign. He began with this problem involving families of curves: Given a family of logarithmic curves $y(x, a) = a \ln x$, all of which pass through $(1, 0)$, find the locus of points x such that $s(x, a) = \int_1^x \frac{dt}{t} \sqrt{t^2 + a^2}$ is a constant. The quantity $s(x, a)$ is the arclength from 1 to x along the curve $y(x, a) = a \ln x$. Leibniz was led to consider the expression $d_a \int p(x, a) \, dx$, the differential with respect to a of $\int p(x, a) \, dx$. Recall that Leibniz always conceived of integrals as sums and differentials as differences; since these operations commute when the initial value is properly chosen, he arrived at the formula

$$d_a \int p(x, a) \, dx = \int d_a p(x, a) \, dx.$$

For various reasons, Leibniz and Bernoulli did not follow up on these discoveries; for almost two decades, this subject lay fallow, until Niklaus I Bernoulli (1687–1759) took it up and developed the basic ideas of partial differentiation. Niklaus I studied mathematics with his uncle Jakob at Basel and earned his masters degree in 1704 with a thesis on infinite series. He was familiar with the preliminary drafts of Jakob's famous *Ars Conjectandi*, and he extended some of its results by applying probability theory to questions in jurisprudence. For this work he was awarded a doctorate of law in 1709; in 1713, he wrote an introductory note to Jakob's *Ars Conjectandi*. Niklaus I published very little; he became known through his correspondence with his contemporaries as a mathematician with a deep concern for the logical foundations of his subject. In 1716, he took up the orthogonal trajectories problem and this led him to the basic concepts of partial differentiation. He defined both partial and complete differentials of variables depending upon a number of other variables. He proved the equality of mixed second-order differentials: If a variable S depends on y and a, then $d_y d_a S = d_a d_y S$, where $d_y S$ denotes the differential of S, with a constant. Niklaus I published his work on orthogonal trajectories, but not the major part of his results on partial differentiation. However, in 1743 he communicated with Euler on this topic.

In the 1730s, Euler wrote articles in which he rederived the results of Leibniz and N. Bernoulli. One of his new and important results was that if $P(x, a)$ was a homogeneous function of degree n in x and a, and $dP = Q dx + R da$, then $nP = Qx + Ra$. Euler applied this formula to the problems of orthogonal trajectories and equal area trajectories. He very quickly perceived partial differentiation as a topic in the calculus of functions of many variables, rather than as merely a procedure necessary for the study of families of curves. After giving the argument for $d_a d_x P = d_x d_a P$ in terms of a family of curves, he wrote that he preferred to give a proof from the very nature of differentiation. He set $P(x + dx, a) = Q$, $P(x, a + da) = R$ and $P(x + dx, a + da) = S$. So $d_x P = Q - P$, and hence $d_a d_x P = (S - R) - (Q - P)$. Similarly, $d_a P = R - P$, so that $d_x d_a P = (S - Q) - (R - P)$. Clearly, it followed that $d_x d_a P = d_a d_x P$.

It was only after Bolzano and Cauchy gave local definitions of continuity and differentiability that nineteenth-century mathematicians began to understand that, even if a result was generally correct, there could be exceptions at certain points. However, this understanding came very slowly. In 1867, the Finnish mathematician Lorenz Lindelöf (1827–1908) presented a counterexample to show the proofs that the mixed second derivatives were independent of their order, published by Schlömilch in 1862 and by Bertrand in 1864, to be invalid. L. Lindelöf's review motivated Weierstrass's student, H. A. Schwarz, in 1873 to give a proof of the equality of the mixed derivatives with fairly strong conditions on the partial derivatives. In order to show the necessity for some conditions on the partial derivatives, Schwarz gave, as a counterexample, the function

$$f(x, y) = x^2 \arctan(y/x) - y^2 \arctan(x/y), \quad f(0,0) = 0,$$

where

$$\partial^2 f/\partial x \partial y = -1 \quad \text{and} \quad \partial^2 f/\partial y \partial x = +1 \quad \text{at} \quad (0,0).$$

At all $(x, y) \neq (0, 0)$,

$$\partial^2 f/\partial x \partial y = (x^2 - y^2)/(x^2 + y^2) = \partial^2 f/\partial y \partial x,$$

so that that Schwarz's function would not have been regarded as a counterexample by eighteenth-century mathematicians. Peano, Stoltz, Hobson, Young, and others succeeded in weakening the conditions for Schwarz's theorem. We note that it is easy to prove that if $f \in C^2$, that is, if f and all its partial derivatives up to the second order are continuous in some open region, then the mixed partial derivatives are equal at all points of that region.

Like partial derivatives, the theory of multiple integrals originated in the study of families of curves. Leibniz made a deeper analysis of the integral $d_a \int_1^x \frac{dt}{t} \sqrt{t^2 + a^2}$ in an unpublished article of 1697. In modern notation, he had the result

$$\int_{a_0}^{a_1} d_a \int_1^x \frac{dt}{t} \sqrt{t^2 + a^2} = \int_{a_0}^{a_1} \int_1^x \frac{a \, da \, dt}{t \sqrt{t^2 + a^2}}.$$

Since the integral $\int_{a_0}^{a_1}$ and d_a cancel out, we have

$$\int_1^x \frac{dt}{t} \sqrt{t^2 + a_1^2} - \int_1^x \frac{dt}{t} \sqrt{t^2 + a_0^2} = \int_{a_0}^{a_1} \int_1^x \frac{a \, da \, dt}{t \sqrt{t^2 + a^2}}.$$

The left-hand side represents the difference in the arclengths along two logarithmic curves. Leibniz was excited by his result; he noted that this was the first time double integrals had appeared in mathematics and that with his work it was possible to integrate and differentiate with respect to more than one variable. Unfortunately, the theory of double integrals did not begin its development until more than half a century later. From the 1740s onward, d'Alembert, Euler, and Lagrange took up the investigation of partial differential equations, requiring differentiation of functions of many variables.

19.1 Preliminary Remarks

Integration in several variables was the next problem to be tackled; in 1769, Euler published a paper on the change of variables formula for double integrals. Four years later, Lagrange extended this to triple integrals. Both Euler and Lagrange gave formal derivations, taking an algebraic approach to calculus, in contrast with Gauss's later method of taking an infinitesimal area in the uv-plane and calculating the change under the transformation $x = x(u, v)$, $y = y(u, v)$.

In 1836, the Russian mathematician Mikhail Ostrogradski (1801–1862) became the first mathematician to present the change of variables formula for the general n-dimensional case. Ostrogradski studied at Kharkov University, and he refused his doctoral degree there to protest religious discrimination against his teacher T. Osipovsky. While at the Sorbonne from 1824 to 1827, he studied with Cauchy and presented a number of important papers to the Paris Academy. His later election to the Petersburg Academy contributed to its return to brilliance. Ostrogradski participated in the work of introducing the Gregorian calendar and the decimal system of measurement in Russia; he also did a great deal to improve mathematics instruction in Russian universities. In 1838, he published a paper criticizing the formal methods of Euler and Lagrange, although his own earlier work had used the same approach. In the same paper, Ostrogradski carefully explained how the area element should be computed under a change of variables.

In 1833, Jacobi gave an evaluation of Euler's beta integral by a simultaneous change of variables, in contrast with Poisson's earlier method of changing the variables one at a time. Subsequently, Jacobi used change of variables in various special cases. Then in 1841, he published his definitive work on functional determinants, including the multiplication rule for the composition of several changes of variables.

Investigations in physics were a motivating factor in the development of the theory of multiple integrals, just as in the theory of partial differential calculus. The integral theorems named after Gauss, Green, and Stokes were developed in the context of studies in fluid mechanics, heat conduction, electricity, and magnetism. In 1813, Gauss, against the backdrop of his studies in magnetism, stated and proved a particular case of the divergence theorem:

$$\int \left(\frac{\partial p}{\partial x} + \frac{\partial q}{\partial y} + \frac{\partial r}{\partial z} \right) w = \int (p \cos \alpha + q \cos \beta + r \cos \gamma) \epsilon,$$

where w and ϵ denoted elements of volume and surface area, the left integral was taken over a solid V with surface boundary S, and α, β, and γ were angles made by the outward normal with the positive direction of the x, y, and z axes, respectively. In 1826, Ostrogradski presented to the Paris Academy a paper on the theory of heat, in which he stated and proved this theorem; this paper was actually published in 1831 by the Petersburg Academy.

In 1828, the British mathematician George Green (1793–1841) published a variant of the divergence theorem in *An Essay on the Application of Mathematical Analysis to the Theories of Electricity and Magnetism*. Green printed this groundbreaking essay at his own expense. In it, he applied many-variable calculus, including some original results, to intricate and carefully described problems in electricity and magnetism. Green, a miller by trade, was almost entirely self-educated, having received only four terms of schooling when he was very small. Yet he produced results of fundamental importance in mathematical physics with references to Laplace, Poisson, Cauchy, and Fourier, although the works of those French scientists were hardly mentioned in the British universities during that period. Green's work is a testament to the degree of general interest in science and mathematics in Britain at that time. In fact, about a century earlier, de Moivre and Simpson gave lessons in mathematics and its applications to non-academics.

Green's biographer, D. M. Cannell, conjectures that Challand Forrest and John Toplis, headmasters of the grammar school, may have tutored Green after he had exhibited great ability and interest. Both men had studied at Cambridge and in 1812 the latter published a translation of the first volume of Laplace's *Mécanique céleste*. Green's remarkable paper earned him the support of E. F. Bromhead of Gonville and Caius College; at the age of forty, Green became a student at Cambridge. Ironically, though his paper brought him such success, it remained largely unknown until the 1850s when William Thomson (Lord Kelvin) had it published in *Crelle's Journal*. The formula known as Green's theorem, though it is not explicitly given in Green's paper, states:

$$\int_C (P\,dx + Q\,dy) = \int\int_D \left(\frac{\partial Q}{\partial x} - \frac{\partial P}{\partial y}\right) dx\,dy,$$

where D is a planar region oriented such that D remains on the left as the boundary C, consisting of a finite number of smooth simple closed curves, is traversed. In an 1829 paper on fluids in equilibrium, Gauss stated this formula in the case where D was a surface in three dimensions. Green's theorem follows by taking D to be a planar region; this case was explicitly stated in an 1831 work by Poisson. In 1846, Cauchy stated Green's theorem without proof and used it to derive his famous theorem in complex analysis. Riemann gave a proof of Green's theorem in his 1851 doctoral dissertation.

Stokes set as an 1854 Cambridge examination problem a generalization of this theorem to a surface in three-dimensional space. The latter result is now known as Stokes's theorem, though it was actually communicated to Stokes by William Thomson (1824–1907), also known as Lord Kelvin. These integral theorems played a significant role in the development of nineteenth-century physics. Thomson and Peter Guthrie Tait (1831–1901) gave a thorough treatment of the subject in their 1867 *Treatise on Natural Philosophy* and Maxwell focused on these integral formulas in the mathematical preliminaries of his great *Treatise on Electricity and Magnetism*.

Mathematical physicists did not find it particularly useful to search for the most general forms of these theorems, but Élie Cartan (1869–1951) and other mathematicians developed generalizations and refinements of the integral theorems, leading to fruitful mathematical theories. In fact, discussions between André Weil and Élie Cartan's son, Henri Cartan, concerning the teaching of Stokes's theorem provided the impetus for the formation of the Bourbaki group. In the mid-1930s, these two French mathematicians were teaching differential and integral calculus when Henri Cartan raised the question of the extent to which the formulas should be generalized for the students. On this point, Weil wrote, "In his book on invariant integrals, Élie Cartan, following Poincaré in emphasizing the importance of this [Stokes's] formula, proposed to extend its domain of validity. Mathematically speaking, the question was of a depth that far exceeded what we were in a position to suspect. Not only did it bring into play the homology theory, along with de Rham's theorems, the importance of which was just becoming apparent; but this question is also what eventually opened the door to the theory of distributions and currents and also to that of sheaves." Thus, Weil, Cartan, and their capable mathematical friends gathered to write a book on the topics of the university calculus course. Weil wrote, "Little did I know that at that moment, Bourbaki was born."

19.2 Homogeneous Functions

During the 1730s, Euler, Clairaut, and Fontaine made independent efforts to work out the differential calculus of several variables. Their work grew out of their studies in mechanics. Alexis Fontaine (1704–1771) created a form of partial differential calculus in his 1732 study of the brachistochrone problem and then extended its scope in his work on the tautochrones of bodies moving in resistant mediums. It is not surprising that he independently found some results also discovered by Euler in a different context, such as their pretty result on homogeneous functions. When dealing with general functions of two independent variables x and y, Euler, Clairaut, and Fontaine generally worked with expressions of the form

$$ax^m y^n + bx^p y^q + cx^r y^s + \cdots. \qquad (19.1)$$

Such a function is called homogeneous of degree $m+n$ when each term is of the same degree, that is, $m+n = p+q = r+s = \cdots$. More generally, a homogeneous function ϕ of degree n satisfies the relation

$$\phi(x, y, z, \ldots) = x^n F\left(\frac{y}{x}, \frac{z}{x}, \ldots\right).$$

If we denote the expression (19.1) by ϕ, then

$$x\frac{\partial \phi}{\partial x} + y\frac{\partial \phi}{\partial y} = (m+n)ax^m y^n + (p+q)bx^p y^q + (r+s)cx^r y^s + \cdots$$
$$= (m+n)\phi. \qquad (19.2)$$

The result in (19.2) is an example of Euler's theorem for homogeneous functions, stating that for a differentiable homogeneous function ϕ of degree n,

$$x\frac{\partial \phi}{\partial x} + y\frac{\partial \phi}{\partial y} + z\frac{\partial \phi}{\partial z} + \cdots = n\phi. \qquad (19.3)$$

In 1740, Euler pointed out in the first letter he wrote to Clairaut that he had stated this theorem for functions of two variables in his 1736 book on mechanics and had given its general form in a paper on partial differential equations presented to the Petersburg Academy in 1734–5, but published in 1740.

In his 1740 paper, Clairaut gave a proof for the case of three variables, generalizable to any number of variables, and attributed it to Fontaine. He supposed ϕ to be a homogeneous function of degree $m+1$ in x, y, z, and he denoted the differential of

ϕ by $M\,dx + N\,dy + P\,dz$, where M, N, P were homogeneous of degree $m \neq -1$ in x, y, z. He set

$$y = xu, \ z = xt, \ dy = x\,du + u\,dx, \ dz = x\,dt + t\,dx,$$

$$M = x^m F, \ N = x^m G, \ P = x^m H.$$

Then he could write

$$d\phi = x^m(F + Gu + Ht)\,dx + x^{m+1}G\,du + x^{m+1}H\,dt.$$

Integrating the differential involving x, he obtained

$$\phi = \int x^m F(F + Gu + Ht)\,dx + T(u, t)$$

$$= \frac{x^{m+1}}{m+1}(F + Gu + Ht) + T(u, t). \tag{19.4}$$

The term $T(u, t)$ was the constant of integration, and it had to involve u and t because the integration was with respect to x. Clairaut very quickly concluded from this equation for ϕ that $T(u, t)$ vanished. Observe that since

$$x^{m+1}(F + gu + Ht) = xM + yN + zP$$

and ϕ are both homogeneous functions of degree $m + 1$, it follows from (19.4) that the constant $T(u, t)$ must be zero. Hence, $xM + yN + zP = (m+1)\phi$ and Euler's theorem, attributed by Clairaut to Fontaine, is proved.

19.3 Cauchy: Taylor Series in Several Variables

In his 1829 work, Cauchy devoted eight of twenty-three differential calculus lectures to several variables. As early researchers on partial differential calculus, Euler, Fontaine, and Clairaut were a little vague about the idea of a differential. Cauchy attempted to provide a more precise definition. He first defined partial derivatives of $u = f(x, y, z, \ldots)$ denoted by

$$\phi(x, y, z, \ldots), \quad \chi(x, y, z, \ldots), \quad \psi(x, y, z, \ldots), \quad \text{etc.,}$$

as the limits (assumed to exist) of the ratios

$$\frac{f(x+i, y, z, \ldots) - f(x, y, z, \ldots)}{i}, \quad \frac{f(x, y+i, z, \ldots) - f(x, y, z, \ldots)}{i},$$

$$\frac{f(x, y, z+i, \ldots) - f(x, y, z, \ldots)}{i},$$

as i approached zero. Then, supposing $\Delta x, \Delta y, \Delta z, \ldots$ to be finite changes in $x, y, z, \ldots$, Cauchy set

$$\Delta u = f(x + \Delta x, y + \Delta y, z + \Delta z, \ldots) - f(x, y, z, \ldots).$$

19.3 Cauchy: Taylor Series in Several Variables

Next, to define differentials, he observed that as $\Delta x, \Delta y, \Delta z, \ldots, \Delta u$ approached zero, there were quantities $dx, dy, dz, \ldots, du$ such that if $\Delta x/dx$ was an infinitesimal α, then the ratios $\Delta y/dy, \ldots, \Delta u/du$ differed from α by a very small value. He set

$$dx = \lim \frac{\Delta x}{\alpha}, \quad dy = \lim \frac{\Delta y}{\alpha}, \quad \ldots, \quad du = \lim \frac{\Delta u}{\alpha}.$$

To evaluate du, Cauchy applied the mean value theorem to obtain the relations

$$f(x+\Delta x, y, z, \ldots) - f(x, y, z, \ldots) = \Delta x \, \phi(x+\theta_1 \Delta x, y, z, \ldots),$$
$$f(x+\Delta x, y+\Delta y, z, \ldots) - f(x+\Delta x, y, z, \ldots) = \Delta y \, \chi(x+\Delta x, y+\theta_2 \Delta y, z, \ldots),$$
$$f(x+\Delta x, y+\Delta y, z+\Delta z, \ldots) - f(x+\Delta x, y+\Delta y, z, \ldots)$$
$$= \Delta z \, \psi(x+\Delta x, y+\Delta y, z+\theta_3 \Delta z, \ldots),$$

etc.,

where $\theta_1, \theta_2, \theta_3, \ldots$ were numbers between 0 and 1. By adding these equations, he obtained

$$f(x+\Delta x, y+\Delta y, z+\Delta z, \ldots) - f(x, y, z, \ldots)$$
$$= \Delta x \, \phi(x+\theta_1 \Delta x, y, z, \ldots) + \Delta y \, \chi(x+\Delta x, y+\theta_2 \Delta y, \ldots) + \text{ etc.}$$

Cauchy then divided by α and let α tend to 0 to find

$$du = \phi(x, y, z, \ldots)dx + \chi(x, y, z, \ldots)dy + \psi(x, y, z, \ldots)dz + \cdots.$$

He then introduced the notation $\phi(x, y, z, \ldots)dx = d_x u$ or $\phi(x, y, z, \ldots) = d_x u/dx$ to write

$$du = d_x u + d_y u + d_z u + \cdots,$$
$$du = \frac{d_x u}{dx}dx + \frac{d_y u}{dy}dy + \frac{d_z u}{dz}dz + \cdots. \tag{19.5}$$

As an application, he deduced the Euler and Fontaine theorem on homogeneous functions, first defining $f(x, y, z, \ldots)$ to be homogeneous of degree a if

$$f(tx, ty, tz, \ldots) = t^a f(x, y, z, \ldots).$$

Taking the derivative with respect to t, he applied (19.5) to obtain

$$\phi(tx, ty, tz, \ldots)x + \chi(tx, ty, tz, \ldots)y + \psi(tx, ty, tz, \ldots)z + \cdots = at^{a-1} f(x, y, \ldots).$$

He then got the required result by setting $t = 1$.

Cauchy also stated and proved Taylor's formula in several variables. This result had been well known since the second half of the eighteenth century. Lagrange gave a detailed discussion of it in his famous 1772 paper on symbolic calculus, where he stated it as

$$f(x+\xi, y+\psi, z+\zeta, \ldots) = e^{\frac{df}{dx}\xi + \frac{df}{dy}\psi + \frac{df}{dz}\zeta + \cdots}. \tag{19.6}$$

Cauchy obtained Taylor's formula by writing $u = f(x, y, z, \ldots)$ and

$$F(\alpha) = f(x + \alpha dx, y + \alpha dy, z + \alpha dz, \ldots),$$

observing that

$$F'(0) = du = d_x u + d_y u + d_z u + \cdots,$$
$$F''(0) = d^2 u = (d_x + d_y + d_z + \cdots)^2 u,$$
$$F'''(0) = d^3 u = (d_x + d_y + d_z + \cdots)^3 u,$$
etc.

The Taylor series for one variable then gave him the series for several variables:

$$f(x + dx, y + dy, z + dz, \ldots) = u + \frac{du}{1} + \frac{d^2 u}{1 \cdot 2} + \frac{d^3 u}{1 \cdot 2 \cdot 3} + \quad \text{etc.}$$

In addition, Cauchy stated Taylor's formula for several variables with remainder, also derived from the corresponding one variable result.

19.4 Clairaut: Exact Differentials and Line Integrals

Alexis Clairaut (1713–1765) studied mathematics with his father, a mathematics teacher in Paris. Clairaut read l'Hôpital's differential calculus book at the age of 10 and was elected to the Paris Academy at 18. In the mid-1730s, Clairaut took part in an expedition to Lapland to gather experimental evidence to test the Newton–Huygens theory that the Earth was flattened at the poles. In his 1743 book on the shape of the Earth, he made use of his research on line integrals and differentials in many variables to confirm the Newton–Huygens theory. In 1739 and 1740, Clairaut presented two papers to the Paris Academy on the line integrals of the differentials $Pdx + Qdy$, $Pdx + Qdy + Rdz$, attempting to define conditions under which these differentials were exact. He was one of the earliest mathematicians to study such questions. Recall that these differentials play a key role in the formulas of Green and Stokes. Note that either of these differentials is exact if there exists a function f of two variables x and y (or three variables x, y, z) such that $df = Pdx + Qdy$ (or $df = Pdx + Qdy + Rdz$). Clairaut thought he had given necessary and sufficient conditions for the exactness of these differentials. It turned out that his conditions were necessary but not sufficient because he mistakenly assumed that differentiation and integration along a curve were invertible processes. But in the 1760s, the French mathematician d'Alembert showed by an example that integration along curves in two dimensions could produce multivalued functions. He did this work with the knowledge of Euler's discovery, around 1750, of the multivalued nature of the logarithmic function.

Clairaut proved the theorem that if $Adx + Bdy$ was the differential of a function of x and y, then $\frac{\partial A}{\partial y} = \frac{\partial B}{\partial x}$, and he mistakenly believed that he had also proved the converse. He did not develop a special notation for partial derivatives, but he clarified by examples that he was keeping x constant and varying y when he wrote $\frac{dA}{dy}$ and similarly for other

19.4 Clairaut: Exact Differentials and Line Integrals

such expressions. He argued that if A was a derivative of some function f with respect to x, with y a constant, then f was obtained after integrating with respect to x. Thus, f would have the form $\int A\,dx + Y$, where Y was a constant of integration and hence purely a function of y. Similarly, f also took the form $\int B\,dy + X$, where X was a function of only x. Thus,

$$\int B\,dy + X = \int A\,dx + Y.$$

Taking the derivative of this equation with respect to y and keeping x a constant, he obtained

$$B = \int \frac{\partial A}{\partial y}\,dx + \frac{dY}{dy}.$$

The partial derivative of this equation with respect to x gave Clairaut

$$\frac{\partial B}{\partial x} = \frac{\partial A}{\partial y},$$

completing his proof of the theorem.

In his famous book of 1743 on the shape of the Earth, Clairaut wrote that the integral $\int (A\,dx + B\,dy)$ could be evaluated without knowing the equation of the curve along which the integral was to be computed, provided $\frac{\partial A}{\partial y} = \frac{\partial B}{\partial x}$. He incorrectly argued that the latter condition was also sufficient for $A\,dx + B\,dy$ to be of the form df for some function f. From this it would follow that

$$\int A\,dx + B\,dy = \int df = f(x, y) - f(x_0, y_0)$$

for any path joining (x_0, y_0) to (x, y). In 1768, d'Alembert pointed out that this was false, using an example of a differential given by Clairaut himself in his 1740 paper:

$$\frac{y\,dx - x\,dy}{xx + yy}.$$

Clairaut stated in his 1740 paper that this differential satisfied $\frac{\partial A}{\partial y} = \frac{\partial B}{\partial x}$, but d'Alembert showed that if this differential were integrated over the perimeter of a circle with center at the origin, the value of the integral would be -2π. Hence the integral was path-dependent. Clairaut's 1740 paper also considered the problem of finding an integrating factor $\mu(x, y)$ such that $\mu M\,dx + \mu N\,dy$ would be exact when $M\,dx + N\,dy$ was not. Clairaut studied differentials in three variables, where he applied the result he obtained for two variables. He observed that if $M\,dx + N\,dy + P\,dz$ was an exact differential of a function of three variables x, y, and z, then

$$\frac{\partial M}{\partial y} = \frac{\partial N}{\partial x}, \quad \frac{\partial M}{\partial z} = \frac{\partial P}{\partial x}, \quad \text{and} \quad \frac{\partial N}{\partial z} = \frac{\partial P}{\partial y}.$$

To prove this, he noted that if he took z to be a constant, then the differential reduced to $M\,dx + N\,dy$, also exact. The first equation given above followed from this; the other two conditions were obtained by taking y constant and then x constant. Clairaut gave the necessary condition for the existence of an integrating factor, making $M\,dx + N\,dy + P\,dz$ exact:

$$N\frac{\partial P}{\partial x} - P\frac{\partial N}{\partial x} + M\frac{\partial N}{\partial z} - N\frac{\partial M}{\partial z} + P\frac{\partial M}{\partial y} - M\frac{\partial P}{\partial y} = 0. \tag{19.7}$$

In his proof, he assumed $\mu(x, y, z)$ to be a function such that $\mu M\,dx + \mu N\,dy + \mu P\,dz$ was exact. Then, by his result in differentials of three variables,

$$\frac{\partial(\mu M)}{\partial y} = \frac{\partial(\mu N)}{\partial x}, \quad \frac{\partial(\mu M)}{\partial z} = \frac{\partial(\mu P)}{\partial x}, \quad \frac{\partial(\mu N)}{\partial z} = \frac{\partial(\mu P)}{\partial y}.$$

After applying the product rule for derivatives and by rearranging, the three equations could be written as

$$\mu\left(\frac{\partial M}{\partial y} - \frac{\partial N}{\partial x}\right) = N\frac{\partial \mu}{\partial x} - M\frac{\partial \mu}{\partial y},$$

$$\mu\left(\frac{\partial N}{\partial z} - \frac{\partial P}{\partial y}\right) = P\frac{\partial \mu}{\partial y} - N\frac{\partial \mu}{\partial z},$$

$$\mu\left(\frac{\partial P}{\partial x} - \frac{\partial M}{\partial z}\right) = M\frac{\partial \mu}{\partial z} - P\frac{\partial \mu}{\partial x}.$$

Multiplying the first equation by P, the second by M, and the third by N and adding, Clairaut obtained equation (19.7) when the terms on the right canceled.

19.5 Euler: Double Integrals

Euler presented the change of variables formula for double integrals in his 1769 paper, "De Formulis Integralibus Duplicatis." He viewed the double integral $\iint Z\,dx dy$ as an interated integral, where he integrated with respect to one variable, keeping the other fixed. In the beginning sections of his paper, he computed some examples of double integrals, starting with

$$\iint \frac{dx dy}{xx + yy} = \int dx \int \frac{dy}{xx + yy}.$$

He noted that

$$\int \frac{dy}{xx + yy} = \frac{1}{x}\arctan\frac{y}{x} + \frac{dX}{dx},$$

where dX/dx was some function of x, and therefore

$$\iint \frac{dx dy}{xx + yy} = \int \frac{dx}{x}\arctan\frac{y}{x} + X.$$

Similarly,

$$\iint \frac{dx dy}{xx + yy} = \int \frac{dy}{y}\arctan\frac{y}{x} + Y.$$

Euler then evaluated the single integrals as series after noting that

$$\arctan(x/y) = \pi/2 - \arctan(y/x)$$
$$= \frac{\pi}{2} - \left(\frac{y}{x} - \frac{y^3}{3x^3} + \frac{y^5}{5x^5} - \frac{y^7}{7x^7} + \frac{y^9}{9x^9} - \text{etc.}\right).$$

Thus, $\int \dfrac{dx}{x} \arctan(y/x) = -\dfrac{y}{x} + \dfrac{y^3}{9x^3} - \dfrac{y^5}{25x^5} + \dfrac{y^7}{49x^7} - \text{etc.} + f(y);$

$\int \dfrac{dy}{y} \arctan(x/y) = \dfrac{\pi}{2} \ln y - \dfrac{y}{x} + \dfrac{y^3}{9x^3} - \dfrac{y^5}{25x^5} + \dfrac{y^7}{49x^7} - \text{etc.} + g(x);$

and finally $\iint \dfrac{dx\,dy}{xx+yy} = X + Y - \dfrac{y}{x} + \dfrac{y^3}{9x^3} - \dfrac{y^5}{25x^5} + \dfrac{y^7}{49x^7} - \text{etc.}$

Euler next considered the problem of changing variables. He took x and y to be functions of u and t so that

$$dx = R\,dt + S\,du, \text{ and } dy = T\,dt + V\,du.$$

His problem was to show how the area element $dx\,dy$ changed when written in terms of u and t. To explain that he could not merely multiply the differentials dx and dy, he considered the case in which the new orthogonal coordinates were obtained by a translation, a rotation, and then a reflection about the x-axis. He computed $dx\,dy$ for this transformation by multiplying the expression for dx with that for dy; the result was not $du\,dt$. He also ended up with a number of meaningless terms in his expression. Euler was aware, however, that under this orthogonal transformation the area should not change.

To derive the necessary change of variables formula, he started with the double integral $\iint dx\,dy$. He took y to be a function of x and u so that $dy = P\,dx + Q\,du$. He noted that x was a constant when integration was performed first with respect to y, so that $dy = Q\,du$ and $\iint dx\,dy = \int dx \int Q\,du$. Now Q was a function of x and u, so he changed the order of integration to write $\iint dx\,dy = \int du \int Q\,dx$. Since $dx = R\,dt + S\,du$, he could write the right-hand side as $\int du \int QR\,dt$, because when integrating with respect to x, u was a constant. Thus, he had $\iint dx\,dy = \iint QR\,du\,dt$. However, Q was a function in x and u, not directly a function of u and t. So Euler observed that the equations

$$dy = T\,dt + V\,du, \text{ and}$$
$$dy = P\,dx + Q\,du = P(R\,dt + S\,du) + Q\,du = PR\,dt + (PS+Q)\,du$$

gave $PRS + QR = VR$ and $PR = T$. Therefore, $QR = VR - TS$, and the change of variables formula emerged as $\iint dx\,dy = \iint (VR - TS)\,du\,dt$. Thus, when variables were changed from x, y to u, t, the product of the differential $du\,dt$ had to be multiplied by the determinant of the equations

$$dx = R\,dt + S\,du \quad \text{and} \quad dy = T\,dt + V\,du.$$

Euler observed that if x and y were interchanged, the result would be $(TS-VR)\,du\,dt$, the negative of the previous result, so that, since the area was positive, he had to take the absolute value of the factor $(VR-TS)$.

19.6 Lagrange's Change of Variables Formula

Lagrange found the need for the change of variables formula for triple integrals in the course of his work on the attraction of ellipsoids. He specifically required the change from Cartesian to spherical coordinates, but he proposed and solved the problem in the general case. Suppose that x, y, z are expressible as functions of p, q, r and conversely. The problem is to express $dx\,dy\,dz$ in terms of $dp\,dq\,dr$. Lagrange started with the chain rule

$$dx = A\,dp + B\,dq + C\,dr; \quad dy = D\,dp + E\,dq + F\,dr; \quad dz = G\,dp + H\,dq + I\,dr, \quad (19.8)$$

where $A, B, C, \ldots$ were functions of p, q, r. He observed that the expressions for dx, dy and dz could not simply be multiplied together because the product would contain cubes and squares of dp, dq and dr and these did not make sense in the triple integral. Following the procedure for evaluating a triple integral, Lagrange proposed to first fix x and y so that $dx = 0, dy = 0$ and vary z; then to fix x and vary y; and finally to vary x. Thus he had three steps:

1. Set $dx = 0$ and $dy = 0$ so that $A\,dp + B\,dq + C\,dr = 0$, and $D\,dp + E\,dq + F\,dr = 0$. Solve for dp and dq in terms of dr to get

$$dp = \frac{BF - CE}{AE - BD}\,dr, \quad dq = \frac{CD - AF}{AE - BD}\,dr;$$

substitute these expressions in the third equation in (19.8) to find

$$dz = \frac{G(BF - CE) + H(CD - AF) + I(AE - BD)}{AE - BD}\,dr. \qquad (19.9)$$

2. To obtain dy, set $dx = 0$ and $dz = 0$. By (19.9) $dz = 0$ and thus $dr = 0$. Hence $A\,dp + B\,dq = 0$, or $dp = -\frac{B}{A}\,dq$. When this was substituted in the equation for dy in (19.8), Lagrange got the required expression

$$dy = \frac{AE - BD}{A}\,dq. \qquad (19.10)$$

3. Finally, to find dx, set $dy = 0$ and $dz = 0$ so that by (19.9) and (19.10) $dq = 0$ and $dr = 0$. Hence

$$dx = A\,dp. \qquad (19.11)$$

Multiplying (19.9), (19.10), and (19.11) would yield the necessary result

$$dx\,dy\,dz = [G(BF - CE) + H(CD - AF) + I(AE - BD)]\,dp\,dq\,dr.$$

Lagrange also noted that one should use the absolute value of the quantity in brackets, since $dx\,dy\,dz$ and $dp\,dq\,dr$ were understood to be positive. He then applied his formula to spherical coordinates p, q, r so that

$$x = r\sin p\cos q, \quad y = r\sin p\sin q, \quad z = r\cos p. \tag{19.12}$$

In this case

$$A = r\cos p\cos q, \qquad B = -r\sin p\sin q, \qquad C = \sin p\cos q,$$
$$D = r\cos p\sin q, \qquad E = r\sin p\cos q, \qquad F = \sin p\sin q,$$
$$G = -r\sin p, \qquad H = 0, \qquad I = \cos p;$$

and therefore $dx\,dy\,dz = r^2 \sin p\,dp\,dq\,dr$. Note that $r^2 \sin p \geq 0$, since p varies from 0 to π.

19.7 Green's Integral Identities

In his 1828 paper, George Green gave statements and derivations of his integral formulas and then applied them to electric and magnetic phenomena. His basic result was the formula

$$\int\!\!\int\!\!\int U\nabla^2 V\,dx\,dy\,dz + \int\!\!\int U\frac{\partial V}{\partial n}\,d\sigma = \int\!\!\int\!\!\int V\nabla^2 U\,dx\,dy\,dz + \int\!\!\int V\frac{\partial U}{\partial n}\,d\sigma, \tag{19.13}$$

where U and V were twice continuously differentiable in a solid three-dimensional region. We note that Green denoted $\nabla^2 V$ by δV and wrote $\frac{d}{dw}$ for $\partial/\partial n$ (the derivative in the normal direction at a point on the surface). Moreover, he denoted the triple integral on the left as $\int dx\,dy\,dz\,U\delta V$. Green proved (19.13) by showing that both sides of the equation were equal to the triple integral

$$\int\!\!\int\!\!\int \left(\frac{\partial V}{\partial x}\frac{\partial U}{\partial x} + \frac{\partial V}{\partial y}\frac{\partial U}{\partial y} + \frac{\partial V}{\partial z}\frac{\partial U}{\partial z}\right) dx\,dy\,dz. \tag{19.14}$$

He noted that integration by parts gave

$$\int\!\!\int\!\!\int \frac{\partial V}{\partial x}\frac{\partial U}{\partial x}\,dx\,dy\,dz = \int\!\!\int \left(V''\frac{\partial U''}{\partial x} - V'\frac{\partial U'}{\partial x}\right) dy\,dz$$
$$- \int\!\!\int\!\!\int V\frac{\partial^2 U}{\partial x^2}\,dx\,dy\,dz, \tag{19.15}$$

where the accents were used to denote the values of those quantities at the limits of the integral, on the surface of the region. Similar results hold for the other two terms in (19.14). Turning to $\int\!\!\int V''\partial U''/\partial x\,dy\,dz$, Green let $d\sigma''$ be the element of the surface corresponding to $dydz$. Since n was normal to the surface he had

$$dydz = -\frac{\partial x}{\partial n}d\sigma'', \text{ and by substitution}$$

$$\iint V''\frac{\partial U''}{\partial x}dydz = -\int V''\frac{\partial U''}{\partial x}\frac{\partial x}{\partial n}d\sigma''.$$

Similarly, the integral corresponding to the smaller value of x on the surface was

$$-\iint V'\frac{\partial U'}{\partial x}dydz = -\int V'\frac{\partial U'}{\partial x}\frac{\partial x}{\partial n}d\sigma'.$$

Green observed that since the sums of the elements denoted by $d\sigma'$ combined with those denoted by $d\sigma''$ formed the complete surface, the double integral in (19.15) could be written as

$$\iint \left(V''\frac{\partial U''}{\partial x} - V'\frac{\partial U'}{\partial x}\right)dydz = -\iint V\frac{\partial U}{\partial x}\frac{\partial x}{\partial n}d\sigma,$$

where the integral on the right-hand was taken over the whole surface. Applying this analysis to all three terms of the integral in (19.14), Green obtained

$$-\int V\left(\frac{\partial u}{\partial x}\frac{\partial x}{\partial n} + \frac{\partial U}{\partial y}\frac{\partial y}{\partial n} + \frac{\partial U}{\partial z}\frac{\partial z}{\partial n}\right)d\sigma - \iiint V\nabla^2 U\,dxdydz$$

$$= -\iint V\frac{\partial U}{\partial n}d\sigma - \iiint V\nabla^2 U\,dxdydz.$$

Since (19.14) was symmetrical in U and V, Green's identity (19.13) followed.

At the end of this derivation, Green noted that the continuity of the derivatives of U and V was essential and showed how the formula would be modified if U had a singularity at a point p' such that U behaved like $1/r$ near p', where r was the distance of p' from the element. We must here understand the element to indicate the point where U was evaluated. Green took an infinitely small sphere of radius a at point p'. He observed that formula (19.13) continued to hold for the region outside the sphere. Moreover, since $\nabla^2 U = \nabla^2 1/r = 0$, the triple integral on the right-hand side of (19.13) vanished for the spherical part, while the triple integral on the left-hand side was of the order of a^2. Similarly, the double integral on the left-hand was of order a on the sphere. Then, since

$$\frac{\partial U}{\partial n} = \frac{\partial U}{\partial r} = -\frac{1}{r^2} = -\frac{1}{a^2},$$

the double integral on the right could be written $-4\pi V' = -4\pi V(p')$, as $a \to 0$. Hence, (19.13) took the form

$$\iiint U\nabla^2 V\,dxdydz + \iint U\frac{\partial V}{\partial n}d\sigma$$

$$= \iiint V\nabla^2 U\,dxdydz + \iint \frac{\partial U}{\partial n}d\sigma - 4\pi V'. \qquad (19.16)$$

19.8 Riemann's Proof of Green's Formula

The result known as Green's formula was apparently first proved in Riemann's inaugural and highly original dissertation. This was presented to the University of Göttingen in 1851 and is famous for its geometric approach to complex analysis, culminating in the Riemann mapping theorem. Riemann proved Green's formula fairly early in his thesis as a basic result needed in his theory of complex analytic functions. He stated the formula in the form

$$\int \left(\frac{\partial X}{\partial x} + \frac{\partial Y}{\partial y}\right) dT = -\int (X\cos\xi + Y\cos\eta) ds, \tag{19.17}$$

where X and Y were functions defined on a region A; dT denoted an area element of the region; and ds an arc element of the boundary. Then, ξ and η were the angles made by the normal to the boundary curve with the x- and y-axes, respectively. Riemann was familiar with Cauchy's work in complex analysis, so he may have seen Green's formula in Cauchy's 1846 paper where it appeared without proof. Cauchy applied (19.17) to prove his integral formula, now known as the Cauchy integral formula. In any case, although Cauchy promised a proof of Green's formula, he never published it, and Riemann provided a simple proof using the fundamental theorem of calculus. It is interesting to note that the referee of Riemann's dissertation, Gauss, had proved Cauchy's integral formula in 1811 but did not publish it, though he communicated the result in a letter to his friend Bessel. Gauss was also aware of Green's formula in a more general form.

To prove (19.17), Riemann divided the region into thin strips parallel to the x-axis. He integrated along a line parallel to the x-axis to obtain

$$\int \frac{\partial X}{\partial x} dx\, dy = dy \int \frac{\partial X}{\partial x} dx = (-X_{\prime} + X'),$$

where X' and $X_{\prime}$ denoted the values of X at the upper and lower end points on the boundary. Riemann then observed that with ξ and η defined as before, dy took the value

$$\cos\xi_{\prime}\, ds_{\prime}$$

on the boundary with the lower value of x, and took the value

$$-\cos\xi'\, ds'$$

on the upper boundary. Thus, after integrating with respect to y, he concluded that

$$\int \frac{\partial X}{\partial x} dT = -\int X\cos\xi\, dx.$$

Similarly, he observed

$$\int \frac{\partial Y}{\partial y} dT = -\int Y\cos\eta\, dx$$

and the result followed.

19.9 Stokes's Theorem

The result known as Stokes's theorem first appeared in print in February 1854 as the eighth problem on the Smith's Prize Exam at Cambridge. The question paper was set by G. G. Stokes; the students were to prove

$$\iint \left(l \left(\frac{\partial Z}{\partial y} - \frac{\partial Y}{\partial z} \right) + m \left(\frac{\partial X}{\partial z} - \frac{\partial Z}{\partial x} \right) + n \left(\frac{\partial Y}{\partial x} - \frac{\partial X}{\partial y} \right) \right) dS$$

$$= \int \left(X \frac{dx}{ds} + Y \frac{dy}{ds} + Z \frac{dz}{ds} \right) ds,$$

where X, Y, Z were functions of x, y, z; dS was an element of a bounded surface; l, m, n were the cosines of the inclination of the normal at dS to the axes; and ds was an element of the boundary line. The double integral was taken over the surface and the single integral over the perimeter of the surface. This theorem had been communicated by William Thomson to Stokes in a letter of July 2, 1850, though the left-hand side of the formula had appeared in earlier papers of Stokes.

Hermann Hankel (1839–1873) gave a proof of Stokes's theorem for the case where the surface could be parameterized by means of two variables. Hankel entered the University of Leipzig in 1857, studying with his father and Möbius. He spent the year 1860 with Riemann in Göttingen and the following year with Weierstrass and Kronecker in Berlin. His refinement of Riemann's theory of integration was an important step toward the measure theoretic integral of Lebesgue, Young, and Borel. In an 1861 monograph on fluid mechanics, Hankel gave his proof without a reference to Stokes; he was probably unaware that the formula was already known. Hankel referred to and used Riemann's Green's theorem for surfaces expressible in the form $z = z(x, y)$. We present a summary of Hankel's proof: Taking

$$dz = \frac{\partial z}{\partial x} dx + \frac{\partial z}{\partial y} dy,$$

rewrite the line integral as

$$\int \left(\overline{X} + \frac{\partial z}{\partial x} \overline{Z} \right) dx + \left(\overline{Y} + \frac{\partial z}{\partial y} \overline{Z} \right) dy.$$

Here $\overline{X}(x, y) = X(x, y, z(x, y))$, and similar expressions hold for $\overline{Y}$ and $\overline{Z}$. By Green's formula, this expression is equal to the double integral

$$\iint \left(\frac{\partial}{\partial x} \left(\overline{Y} + \frac{\partial z}{\partial y} \overline{Z} \right) - \frac{\partial}{\partial y} \left(\overline{X} + \frac{\partial z}{\partial x} \overline{Z} \right) \right) dx dy.$$

Now we can write

$$\frac{\partial \overline{X}}{\partial y} = \frac{\partial X}{\partial y} + \frac{\partial X}{\partial z} \frac{\partial z}{\partial x},$$

and similar equations hold for the partial derivatives of $\overline{Y}$ and $\overline{Z}$. When these are substituted for the integrand in the double integral, the result is

$$-\left[\frac{\partial X}{\partial y} + \frac{\partial X}{\partial z}\frac{\partial z}{\partial y} + \frac{\partial^2 z}{\partial y \partial x}\overline{Z} + \frac{\partial z}{\partial x}\left(\frac{\partial Z}{\partial y} + \frac{\partial Z}{\partial z}\frac{\partial z}{\partial y}\right) - \frac{\partial Y}{\partial x}\right.$$

$$\left. - \frac{\partial Y}{\partial z}\frac{\partial z}{\partial x} - \frac{\partial^2 z}{\partial x \partial y}\overline{Z} - \frac{\partial z}{\partial y}\left(\frac{\partial Z}{\partial x} + \frac{\partial Z}{\partial z}\frac{\partial z}{\partial x}\right)\right].$$

This simplifies to

$$-\left[\left(\frac{\partial X}{\partial y} - \frac{\partial Y}{\partial x}\right) + \left(\frac{\partial Z}{\partial y} - \frac{\partial Y}{\partial z}\right)\frac{\partial z}{\partial x} + \left(\frac{\partial X}{\partial z} - \frac{\partial Z}{\partial x}\right)\frac{\partial z}{\partial y}\right].$$

The normal vector to the surface $z = z(x, y)$ is given by $(\partial z/\partial x, -\partial z/\partial y, 1)$. It then follows that $\partial z/\partial x = -l/n$, $\partial z/\partial y = -m/n$ and $dS = dxdy/n$. This proves Stokes's theorem.

Thomson and Tait proved Stokes's theorem in the beginning of their 1867 work on mathematical physics, as part of a list of results they would need later in the book. Following Thomson and Tait, we replace X, Y, Z, by P, Q, R. They began their proof evaluating the double integral

$$\iint \left(m\frac{\partial P}{\partial z} - n\frac{\partial P}{\partial y}\right) dS. \tag{19.18}$$

We reproduce their succinct argument. First, they divided the surface S into bands by means of planes parallel to the xy-plane and divided each of these bands into rectangles. The breadth of the band between the planes at $x - \frac{1}{2}dx$ and $x + \frac{1}{2}dx$ was $\frac{dx}{\sin\theta}$, with θ denoting the inclination of the tangent plane of S to the plane through x. Hence, if ds denoted the curve in which the plane at x intersected the surface S, they had

$$dS = \frac{1}{\sin\theta} dxds.$$

Since $l = \cos\theta$, they could express the other direction-cosines m and n as

$$m = \sin\theta\cos\phi \quad \text{and} \quad n = \sin\theta\sin\phi, \quad \text{for some } \phi.$$

The double integral (19.18) then took the form

$$\iint dxds \left(\cos\phi\frac{\partial P}{\partial z} - \sin\phi\frac{\partial P}{\partial y}\right) = \iint dxds\frac{dP}{ds} = \int P\,dx.$$

They noted that the terms containing Q and R could be similarly worked out. In a footnote to this theorem, Thomson and Tait commented in their second edition:

> This theorem was given by Stokes in his Smith's prize paper for 1854 (*Cambridge University Calendar*, 1854). The demonstration in the text is an expansion of that indicated in our first edition. A more synthetical proof is given in §69 (q) of Sir W. Thomson's paper on "Vortex Motion," *Trans. R. S. E.* 1869. A thoroughly analytical proof is given by Prof. Clerk Maxwell in his *Electricity and Magnetism* (§24).

The edition quoted above was published in 1890. Thomson did not mention that he had communicated the Stokes formula to Stokes.

The Scottish physicist James Clerk Maxwell (1831–1879) published his book in 1873. In his proof of Stokes's theorem, he assumed that the surface was defined by two parameters, α and β. He supposed that the curves for which α was a constant formed closed curves around a point T on the surface where α took the least value, α_0. The largest value, $\alpha = \alpha_1$, corresponded to the closed curve s. Moreover, the curves for which β was constant formed lines drawn from T to the closed curve s, such that the initial value $\beta = \beta_0$ and the final $\beta = \beta_1$ produced the same line. Using the change of variables

$$dydz = \left(\frac{\partial y}{\partial \alpha}\frac{\partial z}{\partial \beta} - \frac{\partial y}{\partial \beta}\frac{\partial z}{\partial \alpha}\right)d\alpha d\beta,$$

the double integral was transformed to

$$\int\int \left(\frac{\partial P}{\partial \alpha}\frac{\partial x}{\partial \beta} - \frac{\partial P}{\partial \beta}\frac{\partial x}{\partial \alpha}\right)d\alpha d\beta + \cdots + \cdots. \tag{19.19}$$

The other two double integrals involved Q and R. Using integration by parts, Maxwell integrated the first term in the double integral with respect to α to get

$$\int_{\beta_0}^{\beta_1}\left(\left(P\frac{\partial x}{\partial \beta}\right)_{\alpha=\alpha_1} - \left(P\frac{\partial x}{\partial \beta}\right)_{\alpha=\alpha_1}\right)d\beta - \int\int P\frac{\partial^2 x}{\partial \alpha \partial \beta}d\alpha d\beta. \tag{19.20}$$

He similarly integrated the second term with respect to β. The resulting double integral canceled the one in (19.20); thus, the double integral for P in (19.19) reduced to

$$\int_{\beta_0}^{\beta}\left(P\frac{\partial x}{\partial \beta}\right)_{\alpha=\alpha_1}d\beta - \int_{\beta_0}^{\beta}\left(P\frac{\partial x}{\partial \beta}\right)_{\alpha=\alpha_0}d\beta$$

$$-\int_{\alpha_0}^{\alpha_1}\left(P\frac{\partial x}{\partial \alpha}\right)_{\beta=\beta_1}d\alpha + \int_{\alpha_0}^{\alpha_1}\left(P\frac{\partial x}{\partial \alpha}\right)_{\beta=\beta_0}d\alpha.$$

The third and fourth integrals canceled because (α, β_0) and (α, β_1) denoted the same point. The second integral vanished because $\alpha = \alpha_0$ consisted of only one point. Since $\alpha = \alpha_1$ was the curve s, the first integral could be expressed as $\int P\frac{\partial x}{\partial s}dx$. Hence, the double integral (19.19) could be written as

$$\int\int \left(\cos\theta \frac{\partial P}{\partial z} - \sin\theta \frac{\partial P}{\partial y}\right)dxds$$

$$= \int\int \frac{\partial P}{\partial s}dsdx = \int P\,dx.$$

This line integral was taken around the perimeter of the curve bounding the surface S. The proof of Stokes's theorem could then be completed by applying the same procedure to the other two double integrals involving the functions Q and R.

19.10 Exercises

1. Solve the equation $(ix + ky) dx + (lx + my + n) dy = 0$. See Clairaut (1740), p. 302.
2. Prove d'Alembert's assertion that the value of the integral of the differential $(y\,dx - x\,dy)/(x^2 + y^2)$ taken over a circle with the center at the origin is not zero. See d'Alembert (1761–1780), vol. 5, pp. 1–40.
3. Use a double integral to find the volume of the sphere. See Eu. I-17, pp. 293–294.
4. Show that if $AC - B^2 > 0$, then

$$u = Ax^2 + 2Bxy + Cy^2 + 2Dx + 2Ey + F$$

   must have a maximum or minimum. See Cauchy (1829), p. 236.
5. Prove that

$$\iiint \alpha^2 \left(\frac{dU}{dx}\frac{dU'}{dx} + \frac{dU}{dy}\frac{dU'}{dy} + \frac{dU}{dz}\frac{dU'}{dz} \right) dx\,dy\,dz = \quad (19.21)$$

$$\iint U\alpha^2 \delta U'\,dS - \iiint U \left(\frac{d(\alpha^2 \frac{dU'}{dx})}{dx} + \frac{d(\alpha^2 \frac{dU'}{dy})}{dy} + \frac{d(\alpha^2 \frac{dU'}{dz})}{dz} \right) dx\,dy\,dz, \quad (19.22)$$

   where δ denotes the normal derivative and d/dx, d/dy, d/dz are the partial derivatives with respect to x, y, z. See Thomson and Tait (1890), p. 168.
6. Show that if X, Y, Z are continuous and finite within a closed surface S, then the total surface integral of the vector $R = (X, Y, Z)$ would be

$$\iint R \cos \epsilon \, dS = \iiint \left(\frac{dX}{dx} + \frac{dY}{dy} + \frac{dZ}{dz} \right) dx\,dy\,dz,$$

   where the triple integral is extended over the whole space within S, and ϵ is the angle made by the positive normal to the surface with the vector R. See Maxwell (1873), vol. I, pp. 19–20.

19.11 Notes on the Literature

Clairaut (1739) and (1740) contain the results discussed in the text. Cauchy (1829) consists of his course on differential calculus; see pp. 225–228 and 255–257 for the material on differentials and on Taylor series in many variables. For Euler's paper on double integrals, see Eu. I-17, pp. 289–315; for triple integrals, see Lagrange (1867–92), vol. 3, pp. 619–658. See Green (1970), pp. 23–41, for his results in integral calculus, especially pp. 23–27 for the theorems we mention. Riemann's proof of Green's formula can be found on pp. 44–46 of Riemann (1990); Thomson and Tait (1890), p. 143, has their derivation of Stokes's theorem. The quotation about Bourbaki is in Weil (1992), p. 100.

For a historical account of the development of multiple integrals and Stokes's theorem, the reader may wish to see Katz (1979), (1982), and (1985). Engelsman (1984) is an insightful history of the early work on several variables calculus; he discusses the work of Leibniz, Johann and Niklaus I Bernoulli, Euler, and others. A detailed account of the development of calculus in connection with the theories of the earth's shape is presented in Greenberg (1995). This work also provides interesting information on the mathematicians involved and their interactions, as well as a fair amount of mathematical detail. For example, on p. 383, he gives the details omitted by Clairaut in his proof of the result on homogeneous functions mentioned in our text. For a discussion of Gauss's work on multiple integrals, Green's formula, and variational methods, see the article by O. Bolza in Gauss (1863–1927), vol. 10/II.

20

Algebraic Analysis: The Calculus of Operations

20.1 Preliminary Remarks

The operator or operational calculus, the method of treating differential operators as algebraic objects, was once thought to have originated with the English physicist and electrical engineer Oliver Heaviside (1850–1925). Indeed, Heaviside revived and brilliantly applied this method to problems in mathematical physics. But the basic ideas can actually be traced back to Leibniz and Lagrange who must be given credit as the founders of the operational method. With his notation for the differential and integral, Leibniz was able to regard some results on derivatives and integrals as analogs of algebraic results. The later insight of Lagrange was to extend this analogy to infinite series of differentials so that, in particular, he could write the Taylor expansion as an exponential function of a differential operator. In fact, this formal approach to infinite series appeared in the work of Newton himself. For Newton, infinite series in algebra served a purpose analogous to infinite decimals in arithmetic: They were necessary to carry out the algebraic operations to their completion. Newton's insightful algorithms using formal power series were of very wide applicability in analysis, algebra, and algebraic geometry; their power lay precisely in their formal nature. Thus, the algebraic analysis of the eighteenth century can trace its origins to Newton's genius. A branch of algebraic analysis focusing on the combinatorial aspects of power series was developed by the eighteenth-century German combinatorial school.

In a letter of May 1695 to Johann Bernoulli, Leibniz pointed out the formal resemblance between the expression for the nth derivative of a product xy and the binomial expansion of $(x+y)^n$. For $n=2$, for example, Leibniz wrote

$$(x+y)^2 = 1x^2 + 2xy + 1y^2, \quad d^2, xy = 1y\,ddx + 2\,dy\,dx + 1x\,ddy. \qquad (20.1)$$

He made similar remarks in a September 1695 letter to l'Hôpital, in which he used the symbol p for the power (or exponent) so that the analogy would be even more evident. Thus, he denoted x^n by $p^n x$, so that he could write

$$p^e(x+y) = p^e x \cdot p^0 y + \frac{e}{1}p^{e-1}x \cdot p^1 y + \frac{e \cdot e - 1}{1 \cdot 2}p^{e-2}x \cdot p^2 y + \cdots;$$

$$d^e(xy) = d^e x d^0 y + \frac{e}{1}d^{e-1}x \cdot d^1 y + \frac{e \cdot e - 1}{1 \cdot 2}d^{e-2}x \cdot d^2 y + \cdots.$$

The exponent e could be a positive or negative integer; when $e = -n$ was negative, d^{-n} denoted an n-fold integral. He mentioned that the $e = -1$ case was also noted by Johann Bernoulli. In fact, the formula in that case would be equivalent to Taylor's formula. Finally in 1710, Leibniz published a paper on this symbolic analogy. Later, Lagrange, inspired by this paper, extended the scope of this analogy by treating the symbol d, denoting the differential operator, as an algebraic object. In a paper of 1772 presented to the Berlin Academy, he gave the Taylor series formula as

$$u(x+\xi) = u + \frac{du}{dx}\xi + \frac{d^2u}{dx^2}\frac{\xi^2}{2} + \frac{d^3u}{dx^3}\frac{\xi^3}{2 \cdot 3} + \cdots = e^{\frac{du}{dx}\xi}, \tag{20.2}$$

where $\left(\frac{du}{dx}\xi\right)^n$ in the expansion of the exponential was understood to be $\frac{d^n u}{dx^n}\xi^n$. We observe that this point of view was not foreign to Euler. In a brilliant paper of 1750, he suggested replacing the nth derivative by z^n in order to solve a differential equation of infinite order. By this method, he obtained a result from which one could immediately derive the Fourier series of an arbitrary function. Unfortunately, Euler does not seem to have made use of this remarkable result.

The generation of mathematicians after Lagrange chose for clarity, to separate the symbol d/dx from the function u upon which it acted. Lacroix, for example, in his influential work summarizing the eighteenth-century discoveries in calculus, wrote $e^{\frac{du}{dx}\xi}$ as $e^{\xi\frac{d}{dx}}u$. It appears that the French mathematician L. F. Arbogast (1759–1803), the collector and preserver of important mathematical works, was the first to separate the operator from the object on which it operated. Arbogast's method, published in 1800, so impressed the English mathematician Charles Babbage that he wrote:

> Arbogast, in the 6th article of his, "Calcul des derivations," where, by a peculiarly elegant mode of separating the symbols of operation from those of quantity, and operating *upon them* as upon analytical symbols; he derives not only these, but many other much more general theorems with unparalleled conciseness.

Returning to Lagrange's paper, we note that he observed that the difference operator could be expressed as

$$\Delta u(x) = u(x+\xi) - u(x) = e^{\xi\frac{du}{dx}} - 1. \tag{20.3}$$

In Arbogast's notation, write $\Delta u = \left(e^{\xi\frac{d}{dx}} - 1\right)u$. Lagrange applied this formula to obtain a formal though very simple derivation of the Euler–Maclaurin summation, and he extended this to situations involving sums of sums. Laplace used Lagrange's operational method in his work on difference equations; he also attempted to give a rigorous derivation of Lagrange's formulas. In his work of 1800, Arbogast applied this technique to numerous problems, including the solutions of differential equations.

Then in 1808, Barnabé Brisson (1777–1828) independently applied this method to differential equations. A graduate of the École Polytechnique, Brisson published his paper in its journal. In the 1810s, other French mathematicians such as F. J. Servois (1768–1847) and J. F. Français (1775–1833) applied the methods of Lagrange and Arbogast to obtain some results on series. Servois also considered the logical foundation of the operator method. However, after the 1821 publication of Cauchy's *Analyse algébrique*, effectively establishing the limit concept at the foundation of analysis, the operational methods ceased to be developed in France. But in 1826, Cauchy presented a justification of the operational calculus using Fourier transforms. Interestingly, in the early twentieth century, integral transforms were applied to rigorize the operator methods employed by the physicist Heaviside. In 1926, Norbert Wiener created generalized harmonic analysis and one of his motivations was to provide rigorous underpinnings for the operational method.

During the 1830s and 1840s, important work in the operational calculus was done in Britain. Robert Murphy (1806–1843), Duncan Gregory (1813–1844), and George Boole (1815–1864) applied the methods to somewhat more difficult problems than those considered by Français and Servois. Much of the British work was done without full awareness of the earlier Continental work, so that even as late as 1851, William F. Donkin (1814–1869) published a paper in the *Cambridge and Dublin Mathematical Journal* giving an exposition of Arbogast's method of derivations. Thus, the British work was not a direct continuation of the work of Arbogast, Français, and Servois; its origins and motivations lay in a more formal and/or symbolic mathematical approach.

To understand the historical background of the British operational calculus, note that Britain produced a number of outstanding mathematicians in the first half of the eighteenth century, including Cotes, de Moivre, Taylor, Stirling, and Maclaurin. A large part of their work elaborated on or continued the study of topics opened up by Newton. There were also some good textbook writers such as Thomas Simpson and Edmund Stone who explained these developments to a larger audience. In the second half of the century, there was a swift decline in the development of mathematics in Britain. Mathematics was sustained at Cambridge by the almost solitary figure of Edward Waring, whose main interests were algebra and combinatorics, but he had few followers or students and little influence. Also, John Landen did interesting work in analysis, making a significant contribution to elliptic integrals.

British mathematicians had long paid scant attention to the major mathematical advances in continental Europe: the calculus of several variables and its applications to problems of mathematical physics developed by Euler, Fontaine, Clairaut, d'Alembert, Lagrange, and Laplace; major works in algebra produced by Euler, Lagrange, Vandermonde, and Ruffini; and also the brilliant progress in number theory made by Euler and Lagrange. In the early nineteenth century, Robert Woodhouse (1773–1827) appears to be one of the first British mathematicians to attempt to expand the focus of mathematics at Cambridge. He leaned strongly toward a formal or symbolic approach and his main interests lay in the foundations of calculus and the appropriate notation for its development. He also wrote expository works in subjects such as the calculus of variations and gravitation, and his efforts brought this continental work to the notice of the British. In 1803, Woodhouse wrote *The Principles of Analytical Calculation*, a polemical work on the foundation of calculus. He reviewed the foundational ideas of his predecessors: Newton, Leibniz, d'Alembert, Landen, and Lagrange. He rejected the limits of Newton and d'Alembert as well as the infinitesimals of Leibniz as inconsistent and inadequate, advocating instead the algebraic approach of Lagrange and Arbogast, though he disputed specific details. In the preface of his book, he wrote:

> I regard the rule for the multiplication of algebraic symbols, by which addition is compendiously exhibited, as the true and original basis of that calculus, which is equivalent to the fluxionary or differential calculus; on the direct operations of multiplication, are founded the reverse operations of division and extraction of roots, ... they are still farther comprehended under a general formula, called the expansion, or development of a function: from the second term of this expansion, the fluxion or differential of a quantity may immediately be deduced, and in a particular application, it appears to represent the velocity of a body in a motion.

Concerning the equal sign, $=$, Woodhouse maintained that in the context of series, this sign did not denote numerical equality but the result of an operation. So if $1/(1+x)$ denoted the series obtained by dividing 1 by $1+x$, then

$$\frac{1}{1+x} = 1 - x + x^2 - x^3 + \cdots.$$

On the other hand, if $1/(x+1)$ represented the series obtained when 1 was divided by $x+1$, then

$$\frac{1}{x+1} = \frac{1}{x} - \frac{1}{x^2} + \frac{1}{x^3} - \frac{1}{x^4} + \cdots.$$

Woodhouse remarked with reference to the two series that the equality $1/(1+x) = 1/(x+1)$ could not be affirmed.

Woodhouse wrote other articles and books advocating his formal point of view. In his 1809 textbook, *A Treatise on Plane and Spherical Trigonometry*, he defined the trigonometric functions by their series expansions and showed the advantages of the analytic approach over the geometric approach of Newton's *Principia*. Though his 1809 treatise acquired some popularity and went into several editions, Woodhouse was unable to convert the Cambridge dons. Progress in introducing the analytic approach into the curriculum was achieved mainly through the efforts of his students: Edward Ffrench Bromhead (1789–1855), Charles Babbage (1791–1871), George Peacock (1791–1858), and John Herschel (1792–1871). As students at Cambridge, they formed the Analytical Society in 1812 to promote and practice analytical mathematics; they decided to publish a journal called the *Memoirs of the Analytical Society*, though Babbage had wished to name it *The Principles of Pure D-ism* in opposition to the Dot-age of the university. Only one volume was published; it appeared in 1813 and contained one article by Babbage on functional equations and two by Herschel, on trigonometric series and on finite difference equations. As the members scattered, the Analytical Society ceased to meet, though many of its members became fellows or professors at Cambridge. In any case, Babbage, Peacock, and Herschel remained friends. They translated an elementary text by Lacroix on differential and integral calculus and in 1820 published a supplementary collection of examples on calculus, difference equations, and functional equations. Their efforts gradually influenced the teaching of mathematics at Cambridge, leading to the acceptance of Continental methods.

Even though Babbage succeeded Woodhouse as Lucasian Professor of Mathematics in 1828, he spent very little time at Cambridge. Consistent with his formal approach, he became interested in the mechanization of computation and spent the rest of his life on the problems associated with that. He first developed plans to construct a "difference engine," complete with printing device; he hoped it would eventually compute up to twenty decimal places using sixth-order differences. A Swedish engineer, Georg Schentz, used Babbage's description to build a machine with a printer capable of computing eight decimal places using fourth-order differences. Instead of actually building a machine, Babbage himself went on to design a more elaborate computer, called the "analytical engine," inspired by the study of Jacquard's punched cards weaving machine.

John Herschel lost interest in pure mathematics and became a professor of astronomy at Cambridge. So that left George Peacock to carry out the reform or modernization of the teaching of mathematics at Cambridge and more generally in England. In 1832, he published an algebra textbook in which he attempted to put the theory of negative and complex numbers on a firm foundation by dividing algebra into two parts, arithmetical and symbolical. The symbols of arithmetical algebra represented positive numbers, whereas the domain of the symbols in symbolical algebra was extended by the principle of the permanence of equivalent forms. This abstract principle implied, according to Peacock, that any formula in symbolical algebra would yield a formula in arithmetical algebra if the variables were properly chosen. Note that this approach excluded the possibility of a noncommutative algebra.

Ironically, this algebraic approach to calculus taken by the British mathematicians of the 1820s and 1830s stood in contrast with the rigorous methods contemporaneously introduced in Europe by Gauss, Cauchy, Abel, and Dirichlet. The next generation of British mathematicians, including Duncan Gregory, Robert Murphy, George Boole, Leslie Ellis (1817–1859) and others, were aware of the continental approach and yet they felt that their own methods had legitimacy. Early death prevented the talented Duncan Gregory from preparing a new foundation for this method. However, the symbolic method, even if lacking in rigor, had significant influence. The origin of some aspects of modern operational calculus and of the theory of distributions can be seen in the symbolic methods of Gregory and Boole. Moreover, some of the methods themselves were put on a more solid foundation through G. C. Rota's twentieth-century work on umbral calculus.

The British symbolic approach served as the starting point for some significant developments: the symbolic logic of Boole and Augustus De Morgan (1806–1871) and the invariant theory of Boole, Cayley, and Sylvester. Consider, for example, Boole's remarks in the introduction to his 1847 work, *The Mathematical Analysis of Logic*:

> They who are acquainted with the present state of the theory of Symbolical Algebra, are aware, that the validity of the processes of analysis does not depend upon the interpretation of the symbols which are employed, but solely upon the laws of their combination. Every system of interpretation which does not affect the truth of the relations supposed, is equally admissible.

G. H. Hardy wrote in his book *Divergent Series* that the British symbolical mathematicians had the spirit but not the accuracy of the twentieth-century algebraists.

Nevertheless, there is at least one example of abstract algebraic work consistent with the standards of today: Hamilton's theory of couples and of quaternions. The former laid a rigorous algebraic basis for complex numbers. And Hamilton reported that his 1843 discovery of quaternions was guided by a determination for consistency, so that he left open the possibility of an algebra with zero divisors or with noncommutativity. It is noteworthy that around 1819, Gauss composed a multiplication table for quaternions, though apparently he did not develop this further.

Papers by Murphy, Duncan Gregory, and Boole published between 1835 and 1845 provided important steps toward the creation of concepts laying the groundwork for the eventual construction of abstract algebraic theories. Murphy, the son of a shoemaker-parish clerk in Cork County, Ireland, studied mathematics on his own, and his talent soon became known. In 1819, Mr. Mulcahy, a teacher in Cork County, published mathematical problems in the local newspaper; he soon began to receive original solutions from an anonymous reader. He was surprised to discover that his correspondent was a boy of 13. After this, Murphy began to receive encouragement and financial assistance to continue his studies. In 1825, some of his work was brought to the attention of Woodhouse; consequently, Murphy was admitted to Gonville and Caius College, Cambridge, from which he graduated in 1828. In an 1835 paper on definite integrals, Murphy introduced the idea of orthogonal functions, giving them the name reciprocal functions. In his 1837 paper, "First Memoir on the Theory of Analytical Operations", he defined what he called linear operations and showed that their sums and products, obtained by composition, were also linear operations, though the products were not necessarily commutative. He stated a binomial theorem for noncommutative operations, and went on to consider inverses of operations, proving that the inverse of the product of two operations A and B was $B^{-1}A^{-1}$. Murphy also defined the kernel of an operation, naming it the appendage of the operation. He applied his theory mainly to three operations: the differential operator, the difference operator Δ, and the operator transforming a function $f(x)$ to $f(x+h)$. Thus, in his paper Murphy isolated and defined some basic ideas of a system of abstract algebra.

Around this time, Duncan F. Gregory, descendent of the great James Gregory, also began to develop his mathematical ideas. Gregory was born at Edinburgh, Scotland, and graduated from Trinity College, Cambridge, in 1837. Even as a student, Gregory was interested in mathematical research and in encouraging British mathematicians to take up this activity. As a step in this direction, in 1837 he helped found the *Cambridge Mathematical Journal*, of which he was the editor until a few months before his premature death in February 1844. R. Leslie Ellis, who served as editor after this, wrote that Gregory was particularly well qualified for this position for "his acquaintance with mathematical literature was very extensive, while his interest in all subjects connected with it was not only very strong, but also singularly free from the least tinge of jealous or personal feeling. That which another had done or was about to do, seemed to give him as much pleasure as if he himself had been the author of it, and this even when it related to some subject which his own researches might seem to have appropriated." In addition, D. F. Gregory encouraged undergraduates to publish and permitted authors to publish anonymously so that they need not fear for their reputations. This journal was later renamed *Cambridge and Dublin Mathematical Journal*; it then evolved into the

Quarterly Journal of Pure and Applied Mathematics, with editors including William Thomson and J. W. L. Glaisher. In fact, most British mathematicians of the period contributed papers to the *CMJ*, including Augustus De Morgan, J. J. Sylvester, George Gabriel Stokes, Arthur Cayley, George Boole, and William Thomson.

By about 1845, research in operational calculus was no longer widely pursued. In the 1890s, however, Heaviside revived operational methods in order to solve differential equations occurring in electrical engineering problems. Heaviside may or may not have independently devised these methods, but he made at least one important new contribution: The use of the step function $H(t) = 0$ with t negative, and $H(t) = 1$ with t nonnegative. By taking the derivative of this function, he obtained the Dirac delta function. In some situations, Heaviside also used the derivative of the delta function. Because these methods were so successful in solving problems in electrical engineering, mathematicians such as Wiener, Carson, Doetsch, and van der Pol made successful efforts toward putting them on a rigorous footing.

Toward the end of the eighteenth century, algebraic analysis was taken in a different direction by the German combinatorial school founded by C. F. Hindenburg (1741–1808), Professor at Leipzig. Since combinatorial considerations were important in probability computations as well as in deriving formulas for higher derivatives of products of functions and of compositions of functions, Hindenburg saw that he could find relations between/among series through the use of combinatorial concepts. This school took as its starting point and inspiration Euler's extensive use of series to tackle various mathematical problems, as set forth in his *Introductio in Analysin Infinitorum* of 1748. The combinatorial school played a significant role in the overall development of mathematics in Germany; it included among its early members Christian Kramp (1760–1826), Gauss's thesis supervisor J. F. Pfaff (1765–1825), and H. A. Rothe (1773–1842). Later, H. F. Scherk (1798–1885), Franz Ferdinand Schweins (1780–1856), August Leopold Crelle (1780–1855), Weierstrass's teacher Christoph Gudermann (1798–1852), and Moritz A. Stern (1807–1894) made contributions to this tradition, and many of them were active in instituting educational reforms in Germany. In fact, it is very likely that Weierstrass chose to make power series the fundamental object in his study of analysis because of his early contact with Gudermann. Also, Riemann's earliest research on fractional derivatives and infinite series was done while he was a student under Stern, though Riemann eventually took a completely different route as a result of his later association with Dirichlet and Gauss. The combinatorial school produced some interesting results useful even today, and their approach to infinite series is not without significance in modern mathematical research.

Hindenburg believed, and his colleagues agreed, that his most important work was the polynomial formula he proved in 1779. A power series raised to an exponent is another power series:

$$(1 + a_1 x + a_2 x^2 + a_3 x^3 + \cdots)^m = 1 + A_1 x + A_2 x^2 + A_3 x^3 + \cdots.$$

Hindenburg's formula expressed A_n in terms of $a_1, a_2, \ldots, a_n$. De Moivre had already done this with m a positive integer in a paper of 1697. Leibniz and Johann Bernoulli also considered this case in letters exchanged in 1695. This particular case is quite useful; for example, it can be applied to give a short proof of Faà di Bruno's formula, giving the nth derivative of a composition of two functions. Faà di Bruno stated this in the mid-nineteenth century without referring to the earlier proofs; Arbogast offered a proof in 1800.

Using Newton's binomial theorem, Hindenburg extended his formula to fractional and negative m by expanding $(1+y)^m$, where $y = a_1 x + a_2 x^2 + a_3 x^3 + \cdots$, and y^n was obtained from the polynomial theorem for positive integral n. Part of Hindenburg's achievement was to clarify the combinatorial content of the formula. De Moivre had given only the recursive rule for the calculation of A_{n+1} from A_n. In the notation presented by B. F. Thibaut in his 1809 textbook *Grundriss der Allgemeinen Arithmetik*, Hindenburg's formula would be expressed as

$$A_n = \sum_{h=1}^{n} \binom{m}{h} p^{n^h} C.$$

The symbol $^{n^h}C$ represented the sum of all products of h factors taken from $a_1, a_2, \ldots, a_n$, so that the sum of the indices in each summand was n. The symbol p stood for the coefficient associated with each summand, each summand consisting of h factors, and this coefficient gave the number of different permutations of the h factors. Thus, $^{6^3}C$ stood for $a_1^2 a_4 + a_1 a_2 a_3 + a_2^3$ and the number of terms in the sum was the number of partitions of 6 with exactly three parts. Therefore, $p\ ^{6^3}C$ represented

$$\frac{3!}{2!1!} a_1^2 a_4 + \frac{3!}{(1!)^3} a_1 a_2 a_3 + a_2^3.$$

The combinatorial school set great importance on this formula, overestimating its potential. Still, Hindenburg's formula is useful in power series manipulation.

In 1793, H. A. Rothe used Hindenburg's formula to state the reversion of series formula as a combinatorial relation. Two years later, Rothe and Pfaff showed the equivalence of Rothe's formula with the Lagrange inversion formula. In modern times, Lagrange's formula has been regarded as more combinatorial than analytic in nature; in this respect, the combinatorialists were on the right track. Rothe also found one important terminating version of the q-binomial theorem, published in the preface of his 1811 book. In this formula, the coefficients of the powers of x are q-extensions of the binomial coefficients, now called Gaussian polynomials. It is possible that Rothe may have discovered these polynomials even before Gauss's work of 1805, published in 1811. It would be nice to know Rothe's combinatorial interpretation of these polynomials; he gave no proof or comment. In order to get an insight into the combinatorialists' mathematical style, consider comment given by Thomas Muir in his monumental *The Theory of Determinants in the Historical Order of Development*:

> Rothe was a follower of Hindenburg, knew Hindenburg's preface to Rüdiger's *Specimen Analyticum*, and was familiar with what had been done by Cramer and Bézout.... His memoir is very explicit and formal, proposition following definition, and corollary following proposition, in the most methodical manner.

Christian Kramp taught mathematics, chemistry, and experimental physics at École Centrale in Cologne and in 1809 he became professor of mathematics and dean of the faculty of science at Strasbourg. He was a follower of Hindenburg and contributed articles to various journals edited by Hindenburg. In a paper of 1794, he derived some interesting properties of Stirling numbers. One object of interest for him was what he termed a factorial:

$$a^{n|d} = a(a+d)(a+2d)\cdots(a+(n-1)d).$$

He expanded this as a polynomial in a and d, and obtained formulas for the coefficients involving Stirling numbers. Denoting the Stirling numbers of the first and second kinds by $s(n,k)$ and $S(n,k)$, Kramp proved that

$$|s(n+1, n+1-k)| = \sum \binom{n+1}{k+l} \frac{(k+l)!}{j_1! 2^{j_1} j_2! 3^{j_2} j_3! 4^{j_3} \cdots},$$

$$S(n+k, n) = \sum \binom{n+k}{k+l} \frac{(k+l)!}{j_1!(2!)^{j_1} j_2!(3!)^{j_2} j_3!(4!)^{j_3} \cdots},$$

where the sums were over all nonnegative j such that $j_1 + 2j_2 + 3j_3 + \cdots = k$, and where $l = j_1 + j_2 + j_3 + \cdots$. Kramp also introduced the factorial notation, $n!$.

20.2 Lagrange's Extension of the Euler–Maclaurin Formula

In his 1772 "Sur une nouvelle espèce de calcul," Lagrange set out to create a new symbolic method in calculus. As a first step, he expressed the Taylor series of a function $u(x, y, z, \ldots)$ of several variables as

$$u(x+\xi, y+\psi, z+\zeta, \ldots) = e^{\xi \frac{du}{dx} + \psi \frac{du}{dy} + \zeta \frac{du}{dz} + \cdots}.$$

In his symbolic notation, Lagrange understood the numerator of the nth term of $e^{\frac{du}{dx}\xi + \cdots}$ to represent $(\xi \frac{d}{dx} + \psi \frac{d}{dy} + \cdots)^n u$ rather than $(\xi \frac{du}{dx} + \psi \frac{du}{dy} + \cdots)^n$. In effect, Lagrange was treating the derivative operator as an algebraic quantity. The later notation of Arbogast makes this approach clearer, allowing us to write $e^{\xi \frac{d}{dx}} u$ for Lagrange's $e^{\frac{du}{dx}\xi}$ and $\left(e^{\xi \frac{d}{dx}} - 1\right)^\lambda u$ for his $\left(e^{\frac{du}{dx}\xi} - 1\right)^\lambda$. It is easy to see that the last expression is the symbolic form of the λth difference $\Delta^\lambda u$, since we may write

$$\Delta u = u(x+\xi) - u(x) = \left(e^{\xi \frac{d}{dx}} - 1\right) u. \tag{20.4}$$

It follows that the difference operator Δ can be identified with the operator $e^{\xi \frac{d}{dx}} - 1$ and the repeated application of these operations yields

$$\Delta^\lambda u = \left(e^{\xi \frac{d}{dx}} - 1\right)^\lambda u. \tag{20.5}$$

Following Leibniz, Lagrange noted that, given the derivative operator d, he could write

$$d^{-1} = \int, \quad d^{-2} = \int^2, \ldots \text{ and}$$

$$\Delta^{-1} = \Sigma, \quad \Delta^{-2} = \Sigma^2, \ldots.$$

Here $\int^2$ stood for an iterated integral and Σ^2 for an iterated sum. Lagrange applied his symbolic method to a generalization of the Euler–Maclaurin summation by expanding the expression in (20.5). He assumed the series expansion

$$(e^w - 1)^\lambda = w^\lambda (1 + Aw + Bw^2 + Cw^3 + Dw^4 + \cdots)$$

and took the logarithm to obtain

$$\lambda \ln(e^w - 1) - \lambda \ln w = \ln(1 + Aw + Bw^2 + Cw^3 + Dw^4 + \cdots).$$

By differentiation, he found

$$\lambda \left(\frac{e^w}{e^w - 1} - \frac{1}{w} \right) = \frac{A + 2Bw + 3Cw^2 + 4Dw^3 + \cdots}{1 + Aw + Bw^2 + Cw^3 + Dw^4 + \cdots}.$$

Since

$$\frac{e^w}{e^w - 1} = \frac{1}{1 - e^{-w}} = \frac{1}{w - \frac{w^2}{2} + \frac{w^3}{2 \cdot 3} - \frac{w^4}{2 \cdot 3 \cdot 4} + \cdots},$$

he obtained the equation

$$\lambda \left(\frac{1}{2} - \frac{w}{2 \cdot 3} + \frac{w^2}{2 \cdot 3 \cdot 4} - \cdots \right) (1 + Aw + Bw^2 + Cw^3 + \cdots)$$
$$= \left(1 - \frac{w}{2} + \frac{w^2}{2 \cdot 3} - \frac{w^3}{2 \cdot 3 \cdot 4} + \cdots \right) (A + 2Bw + 3Cw^2 + \cdots).$$

Finally, by equating the coefficients of the powers of w, Lagrange found that

$$A = \frac{\lambda}{2}, \quad 2B = \frac{(\lambda + 1)A}{2} - \frac{\lambda}{2 \cdot 3}, \quad 3C = \frac{(\lambda + 2)}{2} B - \frac{(\lambda + 1)}{2 \cdot 3} C + \frac{\lambda}{2 \cdot 3 \cdot 4},$$

$$4D = \frac{(\lambda + 3)}{2} C - \frac{(\lambda + 2)}{2 \cdot 3} B + \frac{(\lambda + 1)}{2 \cdot 3 \cdot 4} A - \frac{\lambda}{2 \cdot 3 \cdot 4 \cdot 5} \text{ etc.} \quad (20.6)$$

Now because $\Delta^{-\lambda} = \Sigma^\lambda$, Lagrange could replace λ by $-\lambda$ in (20.5) to get

$$\Sigma^\lambda u = \frac{\int^\lambda u \, dx^\lambda}{\xi^\lambda} + \alpha \frac{\int^{\lambda-1} u \, dx^{\lambda-1}}{\xi^{\lambda-1}} + \beta \frac{\int^{\lambda-2} u \, dx^{\lambda-2}}{\xi^{\lambda-2}} + \cdots. \quad (20.7)$$

We observe that when λ was changed to $-\lambda$ in (20.6), Lagrange denoted the changed values of A, B, C, ... by α, β, γ, ... The case $\lambda = 1$ in (20.7) is immediately distinguishable as the Euler–Maclaurin formula. In the same paper, Lagrange then proceeded to derive a formula for repeated integrals in terms of sums. He rewrote (20.4) as

$$\xi \frac{du}{dx} = \ln(1 + \Delta) u$$

or in Arbogast's notation,

$$\xi \frac{d}{dx} = \ln(1 + \Delta) \quad (20.8)$$

20.2 Lagrange's Extension of the Euler–Maclaurin Formula

and more generally

$$\xi^\lambda \frac{d^\lambda}{dx^\lambda} = [\ln(1+\Delta)]^\lambda. \tag{20.9}$$

By expanding the right-side of (20.9) as a series in powers of Δ, Lagrange obtained the coefficients of the expansion by a method similar to the one that gave him (20.6). Once again, he replaced λ by $-\lambda$ to obtain

$$\frac{\int^\lambda u\,dx^\lambda}{\xi} = \sum^\lambda u + \mu \sum^{\lambda-1} u + \nu \sum^{\lambda-2} u + w \sum^{\lambda-3} u + \cdots, \quad \text{where} \tag{20.10}$$

$$\mu = \frac{\lambda}{2},\ 2\nu = \frac{(\lambda-1)\mu}{2} - \frac{\lambda}{2\cdot 3},\ 3w = \frac{(\lambda-2)\nu}{2} - \frac{(\lambda-1)\mu}{2\cdot 3} + \frac{\lambda}{3\cdot 4},$$

$$4\chi = \frac{(\lambda-3)w}{2} - \frac{(\lambda-2)\nu}{2\cdot 3} + \frac{(\lambda-1)\mu}{3\cdot 4} - \frac{\lambda}{4\cdot 5},\ \text{etc.}$$

When $\lambda = 1$ in (20.10), Lagrange got the value of the integral in terms a sum involving finite differences:

$$\frac{\int u\,dx}{\xi} = \sum u + \mu u + \nu \Delta u + w \Delta^2 u + \chi \Delta^3 u + \cdots. \tag{20.11}$$

This is exactly the formula communicated by James Gregory to Collins in November 1670. Recall that Gregory most probably discovered this formula by integrating the Gregory–Newton interpolation formula. Lagrange may not have been aware of Gregory's work, but he referred to Cotes, Stirling, and others who used similar, though not identical, results. Laplace may have found this result independently; he used it in his astronomical work. Indeed, this formula was sometimes attributed to Laplace, in particular by Poisson.

Lagrange employed (20.11) to derive an inverse factorial series for $\ln(1+1/x)$, taking $u = 1/x$ and $\xi = 1$ to obtain

$$\ln x = \sum \frac{1}{x} + \frac{\mu}{x} + \nu \Delta \frac{1}{x} + w \Delta^2 \frac{1}{x} + \chi \Delta^3 \frac{1}{x} + \cdots, \quad \text{where}$$

$$\sum \frac{1}{x} = \frac{1}{x-1} + \frac{1}{x-2} + \frac{1}{x-3} + \cdots,$$

$$\Delta \frac{1}{x} = \frac{1}{x+1} - \frac{1}{x} = -\frac{1}{x(x+1)},$$

$$\Delta^2 \frac{1}{x} = \frac{2}{x(x+1)(x+2)},$$

$$\Delta^3 \frac{1}{x} = -\frac{2\cdot 3}{x(x+1)(x+2)(x+3)},\ \text{etc.}$$

Changing x to $x+1$, Lagrange obtained a similar series for $\ln(x+1)$. He then subtracted the series for $\ln x$ to obtain the desired result.

Lagrange's heuristic method was immediately welcomed as a powerful tool in discovering interesting and useful formulas. Laplace's papers on finite differences in the 1770s discussed and used Lagrange's symbolic method. Laplace thought that Lagrange's formula $\Delta^n u = \left(e^{h\frac{d}{dx}} - 1\right)^n u$ could be rigorously established by the use of formal power series. He observed that

$$\Delta^n u = \frac{d^n u}{dx^n} h^n + A_1 \frac{d^{n+1} u}{dx^{n+1}} h^{n+1} + A_2 \frac{d^{n+2} u}{dx^{n+2}} h^{n+2} + \cdots$$

for constant $A_1, A_2, \ldots$. He believed the problem could be reduced to proving that these coefficients were identical with the coefficients of the powers of h in the expansion of $\left(e^h - 1\right)^n$. He noted that the constants $A_1, A_2, \ldots$ were the same for all functions u, so he took $u = e^x$. Then $\Delta e^x = e^{x+h} - e^x = e^x \left(e^h - 1\right)$ and more generally $\Delta^n e^x = e^x \left(e^h - 1\right)^n$. Thus,

$$\left(e^h - 1\right)^n = h^n + A_1 h^{n+1} + A_2 h^{n+2} + \cdots,$$

completing the clever proof.

In 1807, John Brinkley (1766–1835), professor of mathematics at the University of Dublin and a mentor to Hamilton, presented in the *Philosophical Transactions* an interesting expression for the constants $A_1, A_2, \ldots$. He noted that

$$\left(e^h - 1\right)^n = e^{nh} - \binom{n}{1} e^{(n-1)h} + \binom{n}{2} e^{(n-2)h} - \cdots$$

$$= \left(1 - \binom{n}{1} + \binom{n}{2} - \text{etc.}\right)$$

$$+ \left(n - \binom{n}{1}(n-1) + \binom{n}{2}(n-2) - \cdots\right) \frac{h}{1!}$$

$$+ \left(n^2 - \binom{n}{1}(n-1)^2 + \binom{n}{2}(n-2)^2 - \cdots\right) \frac{h^2}{2!} + \cdots$$

$$+ \left(n^m - \binom{n}{1}(n-1)^m + \binom{n}{2}(n-2)^m - \cdots\right) \frac{h^m}{m!} + \cdots.$$

Further, it was clear from the formula

$$\Delta^n f(0) = f(n) - \binom{n}{1} f(n-1) + \binom{n}{2} f(n-2) - \cdots$$

that the coefficient of $h^m/m!$ was $\Delta^n 0^m$. Thus, Brinkley had $A_k = \frac{\Delta^n 0^{n+k}}{(n+k)!}$. Note that these numbers are related to Stirling numbers of the second kind: $(n+k)! A_k = n! s(n+k, n)$.

Brinkley (c. 1763–1835) studied at Cambridge, graduating senior wrangler in 1788. He was the first Royal Astronomer of Ireland and later became Bishop of Cloyne. It was to Brinkley that the 17-year-old Hamilton communicated his work on geometrical optics. Brinkley encouraged Hamilton by presenting his work to the Irish

20.3 Français's Method of Solving Differential Equations

Jacques F. Français, whose mathematical work incorporated some results from the notebooks of his late brother François, based his solution of ordinary differential equations with constant coefficients on the relation between the Arbogast operator E and the differential operator D:

$$E\phi(x) = \phi(x+1) = e^D \phi, \tag{20.12}$$

where $D = d/dx$. We note that Français's notation had δ for D. Now if ϕ were a solution of the equation

$$\frac{d\phi}{dx} - a\phi = 0, \quad \text{or} \quad (D-a)\phi = 0, \tag{20.13}$$

then Français had $D - a = 0$ by the separation of the operator. By (20.12) he had $E = e^D = e^a$ or $E^k = e^{ak}$ and hence $1 = e^{ak} E^{-k}$. He then used this relation to solve (20.13):

$$\phi(x) = 1\phi(x) = e^{ak} E^{-k} \phi(x) = e^{ak} \phi(x-k) \quad \text{or}$$
$$\phi(k) = \phi(0) e^{ak}.$$

Thus, $\phi(x) = Ce^{ax}$, where C was a constant, and the differential equation (20.13) was solved. To solve the general homogeneous differential with constant coefficients

$$D^n \phi + a_1 D^{n-1} \phi + \cdots + a_n \phi = 0,$$

Français separated the operator and factored the nth degree polynomial in D to obtain

$$(D - \alpha_1)(D - \alpha_2) \cdots (D - \alpha_n) = 0.$$

This gave him the n equations:

$$D - \alpha_1 = 0, \ D - \alpha_2 = 0, \ldots, \ D - \alpha_n = 0$$

whose solutions he expressed as $e^{\alpha_1 x}, e^{\alpha_2 x}, \ldots, e^{\alpha_n x}$. Note that this is an exhaustive list of the independent solutions, under the condition that $\alpha_1, \alpha_2, \ldots, \alpha_n$ are all distinct. Français also applied the operational method for the summation of series. He found a series for π, reminiscent of a result obtained by the Kerala school. In a paper of 1811, he started with Euler's series

$$\frac{\pi}{4} \alpha = \sin \alpha - \frac{1}{3^2} \sin 3\alpha + \frac{1}{5^2} \sin 5\alpha - \frac{1}{7^2} \sin 7\alpha + \cdots.$$

He rewrote this as

$$\frac{\pi}{2}\alpha\sqrt{-1} = \left(e^{\alpha\sqrt{-1}} - e^{-\alpha\sqrt{-1}}\right) - \frac{1}{3^2}\left(e^{3\alpha\sqrt{-1}} - e^{-3\alpha\sqrt{-1}}\right) + \frac{1}{5^2}\left(e^{5\alpha\sqrt{-1}} - e^{-5\alpha\sqrt{-1}}\right) - \cdots.$$

He then set $\alpha\sqrt{-1} = D$, $e^D = E$ to obtain

$$\frac{\pi}{2}D = (E - E^{-1}) - \frac{1}{3^2}(E^3 - E^{-3}) + \frac{1}{5^2}(E^5 - E^{-5}) - \cdots.$$

Français next applied this operator equation to $\phi(x)$, so that

$$\frac{\pi}{2}\phi'(x) = \phi(x+1) - \phi(x-1) - \frac{1}{3^2}(\phi(x+3) - \phi(x-3)) + \cdots.$$

Recall that $E\phi(x) = \phi(x+1)$. Taking $\phi(x) = x$, he obtained Leibniz's formula. For $\phi(x) = 1/x$, he found

$$\frac{\pi}{4} \cdot \frac{1}{x^2} = \left(\frac{1}{x^2 - 1}\right) - \frac{1}{3}\left(\frac{1}{x^2 - 3^2}\right) + \frac{1}{5}\left(\frac{1}{x^2 - 5^2}\right) - \frac{1}{7}\left(\frac{1}{x^2 - 7^2}\right) + \cdots. \quad (20.14)$$

Putting $\frac{1}{x^2} = -a$, he could rewrite the equation as

$$\frac{\pi}{4} = \frac{1}{1+a} - \frac{1}{3} \cdot \frac{1}{1+3^2a} + \frac{1}{5} \cdot \frac{1}{1+5^2a} - \frac{1}{7} \cdot \frac{1}{1+7^2a} + \cdots.$$

Then again, by taking $\phi(x) = \ln x$, Français obtained the formula

$$\frac{\pi}{2} \cdot \frac{1}{x} = \ln\left(\frac{x+1}{x-1}\right) - \frac{1}{3^2}\ln\left(\frac{x+3}{x-3}\right) + \frac{1}{5^2}\ln\left(\frac{x+5}{x-5}\right) - \cdots.$$

Finally, to derive another interesting series, he set $a\sqrt{-1} = 1/x$ in (20.14) and then integrated, obtaining

$$\frac{\pi}{4}a = \arctan a - \frac{1}{3^2}\arctan 3a + \frac{1}{5^2}\arctan 5a - \cdots.$$

20.4 Herschel: Calculus of Finite Differences

In the appendix to their English translation of Lacroix's book, Babbage and Herschel included a large number of examples on functional and difference equations, some of which were original. Like his followers, Herschel showed much manipulative ability of an algebraic kind. To see some of Herschel's work on difference equations, consider the equation

$$u_{x+1}u_x - a(u_{x+1} - u_x) + 1 = 0,$$

first given in a paper of Laplace but for which Herschel found a new solution. He differentiated and rewrote it as

$$(a + u_{x+1})\frac{du_x}{dx} - (a - u_x)\frac{du_{x+1}}{dx} = 0.$$

He solved for a in the original equation and used this value in the second equation to find, after simplification

$$\frac{du_{x+1}}{1 + u_{x+1}^2} - \frac{du_x}{1 + u_x^2} = 0, \text{ or } \Delta\int\frac{du_x}{1 + u_x^2} = A,$$

where A was a constant depending on a. Solving this simple difference equation, he had

$$\int\frac{du_x}{1 + u_x^2} = Ax + C,$$

where C was an arbitrary constant. After computing the integral, he got $u_x = \tan(Ax + C)$ and therefore

$$u_{x+1} = \tan(Ax + C + A) = \frac{u_x + \tan A}{1 - u_x \tan A}.$$

At this point he rewrote the original difference equation as

$$u_{x+1} = \frac{u_x + \frac{1}{a}}{1 - u_x \cdot \frac{1}{a}},$$

so he could conclude that $\tan A = \frac{1}{a}$ or $A = \tan^{-1}\frac{1}{a}$. Thus, Herschel obtained the result

$$u_x = \tan\left(x\tan^{-1}\frac{1}{a} + C\right).$$

To see an example of Herschel's symbolic approach, take an analytic function $f(x)$, and let

$$f(e^t) = A_0 + A_1 t + A_2 t^2 + \cdots + A_x t^x + \cdots.$$

Herschel wished to find an expression for A_x; he started with the Taylor expansion

$$f(e^t) = f(1) + \frac{f'(1)}{1}(e^t - 1) + \frac{f''(1)}{1 \cdot 2}(e^t - 1)^2 + \cdots,$$

and noted that for $x \geq 1$, the coefficient of t^x in $f(1)$ was 0; in

$$\frac{f'(1)}{1}(e^t - 1), \quad \text{it was} \quad \frac{f'(1)}{1} \cdot \frac{1}{1 \cdot 2 \cdots x} = \frac{f'(1)}{1} \cdot \frac{\Delta 0^x}{1 \cdot 2 \cdots x};$$

and in

$$\frac{f''(1)}{1 \cdot 2}(e^t - 1)^2, \quad \text{it was} \quad \frac{f''(1)}{1 \cdot 2} \cdot \frac{\Delta^2 0^x}{1 \cdot 2 \cdots x},$$

and so on. He could conclude that

$$A_x = \frac{1}{1 \cdot 2 \cdots x}\left(f(1) \cdot 0^x + \frac{f'(1)}{1}\Delta 0^x + \frac{f''(1)}{1 \cdot 2}\Delta^2 0^x + \cdots\right).$$

He then wrote, "let the symbols of operation be separated from those of quantity, and we get

$$A_x = \frac{1}{1 \cdot 2 \cdots x}\left(f(1) + \frac{f'(1)}{1}\Delta + \frac{f''(1)}{1 \cdot 2}\Delta^2 + \cdots\right)0^x = \frac{f(1+\Delta)0^x}{1 \cdot 2 \cdots x}."$$

Herschel apparently saw that taking a function of an operator was somewhat problematic, commenting that it should be "understood to have no other meaning than its development, of which it is a mere abbreviated expression."

20.5 Murphy's Theory of Analytical Operations

Murphy began his 1837 paper on analytical operations, "The elements of which every distinct analytical process is composed are three, namely, first the *Subject*, that is, the symbol on which a certain notified operation is to be performed; secondly, the *Operation* itself, represented by its own symbol; thirdly, the *Result*, which may be connected with the former two by the algebraic sign of equality." He defined several operations. For example, he denoted by Ψ the operation changing x to $x+h$, and by Δ the operation subtracting the subject from the result of changing x to $x+h$ in the subject. He wrote these operations as

$$[f(x)]\Psi = f(x+h), \quad [f(x)]\Delta = f(x+h) - f(x).$$

The operations themselves could be algebraically combined. Thus, $\Psi = \Delta + 1$, where 1 was the operation under which the subject remained the same. Murphy defined the linearity of an operation:

$$[f(x) + \phi(x)]\Psi = [f(x)]\Psi + [\phi(x)]\Psi.$$

He called two operations fixed or free, depending on whether they were noncommutative or commutative in the given situation. Thus,

$$[x^n]x\Psi = [x^{n+1}]\Psi = (x+h)^{n+1},$$
$$[x^n]\Psi x = [(x+h)^n]x = x(x+h)^n,$$

so that $x\Psi \neq \Psi x$; but for a constant a

$$[x^n]a\Psi = [ax^n]\Psi = a(x+h)^n,$$
$$[x^n]\Psi a = [(x+h)^n]a = a(x+h)^n,$$

so that $a\Psi = \Psi a$.

He also stated a noncommutative binomial theorem: When θ and θ' were fixed operations,

$$(\theta + \theta')^n = \theta^{(n)} + \theta^{(n-1)}\theta' + \theta^{(n-2)}\theta'^{(2)} + \cdots + \theta\theta'^{(n-1)} + \theta'^{(n)}.$$

Here the term $\theta^{(n-1)}\theta'$ represented the sum of n terms formed by placing θ' at the beginning, at the end, and in all the $n-2$ intermediate positions of the expression $\theta \cdot \theta \cdots \theta = \theta^{n-1}$. Similarly, $\theta^{(n-2)}\theta'^{(2)}$ signified a similar sum of $n(n-1)/2$ terms and so on. Murphy carefully defined some important algebraic concepts, such as the inverse and the kernel. Concerning the inverse: "Suppose θ to represent any operation which performed on a subject $[u]$ gives y as the result, then the inverse operation is denoted by θ^{-1}, and is such that when $[y]$ is made the subject u becomes the result." The kernel was called the appendage, denoted by $[0]\theta^{-1}$. Murphy showed, for example, that if d_x denoted the derivative with respect to x, then $[0]d_x^{-1}$ consisted of all the constants. To prove this, he took $[0]d_x^{-1} = \phi(x)$, implying that $[\phi(x)]d_x = 0$, and hence $[\phi(x)]d_x^2 = 0, [\phi(x)]d_x^3 = 0, \ldots$. Murphy then employed Taylor's theorem,

$$\phi(x+h) = \phi(x) + h\frac{d\phi}{dx} + \frac{h^2}{1\cdot 2}\frac{d^2\phi}{dx^2} + \cdots,$$

to obtain $\phi(x+h) = \phi(x)$, meaning $\phi(x)$ was a constant. Here Murphy assumed without comment that ϕ was analytic. To find the kernel of d_x^{-n}, he observed that

$$[0]d_x^{-2} = [0]d_x^{-1}d_x^{-1} = [c]d_x^{-1} + [0]d_x^{-1} = cx + c',$$

and, more generally,

$$[0]d_x^{-n} = A_1 x^{n-1} + A_2 x^{n-2} + \cdots + A_n.$$

In another interesting example, Murphy let the subject be $f(x+y)$; let Ψ_x be the operation under which x received an increment h; and let Ψ_y be the operation under which y received an increment of h. Then, obviously, $[f(x+y)](\Psi_x - \Psi_y) = 0$, and therefore $f(x+y)(\Delta_y - \Delta_x) = 0$, so that $f(x+y)$ was a value in $[0](\Delta_y - \Delta_x)^{-1}$. He explained that $(\Delta_y - \Delta_x)^{-1}$ could be expanded as

$$(\Delta_y - \Delta_x)^{-1} = \Delta_y^{-1} + \Delta_y^{-2}\Delta_x + \Delta_y^{-3}\Delta_x^2 + \Delta_y^{-4}\Delta_x^3 + \cdots,$$

so that $\quad [0](\Delta_y - \Delta_x)^{-1} = [0]\left(\Delta_y^{-1} + \Delta_y^{-2}\Delta_x + \Delta_y^{-3}\Delta_x^2 + \cdots\right).$

Murphy then explained how to derive the Gregory–Newton interpolation formula from this last equation. He noted that $[0]\Delta_y^{-1}$ was a function independent of y, so he had $[0]\Delta_y^{-1} = \phi(x)$, where ϕ was an arbitrary function of x. Similarly, $[0]\Delta_y^{-2} = \phi(x) \cdot \frac{y}{h}$, where the appendage was omitted without loss of generality. Then again,

$$[0]\Delta_y^{-3} = \phi(x) \cdot \frac{y(y-h)}{1 \cdot 2 \cdot h^2} \quad \text{for} \quad [y(y-h)]\Delta_y = (y+h)y - y(y-h) = 2hy.$$

By a similar argument,

$$[0]\Delta_y^{-4} = \phi(x) \cdot \frac{y(y-h)(y-2h)}{1 \cdot 2 \cdot 3 \cdot h^3},$$

and so on. In this way, Murphy obtained the relation

$$[0](\Delta_y - \Delta_x)^{-1} = \phi(x) + \frac{y}{h} \cdot \Delta\phi(x) + \frac{y(y-h)}{1 \cdot 2 \cdot \cdot h^2} \cdot \Delta^2\phi(x)$$
$$+ \frac{y(y-h)(y-2h)}{1 \cdot 2 \cdot 3 \cdot h^3} \cdot \Delta^3\phi(x) + \cdots,$$

and "since $f(x+h)$ is included in this general expression, the particular form to be assigned to the arbitrary $\phi(x)$ is known by making $y = 0$, which gives $\phi(x) = f(x)$." Thus, Murphy had the Gregory–Newton interpolation formula,

$$f(x+y) = f(x) + \frac{y}{h}\Delta f(x) + \frac{y(y-h)}{1 \cdot 2 \cdot h^2}\Delta^2 f(x) + \frac{y(y-h)(y-2h)}{1 \cdot 2 \cdot 3 \cdot h^3}\Delta^3 f(x) + \cdots.$$

Note that this is equivalent to (10.3). From this he derived the binomial theorem as Cotes had done, and perhaps James Gregory before him. Murphy took $f(x) = (1+b)^x, h=1$ and observed that $\Delta^n f(x) = (1+b)^x b^n$ so that the binomial theorem followed after dividing both sides of the equation by $(1+b)^x$:

$$(1+b)^y = 1 + yb + \frac{y(y-1)}{1 \cdot 2}b^2 + \frac{y(y-1)(y-2)}{1 \cdot 2 \cdot 3}b^3 + \cdots.$$

20.6 Duncan Gregory's Operational Calculus

Duncan Gregory published many papers on operational calculus, illustrating the power of the method by elegant derivations of known results. Gregory's proof of Leibniz's formula for the nth derivative of a product of two functions began with the observation that Euler's proof of the binomial theorem

$$(a+b)^n = a^n + na^{n-1}b + \frac{n(n-1)}{1 \cdot 2}a^{n-2}b^2 + \frac{n(n-1)(n-2)}{1 \cdot 2 \cdot 3}a^{n-3}b^3 + \cdots$$

required that n was a fraction. More importantly, Gregory wrote that a, b should satisfy the laws

(1) The commutative, $ab = ba$,

(2) The distributive, $c(a+b) = ca + cb$,

(3) The index law, $a^m \cdot (a^n) = a^{m+n}$.

Gregory added, "Now, since it can be shown that the operations both in the Differential Calculus and the Calculus of Finite Differences are subject to these laws, the Binomial Theorem may be at once assumed as true with respect to them, so that it is not necessary to repeat the demonstration of it for each case." To prove Leibniz's theorem, Gregory observed that

$$\frac{d}{dx}(uv) = u\frac{dv}{dx} + v\frac{du}{dx}.$$

He then rewrote this equation, as had Arbogast:

$$\frac{d}{dx}(uv) = \left(\frac{d'}{dx} + \frac{d}{dx}\right)uv,$$

where $\frac{d'}{dx}$ acted on v but not on u and $\frac{d}{dx}$ acted on u but not on v. Since these operations were independent of each other, they commuted, so

$$\left(\frac{d}{dx}\right)^n (uv) = \left(\frac{d'}{dx} + \frac{d}{dx}\right)^n uv$$

$$= \left(\left(\frac{d'}{dx}\right)^n + n\left(\frac{d'}{dx}\right)^{n-1}\frac{d}{dx} + \cdots\right)uv$$

$$= u\frac{d^n v}{dx^n} + n\frac{d^{n-1}v}{dx^{n-1}}\frac{du}{dx} + \frac{n(n-1)}{1\cdot 2}\frac{d^2 u}{dx^2}\frac{d^{n-2}v}{dx^{n-2}} + \cdots.$$

Gregory remarked that this result was true with n negative or fractional, or "in the cases of integration and general differentiation." He then took $v = 1$ and $n = -1$ to obtain Bernoulli's formula,

$$\int u\, dx = xu - \frac{x^2}{1\cdot 2}\frac{du}{dx} + \frac{x^3}{1\cdot 2\cdot 3}\frac{d^3 u}{dx^3} - \cdots.$$

Using Arbogast's E operator, Gregory derived a proof of the Newton–Montmort transformation, given by Euler in his 1755 differential calculus book. Suppose

$$S = ax + a_1 x^2 + a_2 x^3 + a_3 x^4 + \cdots$$

and $a_1 = Ea, a_2 = E^2 a, a_3 = E^3 a, \ldots$. We write E instead of Gregory's D, since D might be confused with the derivative. Recall that $E = 1 + \Delta$, where Δ is the difference operator; thus, Gregory derived the Newton–Montmort transformation:

$$S = (x + x^2 E + x^3 E^2 + \cdots)a$$

$$= x(1 - xE)^{-1}a = x(1 - x - x\Delta)^{-1}a$$

$$= \frac{x}{1-x}\left(1 - \frac{x}{1-x}\Delta\right)^{-1}a$$

$$= \frac{x}{1-x}\left(1 + \frac{x}{1-x}\Delta + \frac{x^2}{(1-x)^2}\Delta^2 + \cdots\right)a$$

$$= \frac{ax}{1-x} + \Delta a\left(\frac{x}{1-x}\right)^2 + \Delta^2 a\left(\frac{x}{1-x}\right)^3 + \cdots.$$

Recall that we have discussed this formula earlier, as (11.3).

Gregory also found an operational method for solving linear ordinary differential equations with constant coefficients. He began with the equation

$$\frac{d^n y}{dx^n} + A\frac{d^{n-1}y}{dx^{n-1}} + B\frac{d^{n-2}y}{dx^{n-2}} + \cdots + R\frac{dy}{dx} + Sy = X,$$

where X was a function of x. After separating "the signs of operation from those of quantity," the equation became

$$\left(\frac{d^n}{dx^n} + A\frac{d^{n-1}}{dx^{n-1}} + B\frac{d^{n-2}}{dx^{n-2}} + \cdots + R\frac{d}{dx} + S\right)y = X.$$

Note that this can also be written as

$$f\left(\frac{d}{dx}\right)y = X,$$

where f is a polynomial. Gregory's problem was to find

$$y = \left\{f\left(\frac{d}{dx}\right)\right\}^{-1} X$$

and he first worked out the simplest case, where $f(x) = 1+x$ and $X = 0$. Gregory calculated

$$y = \left(1 + \frac{d}{dx}\right)^{-1} 0$$

$$= \left(1 + \frac{d^{-1}}{dx^{-1}}\right)^{-1} \frac{d^{-1}}{dx^{-1}} 0 = \left(1 + \frac{d^{-1}}{dx^{-1}}\right)^{-1} C$$

$$= \left(1 - \frac{d^{-1}}{dx^{-1}} + \frac{d^{-2}}{dx^{-2}} - \cdots\right) C$$

$$= C\left(1 - x + \frac{x^2}{1 \cdot 2} - \frac{x^3}{1 \cdot 2 \cdot 3} + \cdots\right) = Ce^{-x}.$$

He noted that $\frac{d^{-1}}{dx^{-1}} = \int dx$. Now note that if $f(x) = a + x$, one would get $y = ce^{-ax}$. Gregory then observed that

$$\left(\frac{d}{dx} \pm a\right)^n X = e^{\mp ax}\left(\frac{d}{dx}\right)^n e^{\pm ax} X,$$

provable by means of the binomial theorem. Gregory finally considered the general case:

$$\left(\frac{d}{dx} - a_1\right)\left(\frac{d}{dx} - a_2\right)\left(\frac{d}{dx} - a_3\right) \cdots \left(\frac{d}{dx} - a_n\right)y = X.$$

He applied $\left(\frac{d}{dx} - a_1\right)^{-1}$ to both sides of the equation to find

$$\left(\frac{d}{dx} - a_2\right)\left(\frac{d}{dx} - a_3\right) \cdots \left(\frac{d}{dx} - a_n\right)y = \left(\frac{d}{dx} - a_1\right)^{-1} X$$

$$= e^{a_1 x} \int e^{-a_1 x} X \, dx.$$

Similarly,

$$\left(\frac{d}{dx} - a_3\right) \cdots \left(\frac{d}{dx} - a_n\right) y = \left(\frac{d}{dx} - a_2\right)^{-1} e^{a_1 x} \int e^{-a_1 x} X dx$$

$$= e^{a_2 x} \int e^{(a_1 - a_2)x} \left(\int e^{-a_1 x} X dx\right) dx$$

$$= \frac{e^{a_1 x} \int e^{-a_1 x} X dx}{a_1 - a_2} + \frac{e^{a_2 x} \int e^{-a_2 x} X dx}{a_2 - a_1},$$

using integration by parts in the last step. Thus, Gregory's final formula took the form

$$y = \frac{e^{a_1 x} \int e^{-a_1 x} X dx}{(a_1 - a_2)(a_1 - a_3) \cdots (a_1 - a_n)} + \cdots + \frac{e^{a_n x} \int e^{-a_n x} X dx}{(a_n - a_1)(a_n - a_2) \cdots (a_n - a_{n-1})}.$$

In 1811, Français had used the same method, going a step beyond Gregory by using partial fractions to decompose $\left\{f\left(\frac{d}{dx}\right)\right\}^{-1}$. By this technique,

$$y = \left\{f\left(\frac{d}{dx}\right)\right\}^{-1} X = \sum_{i=1}^{n} \frac{N_i}{\frac{d}{dx} - a_i} X,$$

where Français assumed that the roots $a_1, a_2, \ldots, a_n$ of $f(x) = 0$ were distinct. When the value of N_i was substituted, the result was the same as Gregory's. Français also showed that his method could be extended to the case of repeated roots.

20.7 Boole's Operational Calculus

In his 1844 paper "On a General Method in Analysis," Boole extended Murphy and Gregory's symbolic method to treat problems on linear ordinary and partial differential equations with variable coefficients, linear difference equations, summation of series, and the computation of multiple integrals. He started his paper by stating several general propositions on functions of commutative and noncommutative operators. He made frequent use of some special cases and noted that they were already known: Let $x = e^\theta$. Then $x \frac{d}{dx} = \frac{d}{d\theta} = D$, so

$$f(D) e^{m\theta} u = e^{m\theta} f(D + m) u, \tag{20.15}$$

$$f(D) e^{m\theta} = f(m) e^{m\theta}, \tag{20.16}$$

$$D(D-1) \cdots (D-n+1) u = x^n \left(\frac{d}{dx}\right)^n u. \tag{20.17}$$

Though Boole did not explicitly say so, $f(x)$ is a function expandable as a series. Relation (20.15) can be verified from the particular case $f(D) = D^n$. In this case, by Leibniz's formula for the nth derivative of a product, we can see that

$$\frac{d^n}{d\theta^n}(e^{m\theta} u) = e^{m\theta} \left(\frac{d}{d\theta} + m\right)^n u.$$

The other two cases can be easily verified. Boole's fundamental theorem of development was given by the formula

$$f_0(D)u + f_1(D)e^\theta u + f_2(D)e^{2\theta} u + \cdots$$
$$= \sum \{(f_0(m)u_m + f_1(m)u_{m-1} + f_2(m)u_{m-2} + \cdots)e^{m\theta}\}, \qquad (20.18)$$

where $u = \sum u_m e^{m\theta}$. He verified this by substituting the series for u on the left-hand side of the equation and applying (20.15). Boole applied his development theorem to the summation of series, noting that if the coefficients of u satisfied a linear recurrence relation

$$f_0(m)u_m + f_1(m)u_{m-1} + \cdots = 0,$$

then (20.18) yielded a differential equation satisfied by u. In cases where this differential equation could be solved in closed form, he had the sum of the series $\sum u_n x^n$. This method allowed him to use the recurrence relation satisfied by the coefficients of the series in order to quickly find the differential equation satisfied by the series. As an example of a series summation, Boole considered for any real n,

$$u = 1 - \frac{n^2}{1\cdot 2}x^2 + \frac{n^2(n^2-2^2)}{1\cdot 2\cdot 3\cdot 4}x^4 - \frac{n^2(n^2-2^2)(n^2-4^2)}{1\cdot 2\cdots 6}x^6 + \cdots. \qquad (20.19)$$

In this case,

$$u_m = -\frac{n^2 - (m-2)^2}{m(m-1)}u_{m-2}.$$

He could then immediately write the differential equation satisfied by the series as

$$u - \frac{(D-2)^2 - n^2}{D(D-1)}e^{2\theta}u = 1,$$

$$\text{or} \quad D(D-1)u - \left((D-2)^2 - n^2\right)e^{2\theta}u = 0. \qquad (20.20)$$

Applying (20.15) and (20.17),

$$(D-2)^2 e^{2\theta} u = e^{2\theta} D^2 u = e^{2\theta}(D(D-1) + D)u$$
$$= x^2\left(x^2 \frac{d^2}{dx^2} + x\frac{d}{dx}\right)u.$$

Thus, (20.20) was simplified to

$$(1-x^2)\frac{d^2u}{dx^2} - x\frac{du}{dx} + n^2 u = 0. \qquad (20.21)$$

Boole then substituted $\sqrt{1-x^2}\frac{d}{dx} = \frac{d}{dy}$, or $y = \sin^{-1}x$, to convert the differential equation (20.21) to $\frac{d^2u}{dy^2} + n^2 u = 0$. This gave Boole the solution $u = c_1 \cos(ny) + c_2 \sin(ny) = c_1 \cos(n\sin^{-1}x) + c_2 \sin(n\sin^{-1}x)$ with the constants c_1 and c_2 equal to 1 and 0, respectively. He therefore had

$$\cos(n\sin^{-1}x) = 1 - \frac{n^2}{1\cdot 2}x^2 + \frac{n^2(n^2-2^2)}{1\cdot 2\cdot 3\cdot 4}x^4 - \cdots \quad \text{or}$$

$$\cos(ny) = 1 - \frac{n^2}{2!}\sin^2 y + \frac{n^2(n^2-2^2)}{4!}\sin^4 y - \frac{n^2(n^2-2^2)(n^2-4^2)}{6!}\sin^6 y + \cdots.$$

Similarly, Boole noted

$$\sin(ny) = n\sin y - \frac{n(n^2-1^2)}{3!}\sin^3 y + \frac{n(n^2-1^2)(n^2-3^2)}{5!}\sin^5 y + \cdots.$$

Recall that Newton discovered this series (9.16) and communicated it to Leibniz in his first letter of 1676. The series was afterwards employed by Gauss to prove that $\Gamma(x)\Gamma(1-x) = \pi/\sin\pi x$. Boole also used his method to solve some linear differential equations with variable coefficients and considered even more complex equations requiring a somewhat more elaborate technique. A simple example may explain his basic method. Boole set out to solve a differential equation with variable coefficients, commenting that it occurred in the theory of the "Earth's figure":

$$\frac{d^2u}{dx^2} + q^2u - \frac{6u}{x^2} = 0. \tag{20.22}$$

In his solution, Boole employed the general proposition:

The equation $u + \phi(D)e^{r\theta}u = U$ will be converted into the form $v + \psi(D)e^{r\theta}v = V$, by the relations

$$u = P_r\frac{\phi(D)}{\psi(D)}v, \quad U = P_r\frac{\phi(D)}{\psi(D)}V, \tag{20.23}$$

wherein $P_r\frac{\phi(D)}{\psi(D)}$ denotes the infinite symbolical product $\frac{\phi(D)\phi(D-r)\phi(D-2r)\cdots}{\psi(D)\psi(D-r)\psi(D-2r)\cdots}$.

Boole proved this by assuming $u = f(D)v$ and substituting in the first equation to get

$$f(D)v + \phi(D)e^{r\theta}f(D)v = U.$$

By (20.17) this became

$$f(D)v + \phi(D)f(D-r)e^{r\theta}v = U \quad \text{or}$$

$$v + \frac{\phi(D)f(D-r)}{f(D)}e^{r\theta}v = (f(D))^{-1}U.$$

So $\quad \psi(D) = \frac{\phi(D)f(D-r)}{f(D)} \quad$ or

$$f(D) = \frac{\phi(D)}{\psi(D)}f(D-r) = \frac{\phi(D)\phi(D-r)}{\psi(D)\psi(D-r)}f(D-2r) = \cdots.$$

In general, Boole attempted to choose v such that it satisfied the equation $\frac{d^n v}{dx^n} \pm q^n v = X$. Boole applied his general proposition to rewrite (20.22) as

$$u + \frac{q^2}{(D+2)(D-3)} e^{2\theta} u = 0.$$

Now Boole required the equation for v to be $\frac{d^2 v}{dx^2} + q^2 v = X$, or

$$v + \frac{q^2}{D(D-1)} e^{2\theta} v = V.$$

Here

$$\phi(D) = \frac{q^2}{(D+2)(D-3)}, \quad \psi(D) = \frac{q^2}{D(D-1)},$$

so that, by the general proposition,

$$P_2 \frac{\phi(D)}{\psi(D)} = \frac{D-1}{D+2}.$$

Thus, $u = \frac{D-1}{D+2} v$ and $0 = \frac{D-1}{D+2} V$. Boole could then take $V = 0$ and $v = c \sin(qx + c_1)$. Finally,

$$u = \frac{D-1}{D+2} v = \left(1 - 3(D+2)^{-1}\right) c \sin(qx + c_1)$$

$$= c \left(1 - 3e^{-2\theta} D^{-1} e^{2\theta}\right) \sin(qx + c_1)$$

$$= c \left(1 - \frac{3}{x^2} \left(x \frac{d}{dx}\right)^{-1} x^2\right) \sin(qx + c_1)$$

$$= c \sin(qx + c_1) - \frac{3}{x^2} \int dx\, x \sin(qx + c_1)$$

$$= c \left(\left(1 - \frac{3}{q^2 x^2}\right) \sin(qx + c_1) + \frac{3}{qx} \cos(qx + c_1)\right).$$

20.8 Jacobi and the Symbolic Method

In 1847, Jacobi wrote an interesting paper using the operational method to derive two results on transformations of series. He applied the second of these to the derivation of a result in the theory of hypergeometric series knows as Pfaff's transformation. Jacobi apparently wished to bring attention to Pfaff's important result. This wish was finally fulfilled in about 1970, when Richard Askey read Jacobi's paper and made Pfaff's work known to the mathematical community. Jacobi did not explain why he chose to explore the operational method. The work of the British mathematicians may have appealed to his algorithmic style; note that he had visited Britain in 1842. Both of these transformations had been earlier presented in Euler's differential calculus book. Euler's second formula (11.15) stated that if

$$f(x) = a + bx + cx^2 + dx^3 + \cdots \quad \text{then}$$

20.8 Jacobi and the Symbolic Method

$$aA_0 + bA_1x + cA_2x^2 + dA_3x^3 + \cdots$$
$$= A_0 f(x) + \Delta A_0 x \frac{df}{dx} + \frac{\Delta^2 A_0}{1 \cdot 2} x^2 \frac{d^2 f}{dx^2} + \frac{\Delta^3 A_0}{1 \cdot 2 \cdot 3} x^3 \frac{d^3 f}{dx^3} + \cdots.$$

Jacobi's proof of this formula was similar to Duncan Gregory's proof of the Newton–Montmort formula, discussed earlier. Recall the Arbogast operator E used by Gregory: $E^k A_0 = A_k$. The formal steps of the argument were then

$$aA_0 + bA_1x + cA_2x^2 + dA_3x^3 + \cdots$$
$$= (a + bxE + cx^2E^2 + dx^3E^3 + \cdots)A_0$$
$$= f(xE)A_0 = f(x + x(E-1))A_0 = f(x + x\Delta)A_0$$
$$= \left(f(x) + f'(x)x\Delta + \frac{f''(x)}{1 \cdot 2}x^2\Delta^2 + \cdots\right)A_0$$
$$= A_0 f(x) + \Delta A_0 x \frac{df}{dx} + \frac{\Delta^2 A_0}{1 \cdot 2} x^2 \frac{d^2 f}{dx^2} + \cdots.$$

To obtain Pfaff's transformation, Jacobi specialized the sequence $A_0, A_1, A_2, A_3, \ldots$ to

$$1, \frac{\beta}{\gamma}, \frac{\beta(\beta+1)}{\gamma(\gamma+1)}, \frac{\beta(\beta+1)(\beta+2)}{\gamma(\gamma+1)(\gamma+2)}, \ldots$$

and noted that the first and second differences were

$$\frac{\beta-\gamma}{\gamma}, \frac{\beta-\gamma}{\gamma}\frac{\beta}{\gamma+1}, \frac{\beta-\gamma}{\gamma}\frac{\beta(\beta+1)}{(\gamma+1)(\gamma+2)}, \frac{\beta-\gamma}{\gamma}\frac{\beta(\beta+1)(\beta+2)}{(\gamma+1)(\gamma+2)(\gamma+3)}, \ldots$$
$$\frac{(\beta-\gamma)(\beta-\gamma-1)}{\gamma(\gamma+1)}, \frac{(\beta-\gamma)(\beta-\gamma-1)}{\gamma(\gamma+1)}\frac{\beta}{\gamma+2}, \frac{(\beta-\gamma)(\beta-\gamma-1)}{\gamma(\gamma+1)}\frac{\beta(\beta+1)}{(\gamma+2)(\gamma+3)}, \ldots$$

In general, he observed,

$$\Delta^m A_n = \frac{(\beta-\gamma)(\beta-\gamma-1)\cdots(\beta-\gamma-m+1) \cdot \beta(\beta+1)\cdots(\beta+n-1)}{\gamma(\gamma+1)(\gamma+2)\cdots(\gamma+m+n-1)}.$$

In particular, when $n = 0$, he got

$$\Delta^m A_0 = \frac{(\beta-\gamma)(\beta-\gamma-1)\cdots(\beta-\gamma-m+1)}{\gamma(\gamma+1)\cdots(\gamma+m-1)}.$$

Jacobi then set

$$f(x) = (1-x)^{-\alpha} = 1 + \alpha x + \frac{\alpha(\alpha+1)}{1 \cdot 2}x^2 + \frac{\alpha(\alpha+1)(\alpha+2)}{1 \cdot 2 \cdot 3}x^3 + \cdots$$

so that Euler's transformation reduced to Pfaff's transformation:

$$1 + \frac{\alpha \cdot \beta}{1 \cdot \gamma} x + \frac{\alpha(\alpha+1) \cdot \beta(\beta+1)}{1 \cdot 2 \cdot \gamma(\gamma+1)} x^2 + \frac{\alpha(\alpha+1)(\alpha+2) \cdot \beta(\beta+1)(\beta+2)}{1 \cdot 2 \cdot 3 \cdot \gamma(\gamma+1)(\gamma+2)} x^3 + \cdots$$

$$= \frac{1}{(1-x)^\alpha} \left(1 + \frac{\alpha(\beta-\gamma)}{1 \cdot \gamma} \frac{x}{1-x} + \frac{\alpha(\alpha+1) \cdot (\beta-\gamma)(\beta-\gamma-1)}{1 \cdot 2 \cdot \gamma(\gamma+1)} \frac{x^2}{(1-x)^2} + \cdots \right).$$

20.9 Cartier: Gregory's Proof of Leibniz's Rule

In his 2000 preprint "Mathemagics," Pierre Cartier gives a rigorous version of Gregory's argument for Leibniz's rule. Cartier makes use of the tensor product $V \otimes W$ of vector spaces V and W. The vector space $V \otimes W$ consists of all finite sums $\sum \lambda_i (v_i \otimes w_i)$, where λ_i are scalars, $v_i \in V$ and $w_i \in W$, and where $v \otimes w$ is bilinear in v and w. For the purpose at hand, let I be an interval on the real line and $C^\infty(I)$ be the vector space of infinitely differentiable functions on I. Define the operators D_1 and D_2 on $C^\infty(I) \otimes C^\infty(I)$ by

$$D_1(f \otimes g) = Df \otimes g, \quad D_2(f \otimes g) = f \otimes Dg.$$

The two operators commute, that is, $D_1 D_2 = D_2 D_1$. Now define

$$\overline{D}(f \otimes g) = Df \otimes g + f \otimes Dg,$$

so that $\overline{D} = D_1 + D_2$; we can then conclude that

$$\overline{D}^n (f \otimes g) = \sum_{k=0}^{n} \binom{n}{k} D^k f \otimes D^{n-k} g.$$

We can convert the tensor product to an ordinary product by observing that $f \cdot g$ is bilinear in f and g and hence there is a linear map

$$\mu : C^\infty(I) \otimes C^\infty(I) \to C^\infty(I)$$

such that $\mu(f \otimes g) = f \cdot g$. The proof can now be completed:

$$D^n(fg) = D^n(\mu(f \otimes g)) = \mu(\overline{D}^n(f \otimes g))$$

$$= \mu \left(\sum_{k=0}^{n} \binom{n}{k} D^k f \otimes D^{n-k} g \right)$$

$$= \sum_{k=0}^{n} \binom{n}{k} \mu(D^k f \otimes D^{n-k} g)$$

$$= \sum_{k=0}^{n} \binom{n}{k} D^k f \cdot D^{n-k} g.$$

Observe that Cartier succeeds in resolving the problem in Gregory's presentation, that the operators D_1 and D_2 do not apply to both f and g.

20.10 Hamilton's Algebra of Complex Numbers and Quaternions

We have noted a strong algebraic spirit in the work of the British mathematicians of 1830–1850 and have observed important modern algebraic concepts in the work of Murphy. However, William R. Hamilton's (1805–1865) algebraic work was thoroughly modern in its structure and presentation. In 1826, Hamilton's friend J. T. Graves communicated to him some results on imaginary logarithms. This led Hamilton to formulate the theory of algebraic couples as the proper logical foundation for complex numbers. He finally presented this to the British Association in 1834. Gauss also got these ideas around the same time. Hamilton defined complex numbers as a set of pairs of real numbers, called couples, with addition and multiplication defined in a special way. More generally, he determined the necessary and sufficient conditions for a set of couples to form a commutative and associative division algebra. Hamilton first defined the sum and scalar multiplication of couples:

$$(b_1, b_2) + (a_1, a_2) = (b_1 + a_1, b_2 + a_2);$$
$$a \times (a_1, a_2) = (aa_1, aa_2).$$

He took the last equation as the first step toward the definition of the product of two couples by identifying the real number with a couple $(a, 0)$ to get

$$(a, 0) \times (a_1, a_2) = (a, 0)(a_1, a_2) = (a_1, a_2)(a, 0) = (aa_1, aa_2).$$

His aim was to define multiplication in order to satisfy the two conditions

$$(b_1 + a_1, b_2 + a_2)(c_1, c_2) = (b_1, b_2)(c_1, c_2) + (a_1, a_2)(c_1, c_2), \qquad (20.24)$$

$$(c_1, c_2)(b_1 + a_1, b_2 + a_2) = (c_1, c_2)(b_1, b_2) + (c_1, c_2)(a_1, a_2). \qquad (20.25)$$

Now for this type of multiplication to be possible, he had to have

$$(c_1, c_2)(a_1, a_2) = (c_1, 0)(a_1, a_2) + (0, c_2)(a_1, a_2)$$
$$= (c_1 a_1, c_1 a_2) + (0, c_2)(a_1, 0) + (0, c_2)(0, a_2)$$
$$= (c_1 a_1, c_1 a_2) + (0, c_2 a_1) + (0, c_2)(0, a_2)$$
$$= (c_1 a_1, c_1 a_2 + c_2 a_1) + (0, c_2)(0, a_2). \qquad (20.26)$$

It remained to define the product $(0, c_2)(0, a_2) = c_2 a_2 (0, 1)(0, 1)$ contained in the last step. Hamilton set

$$(0, 1)(0, 1) = (\gamma_1, \gamma_2) \qquad (20.27)$$

and determined the necessary and sufficient condition on γ_1 and γ_2 so that the two conditions (20.24) and (20.25) would hold: He supposed (b_1, b_2) to be the result of the product on the left-hand side of (20.26); then by equation (20.27) he had

$$b_1 = c_1 a_1 + \gamma_1 a_2 c_2,$$
$$b_2 = c_1 a_2 + c_2 a_1 + \gamma_2 a_2 c_2.$$

Now to be able to solve these equations for a_1, a_2 when $c_1 c_2 \neq 0$, the necessary and sufficient condition was the nonvanishing of the determinant, where the determinant was given as

$$c_1(c_1 + \gamma_2 c_2) - \gamma_1 c_2^2 = \left(c_1 + \frac{1}{2}\gamma_2 c_2\right)^2 - \left(\gamma_1 + \frac{1}{4}\gamma_2^2\right)c_2^2.$$

This expression was nonvanishing for all $c_1 c_2 \neq 0$ if

$$\gamma_1 + \frac{1}{4}\gamma_2^2 < 0.$$

The case in which $\gamma_1 = -1, \gamma_2 = 0$ gave Hamilton the usual multiplication rule for complex numbers:

$$(b_1, b_2)(a_1, a_2) = (b_1, b_2) \times (a_1, a_2) = (b_1 a_1 - b_2 a_2, b_2 a_1 + b_1 a_2).$$

Further developing the theory of complex numbers, Hamilton showed that the principal square root of $(-1, 0)$ was $(0, 1)$, and since $(-1, 0)$ could be replaced by -1 for brevity, he obtained

$$\sqrt{-1} = (0, 1).$$

He then wrote

> In the THEORY OF SINGLE NUMBERS, the symbol $\sqrt{-1}$ is *absurd*, and denotes an IMPOSSIBLE EXTRACTION, or a merely IMAGINARY NUMBER; but in the THEORY OF COUPLES, the same symbol $\sqrt{-1}$ is *significant*, and denotes a POSSIBLE EXTRACTION, or a REAL COUPLE, namely (as we have just now seen) the *principal square-root of the couple* $(-1, 0)$. In the latter theory, therefore, though not in the former, this sign $\sqrt{-1}$ may properly be employed; and we may write, if we choose, for any couple (a_1, a_2) whatever,
>
> $$(a_1, a_2) = a_1 + a_2 \sqrt{-1}, \ldots.$$

Hamilton next attempted to extend his work to triples, or triplets. His motivation was to obtain an algebra applicable to three-dimensional geometry and physics. He was well aware of the geometrical interpretation of complex numbers as vectors in two dimensions. Under this interpretation, the parallelogram law determined addition; moreover, the length (or modulus) of the product of two complex numbers turned out to be the product of the lengths of the two numbers. In October 1843, Hamilton described his train of thought as he worked toward his October 16 discovery of quaternions. He explained that he considered triplets of the form $x + iy + jz$ representing points (x, y, z) in space. Here j was "another sort of $\sqrt{-1}$, perpendicular to the plane itself." Addition and subtraction of triplets was a simple matter, but multiplication turned out to be a challenge: In a letter of 1865 to his son, Hamilton recalled:

> Every morning in the early part of the above-cited month, on my coming down to breakfast, your brother William Edwin and yourself used to ask me, 'Well, Papa, can you multiply triplets?' Whereto I was always obliged to reply, with a sad shake of the head, 'No, I can only add and subtract them.'

20.10 Hamilton's Algebra of Complex Numbers and Quaternions

In his 1843 description of his discovery, Hamilton recounted his dilemma: In order to multiply triplets, term-by-term multiplication had to be possible and the modulus of the product was required to equal the product of the moduli. He observed that

$$(a+iy+jz)(x+iy+jz) = ax - y^2 - z^2 + i(a+x)y + j(a+x)z + (ij+ji)yz \tag{20.28}$$

and that

$$(a^2+y^2+z^2)(x^2+y^2+z^2) = (ax - y^2 - z^2)^2 + (a+x)^2(y^2+z^2).$$

So the rule for the moduli implied that the last term in (20.28) should be zero, or, $ij + ji = 0$. He was sufficiently audacious to consider the possibility that $ij = ji = 0$; in modern terminology, this meant that i and j would be zero divisors. However, when he examined the general case

$$(a+ib+jc)(x+iy+jz) = ax - by - cz + i(ay+bx) + j(az+cx) + ij(bz-cy)$$

and the corresponding formula for the moduli

$$(a^2+b^2+c^2)(x^2+y^2+z^2) = (ax-by-cz)^2 + (ay+bx)^2 + (az+cx)^2 + (bz-cy)^2,$$

he saw that the coefficient of ij, $bz - cy$, could not be dropped and hence ij could not be zero. Put even more simply, the moduli of the product ij had to be 1 and not 0.

When he reached this result, it dawned on Hamilton that to multiply triplets, he must admit in some sense a fourth dimension, and he described this realization in a letter to his friend J. T. Graves, written the day after he discovered quaternions. By a remarkable coincidence, after completing this letter he came across the May 1843 issue of the *Cambridge Mathematical Journal* containing a paper by Cayley on analytical geometry of n dimensions. In a postscript to his letter to Graves, Hamilton noted that he did not yet know whether or not his ideas were similar to Cayley's. Continuing his description, Hamilton saw that he had to introduce a new imaginary k such that $ij = k$. Thus, he discovered quaternions! Moreover, $ji = -ij = -k$. He wondered whether $k^2 = 1$. But this produced the equation

$$(a+ib+jc+kd)(\alpha+i\beta+j\gamma+k\delta)$$
$$= a\alpha - b\beta - c\gamma + d\delta + i(a\beta + \cdots) + j(a\gamma + \cdots) + k(a\delta + d\alpha + \cdots),$$

implying that

$$(a^2+b^2+c^2+c^2)(\alpha^2+\beta^2+\gamma^2+\delta^2)$$
$$= (a\alpha - b\beta - c\gamma + d\delta)^2 + (a\beta + \cdots)^2 + (a\gamma + \cdots)^2 + (a\delta + d\alpha + \cdots)^2.$$

Of course, this relation could not possibly hold, because the term $2a\alpha d\delta$ in the first square on the right-hand side would not cancel the same term in the last square. So Hamilton took $k^2 = -1$. He then supposed that associativity would probably hold true and hence $-j = (ii)j = i(ij) = ik$. Similarly, $j(ii) = (ji)i = -ki$ or $j = ki$ and so $ik = -ki$. In this way he obtained the basic relations:

$$i^2 = j^2 = k^2 = -1, \; ij = k, \; jk = i, \; ki = j, \; ji = -k, \; kj = -i, \; ik = -j.$$

The product of two quaternions then emerged as

$$(a+ib+jc+kd)(\alpha+i\beta+j\gamma+k\delta)$$
$$= a\alpha - b\beta - c\gamma - d\delta + i(a\beta + b\alpha + c\delta - d\gamma)$$
$$+ j(a\gamma - b\delta + c\alpha + d\beta) + k(a\delta + b\gamma - c\beta + d\alpha).$$

And of course, with this definition,

$$(a^2+b^2+c^2+d^2)(\alpha^2+\beta^2+\gamma^2+\delta^2)$$
$$= (a\alpha - b\beta - c\gamma - d\delta)^2 + (a\beta + b\alpha + c\delta - d\gamma)^2$$
$$+ (a\gamma - b\delta + c\alpha + d\beta)^2 + (a\delta + b\gamma - c\beta + d\alpha)^2.$$

Thus, the modulus of the product equaled the product of the moduli! Interestingly, Euler also knew this equation, in connection with the representation of a number as a sum of four squares. Hamilton concluded his description of this discovery:

Hence we may write, on the plan of my theory of couples,

$$(a,b,c,d)(\alpha,\beta,\gamma,\delta) =$$

$$(a\alpha - b\beta - c\gamma - d\delta, a\beta + b\alpha + c\delta - d\gamma, a\gamma - b\delta + c\alpha + d\beta, a\delta + by - c\beta + d\alpha).$$

Hence $(a,b,c,d)^2 = (a^2-b^2-c^2-d^2, 2ab, 2ac, 2ad)$.

Thus
$$(0,x,y,z)^2 = -(x^2+y^2+z^2);\ (0,x,y,z)^3 = -(x^2+y^2+z^2)(0,x,y,z);$$
$$(0,x,y,z)^4 = +(x^2+y^2+z^2)^2;\ \&c.$$

Therefore

$$e^{(0,x,y,z)} = e^{(ix+jy+kz)} = 1 + \frac{ix+jy+kz}{1} - \frac{x^2+y^2+z^2}{1\cdot 2} - \&c;$$
$$= \cos\sqrt{x^2+y^2+z^2} + \frac{ix+jy+kz}{\sqrt{x^2+y^2+z^2}}\sin\sqrt{x^2+y^2+z^2}$$

and the *modulus* of $e^{(o,x,y,z)} = 1$. [Like the modulus of $e^{(0,x)}$ or $e^{\sqrt{-1}x}$] Let $\sqrt{x^2+y^2+z^2} = \rho$, $x = \rho\cos\phi$, $y = \rho\sin\phi\cos\psi$, $z = \rho\sin\phi\sin\psi$; then $e^{\rho(i\cos\phi+j\sin\phi\cos\psi+k\sin\phi\sin\psi)} = \cos\rho + (i\cos\phi + j\sin\phi\cos\psi + k\sin\phi\sin\psi)\sin\rho$; a theorem, which when $\phi = 0$, becomes the well-known equation

$$e^{i\rho} = \cos\rho + i\sin\rho,\ i = \sqrt{-1}.$$

Hamilton's letter led John T. Graves in December 1843 to produce an eight-dimensional division algebra, the algebra of octaves or octonions. The law of moduli was maintained within this system, so that

$$(a_1^2+a_2^2+\cdots+a_8^2)(b_1^2+b_2^2+\cdots+b_8^2) = c_1^2+c_2^2+\cdots+c_8^2.$$

Hamilton observed that while associativity held for quaternions, it failed to hold for octonions. Graves did not publish his work, though Cayley rediscovered and published it in 1845. Octonions are therefore called Cayley numbers.

The German mathematician A. Hurwitz wrote that Hamilton's two requirements, that term-by-term multiplication be valid and that the product of the moduli be equal to the moduli of the product of n-tuples $(x_1, x_2, \ldots, x_n)$, in fact held only for $n = 1, 2, 4, 8$. This explains why Hamilton was unable to discover a way of multiplying triplets. In the 1870s, C. S. Peirce and Frobenius gave another explanation for Hamilton's failure to work out a three-dimensional division algebra, i.e., an algebra of triplets. They proved that the only real finite-dimensional associative division algebras were: the real numbers, the complex numbers, and the quaternions.

We have seen that Hamilton was initially hesitant to move to the fourth dimension and was struck by Cayley's work outlining a geometry of n dimensions. Note Felix Klein's telling remark on George Green's 1835 paper concerning the attraction of an ellipsoid, "This investigation merits special mathematical interest ... because it is carried out for n dimensions, long before the development of n-dimensional geometry in Germany began." Such was the influence of the formal algebraic approach taken by British mathematicians of the early 1800s, that even the applied mathematician George Green was willing to consider the novel concept of an n-dimensional space.

20.11 Exercises

1. Prove Faà di Bruno's formula for the mth derivative of a composition of two functions:

$$\frac{d^m}{dt^m} g(f(t))$$

$$= \sum \frac{m!}{b_1! b_2! \cdots b_m!} g^{(k)}(f(t)) \left(\frac{f'(t)}{1!}\right)^{b_1} \left(\frac{f''(t)}{2!}\right)^{b_2} \cdots \left(\frac{f^{(m)}(t)}{m!}\right)^{b_m},$$

where the sum is over different solutions of $b_1 + 2b_2 + \cdots + mb_m = m$ and $k = b_1 + b_2 + \cdots + b_m$. Faà di Bruno gave the right-hand side in the form of a determinant:

$$\begin{vmatrix} \binom{m-1}{0} f' g & \binom{m-1}{1} f'' g & \binom{m-1}{2} f''' g & \cdots & \binom{m-1}{m-2} f^{(m-1)} g & \binom{m-1}{m-1} f^{(m)} g \\ -1 & \binom{m-2}{0} f' g & \binom{m-2}{1} f'' g & \cdots & \binom{m-2}{m-3} f^{(m-2)} g & \binom{m-2}{m-2} f^{(m-1)} g \\ 0 & -1 & \binom{m-3}{0} f' g & \cdots & \binom{m-3}{m-4} f^{(m-3)} g & \binom{m-3}{m-3} f^{(m-2)} g \\ . & . & -1 & \cdots & . & . \\ . & . & . & \cdots & . & . \\ . & . & . & \cdots & . & . \\ 0 & 0 & 0 & \cdots & \binom{1}{0} f' g & \binom{1}{1} f'' g \\ 0 & 0 & 0 & \cdots & -1 & \binom{0}{0} f' g \end{vmatrix}$$

where $f^{(i)} \equiv f^{(i)}(t)$ and $g^k = g^{(k)}(f(t))$. Faà di Bruno published this formula without proof or reference in 1855 and then again in 1857. The formulation as a determinant appears to be original with Faà di Bruno, who may also be the only mathematician to be beatified by the Catholic Church. The papers of Craik (2005) and Johnson (2002) contain a detailed history of Faà di Bruno's formula.

2. Solve the difference equation

$$u_{x+1}^2 - 4u_x^2(u_x^2 + 1) = 0.$$

Herschel's hint for the solution is to set $u_x = \sqrt{-1} \sin v_x$.

3. Solve the difference equation

$$u_{x+1}u_x - a_x(u_{x+1} - u_x) + 1 = 0.$$

Herschel remarked that this was a slight generalization of the equation worked out in our text.

4. Sum the series $u = \frac{4x^2}{1\cdot 2\cdot 3} + \frac{5x^4}{2\cdot 3\cdot 4} + \frac{6x^5}{3\cdot 4\cdot 5} + \cdots$. See Boole (1844b), p. 264.

5. Using Boole's notation given in the text, prove his proposition: The equation

$$u + \phi(D)e^{r\theta}u = U$$

will be converted to the form

$$v + \psi(D)e^{r\theta}v = V,$$

by the relations $u = e^{n\theta}v$ and $U = e^{n\theta}V$. See Boole (1844b), p. 247.

6. Let d_x denote the derivative with respect to x. Note that $[f(x+y)](d_y - d_x) = 0$. Apply Murphy's method for the difference operator to the differential in order to obtain the Taylor series for $f(x+y)$. See Murphy (1837), p. 196.

7. Sum the series

$$\sum_{n=1}^{\infty} \arctan\left(1/(1+n+n^2)\right).$$

See Herschel (1820), p. 57.

20.12 Notes on the Literature

See Lagrange (1867–1892), vol. 3, pp. 441–476 for his work on symbolic calculus. Jacobi's paper, "De Seriebus ac Differentiis Observatiunculae," presenting his contribution to the calculus of operations, is given in Jacobi (1969), vol. VI, pp. 174–182. Friedelmeyer (1994) gives an extensive discussion of Arbogast. The quote from Babbage may be found in Babbage and Herschel (1813), p. xi. For the work of Français, see Français (1812–13). Morrison and Morrison (1961) contains papers of Babbage, including one discussing the Analytical Society. Enros (1983) is a nice discussion of the Analytical Society. Herschel (1820), pp. 34–36, contains his treatment of the difference equation of Laplace; our discussion of his evaluation of the coefficient A_n is from pp. 67–68. For the early nineteenth century work on operational calculus in Britain, see Allaire and Bradley (2002). Duncan Gregory's extensive work on the operational method can be found in the early volumes of the *Cambridge Mathematical Journal* or in Gregory (1865); for his work discussed in the text, see pp. 14–27 and 108–123 of the latter. See the interesting article on Gregory by Leslie Ellis (1845), p. 149, for the quote on Gregory.

See Murphy (1837) and Boole (1844b) for their works. The quote from Boole may be seen in Boole (1847), p. 3. The remark attributed to Brinkley about Hamilton appears in many places, including Robert Perceval Graves's article about Hamilton in the *Dublin University Magazine* of 1842, vol. 19, pp. 94–110. Robert Graves, brother of John and Charles Graves, later wrote the three-volume biography of Hamilton. Hamilton's 1865 letter to his son is given in R. P. Graves' biography, Graves (1885), pp. 434–435. See Hamilton (1835) for his work on complex numbers and pp. 127–128 for the quotation on the use of $\sqrt{-1}$. See Hamilton (1945) for quaternions. See also articles by E. L. Ortiz and of S. E. Despeaux in Gray and Parshall (2007) and the paper of Koppelman (1971) for the role of the operational method in the development of abstract algebra. A fascinating discussion of the notation and history of the Stirling numbers, including Kramp's formula, is given by Knuth (2003), pp. 15–44. The work of the German combinatorial school is well discussed in Jahnke (1993). For the quote on Rothe, see Muir (1960), p. 55. For Klein's remark on Green's work, see Klein (1979), p. 217. For remarks on the influence of the German combinatorial school on Gudermann and Weierstrass, see Manning (1975). Becher (1980) is an interesting article on Woodhouse, Babbage, and Peacock.

21

Fourier Series

21.1 Preliminary Remarks

The problem of representing functions by trigonometric series has played as significant a role in the development of mathematics and mathematical physics as that of representing functions as power series. Trigonometric series take the form

$$\frac{1}{2}a_0 + a_1 \cos x + b_1 \sin x + a_2 \cos 2x + b_2 \sin 2x + \cdots, \qquad (21.1)$$

and these series naturally made their appearance in eighteenth-century works on astronomy, a subject dealing with periodic phenomena. Now series (21.1) is called a Fourier series if, for some function $f(x)$ defined on $(0, \pi)$, the coefficients a_n and b_n can be expressed as

$$a_n = \frac{1}{\pi} \int_0^{2\pi} f(t) \cos nt \, dt; \quad b_n = \frac{1}{\pi} \int_0^{2\pi} f(t) \sin nt \, dt. \qquad (21.2)$$

Moreover, if (21.1) converges to some integrable function $f(x)$ and can be integrated term by term, then the coeffieients a_n and b_n will take the form (21.2). Thus, Fourier series have very wide applicability. The Fourier series first occurred explicitly in the 1750 work of Euler, published in 1753, in which he gave the general solution of the difference equation

$$f(x) = f(x-1) + X(x) \qquad (21.3)$$

in the form

$$f(x) = \int X(\xi) \, d\xi + 2 \sum_{n=1}^{\infty} \cos 2n\pi x \int X(\xi) \cos 2n\pi \xi \, d\xi \\ + 2 \sum_{n=1}^{\infty} \sin 2n\pi x \int X(\xi) \sin 2n\pi \xi \, d\xi. \qquad (21.4)$$

Note that the Fourier series for $X(x)$ can readily be derived from this. Earlier, in connection with investigations on the vibratory motion of a stretched string, trigonometric

21.1 Preliminary Remarks

series of this type were used, although the coefficients were not explicitly written as integrals. These researches led to controversy among the principal investigators, d'Alembert, Euler, Bernoulli, and Lagrange, as to whether an 'arbritrary' function could be represented by such series. This dispute began with d'Alembert's 1746 discovery of the wave equation describing the motion of the vibrating string:

$$\sigma \frac{\partial^2 y}{\partial t^2} = T \frac{\partial^2 y}{\partial x^2} \quad \text{or} \quad \frac{1}{c^2} \frac{\partial^2 y}{\partial t^2} = \frac{\partial^2 y}{\partial x^2}, \quad c^2 \equiv \frac{T}{\sigma}, \tag{21.5}$$

where σ and T were constants and y was the displacement of the string. The derivation was based on the work of Taylor dating from 1715. D'Alembert showed that the general solution of equation (21.5) would be of the form

$$y = \Phi(ct+x) + \Psi(ct-x),$$

but the initial and boundary conditions implied a relation between Φ and Ψ. For example, at $x = 0$ and $x = l$, the string would be fixed and hence $y = 0$ for all t at these points. This implied that for all u

$$0 = \Phi(u) + \Psi(u), \quad \text{or} \quad \Psi(u) = -\Phi(u) \quad \text{and} \tag{21.6}$$

$$0 = \Phi(u+l) + \Psi(u-l). \tag{21.7}$$

By (21.6), the general solution took the form $y = \Psi(ct+x) - \Psi(ct-x)$, and by (21.7), Ψ was periodic: $\Psi(u+2l) = \Psi(u)$. Interestingly, d'Alembert's paper also gave the first instance of the use of separation of variables to solve partial differential equations. He set

$$\Psi(ct+x) - \Psi(ct-x) = f(t)g(x) \tag{21.8}$$

and by differentiation obtained the relation

$$\frac{1}{c^2} \frac{f''}{f} = \frac{g''}{g} = A.$$

Since f was independent of x and g of t, A was a constant and the expressions for f and g could be obtained from their differential equations. Note that from the boundary conditions, it can be shown that f and g are sine and cosine functions. D'Alembert, however, saw these solutions as special cases of the general solution.

Euler reacted to d'Alembert's work by publishing his ideas on the matter within a few months. Essentially, he and d'Alembert disagreed on the meaning of the function $\Phi(u)$. D'Alembert thought that Φ had to be an analytic expression, whereas Euler was of the view that Φ was an arbitrary graph defined only by the periodicity condition

$$\Phi(u+2l) = \Phi(u).$$

On this view, Φ could be defined by different expressions in different intervals; in our terms, Φ would be continuous but its derivative could be piecewise continuous. The functions allowed by Euler as solutions of Φ would now be called weak solutions of the equation, while d'Alembert required the solutions to be twice differentiable. And while Euler allowed all possible initial conditions on Φ, d'Alembert ruled out certain initial conditions.

Euler also criticized Taylor's contention that an arbitrary initial vibration would eventually settle into a sinusoidal one. He argued from the equation of motion that higher frequencies would also be involved and that the solution could have the form

$$\sum A_n \sin(n\pi x/l) \cos(n\pi ct/l)$$

with the initial shape given by $\sum A_n \sin(n\pi x)/l$. According to Truesdell, Euler was therefore "*the first to publish formulae for the simple modes of a string* and to observe that they can be combined simultaneously with arbitrary amplitudes." However, Euler did not regard these trigonometric series as the most general solutions of the problem.

At this point, Daniel Bernoulli entered the discussion by presenting in 1748 two memoirs to the Petersburg Academy, in which he explained on physical grounds that the trigonometric solutions found by Euler were in fact the most general possible. Bernoulli wrote that his ideas were based on the work of Taylor, who had observed that the basic shapes of the vibrating string of length a were given by the functions

$$\sin\frac{\pi x}{a}, \sin\frac{2\pi x}{a}, \sin\frac{3\pi x}{a}, \ldots$$

Bernoulli argued that the general form of the curve for the string would be obtained by linear superposition:

$$y = \alpha \sin\frac{\pi x}{a} + \beta \sin\frac{2\pi x}{a} + \gamma \sin\frac{3\pi x}{a} + \delta \sin\frac{4\pi x}{a} + \text{etc.}$$

Unlike Euler, Bernoulli thought that all possible curves assumed by the vibrating string could be obtained in this way. It is interesting to note that in 1728 Bernoulli solved a linear difference equation by taking linear combinations of certain special solutions. However, he was unable to extend this idea to the solutions of ordinary linear differential equations; Euler did this around 1740. Finally in 1748, Bernoulli once again proposed this idea to solve a linear partial differential equation, but this time he gave a physical argument. He apparently saw no need here for the differential equation and thought that the mathematics only obscured the main ideas. This led to further discussion and controversy, mainly involving d'Alembert and Bernoulli.

It seems that these discussions led Euler to further ponder on the problem of expanding functions in terms of trigonometric series. In a paper written around 1752, Euler started with the divergent series

$$\cos x + \cos 2x + \cos 3x + \cdots = -\frac{1}{2}$$

and after integration obtained the formulas

$$\frac{1}{2}x = \sin x - \frac{1}{2}\sin 2x + \frac{1}{3}\sin 3x - \frac{1}{4}\sin 4x + \cdots \quad \text{and} \quad (21.9)$$

$$\frac{1}{12}\pi^2 - \frac{1}{4}x^2 = \cos x - \frac{1}{2^2}\cos 2x + \frac{1}{3^2}\cos 3x - \frac{1}{4^2}\cos 4x + \cdots.$$

21.1 Preliminary Remarks

He gave no range of validity for the formulas, but two decades later D. Bernoulli observed that these results were true only in the interval $-\pi < x < \pi$. Euler also continued the integration process to obtain similar formulas with polynomials of degree 3, 4 and 5; obviously, the process could be continued. The polynomials occurring in this situation were the (Jakob) Bernoulli polynomials. Neither Bernoulli nor Euler seems to have noticed this. It appears that the Swiss mathematician Joseph Raabe (1801–1859) was the first to show this explicitly, around 1850. Recall that the Fourier expansion of the Bernoulli polynomials also follows when Poisson's remainder in the Euler–Maclaurin formula, dating from the 1820s, is set equal to the remainder derived by Jacobi in the 1830s.

In 1759, Lagrange wrote a paper on the vibrating string problem in which he attempted to obtain Euler's general solution with arbitrary functions by first finding the explicit solution for the loaded string and then taking the limit. The equations of motion in the latter case were

$$M\frac{d^2 y_k}{dt^2} = c^2 (y_{k+1} - 2y_k + y_{k-1}), \qquad k = 1, 2, \ldots, n, \tag{21.10}$$

and were first obtained by Johann Bernoulli in 1727. Euler studied them in a slightly different context in 1748 and obtained solutions by setting

$$y_k = A_k \cos \frac{2c}{\sqrt{M}} pt$$

and finding

$$p = \sin \frac{r\pi}{2(n+1)}, \qquad r = 1, \ldots, n$$

and the value of A_k from the corresponding second-order difference equation. Lagrange solved (21.10) by writing the equations as the first-order system

$$\frac{dy_k}{dt} = v_k, \quad \frac{dv_k}{dt} = c^2 (y_{k+1} - 2y_k + y_{k-1}), \quad k = 1, 2, \ldots. \tag{21.11}$$

In the course of this work, Lagrange came close to deriving the Fourier coefficients in the expansion of a function as a series of sines. Instead, he took a different course, since his aim was to derive Euler's general solution rather than a trigonometric series.

Surprisingly, Alexis–Claude Clairaut gave the Fourier coefficients in the case of a cosine series expansion as early as 1757. While studying the perturbations of the sun, he viewed the question of finding the coefficients $A_0, A_1, A_2, \ldots$ in

$$f(x) = A_0 + 2\sum_{m=1}^{\infty} A_m \cos mx \tag{21.12}$$

as an interpolation problem, given that values of f were known at $x = 2\pi/k, 4\pi/k, 6\pi/k, \ldots$. He found

$$A_0 = \frac{1}{k}\sum_{m=1}^{\infty} f\left(\frac{2m\pi}{k}\right), \quad A_n = \frac{1}{k}\sum_{m=1}^{\infty} f\left(\frac{2m\pi}{k}\right) \cos \frac{2mn\pi}{k}$$

and then let $k \to \infty$, to get

$$A_n = \frac{1}{2\pi} \int_0^{2\pi} f(x) \cos nx \, dx. \tag{21.13}$$

Twenty years later, Euler derived (21.13) directly by multiplying (21.12) by $\cos nx$ and using the orthogonality of the cosine function.

Joseph Fourier (1768–1830) lost his parents as a child; he was then sent by the bishop of Auxerre to a military college run by the Benedictines. Fourier's earliest researches were in algebraic equations, and he went to Paris in 1789 to present his results to the Academy. He soon became involved in revolutionary activities and gained a reputation as an orator. In 1795, Fourier began studying with Gaspard Monge; he soon published his first paper and announced plans to present a series of papers on algebraic equations. But Monge selected him to join Napoleon's scientific expedition to Egypt. When Fourier returned to France in 1801, Napoleon appointed him an administrator in Isère. Fourier ably executed his duties, but found time to successfully carry out his difficult researches in heat conduction, presented to the Academy in 1807. His work was reviewed by Lagrange, Laplace, Lacroix, and Monge; Lagrange opposed its publication. Perhaps to make up for this, the Academy then set a prize problem in the conduction of heat, won by Fourier in 1812.

It is not clear whether Euler thought that $f(x)$ in (21.13) was an arbitrary function. But in his 1807 work on heat conduction, Fourier took this view explicitly. He translated a physics problem into the mathematical one of finding a function v such that

$$\frac{d^2v}{dx^2} + \frac{d^2v}{dy^2} = 0,$$

and $v(0, y) = v(r, y) = 0$, $v(x, 0) = f(x)$. By a separation of variables, Fourier found v to be given by the series

$$v = a_1 e^{-\pi y/r} \sin \frac{\pi x}{r} + a_2 e^{-2\pi y/r} \sin \frac{2\pi x}{r} + a_3 e^{-3\pi y/r} \sin \frac{3\pi x}{r} + \cdots,$$

with the coefficients $a_1, a_2, a_3, \ldots$ to be obtained from

$$f(x) = a_1 \sin \frac{\pi x}{r} + a_2 \sin \frac{2\pi x}{r} + a_3 \sin \frac{3\pi x}{r} + \cdots. \tag{21.14}$$

Fourier discussed three methods for deriving these coefficients. In one approach, he converted equation (21.14) into a system of infinitely many equations in infinitely many unknowns $a_1, a_2, a_3, \ldots$. He also considered problems that reduced to cosine series and to series with sines as well as cosines. In his 1913 monograph on such systems, *Les systèmes d'équations linéaires* the Hungarian mathematician Frigyes Riesz (1880–1956) wrote that Fourier was the first to deal with linear equations in infinitely many unknowns. Fourier gave two other methods for determining the coefficients. One method depended on the discrete orthogonality of the sine function, and the other on its continuous orthogonality. Dirichlet later gave a brief exposition of Fourier's discrete orthogonality method, explaining why the integral representation for a_n was plausible.

Fourier regarded the use of an infinite system of equations in infinitely many unknowns as important enough to first discuss a particular case. He expanded a constant function as an infinite series of cosines:

$$1 = a\cos y + b\cos 3y + c\cos 5y + d\cos 7y + \text{ etc.} \tag{21.15}$$

To see briefly how he determined the coefficients $a, b, c, d, \ldots$, we write the equation in the form

$$1 = \sum_{m=1}^{\infty} a_m \cos(2m-1)y.$$

He took derivatives of all orders of this equation and set $y = 0$ to obtain

$$1 = \sum_{m=1}^{\infty} a_m, \quad 0 = \sum_{m=1}^{\infty} (2m-1)^2 a_m,$$

$$0 = \sum_{m=1}^{\infty} (2m-1)^4 a_m, \quad \text{etc.}$$

He considered the first n equations with $n = 1, 2, 3, \ldots$, and replaced these n equations with a new set of n equations, taking $a_m = 0$ for $m > n$. This new system could be regarded as n equations in the n unknowns $a_1^{(n)}, a_2^{(n)}, \ldots, a_n^{(n)}$. By using the well-known formula now known as Cramer's rule, Fourier calculated the Vandermonde determinants appearing in this situation to find that

$$a_1^{(n)} = \frac{3 \cdot 3}{2 \cdot 4} \cdot \frac{5 \cdot 5}{4 \cdot 6} \cdots \frac{(2n-1)(2n-1)}{(2n-2)(2n)},$$

with a similar formula for $a_m^{(n)}$. He assumed that $a_m^{(n)} \to a_m$ as $n \to \infty$. This, by Wallis's formula, gave him $a_1 = 4/\pi$ and in general $a_m = (-1)^{m-1}4/((2m-1)\pi)$. By substituting these a_m back in (21.15), he obtained

$$\frac{\pi}{4} = \cos x - \frac{1}{3}\cos 3x + \frac{1}{5}\cos 5x - \frac{1}{7}\cos 7x + \frac{1}{9}\cos 9x - \text{ etc.}$$

Fourier did not discuss the validity of his method. Interestingly, according to Riesz, the question of the justification of this process was first considered by Henri Poincaré (1854–1912) in 1885. Poincaré's attention was drawn to this problem by a paper of Paul Appell in which Appell applied Fourier's method to obtain the coefficients of a cosine expansion of an elliptic function. Poincaré gave a simple theorem justifying Appell's calculations. A year later he wrote another paper on the subject, "Sur les Determinants d'ordre Infini."

The term infinite determinant was introduced by the American astronomer and mathematician G. W. Hill (1838–1914) in an 1877 paper on lunar theory. In this paper, he solved the equation

$$\frac{d^2w}{dt^2} + \left(\sum_{n=-\infty}^{\infty} \theta_n e^{int}\right) w = 0$$

by making the substitution

$$w = \sum_{n=-\infty}^{\infty} b_n e^{i(n+c)t}$$

and determining b_n from the infinite system of equations

$$\sum_{k=-\infty}^{\infty} \theta_{n-k} b_k - (n+c)^2 b_n = 0, \quad n = -\infty, \ldots, +\infty.$$

Hill employed a procedure similar to that of Fourier and once again Poincaré developed the necessary theorems to justify Hill's result. Poincaré's work was generalized a decade later by the Swedish mathematician Niels Helge von Koch (1870–1924) and this was the starting point for his countryman Ivar Fredholm's (1866–1927) theory of integral equations. Fredholm in turn provided the basis for the pioneering work in the development of functional analysis by David Hilbert and then Riesz, with significant contributions from others such as Erhard Schmidt (1876–1959). They created the ideas and techniques by which linear equations in infinitely many variables could be treated by general methods. The valuable 1913 book by Riesz, one of the earliest monographs on functional analysis, contains an interesting history of the topic. Surely Fourier could not have foreseen that his idea would see such beautiful development. On the other hand, he must have considered it worthy of attention, since he included the long derivation of the formula for the Fourier coefficients by this method when he was well aware of the much shorter method using term-by-term integration.

21.2 Euler: Trigonometric Expansion of a Function

In a very interesting paper of 1750, "De Serierum Determinatione seu Nova Methodus Inveniendi Terminos Generales Serierum," Euler used symbolic calculus to expand a function as a trigonometric series. He also applied the discoveries he had made a decade earlier on solving differential equations with constant coefficients. Given a function X, his problem was to determine $y(x)$ such that

$$y(x) - y(x-1) = X(x).$$

He viewed this as a differential equation of infinite order:

$$\frac{dy}{dx} - \frac{1}{1 \cdot 2}\frac{d^2y}{dx^2} + \frac{1}{1 \cdot 2 \cdot 3}\frac{d^3y}{dx^3} - \cdots = X(x).$$

He noted that if $d^n y/dx^n$ was replaced by z^n, then the left-hand side could be expressed as

$$z - \frac{z^2}{1 \cdot 2} + \frac{z^3}{1 \cdot 2 \cdot 3} - \cdots = 1 - e^{-z}.$$

He observed that the factors of $1 - e^{-z}$ were z and $z^2 + 4kk\pi\pi$ for $k = 1, 2, 3, \ldots$. Hence, dy/dx and $d^2y/dx^2 + 4k^2\pi^2 y$ were factors of the differential equation. The solution of the differential equation corresponding to dy/dx was given by $y = \int X\,dx$, while the solution corresponding to $d^2y/dx^2 + 4kk\pi\pi$ was given by

$$y = 2(\cos 2k\pi \cos 2k\pi x - \sin 2k\pi \sin 2k\pi x) \int X \cos 2k\pi x\,dx$$

$$+ 2(\cos 2k\pi \sin 2k\pi x + \sin 2k\pi \cos 2k\pi x) \int X \sin 2k\pi x\,dx,$$

and since $\sin 2k\pi = 0$, $\cos 2k\pi = 1$, Euler could write the complete solution

$$y = \int X\,dx + 2\cos 2\pi x \int X \cos 2\pi x\,dx + 2\cos 4\pi x \int X \cos 4\pi x\,dx + \cdots$$

$$+ 2\sin 2\pi x \int X \sin 2\pi x\,dx + 2\sin 4\pi x \int X \sin 4\pi x\,dx + \cdots.$$

As an application of this result, Euler gave the trigonometric expansion for

$$y(x) = \ln \Gamma(x+1),$$

though in this case his result was incomplete; he was unable to obtain the asymptotic expansion for $y(x)$. Note that in this case

$$X = y(x) - y(x-1) = \ln x.$$

21.3 Lagrange on the Longitudinal Motion of the Loaded Elastic String

In his study of the vibrating string, Lagrange considered the situation in which the masses were assumed to be at a discrete set of points so that he could express the rate of change with respect to x in terms of finite differences. He wrote the equations in the form

$$\frac{dy_k}{dt} = v_k, \quad \frac{dv_k}{dt} = C^2(y_{k+1} - 2y_k + y_{k-1}), \tag{21.16}$$

where $k = 1, 2, \ldots, m-1$ and $y_0 \equiv y_m \equiv 0$. His idea was to determine constants M_k, N_k and R such that

$$\sum_{k=1}^{m-1}(M_k\,dv_k + N_k\,dy_k) = \sum_{k=1}^{m-1}\left(N_k v_k + C^2 M_k(y_{k+1} - 2y_k + y_{k-1})\right) dt$$

would be reduced to $dz = Rz\,dt$. This required that

$$R(M_k v_k + N_k y_k) = N_k v_k + C^2 M_k (y_{k+1} - 2y_k + y_{k-1}), \quad k = 1, 2, \ldots, m-1, \tag{21.17}$$

or $\quad RM_k = N_k, \quad RN_k = C^2(M_{k+1} - 2M_k + M_{k-1}).$

This meant that M_k satisfied the equation

$$M_{k+1} - \left(\frac{R^2}{C^2} + 2\right) M_k + M_{k-1} = 0. \tag{21.18}$$

Lagrange set $M_k = Aa^k + Bb^k$, so that a and b were roots of

$$x^2 - \left(\frac{R^2}{C^2} + 2\right)x + 1 = 0.$$

Thus, $ab = 1$ and $a + b = \dfrac{R^2}{C^2} + 2$.

Note that because of the restriction on y_0 and y_m, Lagrange could assume without loss of generality that $M_0 \equiv M_m \equiv 0$. He also set $M_1 \equiv 1$. From this it followed that $A + B = 0$ and $Aa + Bb = 1$. With these initial values, he could find the constants A, B in M_k. Thus, he could write

$$M_k = \frac{a^k - b^k}{a - b} \quad \text{and} \quad \frac{a^m - b^m}{a - b} = 0.$$

The last equation gave him $m - 1$ pairs of values $a_n = e^{n\pi i/m}$ and $b_n = e^{-n\pi i/m}$ for $n = 1, 2, \ldots, m - 1$. Corresponding to these were $m - 1$ values of M and R:

$$M_{kn} = \frac{e^{kn\pi i/m} - e^{-kn\pi i/m}}{e^{n\pi i/m} - e^{-n\pi i/m}} = \frac{\sin(kn\pi/m)}{\sin(n\pi/m)}, \tag{21.19}$$

$$R_n = \pm 2iC \sin(n\pi/2m), \quad n = 1, 2, \ldots, m - 1. \tag{21.20}$$

For these values he had the corresponding equations

$$dz_n = R_n z_n \, dt \quad \text{where} \quad z_n = \sum_{k=1}^{m-1} (M_{kn} v_k + R_n M_{kn} y_k). \tag{21.21}$$

The solution of the differential equation for z_n yielded

$$z_n = F_n e^{R_n t}, \quad \text{with } F_n \text{ a constant.}$$

Next he set

$$Z_n = \sum_{k=1}^{m-1} M_{kn} y_k, \tag{21.22}$$

so that $dy_k/dt = v_k$ implied that

$$\frac{dZ_n}{dt} + R_n Z_n = F_n e^{R_n t}. \tag{21.23}$$

Lagrange expressed the constant F_n as $2R_n K_n$ so that he could solve this differential equation in the form

$$Z_n = K_n e^{R_n t} + L_n e^{-R_n t}, \tag{21.24}$$

where L_n was a constant of integration. Here recall that (21.23) can be solved by multiplying it by the integrating factor $e^{R_n t}$. By substituting the value of R_n from (21.20), he could write Z_n in terms of the sine and cosine functions

$$Z_n = P_n \cos(2Ct \sin(n\pi/2m)) + Q_n \frac{\sin(2Ct \sin(n\pi/2m))}{2C \sin(n\pi/2m)}. \tag{21.25}$$

21.3 Lagrange on the Longitudinal Motion of the Loaded Elastic String

Then, Z_n being known, the problem was to determine y_k from (21.22); after substituting the value of M_{kn} from (21.19), (21.22) took the form

$$Z_n \sin \frac{n\pi}{m} = \sum_{k=1}^{m-1} y_k \sin \frac{kn\pi}{m}, \qquad m = 1, 2, \ldots, m-1. \qquad (21.26)$$

The next step was to obtain the $m-1$ unknowns $y_1, y_2, \ldots, y_{m-1}$ from these $m-1$ equations. Several years before Lagrange, in 1748, Euler had encountered this system of equations in his study of the loaded elastic cord. He was able to write the solution in general after studying the special cases where $m \leq 7$. He saw that the result followed from the discrete orthogonality relation for the sine function:

$$\sum_{k=1}^{m-1} \sin \frac{kn\pi}{m} \sin \frac{kp\pi}{m} = \frac{1}{2} m \delta_{np}. \qquad (21.27)$$

He was unable to provide a complete proof. Lagrange gave an ingenious proof of (21.27) and obtained

$$y_j = \frac{2}{m} \sum_{n=1}^{m-1} Z_n \sin \frac{n\pi}{m} \sin \frac{nj\pi}{m}, \qquad (21.28)$$

by multiplying (21.26) by $\sin(nj\pi/m)$, summing over n and applying (21.27). In a later paper, Lagrange observed that the analysis involved in moving from (21.26) to (21.28) also solved an interpolation problem related to trigonometric polynomials. Specifically, given the $m-1$ values

$$f\left(\frac{\pi}{m}\right), f\left(\frac{2\pi}{m}\right), \ldots, f\left(\frac{(m-1)\pi}{m}\right)$$

of a function $f(x)$, the problem was to find a trigonometric polynomial

$$a_1 \sin x + a_2 \sin 2x + \cdots + a_{m-1} \sin(m-1)x \qquad (21.29)$$

passing through $m-1$ points $\left(\frac{k\pi}{m}, f\left(\frac{k\pi}{m}\right)\right), k = 1, 2, \ldots, m-1$. By an application of (21.27), it was clear that

$$ma_n = 2\sin \frac{n\pi}{m} f\left(\frac{\pi}{m}\right) + 2\sin \frac{2n\pi}{m} f\left(\frac{2\pi}{m}\right) + \cdots + 2\sin \frac{(m-1)n\pi}{m} f\left(\frac{(m-1)\pi}{m}\right), \qquad (21.30)$$

and one obtained the coefficients of the trigonometric polynomial interpolating $f(x)$.

We reproduce Dirichlet's 1837 proof of the orthogonality relation (21.27), since it is more illuminating than Lagrange's complicated though clever proof. The same method was clearly described in Fourier's 1822 book on heat. The idea was to apply the addition formula for the sine function; note that this addition formula is also used for the integral analog of (21.27). First note that by the addition formula

$$2\sin \frac{kn\pi}{m} \sin \frac{kp\pi}{m} = \cos \frac{k(n-p)\pi}{m} - \cos \frac{k(n+p)\pi}{m}.$$

So when $n \neq p$, twice the sum in (21.27) is given by

$$\sum_{k=1}^{m-1} \left(\cos \frac{k(n-p)\pi}{m} - \cos \frac{k(n+p)\pi}{m} \right) \qquad (21.31)$$
$$= \frac{\sin(m-\frac{1}{2})(n-p)\frac{\pi}{m}}{2\sin(n-p)\frac{\pi}{2m}} - \frac{\sin(m-\frac{1}{2})(n+p)\frac{\pi}{m}}{2\sin(n+p)\frac{\pi}{2m}} = 0,$$

since each expression is either $-1/2$ or 0, according as $n - p$ is even or odd. To sum the series in (21.31) Dirichlet employed the formula

$$1 + 2\cos 2\theta + 2\cos 4\theta + \cdots + 2\cos 2s\theta = \frac{\sin(2s+1)\theta}{\sin\theta}.$$

Note that this too can be proved by the addition formula for the sine function. Dirichlet pointed out that (21.29) and (21.30) strongly suggested that a function $f(x)$ could be expanded as a Fourier series. Observe that a_n can be expressed as

$$\frac{2}{\pi}\left[\frac{\pi}{m}\sin\frac{0n\pi}{m}f\left(\frac{0\pi}{m}\right) + \frac{\pi}{m}\sin\frac{n\pi}{m}f\left(\frac{\pi}{m}\right) + \cdots + \frac{\pi}{m}\sin\frac{(m-1)n\pi}{m}f\left(\frac{(m-1)\pi}{m}\right)\right],$$

and when $m \to \infty$, the right-hand side tends to

$$\frac{2}{\pi}\int_0^\pi \sin nx\, f(x)\, dx.$$

Thus, $f(x) = a_1 \sin x + a_2 \sin 2x + \cdots + a_n \sin nx + \cdots$, with

$$a_n = \frac{2}{\pi}\int_0^\pi \sin nx\, f(x)\, dx.$$

Observe here that Lagrange missed this opportunity to discover Fourier series, partly because he was focused on obtaining the results of d'Alembert and Euler and partly because he did not think that functions could be represented by such series.

21.4 Euler on Fourier Series

In 1777, Euler submitted a paper to the Petersburg Academy containing a derivation of the Fourier coefficients of a cosine series. This was the first derivation of the coefficients using the orthogonality of the sequence of functions $\cos nx$, $n = 1, 2, \ldots$. Euler's paper was published in 1798, but its contents did not become generally known until much later; a half-century afterwards, Riemann thought that Fourier was the first to give such a derivation. Euler expanded a function Φ as a trigonometric series, $\Phi = A + B\cos\phi + C\cos 2\phi + \cdots$ and gave the coefficients as

$$A = \frac{1}{\pi}\int_0^\pi \Phi\, d\phi, \quad B = \frac{2}{\pi}\int_0^\pi \Phi\, d\phi \cos\phi, \quad C = \frac{2}{\pi}\int_0^\pi \Phi\, d\phi \cos 2\phi, \ldots.$$

21.4 Euler on Fourier Series

His argument was that since $\int d\phi \cos i\phi = \frac{1}{i} \sin i\phi = 0$, on integration from $\phi = 0$ to $\phi = \pi$, he would get

$$\int_0^\pi \Phi \, d\phi = A\pi.$$

Next, by the addition formula for the cosine function,

$$d\phi \cos i\phi \cos \lambda\phi = \frac{1}{2} d\phi \left(\cos (i-\lambda)\phi + \cos (i+\lambda)\phi \right),$$

$$\int_0^\pi d\phi \cos i\phi \cos \lambda\phi = \frac{\sin (i-\lambda)\phi}{2(i-\lambda)} + \frac{\sin (i+\lambda)\phi}{2(i+(\lambda))} \Big]_0^\pi$$

$$= 0, \text{ when } i \neq \lambda. \text{ And when } i = \lambda,$$

$$\int_0^\pi d\phi (\cos i\phi)^2 = \frac{1}{2}\phi + \frac{1}{4i} \sin (2i\phi) \Big]_0^\pi = \frac{1}{2}\pi;$$

$$\text{for} \quad (\cos i\phi)^2 = \frac{1}{2} + \frac{1}{2} \cos (2i\phi).$$

Hence the coefficients $A, B, C, D, \ldots$ were as given earlier.

In this paper, Euler included a proof of the well-known recurrence relation

$$\int_0^\pi d\phi (\cos \phi)^\lambda = \frac{\lambda - 1}{\lambda} \int_0^\pi d\phi (\cos \phi)^{\lambda - 2}.$$

We mention that Euler wrote $\cos \phi^\lambda$ for $(\cos \phi)^\lambda$. Observe that, though he was well aware of the integration by parts formula in the standard form $\int P \, dQ = PQ - \int Q \, dp$, he usually worked it out in a slightly different way. For example, to prove the recurrence formula, Euler started with

$$\int d\phi \cos^\lambda \phi = f \sin \phi \cos^{\lambda - 1} \phi + g \int d\phi \cos^{\lambda - 2} \phi,$$

where f and g had to be determined. He differentiated to get

$$\cos^\lambda \phi = f \cos^\lambda \phi - f(\lambda - 1) \sin^2 \phi \cos^{\lambda - 2} \phi + g \cos^{\lambda - 2} \phi.$$

Since $\sin^2 \phi = 1 - \cos^2 \phi$,

$$\cos^\lambda \phi = \lambda f \cos^\lambda \phi - f(\lambda - 1) \cos^{\lambda - 2} \phi + g \cos^{\lambda - 2} \phi.$$

For this relation to hold, Euler had to have $f = 1/\lambda$ and $g = f(\lambda - 1)$ or $g = (\lambda - 1)/\lambda$. Hence

$$\int d\phi \cos^\lambda \phi = \frac{1}{\lambda} \sin \phi \cos^{\lambda - 1} \phi + \frac{\lambda - 1}{\lambda} \int d\lambda \cos^{\lambda - 2} \phi.$$

Finally, he took the integral from 0 to π. In this paper, as in some others, Euler used the notation $\partial \phi$ instead of $d\phi$.

21.5 Fourier: Linear Equations in Infinitely Many Unknowns

Fourier rediscovered Euler's derivation of the Fourier coefficients. Nevertheless, Fourier sought alternative derivations to convince the mathematical community of the correctness of the Fourier expansion. He also presented several derivations of Fourier expansions of specific functions. In spite of these efforts, his theory encountered a certain amount of opposition, mainly from the older generation. In section six of the third chapter of his famous book, *Théorie analytique de la chaleur*, Fourier considered the problem of determining the coefficients in the sine expansion of an odd function. He reduced this problem to that of solving an infinite system of equations in infinitely many unknowns. It is interesting to see how beautifully Fourier carried out the computations; he started with a sine series expansion of an odd function

$$\phi(x) = a\sin x + b\sin 2x + c\sin 3x + d\sin 4x + \cdots. \tag{21.32}$$

He let $A = \phi'(0)$, $B = -\phi'''(0)$, $C = \phi^{(5)}(0)$, $D = -\phi^{(7)}(0)$, ..., so that by repeatedly differentiating (21.32), he had

$$\begin{aligned}
A &= a + 2b + 3c + 4d + 5e + \cdots, \\
B &= a + 2^3 b + 3^3 c + 4^3 d + 5^3 e + \cdots, \\
C &= a + 2^5 b + 3^5 c + 4^5 d + 5^5 e + \cdots, \\
D &= a + 2^7 b + 3^7 c + 4^7 d + 5^7 e + \cdots, \\
E &= a + 2^9 b + 3^9 c + 4^9 d + 5^9 e + \cdots,
\end{aligned} \tag{21.33}$$

and so on. He broke up this system into the subsystems

$$a_1 = A_1 \qquad \begin{aligned} a_2 + 2b_2 &= A_2, \\ a_2 + 2^3 b_2 &= B_2, \end{aligned} \qquad \begin{aligned} a_3 + 2b_3 + 3c_3 &= A_3, \\ a_3 + 2^3 b_3 + 3^3 c_3 &= B_3, \\ a_3 + 2^5 b_3 + 3^5 c_3 &= C_3, \end{aligned} \tag{21.34}$$

$$\begin{aligned}
a_4 + 2b_4 + 3c_4 + 4d_4 &= A_4, \\
a_4 + 2^3 b_4 + 3^3 c_4 + 4^3 d_4 &= B_4, \\
a_4 + 2^5 b_4 + 3^5 c_4 + 4^5 d_4 &= C_4, \\
a_4 + 2^7 b_4 + 3^7 c_4 + 4^7 d_4 &= D_4,
\end{aligned} \tag{21.35}$$

$$\begin{aligned}
a_5 + 2b_5 + 3c_5 + 4d_5 + 5e_5 &= A_5, \\
a_5 + 2^3 b_5 + 3^3 c_5 + 4^3 d_5 + 5^3 e_5 &= B_5, \\
a_5 + 2^5 b_5 + 3^5 c_5 + 4^5 d_5 + 5^5 e_5 &= C_5, \\
a_5 + 2^7 b_5 + 3^7 c_5 + 4^7 d_5 + 5^7 e_5 &= D_5, \\
a_5 + 2^9 b_5 + 3^9 c_5 + 4^9 d_5 + 5^9 e_5 &= E_5,
\end{aligned} \tag{21.36}$$

and so on. Fourier's strategy was to solve the first equation for a_1, the second for $2b_2$, the third for $3c_3$, and so on. He wrote that the equations could be solved by inspection,

21.5 Fourier: Linear Equations in Infinitely Many Unknowns

meaning that they could be obtained by Cramer's rule, since the determinants in the equations were Vandermonde determinants. He also established the recursive relations connecting a_{j-1} with a_j, b_{j-1} with b_j, c_{j-1} with c_j, etc., and similarly with the right-hand sides of the equations, A_j, B_j, C_j, etc. He assumed that as $j \to \infty$, $a_j \to a$, $b_j \to b$, $c_j \to c$, etc., and that $A_j \to A$, $B_j \to B$, $C_j \to C$. To find the recursive relations, Fourier eliminated e_5 from the last five equations to get

$$\begin{aligned}
a_5(5^2 - 1^2) + 2b_5(5^2 - 2^2) + 3c_5(5^2 - 3^2) + 4d_5(5^2 - 4^2) &= 5^2 A_5 - B_5, \\
a_5(5^2 - 1^2) + 2^3 b_5(5^2 - 2^2) + 3^3 c_5(5^2 - 3^2) + 4^3 d_5(5^2 - 4^2) &= 5^2 B_5 - C_5, \\
a_5(5^2 - 1^2) + 2^5 b_5(5^2 - 2^2) + 3^5 c_5(5^2 - 3^2) + 4^5 d_5(5^2 - 4^2) &= 5^2 C_5 - D_5, \\
a_5(5^2 - 1^2) + 2^7 b_5(5^2 - 2^2) + 3^7 c_5(5^2 - 3^2) + 4^7 d_5(5^2 - 4^2) &= 5^2 D_5 - E_5.
\end{aligned} \quad (21.37)$$

Fourier then argued that for this system to coincide with the system of four equations in (21.35), he must have

$$a_4 = (5^2 - 1^2)a_5, \quad b_4 = (5^2 - 2^2)b_5, \quad c_4 = (5^2 - 3^2)c_5, \, d_4 = (5^2 - 4^2)d_5, \quad (21.38)$$

$$A_4 = 5^2 A_5 - B_5, \quad B_4 = 5^2 B_5 - C_5, \quad C_4 = 5^2 C_5 - D_5, \quad D_4 = 5^2 D_5 - E_5. \quad (21.39)$$

We remark that he wrote out all his equations in this manner, noting that this reasoning would apply in general to the $m \times m$ system of equations. We now write his formulas in shorter form. From the above relations, it is evident that

$$\begin{aligned}
a_{j-1} &= a_j(j^2 - 1^2), \quad j = 2, 3, 4, \ldots, \\
b_{j-1} &= b_j(j^2 - 2^2), \quad j = 3, 4, 5, \ldots, \\
c_{j-1} &= c_j(j^2 - 3^2), \quad j = 4, 5, 6, \ldots, \\
d_{j-1} &= d_j(j^2 - 4^2), \quad j = 5, 6, 7, \ldots.
\end{aligned} \quad (21.40)$$

Moreover,

$$\begin{aligned}
A_{j-1} &= j^2 A_j - B_j, \quad j = 2, 3, 4, \ldots, \\
B_{j-1} &= j^2 B_j - C_j, \quad j = 3, 4, 5, \ldots, \\
C_{j-1} &= j^2 C_j - D_j, \quad j = 4, 5, 6, \ldots, \\
D_{j-1} &= j^2 D_j - E_j, \quad j = 5, 6, 7, \ldots.
\end{aligned} \quad (21.41)$$

As we noted before, Fourier assumed that $a_j \to a, b_j \to b, \ldots, A_j \to A, \ldots$ as $j \to \infty$. So from (21.40), he could conclude that

$$\begin{aligned}
a &= \frac{a_1}{(2^2 - 1^2)(3^2 - 1^2)(4^2 - 1^2)\cdots}, \\
b &= \frac{b_2}{(3^2 - 2^2)(4^2 - 2^2)(5^2 - 2^2)\cdots}, \\
c &= \frac{c_3}{(4^2 - 3^2)(5^2 - 3^2)(6^2 - 3^2)\cdots},
\end{aligned} \quad (21.42)$$

and so on. Similarly, a repeated application of (21.41) gave him

$$A_1 = A_2 2^2 - B_2, \quad A_1 = A_3 2^2 \cdot 3^2 - B_3(2^2 + 3^2) + C_3,$$
$$A_1 = A_4 2^2 \cdot 3^2 \cdot 4^2 - B_4(2^2 \cdot 3^2 + 2^2 \cdot 4^2 + 3^2 \cdot 4^2) + C_4(2^2 + 3^2 + 4^2) - D_4, \text{ etc.}$$

To understand Fourier's next step, one may divide the first value of A_1 by 2^2, the second value of A_1 by $2^2 \cdot 3^2$, the third by $2^2 \cdot 3^2 \cdot 4^2$, and consider the form of the right-hand side. So by dividing the ultimate equation A_1 by $2^2 \cdot 3^2 \cdot 4^2 \cdot 5^2 \cdots$, Fourier obtained by equations (21.34) and (21.42)

$$\frac{A_1(=a_1)}{2^2 \cdot 3^2 \cdot 4^2 \cdot 5^2 \cdots} = \frac{a(2^2-1)(3^2-1)(4^2-1)(5^2-1)\cdots}{2^2 \cdot 3^2 \cdot 4^2 \cdot 5^2 \cdots}$$
$$= A - B\left(\frac{1}{2^2} + \frac{1}{3^2} + \frac{1}{4^2} + \cdots\right) + C\left(\frac{1}{2^2 \cdot 3^2} + \frac{1}{2^2 \cdot 4^2} + \frac{1}{3^2 \cdot 4^2} + \cdots\right) \quad (21.43)$$
$$- D\left(\frac{1}{2^2 \cdot 3^2 \cdot 4^2} + \frac{1}{2^2 \cdot 3^2 \cdot 5^2} + \frac{1}{3^2 \cdot 4^2 \cdot 5^2} + \cdots\right) + \cdots$$
$$= A - BP_1 + CQ_1 - DR_1 + ES_1 - \cdots.$$

We note that by $P_1, Q_1, R_1, \ldots$ Fourier meant the sums of products of $\frac{1}{2^2}, \frac{1}{3^2}, \frac{1}{4^2}, \cdots$ taken one, two, three, ... at a time. This gave him the value of a in terms of A, B, C, D etc. To find the values of $b, c, d, \ldots$ in a similar manner, Fourier solved the second and third systems in (21.34) to find $2b_2$ and $3b_3$. Similarly, he solved (21.35) for $4b_4$ etc. to arrive at the solutions:

$$A - BP_2 + CQ_2 - DR_2 + \cdots = 2b_2 \frac{(1^2 - 2^2)}{1^2 \cdot 3^2 \cdot 4^2 \cdot 5^2 \cdots},$$
$$A - BP_3 + CQ_3 - DR_3 + \cdots = 3c_3 \frac{(1^2 - 3^2)(2^2 - 3^2)}{1^2 \cdot 2^2 \cdot 4^2 \cdot 5^2 \cdot 6^2 \cdots}, \quad (21.44)$$
$$A - BP_4 + CQ_4 - DR_4 + \cdots = 4d_4 \frac{(1^2 - 4^2)(2^2 - 4^2)(3^2 - 4^2)}{1^2 \cdot 2^2 \cdot 3^2 \cdot 5^2 \cdot 6^2 \cdots},$$

and so on. The starting points for deriving these equations were

$$2b_2(1^2 - 2^2) = A_2 1^2 - B_2,$$
$$3c_3(1^2 - 3^2)(2^2 - 3^2) = A_3 1^2 \cdot 2^2 - B_3(1^2 + 2^2) + C_3,$$
$$4d_4(1^2 - 4^2)(2^2 - 4^2)(3^2 - 4^2) \quad (21.45)$$
$$= A_4 1^2 \cdot 2^2 \cdot 3^2 - B_4(1^2 \cdot 2^2 + 1^2 \cdot 3^2 + 2^2 \cdot 3^2) + C_4(1^2 + 2^2 + 3^2) - D_4,$$

and so on. As before, these relations were continued by repeated use of (21.41). Once again Fourier applied (21.40) to express $b_2, c_3, d_4, \ldots$ in terms of $b, c, d, \ldots$. Recall

21.5 Fourier: Linear Equations in Infinitely Many Unknowns

that $b_j \to b$, $c_j \to c$, $c_j \to d$, etc. He then had

$$A - BP_1 + CQ_1 - DR_1 + \cdots = a\left(1 - \frac{1^2}{2^2}\right)\left(1 - \frac{1^2}{3^2}\right)\left(1 - \frac{1^2}{4^2}\right)\cdots,$$

$$A - BP_2 + CQ_2 - DR_2 + \cdots = 2b\frac{(1^2 - 2^2)(3^2 - 2^2)(4^2 - 2^2)\cdots}{1^2 \cdot 3^2 \cdot 4^3 \cdot 5^2 \cdots}$$

$$= 2b\left(1 - \frac{2^2}{1^2}\right)\left(1 - \frac{2^2}{3^2}\right)\left(1 - \frac{2^2}{4^2}\right)\cdots,$$

$$A - BP_3 + CQ_3 - DR_3 + \cdots = 3c\left(1 - \frac{3^2}{1^2}\right)\left(1 - \frac{3^2}{2^2}\right)\left(1 - \frac{3^2}{4^2}\right)\cdots,$$

$$A - BP_4 + CQ_4 - DR_4 + \cdots = 4d\left(1 - \frac{4^2}{1^2}\right)\left(1 - \frac{4^2}{2^2}\right)\left(1 - \frac{4^2}{3^2}\right)\left(1 - \frac{4^2}{5^2}\right)\cdots,$$

(21.46)

and so on. To compute the values of the products on the right-hand side of (21.46), and the values of $P_j, Q_j, R_j, S_j, \ldots$, observe that

$$\frac{\sin \pi x}{\pi x} = \left(1 - \frac{x^2}{1^2}\right)\left(1 - \frac{x^2}{2^2}\right)\left(1 - \frac{x^2}{3^2}\right)\cdots. \tag{21.47}$$

Fourier did not write down the details of the evaluations of the products on the right-hand sides of (21.46), but they are fairly simple. Note that the first product has a factor $(1 - 1^2/1^2)$ missing, the second $(1 - 2^2/2^2)$, the third $(1 - 3^2/3^2)$, etc. Now the value of the product with a missing $1 - j^2/j^2$ can be evaluated by (21.47) to be

$$\lim_{\epsilon \to 0} \frac{\sin \pi(j+\epsilon)}{\pi(j+\epsilon)\left(1 - \frac{(j+\epsilon)^2}{j^2}\right)} = \lim_{\epsilon \to 0} \frac{j^2(-1)^{j-1}\sin \pi\epsilon}{\pi(j+\epsilon)(2j+\epsilon)\epsilon} = \frac{(-1)^{j-1}}{2}. \tag{21.48}$$

To find $P_j, Q_j, R_j, \ldots$, he expanded the product on the right-hand side as a series

$$1 - Px^2 + Qx^4 - Rx^6 + \cdots, \tag{21.49}$$

so that $P, Q, R, \ldots$ were sums of products of $1, \frac{1}{2^2}, \frac{1}{3^2}, \ldots$ taken one, two, three, $\ldots$ (respectively) at a time. He could then equate this series with the known power series for $\sin(\pi x)/\pi x$:

$$1 - \frac{x^2\pi^2}{3!} + \frac{x^4\pi^4}{5!} - \frac{x^6\pi^6}{7!} + \cdots$$

to get

$$P = \pi^2/3!, \quad Q = \pi^4/5!, \quad R = \pi^6/7!, \quad \ldots. \tag{21.50}$$

Moreover, it is easy to see that

$$\left(1-\frac{y}{j^2}\right)(1-P_j y + Q_j y^2 - R_j y^3 + \cdots) = 1 - Py + Qy^2 - Ry^3 + \cdots,$$

$$P_j + \frac{1}{j^2} = P, \quad Q_j + \frac{1}{j^2} P_j = Q, \quad R_j + \frac{1}{j^2} Q_j = R, \ldots.$$

Hence $P_j = P - \dfrac{1}{j^2} = \dfrac{\pi^2}{3!} - \dfrac{1}{j^2}, \quad Q_j = \dfrac{\pi^4}{5!} - \dfrac{1}{j^2}\dfrac{\pi^2}{3!} + \dfrac{1}{j^4}, \ldots.$ (21.51)

Using these expressions in (21.46), he obtained the relations

$$\frac{1}{2}a = A - B\left(\frac{\pi^2}{3!} - \frac{1}{1^2}\right) + C\left(\frac{\pi^4}{5!} - \frac{1}{1^2}\frac{\pi^2}{3!} + \frac{1}{1^4}\right)$$

$$- D\left(\frac{\pi^6}{7!} - \frac{1}{1^2}\frac{\pi^4}{5!} + \frac{1}{1^4}\frac{\pi^2}{3!} - \frac{1}{1^6}\right) + \cdots,$$

$$-\frac{1}{2}2b = A - B\left(\frac{\pi^2}{3!} - \frac{1}{2^2}\right) + C\left(\frac{\pi^4}{5!} - \frac{1}{2^2}\frac{\pi^2}{3!} + \frac{1}{2^4}\right) \quad (21.52)$$

$$- D\left(\frac{\pi^6}{7!} - \frac{1}{2^2}\frac{\pi^4}{5!} + \frac{1}{2^4}\frac{\pi^2}{3!} - \frac{1}{2^6}\right) + \cdots,$$

$$\frac{1}{2}3c = A - B\left(\frac{\pi^2}{3!} - \frac{1}{3^2}\right) + C\left(\frac{\pi^4}{5!} - \frac{1}{3^2}\frac{\pi^2}{3!} + \frac{1}{3^4}\right) - \cdots,$$

etc. Now recall that $A = \phi'(0), -B = \phi'''(0), C = \phi^{(5)}(0), \ldots$; one may use the expressions for $a, b, c, \ldots$ in equation (21.42) to get

$$\frac{1}{2}\phi(x) = \sin x \left\{\phi'(0) + \phi'''(0)\left(\frac{\pi^2}{3!} - \frac{1}{1^2}\right) + \phi^{(5)}(0)\left(\frac{\pi^4}{5!} - \frac{1}{1^2}\frac{\pi^2}{3!} + \frac{1}{1^4}\right) + \cdots\right\}$$

$$- \frac{1}{2}\sin 2x \left\{\phi'(0) + \phi'''(0)\left(\frac{\pi^2}{3!} - \frac{1}{2^2}\right) + \phi^{(5)}(0)\left(\frac{\pi^4}{5!} - \frac{1}{2^2}\frac{\pi^2}{3!} + \frac{1}{2^4}\right) + \cdots\right\}$$

$$+ \frac{1}{3}\sin 3x \left\{\phi'(0) + \phi'''(0)\left(\frac{\pi^2}{3!} - \frac{1}{3^2}\right) + \phi^{(5)}(0)\left(\frac{\pi^4}{5!} - \frac{1}{3^2}\frac{\pi^2}{3!} + \frac{1}{3^4}\right) + \cdots\right\}$$

$$\cdots\cdots$$

Fourier noted that the expression in the first set of chain brackets was the Maclaurin series for

$$\frac{1}{\pi}\left\{\phi(\pi) - \frac{1}{1^2}\phi''(\pi) + \frac{1}{1^4}\phi^{(4)}(\pi) - \frac{1}{1^6}\phi^{(6)}(\pi) + \cdots\right\}.$$

Similarly, the expressions in the second and third brackets were

$$\frac{1}{\pi}\left\{\phi(\pi) - \frac{1}{2^2}\phi''(\pi) + \frac{1}{2^4}\phi^{(4)}(\pi) - \frac{1}{2^6}\phi^{(6)}(\pi) + \cdots\right\};$$

$$\frac{1}{\pi}\left\{\phi(\pi) - \frac{1}{3^2}\phi''(\pi) + \frac{1}{3^4}\phi^{(4)}(\pi) - \frac{1}{3^6}\phi^{(6)}(\pi) + \cdots\right\}.$$

To sum the expressions in chain brackets, Fourier observed that

$$s(x) = \phi(x) - \frac{1}{m^2}\phi''(x) + \frac{1}{m^4}\phi^{(4)}(x) - \cdots$$

satisfied the differential equation

$$\frac{1}{m^2}s''(x) + s(x) = \phi(x).$$

He noted that the general solution of this differential equation was

$$s(x) = C_1 \cos mx + C_2 \sin mx + m \sin mx \int_0^x \phi(t) \cos mt\, dt$$
$$- m \cos mx \int_0^x \phi(t) \sin mt\, dt.$$

Because $\phi(x)$ was an odd function, its even-order derivatives were also odd functions, making $s(x)$ an odd function. This meant that $C_1 = 0$. Hence,

$$s(\pi) = (-1)^{m+1} m \int_0^\pi \phi(t) \sin mt\, dt,$$

and this in turn implied

$$a_m = \frac{2}{\pi} \int_0^\pi \phi(t) \sin mt\, dt.$$

Thus, Fourier found the "Fourier" coefficients.

21.6 Dirichlet's Proof of Fourier's Theorem

Fourier's work clearly demonstrated the tremendous significance of trigonometric series in the study of heat conduction and more generally in solving partial differential equations with boundary conditions. As we have seen, Fourier offered many arguments for the validity of his methods. But when the work of Gauss, Cauchy, and Abel on convergence of series became known in the 1820s, Fourier's methods were perceived to be nonrigorous. Lejeune Dirichlet (1805–1859) studied in France with Fourier and Poisson, who introduced him to problems in mathematical physics. At the same time, Dirichlet became familiar with the latest ideas on the rigorous treatment of infinite power series. Dirichlet's first great achievement was to treat infinite trigonometric series with equal rigor, thereby vindicating Fourier, who had befriended him.

In 1829, Dirichlet published his famous paper on Fourier series, "Sur la convergence des séries trigonométriques qui servent a représenter une fonction arbitraire entre des limites données" in the newly founded *Crelle's Journal*. Eight years later he published the same paper in the Berlin Academy journal with further computational details and a more careful analysis of convergence. We follow the 1829 paper, whose title indicates that Dirichlet's aim was to obtain conditions on an arbitrary function so that the corresponding Fourier series would converge to the function. Dirichlet started his paper by observing that Fourier began a new era in analysis by applying trigonometric series in his researches on heat. However, he noted that only one paper, published by Cauchy in 1823, had discussed the validity of this method. Dirichlet noted, moreover, that the results of Cauchy's paper were inconclusive because they were based on the false premise that if the series with nth term $v_n = (A \sin nx)/n$ converged, the series $\sum u_n$ also converged when u_n/v_n had 1 as a limit. Dirichlet produced examples of two series with nth terms

$$(-1)^n/\sqrt{n} \quad \text{and} \quad (1+(-1)^n/\sqrt{n})(-1)^n/\sqrt{n}.$$

Dirichlet pointed out that the ratio of the nth terms approached 1 as n tended to infinity, but the first series converged and the second diverged.

Now in article 235 of his book on heat, Fourier gave the formula

$$f(x) = \frac{1}{\pi} \int f(\alpha) d\alpha \left(\frac{1}{2} + \sum \cos i(x-\alpha) \right).$$

Dirichlet analyzed the partial sums of this series under the assumption that the function $f(x)$ was piecewise monotonic. Taking $n+1$ terms of the series and using

$$\frac{1}{2} + \cos(\alpha-x) + \cos 2(\alpha-x) + \cdots + \cos n(\alpha-x) = \frac{\sin\left(n+\frac{1}{2}\right)(\alpha-x)}{2\sin\frac{1}{2}(\alpha-x)},$$

Dirichlet represented the partial sum by

$$s_n(x) = \frac{1}{\pi} \int_{-\pi}^{\pi} f(\alpha) \frac{\sin\left(n+\frac{1}{2}\right)(\alpha-x)}{2\sin\frac{1}{2}(\alpha-x)} d\alpha.$$

He proved that this integral converged to

$$\frac{f(x+0) + f(x-0)}{2},$$

when f satisfied certain conditions. For this purpose, he first demonstrated the theorem: For any function $f(\beta)$, continuous and monotonic in the interval (g, h), where $0 \le g < h \le \pi/2$, the integral (for $0 \le g < h$)

$$\int_g^h f(\beta) \frac{\sin i\beta}{\sin \beta} d\beta$$

converges to a limit as i tends to infinity. The limit is zero except when $g = 0$, in which case the limit is $\frac{\pi}{2} f(0)$. We present Dirichlet's argument in a slightly condensed form, for the most part using his notation. First note that we can write

$$\int_0^\infty (\sin x)/x \, dx = \pi/2$$

$$\int_0^\pi + \int_\pi^{2\pi} + \cdots + \int_{n\pi}^{(n+1)\pi} + \cdots \frac{\sin x}{x} dx = \frac{\pi}{2}.$$

Since $\sin x$ changes signs in the successive intervals $[0, \pi]$, $[\pi, 2\pi]$, ... and the integrand is decreasing, we can write the sum as

$$k_1 - k_2 + k_3 - \cdots + (-1)^{n-1} k_n + \cdots = \frac{\pi}{2}, \qquad (21.53)$$

$$\text{where} \quad k_{n+1} = \left| \int_{n\pi}^{(n+1)\pi} \frac{\sin x}{x} dx \right|.$$

The series converges and hence $k_n \to 0$ as $n \to \infty$. Now consider the integral

$$I = \int_0^h \frac{\sin i\beta}{\sin \beta} f(\beta) d\beta,$$

where $f(\beta)$ is decreasing and positive. Divide the interval $[0, h]$ by the points

$$0 < \frac{\pi}{i} < \frac{2\pi}{i} < \cdots < \frac{r\pi}{i} < h,$$

where r is the largest integer for which the last inequality holds. I is the sum of the integrals on these $r+1$ subintervals. On comparing two of the consecutive subintervals, we see that for $\upsilon < r$

$$\left| I_\upsilon = \int_{(\upsilon-1)\frac{\pi}{i}}^{\upsilon\frac{\pi}{i}} \frac{\sin i\beta}{\sin \beta} f(\beta) d\beta \right| \geq \left| I_{\upsilon+1} = \int_{\upsilon\frac{\pi}{i}}^{(\upsilon+1)\frac{\pi}{i}} \frac{\sin i\beta}{\sin \beta} f(\beta) d\beta \right|.$$

Verify this by changing β to $\frac{\pi}{i} + \beta$, so that the second integral can be written as

$$-\int_{(\upsilon-1)\frac{\pi}{i}}^{\upsilon\frac{\pi}{i}} \frac{\sin i\beta}{\sin(\beta + \frac{\pi}{i})} f\left(\beta + \frac{\pi}{i}\right) d\beta.$$

Also, $f(\beta)$ is decreasing so that

$$\frac{f(\beta)}{\sin \beta} > \frac{f\left(\beta + \frac{\pi}{i}\right)}{\sin\left(\beta + \frac{\pi}{i}\right)}.$$

Thus,

$$I = I_1 - I_2 + I_3 - I_4 + \cdots \pm I_r \mp I_h,$$

where I_h is defined over the interval $(r\pi/i, h)$ so that the I_j are positive and decreasing right up to the last term I_h. Next, let

$$K_\upsilon = \left| \int_{(\upsilon-1)\pi/i}^{\upsilon\pi/i} \frac{\sin i\beta}{\sin \beta} d\beta \right|$$

$$= \left| \int_{(\upsilon-1)\pi}^{\upsilon\pi} \frac{\sin \gamma}{i \sin(\gamma/i)} d\gamma \right|.$$

Observe that the last integral is obtained by the change of variables $\gamma = i\beta$. As $i \to \infty$, this integral tends to $\int_{(\upsilon-1)\pi}^{\upsilon\pi} (\sin \gamma)/\gamma \, d\gamma = k_\upsilon$. Next fix a number m, assumed for convenience to be even, and let r be greater than m. Let ρ_υ be such that

$$f\left(\frac{(\upsilon-1)\pi}{i}\right) \leq \rho_\upsilon \leq f\left(\frac{\upsilon\pi}{i}\right) \text{ and } I_\upsilon = \rho_\upsilon K_\upsilon.$$

Then

$$I = (K_1\rho_1 - K_2\rho_2 + K_3\rho_3 - \cdots - K_m\rho_m)$$
$$+ (K_{m+1}\rho_{m+1} - K_{m+2}\rho_{m+2} + \cdots) = I(m) + I',$$

where $I(m)$ consists of the m terms inside the first set of parentheses and I' represents the remaining terms, inside the second set of parentheses. Therefore, the sum $I(m)$ as $i \to \infty$ converges to

$$f(0)(k_1 - k_2 + k_3 - \cdots - k_m) = s_m f(0).$$

This means that the sums $I(m)$ and $s_m f(0)$ can be made less than a positive number w no matter how small. The sum I' is an alternating series with decreasing terms and hence is less than $K_{m+1}\rho_{m+1}$; note that this converges to $k_{m+1} f(0)$. Thus, by (21.53), $|I'| < k_{m+1} f(0)| + w'$, where w' can be made arbitrarily small. Moreover,

$$\left|\frac{\pi}{2} - s_m\right| < k_{m+1} \text{ and so } \left|I - \frac{\pi}{2} f(0)\right| < w + w' + 2f(0)k_{m+1}. \tag{21.54}$$

This proves the theorem for f positive and $g = 0$. If $g > 0$, then

$$\int_g^h \frac{\sin i\beta}{\sin \beta} f\beta \, d\beta = \int_0^h \frac{\sin i\beta}{\sin \beta} f\beta \, d\beta - \int_0^g \frac{\sin i\beta}{\sin \beta} f\beta \, d\beta.$$

At this point, one may conclude that both these integrals tend to $\frac{\pi}{2} f(0)$ as $i \to \infty$. So $I_g \to 0$ as $i \to \infty$. This proves the theorem for positive decreasing f. If f also assumes negative values, then choose a constant C large enough that $C + f$ is positive. If f is increasing, $-f$ is decreasing, taking care of that case, and the theorem is proved.

Dirichlet noted that if f was discontinuous at 0, then by the previous argument, $f(0)$ could be replaced by $f(\epsilon)$ where ϵ was an infinitely small positive number. In his 1837 paper, he denoted $f(x + \epsilon)$ by $f(x + 0)$, the right-hand limit of $f(t)$ as $t \to x$. This is now standard notation.

To prove Fourier's theorem, break up the integral for $s_n(x)$ into two parts, one taken from $-\pi$ to x and the other from x to π. If α is replaced by $x - 2\beta$ in the first integral and by $x + 2\beta$ in the second, then we have

$$s_n(x) = \int_0^{(\pi+x)/2} \frac{\sin(2n+1)\beta}{\sin \beta} f(x - 2\beta) \, d\beta + \int_0^{(\pi-x)/2} \frac{\sin(2n+1)\beta}{\sin \beta} f(x + 2\beta) \, d\beta.$$

Suppose $x \neq -\pi$ or π and $\beta - x < \pi$. The function $f(x + 2\beta)$ in the second integral may be discontinuous at several points between $\beta = 0$ and $\beta = (\pi - x)/2$, and it may also have several external points in this interval. Denote these points by $l, l', l'', \ldots, l^\upsilon$ in ascending order, and decompose the second integral over the intervals $(0, l), (l, l'), \ldots$. By the theorem, the first of these $\upsilon + 1$ integrals has the limit $f(x + \epsilon)\frac{\pi}{2}$ (i.e., $f(x + 0)\frac{\pi}{2}$) and the others have the limit zero as $n \to \infty$. If in the first integral for $s_n(x)$ we have $\beta + x \geq \pi$, then write it as

$$\int_0^{\pi/2} \frac{\sin(2n+1)\beta}{\sin \beta} f(x - 2\beta) \, d\beta + \int_{\pi/2}^{(\beta+x)/2} \frac{\sin(2n+1)\beta}{\sin \beta} f(x - 2\beta) \, d\beta.$$

The first integral tends to $f(x-\epsilon)\frac{\pi}{2}$ as $n \to \infty$. A similar argument shows that

$$\int_0^{(\pi+x)/2} \frac{\sin(2n+1)\beta}{\sin \beta} f(x-2\beta)\, d\beta \text{ tends to } f(x+\epsilon)\frac{\pi}{2}.$$

21.7 Dirichlet: On the Evaluation of Gauss Sums

In 1835, Dirichlet presented a paper to the Berlin Academy explaining how the definite integral $\int_{-\infty}^{\infty} e^{ix^2}\, dx$ could be applied to evaluate the Gauss sum $\sum_{k=0}^{n-1} e^{2\pi i k^2/n}$. He gave a slightly simpler version of this proof in his famous 1840 paper in *Crelle's Journal* on the applications of infinitesimal analysis to the theory of numbers. In this paper, Dirichlet also derived the class number formula for quadratic forms and proved his well-known theorem on primes in arithmetic progressions. In his paper of 1837, Dirichlet gave an even more careful analysis of convergence.

He started his evaluation of the Gauss sum by first proving a finite form of the Poisson summation formula. Note that in 1826, Poisson had used such a finite formula to deduce the Euler–Maclaurin summation. Dirichlet began with a continuous function $g(x)$ in $[0, \pi]$ expandable as a Fourier series:

$$\pi g(x) = c_0 + 2 \sum_{s=1}^{\infty} c_s \cos sx, \tag{21.55}$$

$$\text{where } c_s = \int_0^{\pi} g(x) \cos sx\, dx. \tag{21.56}$$

It followed for $x = 0$ that

$$c_0 + 2 \sum_{s=1}^{\infty} c_s = \pi g(0). \tag{21.57}$$

He then set

$$g(x) = f(x) + f(2\pi - x) + f(2\pi + x) + \cdots + f(2(h-1)\pi + x) + f(2h\pi - x), \tag{21.58}$$

where $f(x)$ was continuous on $[0, 2h\pi]$. He observed that

$$c_s = \int_0^{\pi} g(x) \cos sx\, dx = \int_0^{2h\pi} f(x) \cos sx\, dx. \tag{21.59}$$

By using the value of $g(0)$ from (21.58), he could rewrite (21.57) in the form

$$c_0 + 2 \sum_{s=1}^{\infty} c_s = \pi \left(f(0) + f(2n\pi) + 2 \sum_{s=1}^{h-1} f(2s\pi) \right), \tag{21.60}$$

where c_s was given by (21.59). This was the finite form of the Poisson summation formula employed by Dirichlet. He then considered the integral

$$\int_{-\infty}^{\infty} \cos x^2\, dx = a,$$

where a was some number to be determined. Since Euler had evaluated this integral, Dirichlet knew its exact value. However, Dirichlet's method of evaluating Gauss sums was such that it also determined the value of a. He set

$$x = \frac{z}{2}\sqrt{\frac{n}{2\pi}},$$

where n was a positive integer divisible by 4, transforming the integral to

$$\int_{-\infty}^{\infty} \cos\left(\frac{n}{8\pi}z^2\right) dz = 2a\sqrt{\frac{2\pi}{n}}. \tag{21.61}$$

Dirichlet then rewrote the last integral as a sum:

$$\sum_{s=-\infty}^{\infty} \int_{2s\pi}^{2(s+1)\pi} \cos\left(\frac{n}{8\pi}z^2\right) dz = \sum_{s=-\infty}^{\infty} \int_0^{2\pi} \cos\frac{n}{8\pi}(2s\pi + z)^2 dz. \tag{21.62}$$

He observed that since n was divisible by 4,

$$\cos\frac{n}{8\pi}(2s\pi + z)^2 = \cos\frac{n}{8\pi}(4s^2\pi^2 + 4s\pi z + z^2) = \cos\left(\frac{snz}{2} + \frac{n}{8\pi}z^2\right).$$

Then, by the addition formula, he had

$$\cos\left(\frac{snz}{2} + \frac{n}{8\pi}z^2\right) + \cos\left(-\frac{snz}{2} + \frac{n}{8\pi}z^2\right) = 2\cos\left(\frac{snz}{2}\right)\cos\left(\frac{n}{8\pi}z^2\right).$$

Hence, (21.61) and (21.62) could be expressed as

$$\int_0^{2\pi} \cos\left(\frac{n}{8\pi}z^2\right) dz + 2\sum_{s=1}^{\infty} \int_0^{2\pi} \cos\left(\frac{n}{8\pi}z^2\right) \cos\left(s\frac{nz}{2}\right) dz = 2a\sqrt{\frac{2\pi}{n}}.$$

Dirichlet substituted $nz = 2x$ in this formula to obtain

$$\int_0^{n\pi} \cos\left(\frac{x^2}{2n\pi}\right) dx + 2\sum_{s=1}^{\infty} \int_0^{n\pi} \cos\left(\frac{x^2}{2n\pi}\right) \cos sx\, dx = a\sqrt{2n\pi}. \tag{21.63}$$

Since n was an even number expressible as $2h$, the sum on the left-hand side coincided with the sum on the left-hand side of (21.60) when $f(x) = \cos(x^2/2n\pi)$. By combining (21.60) and (21.63), Dirichlet arrived at the formula

$$\cos 0 + \cos\left(\frac{n}{2}\right)^2 \frac{2\pi}{n} + 2\sum_{s=1}^{\frac{n}{2}-1} \cos s^2 \frac{2\pi}{n} = a\sqrt{\frac{2n}{\pi}}. \tag{21.64}$$

He then observed that

$$\cos s^2 \frac{2\pi}{n} = \cos(n-s)^2 \frac{2\pi}{n},$$

and therefore (21.64) could be expressed in the simpler form

$$\sum_{s=0}^{n-1} \cos s^2 \frac{2\pi}{n} = a\sqrt{\frac{2n}{\pi}}. \tag{21.65}$$

21.7 Dirichlet: On the Evaluation of Gauss Sums

Dirichlet next remarked that the value of a was independent of n so that by choosing $n = 4$, he could write

$$2 = 2a\sqrt{\frac{2}{\pi}}, \quad \text{or} \quad a = \sqrt{\pi/2},$$

and therefore he could express the Gauss sum as

$$\sum_{s=0}^{n-1} \cos \frac{s^2 2\pi}{n} = \sqrt{n}. \tag{21.66}$$

Operating in the same manner with $\int_{-\infty}^{\infty} \sin x^2 \, dx$, he arrived at

$$\sum_{s=0}^{n-1} \sin \frac{s^2 2\pi}{n} = \sqrt{n}. \tag{21.67}$$

Dirichlet pointed out that the sums (21.66) and (21.67) could be similarly evaluated for n of the form $4\mu + 1$, $4\mu + 2$, and $4\mu + 3$. However, it was possible to obtain these sums in a different way. For that purpose he defined, for positive integers m and n,

$$\sum_{s=0}^{n-1} e^{2ms^2 \pi i/n} = \phi(m,n).$$

He then wrote

$$\phi(m,n) = \phi(m',n) \quad \text{when } m \equiv m' \pmod{n}; \tag{21.68}$$

$$\phi(m,n) = \phi(c^2 m, n) \quad \text{when } c \text{ was prime to } n; \tag{21.69}$$

$$\phi(m,n)\phi(n,m) = \phi(1,mn) \quad \text{when } m \text{ and } n \text{ were coprime}. \tag{21.70}$$

Dirichlet proved the third equation by observing that

$$\phi(m,n)\phi(n,m) = \sum_{s=0}^{n-1} \sum_{t=0}^{m-1} e^{2ms^2 \pi i/n} e^{2nt^2 \pi i/m}$$

$$= \sum_{s=0}^{n-1} \sum_{t=0}^{m-1} e^{(m^2 s^2 + n^2 t^2) 2\pi i/(mn)}$$

$$= \sum_{s=0}^{n-1} \sum_{t=0}^{m-1} e^{(ms+nt)^2 2\pi i/(mn)}.$$

Since m and n were chosen relatively prime, Dirichlet argued that $ms + nt$ assumed all the residues (mod mn) as s and t ranged over the values $0, 1, \ldots, n-1$ and $0, 1, \ldots, m-1$, respectively. Therefore,

$$\phi(m,n)\phi(n,m) = \sum_{s=0}^{mn-1} e^{2\pi s^2 i/(mn)}, \tag{21.71}$$

and Dirichlet's proof of (21.70) was complete. Note that Gauss gave a similar argument in 1801. Dirichlet then observed that for n, a multiple of 4, (21.66) and (21.67) implied

$$\phi(1,n) = (1+i)\sqrt{n}. \tag{21.72}$$

And for odd n, (21.70) and (21.72) gave

$$\phi(4,n)\phi(n,4) = \phi(1,4n) = 2(1+i)\sqrt{n}. \tag{21.73}$$

Moreover, by (21.69), $\phi(4,n) = \phi(1,n)$ when n was odd, and by (21.68) $\phi(n,4) = \phi(1,4)$ or $\phi(3,4)$, depending on whether $n = 4\mu + 1$ or $n = 4\mu + 3$. Since

$$\phi(1,4) = 2(1+i), \quad \text{and} \quad \phi(3,4) = 2(1-i),$$

Dirichlet could conclude that

$$\phi(1,n) = \sqrt{n}, \; n = 4\mu + 1; \; \phi(1,n) = i\sqrt{n}, \; n = 4\mu + 3. \tag{21.74}$$

Finally, when $n = 4\mu + 2$, he argued that $n/2$ and 2 were relatively prime, so that by (21.70)

$$\phi(2,n/2)\phi(n/2,2) = \phi(1,n) \quad \text{and} \quad \phi(n/2,2) = \phi(1,2) = 0.$$

$$\text{Thus,} \quad \phi(1,n) = 0, \quad n = 4\mu + 2.$$

Gauss gave a proof of the quadratic reciprocity theorem by using the values of the Gauss sums. Dirichlet repeated these arguments in his papers, though in a simpler form. In fact, in his papers and lectures Dirichlet presented many number theoretic ideas of Gauss within an easily understandable approach. For example, he published a one-page proof of a theorem of Gauss on the biquadratic character of 2. It is interesting to note that the British number theorist, H. J. S. Smith (1826–1883), presented this result in the first part of his report on number theory published in 1859; he wrote in a footnote:

> The death of this eminent geometer in the present year (May 5, 1859) is an irreparable loss to the science of arithmetic. His original investigations have probably contributed more to its advancement than those of any other writer since the time of Gauss; if, at least, we estimate results rather by their importance than by their number. He has also applied himself (in several of his memoirs) to give an elementary character to arithmetical theories which, as they appear in the work of Gauss, are tedious and obscure; and he has thus done much to *popularize* the theory of numbers among mathematicians – a service which it is impossible to appreciate too highly.

Noting Smith's remark on the importance, rather than the number, of Dirichlet's results, we observe that Gauss made a similar comment when he recommended Dirichlet for the order *pour le mérite* in 1845: "The same [Dirichlet] has – as far as I know – not yet published a big work, and also his individual memoirs do not yet comprise a big volume. But they are jewels, and one does not weigh jewels on a grocer's scales."

21.8 Exercises

1. Solve $y(x) - y(x-1) = \ln x$ by Euler's trigonometric series method. See Eu. I-14, pp. 513–515.

2. Solve the equation $\frac{d^2v}{dx^2} + \frac{d^2v}{dy^2} = 0$ by assuming $v = F(x)f(y)$ to obtain $F(x) = e^{-mx}$, $f(y) = \cos my$. Let $v = \phi(x, y)$ and assume the boundary conditions $\phi(x, \pm\pi/2) = 0$ and $\phi(0, y) = 1$. Show that

$$\phi(x, y) = \frac{4}{\pi}(e^{-x}\cos y - \frac{1}{3}e^{-3x}\cos 3y + \frac{1}{5}e^{-5x}\cos 5y - \cdots).$$

See Fourier (1955), pp. 134–144.

3. Show that $\pi/2 = \arctan u + \arctan(1/u)$. Let $u = e^{ix}$ and expand $\arctan u$ and $\arctan(1/u)$ as series to obtain

$$\frac{\pi}{4} = \cos x - \frac{1}{3}\cos 3x + \frac{1}{5}\cos 5x - \cdots.$$

See Fourier (1955), p. 154.

4. Let a denote a quadratic residue modulo p, a prime, and let b denote a quadratic nonresidue. Show that

$$\phi(1, p) = 1 + 2\sum e^{a2\pi i/p} = i^{(p-1)^2/4}\sqrt{p},$$

where the sum is over all the residues a. Show also that

$$\phi(m, p) = \left(\frac{m}{p}\right)\phi(1, p) = 1 + 2\sum e^{a2m\pi i/p},$$

where $\left(\frac{m}{p}\right)$ denotes the Legendre symbol. Deduce that

$$\sum e^{2m\pi ia/p} - \sum e^{2m\pi ib/p} = \left(\frac{m}{p}\right)i^{(p-1)^2/4}\sqrt{p}.$$

See Dirichlet (1969), pp. 478–479.

5. Suppose p and q are primes. Use $\phi(p, q)\phi(q, p) = \phi(1, pq)$ and the results in the previous exercise to prove the law of quadratic reciprocity:

$$\left(\frac{p}{q}\right)\left(\frac{q}{p}\right) = (-1)^{\frac{p-1}{2}\cdot\frac{q-1}{2}}.$$

This proof originates with Gauss's 1808 paper, "Summatio Quarundam Serierum Singularium." For the derivation discussed in this exercise, see Dirichlet and Dedekind (1999), p. 206–207.

21.9 Notes on the Literature

For Lagrange's 1859 paper discussed in the text, see Lagrange (1867–1892), vol. 1, pp. 72–90. This consists of a part of his paper, "Recherches sur la nature et la propagation du son." See Eu. I-14, pp. 463–515 for the difference equation $f(x) - f(x-1) = X$ and Eu. I-16/2, pp. 333–41, for the derivation of the Fourier coefficients. For Wiener's treatment of Euler's difference equation, see Wiener (1979), vol. 2, pp. 443–453. Wiener's paper contributed to the effort to make operational calculus rigorous. Fourier (1955), pp. 168–185 contains his derivation of the Fourier coefficients by means of infinitely many linear equations in infinitely many unknowns. Riesz (1913) contains a valuable commentary on this work of Fourier. See Dirichlet (1969), vol. 1, pp. 117–132, for his 1829 paper. The more detailed paper of 1837, proving the same result, is on pp. 133–160, and his evaluation of the Gauss sum can be found on pp. 473–479 of the same volume. The deduction of quadratic reciprocity appears in Dirichlet and Dedekind (1999), pp. 208–209. Smith (1965b), p. 72, has the quote on Dirichlet. See Duke and Tschinkel (2005), p. 18, for Gauss's comment on Dirichlet. The article by J. Elstrodt in Duke and Tschinkel gives a good summary of Dirichlet's mathematical achievements. Truesdell (1960) offers an excellent account of the mathematical contributions to the subject of mechanics, related to topics mentioned in our text, by Euler, d'Alembert, D. Bernoulli, and Lagrange. Truesdell's quote concerning Euler's formulas for simple modes of an oscillating string may be found in Truesdell (1984), p. 250. See also Bottazzini (1986) and Yushkevich (1971) for the development of the concept of a function in connection with Fourier series.

22

Trigonometric Series after 1830

22.1 Preliminary Remarks

At the end of his 1829 paper on Fourier series, Dirichlet pointed out that the concept of the definite integral required further investigation if the theory of Fourier series were to include functions with an infinite number of discontinuities. In this connection he gave the example of a function $\phi(x)$ defined as a fixed constant for rational x and another fixed constant for irrational x. Such a function could not be integrated by Cauchy's definition of an integral. Dirichlet stated his plan to publish a paper on this topic at the foundation of analysis, but he never presented any results on it, though he gave important applications of Fourier series to number theory.

Bernhard Riemann (1826–1866), a student of Dirichlet, took up this question as he discussed trigonometric series in his *Habilitation* paper of 1853. The first part of the paper gave a brief history of Fourier series, a topic Riemann studied with Dirichlet's help. In the later portion, Riemann briefly considered a new definition of the integral and then went on to study general trigonometric series of the form

$$\frac{1}{2}a_0 + \sum_{n=1}^{\infty}(a_n \cos nx + b_n \sin nx),$$

where the coefficients a_n and b_n were not necessarily defined by the Euler–Fourier integrals. Using these series, he could represent nonintegrable functions in terms of trigonometric series. Here he introduced methods still not superseded, though they have been further developed. He associated with the trigonometric series a continuous function $F(x)$ obtained by twice formally integrating the series. Riemann then defined the generalized second, or Riemann–Schwarz, derivative of $F(x)$ as $\lim_{h \to 0} \left(\Delta^2 F(x-h) \right) / h^2$ and proved that if the trigonometric series converged to some $f(x)$, then the Riemann–Schwarz derivative was equal to $f(x)$. In addition, he proved that if a_n and b_n tended to zero, then

$$\lim_{h \to 0} \frac{\Delta^2 F(x-h)}{h} = 0. \tag{22.1}$$

Riemann's paper also contained a number of very interesting examples, raising important questions. Perhaps he did not publish the paper because he was unable to answer these questions; Dedekind had it published in 1867 after Riemann's premature death.

The publication of this paper by Riemann led Heinrich Heine (1821–1881) to ask whether more than one trigonometric series could represent the same function. He applied Weierstrass's result that when a series converged uniformly, term-by-term integration was possible. Weierstrass taught this theorem in his lectures in Berlin starting in the early 1860s, though he had discovered it two decades earlier. From this result, Heine concluded that a uniformly convergent trigonometric series was a Fourier series; he defined a generally uniformly convergent series to cover the case of the series of continuous functions converging to a discontinuous function. Such series converged uniformly on the intervals obtained after small neighborhoods around the discontinuities had been removed. In a paper of 1870, Heine stated and proved that a function could not be represented by more than one generally uniformly convergent trigonometric series.

When Georg Cantor (1845–1918) joined Heine at the University of Halle in 1869, Heine awakened his interest in this uniqueness question. Cantor had studied at the University of Berlin under Kummer, Kronecker, and Weierstrass and wrote his thesis on quadratic forms under Kummer. At Heine's suggestion, Cantor studied Riemann's paper containing the observation, without proof, that if

$$a_n \cos nx + b_n \sin nx \to 0 \quad \text{as} \quad n \to \infty$$

for all x in an interval, then $a_n \to 0$ and $b_n \to 0$ as $n \to \infty$. Cantor's first paper on trigonometric series, published in 1870, provided a proof of this important assertion, now known as the Cantor-Lebesgue theorem. This was the first step in Cantor's proof of the uniqueness theorem that if two trigonometric series converge to the same sum in $(0, 2\pi)$ except for a finite number of points, then the series are identical. Note that Henri Lebesgue (1875–1941) later proved the theorem in a more general context.

To prove his theorem, Cantor needed to show that if the generalized second, or Riemann–Schwarz, derivative of a continuous function was zero in an interval, then the function was linear in that interval. So in a letter of February 17, 1870, he asked his friend Hermann Schwarz for a proof of this result. Schwarz had received his doctoral degree a few years before Cantor, but they had both studied under Kummer and Weierstrass at Berlin. Schwarz left the University of Halle in 1869 and went to Zurich, but they corresponded often. In fact, Schwarz wrote to Cantor on February 25, 1870, "The fact that I wrote to you at length yesterday is no reason why I should not write again today." In this letter Schwarz gave what he said was the first rigorous proof of the theorem that if a function had a zero derivative at every value in an interval, then the function was a constant in that interval.

Schwarz provided a proof of the result Cantor needed for his uniqueness theorem. Cantor next studied the case with exceptional points, at which the series was not known to converge to zero. Was the value of every coefficient still zero? He supposed c to be an exceptional point in an interval (a, b) so that the series converged to zero in the intervals (a, c) and (c, b). Now Riemann's second theorem, given by (22.1), implied that the slopes of the two lines had to be the same, and hence $F(x)$ was linear in (a, b), and the uniqueness theorem followed. Clearly, the argument could be extended to a finite number of exceptional points. When Cantor realized this, he asked whether there could be an infinite number of exceptional points; he soon understood that even if the exceptional points were infinite in number, as long as they had only a finite number of limit points, his basic argument would still be effective.

Leopold Kronecker (1823–1891) was initially quite interested in the work of Cantor on the uniqueness of trigonometric series. After the publication of Cantor's first paper, Kronecker explained to him that the proof of the Cantor–Lebesgue theorem could be simplified by means of an idea contained in Riemann's paper. However, as Cantor's work progressed and he began to use increasingly intricate infinite sets, Kronecker lost sympathy with Cantor's ideas and became a passionate critic of the theory of infinite sets. Cantor, on the other hand, abandoned the study of trigonometric series and after 1872 became more and more intrigued by infinite sets, at that time completely unexplored territory. Luckily, Cantor found an understanding and kindred spirit in Dedekind, who had himself done some work on infinite sets. Cantor started a correspondence with Dedekind in 1872 that continued off and on for several years. Dedekind helped Cantor write up a concise proof of the countability of the set of algebraic numbers, and in 1874 this theorem appeared in Cantor's first paper on infinite sets.

Though there was some opposition to Cantor's theory, it was directly and indirectly successful as sets became basic objects in the language of mathematics. Without this concept, such early twentieth-century innovations as measure theory and the Lebesgue integral would hardly have been possible. These advances in turn had consequences for the theory of trigonometric series and the theory of uniqueness of such series. As an example, consider the noteworthy theorem of W. H. Young from a 1906 paper: "If the values of a function be assigned at all but a countable set of points, it can be expressed as a trigonometric series in at most one way."

22.2 The Riemann Integral

In his 1853 paper on trigonometric series, Riemann observed that since the Euler–Fourier coefficients were defined by integrals, he would begin his study of Fourier series with a clarification of the concept of an integral. To understand his definition, let f be a bounded function defined on an interval (a, b) and let $a = x_0 < x_1 < x_2 < \cdots < x_{n-1} < x_n = b$. Denote the length of the subinterval $x_k - x_{k-1}$ by δ_k where $k = 1, 2, \ldots, n$. Let $0 \leq \epsilon_k \leq 1$ and set

$$s = \delta_1 f(a + \epsilon_1 \delta_1) + \delta_2 f(x_1 + \epsilon_2 \delta_2) + \delta_3 f(x_2 + \epsilon_3 \delta_3) + \cdots + \delta_n f(x_{n-1} + \epsilon_n \delta_n).$$

Riemann noted that the value of the sum s depended on δ_k and ϵ_k, but if it approached infinitely close to a fixed limit A as all the δs became infinitely small, then this limit would be denoted by $\int_a^b f(x)\,dx$. On the other hand, if the sum s did not have this property, then $\int_a^b f(x)\,dx$ had no meaning.

Riemann then extended the definition of an integral to include unbounded functions, as Cauchy had done. Thus, if $f(x)$ was infinitely large at a point c in (a,b), then

$$\int_a^b f(x)\,dx = \lim_{\alpha_1 \to 0} \int_a^{c-\alpha_1} f(x)\,dx + \lim_{\alpha_2 \to 0} \int_{c+\alpha_2}^b f(x)\,dx.$$

Riemann next raised the question: When was a function integrable? He gave his answer in terms of the variations of the function within subintervals. He let D_k denote the difference between the largest and the smallest values of the function in the interval (x_{k-1}, x_k) for $k = 1, 2, \ldots, n$. He then argued that if the function was integrable, the sum

$$\sum = \delta_1 D_1 + \delta_2 D_2 + \cdots + \delta_n D_n$$

must become infinitely small as the values of δ became small. He next observed that for $\delta_k \leq d$ ($k = 1, \ldots, n$), this sum would have a largest value, $\Delta(d)$. Moreover, $\Delta(d)$ decreased with d and $\Delta(d) \to 0$ as $d \to 0$. He noted that if s were the total length of those intervals in which the function varied more than some value σ, then the contribution of those intervals to $\sum$ was $\geq \sigma s$. Thus, he arrived at

$$\sigma s \leq \delta_1 D_1 + \delta_2 D_2 + \cdots + \delta_n D_n \leq \Delta$$
$$\text{or} \quad s \leq \Delta/\sigma. \tag{22.2}$$

From this inequality, he concluded that for a given σ, Δ/σ could be made arbitrarily small by a suitable choice of d and hence the same was true for s. Riemann could then state that a bounded function $f(x)$ was integrable only if the total length of the intervals in which the variations of $f(x)$ were $> \sigma$ could be made arbitrarily small by a suitable choice of d. He also gave a short argument proving the converse. Riemann's proof omitted some details necessary to make it completely convincing. In fact, in an 1875 paper presented to the London Mathematical Society, H. J. S. Smith formulated a clearer definition of integrability and a modified form of Riemann's theorem.

Riemann gave a number of interesting examples of applications of this theorem, remarking that they were quite novel. For instance, he considered the function defined by the series

$$f(x) = \frac{(x)}{1} + \frac{(2x)}{4} + \frac{(3x)}{9} + \cdots = \sum_{n=1}^{\infty} \frac{(nx)}{n^2}, \tag{22.3}$$

where (x) was the difference between x and the closest integer; in the ambiguous case when x was at the midpoint between two successive integers, (x) was taken to be zero. He showed that for $x = p/2n$ where p and n were relatively prime,

$$f(x+0) = f(x) - \frac{1}{2nn}\left(1 + \frac{1}{9} + \frac{1}{25} + \cdots\right) = f(x) - \frac{\pi\pi}{16nn},$$

$$f(x-0) = f(x) + \frac{1}{2nn}\left(1 + \frac{1}{9} + \frac{1}{25} + \cdots\right) = f(x) + \frac{\pi\pi}{16nn};$$

22.3 Smith: Revision of Riemann and Discovery of the Cantor Set

at all other values of x, $f(x)$ was continuous. Riemann applied his theorem to show that, although (22.3) had an infinite number of discontinuities, it was integrable over $(0, 1)$.

22.3 Smith: Revision of Riemann and Discovery of the Cantor Set

Henry Smith did his most notable work in number theory and elliptic functions, but his 1875 paper "On the Integration of Discontinuous Functions" also obtained some important results later found by Cantor. Though continental mathematicians did not notice this paper, it anticipated by eight years Cantor's construction of a ternary set. In order to reformulate Riemann's definition and theorem on integrability, Smith efficiently set up the modern definition of the Riemann integral in terms of the upper and lower Riemann sums, politely pointing out the gap in Riemann's work:

> Riemann, in his Memoir ..., has given an important theorem which serves to determine whether a function $f(x)$ which is discontinuous, but not infinite, between the finite limits a and b, does or does not admit of integration between those limits, the variable x, as well as the limits a and b, being supposed real. Some further discussion of this theorem would seem to be desirable, partly because, in one particular at least, Riemann's demonstration is wanting in formal accuracy, and partly because the theorem itself appears to have been misunderstood, and to have been made the basis of erroneous inferences.
>
> Let d be any given positive quantity, and let the interval $b-a$ be divided into any *segments* whatever, $\delta_1 = x_1 - a$, $\delta_2 = x_2 - x_1$, ..., $\delta_n = b - x_{n-1}$, subject only to the condition that none of these segments surpasses d. We may term d the *norm* of the division; it is evident that there is an infinite number of different divisions having a given norm; and that a division appertaining to any given norm, appertains also to every greater norm. Let $\epsilon_1, \epsilon_2, \ldots, \epsilon_n$ be positive proper fractions; if, when the norm d is diminished indefinitely, the sum
>
> $$S = \delta_1 f(a + \epsilon_1 \delta_1) + \delta_2 f(x_1 + \epsilon_2 \delta_2) + \cdots + \delta_n f(x_{n-1} + \epsilon_n \delta_n)$$
>
> converges to a definite limit, whatever be the mode of division, and whatever be the fractions $\epsilon_1, \epsilon_2, \ldots, \epsilon_n$, that limit is represented by the symbol $\int_a^b f(x) dx$, and the function $f(x)$ is said to admit of integration between the limits a and b. We shall call the values of $f(x)$ corresponding to the points of any segment the *ordinates* of that segment; by the *ordinate difference* of a segment we shall understand the difference between the greatest and least ordinates of the segment. For any given division $\delta_1, \delta_2, \ldots, \delta_n$, the greatest value of S is obtained by taking the maximum ordinate of each segment, and the least value of S by taking the minimum ordinate of each segment; if D_i is the ordinate difference of the segment d_i, the difference θ between those two values of S is
>
> $$\theta = \delta_1 D_1 + \delta_2 D_2 + \cdots + \delta_n D_n.$$
>
> But, for a given norm d, the greatest value of S, and the least value of S, will in general result, not from one and the same division, but from two different divisions, each of them having the given norm. Hence the difference Θ between the greatest and least values that S can acquire for a given norm, is, in general, greater than the greatest of the differences θ. To satisfy ourselves, in any given case, that S converges to a definite limit, when d is diminished without limit, we must be sure that Θ diminishes without limit; and it is not enough to show (as the form of Riemann's proof would seem to imply) that θ diminishes without limit, even if this should be shown for every division having the norm d.

With this revised definition of the integral, Smith was in a position to restate Riemann's condition for integrability: "Let σ be any given quantity, however small; if, in every division of norm d, the sum of the segments, of which the ordinate differences surpass σ, diminishes without limit, as d diminishes without limit, the function admits of integration; and, *vice versa*, if the function admits of integration, the sum of these segments diminishes without limit with d."

Recall that Cantor was led to his theory of infinite sets through his researches in trigonometric series, and these in turn had their origins in Riemann's paper. This paper also inspired mathematicians to investigate the possibility of other peculiar or pathological functions and to construct infinite sets with apparently strange properties. In 1870 Hermann Hankel, a student of Riemann, constructed infinite nowhere-dense sets and he gave a flawed proof that a function with discontinuities only on a nowhere dense set was integrable. However, Hankel succeeded in proving that the set of points of continuity of an integrable function was dense.

Smith was the first to notice the mistake in Hankel's proof; to begin to tackle this problem, he divided the interval $(0, 1)$ into $m \geq 2$ equal parts where the last segment was not further divided. The remaining $m - 1$ segments were again divided into m equal parts with the last segments of each left undivided. This process was continued ad infinitum to obtain the set P of division points. Smith proved that P was nowhere dense; he called them points "in loose order." The union of the set P and its limit points is now called a Cantor set since in 1883 Cantor constructed such a set with $m = 3$. Smith showed that after k steps, the total length of the divided segments was $(1 - 1/m)^k$; so that as k increased indefinitely, the points of P were located on segments occupying only an infinitesimal portion of the interval $(0, 1)$. He then applied Riemann's criterion for integrability to show that bounded functions with discontinuities only at P would be integrable.

With a slight modification of this construction, Smith showed that there existed nowhere dense sets of positive measure. The first step in his modification was the same as before. In the second step he divided the $m - 1$ divided segments into m^2 parts, but did not further divide the last segment of each of these. The $(m-1)(m^2-1)$ remaining segments were divided into m^3 parts, and so on. After k steps, Smith found the total length of the divided segments to be $(1-1/m)(1-1/m^2)\cdots(1-1/m^k)$. He noted that the limit $\prod_{k=1}^{\infty}(1-1/m^k)$ was not equal to zero. He again proved that the set of division points Q was nowhere dense but that in this case a function with discontinuities at the points in Q was not integrable. Smith then noted, "The result obtained in the last example deserves attention, because it is opposed to a theory of discontinuous functions, which has received the sanction of an eminent geometer, Dr. Hermann Hankel, whose recent death at an early age is a great loss to mathematical science."

In 1902, Lebesgue proved that a bounded function was Riemann integrable if and only if the set of its discontinuities was of measure zero. Smith would perhaps not have been surprised at this result.

22.4 Riemann's Theorems on Trigonometric Series

After defining the integral, Riemann also investigated the question of whether a function could be represented by a trigonometric series without assuming any specific properties

22.4 Riemann's Theorems on Trigonometric Series

of the function, such as whether the function was integrable. Of course, if a function is not integrable it cannot have a Fourier series. Thus, Riemann focused on series of the form

$$\Omega = \frac{1}{2}a_0 + (a_1 \cos x + b_1 \sin x) + (a_2 \cos 2x + b_2 \sin 2x) + \cdots$$
$$= A_0 + A_1 + A_2 + \cdots,$$

where $A_0 = a_0/2$ and for $n > 0$

$$A_n = a_n \cos nx + b_n \sin nx.$$

He assumed that $A_n \to 0$ as $n \to \infty$ and he associated with Ω a function $F(x)$ obtained by twice formally integrating the series Ω. Thus, he set

$$C + C'x + A_0 \frac{xx}{2} - A_1 - \frac{A_2}{4} - \frac{A_3}{9} - \cdots = F(x) \tag{22.4}$$

and proved $F(x)$ continuous by showing that the series was uniformly convergent, though he did not use this terminology. He then stated his first theorem on $F(x)$:

If the series Ω converges,

$$\frac{F(x+\alpha+\beta) - F(x+\alpha-\beta) - F(x-\alpha+\beta) + F(x-\alpha-\beta)}{4\alpha\beta}, \tag{22.5}$$

converges to the same value as the series if α and β become infinitely small in such a way that their ratio remains finite (bounded).

By using the addition formula for sine and cosine, Riemann saw that expression (22.5) reduced to

$$A_0 + A_1 \frac{\sin \alpha}{\alpha} \frac{\sin \beta}{\beta} + A_2 \frac{\sin 2\alpha}{2\alpha} \frac{\sin 2\beta}{2\beta} + A_2 \frac{\sin 3\alpha}{3\alpha} \frac{\sin 3\beta}{3\beta} + \cdots.$$

When $\alpha = \beta$, he had the equation

$$\frac{F(x+2\alpha) - 2F(x) + F(x-2\alpha)}{4\alpha\alpha} = A_0 + A_1 \left(\frac{\sin \alpha}{\alpha}\right)^2 + A_2 \left(\frac{\sin 2\alpha}{2\alpha}\right)^2 + \cdots. \tag{22.6}$$

Riemann first proved this theorem for the $\alpha = \beta$ case and then deduced the general case. Observe that as $\alpha \to 0$, the series (22.6) converges termwise to Ω. Thus, Riemann's task was essentially to show that (22.6) converged uniformly with respect to α. We follow Riemann in detail, keeping in mind that Riemann did not use absolute values as we would today. Suppose that the series Ω converges to a function $f(x)$. Write

$$A_0 + A_1 + \cdots + A_{n-1} = f(x) + \epsilon_n \tag{22.7}$$

so that $\quad A_0 = f(x) + \epsilon_1 \quad$ and $\quad A_n = \epsilon_{n+1} - \epsilon_n. \tag{22.8}$

Riemann noted that, because of convergence, for any positive number δ, there existed an integer m such that $\epsilon_n < \delta$ for $n > m$. By (22.8) and using summation by parts, he concluded that

$$\sum_{n=0}^{\infty} A_n \left(\frac{\sin n\alpha}{n\alpha}\right)^2 = f(x) + \sum_{n=1}^{\infty} \epsilon_n \left(\left(\frac{\sin(n-1)\alpha}{(n-1)\alpha}\right)^2 - \left(\frac{\sin n\alpha}{n\alpha}\right)^2\right). \tag{22.9}$$

He then took α to be sufficiently small, so that $m\alpha < \pi$ and let s be the largest integer in π/α. He divided the last sum into three parts:

$$\sum_{n=1}^{m} + \sum_{n=m+1}^{s} + \sum_{n=s+1}^{\infty}.$$

The first sum was a finite sum of continuous functions, and it could be made arbitrarily small by taking α sufficiently small. In the second sum, the factor multiplying ϵ_n was positive and hence the sum could be written

$$< \delta \left(\left(\frac{\sin m\alpha}{m\alpha}\right)^2 - \left(\frac{\sin s\alpha}{s\alpha}\right)^2\right).$$

Note that Riemann assumed $\pi/2 \leq \alpha \leq \pi$ for any n in the second sum, although he did not explicitly mention this. To show that the third sum could be made arbitrarily small, he rewrote the general term as the sum of

$$\epsilon_n \left(\left(\frac{\sin(n-1)\alpha}{(n-1)\alpha}\right)^2 - \left(\frac{\sin(n-1)\alpha}{n\alpha}\right)^2\right) \quad \text{and}$$

$$\epsilon_n \left(\left(\frac{\sin(n-1)\alpha}{n\alpha}\right)^2 - \left(\frac{\sin n\alpha}{n\alpha}\right)^2\right) = -\epsilon_n \frac{\sin(2n-1)\alpha \sin \alpha}{(n\alpha)^2}.$$

It was then clear that the general term in the third sum was less than

$$\delta \left(\frac{1}{(n-1)^2 \alpha\alpha} - \frac{1}{nn\alpha\alpha}\right) + \delta \frac{1}{nn\alpha}.$$

Thus, the third sum was less than

$$\delta \left(\frac{1}{(s\alpha)^2} + \frac{1}{s\alpha}\right).$$

Then, for infinitely small α, this expression became

$$\delta \left(\frac{1}{\pi\pi} + \frac{1}{\pi}\right).$$

Riemann concluded that the infinite series on the right-hand side of (22.9) could not be greater than

$$\delta(1 + 1/\pi + 1/\pi^2),$$

so that the theorem was proved. Riemann's argument can be shortened by observing that the second and third sums, in absolute value, are together less than

$$\delta \sum_{n=m+1}^{\infty} \int_{(n-1)\alpha}^{n\alpha} \left| \frac{d}{dt}\left(\frac{\sin^2 t}{t^2}\right) \right| dt < \delta \int_0^{\infty} \left| \frac{d}{dt}\left(\frac{\sin^2 t}{t^2}\right) \right| dt.$$

Since the last integral is convergent, the result follows. To prove the general result, when $\alpha \neq \beta$, Riemann set

$$F(x+\alpha+\beta) - 2F(x) + F(x-\alpha-\beta) = (\alpha+\beta)^2(f(x)+\delta_1),$$
$$F(x+\alpha-\beta) - 2F(x) + F(x-\alpha+\beta) = (\alpha-\beta)^2(f(x)+\delta_2),$$

so that

$$\big(F(x+\alpha+\beta) - F(x+\alpha-\beta) - F(x-\alpha+\beta) + F(x-\alpha-\beta)\big)/4\alpha\beta$$
$$= f(x) + \frac{(\alpha+\beta)^2}{4\alpha\beta}\delta_1 - \frac{(\alpha-\beta)^2}{4\alpha\beta}\delta_2.$$

The special case $\alpha = \beta$ implied that δ_1 and δ_2 became small as α and β got small. Moreover, the factors $(\alpha+\beta)^2/4\alpha\beta$ and $(\alpha-\beta)^2/4\alpha\beta$ remained bounded when β/α was bounded. This proved the general case. Observe that the limit

$$\lim_{h \to 0} \frac{F(x+h) + F(x-h) - 2F(x)}{h^2}$$

is called the Schwarz, or Riemann–Schwarz, derivative of F. Riemann called this the "second differential quotient." Note that

$$F(x+h) + F(x-h) - 2F(x) = \Delta^2 F(x-h), \text{ where } \Delta F(x-h) = F(x) - F(x-h).$$

In general, $F(x)$ is continuous, as Riemann proved, but not necessarily differentiable. So here we have an instance of a generalized second derivative, although Riemann did not express himself in those terms.

Riemann's second theorem stated that when $A_n \to 0$ as $n \to \infty$, then

$$\frac{F(x+2\alpha) + F(x-2\alpha) - 2F(x)}{2\alpha}$$

tends to 0 as α tends to 0. In his terse style, Riemann gave a succinct argument for this, along lines similar to his proof of his first theorem. In his *The Apprenticeship of a Mathematician*, André Weil wrote that both he and his sister Simone found great value in the works of great minds and that he was very lucky to start off his mathematical reading of the greats with Riemann; he found that Riemann's works "are not hard to read, as long as one realizes that every word is loaded with meaning; there is perhaps no other mathematician whose writing matches Riemann's for density."

22.5 The Riemann–Lebesgue Lemma

The Riemann–Lebesgue lemma states that if $f(x)$ is integrable over (a,b), then as $t \to \infty$,

$$\int_a^b f(x)\cos tx\, dx \to 0, \quad \text{and} \quad \int_a^b f(x)\sin tx\, dx \to 0.$$

Note that this result implies that the nth Fourier coefficients of an integrable function tend to zero as $n \to \infty$. Riemann derived his lemma from his integrability condition in an interesting way. He began by writing

$$\int_0^{2\pi} f(x)\sin nx\, dx = \sum_{k=1}^n \int_{2(k-1)\pi/n}^{2k\pi/n} f(x)\sin nx\, dx.$$

He noted that $\sin nx$ was positive in the first half of the subinterval $\left(\frac{2(k-1)\pi}{n}, \frac{2k\pi}{n}\right)$ and negative in the second half. He supposed that in the whole subinterval he had $m_k \le f(x) \le M_k$, where M_k was taken to be the largest value of $f(x)$ in the subinterval and m_k the least. We may assume these to be the least upper bound and greatest lower bound, respectively. Thus, in the first half of the subinterval,

$$\int_{2(k-1)\pi/n}^{(2k-1)\pi/n} f(x)\sin nx\, dx \le M_k \int_{2(k-1)\pi/n}^{(2k-1)\pi/n} \sin nx\, dx = 2M_k/n.$$

Similarly, in the second half of the subinterval, the integral would be less than $-2m_k/n$. It followed that

$$\int_{2(k-1)\pi/n}^{2k\pi/n} f(x)\sin nx\, dx \le \frac{2}{n}(M_k - m_k)$$

and hence

$$\left|\int_0^{2\pi} f(x)\sin nx\, dx\right| \le \sum_{k=1}^n \frac{2}{n}(M_k - m_k) = \frac{1}{\pi}\sum_{k=1}^n \delta_k D_k,$$

where δ_k was the length of the kth subinterval and D_k was the variation of $f(x)$ on that interval. By his own definition of integrablility of $f(x)$, the sum $\sum \delta_k D_k$ had to become infinitely small as n became infinitely large. This proved the theorem. Observe that the definition of integrablility was perfect for obtaining this result on the Fourier coefficients, leading some to speculate that Riemann fashioned the definition with this result in mind.

22.6 Schwarz's Lemma on Generalized Derivatives

Recall that in connection with his work on trigonometric series, Cantor in 1870 asked Schwarz whether the following result was true: If $F(x)$ is continuous in an interval

$a \leq x \leq b$ and

$$\lim_{\alpha \to 0} \frac{F(x+\alpha) - 2F(x) + F(x-\alpha)}{\alpha\alpha} = 0 \qquad (22.10)$$

for all x in the interval, then $F(x)$ is a linear function. Schwarz replied that this was indeed a theorem and provided a proof published twenty years later in his collected mathematical works. Cantor used the theorem and gave the proof, credited to Schwarz, in his 1870 paper.

Now note that if F is twice differentiable, then its second derivative and generalized second derivative are identical; moreover, by (22.10), $F(x)$ is linear. Briefly, Schwarz's proof of the general case began by setting

$$\phi(x) = \left| F(x) - F(a) - \frac{x-a}{b-a}(F(a) - F(b)) \right| - \frac{1}{2}k(x-a)(b-x), \qquad (22.11)$$

where k was a positive quantity to be chosen later. Schwarz did not employ the absolute value sign, instead using an ϵ, equal to plus or minus 1, as a factor to maintain a positive value. Observe that $\phi(a) = 0$ and $\phi(b) = 0$. If the expression inside the absolute value sign in (22.11) is zero for all x in $a \leq x \leq b$, then $F(x)$ is a linear function. Suppose the value of the expression is not zero. Since $\phi(x)$ is continuous, it has a maximum at some point x_0. Take k sufficiently small that the value of $\phi(x_0)$ is positive. By the definition of maximum,

$$\phi(x_0 + \alpha) - \phi(x_0) \leq 0 \quad \text{and} \quad \phi(x_0 - \alpha) - \phi(x_0) \leq 0;$$

thus,

$$\phi(x_0 + \alpha) - 2\phi(x_0) + \phi(x_0 - \alpha) \leq 0.$$

But

$$\lim_{\alpha \to 0} \frac{\phi(x_0 + \alpha) - 2\phi(x_0) + \phi(x_0 - \alpha)}{\alpha\alpha}$$

$$= \lim_{\alpha \to 0} \left(\frac{F(x_0 + \alpha) - 2F(x_0) + F(x_0 + \alpha)}{\alpha\alpha} + k \right) = k > 0.$$

This contradiction implies that $F(x)$ is a linear function. Note that Weierstrass is credited with the 1841 invention of the absolute value sign we use today.

22.7 Cantor's Uniqueness Theorem

Cantor first stated his uniqueness theorem in 1870, though he later gave generalizations. His first theorem stated that if a trigonometric series

$$\frac{1}{2}a_0 + \sum_{n=1}^{\infty}(a_n \cos nx + b_n \sin nx)$$

converged to zero at every point of the interval $(-\pi, \pi)$, then $a_0 = 0$ and $a_n = b_n = 0$ for $n \geq 1$. To prove this, Cantor first used the convergence of the trigonometric series to produce a tedious proof that $a_n \to 0$ and $b_n \to 0$ as $n \to \infty$. Later on, Kronecker helped Cantor realize that he could greatly streamline his proof by working with a different series. But we continue with Cantor's original proof based on this result. Observe that he could apply Riemann's second theorem so that the second Riemann–Schwarz derivative of

$$F(x) = \frac{1}{4} a_0 x^2 - \sum_{n=1}^{\infty} \frac{1}{n^2} (a_n \cos nx + b_n \sin nx)$$

was zero in $(-\pi, \pi)$. By Schwarz's lemma, $F(x)$ was a linear function $ax + b$, and he had

$$\sum_{n=1}^{\infty} \frac{1}{n^2} (a_n \cos nx + b_n \sin nx) = \frac{1}{4} a_0 x^2 - ax - b.$$

Since the left-hand side was periodic, a_0 and a had to be zero. Because the series was uniformly convergent, Cantor could multiply by $\cos mx$ and $\sin mx$ and integrate term by term to obtain

$$\frac{\pi a_m}{m^2} = -b \int_{-\pi}^{\pi} \cos mx \, dx = 0, \quad \frac{\pi b_m}{m^2} = -b \int_{-\pi}^{\pi} \sin mx \, dx = 0,$$

for $m \geq 1$. This concludes Cantor's original proof. Observe that as a student of Weierstrass, he was quite familiar with uniform convergence and its connection with integration, but at that time the concept of uniform convergence was not well known.

To take care of the first step concerning a_n and b_n, Kronecker pointed out that it was not necessary to prove that these coefficients tended to zero. Instead, he called the trigonometric series in the theorem $f(x)$ and defined a new function in terms of u:

$$g(u) = \frac{1}{2}\big(f(x+u) + f(x-u)\big)$$
$$= \frac{1}{2} a_0 + \sum_{n=1}^{\infty} (a_n \cos nx + b_n \sin nx) \cos nu = \frac{1}{2} a_0 + \sum_{n=1}^{\infty} A_n \cos nu.$$

Since the series $f(x)$ converged, $g(u)$ also converged and therefore $A_n = a_n \cos nx + b_n \sin nx \to 0$ as $n \to \infty$ for all x in $(-\pi, \pi)$. With this new first step, Riemann's second theorem could then be applied, using $g(u)$ instead of $f(x)$, yielding $A_n = 0$ for $n \geq 1$. Thus, $a_n = 0$ and $b_n = 0$ for $n \geq 1$, so he also had $a_0 = 0$. Though Kronecker assisted Cantor with this argument, the germ of the idea was already in Riemann's paper.

Cantor extended the uniqueness theorem in an 1871 paper by requiring convergence to zero of $\frac{1}{2} a_0 + \sum_{n=1}^{\infty} A_n$ at all but a finite number of points in $(-\pi, \pi)$. He supposed x_ν to be a point at which the series did not converge. Now by Cantor's first proof, on the left-hand side of x_ν, $F(x) = k_\nu x + l_\nu$ for some constants k_ν and l_ν, whereas on the right-hand side, $F(x) = k_{\nu+1} x + l_{\nu+1}$. Now because $F(x)$ was continuous, $k_\nu x_\nu + l_\nu = k_{\nu+1} x_\nu + l_\nu$

and by Riemann's second theorem

$$\lim_{\alpha \to 0} \frac{F(x_\nu + \alpha) - 2F(x_\nu) + F(x_\nu - \alpha)}{\alpha}$$
$$= \lim_{\alpha \to 0} \frac{x_\nu(k_{\nu+1} - k_\nu) + l_{\nu+1} - l_\nu + \alpha(k_{\nu+1} - k_\nu)}{\alpha} = 0.$$

This implied that $k_{\nu+1} = k_\nu$ and $l_{\nu+1} = l_\nu$; therefore, $F(x)$ was defined by the same linear function in the whole interval $(-\pi, \pi)$.

Cantor then extended the argument to an infinite set with a finite number of limit points. Summarizing his argument, suppose $x_1, x_2, x_3, \ldots$ to be a sequence with one limit point x. Then, by the previous argument, the isolated points $x_1, x_2, x_3, \ldots$ can be removed, and then finally, after an infinite number of steps, x is isolated and can be removed. Kronecker was horrified at this mode of argument, involving the completion of an infinite number of steps; he suggested to Cantor that he refrain from publishing his paper. But to Cantor's way of thinking, this kind of reasoning was quite legitimate, since he subscribed to the concept of a completed infinity. Cantor gave further extensions of his uniqueness theorem to more general infinite sets. The enterprise led him to turn his attention toward set theory rather than analysis, and he spent the rest of his life creating and developing the theory of infinite sets.

22.8 Exercises

1. A solution of an equation $a_0 x^n + a_1 x^{n-1} + \cdots + a_n = 0$ where $a_0, a_1, \ldots, a_n$ are integers is called an algebraic number. Let $|a_0| + |a_1| + \cdots + |a_n| + n$ be the height of the equation. Show that there exist only a finite number of equations of a given height. Use this theorem to prove Dedekind's result that the set of algebraic numbers is countable, that is, the set can be put in one-to-one correspondence with the set of natural numbers. This theorem and this proof appeared in print in an 1874 paper of Cantor. Dedekind had communicated the proof to Cantor in November 1873. Uncharacteristically, Cantor did not mention Dedekind's contribution. See Ferreirós (1993) for a possible explanation.
2. Read Wilbraham (1848); this paper contains the first discussion of the Gibbs phenomenon, dealing with overshoot in the convergence of the partial sums of certain Fourier series in the neighborhood of a discontinuity of the function. See Hewitt and Hewitt (1980) for a detailed discussion and history of the topic.
3. In his paper, Riemann gave the function $f(x) = \frac{d}{dx}(x^\nu \cos 1/x)$, where $0 < \nu < 1/2$ as an example of an integrable function, not representable as a Fourier series and having an infinite number of maxima and minima. Analyze this claim.
4. Show that the series

$$\sum_{n=1}^{\infty} \frac{\sin n^2 x}{n^2}$$

converges to a continuous function. Prove that the function does not have a derivative at $\zeta \pi$ if ζ is irrational; prove the same if $\zeta = 2A/(4B+1)$ or $(2A+1)/2B$ for integers A and B. Show that when $\zeta = (2A+1)/(2B+1)$, the function has a derivative $= -1/2$. In 1916, Hardy proved the nondifferentiability portion of this above result; in 1970, J. Gerver proved the differentiability portion. In his lectures, Riemann discussed this series, apparently without stating the theorem. Weierstrass was of the opinion that Riemann may have intended this to be an example of a continuous but nondifferentiable function. Unable to prove this, Weierstrass constructed a different example, given in the next exercise. See Segal (1978).

5. Show that the function $f(x) = \sum_{n=0}^{\infty} b^n \cos(a^n \pi x)$, where $0 < b < 1$; a is an odd integer; $ab > 1 + 3\pi/2$, and the function

$$g(x) = \sum_{n=1}^{\infty} \frac{\cos(n!x)}{n!}$$

are both continuous and everywhere nondifferentiable. Weierstrass presented the first example in his lectures, and Paul du Bois-Reymond published it in 1875. G. Darboux published the second example in 1879. See Weierstrass (1894–1927), vol. 2, pp. 71–74.

6. Let $t = \sum c_n/2^n$, with $c_n = 0$ or 1, be the binary expansion of $0 \leq t \leq 1$. Set $f(t) = \sum a_n/2^n$, where a_n denotes the number of zeros among $c_1, c_2, \ldots, c_n$ if $c_0 = 0$; if $c_0 = 1$, then a_n denotes the number of ones. Prove that $f(t)$ is continuous and single-valued for $0 \leq t \leq 1$ and that $f(t)$ is not differentiable for any t. See Takagi (1990), pp. 5–6. Teiji Takagi (1875–1960) graduated from the University of Tokyo and then studied under Schwarz, Frobenius, and Hilbert in Berlin and Göttingen 1898–1901. Even before going to Germany, Takagi studied Hilbert's 1897 *Zahlbericht*. His thesis proved the statement from Kronecker's *Jugendtraum* that all the abelian extensions of the number field $Q(\sqrt{-1})$ can be obtained by the division of the lemniscate. Takagi did his most outstanding work in class field theory; he was one of the first Japanese mathematicians to begin his career after the transition to Western mathematics in Japan, and he was instrumental in establishing a tradition of algebraic number theory there. See Miyake (1994) and Sasaki (1994). These two papers, along with other papers of interest, are contained in Sasaki, Sugiura, and Dauben (1994).

7. For Bolzano's example of a continuous nowhere differentiable function, dating from about 1830, read Strichartz (1995), pp. 403–406. He gives a graphical presentation and points out that it has close connections with fractals.

8. Show that the series

$$\sum_{n=2}^{\infty} \frac{\sin nx}{\ln n}$$

converges to a function not integrable in any interval containing the origin. Then derive the conclusion that this trigonometric series is not a Fourier series. This example is due to P. Fatou and is referred to in Lebesgue (1906), p. 124.

9. Prove W. H. Young's theorem that if $q_0 \geq q_1 \geq \cdots$ form a monotone descending sequence with zero as limit, and their decrements also form a monotone

descending sequence, viz., $q_0 - q_1 \geq q_1 - q_2 \geq \cdots$, then the trigonometric series

$$\frac{1}{2} q_0 + \sum_{n=1}^{\infty} q_n \cos nx$$

is the Fourier series of a positive summable function. Use this to prove that

$$\sum_{n=2}^{\infty} (\cos nx)/(\ln n)^c,$$

where $c > 0$ is a Fourier series. For this and the next exercise, see G. C. Young and W. H. Young (2000), pp. 449–478.

10. Prove that if $q_1 \geq q_2 \geq \cdots$ form a monotone descending sequence of constants with zero as limit and $\sum_{n=1}^{\infty} n^{-1} q_n$ converges, then

$$\sum_{n=1}^{\infty} q_n \sin nx$$

is the Fourier series of a summable function bounded below for positive values of x and bounded above for negative values of x. See exercise 9.

11. Prove that if $f \in L^1(-\pi, \pi)$, then the Poisson integral

$$\frac{1}{2\pi} \int_{-\pi}^{\pi} f(t) \frac{1-r^2}{1-2r\cos(t-x)+r^2} dt$$

converges almost everywhere (a.e.) to $f(x)$ as $r \to 1^-$. See Fatou (1906).

12. Given a series $\sum_{n=1}^{\infty} A_n$, $A_n = a_n \cos nx + b_n \sin nx$, define its conjugate as the series $\sum_{n=1}^{\infty} B_n$, where $B_n = -b_n \cos nx + a_n \sin nx$. Suppose then that

$$\sum_{n=1}^{\infty} (a_n^2 + b_n^2) < \infty.$$

Prove the Riesz–Fischer theorem that there exist functions $f, g \in L^2(-\pi, \pi)$ such that $f \sim \sum A_n$ and $g \sim \sum B_n$. Show also Lusin's result that

$$\frac{1}{2\pi} \int_{-\pi}^{\pi} f(t) \frac{1-r^2}{1-2r\cos(t-x)+r^2} dt$$
$$= \frac{1}{\pi} \int_{-\pi}^{\pi} g(t) \frac{r \sin(t-x)}{1-2r\cos(t-x)+r^2} dt = f(x) \text{ a.e.}$$

Next, deduce the formula for the Cauchy principal value integral:

$$\lim_{\epsilon \to 0^+} \frac{1}{\pi} \int_{\epsilon \leq |t| \leq \pi} g(x+t) \frac{dt}{2\tan(t/2)} = f(x) \text{ a.e.} \qquad (22.12)$$

For an arbitrary function g, the conjugate $\tilde{g}$ is defined by the negative of the principal value integral in (22.12). If $g \in L^1$, then in general $\tilde{g}$ might or might not be in L^1, but $\tilde{g} \in L^p$ for $0 < p < 1$. If $g \in L^p$, for $p > 1$, then $\tilde{g} \in L^p$. Note that Lusin proved the last result when $p = 2$. See Lusin (1913). Nikolai Lusin (1883–1950) was a student of Dmitri Egorov (1869–1931) at Moscow University and he founded an important school of mathematics there with students such as Kolmogorov, Menshov, and Privalov. They developed what is now called the complex method in Fourier analysis.

13. Concerning $S_n f$, the nth partial sum of the Fourier series of f, given by $\sum_{k=1}^{n} A_k$, show that

$$S_n f(x) = \sum_{k=1}^{n} (a_k \cos kx + b_k \sin kx)$$
$$= \frac{1}{\pi} \int_{-\pi}^{\pi} g(x+t) \left(\frac{1}{2\tan(t/2)} - \frac{\cos(n+1/2)t}{2\sin(t/2)} \right) dt,$$

where the two integrals on the right-hand side should be taken as Cauchy principal values. Combine this with the result in exercise 12 to show that $\lim_{n \to \infty} S_n f(x) = f(x)$ a.e. if and only if the principal value integral satisfies

$$\lim_{n \to \infty} \int_{-\pi}^{\pi} g(x+t) \frac{\cos nt}{t} dt = 0 \text{ a.e.}$$

From (22.12) it follows that the principal value integral $\int_{-\pi}^{\pi} \frac{g(x+t)}{t} dt$ exists a.e. for $g \in L^2$. Lusin also had an example of a continuous function g with $\int_{-\pi}^{\pi} \left| \frac{g(x+t)}{t} \right| dt = \infty$ on a set of positive measure. Note that in order for this principal value integral to converge, there must have been a good deal of cancellation. Lusin conjectured the almost everywhere convergence of the Fourier series of square integrable functions because he thought that the cancellation in the principal value integral was the reason for the convergence of the series. Kolmogorov (1923) contains an example of an integrable, but not square-integrable, function whose Fourier series diverged everywhere. Lennart Carleson proved Lusin's conjecture in 1966, and Richard Hunt soon extended Carleson's theorem to L^p functions with $p > 1$. One of the important concepts needed in the Carleson and Hunt proofs was that of maximal functions. For a locally integrable function f, the Hardy–Littlewood maximal function is defined by

$$Mf(x) = \sup_{h>0} \frac{1}{h} \int_{x-h}^{x+h} |f(t)| dt.$$

14. Prove that if $f \in L^p(-\pi, \pi)$ for $1 < p < \infty$, then $\tilde{f} \in L^p$ and $||\tilde{f}||_p \leq C_p ||f||_p$. This theorem is due to Marcel Riesz (1928). Also deduce that $||S_n f||_p \leq C_p ||f||_p$.

15. Show that if $f \in L^p$ for $1 < p < \infty$, then

$$||Mf||_p \leq C_p ||f||_p.$$

Show also that if $f(r, x)$ is the Poisson integral of f, then

$$\sup_{0 \le r < 1} |f(r, x)| \le C M f(x),$$

and hence $\left\| \sup_{0 \le r < 1} |f(r, x)| \right\|_p \le C_p \|f\|_p.$

These results were published in 1930 by Hardy and Littlewood; see Hardy (1966–1979), p. 509–544, especially pp. 530–538.

22.9 Notes on the Literature

For Carleson's proof of the convergence theorem, see Carleson (1966). Hunt's extension can be found in Haimo (1968), pp. 235–255. For historical background on the convergence of Fourier series, see Hunt's paper in Butzer and Sz.-Nagy (1974). Riemann's paper, "Ueber die Darstellbarkeit einer Function durch eine trigonometrische Reihe," was published by Dedekind in 1867. The easiest place to find the paper is in the third edition of Riemann's collected papers edited by R. Narasimhan, Riemann (1990). This contains all the material from the previous editions together with some new material, including essays by outstanding mathematicians on particular aspects of Riemann's work. An English translation of Riemann's paper is in S. Hawking (2005). Laugwitz (1999) presents a lively account of Riemann's life and mathematical work, including trigonometric series and complex variables.

See Smith (1965a), vol. 2, pp. 86–89 for the quotations from his paper. His construction of the Cantor set appears on pp. 94–95 and the reference to Hankel is on p. 95. Schwarz (1972), vol. 2, pp. 341–343 gives the proof of the theorem on the second difference quotient, used by Cantor to prove his uniqueness theorem. Cantor (1932) contains his work on the uniqueness of trigonometric series. See Dauben (1979) for a discussion of the development of Cantor's mathematical thought.

Cooke (1993) is an interesting history of the work on the uniqueness of trigonometric series, and it also surveys recent contributions. The article by Zygmund in Ash (1976) contains some insightful remarks on the development of Fourier series. Hawkins (1975) presents a detailed but very readable account of the development of integration theory from Riemann to Lebesgue. See also Bressoud (2008) for more on Lebesgue. Meschkowski (1964) gives an English translation of Schwarz's February 25, 1870 letter to Cantor. This letter contains a proof of the theorem that a function whose derivative vanishes in an interval must be constant in that interval. Schwarz explained that his proof made use of ideas from Weierstrass's 1861 lecture at the Technical University in Berlin. For Weil's remarks on Riemann, see Weil (1992), p. 40. For a modern discussion of Riemann integrability, see Bressoud (2007), p. 251.

23

The Gamma Function

23.1 Preliminary Remarks

The problem of interpolating the sequence of factorials 0!, 1!, 2!, 3!, ... appeared in Wallis's 1655 book on the quadrature of a circle. Specifically, the problem would be to find a function $f(x)$ of a positive real variable x, such that $f(x+1) = x f(x)$ and $f(n) = (n-1)!$, when n is a positive integer. One might wish to have $f(n) = n!$ but we will identify $f(x)$ with the gamma function $\Gamma(x)$, so we shift the function $f(n) = n!$ to the right by a unit. In his book, Wallis solved this problem for half-integral values of x but did not work out the details of the general case. Strangely, for seventy years after this, mathematicians did not take up this subject.

Euler and Stirling made significant contributions to this problem starting in the late 1720s. They worked independently, Euler in Russia and Stirling in Scotland; their approaches and aims were also distinct. In the mid-1730s, they came to know of each other's work and had a brief correspondence. Always the algorist, Euler was interested in obtaining analytic expressions for the interpolating function $f(x)$. His first paper on the subject, written in 1730, gives two different representations of $f(x)$, one as an infinite product and the other as a definite integral. On the other hand, Stirling was a numerical analyst interested in finding efficient methods for computing $f(x)$. He was undoubtedly extremely experienced in computation and demonstrated that he knew the values of many mathematical constants to several decimal places. Without giving an explicit analytic formula for $f(x)$, but making use of Newton's method of interpolation, he calculated the value of $(1/2)! = f(3/2)$ as 0.8862269251. He recognized this as $\sqrt{\pi}/2$ and, indeed, this is the correct value of $\Gamma(3/2)$.

There was a common feature in the thinking of Euler and Stirling: They both believed that there was only one reasonable or logical interpolating function $f(x)$. Thus, Euler did not prove in his first paper that the integral and infinite product representations of $f(x)$ were equal for all $x > 0$, but merely that they were equal for positive integral values of x. Similarly, Stirling thought that his numerical methods gave the value of the unique interpolating function $f(x)$. The later work of H. Bohr and J. Mollerup from around 1920 showed that to obtain uniqueness of the interpolating function one must

23.1 Preliminary Remarks

assume the convexity of $\ln f(x)$, in addition to the two above-mentioned properties $f(x+1) = x f(x)$ and $f(1) = 1$.

Leonhard Euler (1707–1783) was born in Basel, Switzerland, and studied at the University of Basel from 1720–1724. After this he studied independently, concentrating on mathematics, physics, and astronomy, under the guidance of Johann Bernoulli, with whom he met once a week. Adhering to this regime, Euler quickly became an excellent mathematician and by 1725 he began seeking a position. Failing to find one in Switzerland, he moved to Russia in 1727 to join his friends Daniel and Niklaus (also Nicolaus) II Bernoulli at the newly founded Petersburg Academy. The Bernoulli brothers received their appointments when their father Johann declined a position and persuaded the Academy to employ his sons. Euler was originally appointed to a position in medicine, prompting him to brush up on his anatomy, but he ended up getting a situation in mathematics when Niklaus II died unexpectedly before Euler arrived in St. Petersburg. Euler enjoyed a very stimulating scientific collaboration with Daniel until the latter returned to Basel in 1834. Euler also developed a friendship with Christian Goldbach (1690–1764) from Prussia, whom Clifford Truesdell described as "an energetic and intelligent Prussian for whom mathematics was a hobby, the entire realm of letters an occupation, and espionage a livelihood." Euler and Goldbach corresponded extensively with each other, and Goldbach sometimes suggested problems, stimulating Euler to important mathematical discoveries. Euler spent 1741–66 in Berlin and then returned to St. Petersburg where he died, mathematically active until the end.

Euler became interested in the interpolation problem when it appeared in a 1728 paper presented by Goldbach to the St. Petersburg Academy. Goldbach also mentioned the problem in his letters to Daniel Bernoulli who may have discussed the matter with Euler. Bernoulli outlined a solution in a postscript to a letter to Goldbach dated October 6, 1729. He let A stand for an infinite number. Then the general xth term of the factorial sequence was given by

$$\left(A + \frac{x}{2}\right)^{x-1} \left(\frac{2}{1+x} \cdot \frac{3}{2+x} \cdot \frac{4}{3+x} \cdots \frac{A}{A-1+x}\right).$$

He noted that when $x = 3/2$ and $A = 8$, the value of the preceding expression was approximately 1.3005. He had made a computational error here, and the value should have been 1.329, as he observed in a letter two weeks later. This value of $(3/2)!$ is correct to three decimal places. Even at this early stage of his career, Daniel Bernoulli did not pursue this problem any further, and it was left to Euler to initiate and develop the theory of the gamma function. In fact, Daniel was primarily a mathematical physicist and after middle age, his interest in pure mathematical questions waned.

Euler's October 15, 1729 letter to Goldbach gave the value of $\Gamma(m+1)$, representing the mth term of the factorial sequence, to be

$$\Gamma(m+1) = \frac{1 \cdot 2^m}{1+m} \cdot \frac{2^{1-m} \cdot 3^m}{2+m} \cdot \frac{3^{1-m} \cdot 4^m}{3+m} \cdot \frac{4^{1-m} \cdot 5^m}{4+m} \cdots. \qquad (23.1)$$

He observed that the infinite product reduced to $m!$ when m was a positive integer, though he verified this only for $m = 2$ and $m = 3$. He also noted in the letter that the infinite product (23.1), when terminated after n terms, could be written as

$$\frac{1 \cdot 2 \cdot 3 \cdots n \cdot (n+1)^m}{(1+m)(2+m)\cdots(n+m)}.$$

This implied

$$\Pi(m) = \Gamma(m+1) = \lim_{n\to\infty} \frac{n!n^m}{(m+1)(m+2)\cdots(m+n)}. \qquad (23.2)$$

In 1812, Gauss gave the definition of the gamma function as (23.2), often called Gauss's definition of the gamma function, although Euler found it in 1729 and published it in 1776.

Euler recognized that the gamma function was important, but he did not introduce a notation for it. Legendre employed the symbol Γ, while Gauss used Π. In general, Euler simply wrote $1 \cdot 2 \cdot 3 \cdots x$ to mean $\Gamma(x+1)$. He once wrote $[x]$ for $\Gamma(x+1)$, but this was a temporary device and he used the square brackets to stand for other things as well. Soon after he wrote this letter to Goldbach, Euler presented to the Academy a long paper on the subject, although the Academy did not publish it until 1738. Euler made it clear that the source of his inspiration was Wallis's *Arithmetica Infinitorum*. In fact, a large part of the paper was a reworking of the results of Wallis, who had also grappled with the problem of interpolating the factorial sequence. Recall two important results of Wallis from the the first two sections of chapter 3. First,

$$\int_0^1 (1-x^{1/p})^q dx = \frac{p!q!}{(p+q)!} = \frac{p!}{(q+1)\cdots(q+p)} = \frac{q!}{(p+1)\cdots(p+q)}. \qquad (23.3)$$

From this relation Wallis could compute the integral when either p or q was a positive integer. The second result took $p = q = 1/2$ in (23.3). Then, since the integral represented the area under the circle $y = \sqrt{1-x^2}$ on the interval $(0, 1)$, Wallis's formula for π gave

$$\frac{\pi}{4} = ((1/2)!)^2 = \frac{2}{3}\cdot\frac{4}{3}\cdot\frac{4}{5}\cdot\frac{6}{5}\cdots\frac{2n}{2n+1}\cdot\frac{2n+1}{2n+2}\cdots,$$

$$\text{or } \frac{\sqrt{\pi}}{2} = (1/2)! = \lim_{n\to\infty} \frac{n!(n+1)^{1/2}}{\left(\frac{1}{2}+1\right)\left(\frac{1}{2}+2\right)\cdots\left(\frac{1}{2}+n\right)}. \qquad (23.4)$$

This was the $m = 1/2$ case of Euler's formula (23.2).

The new feature in Euler's paper was his integral representation

$$\Gamma(m+1) = m! = \int_0^1 (-\ln x)^m dx;$$

he verified this only for the case when m was a nonnegative integer. He assumed that the result continued to hold for all real m, so that Wallis's formula (23.4) gave him

$$\int_0^1 (-\ln x)^{1/2} dx = \frac{\sqrt{\pi}}{2}. \qquad (23.5)$$

Note that a change of variables $x = e^{-t^2}$, followed by integration by parts, gives the probability integral

$$\int_0^\infty e^{-t^2} dt = \frac{\sqrt{\pi}}{2}. \tag{23.6}$$

Thus, Euler had actually found the probability integral in the form (23.5). At the end of the paper, he explained how the gamma function could be used to define fractional derivatives. This was a problem already raised by Leibniz, as Euler may have been aware. He observed that when n was a nonnegative integer,

$$\frac{d^n z^e}{dz^n} = e(e-1)\cdots(e-n+1)z^{e-n};$$

and so one could define, for any positive real number n,

$$\frac{d^n z^e}{dz^n} = \frac{\int_0^1 (-\ln x)^e dx}{\int_0^1 (\ln x)^{e-n} dx} z^{e-n}. \tag{23.7}$$

By taking $e = 1$ and $n = 1/2$, he had

$$\frac{d^{1/2} z}{dz^{1/2}} = 2\sqrt{\frac{z}{\pi}}. \tag{23.8}$$

Euler did not do any more with this concept, later rediscovered by Abel who in 1823 applied it to the solution of an integral equation.

In his second paper of 1739, Euler gave a proof of (23.3), extended to the case when both p and q were not integers. Interestingly, his proof followed that of Wallis for (23.4), with a simplification. We note that formula (23.3) is usually written as

$$\int_0^1 t^{p-1}(1-t)^{q-1} dt = \frac{\Gamma(p)\Gamma(q)}{\Gamma(p+q)}. \tag{23.9}$$

Sometime in the 1740s or perhaps earlier, Euler found a connection between the gamma function and the trigonometric functions, in the form of his reflection formula:

$$\Gamma(x)\Gamma(1-x) = \pi/\sin \pi x. \tag{23.10}$$

He explicitly stated this result in an important paper, written in 1749 but published in 1768, giving the functional relation for the zeta function.

Euler made a curious observation in his 1729 letter to Goldbach. He wrote that the value of (23.1) when $m = 1/2$ was $\frac{1}{2}\sqrt{\sqrt{-1}\log(-1)}$, equal to the square root of the area of a circle with diameter 1. This amounted to $\sqrt{-1}\ln(-1) = \pi$. At that time, mathematicians did not have a clear idea about how the logarithm of a negative number should be defined. Leibniz and Johann Bernoulli had some correspondence on this point in the 1710s, but these discussions brought forth nothing of real value. Eventually, Euler produced a complete definition of the logarithm of a complex number, including its property of being multivalued. The question is, did Euler have a good understanding of this definition in 1729? Perhaps he did not. Roger Cotes's posthumous work, *Harmonia Mensurarum*, published in 1722, had the formula

$$\sqrt{-1}\log(\cos\theta + i\sin\theta) = \theta$$

and Euler's formula covered the particular case when $\theta = \pi$. Moreover, Cotes's result had an error in sign and this error reappeared in Euler, if we take the principal value of $\log(-1)$ to be $i\pi$. It seems reasonable to draw the conclusion that Euler got his result from Cotes. However, the formula of Cotes set Euler on the right track toward his own more conclusive results, finally written up in the 1740s.

Euler did not deal with the question of the convergence of the infinite product (23.1). It was not the practice among mathematicians of the eighteenth century to go into the details of convergence problems. However, the manner of Euler's expression in some cases leads us to believe that he had clear ideas about what was meant by convergence. For example, in (23.1) Euler did not cancel the factors, showing us that here he was not unaware of convergence issues. One may easily check that the nth term of the product is

$$\frac{n^{1-m} \cdot (n+1)^m}{n+m} = \left(1 + \frac{1}{n}\right)^m \left(1 + \frac{m}{n}\right)^{-1} = 1 + \frac{m(m-1)}{2n^2} + O\left(\frac{1}{n^3}\right),$$

and thus that the infinite product converges.

The eighteenth-century mathematicians produced an enormous body of analytical results without a substantial discussion of convergence. The first mathematician to seriously think about convergence issues was Carl Friedrich Gauss (1777–1855). Like Euler, he had an extremely broad range of interests, covering almost every area of pure and applied mathematics. His paper on the gamma function was a part of a larger work on hypergeometric series published in 1813. He founded his study of convergence on the theory of limits of sequences. In an unpublished early work, he discussed concepts such as the upper and lower limits of sequences. It is difficult to determine the influences informing Gauss's work. Of course, he was extremely well read and was very familiar with the works of his great predecessors. But he appeared to prefer to work in isolation. So it is not clear what motivated him to study convergence of infinite series and products, besides a desire for greater mathematical rigor. Thus, in the 1813 paper mentioned earlier, Gauss showed that the limit in (23.2) existed. He also gave a new method of deriving Euler's results (23.5), (23.9), and (23.10). At the heart of Gauss's new method was the summation formula

$$1 + \frac{a \cdot b}{1 \cdot c} + \frac{a(a+1) \cdot b(b+1)}{1 \cdot 2 \cdot c(c+1)} + \frac{a(a+1)(a+2) \cdot b(b+1)(b+2)}{1 \cdot 2 \cdot 3 \cdot c(c+1)(c+2)} + \cdots$$
$$= \frac{\Gamma(c)\Gamma(c-a-b)}{\Gamma(c-a)\Gamma(c-b)}, \quad (23.11)$$

where a, b, c were complex numbers with $\operatorname{Re}(c - a - b) > 0$. He gave a completely satisfactory proof of this formula, given in our chapter on hypergeometric functions.

23.1 Preliminary Remarks

Gauss also found the multiplication formula for the gamma function:

$$n^{nz-1/2}\Gamma(z)\Gamma\left(z+\frac{1}{n}\right)\Gamma\left(z+\frac{2}{n}\right)\cdots\Gamma\left(z+\frac{n-1}{n}\right)$$
$$= (2\pi)^{(n-1)/2}\Gamma(nz), \tag{23.12}$$

where n was a positive integer. The reflection formula (23.10) suggests that the inspiration for this must have been the similar formula for $\sin n\pi z$ discovered and published by Euler in his 1748 *Introductio in Analysin Infinitorum*. In slightly modified form, Euler's formula was

$$\sin n\pi z = 2^{n-1}\sin \pi z \sin \pi\left(z+\frac{1}{n}\right)\sin \pi\left(z+\frac{2}{n}\right)\cdots\sin \pi\left(z+\frac{n-1}{n}\right). \tag{23.13}$$

Euler also gave a special case of (23.12):

$$\sqrt{n}\,\Gamma\left(\frac{1}{n}\right)\Gamma\left(\frac{2}{n}\right)\cdots\Gamma\left(\frac{n-1}{n}\right) = (2\pi)^{(n-1)/2}. \tag{23.14}$$

Slightly before Gauss's paper was published, Legendre discovered the duplication formula, the $n = 2$ case of Gauss's formula (23.12). Legendre's proof employed the integral representation of the gamma function, and this in turn suggested the problem of deriving the properties of the gamma function using definite integrals. At that time, definite integrals were appearing in many areas of mathematics and its applications. For Euler, this topic was a life-long interest; he had already evaluated several definite integrals by means of a variety of techniques. S. P. Laplace and Legendre also pursued the study of definite integrals, of great usefulness in solving problems in probability theory and mechanics. The method of Fourier transforms, originated by Fourier in his work on heat conduction and its applications to wave phenomena, also produced numerous definite integrals.

By 1810, several French mathematicians had published papers whose aim was to evaluate classes of definite integrals. In 1814, Cauchy wrote a long memoir on definite integrals, the first of his many contributions to what would become complex function theory. A decade later, Cauchy gave a precise definition of a definite integral in his lectures at the École Polytechnique; he then proceeded to define improper integrals and their convergence.

Dirichlet, though a Prussian, studied in Paris in the mid 1820s. He mastered Cauchy's ideas on rigor and applied them to the series introduced into mathematics and mathematical physics by his friend Fourier. Even in his first paper on Fourier series, Dirichlet recognized the importance of extending the definite integral to include highly discontinuous functions. He even made use of improper integrals in his number theoretic work. He employed the integrals $\int_0^\infty \cos x^2\,dx$ and $\int_0^\infty \sin x^2\,dx$, closely related to the gamma function, to obtain a remarkable evaluation of the quadratic Gauss sum. Recall that we discussed this in chapter 21, section 7. In his famous work on primes in arithmetic progressions, Dirichlet used Euler's integral formula for the gamma function in the form

$$\int_0^1 x^{n-1} (\ln(1/x))^{s-1} dx = \frac{\Gamma(s)}{n^s} \qquad (23.15)$$

to represent as integrals certain Dirichlet series called L-functions. We discuss this in detail in chapter 32. Dirichlet's number theoretic work motivated him to further investigate the gamma function within the theory of definite integrals. He wrote several papers on the topic, including one dealing with a multidimensional generalization of Euler's beta integral (23.9).

Dirichlet's interest in definite integrals, expressed through his publications and his lectures at Berlin University, created an interest in this topic among German mathematicians. Thus, in 1852, Richard Dedekind, who did great work in number theory, wrote his Göttingen doctoral thesis on Eulerian integrals. Riemann, greatly influenced by Dirichlet, made brilliant use of definite integrals, and the gamma integral in particular, in his great 1859 paper on the distribution of primes. In this paper, he expressed the zeta function as a contour integral from which he derived the functional equation for the zeta function. This work of Riemann inspired his student Hermann Hankel (1839–1873) to find in 1863 a contour integral representation for $\Gamma(z)$, valid for all complex z except the negative integers.

The gamma function also played a significant role in the development of the theory of infinite products. In an 1848 paper, the English mathematician F. W. Newman explained how an exponential factor $e^{-x/n}$ in $(1+x/n)e^{-x/n}$ ensured the convergence of the product $\prod_{n=1}^{\infty}(1+x/n)e^{-x/n}$. Using this product, he obtained a new representation for the gamma function. Oscar Schlömilch (1823–1901), a student of Dirichlet, published this result in 1843, taking an integral of Dirichlet as a starting point of the proof. Schlömilch's work was based on the evaluation of definite integrals. In 1856, Weierstrass gave a foundation to the theory of the gamma function by defining it in terms of an infinite product. In fact, the ideas of this paper inspired him to construct entire functions with a prescribed sequence of zeros.

The gamma function is one of the basic special functions, cropping up again and again. Consequently, mathematicians have tried to derive its properties from several different points of view. In 1930, Emil Artin observed that the concept of logarithmic convexity, used by Bohr and Mollerup to prove the equivalence of the product and integral representations of the gamma function, could be employed to characterize and develop the properties of this function. While Artin worked with real variables, in 1939 Helmut Wielandt gave a complex analytic characterization. The defining property other than the obvious $f(z+1) = zf(z)$ was that $f(z)$ was bounded in the vertical strip $1 \leq \text{Re}\, z < 2$.

23.2 Stirling: $\Gamma(1/2)$ by Newton–Bessel Interpolation

James Stirling gave a remarkable numerical evaluation of $\Gamma(1/2)$. He tabulated the base 10 logarithms of the twelve numbers $5!, 6!, \ldots, 16!$ and then applied the Newton-Bessel interpolation formula to obtain the middle value $(21/2)!$. Then by successive

23.2 Stirling: $\Gamma(1/2)$ by Newton–Bessel Interpolation

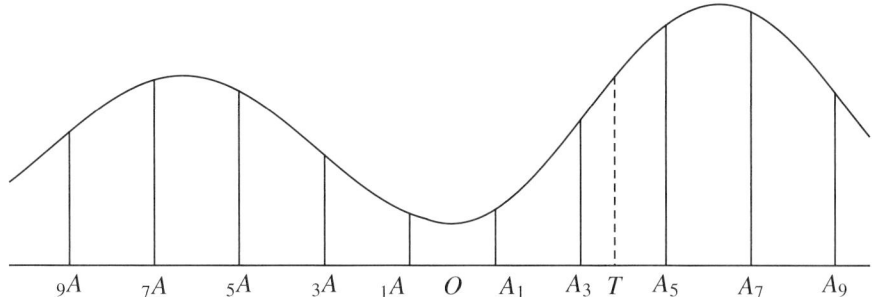

Figure 23.1. Stirling: gamma values by interpolation.

division, he computed $\Gamma(1/2) = (-1/2)!$ to ten decimal places and recognized it to be $\sqrt{\pi}$. In the *Methodus Differentialis*, proposition 20, Stirling described the interpolation formula:

He first supposed an even number of equidistant ordinates and, using a diagram similar to Figure 23.1, he denoted them $_9A, _7A, _5A, \ldots, A_5, A_7, A_9$. Note that these values refer to heights of the line segments. He called $_1A$ and A_1 the middle values and set A to be their sum. Thus, he had nine differences of these ten numbers, such as $_7A - _9A, _5A - _7A, \ldots, A_9 - A_7$. He called the middle difference a. Taking the eight differences of these nine differences, he denoted the sum of the middle two terms as B. Next, Stirling called the middle term of the seven differences of the eight second differences b, and so on. He took O to be the midpoint of $_1A$ and A_1 and let T be an arbitrary ordinate; he let $z/2$ be the ratio of the distance between O and that point whose ordinate was T, and the distance between $_1A$ and A_1. Note that this last distance is also the distance between any two successive ordinates. Stirling wrote the formula as

$$T = \frac{A+az}{2} + \frac{3B+bz}{2} \times \frac{z^2-1}{4 \cdot 6} + \frac{5C+cz}{2} \times \frac{z^2-1}{4 \cdot 6} \times \frac{z^2-9}{8 \cdot 10}$$
$$+ \frac{7D+dz}{2} \times \frac{z^2-1}{4 \cdot 6} \times \frac{z^2-9}{8 \cdot 10} \times \frac{z^2-25}{12 \cdot 14}$$
$$+ \frac{9E+ez}{2} \times \frac{z^2-1}{4 \cdot 6} \times \frac{z^2-9}{8 \cdot 10} \times \frac{z^2-25}{12 \cdot 14} \times \frac{z^2-49}{16 \cdot 18} + \cdots. \qquad (23.16)$$

Stirling also noted that z was positive when T was on the right-hand side of the middle point O, as in Figure 23.1, and negative when it lay on the left-hand side. To write this in modern form, set

$$_1A = f\left(s - \frac{h}{2}\right), \ A_1 = f\left(s + \frac{h}{2}\right), \ _2A = f\left(s - \frac{3}{2}h\right), \ A_2 = f\left(s + \frac{3}{2}h\right), \text{ etc.}$$

Then the Newton–Bessel formula is given by

$$f\left(s+\left(x-\frac{1}{2}\right)h\right) = \frac{1}{2}\left(f\left(s-\frac{h}{2}\right)+f\left(s+\frac{h}{2}\right)\right)+\left(x-\frac{1}{2}\right)\Delta f\left(s-\frac{h}{2}\right)$$
$$+\frac{x(x-1)}{2!}\frac{1}{2}\left(\Delta^2 f\left(s-\frac{3h}{2}\right)+\Delta^2 f\left(s-\frac{h}{2}\right)\right)$$
$$+\frac{x(x-1)(x-1/2)}{3!}\Delta^3 f\left(s-\frac{3h}{2}\right)+\cdots. \qquad (23.17)$$

Stirling's expression is obtained by setting $z = 2x - 1$ and combining pairs of terms in (23.17).

In Example 2 of Proposition 21, he explained his approach to the evaluation of $\Gamma(x)$ for $x > 0$:

> Let the series [sequence] to be interpolated be 1, 1, 2, 6, 24, 120, 720, etc. whose terms are generated by repeated multiplication of the numbers 1, 2, 3, 4, 5, etc. Since these terms increase very rapidly, their differences will form a divergent progression, as a result of which the ordinate of the parabola does not approach the true value. Therefore, in this and similar cases I interpolate the logarithms of the terms, whose differences can in fact form a rapidly convergent series, even if the terms themselves increase very rapidly as in the present example.

Stirling interpolated the sequence

$$\ln 0!, \ln 1!, \ln 2!, \ldots, \ln n!, \ln(n+1)!, \ldots.$$

The first difference would be

$$\ln(n+1)! - \ln n! = \ln(n+1);$$

the second difference would be written as

$$\ln(n+1) - \ln n = \ln\left(1+\frac{1}{n}\right) \approx \frac{1}{n};$$

and the third difference would be written as

$$\ln(n+1) - 2\ln(n) + \ln(n-1) = \ln\left(1-\frac{1}{n^2}\right) \approx -\frac{1}{n^2}.$$

Thus, we can see that the successive differences get small rapidly when n is large enough, though $\ln n!$ increases. This means that if one desires to find the value of $\Gamma(n+1)$ for a small value, $n = \frac{1}{2}$, then one must first compute $\ln \Gamma(n+1)$ for a larger value of n by the method of differences and then apply the functional equation $\Gamma(n+1) = n\Gamma(n)$. Stirling thought that the interpolating function must satisfy the same functional relation as the successive terms of the original sequence. In fact, in Proposition 16 of his book, he attempted to explain why this must be so. Stirling then continued:

> Now I propose to find the term which stands in the middle between the first two 1 and 1. And since the logarithms of the initial terms have slowly convergent differences, I first seek the term

standing in the middle between two terms which are sufficiently far removed from the beginning, for example, that between the eleventh term 3628800 and the twelfth term 39916800: and when this is given, I may go back to the term sought by means of Proposition 16. And since there are some terms located on both sides of the intermediate term which is to be determined first, I set up the operation by means of the second case of Proposition 20.

Actually, Stirling worked with $\log_{10}$ since the logarithmic tables were often in base 10. He therefore took the twelve known ordinates to be $\log_{10} \Gamma(\frac{1}{2}(z+23))$ for $z = \pm 1, \pm 3, \pm 5, \pm 7, \pm 9$, and ± 11 and used Newton-Bessel to find the value at $z = 0$. From this he found $\Gamma(11.5)$ and after successive division by $10.5, 9.5, \ldots$, down to 1.5, he computed $\Gamma(\frac{3}{2})$ to ten decimal places. After this amazing calculation Stirling commented, "From this is established that the term between 1 and 1 [referring to 0! and 1!] is .8862269251, whose square is .7853... etc., namely, the area of a circle whose diameter is one. And twice that, 1.7724538502, namely the term which stands before the first principal term by half the common interval, is equal to the square root of the number 3.1415926... etc., which denotes the circumference of a circle whose diameter is one."

Here Stirling gave the value of $\Gamma(\frac{1}{2})$, obtained from $\Gamma(\frac{3}{2}) = \frac{1}{2}\Gamma(\frac{1}{2})$. His value for $\Gamma(\frac{3}{2})$ is incorrect only in the tenth decimal place; when rounded, the tenth place should be 5 instead of 1.

23.3 Euler's Evaluation of the Beta Integral

Euler followed Wallis quite closely in his paper of 1730. He computed the integral below by expanding $(1-x)^n$ as a series and integrating term by term:

$$\int_0^1 x^e (1-x)^n dx = \frac{1}{e+1} - \frac{n}{1 \cdot (e+2)} + \frac{n(n-1)}{1 \cdot 2(e+3)} - \frac{n(n-1)(n-2)}{1 \cdot 2 \cdot 3(e+4)} + \text{etc.}$$

He checked that for $n = 0, 1, 2, 3$ the results were

$$\frac{1}{e+1}, \frac{1}{(e+1)(e+2)}, \frac{1 \cdot 2}{(e+1)(e+2)(e+3)}, \frac{1 \cdot 2 \cdot 3}{(e+1)(e+2)(e+3)(e+4)}.$$

From this he concluded that the value of the integral in the general case was

$$\frac{1 \cdot 2 \cdot 3 \cdots n}{(e+1)(e+2) \cdots (e+n+1)}.$$

When $e = f/g$, he wrote

$$\frac{f+(n+1)g}{g^{n+1}} \int_0^1 x^{f/g} (1-x)^n dx = \frac{1 \cdot 2 \cdot 3 \cdots n}{(f+g)(f+2g) \cdots (f+ng)}. \tag{23.18}$$

Euler then observed that l'Hôpital's rule, or rather their teacher Johann Bernoulli's rule, gave

$$\lim_{x \to 0} \frac{1-x^g}{g} = -\ln x,$$

and he wrote this as $\frac{1-x^0}{0} = -lx$. He changed x to x^g in (23.18) to get

$$(f+(n+1)g)\int_0^1 x^{f+g-1}\left(\frac{1-x^g}{g}\right)^n dx = \frac{1\cdot 2\cdot 3\cdots n}{(f+g)\cdots(f+ng)}.$$

Next he set $f=1$ and took the limit as $g \to 0$, or as he wrote, set $f=1$ and $g=0$ to obtain

$$\int_0^1 (-\ln x)^n dx = 1\cdot 2\cdot 3\cdots n. \tag{23.19}$$

Euler then concluded that when $n=\frac{1}{2}$, Wallis's result gave the value of the integral to be $\sqrt{\pi}/2$. In his first paper, he used Wallis's result to evaluate $\int_0^1 (-\ln x)^{\frac{1}{2}} dx$, but gave no independent derivation.

In all of the eighteenth century, there was no greater master than Euler in this area of mathematics. His method of evaluating the definite integral (23.19) by taking a limit was completely new. In the course of his career, Euler gave several methods for computing definite integrals. Of course, it was only in the nineteenth century that his methods were fully justified.

Several proofs of (23.10) follow immediately from formulas appearing in Euler's papers. He did not explicitly state this result very often, perhaps because he had not developed a convenient notation for $\Gamma(x)$. In his simple proof published in 1772, Euler used the products for $\Gamma(x)$ and $\sin \pi x$: He let

$$[m] = \frac{1\cdot 2^m}{1+m}\cdot\frac{2^{1-m}3^m}{2+m}\cdot\frac{3^{1-m}\cdot 4^m}{3+m}\text{etc.}$$

Then

$$[-m] = \frac{1\cdot 2^{-m}}{1-m}\cdot\frac{2^{1+m}3^{-m}}{2-m}\cdot\frac{3^{1+m}\cdot 4^{-m}}{3-m}\text{etc.},$$

$$[m][-m] = \frac{1}{1-m^2}\cdot\frac{2^2}{2^2-m^2}\cdot\frac{3^2}{3^2-m^2}\cdot\text{etc.} = \frac{\pi m}{\sin \pi m}. \tag{23.20}$$

The last equation followed from the infinite product for $\sin x$. Here Euler used the symbol $[m]$ for $\Gamma(m+1)$ but its use was merely provisional.

Euler did not prove (23.9) in his first paper on the gamma function, but the second paper, presented to the Petersburg Academy in 1739, contained an argument upon which a proof can be worked out. He started with the observation that

$$\int_0^1 x^{m-1}(1-x^{nq})^{\frac{p}{q}} dx = \frac{m+(p+q)n}{m}\int_0^1 x^{m+nq-1}(1-x^{nq})^{\frac{p}{q}} dx. \tag{23.21}$$

Wallis had stated a similar functional relation without proof, but with the development of calculus, this could easily be proved by using integration by parts. In this paper, Euler wrote that the result was easy to prove, and when he returned to the subject over three decades later, he provided the details, revealing his explanation of the technique

23.3 Euler's Evaluation of the Beta Integral

of integration by parts. Euler supposed

$$\int x^{f-1}(1-x^g)^m dx = A \int x^{f-1}(1-x^g)^{m-1} dx + B x^f (1-x^g)^m. \qquad (23.22)$$

He took the derivative to get

$$x^{f-1}(1-x^g)^m = A x^{f-1}(1-x^g)^{m-1} - Bmg x^{f+g-1}(1-x^g)^{m-1} + Bf x^{f-1}(1-x^g)^m,$$

or

$$1 - x^g = A - Bmg x^g + Bf(1-x^g)$$
$$= A - Bmg + B(f+gm)(1-x^g).$$

Thus,

$$A - Bmg = 0 \text{ and } B(f+mg) = 1, \text{ or}$$

$$A = \frac{mg}{f+gm}, \text{ and } B = \frac{1}{f+mg}.$$

Next, equation (23.21) followed from (23.22) by choosing the parameters appropriately. Euler applied the functional relation infinitely often to arrive at

$$\int_0^1 x^{m-1}(1-x^{nq})^{\frac{p}{q}} dx =$$
$$\frac{(m+(p+q)n)(m+(p+2q)n)\cdots(m+(p+\infty q)n)}{m(m+nq)(m+2nq)\cdots(m+\infty nq)}$$
$$\times \int_0^1 x^{m+\infty nq-1}(1-x^{nq})^{\frac{p}{q}} dx. \qquad (23.23)$$

We note that the infinite product diverges and the integral on the right-hand side vanishes. We can, however, define the right-hand side as a limit. Let us continue to go along with Euler, who followed Wallis again by taking another integral similar to the one on the left-hand side of (23.23) but which could be exactly evaluated. He took $m = nq$ to obtain

$$\frac{1}{(p+q)n} = \int_0^1 x^{nq-1}(1-x^{nq})^{\frac{p}{q}} dx$$
$$= \frac{(nq+(p+q)n)(nq+(p+2q)n)\cdots(nq+(p+\infty q)n)}{nq(2nq)(3nq)\cdots(\infty nq)}$$
$$\times \int_0^1 x^{nq+\infty nq-1}(1-x^{nq})^{\frac{p}{q}} dx. \qquad (23.24)$$

Euler then observed that if k was an infinite number and α finite, then

$$\int_0^1 x^k (1-x^{nq})^{\frac{p}{q}} dx = \int_0^1 x^{k+\alpha}(1-x^{nq})^{\frac{p}{q}} dx. \qquad (23.25)$$

456   The Gamma Function

So, dividing equation (23.23) by equation (23.24), the integrals on the right-hand side canceled, and the result was

$$\int_0^1 x^{m-1}(1-x^{nq})^{\frac{p}{q}}dx$$
$$= \frac{1}{(p+q)n} \cdot \frac{nq(m+(p+q)n) \cdot 2nq(m+(p+2q)n) \cdot 3nq(m+(p+3q)n)\cdots}{m(p+2q)n(m+nq)(p+3q)n(m+2nq)\cdots}.$$

Replacing n by n/q, the form of the relation became

$$\int_0^1 x^{m-1}(1-x^n)^{\frac{p}{q}}dx$$
$$= \frac{1}{(p+q)n} \cdot \frac{1(mq+(p+q)n)}{m(p+2q)} \cdot \frac{2(mq+(p+2q)n)}{(m+n)(p+3q)} \cdot \frac{3(mq+(p+3q)n)}{(m+2n)(p+4q)}\cdots. \quad (23.26)$$

This was Euler's final result, equivalent to (23.9). Observe that if we set $n=q=1$ and then replace p by $n-1$, we get

$$\int_0^1 x^{m-1}(1-x)^{n-1}dx$$
$$= \frac{1 \cdot (m+n) \cdot 2 \cdot (m+n+1) \cdot 3 \cdot (m+n+2)\cdots}{n \cdot m \cdot (n+1) \cdot (m+1) \cdot (n+2) \cdot (m+2)\cdots}$$
$$= \lim_{s \to \infty} \left(\frac{s!s^n}{n(n+1)\cdots(n+s)} \cdot \frac{s!s^m}{m(m+1)\cdots(m+s)} \cdot \frac{(m+n)\cdots(m+n+s)}{s!s^{m+n}} \right)$$
$$= \frac{\Gamma(n)\Gamma(m)}{\Gamma(m+n)}. \quad (23.27)$$

Also note that, since both sides of (23.25) are zero, we have to understand Euler to mean that the ratio of the two sides tends to 1 as $k \to \infty$.

By looking at the infinite product in the last equation from another viewpoint, we get a different infinite product for the gamma function. In 1848, more than a century after Euler, the English scholar and mathematician Francis W. Newman (1805–1897) presented this result:

$$\Gamma(m) = \frac{e^{-\gamma m}}{m} \prod_{k=1}^{\infty} \left(1+\frac{m}{k}\right)^{-1} e^{m/k}, \quad (23.28)$$

where γ was Euler's constant defined by

$$\gamma = \lim_{k \to \infty} (1+\frac{1}{2}+\frac{1}{3}+\cdots+\frac{1}{k}-\ln k). \quad (23.29)$$

This representation of the gamma function is sometimes attributed to Weierstrass, who defined the function by a similar infinite product in 1856. In fact, the German mathematician Schlömilch discovered (23.28) even earlier, in 1843; his proof appears in the exercises. Newman wrote the product in (23.27) as

$$\frac{m+n}{mn} \cdot \frac{1+(m+n)}{1+m \cdot 1+n} \cdot \frac{1+\frac{1}{2}(m+n)}{1+\frac{1}{2}m \cdot 1+\frac{1}{2}n} \cdot \frac{1+\frac{1}{3}(m+n)}{1+\frac{1}{3}m \cdot 1+\frac{1}{3}n} \cdots$$

and then observed that the product

$$m(1+m)(1+\frac{1}{2}m)(1+\frac{1}{3}m)\cdots$$

was divergent because its logarithm was

$$\ln m + (1+\frac{1}{2}+\frac{1}{3}+\cdots)m - \frac{1}{2}(1+\frac{1}{2^2}+\frac{1}{3^2}+\cdots)m^2 + \frac{1}{3}(1+\frac{1}{2^3}+\cdots)m^3 - \cdots$$

and the coefficient of m was a divergent series. To remove this defect, he defined a new product,

$$m \cdot \frac{1+m}{e^m} \cdot \frac{1+\frac{1}{2}m}{e^{\frac{1}{2}m}} \cdot \frac{1+\frac{1}{3}m}{e^{\frac{1}{3}m}} \cdots, \qquad (23.30)$$

whose logarithm did not have this divergent part, so that the product converged. He also observed that (23.30) was meaningful when m was a complex number. He next defined $\Gamma(m)$ by equation (23.28). It then followed immediately from (23.27) and this definition that

$$\int_0^1 x^{m-1}(1-x)^{n-1}dx = \frac{\Gamma(m)\Gamma(n)}{\Gamma(m+n)},$$

where Re $m > 0$ and Re $n > 0$ would be required for the convergence of the integral. The Newman-Schlömilch product (23.30) is not difficult to derive from Euler's product (23.1).

23.4 Gauss's Theory of the Gamma Function

Gauss's work on the gamma function was marked by his systematic approach and a greater sense of rigor than was earlier practiced. He discussed the convergence of series and products but did not justify changing the order of limits, or term-by-term integration of infinite series. He started with the finite product

$$\Pi(k,z) = \frac{1 \cdot 2 \cdot 3 \cdots k}{(z+1)(z+2)\cdots(z+k)} k^z, \qquad (23.31)$$

where k was a positive integer and z a complex number not equal to a negative integer. He first proved that the limit as $k \to \infty$ existed. For this purpose, he noted that

$$\Pi(k+n, z) = \Pi(k, z) \frac{\left(1 - \frac{1}{k+1}\right)^{-z}}{1 + \frac{z}{k+1}} \cdots \frac{\left(1 - \frac{1}{k+n}\right)^{-z}}{1 + \frac{z}{k+n}}, \qquad (23.32)$$

and that the logarithm of the product written after $\Pi(k, z)$ remained finite as $n \to \infty$. This proved that the limit existed.

Wallis and Euler had perceived the significance of the gamma function for the evaluation of certain definite integrals. Gauss's important contribution here was to use the gamma function to sum series; some early hints of this also appeared in the works of Stirling, Euler, and Pfaff. Gauss's insight opened up the subject of summation of series of the hypergeometric type. Moreover, Gauss used (23.11) to establish the basic results on the gamma function. Interestingly, in this connection Gauss made use of two series discovered by Newton while he was a student at Cambridge:

$$\arcsin x = x + \frac{\frac{1}{2}}{1!} \frac{x^3}{3} + \frac{\frac{1}{2} \cdot \frac{3}{2}}{2!} \frac{x^5}{5} + \frac{\frac{1}{2} \cdot \frac{3}{2} \cdot \frac{5}{2}}{3!} \frac{x^7}{7} + \cdots \quad \text{and}$$

$$\sin n\theta = n \sin \theta - \frac{n(n^2 - 1^2)}{3!} \sin^3 \theta + \frac{n(n^2 - 1^2)(n^2 - 3^2)}{5!} \sin^5 \theta - \cdots,$$

where n was not necessarily an integer. Gauss's analysis of the convergence of the hypergeometric series showed that the first formula was true for $|x| \leq 1$ and the second for $|\theta| \leq \pi/2$. By contrast, Newton took a much more cavalier approach toward convergence questions. He is known to have discussed the convergence of the geometric series on one occasion, but his remarks contained no new insights. To write (23.11) in a compact form, we employ the modern notation for a shifted factorial:

$$\begin{aligned} (a)_n &= a(a+1)(a+2) \cdots (a+n-1), & \text{for } n > 0, \\ &= 1, & \text{for } n = 0. \end{aligned} \qquad (23.33)$$

We can now write

$$\sum_{n=0}^{\infty} \frac{(a)_n (b)_n}{n! (c)_n} = \frac{\Gamma(c) \Gamma(c-a-b)}{\Gamma(c-a) \Gamma(c-b)}, \qquad (23.34)$$

when $\operatorname{Re}(c - a - b) > 0$. To obtain the value of $\Gamma(1/2)$, Gauss took $x = 1$ in Newton's series for $\arcsin x$ to get

$$\frac{\pi}{2} = \sum_{n=0}^{\infty} \frac{(1/2)_n (1/2)_n}{n! (3/2)_n} = \frac{\Gamma(3/2) \Gamma(1/2)}{\Gamma(1) \Gamma(1)} = 1/2 \left(\Gamma(1/2)\right)^2 \qquad (23.35)$$

$$\text{or} \quad \Gamma(1/2) = \sqrt{\pi}. \qquad (23.36)$$

23.4 Gauss's Theory of the Gamma Function

He then derived Euler's reflection formula (23.10) by taking $\theta = \frac{\pi}{2}$ in Newton's series for $\sin n\theta$ where n was any real number. In that case,

$$\sin\frac{n\pi}{2} = n - \frac{n(n-1)(n+1)}{2\cdot 3} + \frac{n(n-1)(n+1)(n-3)(n+3)}{2\cdot 3\cdot 4\cdot 5}\cdots$$

$$= n\sum_{k=0}^{\infty}\frac{(\frac{-n+1}{2})_k(\frac{n+1}{2})_k}{k!(3/2)_k}$$

$$= n\frac{\Gamma(3/2)\Gamma(1/2)}{\Gamma(3/2+n/2-1/2)\Gamma(3/2-n/2-1/2)}$$

$$= \frac{n\pi}{2\Gamma(1+n/2)\Gamma(1-n/2)}. \qquad (23.37)$$

Note that in the last two steps, (23.36) and (23.34) were employed. Gauss next set $x = \frac{n}{2}$ in (23.37) to obtain

$$\Gamma(1+x)\Gamma(1-x) = \frac{\pi x}{\sin\pi x} \quad \text{or} \qquad (23.38)$$

$$\Gamma(x)\Gamma(1-x) = \frac{\pi}{\sin\pi x}. \qquad (23.39)$$

Finally, to derive (23.9), he wrote

$$\int_0^1 x^{m-1}(1-x)^{n-1}dx$$

$$= \int_0^1 x^{m-1}\left(1-(n-1)x+\frac{(n-1)(n-2)}{2!}x^2-\cdots\right)dx$$

$$= \frac{1}{m}-\frac{n-1}{m+1}+\frac{(n-1)(n-2)}{2!(m+2)}-\frac{(n-1)(n-2)(n-3)}{3!(m+3)}+\cdots$$

$$= \frac{1}{m}\left[1+\frac{(-n+1)m}{m+1}+\frac{(-n+1)(-n+2)m(m+1)}{2!(m+1)(m+2)}+\cdots\right]$$

$$= \frac{1}{m}\cdot\frac{\Gamma(m+1+n-1-m)\Gamma(m+1)}{\Gamma(m+1+n-1)\Gamma(m+1-m)} = \frac{\Gamma(m)\Gamma(n)}{\Gamma(m+n)}. \qquad (23.40)$$

Gauss derived the integral representation for the gamma function by setting $y = nx$ in (23.40), where n was an integer, to obtain

$$\int_0^n y^{m-1}\left(1-\frac{y}{n}\right)^n dy = \frac{n!n^{m-1}}{m(m+1)\cdots(m+n-1)}.$$

The limit as $n \to \infty$ gave

$$\int_0^\infty y^{m-1}e^{-y}dy = \Gamma(m),$$

a result also due to Euler. Gauss did not justify that $\lim_{n\to\infty}\int = \int\lim_{n\to\infty}$ in this situation.

Gauss also defined a new function $\Psi(z)$, given by

$$\Psi(z) = \frac{d}{dz} \ln \Gamma(z+1) = \frac{\Gamma'(z+1)}{\Gamma(z+1)}. \qquad (23.41)$$

He observed that $\Psi(z)$, the digamma function, was almost as remarkable a function as $\Gamma(z)$ and noted some of its more important properties. See the exercises for some of these properties.

23.5 Poisson, Jacobi, and Dirichlet: Beta Integrals

The early nineteenth-century mathematicians, striving to better understand how to manipulate definite integrals, used them to give new proofs of already known properties of the gamma and beta integrals. Poisson, Jacobi, and Dirichlet showed how double integrals could be employed to evaluate Euler's beta integral (23.9). It is interesting to see that Poisson made a change of variables in the double integral one variable at a time, while Jacobi changed from one pair of variables to another. Poisson's derivation appeared in papers of 1811 and 1823; Jacobi's proof was published in 1833. In 1841, Jacobi wrote his important paper on functional determinants or Jacobians (so called by J. J. Sylvester) from which arose the change of variables formula for n-dimensional integrals. Around 1835, the Russian mathematician M. Ostrogradski gave a formal derivation of the general change of variables formula.

In Poisson's evaluation of the beta integral, we first observe that the substitution $t = 1/(1+s)$ gives

$$\int_0^1 t^{p-1}(1-t)^{q-1} dt = \int_0^\infty \frac{s^{q-1} ds}{(1+s)^{p+q}}. \qquad (23.42)$$

Euler knew this, but Poisson also noted that the integrals converged if p and q were real and positive, or if p and q were complex with positive real parts. Poisson started by multiplying the integrals for $\Gamma(p)$ and $\Gamma(q)$ to get

$$\Gamma(p)\Gamma(q) = \int_0^\infty \int_0^\infty e^{-x} e^{-y} x^{p-1} y^{q-1} dx dy. \qquad (23.43)$$

He then substituted xy and xdy for y and dy, followed by $x/(1+y)$ and $dx/(1+y)$ in place of x and dx, to obtain

$$\begin{aligned}
\Gamma(p)\Gamma(q) &= \int_0^\infty \int_0^\infty \frac{e^{-x} x^{p+q-1} y^{q-1}}{(1+y)^{p+q}} dx dy \\
&= \int_0^\infty e^{-x} x^{p+q-1} dx \int_0^\infty \frac{y^{q-1} dy}{(1+y)^{p+q}} \\
&= \Gamma(p+q) \int_0^\infty \frac{y^{q-1} dy}{(1+y)^{p+q}}.
\end{aligned}$$

This proved (23.9). In 1833 Jacobi gave a different substitution. He set $x + y = r$, $x = rw$, so that r ranged from 0 to ∞ and w from 0 to 1. He then noted that

$$dx\,dy = r\,dr\,dw, \tag{23.44}$$

so that the change of variables from x, y to r, w in (23.43) gave

$$\Gamma(p)\Gamma(q) = \int_0^\infty e^{-r} r^{p+q-1}\,dr \int_0^1 w^{p-1}(1-w)^{q-1}\,dw.$$

He did not explain how he obtained (23.44), probably because he took it to be well known.

We note that Poisson gave the conditions for convergence of the integrals, reflecting the increasing awareness among mathematicians that rigor was important. In fact, the works of Gauss, Cauchy, and Abel on infinite series contain the first significant expressions of this rigor. And Dirichlet was particularly attentive to questions of rigor in his important work on integrals, both in papers and in lectures. For example, he discussed the conditions, such as absolute convergence, for changing the order of integration in a double integral.

Dirichlet's evaluation of Euler's integral and his proof of Gauss's multiplication formula using integrals are both given in the exercises at the end of this chapter. Here we mention a multivariable extension of Euler's integral presented by Dirichlet to the Berlin Academy in 1839:

$$\int\cdots\int x_1^{\alpha_1-1} x_2^{\alpha_2-1}\cdots x_n^{\alpha_n-1}dx_1 dx_2\cdots dx_n = \frac{\Gamma(\alpha_1)\Gamma(\alpha_2)\cdots\Gamma(\alpha_n)}{\Gamma(1+\alpha_1+\cdots+\alpha_n)}, \tag{23.45}$$

where $\alpha_i > 0$ and the integral is taken over the region $\sum_{i=1}^n x_i \leq 1$, $x_i \geq 0$, $i = 1, 2, \ldots n$. Note that this formula is an iterated form of Euler's beta integral; in 1941, Selberg discovered a genuine multidimensional generalization of Euler's beta integral. Also note that the use of Euler's beta integral yields a new proof of Euler's reflection formula. For this purpose, take $q = 1 - p$ in the infinite integral (23.42). Then we have

$$\Gamma(p)\Gamma(1-p) = \int_0^\infty \frac{s^{p-1}}{1+s}ds.$$

So if we prove that the value of the integral is $\pi/\sin p\pi$, we have our proof. Euler himself evaluated this integral in 1738 for p a rational number. Dedekind included an improved and streamlined version of this method in his doctoral thesis of 1852. He let

$$B = \int_0^\infty \frac{x^{\frac{m}{n}-1}}{x+1}dx = n\int_0^\infty \frac{x^{m-1}}{x^n+1}dx. \tag{23.46}$$

Then $x^n + 1 = (x - \zeta)(x - \zeta^3)\cdots(x - \zeta^{2n-1})$,

where $\zeta = e^{\pi i/n}$. By a partial fractions expansion

$$\frac{x^{m-1}}{x^n+1} = \frac{-1}{n}\sum_{k=1}^n \frac{\zeta^{(2k-1)m}}{x - \zeta^{2k-1}}.$$

From this, Dedekind could conclude

$$n\int \frac{x^{m-1}}{x^n+1}dx = -\sum_{k=1}^{n}\zeta^{m(2k-1)}\log(\zeta^{2k-1}-x).$$

The last expression was easy to evaluate at $x=0$ but not at $x=\infty$. So Dedekind rewrote this expression as

$$-\sum_{k=1}^{n}\zeta^{m(2k-1)}\log\left(\frac{\zeta^{2k-1}}{x}-1\right) - \log x \sum_{k=1}^{n}\zeta^{m(2k-1)}. \qquad (23.47)$$

This last sum was zero because

$$\sum \zeta^{m(2k-1)} = \zeta^m \sum (\zeta^{2m})^{k-1} = \zeta^m \frac{\zeta^{2mn}-1}{\zeta^{2m}-1} = 0.$$

Thus,

$$n\int \frac{x^{m-1}}{x^n+1}dx = -\sum_{k=1}^{n}\zeta^{m(2k-1)}\log\left(\frac{\zeta^{2k-1}}{x}-1\right). \qquad (23.48)$$

He next used this relation at $x=\infty$ and the previous one at $x=0$ to get

$$B = \sum_{k=1}^{n}\zeta^{m(2k-1)}\log(\zeta^{2k-1}) = \frac{\pi i}{n}\sum_{k=1}^{n}(2k-1)\zeta^{m(2k-1)} = \frac{\pi}{\sin(m\pi/n)}.$$

Dedekind remarked that his use of the second relation to evaluate the integral at ∞ made his proof shorter than the ones found in integral calculus textbooks. It is interesting to compare Dedekind's derivation with that of Euler, (13.38). Dedekind gave three other evaluations of this integral. One of these used Cauchy's new calculus of residues. Another proof, included in the exercises, employed differential equations, and was an original contribution of Dedekind. In the third proof, given earlier by Euler, Dedekind expressed the integral as a partial fractions expansion.

23.6 Bohr, Mollerup, and Artin on the Gamma Function

In 1922, the important Danish language textbook on analysis was published, written by colleagues at the Polytecknisk Lareanstat, Harald Bohr (1887–1951) and Johannes Mollerup (1872–1937). Bohr gained early fame in 1908 as a member of his country's silver-medal Olympic soccer team; he worked on the Riemann zeta function and did his most original mathematical work in creating the theory of almost periodic functions. Bohr and Mollerup's four-volume work, *Laerebog i Matematisk Analyse*, published in various editions, had a profound effect on the teaching of analysis in Denmark, greatly raising the standards. In this work, they applied the idea of logarithmic convexity to prove that the gamma integral equaled the infinite product, that is,

$$\int_0^\infty e^{-x}x^{m-1}dx = \lim_{n\to\infty}\frac{n!n^{m-1}}{m(m+1)\cdots(m+n-1)}. \qquad (23.49)$$

23.6 Bohr, Mollerup, and Artin on the Gamma Function

They started with the right-hand side of (23.49) as the definition of $\Gamma(x)$. They relied on a definition of the Danish mathematician Johan Jensen of a convex function as a function $\phi(x)$ with the property that for every pair of numbers $x_1 < x_3$ and $x_2 = (x_1 + x_3)/2$

$$\phi(x_2) \leq (\phi(x_1) + \phi(x_3))/2. \tag{23.50}$$

When ϕ was continuous, Jensen noted that (23.50) was equivalent to

$$\phi(tx_1 + (1-t)x_3) \leq t\phi(x_1) + (1-t)\phi(x_3), \ 0 < t < 1 \tag{23.51}$$

for all pairs of numbers $x_1 < x_3$. We note that Jensen's definition of convexity arose from a study of inequalities.

In proving (23.49), we follow closely the Bohr–Mollerup notation and argument; by contrast, textbooks usually follow the treatment of Artin. The result we now refer to as the Bohr–Mollerup theorem was not explicitly stated by Bohr and Mollerup but follows from their argument. Bohr and Mollerup denoted the integral in (23.49) by $\Lambda(m)$ and then observed that $\Lambda(x+1) = x\Lambda(x)$ and $\Lambda(1) = 1$. Moreover, by the Cauchy-Schwarz inequality

$$\left(\Lambda\left(\frac{x_1+x_3}{2}\right)\right)^2 \leq \Lambda(x_1)\Lambda(x_3), \quad 0 < x_1 < x_3. \tag{23.52}$$

Observe that this result is equivalent to the logarithmic convexity of $\Lambda(x)$ because $\Lambda(x)$ is continuous and (23.52) implies

$$\ln \Lambda\left(\frac{x_1+x_3}{2}\right) \leq \frac{1}{2}\left(\ln \Lambda(x_1) + \ln \Lambda(x_3)\right).$$

Following Bohr and Mollerup, write $x_2 = (x_1 + x_3)/2$. They then set

$$P(x) = \Lambda(x)/\Gamma(x) \tag{23.53}$$

where $\Gamma(x)$ was defined by (23.2). It followed that $P(1) = 1$ and

$$P(x+1) = P(x) \quad \text{for } x > 0. \tag{23.54}$$

Bohr and Mollerup then used (23.52) to show that $P(x) \equiv 1$. For this purpose, they noted that when n was an integer,

$$\lim_{n \to \infty} \frac{\Gamma(x_1+n)\Gamma(x_3+n)}{(\Gamma(x_2+n))^2} = 1, \tag{23.55}$$

because

$$\lim_{n \to \infty} \frac{\Gamma(n+x)}{n^{x-1}n!} = \lim_{n \to \infty} \frac{(n-1+x)(n-2+x)\cdots x\Gamma(x)}{n^{x-1}n!} = 1.$$

Now by the periodicity of $P(x)$ given in (23.54), they had

$$\frac{P(x_1)P(x_3)}{(P(x_2))^2} = \frac{P(x_1+n)P(x_3+n)}{(P(x_2+n))^2}.$$

By letting $n \to \infty$ in this equation and using (23.52) and (23.55), they could see that

$$1 \leq \frac{P(x_1)P(x_3)}{(P(x_2))^2}. \tag{23.56}$$

Next, they supposed $P(x)$ was not a constant. Then, because $P(x+1) = P(x)$, it was possible to choose $x_1 < x_2$ such that $P(x_1) < P(x_2)$. They took the sequence $x_1 < x_2 < x_3 < x_4 < \cdots$ such that the difference between two consecutive numbers was always the same, that is, equal to $x_2 - x_1$. This meant that if $x_{n-1} < x_n < x_{n+1}$ was a part of the above sequence, then $x_n = (x_{n-1} + x_{n+1})/2$. By (23.56), they obtained

$$1 < \frac{P(x_2)}{P(x_1)} \leq \frac{P(x_3)}{P(x_2)} \leq \frac{P(x_4)}{P(x_3)} \leq \cdots. \tag{23.57}$$

These inequalities implied by induction that $P(x_n)/P(x_1) \geq (P(x_2)/P(x_1))^n$. They could conclude that $P(x_n) \to \infty$ as $n \to \infty$. They noted that $P(x)$ was continuous on $[1,2]$ and hence bounded on that interval. So they got a contradiction by the periodicity of P. Thus, $P(x)$ was a constant, necessarily equal to 1, and their proof was complete.

Emil Artin (1898–1962) saw that the Bohr–Mollerup proof of (23.49) could be simplified if (23.51) instead of (23.50) were used for convexity. Artin's argument applied Hölder's inequality to show that

$$\ln(\Lambda(tx_1 + (1-t)x_3)) \leq t \ln \Lambda(x_1) + (1-t)\ln \Lambda(x_3), \text{ for } 0 < t < 1.$$

He then proved more generally that if $f(0) = 1$, $f(x+1) = xf(x)$, and $\ln f(x)$ satisfied (23.51), then

$$f(x) = \lim_{n \to \infty} \frac{n! n^{x-1}}{x(x+1)\cdots(x+n-1)}.$$

Artin's proof was quite short. Note that by the first two conditions, $\ln f(n) = \ln(n-1)!$, when n was a positive integer. Next, let $0 < x \leq 1$. With $x_1 = n$, $x_3 = x + n + 1$ and $t = x/(1+x)$ in (23.51) Artin had

$$(x+1)\ln n! \leq x \ln(n-1)! + \ln f(n+1+x).$$

This simplified to

$$\frac{n! n^{x-1}}{x(x+1)\cdots(x+n-1)} \cdot \frac{n}{n+x} \leq f(x).$$

Similarly, with $x_1 = n+1$, $x_3 = n+2$ and $t = 1-x$, he had, after simplification,

$$f(x) \leq \frac{n!(n+1)^x}{x(x+1)\cdots(x+n)} = \frac{n! n^{x-1}}{x(x+1)\cdots(x+n-1)} \cdot \left(\frac{n+1}{n}\right)^{x-1} \cdot \frac{n+1}{n+x}.$$

The two inequalities yielded the required formula when $n \to \infty$. Artin was a number theorist and algebraist. In algebra, he was a disciple of Emmy Noether (1882–1935) and advocated a very abstract point of view. It is therefore interesting to see him make this contribution to special functions. Some of his other results in this area are mentioned in the exercises. In addition, the reader may refer to chapter 3, sections 2 and 4, for the the use of logarithmic convexity by Wallis and Stieltjes.

23.7 Kummer's Fourier Series for $\ln \Gamma(x)$

By an interesting application of definite integrals, in 1847 Kummer derived the Fourier series for $\ln \Gamma(x)$. This formula is important in number theory, although Kummer's purpose was to obtain a new derivation for Gauss's multiplication formula for the gamma function. The latter can be written as

$$\sum_{k=0}^{n-1} \ln \Gamma\left(x + \frac{k}{n}\right) = \frac{1}{2}(n-1)\ln 2\pi + \frac{1}{2}(1-2nx)\ln n + \ln \Gamma(nx). \quad (23.58)$$

Kummer explained why he thought of the Fourier series in this connection. Suppose

$$f(x) = A_0 + 2\sum_{k=1}^{\infty} A_k \cos 2k\pi x + 2\sum_{k=1}^{\infty} B_k \sin 2k\pi x, \quad \text{for } 0 < x < 1, \quad (23.59)$$

where $\quad A_k = \int_0^1 f(x) \cos 2k\pi x \, dx, \quad B_k = \int_0^1 f(x) \sin 2k\pi x \, dx. \quad (23.60)$

Then

$$\sum_{k=0}^{n-1} f\left(x + \frac{k}{n}\right) = n\left(A_0 + 2\sum_{k=1}^{\infty} A_{nk} \cos 2kn\pi x + 2\sum_{k=1}^{\infty} B_{nk} \sin 2kn\pi x\right). \quad (23.61)$$

Moreover, by denoting

$$F(x) = A_0 + 2\sum_{k=1}^{\infty} A_{nk} \cos 2k\pi x + 2\sum_{k=1}^{\infty} B_{nk} \sin 2k\pi x,$$

the right-hand side of (23.61) was $nF(nx)$. Thus, equation (23.61) was suggestive of Gauss's formula (23.58). So Kummer took $f(x) = \ln \Gamma(x)$ in (23.59). Then by Euler's reflection formula

$$\ln \Gamma(x) + \ln \Gamma(1-x) = \ln 2\pi - \ln(2\sin \pi x)$$

$$= \ln 2\pi - \sum_{k=1}^{\infty} \frac{\cos 2k\pi x}{k}, \quad (23.62)$$

where the last relation followed from Euler's Fourier series for $\ln(2\sin \pi x)$. By (23.59),

$$\ln \Gamma(x) + \ln \Gamma(1-x) = 2A_0 + \sum_{k=1}^{\infty} 4A_k \cos 2k\pi x$$

and hence, by (23.62),

$$A_0 = \frac{1}{2}\ln 2\pi, \quad A_k = \frac{1}{4k}.$$

Kummer had to work harder to find B_k. He started with Plana's formula

$$\ln \Gamma(x) = \int_0^1 \left(\frac{1 - t^{x-1}}{1-t} - x + 1\right) \frac{dt}{\ln t}, \quad x > 0. \quad (23.63)$$

This was an integrated form of Gauss's formula for $\Psi(x)$. So he had

$$B_k = \int_0^1 \int_0^1 \left(\frac{1-t^{x-1}}{1-t} - x + 1\right) \frac{\sin 2k\pi x}{\ln t} dt\, dx. \quad (23.64)$$

Since $\int_0^1 \sin 2k\pi x\, dx = 0$, $\quad \int_0^1 x \sin 2k\pi x\, dx = -\frac{1}{2k\pi}$

and $\int_0^1 t^{x-1} \sin 2kx\, dx = \frac{(1-t)2k\pi}{t((\ln t)^2 + 4k^2\pi^2)}$,

Kummer reduced (23.64) to

$$B_k = \int_0^1 \left(\frac{-2k\pi}{t((\ln t)^2 + 4k^2\pi^2)} + \frac{1}{2k\pi}\right) \frac{dt}{\ln t}.$$

Then, with $t = e^{-2k\pi u}$,

$$B_k = \frac{1}{2k\pi} \int_0^\infty \left(\frac{1}{1+u^2} - e^{-2k\pi u}\right) \frac{du}{u}.$$

When $k = 1$,

$$B_1 = \frac{1}{2\pi} \int_0^\infty \left(\frac{1}{1+u^2} - e^{-2\pi u}\right) \frac{du}{u}.$$

Kummer then employed a result of Dirichlet; see exercise 3(b):

$$-\frac{\gamma}{2\pi} = \frac{1}{2\pi} \int_0^\infty \left(e^{-u} - \frac{1}{1+u}\right) \frac{du}{u},$$

where γ was Euler's constant. Therefore,

$$B_1 - \frac{\gamma}{2\pi} = \frac{1}{2\pi} \int_0^\infty \frac{e^{-u} - e^{-2\pi u}}{u} du + \frac{1}{2\pi} \int_0^\infty \left(\frac{1}{1+u^2} - \frac{1}{1+u}\right) \frac{du}{u}.$$

The first integral equaled $\ln 2\pi$ and a change of variables t to $1/t$ showed that the value of the second integral was 0. Thus,

$$B_1 = \frac{\gamma}{2\pi} + \frac{1}{2\pi} \ln 2\pi.$$

To find B_k, he observed that

$$kB_k - B_1 = \frac{1}{2\pi} \int_0^\infty \frac{e^{-2\pi u} - e^{-2k\pi u}}{u} du = \frac{1}{2\pi} \ln k.$$

Thus, $B_k = \frac{1}{2k\pi}(\gamma + \ln 2k\pi), \quad k = 1, 2, 3, \ldots,$

and Kummer got his Fourier expansion:

$$\ln \Gamma(x) = \frac{1}{2} \ln 2\pi + \sum_{k=1}^\infty \frac{\cos 2\pi kx}{2k} + \frac{1}{\pi} \sum_{k=1}^\infty \frac{\gamma + \ln 2\pi + 2\ln k}{2k} \sin 2k\pi x.$$

Note that Kummer's formula is a particular case of the functional equation for the Hurwitz zeta function.

23.8 Exercises

1. Show that by taking $k = p + q\sqrt{-1}$, $p > 0$ in

$$\int_0^\infty x^{n-1} e^{-kx} dx = \frac{\Gamma(n)}{k^n},$$

we get

$$\int_0^\infty x^{n-1} e^{-px} \cos qx \, dx = \frac{\Gamma(n) \cos n\theta}{f^n} \quad \text{and}$$

$$\int_0^\infty x^{n-1} e^{-px} \sin qx \, dx = \frac{\Gamma(n) \sin n\theta}{f^n},$$

where $f = (p^2 + q^2)^{1/2}$ and $\tan \theta = \frac{q}{p}$.

Deduce that

$$\int_0^\infty e^{-px} \frac{\cos qx}{\sqrt{x}} dx = \frac{\sqrt{\pi}}{f} \sqrt{\frac{f-p}{2}},$$

$$\int_0^\infty e^{-px} \frac{\sin qx}{\sqrt{x}} dx = \frac{\sqrt{\pi}}{f} \sqrt{\frac{f+p}{2}},$$

$$\int_0^\infty \frac{\cos x}{\sqrt{x}} dx = \sqrt{\frac{\pi}{2}}, \quad \int_0^\infty \frac{\sin x}{\sqrt{x}} dx = \sqrt{\frac{\pi}{2}},$$

$$\int_0^\infty e^{-px} \frac{\sin qx}{x} dx = \theta, \quad \text{and} \quad \int_0^\infty \frac{\sin x}{x} dx = \frac{\pi}{2}.$$

All these definite integrals appeared in Euler's paper of 1781, though he had evaluated some of them earlier by other methods. See Eu I-19, pp. 217–227. Euler's deductions were formal and he was the first to make use of complex parameters in this way. Although he initially assumed the parameter $p > 0$, he let $p \to 0$ to obtain the later integrals. He expressed this by setting $p = 0$. He had some reservations about presenting the final integral for $(\sin x)/x$ but numerical computation convinced him of its correctness. This paper was influential in the development of complex analysis by Cauchy. Legendre and Laplace referred to it when extending its methods to evaluate other integrals and these results motivated Cauchy to begin his work on complex integrals in 1814.

2. If Euler found a formula interesting, he often evaluated it in more than one way. Complete the details and verify the steps of the three methods he gave to show that the integral $\int_0^{\pi/2} \ln \sin \phi \, d\phi = -\frac{\pi}{2} \ln 2$.

The Gamma Function

(a) Euler began by setting $x = \sin\phi$ to get

$$\int_0^1 \frac{\ln x}{\sqrt{1-x^2}} dx = \int_0^1 \frac{\ln\sqrt{(1-y^2)}}{\sqrt{1-y^2}} dy$$

$$= -\int_0^1 \frac{\left(\frac{y^2}{2} + \frac{y^4}{4} + \frac{y^6}{6} + \cdots\right)}{\sqrt{1-y^2}} dy$$

$$= -\frac{\pi}{2}\left(\frac{1}{2^2} + \frac{1 \cdot 3}{2 \cdot 4^2} + \frac{1 \cdot 3 \cdot 5}{2 \cdot 4 \cdot 6^2} + \frac{1 \cdot 3 \cdot 5 \cdot 7}{2 \cdot 4 \cdot 6 \cdot 8^2} + \cdots\right).$$

Euler showed that the sum of the series was $\ln 2$. For this purpose, he noted that by the binomial expansion

$$\int_0^x \frac{1}{z}\left(\frac{1}{\sqrt{1-z^2}} - 1\right) dz = \frac{1}{2^2}x^2 + \frac{1 \cdot 3}{2 \cdot 4^2}x^4 + \frac{1 \cdot 3 \cdot 5}{2 \cdot 4 \cdot 6^2}x^6 + \text{etc.}$$

He applied the substitution $v = \sqrt{1-z^2}$ to evaluate

$$\int \frac{1}{z\sqrt{1-z^2}} dz = -\ln\frac{1 + \sqrt{1-z^2}}{z} + C.$$

Hence, $\int_0^x \frac{1}{z}\left(\frac{1}{\sqrt{1-z^2}} - 1\right) dz = \ln\frac{2}{1+\sqrt{1-x^2}}.$

(b) In his second method, Euler started with a divergent series. He applied the addition formula $2\sin n\theta \sin\theta = \cos(n-1)\theta - \cos(n+1)\theta$ to get

$$\frac{\cos\theta}{\sin\theta} = 2\sin 2\theta + 2\sin 4\theta + 2\sin 6\theta + 2\sin 8\theta + \text{etc.}$$

He integrated to obtain

$$\ln\sin\theta = C - \cos 2\theta - \frac{1}{2}\cos 4\theta - \frac{1}{3}\cos 6\theta - \frac{1}{4}\cos 8\theta - \text{etc.} \quad (23.65)$$

Then $\theta = \pi/2$ gave $C = -\ln 2$. Euler integrated again to obtain the required result.

(c) Euler proved the more general formula

$$\int_0^1 x^{p-1} X \ln x \, dx = \int_0^1 x^{p-1} X \, dx \int_0^1 \frac{x^{p-1}(x^m - 1)}{1 - x^n} dx, \quad (23.66)$$

where $X = (1-x^n)^{\frac{m-n}{n}}$. The result in (a) and (b) would be obtained by setting $n = 2$, $m = p = 1$. To prove (23.66), Euler set $P = \int_0^1 x^{p-1} X \, dx$. By an argument which gives (23.26), show that

$$P = \frac{n}{m} \cdot \frac{2n}{m+n} \cdot \frac{3n}{m+2n} \cdots \times \frac{p+m}{p} \cdot \frac{p+m+n}{p+n} \cdot \frac{p+m+2n}{p+2n} \cdot \text{etc.}$$

He then let p be the variable and m, n be constants to obtain

$$\frac{dP/dp}{p} = \frac{1}{m+p} - \frac{1}{p} + \frac{1}{m+p+n} - \frac{1}{p+n} + \frac{1}{m+p+2n}$$
$$- \frac{1}{p+n} + \text{etc.}$$

Prove the result by showing how that this partial fractions expansion equals

$$\int_0^1 \frac{x^{p-1}(x^m - 1)}{1 - x^n} dx.$$

Euler actually worked with a product for P/Q where Q was an integral similar to P. Lacroix (1819) gave the preceding simplification on p. 437. These results are contained in Eu. I-18, pp. 23–50. This volume is full of ingenious evaluations of definite integrals. Observe that (23.65) is the Fourier series expansion of $\ln \sin \theta$ used by Kummer to obtain (23.62). Also, Euler could have derived (23.65) without using divergent series by expanding $\ln(1 - e^{2i\theta})$. But Euler treated divergent series as very much a part of mathematics, a view validated only in the twentieth century.

3. The following derivation of Gauss's multiplication formula is due to Dirichlet (1969), pp. 274–276. Verify the successive steps for Re $a > 0$:

(a) $\int_0^\infty (e^{-y} - e^{-sy}) \frac{dy}{y} = \ln s$.

(b)

$$\Gamma'(a) = \int_0^\infty e^{-s} s^{a-1} \ln s \, ds$$
$$= \int_0^\infty \frac{dy}{y} \left(e^{-y} \int_0^\infty e^{-s} s^{a-1} ds - \int_0^\infty e^{-(1+y)s} s^{a-1} ds \right)$$
$$= \Gamma(a) \int_0^\infty \frac{dy}{y} \left(e^{-y} - \frac{1}{(1+y)^a} \right). \qquad (23.67)$$

(c) $\frac{d}{da} \ln \Gamma(a) = \int_0^1 \left(e^{1-1/x} - x^a \right) \frac{dx}{x(1-x)}$.

(d) Let

$$S = n \int_0^1 \left(\frac{n e^{1-1/x^n}}{1 - x^n} - \frac{x^{na}}{1 - x} \right) \frac{dx}{x}. \quad \text{Then}$$

$$S = \sum_{k=0}^{n-1} \frac{d}{da} \ln \Gamma\left(a + \frac{k}{n} \right).$$

(e) Change a to na in (c) and subtract the result from (d) to see that $S - \frac{d}{da} \ln \Gamma(na)$ is independent of a. Denote this quantity by p and integrate to get

$$\prod_{k=0}^{n-1} \Gamma\left(a + \frac{k}{n}\right) = qp^a \Gamma(na).$$

(f) Change $a \to a + 1/n$ in (e) to deduce that $p = n^{-n}$.

(g) Euler's formula (23.14) implies that $q = (2\pi)^{\frac{n-1}{2}} \sqrt{n}$.

(h) Show that Euler's formula (23.14) is obtained by applying $\Gamma(x)\Gamma(1-x) = \pi/\sin \pi x$.

4. Show that for suitably chosen constants a, b, c, and k,

(a) $\int_0^\infty e^{-(c+z)y} y^{a-1} dy = \frac{\Gamma(a)}{(c+z)^a}$.

(b) $\int_0^\infty e^{-cy} y^{a-1} \left(\int_0^\infty e^{-(k+y)z} z^{b-1} dz\right) dy = \Gamma(a) \int_0^\infty \frac{e^{-kz} z^{b-1}}{(c+z)^a} dz$.

(c) $\Gamma(b) \int_0^\infty \frac{e^{-cy} y^{a-1}}{(k+y)^b} dy = \Gamma(a) \int_0^\infty \frac{e^{-kz} z^{a-1}}{(c+z)^a} dz$.

(d) $\int_0^\infty \frac{y^{a-1}}{(1+y)^{a+b}} dy = \frac{\Gamma(a)\Gamma(b)}{\Gamma(a+b)}$.

See Dirichlet (1969), vol. I, p. 278.

5. Use Dirichlet's integral formula for $\Gamma'(a)/\Gamma(a)$ (23.67) to show that for $a > 0$

$$\gamma + \frac{d}{da} \ln \Gamma(a+1) = \int_0^1 \frac{1 - y^a}{1 - y} dy$$
$$= \left(1 - \frac{1}{a+1}\right) + \left(\frac{1}{2} - \frac{1}{a+2}\right) + \left(\frac{1}{3} - \frac{1}{a+3}\right) + \cdots.$$

Deduce the infinite product for $\Gamma(a)$:

$$\Gamma(a+1) = a\Gamma(a) = e^{-\gamma a} \cdot \frac{e^a}{1+a} \cdot \frac{e^{a/2}}{1+a/2} \cdot \frac{e^{a/3}}{1+a/3} \cdots.$$

For details, see Schlömilch (1843).

6. Show that Dirichlet's integral formula, (23.67) can be obtained from Gauss's (23.72).

7. For $0 < b < 1$, set $B(b) = \int_0^\infty \frac{t^{b-1}}{1+t} dt$. Show that

$$\int_0^\infty \frac{t^{b-1}}{st+1} dt = Bs^{-b} \quad \text{and}$$

$$\int_0^\infty \frac{t^{b-1}}{t+s} dt = Bs^{b-1}.$$

Deduce that

$$B \frac{(s^{b-1} - s^{-b})}{s-1} = \int_0^\infty \frac{t^{b-1}(t-1)}{(st+1)(t+s)} dt. \qquad (23.68)$$

Observe that

$$B^2 = \int_0^\infty \frac{1}{s+1}\left(\int_0^\infty \frac{t^{b-1}}{t+s}dt\right)ds$$
$$= \int_0^\infty \frac{t^{b-1}\ln t}{t-1}dt.$$

Deduce that

$$\int_{1-y}^y B^2 dt = \int_0^\infty \frac{t^{y-1}-t^{-y}}{t-1}dt.$$

From (23.68) obtain

$$B(b)\int_{1-b}^b [B(t)]^2 dt = 2\int_0^\infty \frac{t^{b-1}\ln t}{1+t}dt = 2B'(b).$$

Observe that $B(b) = B(1-b)$ and deduce that $B'(1/2) = 0$,

$$\int_{1-b}^b [B(t)]^2 dt = 2\int_{1/2}^b [B(t)]^2 dt$$
$$\text{and}\quad B(b)\int_{1/2}^b [B(t)]^2 dt = B'(b).$$

Now show that B satisfies the differential equation $BB'' - (B')^2 = B^4$. Solve the differential equation with initial conditions $B(1/2) = \pi$ and $B'(1/2) = 0$ to obtain $B = \pi \csc \pi b$. This is Dedekind's evaluation of the Eulerian integral B, a part of his doctoral dissertation. See Dedekind (1930), vol. 1, pp. 19–22 and 29–31.

8. Let $c_1, c_2, \ldots, c_n$ be positive constants and set $f(x) = (x+c_1)(x+c_2)\cdots(x+c_n)$. Show that for $0 < b < n$

$$\int_0^\infty \frac{x^{b-1}}{f(s)}dx = \frac{\pi}{\sin b\pi}\sum_{k=1}^n \frac{c_k^{b-1}}{f'(-c_k)},$$

where f' denotes the derivative of f. See Dedekind (1930), vol. 1, p. 24.

9. (a) Suppose that $\phi(x)$ is positive and twice continuously differentiable on $0 < x < \infty$ and satisfies (i) $\phi(x+1) = \phi(x)$ and (ii) $\phi\left(\frac{x}{2}\right)\phi\left(\frac{x+1}{2}\right) = d\phi(x)$, where d is a constant. Prove that ϕ is a constant.
   (b) Show that $\Gamma(x)\Gamma(1-x)\sin \pi x$ satisfies the conditions of the first part of the problem. Deduce Euler's formula (23.10). This proof of Euler's reflection formula (23.10) is due to Artin (1964), chapter 4.

10. Suppose that f is a positive and twice continuously differentiable function on $0 < x < \infty$ and satisfies $f(x+1) = xf(x)$ and $2^{2x-1}f(x)f(x+\frac{1}{2}) = \sqrt{\pi}f(2x)$. Show that $f(x) = \Gamma(x)$. See Artin (1964).

11. Prove the following results of Gauss on the digamma function:

(a) For a positive integer n,

$$\Psi(z+n) = \Psi(z) + \frac{1}{z+1} + \frac{1}{z+2} + \cdots + \frac{1}{z+n}.$$

(b) $\Psi(0) = \Gamma'(1) = -\gamma = -0.57721566490153286060653$. Euler computed the constant γ correctly to fifteen decimal places by an application of the Euler–Maclaurin summation formula. About twenty years later, in 1790, the Italian mathematician, Lorenzo Mascheroni (1750–1800) computed γ to thirty-two decimal places by the same method. To compute γ, Gauss gave two asymptotic series for $\Psi(z)$, obtained by taking the derivatives of the de Moivre and Stirling asymptotic series for $\ln(z+1)$. His value differed from Mascheroni's in the twentieth place and so he persuaded F. B. G. Nicolai, a calculating prodigy, to repeat the computation and to extend it further. Nicolai calculated to forty places, given by Gauss in a footnote, and verified that Gauss's computation was correct.

(c)
$$\Psi(-z) - \Psi(z-1) = \pi \cot \pi z. \tag{23.69}$$

(d)
$$\Psi(x) - \Psi(y) = -\frac{1}{x+1} + \frac{1}{y+1} - \frac{1}{x+2} + \frac{1}{y+2} - \frac{1}{x+3} + \text{etc.}$$

(e)
$$\Psi(z) + \Psi\left(z - \frac{1}{n}\right) + \Psi\left(z - \frac{2}{n}\right) + \cdots + \Psi\left(z - \frac{n-1}{n}\right)$$
$$= n\Psi(nz) - n \ln n.$$

(f)
$$\Psi\left(-\frac{1}{n}\right) + \Psi\left(-\frac{2}{n}\right) + \cdots + \Psi\left(-\frac{n-1}{n}\right) = -(n-1)\gamma - n \ln n.$$

(g) For n an odd integer and m a positive integer less than n,

$$\Psi\left(-\frac{m}{n}\right) = -\gamma + \frac{1}{2}\pi \cot \frac{m\pi}{n} - \ln n$$
$$+ \sum_{k=1}^{n-1} \cos \frac{2km\pi}{n} \ln\left(2 - 2\cos \frac{k\pi}{n}\right). \tag{23.70}$$

(h) For n even,

$$\Psi\left(-\frac{m}{n}\right) = \pm \ln 2 - \gamma + \frac{1}{2}\pi \cos\frac{m\pi}{n} - \ln n$$
$$+ \sum_{k=1}^{n-2} \cos\frac{2km\pi}{n} \ln\left(2 - 2\cos\frac{k\pi}{n}\right), \qquad (23.71)$$

where the upper sign is taken for m even, and the lower for m odd.

(i)
$$\Psi(t) = \int_0^1 \left(-\frac{1}{\ln x} - \frac{x^t}{1-x}\right) dx, \quad t > -1. \qquad (23.72)$$

In unpublished work, Gauss explained how $\Psi(z)$ could be used to express the second independent solution of the hypergeometric equation in certain special circumstances. See Gauss (1868–1927), vol. 3, pp. 154–160.

12. Prove that

$$\frac{1}{2\pi i}\int_{-i\infty}^{i\infty}\Gamma(a+s)\Gamma(b+s)\Gamma(c-s)\Gamma(d-s)ds$$
$$=\frac{\Gamma(a+c)\Gamma(a+d)\Gamma(b+c)\Gamma(b+d)}{\Gamma(a+b+c+d)},$$

where the path of integration is curved so that the poles of $\Gamma(c-s)\Gamma(d-s)$ lie on the right of the path and the poles of $\Gamma(a+s)\Gamma(b+s)$ lie on the left. This formula is due to Barnes (1908); it played an important role in his theory of the hypergeometric function. It is an extension of Euler's beta integral formula (23.9), as pointed out by Askey. This can be seen by replacing b and d by $b-it$ and $d+it$, respectively, and then setting $s=tx$ and letting $t\to\infty$.

13. Show that if n is a positive integer and α,β,γ are complex numbers such that Re $\alpha > 0$, Re $\beta > 0$, and Re $\gamma > -\min(1/n, (\text{Re }\alpha)/(n-1), (\text{Re }\beta)/(n-1))$, then

$$\int_0^1\cdots\int_0^1 \prod_{i=1}^n \left(x_i^{\alpha-1}(1-x_i)^{\beta-1}\right)|\Delta(x)|^{2\gamma} dx_1 dx_2\cdots dx_n$$
$$= \prod_{j=1}^n \frac{\Gamma(\alpha+(j-1)\gamma)\Gamma(\beta+(j-1)\gamma)\Gamma(1+j\gamma)}{\Gamma(\alpha+\beta+(n+j-2)\gamma)\Gamma(1+\gamma)}$$

where $\displaystyle \Delta(x) = \prod_{1\le i<j\le n}(x_i - x_j)$.

This multidimensional generalization of Euler's beta integral was discovered by Selberg around 1940; see Selberg (1989), vol. 1, pp. 204–213. He used it to prove a generalization of some theorems of Hardy, Pólya, and Gelfond on entire functions. Since Selberg's formula did not become well known until the 1970s, the theoretical physicists M. L. Mehta and F. J. Dyson conjectured and used the following limiting case of Selberg's formula: For Re $\gamma > -1/n$,

$$\int_{-\infty}^{\infty}\cdots\int_{-\infty}^{\infty}\exp\left(-\frac{1}{2}\sum_{i=1}^{n}x_i^2\right)\prod_{1\le i<j\le n}|x_i-x_j|^{2\gamma}dx_1dx_2\cdots dx_n$$
$$=(2\pi)^{n/2}\prod_{j=1}^{n}\frac{\Gamma(\gamma j+1)}{\Gamma(\gamma+1)}.$$

For other applications of the Selberg integral, see Forrester and Warnaar (2008).

14. Suppose F is a holomorphic function in the right half complex plane Re $z > 0$. Suppose also that $F(1) = 1$, $F(z+1) = zF(z)$ and that $F(z)$ is bounded in the vertical strip $1 \le$ Re $z < 2$. Then $F(z) = \Gamma(z)$ for Re $z > 0$. This uniqueness theorem, useful for giving short proofs of several basic results on the gamma function, was proved by Helmut Wielandt in 1939 and published by Konrad Knopp in 1941. See Remmert (1996), who quotes a letter of Wielandt explaining this and gives references.

15. Prove Dirichlet's formula: Suppose $c_1, c_2, c_3, \ldots$ is a sequence of complex numbers which satisfy $c_{n+k} = c_n$. Suppose $\sum_{n=1}^{\infty}\frac{c_n}{n^s}$ converges absolutely. Then

$$\sum_{n=1}^{\infty}\frac{c_n}{n^s} = \frac{1}{\Gamma(s)}\int_0^1 \frac{\sum_{n=1}^{k}c_n x^{n-1}}{1-x^k}\ln^{s-1}\left(\frac{1}{x}\right)dx.$$

This is the formula Dirichlet applied to the problem of primes in arithmetic progressions. See Dirichlet (1969), vol. I, pp. 421–422.

16. Observe that for $a < 1/2$

$$\int_0^1 t^{\frac{1}{2}-a}(1-t)^{\frac{1}{2}-a}dt = (\Gamma(1/2-a))^2/\Gamma(1-2a).$$

Write the integrand as $2^{2a-1}(1-(1-2t)^2)^{a-1/2}$, apply a change of variables, and then use Euler's reflection formula to obtain

$$\sqrt{\pi}\Gamma(a) = 2^{1-2a}\cos(a\pi)\Gamma(2a)\cdot\Gamma\left(\frac{1}{2}-a\right).$$

This proof of the duplication formula is Legendre's (1811–1817), vol. 1, p. 284.

23.9 Notes on the Literature

For Stirling's work, see the English translation of Stirling's *Methodus Differentialis* by Tweddle (2003); see especially pp. 124–127 for the quoted passages. Tweddle has added 120 pages of notes to clarify and explain Stirling's propositions in modern terms. See Truesdell (1984), p. 345, for the quote concerning Goldbach. For the correspondence between Euler and Goldbach, see vol. 1 of Fuss (1968). The second volume includes the letters of D. Bernoulli, mentioned in the text. The integral formula for the gamma function (23.16) as well as the fractional derivatives (23.8) and (23.9) appear in Euler's 1730–31 paper "De Progressionibus Transcendentibus seu Quarum Termini Generales Algebraice dari Nequeunt." See Eu. I-14, pp. 1–24, particularly pp. 12 and 23. Euler treated the beta integral in several papers, some written in the 1730s and some when he later returned to this subject, in the 1760s and 1770s. His integral calculus book of 1768, contained in Eu. I-11–13, presents a thorough treatment of this whole topic. The evaluation of the beta integral as an infinite product, in the form given in the text, can be found in his 1739 paper "De Productis ex Infinitis Factoribus Ortis." See Eu. I-14, pp. 260–90, especially pp. 282–84. For Euler's reflection formula (23.20), refer to Eu. I-17, pp. 316–357, particularly § 43. For Dedekind's evaluation of the reflection formula by means of an integration of a rational function, see Dedekind (1930), vol. I, pp. 8–9. Euler included sections on the gamma function in his books on the differential and integral calculus. His differential calculus (reprinted in Eu. I-10) has a chapter titled "De Interpolatione Serierum" discussing infinite products related to the gamma function. The evaluation of the beta integral is carried out in even more detail than in the 1739 paper in his integral calculus (Eu. I-12). See also vol. 17.

The product for the gamma function can be found in Schlömilch (1843) and Newman (1848). The journal containing Newman's paper was a continuation of the *Cambridge Mathematical Journal* and in 1848, the journal was edited by the mathematical physicist William Thomson.

Gauss's work on the gamma function appears in his paper on the hypergeometric series, "Disquisitiones Generales Circa Seriem Infinitam . . . ." See Gauss (1863–1927), vol. 3, pp. 123–162, especially pp. 144–152. For evaluations of the beta integrals by using double integrals, see Poisson (1823), especially pp. 477–478 and Jacobi (1969), vol. 6, pp. 62–63. For the multivariable iterated form of the beta integral (23.45), see Dirichlet (1969), vol. I, p. 389; see Selberg (1989), vol. 1, pp. 204–213 for his multidimensional generalization of the beta integral.

For a discussion of Kummer's Fourier series expansion of $\ln \Gamma(x)$, see Andrews, Askey, and Roy (1999), pp. 29–32. For the Bohr–Mollerup derivation of the gamma integral as a product, see Bohr and Mollerup (1922), vol. III, pp. 149–164. Also see Artin (1964), an English translation of his 1930 monograph. For more history of the gamma function, see Davis (1959) and Dutka (1991).

24

The Asymptotic Series for $\ln \Gamma(x)$

24.1 Preliminary Remarks

Since the time of Newton, expressions involving the factorial appeared in the solutions of combinatorial problems and as coefficients of some important power series. However, good approximations for $m!$ were obtained only around 1730 by the joint efforts of Abraham de Moivre (1667–1754) and James Stirling (1692–1770), both working in Britain. The story of their cooperation is fascinating. Although he was born in France, de Moivre was a victim of religious discrimination there; he relocated to Britain as a young man and worked there for the rest of his life. In 1721, de Moivre found a method for converting the sum $\ln m! = \sum_{k=1}^{m} \ln k$ into an asymptotic series. Later in the 1720s, Stirling introduced an improvement in de Moivre's series and then gave a different method. De Moivre's work appeared in the *Supplement* to his *Miscellanea Analytica* in 1730; Stirling's series was published the same year in his *Methodus Differentialis*. Stirling presented the formula, for $s = m + 1/2$,

$$\sum_{k=1}^{m} \ln k = \ln m! = s \ln s + \frac{1}{2} \ln 2\pi - s - \frac{1}{24s} + \frac{7}{2880s^3} - \text{etc.} \qquad (24.1)$$

But de Moivre gave a slightly different form:

$$\ln m! = \left(m + \frac{1}{2}\right) \ln m + \frac{1}{2} \ln 2\pi - m + \frac{1}{12m} - \frac{1}{360m^3} + \frac{1}{1260m^5} - \frac{1}{1680m^7} + \text{etc.} \qquad (24.2)$$

The approximation $m! \sim \sqrt{2\pi} \, m^{m+1/2} e^{-m}$ is now known as Stirling's approximation, though it is a consequence of de Moivre's series. Stirling's series actually suggests the approximation

$$m! \sim \sqrt{2\pi} \left(\frac{m+1/2}{e}\right)^{m+1/2}. \qquad (24.3)$$

24.1 Preliminary Remarks

In fact, neither de Moivre nor Stirling explicitly stated either of these approximations. It seems that the first appearance of the result called Stirling's approximation occurred in a letter from Euler to Goldbach dated June 23, 1744.

De Moivre's motivation in developing the series (24.2) arose from his interest in probability theory, a subject he started cultivating in 1707 at the age of 40. He became familiar with the works of Jakob Bernoulli, Niklaus I Bernoulli, and Pierre Montmort and went on to make very important contributions of his own. He made a living as a consultant to gamblers, speculators, and rich patrons, helping them solve problems related to games of chance or the calculation of annuities. He published the first edition of his *Doctrine of Chances* in 1718. This may have led Sir Alexander Cuming (1690–1775) to consult de Moivre in 1721 about a problem arising in the context of gambling. Cuming lived an eventful and long life, as an alchemist, a member of the Scottish bar, a Fellow of the Royal Society of London, a Cherokee chief, and a baronet - and yet he died in poverty. Mathematically, the problem he posed to de Moivre reduced to: A coin tossed n times had p as the probability of getting heads. The probability of x heads in n tosses would then be

$$b(x,n,p) = \binom{n}{x} p^x (1-p)^{n-x}. \tag{24.4}$$

The problem was to compute $\sum_{x=0}^{n} |x-p|\, b(x,n,p)$. For $p = 1/2$, de Moivre found the value to be

$$\frac{n}{2^{n+1}} \binom{n}{\lfloor n/2 \rfloor}.$$

With n an even number $n = 2m$, the expression reduced to

$$\frac{m}{2^{2m}} \binom{2m}{m}. \tag{24.5}$$

So de Moivre's problem was to obtain a reasonable approximation for (24.5), easily usable by a person such as Cuming. De Moivre saw that he could use the power series for $\ln(1+x)$, found by Newton and Mercator in the 1660s, and Bernoulli's formula for the sums of powers of integers published in 1713, to show that

$$\ln\left(\frac{1}{2^{2m}} \binom{2m}{m}\right) \sim \left(2m - \frac{1}{2}\right) \ln(2m-1) - 2m\ln(2m) + \ln 2$$
$$+ \frac{1}{12} - \frac{1}{360} + \frac{1}{1260} - \frac{1}{1680} + \cdots. \tag{24.6}$$

The series of constants was infinite but de Moivre took only the first four terms to get 0.7739 as the value of the constant. In his *Supplement*, he wrote that the terms after the fourth constant do not decrease, and in the 1756 edition of the *Doctrine of Chances*, he remarked that the series converged, but slowly. In fact, the series diverges but the first four terms give a good approximation for the expression on the left-hand side of (24.6). Note that this happens because the series in (24.1), (24.2), and (24.6) are asymptotic series. Their true nature was not understood until the nineteenth century, but eighteenth-century mathematicians such as de Moivre, Stirling, Euler, Maclaurin, Lagrange, and Laplace had an intuitive understanding, allowing them to make effective use such series. It is very likely that de Moivre got a pretty good idea of the constant by taking some particular values of m in (24.6). Thus, he had

$$\frac{1}{2^{2m}}\binom{2m}{m} \sim \frac{2.168}{\sqrt{2m-1}}\left(1-\frac{1}{2m}\right)^{2m}. \tag{24.7}$$

Observe that for large m, $(1-1/2m)^{2m}$ is approximately $1/e$. By contrast, Stirling's approximation yields

$$\frac{1}{2^{2m}}\binom{2m}{m} \sim \frac{1}{\sqrt{\pi m}}.$$

The role of π was perceived by Stirling, whose letter of June 1729 to de Moivre again brought Cuming into the picture.

> About four years ago, when I informed Mr. *Alex. Cuming* that problems concerning the Interpolation and Summation of series and others of this type which are not susceptible to the commonly accepted analysis, can be solved by Newton's Method of Differences, the most illustrious man replied that he doubted if the problem solved by you some years before about finding the middle coefficient in an arbitrary power of the binomial could be solved by differences. Then, led by curiosity and confident that I would be doing a favour to a most deserving man of Mathematics, I took it up willingly: and I admit that difficulties arose which prevented me from arriving at the desired conclusion rapidly, but I do not regret the labour, if I have in fact finally achieved a solution which is so acceptable to you that you consider it worthy of inclusion in your own writings.

Stirling then gave two series, one for the square of $b(m, 2m, 1/2)$ in (24.4), and the other for its reciprocal:

$$\left(\frac{1}{2^{2m}}\binom{2m}{m}\right)^{-2}$$
$$= \pi m\left(1 + \frac{1}{2^2(m+1)} + \frac{(1\cdot 3)^2}{2^4 2!(m+1)(m+2)} + \frac{(1\cdot 3\cdot 5)^2}{2^6 3!(m+1)(m+2)(m+3)} + \cdots\right) \tag{24.8}$$

and

$$\left(\frac{1}{2^{2m}}\binom{2m}{m}\right)^2 = \frac{2}{\pi(2m+1)}$$
$$\times \left(1 + \frac{1}{2^2(m+\frac{3}{2})} + \frac{(1\cdot 3)^2}{2^4 2!(m+\frac{3}{2})(m+\frac{5}{2})} + \frac{(1\cdot 3\cdot 5)^2}{2^6 3!(m+\frac{3}{2})(m+\frac{5}{2})(m+\frac{7}{2})} + \cdots\right). \tag{24.9}$$

We may now recognize these series as hypergeometric; they can be computed by the summation formula by means of which Gauss proved his fundamental results for the gamma function. Stirling may have initially come upon this result by an interpolation technique described in his book. He also gave a different method, mentioned in his letter, using difference equations. When de Moivre received Stirling's communication,

24.1 Preliminary Remarks

he was astonished to see the appearance of π in this context. He searched the literature for a result that would explain this and found that Wallis's formula was just what he needed. In the *Miscellanea Analytica* de Moivre restated (24.6). His result in modern notation is

$$\ln\left(\frac{1}{2^{2m}}\binom{2m}{m}\right) = \left(2m - \frac{1}{2}\right)\ln(2m-1) - 2m\ln(2m) + \ln 2 - \frac{1}{2}\ln(2\pi)$$

$$+ 1 - \sum_{k=1}^{\infty} \frac{B_{2k}}{(2k-1)2k}\left(\frac{2}{m^{2m-1}} - \frac{1}{(2m-1)^{2m-1}}\right). \quad (24.10)$$

Stirling then published his series (24.2) for $\ln m!$. After that, de Moivre published the *Supplement*, presenting his own proof of (24.1). Just as in the case of (24.10), it was an application of three formulas: the power series for $\ln(1+x)$, the formula for $\sum_{k=1}^{n} k^m$, and Wallis's formula for π.

Soon after this, Euler and Maclaurin discovered a very general formula from which (24.1) and (24.2) could be easily derived. In his paper on the gamma function, Gauss referred to Euler's derivation. He noted that though Euler stated the result for $\ln \Gamma(x)$ when x was a positive integer, his method applied to the general case so that

$$\ln \Gamma(x+1) = \left(x + \frac{1}{2}\right)\ln x - x + \frac{1}{2}\ln 2\pi + \frac{B_2}{1\cdot 2x} + \frac{B_4}{3\cdot 4x^3} + \frac{B_6}{5\cdot 6x^5} + \text{etc.} \quad (24.11)$$

Note that Gauss denoted the absolute values of the Bernoulli numbers by $A, B, C, D, \ldots$ and used $\Pi(x)$ to mean $\Gamma(x+1)$. Euler also proved that (24.1), (24.2) and more generally (24.11) were divergent. This followed from his beautiful formula for Bernoulli numbers

$$B_{2k} = \frac{(-1)^{k-1}(2k)!}{2^{2k-1}\pi^{2k}}\left(1 + \frac{1}{2^{2k}} + \frac{1}{3^{2k}} + \cdots\right) = O\left(\left(\frac{k}{\pi}\right)^{2k}\sqrt{k}\right). \quad (24.12)$$

Gauss, with his desire for complete mathematical rigor, declared that it was important to determine why a divergent series gave an excellent approximation when only the first few terms were used. He also pointed out that the series (24.1) and (24.2) of de Moivre and Stirling, in their more general form for $\ln \Gamma(x+1)$, could be obtained from each other by the duplication formula for the gamma function.

In 1843, Cauchy gave an explanation of the peculiar nature of the series (24.11). He proved that

$$\mu(x) = \ln \Gamma(x) - (x - 1/2)\ln x + x - \frac{1}{2}\ln 2\pi$$

$$= \sum_{k=1}^{m} \frac{B_{2k}}{(2k-1)2kx^{2k-1}} + \frac{\theta B_{2m+2}}{(2m+1)(2m+2)x^{2m+1}}, \text{ where } 0 < \theta < 1. \quad (24.13)$$

This means that when m terms of the series (24.11) are used, the error is less than the $(m+1)$th term and has the same sign as that term, determined by the sign of B_{2m+2}. Thus, the eighteenth-century mathematicians had the judgment to choose an ideal stopping point in the series for their numerical calculations. Although Poisson and Jacobi had already proved Cauchy's result in a more general situation, they did not specifically note this important particular case. It is also possible that Cauchy wished to show how an integral representation for $\mu(x)$ in (24.13), due to his friend Binet, could be used for the proof of (24.13).

Jacques Binet (1786–1856) studied at the École Polytechnique from 1804 to 1806 and returned to the institution as an instructor in 1807. His main interests were astronomy and optics, though he contributed some important papers in mathematics. He was a good friend of Cauchy, and in 1812 the two generalized some results on determinants and took the subject to a higher level of generality. In particular, Binet stated the multiplication theorem in more general terms, so that his work can be taken as an early discussion of the product of two rectangular matrices. In an 1839 paper of over 200 pages, Binet gave two integral representations for $\mu(x)$ in (24.13). These are now called Binet's formulas. In applying integrals to study the gamma function, Binet was following the trend of the 1830s. Thus, he used Euler's formula for the beta integral $\int_0^1 x^{m-1}(1-x)^{n-1} dx$ to prove Stirling's formulas (24.8) and (24.9).

Although the asymptotic series (24.11) and similar series were used frequently after 1850, it was in 1886 that Henri Poincaré gave a formal definition. He noted that it was well known to geometers that if S_n denoted the terms of the series for $\ln \Gamma(x+1)$ up to and including $(B_{2n}/2n(2n-1))1/x^{2n}$, then the expression $x^{2n+1}(\ln \Gamma(x+1) - S_n)$ tended to 0 when x increased indefinitely. He then defined an asymptotic series:

I say that a *divergent series*

$$A_0 + \frac{A_1}{x} + \frac{A_2}{x^2} + \cdots + \frac{A_n}{x^n} + \cdots,$$

where the sum of the first $n+1$ terms is S_n, *asymptotically* represents a function $J(x)$ if the expression $x^n(J - S_n)$ tends to 0 when x increases indefinitely.

He showed that asymptotic series behaved well under the algebraic operations of addition, subtraction, multiplication, and division. Term-by-term integration of an asymptotic series also worked, but not differentiation. Poincaré noted that the theory remained unchanged if one supposed that x tended to infinity radially (in the complex plane) with a fixed nonzero argument. However, a divergent series could not represent one and the same function J in all directions of radial approach to infinity. He also observed that the same series could represent more than one function asymptotically. Poincaré applied his theory of asymptotic series to the solution of differential equations, though the British mathematician George Stokes developed some of these ideas earlier, in the 1850s and 1860s, in connection with Bessel's equation.

The Dutch mathematician Thomas Joannes Stieltjes (1856–1894) also developed a theory of asymptotic series; following Legendre, he labeled it semiconvergent series. Stieltjes's paper appeared in the same year as Poincaré's, 1886. Then in 1889, Stieltjes extended formula (24.2) to the slit complex plane $\mathbb{C}^- = \mathbb{C} \setminus (-\infty, 0]$. Until then, the formula was known to hold only in the right half-plane. He accomplished this extension by a systematic use of the formula

$$\mu(z) = \int_0^\infty \frac{t - [t] - 1/2}{z + t} dt, \ z \in \mathbb{C}^-, \qquad (24.14)$$

where μ was defined by (24.13).

24.2 De Moivre's Asymptotic Series

De Moivre's derivation of (24.1) in the *Supplement* started with

$$\ln \frac{m^{m-1}}{(m-1)!} = \ln \prod_{k=1}^{m-1} \left(1 - \frac{k}{m}\right)^{-1} = -\sum_{k=1}^{m-1} \ln\left(1 - \frac{k}{m}\right)$$

$$= \sum_{k=1}^{m-1} \sum_{n=1}^\infty \frac{k^n}{nm^n} = \sum_{n=1}^\infty \frac{1}{nm^n} \sum_{k=1}^{m-1} k^n. \qquad (24.15)$$

We remark that de Moivre did not use the summation or factorial notation. He effected the change in the order of summation by writing the series for $\ln(1 - \frac{k}{m})$ in rows, for some values of k, and then summing the columns. He then reproduced Jakob Bernoulli's table for the sums of powers of integers and applied it to the inner sum in the last expression of (24.15). In modern notation, Bernoulli's formula is

$$\sum_{k=1}^{m-1} k^n = \frac{(m-1)^{n+1}}{n+1} + \frac{1}{2}(m-1)^n$$

$$+ \binom{n}{1} \frac{B_2}{2}(m-1)^{n-1} + \binom{n}{3} \frac{B_4}{4}(m-1)^{n-3} + \cdots. \qquad (24.16)$$

Thus, de Moivre had the equation

$$\ln \frac{m^{m-1}}{(m-1)!} = \frac{1}{m}\left(\frac{(m-1)^2}{2} + \frac{m-1}{2}\right) + \frac{1}{2mm}\left(\frac{(m-1)^3}{3} + \frac{(m-1)^2}{2} + \frac{m-1}{6}\right)$$

$$+ \frac{1}{3m^3}\left(\frac{(m-1)^4}{4} + \frac{(m-1)^3}{2} + 3B_2\frac{(m-1)^2}{2}\right)$$

$$+ \frac{1}{4m^4}\left(\frac{(m-1)^5}{5} + \frac{(m-1)^4}{2} + 4B_2\frac{(m-1)^3}{2} + 4B_4\frac{(m-1)}{4}\right) + \cdots.$$

He then set $x = \frac{m-1}{m}$ and changed the order of summation to get

$$\ln \frac{m^{m-1}}{(m-1)!} = m\left(\frac{x^2}{2} + \frac{x^3}{6} + \frac{x^4}{12} + \cdots\right) + \frac{1}{2}\left(x + \frac{x^2}{2} + \frac{x^3}{3} + \cdots\right)$$

$$+ \frac{B_2}{2m}\left(x + x^2 + x^3 + \cdots\right) + \frac{B_4}{4m^3}\left(x + \frac{4}{2}x^2 + \frac{5 \cdot 4}{3 \cdot 2}x^3 + \cdots\right) + \cdots. \qquad (24.17)$$

The general term, left unexpressed by de Moivre, as was the common practice, would have been

$$\frac{B_{2r}}{2rm^{2r-1}}\left(\binom{2r}{1}\frac{x}{2r} + \binom{2r+1}{2}\frac{x^2}{2r+1} + \binom{2r+2}{3}\frac{x^3}{2r+2} + \cdots\right)$$

$$= \frac{B_{2r}}{2r(2r-1)m^{2r-1}}\left(\binom{2r-1}{1}x + \binom{2r}{2}x^2 + \binom{2r+1}{3}x^3 + \cdots\right)$$

$$= \frac{B_{2r}}{2r(2r-1)m^{2r-1}}\left((1-x)^{-2r+1} - 1\right)$$

$$= \frac{B_{2r}}{2r(2r-1)}\left(1 - \frac{1}{m^{2r-1}}\right). \tag{24.18}$$

The second series in (24.17) summed to

$$-\ln(1-x) = -\ln\left(1 - \frac{m-1}{m}\right) = \ln m,$$

and the first series turned out to be the integral of this series, equal to

$$(1-x)\ln(1-x) + x = \frac{m - 1 - \ln m}{m}.$$

This computation involves integration by parts, and it is interesting to see how de Moivre handled it. He used Newton's notation for fluxions, as was only natural since de Moivre worked in England and was Newton's friend. He set

$$v = \ln\frac{1}{1-x} = x + \frac{x^2}{2} + \frac{x^3}{3} + \frac{x^4}{4} + \cdots.$$

Then $v\dot{x} = x\dot{x} + \frac{1}{2}x^2\dot{x} + \frac{1}{3}x^3\dot{x} + \frac{1}{4}x^4\dot{x} + \cdots$

$F.v\dot{x} = \frac{1}{2}x^2 + \frac{1}{6}x^3 + \frac{1}{12}x^4 + \frac{1}{20}x^5 + \cdots.$ (F. = fluent = integral).

He then set

$$q = vx - F.v\dot{x} = x^2 + \frac{1}{2}x^3 + \frac{1}{3}x^4 + \frac{1}{4}x^5 + \cdots - \left(\frac{1}{2}x^2 + \frac{1}{6}x^3 + \frac{1}{12}x^4 + \cdots\right)$$

so that

$$\dot{q} = \dot{x}\left(2x + \frac{3}{2}x^2 + \frac{4}{3}x^3 + \frac{5}{4}x^4 + \cdots\right) - \dot{x}\left(x + \frac{1}{2}x^2 + \frac{1}{3}x^3 + \frac{1}{4}x^4 + \cdots\right)$$

$$= \frac{\dot{x}x}{1-x} = -\dot{x} + \dot{v}.$$

Therefore $q = -x + v$ and $F.v\dot{x} = vx - q = \frac{m-1-\ln m}{m}$. Using the above simplifications, (24.17) became

$$\ln \frac{m^{m-1}}{(m-1)!} = (m-1) - \ln m + \frac{1}{2} \ln m$$
$$+ \frac{B_2}{2}\left(1 - \frac{1}{m}\right) + \frac{B_4}{3 \cdot 4}\left(1 - \frac{1}{m^3}\right) + \frac{B_6}{5 \cdot 6}\left(1 - \frac{1}{m^5}\right) + \cdots$$
$$= (m-1) - \frac{1}{2}\ln m + \frac{1}{12} - \frac{1}{360} + \frac{1}{1260} - \frac{1}{1680} + \cdots$$
$$- \frac{1}{12m} + \frac{1}{360m^3} - \frac{1}{1260m^5} + \frac{1}{1680m^7} - \cdots.$$

After adding $\ln m$ to each side and rearranging terms, de Moivre had

$$\ln m! = \left(m + \frac{1}{2}\right)\ln m - m + 1 - \frac{1}{12} + \frac{1}{360} - \frac{1}{1260} + \frac{1}{1680} - \cdots$$
$$+ \frac{1}{12m} - \frac{1}{360m^3} + \frac{1}{1260m^5} - \frac{1}{1680m^7} + \cdots. \quad (24.19)$$

De Moivre remarked that the constant in this equation could be quickly computed by taking $m = 2$. In that case,

$$C = 1 - \frac{1}{12} + \frac{1}{360} - \frac{1}{1260} + \frac{1}{1680} - \cdots$$
$$= 2 - \frac{3}{2}\ln 2 - \frac{1}{12 \times 2} + \frac{1}{360 \times 8} - \frac{1}{1260 \times 32} + \frac{1}{1680 \times 128} - \cdots. \quad (24.20)$$

As we have noted before, the two series here are divergent but the terms as written down by de Moivre gave a good approximation for C. After learning of Stirling's result on the asymptotic value of $\frac{1}{2^{2m}}\binom{2m}{m}$, de Moivre realized that $C = \frac{1}{2}\ln(2\pi)$, and he proved it using Wallis's formula. Stirling and de Moivre's derivations for the value of C were identical; we present the details in the next section.

24.3 Stirling's Asymptotic Series

Stirling's *Methodus Differentialis* gave several ingenious applications of difference equations. His derivations of (24.8) and (24.9) were probably his most imaginative use of difference equations. It is obvious from his letter to de Moivre that he was quite proud of his solutions. We give details of this work and derive equation (24.2). Proposition 23 of Stirling's book states the problem: to find the ratio of the middle coefficient to the sum of all coefficients in any power of the binomial. Stirling observed that the sequence $1, 2, \frac{8}{3}, \frac{16}{5}, \frac{128}{35}, \ldots$ or $2^{2m} \div \binom{2m}{m}$, $m = 0, 1, 2, \ldots$ satisfied the relation $T' = (n+2)T/(n+1)$, where $n = 2m = 0, 2, 4, \ldots$ and T' denoted the term after T.

So if T was the nth term, T' would be obtained by changing n to $n+2$ in T. In modern notation, $T_{n+2} = \frac{n+2}{n+1} T_n$. Stirling rewrote this relation after squaring it:

$$2T'^2 + (n+2)(T^2 - T'^2) - \frac{T'^2}{n+2} = 0. \tag{24.21}$$

This is the difference equation into which Stirling substituted an inverse factorial series to solve his problem. Since he had so many difference equations from which to choose, it is hard to discern how Stirling was guided to this one; it worked very successfully. Stirling first took

$$T^2 = An + \frac{Bn}{n+2} + \frac{Cn}{(n+2)(n+4)} + \frac{Dn}{(n+2)(n+4)(n+6)} + \cdots$$
$$= An + B + \frac{C - 2B}{n+2} + \frac{D - 4C}{(n+2)(n+4)} + \cdots. \tag{24.22}$$

Then

$$T'^2 = A(n+2) + B + \frac{C - 2B}{n+4} + \frac{D - 4C}{(n+4)(n+6)} + \cdots, \tag{24.23}$$

so that

$$(n+2)(T^2 - T'^2) = -2A(n+2) + \frac{2C - 4B}{n+4} + \frac{4D - 16C}{(n+4)(n+6)} + \cdots. \tag{24.24}$$

It followed from (24.22) after replacing T by T' and n by $n+2$ that

$$\frac{T'^2}{n+2} = A + \frac{B}{n+4} + \frac{C}{(n+4)(n+6)} + \frac{D}{(n+4)(n+6)(n+8)} + \cdots.$$

He used the three series (24.22), (24.23), and (24.24) in (24.21), and the result was

$$2B - A + \frac{4C - 9B}{n+4} + \frac{6D - 25C}{(n+4)(n+6)} + \frac{8E - 49D}{(n+4)(n+6)(n+8)} + \cdots = 0.$$

This implied the relations

$$2B - A = 0, \quad 4C - 9B = 0, \quad 6D - 25C = 0, \quad 8E - 49D = 0,$$

and these in turn implied the series in (24.8) except for the value of A. As was the practice among the eighteenth-century mathematicians, Stirling computed only the first few coefficients $B, C, D, \ldots$ and gave no expression for the general term. To find A, he argued that by (24.22) for large n, $T^2 = An$. Stirling then made an application of Wallis's formula, to which he referred in his exposition. For $n = 2m$,

$$\frac{T^2}{n} = \frac{1}{2m} \left(\frac{1}{2^{2m}} \binom{2m}{m} \right)^{-2}$$
$$= \frac{2^2 \times 4^2 \times \cdots \times (2m)^2}{3^2 \times 5^2 \times \cdots \times (2m-1)^2 \times 2m+1} \cdot \frac{2m+1}{2m} \to \frac{\pi}{2} \text{ as } n \to \infty.$$

24.3 Stirling's Asymptotic Series

First in his *Miscellanea* and then again in the 1756 edition of his *Doctrine of Chances*, de Moivre praised Stirling for his introduction of π in the asymptotic series for $\ln n!$. In the latter work he wrote, "I own with pleasure that this discovery, besides that it saved trouble, has spread a singular Elegancy on the Solution."

Stirling's series (24.2) for $\ln m!$ was a corollary of his main result, contained in proposition 28 of his book. The purpose of the proposition was to find the sum of any number of logarithms, whose arguments were in arithmetic progression. He denoted the progression by $x+n, x+2n, x+3n, \ldots, z-n$, where the logarithms were taken base 10. Since $\log_{10} x = \ln x / \ln 10$, he defined the number $a = 1/\ln 10$ and gave the approximate value of a to be 0.43429,44819,03252. To state his result, Stirling began with the series

$$f(z) = \frac{z}{2n} \log_{10} z - \frac{a}{2n} z + a A_1 \frac{n}{z} + a A_2 \frac{n^3}{z^3} + a A_3 \frac{n^5}{z^5} + a A_4 \frac{n^7}{z^7} + \cdots, \quad (24.25)$$

where the numbers $A_1, A_2, A_3, \ldots$ were such that

$$\sum_{k=1}^{m} \binom{2m-1}{2k-2} A_k = -\frac{1}{4m(2m+1)}. \quad (24.26)$$

He had the values $A_1 = -\frac{1}{12}$, $A_2 = \frac{7}{360}$, $A_3 = -\frac{31}{1260}$, $A_4 = \frac{127}{1680}$, and $A_5 = -\frac{511}{1188}$. In fact, one can show that

$$A_k = -\frac{(2^{2k-1}-1)B_{2k}}{2k(2k-1)}.$$

Stirling may not have recognized this connection with Bernoulli numbers when he discovered his result on $\ln m!$. Later, after reading de Moivre's book, where Bernoulli numbers were explicitly mentioned, Stirling investigated properties of these numbers and discussed them in some unpublished notes. Stirling's main theorem was that

$$\log_{10}((x+n)(x+3n)(x+5n) \cdots (z-n)) = f(z) - f(x), \quad (24.27)$$

where $f(z)$ was the series defined in (24.25). His proof consisted in observing that

$$f(z) - f(z-2n) = \log_{10} z - a \left(\frac{n}{z} + \frac{1}{2} \left(\frac{n}{z} \right)^2 + \frac{1}{3} \left(\frac{n}{z} \right)^3 + \cdots \right)$$

$$= \log_{10} z - \log_{10} \left(1 - \frac{n}{z} \right) = \log_{10}(z-n). \quad (24.28)$$

Stirling apparently left the verification of this equation to the reader; an outline of the proof appears in the exercises. He made the remark that the terms in $f(z)$ and $f(z-2n)$ had first to be reduced to the same form. The theorem follows immediately from (24.28):

$$f(z) - f(x) = (f(z) - f(z-2n))$$
$$+ (f(z-2n) - f(z-4n)) + \cdots + (f(x+2n) - f(x))$$
$$= \log_{10}(z-n) + \log_{10}(z-3n) + \cdots + \log_{10}(x+n).$$

Stirling applied his theorem to derive his series for $\log_{10} m!$ by taking $x = 1/2$, $n = 1/2$, and $z = m + 1/2$. From this he had

$$f(m+1/2) = (m+\frac{1}{2})\log_{10}(m+\frac{1}{2}) - a(m+\frac{1}{2}) - \frac{a}{24(m+\frac{1}{2})} + \frac{7a}{2880(m+\frac{1}{2})^3} - \text{etc.}$$

Next, by (24.27), $\log_{10} m! = f(m+1/2) - f(1/2)$. Stirling wrote that

$$-f(1/2) = \frac{1}{2}\log_{10} 2\pi \approx 0.39908, 99341, 79,$$

but he did not explain how he arrived at $\frac{1}{2}\log_{10} 2\pi$. Perhaps he numerically computed $\log_{10} m! - f(m+1/2)$ for a large enough value of m and noticed that he had half the value of $\log_{10} 2\pi$; he must have been very familiar with this value, based on his extensive numerical calculations. Recall that he had recognized $\sqrt{\pi}$ from its numerical value when computing $(1/2)!$. Of course, he could also have provided a proof using Wallis's formula, as he did in the situation discussed above.

24.4 Binet's Integrals for $\ln \Gamma(x)$

Binet knew that Stirling's two series (24.8) and (24.9) could be derived by Gauss's summation formula. In addition, he gave an interesting proof of (24.8) using integrals:

$$B\left(m, \frac{1}{2}\right) = \int_0^1 x^{m-1}(1-x)^{-1/2} dx = \int_0^1 x^{m-\frac{1}{2}}(1-x)^{-1/2}(1-(1-x))^{-1/2} dx$$

$$= \int_0^1 x^{m-1/2}(1-x)^{-1/2}\left(1 + \frac{\frac{1}{2}}{1!}(1-x) + \frac{\frac{1}{2} \cdot \frac{3}{2}}{2!}(1-x)^2 + \cdots\right) dx$$

$$= B\left(m+\frac{1}{2}, \frac{1}{2}\right) + \frac{\frac{1}{2}}{1!}B\left(m+\frac{1}{2}, \frac{3}{2}\right) + \frac{\frac{1}{2} \cdot \frac{3}{2}}{2!}B\left(m+\frac{1}{2}, \frac{5}{2}\right) + \cdots. \tag{24.29}$$

When Euler's formula, $B(x, y) = \Gamma(x)\Gamma(y)/\Gamma(x+y)$, was applied in (24.29), Stirling's formula (24.8) followed. Binet's proof of (24.9) ran along similar lines. He also gave two integral representations for

$$\mu(x) = \ln \Gamma(x) - \left(x - \frac{1}{2}\right)\ln x + x - \frac{1}{2}\ln 2\pi. \tag{24.30}$$

These useful representations were:

$$\mu(x) = \int_0^\infty \left(\frac{1}{2} - \frac{1}{t} + \frac{1}{e^t - 1}\right)\frac{e^{-xt}}{t} dt, \tag{24.31}$$

$$\mu(x) = 2\int_0^\infty \frac{\arctan(t/x)}{e^{2\pi t} - 1} dt. \tag{24.32}$$

24.4 Binet's Integrals for $\ln \Gamma(x)$

Binet demonstrated the equality of the two expressions for $\mu(x)$ by using the two formulas:

$$\int_0^\infty e^{-sy} \sin(ty)\, dy = \frac{t}{t^2 + s^2}, \tag{24.33}$$

$$4\int_0^\infty \frac{\sin(ty)}{e^{2\pi y} - 1}\, dy = \frac{e^t + 1}{e^t - 1} - \frac{2}{t}. \tag{24.34}$$

He attributed the first of these to Euler and the second to Poisson. He multiplied the second equation by $e^{-st}\, dt$, integrated over $(0, \infty)$, and then used the first integral to get

$$\int_0^\infty e^{-st}\left(\frac{e^t + 1}{e^t - 1} - \frac{2}{t}\right) dt = 4\int_0^\infty \frac{t\, dt}{(t^2 + s^2)(e^{2\pi t} - 1)}.$$

Binet then integrated both sides of this equation with respect to s over the interval (x, ∞), and changed the order of integration to obtain

$$\int_0^\infty \left(\frac{1}{2} - \frac{1}{t} + \frac{1}{e^t - 1}\right) \frac{e^{-xt}}{t}\, dt = 2\int_0^\infty \frac{\arctan(t/x)}{e^{2\pi t} - 1}\, dt.$$

Thus, it was sufficient to prove one of the formulas for $\mu(x)$, and Binet proved the first one, starting from the definition of $\mu(x)$:

$$\mu(x+1) - \mu(x) = \left(x + \frac{1}{2}\right)\ln\left(1 - \frac{1}{x+1}\right) + 1$$

$$= -\sum \frac{(n-1)}{2n(n+1)(x+1)^n},$$

or $2\mu(x) = 2\mu(x+1) + \sum \frac{n-1}{n(n+1)(x+1)^n}.$

By Stirling's approximation, he had $\mu(x) \to 0$ as $x \to \infty$, and hence

$$2\mu(x) = \frac{1}{2\cdot 3}\sum_{k=1}^\infty \frac{1}{(x+k)^2} + \frac{2}{3\cdot 4}\sum_{k=1}^\infty \frac{1}{(x+k)^3} + \frac{3}{4\cdot 5}\sum_{k=1}^\infty \frac{1}{(x+k)^4} + \text{etc.}$$

By Euler's gamma integral

$$\frac{\Gamma(n+1)}{(k+x)^{n+1}} = \int_0^\infty t^n e^{-t(k+x)}\, dt, \tag{24.35}$$

and therefore

$$\Gamma(n+1) \sum_{k=1}^\infty \frac{1}{(x+k)^{n+1}} = \int_0^\infty t^n \left(e^{-t(x+1)} + e^{-t(x+2)} + e^{-t(x+3)} + \cdots\right) dt$$

$$= \int_0^\infty \frac{t^n e^{-xt}}{e^t - 1}\, dt.$$

He then wrote

$$2\mu(x) = \int_0^\infty \frac{e^{-xt}}{e^t - 1} \left(\frac{t}{2 \cdot 3} + \frac{2t^2}{2 \cdot 3 \cdot 4} + \frac{3t^3}{2 \cdot 3 \cdot 4 \cdot 5} + \cdots \right) dt$$

and an easy calculation showed that the sum of the series inside the parentheses would be

$$(e^t - 1) \left(\frac{1}{t} - \frac{2}{t^2} \right) + \frac{2}{t}.$$

This completes Binet's ingenious proof of his formulas.

De Moivre's form of the asymptotic series for $\ln \Gamma(x)$ can be obtained from Binet's integrals. Start with Euler's generating function for Bernoulli numbers,

$$\frac{t}{e^t - 1} = 1 - \frac{1}{2}t + \sum_{n=1}^\infty \frac{B_{2n}}{(2n)!} t^{2n}.$$

It follows easily that the integrand in the first integral for $\mu(x)$ is

$$\frac{1}{t} \left(\frac{1}{e^t - 1} - \frac{1}{t} + \frac{1}{2} \right) = \sum_{n=0}^\infty \frac{B_{2n+2}}{(2n+2)!} t^{2n}. \tag{24.36}$$

We substitute this series in the integrand and integrate term by term; an application of (24.35) then yields de Moivre's asymptotic series. Unfortunately, however, this last operation is invalid because the series (24.36) is convergent only for $|t| < 2\pi$, whereas we are integrating on $(0, \infty)$.

24.5 Cauchy's Proof of the Asymptotic Character of de Moivre's Series

In an 1843 paper in the *Comptes Rendus*, Cauchy proved that de Moivre's series, though divergent, was asymptotic and hence useful for computation. Cauchy eliminated the convergence difficulty of the infinite series (24.36) by deriving its finite form:

$$\frac{1}{t} \left(\frac{1}{e^t - 1} - \frac{1}{t} + \frac{1}{2} \right) = \sum_{n=0}^{m-1} \frac{B_{2n+2}}{(2n+2)!} t^{2n} + \frac{\theta B_{2m+2}}{(2m+2)!} t^{2m}, \quad 0 < \theta < 1. \tag{24.37}$$

Cauchy's important formula (24.13) followed immediately when (24.37) was applied in Binet's integral for $\mu(x)$. This proved that de Moivre's series for $\ln \Gamma(x)$ was an asymptotic series. Cauchy first proved that the left-hand side of (24.37) had the partial fractions expansion

$$\frac{1}{t} \left(\frac{1}{e^t - 1} - \frac{1}{t} + \frac{1}{2} \right)$$
$$= 2 \left(\frac{1}{t^2 + (2\pi)^2} + \frac{1}{t^2 + (4\pi)^2} + \frac{1}{t^2 + (6\pi)^2} + \cdots \right). \tag{24.38}$$

He indicated a proof of this, although Euler had already found and proved the result. Cauchy then noted that for $u > 0$,

$$\frac{1}{1+u} = 1 - u + u^2 - u^3 + \cdots \mp u^{m-1} \pm \frac{u^m}{1+u}$$

$$= 1 - u + u^2 - u^3 + \cdots \mp u^{m-1} \pm \theta u^m, \ 0 < \theta < 1.$$

Therefore,

$$\frac{1}{t^2 + (2k\pi)^2} = \frac{1}{(2k\pi)^2} \cdot \frac{1}{1+(t/2k\pi)^2} = \sum_{n=0}^{m-1} \frac{(-t^2)^n}{(2k\pi)^{2n+2}} + \frac{(-1)^m \theta t^{2m}}{(2k\pi)^{2m+2}}.$$

Cauchy then substituted this in (24.38) and applied Euler's formula for Bernoulli numbers (24.12) to get (24.37). When the latter series was used in Binet's formula (24.31), the asymptotic character of de Moivre's series became evident.

24.6 Exercises

1. Prove that if $S_m = \sqrt{2\pi} \left(\frac{m+1/2}{e}\right)^{m+1/2}$ and $D_m = \sqrt{2\pi m} \left(\frac{m}{e}\right)^m$, then

$$\lim_{m \to \infty} \frac{S_m - m!}{m! - D_m} = \frac{1}{2}.$$

This shows that (24.3), implied by Stirling's series, is a better approximation for $m!$ than D_m, resulting from de Moivre's series but called Stirling's approximation. Note that S_m gives values larger than $m!$, while D_m underestimates $m!$. See Tweddle (1984).

2. Prove that

$$\int_0^1 \ln \Gamma(x+u) \, du = x \ln x - x + \frac{1}{2} \ln 2\pi \tag{24.39}$$

by the following methods:

(a) Observe that the integral is equal to

$$\lim_{n \to \infty} \frac{1}{n} \left(\ln \Gamma(x) + \ln \Gamma\left(x + \frac{1}{n}\right) + \cdots + \ln \Gamma\left(x + \frac{n-1}{n}\right) \right).$$

Apply Gauss's multiplication formula and Stirling's approximation.

(b) Apply Euler's reflection formula $\Gamma(x)\Gamma(1-x) = \frac{\pi}{\sin \pi x}$ to compute the limit

$$\int_0^1 \ln \Gamma(u) \, du = \lim_{n \to \infty} \frac{1}{n} \left(\ln \Gamma\left(\frac{1}{n}\right) + \ln \Gamma\left(\frac{2}{n}\right) + \cdots + \ln \Gamma\left(\frac{n-1}{n}\right) \right)$$

$$= \frac{1}{2} \ln 2\pi. \tag{24.40}$$

Take the derivative of (24.39) with respect to x to show that

$$\int_0^1 \ln\Gamma(u+x)\,du = x\ln x - x + \int_0^1 \ln\Gamma(u)\,du.$$

(c) Denote the integral in (24.40) by C and show that

$$2C = -\int_0^1 f(u)\,du \quad \text{where} \quad f(x) = \ln\left(\frac{\sin\pi u}{u}\right).$$

Show that $C = \frac{1}{2}\ln 2\pi$ by proving (i) $\int_0^1 f(u)\,du = \int_0^1 f\left(\frac{u}{2}\right)\,du$ and (ii) $f(u) = f\left(\frac{u}{2}\right) + f\left(\frac{1-u}{2}\right) + \ln 2\pi$.

The proofs in (a) and (b) were published by Stieltjes in 1878. See Stieltjes (1993), vol. 1, pp. 114–18. The proof in (c) was attributed to Mathias Lerch by Hermite in his 1891 lectures at the École Normale. See Hermite (1891).

3. Integrate Gauss's formula for $\frac{\Gamma'(1+y)}{\Gamma(1+y)}$ to obtain Plana's formula

$$\ln\Gamma(u) = \int_0^\infty \left(\frac{1-e^{(1-u)x}}{e^x - 1} + (u-1)e^{-x}\right)\frac{dx}{x}.$$

Then, by another integration, show that

$$J \equiv \int_a^{a+1} \ln\Gamma(u)\,du = \int_0^\infty \left(\frac{e^{-ax}}{x} + \frac{e^{-x}}{e^{-x}-1} - \left(a - \frac{1}{2}\right)e^{-x}\right)\frac{dx}{x}.$$

Deduce that

$$\ln\Gamma(a) = J - \frac{1}{2}\ln a + \int_0^\infty \left(\frac{1}{2} - \frac{1}{x} + \frac{1}{e^x - 1}\right)\frac{e^{-ax}}{x}\,dx.$$

This is Binet's formula (24.31) after the value of J is substituted from exercise 2. This proof of Binet's formula is from Hermite (1891).

4. Prove the formulas used by Binet:

$$\int_0^\infty e^{-xy}\sin(ty)\,dy = \frac{t}{t^2 + x^2}$$

$$4\int_0^\infty \frac{\sin(ty)}{e^{2\pi} - 1}\,dy = \frac{e^t + 1}{e^t - 1} - \frac{2}{t}.$$

5. Let $n = 2m$ and $Y_n = \left(\frac{1}{2^{2m}}\binom{2m}{m}\right)^2$. Prove the recurrence relation

$$(n+1)(n+3)(Y_n - Y_{n+2}) - 2(n+1)Y_n - Y_{n+2} = 0, \ n = 0, 2, 4, 6, \ldots.$$

Assume

$$Y_n = \frac{\alpha_1}{n+1} + \frac{\alpha_2}{(n+1)(n+3)} + \frac{\alpha_3}{(n+1)(n+3)(n+5)} + \cdots.$$

Then employ the recurrence relation to prove that $(2k-2)\alpha_k = (2k-3)^2\alpha_{k-1}$, $k = 2, 3, 4, \ldots$. Finally, use Wallis's formula to show that $\alpha_1 = 1/\pi$. This is Stirling's formal proof of (24.9) and is very similar to his proof of (24.8) given in the text.

6. Obtain Binet's proof of (24.9) by observing that

$$B\left(m+\frac{1}{2},\frac{1}{2}\right) = \int_0^1 x^m(1-x)^{-1/2}(1-(1-x))^{-1/2}dx,$$

and following his argument for (24.8), given in the text.

7. Note that

$$\ln\left(\frac{1}{2^{2m}}\binom{2m}{m}\right) = (-2m+1)\ln 2 + \sum_{k=1}^{m-1}\ln\frac{1+k/m}{1-k/m}.$$

Now apply de Moivre's method from the *Miscellanea Analytica*, given in the text, to obtain (24.10).

8. Prove the formula (24.38) used by Cauchy in his derivation of the remainder in the series for $\mu(n)$. Cauchy started with the infinite product for $\sinh(t/2)$, due to Euler, and took the logarithmic derivative. In fact, Euler was aware of this result and this proof.

9. In the *Methodus*, Stirling gave two more formulas for $b_m = \binom{2m}{m}$:

$$\left(\frac{2^{2m}}{b_m}\right)^2 = \frac{\pi}{2}(2m+1)\left(1 - \frac{1^2}{2(2m-3)} + \frac{1^2 \cdot 3^2}{2 \cdot 4(2m-3)(2m-5)} - \text{etc.}\right),$$

$$\left(\frac{b_m}{2^{2m}}\right)^2 = \frac{1}{\pi m}\left(1 - \frac{1^2}{2(2m-2)} + \frac{1^2 \cdot 3^2}{2 \cdot 4(2m-2)(2m-4)} - \text{etc.}\right).$$

In his analysis of Stirling's work, Binet pointed out that these formulas were incorrect and should be replaced by

$$\left(\frac{2^{2m}}{b_m}\right)^2 = \frac{\pi}{2}(2m+1)\sum_{k=0}^{\infty}\frac{\left(\frac{1}{2}\right)_k\left(-\frac{1}{2}\right)_k}{k!(m+1)_k}$$

$$\left(\frac{b_m}{2^{2m}}\right)^2 = \frac{1}{\pi m}\sum_{k=0}^{\infty}\frac{\left(\frac{1}{2}\right)_k\left(-\frac{1}{2}\right)_k}{k!\left(m+\frac{1}{2}\right)_k},$$

where $(a)_k = a(a+1)\cdots(a+k-1)$. Prove Binet's formulas. See Binet (1839), pp. 319–320. For an analysis of Stirling's results, see Tweddle (2003).

10. From de Moivre's version (24.11), use Legendre's duplication formula $\sqrt{\pi}\,\Gamma(2x) = 2^{2x-1}\Gamma(x)\Gamma(x+1/2)$ to obtain Stirling's series

$$\ln\Gamma(x+1) = \left(x+\frac{1}{2}\right)\ln\left(x+\frac{1}{2}\right) - \left(x+\frac{1}{2}\right)$$
$$+ \frac{1}{2}\ln 2\pi - \frac{B_2}{4\left(x+\frac{1}{2}\right)} + \frac{7B_4}{96\left(x+\frac{1}{2}\right)^3} - \text{etc.} \quad (24.41)$$

See Gauss (1863–1927), vol. III, p. 152.

11. Prove that if $S(n,k)$ denotes the Stirling numbers of the second kind, then

$$\sum_{k=1}^{n} \frac{(-1)^k k!}{k+1} S(n,k) = B_n, \ n = 1, 2, 3, \ldots.$$

See Tweddle (1988), p. 16, for the reference to unpublished work of Stirling containing this result.

12. Prove that if $A_1, A_2, A_3, \ldots$ is a sequence of rational numbers satisfying

$$\sum_{k=1}^{n} \binom{2n-1}{2k-2} A_k = -\frac{1}{4n(2n+1)}, \ n = 1, 2, 3, \ldots,$$

$$\text{then} \quad A_k = -\frac{(2^{2k-1}-1)B_{2k}}{2k(2k-1)}.$$

See Tweddle (1988).

13. Prove Stirling's result (24.28) as follows: First set

$$f(z) = \frac{z}{2n}\log_{10} z - \frac{az}{2n} + a\sum_{k=1}^{\infty} A_k \left(\frac{n}{z}\right)^{2k-1}.$$

Then one obtains

$$f(z) - f(z-2n) = \frac{z\log_{10} z}{2n} - \frac{(z-2n)\log_{10}(z-2n)}{2n} - a$$
$$+ a\sum_{k=1}^{\infty} A_k\left(\left(\frac{n}{z}\right)^{2k-1} - \left(\frac{n}{z-2n}\right)^{2k-1}\right). \quad (24.42)$$

Note that

$$\log_{10} z = \log_{10}\left((z-n)\left(1+\frac{n}{z-n}\right)\right) \text{ and}$$

$$\log_{10}(z-2n) = \log_{10}\left((z-n)\left(1-\frac{n}{z-n}\right)\right),$$

and show that

$$\frac{z\log_{10} z}{2n} - \frac{(z-2n)\log_{10}(z-2n)}{2n} - a$$
$$= \log_{10}(z-n) - a \sum_{k=1}^{\infty} \frac{1}{2k(2k+1)} \left(\frac{n}{z-n}\right)^{2k}. \quad (24.43)$$

Next, take $\frac{n}{z} = \frac{n}{z-n}\left(1 + \frac{n}{z-n}\right)^{-1}$ and $\frac{n}{z-2n} = \frac{n}{z-n}\left(1 - \frac{n}{z-n}\right)^{-1}$ in (24.42) and use

$$\sum_{s=1}^{k} \binom{2k-1}{2s-2} A_s = -\frac{1}{4k(2k+1)} \quad (24.44)$$

to obtain (24.28). See Tweddle (2003), pp. 269–270.

24.7 Notes on the Literature

The excerpt from the English translation of Stirling's Latin letter to de Moivre is from Tweddle (2003), pp. 285–287. The Latin original is given in Tweedie (1922), p. 46. De Moivre's praise of Stirling can be found in de Moivre (1967), p. 244. For Binet's results, see pp. 239–241 and 321–323 of his very long paper, Binet (1839). For Stieltjes's work on $\mu(z)$, see Stieltjes (1993), vol. 2, pp. 6–62. Schneider (1968) gives a thorough analysis of de Moivre's work. Also see Hald (1990), pp. 480–489 for an excellent treatment of de Moivre and Stirling. Cauchy (1843) gives his work on the legitimacy asymptotic series. Poincaré (1886) contains his definition of an asymptotic series.

25

The Euler–Maclaurin Summation Formula

25.1 Preliminary Remarks

The Euler–Maclaurin summation formula is among the most useful and important formulas in all of mathematics, independently discovered by Euler and Maclaurin in the early 1730s. In modern form, the formula is given by

$$\sum_{k=m}^{n} f(k) = \int_{m}^{n} f(x)\,dx + \frac{1}{2}(f(m) + f(n))$$
$$+ \sum_{s=1}^{q} \frac{B_{2s}}{(2s)!} \left(f^{(2s-1)}(n) - f^{(2s-1)}(m) \right) + R_q(f), \qquad (25.1)$$

where

$$R_q(f) = \frac{-1}{(2q)!} \int_{m}^{n} B_{2q}(x - [x]) f^{(2q)}(x)\,dx. \qquad (25.2)$$

The B_{2s} are the Bernoulli numbers; the Bernoulli polynomial $B_q(t)$ is defined by

$$B_q(t) = \sum_{k=0}^{q} \binom{q}{k} B_k t^k. \qquad (25.3)$$

Note that since $B_{2k+1} = 0$ for $k \geq 1$, only odd order derivatives appear in the sum (25.1). It can therefore be shown, applying integration by parts, that changing every $2q$ to $2q+1$, also changing the $-$ to $+$, does not effect a change in R_q.

The Euler–Maclaurin summation formula arose out of efforts to find approximate values for finite and infinite series. During the 1720s, the series $\zeta(2) = \sum_{n=1}^{\infty} 1/n^2$ received a good deal of attention. Since the exact evaluation of this series appeared to be out of reach at that time, several mathematicians devised methods to compute approximations for this series. Stirling found some ingenious methods for transforming this and similar slowly convergent series to more rapidly convergent series. In his *Methodus Differentialis* of 1730, Stirling computed $\zeta(2)$ by three different methods, one of which gave the correct value to sixteen decimal places. Around 1727, Daniel Bernoulli and Goldbach also showed a passing interest in the problem by computing $\zeta(2)$ to a few decimal places. This may have caused Euler, their colleague at the St. Petersburg Academy, to study this problem. In a paper of 1729, Euler used integration to derive the formula

$$\sum_{n=1}^{\infty} \frac{1}{n^2} = \sum_{n=0}^{\infty} \frac{1}{2^n} \cdot \frac{1}{(n+1)^2} + (\ln 2)^2. \tag{25.4}$$

The series on the right-hand side was evidently much more rapidly convergent than the original series for $\zeta(2)$, and Euler determined that $\zeta(2) \approx 1.644934$. In fact, Euler had a result more general than (25.4); this involved the dilogarithmic function defined by the series $\sum_{n=1}^{\infty} x^n/n^2$. Perhaps Euler's work on the dilogarithm led him to apply calculus to the problem of the summation of the general series $\sum_{k=1}^{n} f(k)$. The result was a paper he presented to the Academy in 1732 (published in 1738) in which he briefly mentioned the Euler–Maclaurin formula in the form

$$\sum_{k=1}^{n} t(k) = \int t \, dn + \alpha t + \beta \frac{dt}{dn} + \gamma \frac{d^2 t}{dn^2} + \delta \frac{d^3 t}{dn^3} + \cdots, \tag{25.5}$$

where $\alpha, \beta, \gamma, \ldots$ were computed from the equations

$$\alpha = \frac{1}{2}, \ \beta = \frac{1}{1 \cdot 2}\alpha - \frac{1}{1 \cdot 2 \cdot 3}, \ \gamma = \frac{1}{1 \cdot 2}\beta - \frac{1}{1 \cdot 2 \cdot 3}\alpha + \frac{1}{1 \cdot 2 \cdot 3 \cdot 4};$$

$$\delta = \frac{1}{1 \cdot 2}\gamma - \frac{1}{1 \cdot 2 \cdot 3}\beta + \frac{1}{1 \cdot 2 \cdot 3 \cdot 4}\alpha - \frac{1}{1 \cdot 2 \cdot 3 \cdot 4 \cdot 5}, \ldots.$$

He gave an application of (25.5) to the summation of the very simple example, $\sum_{k=1}^{n}(k^2+2k)$, and then proceeded to discuss other types of series. He wrote a longer paper on the subject four years later, in 1736, in which he explicitly evaluated $\sum_{k=1}^{n} k^r$ as polynomials in n for $r = 1, 2, \ldots, 16$. This should have alerted Euler to the fact that $\alpha, \beta, \gamma, \delta, \ldots$ were closely related to the Bernoulli numbers. Jakob Bernoulli defined his numbers in exactly this context, except that in his published work he gave the polynomials up to $r = 10$. Euler was perhaps not aware of Bernoulli's work at this stage. In a highly interesting paper, written in 1740, Euler explained that the generating function for the numbers $\alpha, \beta, \gamma, \ldots$ was given by

$$S = 1 + \alpha z + \beta z^2 + \gamma z^3 + \cdots = \frac{1}{1 - \frac{z}{1 \cdot 2} + \frac{z^2}{1 \cdot 2 \cdot 3} - \frac{z^3}{1 \cdot 2 \cdot 3 \cdot 4} + \frac{z^4}{1 \cdot 2 \cdot 3 \cdot 4 \cdot 5} - \cdots}$$

$$= \frac{ze^z}{e^z - 1}. \tag{25.6}$$

Euler also offered an explanation for the appearance of the Bernoulli numbers in two such very different situations: in the values of $\zeta(2n), n = 1, 2, 3, \ldots$ and in the Euler–Maclaurin formula. In rough terms, his explanation was that the generating function for both cases was the same. In his 1736 paper, he also computed $\zeta(n) = \sum_{k=1}^{\infty} 1/k^n$ to fifteen decimal places for $n = 2, 3, 4$, making use of (25.5). The series on the right-hand side of (25.5) in these cases were asymptotic series and Euler manipulated them exactly as Stirling and de Moivre had done in a different context, using the first few terms of the asymptotic series, up to the point where the terms started getting large.

Maclaurin's results related to (25.1) appeared in his influential book *Treatise of Fluxions*. This work was published in 1742 in two volumes, although the first volume, containing the statements of the Euler–Maclaurin formula, was already typeset in 1737. Colin Maclaurin (1698–1746) studied at the University of Glasgow, Scotland, but the mathematician who had the greatest formative influence on him was Newton, whom he met in 1719. Much of Maclaurin's work on algebra, calculus, and dynamics arose directly from topics on which Newton had published results. Maclaurin was professor of mathematics at the University of Edinburgh from 1726 to 1746, having been recommended to the position by Newton. Maclaurin was probably inspired to discover the Euler–Maclaurin formula by the results of de Moivre and Stirling on the asymptotic series for $\sum_{k=1}^{n} \ln k$. Newton had a result on the sum $\sum_{k=1}^{n} 1/(a+kb)$, giving the first terms of the Euler–Maclaurin formula for this particular case. Newton gave this result in a letter of July 20, 1671, to Collins, but Maclaurin was most likely unaware of it.

It is a curious fact that Euler and Maclaurin learned of each other's works even before they were published. This was a result of the brief correspondence between Euler and Stirling. In June 1736, Euler wrote Stirling about his formula and mentioned applications to the summation of $\sum_{k=1}^{\infty} 1/k^2$ and $\sum_{k=1}^{n} 1/k$. He wrote the latter result as

$$1 + \frac{1}{2} + \frac{1}{3} + \cdots + \frac{1}{x} = C + \ln x + \frac{1}{2x} - \frac{1}{12x^2} + \frac{1}{120x^4} - \frac{1}{252x^6} + \frac{1}{240x^8} - \frac{1}{132x^{10}} + \frac{691}{32760x^{12}} - \text{etc.} \qquad (25.7)$$

We can see that the value of C, Euler's constant γ, would be $\lim_{x \to \infty} \left(\sum_{k=1}^{x} \frac{1}{k} - \ln x \right)$, and Euler gave this value as 0.5772156649015329 in his 1736 paper and in his letter.

Then in 1737, Stirling received from Maclaurin the galley proofs of some portions of the first volume of Maclaurin's treatise, containing two formulations of the Euler–Maclaurin formula. Because of some business preoccupations, Stirling did not reply to Euler's letter until April 1738. He then informed Euler about Maclaurin's work and about his communications with Maclaurin on Euler's work. Stirling also told Euler that Maclaurin had promised to acknowledge Euler's work in his book. And indeed Maclaurin did so. Concerning this point, Euler wrote in his 27 July 1738 reply to Stirling:

> But in this matter I have very little desire for anything to be detracted from the fame of the celebrated Mr Maclaurin since he probably came upon the same theorem for summing series before me, and consequently deserved to be named as its first discoverer. For I found that theorem about four years ago, at which time I also described its proof and application in greater detail to our Academy.

Unfortunately, Euler forgot to mention Maclaurin in his differential calculus book of 1755, where he discussed this formula.

Maclaurin presented four formulas and he understood these to be variations of the same result. In modern notation, two of these can be given as

$$\sum_{k=0}^{n-1} f(a+k) = \int_a^{a+n} f(x)\,dx + \frac{1}{2}(f(a) - f(a+n)) + \frac{1}{12}(f'(a) - f'(a+n))$$
$$- \frac{1}{720}(f'''(a) - f'''(a+n)) + \frac{1}{30240}(f^v(a) - f^v(a+n)) - \cdots, \tag{25.8}$$

$$\sum_{k=0}^{n} f(a+k) = \int_{a-1/2}^{a-1/2+n} f(x)\,dx + \frac{1}{24}(f'(a-1/2) - f'(a-1/2+n))$$
$$- \frac{7}{5760}(f'''(a-1/2) - f'''(a-1/2+n))$$
$$+ \frac{31}{967680}(f^v(a-1/2) - f^v(a-1/2+n)) - \cdots. \tag{25.9}$$

The remaining two formulas were for cases where the series was infinite, and it was assumed that $f(x)$ and its derivatives tended to zero as $x \to \infty$. Maclaurin derived the de Moivre and Stirling forms of the approximations for $n!$ by taking $f(x) = \ln x$ in (25.8) and (25.9), respectively. He also applied his results to obtain Jakob Bernoulli's formula for sums of powers of integers as well as approximations of $\zeta(n)$ for some values of n.

In 1772, Lagrange gave a formal expression for the Taylor series as a basis for an interesting derivation of the Euler–Maclaurin summation formula and of some extensions involving sums of sums. Suggestive of important analytical applications, Lagrange's formula for the Taylor series was

$$f(x+h) = f(x) + hf'(x) + \frac{h^2}{2!}f''(x) + \cdots$$
$$= \left(1 + hD + \frac{h^2 D^2}{2!} + \cdots\right) f(x) = e^{hD} f(x). \tag{25.10}$$

Here D represents the differential operator d/dx.

Clearly, Lagrange charted out a new approach with his algebraic conception of the derivative, and yet this algebraic perspective can be traced back to the work of Leibniz. Leibniz had been struck by the formal analogy between the differential operator and algebraic quantities; Lagrange implemented this idea by going a step further and identifying the derivative operator with an algebraic quantity. This formal method was used by some French and British mathematicians of the first half of the nineteenth century, leading to significant mathematical developments.

The eighteenth-century mathematicians made very effective use of the Euler–Maclaurin formula but did not seem too concerned about the reasons for this effectiveness, especially where divergent asymptotic series were involved. Gauss, with his interest in rigor, was the first mathematician to express the need for an investigation into this question. He did this in his 1813 paper on the gamma function and hypergeometric series and then again in his 1816 paper on the fundamental theorem of algebra. Interestingly, the rigor needed for the careful discussion of the Euler–Maclaurin summation formula was provided by the French mathematical physicist, S. D. Poisson.

Siméon Denis Poisson (1781–1840) studied at the École Polytechnique in Paris where he came under the influence of Laplace and Lagrange. The latter lectured on analytic functions at the Polytechnique, where he introduced the remainder term for the Taylor series. In the 1820s, Cauchy lectured at the Polytechnique on the application of this remainder term to a rigorous discussion of the power series representation of functions. Poisson's contribution was to derive the remainder term for the Euler–Maclaurin series. His motivation for this 1826 work was to explain an apparent paradox in Legendre's use of the Euler–Maclaurin formula to numerically evaluate the elliptic integral

$$\int_0^{\pi/2} \sqrt{1 - k^2 \sin^2 \theta} \, d\theta. \tag{25.11}$$

The sum on the left-hand side of (25.8), after a small modification to allow for non-integer division points of the interval $(a, a+n)$, can be used to approximate the integral on the right-hand side. Now the integrand in (25.11), $f(x) = \sqrt{1 - k^2 \sin^2 x}$, is such that its odd order derivatives vanish at 0 and $\pi/2$. Thus, the series on the right-hand side of (25.8) vanishes, since it involves only the odd order derivatives. This implies the absurd result that the sum on the left-hand side remains unchanged no matter how many division points are chosen in the interval. It was to resolve this paradox, rather than to explain the effectiveness of the asymptotic series, that Poisson developed the remainder term for the Euler–Maclaurin formula.

Another peculiar feature of Poisson's work was that he used Fourier series instead of Taylor series to find the remainder term. Poisson learned the technique of Fourier series from Fourier's long 1807 paper on heat conduction. From 1811 on, Poisson published several papers on Fourier series and was very familiar with its techniques. In particular, he applied the result now known as the Poisson summation formula (originally due to Cauchy) to several problems, including the present one. He obtained the remainder after q terms as an integral whose integrand had the form

$$\left(\sum_{n=1}^{\infty} \frac{1}{n^{2q}} \cos 2\pi n x \right) f^{(2q)}(x). \tag{25.12}$$

In his 1826 paper, Poisson also applied the Euler–Maclaurin formula to the derivation of a result he attributed to Laplace. Laplace arrived at his result as he attempted to approximate an integral by a sum during his study of the variations of the elements of the orbit of a comet. It is a remarkable fact that James Gregory communicated just this result to Collins in a letter dated November 23, 1670.

Jacobi, who was aware of Poisson's paper, gave a different derivation of Euler–Maclaurin using Taylor series and his remainder was essentially the expression in equation (25.2). Jacobi defined, but did not name, the even Bernoulli polynomials $B_{2n}(x)$ and gave their generating function. In the 1840s, Raabe gave them the name Bernoulli polynomials. Jacobi also proved the important result that $B_{4m+2}(x) - B_{4m+2}$

was positive while $B_{4m}(x) - B_{4m}$ was negative in the interval $(0, 1)$. From this he was able to give sufficient conditions on f that the remainder term had the same sign and the magnitude of at most the first omitted term in the series on the right-hand side of (25.8). One set of sufficient conditions was that the sign of $f^{(2m)}(x)$ did not change for $x > a$ and that the product $f^{(2m)}(x) f^{(2m+2)}(x)$ was positive. Since this was clearly true for $f(x) = \ln x$, Jacobi's result actually implied that de Moivre's and Stirling's series were asymptotic. Thus, though Jacobi did not explicitly mention it, he had resolved the problem raised by Gauss.

The papers of Poisson and Jacobi show that the Euler–Maclaurin formula follows from the Poisson summation formula. Thus, these two extremely important formulas are essentially equivalent. Moreover, by comparing the remainders in the formulas of Poisson and Jacobi, we observe that the Bernoulli polynomials $B_n(x)$ restricted to $0 \leq x \leq 1$ have Fourier series expansions. Surprisingly, Euler and D. Bernoulli were aware of this fact and in the 1770s, Euler gave a very interesting derivation of this result by starting with a divergent series.

In 1823, the Norwegian mathematician Niels Abel (1802–1829) found another summation formula. There was little mathematical instruction at Abel's alma mater, University of Christiania, so he independently studied the works of Euler, Lagrange, and Laplace. Before doing his great work on algebraic equations and elliptic and Abelian functions, Abel made some interesting discoveries as a student. For example, he found an integral representation for the Bernoulli numbers, and he substituted this in the Euler–Maclaurin formula to write:

$$\sum \phi(x) = \int \phi(x)\,dx - \frac{1}{2}\phi(x) + \int_0^\infty \frac{\phi\left(x + \frac{t}{2}\sqrt{-1}\right) - \phi\left(x - \frac{t}{2}\sqrt{-1}\right)}{2\sqrt{-1}} \frac{dt}{e^{\pi t} - 1}.$$

Interestingly, the Italian astronomer and mathematician Giovanni Plana (1781–1864) discovered this result three years before Abel. Plana studied with Lagrange at the École Polytechnique; both Lagrange and Fourier supported Plana in the course of his long and illustrious career. It appears that in 1889 Kronecker used complex analytic methods to offer the first rigorous proof of this formula.

25.2 Euler on the Euler–Maclaurin Formula

Euler's problem was to sum $\sum_{k=1}^{x} t(k) = S(x)$, where he assumed that $t(x)$ and $S(x)$ were analytic functions for $x > 0$. Naturally, he did not state such conditions, but his calculations imply them. His procedure for solving this problem was almost reckless. He expanded $S(x - 1)$ as a Taylor series:

$$S(x - 1) = S(x) - S'(x) + \frac{1}{2!}S''(x) - \frac{1}{3!}S'''(x) + \cdots.$$

Following his notation except for the factorials, Euler then had

$$t(n) = S(n) - S(n-1) = \frac{dS}{dn} - \frac{1}{2!}\frac{d^2 S}{dn^2} + \frac{1}{3!}\frac{d^3 S}{dn^3} - \frac{1}{4!}\frac{d^4 S}{dn^4} + \cdots. \tag{25.13}$$

To determine S from this equation, he assumed

$$S = \int t\,dn + \alpha t + \beta \frac{dt}{dn} + \gamma \frac{d^2t}{dn^2} + \delta \frac{d^3t}{dn^3} + \cdots. \tag{25.14}$$

Next, he substituted this series for S on the right-hand side of (25.13) and equated coefficients. He called this a well-known method, probably referring to the method of undetermined coefficients. In his first paper on this topic, he merely noted the values of $\alpha, \beta, \gamma, \delta, \cdots$ obtained when this substitution was carried out. In the second paper he observed that he got

$$t = \left(t + \alpha \frac{dt}{dn} + \cdots\right) - \frac{1}{2!}\left(\frac{dt}{dn} + \alpha \frac{d^2t}{dn^2} + \cdots\right) + \frac{1}{3!}\left(\frac{d^2t}{dn^2} + \alpha \frac{d^3t}{dn^3} + \cdots\right) - \cdots.$$

The term t on both sides canceled, and thus the coefficients of $\frac{dt}{dn}, \frac{d^2t}{dn^2}, \ldots$ had to be zero. This gave him

$$\alpha = \frac{1}{2}, \quad \beta = \frac{\alpha}{2} - \frac{1}{6}, \quad \gamma = \frac{\beta}{2} - \frac{\alpha}{6} + \frac{1}{24}, \ldots,$$

so that $\quad \alpha = \frac{1}{2}, \quad \beta = \frac{1}{12}, \quad \gamma = 0, \quad \delta = -\frac{1}{720}.$

Finally, Euler could write the Euler–Maclaurin formula as

$$S = \int t\,dn + \frac{1}{2}t + \frac{1}{12}\frac{dt}{dn} - \frac{1}{720}\frac{d^3t}{dn^3} + \frac{1}{30240}\frac{d^5t}{dn^5} - \cdots.$$

In fact, he calculated the terms up to the fifteenth derivative. Euler applied this formula to the approximate summation of $\sum_{k=1}^{\infty} 1/k^2$. He let the general term be $X = 1/x^2$. Then, with Euler's use of Const. for the constant term,

$$\int X\,dx = \text{Const.} - \frac{1}{x} \quad \text{and}$$

$$\frac{dX}{dx} = -\frac{2}{x^3}, \quad \frac{d^3X}{dx^3} = -\frac{2\cdot 3\cdot 4}{x^5}, \quad \frac{d^5X}{dx^5} = -\frac{2\cdot 3\cdot 4\cdot 5\cdot 6}{x^7} \quad \text{etc.}$$

Hence, by the Euler–Maclaurin formula,

$$1 + \frac{1}{4} + \frac{1}{9} + \cdots + \frac{1}{x^2} = S$$

$$= \text{Const.} - \frac{1}{x} + \frac{1}{2x^2} - \frac{1}{6x^3} + \frac{1}{30x^5} - \frac{1}{42x^7} + \frac{1}{30x^9} - \frac{5}{66x^{11}} + \frac{691}{2730x^{13}} - \frac{7}{6x^{15}} + \text{etc.} \tag{25.15}$$

Euler took $x = 10$ and determined that $\sum_{k=1}^{10} 1/k^2 = 1.549767731166540$. He then summed the nine terms on the right-hand side of equation (25.15). These calculations gave him the value of Const. $= 1.644934066848226 43647$. When $x = \infty$, the terms $1/x, 1/2x^2, \ldots$ vanished and hence $\sum_{k=1}^{\infty} 1/k^2 = \text{Const.}$ In this manner, Euler also computed $\sum_{k=1}^{\infty} 1/k^p$ for $p = 3$ and $p = 4$.

25.3 Maclaurin's Derivation of the Euler–Maclaurin Formula

Maclaurin's proof of the Euler–Maclaurin formula is similar to that of Euler, though the procedure appears to be more rigorous. Maclaurin described his results in geometric terms, but his arguments were mostly analytic. However, to enter Maclaurin's geometric mode of thought, we start with his proof of the integral test, usually attributed to Cauchy, who proved it in his lectures of 1828. The Euler–Maclaurin formula may be viewed as a refinement of the integral test.

Referring to Figure 25.1, in section 350 of his treatise, Maclaurin wrote:

> Let the terms of any progression be represented by the perpendiculars AF, BE, CK, HL, &c. that stand upon the base AD at equal distances; and let PN be any ordinate of the curve FNe that passes through the extremities of those perpendiculars. Suppose AP to be produced; and according as the area $APNF$ has a limit which it never amounts to, or may be produced till it exceed any give space, there is a limit which the sum of the progression never amounts to, or it may be continued till its sum exceed any given number. For let the rectangles FB, EC, KH, LI, &c. be completed, and, the area $APNF$ being continued over the same base, it is always less than the sum of all those rectangles, but greater than the sum of all the rectangles after the first. Therefore the area $APNF$ and the sum of those rectangles either both have limits, or both have none; and it is obvious, that the same is to be said of the sum of the ordinates AF, BE, CK, HL, &c. and of the sum of the terms of the progression that are represented by them.

Maclaurin's derivation of the Euler–Maclaurin formula followed a slightly less dangerous path than Euler's. His initial description was geometric, but once he had defined his terms with the help of a picture, his argument was analytic. The Maclaurin series for $f(x) = \int_0^x y(t)\,dt$ was

$$f(x) = y(0) + \frac{x^2}{2!}y'(0) + \frac{x^3}{3!}y''(0) + \cdots.$$

Thus,

$$\int_0^1 y\,dx = y(0) + \frac{1}{2!}y'(0) + \frac{1}{3!}y''(0) + \frac{1}{4!}y'''(0) + \cdots,$$

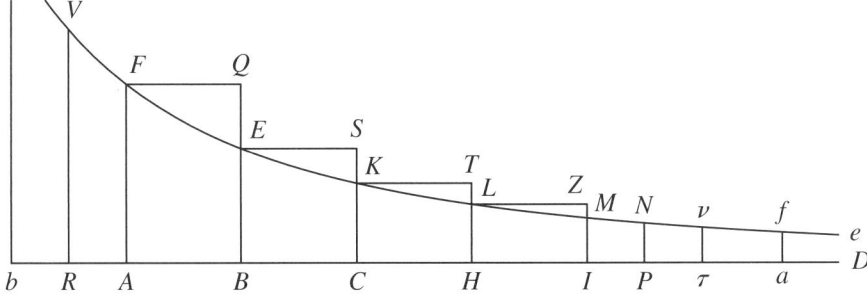

Figure 25.1. Maclaurin's geometric statement of his formula.

or

$$y(0) = \int_0^1 y\,dx - \frac{1}{2!}y'(0) - \frac{1}{3!}y''(0) - \frac{1}{4!}y'''(0) - \cdots. \qquad (25.16)$$

Similarly

$$y'(0) = \int_0^1 y'\,dx - \frac{1}{2!}y''(0) - \frac{1}{3!}y'''(0) - \frac{1}{4!}y^{iv}(0) - \cdots,$$

$$y''(0) = \int_0^1 y''\,dx - \frac{1}{2!}y'''(0) - \frac{1}{3!}y^{iv}(0) - \cdots,$$

$$y'''(0) = \int_0^1 y'''\,dx - \frac{1}{2!}y^{iv}(0) - \cdots.$$

Maclaurin used these equations to eliminate $y'(0), y''(0), \ldots$ in (25.16), obtaining

$$y(0) = \int_0^1 y\,dx - \frac{1}{2}\int_0^1 y'\,dx + \frac{1}{12}\int_0^1 y''\,dx - \frac{1}{720}\int_0^1 y^{(4)}\,dx + \cdots. \qquad (25.17)$$

Thus, he obtained another form of the Euler–Maclaurin formula:

$$y(0) + y(1) + \cdots + y(n-1)$$
$$= \int_0^n y\,dx - \frac{1}{2}\int_0^n y'\,dx + \frac{1}{12}\int_0^n y''\,dx - \frac{1}{720}\int_0^n y^{iv}\,dx + \cdots. \qquad (25.18)$$

Maclaurin also explained how the coefficients were obtained. The reader should now have little trouble in following Maclaurin's language and argument from section 828 of his book, while referring to Figure 25.2.

> Suppose the base $AP = z$, the ordinate $PM = y$, and, the base being supposed to flow uniformly, let $\dot{z} = 1$. Let the first ordinate AF be represented by a, $AB = 1$, and the area $ABEF = A$. As A is the area generated by the ordinate y, so let B, C, D, E, F, &c. represent the areas upon the same base AB generated by the respective ordinates $\dot{y}, \ddot{y}, \dddot{y}, \ddddot{y}$, &c. Then $AF = a = A - \frac{B}{2} + \frac{C}{12} - \frac{E}{720} + \frac{G}{30240} - $ &c. For, by art. 752, $A = a + \frac{\dot{a}}{2} + \frac{\ddot{a}}{6} + \frac{\dddot{a}}{24} + \frac{\ddddot{a}}{120} + $ &c. whence we have the equation $(Q)\, a = A - \frac{\dot{a}}{2} - \frac{\ddot{a}}{6} - \frac{\dddot{a}}{24} - \frac{\ddddot{a}}{120} - $ &c. In like manner, $\dot{a} = B - \frac{\ddot{a}}{2} - \frac{\dddot{a}}{6} - \frac{\ddddot{a}}{24} - $ &c. $\ddot{a} = C - \frac{\dddot{a}}{2} - \frac{\ddddot{a}}{2} - $ &c. $\dddot{a} =$

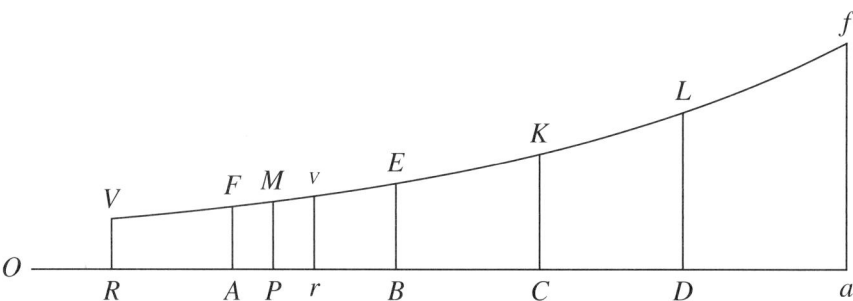

Figure 25.2. Maclaurin's representation of E-M formula.

$D - \frac{\dddot{a}}{2} - \&c.\ddot{a} = E - \&c.$ by which latter equations, if we exterminate $\dot{a}, \ddot{a}, \dddot{a}, \ddddot{a}, \&c.$ from the value of a in the equation Q, we find that $a = A - \frac{B}{2} + \frac{C}{12} - \frac{E}{720} + \&c.$ The coefficients are continued thus: let $k, l, m, n, \&c.$ denote the respective coefficients of $\dot{a}, \ddot{a}, \dddot{a}, \&c.$ in the equation Q; that is, let $k = \frac{1}{2}, l = \frac{1}{6}, m = \frac{1}{120}, \&c.$; suppose $K = k = \frac{1}{2}, L = kK - l = \frac{1}{12}, M = kL - KK + m = 0, N = kM - lL + mK - n = -\frac{1}{720}$, and so on; then $a = A - KB + LC - MD + NE - \&c.$ where the coefficients of the alternate areas $D, F, H, \&c.$ vanish.

25.4 Poisson's Remainder Term

Poisson entitled his 1826 paper on the Euler–Maclaurin formula "Sur le calcul numérique des intégrales définies." This paper applied Fourier series and the Poisson summation formula, in particular, to a variety of problems. It began with a brief sketch of the proof given in an earlier paper that the Abel means of a Fourier series of a given function converged to the function at a point of continuity. Poisson appears to assume that this was a proof that the Fourier series of a function converged to the function.

Poisson partitioned $[-a, a]$ into $2n$ equal parts with $a = nw$. He then wrote

$$\int_{-a}^{a} f(x)\,dx = wP_n + Q_n, \tag{25.19}$$

where

$$P_n = \frac{1}{2} f(-nw) + f(-nw + w) + f(-nw + 2w) + \cdots$$
$$+ f(nw - 2w) + f(nw - w) + \frac{1}{2} f(nw), \tag{25.20}$$

and Q_n was the error or remainder. The expression P_n was obtained by applying the trapezoidal rule to the given set of division points. To find the remainder, he started with Fourier's formula:

$$f(x) = \frac{1}{2a} \int_{-a}^{a} f(t)\,dt + \frac{1}{a} \int_{-a}^{a} \sum_{k=1}^{\infty} \cos \frac{k\pi(x-t)}{a} f(t)\,dt, \tag{25.21}$$

when $-a < x < a$. When $x = a$, the left-hand side was replaced by $\frac{1}{2}(f(a) + f(-a))$. He took $x = nw, (n-1)w, \ldots, 0, -w, -2w, \ldots, -(n-1)w$ successively in (25.21) and added to get

$$P_n = \frac{2n}{2nw} \int_{-a}^{a} f(t)\,dt + \frac{1}{nw} \int_{-a}^{a} \sum_{j=-(n-1)}^{n} \sum_{k=1}^{\infty} \cos \frac{k\pi(jw-t)}{nw} f(t)\,dt. \tag{25.22}$$

It is easy to check that the inner sum in (25.22), after changing the order of summation, would be

$$\sum_{j=-(n-1)}^{n} \cos \frac{k\pi(jw-t)}{nw} = \begin{cases} 2n \cos \frac{k\pi t}{nw} & \text{when } k = 2nl, \\ 0 & \text{otherwise.} \end{cases}$$

So he obtained

$$wP_n = \int_{-a}^{a} f(t)\,dt + 2\int_{-a}^{a} \sum_{l=1}^{\infty} \cos\frac{2l\pi t}{w} f(t)\,dt$$
$$= \int_{-a}^{a} f(t)\,dt - Q_n. \qquad (25.23)$$

To find another expression for Q_n, Poisson applied integration by parts repeatedly to get

$$\int_{-a}^{a} \cos\frac{2l\pi t}{w} f(t)\,dt = \frac{w}{2\pi l}\sin\frac{2l\pi t}{w} f(t)\Big]_{-a}^{a} - \int_{-a}^{a} \frac{w}{2\pi l}\sin\frac{2l\pi t}{w} f'(t)\,dt$$
$$= \frac{w^2}{4\pi^2 l^2}\cos\frac{2l\pi t}{w} f'(t)\Big]_{-a}^{a} - \int_{-a}^{a} \frac{w^2}{4\pi^2 l^2}\cos\frac{2l\pi t}{w} f''(t)\,dt$$
$$= \frac{w^2}{4\pi^2 l^2}\left(f'(a) - f'(-a)\right) - \frac{w^2}{4\pi^2}\int_{-a}^{a} \frac{1}{l^2}\cos\frac{2l\pi t}{w} f''(t)\,dt.$$

Thus,

$$Q_n = -\frac{2w^2}{4\pi^2}\sum_{l=1}^{\infty}\frac{1}{l^2}\left(f'(a) - f'(-a)\right) + \frac{2w^4}{(2\pi)^4}\sum_{l=1}^{\infty}\frac{1}{l^4}\left(f'''(a) - f'''(-a)\right)$$
$$-\frac{2w^6}{(2\pi)^6}\sum_{l=1}^{\infty}\frac{1}{l^6}(f^v(a) - f^v(-a)) + \cdots$$
$$+ \frac{2w^{2m}}{(2\pi)^{2m}}\sum_{l=1}^{\infty}\frac{1}{l^{2m}}\left(f^{(2m-1)}(a) - f^{(2m-1)}(-a)\right) + R_m, \qquad (25.24)$$

where

$$R_m = -2(-1)^m \left(\frac{w}{2\pi}\right)^{2m}\int_{-a}^{a}\sum_{l=1}^{\infty}\frac{1}{l^{2m}}\cos\frac{2l\pi t}{w} f^{(2m)}(t)\,dt. \qquad (25.25)$$

Poisson then showed that

$$\sum_{l=1}^{\infty}\frac{1}{l^{2m}} = \frac{(-1)^{m-1} 2^{2m-1}\pi^{2m}}{(2m)!} B_{2m} \qquad (25.26)$$

by an argument identical with one of several given by Euler. With this formula, Poisson had

$$Q_n = \frac{B_2}{2!}\left(f'(a) - f'(-a)\right)w^2 + \frac{B_4}{4!}\left(f'''(a) - f'''(-a)\right)w^4 + \cdots$$
$$+ \frac{B_{2m}}{(2m)!}\left(f^{(2m-1)}(a) - f^{(2m-1)}(-a)\right)w^{2m} + R_m. \qquad (25.27)$$

By substituting this expression for Q_n in (25.19), Poisson obtained the Euler–Maclaurin formula with remainder given by (25.25).

Poisson then took the interval of integration to be $[0, c]$ by a change of variables and gave a number of applications. For instance, he applied this to the derivation of Gregory's formula

$$\int_0^c f(x)\,dx = wP_n + \frac{w}{12}\Delta F(0) - \frac{w}{24}\Delta^2 F(0) + \frac{19w}{720}\Delta^3 F(0) - \frac{3w}{160}\Delta^4 F(0) + \text{etc},$$

where $c = nw$ and $F(z) = f(z) + f(c-z)$. In order to obtain Gregory's result, Poisson used the Gregory–Newton difference formula to determine the derivatives in the Euler–Maclaurin formula. Note that one can find the derivatives in terms of the difference operator Δ by using Lagrange's symbolic method:

$$e^{wD} = 1 + \Delta \quad \text{or} \quad wD = \ln(1+\Delta) = \Delta - \frac{1}{2}\Delta^2 + \frac{1}{3}\Delta^3 - \cdots,$$

$$w^3 D^3 = \left(\Delta - \frac{1}{2}\Delta^2 + \frac{1}{3}\Delta^3 - \cdots\right)^3 = \Delta^3 - \frac{3}{2}\Delta^4 + \frac{7}{4}\Delta^5 - \cdots \text{ etc.}$$

Poisson also observed that if the derivatives of f were bounded at 0 and c, then the remainder R_m was of the order w^{2m+1} as $w \to 0$. But it was still possible, for some f, that the remainder would tend to ∞ as $m \to \infty$ for a fixed w. Thus, by means of his remainder term, Poisson was able to clearly explain the difference between an asymptotic series and a convergent one, a distinction only intuitively understood in the eighteenth century.

25.5 Jacobi's Remainder Term

Jacobi used Taylor series instead of Fourier series to obtain the remainder term. He expressed the remainder by means of Bernoulli polynomials, and in the course of his proof he found their generating function. Jacobi began his proof by presenting some results on Bernoulli numbers and polynomials:

$$\frac{1}{(2m+1)!} + B_1 \frac{1}{(2m)!} + \frac{B_2}{2!}\frac{1}{(2m-1)!} + \frac{B_4}{4!}\frac{1}{(2m-3)!} + \cdots + \frac{B_{2m}}{(2m)!} = 0, \tag{25.28}$$

$$\frac{1}{(2m+2)!} + B_1 \frac{1}{(2m+1)!} + \frac{B_2}{2!}\frac{1}{(2m)!} + \frac{B_4}{4!}\frac{1}{(2m-2)!} + \cdots + \frac{B_{2m}}{(2m)!2!} = 0, \tag{25.29}$$

and

$$\frac{1}{2}\left(\frac{1-e^{xz}}{1-e^z} - \frac{1-e^{-xz}}{1-e^{-z}}\right)$$
$$= \frac{B_2(x) - B_2}{2!}z + \frac{B_4(x) - B_4}{4!}z^3 + \frac{B_6(x) - B_6}{6!}z^5 + \cdots. \tag{25.30}$$

The first two relations were found by Jakob Bernoulli and the third was the form of the generating function used by Jacobi to prove some properties of Bernoulli polynomials.

Jacobi proceeded to define the functions

$$\Psi(x) = \int_a^x f(t)\,dt, \quad \text{and} \quad \Psi(x) - \Psi(x-h) = \Theta(x). \tag{25.31}$$

Now letting $x - a$ be a multiple of h, set Jacobi's notation as

$$\sum_a^x \Theta(x) = \Theta(a+h) + \Theta(a+2h) + \cdots + \Theta(x) = \Psi(x) - \Psi(a) = \Psi(x). \tag{25.32}$$

Since $\Theta(x) = \Psi(x) - \Psi(x-h)$, Taylor's theorem with integral remainder gives

$$\Theta(x) = \Psi'(x)h - \Psi''(x)\frac{h^2}{2!} + \cdots + (-1)^{n-1}\Psi^{(n)}(x)\frac{h^n}{n!}$$

$$+ (-1)^n \int_0^h \frac{(h-t)^n}{n!} \Psi^{(n+1)}(x-t)\,dt.$$

Because $\Psi'(x) = f(x)$, $\Psi^{(m+1)}(x) = f^{(m)}(x)$, Jacobi wrote

$$\sum_a^x \frac{\Theta(x)}{h} = \int_a^x \frac{f(t)}{h}\,dt$$

$$= \sum_a^x \left(f(x) - f'(x)\frac{h}{2} + f''(x)\frac{h^2}{2\cdot 3} - \cdots + (-1)^{n-1} f^{(n-1)}(x)\frac{h^{n-1}}{n!} \right)$$

$$+ (-1)^n \int_0^h \frac{(h-t)^n}{hn!} \sum_a^x f^{(n)}(x-t)\,dt.$$

He eliminated the terms involving h in the sum on the right by using (25.28) and (25.29) in an elegant way. He replaced $f(x)$ successively by

$$B_1 f'(x)h, \; \frac{B_2}{2!} f''(x)h^2, \; -\frac{B_4}{4!} f^{iv}(x)h^4, \ldots, (-1)^{m+1}\frac{B_{2m}}{(2m)!} f^{(2m)}(x)h^{2m}.$$

As he replaced $f(x)$ by each successive expression, he replaced n by $n-1, n-2, n-4, \ldots, n-2m$, respectively. He then added all the equations to get

$$\int_a^x \left(\frac{f(t)}{h} + \frac{B_1}{1!} f'(t) + \frac{B_2}{2!} f''(t)h + \frac{B_4}{4!} f^{iv}(t)h^3 + \cdots + \frac{B_{2m}}{(2m)!} f^{(2m)}(t)h^{2m-1} \right) dt$$

$$= \sum_a^x f(x) + \int_0^h \frac{h^{2m+1}}{(2m+2)!} \left(B_{2m+2}\left(\frac{t}{h}\right) - B_{2m+2} \right) \sum_a^x f^{(2m+2)}(x-t)\,dt.$$

This is the form in which Jacobi gave the remainder, but we can easily convert it to the modern form. Take $h = 1$ so that the remainder, without $(2m+2)!$ in the denominator, takes the form

$$\int_0^1 B_{2m+2}(t) \left(f^{(2m+2)}(a+1-t) + f^{(2m+2)}(a+2-t) + \cdots + f^{(2m+2)}(x-t) \right) dt.$$

When t is changed to $1-t$, the integral is

$$(-1)^m \left(\int_0^1 B_{2m+2}(t) f^{(2m+2)}(a+t)\, dt + \int_0^1 B_{2m+2}(t) f(a+1+t)\, dt + \cdots \right)$$

$$= (-1)^m \left(\int_a^{a+1} B_{2m+2}(t-[t]) f^{(2m+2)}(t)\, dt + \int_{a+1}^{a+2} B_{2m+2}(t-[t]) f^{2m+2}(t)\, dt + \cdots \right)$$

$$= (-1)^m \int_a^x B_{2m+2}(t-[t]) f^{(2m+2)}(t)\, dt.$$

This is the remainder in modern form. A comparison of this remainder with that of Poisson shows that the Bernoulli polynomials restricted to $0 \leq x < 1$ have very nice Fourier expansions of which Euler and Daniel Bernoulli were aware.

25.6 Euler on the Fourier Expansions of Bernoulli Polynomials

Euler and Daniel Bernoulli obtained the Fourier series of the Bernoulli polynomials by repeated integration of the equation

$$\cos u + \cos 2u + \cos 3u + \cdots = -\frac{1}{2}. \tag{25.33}$$

Euler justified this relation by setting $x = e^{iu}$ in $1 + x + x^2 + x^3 + \cdots = \frac{1}{1-x}$. He knew that the series (25.33) was divergent. Nevertheless, he integrated it to get

$$\sum_{n=1}^{\infty} \frac{\sin nu}{n} = A - \frac{1}{2} u, \tag{25.34}$$

where A was the constant of integration. To find A, he observed that he must not set $u = 0$; he also saw that for small u, $(\sin nu)/n = u$, and the series on the left in (25.34) took the form $u + u + u + u + \cdots$. The sum of this series was infinite. So he let $u = \pi + w$ with w small to get

$$-w + w - w + w - \cdots = A - \frac{1}{2}(\pi + w).$$

When $w \to 0$, the left-hand side vanished and hence $A = \frac{\pi}{2}$. Thus,

$$\sum_{n=1}^{\infty} \frac{\sin nu}{n} = \frac{1}{2}(\pi - u). \tag{25.35}$$

This formula turned out to be quite correct; Euler obtained it by integrating a divergent series! Euler attributed this method of finding A to Daniel Bernoulli. Integration of (25.35) gave

$$\sum_{n=1}^{\infty} \frac{\cos nu}{n^2} = B - \frac{\pi u}{2} + \frac{u^2}{4}. \tag{25.36}$$

Next, $u = 0$ and $u = \pi$ produced the two equations

$$\sum_{n=1}^{\infty} \frac{1}{n^2} = B \quad \text{and} \quad \sum_{n=1}^{\infty} \frac{(-1)^n}{n^2} = B - \frac{\pi^2}{4}.$$

Euler added these equations to obtain

$$\sum_{n=1}^{\infty} \frac{2}{(2n)^2} = 2B - \frac{\pi^2}{4}.$$

The left side of this equation could be expressed as

$$\frac{1}{2}\sum_{n=1}^{\infty} \frac{1}{n^2} = \frac{1}{2}B \quad \text{and thus} \quad B = \frac{\pi^2}{6}.$$

By (25.36) Euler obtained

$$\sum_{n=1}^{\infty} \frac{\cos nu}{n^2} = \frac{\pi^2}{6} - \frac{\pi u}{2} + \frac{u^2}{4}. \tag{25.37}$$

Euler continued by repeated integration to write

$$\sum_{n=1}^{\infty} \frac{\sin nu}{n^3} = \frac{\pi^2 u}{6} - \frac{\pi u^2}{4} + \frac{u^3}{12},$$

$$\sum_{n=1}^{\infty} \frac{\cos nu}{n^4} = \frac{\pi^4}{90} - \frac{\pi^2 u^2}{12} + \frac{\pi u^3}{12} - \frac{u^4}{48},$$

$$\sum_{n=1}^{\infty} \frac{\sin nu}{n^5} = \frac{\pi^4 u}{90} - \frac{\pi^2 u^3}{36} + \frac{\pi u^4}{48} - \frac{u^5}{240},$$

$$\sum_{n=1}^{\infty} \frac{\cos nu}{n^6} = \frac{\pi^6}{945} - \frac{\pi^4}{90}\cdot\frac{u^2}{2} + \frac{\pi^2}{6}\cdot\frac{u^4}{24} - \frac{\pi}{2}\cdot\frac{u^5}{120} + \frac{1}{2}\cdot\frac{u^6}{720}.$$

In particular, he had a new method for computing the exact value of $\sum_{n=1}^{\infty} \frac{1}{n^{2k}}$. To see the connection with Bernoulli polynomials, write the last equation as

$$2 \times 6! \sum_{n=1}^{\infty} \frac{\cos 2n\pi t}{(2\pi n)^6} = t^6 - 3t^5 + \frac{5}{2}t^4 - \frac{1}{2}t^2 + \frac{1}{42}.$$

Note that this polynomial is, in fact, $B_6(t)$. Euler did not mention the connection between the polynomials appearing here and the polynomials obtained when $\sum_{k=1}^{n} k^m$, $m = 1, 2, 3, \ldots$ were expressed as polynomials in n.

25.7 Abel's Derivation of the Plana–Abel Formula

In 1823, while he was still a student in Norway, Abel published a paper in which he gave a number of applications of definite integrals. In one of these, he expressed Bernoulli numbers as a definite integral. Euler was already aware of this formula, but Abel used it in the Euler–Maclaurin summation formula to obtain the Plana–Abel formula. Since he was dealing with divergent series, his proof was not rigorous. A rigorous proof depends on the complex analytic methods being developed around that time by Cauchy. Abel observed that

$$A_n = \frac{2n}{2^{2n-1}} \int_0^\infty \frac{t^{2n-1}}{e^{\pi t} - 1} dt, \qquad (25.38)$$

where $A_n = (-1)^{n-1} B_{2n}$.

He obtained this from the expansion

$$\frac{1}{e^{\pi t} - 1} = e^{-\pi t} + e^{-2\pi t} + e^{-3\pi t} + \cdots$$

and term-by-term integration. He wrote the Euler–Maclaurin formula in the form

$$\sum \phi(x) = \int \phi(x) \, dx - \frac{1}{2}\phi(x) + A_1 \frac{\phi'(x)}{2!} - A_2 \frac{\phi'''(x)}{4!} + \cdots$$

$$= \int \phi(x) \, dx - \frac{1}{2}\phi(x) + \frac{\phi'(x)}{2!} \int_0^\infty \frac{t \, dt}{e^{\pi t} - 1} - \frac{\phi'''(x)}{3! 2^3} \int_0^\infty \frac{t^3 \, dt}{e^{\pi t} - 1} + \cdots$$

$$= \int \phi(x) \, dx - \frac{1}{2}\phi(x) + \int_0^\infty \frac{dt}{e^{\pi t} - 1} \left(\phi'(x) \frac{t}{2} - \frac{\phi'''(x)}{3!} \frac{t^3}{2^3} + \cdots \right)$$

$$= \int \phi(x) \, dx - \frac{1}{2}\phi(x) + \frac{1}{2\sqrt{-1}} \int_0^\infty \frac{\phi\left(x + \frac{t}{2}\sqrt{-1}\right) - \phi\left(x - \frac{t}{2}\sqrt{-1}\right)}{e^{\pi t} - 1} dt.$$

This is the Plana–Abel formula. The last step followed from the Taylor expansions of the functions in the numerator of the integrand. After studying Cauchy's lectures on analysis, Abel came to realize that his derivation was invalid. In an 1826 letter to his friend Holmboe, he wondered why the use of divergent series could often lead to correct results. It was not until the early years of the twentieth century that this puzzle was fully understood.

25.8 Exercises

1. In his 1671 letter to Collins, Newton expressed his result on $\sum_{k=1}^{p} \frac{a}{b+kc}$ as follows:

    Any musical progression $\frac{a}{b} \cdot \frac{a}{b+c} \cdot \frac{a}{b+2c} \cdot \frac{a}{b+3c} \cdot \frac{a}{b+4c}$ etc. being propounded whose last term is $\frac{a}{d}$: for ye following operation choose any convenient number e (whither whole broken or surd) which intercedes these limits $\frac{2mn}{b+d}$ and $\sqrt{mn}$; supposing $b - \frac{1}{2}c$ to bee m, and $b + \frac{1}{2}c$ to bee n. And this proportion will give you the aggregate of the terms very near the truth.
    As ye Logarithm of $\frac{e+\frac{1}{2}c}{e-\frac{1}{2}c}$ to ye Logarithm of $\frac{n}{m}$, so is $\frac{a}{e}$ to ye desired summe.

Verify Newton's approximation, stated in modern terminology: Let $m = b - \frac{1}{2}c$ and $n = d + \frac{1}{2}c$ and $d = b + (p-1)c$. Then

$$\sum_{k=1}^{p} \frac{a}{b+kc} \approx \frac{a \ln(n/m)}{e \ln(e + \frac{1}{2}c)/(c - \frac{1}{2}c))},$$

where $\frac{2mn}{m+n} < e < \sqrt{mn}$. See Newton (1959–1960), pp. 68–70.

2. Suppose $p = \int \frac{\partial x}{x} \ln y$ and $q = \int \frac{\partial y}{y} \ln x$, where the symbols ∂x, ∂y denote partial differentiation.

   (a) Show that $p + q = \ln x \ln y + C$.
   (b) Take $y = 1 - x$, $0 < x < 1$. Show that

   $$p = -\sum_{n=1}^{\infty} \frac{x^n}{n^2} \quad \text{and} \quad q = -\sum_{n=1}^{\infty} \frac{y^n}{n^2}.$$

   Let $x \to 1$ to get $C = -\sum_{n=1}^{\infty} 1/n^2 = -\pi^2/6$. Thus,

   $$\text{Li}_2(x) + \text{Li}_2(y) = \frac{\pi^2}{6} - \ln x \ln y, \tag{25.39}$$

   where $\text{Li}_2(x) = \sum_{n=1}^{\infty} x^n/n^2$.

   (c) Take $y = x - 1$. Observe that

   $$\ln y = \ln x + \ln\left(1 - \frac{1}{x}\right) = \ln x - \frac{1}{x} - \frac{1}{2x^2} - \frac{1}{3x^3} - \frac{1}{4x^4} - \text{etc.}$$

   Show that

   $$p = \frac{1}{2}(\ln x)^2 + \text{Li}_2(1/x) \text{ and } q = -\text{Li}_2(-y).$$

   Deduce that

   $$p + q = \frac{\pi^2}{6} + \ln x \ln \frac{y}{\sqrt{x}}.$$

   (d) Deduce from (c) that for $a = (\sqrt{5} - 1)/2$

   $$\text{Li}_2(a) - \text{Li}_2(-a) = \frac{\pi^2}{6} - \ln a \ln(a\sqrt{a}).$$

The results and methods above are from Euler's paper, written in 1779 and published in 1811, devoted entirely to the topic of the dilogarithm. By using partial derivatives, he made the proof of (25.39) somewhat shorter than his 1729 proof of the same result. In 1735 Euler had also been able to evaluate $\zeta(2)$; he made use of this in his 1779 paper. See Eu. I-16-2, pp. 117–138.

3. Let $s = 1^n + 2^n + 3^n + \cdots + x^n$. Use Euler–Maclaurin summation to prove

$$s = \frac{x^{n+1}}{n+1} + \frac{x^n}{2} + \frac{1}{2}\binom{n}{1}\frac{x^{n-1}}{6} - \frac{1}{4}\binom{n}{3}\frac{x^{n-3}}{30} + \frac{1}{6}\binom{n}{5}\frac{x^{n-5}}{42}$$
$$- \frac{1}{8}\binom{n}{7}\frac{x^{n-7}}{30} + \frac{1}{10}\binom{n}{9}\frac{5x^{n-9}}{66} - \frac{1}{12}\binom{n}{11}\frac{691x^{n-11}}{2730}$$
$$+ \frac{1}{14}\binom{n}{13}\frac{7x^{n-13}}{6} - \frac{1}{16}\binom{n}{15}\frac{3617x^{n-15}}{510} + \text{etc.}$$

Euler gave this formula in his 1736 paper (Eu. I-14, 108–123) and specifically listed the sums for $n = 1, 2, \ldots, 16$. Note the explicit appearance of the Bernoulli numbers, $\frac{1}{6}, -\frac{1}{30}, \frac{1}{42}, \ldots, -\frac{3617}{510}$ in Euler's presentation of the formula. Naturally, Euler wrote

$$\frac{1}{k+1}\binom{n}{k} \quad \text{as} \quad \frac{n(n-1)\cdots(n-k+1)}{1 \cdot 2 \cdots k(k+1)}.$$

4. Verify Euler's computations leading to the value of γ to fifteen decimal places. He took $n = 10$ in (25.7) and determined that

$$1 + \frac{1}{2} + \frac{1}{3} + \cdots + \frac{1}{10} = 2.9289682539682539.$$

He also knew that $\ln 10 = 2.302585092994045684$. He used precisely the terms in (25.7) so that he had to calculate

$$\frac{1}{20} - \frac{1}{1200} + \frac{1}{1200000} - \cdots + \frac{691}{32760 \times 10^{12}} - \frac{1}{12 \times 10^{14}}.$$

From this he obtained

$$\text{Const.} = \gamma = \lim_{n \to \infty}\left(\sum_{k=1}^{n}\frac{1}{k} - \ln n\right) = 0.5772156649015329.$$

This is just one example of the kind of numerical calculation that Euler undertook on a regular, if not daily, basis.

In this connection, also prove the inequalities of Mengoli:

$$\frac{1}{n+1} + \frac{1}{n+2} + \cdots + \frac{1}{np} < \ln p < \frac{1}{n} + \frac{1}{n+1} + \cdots + \frac{1}{np-1}. \quad (25.40)$$

Mengoli proved these in his *Geometria Speciosa* of 1659. See Hofmann's article on the Euler–Maclaurin formula, in Hofmann (1990), vol. 1, pp. 233–40, especially p. 237.

5. Use the Euler–Maclaurin formula to show that

$$\sum_{k=1}^{\infty} \frac{1}{k^3} = 1.202056903159594$$

$$\sum_{k=1}^{\infty} \frac{1}{k^4} = 1.0823232337110824.$$

These results and the evaluation of γ in the previous exercise are in Euler's 1736 paper (Eu. I-14, pp. 118–21).

6. Verify Euler's formal computations to obtain a formula for the alternating series

$$s(x) = f(x) - f(x+b) + f(x+2b) - \text{etc:}$$

From $s(x+2b) - s(x) = -f(x) + f(x+b)$, deduce that

$$f(x+b) - f(x) = \frac{2b}{1!} \frac{ds}{dx} + \frac{4b^2}{2!} \frac{d^2s}{dx^2} + \text{etc.}$$

Thus,

$$\frac{b}{1!} \frac{df}{dx} + \frac{b^2}{2!} \frac{d^2f}{dx^2} + \text{etc.} = \frac{2b}{1!} \frac{ds}{dx} + \frac{4b^2}{2!} \frac{d^2s}{dx^2} + \text{etc.} \qquad (25.41)$$

Assume with Euler that

$$\frac{ds}{dx} = \alpha \frac{df}{dx} + \beta \frac{d^2f}{dx^2} + \gamma \frac{d^3f}{dx^3} + \text{etc.}$$

and substitute in (25.41), equating coefficients, to get

$$2s = \text{Const.} + f(x) + \frac{b}{2} \frac{df}{dx} - \frac{b^3}{4!} \frac{d^3f}{dx^3} + \frac{3b^5}{6!} \frac{d^5f}{dx^5} - \frac{17b^7}{8!} \frac{d^7f}{dx^7}$$
$$+ \frac{155b^9}{10!} \frac{d^9f}{dx^9} - \frac{2073b^{11}}{12!} \frac{d^{11}f}{dx^{11}} + \frac{38227b^{13}}{14!} \frac{d^{13}f}{dx^{13}} - \text{etc.} \qquad (25.42)$$

Euler used this formula to compute the series

$$\frac{1}{x} - \frac{1}{x+b} + \frac{1}{x+2b} - \frac{1}{x+3b} + \text{etc.}$$

He applied it to

$$\frac{\pi}{4} = 1 - \frac{1}{3} + \frac{1}{5} - \frac{1}{7} + \cdots$$

by taking $x = 25$, $b = 2$ and then computing $1 - \frac{1}{3} + \cdots - \frac{1}{23}$ separately. In this way he obtained a value of $\pi/4$ correct to eleven decimal places. The summation formula (25.42) is often called the Boole summation formula, though Euler had the result a hundred years before Boole. This work is in Eu. I-14, pp. 128–130.

7. Show that

$$f(x) - f(x+1) + f(x+2) - \cdots \qquad (25.43)$$

$$= \frac{1}{2}f(x) + \sum_{n=1}^{\infty} \frac{(1-2^{2n})B_{2n}}{(2n)!} \frac{d^{2n-1}f}{dx^{2n-1}}. \qquad (25.44)$$

Although this is not Euler's notation, he pointed out the connection between the terms of the series and the Bernoulli numbers. The paper appeared in 1788, though it was presented to the St. Petersburg Academy in 1776. See Eu. I-16-1, p. 57.

8. Show by a formal calculation that when m is a positive integer

$$a^m - (a+b)^m + (a+2b)^m - (a+3b)^m + \cdots$$

$$= \frac{a^m}{2} - \frac{(2^2-1)B_2}{2}\binom{m}{1}a^{m-1}b - \frac{(2^4-1)B_4}{4}\binom{m}{3}a^{m-3}b^3 - \cdots$$

$$- \frac{(2^{m+1}-1)B_{m+1}}{m+1}b^m.$$

For $a = 0, b = 1$ this gives

$$0^m - 1^m + 2^m - 3^m + 4^m - \cdots = -\frac{B_{m+1}(2^{m+1}-1)}{m+1}.$$

This also follows from the formula (25.44) above. Euler used this obviously divergent series to prove the functional relation for the zeta function. See Eu. I-15, p. 76.

25.9 Notes on the Literature

The Euler–Maclaurin was one of Euler's favorite formulas; he discussed or used it in numerous papers. He first stated it in "Methodus Generalis Summandi Progressiones," Eu. I-14, p. 43. Although this paper dealt with a different topic, Euler had apparently discovered this result at just that time and wished to announce it quickly. He investigated the subject more thoroughly in the 1736 paper "Inventio Summae cuiusque Seriei ex Dato Termino Generali," Eu. I-14, pp. 108–123. There he gave approximations for $\zeta(n)$, $n = 2, 3, 4$ and for Euler's constant γ. He also obtained formulas for the sums of powers of consecutive integers for pairs up to sixteen. Euler then extended his formula to the summation of alternating series in his "Methodus Universalis Series Summandi Ulterius Promota," also published in 1736. See Eu. I-14, pp. 124–137. His paper written in 1740, published ten years after that, "De Seriebus Quibusdam Considerationes," contains the generating function for the Bernoulli numbers. See Eu. I-14, p. 436. Euler included a full treatment of the Euler–Maclaurin in chapter 5 of part two of his 1755 differential calculus book. See Eu. I-10. The other chapters contain much interesting work on series. For the quote about Maclaurin from Euler's letter, see Tweddle (1988), p. 146.

For the derivations of the remainder term, see Poisson (1823) and Jacobi (1969), vol. 6, pp. 64–75. Incidentally, Poisson refers to Euler for the Euler–Maclaurin formula, while the title of Jacobi's paper is "De Usu Legitimo Formulae Summatoriae Maclaurinianae." Abel's 1823 paper presenting the Plana–Abel formula appeared in Norwegian in the Norwegian journal *Magazin for Naturvidenskaberne*; it was reprinted in French as "Solution de quelques problèmes à l'aide d'intégrales définies." See Abel (1965), vol. I, pp. 11–27, especially p. 23. Hardy (1949) gives an excellent treatment of the work of Jacobi and Poisson on the Euler–Maclaurin and Cohen (2007) presents a modern treatment of the E-M and its extensions.

26

L-Series

26.1 Preliminary Remarks

One of the most difficult and outstanding mathematical problems of the early eighteenth century was the summation of the series $\sum_{n=0}^{\infty} 1/n^2$. In his 1650 book *Novae Quadraturae Arithmeticae Seu De Additione Fractionum*, Pietro Mengoli (1626–1686) considered the sum of the reciprocals of figurate numbers: natural numbers, triangular numbers, square numbers, and so on. For the natural numbers, Mengoli showed that the sum of their reciprocals diverged, or that the harmonic series was divergent. For triangular numbers, he showed how the reciprocal of each triangular number could be written as a difference of two fractions, thus summing the series. But in the next step, square numbers, Mengoli posed the problem of summing their reciprocals, but could not solve it. He expressed surprise that the series of triangular reciprocals could be more easily summed than the series of square reciprocals, saying that a "richer intellect" would be required to solve this problem. Leibniz, Jakob and Johann Bernoulli, and James Stirling all later attempted to sum this series. In fact, this question became known as the Basel problem because it frustrated the very best efforts of Jakob Bernoulli of Basel, who wrote that he would be greatly indebted to anyone who would send him a solution. Unfortunately, the solution was not found until thirty years after Bernoulli's death; in 1735, Euler became the first to sum this series. With characteristic brilliance, Euler made use of the formulas relating the roots to the coefficients of algebraic equations and he boldly applied them to equations of infinite degree. When Euler communicated his results without proof to Stirling in 1736, the latter wrote in response that Euler must have tapped a new source, since the old methods were insufficient. Euler had indeed found a new source, beginning with the equation

$$0 = 1 - \frac{1}{y} \sin x = 1 - \frac{1}{y}\left(x - \frac{x^3}{3!} + \frac{x^5}{5!} - \cdots\right) \qquad (26.1)$$

where y was a constant between -1 and 1. He argued that $2n\pi + A$ and $(2n+1)\pi - A$ gave a complete list of the roots of (26.1), provided that $\sin A = y$ and that n assumed all possible integer values. Hence, he factored (26.1) as

$$\left(1 - \frac{x}{A_1}\right)\left(1 - \frac{x}{A_2}\right)\left(1 - \frac{x}{A_3}\right)\cdots, \tag{26.2}$$

where $A_1, A_2, A_3, \ldots$ were all the roots. He equated the coefficients of x to obtain the sum of the reciprocals of the roots:

$$\frac{1}{A_1} + \frac{1}{A_2} + \frac{1}{A_3} + \cdots = \frac{1}{y}. \tag{26.3}$$

To obtain the sums of the squares, cubes, fourth powers, etc., of the reciprocals of the roots, he applied the Girard–Newton formulas (18.3). Thus, when $y = 1$ and $A = \frac{\pi}{2}$, he had the formula

$$\frac{4}{\pi} - \frac{4}{3\pi} + \frac{4}{5\pi} + \cdots = 1 \quad \text{or} \tag{26.4}$$

$$1 - \frac{1}{3} + \frac{1}{5} + \cdots = \frac{\pi}{4}. \tag{26.5}$$

From the Girard–Newton formulas he obtained

$$1 + \frac{1}{3^2} + \frac{1}{5^2} + \cdots = \frac{\pi^2}{8}, \tag{26.6}$$

$$1 - \frac{1}{3^3} + \frac{1}{5^3} - \cdots = \frac{\pi^3}{32}, \tag{26.7}$$

$$1 + \frac{1}{3^4} + \frac{1}{5^4} + \cdots = \frac{\pi^4}{96}, \tag{26.8}$$

etc. He then observed, as Jakob Bernoulli had also seen, that

$$1 + \frac{1}{2^2} + \frac{1}{3^2} + \frac{1}{4^2} + \cdots = 1 + \frac{1}{3^2} + \frac{1}{5^2} + \cdots + \frac{1}{4}\left(1 + \frac{1}{2^2} + \frac{1}{3^2} + \cdots\right)$$

and hence by (26.6)

$$\sum_{n=1}^{\infty} \frac{1}{n^2} = \frac{\pi^2}{6}. \tag{26.9}$$

Similarly, Euler found

$$\sum_{n=1}^{\infty} \frac{1}{n^4} = \frac{\pi^4}{90}, \tag{26.10}$$

and so on. To derive other series, he took y as constants other than one. For example, for $y = 1/\sqrt{2}$ and $y = \sqrt{3}/2$ he had, respectively,

$$1 + \frac{1}{3} - \frac{1}{5} - \frac{1}{7} + \frac{1}{9} + \frac{1}{11} - \frac{1}{13} - \frac{1}{15} + \cdots = \frac{\pi}{2\sqrt{2}} \quad \text{and} \tag{26.11}$$

$$1 + \frac{1}{2} - \frac{1}{4} - \frac{1}{5} + \frac{1}{7} + \frac{1}{8} - \frac{1}{10} - \frac{1}{11} + \cdots = \frac{2\pi}{3\sqrt{3}}. \tag{26.12}$$

26.1 Preliminary Remarks

Recall that Newton had earlier proved (26.11) by the integration of the rational function $(1+x^2)/(1+x^4)$ over $[0, 1]$. Indeed, Euler credited Newton with this result. Of course, Madhava and Leibniz found (26.5) by integrating the rational function $1/(1+x^2)$. In the same manner, (26.12) can by obtained by the integration of $(1+x)/(1+x^3)$. Thus, integration of rational functions is a powerful method for evaluating many series that sum to multiples of π or to logarithms of numbers. However, this method is not as effective for series such as (26.6) through (26.10). Euler, however, provided a new insight, so that one could efficiently sum up many series. When Euler communicated his method to some of his mathematical correspondents, there were objections to his procedure. How did he know, for example, that $n\pi$ were the only roots of $\sin x = 0$? There could be complex roots! How could he employ the Girard–Newton formulas, applicable to polynomials, for equations of infinite degree? In addition, there were also convergence questions concerning some of Euler's series.

Euler was well aware of these inevitable objections, but he believed in the correctness of his formulas. His methods had succeeded in rederiving known formulas and, moreover, numerical methods such as the Euler–Maclaurin formula showed him that his results were correct to many decimal places. So Euler made great efforts to resolve any doubts about his method, as well as to prove his formulas using alternative procedures. For example, by proving the product formulas for $\sin x$ and $\cos x$, (16.20) and (16.21), he showed that these functions had only the well-known zeros and no others. And in 1737 Euler gave an ingenious method for deriving (26.6) and (26.9) by computing $\int_0^1 \arcsin x / \sqrt{1-x^2}\, dx$ in two different ways. Unfortunately, this method did not extend to formulas such as (26.7), (26.8), and (26.10) where the powers of the denominators were greater than two. Meanwhile, in 1738, Niklaus I Bernoulli gave an extremely clever proof of (26.6) by squaring the series (26.5). Euler soon simplified the argument and communicated his result to his friend Goldbach, who suggested considering even more general series, now known as double-zeta values; Euler found some very interesting results about them. This laid the foundation for the fertile modern topic of multizeta values.

In a paper of 1743 published in Berlin, Euler used the partial fractions expansions of $\cot \pi x$ and $\csc \pi x$ and their derivatives to find the sum of $\sum 1/n^{2k}$ and related series. This is essentially the method often used in modern textbooks, although Euler, Daniel Bernoulli, and Landen found other significant proofs. In 1754, Euler discovered the "Fourier" expansion

$$\sin \phi - \frac{1}{2}\sin 2\phi + \frac{1}{3}\sin 3\phi - \frac{1}{4}\sin 4\phi + \frac{1}{5}\sin 5\phi - \text{etc.} = \frac{\phi}{2}, \tag{26.13}$$

yielding (26.5) when $\phi = \frac{\pi}{2}$. Integration of (26.13) gave him

$$\cos \phi - \frac{1}{2^2}\cos 2\phi + \frac{1}{3^2}\cos 3\phi - \frac{1}{4^2}\cos 4\phi + \text{etc.} = C - \frac{\phi^2}{4}. \tag{26.14}$$

We have seen in the previous chapter that in 1773 Daniel Bernoulli showed that the values of $\sum 1/n^{2k}$ and $\sum (-1)^n/(2n+1)^{2k+1}$ could be obtained from (26.14). Euler further improved on these results, and their joint work produced the Fourier expansions for Bernoulli polynomials.

Recall that in 1758 Landen evaluated the sums $\sum 1/n^{2k}$ and $\sum (-1)^n/(2n+1)^{2k+1}$ by means of the polylogarithmic functions. To do this, he employed complex numbers through the use of $\log(-1)$. In the 1770s, Euler made further strides in this area, though his work had a few errors, especially where complex numbers were used. Thus, by the 1770s, Euler had worked out many ways of evaluating $\sum 1/n^2$. In his papers written during that period, he described his methods and even added a new one, employing integration and differentiation under the integral sign.

From the start of his work on what we would now call zeta values, Euler observed that the numbers appearing in the values of $\sum 1/n^{2k}$, $k = 1, 2, 3, \ldots$ also presented themselves as coefficients in the Euler–Maclaurin summation formula. He was intrigued by this puzzle and wrote in his 1738 letter to Stirling that an explanation of this would be a significant advancement. In a 1740 paper, Euler began to understand this, as he used differential equations to obtain the Taylor series expansions of $\cot \pi x$ and $xe^x/(e^x - 1)$. The coefficients of the series for $\cot \pi x$ involved the sums $\sum 1/n^{2k}$; on the other hand, the coefficients of $xe^x/(e^x - 1)$ involved the Bernoulli numbers and were the coefficients in the Euler–Maclaurin series. Thus, Euler made his own "significant advancement." It was around 1740 that Euler more precisely understood the relation between the trigonometric functions and the exponential function, noting that

$$\cot x = i\left(1 + \frac{2}{e^{2ix} - 1}\right), \quad \text{and} \quad \frac{xe^x}{e^x - 1} = x\left(1 + \frac{2}{e^x - 1}\right) \tag{26.15}$$

gave a simpler reason for the appearance of Bernoulli numbers in the summation of $\sum 1/n^{2k}$. He gave details in his differential calculus book of 1755.

Euler was also mystified by the fact that, even though he could sum

$$\sum 1/n^{2k} \quad \text{or} \quad \sum (-1)^{n+1}/n^{2k}$$

in various ways, he was unable to find the sum of the series with odd powers, $\sum 1/n^{2k+1}$. Of course, this is a problem that has baffled mathematicians to the present day. To shed some light on this question, Euler considered the divergent series $\sum (-1)^{n+1} n^k$. In a 1740 paper, he noted that

$$1 - 1 + 1 - 1 + 1 - \text{etc.} = \frac{1}{2}, \tag{26.16}$$

$$1 - 2^{2k} + 3^{2k} - 4^{2k} + \text{etc.} = 0, \tag{26.17}$$

$$1 - 2^{2k-1} + 3^{2k-1} - 4^{2k-1} + \text{etc.}$$
$$= (-1)^{k-1} 2 \cdot \frac{1 \cdot 2 \cdots (2k-1)}{\pi^{2k}} \left(1 + \frac{1}{3^{2k}} + \frac{1}{5^{2k}} + \cdots\right), \tag{26.18}$$

$k = 1, 2, 3, \ldots$. Note that the series in parentheses in (26.18) can also be written as

$$\frac{2^{2k} - 1}{2^{2k} - 2}\left(1 - \frac{1}{2^{2k}} + \frac{1}{3^{2k}} - \frac{1}{4^{2k}} + \text{etc.}\right). \tag{26.19}$$

Thus, the series with even/odd powers in the denominator were related to the series with odd/even powers in the numerator; however, the series with even powers in the numerator summed to zero (except for $k = 0$), and hence gave no information about the odd series $\sum (-1)^{n+1}/n^{2k+1}$, $k = 1, 2, 3, \ldots$. In the exceptional case, for $k = 0$, one gets $\sum (-1)^{n+1}/n^{2k+1} = \sum (-1)^{n-1}/n = \ln 2$. We observe that Euler was well aware that the series on the left-hand side were divergent, but he plunged right in anyway, since this work was yielding him insight into a very challenging problem. Indeed, he was justified in his audacity, since this approach led him to the functional equation for the zeta function.

It appears that at some point in the 1740s, Euler started thinking of the series $\sum 1/n^{2k}$ and other related series as particular values of the functions defined by $\zeta(s) = \sum 1/n^s$, etc. Here note, however, that the label $\zeta(s)$ was later given by Riemann. In a paper presented to the Berlin Academy in 1749 but published in 1768, Euler drew a relation between $\zeta(s)$ and $\zeta(1-s)$, using the equation he wrote as

$$\frac{1 - 2^{-m} + 3^{-m} - 4^{-m} + 5^{-m} - 6^{-m} + \text{etc.}}{1 - 2^{m-1} + 3^{m-1} - 4^{m-1} + 5^{m-1} - 6^{m-1} + \text{etc.}}$$
$$= -\frac{1 \cdot 2 \cdot 3 \cdots (m-1)(2^m - 1)}{(2^{m-1} - 1)\pi^m} \cos \frac{m\pi}{2}, \quad (26.20)$$

where m was a real number. In this way, Euler found a generalization of (26.18).

In 1826, Abel wrote his friend Holmboe that equation (26.17) was a laughable equation to write. Abel's early training had been in the formal mathematical tradition of which Euler was considered the model. After he studied Cauchy's lectures on analysis, Abel changed his view of mathematics; he believed it illegitimate to use divergent series at all and therefore wished to abolish formulas such as (26.17), (26.18), and (26.20). However, Euler had a very clear idea that the definition of the divergent series in these formulas amounted to a limit:

$$1 - 2^n + 3^n - 4^n + \cdots = \lim_{x \to 1^-} \left(1 - 2^n x + 3^n x^2 - 4^n x^3 + \cdots \right). \quad (26.21)$$

He found the values of these limits, for nonnegative integer values of n, by repeated multiplication by x followed by differentiation of the geometric series formula

$$1 - x + x^2 - x^3 + x^4 - \cdots = \frac{1}{1+x}. \quad (26.22)$$

Euler's technique of summing (26.21) became an important summability method in the theory of divergent series developed after 1890. Ironically, it is called the Abel sum.

Euler verified (26.20) for all integer values of m and for two fractional values, $m = 1/2$ and $m = 3/2$. In general, he meant $1 \cdot 2 \cdot 3 \cdots (m-1)$ to stand for the gamma function $\Gamma(m)$. For positive integer values of m, he made illegitimate but successful use of the Euler–Maclaurin summation formula to sum the divergent series

$$x^m - (x+1)^m + (x+2)^m - (x+3)^m + \cdots. \quad (26.23)$$

By doing this, he could bring into play the Bernoulli numbers that appear in the Euler–Maclaurin summation, using them to evaluate the sums on the left-hand side of (26.21). In fact, he could have done this without using Euler–Maclaurin, had he first applied the change of variables $x = e^{-y}$ in (26.22). Euler believed that a divergent series, especially an alternating series, had a definite value, obtainable by varying methods.

In order to verify (26.20) for $m = 0$, Euler used (26.16) and the series for $\ln 2$. For negative integers, he noted that under the transformation $m \to 1 - m$, both sides were converted into their reciprocals. For the right-hand side, he required the reflection formula for the gamma function $\Gamma(m)\Gamma(1-m) = \pi/\sin \pi m$, and it appears that Euler explicitly stated this formula for the first time in this paper. For $m = 1/2$, Euler used the value $\Gamma(1/2) = \sqrt{\pi}$; he had known this value since 1729, but it also followed immediately from his reflection formula. Finally, for $m = 3/2$, Euler computed both sides to several decimal places, checking that the results were identical. To do this, he applied the Euler–Maclaurin formula to sum the divergent series $1 - \sqrt{2} + \sqrt{3} - \sqrt{4} + \cdots = 0.380129$, as well as the convergent series

$$1 - \frac{1}{2\sqrt{2}} + \frac{1}{3\sqrt{3}} - \frac{1}{4\sqrt{4}} + \cdots = 0.765158.$$

G. Faber, the editor of vol. 15 of Euler's *Opera Omnia* in which this paper appeared, noted that the values could be expressed more exactly as 0.380105 and 0.765147.

At the end of his 1749 paper, Euler mentioned without proof the functional equation for a special L-function:

$$\frac{1 - 3^{n-1} + 5^{n-1} - 7^{n-1} + \text{etc.}}{1 - 3^{-n} + 5^{-n} - 7^{-n} + \text{etc.}} = \frac{1 \cdot 2 \cdot 3 \cdots (n-1) 2^n}{\pi^n} \sin \frac{n\pi}{2}. \tag{26.24}$$

These results on the functional relations for the zeta and L-functions went unnoticed. In the 1840s, the functional equation (26.20) was given a complete proof for $0 < m < 1$ by the Swedish mathematician Carl Johan Malmsten (1814–1886), who mentioned seeing (26.20) somewhere in Euler. Oscar Schlömilch independently found a proof and stated it as a problem in a journal. A solution was published by Thomas Clausen (1801–1885) in 1858 and another was noted by Eisenstein on the last blank page in his copy of the *Disquisitiones* in the French translation of 1807. André Weil conjectured that Eisenstein discussed this topic with Riemann, providing the impetus for Riemann's well-known paper of 1859 on the zeta function. Indeed, Riemann and Eisenstein had been close friends in Berlin, and Eisenstein's note is dated April 1849, just before Riemann left Berlin for Göttingen.

In 1737, Euler discovered another important result on the zeta function, that $\sum_{n=1}^{\infty} 1/n^s$ could be expressed as an infinite product involving prime numbers:

$$\sum_{n=1}^{\infty} 1/n^s = \prod_p (1 - p^{-s})^{-1}. \tag{26.25}$$

This formula has important consequences, just as the product formulas for the trigonometric functions have significant corollaries. Euler himself almost immediately saw that he could derive the infinitude of the primes from this formula. He later extended the product formula to some L-series. For example,

$$1 - \frac{1}{3^s} + \frac{1}{5^s} - \frac{1}{7^s} + \cdots = \prod (1 - p^{-s})^{-1} (1 + q^{-s})^{-1},$$

where p and q are primes of the form $4n+1$ and $4n+3$, respectively. He employed this formula to show that there were infinitely many primes of the form $4n+1$, and infinitely many of the form $4n+3$.

In order to extend Euler's result to primes in arithmetic progressions, Dirichlet defined the concept of a character; in modern terms, this would be a homomorphism from the residue classes of integers, modulo m, to the complex numbers of magnitude one:

$$\chi : \mathbb{Z}_m^\times \to \mathbb{C}^\times, \quad \chi(ab) = \chi(a)\chi(b). \tag{26.26}$$

Note that $\chi(a)$ is automatically a $\phi(m)$th root of unity, where ϕ is the Euler totient function. Dirichlet constructed series corresponding to these characters:

$$L(\chi, x) = \sum_{n=1}^{\infty} \frac{\chi(n)}{n^s}. \tag{26.27}$$

He noted that terms with n not relatively prime to m were omitted. In modern notation, we set $\chi(n) = 0$ if the greatest common divisor (gcd) of n and m is greater than one. For example, for $m = 4$, $\mathbb{Z}_4^\times$ consists of two residue classes relatively prime to 4, represented by 1 and 3. Then $\chi(1) = 1$, $\chi(3) = -1$ is a character defined on $\mathbb{Z}_4^\times$ and the corresponding series would be

$$1 - \frac{1}{3^s} + \frac{1}{5^s} - \frac{1}{7^s} + \cdots .$$

In this way, Dirichlet systematized and generalized Euler's results. In particular, he had

$$L(\chi, s) = \prod_p (1 - \chi(p)/p^s)^{-1},$$

where the product was taken over the primes not dividing m.

26.2 Euler's First Evaluation of $\sum 1/n^{2k}$

Euler's evaluation was based on the factorization given by (26.1) and (26.2):

$$1 - \frac{1}{y}\left(\frac{x}{1} - \frac{x^3}{1 \cdot 2 \cdot 3} + \frac{x^5}{1 \cdot 2 \cdot 3 \cdot 5} - \cdots\right) = \left(1 - \frac{x}{A_1}\right)\left(1 - \frac{x}{A_2}\right)\left(1 - \frac{x}{A_3}\right)\cdots . \tag{26.28}$$

Note that Euler wrote $A, B, C, D, \ldots$ instead of $A_1, A_2, A_3, A_4, \ldots$. He observed that the coefficients of the powers of x were the elementary symmetric functions of the infinitely many variables $1/A_1, 1/A_2, 1/A_3, \ldots$. Therefore, by equating coefficients, he had

$$\alpha \equiv \sum \frac{1}{A_i} = \frac{1}{y}, \quad \beta \equiv \sum_{i<j} \frac{1}{A_i A_j} = 0, \quad \gamma \equiv \sum_{i<j<k} \frac{1}{A_i A_j A_k} = \frac{1}{6y} \text{ etc.,} \tag{26.29}$$

where α, β, γ, δ, etc., denoted the elementary symmetric functions. Euler then applied the Girard–Newton formulas (18.3) connecting the sums of the squares, cubes, fourth powers, etc., of $1/A_1$, $1/A_2$, $1/A_3$, ... with the symmetric functions α, β, γ, etc. Thus, he got

$$\sum \frac{1}{A_i^2} = \alpha^2 - 2\beta, \quad \sum \frac{1}{A_i^3} = \alpha^3 - 3\alpha\beta + 3\gamma,$$

$$\sum \frac{1}{A_i^4} = \alpha^4 - 4\alpha^2\beta + 4\alpha\gamma + 2\beta^2 - 4\delta, \text{ etc.}$$

For $y = 1$, the roots A_1, A_2, A_3, ... of the equation $\sin x = 1$ were

$$\frac{\pi}{2}, \frac{\pi}{2}, -\frac{3\pi}{2}, -\frac{3\pi}{2}, \frac{5\pi}{2}, \frac{5\pi}{2}, \ldots,$$

since $\sin x - 1 = 0$ had double roots. Thus, Euler obtained equations (26.6), (26.7), and (26.8). Clearly, he could continue the calculations to arbitrarily large powers of the reciprocals of the roots. Euler explicitly wrote the values of $\sum 1/n^{2k}$ for $k = 1, 2, \ldots, 6$, and the last of these turned out to be

$$1 + \frac{1}{2^{12}} + \frac{1}{3^{12}} + \frac{1}{4^{12}} + \cdots = \frac{691\pi^{12}}{6825 \cdot 93555}. \tag{26.30}$$

The appearance of the fairly large prime 691 may have alerted Euler to the connection of zeta values with Bernoulli numbers. Recall that this prime had already appeared in the Euler–Maclaurin series he had found only two or three years earlier, and at the time he discovered (26.30), he was still intensely studying the Euler–Maclaurin summation.

26.3 Euler: Bernoulli Numbers and $\sum 1/n^{2k}$

In a paper of 1740, Euler explained the connection between the Bernoulli numbers appearing in the Euler–Maclaurin formula and the sums $\sum 1/n^{2k}$. A year earlier, he had found the partial fractions expansion of $\cot x$ and he made use of this in his explanation. Euler started with a generating function for the sums $\sum 1/n^{2k}$ and changed the order of summation to obtain a partial fractions expansion that he could recognize as a cotangent function. Denoting the generating function by S, he had by a rearrangement of terms

$$S = \left(\sum_{n=1}^{\infty} \frac{1}{n^2}\right)x^2 + \left(\sum_{n=1}^{\infty} \frac{1}{n^4}\right)x^4 + \left(\sum_{n=1}^{\infty} \frac{1}{n^6}\right)x^6 + \cdots$$

$$= x^2 + x^4 + x^6 + \cdots + \frac{x^2}{2^2} + \frac{x^4}{2^4} + \frac{x^6}{2^6} + \cdots + \frac{x^2}{3^2} + \frac{x^4}{3^4} + \frac{x^6}{3^6} + \cdots$$

$$= \frac{x^2}{1-x^2} + \frac{x^2}{2^2 - x^2} + \frac{x^2}{3^2 - x^2} + \cdots = \frac{1}{2} - \frac{\pi x}{2 \tan \pi x}. \tag{26.31}$$

At this point, Euler could have used equation (26.15) to derive the value of $\sum 1/n^{2k}$ in terms of Bernoulli numbers. By expressing $\tan \pi x$ in terms of the exponential function, he would have obtained

$$1 - \frac{\pi x}{\tan \pi x} = 1 + \pi i x - \frac{2\pi i x e^{2\pi i x}}{e^{2\pi i x} - 1}. \tag{26.32}$$

26.3 Euler: Bernoulli Numbers and $\sum 1/n^{2k}$

He next could have used his generating function for the Bernoulli numbers appearing in the Euler–Maclaurin formula to express the right-hand side as

$$\frac{1}{2}\sum_{n=1}^{\infty}(-1)^{n-1}\frac{B_{2n}}{(2n)!}(2\pi x)^{2n}.$$

Now, in 1740 Euler was just beginning to delve into the connection between the circular and exponential functions; he was not yet ready to make full use of it. For example, in a letter to Johann Bernoulli written during this period, he explained the equality of $2\cos x$ and $e^{ix}+e^{-ix}$ by means of differential equations. Similarly, in his 1740 paper, he proved (26.32) through the use of differential equations. Thus, Euler continued his argument, proceeding to define $A, B, C, \ldots$ by $A = \frac{1}{\pi^2}\sum 1/n^2$, $B = \frac{1}{\pi^4}\sum 1/n^4$, etc. Since

$$S = \frac{1}{2}\left(1 - \frac{\pi x}{\tan \pi x}\right), \quad u = \arctan\frac{u}{1-2S},$$

where $u = \pi x$, a simple calculation showed Euler that S satisfied

$$2u\frac{dS}{du} + 2S = u^2 + 4S^2.$$

He substituted the series $S = Au^2 + Bu^4 + Cu^6 + Du^8 + \cdots$ into the differential equation and determined that

$$A = \frac{1}{6}, \quad B = \frac{2A^2}{5}, \quad C = \frac{4AB}{7},$$

$$D = \frac{AC + 2B^2}{9}, \quad E = \frac{4AD + BC}{11}, \text{ etc.} \quad (26.33)$$

He then observed that the coefficients in the Euler–Maclaurin series were generated by

$$s = \frac{xe^x}{e^x - 1} \equiv 1 + \alpha x + \beta x^2 + \gamma x^3 + \delta x^4 + \text{ etc.}$$

and saw that s satisfied the differential equation

$$x\frac{ds}{dx} - s - sx + s^2 = 0.$$

By substituting the series for s, he obtained relations for the coefficients $\alpha, \beta, \gamma, \delta$, etc. He noted that except for $\alpha = 1/2$, the coefficients of the odd powers were zero. To see this more easily, the reader may consider the fact that $-x/2 + xe^x/(e^x - 1)$ is an even function. Euler next set $\beta = A/2$, $\delta = -B/2^3$, $\zeta = C/2^5$, $\theta = -D/2^7$, $\chi = E/2^9$, etc., where ζ, θ, χ denoted the coefficients of x^6, x^8, x^{10}, respectively. He then showed that these $A, B, C, \ldots$ also satisfied the relations (26.33). Thus, Euler had the formula we now write as

$$\zeta(2n) = 1 + \frac{1}{2^{2n}} + \frac{1}{3^{2n}} + \frac{1}{4^{2n}} + \cdots = (-1)^{n-1}\frac{2^{2n-1}\pi^{2n}}{(2n)!}B_{2n}. \quad (26.34)$$

26.4 Euler's Evaluation of Some L-Series Values by Partial Fractions

Euler's essential idea in the derivation of his famous zeta value formula, proved in the last section, was the partial fractions expansion of $\pi/\tan \pi x$. After his move to Berlin in 1741, Euler followed this up with a paper in the Berlin Academy journal of 1743. There he showed how the same partial fractions could also be applied to the derivation of several L-series values. He started with

$$\frac{\pi}{\sin s\pi} = \frac{1}{s} + \frac{1}{1-s} - \frac{1}{1+s} - \frac{1}{2-s} + \frac{1}{2+s} + \frac{1}{3-s} - \frac{1}{3+s} - \text{etc.} \qquad (26.35)$$

$$\frac{\pi}{\tan s\pi} = \frac{1}{s} - \frac{1}{1-s} + \frac{1}{1+s} - \frac{1}{2-s} + \frac{1}{2+s} - \frac{1}{3-s} + \frac{1}{3+s} - \text{etc,} \qquad (26.36)$$

where he took s to be a rational number $s = p/q$. He assigned specific integer values to p and q and evaluated several series, including (26.5), (26.11), and (26.12).

To get the series for the squares of the partial fractions, Euler took the derivatives of (26.35) and (26.36) to get

$$\frac{\pi^2 \cos \pi s}{(\sin \pi s)^2} = \frac{1}{s^2} - \frac{1}{(1-s)^2} - \frac{1}{(1+s)^2} + \frac{1}{(2-s)^2} + \frac{1}{(2+s)^2} - \frac{1}{(3-s)^2} - \text{etc.,} \qquad (26.37)$$

$$\frac{\pi^2}{(\sin \pi x)^2} = \frac{1}{s^2} + \frac{1}{(1-s)^2} + \frac{1}{(1+s)^2} + \frac{1}{(2-s)^2} + \frac{1}{(2+s)^2} + \frac{1}{(3-s)^2} + \text{etc.} \qquad (26.38)$$

Among examples of these relations, for $s = 1/3$ Euler gave

$$\frac{2\pi^2}{27} = 1 - \frac{1}{2^2} - \frac{1}{4^2} + \frac{1}{5^2} + \frac{1}{7^2} - \frac{1}{8^2} - \frac{1}{10^2} + \text{etc.,} \qquad (26.39)$$

$$\frac{4\pi^2}{27} = 1 + \frac{1}{2^2} + \frac{1}{4^2} + \frac{1}{5^2} + \frac{1}{7^2} + \frac{1}{8^2} + \frac{1}{10^2} + \text{etc.,} \qquad (26.40)$$

and for $s = 1/4$ in (26.37) he obtained

$$\frac{\pi^2}{8\sqrt{2}} = 1 - \frac{1}{3^2} - \frac{1}{5^2} + \frac{1}{7^2} + \frac{1}{9^2} - \frac{1}{11^2} - \frac{1}{13^2} + \text{etc.} \qquad (26.41)$$

He knew that he could obtain (26.40) directly from (26.9). Near the end of the paper he noted that if P and Q denoted the left-hand sides of (26.35) and (26.36), respectively, then

$$\frac{(-1)^{n-1}}{(n-1)!} \frac{d^{n-1}P}{ds^{n-1}} = \frac{1}{s^n} + (-1)^{n-1} \frac{1}{(1-s)^n} - \frac{1}{(1+s)^n} + (-1)^n \frac{1}{(2-s)^n} \\ + \frac{1}{(2+s)^n} + (-1)^{n-1} \frac{1}{(3-s)^n} - \text{etc.} \qquad (26.42)$$

$$\frac{(-1)^{n-1}}{(n-1)!} \frac{d^{n-1}Q}{ds^{n-1}} = \frac{1}{s^n} + (-1)^{n-1} \frac{1}{(1-s)^n} + \frac{1}{(1+s)^n} + (-1)^{n-1} \frac{1}{(2-s)^n} \\ + \frac{1}{(2+s)^n} + (-1)^{n-1} \frac{1}{(3-s)^n} + \text{etc.} \qquad (26.43)$$

With these expansions, it was possible to obtain more L-series values. It may be helpful to observe that in Dirichlet's terms, the series (26.39) and (26.40) corresponded to L-series with character modulo 3. In (26.27), put $\chi(3m \pm 1) = \pm 1$ to obtain the first series and put $\chi(3m \pm 1) = 1$ to obtain the second series. On the other hand, series (26.41) would be defined by the character modulo 8, such that $\chi(8m \pm 1) = 1$ and $\chi(8m \pm 3) = -1$.

26.5 Euler's Evaluation of $\sum 1/n^2$ by Integration

Because some mathematicians raised objections to his first evaluation of $\sum 1/n^2$, Euler looked for other methods. Now his evaluations by partial fractions were immune to these objections, but even before finding the partial fractions method, Euler discovered an ingenious technique using integration. In 1737, Euler worked out and communicated to Johann Bernoulli his integration method. But the paper containing this method appeared in 1743 in the *Journal littéraire d'Allemagne*. It was unusual for Euler to use this journal; consequently, the paper received very scant notice. It was finally reprinted in the 1907–1908 *Bibliotheca Mathematica* as a forgotten work of Euler. However, Euler's evaluation of $\sum 1/(2n+1)^2$, taken from this paper, was reproduced without attribution in a 1750s calculus book by Simpson.

Briefly (and in modern notation), Euler started with Newton's series

$$\arcsin x = x + \frac{1}{2}\frac{x^3}{3} + \frac{1 \cdot 3}{2 \cdot 4}\frac{x^5}{5} + \frac{1 \cdot 3 \cdot 5}{2 \cdot 4 \cdot 6}\frac{x^7}{7} + \cdots$$

to get

$$\frac{1}{2}(\arcsin x)^2 = \int_0^x \frac{\arcsin t}{\sqrt{1-t^2}} dt = \int_0^x \left(t + \frac{1}{2}\frac{t^3}{3} + \cdots\right)\frac{dt}{\sqrt{1-t^2}}. \quad (26.44)$$

Assuming n was odd, since only odd powers appeared in the series, integration by parts gave him

$$\int_0^1 \frac{x^{n+2}}{\sqrt{1-x^2}} dx = \frac{n+1}{n+2} \int_0^1 \frac{x^n}{\sqrt{1-x^2}} dx$$
$$= \frac{(n+1)(n-1)(n-3)\cdots 2}{(n+2)(n)(n-2)\cdots 3}. \quad (26.45)$$

Euler applied this to (26.44) with $x = 1$ to obtain

$$\frac{\pi^2}{8} = 1 + \frac{1}{3^2} + \frac{1}{5^2} + \frac{1}{7^2} + \cdots.$$

Unfortunately, this method could not be extended to $\sum 1/n^{2k}$ for $k = 2, 3, \ldots$, though Euler attempted it. For example, he considered the series for $(\arcsin x)^2$ divided by $\sqrt{1-x^2}$ and integrated over $(0,1)$. After some similar calculations, he obtained

$$\frac{\pi^3}{48} = \frac{1}{2^2}\cdot\frac{\pi}{2} + \frac{1}{4^2}\cdot\frac{\pi}{2} + \frac{1}{6^2}\cdot\frac{\pi}{2} + \frac{1}{8^2}\cdot\frac{\pi}{2} + \text{etc.,}$$

a result equivalent to $\pi^2/6 = \sum 1/n^2$.

In a letter dated August 27, 1737 of the Russian or Julian calendar, Euler communicated to his former teacher, Johann Bernoulli, the evaluation of $\sum 1/n^2$ by means of (26.44). In his reply of November 6 (Gregorian), Bernoulli expressed his admiration for this method and observed that it had led him to find a new series for $\pi^2/8$:

$$\frac{1}{1\cdot 2} + \frac{2}{1\cdot 3\cdot 4} + \frac{2\cdot 4}{1\cdot 3\cdot 5\cdot 6} + \frac{2\cdot 4\cdot 6}{1\cdot 3\cdot 5\cdot 7\cdot 8} + \frac{2\cdot 4\cdot 6\cdot 8}{1\cdot 3\cdot 5\cdot 7\cdot 9\cdot 10} + \cdots = \frac{\pi^2}{8}. \quad (26.46)$$

We note that Bernoulli wrote C instead of π. Since the Russian calendar at that time was about ten days behind the Gregorian calendar, keeping track of correspondence can be challenging.

Bernoulli gave no further details in his letter, but in 1742 he offered an explanation in the fourth volume of his *Opera Omnia*. His method was to divide Newton's transformed series for $\arctan t$, written up in the "De Computo," by $1+t^2$ and then integrate. Recall that Newton had not published his work, so the alternative series for $\arctan t$ in powers of $t/(1+t^2)$ was Bernoulli's rediscovery. Thus, Bernoulli had

$$\frac{(\arctan t)^2}{2} = \int \frac{\arctan t}{1+t^2}\,dt$$

$$= \int \left(\frac{t}{(1+t^2)^2} + \frac{2t^3}{1\cdot 3\cdot (1+t^2)^3} + \frac{2\cdot 4t^5}{1\cdot 3\cdot 5(1+t^2)^4} + \cdots\right)dt. \quad (26.47)$$

He obtained (26.46) by integrating this formula over $(0,\infty)$. Concerning (26.46), Euler noted in a letter dated December 10, 1737 (Julian), that he had found the more general series

$$\frac{1}{2}(\arcsin x)^2 = \frac{x^2}{1\cdot 2} + \frac{2\cdot x^4}{1\cdot 3\cdot 4} + \frac{2\cdot 4\cdot x^6}{1\cdot 3\cdot 5\cdot 6} + \frac{2\cdot 4\cdot 6\cdot x^8}{1\cdot 3\cdot 5\cdot 7\cdot 8} + \cdots. \quad (26.48)$$

Euler went on to remark that Bernoulli's formula followed from this by taking $x=1$. Moreover, he observed that other interesting series would result by taking $x = 1/2, 1/\sqrt{2}$, or $\sqrt{3}/2$.

In his 1743 paper, referred to earlier, Euler gave a derivation of (26.48) by observing that $(\arcsin x)^2$ satisfied the second-order linear differential equation

$$(1-x^2)\frac{d^2 y}{dx^2} - x\frac{dy}{dx} - 2 = 0. \quad (26.49)$$

He then solved this equation by infinite series to prove (26.48). Observe, however, that Bernoulli could have obtained Euler's formula (26.48) from his own (26.47): Rewrite (26.47) as

$$\frac{1}{2}(\arctan z)^2 = \int_0^z \frac{\arctan t}{1+t^2}\,dt$$

$$= \sum_{n=0}^\infty \frac{2^{2n}(n!)^2}{(2n+1)!}\int_0^z \left(\frac{t^2}{1+t^2}\right)^n \frac{t\,dt}{(1+t^2)^2}$$

$$= \sum_{n=0}^{\infty} \frac{2^{2n}(n!)^2}{2(2n+1)!} \int_0^{z^2/(1+z^2)} u^n \, du$$

$$= \sum_{n=0}^{\infty} \frac{2^{2n}(n!)^2}{(2n+2)!} \left(\frac{z^2}{1+z^2}\right)^{n+1}. \tag{26.50}$$

Now set $x^2 = z^2/(1+z^2)$ so that $z = \tan(\arcsin x)$ and (26.48) follows.

We here note the remarkable fact that the Japanese mathematician Takebe Katahiro (1664–1739) published both Bernoulli's (26.46) and Euler's (26.48) series in his 1722 treatise, *Tetsujutsu Sankei*. Takebe's approach was very different, since he apparently made his discoveries of these series on the basis of a considerable amount of numerical work. He related the length of an arc of a circle determined by a chord to the height of the chord. The latter would be the distance between the midpoint of the arc and the midpoint of the chord. After finding the series, Takebe sought an analytic justification for it. The *Tetsujutsu* exerted great influence on the development of mathematics in eighteenth-century Japan, spurring Japanese mathematicians to discover other series for π.

26.6 N. Bernoulli's Evaluation of $\sum 1/(2n+1)^2$

Euler was eager to find many different evaluations of $\sum 1/n^2$ and in this he was assisted by Niklaus I Bernoulli who in 1738 published a very interesting method by squaring the Madhava–Leibniz series for $\pi/4$, given by (26.5). Bernoulli's derivation involved many transformations of series but in a July 1738 letter to Johann Bernoulli, Euler gave a shorter proof by greatly simplifying the second portion. We present this simplified proof, whose fundamental idea remained Bernoulli's.

Bernoulli first observed that by squaring equation (26.5) he had

$$\frac{\pi^2}{16} = \left(1 - \frac{1}{3} + \frac{1}{5} - \frac{1}{7} + \cdots\right)^2 = \sum_{n=0}^{\infty} \frac{1}{(2n+1)^2} - 2\sum_{n=0}^{\infty} \frac{1}{2n+1} \cdot \frac{1}{2n+3}$$
$$+ 2\sum_{n=0}^{\infty} \frac{1}{2n+1} \cdot \frac{1}{2n+5} - 2\sum_{n=0}^{\infty} \frac{1}{2n+1} \cdot \frac{1}{2n+7} + \cdots. \tag{26.51}$$

The first series on the right was the sum of the squares of 1, 1/3, 1/5, …. The other series were the sums of the mixed terms obtained by squaring. He then noted that

$$2\sum_{n=0}^{\infty} \frac{1}{2n+1} \cdot \frac{1}{2n+3} = \sum_{n=0}^{\infty} \left(\frac{1}{2n+1} - \frac{1}{2n+3}\right) = 1,$$

$$2\sum_{n=0}^{\infty} \frac{1}{2n+1} \cdot \frac{1}{2n+5} = \frac{1}{2}\sum_{n=0}^{\infty} \left(\frac{1}{2n+1} - \frac{1}{2n+5}\right) = \frac{1}{2}\left(1 + \frac{1}{3}\right),$$

$$2\sum_{n=0}^{\infty} \frac{1}{2n+1} \cdot \frac{1}{2n+7} = \frac{1}{3}\sum_{n=0}^{\infty} \left(\frac{1}{2n+1} - \frac{1}{2n+7}\right) = \frac{1}{3}\left(1 + \frac{1}{3} + \frac{1}{5}\right),$$

and so on. Hence,

$$\frac{\pi^2}{16} = \sum_{n=0}^{\infty} \frac{1}{(2n+1)^2} - \left(1 - \frac{1}{2}\left(1 + \frac{1}{3}\right) + \frac{1}{3}\left(1 + \frac{1}{3} + \frac{1}{5}\right) - \cdots\right). \qquad (26.52)$$

Euler's simplification took effect at this point. To sum the series within the parentheses in (26.52), he observed that

$$\begin{aligned}\frac{\arctan x}{1+x^2} &= (1 - x^2 + x^4 - x^6 + \cdots)\left(x - \frac{x^3}{3} + \frac{x^5}{5} - \frac{x^7}{7} + \cdots\right) \\ &= x - x^3\left(1 + \frac{1}{3}\right) + x^5\left(1 + \frac{1}{3} + \frac{1}{5}\right) - x^7\left(1 + \frac{1}{3} + \frac{1}{5} + \frac{1}{7}\right) + \cdots.\end{aligned} \qquad (26.53)$$

Euler then integrated over (0,1) to obtain

$$\frac{1}{2}(\arctan 1)^2 = \frac{1}{2}\left(\frac{\pi}{4}\right)^2 = \frac{1}{2} \cdot \frac{1}{4}\left(1 + \frac{1}{3}\right) + \frac{1}{6}\left(1 + \frac{1}{3} + \frac{1}{5}\right) - \cdots.$$

Thus, he showed that the series within parentheses in (26.52) was equal to $\pi^2/16$; hence, $\sum 1/(2n+1)^2 = \pi^2/8$, as was required. Not only did Euler simplify N. Bernoulli's proof but he also obtained a more general result. This result and the inspiration and assistance of his friend Christian Goldbach eventually lead him to a fruitful study of double zeta values.

26.7 Euler and Goldbach: Double Zeta Values

The route toward the consideration of double zeta values started with a theorem communicated by Euler to Goldbach on August 28, 1742. Note that this is a generalization of the last equation, mentioned above, in N. Bernoulli's evaluation. If

$$s = 1 + \frac{a}{n+1} + \frac{aa}{2n+1} + \frac{a^3}{3n+1} + \frac{a^4}{4n+1} + \cdots, \quad \text{then} \qquad (26.54)$$

$$\begin{aligned}\frac{ss}{2} &= \frac{1}{2} + \frac{a}{n+2}\left(1 + \frac{1}{n+1}\right) + \frac{aa}{2n+2}\left(1 + \frac{1}{n+1} + \frac{1}{2n+1}\right) \\ &\quad + \frac{a^3}{3n+2}\left(1 + \frac{1}{n+1} + \frac{1}{2n+1} + \frac{1}{3n+1}\right) \\ &\quad + \frac{a^4}{4n+2}\left(1 + \frac{1}{n+1} + \frac{1}{2n+1} + \frac{1}{3n+1} + \frac{1}{4n+1}\right) + \text{etc.} \qquad (26.55)\end{aligned}$$

Goldbach was intrigued by this result and raised some questions about it in a letter dated October 1, 1742. Euler responded by explaining how Niklaus I Bernoulli's result, discussed in the previous section, could be obtained from the theorem. This led Goldbach to consider the series now known as double zeta values. In a letter of December 24,

26.7 Euler and Goldbach: Double Zeta Values

1742 to Euler, Goldbach wrote that the type of series in Euler's theorem suggested the study of the series

$$1 + \frac{1}{2^n}\left(1 + \frac{1}{2^m}\right) + \frac{1}{3^n}\left(1 + \frac{1}{2^m} + \frac{1}{3^m}\right) + \frac{1}{4^n}\left(1 + \frac{1}{2^m} + \frac{1}{3^m} + \frac{1}{4^m}\right) + \cdots. \quad (26.56)$$

Let us denote this series by $\zeta_G(n, m)$. In modern terminology, this is almost the double zeta value defined by

$$\zeta(n, m) = 1 + \frac{1}{2^n} + \frac{1}{3^n}\left(1 + \frac{1}{2^m}\right) + \frac{1}{4^n}\left(1 + \frac{1}{2^m} + \frac{1}{3^m}\right) + \cdots$$

$$= \zeta_G(n, m) - \zeta(m + n). \quad (26.57)$$

Goldbach further wrote that he had found

$$\zeta_G(3, 1) = \pi^4/72, \quad \text{and} \quad 2\zeta_G(5, 1) + \zeta_G(4, 2) = 19\pi^6/(2 \cdot 5 \cdot 7 \cdot 3^4). \quad (26.58)$$

He also mentioned that, while he could not evaluate $\zeta_G(n, m)$, he could handle $\zeta_G(n, m) + \zeta_G(m, n)$. Euler must have been greatly fascinated by these series, for on January 5, 1743, he responded with details of the proof of his theorem and then two weeks later gave a number of evaluations of particular cases of Goldbach's series. Thus, he had

$$\zeta_G(3, 1) = \frac{1}{2}(\zeta(2))^2; \quad \zeta_G(5, 1) = \zeta(2)\zeta(4) - \frac{1}{2}(\zeta(3))^2;$$

$$\zeta_G(7, 1) = \zeta(2)\zeta(6) - \zeta(3)\zeta(5) + \frac{1}{2}(\zeta(4))^2;$$

$$\zeta_G(9, 1) = \zeta(2)\zeta(8) - \zeta(3)\zeta(7) + \zeta(4)\zeta(6) - \frac{1}{2}(\zeta(5))^2;$$

$$\zeta_G(2, 2) = \frac{1}{2}(\zeta(2))^2 + \frac{1}{2}\zeta(4); \quad \zeta_G(4, 2) = (\zeta(3))^2 - \frac{1}{3}\zeta(6);$$

$$\zeta_G(6, 2) = 2\zeta(3)\zeta(5) - \frac{3}{2}(\zeta(4))^2 + \frac{1}{4}\zeta(8);$$

$$\zeta_G(8, 2) = 2\zeta(3)\zeta(7) - 3\zeta(4)\zeta(6) + \frac{4}{2}(\zeta(5))^2 - \frac{1}{5}\zeta(10);$$

$$\zeta_G(3, 3) = \frac{1}{2}(\zeta(3))^2 + \frac{1}{2}\zeta(6);$$

$$\zeta_G(5, 3) = \frac{3}{2}(\zeta(4))^2 - \frac{5}{8}\zeta(8) = \frac{\pi^8}{16 \cdot 3 \cdot 25 \cdot 7}. \quad (26.59)$$

Euler showed how Goldbach's results (26.58) could be derived from these values and, concerning Goldbach's remark on $\zeta_G(m, n) + \zeta_G(n, m)$, he observed that

$$\zeta(m)\zeta(n) + \zeta(m + n) = \zeta_G(m, n) + \zeta_G(n, m). \quad (26.60)$$

Presumably, Goldbach had also found this elementary but basic formula now written as

$$\zeta(m)\zeta(n) - \zeta(m + n) = \zeta(m, n) + \zeta(n, m). \quad (26.61)$$

In a letter of February 26, 1743, Euler explained that his results depended on the partial fractions identity:

$$\frac{1}{x^m(x+a)^n} = \frac{A_0}{x^m} + \frac{A_1}{x^{m-1}} + \cdots + \frac{A_{m-1}}{x} + \frac{B_0}{(x+a)^n} + \frac{B_1}{(x+a)^{n-1}} + \cdots + \frac{B_{n-1}}{x+a} \quad (26.62)$$

where

$$A_k = \frac{(-1)^k n(n+1)\cdots(n+k-1)}{k!\, a^{n+k}}; \quad B_k = \frac{(-1)^m m(m+1)\cdots(m+k-1)}{k!\, a^{m+k}}. \quad (26.63)$$

This identity was needed when he considered the series

$$\zeta(m) \times \zeta(n) - \zeta(m+n) = \sum_{x=1}^{\infty} \frac{1}{x^m} \sum_{a=1}^{\infty} \frac{1}{(x+a)^n} + \sum_{x=1}^{\infty} \frac{1}{x^n} \sum_{a=1}^{\infty} \frac{1}{(x+a)^m}. \quad (26.64)$$

Needless to say, Euler did not use this notation in his letter or in the paper he wrote up more than three decades later, in 1775. He wrote several terms of the expressions in (26.62) and (26.64) to make it clear how the series progressed. In his 1775 paper, Euler noted that (26.62) could be obtained by the method he had presented in his *Introductio* of 1748. Here note that Euler's notation for $\zeta(s)$ and $\zeta_G(m,n)$ used the integral sign for summation:

$$\zeta(s) = \int \frac{1}{x^s}, \quad \zeta_G(m,n) = \int \frac{1}{z^m}\left(\frac{1}{y^n}\right). \quad (26.65)$$

To evaluate (26.64), Euler started with the simple observation that

$$\sum_{x=1}^{\infty} \frac{1}{(x+a)^t} = \sum_{x=1}^{\infty} \frac{1}{x^t} - \sum_{k=1}^{a} \frac{1}{k^t}. \quad (26.66)$$

He then applied (26.62) and (26.63) to transform the series s:

$$\begin{aligned}
s &= \sum_{x=1}^{\infty} \frac{1}{x^m(x+a)^n} \\
&= \frac{1}{a^n} \sum_{z=1}^{\infty} \frac{1}{z^m} - \frac{n}{a^{n+1}} \sum_{z=1}^{\infty} \frac{1}{z^{m-1}} + \frac{n(n+1)}{2!\, a^{n+2}} \sum_{z=1}^{\infty} \frac{1}{z^{m-2}} - \cdots \\
&\quad + (-1)^m \left(\frac{1}{a^m} \sum_{z=1}^{\infty} \frac{1}{z^n} + \frac{m}{a^{m+1}} \sum_{z=1}^{\infty} \frac{1}{z^{n-1}} + \frac{m(m+1)}{2!\, a^{m+2}} \sum_{z=1}^{\infty} \frac{1}{z^{n-2}} + \cdots \right) \\
&\quad + (-1)^{m+1} \left(\frac{1}{a^m} \sum_{k=1}^{a} \frac{1}{k^n} + \frac{m}{a^{m+1}} \sum_{k=1}^{a} \frac{1}{k^{n-1}} + \frac{m(m+1)}{2!\, a^{m+2}} \sum_{k=1}^{a} \frac{1}{k^{n-2}} + \cdots \right).
\end{aligned} \quad (26.67)$$

26.7 Euler and Goldbach: Double Zeta Values

By summing over a, he then obtained an expression for the first series on the right-hand side of (26.64):

$$\sum_{x=1}^{\infty} \frac{1}{x^m} \sum_{a=1}^{\infty} \frac{1}{(x+a)^n} = \zeta(n)\zeta(m) - n\zeta(n+1)\zeta(m-1) + \frac{n(n+1)}{2!}\zeta(n+2)\zeta(m-2) - \cdots$$

$$+ (-1)^m \left(\zeta(m)\zeta(n) + m\zeta(m+1)\zeta(n-1) + \frac{m(m+1)}{2!}\zeta(m+2)\zeta(n-2) + \cdots \right)$$

$$+ (-1)^{m-1} \left(\zeta_G(m,n) + m\zeta_G(m+1,n-1) + \frac{m(m+1)}{2!}\zeta_G(m+2,n-2) + \cdots \right). \tag{26.68}$$

By interchanging m and n, Euler immediately got the formula for $\sum \frac{1}{x^n} \sum \frac{1}{(x+a)^m}$. Thus, he had established all the formulas necessary to evaluate the results in (26.59).

Euler evaluated several specific examples in his paper. He took $m+n=3$ with $m=2, n=1$ and by applying (26.68) and using an analogous formula with m and n interchanged, he got

$$\zeta(2)\zeta(1) - \zeta(3)$$
$$= 2\zeta(2)\zeta(1) - \zeta_G(2,1) - 2\zeta(2)\zeta(1) + \zeta_G(1,2) + \zeta_G(2,1)$$
$$= \zeta_G(1,2). \tag{26.69}$$

Then by (26.60), he obtained

$$\zeta(2)\zeta(1) + \zeta(3) = \zeta_G(2,1) + \zeta_G(1,2). \tag{26.70}$$

Again, by subtracting (26.69) from (26.70), Euler could conclude that

$$\zeta_G(2,1) = 2\zeta(3), \tag{26.71}$$

or in modern notation for the double-zeta function:

$$\zeta(2,1) = \zeta(3). \tag{26.72}$$

Observe that series $\zeta(1)$ and $\zeta(1,2)$ are both logarithmically divergent, though it is possible to suitably modify Euler's argument to avoid the use of divergent series. To compute $\zeta_G(4,1)$, Euler took $m=4, n=1$ to get, after simplification,

$$\zeta_G(2,3) + \zeta_G(3,2) - \zeta_G(4,1) = 2\zeta(2)\zeta(3) - 2\zeta(5).$$

He then took $m=2, n=3$ in (26.60) to obtain

$$\zeta(2)\zeta(3) + \zeta(5) = \zeta_G(2,3) + \zeta_G(3,2). \tag{26.73}$$

Combined with the previous equation, this gave $\zeta_G(4,1) = 3\zeta(5) - \zeta(2)\zeta(3)$. Euler noted that he was unable to obtain $\zeta_G(2,3)$ and $\zeta_G(3,2)$ by this method. When he set

$m = 3, n = 2$ in (26.68), the result was once again (26.73). Euler effectively remedied this drawback by developing a new algebraic method in the later part of the paper.

The Euler–Goldbach double-zeta values can be generalized to multizeta values, defined as

$$\zeta(s_1, s_2, \ldots, s_k) = \sum_{n_1 > n_2 > \cdots n_k > 0} \frac{1}{n_1^{s_1} n_2^{s_2} \cdots n_k^{s_k}}, \qquad (26.74)$$

where $s_1, s_2 \ldots, s_k$ are natural numbers. These values have been found to have connections with basic objects in number theory, algebraic geometry, and topology. They have been studied by the use of methods from combinatorics, real and complex analysis, algebra, and number theory, creating great current interest in this topic.

Christian Goldbach, who was the first to see the potential for double-zeta series, was a Prussian but moved to Russia in the 1720s and remained there until his death. In 1725 he became secretary of conferences at the Petersburg Academy. He was a man of diverse talents and his chief hobbies were languages, number theory, differential calculus, and infinite series. He was one of the educators of the young Tsar Peter (1715–1730). As we have seen, Euler frequently communicated important results to his friend Goldbach, who proposed problems of possible interest to Euler and made helpful comments. Goldbach informed Euler of Fermat's unproved theorems and proposed the well-known Goldbach conjecture. Thus, he succeeded in directing Euler's extraordinary talents toward the field of number theory, in which Euler's other colleagues had minimal interest before Lagrange entered the scene. The first volume of P. Fuss's *Correspondance mathématique et physique de quelques célèbres géomètres du XVIIIème siècle* contains 176 letters written by Euler and Goldbach to each other over more than thirty-four years. Euler apparently regarded Goldbach as his close friend, writing him an urgent letter in 1738 when his eyesight was threatened. Goldbach then made unsuccessful attempts to relieve his friend of his burdensome responsibilities in geographical studies. We note that it was in 1766, when Euler returned to St. Petersburg after a twenty-five year stay in Berlin, that he became blind for all practical purposes.

26.8 Dirichlet's Summation of $L(1, \chi)$

The series (26.5) gives the value of $L(1, \chi)$ for the nontrivial character modulo 4, defined by $\chi(4n \pm 1) = \pm 1$. Euler employed this series to prove that primes of the form $4n + 1$ and of the form $4n - 1$ were both infinite in number. In the 1830s, J. P. G. Lejeune Dirichlet went further, giving a general evaluation of $L(1, \chi)$ to prove his results on quadratic forms and on primes in arithmetic progressions. Dirichlet first examined the case in which χ was a character modulo p, where p was a prime. We limit our discussion to this simple case. Dirichlet defined this character by taking any generator g of the cyclic group consisting of the integers modulo p without the zero element. Next, he let w be any $(p-1)$th root of unity. For n not divisible by p, he set

$$\chi(n) = w^{\gamma_n} \quad \text{where} \quad g^{\gamma_n} \equiv n \pmod{p}.$$

26.8 Dirichlet's Summation of $L(1, \chi)$

By convention, $\chi(mp) = 0$. More details on Dirichlet's theory of characters are given later in this chapter. To evaluate the L-series $\sum_{n=1}^{\infty} \frac{w^{\gamma n}}{n^s}$ at $s = 1$ when $w \neq 1$, Dirichlet first expressed the series as an integral. Note that the terms in which n was a multiple of p were taken to be 0. In a paper of 1769 on series related to the zeta function, Euler had used the idea of expressing this type of series as an integral. Like Euler, Dirichlet started with

$$\int_0^1 x^{n-1} (\log(1/x))^{s-1} dx = \frac{\Gamma(s)}{n^s}. \tag{26.75}$$

From the periodicity of the character $w^{\gamma n}$, he had

$$L_w(s) = \sum_{n=1}^{\infty} \frac{w^{\gamma n}}{n^s} = \sum_{k=1}^{p-1} w^{\gamma k} \left(\frac{1}{k^s} + \frac{1}{(k+p)^s} + \frac{1}{(k+2p)^s} + \cdots \right)$$

$$= \frac{1}{\Gamma(s)} \sum_{k=1}^{p-1} w^{\gamma k} \sum_{d=0}^{\infty} \int_0^1 x^{k+dp-1} (\log(1/x))^{s-1} dx$$

$$= \frac{1}{\Gamma(s)} \sum_{k=1}^{p-1} w^{\gamma k} \int_0^1 \frac{x^{k-1}}{1-x^p} (\log(1/x))^{s-1} dx$$

$$= \frac{1}{\Gamma(s)} \int_0^1 \frac{x^{-1} f(x)}{1-x^p} (\log(1/x))^{s-1} dx \tag{26.76}$$

$$\text{where} \quad f(x) = \sum_{k=1}^{p-1} w^{\gamma k} x^k. \tag{26.77}$$

Unlike most eighteenth-century mathematicians, Dirichlet dealt carefully with convergence and term-by-term integration, so he summed only the first $(p-1)h$ terms of the series and then showed that this sum differed from the integral (26.76) by an integral with the limit zero as $h \to \infty$. When $s = 1$, the logarithmic term in the integral (26.76) vanished. Thus, Dirichlet observed that $L_w(1)$, an integral of a rational function, could be computed in terms of logarithms and circular functions.

Dirichlet pointed out that the factors of $x^p - 1$ were of the form $x - e^{2m\pi i/p}$; hence,

$$\frac{x^{-1} f(x)}{x^p - 1} = \sum_{m=1}^{p-1} \frac{A_m}{x - e^{2m\pi i/p}},$$

$$\text{where} \quad A_m = \lim_{x \to e^{2m\pi i/p}} \frac{(x - e^{2m\pi i/p}) x^{-1} f(x)}{x^p - 1}.$$

This limit was the value of $\frac{x^{-1} f(x)}{p x^{p-1}}$ at $x = e^{2m\pi i/p}$; thus, he found

$$A_m = \frac{1}{p} f\left(e^{2m\pi i/p}\right) = \frac{1}{p} \sum_{k=1}^{p-1} w^{\gamma k} e^{2km\pi i/p}. \tag{26.78}$$

Next, Dirichlet set $km \equiv h \pmod{p}$ so that $w^{\gamma_k} = w^{-\gamma_m} w^{\gamma_h}$ and

$$f\left(e^{2m\pi i/p}\right) = w^{-\gamma_m} \sum_{h=1}^{p-1} w^{\gamma_h} e^{2h\pi i/p} = w^{-\gamma_m} f(e^{2\pi i/p}).$$

In this manner, Dirichlet arrived at

$$L_w(1) = \sum_{n=1}^{\infty} \frac{w^{\gamma_n}}{n} = -\frac{1}{p} f(e^{2\pi i/p}) \sum_{m=1}^{p-1} w^{-\gamma_m} \int_0^1 \frac{dx}{x - e^{2m\pi i/p}}. \tag{26.79}$$

He next noted that the last integral could be expressed as

$$\log(1 - e^{-2m\pi i/p}) = \log\left(2\sin\frac{m\pi}{p}\right) + i\frac{\pi}{2}\left(1 - \frac{2m}{p}\right),$$

so that

$$\sum_{n=1}^{\infty} \frac{w^{\gamma_n}}{n} = -\frac{1}{p} f\left(e^{2\pi i/p}\right) \sum_{m=1}^{p-1} w^{-\gamma_m} \left(\log\left(2\sin\frac{m\pi}{p}\right) + i\frac{\pi}{2}\left(1 - \frac{2m}{p}\right)\right). \tag{26.80}$$

Dirichlet further observed that this formula took a much simpler form when $w = -1$. This corresponded to the quadratic character $(-1)^{\gamma_n} = \left(\frac{n}{p}\right)$; he then had

$$\sum_{n=1}^{\infty} \left(\frac{n}{p}\right)\frac{1}{n} = -\frac{1}{p} f(e^{2\pi i/p}) \sum_{m=1}^{p-1} \left(\frac{m}{p}\right)\left(\log\left(2\sin\frac{m\pi}{p}\right) + i\frac{\pi}{2}\left(1 - \frac{2m}{p}\right)\right).$$

Since $\sum_{m=1}^{p-1} \left(\frac{m}{p}\right) = \sum_{m=1}^{p-1}(-1)^{\gamma_m} = 0$, he could simplify to obtain

$$\sum_{n=1}^{\infty} \left(\frac{n}{p}\right)\frac{1}{n} = -\frac{1}{p} f(e^{2\pi i/p}) \sum_{m=1}^{p-1} \left(\frac{m}{p}\right)\left(\log\left(2\sin\frac{m\pi}{p}\right) - i\frac{m\pi}{p}\right). \tag{26.81}$$

He then noted that

$$\left(\frac{p-m}{p}\right) = \left(\frac{-m}{p}\right) = \left(\frac{-1}{p}\right)\left(\frac{m}{p}\right) = (-1)^{\frac{p-1}{2}}\left(\frac{m}{p}\right) = \pm\left(\frac{m}{p}\right),$$

using plus if p took the form $4n+1$ and minus if p took the form $4n+3$. Note that when p is of the form $4n+1$, the imaginary part of the sum vanishes because $\sum m\left(\frac{m}{p}\right) = 0$ when $\left(\frac{p}{m}\right) = \left(\frac{p-m}{m}\right)$. Dirichlet could then conclude that

$$\sum_{n=1}^{\infty} \left(\frac{n}{p}\right)\frac{1}{p} = \frac{1}{p} f(e^{2\pi i/p}) \log \frac{\Pi \sin\frac{b\pi}{p}}{\Pi \sin\frac{a\pi}{p}}, \tag{26.82}$$

where a represented quadratic residues (mod p) and b nonresidues. Observe that for the case $p = 4n+3$,

$$\left(\frac{m}{p}\right)\log\left(2\sin\frac{m\pi}{p}\right) = -\left(\frac{p-m}{p}\right)\log\left(2\sin\frac{(m-p)\pi}{p}\right),$$

and hence the sum of these terms is zero and

$$\sum_{n=1}^{\infty}\left(\frac{n}{p}\right)\frac{1}{n} = \frac{\pi}{p} f(e^{2\pi i/p})\left(\sum a - \sum b\right)\sqrt{-1}. \tag{26.83}$$

Moreover, the term $f(e^{2\pi i/p})$ is the quadratic Gauss sum

$$\sum_{k=1}^{p-1}(-1)^{\gamma_k} e^{2k\pi i/p} = \sum_{k=1}^{p-1}\left(\frac{k}{p}\right) e^{2k\pi i/p} = \begin{cases} \sqrt{p}, & p = 4n+1, \\ i\sqrt{p}, & p = 4n+3 \end{cases}.$$

In this way, we obtain Dirichlet's final formulas

$$\sum_{n=1}^{\infty}\left(\frac{n}{p}\right)\frac{1}{n} = \frac{1}{\sqrt{p}} \log \frac{\Pi \sin \frac{b\pi}{p}}{\Pi \sin \frac{a\pi}{p}}, \quad p \equiv 1 \pmod{4}, \tag{26.84}$$

$$\sum_{n=1}^{\infty}\left(\frac{n}{p}\right)\frac{1}{n} = \frac{\pi}{p\sqrt{p}}\left(\sum b - \sum a\right) \quad p \equiv 3 \pmod{4}. \tag{26.85}$$

Dirichlet wrote that the last formula implied that for primes of the form $4n+3$, $\sum b > \sum a$, that is, the sum of the quadratic nonresidues was greater than the sum of the quadratic residues, and that it would be difficult to prove this in a different way.

26.9 Eisenstein's Proof of the Functional Equation

The discovery of Eisenstein's proof of the functional equation began in 1964 when B. Artmann came across Eisenstein's old copy of Gauss's *Disquisitiones* in the Giessen University Mathematical Institute Library. This book had belonged to Ferdinand Eisenstein (1823–1852) and then to Eugen Netto (1848–1919), student of Weierstrass and Kummer, before arriving at the Library. The proof, in Eisenstein's hand and dated 1849, appeared on the last blank page of the book; with the help of Artmann and the librarian, André Weil was able to examine it and to publish it in a paper of 1989. Eisenstein's proof started with the formula

$$\int_0^{\infty} e^{\sigma \psi i} \psi^{q-1} d\psi = \frac{\Gamma(q)}{(\pm \sigma)^q} e^{\pm q\pi i/2}. \tag{26.86}$$

This is in fact the Fourier transform of the function

$$f(\psi) = \begin{cases} \psi^{q-1} & \text{for } \psi > 0, \quad 0 < q < 1, \\ 0 & \text{for } \psi < 0. \end{cases}$$

For this formula, Eisenstein referred to a 1836 paper by Dirichlet on definite integrals. In that paper, Dirichlet noted that the formula was first found by Euler but that Poisson gave the proof, with the convergence condition $0 < q < 1$. Eisenstein then applied the Poisson summation formula

$$\sum_{n=-\infty}^{\infty} \phi(n) = \sum_{m=-\infty}^{\infty} \hat{\phi}(m), \tag{26.87}$$

where $\hat{\phi}$ was the Fourier transformation of ϕ, to the function

$$\phi(x) = \begin{cases} e^{2\pi\alpha(x-\beta)i}(x-\beta)^{q-1} & \text{for } x > \beta,\, 0 < \alpha < 1,\, 0 < \beta < 1, \\ 0 & \text{for } x < \beta. \end{cases}$$

He then had

$$\frac{e^{2\pi\alpha(1-\beta)i}}{(1-\beta)^{1-q}} + \frac{e^{2\pi\alpha(2-\beta)i}}{(2-\beta)^{1-q}} + \frac{e^{2\pi\alpha(3-\beta)i}}{(3-\beta)^{1-q}} + \cdots$$

$$= \sum_{\sigma=-\infty}^{\infty} \int_{\beta}^{\infty} e^{2\pi\alpha(\lambda-\beta)i}(\lambda-\beta)^{q-1} e^{2\pi i\sigma\lambda} \, d\lambda$$

$$= \sum_{\sigma=-\infty}^{\infty} \int_{0}^{\infty} e^{2\pi(\alpha+\sigma)\lambda i} e^{2\pi i\lambda\beta} \lambda^{q-1} \, d\lambda \qquad (26.88)$$

$$= \frac{\Gamma(q)}{(2\pi)^q} e^{q\pi i/2} \sum_{\sigma=0}^{\infty} \frac{e^{2\pi i\sigma\beta}}{(\sigma+\alpha)^q} + \frac{\Gamma(q)}{(2\pi)^q} e^{-q\pi i/2} \sum_{\sigma=1}^{\infty} \frac{e^{2\pi i\sigma\beta}}{(\sigma-\alpha)^q}$$

where the last step followed from (26.86). By taking $\alpha = \beta = 1/2$, he obtained the functional equation for the L-function

$$1 - \frac{1}{3^{1-q}} + \frac{1}{5^{1-q}} - \frac{1}{7^{1-q}} + \cdots = \frac{2^q \Gamma(q)}{\pi^q} \sin\frac{q\pi}{2} \left(1 - \frac{1}{3^q} + \frac{1}{5^q} + \frac{1}{7^q} + \cdots\right). \qquad (26.89)$$

Eisenstein also observed at this point that when q was replaced by $1-q$ and the two formulas were multiplied, he got another proof of Euler's reflection formula

$$\Gamma(q)\Gamma(1-q) = \frac{\pi}{\sin\pi q}.$$

26.10 Riemann's Derivations of the Functional Equation

Eisenstein may have discussed his proof of the functional equation with Riemann, perhaps inspiring Riemann's 1859 paper on the number of primes less than a given number. In this paper, Riemann used complex analysis to give two new proofs of the functional equation. One proof made use of contour integration and the second, deeper proof employed the transformation of a theta function. The latter method presaged a connection between modular forms and the corresponding Dirichlet series obtained by applying the Mellin transform.

The proof by contour integration started with two formulas due to Euler, though Riemann did not attribute them to anyone, perhaps regarding them as well known: For

26.10 Riemann's Derivations of the Functional Equation

Re $s > 0$,

$$\int_0^\infty e^{-nx} x^{s-1} dx = \frac{\Gamma(s)}{n^s}, \tag{26.90}$$

$$\Gamma(s)\zeta(s) = \int_0^\infty \frac{x^{s-1}}{e^x - 1} dx. \tag{26.91}$$

We here mention that Riemann used Gauss's notation for the gamma function: $\Pi(s-1)$. Observe that the second formula follows from the first, using the geometric series expansion

$$\frac{1}{e^x - 1} = e^{-x} + e^{-2x} + e^{-3x} + \cdots.$$

Riemann next considered the integral

$$\int \frac{(-x)^{s-1}}{e^x - 1} dx \tag{26.92}$$

over a contour from $+\infty$ to $+\infty$ in the positive sense around the boundary of a region containing in its interior 0 but no other singularities of the integrand. He noted that this integral simplified to

$$\left(e^{-\pi s i} - e^{\pi s i}\right) \int_0^\infty \frac{x^{s-1}}{e^x - 1} dx, \tag{26.93}$$

provided that one used the branch of the many-valued function $(-x)^{s-1} = e^{(s-1)\log(-x)}$ for which $\log(-x)$ was real for negative values of x. From (26.91), (26.92), and (26.93), he concluded that

$$2 \sin s\pi \, \Gamma(s) \zeta(s) = i \int_\infty^\infty \frac{(-x)^{s-1}}{e^x - 1} dx. \tag{26.94}$$

Riemann pointed out that this integral defined $\zeta(s)$ as an analytic function of s with a singularity at $s = 1$. In addition, we note that since

$$\frac{x}{e^x - 1} = 1 - \frac{1}{2}x + B_2 \frac{x^2}{2!} - B_4 \frac{x^4}{4!} + B_6 \frac{x^6}{6!} - \cdots,$$

two of Euler's famous formulas are immediate corollaries, though Riemann noted only the first one:

$$\zeta(-2n) = 0, \quad \text{and} \quad \zeta(1-2n) = \frac{(-1)^n B_{2n}}{2n}, \quad n = 1, 2, \ldots. \tag{26.95}$$

To obtain the functional equation, Riemann remarked at this point that for Re $s < 0$, the contour for the integral in (26.94) could be viewed as if defined (with a negative orientation) as the boundary of the complementary region containing the singularities $\pm 2n\pi i$, $n > 0$ of the integrand. Since the residue at $2n\pi i$ was $(-n2\pi i)^{s-1}(-2\pi i)$, he obtained the equation

$$2 \sin s\pi \, \Gamma(s) \zeta(s) = (2\pi)^s \sum n^{s-1} \left((-i)^{s-1} + i^{s-1}\right). \tag{26.96}$$

Riemann noted that by using the known properties of the gamma function, (26.96) could be seen as equivalent to the statement that:

$$\Gamma(s/2)\pi^{-s/2}\zeta(s) \tag{26.97}$$

was invariant under the transformation $s \to 1-s$. And this was the functional equation for $\zeta(s)$. In his 1859 paper, Riemann noted only the first equation in (26.95), though he clearly knew the second one as well; when combined with the functional equation, this yields a new proof of Euler's formula

$$\zeta(2n) = \frac{(-1)^{n-1}2^{2n-1}\pi^{2n}B_{2n}}{(2n)!}.$$

Riemann wrote that the expression in (26.97) and its invariance led him to consider the integral for $\Gamma(s/2)$ and thus directed him to another important derivation of the functional equation. Since

$$\Gamma(s/2)\pi^{-s/2}\frac{1}{n^s} = \int_0^\infty x^{(s/2)-1}e^{-n^2\pi x}\,dx,$$

Riemann used term-by-term integration to find that

$$\Gamma(s/2)\pi^{-s/2}\zeta(s) = \int_0^\infty x^{(s/2)-1}\sum_{n=1}^\infty e^{-n^2\pi x}\,dx. \tag{26.98}$$

It was proved by Cauchy and Poisson, and a little later by Jacobi, that

$$\sum_{n=-\infty}^\infty e^{-n^2\pi x} = \frac{1}{\sqrt{x}}\sum_{n=-\infty}^\infty e^{-n^2\pi/x}. \tag{26.99}$$

In Riemann's notation, this was equivalent to

$$2\psi(x) + 1 = x^{-1/2}\left(2\psi(1/x) + 1\right),$$

$$\text{where } \psi(x) = \sum_{n=1}^\infty e^{-n^2\pi x}.$$

Riemann referred to Jacobi's *Fundamenta Nova* for (26.99). We note that Jacobi's proof of (26.99) used elliptic functions, while Cauchy and Poisson employed Fourier analysis. See chapter 37 for Jacobi's proof. Next Riemann rewrote (26.98) as

$$\begin{aligned}\Gamma(s/2)\pi^{s/2}\zeta(s) &= \int_1^\infty \psi(x)x^{(s/2)-1}dx + \int_0^1 \psi(1/x)x^{(s-3)/2}dx \\ &\quad + \frac{1}{2}\int_0^1 \left(x^{(s-3)/2} - x^{(s/2)-1}\right)dx \\ &= \frac{1}{s(s-1)} + \int_1^\infty \psi(x)\left(x^{(s/2)-1} + x^{-(1+s)/2}\right)dx.\end{aligned} \tag{26.100}$$

This reproved the functional equation because the right-hand side was invariant under $s \to 1-s$. Moreover, $\zeta(s)$ was once again defined for all complex $s \neq 1$. To emphasize the significance of the line $\operatorname{Re} s = \frac{1}{2}$, Riemann set $s = \frac{1}{2} + it$ and denoted the left-hand side of (26.100) as $\xi(t)$, so that he had

$$\xi(t) = \frac{1}{2} - \left(t^2 + \frac{1}{4}\right) \int_1^\infty \psi(x) x^{-3/4} \cos\left(\frac{1}{2} t \log x\right) dx, \qquad (26.101)$$

$$\xi(t) = 4 \int_1^\infty \frac{d}{dx}(x^{3/2} \psi'(x)) x^{-1/4} \cos\left(\frac{1}{2} t \log x\right) dx. \qquad (26.102)$$

26.11 Euler's Product for $\sum 1/n^s$

In his 1859 paper giving the formula for the number of primes less than a given number, Riemann remarked that he had taken Euler's infinite product for the zeta function as the starting point for his investigations. Indeed, it was Euler's product representation for the zeta function that made it possible to perceive the connection between the zeta function and prime numbers.

In a 1737 paper, Euler showed how to convert the series for the zeta function, $\sum_{k=1}^\infty 1/k^n$, into a product. Euler's insightful argument, amounting to an application of the fundamental theorem of arithmetic, is here presented in its original form. Euler let

$$x = 1 + \frac{1}{2^n} + \frac{1}{3^n} + \frac{1}{4^n} + \frac{1}{5^n} + \frac{1}{6^n} + \text{etc.}$$

Then $\quad \dfrac{1}{2^n} x = \dfrac{1}{2^n} + \dfrac{1}{4^n} + \dfrac{1}{6^n} + \dfrac{1}{8^n} + \text{etc.}$

Removing all even numbers by subtraction, he got

$$\frac{2^n - 1}{2^n} x = 1 + \frac{1}{3^n} + \frac{1}{5^n} + \frac{1}{7^n} + \frac{1}{9^n} + \text{etc.}$$

He multiplied this by $\frac{1}{3^n}$ to obtain

$$\frac{2^n - 1}{2^n} \cdot \frac{1}{3^n} x = \frac{1}{3^n} + \frac{1}{9^n} + \frac{1}{15^n} + \text{etc.}$$

Again, Euler removed by subtraction all multiples of 3 so that

$$\frac{2^n - 1}{2^n} \cdot \frac{3^n - 1}{3^n} x = 1 + \frac{1}{5^n} + \frac{1}{7^n} + \text{etc.}$$

By continuing this process with each of the prime numbers, all numbers on the right-hand side except one were eliminated, yielding

$$x \cdot \frac{2^n - 1}{2^n} \cdot \frac{3^n - 1}{3^n} \cdot \frac{5^n - 1}{5^n} \cdot \text{etc.} = 1,$$

or $\quad 1 + \dfrac{1}{2^n} + \dfrac{1}{3^n} + \dfrac{1}{4^n} + \text{etc.} = \dfrac{2^n}{2^n - 1} \cdot \dfrac{3^n}{3^n - 1} \cdot \dfrac{5^n}{5^n - 1} \cdot \text{etc.}$

Note that this was in essence the fundamental theorem of arithmetic in analytic form. Euler's 1748 book *Introductio in Analysin Infinitorum* made this connection more clear, as he expressed this infinite product in almost modern form:

$$\frac{1}{\left(1-\frac{1}{2^n}\right)\left(1-\frac{1}{3^n}\right)\left(1-\frac{1}{5^n}\right) \text{ etc.}}$$

To see the unique factorization theorem here, simply expand these fractions using the geometric series.

In 1837, Dirichlet defined L-functions for which he found an analogous infinite product. For example, in the case of characters modulo a prime p, he stated the result as

$$\prod \frac{1}{1 - w^{\gamma_q}\frac{1}{q^s}} = \sum w^{\gamma_n} \cdot \frac{1}{n^s}.$$

The product was defined over all primes other than p, while w was a $(p-1)$th root of unity. Dirichlet used this formula in his proof of his famous theorem on primes in arithmetic progressions. Note also that the product formula shows that the series on the left-hand side of (26.85) has to be positive, justifying Dirichlet's remark on it.

26.12 Dirichlet Characters

Dirichlet's construction of characters was based on a theorem first observed by Euler and later completely proved by Gauss in his 1801 *Disquisitiones Arithmeticae*. Gauss showed that for any prime p, the multiplicative group modulo p, whose elements could be represented by the integers $1, 2, \ldots, p-1$, was a cyclic group. This means that there is at least one g among these $p-1$ integers such that for any n, not a multiple of p, there exists an integer γ_n such that

$$g^{\gamma_n} \equiv n \pmod{p}. \tag{26.103}$$

This equation implies that for positive integers m and n not multiples of p,

$$g^{\gamma_{mn}} \equiv mn \equiv g^{\gamma_m} g^{\gamma_n} \equiv g^{\gamma_m + \gamma_n} \pmod{p},$$

and hence

$$\gamma_{mn} \equiv \gamma_m + \gamma_n \pmod{(p-1)}. \tag{26.104}$$

So if w is a $(p-1)$th root of unity, we have

$$w^{\gamma_{mn}} = w^{\gamma_m + \gamma_n} = w^{\gamma_m} w^{\gamma_n}. \tag{26.105}$$

The complex number w can be written as $e^{2\pi i k/(p-1)}$ for $k = 1, 2, \ldots, p-1$. For any one of these $p-1$ complex numbers w, Dirichlet defined a character with values w^{γ_1}, $w^{\gamma_2}, \ldots, w^{\gamma_{p-1}}$ and with the property

$$\frac{w^{\gamma_n}}{n^s} \cdot \frac{w^{\gamma_m}}{m^s} = \frac{w^{\gamma_{mn}}}{(mn)^s}. \tag{26.106}$$

26.12 Dirichlet Characters

He observed that by (26.105)

$$\frac{1}{1 - w^{\gamma_q}\frac{1}{q^s}} = 1 + w^{\gamma_q} \cdot \frac{1}{q^s} + w^{\gamma_{q^2}} \cdot \frac{1}{q^{2s}} + \cdots$$

for $s > 1$. Then, by the unique factorization theorem,

$$\prod \frac{1}{1 - w^{\gamma_q}\frac{1}{q^s}} = \sum w^{\gamma_n} \cdot \frac{1}{n^s}. \tag{26.107}$$

The infinite product was defined over all primes not equal to p, and the sum was taken over all positive integers not divisible by p. Note that this sum can be taken over all positive integers with the convention that $w^{\gamma_n} = 0$ when n is a multiple of p. When $w = -1$, we have $w^{\gamma_n} = \pm 1$, depending on whether γ_n is even or odd. If it is even, we can write the left-hand side of (26.103) as a square and hence n is a square modulo p, or rather, n is a quadratic residue. We can therefore write

$$(-1)^{\gamma_n} = \left(\frac{n}{p}\right), \tag{26.108}$$

where $\left(\frac{n}{p}\right)$ is the Legendre symbol; it is $+1$ when n is a quadratic residue (mod p) and -1 when n is a quadratic nonresidue. For this character, we can write (26.107) as

$$\prod_q \frac{1}{1 - \left(\frac{q}{p}\right)q^{-s}} = \sum_{n=1}^{\infty} \frac{\left(\frac{n}{p}\right)}{n^s},$$

where $\left(\frac{n}{p}\right) = 0$ when n is a multiple of p and the product is taken over all primes not equal to p.

In his 1837 paper on primes within any arithmetic progression, Dirichlet also defined characters modulo any positive integer m. For this purpose, he employed a result from Gauss's *Disquisitiones*: For any odd prime p and positive integer k, the multiplicative group modulo p^k, that is, the integers relatively prime to p and represented by integers less than p^k, is a cyclic group. This theorem enabled Dirichlet to define $\phi(p^k) = p^k - p^{k-1}$ different characters corresponding to the $\phi(p^k)$ values $e^{2\pi i m/\phi(p^k)}$, $m = 1, 2, \ldots, \phi(p^k)$. Letting w denote any one of these values, the value of the corresponding character at n where n was not divisible by p, would be w^{γ_n}. As before, γ_n was defined as in (26.103), with respect to a generator g of the multiplicative group modulo p^k.

For powers of 2, the situation was slightly more complex. Clearly, the multiplicative groups mod 2 and mod 4 are cyclic. Another result from the *Disquisitiones* stated that every relatively prime residue class mod 2^k, where $k \geq 3$, could be represented uniquely as $(-1)^{\gamma} 5^{\gamma'}$, where γ was defined to the modulus 2 and γ' to the modulus $\frac{1}{2}\phi(2^k) = 2^{k-2}$. Again, Dirichlet used Gauss's result to define the characters modulo powers of 2 by

$$w^{\gamma}(w')^{\gamma'}, \quad \text{where} \quad w^2 = 1 \quad \text{and} \quad (w')^{2^{k-2}} = 1.$$

Dirichlet noted that the number of such characters was

$$2^{k-1} = \phi(2^k).$$

Next, Dirichlet defined characters modulo

$$m = 2^k p_1^{k_1} p_2^{k_2} \cdots p_l^{k_l}.$$

He considered an integer n relatively prime to m and assumed

$$n = (-1)^\gamma 5^{\gamma'} \pmod{2^k} \quad \text{and} \quad n = g_j^{\gamma_{n,j}} \pmod{p_j^{k_j}}$$

where g_j was the generator of the relatively prime residue classes modulo $p_j^{k_j}$. Then he gave the value of an arbitrary character at n modulo m as

$$w^\gamma (w')^{\gamma'} w_1^{\gamma_{n,1}} w_2^{\gamma_{n,2}} \cdots w_l^{\gamma_{n,l}}. \tag{26.109}$$

Here w_j was a root of $w_j^{(p-1)p^{j-1}} - 1 = 0$ and there were

$$\phi(m) = m \prod_{p|m} \left(1 - \frac{1}{p}\right)$$

such characters. Dirichlet showed that with this general definition of a character, the product formula (26.107) would continue to hold. The operative idea behind the product formula was the multiplicative property of characters.

26.13 Exercises

1. Show that

$$1 - \frac{1}{3^5} + \frac{1}{5^5} - \frac{1}{7^5} + \frac{1}{9^5} - \text{etc.} = \frac{5\pi^5}{1536},$$

$$1 + \frac{1}{3^6} + \frac{1}{5^6} + \frac{1}{7^6} + \frac{1}{9^6} + \text{etc.} = \frac{\pi^6}{960},$$

$$1 - \frac{1}{3^7} + \frac{1}{5^7} - \frac{1}{7^7} + \frac{1}{9^7} - \text{etc.} = \frac{61\pi^7}{184320},$$

$$1 + \frac{1}{3^8} + \frac{1}{5^8} + \frac{1}{7^8} + \frac{1}{9^8} + \text{etc.} = \frac{17\pi^8}{161280}.$$

See Eu. I-14, p. 81.

2. Express Newton's series (26.11) as a Dirichlet L-series. Do the same with Euler's series (26.39) and (26.41).
3. Divide Takebe's series

$$\frac{1}{2}(\arcsin x)^2 = \frac{x^2}{2} + \frac{2}{3} \cdot \frac{x^4}{4} + \frac{2 \cdot 4}{3 \cdot 5} \cdot \frac{x^6}{6} + \frac{2 \cdot 4 \cdot 6}{3 \cdot 5 \cdot 7} \cdot \frac{x^8}{8} + \cdots$$

by $\sqrt{1-x^2}$ and integrate term by term over (0,1) to obtain

$$\frac{\pi^2}{6} = 1 + \frac{1}{2^2} + \frac{1}{3^2} + \frac{1}{4^2} + \cdots.$$

See Eu. I-14, p. 184.

4. Show that

$$1 + \frac{1}{2^{24}} + \frac{1}{3^{24}} + \frac{1}{4^{24}} + \frac{1}{5^{24}} + \text{etc.} = \frac{2^{23}}{1 \cdot 2 \cdot 3 \cdots 25} \cdot \frac{1181820455}{546} \pi^{24},$$

$$1 + \frac{1}{2^{26}} + \frac{1}{3^{26}} + \frac{1}{4^{26}} + \frac{1}{5^{26}} + \text{etc.} = \frac{2^{25}}{1 \cdot 2 \cdot 3 \cdots 27} \cdot \frac{76977927}{2} \pi^{26}.$$

See Eu. I-14, p. 185.

5. Prove Eisenstein's formula:

$$e^{\beta \pi i} \sum_{\sigma=0}^{\infty} \frac{(-1)^\sigma}{(\sigma+\beta)^{1-q}} = \frac{\Gamma(q)}{(2\pi)^q} \left(e^{q\pi i/2} + e^{-2\beta\pi i - q\pi i/2} \right) \sum_{\sigma=0}^{\infty} \frac{e^{-\sigma\beta \cdot 2\pi i}}{(\sigma+\frac{1}{2})^q}.$$

See Weil (1989a).

6. Prove that

$$1 - \frac{\pi x}{\tan \pi x} = \sum_{n=1}^{\infty} (-1)^{n-1} \frac{2^{2n} B_{2n} \pi^{2n}}{(2n)!} x^{2n}.$$

Combine this with (26.31) to derive Euler's formula (26.34). See Eu. I-10, p. 325.

7. Prove that for $|a| < \pi$ and $0 < s < 1$,

$$\sum_{k=0}^{\infty} (-1)^k \left(\frac{1}{((2k+1)\pi + a)^s} - \frac{1}{((2k+1)\pi - a)^s} \right)$$

$$= \frac{1}{\Gamma(s) \sin \frac{s\pi}{2}} \sum_{k=1}^{\infty} \frac{(-1)^{k-1} \sin ka}{k^{1-s}}.$$

Deduce the functional equation for $L(s) = \sum_{k=0}^{\infty} (-1)^k (2k+1)^{-s}$.

See Malmsten (1849). Carl Malmsten became professor of mathematics in Uppsala in 1841; during his career, he made significant contributions to the development of the Swedish mathematical tradition. See Gårding (1994).

8. Prove Goldbach's formula

$$1 + \frac{1}{2^3}(1 + \frac{1}{2}) + \frac{1}{3^3}(1 + \frac{1}{2} + \frac{1}{3}) + \cdots = \frac{\pi^4}{72}.$$

See Fuss (1968), p. 197.

9. Prove Euler's formula

$$1 + \frac{1}{2^5}(1 + \frac{1}{2^3}) + \frac{1}{3^5}(1 + \frac{1}{2^3} + \frac{1}{3^3}) + \cdots = \frac{\pi^8}{16 \cdot 3 \cdot 25 \cdot 7}.$$

See Fuss (1968), p. 190.

10. Prove Euler's formula (26.55). See Fuss (1968), pp. 181–182.
11. Use the generating function for Bernoulli numbers to show that

    (a) $\cot u = \sum_{k=0}^{\infty}(-1)^k 2^{2k} \frac{B_{2k}}{(2k)!} u^{2k-1}$,

    (b) $\tan u = \sum_{k=1}^{\infty}(-1)^{k-1} 2^{2k}(2^{2k}-1) \frac{B_{2k}}{(2k)!} u^{2k-1}$,

    (c) $\sec u = \sum_{k=0}^{\infty} E_{2k} \frac{u^{2k}}{(2k)!}$,

    where $E_{2k} = \frac{2^{2k+2}(2k)!}{\pi^{2k+1}} \left(1 - \frac{1}{3^{2k+1}} + \frac{1}{5^{2k+1}} - \frac{1}{7^{2k+1}} + \cdots\right)$,

    (d) $\csc u = \frac{1}{u} + \sum_{k=1}^{\infty}(-1)^{k-1} 2(2^{2k-1}-1) \frac{B_{2k}}{(2k)!} u^{2k-1}$.

    Note that the E_{2k} are called Euler numbers. The results in (a) and (d) were explicitly used by Euler in several papers; (b) and (c) are implicitly contained in Eu. I-17, pp. 384–420 (published 1775).

12. Show that for $t = s - 1/2$ and $0 < s < 1$,

    $$\int_0^1 \frac{x^{s-1} - x^{-s}}{1-x} dx = -\pi \tan \pi t.$$

    Euler took successive derivatives of both sides with respect to s and set $s = 1/2$ or $t = 0$. Verify that after taking the first derivative and setting $s = 1/2$, the result is

    $$\int_0^1 \frac{2\ln x}{1-x} \frac{dx}{x^{1/2}} = -\pi^2,$$

    or $$\int_0^1 \frac{\ln y}{1-y^2} dy = -\frac{\pi^2}{8}.$$

    More generally show that

    $$\int_0^1 \frac{(\ln y)^{2k-1}}{1-y^2} dy = (-1)^k (2^{2k}-1) \frac{B_{2k}}{4k} \pi^{2k}. \qquad (26.110)$$

    Euler wrote down the formulas for $k = 1, 2$ and 3. See Eu. I-17, p. 406.

13. Show that for $t = s - \frac{1}{2}$ and $0 < s < 1$

    $$\int_0^1 \frac{x^{s-1} + x^{-s}}{1+x} dx = \pi \sec \pi t.$$

    Use the method of the previous problem and the series for $\sec u$ in exercise 11 to prove the formula for Euler numbers

    $$\int_0^1 \frac{(\ln y)^{2k}}{1+y^2} dy = \frac{E_{2k} \pi^{2k+1}}{2^{2k+2}}.$$

    Euler computed the Euler numbers E_{2k} for $k = 0, 1, 2, 3, 4$ to obtain $1, 1, 5, 61, 1385$, respectively. See Eu. I-17, pp. 401, 405.

14. Let $P = \int_0^1 \frac{y(\ln y)^{2k-1}}{1-y^2} dy$, and let Q denote the integral in (26.110). Observe that

$$Q \pm P = \int_0^1 \frac{(\ln y)^{2k-1}}{1 \mp y} dy,$$

$$P = \frac{1}{2^{2k}} \int_0^1 \frac{(\ln y)^{2k-1}}{1-y} dy.$$

Deduce that

$$\int_0^1 \frac{(\ln y)^{2k-1}}{1-y} dy = (-1)^k \frac{B_{2k}}{4k} (2\pi)^{2k},$$

$$\int_0^1 \frac{(\ln y)^{2k-1}}{1+y} dy = (-1)^k (2^{2k-1} - 1) \frac{B_{2k}}{2k} \pi^{2k}.$$

Euler gave this argument in Eu. I-17, pp. 406–407.

15. Show that

$$\int_0^1 y^m (\ln y)^{2k-1} dy = \frac{\Gamma(2k)}{m^{2k}}.$$

From this, compute the ζ and L-series values $\sum_{n=1}^{\infty} 1/n^{2k}$, $\sum_{n=1}^{\infty} (-1)^n/n^{2k}$, and $\sum_{n=1}^{\infty} (-1)^{n-1}/(2n-1)^{2k-1}$. In 1737 Euler used integration to exactly evaluate $\sum_{n=1}^{\infty} 1/n^2$ but he regretted that the method did not extend to $k \geq 2$. In 1774, he finally found what he was looking for. See Eu. I-17, pp. 428–451.

26.14 Notes on the Literature

Euler's early papers on $\sum 1/n^2$ are in Eu. I-14. Some of the most important are: "De Summis Serierum Reciprocarum," 1734–1735, "De Summis Serierum Reciprocarum ex Potestatibus Numerorum Naturalicum Ortarum Dissertatio Altera, ...," 1743, "Demonstration de la somme de cette suite $1 + \frac{1}{4} + \frac{1}{9} + \frac{1}{16} + \frac{1}{25} + \frac{1}{36} +$ etc.," 1743, and "De Seriebus quibusdam Considerationes," 1740. In the first paper listed here, Euler summed the series given in the equations (26.4) through (26.12) and several more by using equation (26.1). By the time he wrote the second paper, he had found proofs of the product formulas for $\sin x$ and $\cos x$ as well as the resulting partial fractions expansions; he used these to sum $\zeta(2n)$. Euler used the integral calculus to sum $\sum 1/n^2$ in the third paper. This paper was republished in the *Bibliotheca Mathematica* (1907–1908) with a commentary by Paul Stäckel. Stäckel gave an account of Euler's work on this problem and his communications with the Bernoullis on this and related questions, included in Eu. I-14. In the fourth paper, Euler worked out the connection between $\sum 1/n^{2k}$ and the Bernoulli number B_{2k}. The infinite product for $\zeta(n)$ is on pp. 243–244 of Eu. I-14. Goldbach's correspondence with Euler on double-zeta values can be found on pp. 160–208 of Fuss (1968).

N. Bernoulli's paper on $\sum 1/n^2$ appeared in the Petersburg Academy publication in 1747, though it was received in 1738 and had that publication date; see N. Bernoulli (1738). Euler's letter of July 30, 1738, with the improvement on N. Bernoulli's derivation, is in Eu. 4A-2, pp. 230–236. Dirichlet's evaluation of $L(\chi, 1)$, and his theory of characters is given on pp. 316–329 of Dirichlet (1969). Riemann's derivation of the functional equation for the zeta function is contained in his paper on prime numbers, "Ueber die Anzahl der Primzahlen unter einer gegebenen Grösse" in Riemann (1990), pp. 177–185. An English translation of this paper appears in Edwards (2001), pp. 299–305. For a discussion of some recent work on multizeta values, see Borwein, Bailey, and Girgensohn (2004), Eie (2009), and Varadarajan (2006). Horiuchi (1994) discusses Takebe's *Tetsujutsu Sankei* of 1722. For Eisenstein's derivation of the functional equation, see Weil (1989a).

27

The Hypergeometric Series

27.1 Preliminary Remarks

The hypergeometric series and associated functions are among the most important in mathematics, partly because they cover a large class of valuable special functions as either particular cases or as limiting cases. More importantly, because they have the appropriate degree of generality, very useful transformation formulas and other relations can be proved about them. The hypergeometric series is defined by

$$F(a,b,c,x) = {}_2F_1\left(\begin{matrix}a,b\\c\end{matrix}; x\right) = 1 + \frac{a \cdot b}{1 \cdot c}x + \frac{a(a+1) \cdot b(b+1)}{1 \cdot 2 \cdot c(c+1)}x^2 + \cdots. \quad (27.1)$$

The expressions involved can be written more briefly if we adopt the modern notation for the shifted factorial:

$$(a)_n = a(a+1)\cdots(a+n-1) \quad \text{for } n \geq 1, \ (a)_0 = 1. \quad (27.2)$$

Thus, $\quad F(a,b,c,x) = {}_2F_1\left(\begin{matrix}a,b\\c\end{matrix}; x\right) = \sum_{n=0}^{\infty} \frac{(a)_n(b)_n}{n!(c)_n} x^n. \quad (27.3)$

The subscript notation in F was introduced in the twentieth century when similar series with varying numbers of parameters, such as a, b, c, were considered. Note the following examples of hypergeometric series in Gauss's notation:

$$(1-x)^{-\alpha} = F(\alpha, 1, 1, x); \quad \log\frac{1+x}{1-x} = 2x\, F(1/2, 1, 3/2, x^2);$$

$$e^x = \lim_{a\to\infty} F(1,1,1,x/a); \quad J_\alpha(x) = \frac{(x/2)^\alpha}{\Gamma(\alpha+1)} \lim_{a,b\to\infty} F(a,b,\alpha+1,-x^2/4ab).$$

Historically, hypergeometric series occurred not only in the study of power series but also as inverse factorial series in finite difference theory. James Stirling, in particular, employed them in the approximate summation of series and in this connection also discovered special cases of important transformation formulas. However, in 1778, Euler first introduced the hypergeometric series in the form (27.1). He proved that the series satisfied the second-order differential equation:

$$x(1-x)\frac{d^2F}{dx^2} + (c-(a+b+1)x)\frac{dF}{dx} - abF = 0, \tag{27.4}$$

and then used this equation to prove an important transformation formula:

$$F(a,b,c,x) = (1-x)^{c-a-b}F(c-a,c-b,c,x). \tag{27.5}$$

The binomial factor can be moved to the left-hand side, as $(1-x)^{a+b-c}$. When this is expanded as a series and multiplied by the hypergeometric function on the left-hand side, the coefficients of x^n on the two sides give the identity

$$\sum_{k=0}^{n} \frac{(a)_k(b)_k(a+b-c)_{n-k}}{k!(c)_k(n-k)!} = \frac{(c-a)_n(c-b)_n}{n!(c)_n} \quad \text{or} \tag{27.6}$$

$$\sum_{k=0}^{n} \frac{(-n)_k(a)_k(b)_k}{k!(c)_k(1+a+b-c-n)_k} = \frac{(c-a)_n(c-b)_n}{(c)_n(c-a-b)_n}, \tag{27.7}$$

or in the following modern notation, whose meaning is obvious from (27.7):

$$_3F_2\left(\begin{matrix}-n,a,b\\c,1+a+b-c-n\end{matrix};1\right) = \frac{(c-a)_n(c-b)_n}{(c)_n(c-a-b)_n}. \tag{27.8}$$

Observe that this identity is formally equivalent to (27.5).

In 1797, Johann Friedrich Pfaff (1765–1825) proved Euler's transformation (27.5) by giving an inductive proof of (27.8). Pfaff was among the leading mathematicians in Germany during the late eighteenth and early nineteenth centuries; he was the formal thesis advisor for Gauss. His results on second-order differential equations were inspired by Euler, whose work on this topic appeared in his three volumes on the integral calculus. Euler's work on series provided the starting point for the German combinatorial school founded by C. F. Hindenburg (1741–1808), of which Pfaff was a member. Pfaff's formula (27.8) is very useful for evaluating certain types of sums of products of binomial coefficients occurring in combinatorial problems. In order to save this identity and some other of Pfaff's results from oblivion, Jacobi referred to it in a paper of 1845. We remark that Jacobi was interested in the history of mathematics and consistently attempted to give credit to the original discoverer of a concept or formula. In spite of Jacobi's efforts, this identity was forgotten for many years. In 1890, it was finally rediscovered and published by L. Saalschütz, with whose name it was associated for many years. In the 1970s, Askey noticed Jacobi's reference and renamed it the Pfaff-Saalschütz identity. Pfaff could not have foreseen that in the 1990s, his method of proving (27.8) would become the foundation of George Andrews's general method for proving hypergeometric identities useful in computer algebra systems. Pfaff also found the terminating form of another important hypergeometric transformation:

$$F(a,b,c,x) = (1-x)^{-a}F(a,c-b,c,x/(x-1)). \tag{27.9}$$

Note that Pfaff took the parameter a to be a negative integer so that the series on both sides were finite. Pfaff derived this formula from a study of the differential equation

$$x^2(a+bx^n)\frac{d^2y}{dx^2} + x(c+ex^n)\frac{dy}{dx} + (f+gx^n)y = X,$$

where X was a function of x. Euler earlier discussed the homogeneous form of this equation in his book on the integral calculus. Note that Newton's transformation (11.4) is a particular case of (27.9), obtained by taking $a = 1$, $b = 1/2$, $c = 3/2$, and $x = -t^2$. Stirling's formula (11.13), obtained by equating the series in (11.31) and (11.32), can also be derived from (27.9) by taking $a = -1$ and $x = 1/m$. It is possible that Pfaff was motivated to study the series in (27.9) by Hindenburg's 1781 work on the following problem: For given numbers α and β, transform a series $ay + by^2 + cy^3 + \cdots$ to a series of the form

$$\frac{Ay}{\alpha+\beta y} + \frac{By^2}{(\alpha+\beta y)^2} + \frac{Cy^3}{(\alpha+\beta y)^3} + \cdots;$$

thus, determine $A, B, C, \ldots$ in terms of $a, b, c, \ldots$.

Gauss was the first mathematician to undertake a systematic and thorough study of the hypergeometric function. His treatment of the subject appeared in a paper of 1812. It is possible that Gauss was introduced to the topic when he visited Helmstedt in 1799 to use the university library and rented a room in Pfaff's home. One imagines this to be very likely, since Gauss and Pfaff took walks together every evening and discussed mathematics. Gauss does not refer to earlier work on hypergeometric series so it is hard to determine what he had learned from others. The two most notable features of Gauss's contributions to hypergeometric series were his use of contiguous relations to derive the basic formulas and his determination of the conditions for the convergence of the series. Some of his unpublished work shows that he wanted to build the foundation of analysis on a rigorous theory of limits, for which purpose he carefully defined the concepts of superior and inferior limits of sequences.

Gauss defined functions contiguous to $F(a,b,c,x)$ as those functions arising from it when the first, second or third parameter a,b,c was increased or diminished by one while the other three remained the same. Gauss may have seen the importance of contiguous functions by reading Stirling's 1730 *Methodus*. He found that there was a linear relation between $F(a,b,c,x)$ and any two contiguous functions; such an equation is now called a contiguous relation. Clearly there would be $\binom{6}{2} = 15$ such relations, and Gauss listed all of them in the first section of his 1812 paper. From these relations he derived continued fractions expansions of ratios of hypergeometric functions, his fundamental summation formula for $F(a,b,c,1)$, and the differential equation for $F(a,b,c,x)$. He derived the latter in the second (unpublished) part of his paper. In this part, Gauss derived transformation formulas in the same manner as Euler before him, except that he also gave examples of quadratic transformations. For example:

$$F\left(a,b,a+b+1/2, 4x-4x^2\right) = F(2a, 2b, a+b+1/2, x). \tag{27.10}$$

Gauss treated a, b, c, and x as complex variables and in this connection he pointed out that it was necessary to exercise care when dealing with values of x outside the circle of convergence of the series. Thus, when x was changed to $1-x$ in (27.10), the left-hand side would remain unchanged, leading to the evidently contradictory result that

$$F(2a, 2b, a+b+1/2, x) = F(2a, 2b, a+b+1/2, 1-x). \qquad (27.11)$$

Gauss called this result a paradox and his explanation, from the unpublished portion of his paper, is highly interesting, showing that as early as 1812 he was thinking of analytic continuation of functions:

> To explain this, it ought to be remembered that proper distinction should be made between the two significations of the symbol F, viz., whether it represents the function whose nature is expressed by the differential equation [(27.4)], or simply the sum of an infinite series. The latter is always a perfectly determinate quantity so long as the fourth element lies between -1 and $+1$, and care must be taken not to exceed these limits for otherwise it is entirely without any meaning. On the other hand, according to the former signification, it [F] represents a general function which always varies subject to the law of continuity if the fourth element vary continuously whether you attribute real values or imaginary values to it, provided you always avoid the values 0 and 1. Hence it is evident that in the latter sense, the function may for equal values of the fourth element (the passage or rather the return being made through imaginary quantities) attain unequal values of which that which the *series* F represents is only one, so that it is not at all contradictory that while some *one* value of the function $F(a, b, a+b+1/2, 4y - 4yy)$ is equal to $F(2a, 2b, a+b+1/2, y)$ the *other* value should be equal to $F(2a, 2b, a+b+1/2, 1-y)$ and it would be just as absurd to deduce thence the equality of these values as it would be to conclude, that since Arc. sin $\frac{1}{2} = 30°$, Arc. sin $\frac{1}{2} = 150°$, $30° = 150°$. – But if we take F in the less general sense, viz. simply as the sum of the series F, the arguments by which we have deduced (27.10), necessarily suppose y to increase from the value 0 only up to the point when $x[= 4y - 4yy]$ becomes $= 1$, i.e. up to $y = 1/2$. At this point, indeed, the *continuity* of the series $P = F(a, b, a+b+1/2, 4y - 4yy)$ is interrupted, for evidently $\frac{dP}{dy}$ jumps suddenly from a positive (finite) value to a negative. Thus in this sense equation (27.10) does not admit of being extended outside the limits $y = 1/2 - \sqrt{1/2}$ up to $y = 1/2$. If preferred, the same equation can also be put thus: –
>
> $$F\left(a, b, a+b+\frac{1}{2}, x\right) = F\left(2a, 2b, a+b+\frac{1}{2}, \frac{1-\sqrt{1-x}}{2}\right).$$

Again, Gauss's letter of December 18, 1811, to his friend F. W. Bessel (1784–1846) shows how far he had advanced in developing a theory of functions of complex variables:

> What should we make of $\int \phi x.dx$ for $x = a + bi$? Obviously, if we're to proceed from clear concepts, we have to assume that x passes, via infinitely small increments (each of the form $\alpha + i\beta$), from that value at which the integral is supposed to be 0, to $x = a + bi$ and that then all the $\phi x.dx$ are summed up. In this way the meaning is made precise. But the progression of x values can take place in infinitely many ways: Just as we think of the realm of all real magnitudes as an infinite straight line, so we can envision the realm of all magnitudes, real and imaginary, as an infinite plane wherein every point which is determined by an abscissa a and an ordinate b represents as well the magnitude $a + bi$. The continuous passage from one value of x to another $a + bi$ accordingly occurs along a curve and is consequently possible in infinitely many ways. But I maintain that the integral $\int \phi x.dx$ computed via two different such passages always gets the same value as long as $\phi x = \infty$ never occurs in the region of the plane enclosed by the curves describing these two passages. This is a very beautiful theorem, whose not-so-difficult proof I will give when an appropriate occasion comes up. It is closely related to other beautiful truths having to do with developing functions in series. The passage from point to point can always be carried

out without ever touching one where $\phi x = \infty$. However, I demand that those points be avoided lest the original basic conception of $\int \phi x . dx$ lose its clarity and lead to contradictions. Moreover it is also clear from this how a function generated by $\int \phi x . dx$ could have several values for the same values of x, depending on whether a point where $\phi x = \infty$ is gone around not at all, once, or several times. If, for example, we define $\log x$ via $\int \frac{1}{x} dx$ starting at $x = 1$, then arrive at $\log x$ having gone around the point $x = 0$ one or more times or not at all, every circuit adds the constant $+2\pi i$ or $-2\pi i$; thus the fact that every number has multiple logarithms becomes quite clear.

Thus in 1811, Gauss had a clear conception of complex integration and had discovered Cauchy's integral theorem, published by Cauchy in 1825. He had also begun to understand the reason for a function being multivalued; this understanding informed Gauss's comments on (27.11). It is possible that Gauss was motivated to study quadratic transformations by his discovery during the mid-1790s of the connection between the arithmetic-geometric mean and the complete elliptic integral. This integral is defined by

$$K(k) = \int_0^{\frac{\pi}{2}} \frac{d\theta}{\sqrt{1 - k^2 \sin^2 \theta}} = \frac{\pi}{2} F\left(\frac{1}{2}, \frac{1}{2}, 1, k^2\right).$$

In his unpublished paper, Gauss also computed two independent solutions of the hypergeometric equation in the neighborhood of 0, 1, and ∞. He obtained explicit formulas linearly relating a solution in the neighborhood of one of these points with two independent solutions in the neighborhood of another point. As an example, consider Gauss's result

$$F(a, b, c, x) = \frac{\Gamma(c)\Gamma(b-a)}{\Gamma(c-a)\Gamma(b)} (-x)^{-a} F(a, a+1-c, a+1-b, 1/x)$$
$$+ \frac{\Gamma(c)\Gamma(a-b)}{\Gamma(a)\Gamma(c-b)} (-x)^{-b} F(b, b+1-c, b+a-a, 1/x).$$

The functions on the right-hand side were solutions in the neighborhood of infinity. Gauss also considered the case where the parameter c was an integer so that the second independent solution involved a logarithmic term. Euler was also aware of this situation. Gauss went further by showing that the digamma function, $\psi(x) \equiv \Gamma'(x)/\Gamma(x)$, defined in the first part of his paper, could be employed to obtain an expression for the second solution.

Gauss's paper was quite influential, especially among German mathematicians, who produced much important research on this topic in the next three or four decades. In 1833, as part of his doctoral dissertation, P. C. O. Vorsselman de Heer gave the integral representation

$$F(a, b, c, x) = \frac{\int_0^1 t^{b-1}(1-t)^{c-b-1}(1-xt)^{-a} dt}{\int_0^1 t^{b-1}(1-t)^{c-b-1} dt}. \tag{27.12}$$

Note that the integral in the denominator is the beta integral, evaluated by Euler, equal to $\Gamma(b)\Gamma(c-b)/\Gamma(c)$. This integral representation of $F(a, b, c, x)$ was independently found by Kummer and published a few years later in his long memoir on hypergeometric functions. However, in a posthumous paper, Jacobi attributed this formula to Euler, though it seems that it does not appear explicitly in Euler's work. However, Euler did give an integral representation of a solution of a differential equation closely related to the hypergeometric equation; this may have been Jacobi's reason for the attribution.

In 1828, the Danish mathematician Thomas Clausen (1801–1885) obtained a significant result of a different kind. Clausen was born to poor farming people and did not learn to read or write until the age of 12. He encountered many difficulties due to his humble origins. But Gauss thought highly of him, and Clausen's abundant mathematical talent was eventually recognized. He considered the square of a hypergeometric series and found for $c = a + b + 1/2$

$$(F(a,b,c,x))^2 \equiv \left({}_2F_1\left(\begin{matrix} a,b \\ c \end{matrix} ; x \right) \right)^2 = {}_3F_2\left(\begin{matrix} 2a, 2b, a+b \\ 2a+2b, a+b+\frac{1}{2} \end{matrix} ; x \right), \qquad (27.13)$$

$$\text{where} \quad \sum_{n=0}^{\infty} \frac{(a)_n (b)_n (c)_n}{n!(d)_n (e)_n} x^n \equiv {}_3F_2\left(\begin{matrix} a,b,c \\ d,c \end{matrix} ; x \right).$$

In 1836, Ernst Kummer (1810–1893) published the first major work on hypergeometric functions after Gauss. He rediscovered much of the material in the unpublished portion of Gauss's paper, including quadratic transformations. In fact, these transformations are implicitly contained in Gauss's published paper. Kummer also found some results for ${}_3F_2$ functions, including the existence of three-term contiguous relations when $x = 1$. Kummer was trained as a high school teacher; he taught at that level 1831–1841. In 1834, while serving a year in the army, he communicated some papers in analysis to Jacobi who is reported by E. Lampe to have commented: "There we are; now the Prussian musketeers even enter into competition with the professors by way of mathematical works." However, Jacobi was impressed with the work done by Kummer under difficult circumstances and wrote in his reply, "If you think that I could be of any help with obtaining an academic position, I would be happy to offer my humble services – less because I think that you would need them, or that they would be significant, but as a token of my great respect for your talent and your works." Dirichlet and Jacobi worked to find Kummer a university position. He became a professor at Breslau in 1842 and moved to Berlin in 1855, when Dirichlet vacated his chair there to take up the position at Göttingen left open by Gauss's death.

In the 1840s, Jacobi wrote some interesting results on hypergeometric series. In the posthumous paper mentioned earlier, he showed that the sequence of hypergeometric polynomials $F(-n, b, c, x)$ where $n = 0, 1, 2, \ldots$, were orthogonal with respect to a suitable distribution. Following Euler, he also worked out how definite integrals could be employed to study solutions of the hypergeometric equation. In another paper, he applied the symbolic method to obtain some known transformation formulas for hypergeometric functions.

In a paper of 1857, Bernhard Riemann took a very different approach to hypergeometric functions as part of his new theory of functions of a complex variable. Riemann gave the foundation of this theory in his famous doctoral dissertation of 1851. An important idea first given in this work and later applied to the theory of abelian functions, hypergeometric functions, and the zeta function was that a complex analytic function was to a large extent determined by the nature and location of its singularities. The

27.1 Preliminary Remarks

singularities of the hypergeometric equation are at 0, 1, and ∞. In his 1857 paper, Riemann considered the more general case where the singularities of a function were at three distinct values a, b, and c. He axiomatically defined a set of functions, called P functions, satisfying certain properties in the neighborhood of the three singularities, but without reference to the hypergeometric function or equation. Riemann showed that P functions were solutions of a second-order differential equation reducible to the hypergeometric equation when the singular points were 0, 1, and ∞. He also developed a very simple transformation theory for P functions by means of which one could derive a large number of relations among hypergeometric functions with little calculation.

We have seen that Gauss emphasized the fact that the hypergeometric series represented a hypergeometric function in only a small part of the domain of definition of the function. Moreover, the function was multivalued. Perhaps unable to develop a theory of complex variables to treat the hypergeometric function to his satisfaction, Gauss held back publication of the second part of his paper on the subject. Riemann saw Gauss's full paper in 1855, after Gauss's death. Surely this problem left pending by Gauss provided Riemann with great motivation for his landmark 1857 paper. Riemann also had a strong interest in mathematical physics; as he mentioned in the introduction to his paper, the hypergeometric function had numerous applications in physical and astronomical researches. After 1857, Riemann continued his investigations on the theory of ordinary differential equations with algebraic coefficients. His lectures and writings on the topic were published posthumously and eventually led to the formulation of what is now known as the Riemann–Hilbert problem.

Felix Klein (1849–1925) was one of the earliest mathematicians to understand and propagate the ideas of Riemann. In 1893, he gave a course of lectures on Riemann's theory of hypergeometric functions. Interestingly, a decade later, the English mathematician E. W. Barnes (1874–1953) presented an alternative development of the hypergeometric function, based on the complex analytic technique of the Mellin transform, making use of Cauchy's calculus of residues.

R. H. Mellin (1854–1935) was a Finnish mathematician who studied analysis first under Mittag–Leffler in Stockholm and then with Weierstrass in Berlin. He started teaching in 1884 at what was later named the Technical University of Finland. He founded a tradition of research in complex function theory in Finland, continued by mathematicians such as Ernst Lindelöf, Frithiof and Rolf Nevanlinna, and Lars V. Ahlfors. Mellin gave a general formulation of the Mellin transform in an 1895 treatise on the gamma and hypergeometric functions. For a function $f(x)$ integrable on $(0, \infty)$, the Mellin transform is defined by

$$F(s) = \int_0^\infty x^{s-1} f(x) dx. \qquad (27.14)$$

If $f(x) = O(x^{-a+\epsilon})$ as $x \to 0+$ and $f(x) = O(x^{b-\epsilon})$ as $x \to +\infty$, for $\epsilon > 0$ and $a < b$, then the integral converges absolutely and defines an analytic function in the strip $a < \operatorname{Re} s < b$. Mellin gave the inversion formula:

$$f(x) = \frac{1}{2\pi i} \int_{c-\infty i}^{c+\infty i} x^{-s} F(s) ds, \ a < c < b. \qquad (27.15)$$

In particular, we have the pair of formulas (stated without convergence conditions) very useful in analytic number theory:

$$\Gamma(s) = \int_0^\infty x^{s-1} e^{-x} dx \text{ and } e^{-x} = \frac{1}{2\pi i} \int_{c-\infty i}^{c+\infty i} \Gamma(s) x^{-s} ds. \qquad (27.16)$$

In fact, Riemann had already used the Mellin transform in his famous paper on the distribution of primes. Other particular cases of the transform were derived by others, including Mellin himself, before he stated the general formula. The second formula in (27.16) was apparently first discovered by the French mathematician Eugène Cahen in 1893. His thesis on the Riemann zeta function and its analogs contains several interesting results on Dirichlet series, though some of these were not rigorously proved until more than a decade later. Cahen followed Riemann in taking the Mellin transforms of a function analogous to the theta function to obtain functional equations for the corresponding Dirichlet series. He considered some analogs of the theta function:

$$\sum_{n=1}^\infty \left(\frac{n}{p}\right) e^{-n^2 \pi x/p}, \; \sum_{n=1}^\infty n\left(\frac{n}{p}\right) e^{-n^2 \pi x/p}, \; \sum_{n=1}^\infty \frac{\sigma_1(n)}{n} e^{-2n\pi x}$$

where $\left(\frac{n}{p}\right)$ denoted the Legendre symbol and $\sigma_1(n)$ the sum of the divisors of n. Cahen employed the first sum when $p \equiv 1 \pmod 4$, and the second when $p \equiv 3 \pmod 4$.

E. W. Barnes studied at Trinity College, Cambridge, from 1893 to 1896. Most of his mathematical work was done in the period 1897–1910 on the double gamma function, hypergeometric functions and Mellin transforms, and the theory of entire functions. In 1915 Barnes left Cambridge to pursue his second career. He was ordained in 1922 and appointed to the Bishopric of Birmingham in 1924, an office he held until 1952.

Barnes's starting point was the observation that from Euler's integral representation, and by expanding $(1-xt)^{-a}$ as a series, the Mellin transform of the hypergeometric function would be

$$\int_0^\infty x^{s-1} F(a,b,c,-x) dx = \frac{\Gamma(c)}{\Gamma(a)\Gamma(b)} \frac{\Gamma(s)\Gamma(a-s)\Gamma(b-s)}{\Gamma(c-s)}, \qquad (27.17)$$

for $\min(\operatorname{Re} a, \operatorname{Re} b) > \operatorname{Re} s > 0$. This suggested the integral representation for the hypergeometric function:

$$\frac{\Gamma(a)\Gamma(b)}{\Gamma(c)} F(a,b,c,x) = \frac{1}{2\pi i} \int_{k-\infty i}^{k+\infty i} \frac{\Gamma(s)\Gamma(a-s)\Gamma(b-s)}{\Gamma(c-s)} (-x)^{-s} ds, \qquad (27.18)$$

where $\min(\operatorname{Re} a, \operatorname{Re} b) > k > 0$ and $c \neq 0, -1, -2, \ldots$. This is Barnes's integral for the hypergeometric function and provides the basis for an alternative development of these functions. A precise statement of the integral formula requires conditions on the path of integration.

27.2 Euler's Derivation of the Hypergeometric Equation

We follow Euler's notation as it is easy to understand and his derivation is quite short and straightforward. Euler let s denote the hypergeometric series (27.1). Then

$$\partial(x^c \partial s) = ab x^{c-1} + \frac{ab}{1 \cdot c}(a+1)(b+1)x^c + \cdots$$

$$\partial(x^a s) = ax^{a-1} + \frac{ab}{1 \cdot c}(a+1)x^a + \cdots.$$

Note that, for the sake for brevity, he frequently suppressed ∂x. Now

$$\partial(x^{b+1-a}\partial(x^a s)) = abx^{b-1} + \frac{ab}{1 \cdot c}(a+1)(b+1)x^b + \cdots$$
$$= x^{b-c}\partial(x^c \partial s),$$

or $\partial(ax^b s + x^{b+1}\partial s) = x^{b-c}(cx^{c-1}\partial s + x^c \partial\partial s)$

or $a(bx^{b-1}s + x^b \partial s) + (b+1)x^b \partial s + x^{b+1}\partial\partial s = cx^{b-1}\partial s + x^b \partial\partial s.$

Dividing by x^{b-1}, he got the hypergeometric equation

$$x(1-x)\partial\partial s + (c - (a+b+1)x)\partial s - abs = 0. \tag{27.19}$$

Euler gave an equally simple proof of the transformation formula. He showed that $s = (1-x)^n z$ also satisfied a second-order differential equation with the hypergeometric form when $n = c - a - b$. He started by taking the logarithmic derivative of s to obtain

$$\frac{\partial s}{s} = \frac{\partial z}{z} - \frac{n\partial x}{1-x}. \tag{27.20}$$

The derivative of this equation was

$$\frac{\partial\partial s}{s} - \frac{(\partial s)^2}{s^2} = \frac{\partial\partial z}{z} - \frac{(\partial z)^2}{zz} - \frac{n(\partial x)^2}{(1-x)^2}. \tag{27.21}$$

We remark that Euler wrote ∂s^2 for $(\partial s)^2$. He then squared (27.20) to get

$$\frac{(\partial s)^2}{s^2} = \frac{(\partial z)^2}{z^2} - \frac{2n\partial x \partial z}{z(1-x)} + \frac{nn(\partial x)^2}{(1-x)^2}.$$

He added this equation to (27.21) to get

$$\frac{\partial\partial s}{s} = \frac{\partial\partial z}{z} - \frac{2n\partial x \partial z}{z(1-x)} + \frac{n(n-1)(\partial x)^2}{(1-x)^2}. \tag{27.22}$$

When (27.20) and (27.22) were applied to the hypergeometric equation (27.19), he could write

$$x(1-x)\frac{\partial\partial z}{z} - \frac{2nx\partial x \partial z}{z} + (c-(a+b+1)x)\frac{\partial z}{z}$$
$$+ \frac{n(n-1)x(\partial x)^2}{1-x} - \frac{n(c-(a+b+1)x)\partial x}{1-x} - ab = 0. \tag{27.23}$$

Next, the two terms with $1-x$ in the denominator, the second of which had a suppressed ∂x, combined to form

$$\frac{n((n+a+b)x-c)}{1-x}.$$

When $n+a+b=c$, the factor $1-x$ canceled. For this n, (27.23) was reduced to

$$x(1-x)\partial\partial z + [c+(a+b-2c-1)x]\partial z - (c-a)(c-b)z = 0, \tag{27.24}$$

an equation of the hypergeometric type. Thus,

$$z = F(c-a, c-b, c, x) = (1-x)^{a+b-c} F(a, b, c, x). \tag{27.25}$$

This proved Euler's transformation (27.5).

27.3 Pfaff's Derivation of the $_3F_2$ Identity

We have already noted that equation (27.25) is equivalent to Pfaff's identity (27.7). Pfaff gave a very interesting proof of this, given here in modern notation using shifted factorials. Let

$$S_n(a,b,c) = \sum_{j=0}^n \frac{(-n)_j (a)_j (b)_j}{j!(c)_j (1-n+a+b-c)_j}.$$

Then, by a simple calculation,

$$S_n(a,b,c) - S_{n-1}(a,b,c)$$

$$= \sum_{j=0}^n \left(\frac{(-n)_j (a)_j (b)_j}{j!(c)_j (1-n+a+b-c)_j} - \frac{(1-n)_j (a)_j (b)_j}{j!(c)_j (2-n+a+b-c)_j} \right)$$

$$= \frac{-(1+a+b-c)ab}{c(1+a+b-c-n)(2+a+b-c-n)} S_{n-1}(a+1, b+1, c+1). \tag{27.26}$$

By induction, the recurrence (27.26), combined with the initial value $S_0(a,b,c)=1$, uniquely determines $S_n(a,b,c)$. Pfaff could easy verify that

$$\sigma_n(a,b,c) = \frac{(c-a)_n (c-b)_n}{(c)_n (c-a-b)_n}$$

satisfied the same recurrence relation and initial condition, proving his formula (27.7).

This formula is quite useful and important, though this does not seem to have been realized until the twentieth century when it found applications to the evaluation of combinatorial sums of products of binomial coefficients. In this connection, the Chinese mathematician Li Shanlan (1811–1882) is of historical interest. He was trained in the Chinese mathematical tradition, though later in life he came to learn about Western works on algebra, analytic geometry, and calculus. At the age of 8, he studied the ancient Chinese text *Jiuzhang Suanshu*, and six years later he read a Chinese translation of the first six books of Euclid's *Elements*. Soon after that, he studied Chinese works on algebra and trigonometry. Eventually he became interested in the summation of finite series. He made some interesting discoveries involving Stirling numbers, Euler numbers and other numbers and series of combinatorial significance, contained in his work *Duoji Bilei*. This may be translated as "Heaps Summed Using Analogies"; heaps refer to finite sums. In this work, Li Shanlan developed and generalized the concepts and formulas of earlier researchers such as Wang Lai (1768–1813) and Dong Youcheng (1791–1823). Li Shanlan presented the following summation formula:

$$\sum_{j=0}^{k} \binom{k}{j}^2 \binom{n+2k-j}{2k} = \binom{n+k}{k}^2. \tag{27.27}$$

This formula was brought to the notice of the Hungarian mathematician Paul Turán (1910–1976) in 1937. He gave a proof using Legendre polynomials, published in 1954. This aroused the curiosity of other mathematicians, and it was established that the combinatorial sum (27.27) could be written as

$$\binom{n+2k}{2k} {}_3F_2\left(\begin{matrix}-k,-k,-n\\1,-n-2k\end{matrix};1\right),$$

and therefore (27.27) could be derived from Pfaff's formula. Jacobi's perceptive effort to prevent this formula from being forgotten provides further evidence of his insight into formulas and his stature as an algorist. As another application of Pfaff's identity, note that it can be written as

$$\sum_{k=0}^{n} \frac{(-n)_k (a)_k (b)_k}{k!(c)_k(-n+1+a+b-c)_k} = \frac{(c-a)_n}{n!n^{c-a-1}} \cdot \frac{(c-b)_n}{n!n^{c-b-1}} \cdot \frac{n!n^{c-1}}{(c)_n} \cdot \frac{n!n^{c-a-b-1}}{(c-a-b)_n}.$$

When $n \to \infty$ and $\text{Re}(c-a-b) > 0$, by (23.2), we obtain Gauss's ${}_2F_1$ summation mentioned earlier as (23.11):

$$F(a,b,c,1) = \frac{\Gamma(c)\Gamma(c-a-b)}{\Gamma(c-a)\Gamma(c-b)}, \tag{27.28}$$

though we do not know whether Gauss was aware of this derivation.

27.4 Gauss's Contiguous Relations and Summation Formula

The contiguous relations can be given in compact form if we use the following notation for contiguous functions:

$$F = F(a,b,c,x), \quad F(a+) = F(a+1,b,c,x), \quad \text{etc.}$$

Gauss wrote down all of the fifteen contiguous relations connecting F with two functions contiguous to it. Here we give four examples:

$$(c - 2a - (b - a)x)F + a(1 - x)F(a+) - (c - a)F(a-) = 0, \qquad (27.29)$$

$$(c - a - b)F + a(1 - x)F(a+) - (c - b)F(b-) = 0, \qquad (27.30)$$

$$(c - a - 1)F + aF(a+) - (c - 1)F(c-) = 0, \qquad (27.31)$$

$$c(c - 1 - (2c - a - b - 1)x)F + (c - a)(c - b)xF(c+) - c(c - 1)(1 - x)F(c-) = 0. \qquad (27.32)$$

From the fifteen relations, one may obtain other relations in which more than one parameter is changed by one or more; we give a relation presented by Gauss, where our notation has the obvious meaning.

$$F(b+, c+) - F = \frac{a(c - b)x}{c(c + 1)} F(a+, b+, c + 2). \qquad (27.33)$$

Gauss proved relations (27.30) and (27.31): First, let

$$M = \frac{(a + 1)_{n-1}(b)_{n-1}}{n!(c)_n}.$$

Then the coefficients of x^n in F, $F(b-)$, $F(a+)$, $F(c-)$, and $xF(a+)$ would be

$$a(b + n - 1)M, \ a(b - 1)M, (a + n)(b + n - 1)M,$$

$$\frac{a(b + n - 1)(c + n - 1)M}{c - 1}, \ n(c + n - 1)M,$$

respectively. To obtain (27.31), it was therefore sufficient for him to check that

$$a(c - a - 1)(b + n - 1) + a(a + n)(b + n - 1) - a(b + n - 1)(c + n - 1) = 0.$$

Equation (27.30) can be proved in a similar manner; equation (27.33) can also be proved by the direct method. Gauss found his formula (27.28) for $F(a, b, c, 1)$ by taking $x = 1$ in (27.32) to obtain

$$F(a, b, c, 1) = \frac{(c - a)(c - b)}{c(c - a - b)} F(a, b, c + 1, 1). \qquad (27.34)$$

Note that he proved the convergence of the series for $\mathrm{Re}\,(c - a - b) > 0$; thus, the series on the right-hand side also converged. By repeated application of this equation he got

$$F(a, b, c, 1) = \frac{(c - a)_n(c - b)_n}{(c)_n(c - a - b)_n} F(a, b, c + n, 1). \qquad (27.35)$$

Gauss could then express the right-hand side of the equation in terms of the gamma function, just as we obtained (27.28); he then let $n \to \infty$ to get the result.

27.5 Gauss's Proof of the Convergence of $F(a,b,c,x)$ for $c-a-b>0$

Gauss's proof of this important result was based on the formula

$$(\beta - \alpha - 1)\left(1 + \frac{\alpha}{\beta} + \frac{\alpha(\alpha+1)}{\beta(\beta+1)} + \cdots + \frac{(\alpha)_k}{(\beta)_k}\right) = \beta - 1 - \frac{(\alpha)_{k+1}}{(\beta)_k}. \quad (27.36)$$

This summation formula follows immediately from the following relation; although Gauss did not state it explicitly, he knew it well from his numerous calculations with hypergeometric series.

$$\frac{(\alpha)_k}{(\beta)_{k-1}} - \frac{(\alpha)_{k+1}}{(\beta)_k} = (\beta - \alpha - 1)\frac{(\alpha)_k}{(\beta)_k}. \quad (27.37)$$

A simple algebraic calculation is sufficient to check this relation. The idea was to write a hypergeometric term as a difference of two terms. It is interesting that in 1978, Bill Gosper showed the tremendous effectiveness of this approach in the summation of series of hypergeometric type. Gosper's method is now one of the fundamental algorithms used to sum such series. Now note that the ratio of the $(n+1)$th term over the nth term of the series $F(a,b,c,x)$ (omitting x) is

$$\frac{(a+n)(b+n)}{(1+n)(c+n)} = \frac{n^2 + (a+b)n + ab}{n^2 + (c+1)n + c}. \quad (27.38)$$

We take a, b, c real, though the argument also applies to complex values. Gauss proved, more generally, that if the ratio of the consecutive terms in a series was

$$\frac{n^\lambda + An^{\lambda-1} + Bn^{\lambda-2} + Cn^{\lambda-3} + \cdots}{n^\lambda + an^{\lambda-1} + bn^{\lambda-2} + cn^{\lambda-3} + \cdots} \quad (27.39)$$

and $A - a$ was a negative quantity with absolute value greater than unity, then the series converged. And when this result is applied to the special case of $F(a,b,c,x)$, it follows from (27.38) that the hypergeometric series converges for $c+1-a-b > 1$ or $c-a-b > 0$. To prove the theorem, write the series, for which the ratio of terms is given by (27.39), as $M_1 + M_2 + M_3 + \cdots$. We remark that Gauss did not use subscripts; he wrote the series as $M + M' + M'' + \cdots$. Now since $a > A+1$, there is a sufficiently small number h such that $a - h > A + 1$, or $a - h - 1 > A$. Now observe that if the fraction (27.39) is multiplied by $\frac{n}{n-1-h}$, we have

$$\frac{n}{n-1-h} \frac{M_{n+1}}{M_n} = \frac{n^{\lambda+1} + An^\lambda + \cdots}{n^{\lambda+1} + (a-h-1)n^\lambda + \cdots}.$$

If n is large enough, the last ratio is less than 1. Suppose this true for $n \geq N$. Then

$$|M_{N+1}| < \frac{N-1-h}{N}|M_n|,$$

$$|M_{N+2}| < \frac{N-h}{N+1}|M_{n+1}| < \frac{(N-h-1)(N-h)}{N(N+1)}|M_N|,$$

$$\cdots$$

$$|M_{N+k}| < \frac{(N-h-1)(N-h)\cdots(N-h-1+k-1)}{N(N+1)\cdots(N+k-1)}|M_N|.$$

Hence,

$$|M_N| + |M_{N+1}| + \cdots + |M_{N+k}|$$
$$= |M_N|\left(1 + \frac{N-h-1}{N} + \frac{(N-h-1)(N-h)}{N(N+1)} + \cdots + \frac{(N-h-1)_k}{(N)_k}\right)$$
$$= \frac{|M_N|}{h}\left(N - 1 - \frac{(N-h-1)_{k+1}}{(N)_k}\right),$$

where the last equation follows from (27.36). The term $\frac{(N-h-1)_{k+1}}{(N)_k}$ tends to zero as $k \to \infty$ because

$$\lim_{k \to \infty} \frac{(N-h-1)_{k+1}}{(N)_k} = \lim_{k \to \infty} \left(\frac{k!k^N}{(N)_k}\right)\left(\frac{(N-h)_k}{k!k^{N-h}}\right)\frac{N-h-1}{k^h}.$$

Now the first two expressions in parentheses have the limit $\Gamma(N)/\Gamma(N-h)$, while $\lim_{k \to \infty}(N-h-1)/k^h = 0$. Thus, Gauss proved that

$$\sum_{k=N}^{\infty}|M_k| < \frac{N-1}{h}|M_N|,$$

and the convergence of $\sum_{n=1}^{\infty} M_n$ followed. Observe that Gauss's method leads to a great refinement of the ratio test.

27.6 Gauss's Continued Fraction

Gauss derived an important continued fraction from the contiguous relation (27.33). He set

$$G(a,b,c,x) = \frac{F(a,b+1,c+1,x)}{F(a,b,c,x)}, \quad \text{so that}$$

$$\frac{F(a+1,b,c+1,x)}{F(a,b,c,x)} = \frac{F(b,a+1,c+1,x)}{F(b,a,c,x)} = G(b,a,c,x).$$

Then, dividing (27.33) by $F(a,b+1,c+1,x)$, he obtained

$$1 - \frac{1}{G(a,b,c,x)} = \frac{a(c-b)}{c(c+1)}xG(b+1,a,c+1,x),$$

or

$$G(a,b,c,x) = \frac{1}{1 - \frac{a(c-b)}{c(c+1)} x G(b+1,a,c+1,x)}. \tag{27.40}$$

This process could be continued:

$$G(b+1,a,c+1,x) = \frac{1}{1 - \frac{(b+1)(c+1-a)}{(c+1)(c+2)} x G(a+1,b+1,c+2,x)},$$

and thus
$$G(a,b,c,x) = \frac{1}{1-} \frac{\alpha_0 x}{1-} \frac{\beta_1 x}{1-} \frac{\alpha_1 x}{1-} \frac{\beta_2 x}{1-} \cdots, \tag{27.41}$$

where

$$\alpha_n = \frac{(a+n)(c+n-b)}{(c+2n)(c+2n+1)} \quad \text{and} \quad \beta_n = \frac{(b+n)(c+n-a)}{(c+2n-1)(c+2n)}. \tag{27.42}$$

Gauss mentioned an important particular case: when $b = 0$. In that case,

$$G(a,0,c-1,x) = F(a,1,c,x), \tag{27.43}$$

and the formulas in (27.42) took the form

$$\alpha_n = \frac{(a+n)(c+n-1)}{(c+2n-1)(c+2n)} \quad \text{and} \quad \beta_n = \frac{n(c+n-1-a)}{(c+2n-2)(c+2n-1)}. \tag{27.44}$$

For $a = 1$ and $c = \frac{3}{2}$ and $x = t^2$, Gauss had

$$\log \frac{1+t}{1-t} = \frac{2t}{1-} \frac{\frac{1}{3}t^2}{1-} \frac{\frac{2\cdot 2}{3\cdot 5}t^2}{1-} \frac{\frac{3\cdot 3}{5\cdot 7}t^2}{1-} \cdots.$$

This continued fraction played a fundamental role in Gauss's theory of numerical integration.

27.7 Gauss: Transformations of Hypergeometric Functions

Gauss found solutions of the hypergeometric equation other than $F(\alpha,\beta,\gamma,x)$ and also used the hypergeometric equation to obtain transformation formulas, just as Euler had done. Note that Gauss used the symbols $\alpha, \beta,$ and γ and employed a, b for variables in a different context. We shall follow that practice here. He set $x = 1 - y$ in the hypergeometric equation to get

$$(y - yy)\frac{ddP}{dy^2} + (\alpha + \beta + 1 - \gamma - (\alpha + \beta + 1)y)\frac{dP}{dy} - \alpha\beta P = 0.$$

Clearly, $P = F(\alpha, \beta, \alpha + \beta + 1 - \gamma, y)$ was a solution of this equation and hence $F(\alpha, \beta, \alpha + \beta + 1 - \gamma, 1 - x)$ would be an independent solution of the hypergeometric equation. Gauss noted that any solution of the hypergeometric equation must be a linear combination of these two. He then looked for solutions of the form $P = x^\mu P'$ by substituting this expression for P in the equation. He observed that the equation for P' was of the hypergeometric form when $\mu = 0$ or $\mu = 1 - \gamma$. In the latter case, the equation for P' was

$$(x - xx)\frac{ddP'}{dx^2} + (2 - \gamma - (\alpha + \beta + 3 - 2\gamma)x)\frac{dP'}{dx} - (\alpha + 1 - \gamma)(\beta + 1 - \gamma)P' = 0.$$

Thus,

$$P = x^{1-\gamma} F(\alpha + 1 - \gamma, \beta + 1 - \gamma, 2 - \gamma, x)$$
$$= (1-x)^{\gamma - \alpha - \beta} x^{1-\gamma} F(1 - \alpha, 1 - \beta, 2 - \gamma, x)$$

would be another solution of the original hypergeometric equation. Observe that the last step followed from an application of Euler's transformation (27.5). It then followed that there existed constants M and N such that

$$F(\alpha, \beta, \alpha + \beta + 1 - \gamma, 1 - x) = MF(\alpha, \beta, \gamma, x)$$
$$+ Nx^{1-\gamma}(1-x)^{\gamma - \alpha - \beta} F(1 - \alpha, 1 - \beta, 2 - \gamma, x).$$

Gauss determined after three pages of interesting calculations that

$$M = \frac{\Gamma(\alpha + \beta + 1 - \gamma)\Gamma(1 - \gamma)}{\Gamma(\alpha + 1 - \gamma)\Gamma(\beta + 1 - \gamma)} \text{ and } N = \frac{\Gamma(\alpha + \beta + 1 - \gamma)\Gamma(\gamma - 1)}{\Gamma(\alpha)\Gamma(\beta)}.$$

We observe that the case in which α is a negative integer was given by Pfaff in 1797. In this case, the second term is zero because $\Gamma(\alpha)$ appears in the denominator. Gauss remarked that this formula was useful for computational purposes. Clearly, a series would converge more rapidly for x between 0 and $1/2$ than for x between $1/2$ and 1. A formula of this type could be applied to convert a slowly convergent series to two more rapidly convergent ones. But Gauss cautioned that this formula would not be applicable if the series to be transformed was such that the third parameter minus the sum of the first two turned out to be an integer. He then went on to show that if this occurred, the formula could be modified by the use of his Ψ function and the logarithm. He explicitly worked out the formula for the elliptic integral $F(1/2, 1/2, 1, 1 - x)$.

Gauss also found solutions at infinity. He set $x = 1/y$ and then $P = y^\mu P'$ and observed that P' was hypergeometric when $\mu = \alpha$ or β. Thus, he obtained P as

$$x^{-\alpha} F(\alpha, \alpha + 1 - \gamma, \alpha + 1 - \beta, 1/x) \text{ or } x^{-\beta} F(\beta, \beta + 1 - \gamma, \beta + 1 - \alpha, 1/x).$$

He then expressed $F(\alpha, \beta, \gamma, x)$ as a linear combination of these solutions.

Gauss derived another general transformation formula by taking $x = y/(y - 1)$ in the hypergeometric equation and then $P = (1 - y)^\mu P'$, so that another hypergeometric equation would be obtained when $\mu = \alpha$ or β. This gave the necessary result:

$$F(\alpha, \beta, \gamma, x) = (1 - y)^\alpha F(\alpha, \gamma - \beta, \gamma, y)$$
$$= (1 - x)^{-\alpha} F\left(\alpha, \gamma - \beta, \gamma, \frac{x}{x - 1}\right).$$

27.7 Gauss: Transformations of Hypergeometric Functions

In 1797 Pfaff published this result for the case in which α was a negative integer. Gudermann proved the generalization in 1830. Three years later, P. E. O. Vorsselman de Heer noted in his thesis that Euler's transformation could be obtained when the preceding transformation was applied to itself; Kummer also observed this fact.

In the published part of his 1812 paper, Gauss found the values of the coefficients in the expansion

$$(aa + bb - 2ab\cos\phi)^{-n} = A + 2A'\cos\phi + 2A''\cos 2\phi + 2A'''\cos 3\phi + \cdots \quad (27.45)$$

in terms of hypergeometric series. He noted that

$$
\begin{aligned}
A^{(p)} &= \frac{1}{a^{2n}} \binom{n+p-1}{p} \left(\frac{b}{a}\right)^p F\left(n, n+p, p+1, \frac{bb}{aa}\right) \\
&= \frac{1}{(aa+bb)^n} \binom{n+p-1}{p} \left(\frac{ab}{aa+bb}\right)^p \\
&\quad \times F\left(\frac{n+p}{2}, \frac{n+p+1}{2}, p+1, \frac{4aabb}{(aa+bb)^2}\right) \\
&= \frac{1}{(a\pm b)^{2n}} \binom{n+p-1}{p} \left(\frac{ab}{(a\pm b)^2}\right)^p F\left(n+p, p+\frac{1}{2}, n+\frac{1}{2}, \frac{\pm 4ab}{(a\pm b)^2}\right).
\end{aligned}
\quad (27.46)
$$

Note that Euler studied the series (27.45) in a 1749 memoir on the perturbation of planetary orbits and in 1766 Lagrange found the first series for $A^{(p)}$ in (27.46). This series and its coefficients have been studied intensively, both analytically and numerically, and Gauss's interest in them was evident. If we take $a = 1$ and $x = b^2$, the second equation in (27.46) gives

$$(1+x)^{n+p} F(n, n+p, p+1, x) = F\left(\frac{n+p}{2}, \frac{n+p+1}{2}, p+1, \frac{4x}{(1+x)^2}\right). \quad (27.47)$$

This is an example of a quadratic transformation because the variable on one side is x, or it could be a fractional linear transformation of x, while the variable on the right-hand side involves x^2. It is very likely that equation (27.47) led Gauss to study such transformations in the second (unpublished) part of his paper. He set $x = 4y/(1+y)^2$ in the hypergeometric equation and then $P = (1+y)^{2\alpha} Q$ to find that the equation satisfied by Q was

$$(1+y)(y-y^2)\frac{d^2Q}{dy^2} + \left(\gamma - (4\beta - 2\gamma)y + (\gamma - 4\alpha - 2)y^2\right)\frac{dQ}{dy}$$
$$- 2\alpha(2\beta - \gamma + (2\alpha + 1 - \gamma)y)Q = 0.$$

Now note that when $\beta = \alpha + 1/2$, $1 + y$ is a common factor in this equation. We remark that equation (27.47) guided Gauss in the substitutions for x, P and β. We next have $Q = F(2\alpha, 2\alpha + 1 - \gamma, \gamma, y)$ and finally

$$(1+y)^{2\alpha} F(2\alpha, 2\alpha + 1 - \gamma, \gamma, y) = F\left(\alpha, \alpha + 1/2, \gamma, 4y/(1+y)^2\right). \quad (27.48)$$

27.8 Kummer's 1836 Paper on Hypergeometric Series

Kummer independently rediscovered Gauss's unpublished results on hypergeometric functions, including the quadratic transformations. Of course, he was familiar with Gauss's published paper and with the work of Euler, Pfaff, Jacobi, and Gudermann on this topic. Kummer took a general approach. He set out to determine all functions of z and w of x such that $y = w F(\alpha', \beta', \gamma', z)$ satisfied the equation

$$y'' + (\gamma - (\alpha + \beta + 1)x) y' - \alpha \beta y = 0$$

and α', β', γ' were linear combinations of α, β, γ. He found z to be a fractional linear transformation $(ax+b)/(cx+d)$ and that w could be taken to be

$$x^{1-\gamma}, \ (1-x)^{\gamma-\alpha-\beta}, \ x^{1-\gamma}(1-x)^{\gamma-\alpha-\beta}, \ \text{or } 1.$$

Specifically, z could be any one of the six fractional linear transformations serving to permute the values 0, 1, and ∞. These would be

$$z = x, z = 1-x, z = 1/x, z = 1/(1-x), z = x/(x-1), z = (x-1)/x.$$

When $z = x$, he obtained the four forms

$$F(\alpha, \beta, \gamma, x), \quad (1-x)^{\gamma-\alpha-\beta} F(\gamma-\alpha, \gamma-\beta, \gamma, x),$$

$$x^{1-\gamma} F(\alpha-\gamma+1, \beta-\gamma+1, 2-\gamma, x),$$

$$x^{1-\gamma}(1-x)^{\gamma-\alpha-\beta} F(1-\alpha, 1-\beta, 2-\gamma, x).$$

Thus, he obtained twenty-four solutions of the hypergeometric equation and determined the linear relation among any three of them.

Kummer may have become interested in quadratic transformations after studying Gauss's published equation (27.46). His interest in elliptic integrals may have provided him with further motivation to study these transformations. It was clear to Kummer, as it was to Gauss, that quadratic transformations existed when the parameters α, β, γ in $F(\alpha, \beta, \gamma, x)$ satisfied certain relations. So Kummer considered the linear relations among the parameters leading to such transformations. In this way, he rediscovered Gauss's results as well as new ones. For example, he obtained

$$F(\alpha, \beta, 2\beta, x) = (1-x)^{\beta-\alpha} \left(1 - \frac{x}{2}\right)^{\alpha-2\beta} F\left(\beta - \frac{\alpha}{2}, \frac{2\beta-\alpha+1}{2}, \beta + \frac{1}{2}, \left(\frac{x}{2-x}\right)^2\right).$$

Note that by applying Euler's transformation to the right-hand side, we get the simpler form

$$F(\alpha, \beta, 2\beta, x) = \left(1 - \frac{x}{2}\right)^{-\alpha} F\left(\frac{\alpha}{2}, \frac{\alpha+1}{2}, \beta + \frac{1}{2}, \left(\frac{x}{2-x}\right)^2\right).$$

This and Gauss's transformation (27.48) are the two basic quadratic transformations; from these, the others can be obtained by using fractional linear transformations or the three-term relations among the different solutions of the hypergeometric equation. At the end of his paper, Kummer commented on the more general hypergeometric series

$$1 + \frac{\alpha \cdot \beta \cdot \lambda}{1 \cdot \gamma \cdot \nu} x + \frac{\alpha(\alpha+1) \cdot \beta(\beta+1) \cdot \lambda(\lambda+1)}{1 \cdot 2 \cdot \gamma(\gamma+1) \cdot \nu(\nu+1)} x^2 + \cdots.$$

He wrote that he was unable to obtain general transformation formulas for this function, although he had several for the case $x = 1$. As an example, he presented

$$\sum_{k=0}^{\infty} \frac{(\alpha)_k (\beta)_k (\lambda)_k}{k!(\gamma)_k (\nu)_k} = \frac{\Gamma(\nu)\Gamma(\nu+\gamma-\alpha-\beta-\lambda)}{\Gamma(\nu-\lambda)\Gamma(\nu+\gamma-\alpha-\beta)} \sum_{k=0}^{\infty} \frac{(\gamma-\alpha)_k (\gamma-\beta)_k (\lambda)_k}{k!(\gamma)_k (\nu+\gamma-\alpha-\beta)_k}. \tag{27.49}$$

He observed that, in general, this series could not be summed in terms of the gamma function, but when $\lambda = 1$ and $\nu = 2(\alpha + \beta - \gamma + 1)$, then its value would be

$$\frac{(\alpha+\beta-\gamma+1)(\gamma-1)}{(\alpha-\gamma+1)(\beta-\gamma+1)} \left(\frac{\Gamma(\gamma-1)\Gamma(\alpha+\beta-\gamma+1)}{\Gamma(\alpha)\Gamma(\beta)} - 1 \right). \tag{27.50}$$

Recall that Stirling discovered a particular case of Kummer's transformation where $\lambda = 1$ and $\nu = \beta + 1$; see (11.40).

27.9 Jacobi's Solution by Definite Integrals

Euler gave a method of solving differential equations using definite integrals. He applied it to solve several second-order differential equations, including one related to the hypergeometric equation. Jacobi worked out the specific details of the method for the hypergeometric equation and showed how to obtain the twenty-four solutions of Kummer. Jacobi started with the observation that for

$$V = u^{\beta-1}(1-u)^{\gamma-\beta-1}(1-xu)^{-\alpha}, \tag{27.51}$$

$$x(1-x)\frac{d^2V}{dx^2} + (\gamma - (\alpha+\beta+1)x)\frac{dV}{dx} - \alpha\beta V = -\alpha \frac{d}{du}\left(\frac{u(1-u)}{1-xu}V\right)$$

$$= -\alpha u^{\beta}(1-u)^{\gamma-\beta}(1-xu)^{-\alpha-1}.$$

Hence, $y = \int_0^1 V\, du$ would be a solution of the hypergeometric equation for $\beta > 0$ and $\gamma - \beta > 0$ because

$$-\alpha \int_0^1 d\left(\frac{u(1-u)}{1-xu}V\right) = -\alpha \frac{u(1-u)}{1-xu}V \bigg|_0^1 = -\alpha u^{\beta}(1-u)^{\gamma-\beta}(1-xu)^{-\alpha-1} \bigg|_0^1 = 0.$$

The expression $u^{\beta}(1-u)^{\gamma-\beta}(1-xu)^{-\alpha-1}$ also vanished at $u = \pm\infty$ when $\gamma - \alpha - 1 < 0$. So if g and h were a pair of the values $0, 1, \pm\infty$, then, Jacobi observed, the integral

$y = \int_g^h V\, du$ would be a solution of the hypergeometric equation under suitable conditions on α, β, γ.

Jacobi also considered a solution of the form $y = \int_g^{\epsilon/x} V\, du$ where ϵ was a constant. When this y was substituted in the hypergeometric equation, Jacobi obtained

$$-(\gamma - \beta - 1)\epsilon^\beta (1-\epsilon)^{1-\alpha} x^{1-\gamma}(x-\epsilon)^{\gamma-\beta-2} + \alpha g^\beta (1-g)^{\gamma-\beta}(1-xg)^{-\alpha-1}.$$

The expression involving ϵ vanished for $\epsilon = 1$ when $1 - \alpha > 0$, so for $y = \int_g^{1/x} V\, du$ to be a solution, Jacobi required that $1 - \alpha > 0$. Taking x to be positive, Jacobi had the six solutions:

- $y = \int_0^1 V\, du$, when β and $\gamma - \beta$ were positive;
- $y = \int_0^{-\infty} V\, du$, when β and $\alpha + 1 - \gamma$ were positive;
- $y = \int_1^\infty V\, du$, when $\gamma - \beta$ and $\alpha + 1 - \gamma$ were positive;
- $y = \int_0^{1/x} V\, du$, when β and $1 - \alpha$ were positive;
- $y = \int_{1/x}^\infty V\, du$, when $\alpha + 1 - \gamma$ and $1 - \alpha$ were positive;
- $y = \int_1^{1/x} V\, du$, when $\gamma - \beta$ and $1 - \alpha$ were positive.

Jacobi then noted that the integral $\int_0^1 u^\lambda (1-u)^\mu (1-au)^\nu\, du$ was in fact a constant times the series $F(-\nu, \lambda+1, \lambda+\mu+2, a)$. This series could be derived by expanding $(1-au)^\nu$ by the binomial expansion and then performing term-by-term integration. Also, note that the last five integrals could actually be obtained by a suitable substitution in the first one. For example, to go from $\int_0^1 V\, du$ to $\int_0^{-\infty} V\, du$, set $u = \frac{v-1}{v}$. Then

$$y = \int_0^{-\infty} V\, du = (-1)^\beta x^{-\alpha} \int_0^1 v^{\alpha-\gamma}(1-v)^{\beta-1}\left(1 - v\frac{x-1}{x}\right)^{-\alpha} dv.$$

The corresponding hypergeometric series would be

$$F\left(\alpha, \alpha+1-\gamma, \alpha+\beta+1-\gamma, \frac{x-1}{x}\right).$$

In this manner, Jacobi represented the six integrals as hypergeometric functions:

- $F(\alpha, \beta, \gamma, x)$, substitution $u = v$;
- $x^{-\alpha} F\left(\alpha, \alpha+1-\gamma, \alpha+\beta+1-\gamma, \frac{x-1}{x}\right)$, substitution $u = \frac{v-1}{v}$;
- $x^{-\alpha} F\left(\alpha, \alpha+1-\gamma, \alpha+1-\beta, \frac{1}{x}\right)$, substitution $u = \frac{1}{v}$;
- $x^{-\beta} F\left(\beta, \beta+1-\gamma, \beta+1-\alpha, \frac{1}{x}\right)$, substitution $u = \frac{v}{x}$;
- $x^{1-\gamma} F(\alpha+1-\gamma, \beta+1-\gamma, 2-\gamma, x)$, substitution $u = \frac{1}{xv}$;
- $x^{\alpha-\gamma}(1-x)^{\gamma-\alpha-\beta} F\left(\gamma-\alpha, 1-\alpha, \gamma+1-\alpha-\beta, \frac{x-1}{x}\right)$, substitution $u = \frac{1}{x+(1-x)v}$.

Jacobi then observed that, other than the identity, the fractional linear transformations mapping 0, 1 to itself could be given as

$$u = 1 - v, \quad u = \frac{v}{1-x+vx}, \quad u = \frac{1-v}{1-vx}.$$

Then $V\,du$ was, respectively,

$$(1-x)^{-\alpha}v^{\gamma-\beta-1}(1-v)^{\beta-1}\left(1-\frac{xv}{x-1}\right)^{-\alpha}dv,$$

$$(1-x)^{-\beta}v^{\beta-1}(1-v)^{\gamma-\beta-1}\left(1-\frac{xv}{x-1}\right)^{\alpha-\gamma}dv,$$

$$(1-x)^{\gamma-\alpha-\beta}v^{\gamma-\beta-1}(1-v)^{\beta-1}(1-vx)^{\alpha-\gamma}dv.$$

Observe that we have $y = \int_0^1 V\,du$ as a constant times each of the four expressions:

$$F(\alpha,\beta,\gamma,x) = (1-x)^{-\alpha}F\left(\alpha,\gamma-\beta,\gamma,\frac{x}{x-1}\right)$$

$$= (1-x)^{-\beta}F\left(\gamma-\alpha,\beta,\gamma,\frac{x}{x-1}\right)$$

$$= (1-x)^{\gamma-\alpha-\beta}F(\gamma-\alpha,\gamma-\beta,\gamma,x).$$

Similarly, there are four expressions with each of the six integral solutions, yielding Kummer's twenty-four solutions.

27.10 Riemann's Theory of Hypergeometric Functions

Kummer showed that the twenty-four solutions of the hypergeometric equation could be expressed as hypergeometric series in $x, 1-x, 1/x, 1-1/x, \frac{1}{1-x}, \frac{1}{1-1/x}$ multiplied by suitable powers of x and/or $1-x$. He also gave the relations among any three overlapping solutions. In a paper of 1857, Riemann reversed this process, starting with a set of functions with three properties; these properties in turn uniquely determined the functions up to a constant factor, as well as the differential equation of which these functions were the complete set of solutions. He denoted by

$$P\left\{\begin{matrix} a & b & c \\ \alpha & \beta & \gamma & x \\ \alpha' & \beta' & \gamma' \end{matrix}\right\}$$

any function satisfying the three properties:

- For all values of x except a, b, c, called branch points, P was single valued and finite.
- Between any three branches P', P'', P''' of this function, there was a linear homogeneous relation with constant coefficients,

$$C'P' + C''P'' + C'''P''' = 0.$$

- The function could be written in the form

$$C_\alpha P^\alpha + C_{\alpha'} P^{\alpha'},\ C_\beta P^{(\beta)} + C_{\beta'} P^{(\beta')},\ C_\gamma P^{(\gamma)} + C_{\gamma'} P^{(\gamma')},$$

where C_α, $C_{\alpha'}$, $\cdots C_{\gamma'}$ were constants and

$$(x-a)^{-\alpha} P^{(\alpha)}, \ (x-a)^{-\alpha'} P^{(\alpha')}$$

were single valued near $x = a$ and nonvanishing and finite at $x = a$; a similar requirement would hold for

$$(x-b)^{-\beta} P^{(\beta)}, \ (x-b)^{-\beta'} P^{(\beta')} \text{ at } x=b$$

and for $(x-c)^{-\gamma} P^{(\gamma)}, \ (x-c)^{-\gamma'} P^{(\gamma')}$ at $x=c$.

Moreover, $\alpha - \alpha'$, $\beta - \beta'$, $\gamma - \gamma'$ were not integers and $\alpha + \alpha' + \beta + \beta' + \gamma + \gamma' = 1$.

It follows immediately from the definition of P that if x' is a fractional linear transformation of x mapping a, b, c to a', b', c', then

$$P \left\{ \begin{matrix} a & b & c \\ \alpha & \beta & \gamma & x \\ \alpha' & \beta' & \gamma' \end{matrix} \right\} = P \left\{ \begin{matrix} a' & b' & c' \\ \alpha & \beta & \gamma & x' \\ \alpha' & \beta' & \gamma' \end{matrix} \right\} \tag{27.52}$$

Here recall that every conformal mapping of $\mathbb{C} \cup \{\infty\}$ is of the form

$$x' = \frac{\lambda x + \mu}{\delta x + \nu}, \text{ where } \lambda \nu - \mu \delta = 1.$$

We can therefore choose a', b', c' to be $0, \infty, 1$. It is also clear from the definition of the Riemann P function that

$$\left(\frac{x-a}{x-b} \right)^\delta P \left\{ \begin{matrix} a & b & c \\ \alpha & \beta & \gamma & x \\ \alpha' & \beta' & \gamma' \end{matrix} \right\} = P \left\{ \begin{matrix} a & b & c \\ \alpha+\delta & \beta-\delta & \gamma & x \\ \alpha'+\delta & \beta'-\delta & \gamma' \end{matrix} \right\};$$

$$x^\delta (1-x)^\epsilon P \left\{ \begin{matrix} 0 & \infty & 1 \\ \alpha & \beta & \gamma & x \\ \alpha' & \beta' & \gamma' \end{matrix} \right\}$$

$$= P \left\{ \begin{matrix} 0 & \infty & 1 \\ \alpha+\delta & \beta-\delta-\epsilon & \gamma+\epsilon & x \\ \alpha'+\delta & \beta'-\delta-\epsilon & \gamma'+\epsilon \end{matrix} \right\}. \tag{27.53}$$

Following Riemann, we write $P \left(\begin{matrix} \alpha & \beta & \gamma \\ \alpha' & \beta' & \gamma' \end{matrix} x \right)$ when the first row is $0 \ \infty \ 1$. We may immediately write the relations

$$P\begin{pmatrix} 0 & a & 0 \\ 1-c & b & c-a-b \end{pmatrix} x\end{pmatrix} = (1-x)^{-a} P\begin{pmatrix} 0 & a & 0 & \frac{x}{x-1} \\ 1-c & c-b & b-a & \frac{x}{x-1} \end{pmatrix}$$

$$= x^{-a} P\begin{pmatrix} 0 & a & 0 & \frac{1}{x} \\ b-a & 1-c+b & c-a-b & \frac{1}{x} \end{pmatrix}$$

$$= (1-x)^{c-a-b} P\begin{pmatrix} 0 & c-a & 0 \\ 1-c & c-b & a+b-c \end{pmatrix} x\end{pmatrix}.$$
(27.54)

Riemann also studied contiguous relations satisfied by the P functions. Following Gauss, he used these relations to find the differential equation satisfied by P. In fact, Riemann worked out the details only for the case $\gamma = 0$, sufficient for his purpose. Felix Klein's student Erwin Papperitz (1857–1938) presented the general case in 1889. Riemann found that the equation satisfied by $P\begin{pmatrix} \alpha & \beta & 0 \\ \alpha' & \beta' & \gamma' \end{pmatrix} x\end{pmatrix}$ was

$$(1-x)\frac{d^2 y}{d \log x^2} - (\alpha + \alpha' + (\beta + \beta')x)\frac{dy}{d \log x} + (\alpha\alpha' - \beta\beta' x)y = 0.$$

He showed quite easily from this equation that

$$F(a, b, c, x) = \text{const.} P\begin{pmatrix} 0 & a & 0 \\ 1-c & b & c-a-b \end{pmatrix} x\end{pmatrix}.$$

Moreover, the Pfaff and Euler transformations follow from this equation and (27.54).

Riemann's work on the hypergeometric equation led to important developments in the theory of linear differential equations. Riemann himself foresaw some of these developments, though he did not publish his ideas. In 1904, James Pierpont wrote about this aspect of nineteenth-century mathematics: "A particular class of linear differential equations of great importance is the hypergeometric equation; the results obtained by Gauss, Kummer, Riemann, and Schwarz relating to this equation have had the greatest influence on the development of the general theory. The great extent of the theory of linear differential equations may be estimated when we recall that within its borders it embraces not only almost all the elementary functions, but also the modular and automorphic functions."

27.11 Exercises

1. Verify (27.50).
2. Show that $y = (\arcsin x)^2/2$ satisfies the differential equation

$$(1-x^2)\frac{d^2 y}{dx^2} - x\frac{dy}{dx} - 1 = 0.$$

Deduce Takebe's formula

$$\frac{1}{2}(\arcsin x)^2 = \sum_{n=0}^{\infty} \frac{2^{2n}(n!)^2}{(2n+2)!} x^{2n+2}.$$

Prove Clausen's 1828 observation that this formula is a particular case of his formula (27.13). See Clausen (1828). Also see Eu. 14, pp. 156–186 and the correspondence of Euler and Johann Bernoulli on this topic: Eu. 4A-2, pp.161–262.

3. Prove the following examples mentioned in Gauss's 1812 paper on hypergeometric series.

$$\sin nt = n \sin t \, F\left(\frac{1}{2}n + \frac{1}{2}, -\frac{1}{2}n + \frac{1}{2}, \frac{3}{2}, \sin^2 t\right);$$

$$\sin nt = n \sin t \cos t \, F\left(\frac{1}{2}n + 1, -\frac{1}{2}n + 1, \frac{3}{2}, \sin^2 t\right);$$

$$\cos nt = F\left(\frac{1}{2}n, -\frac{1}{2}n, \frac{1}{2}, \sin^2 t\right);$$

$$\cos nt = \cos t \, F\left(\frac{1}{2}n + \frac{1}{2}, -\frac{1}{2}n + \frac{1}{2}, \frac{1}{2}, \sin^2 t\right).$$

See Gauss (1863–1927), vol. 3, p. 127.

4. Show that

$$e^t = \frac{1}{1-} \frac{t}{1+} \frac{\frac{1}{2}t}{1-} \frac{\frac{1}{6}t}{1+} \frac{\frac{1}{6}t}{1-} \frac{\frac{1}{10}t}{1+} \frac{\frac{1}{10}t}{1-} \cdots$$

$$t = \frac{\sin t \cos t}{1-} \frac{\frac{1\cdot 2}{1\cdot 3}\sin^2 t}{1-} \frac{\frac{1\cdot 2}{5\cdot 7}\sin^2 t}{1-} \frac{\frac{3\cdot 4}{5\cdot 7}\sin^2 t}{1-} \frac{\frac{3\cdot 4}{7\cdot 9}\sin^2 t}{1-}.$$

See Gauss (1863–1927), vol. 3, pp. 136–137.

5. Show that

$$F(2, 4, 9/2, x) = (1-x)^{-\frac{3}{2}} F(5/2, 1/2, 9/2, x).$$

Gauss stated this without proof in the *Ephemeridibus Astronomicis Berolinensibus* 1814, p. 257. He gave a proof in the unpublished second part of his paper. See Gauss (1863–1927), vol. 3, p. 209.

6. Set $x = 1 - y$ in the hypergeometric differential equation and from its form deduce that $F(a, b, a+b+1-c, 1-x)$ is another solution of the hypergeometric equation. See Gauss (1863–1927), vol. 3, p. 208.

7. Show that when $x = y/(y-1)$, the hypergeometric equation changes to

$$(1-y)(y-y^2)\frac{d^2 F}{dy^2} + (1-y)(c + (a+b-c-1)y)\frac{dF}{dy} + abF = 0.$$

In this equation set $F = (1-y)^\mu G$ to show that G satisfies

$$(1-y)(y-y^2)\frac{d^2 G}{dy^2} + (1-y)(c + (a+b-c-1)y - 2\mu y)\frac{dG}{dy}$$
$$+ \left((ab - \mu(a+b-c-1)y) + (\mu^2 - \mu)y\right) G = 0.$$

27.11 Exercises

Show that when $\mu = a$ or $\mu = b$, then the coefficient of G is divisible by $(1-y)$ and thus deduce that

$$F(a,b,c,x) = (1-x)^{-a} F(a, c-b, c, x/(x-1)).$$

This is Gauss's proof of Pfaff's transformation. See Gauss (1863–1927), vol. 3, pp. 217–218.

8. Set $x = 4y - 4y^2$. Show that the hypergeometric equation takes the form

$$(y-y^2)\frac{d^2F}{dy^2} + (c - (4a+4b+2)y + (4a+4b+2)y^2)\frac{1}{1-2y}\frac{dF}{dy} - 4abF = 0.$$

Next, show that the fraction in the middle term is removed by putting $c = a + b + \frac{1}{2}$. Also deduce that

$$F\left(a, b, a+b+\frac{1}{2}, 4y-4y^2\right) = F\left(2a, 2b, a+b+\frac{1}{2}, y\right).$$

It was by this example that Gauss illustrated the multivaluedness of the hypergeometric function. See Gauss (1863–1927), vol. 3, pp. 225–227.

9. Prove Kummer's transformation (27.49) and its corollary (27.50). Kummer stated these formulas without proof at the end of his 1836 paper. See Kummer (1975), vol. 2, pp. 75–166.

10. Prove Euler's continued fraction formula

$$\frac{\beta x}{\gamma} \frac{{}_2F_1(-\alpha, \beta+1; \gamma+1, -x)}{{}_2F_1(-\alpha; \beta; \gamma; -x)} = \frac{\beta x}{\gamma - (\alpha+\beta+1)x} + \frac{(\beta+1)(\alpha+\gamma+1)x}{\gamma+1-(\alpha+\beta+2)x} + \frac{(\beta+2)(\alpha+\gamma+2)x}{\gamma+2-(\alpha+\beta+3)x} + \cdots.$$

See Eu. I-14 pp. 291–349.

11. Prove Ramanujan's integral formula

$$\int_0^\infty x^{s-1}(\phi(0) - \phi(1)x + \phi(2)x^2 - \cdots)dx = \frac{\pi}{\sin s\pi}\phi(-s).$$

See Berndt (1985–1998), part I, pp. 295–307. See also Hardy (1978), pp. 186–190; he relates this formula of Ramanujan with a 1914 interpolation theorem of F. Carlson, useful in proving hypergeometric formulas.

12. Use the following outline to determine when the square of a hypergeometric function

$$y = \sum_{n=0}^\infty \frac{(\alpha)_n(\beta)_n}{n!(\gamma)_n} x^n \text{ takes the form } z = \sum_{n+0}^\infty \frac{(\alpha')_n(\beta')_n(\delta')_n}{n!(\gamma')_n(\epsilon')_n} x^n.$$

(i) Show that when the hypergeometric equation is multiplied by x and then differentiated, the result is

$$(x^3 - x^2)\frac{d^3y}{dx^3} + ((\alpha+\beta+4)x^2 - (\gamma+2)x)\frac{d^2y}{dx^2}$$
$$+ ((2\alpha+2\beta+\alpha\beta+2)x - \gamma)\frac{dy}{dx} + \alpha\beta y = 0.$$

(ii) Show that z satisfies the differential equations

$$(x^3 - x^2)\frac{d^3z}{dx^3} + ((3+\alpha'+\beta'+\delta')x^2 - (1+\gamma'+\epsilon')x)\frac{d^2z}{dx^2}$$
$$+ (1+\alpha'+\beta'+\delta'+\alpha'\beta'+\alpha'\delta'+\beta'\delta')x - \gamma'\epsilon')\frac{dz}{dx} + \alpha'\beta'\delta'z = 0.$$

(iii) Show that if $z = y^2$, then the equation in (ii) becomes

$$(x^3 - x^2)2y\frac{d^3y}{dx^3} + ((3+a')x^2 - (1+d')x)2y\frac{d^2y}{dx^2}$$
$$+ ((1+a'+b')x - e')2y\frac{dy}{dx} + c'y^2$$
$$+ 6(x^3-x^2)\frac{dy}{dx}\cdot\frac{d^2y}{dx^2} + 2((3+a')x^2 - (1+d')x)\left(\frac{dy}{dx}\right)^2 = 0,$$

where $a' = \alpha' + \beta' + \delta'$, $b' = \alpha'\beta' + \alpha'\delta' + \beta'\delta'$, $c' = \alpha'\beta'\delta'$, $d' = \epsilon' + \gamma'$, $e' = \epsilon'\gamma'$.

(iv) Multiply the hypergeometric equation by $\frac{2yA}{x} + B\frac{dy}{dx}$ and equation (i) by $2y$ and add the two equations. Compare the resulting equation with (iii), and deduce that

$$\gamma = \alpha + \beta + 1/2,\ A = 2\alpha + 2\beta - 1,\ B = b',\ a' = 3\alpha + 3\beta,$$
$$b' = 2\alpha^2 + 8\alpha\beta + 2\beta^2,\ c' = 4(\alpha+\beta)\alpha\beta,\ d' = 3\gamma - 1,\ e' = (2\gamma-1)\gamma.$$

(v) Deduce that $\alpha' = 2\alpha$, $\beta' = 2\beta$, $\delta' = \alpha + \beta$, $\gamma' = \gamma$, $\epsilon' = 2\gamma - 1$.

(vi) Conclude that

$$\left({}_2F_1\left(\begin{matrix}\alpha,\beta,\\ \alpha+\beta+1/2\end{matrix}; x\right)\right)^2 = {}_3F_2\left(\begin{matrix}2\alpha,2\beta,\alpha+\beta\\ \alpha+\beta+1/2,2\alpha+2\beta\end{matrix}; x\right).$$

See Clausen (1828).

27.12 Notes on the Literature

The translation of the quotation from Gauss is taken from D. Kikuchi's translation of Gauss's two-part paper, included in *Memoirs on Infinite Series*, an 1891 publication of the Tokio [sic] Mathematical and Physical Society. Robert Burckel's English translation of Gauss's letter to Bessel is from Remmert (1991), pp. 167–168. Euler's paper, "Specimen transformationis singularis serierum," (Eu. I-16-2, pp. 41–55) contains his work on the hypergeometric equation. For Gauss's papers on the hypergeometric series, see Gauss (1863–1927), vol. 3, pp. 125–162 and 206–229. Discussions of Gauss's convergence test, including observations illustrating that Gauss's work on this topic was far ahead of its time, are available in Bressoud (2007) and Knopp (1990).

For Kummer's papers on hypergeometric series, see Kummer (1975), vol. 2, pp. 75–166. Jacobi's musketeer remark appeared in Lampe's obituary of Kummer; see Kummer (1975), vol. 1, p. 18. For more on Jacobi's remark and for his letter, see Pieper's entertaining "A Network of Scientific Philanthropy" (2007). Jacobi's paper, "Untersuchungen über die Differentialgleichung der hypergeometrischen Reihe," was edited and published by Heine in 1859, eight years after Jacobi's death. See Jacobi (1969), vol. 6, pp. 184–202. Jacobi wrote other papers on the hypergeometric series; this one elaborated on Euler's important idea of solving a differential equation by definite integrals.

Riemann's 1857 paper defined the P function and mentioned Gauss in the title; see Riemann (1990), pp. 99–115. In 1859, perhaps influenced by the work of Jacobi mentioned above, Riemann gave a course of lectures in which he defined the P function by means of a complex integral. See Riemann (1990), pp. 667–691. The reader may read more on Li Shanlan and other Chinese mathematicians in Martzloff (1997), pp. 341–350. Also see Turán (1990), vol. 1, pp. 743–747 on Li Shanlan and a proof of his formula. For the quote from his recently republished article, see Pierpont (2000). For a history of the hypergeometric series, see Dutka (1984).

28
Orthogonal Polynomials

28.1 Preliminary Remarks

Orthogonal polynomials played an important role in the nineteenth-century development of continued fractions, hypergeometric series, numerical integration, and approximation theory; in the twentieth century, they additionally contributed to progress in the moment problem and in functional analysis. However, orthogonal polynomials may not have received recognition proportional to their significance, leading Barry Simon to dub them "the Rodney Dangerfield of analysis." Nevertheless, when Paul Nevai edited the proceedings of a 1989 conference on this subject, he stamped on the dedication page, "I love orthogonal polynomials."

A sequence of polynomials $p_n(x)$, $n = 0, 1, 2, \ldots$, is said to be orthogonal with respect to a weight function $w(x)$ over an interval (a, b) where $-\infty \leq a < b \leq \infty$, if

$$\int_a^b p_n(x) p_m(x) w(x) \, dx = A_n \delta_{mn}, \tag{28.1}$$

where $A_n \neq 0$. In a paper on probability written in the early 1770s, Lagrange defined a sequence of polynomials containing as special cases the Legendre polynomials. Denoted by $P_n(x)$, the Legendre polynomials are obtained when $a = -1$, $b = 1$ and $w(x) \equiv 1$, in (28.1). Lagrange gave a three-term recurrence relation for his sequence of polynomials; for the particular case of Legendre polynomials, this recurrence amounted to

$$(2n+1) x P_n(x) = (n+1) P_{n+1}(x) + n P_{n-1}(x) \tag{28.2}$$

$$n = 1, 2, 3, \ldots, \quad P_0(x) = 1, \; P_1(x) = x.$$

In a paper of 1785 on the attraction of spheroids of revolution, Legendre defined the polynomials now bearing his name by the expansion

$$(1 - 2\cos\theta y + y^2)^{-1/2} = 1 + P_1(\cos\theta) y + P_2(\cos\theta) y^2 + P_3(\cos\theta) y^3 + \cdots. \tag{28.3}$$

In this memoir, Legendre needed only the polynomials of even degree and he explicitly presented $P_2(\cos\theta)$, $P_4(\cos\theta)$, $P_6(\cos\theta)$, and $P_8(\cos\theta)$. We note his first two examples:

$$P_2(\cos\theta) = \frac{3}{2}\cos^2\theta - \frac{1}{2}; \quad P_4(\cos\theta) = \frac{5\cdot 7}{2\cdot 4}\cos^4\theta - \frac{3\cdot 5}{2\cdot 4}2\cos^2\theta + \frac{1\cdot 3}{2\cdot 4}. \quad (28.4)$$

In the second volume of his *Exercices de calcul intégral* of 1817, Legendre gave the orthogonality relation and an expression for the general $P_n(x)$. Legendre polynomials played an important role in the celestial mechanics of Laplace, Legendre, and others.

Gauss used Legendre polynomials in his 1814 paper on numerical integration, extending the work of Newton and Cotes. But Gauss did not refer to the earlier work on these polynomials; rather, he conceived of Legendre polynomials as an outgrowth of his work in hypergeometric series. The groundbreaking approach and methodology taken by Gauss in this paper led to important advances in nineteenth-century numerical analysis. Briefly summarizing Gauss, we suppose $\int_c^d y(x)\,dx$ is to be computed. Let points $a(=a_0), a_1, \ldots, a_n$ be chosen in $[c, d]$ and let the corresponding values of y at these points be $y_0, y_1, \ldots, y_n$. Set

$$f(x) = (x-a)(x-a_1)(x-a_2)\ldots(x-a_n). \quad (28.5)$$

Note that the nth degree Lagrange-Waring polynomial

$$z_n(x) = \sum_{k=0}^{n} \frac{f(x) y_k}{f'(a_k)(x - a_k)} \quad (28.6)$$

passes through (a_k, y_k), $k = 0, 1, \ldots, n$, and therefore interpolates $y(x)$; thus, we may write $y(x) = z_n(x) + r_n(y)$. Then

$$\int_c^d y(x)\,dx = \sum_{k=0}^{n} \lambda_k y_k + R_n(y), \quad (28.7)$$

where

$$\lambda_k = \int_c^d \frac{f(x)\,dx}{f'(a_k)(x-a_k)} \quad \text{and} \quad R_n(y) = \int_c^d r_n(y)\,dx. \quad (28.8)$$

It is clear that if $y(x)$ is a polynomial of degree $\leq n$, then $y(x) = z_n(x)$ and hence $R_n(y) = 0$. In the Newton–Cotes scheme, the points $a_0, a_1, \ldots, a_n$ were equally spaced. Gauss considered whether he could prove $R_n(y) = 0$ for a larger class of polynomials by varying the nodes $a, a_1, \ldots, a_n$. Since there were $n+1$ points to be varied, he wanted the class of polynomials, for which $R_n(y) = 0$, to consist of all polynomials of degree $\leq 2n+1$; indeed, he succeeded in proving this. In short, his argument began with the observation that

$$R_n\left(\frac{1}{t-x}\right) = R_n\left(\frac{1}{t} + \frac{x}{t^2} + \frac{x^2}{t^3} + \cdots\right) = \sum_{k=0}^{\infty} \frac{R_n(x^k)}{t^{k+1}}. \quad (28.9)$$

His problem was to choose $a, a_1, \ldots, a_n$ so that $R_n(x^k) = 0$ for $k = 0, 1, \ldots, 2n+1$; then he could write

$$R_n\left(\frac{1}{t-x}\right) = O\left(\frac{1}{t^{2n+3}}\right), \quad t \to \infty. \tag{28.10}$$

When $[c, d] = [-1, 1]$, from his results on hypergeometric functions, Gauss had

$$\int_{-1}^{1} \frac{dx}{t-x} = \ln\frac{1+1/t}{1-1/t} = \frac{2}{t-}\frac{1/3}{t-}\frac{2\cdot 2/(3\cdot 5)}{t-}\frac{3\cdot 3/(5\cdot 7)}{t-}\cdots. \tag{28.11}$$

He also knew that the $(n+1)$th convergent of this continued fraction was a rational function $S_n(t)/P_{n+1}(t)$, where S_n was of degree n and P_{n+1} of degree $n+1$. Moreover, this rational function approximated the continued fraction up to the order t^{-2n-3}. So Gauss factorized $P_{n+1}(t)$ and wrote $S_n(x)/P_{n+1}(x)$ as a sum of partial fractions:

$$\frac{S_n(x)}{P_{n+1}(x)} = \sum_{k=0}^{n} \frac{\lambda_k}{x - a_k}. \tag{28.12}$$

He then easily showed that by using these $a, a_1, \ldots, a_n$ and $\lambda, \lambda_1, \ldots, \lambda_n$, he would obtain the result. Gauss explicitly wrote down the polynomials $P_{n+1}(x)$ for $n = 0, 1, 2, \ldots, 6$; we can see they are Legendre polynomials of degrees 1 to 7, although Gauss did not make this observation. Instead, he gave the hypergeometric representations of the polynomials P_{n+1} and of the remainder

$$\ln\frac{1+1/t}{1-1/t} - \frac{S_n(t)}{P_{n+1}(t)}.$$

At the end of the paper, he computed the zeros of the Legendre polynomials of degree seven and less with the corresponding λ. He used these results to compute the integral $\int dx/\ln x$ over the interval $x = 100000$ to $x = 200000$. Note that Gauss was well aware that $\int_2^x dt/\ln t$ gave a good approximation for the number of primes less than x.

In a paper of 1826, Jacobi pointed out that Gauss's proof ultimately depended on the orthogonality of $P_{n+1}(x)$. To see this, suppose $y(x)$ is a polynomial of degree at most $2n+1$. Then $y(x) = q(x)P_{n+1}(x) + r(x)$, where $q(x)$ and $r(x)$ are polynomials of degree at most n. Next note that

$$\int_{-1}^{1} y(x)dx = \int_{-1}^{1} q(x)P_{n+1}(x)dx + \int_{-1}^{1} r(x)dx.$$

The first integral on the right-hand side vanishes by the orthogonality of $P_{n+1}(x)$, and the second integral can be exactly computed by the Newton–Cotes method because the degree of $r(x)$ is not greater than n. In fact, Jacobi did not start his reasoning process with the Legendre polynomial; at that time he may not have known of the earlier work of Legendre, Laplace, and others on Legendre polynomials. His argument produced these polynomials, their orthogonality, and the byproduct that

$$P_n(x) = \frac{1}{2^n n!}\frac{d^n}{dx^n}(x^2-1)^n. \tag{28.13}$$

28.1 Preliminary Remarks

Interestingly enough, Rodrigues and Ivory had already independently discovered this useful and important formula for Legendre polynomials. In 1808, Olinde Rodrigues enrolled in the Lycée Impérial, later named Lycée Louis-LeGrand and where Galois also studied. After graduating in 1812, he was admitted to the Université de Paris, submitting a doctoral thesis on the attraction of spheroids in 1815. Unfortunately, the haphazard journal in which his memoir on this subject was published produced only three volumes from 1814 to 1816. This partly explains why Rodrigues's work and the formula for Legendre polynomials, in particular, were not noticed. For several decades, the result was referred to as the formula of Ivory and Jacobi. In 1865, Hermite finally pointed out Rodrigues's paper; Cayley referred to it in a different context in 1858.

James Ivory (1765–1842) was an essentially self-taught Scottish mathematician whose interest was mainly in applied areas. He received much recognition, but perhaps suffered from depression, curtailing his career; in a letter to MacVey Napier he declared, "I believe on the whole I am the most unlucky person that ever existed." Most of Ivory's inspiration was drawn from the work of the French mathematicians Laplace, Legendre, and Lagrange; his papers contain several references to Laplace's book on celestial mechanics. Ivory published (28.13) in a 1824 paper on the shape of a revolving homogeneous fluid mass in equilibrium; he derived the formula from a result in his earlier 1812 paper on the attraction of a spheroid. In his 1824 paper, Ivory remarked on the formula, "From this very simple expression, the most remarkable properties of the coefficients of the expansion of $1/f$, are very readily deduced." Here f refers to the expression $(1 - 2\cos\theta y + y^2)^{1/2}$.

The Irish mathematician Robert Murphy (1806–1843), mentioned in chapter 20, had a brief mathematical career during which he published papers on integral equations, operator theory, and algebraic equations. He was perhaps the first to understand the significance of orthogonality; in a series of papers on integral equations in the early 1830s, Murphy considered the following problem: Suppose

$$\phi(x) = \int_0^1 t^x f(t)\,dt, \quad x = 0, 1, 2, \ldots.$$

Determine the function $f(t)$ from the function $\phi(x)$. One of the simplest results he stated in this connection was that if $\phi(x)$ was of the form $\frac{A}{x} + \frac{B}{x^2} + \frac{C}{x^3} + \cdots$, then $f(t)$ would be given by $1/t$ multiplied by the coefficient of $1/x$ in $\phi(x) \cdot t^{-x}$. As an extension of the previously stated problem, Murphy considered the determination of $f(t)$ from a knowledge of $\phi(x)$ for a finite number of values of x, say $x = 0, 1, \ldots, n-1$. The simplest case is when $\phi(x) = 0$ and this leads to the Legendre polynomials as solutions for $f(t)$. Murphy called such functions reciprocal rather than orthogonal. He also considered cases where t was replaced by $\ln t$, and this led him to the Laguerre polynomials

$$T_n(u) = \frac{e^{-u}}{n!} \cdot \frac{d^n}{du^n}(u^n e^u), \quad n = 0, 1, 2, \ldots.$$

He proved their orthogonality and found their generating function by applying the Lagrange inversion formula. Recall that a century before this, D. Bernoulli and Euler had studied these Laguerre polynomials; Bernoulli computed the zeros of several of them by his method of recurrent series.

It may be fair to say that Pafnuty Chebyshev was the creator of the theory of orthogonal polynomials and its applications. In an important paper of 1855, Chebyshev introduced and studied discreet orthogonal polynomials. This and later papers were associated with the areas of continued fractions, least squares approximations, interpolation, and approximate quadrature. Later in his career, Chebyshev's excellent students, including A. A. Markov and E. I. Zolotarev, continued his work in these and other areas.

28.2 Legendre's Proof of the Orthogonality of His Polynomials

In his *Exercices*, Legendre used the generating function for Legendre polynomials to offer an elegant and short proof of their orthogonality. Note that Legendre denoted $P_n(x)$ by X^n, but we use the more modern notation. He started with the generating function

$$(1 - 2xy + y^2)^{-1/2} = \sum_{n=0}^{\infty} P_n(x) y^n,$$

where $|x| \leq 1$ and $|y| < 1$, and considered the integral

$$I = \int_{-1}^{1} \left(\sum_{n=0}^{\infty} P_n(x) r^n y^n \right) \left(\sum_{m=0}^{\infty} P_m(x) y^m / r^m \right) dx \qquad (28.14)$$

$$= \int_{-1}^{1} \frac{dx}{\sqrt{(1 - 2xry + r^2 y^2)(1 - \frac{2xy}{r} + \frac{y^2}{r^2})}}.$$

He set

$$x = \frac{1 + r^2 y^2 - z^2}{2ry}$$

to obtain

$$I = -\frac{1}{2} \int_{1+ry}^{1-ry} \frac{dz}{\sqrt{(z^2 - 1 + r^2 + y^2 - r^2 y^2)}}$$

$$= \ln\left(-z + \sqrt{(z^2 - 1 + r^2 + y^2 - r^2 y^2)}\right) \Big|_{1+ry}^{1-ry}$$

$$= \ln(-1 + ry + r - y) - \frac{1}{2} \ln(-1 - ry + r + y)$$

$$= \ln \frac{1+y}{1-y} = 2 + \frac{2}{3} y^2 + \frac{2}{5} y^4 + \frac{2}{7} y^6 + \cdots + \frac{2}{2n+1} y^{2n} + \cdots.$$

Comparing this expression with the integral (28.14), he obtained orthogonality:

$$\int_{-1}^{1} P_n(x) P_m(x) \, dx = \frac{2}{2n+1} \delta_{mn}. \qquad (28.15)$$

He also used the generating function to obtain

$$P_n(x) = \frac{1 \cdot 3 \cdot 5 \cdots (2n-1)}{1 \cdot 2 \cdot 3 \cdots n} x^n - \frac{1 \cdot 3 \cdots (2n-3)}{1 \cdot 2 \cdot (n-2)} \frac{x^{n-2}}{2} + \cdots. \tag{28.16}$$

28.3 Gauss on Numerical Integration

Gauss started his 1814 paper with a discussion of the Newton–Cotes method for numerical integration. Let $[0, 1]$ be the interval and let $a, a_1, a_2, \ldots, a_n$ be $n+1$ points in that interval. Set

$$f(x) = (x-a)(x-a_1)(x-a_2)\ldots(x-a_n)$$
$$= x^{n+1} + c_1 x^n + c_2 x^{n-1} + \cdots + c_{n+1}. \tag{28.17}$$

Let y be a function to be integrated over $[0, 1]$ and let $y(= y_0), y_1, y_2, \ldots, y_n$ be its values at $a = a_0, a_1, a_2, \ldots, a_n$, respectively. The Lagrange-Waring interpolating polynomial of degree n for y is then given by

$$g(x) = \frac{f(x)y}{f'(a)(x-a)} + \sum_{k=1}^{n} \frac{f(x)y_k}{f'(a_k)(x-a_k)} = \sum_{k=0}^{n} \frac{f(x)y_k}{f'(a_k)(x-a_k)}. \tag{28.18}$$

The Newton–Cotes method then consists in integrating the interpolating polynomial:

$$\int_0^1 y \, dt = \sum_{k=0}^{n} \lambda_k y_k + R_n(y) \tag{28.19}$$

$$\text{where} \quad \lambda_k = \int_0^1 \frac{f(x)\,dx}{f'(a_k)(x-a_k)}, \quad k = 0, 1, \ldots n, \tag{28.20}$$

and $R_n(y)$ is the remainder. This remainder is zero when y is a polynomial of degree at most n. Gauss asked whether it was possible to choose $a, a_1, \ldots, a_n$ in such a way that the remainder would be zero for polynomials of degree at most $2n+1$, with the points $a, a_1, \ldots, a_n$ no longer equally spaced as in the Newton–Cotes procedure. Gauss observed that since $f(a) = 0$, he had

$$\frac{f(x)}{x-a} = \frac{x^{n+1} - a^{n+1} + c_1(x^n - a^n) + \cdots + c_n(x-a)}{x-a}$$
$$= x^n + x^{n-1}a + x^{n-2}a^2 + \cdots + a^n$$
$$\quad + c_1 x^{n-1} + c_1 x^{n-2}a + \cdots + c_1 a^{n-1}$$
$$\quad + \cdots.$$

Hence, after rearranging terms,

$$\int_0^1 \frac{f(x)}{x-a}dx = a^n + c_1 a^{n-1} + c_2 a^{n-2} + \cdots + c_n$$
$$+ \frac{1}{2}(a^{n-1} + c_1 a^{n-2} + \cdots + c_{n-1})$$
$$+ \frac{1}{3}(a^{n-2} + c_1 a^{n-3} + \cdots + c_{n-2})$$
$$+ \cdots$$
$$+ \frac{1}{n}(a + c_1)$$
$$+ \frac{1}{n+1}. \qquad (28.21)$$

Here Gauss noted that the nonnegative powers of x in the product

$$(x^{n+1} + c_1 x^n + \cdots + c_{n+1})\left(\frac{1}{x} + \frac{1}{2x^2} + \frac{1}{3x^3} + \cdots\right) = -f(x)\ln\left(1 - \frac{1}{x}\right) \qquad (28.22)$$

gave the terms on the right-hand side of (28.21) when $x = a$. So he could write

$$-f(x)\ln(1 - 1/x) = T_1(x) + T_2(x), \qquad (28.23)$$

where $T_1(x)$ was the polynomial or principal part of $-f(x)\ln(1 - 1/x)$; then, by (28.20),

$$T_1(a_k) = \int_0^1 \frac{f(x)}{x - a_k}dx = \lambda_k f'(a_k), \quad k = 0, 1, \ldots, n. \qquad (28.24)$$

Denoting $R_n(x^m)$ by k_m, Gauss used (28.19) to obtain

$$\sum_{k=0}^n \lambda_k a_k^m = \frac{1}{m+1} - k_m, \quad m = 0, 1, 2, \ldots. \qquad (28.25)$$

It followed that

$$\sum_{k=0}^n \frac{\lambda_k}{x - a_k} = \sum_{k=0}^n \frac{\lambda_k}{x}\left(1 - \frac{a_k}{x}\right)^{-1} = \sum_{k=0}^n \left(\frac{\lambda_k}{x} + \frac{\lambda_k a_k}{x^2} + \frac{\lambda_k a_k^2}{x^3} + \cdots\right)$$
$$= (1 - k_0)\frac{1}{x} + \left(\frac{1}{2} - k_1\right)\frac{1}{x^2} + \left(\frac{1}{3} - k_2\right)\frac{1}{x^3} + \left(\frac{1}{4} - k_3\right)\frac{1}{x^4} + \cdots$$
$$= -\ln\left(1 - \frac{1}{x}\right) - \left(\frac{k_{n+1}}{x^{n+2}} + \frac{k_{n+2}}{x^{n+3}} + \cdots\right). \qquad (28.26)$$

In the last equation, Gauss used the fact that $k_j = 0$ for $j = 0, 1, \ldots, n$; Gauss then had to determine conditions on $f(x)$ so that $k_j = 0$ for $j = n+1, n+2, \ldots, 2n$ as well. By (28.23) and (28.24), he could deduce that

$$T_1(x) = f(x)\sum_{k=0}^n \frac{\lambda_k}{x - a_k}.$$

28.3 Gauss on Numerical Integration

This was possible because both sides were polynomials of degree n, equal for $n+1$ values of x, given by $a_0, a_1, a_2, \ldots, a_n$. Thus, by (28.23) and (28.26), it followed that

$$f(x)\left(\frac{k_{n+1}}{x^{n+2}} + \frac{k_{n+2}}{x^{n+3}} + \frac{k_{n+3}}{x^{n+4}} + \cdots\right) = T_2(x). \tag{28.27}$$

Gauss used this analysis to find $f(x)$ of small degrees. For example, when $n=0$, $f(x) = x + c_1$, he had to consider

$$(x+c_1)\left(\frac{1}{x} + \frac{1}{2x^2} + \frac{1}{3x^3} + \cdots\right).$$

For the coefficient of $\frac{1}{x}$ to be zero, he required that $c_1 + 1/2 = 0$ or $c_1 = -1/2$. For $n=1$, $f(x) = x^2 + c_1 x + c_2$, and the coefficients of $1/x$ and $1/x^2$ in the expansion $-f(x)\ln(1 - 1/x)$ then had to be zero. He could then write the equations for c_1 and c_2:

$$c_2 + \frac{1}{2}c_1 + \frac{1}{3} = 0 \quad \text{and} \quad \frac{1}{2}c_2 + \frac{1}{3}c_1 + \frac{1}{4} = 0.$$

Thus, $c_1 = -1$ and $c_2 = 1/6$ and so the polynomial was $x^2 - x + 1/6$. Gauss then changed the variable so that the interval of integration became $[-1, 1]$. In this case, he had to choose the polynomial $U(x)$ of degree $n+1$ so that

$$\frac{1}{2}U(x)\ln\frac{1+1/x}{1-1/x} = U(x)\left(\frac{1}{x} + \frac{1}{3x^2} + \frac{1}{5x^3} + \cdots\right) = U_1(x) + U_2(x)$$

had appropriate negative powers of x with zero coefficients. In fact, the zeros u of U were related to zeros a of f by $u = 2a - 1$; U_1 and U_2 corresponded to T_1 and T_2 of equation (28.23). With this change of variables, the polynomials for $n = 1, 2, 3$ were x, $x^2 - 1/3$, and $x^3 - 3x/5$. Note that these are the Legendre polynomials of the first three degrees, normalized so that they are monic.

Gauss proceeded to give a method using continued fractions in order to quickly determine the polynomials $f(x)$. From his paper on hypergeometric functions, he had the expression

$$\phi(x) = \frac{1}{2}\ln\frac{x+1}{x-1} = \frac{1}{x-} \frac{1^2/3}{x-} \frac{2^2/3 \cdot 5}{x-} \frac{3^2/5 \cdot 7}{x-} \cdots. \tag{28.28}$$

He then showed that if the nth convergent of his continued fraction was $P_n(x)/Q_n(x)$, then

$$\phi(x) - \frac{P_n(x)}{Q_n(x)} = O\left(\frac{1}{x^{2n+1}}\right).$$

From this he could conclude that if $Q_n(x)$ was monic, then

$$Q_{n+1}(x) = U(x) \quad \text{and} \quad P_{n+1}(x) = U_1(x).$$

In this manner, Gauss completely solved his problem. Observe that the points a_k, $k = 0, 1, \ldots, n$ are the zeros of the Legendre polynomials and that the numbers λ_k

could be obtained from
$$\frac{P_{n+1}(x)}{Q_{n+1}(x)} = \sum_{k=0}^{n} \frac{\lambda_k}{x - a_k},$$
so that $\lambda_k = P_{n+1}(a_k)/Q'_{n+1}(a_k)$.

28.4 Jacobi's Commentary on Gauss

In the introduction to his 1826 paper on Gauss's new method of approximate quadrature Jacobi remarked that the simplicity and elegance of Gauss's results led him to believe that there was a simple and direct way of deriving them. The object of his paper was to present such a derivation, making use of his work from his doctoral dissertation on the Lagrange-Waring interpolation formula. Jacobi proceeded, in his usual lucid style, to show that Gauss's numerical integration method was effective because of its use of orthogonal polynomials. Abbreviating Jacobi's work for convenience, suppose $\phi(x) = \prod_{k=1}^{n}(x - x_k)$, where the x_i are distinct and suppose $f(x)$ is a polynomial of degree $\leq n - 1$. Then

$$\frac{f(x)}{\phi(x)} = \frac{A_1}{x - x_1} + \frac{A_2}{x - x_2} + \cdots + \frac{A_n}{x - x_n},$$

where $\quad a_k = \lim_{x \to x_k} \frac{(x - x_k) f(x)}{\phi(x)} = \lim_{x \to x_k} \frac{(x - x_k) f(x)}{\phi(x) - \phi(x_k)} = \frac{f(x_k)}{\phi'(x_k)}.$

So if $x_1, x_2, \ldots, x_n$ are the interpolation points, then any polynomial $f(x)$ of degree at most $n - 1$ can be expressed by the formula

$$f(x) = \sum_{k=1}^{n} \frac{f(x_k)\phi(x)}{\phi'(x_k)(x - x_k)},$$

attributed by Jacobi to Lagrange. The integral of such a polynomial is given exactly by the Newton–Cotes formula. On the other hand, if the degree of f is greater than $n - 1$, then divide $f(x)$ by $\phi(x)$ to get

$$\frac{f(x)}{\phi(x)} = V(x) + \frac{U(x)}{\phi(x)},$$

where $U(x)$ and $V(x)$ are polynomials and the degree of U is less than or equal to $n - 1$. Now assume with Jacobi that

$$f(x) = a + a_1 x + a_2 x^2 + \cdots + a_n x^n + a_{n+1} x^{n+1} + \cdots + a_{2n} x^{2n} + \cdots \quad \text{and}$$

$$\frac{1}{\phi(x)} = \frac{A_1}{x^n} + \frac{A_2}{x^{n+1}} + \cdots + \frac{A_{n+1}}{x^{2n}} + \cdots.$$

Then

$$V(x) = a_n A_1 + a_{n+1}(A_1 x + A_2) + a_{n+2}(A_1 x^2 + A_2 x + A_3) + \cdots$$
$$+ a_{2n-1}(A_1 x^{n-1} + A_2 x^{n-2} + \cdots + A_n) + \cdots.$$

Jacobi observed that according to Newton's method, to compute $\int_{-1}^{1} f(x)\,dx$, one would substitute $U(x)$ for $f(x)$ and the error would be

$$\Delta = \int f(x)\,dx - \int U\,dx = \int \phi(x) V\,dx.$$

He then noted that the expression for V did not involve $a_1, a_2, \ldots, a_{n-1}$ and hence the error, Δ, would be independent of these coefficients of f. The question was whether ϕ could be chosen so that the error would be independent of a_n, a_{n+1}, etc. Clearly, if $\int \phi x^k = 0$, for $k = 0, 1, \ldots, l$, then Δ would also be independent of $a_n, a_{n+1}, \ldots, a_{n+l-1}$. Since $\int (\phi(x))^2 dx > 0$, the value of l could be at most $n-1$. Thus if, $\int \phi x^k = 0$, for $k = 0, 1, \ldots, n-1$, then $\int f(x)\,dx$ was exact for polynomials of degree $\leq 2n-1$. This meant that $\phi(x)$ should be a constant multiple of the Legendre polynomial of degree n and Jacobi had succeeded in showing that orthogonality lay at the root of the Gaussian method of numerical integration.

28.5 Murphy and Ivory: The Rodrigues Formula

Robert Murphy's discussion of orthogonal polynomials appeared in his two publications on the inverse method of definite integrals of 1833 and 1835, written in 1832 and 1833, and in his 1833 treatise on physics. He considered the integral $\phi(x) = \int_0^1 f(t) t^x dt$ and determined the form of the polynomial $f(t)$ such that $\phi(x)$ was zero for $x = 0, 1, \ldots, n-1$. He let

$$f(t) = 1 + A_1 t + A_2 t^2 + \cdots + A_n t^n, \quad \text{so that}$$

$$\phi(x) = \frac{1}{x+1} + \frac{A_1}{x+2} + \frac{A_2}{x+3} + \cdots + \frac{A_n}{x+n+1} = \frac{P}{Q}, \qquad (28.29)$$

where $Q = (x+1)(x+2) \cdots (x+n+1)$ and P was a polynomial of degree at most n. To find an expression for $f(t)$ when $\phi(x) = 0$ for $x = 0, 1, \ldots, n-1$, Murphy argued that P would have the form $cx(x-1) \cdots (x-n+1)$. Thus,

$$\frac{1}{x+1} + \frac{A_1}{x+2} + \frac{A_2}{x+3} + \cdots + \frac{A_n}{x+n+1} = \frac{cx(x-1) \cdots (x-n+1)}{(x+1)(x+2) \cdots (x+n+1)}.$$

Multiplying both sides by $x + k$ and then setting $x = -k$ for $k = 1, \ldots, n+1$, he got the result

$$c = (-1)^n, \quad A_1 = -\frac{n}{1} \cdot \frac{n+1}{1}, \quad A_2 = \frac{n(n-1)}{1 \cdot 2} \cdot \frac{(n+1)(n+2)}{1 \cdot 2}, \text{ etc.}$$

Hence,

$$f(t) = 1 - \frac{n}{1} \cdot \frac{n+1}{1} t + \frac{n(n-1)}{1 \cdot 2} \cdot \frac{(n+1)(n+2)}{1 \cdot 2} t^2 - \cdots$$

$$= \frac{d^n}{dt^n} \frac{\left(t^n \left(1 - nt + \frac{n(n-1)}{1 \cdot 2} t^2 - \cdots\right)\right)}{1 \cdot 2 \cdot 3 \cdots n}$$

$$= \frac{1}{1 \cdot 2 \cdots n} \frac{d^n}{dt^n} (t(1-t))^n,$$

completing Murphy's proof of the Rodrigues formula.

Now Ivory's proof involved differential equations. He showed that $P_k(x)$, the kth Legendre polynomial, satisfied the equation

$$(k-n)(k+n+1)(1-x^2)^n \frac{d^n P_k}{dx^n} + \frac{d}{dx}\left((1-x^2)^{n+1} \frac{d^{n+1}}{dx^{n+1}} P_k\right) = 0.$$

Ivory presented this result in his 1812 paper. Twelve years later, unaware of Rodrigues's earlier work, he observed that by a repeated use of this equation, he could obtain the Rodrigues formula. He set

$$\phi_n = (1-x^2)^n \frac{d^n}{dx^n} P_k \quad \text{and} \quad \phi_0 = P_k$$

$$\text{so that} \quad \phi_0 + \frac{1}{k(k+1)} \frac{d}{dx} \phi_1 = 0,$$

$$\phi_1 + \frac{1}{(k-1)(k+2)} \frac{d}{dx} \phi_2 = 0,$$

$$\cdots$$

$$\phi_{k-1} + \frac{1}{1 \cdot 2k} \frac{d}{dx} \phi_k = 0.$$

Thus,

$$\phi_0 = \frac{(-1)^k}{1 \cdot 2 \cdot 3 \cdots 2k} \frac{d^k}{dx^k} \left((1-x^2)^k \frac{d^k}{dx^k} P_k\right).$$

Now, from (28.16), he could deduce $\frac{d^k}{dx^k} P_k = 1 \cdot 3 \cdot 5 \cdots (2k-1)$, and therefore he could write the required result,

$$P_k(x) = \frac{(-1)^k}{2 \cdot 4 \cdot 6 \cdots 2k} \frac{d^k}{dx^k} (1-x^2)^k.$$

Also mentioned in chapter 29 on q-series, Olinde Rodrigues employed differential equations to obtain his formula in 1815; in fact, he referred to Ivory's 1812 paper. Still, note that Rodrigues has priority in this matter, since Ivory did not work out his final result until 1824.

28.6 Liouville's Proof of the Rodrigues Formula

In 1837, Ivory and Jacobi published a joint paper in Liouville's journal, containing a proof of the Rodrigues formula, using Lagrange inversion. They were both unaware that the French mathematician Rodrigues had already published his result in 1815, albeit in an obscure journal. This interesting collaboration took place at the suggestion of Jacobi, who wrote to Ivory that, since they had independently obtained the Rodrigues formula, they could publish a joint paper to broadcast this result in France, where it was unknown. In the same issue of his journal, Liouville published an alternate, more transparent proof, in fact similar to one published by Jacobi almost ten years earlier. Liouville started by reproducing Legendre's result that

$$\int_{-1}^{1} x_m x_n \, dx = \frac{2}{2n+1} \delta_{mn},$$

where x_m denoted the Legendre polynomial of degree m. Liouville then observed that x_n was a polynomial of exact degree n, and hence any nth degree polynomial had to be a linear combination of the polynomials $x_0, x_1, \ldots, x_n$. He let y be any polynomial of degree $n-1$, so that for some constants $A_0, A_1, \ldots, A_{n-1}$

$$y = A_0 + A_1 x_1 + A_2 x_2 + \cdots + A_{n-1} x_{n-1}.$$

From this, he had

$$d^n y = 0, \quad \int_{-1}^{1} y x_n \, dx = 0.$$

Since $d^n y = 0$, repeated integration by parts yielded

$$\int_{-1}^{x} y x_n \, dt = y \int_{-1}^{x} x_n \, dt - y' \int_{-1}^{x} \int_{-1}^{t} x_n \, dt_1 \, dt + y'' \int_{-1}^{x} \int_{-1}^{t} \int_{-1}^{t_1} x_n \, dt_2 \, dt_1 \, dt$$

$$+ \cdots + (-1)^{n-1} y^{(n-1)} \int_{-1}^{x} \int_{-1}^{t} \cdots \int_{-1}^{t_{n-2}} x_n \, dt_{n-1} \, dt_{n-2} \cdots dt.$$

Because the left-hand side was zero for $x = 1$, and y was an arbitrary polynomial of degree $n-1$, for $x = 1$ he obtained

$$\int_{-1}^{x} x_n \, dt = 0, \quad \int_{-1}^{x} \int_{-1}^{t} x_n \, dt_1 \, dt = 0, \cdots, \int_{-1}^{x} \int_{-1}^{t} \cdots \int_{-1}^{t_{n-2}} x_n \, dt_{n-1} \, dt_{n-2} \cdots dt = 0.$$

Liouville denoted the polynomial of degree $2n$ in the last equation by $\phi(x)$, or,

$$\phi(x) = \int_{-1}^{x} \int_{-1}^{t} \cdots \int_{-1}^{t_{n-2}} x_n \, dt_{n-1} \, dt_{n-2} \cdots dt.$$

Then $\phi(1) = \phi'(1) = \cdots = \phi^{(n-1)}(1) = 0$. This implied that $(x-1)^n$ was a factor of $\phi(x)$. Since it was obvious that $\phi(-1) = \phi'(-1) = \cdots = \phi^{(n-1)}(-1) = 0$, he could conclude that $(x+1)^n$ was also a factor; hence, $(x^2-1)^n$ was a factor of $\phi(x)$. Also,

$\phi(x)$ was of degree $2n$, so, clearly, $\phi(x) = D(x^2-1)^n$ for a constant D. Therefore, for some constant H_n,

$$x_n = H_n \frac{d^n}{dx^n}(x^2-1)^n$$

$$= H_n \left((x+1)^n \frac{d^n}{dx^n}(x-1)^n + \frac{n}{1} \cdot \frac{d}{dx}(x+1)^n \frac{d^{n-1}}{dx^{n-1}}(x-1)^n + \cdots \right).$$

Observe that Liouville applied Leibniz's formula for the nth derivative of a product. Now note that, except for the first, every term in this expression was zero at $x = 1$, so that he could write

$$x_n(1) = 1 \cdot 2 \cdot 3 \cdots n \cdot 2^n H_n.$$

Note also that, for $x = 1$, the generating function of the Legendre polynomials is

$$(1 - 2xz + z^2)^{-1/2} = (1-z)^{-1} = (1 + z + z^2 + \cdots).$$

So Liouville could conclude that $x_n(1) = 1$, and $H_n = 1/(n!2^n)$, proving the result.

Liouville also proved that a function $f(x)$ could be expanded in terms of x_n. He first set

$$F(x) = \sum_{n=0}^{\infty} \frac{2n+1}{2} \cdot x_n \cdot \int_{-1}^{1} f(x) x_n \, dx.$$

Multiplying both sides by x_n and integrating over $(-1, 1)$, he obtained

$$\int_{-1}^{1} (F(x) - f(x)) x_n \, dx = 0, \quad \text{and therefore}$$

$$\int_{-1}^{1} (F(x) - f(x)) y \, dx = 0$$

for an arbitrary polynomial y. Liouville took $y = x^n$ to get

$$\int_{-1}^{1} (F(x) - f(x)) x^n \, dx = 0, \; n = 0, 1, 2, \ldots.$$

He then concluded that $f(x) = F(x)$ and the result was proved.

To show that his conclusion was justified, Liouville also derived the additional theorem that if $f(x)$ was continuous and finite on $[a, b]$ and $\int_a^b x^n f(x) \, dx = 0$ for $n = 0, 1, 2, \ldots$, then $f(x) = 0$ on $[a, b]$. His proof applied only to those functions having a finite number of changes of sign in the interval $[a, b]$, though he failed to remark on this. He began his proof by assuming the geometrically evident proposition that if $f(x)$ was always nonnegative in $[a, b]$ and $\int_a^b f(x) \, dx = 0$, then $f(x)$ had to be identically zero. We remark that Cauchy's ideas on integrals and continuity from the 1820s can be applied to provide an effective proof of this assumption. Next, Liouville supposed that $f(x)$ changed sign at the values $x_1, x_2, \ldots, x_n$ inside $[a, b]$. He let $\psi(x) = (x - x_1)(x - x_2) \cdots (x - x_n)$, and noted that $f(x)\psi(x)$ would have no changes of sign in $[a, b]$ and that $\int_a^b \psi(x) f(x) \, dx = 0$. He could conclude $f(x)\psi(x) \equiv 0$ and $f(x) \equiv 0$, giving him the required result. A modern proof of the proposition might use the Weierstrass approximation theorem, but that was not stated until some decades later. Moreover, observe that ideas such as uniform convergence had not been discovered in Liouville's time. In fact, he did not even clarify the type of interval he was working with and had to explicitly state that $f(x)$ was finite, or bounded, at all points.

28.7 The Jacobi Polynomials

In his significant 1859 posthumously published paper, "Untersuchungen über die Differentialgleichung der hypergeometrischen Reihe," edited by Heine, Jacobi used the hypergeometric differential equation to derive a Rodrigues-type formula for hypergeometic polynomials. These polynomials are now referred to as Jacobi polynomials, and Jacobi further showed them to be orthogonal with respect to the beta distribution. In this paper, Jacobi also obtained the generating function for Jacobi polynomials by an application of the Lagrange inversion formula. Note that Jacobi polynomials are in fact generalizations of Legendre polynomials. It is hard to determine exactly when Jacobi discovered these polynomials. In the 1840s, he published some papers on hypergeometric functions and related topics, but remarks of Kummer indicate that Jacobi had studied these functions even earlier than that. Jacobi started his investigations with the observation that if y satisfied the hypergeometric differential equation, then

$$x(1-x)y^{(2)} + (c - (a+b+1)x)y' - aby = 0,$$
$$x(1-x)y^{(3)} + (c+1 - (a+b+3)x)y^{(2)} - (a+1)(b+1)y' = 0,$$
$$x(1-x)y^{(4)} + (c+2 - (a+b+5)x)y^{(3)} - (a+2)(b+2)y^{(2)} = 0,$$
$$\cdots$$
$$x(1-x)y^{(n+1)} + \big(c+n-1 - (a+b+2n-1)x\big)y^{(n)}$$
$$-(a+n-1)(b+n-1)y^{(n-1)} = 0. \quad (28.30)$$

To understand why these equations follow one after the other, note that, in Gauss's notation, if

$$y = F(a,b,c,x), \quad \text{then} \quad y' = (ab/c)\, F(a+1, b+1, c+1, x).$$

(In more modern notation, one might write y as ${}_2F_1\begin{pmatrix} a,b \\ c \end{pmatrix}; x$.) The parameters a, b, c change to $a+1, b+1, c+1$, respectively, when one takes the derivative of a hypergeometric function. Following Jacobi, multiply (28.30) by

$$x^{c+n-2}(1-x)^{a+b-c+n-1}$$

and rewrite it as

$$\frac{d}{dx}\big(x^n(1-x)^n M y^{(n)}\big) = (a+n-1)(b+n-1)x^{n-1}(1-x)^{n-1} M y^{(n-1)},$$

where $M = x^{c-1}(1-x)^{a+b-c}$. By iteration, he had

$$\frac{d^n}{dx^n}\left(x^n(1-x)^n M y^{(n)}\right) = a(a+1)\cdots(a+n-1)b(b+1)\cdots(b+n-1)My.$$

Next Jacobi took $b = -n$, so that $y = F(-n, a, c, x)$ would be a polynomial of degree n; then $y^{(n)}$ was a constant and the equation became

$$F(-n, a, c, x) = \frac{x^{1-c}(1-x)^{c+n-a}}{c(c+1)\cdots(c+n-1)} \frac{d^n}{dx^n}[x^{c+n-1}(1-x)^{a-c}].$$

Replacing a by $a + n$, he obtained the Rodrigues-type formula

$$X_n = F(-n, a+n, c, x) = \frac{x^{1-c}(1-x)^{c-a}}{c(c+1)\cdots(c+n-1)} \frac{d^n}{dx^n}[x^{c+n-1}(1-x)^{a+n-c}].$$

Jacobi used the Lagrange inversion formula to find the generating function of the polynomials X_n; for $\xi = 1 - 2x$ and $(c)_n$ denoting the shifted factorial:

$$\sum_{n=0}^{\infty} \frac{(c)_n}{n!} h^n X_n =$$

$$\frac{x^{1-c}(1-x)^{c-a}\left(h - 1 + \sqrt{1 - 2h\xi + h^2}\right)^{c-1}\left(h + 1 - \sqrt{1 - 2h\xi + h^2}\right)^{a-c}}{(2h)^{a-1}\sqrt{1 - 2h\xi + h^2}}.$$

He then used the hypergeometric differential equation to prove the orthogonality relation for X_n when $c > 0$ and $a + 1 - c > 0$. Observe that the latter conditions were necessary for the convergence of the integrals. He then let

$$J_{m,n} = \int_0^1 X_m X_n x^{c-1}(1-x)^{a-c} dx.$$

Since X_n satisfied the differential equation

$$x(1-x)X_n'' + (c - (a+1)x)X_n' = -n(n+a)X_n,$$

he could deduce that

$$-n(n+a)J_{m,n} = \int_0^1 X_m \frac{d}{dx}[x^c(1-x)^{a+1-c}X_n'] dx$$

$$= \int_0^1 X_n \frac{d}{dx}[x^c(1-x)^{a+1-c}X_m'] dx = -m(m+a)J_{m,n}.$$

Thus, taking $m \neq n$, he had $J_{m,n} = 0$. When $m = n$, then integration by parts yielded

$$n(n+a)J_{m,n} = \int_0^1 X_n' X_n' x^c (1-x)^{a+1-c} dx.$$

Since X'_m and X'_n were again hypergeometric polynomials, this relation implied that

$$(n-1)(n+a+1)\int_0^1 X'_m X'_n x^c (1-x)^{a+1-c} dx = \int_0^1 X''_m X''_n x^{c+1}(1-x)^{a+2-c} dx.$$

Now $X_n^{(n)}$ was a constant so a repeated application of this formula finally produced a beta integral, computable in terms of gamma functions. The eventual result was then

$$J_{n,n} = \frac{n!}{a+2n} \frac{(\Gamma(c))^2 \Gamma(a-c+n+1)}{\Gamma(a+n)\Gamma(c+n)}.$$

The polynomials X_n are the Jacobi polynomials, except for a constant factor. In more modern notation, taking $c = \alpha + 1$ and $a = \alpha + \beta + 1$ in the expression for X_n, the Jacobi polynomials may be expressed as

$$P^{(\alpha,\beta)}(\xi) := \frac{(\alpha+1)_n}{n!} F(-n, n+\alpha+\beta+1, \alpha+1, (1-\xi)/2)$$

$$= (-1)^n \frac{(1-\xi)^{-\alpha}(1+\xi)^{-\beta}}{2^n} \frac{d^n}{d\xi^n}[(1-\xi)^{n+\alpha}(1+\xi)^{n+\beta}].$$

Thus, observe that the orthogonality relation would hold over $[-1, 1]$ with respect to the beta distribution $(1-x)^\alpha (1+x)^\beta$ for $\alpha, \beta > -1$.

Jacobi briefly noted that for $x = (1-\xi)/2$,

$$(1 - 2h\xi + h^2)^{-c} = \sum_{n=0}^{\infty} h^n Y_n,$$

where

$$Y_n = \frac{2c(2c+1)\cdots(2c+n-1)}{n!} F\left(-n, 2c+n, \frac{2c+1}{2}, x\right)$$

$$= \frac{4^n c(c+1)\cdots(c+n-1)}{(2c+n)(2c+n+1)\cdots(2c+2n-1)} \frac{[x(1-x)]^{\frac{1}{2}(1-2c)}}{n!} \qquad (28.31)$$

$$\times \frac{d^n}{dx^n}[x(1-x)]^{\frac{1}{2}(2c+2n-1)}.$$

We now designate the Y_n as ultraspherical or Gegenbauer polynomials:

$$Y_n := C_n^c(\xi) = \frac{(2c)_n}{(c+1/2)_n} P_n^{(c-1/2, c-1/2)}(\xi),$$

where $(a)_n$ denotes the shifted factorial $a(a+1)\cdots(a+n-1)$. Gegenbauer polynomials, named after the Austrian mathematician Leopold Gegenbauer (1849–1903), student of Weierstrass and Kronecker, are special cases of Jacobi polynomials. They occur when the parameters α and β are equal, and they are of great independent interest. Note that the above generating function for the ultraspherical polynomials is different from the generating function obtained from the one for the Jacobi polynomials.

28.8 Chebyshev: Discrete Orthogonal Polynomials

P. L. Chebyshev introduced discrete orthogonal polynomials into mathematics in his 1855 article, "Sur les fractions continues." His work in this area, like many of his other efforts, was motivated by practical problems for which he sought effective solutions. Chebyshev made frequent use of orthogonal polynomials, continuous as well as discrete, and he was probably the first mathematician to emphasize their importance and applicability to problems in both pure and applied mathematics. Chebyshev was greatly influenced in this connection by the papers of Gauss and Jacobi on numerical integration. Chebyshev studied at Moscow University from 1837 to 1841 where N. D. Brashman instructed him in practical mechanics, motivating some of Chebyshev's later work. In 1846, Chebyshev wrote a master's thesis on a topic in probability; this subject also became his lifelong interest. Chebyshev's 1855 paper laid the foundation for his work on orthogonal polynomials. In presenting his work, we at times follow the notation given by N. I. Akhiezer in his article on Chebyshev's work. This notation more clearly reveals the dependence of certain quantities on the given variables.

Chebyshev began his paper by stating the problem in rather general and vague terms: Suppose $F(x)$ is approximately known for $n+1$ values $x = x_0, x_1, \ldots, x_n$, and that $F(x)$ can be represented by a polynomial of degree $m \leq n$,

$$a + bx + cx^2 + \cdots + gx^{m-1} + hx^m.$$

Find the value of $F(x)$ at $x = X$ so that the errors in $F(x_0)$, $F(x_1)$, ..., $F(x_n)$ have minimal influence on $F(X)$. From a practical standpoint, the problem makes good sense. For example, the values of some function $y = F(x)$ may be obtained by observation for $x = x_0, x_1, \ldots, x_n$. These values would have experimental errors so that $y_i \simeq F(x_i)$. Thus, $F(x)$ is a polynomial of degree $m \leq n$ and the problem is to determine $F(x)$ in such a way that the errors of observation have the least influence. In more specific terms, Chebyshev stated the problem: Find a polynomial $F(x)$ of the form

$$F(x) = \mu_0 \lambda_0(x) y_0 + \mu_1 \lambda_1(x) y_1 + \cdots + \mu_n \lambda_n(x) y_n, \qquad (28.32)$$

where $\lambda_i(x)$ are unknown polynomials of degree $\leq m$ and $\mu_i > 0$ are weights associated with observed values y_i subject to the following two conditions: The identity

$$f(X) = \mu_0 \lambda_0(X) f(x_0) + \mu_1 \lambda_1(X) f(x_1) + \cdots + \mu_n \lambda_n(X) f(x_n) \qquad (28.33)$$

must hold for any polynomial of degree at most m; and one must minimize the sum

$$W(X) = \mu_0 \big(\lambda_0(X)\big)^2 + \mu_1 \big(\lambda_1(X)\big)^2 + \cdots + \mu_n \big(\lambda_n(X)\big)^2. \qquad (28.34)$$

Thus, $W(X)$ had to be minimized with respect to the constraints of (28.33), equivalent to the $m+1$ conditions

$$X^k = \mu_0 \lambda_0(X) x_0^k + \mu_1 \lambda_1(X) x_1^k + \cdots + \mu_n \lambda_n(X) x_n^k, \qquad (28.35)$$

for $k = 0, 1, \ldots, m$. Then Chebyshev applied the method of Lagrange multipliers with $\lambda_0, \lambda_1, \ldots, \lambda_n$ as the variables and with $l_0(X), l_1(X), \ldots, l_m(X)$ as the $m+1$ multipliers.

28.8 Chebyshev: Discrete Orthogonal Polynomials

This gave him the $n+1$ relations

$$\frac{\partial W}{\partial \lambda_i} - \frac{\partial}{\partial \lambda_i} \sum_{k=0}^{m} l_k(X) \left(\mu_0 \lambda_0 x_0^k + \cdots + \mu_n \lambda_n x_n^k - X^k\right) = 0,$$

$$\text{or} \quad 2\lambda_i(X) = l_0(X) + l_1(X)x_i + \cdots + l_m(X)x_i^m, \tag{28.36}$$

for $i = 0, 1, \ldots, m+1$. Chebyshev wrote that the whole difficulty boiled down to solving this system of equations. He denoted the polynomial on the right-hand side of equation (28.36) by $2K_m(X, x_i)$, obtaining

$$K_m(X, x) = \frac{1}{2} \sum_{k=0}^{m} l_k(X) x^k. \tag{28.37}$$

Thus, Chebyshev's problem was to find an expression for $\lambda_i(X) = K_m(X, x_i)$, $i = 0, 1, \ldots, n$. Note that the constraints (28.35) could be written as

$$\sum_{i=0}^{n} \mu_i K_m(X, x_i) x_i^k = X^k, \quad k = 0, 1, \ldots, m. \tag{28.38}$$

These relations implied that the polynomials $K_m(X, x_i)$ should be such that, for some function $A(X)$,

$$\sum_{i=0}^{n} \frac{\mu_i K_m(X, x_i)}{x - x_i} - \frac{1}{x - X} = \frac{A(X)}{x^{m+2}} + \cdots. \tag{28.39}$$

Note that this relation could be rewritten as

$$K_m(X, x) \sum_{i=0}^{n} \frac{\mu_i}{x - x_i} - N(X, x) - \frac{1}{x - X} = \frac{A(X)}{x^{m+2}} + \cdots \tag{28.40}$$

with $N(X, x)$ a polynomial of degree $m-1$ in x. Of course, this was made possible by the elementary relation that if $g(x)$ is a polynomial of degree m, then there is a polynomial $h(x)$ of degree $m-1$ such that

$$\frac{g(x)}{x - x_i} = h(x) + \frac{g(x_i)}{x - x_i}.$$

Here Chebyshev considered an additional but related problem: Find a polynomial $\psi_m(x)$ of degree m and a polynomial $\pi_m(x)$ of degree at most $m-1$ so that

$$\psi_m(x) \sum_{i=0}^{n} \frac{\mu_i}{x - x_i} - \pi_m(x) = O\left(\frac{1}{x^{m+2}}\right). \tag{28.41}$$

Chebyshev's study of Gauss's paper on numerical integration showed him that the answer lay in the continued fraction expansion

$$\sum_{i=0}^{n} \frac{\mu_i}{x - x_i} = \frac{1}{q_1+} \frac{1}{q_2+} \frac{1}{q_3+} \cdots \frac{1}{q_{n+1}}, \tag{28.42}$$

where $q_m = A_m x + B_m$ were linear functions for $m = 1, 2, \ldots$. In fact, the mth ($m \leq n+1$) convergent was the rational function $\pi_m(x)/\psi_m(x)$, producing the polynomials required in (28.41). Chebyshev proved that

$$\lambda_i(x) = K_m(x, x_i) = (-1)^m \frac{\psi_{m+1}(x)\psi_m(x_i) - \psi_m(x)\psi_{m+1}(x_i)}{x - x_i}. \tag{28.43}$$

He then derived another relation for $\lambda_i(x)$ by using the three-term relation for $\psi_m(x)$ obtained from the continued fraction

$$\psi_{m+1}(x) = q_{m+1}\psi_m(x) + \psi_{m-1}(x)$$
$$= (A_{m+1} x + B_{m+1})\psi_m(x) + \psi_{m-1}(x).$$

When this was substituted in (28.43) he could obtain, after simplification,

$$(-1)^m \lambda_i(x) = \left(A_{m+1}\psi_m(x)\psi_m(x_i) - \frac{\psi_m(x)\psi_{m-1}(x_i) - \psi_m(x_i)\psi_{m-1}(x)}{x - x_i} \right).$$

Repeating this process m times, he got

$$(-1)^m \lambda_i(x) = \sum_{j=0}^{m} (-1)^{m-j} A_{j+1} \psi_j(x_i) \psi_j(x)$$
$$= \frac{\psi_{m+1}(x)\psi_m(x_i) - \psi_m(x)\psi_{m+1}(x_i)}{x - x_i}. \tag{28.44}$$

This important relation is usually called the Christofell-Darboux formula; they obtained it in a similar way, but Chebyshev published the formula more than a decade earlier. When he substituted the value of $\lambda_i(x)$ in (28.44) in (28.32), Chebyshev had

$$F(x) = \sum_{i=0}^{n} \left(\sum_{j=0}^{m} (-1)^j A_{j+1} \psi_j(x_i) \psi_j(x) \right) \mu_i F(x_i). \tag{28.45}$$

He then set $F(x) = \psi_m(x)$ and equated the coefficients of $\psi_k(x)$ on both sides to obtain the orthogonality relation

$$(-1)^k A_{k+1} \sum_{i=0}^{n} \mu_i \psi_k(x_i) \psi_m(x_i) = \delta_{km}, \tag{28.46}$$

and in particular

$$A_{k+1} = \frac{(-1)^k}{\sum_{i=0}^{n} \mu_i \psi_k^2(x_i)}. \tag{28.47}$$

28.8 Chebyshev: Discrete Orthogonal Polynomials

So his final result was expressed as

$$\lambda_i(x) = \frac{\sum_{k=0}^{m} \mu_k \psi_k(x_i) \psi_k(x)}{\sum_{i=0}^{n} \mu_k \psi_k^2(x_i)}. \tag{28.48}$$

Chebyshev concluded his paper by stating and proving two results on least squares. For the first result, he supposed V to be a polynomial of degree m with the coefficient of x^m the same as that of $\psi_m(x)$. He then showed that the sum $\sum_{i=0}^{n} \mu_i V^2(x_i)$ had the least value when $V = \psi_m(x)$. To prove this, Chebyshev set

$$V = A_0 \psi_0(x) + \cdots + A_{m-1} \psi_{m-1}(x) + \psi_m(x) \quad \text{and then}$$

$$\sum_{i=0}^{n} \mu_i V^2(x_i) = \sum_{i=0}^{n} \mu_i \big(A_0 \psi_0(x_i) + \cdots + A_{m-1} \psi_{m-1}(x_i) + \psi_m(x_i)\big)^2.$$

For a minimum, the derivatives with respect to the A_j should be zero. Thus,

$$2 \sum_{i=0}^{n} \mu_i \psi_j(x_i) \big(A_0 \psi_0(x_i) + A_{m-1} \psi_{m-1}(x_i) + \psi_m(x_i)\big) = 0, \quad j = 0, \ldots, m-1.$$

An application of the orthogonality relation (28.46) gave

$$A_j \sum_{i=0}^{n} \mu_i \psi_j^2(x_i) = 0, \quad j = 0, 1, \ldots, m-1.$$

This implied $A_j = 0$, $j = 0, 1, \ldots, m-1$, and hence $V = \psi_m(x)$. In his second result, Chebyshev proved that

$$\sum_{i=0}^{n} \mu_i \left(F(x_i) - \sum_{j=0}^{m} A_j \psi_j(x_i) \right)^2 \tag{28.49}$$

was a minimum when

$$A_j = \frac{\sum_{i=0}^{n} \mu_i \psi_j(x_i) F(x_i)}{\sum_{i=0}^{n} \mu_i \psi_j^2(x_i)}.$$

He once more took the derivatives of (28.49) with respect to A_j, $j = 0, \ldots, m$ to obtain

$$2 \sum_{i=0}^{n} \mu_i \psi_j(x_i) \left(F(x_i) - \sum_{j=0}^{m} A_j \psi_j(x_i) \right) = 0, \quad j = 0, 1, \ldots, m.$$

Again using the orthogonality relation (28.46), these equations reduced to

$$\sum_{i=0}^{n} \mu_i \psi_j(x_i) F(x_i) - A_j \sum_{i=0}^{n} \mu_i \psi_j^2(x_i) = 0, \quad j = 0, 1, \ldots, m,$$

implying the required result.

28.9 Chebyshev and Orthogonal Matrices

In his 1855 paper, before the theory of matrices was formally developed, Chebyshev gave a very interesting construction of an orthogonal matrix, noting that in a paper of 1771, Euler also constructed such squares. However, after 1855, Chebyshev did not develop this topic further. Chebyshev defined the function

$$\Phi_k(x_i) = \sqrt{\alpha_i}\,\psi_k(x_i), \quad i, k = 0, 1, \ldots, n$$

$$\text{where} \quad \alpha_i = \mu_i / \sum_{j=0}^{n} \mu_j \psi_j^2(x_i).$$

He then considered the square tableau

$$\begin{matrix} \Phi_0(x_0) & \Phi_0(x_1) & \cdots & \Phi_0(x_n) \\ \Phi_1(x_0) & \Phi_1(x_1) & \cdots & \Phi_1(x_n) \\ \vdots & \vdots & & \vdots \\ \Phi_n(x_0) & \Phi_n(x_1) & \cdots & \Phi_n(x_n). \end{matrix} \qquad (28.50)$$

From the orthogonality relation (28.46), he deduced that the sum of the squares of the terms in each row and in each column was one. Also, in any two rows or columns, the sum of the products of their corresponding terms would be zero.

28.10 Chebyshev's Discrete Legendre and Jacobi Polynomials

In his 1864 paper "Sur l'interpolation," Chebyshev took $\mu_i = 1$ and $x_i = i$ for $i = 0, 1, \ldots, n-1$ with μ defined as previously. See equations (28.41) and (28.42). Then the polynomials $\psi_0(x), \psi_1(x), \psi_2(x), \ldots$ were the denominators in the continued fraction expansion of

$$\frac{1}{x} + \frac{1}{x-1} + \frac{1}{x-2} + \cdots + \frac{1}{x-n+1}.$$

By (28.46), these in turn satisfied the relations

$$\sum_{i=0}^{n} \psi_l(i)\psi_m(i) = 0 \quad \text{for} \quad m < l. \qquad (28.51)$$

Chebyshev found a Rodrigues-type formula for $\psi_k(x)$, where the differential operator was replaced by the finite difference operator. His two-step approach was exactly the discrete analog of the method employed by Jacobi in his paper on numerical integration. In the first step, Chebyshev proved that if there was a polynomial $f(x)$ of degree l such that

$$\sum_{i=0}^{n} f(i)\psi_m(i) = 0 \quad \text{for} \quad m < l,$$

then there existed a constant C such that $f(x) = C\psi_l(x)$. For the second step, he showed that the polynomial of degree l given by

$$f(x) = \Delta^l x(x-1)\cdots(x-l+1)(x-n)(x-n-1)\cdots(x-n-l+1)$$

28.10 Chebyshev's Discrete Legendre and Jacobi Polynomials

satisfied the required condition. Thus, he had the Rodrigues-type formula $\psi_l(x) = C_l f(x)$, where C_l was a constant.

Chebyshev also gave an interesting interpolation formula in terms of $\psi_l(x)$. He supposed $u_0, u_1, \ldots, u_{n-1}$ to be the values of a function u at $x = 0, 1, \ldots, n-1$. The interpolation formula would then be expressed as

$$u = \frac{\sum_{i=0}^{n-1} u_i}{n} + \frac{3 \sum_{i=0}^{n-1} (i+1)(n-i-1) \Delta u_i}{1^2 n(n^2-1^2)} \Delta x(x-n)$$
$$+ 5 \sum_{i=0}^{n-1} \frac{(i+1)(i+2)(n-i-1)(n-i-2) \Delta^2 u_i}{(2!)^2 n(n^2-1^2)(n^2-2^2)} \quad (28.52)$$
$$\times \Delta^2 x(x-1)(x-n)(x-n-1) + \cdots.$$

Interestingly, in an 1858 paper, "Sur une nouvelle série," Chebyshev took the points as $x_1 = h$, $x_2 = 2h$, $\ldots$, $x_n = nh$ such that the orthogonal polynomials could be expressed in the form

$$\psi_l(x) = C_l \Delta^l (x-h)(x-2h) \cdots (x-lh)(x-nh-h) \cdots (x-nh-lh).$$

The formula corresponding to (28.52) would then be written

$$u = \frac{1}{n} \sum u_i + \frac{3 \sum i(n-i) \Delta u_i}{1^2 n(n^2-1^2) h^2} \Delta(x-h)(x-nh-h)$$
$$+ 5 \frac{\sum i(i+1)(n-i)(n-i-1) \Delta^2 u_i}{(2!)^2 n(n^2-1^2)(n^2-2^2) h^4} \quad (28.53)$$
$$\times \Delta(x-h)(x-2h)(x-nh-h)(x-nh-2h) + \cdots.$$

Chebyshev observed that if he set $h = 1/n$ in his interpolation formula (28.53) and let $n \to \infty$, he obtained a series in terms of Legendre polynomials. On the other hand, if he set $h = 1/n^2$ and let $n \to \infty$, he arrived at the Maclaurin series expansion! At the end of the paper, Chebyshev made the insightful remark that one might use discrete orthogonal polynomials to approximate the sum $\sum_{i=1}^{n} F(ih)$, just as Gauss had used Legendre polynomials for numerical integration.

Chebyshev appears to have independently discovered the Jacobi polynomials. He gave the generating function and proved orthogonality for these polynomials in an 1870 paper based on the work of Legendre. Chebyshev there proved that if

$$F(s,x) = \frac{(1+s+\sqrt{1-2sx+s^2})^{-\gamma}(1-s+\sqrt{1-2sx+s^2})^{-\mu}}{\sqrt{1-2sx+x^2}}$$
$$= \sum_{n=0}^{\infty} T_n(x) s^n, \quad \text{then}$$

$$\int_{-1}^{1} F(s,x) F(t,x) (1-x)^{\mu} (1+x)^{\gamma} dx$$

was purely a function of st. This gave the orthogonality of $T_n(x)$ with respect to the beta distribution $(1-x)^\mu (1+x)^\nu$. Note that the $T_n(x)$ are the Jacobi polynomials, generalizing the Legendre polynomials. On the basis of this work, in 1875 Chebyshev defined the discrete Jacobi polynomials. He also showed that his interpolation formulas could be applied to problems in ballistics.

28.11 Exercises

1. Show that
$$\int_{-1}^{1} \frac{1}{x-u} \frac{du}{\sqrt{1-u^2}} = \frac{\pi}{\sqrt{x^2-1}} = \frac{\pi}{x-} \frac{1}{2x-} \frac{1}{2x-} \cdots.$$

   Show also that the denominators of the convergents of the continued fractions are $\cos\phi$, $\cos 2\phi$, ..., where $x = \cos\phi$. See Chebyshev (1899–1907), vol. 1, pp. 501–508, especially p. 502.

2. Let d be the greatest integer in $n/2$, where n is a positive integer. Show that the ultraspherical polynomials C_n^λ satisfy the relation
$$C_n^\lambda = \sum_{k=0}^{d} \frac{(\lambda)_{n-k}(\lambda-\mu)_k(n+\mu-2k)}{k!\mu(\mu+1)_{n-k}} C_{n-2k}^\mu(x),$$

   where for any quantity a, $(a)_k$ denotes the shifted factorial $a(a+1)\cdots(a+k-1)$. See Gegenbauer (1884).

3. Suppose n unit charges are distributed at $x_1, x_2, \ldots, x_n$ inside the interval $(-1, 1)$ and that there is an extra charge $\alpha > 0$ at -1 and another, $\beta > 0$, at $+1$. The electrostatic energy would be given by
$$L = \sum_{1 \le i < j \le n} \log \frac{1}{|x_i - x_j|} + \alpha \sum_{i=1}^{n} \log \frac{1}{|1+x_i|} + \beta \sum_{i=1}^{n} \log \frac{1}{|1-x_i|}.$$

   Show that L is minimum when $x_1, x_2, \ldots, x_n$ are zeros of a polynomial $\phi(x)$ satisfying the differential equation
$$(1-x^2)\phi'' + 2(\alpha - \beta - (\alpha+\beta)x)\phi'(x) + n(n+2\alpha+2\beta-1)\phi = 0.$$

   Show that ϕ is a constant multiple of the Jacobi polynomial $P^{(\beta,\alpha)}(x)$. See Stieltjes (1993), vol. 2, pp. 79–80. Note that these volumes have two sets of page numbers; we refer to the bottom numbers. See also the interesting commentary of Walter Van Assche in vol. 1, particularly pp. 13–16.

4. Show that
$$\int_x^\infty \frac{e^{-t}}{t} dt = \frac{e^{-x}}{x+1-} \frac{1}{x+3-} \frac{1}{(x+5)/4-} \frac{1/2^2}{(x+7)/9-} \frac{1/3^2}{(x+9)/16-} \cdots.$$

Denote the mth convergent by $e^{-x}\phi_m(x)/f_m(x)$. Then show that

$$xf_n'(x) = nf_n(x) - n^2 f_{n-1}(x),$$
$$f_{n+1}(x) = (x + 2n + 1) f_n(x) - n^2 f_{n-1}(x),$$
$$\int_{-\infty}^0 e^x f_n(x) f_m(x)\,dx = (n!)^2 \delta_{mn}.$$

See Laguerre (1972), vol. 1, pp. 431–35.

5. Show that

$$\frac{1}{(n-k)!} \frac{d^{n-k}}{dx^{n-k}} (x^2 - 1)^n = \frac{(x^2 - 1)^k}{(n+k)!} \frac{d^{n+k}}{dx^{n+k}} (x^2 - 1)^n.$$

See Rodrigues (1816).

6. Show that if $z = \cos x$, then

$$\frac{d^{i-1}(1-z^2)^{(2i-1)/2}}{dz^{i-1}} = (-1)^{i-1} 3 \cdot 5 \cdots (2i-1) \frac{\sin ix}{i};$$

$$\frac{d^i(1-z^2)^{(2i-1)/2}}{dz^i} dz = (-1)^{i-1} 3 \cdot 5 \cdots (2i-1) \cos ix\, dx.$$

See Jacobi (1969), vol. 6, pp. 90–91.

28.12 Notes on the Literature

For the quote from Ivory, see Craik (2000). Gauss (1863–1927), vol. 3, pp. 163–196, contains his paper on numerical integration. Goldstine's (1977) very interesting book on the history of numerical analysis gives a thorough and readable account of Gauss's work; see pp. 224–232. For Legendre's proof of orthogonality, see Legendre (1811–1817), vol. 2, pp. 249–250. See Jacobi (1969), vol. 6, pp. 3–11, for his 1826 commentary on Gauss's paper. Ivory and Jacobi (1837) is their joint paper published in the then newly founded *Liouville's Journal*. This journal gave Liouville the opportunity to review many papers before they appeared in print, and then react to them. For example, the paper of Ivory and Jacobi stimulated Liouville to write his two short notes (1837a) and (1837b) in the same volume of his journal. Ivory (1812) and (1824), taken together, contain his derivation of the Rodrigues formula. Altmann and Ortiz (2005) is completely devoted to the work of Rodrigues in and outside of mathematics. See particularly the articles by Grattan-Guinness and Askey in this book. See Murphy (1833) and (1835) for his two long papers on definite integrals and orthogonal polynomials.

See Chebyshev (1899–1907), vol. 1, pp. 203–230, for his 1855 paper on continued fractions and discrete orthogonal polynomials. See pp. 541–560 for his paper on discrete Legendre polynomials; pp. 1–8 of the second volume present Chebyshev's proof of the orthogonality of the Jacobi polynomials. See also N. I. Akhiezer's readable and insightful commentary on Chebyshev's work on continued fractions in Kolmogorov and

Yushkevich (1998). Steffens (2006) gives a fairly comprehensive history of Chebyshev and his students' contributions to orthogonal polynomials and approximation theory. Simon's remark on orthogonal polynomials is the first sentence of Simon (2005). Nevai (1990) is a collection of interesting papers on orthogonal polynomials and their numerous applications.

29

q-Series

29.1 Preliminary Remarks

The theory of q-series in modern mathematics plays a significant role in partition theory and modular functions as well as in some aspects of Lie algebras and statistical mechanics. This subject began quietly, however, with two combinatorial problems posed in a September 1740 letter from Phillipe Naudé (1684–1747) to Euler. Naudé, a mathematician of French origin working in Berlin, asked how to find the number of ways in which a given number could be expressed as the sum of a fixed number, first of distinct integers and then without the requirement that the integers in the sum be distinct.

As an example of both these problems, 7 can be expressed as a sum of three distinct integers in one way, $1+2+4$; whereas it can be expressed as a sum of three integers in four ways: $1+1+5$, $1+2+4$, $1+3+3$, $2+2+3$. Euler received Naudé's letter in St. Petersburg, just before he moved to Berlin. Within two weeks, in a reply to Naudé, Euler outlined a solution, and soon after that he presented a complete solution to the Petersburg Academy. In 1748, he devoted a whole chapter to this topic in his *Introductio in Analysin Infinitorum*. The essential idea in Euler's solution was that the coefficient of $q^k x^m$ in the series expansion of the infinite product

$$f(q,x) = (1+qx)(1+q^2x)(1+q^3x)\cdots \tag{29.1}$$

gave the number of ways of writing k as a sum of m distinct positive integers. Euler used the functional relation

$$f(q,x) = (1+qx)f(q,qx) \tag{29.2}$$

to prove that

$$f(q,x) = \sum_{m=0}^{\infty} \frac{q^{m(m+1)/2} x^m}{(1-q)(1-q^2)\cdots(1-q^m)}. \tag{29.3}$$

He noted that

$$\frac{1}{(1-q)(1-q^2)\cdots(1-q^m)}$$
$$= (1+q+q^{1+1}+\cdots)(1+q^2+q^{2+2}+\cdots)\cdots(1+q^m+q^{m+m}+\cdots)$$
$$= \sum_{n=0}^{\infty} a_n q^n, \qquad (29.4)$$

where the middle product showed that a_n, the coefficient of q^n, was the number of ways of writing n as a sum of integers chosen from the set $1, 2, \ldots, m$. This implied that the coefficient of $q^k x^m$ on the right-hand side of (29.3) was the number of ways of writing $k - m(m+1)/2$ as a sum of integers from the set $1, 2, \ldots, m$. Thus, Euler stated the theorem: The number of different ways in which the number n can be expressed as a sum of m different numbers is the same as the number of different ways in which $n - m(m+1)/2$ can be expressed as the sums of the numbers $1, 2, 3, \ldots, m$.

For the second problem, Euler used the product

$$g(q,x) = \prod_{n=1}^{\infty} (1-q^n x)^{-1} \qquad (29.5)$$

and obtained the corresponding series and theorem in a similar way. Euler here used functional relations to evaluate the product as a series, just as he earlier employed functional relations to evaluate the beta integral as a product. Of course, this method goes back to Wallis.

Euler also considered the case $x = 1$. In that case (in modern notation), we have

$$\prod_{n=1}^{\infty} (1-q^n)^{-1} = (1+q+q^{1+1}+\cdots)(1+q^2+q^{2+2}+\cdots)(1+q^3+q^{3+3}+\cdots)\cdots$$
$$= \sum_{n=0}^{\infty} p(n) q^n, \qquad (29.6)$$

where $p(n)$ is the number of partitions of n, or the number of ways in which n can be written as a sum of positive integers. For example, $p(4) = 5$ because 4 has the five partitions

$$1+1+1+1, \ 2+1+1, \ 2+2, \ 3+1, \ 4.$$

The product in (29.6) also led Euler to consider its reciprocal, $\prod_{n=1}^{\infty} (1-q^n)$. He attempted to expand this as a series but it took him nine years to completely resolve this difficult problem. In his first attempt, he multiplied a large number of terms of the product to find that

$$\prod_{n=1}^{\infty}(1-q^n) = 1 - q - q^2 + q^5 + q^7 - q^{12} - q^{15} + q^{22} + q^{26} - q^{35} - q^{40} + q^{51} + \cdots.$$
$$(29.7)$$

He quickly found a general expression for the exponents, $m(3m \pm 1)/2$. He most probably did this by considering the differences in the sequence of exponents; note that the sequence of exponents is

$$0, 1, 2, 5, 7, 12, 15, 22, 26, 35, 40, 51, \ldots.$$

Observe that the sequence of differences is then

$$1, 1, 3, 2, 5, 3, 7, 4, 9, 5, 11, \ldots.$$

The pattern of this sequence suggests that one should group the sequence of exponents into two separate sequences, first taking the exponents of the odd-numbered terms and then the exponents of the even-numbered terms. For example, the sequence of exponents of the odd-numbered terms is $0, 2, 7, 15, 26, 40, \ldots$, and their differences are $2, 5, 8, 11, 14, \ldots$. Since the differences of these differences are 3 in every case, we may apply the formula of Zhu Shijie and Montmort (11.23) in order to perceive that the $(n+1)$th term of the sequence of odd-numbered exponents will be given by

$$0 + 2n + 3n(n-1)/2 = n(3n+1)/2.$$

Similarly, the nth term in the sequence of even-term exponents is $n(3n-1)/2$. In the *Introductio*, Euler wrote, "If we consider this sequence with some attention we will note that the only exponents which appear are of the form $(3n^2 \pm n)/2$ and that the sign of the corresponding term is negative when n is odd, and the sign is positive when n is even." Thus, Euler made the conjecture

$$\prod_{n=1}^{\infty}(1-q^n) = \sum_{m=-\infty}^{\infty}(-1)^m q^{m(3m+1)/2} = 1 + \sum_{m=1}^{\infty}(-1)^m q^{m(3m\pm 1)/2}, \quad (29.8)$$

and finally found a proof of this in 1750. He immediately wrote Goldbach about the details of the proof, explaining that it depended on the algebraic identity:

$$(1-\alpha)(1-\beta)(1-\delta) \text{ etc.} = 1 - \alpha - \beta(1-\alpha) - \gamma(1-\alpha)(1-\beta)$$
$$- \delta(1-\alpha)(1-\beta)(1-\gamma) - \text{ etc.}$$

This identity is easy to check, since the first three terms on the right-hand side add up to

$$1 - \alpha - \beta(1-\alpha) = (1-\alpha)(1-\beta),$$

and when this is added to the fourth term, we get

$$(1-\alpha)(1-\beta) - \gamma(1-\alpha)(1-\beta) = (1-\alpha)(1-\beta)(1-\gamma)$$

and so on. An interesting feature of the series (29.8) is that the exponent of q is a quadratic in m, the index of summation. Surprisingly, series of this kind had already appeared in 1690 within Jakob Bernoulli's works on probability theory, but he was unable to do much with them. Over a century later, Gauss initiated a systematic study of these series. Entry 58 of Gauss's mathematical diary, dated February 1797, gives a continued fraction expansion of one of Bernoulli's series:

$$1 - a + a^3 - a^6 + a^{10} - \cdots$$

$$= \cfrac{1}{1 + \cfrac{a}{1 + \cfrac{a^2 - a}{1 + \cfrac{a^3}{1 + \cfrac{a^4 - a^2}{1 + \cfrac{a^5}{1 + \text{etc.}}}}}}} \qquad (29.9)$$

Gauss added the comment, "From this all series where the exponents form a series of the second order are easily transformed." About a year later, he raised the problem of expressing $1 + q + q^3 + q^6 + q^{10} + \cdots$ as an infinite product. Gauss came upon series of this type around 1794 in the context of his work on the arithmetic-geometric mean. This latter work was absorbed into his theory of elliptic functions. Series (29.8) and (29.9) are actually examples of the special kind of q-series called theta functions. Theta functions also arose naturally in Fourier's 1807 study of heat conduction.

Unfortunately, Gauss did not publish any of his work on theta or elliptic functions, and it remained for Abel and Jacobi to independently rediscover much of this work, going beyond Gauss in many respects. Around 1805–1808, Gauss began to view q-series in a different way. For example, his 1811 paper on q-series dealt with a generalization of the binomial coefficient and the binomial series. In particular, he defined the Gaussian polynomial

$$(m, \mu) = \frac{(1 - q^m)(1 - q^{m-1})(1 - q^{m-2}) \cdots (1 - q^{m-\mu+1})}{(1 - q)(1 - q^2)(1 - q^3) \cdots (1 - q^\mu)}. \qquad (29.10)$$

Note that Gauss wrote x instead of q. Observe that as $q \to 1$

$$(m, \mu) \to \binom{m}{\mu}. \qquad (29.11)$$

This work led to an unexpected byproduct: an evaluation of the Gauss sum $\sum_{k=0}^{n-1} e^{2\pi i k^2/n}$ where n was an odd positive integer. This sum had already appeared naturally in Gauss's theory of the cyclotomic equation $x^n - 1 = 0$, to which he had devoted the final chapter of his 1801 *Disquisitiones Arithmeticae*. There Gauss had computed the square of the Gauss sum, but he was unable to determine the correct sign of the square root. Already in 1801, he knew that it was important to find the exact value of the sum; he expended considerable effort over the next four years to compute the Gauss sum, and it was a complete surprise for him when the result dropped out of his work on q-series. In September 1805, he wrote his astronomer friend, Wilhelm Olbers,

> What I wrote there [Disqu. Arith. section 365] ..., I proved rigorously, but I was always annoyed by what was missing, namely, the determination of the sign of the root. This gap spoiled whatever

else I found, and hardly a week may have gone by in the last four years without one or more unsuccessful attempts to unravel this knot - just recently it again occupied me much. But all the brooding, the searching, was to no avail, and I had sadly to lay down my pen again. A few days ago, I finally succeeded - not by my efforts, but by the grace of God, I should say. The mystery was solved the way lightning strikes, I myself could not find the connection between what I knew previously, what I investigated last, and the way it was finally solved.

He recorded these events in his diary:

(May 1801) A method for proving the first fundamental theorem has been found by means of a most elegant theorem in the division of the circle, thus

$$\sum {\sin\atop \cos} {nn\over a} P = + \sqrt{a} \left| {0 \atop \sqrt{a}} \right. \left| {0 \atop +\sqrt{a}} \right. \left| {+\sqrt{a} \atop 0} \right. \tag{29.12}$$

according as $a \equiv 0, 1, 2, 3 \pmod 4$ substituting for n all numbers from 0 to $(a-1)$. (August 1805) The proof of the most charming theorem recorded above, May 1801, which we have sought to prove for 4 years and more with every effort, at last perfected.

Conceptually, this was a major achievement, since it served to connect cyclotomy with the reciprocity law. Gauss may have initially considered the polynomial $\sum_{k=0}^{m}(m,k)x^k$ as a possible analog of the finite binomial series. In any case, he expressed the sum as a finite product when $x = -1$ and when $x = \sqrt{q}$, and these formulas finally yielded the correct value of the Gauss sum. It is interesting to note that the polynomial $\sum_{k=0}^{m}(m,k)x^k$ played a key role in Szegő's theory of orthogonal polynomials on the unit disc.

Gauss found the appropriate q-extension of the terminating binomial theorem, perhaps around 1808, but he did not publish it. In 1811, Heinrich A. Rothe (1773–1841) first published this result in the preface of his *Systematisches Lehrbuch der Arithmetik* as the formula

$$\sum_{k=0}^{m} {1-q^m \over 1-q} \cdot {1-q^{m-1} \over 1-q^2} \cdot \ldots \cdot {1-q^{m-k+1} \over 1-q^k} \cdot q^{k(k+1)/2} x^{m-k} y^k$$
$$= (x+y)(x+qy)\cdots(x+q^{k-1}y). \tag{29.13}$$

Although this was the most important result in the book, Rothe excluded it from the body of text, apparently in order to keep the book within the size required by the publisher. Gauss's paper and Rothe's formula indicated a direction for further research on q-series relating to the extension of the binomial theorem. This path was not pursued until the 1840s, except for Schweins's *Analysis* of 1820. This work presented a q-extension of Vandermonde's identity (29.58).

In the 1820s, Jacobi investigated q-series in connection with his work on theta functions, a byproduct of his researches on elliptic functions. His most remarkable discovery in this area was the triple product identity. Jacobi's famous *Fundamenta*

Nova of 1829 stated the formula as

$$(1+qz)(1+q^3z)(1+q^5z)\cdots\left(1+\frac{q}{z}\right)\left(1+\frac{q^3}{z}\right)\left(1+\frac{q^5}{z}\right)\cdots$$
$$= \frac{1+q\left(z+\frac{1}{z}\right)+q^4\left(z^2+\frac{1}{z^2}\right)+q^9\left(z^3+\frac{1}{z^3}\right)+\cdots}{(1-q^2)(1-q^4)(1-q^6)(1-q^8)\cdots}. \qquad (29.14)$$

Jacobi regarded this identity as his most important formula in pure mathematics. He gave several very important applications. In one of these, he derived an identity, giving the number of representations of an integer as a sum of four squares. In another, he obtained an important series expression for the square root of the period of some elliptic functions, allowing him to find a new derivation of the following transformation of a theta function, originally due to Cauchy and Poisson:

$$1+2\sum_{n=1}^{\infty}e^{-n^2\pi x} = \frac{1}{\sqrt{x}}\left(1+2\sum_{n=1}^{\infty}e^{-n^2\pi/x}\right). \qquad (29.15)$$

Jacobi also published a long paper on those series whose powers are quadratic forms; the triple product identity formed the basis for this. In the 1820s, when Gauss learned of Jacobi's work, he informed Jacobi that he had already found (29.14) in 1808. Legendre, on very friendly terms with Jacobi, refused to believe that Gauss had anticipated his friend. In a letter to Jacobi, Legendre wrote, "Such outrageous impudence is incredible in a man with enough ability of his own that he should not have to take credit for other people's discoveries." Then again, Legendre had had his own priority disputes with Gauss with regard to quadratic reciprocity and the method of least squares.

In the early 1840s, papers on q-series appeared in quick succession by Cauchy in France and Eisenstein, Jacobi, and E. Heine in Germany. As a second-year student at Berlin in 1844, Eisenstein presented twenty-five papers for publication to *Crelle's Journal*. One of them, "Neuer Beweis und Verallgemeinerung des binomischen Lehrsatzes," began with the statement and proof of the Rothe-Gauss theorem; it then applied Euler's approach to the proof of the binomial theorem to obtain a version of the q-binomial theorem. Some details omitted by Euler in his account were treated in Eisenstein's paper.

Jacobi and Cauchy stated and proved the q-binomial theorem in the form

$$1+\frac{v-w}{1-q}z+\frac{(v-w)(v-qw)}{(1-q)(1-q^2)}z^2+\frac{(v-w)(v-qw)(v-q^2w)}{(1-q)(1-q^2)(1-q^3)}z^3+\cdots$$
$$= \frac{(1-wz)(1-qwz)(1-q^2wz)(1-q^3wz)\cdots}{(1-vz)(1-qvz)(1-q^2vz)(1-q^3vz)\cdots}. \qquad (29.16)$$

The idea in this proof was the same as the one used by Euler to prove (29.3), clearly a particular case. Jacobi also went on to give a q-extension of Gauss's $_2F_1$ summation formula. At that time, it was natural for someone to consider the q-extension of a general $_2F_1$ hypergeometric series; E. Heine did just that, and we discuss his work in chapter 30.

29.2 Jakob Bernoulli's Theta Series

It is interesting that the series with quadratic exponents, normally arising in the theory of elliptic functions, occurred in Bernoulli's work in probability. In 1685, he proposed the following two problems in the *Journal des Sçavans*:

> Let there be two players A and B, playing against each other with two dice on the condition that whoever first throws a 7 will win. There are sought their expectations if they play in one of these orders:
>
> 1. A once, B once, A twice, B twice, A three times, B three times, A four times, B four times, etc.
> 2. A once, B twice, A three times, B four times, A five times, etc.

In his *Ars Conjectandi*, Bernoulli wrote that in May 1690, when no solution to this problem had yet appeared, he communicated a solution to *Acta Eruditorum*. In the first case, Bernoulli gave the probability for A to win as

$$1 - m + m^2 - m^4 + m^6 - m^9 + m^{12} - m^{16} + m^{20} - m^{25} + \text{ etc.} \tag{29.17}$$

In the second case, the probability for A to win was

$$1 - m + m^3 - m^6 + m^{10} - m^{15} + m^{21} - m^{28} + m^{36} - m^{45} + \text{ etc.} \tag{29.18}$$

In both cases, $m = 5/6$. To make the quadratic exponents explicit, write the two series as

$$1 + \sum_{n=1}^{\infty} m^{n(n+1)} - \sum_{n=1}^{\infty} m^{n^2} \text{ and } \sum_{n=0}^{\infty} (-1)^n m^{n(n+1)/2}.$$

Bernoulli remarked that the summation of these series was difficult because of the unequal jumps in the powers of m. He noted that numerical approximation to any degree of accuracy was easy and for $m = 5/6$, the value of the second series was 0.52393; we remark that this value is inaccurate by only one in the last decimal place. Jakob Bernoulli was very interested in polygonal and figurate numbers; in fact, he worked out the sum of the reciprocals of triangular numbers. Here he had series with triangular and square numbers as exponents. Gauss discovered a way to express these series as products. Euler found the product expansion of a series with pentagonal numbers as exponents.

29.3 Euler's q-series Identities

In response to the problems of Naudé, Euler proved the two identities:

$$(1+qx)(1+q^2x)(1+q^3x)(1+q^4x)\cdots$$
$$= 1 + \frac{q}{1-q}x + \frac{q^3}{(1-q)(1-q^2)}x^2 + \cdots + \frac{q^{m(m+1)/2}}{(1-q)\cdots(1-q^m)}x^m + \cdots,$$

$$\tag{29.19}$$

$$\frac{1}{(1-qx)(1-q^2x)(1-q^3x)\cdots}$$
$$= 1 + \frac{q}{1-q}x + \frac{q^2}{(1-q)(1-q^2)}x^2 + \cdots + \frac{q^m}{(1-q)\cdots(1-q^m)}x^m + \cdots. \quad (29.20)$$

Euler's argument for the first identity was outlined in the opening remarks of this chapter. His proof of his second identity ran along similar lines. We here follow Euler's presentation from his *Introductio*, noting that Euler wrote x for our q and z for our x. Note also that the term q-series came into use only in the latter half of the nineteenth century, appearing in the works of Cayley, Rogers, and others. Jacobi may possibly have been the first to use the symbol q in this context, though he did not use the term q-series. Euler let Z denote the infinite product on the left of (29.20) and he assumed that Z could be expanded as a series:

$$Z = 1 + Px + Qx^2 + Rx^3 + Sx^4 + \cdots. \quad (29.21)$$

When x was replaced by qx in Z, he got

$$\frac{1}{(1-q^2x)(1-q^3x)(1-q^4x)\cdots} = (1-qx)Z.$$

Making the same substitution in (29.21), he obtained

$$(1-qx)Z = 1 + Pqx + Qq^2x^2 + Rq^3x^3 + Sq^4x^4 + \cdots. \quad (29.22)$$

When the series for Z was substituted in (29.22) and the coefficients of the various powers of x were equated, the result was

$$P = \frac{q}{1-q}, \quad Q = \frac{Pq}{1-q^2}, \quad R = \frac{Qq}{1-q^3}, \quad S = \frac{Rq}{1-q^4}, \text{ etc.}$$

and this proved (29.20).

29.4 Euler's Pentagonal Number Theorem

Pentagonal numbers can be generated by the exponents $m(3m \pm 1)/2$ in Euler's formula (29.8)

$$\prod_{n=1}^{\infty}(1-x^n) = 1 + \sum_{m=1}^{\infty}(-1)^m x^{m(3m\pm 1)/2}. \quad (29.23)$$

This identity is often referred to as the pentagonal number theorem. Euler's proof is elementary and employs simple algebra in an ingenious way; we present it almost exactly as it appeared in Euler's June 1750 letter to Goldbach. He began with the algebraic identity mentioned earlier

$$(1-\alpha)(1-\beta)(1-\delta)(1-\gamma) \text{ etc.}$$
$$= 1 - \alpha - \beta(1-\alpha) - \delta(1-\alpha)(1-\beta) - \gamma(1-\alpha)(1-\beta)(1-\delta) - \cdots.$$

29.4 Euler's Pentagonal Number Theorem

From this he had

$$(1-x)(1-x^2)(1-x^3)(1-x^4)(1-x^5) \text{ etc.} = S$$
$$= 1 - x - x^2(1-x) - x^3(1-x)(1-x^2) - x^4(1-x)(1-x^2)(1-x^3) - \text{etc.}$$

He set $S = 1 - x - Axx$, where

$$A = 1 - x + x(1-x)(1-x^2) + x^2(1-x)(1-x^2)(1-x^3) + \text{etc.}$$

Multiplying out by the factor $1 - x$ in each term, he obtained

$$A = 1 - x \quad\quad - x^2(1-x^2) \quad\quad - x^3(1-x^2)(1-x^3) - \text{etc.}$$
$$+ x(1-x^2) + x^2(1-x^2)(1-x^3) + x^3(1-x^2)(1-x^3)(1-x^4) + \text{etc.}$$
$$= 1 - x^3 \quad\quad - x^5(1-x^2) \quad\quad - x^7(1-x^2)(1-x^3) - \text{etc.}$$

He set $A = 1 - x^3 - Bx^5$, where

$$B = 1 - x^2 + x^2(1-x^2)(1-x^3) + x^4(1-x^2)(1-x^3)(1-x^4) + \text{etc.}$$

After multiplying out by the factor $1 - x^2$, appearing in each term of B, he arrived at

$$B = 1 - x^2 \quad\quad - x^4(1-x^3) \quad\quad - x^6(1-x^3)(1-x^4) - \text{etc.}$$
$$+ x^2(1-x^3) + x^4(1-x^3)(1-x^4) + x^6(1-x^3)(1-x^4)(1-x^5) + \text{etc.}$$
$$= 1 - x^5 \quad\quad - x^8(1-x^3) \quad\quad - x^{11}(1-x^3)(1-x^4) - \text{etc.}$$

Euler then set $B = 1 - x^5 - x^8 C$, where $C = 1 - x^3 + x^3(1-x^3)(1-x^4) + x^6(1-x^3)(1-x^4)(1-x^5) + $ etc. Multiplying out by $1 - x^3$,

$$C = 1 - x^3 \quad\quad - x^6(1-x^4) \quad\quad - x^9(1-x^4)(1-x^5) - \text{etc.}$$
$$+ x^3(1-x^4) + x^6(1-x^4)(1-x^5) + x^9(1-x^4)(1-x^5)(1-x^6) + \text{etc.}$$
$$= 1 - x^7 \quad\quad - x^{11}(1-x^4) \quad\quad - x^{15}(1-x^4)(1-x^5) - \text{etc.}$$

When this process was continued, he got

$$C = 1 - x^7 - x^{11} D, \quad D = 1 - x^9 - x^{14} E, \quad E = 1 - x^{11} - x^{17} F.$$

This completed Euler's proof. To describe it more succinctly, write $S = P_0$, $A = P_1$, $B = P_2$, $C = P_3$ and so on. If he had completed the inductive step, Euler would have shown that

$$P_{n-1} = 1 - x^{2n-1} - x^{3n-1} P_n, \quad \text{where} \tag{29.24}$$

$$P_n = \sum_{k=0}^{\infty} x^{kn}(1-x^n)(1-x^{n+1}) \cdots (1-x^{n+k}). \tag{29.25}$$

Since Euler's method of proving (29.24) is useful in establishing other identities in q-series, we describe it in the general situation. The first step is to break up each term of P_n into two parts

$$x^{kn}(1-x^{n+1})\cdots(1-x^{n+k}) - x^{(k+1)n}(1-x^{n+1})\cdots(1-x^{n+k});$$

in the second step, take the second (negative) part and add it to the first part of the next term of P_n:

$$-x^{(k+1)n}(1-x^{n+1})\cdots(1-x^{n+k}) + x^{(k+1)n}(1-x^{n+1})\cdots(1-x^{n+k+1})$$
$$= -x^{(k+2)n+k+1}(1-x^{n+1})\cdots(1-x^{n+k}).$$

It can now be seen that

$$P_n = 1 - x^{2n+1} - x^{3n+2}\sum_{k=0}^{\infty} x^{k(n+1)}(1-x^{n+1})\cdots(1-x^{n+1+k})$$
$$= 1 - x^{2n+1} - x^{3n+2} P_{n+1},$$

proving (29.24) by induction. Euler's method was used by Gauss, and then in 1884 Cayley applied it to prove an interesting identity of Sylvester; Rogers and Ramanujan independently employed the idea to prove the Rogers–Ramanujan identities. Recently, Andrews has further developed this method.

A repeated application of (29.24) converts the infinite product in (29.23) to the required sum:

$$(1-x)(1-x^2)(1-x^3)(1-x^4)\cdots$$
$$= 1 - x - x^2(1-x^3) + x^{2+5}(1-x^5) - x^{2+5+8}(1-x^7) + \cdots$$
$$+ (-1)^{n-1} x^{2+5+\cdots+3n-4}(1-x^{2n-1}) + (-1)^n x^{2+5+\cdots+3n-1}(1-x^{2n+1}) + \cdots$$
$$= 1 - (x+x^2) + (x^5+x^7) - (x^{12}+x^{15}) + \cdots + (-1)^n \left(x^{\frac{n(3n-1)}{2}} + x^{\frac{n(3n+1)}{2}} \right) + \cdots.$$

As Euler noted, this series can also be written as

$$\cdots + x^{26} - x^{15} + x^7 + x^0 - x^1 + x^5 - x^{12} + \cdots = \sum_{n=-\infty}^{\infty} (-1)^n x^{n(3n-1)/2}.$$

This result of Euler was quite remarkable and its proof ingenious; it made quite an impression on the young Gauss who continued Euler's work in new directions.

29.5 Gauss: Triangular and Square Numbers Theorem

We saw Euler's algebraic virtuosity in his proof of the pentagonal number theorem. It is therefore interesting to see Gauss's extremely skillful performance in his similar evaluation of series with triangular and square numbers as exponents. Gauss too divided each term of an appropriate series into two parts and added the second part of each term to the first part of the next term. These formulas for triangular and square exponents

29.5 Gauss: Triangular and Square Numbers Theorem

are particular cases of the triple product identity. Gauss knew this, but he liked his ingenious calculations enough to make a brief note of his method in a paper published only in his collected works, "Zur Theorie der transscendenten Functionen Gehörig." He gave a proof of the formula

$$\frac{1-x}{1+x} \cdot \frac{1-xx}{1+xx} \cdot \frac{1-x^3}{1+x^3} \cdot \frac{1-x^4}{1+x^4} \cdot \text{etc.} = 1 - 2x + 2x^4 - 2x^9 + 2x^{16} - \text{etc. for } |x| < 1.$$

Gauss started with the series:

$$P = 1 + \sum_{k=1}^{\infty} \frac{x^{kn}}{1+x^n} \cdot \frac{(1-x^{2n+k})(1-x^{n+1})(1-x^{n+2})\cdots(1-x^{n+k-1})}{(1+x^{n+1})(1+x^{n+2})(1+x^{n+3})\cdots(1+x^{n+k})}$$

$$Q = \frac{x^n}{1+x^n} + \frac{x^{2n}}{1+x^n} \cdot \frac{1-x^{n+1}}{1+x^{n+1}} + \frac{x^{2n}}{1+x^n} \cdot \frac{1-x^{n+1}}{1+x^{n+1}} \cdot \frac{1-x^{n+2}}{1+x^{n+2}} + \text{etc.}$$

$$R = P - Q.$$

He evaluated R in two different ways. First, he subtracted the kth term in Q from the kth term in P for each k to get

$$R = \frac{1}{1+x^n} + \sum_{k=1}^{n} \frac{x^{kn}}{1+x^n} \cdot \frac{(1-x^n)(1-x^{n+1})\cdots(1-x^{n+k-1})}{(1+x^{n+1})(1+x^{n+2})\cdots(1+x^{n+k})}. \quad (29.26)$$

He denoted this series for R by $\phi(x,n)$. To find another series for R, Gauss subtracted the kth term in Q from the $(k+1)$th term in P for each k to get

$$R = 1 - \frac{x^{2n+1}}{1+x^{n+1}} - \frac{x^{2n+2}}{1+x^{n+1}} \cdot \frac{1-x^{n+1}}{1+x^{n+2}} - \frac{x^{2n+3}}{1+x^{n+1}} \cdot \frac{1-x^{n+1}}{1+x^{n+2}} \cdot \frac{1-x^{n+2}}{1+x^{n+3}} - \text{etc.}$$

He concluded from this relation that

$$R = 1 - x^{2n+1} \cdot \phi(x, n+1) \quad \text{or}$$

$$\phi(x,n) = 1 - x^{2n+1} \cdot \phi(x, n+1).$$

Gauss noted that the relation was true for $n \geq 1$, and for such n

$$\phi(x,n) = 1 - x^{2n+1} + x^{4n+4} - x^{6n+9} + x^{8n+16} - \text{etc.}$$

Note that the series P, Q, and R with $n \geq 1$ are absolutely convergent and that the terms can be rearranged. When $n = 0$, the series for P and Q are divergent and care must be exercised. It is clear from the definition of $\phi(x,n)$ given by (29.26) that $\phi(x,0) = 1/2$. For clarity, we now employ notation not used by Gauss. Let $p_1, p_2, p_3, \ldots$ and $q_1, q_2, q_3, \ldots$ denote the consecutive terms of P and Q when $n = 0$. Then

$$\phi(x,0) = \lim_{m \to \infty} ((p_1 - q_1) + (p_2 - q_2) + \cdots + (p_m - q_m))$$

$$= \lim_{m \to \infty} (p_1 + (p_2 - q_1) + (p_3 - q_2) + \cdots + (p_m - q_{m-1})) - \lim_{m \to \infty} q_m.$$

Gauss denoted the second limit, $\lim_{m\to\infty} q_m$, by T and called it the last term of the series Q, with $n = 0$. He observed that the first limit could be expressed as $1 - x\phi(x, 1)$. Thus, he had $T = 1 - x\phi(x, 1) - \phi(x, 0)$ or

$$\phi(x, 0) = 1 - x\phi(x, 1) - T = 1 - x + x^4 - x^9 + x^{16} - \cdots - T.$$

From the definition of T, Gauss could see that

$$T = \frac{1}{2} \frac{1-x}{1+x} \cdot \frac{1-xx}{1+xx} \cdot \frac{1-x^3}{1+x^3} \cdots,$$

and since $\phi(x, 0) = 1/2$, Gauss finally had

$$2T = \frac{1-x}{1+x} \cdot \frac{1-xx}{1+xx} \cdot \frac{1-x^3}{1+x^3} \cdots = 1 - 2x + 2x^4 - 2x^9 + 2x^{16} - \cdots.$$

He gave an abbreviated form of the argument for the series with triangular numbers. We reproduce Gauss's calculation exactly:

$$P_1 = \frac{1-x^{2n+2}}{1-x^{n+1}} + \frac{x^n \cdot 1-x^{2n+4} \cdot 1-x^{n+2}}{1-x^{n+1} \cdot 1-x^{n+3}}$$
$$+ \frac{x^{2n} \cdot 1-x^{2n+6} \cdot 1-x^{n+2} \cdot 1-x^{n+4}}{1-x^{n+1} \cdot 1-x^{n+3} \cdot 1-x^{n+5}} + \text{etc.}$$

$$Q_1 = \frac{x^n \cdot 1-x^{n+2}}{1-x^{n+1}} + \frac{x^{2n} \cdot 1-x^{n+2} \cdot 1-x^{n+4}}{1-x^{n+1} \cdot 1-x^{n+3}}$$
$$+ \frac{x^{3n} \cdot 1-x^{n+2} \cdot 1-x^{n+4} \cdot 1-x^{n+6}}{1-x^{n+1} \cdot 1-x^{n+3} \cdot 1-x^{n+5}} + \text{etc.}$$

$$R_1 = \frac{1-x^n}{1-x^{n+1}} + \frac{x^n \cdot 1-x^n \cdot 1-x^{n+2}}{1-x^{n+1} \cdot 1-x^{n+3}} + \frac{x^{2n} \cdot 1-x^n \cdot 1-x^{n+2} \cdot 1-x^{n+4}}{1-x^{n+1} \cdot 1-x^{n+3} \cdot 1-x^{n+5}} + \text{etc.}$$

where R_1 was obtained by subtracting Q_1 termwise from P_1. Gauss denoted this series for R_1 as $\psi(x, n)$. Then, by subtracting the kth term of Q_1 from the $(k+1)$th term of P_1, Gauss had

$$R_1 = 1 + x^{n+1} + \frac{x^{2n+3} \cdot 1-x^{n+2}}{1-x^{n+3}} + \frac{x^{2n+3} \cdot x^{n+2} \cdot 1-x^{n+2} \cdot 1-x^{n+4}}{1-x^{n+3} \cdot 1-x^{n+5}} + \text{etc.}$$
$$= 1 + x^{n+1} + x^{2n+3} \psi(x, n+2) = \psi(x, n),$$

when $n \geq 1$. Therefore,

$$\psi(x, n) = 1 + x^{n+1} + x^{2n+3} + x^{3n+6} + x^{4n+10} + \text{etc.}$$

In the case $n = 0$, $\psi(x, 0) = 0$. Moreover,

$$\psi(x, 0) = 1 + x + x^3 \psi(x, 2) - \frac{1-x^2}{1-x} \cdot \frac{1-x^4}{1-x^3} \cdot \frac{1-x^6}{1-x^5} \text{ etc.}$$

Hence, the required result followed:

$$\frac{1-x^2}{1-x} \cdot \frac{1-x^4}{1-x^3} \cdot \frac{1-x^6}{1-x^5} \cdots = 1 + x + x^3 + x^6 + x^{10} + \text{etc.}$$

29.6 Gauss Polynomials and Gauss Sums

In his paper of 1811, Gauss defined the q-extension of a binomial coefficient by

$$(m,\mu) = \frac{(1-q^m)(1-q^{m-1})(1-q^{m-2})\cdots(1-q^{m-\mu+1})}{(1-q)(1-q^2)(1-q^3)\cdots(1-q^\mu)}. \tag{29.27}$$

He noted the easily verified formula

$$(m,\mu+1) = (m-1,\mu+1) + q^{m-\mu-1}(m-1,\mu). \tag{29.28}$$

Note that it follows from this that (m,μ) is a polynomial in q when m is a positive integer. These polynomials are now called Gaussian polynomials and are extensions of the binomial coefficients $\binom{m}{\mu}$. We remark that we are using the familiar symbol q, although Gauss used x.

Let us now see how Gauss evaluated the polynomial $\sum_{\mu=0}^{m}(m,\mu)x^\mu$ for $x=-1$ and $x=\sqrt{q}$. For $x=-1$, Gauss used (29.28) to show that

$$f(q,m) = 1 - (m,1) + (m,2) - (m,3) + (m,4) - \cdots$$

satisfied the functional relation

$$f(q,m) = (1-q^{m-1})f(q,m-2). \tag{29.29}$$

Since $f(q,0)=1$ and $f(q,1)=0$, Gauss deduced that

$$f(q,m) = (1-q)(1-q^3)\cdots(1-q^{m-1}), \qquad \text{for } m \text{ even,}$$
$$= 0, \qquad \text{for } m \text{ odd.} \tag{29.30}$$

For $x=\sqrt{q}$, Gauss wrote

$$F(q,m) = 1 + q^{1/2}(m,1) + q(m,2) + q^{3/2}(m,3) + \cdots$$
$$= q^{m/2} + q^{(m-1)/2}(m,1) + q^{(m-2)/2}(m,2) + q^{(m-3)/2}(m,3) + \cdots. \tag{29.31}$$

Note that the second (finite) series is identical to the first one, but is in reverse order. Gauss then multiplied the second series by $q^{(m+1)/2}$, and added the result to the first series, yielding

$$(1+q^{(m+1)/2})F(q,m) = 1 + q^{1/2}(m,1) + q(m,2) + q^{3/2}(m,3) + \cdots$$
$$+ q^{1/2}\cdot q^m + q\cdot q^{m-1}(m,1) + q^{3/2}\cdot q^{m-2}(m,2) + \cdots$$
$$= 1 + q^{1/2}(q^m + (m,1)) + q((m,2) + q^{m-1}(m,1))$$
$$+ q^{3/2}((m,3) + q^{m-2}(m,2)) + \cdots.$$

By (29.28), he concluded that $(1+q^{(m+1)/2})F(q,m) = F(q,m+1)$; since $F(q,0)=1$, he had the required result

$$F(q,m) = (1+q^{1/2})(1+q)(1+q^{3/2})\cdots(1+q^{m/2}). \tag{29.32}$$

Gauss used formulas (29.30) and (29.32) to show that

$$\sum_{k=0}^{n-1} e^{2\pi i k^2/n} = \frac{1+i^{-n}}{1+i^{-1}}\sqrt{n}. \tag{29.33}$$

We note that the expression on the left side is called a quadratic Gauss sum. He proved this formula in four separate exhaustive cases: for $n \equiv 0, 1, 2, 3 \pmod 4$. Gauss explained how to convert the expression $F(q,m)$ into a Gauss sum. He set $q^{1/2} = -y^{-1}$ so that

$$F(y^{-2}, m) = 1 - y^{-1}\frac{1-y^{-2m}}{1-y^{-2}} + y^{-2}\frac{(1-y^{-2m})(1-y^{-2m+2})}{(1-y^{-2})(1-y^{-4})} - \cdots. \tag{29.34}$$

He then set $m = n-1$ and took y to be a primitive root of $y^n - 1 = 0$, to get

$$\frac{1-y^{-2m}}{1-y^{-2}} = \frac{1-y^2}{1-y^{-2}} = -y^2\,;\ \frac{1-y^{-2m+2}}{1-y^{-4}} = -y^4\,;\ \frac{1-y^{-2m+4}}{1-y^{-6}} = -y^6\cdots.$$

Thus, he found

$$F(y^{-2}, m) = 1 + y^{-1}\cdot y^2 + y^{-2}\cdot y^2\cdot y^4 + y^{-3}\cdot y^2\cdot y^4\cdot y^6 + \cdots$$
$$= 1 + y + y^4 + y^9 + \cdots + y^{(n-1)^2}. \tag{29.35}$$

Observe that for $y = e^{2\pi i/n}$, the expression (29.35) was the Gauss sum. From (29.32) and (29.35), it followed that

$$1 + y + y^4 + y^9 + \cdots + y^{(n-1)^2} = (1-y^{-1})(1+y^{-2})(1-y^{-3})\cdots(1\pm y^{-n+1}), \tag{29.36}$$

when y was a primitive root of $y^n - 1 = 0$. Gauss showed that when $y = e^{2\pi i/n}$, the product in (29.36) reduced to the expression on the right-hand side of (29.33). The case $n = 4s + 2$ is elementary. In fact, for this case Gauss observed that for any primitive root y, $y^{2s+1} = -1$, so that $y^{(2s+1)^2} = -1$. Moreover, for any integer t,

$$y^{(2s+1+t)^2} = y^{(2s+1)^2 + (4s+2)t + t^2} = -y^{t^2}.$$

Therefore, by cancellation of terms, Gauss found the sum (29.35) to be zero. Turning to the case $n = 4s$, Gauss applied (29.36) to evaluate (29.33). Now $y^{(2s+t)^2} = y^{t^2}$, and hence

$$1 + y + y^4 + \cdots + y^{(n-1)^2} = 2(1 + y + y^4 + \cdots + y^{(2s-1)^2}). \tag{29.37}$$

By taking $m = \frac{1}{2}n - 1 = 2s - 1$ in (29.34) and using the calculations leading up to (29.36), Gauss had

$$1 + y + y^4 + \cdots + y^{(2s-1)^2} = (1-y^{-1})(1+y^{-2})(1-y^{-3})\cdots(1-y^{-2s+1}). \tag{29.38}$$

29.6 Gauss Polynomials and Gauss Sums

Then $y^{2s} = -1$, and hence $1 + y^{-2k} = -y^{2s-2k}(1 - y^{-2s+2k})$. He applied this to the product in (29.38), so that by (29.37)

$$F \equiv 1 + y + y^4 + \cdots + y^{(n-1)^2}$$
$$= 2(-1)^{s-1} y^{s^2-s}(1-y^{-1})(1-y^{-2})\cdots(1-y^{-2s+1}). \tag{29.39}$$

Again, from the fact that $1 - y^{-k} = -y^{-k}(1 - y^{-4s+k})$, Gauss got

$$(1-y^{-1})(1-y^{-2})(1-y^{-3})\cdots(1-y^{-2s+1})$$

$$= (-1)^{2s-1} y^{-2s^2+s}(1-y^{-2s-1})(1-y^{-2s-2})(1-y^{-2s-3})\cdots(1-y^{-4s+1}). \tag{29.40}$$

Therefore, by (29.39) and (29.40)

$$F = 2(-1)^{3s-2} y^{-s^2}(1-y^{-2s-1})(1-y^{-2s-2})\cdots(1-y^{-4s+1}). \tag{29.41}$$

Next, Gauss took the product of (29.39) and (29.41) and multiplied by $1 - y^{-2s}$ to obtain

$$(1-y^{-2s})F^2 = 4(-1)^{4s-3} y^{-s}(1-y^{-1})(1-y^{-2})\cdots(1-y^{-4s+1}). \tag{29.42}$$

So Gauss could conclude that $F^2 = 2y^s n = \pm 2in$, since $y^{2s} = -1$ or $y^s = \pm i$. Note that Gauss also made use of the fact that the product

$$(1-y^{-1})(1-y^{-2})\cdots(1-y^{-4s+1})$$

was equal to n, because $y^{-1}, \ldots, y^{-4s+1}$ were all the nontrivial nth roots of unity. By taking square roots, Gauss obtained

$$F = 1 + y + y^4 + \cdots + y^{(n-1)^2} = \pm(1+i)\sqrt{n}. \tag{29.43}$$

To determine the sign when $y = e^{2\pi i/n}$, Gauss set $y = p^2$ in (29.38) and used $p^n = -1$ to get

$$F = 2(1+p^{n-2})(1+p^{-4})(1+p^{n-6})(1+p^{-8})\cdots(1+p^{-n+4})(1+p^2).$$

He rewrote this equation as

$$F = 2(1+p^2)(1+p^{-4})(1+p^6)\cdots(1+p^{-n+4})(1+p^{n-2})$$

and observed that $1 + p^{\pm 2k} = 2p^{\pm k}\cos(k\pi/n)$, finally concluding that

$$F = 2^{2s} p^s \cos\frac{\pi}{n} \cos\frac{2\pi}{n} \cos\frac{3\pi}{n} \cdots \cos\frac{(2s-1)\pi}{n}.$$

Now $p^s = \cos\frac{\pi}{4} + i\sin\frac{\pi}{4} = (1+i)/\sqrt{2}$ and since all the cosine values were positive, Gauss determined that the sign in (29.43) was positive. This concluded his proof of the case $n = 4s$. The other two cases of (29.33), where n is odd, or $n = 4s + 1$ or $4s + 3$, are the most important because they lead to the proof of the quadratic reciprocity theorem. For these cases, Gauss first gave a detailed derivation using (29.30), although he indicated that (29.32) could also serve the purpose. So it remained for Gauss to prove that the Gauss sum in (29.33) was equal to $\sqrt{n}$ when n was of the form $4m+1$, and equal to $i\sqrt{n}$ when n took the form $4m+3$. He took $n = s+1$ where s was even, x was a primitive root of $x^n - 1 = 0$, and $q = x^{-2}$. Then

$$\frac{1-q^{s-j}}{1-q^{j+1}} = \frac{1-x^{-2(s-j)}}{1-x^{-2j-2}} = \frac{1-x^{-2(n-1-j)}}{1-x^{-2j-2}} = \frac{1-x^{2j+2}}{1-x^{-2j-2}} = -x^{2(j+1)}$$

and the Gaussian polynomial was given by

$$(s,k) = (-1)^k x^{2(1+2+\cdots+k-1)} = (-1)^k x^{k(k-1)}.$$

Using this in (29.30), he had

$$\sum_{k=1}^{n} x^{k(k-1)} = (1-x^{-2})(1-x^{-6})\cdots(1-x^{-2(n-2)})$$

$$= x^{\frac{1}{4}(n-1)^2}(x-x^{-1})(x^3-x^{-3})\cdots(x^{n-2}-x^{-(n-2)}). \qquad (29.44)$$

Since x was an nth root of unity and since

$$\frac{1}{4}(n-1)^2 + k(k-1) = \frac{1}{4}(n^2 - 2n + (2k-1)^2),$$

$$x^{\frac{1}{4}(n-1)^2+k(k-1)} = x^{\frac{1}{4}(n-(2k-1))^2} = x^{(e-k)^2}, \quad \text{where } n+1 = 2e.$$

Thus Gauss could rewrite (29.44) as

$$W \equiv \sum_{k=0}^{n-1} x^{k^2} = (x-x^{-1})(x^3-x^{-3})\cdots(x^{n-2}-x^{-(n-2)}). \qquad (29.45)$$

We note that Gauss also worked out a derivation of this formula using (29.32). Now $x^{n-2} - x^{-(n-2)} = -(x^2 - x^{-2})$ etc. implied that

$$W = (-1)^{\frac{n-1}{2}}(x^2-x^{-2})(x^4-x^{-4})\cdots(x^{n-1}-x^{-n+1}). \qquad (29.46)$$

By multiplying (29.45) and (29.46), he obtained

$$W^2 = (-1)^{\frac{n-1}{2}}(x-x^{-1})(x^2-x^{-2})(x^3-x^{-3})\cdots(x^{n-1}-x^{-n+1}).$$

When n was of the form $4s+1$, the factor $(-1)^{\frac{n-1}{2}}$ became $+1$ and when n was of the form $4s+3$, it became -1. Thus, he obtained

$$W^2 = \pm x^{\frac{1}{2}(n-1)n}(1-x^{-2})(1-x^{-4})\cdots(1-x^{-2(n-1)}).$$

Using an argument similar to the one for (29.43), Gauss concluded that $W = \pm n$, where the $+$ sign applied to $n = 4s+1$ and the $-$ sign to $4s+3$. Note that Gauss had already arrived at this point in 1801, but by a different route. The problem remaining in 1805 was to choose the correct sign for the square root to obtain W. To find that, Gauss set $x = e^{2\pi i/n}$ in (29.45) to get

$$W = (2i)^{\frac{n-1}{2}} \sin \frac{2\pi}{n} \sin \frac{6\pi}{n} \cdots \sin \frac{2(n-2)\pi}{n}.$$

Whether $n = 4s + 1$ or $n = 4s + 3$, Gauss saw that there were clearly s negative factors in the sine product. Thus, Gauss could conclude that $W = \sqrt{n}$ for $n = 4s + 1$ and $W = i\sqrt{n}$ for $n = 4s + 3$.

29.7 Gauss's q-Binomial Theorem and the Triple Product Identity

Gauss wrote a paper "Hundert Theoreme über die neuen Transscendenten," but he did not publish it. In this paper, he derived a form of the terminating q-binomial theorem. He then wrote the result in a symmetric form and by an ingenious argument derived the triple product identity. We follow Gauss's notation and proof: He stated the terminating q-binomial theorem in the form

$$1 + \frac{a^n - 1}{a - 1} t + \frac{a^n - 1 \cdot a^n - a}{a - 1 \cdot aa - 1} tt + \frac{a^n - 1 \cdot a^n - a \cdot a^n - aa}{a - 1 \cdot aa - 1 \cdot a^3 - 1} t^3 + \text{etc.}$$

$$= (1+t)(1+at)(1+aat)\cdots(1+a^{n-1}t). \tag{29.47}$$

Recall that we would write q instead of a. To prove the formula inductively, he denoted the sum as T and multiplied it by $(1 + a^n t)$ to obtain a series of the same form with n changed to $n + 1$. The reader may work out this calculation. Gauss next observed that by taking $T = \theta(n)$, one could see that $T(1 + a^n t) = \theta(n+1)$. Thus, the terminating q-binomial theorem was proved inductively.

To prove the triple-product identity, he wrote his result in a symmetric form. He took n even, set $y = a^{\frac{n-1}{2}} t$ and $x^2 = a$ to transform (29.47) into

$$1 + \frac{1-x^n}{1-x^{n+2}} x \left(y + \frac{1}{y}\right) + \frac{1-x^n}{1-x^{n+2}} \cdot \frac{1-x^{n-2}}{1-x^{n+4}} \cdot x^4 \left(yy + \frac{1}{yy}\right)$$

$$+ \frac{1-x^n}{1-x^{n+2}} \cdot \frac{1-x^{n-2}}{1-x^{n+4}} \cdot \frac{1-x^{n-4}}{1-x^{n+6}} \cdot x^9 \left(y^3 + \frac{1}{y^3}\right) + \cdots$$

$$= \frac{1-xx}{1-x^{n+2}} \cdot \frac{1-x^4}{1-x^{n+4}} \cdot \frac{1-x^6}{1-x^{n+6}} \cdots \cdot \frac{1-x^n}{1-x^{2n}}$$

$$\cdot (1+xy)(1+x^3 y) \cdots \cdot (1+x^{n-1} y) \left(1 + \frac{x}{y}\right)$$

$$\cdot \left(1 + \frac{x^3}{y}\right) \cdots \cdot \left(1 + \frac{x^{n-1}}{y}\right). \tag{29.48}$$

Next, he took $|x| < 1$, so that $x^n \to 0$ as $n \to \infty$. The result was the triple product identity:

$$1 + x(y + y^{-1}) + x^4(yy + y^{-2}) + x^9(y^3 + y^{-3}) + \cdots$$
$$= (1 - xx)(1 - x^4)(1 - x^6) \cdots (1 + xy)(1 + x^3 y)(1 + x^5 y) \qquad (29.49)$$
$$\cdots (1 + xy^{-1})(1 + x^3 y^{-1})(1 + x^5 y^{-1}) \cdots.$$

Now let us examine the algebraic steps Gauss used to get the required symmetric form (29.48). Note that the last term in the left-hand side of (29.47) was $a^{n(n-1)/2} t^n = y^n$. Combining the first and last terms of the sum, then the second to the last but one, and so on, he arrived at

$$1 + y^n + \frac{1-a^n}{1-a} t \left(1 + y^{n-2}\right) + \frac{1-a^n}{1-a} \cdot \frac{1-a^{n-1}}{1-aa} \cdot att \left(1 + y^{n-4}\right) + \cdots$$
$$+ \frac{1-a^n}{1-a} \cdot \frac{1-a^{n-1}}{1-aa} \cdots \frac{1-a^{\frac{1}{2}n+2}}{1-a^{\frac{1}{2}n-1}} \cdot a^{\frac{1}{2}\left(\frac{1}{2}n-1\right)\left(\frac{1}{2}n-2\right)} t^{\frac{1}{2}n-1}(1 + yy)$$
$$+ \frac{1-a^n}{1-a} \cdot \frac{1-a^{n-1}}{1-aa} \cdots \frac{1-a^{\frac{1}{2}n+1}}{1-a^{\frac{1}{2}n}} \cdot a^{\frac{1}{2} \cdot \frac{1}{2}n \cdot \frac{1}{2}n-1} t^{\frac{1}{2}n}.$$

He set $a = x^2$, denoted the last term by A and took it out as a common factor to get

$$A\left(1 + \frac{1-x^n}{1-x^{n+2}} \cdot x(y + y^{-1}) + \frac{1-x^n}{1-x^{n+2}} \cdot \frac{1-x^{n-2}}{1-x^{n+4}} x^4(yy + y^{-2})\right.$$
$$\left. + \frac{1-x^n}{1-x^{n+2}} \cdot \frac{1-x^{n-2}}{1-x^{n+4}} \cdot \frac{1-x^{n-4}}{1-x^{n+6}} \cdot x^9(y^3 + y^{-3}) + \cdots\right),$$

where A could be written as

$$\frac{1-x^{n+2}}{1-xx} \cdot \frac{1-x^{n+4}}{1-x^4} \cdot \frac{1-x^{n+6}}{1-x^6} \cdots \frac{1-x^{2n}}{1-x^n} \cdot \frac{y^{\frac{1}{2}n}}{x^{\frac{1}{4}nn}}.$$

He rewrote the product $(1+t)(1+at)(1+aat) \cdots (1+a^{n-1}t)$ as

$$\left(1 + \frac{y}{x^{n-1}}\right)\left(1 + \frac{y}{x^{n-3}}\right) \cdots \left(1 + \frac{y}{x}\right)(1 + yx)(1 + yx^3) \cdots (1 + yx^{n-1}).$$

To complete the calculations necessary for the symmetric form (29.48), it was sufficient for Gauss to observe that the first half of the product could be rewritten as

$$\frac{y^{\frac{1}{2}n}}{x^{\frac{1}{4}nn}} \left(1 + \frac{x^{n-1}}{y}\right)\left(1 + \frac{x^{n-3}}{y}\right) \cdots \left(1 + \frac{x}{y}\right).$$

It is interesting to note that the triple product formula (29.49) contains a plethora of important special cases. Euler's pentagonal numbers identity follows on taking $x = q^{3/2}$ and $y = -q^{1/2}$. Gauss's formula for triangular numbers, derived earlier, follows by taking $x = q^{1/2}$ and $y = q^{1/2}$.

It is surprising that Gauss did not publish his work related to the triple product. Gauss's 1811 paper correctly noted the significance of the Gaussian polynomial (m, μ) and later work of Rodrigues, P. MacMahon, and others revealed the combinatorial import of the Gaussian polynomial and its generalization. In addition, Gaussian polynomials played an important role in Cayley and Sylvester's development of invariant theory. It remained for Jacobi to rediscover the triple product formula and use it in his theory of elliptic functions.

29.8 Jacobi: Triple Product Identity

In his work in the theory of elliptic functions, Jacobi encountered numerous infinite products, a large number of which were particular cases of the product side of the triple-product identity. And the product side of this identity was composed of two infinite products, first elucidated by Euler, of the form (29.1). Because Jacobi wished to convert his products in elliptic function theory into series, it was only natural for him to start with Euler's formula (29.3). Change q to q^2 and x to z/q to get

$$(1+qz)(1+q^3z)(1+z^5z)(1+q^7z)\cdots$$
$$= 1 + \frac{qz}{1-q^2} + \frac{q^4z^2}{(1-q^2)(1-q^4)} + \frac{q^9z^3}{(1-q^2)(1-q^4)(1-q^6)} + \cdots.$$

Jacobi then multiplied this equation by one in which z was replaced by $1/z$, to obtain

$$(1+qz)(1+q^3z)(1+q^5z)\cdots(1+q/z)(1+q^3/z)(1+q^5/z)\cdots$$
$$= 1 + \frac{qz}{1-q^2} + \frac{q^4z^2}{(1-q^2)(1-q^4)} + \frac{q^9z^3}{(1-q^2)(1-q^4)(1-q^6)} + \cdots$$
$$\times \left(1 + \frac{q}{1-q^2}\frac{1}{z} + \frac{q^4}{(1-q^2)(1-q^4)}\frac{1}{z^2} + \frac{q^9}{(1-q^2)(1-q^4)(1-q^6)}\frac{1}{z^3} + \cdots\right).$$

Jacobi observed that the coefficient of $z^n + 1/z^n$ in the product on the right-hand side was

$$\frac{q^{nn}}{(1-q^2)(1-q^4)\cdots(1-q^{2n})}$$
$$\times (1 + \frac{q^2}{1-q^2}\cdot\frac{q^{2n}}{1-q^{2n+2}} + \frac{q^8}{(1-q^2)(1-q^4)}\cdot\frac{q^{4n}}{(1-q^{2n+2})(1-q^{2n+4})}$$
$$+ \frac{q^{18}}{(1-q^2)(1-q^4)(1-q^6)}\cdot\frac{q^{6n}}{(1-q^{2n+2})(1-q^{2n+4})(1-q^{2n+6})} + \cdots). \quad (29.50)$$

It seems that Jacobi had some trouble simplifying this expression and this delayed him for quite a while. But he succeeded in resolving the problem by proving that

$$\prod_{n=1}^{\infty}(1-q^nz)^{-1} = \sum_{n=1}^{\infty}\frac{q^{n^2}z^n}{(1-q)(1-q^2)\cdots(1-q^n)(1-qz)\cdots(1-q^nz)}. \quad (29.51)$$

He replaced q by q^2 and then set $z = q^{2n}$ to sum the series in (29.50). He thus found the coefficient of $z^n + \frac{1}{z^n}$ to be $q^{nn}/\left((1-q^2)\cdots(1-q^{2n})\right)$. This proved the triple product identity. To prove (29.51), Jacobi assumed that the product on the left-hand side could be expressed as a sum of terms of the form $A_n z^n/((1-qz)\cdots(1-q^n z))$. For A_n, he applied the standard procedure of changing z to qz to get a functional relation. Obviously, the difficult point here was to conceive of that form of the series in which the variable z would also appear in the denominator. Neither Euler nor Gauss came up with such a series.

Jacobi's formula (29.51) is very interesting. Note that the product on the left-hand is the same as the product in Euler's second formula (29.20) but the series on the right, though similar in appearance, has an additional factor in the denominator of each term of the sum. Jacobi may have asked whether it was possible to directly transform one series into the other. This suggests a transformation theory of q-series similar to that for hypergeometric series. Heinrich Eduard Heine (1821–1881) paved the way for the study of transformations of q-series in his 1846 theory of the q-hypergeometric series. We also note that in 1843, Cauchy gave a generalization of (29.51).

29.9 Eisenstein: q-Binomial Theorem

Eisenstein's "Neuer Beweis und Verallgemeinerung des binomischen Lehrsatzes" was one of three papers he submitted to *Crelle's Journal* in May 1844. In this paper, he proved the general q-binomial theorem, although Eisenstein wrote p instead of q, and deduced from it the ordinary binomial theorem. His proof was based on an idea of Euler and employed the multiplication of series. Eisenstein did not refer to Euler, but mentioned Dirichlet and Ohm, who may have discussed Euler's idea in their lectures. Eisenstein first proved the finite case of the q-binomial theorem. For this he defined for a positive integer α,

$$\phi(x, \alpha) = (1+x)(1+qx)(1+q^2 x)\cdots(1+q^{\alpha-1}x). \tag{29.52}$$

He proved Rothe's formula without reference to Rothe:

$$\phi(x, \alpha) = \sum_{t=0}^{\alpha} A_t x^t, \tag{29.53}$$

$$\text{where} \quad A_t = \frac{q^\alpha - 1}{q-1} \cdot \frac{q^{\alpha-1} - 1}{q^2 - 1} \cdots \frac{q^{\alpha-t+1} - 1}{q^t - 1} q^{\frac{1}{2}t(t-1)}. \tag{29.54}$$

Note that this was done in the standard way by using the relation

$$(1 + q^\alpha x)\phi(x, \alpha) = (1+x)\phi(qx, \alpha).$$

Eisenstein stated the general q-binomial theorem in the form

$$\phi(x, \alpha) \equiv \sum_{t=0}^{\infty} A_t x^t = \frac{(1+x)(1+qx)(1+q^2 x)\cdots}{(1+q^\alpha x)(1+q^{\alpha+1}x)(1+q^{\alpha+2}x)\cdots}, \tag{29.55}$$

where $|q| < 1$, and α was any number. To prove this, he first wished to show that

$$\phi(x, \alpha + \beta) = \phi(x, \alpha)\phi(q^\alpha x, \beta). \tag{29.56}$$

For this purpose, he demonstrated that

$$C_t = A_t + A_{t-1}B_1 q^\alpha + A_{t-2}B_2 q^{2\alpha} + \cdots + B_t q^{t\alpha}, \tag{29.57}$$

where B_t and C_t were obtained from (29.54) by replacing α by β and α by $\alpha + \beta$, respectively. He noted that (29.56) and (29.57) were clearly true when α and β were positive integers. Eisenstein then set $u = q^\alpha$ and $v = q^\beta$ and observed that both sides of (29.57) were equal for infinitely many values of u and v, and thus (29.57) was identically true. At this point, Eisenstein noted that the proof could be completed in the usual manner and referred to Dirichlet and Ohm. From chapter 4, one may see that Eisenstein intended to use (29.56) to prove (29.55) for all integers α, and then for all rational numbers, and finally (by continuity) for all real α.

29.10 Jacobi's q-Series Identity

In 1845, Jacobi proved the q-binomial theorem and obtained an extension of the Vandermonde identity, as well as an extension of Gauss's $_2F_1$ summation formula. Recall Gauss's summation formula:

$$\sum_{n=0}^{\infty} \frac{(a)_n (b)_n}{n!(c)_n} = \frac{\Gamma(c)\Gamma(c-a-b)}{\Gamma(c-a)\Gamma(c-b)}$$

when $\text{Re}(c - a - b) > 0$. Note that when $a = -m$, a negative integer, we have Vandermonde's identity

$$\sum_{n=0}^{m} \frac{(-m)_n (b)_n}{n!(c)_n} = \frac{(c-b)_m}{(c)_m}. \tag{29.58}$$

This identity is not difficult and follows immediately from the Gregory–Newton interpolation formula; it can also be obtained by multiplying two binomial series and equating coefficients. As Richard Askey has pointed out, around 1301, the Chinese mathematician Chu Shih-Chieh (also Zhu Shijie) discovered two equations; when they are combined, they yield this identity. Vandermonde found it in 1772.

In Jacobi's notation, the q-binomial theorem was stated as

$$[w, v] \equiv 1 + \frac{v-w}{1-x} z + \frac{(v-w)(v-xw)}{(1-x)(1-x^2)} z^2 + \frac{(v-w)(v-xw)(v-x^2w)}{(1-x)(1-x^2)(1-x^3)} z^3 + \cdots$$

$$= \frac{(1-wz)(1-xwz)(1-x^2wz)(1-x^3wz)\cdots}{(1-vz)(1-xvz)(1-x^2vz)(1-x^3vz)\cdots}. \tag{29.59}$$

Let $\phi(z)$ denote the product. In his proof, Jacobi assumed that

$$\phi(z) = 1 + A_1 z + A_2 z^2 + A_3 z^3 + A_4 z^4 + \cdots$$

and observed that $\phi(z)$ satisfied the functional relation

$$\phi(z) - \phi(xz) = v\phi(z) - w\phi(xz).$$

Thus, the coefficients $A_1, A_2, A_3, \ldots$ satisfied the equations

$$(1-x)A_1 = v - w, \ (1-x^2)A_2 = (v - xw)A_1, \ (1-x^3)A_3 = (v - x^2 w)A_2, \cdots.$$

By induction, this gives the desired result.

Now note that from the product expression for $[w, v]$ it is easy to see that $[w, v][v, 1] = [w, 1]$. If the corresponding series are substituted in this equation and the coefficient of z^p equated on the two sides, then the result is a q-extension of the Chu-Vandermonde identity. Jacobi wrote that he saw this result in Schweins's *Analysis*:

$$\frac{(1-w)(1-xw)(1-x^2 w)\cdots(1-x^{p-1}w)}{(1-x)(1-x^2)(1-x^3)\cdots(1-x^p)}$$
$$= \sum_{k=0}^{p} \frac{(v-w)(v-xw)\cdots(v-x^{k-1}w)}{(1-x)(1-x^2)\cdots(1-x^k)} \cdot \frac{(1-v)(1-xv)\cdots(1-x^{p-k-1}v)}{(1-x)(1-x^2)\cdots(1-x^{p-k})}. \quad (29.60)$$

Note that the empty products occurring in the sum have the value 1. Now, it is possible to prove Gauss's formula from the Chu-Vandermonde identity, but it is not easy, and such a proof was not known in Jacobi's time. But in a beautiful argument, Jacobi used (29.60), the q-extension of Vandermonde, to prove a q-extension of Gauss's formula. He divided both sides of the equation by the first term on the right-hand side, to get (after a change of variables)

$$\frac{(1-u)(1-xu)(1-x^2 u)\cdots(1-x^{p-1}u)}{(1-r)(1-xr)(1-x^2 r)\cdots(1-x^{p-1}r)}$$
$$= 1 + \sum_{k=1}^{p} (-1)^k \frac{(u-r)(u-xr)\cdots(u-x^{k-1}r)}{(1-x)(1-x^2)\cdots(1-x^k)}$$
$$\times \frac{(1-x^p)(1-x^{p-1})\cdots(1-x^{p-k+1})}{(1-r)(1-xr)\cdots(1-x^{k-1}r)} x^{k(k-1)/2}. \quad (29.61)$$

Jacobi stated the extension of Gauss's formula in the form

$$1 + \frac{(1-s)(1-t)}{(1-x)(1-r)} r + \frac{(1-s)(x-s)(1-t)(x-t)}{(1-x)(1-x^2)(1-r)(1-xr)} r^2$$
$$+ \frac{(1-s)(x-s)(x^2-s)(1-t)(x-t)(x^2-t)}{(1-x)(1-x^2)(1-x^3)(1-r)(1-xr)(1-x^2 r)} r^3 + \cdots$$
$$= \frac{(1-sr)(1-tr)}{(1-r)(1-str)} \cdot \frac{(1-xsr)(1-xtr)}{(1-xr)(1-xstr)} \cdot \frac{(1-x^2 sr)(1-x^2 tr)}{(1-x^2 r)(1-x^2 str)} \cdots. \quad (29.62)$$

He showed that when $t = x^p$, and $p = 0, 1, 2, \ldots$, this was reduced to the identity (29.61). Thus, (29.62) was true for an infinite number of values of t and, by symmetry, for an infinite number of values of s. He then observed that

$$\frac{(1-sr)(1-tr)}{(1-r)(1-str)} = 1 + c_1 r + c_2 r^2 + c_3 r^3 + \cdots$$

where $c_1, c_2, c_3, \ldots$ were polynomials in s and t. This implied that the product on the right-hand side of (29.62) was of the form

$$(1 + c_1 r + c_2 r^2 + c_3 r^3 + \cdots) \cdot (1 + c_1 x r + c_2 x^2 r^2 + \cdots) \cdot (1 + c_1 x^2 r + c_2 x^4 r^2 + \cdots) \cdots.$$

This product would then be of the form $1 + b_1 r + b_2 r^2 + b_3 r^3 + \cdots$, where $b_1, b_2, b_3, \ldots$ were polynomials in s and t. To complete the proof, Jacobi wrote the left-hand side of (29.62) in powers of r as $1 + k_1 r + k_2 r^2 + \cdots$, so that $k_1, k_2, \ldots$ were polynomials in s and t. Jacobi concluded that $b_i = k_i$ because it held for an infinite number of values of s and t. This completed the proof of (29.62). Jacobi also observed, without giving a precise definition, that the products on the right-hand sides of (29.62) could be considered q-analogs of the gamma functions in Gauss's formula. Very soon after this, Heine obtained a nearly correct definition.

29.11 Cauchy and Ramanujan: The Extension of the Triple Product

In 1843, Augustin-Louis Cauchy published an important paper containing the first statement and proof of the general q-binomial theorem and an extension of the triple product identity. To be clear and succinct in stating the results of Cauchy and Ramanujan, we introduce the following modern notation: Let

$$(a; q)_n = (1-a)(1-aq) \cdots (1-aq^{n-1}), \qquad \text{for } n \geq 1,$$
$$= 1, \qquad \text{for } n = 0,$$
$$= \frac{1}{(1-q^{-1}a)(1-q^{-2}a) \cdots (1-q^{-n}a)}, \qquad \text{for } n \leq 0.$$

And $(a; q)_\infty = (1-a)(1-qa)(1-q^2 a) \cdots$. Using this notation, the q-binomial theorem can be stated as

$$\sum_{n=0}^{\infty} \frac{(a;q)_n}{(q;q)_n} x^n = \frac{(ax;q)_\infty}{(x;q)_\infty}.$$

For convergence we require $|x| < 1$, $|q| < 1$. Cauchy's extension of the triple product identity can now be stated for $0 < |bx| < 1$:

$$\sum_{n=-\infty}^{\infty} \left(\frac{a}{b}; q\right)_n b^n x^n = \frac{(ax;q)_\infty \left(\frac{q}{ax};q\right)_\infty (q;q)_\infty}{(bx;q)_\infty \left(\frac{bq}{a};q\right)_\infty}.$$

Here, Cauchy failed to find the better result, called the Ramanujan $_1\psi_1$ sum, generalizing the q-binomial theorem as well as the triple product identity. G. H. Hardy found this theorem without proof in Srinivasa Ramanujan's (1887–1920) notebooks and published them in his 1940 lectures on Ramanujan's work. This formula provides the basis for the study of bilateral q-series.

Ramanujan's theorem can be stated as

$$\sum_{n=-\infty}^{\infty} \frac{(a;q)_n}{(b;q)_n} x^n = \frac{(ax;q)_\infty (q/ax;q)_\infty (q;q)_\infty (b/a;q)_\infty}{(x;q)_\infty (b/ax;q)_\infty (b;q)_\infty (q/a;q)_\infty},$$

where $|q| < 1$ and $|b/a| < |x| < 1$.

29.12 Rodrigues and MacMahon: Combinatorics

Olinde Rodrigues and Percy Alexander MacMahon made important contributions to combinatorial problems connected with Gaussian polynomials and their generalizations. Olinde Rodrigues (1794–1851) was a French mathematician whose ancestors most probably left Spain, fleeing the persecution of the Jews. He studied at the Lycée Impérial in Paris and then at the new Université de Paris. He published six mathematical papers during 1813–16, one of which contains his well-known formula for Legendre polynomials. He did not pursue an academic career, perhaps because of religious discrimination. In fact, he apparently gave up mathematical research for over two decades, returning to it in 1838; he then produced papers on combinatorics and an important work on rotations.

Rodrigues's theorem from 1839 gave the generating function for the number of permutations $Z(n,k)$ of n distinct objects with k inversions; this was the number of permutations $a_1, a_2, \ldots, a_n$, of $1, 2, 3, \ldots, n$ with k pairs (a_i, a_j), such that $i < j$ and $a_i > a_j$. The values of k range from 0 to $n(n-1)/2$. To find the generating function of $Z(n,k)$, Rodrigues argued that $Z(n,k)$ was the number of integer solutions of the equation

$$x_0 + x_1 + x_2 + \cdots + x_{n-1} = k,$$

where $0 \leq x_i \leq i$ for $i = 0, 1, \ldots, n-1$. This implied that the $Z(n,k)$ was the coefficient of t^k in the product

$$(1+t)(1+t+t^2)(1+t+t^2+t^3)\cdots(1+t+t^2+\cdots+t^{n-1}).$$

As immediate corollaries, Rodrigues had

$$Z(n,0) + Z(n,1) + \cdots + Z(n,n-1) = n!,$$

$$Z(n,0) - Z(n,1) + Z(n,2) - \cdots + (-1)^{n-1} Z(n,n-1) = 0.$$

The first relation also answered a question posed by Stern on the sum of all the inversions in the permutations of n letters. Note that we can write Rodrigues's result as

$$\sum_{k=0}^{n(n-1)/2} Z(n,k) q^k = \frac{(1-q)(1-q^2)\cdots(1-q^n)}{(1-q)^n}. \tag{29.63}$$

Then, we see that the expression on the right-hand side is the q-extension of $n!$.

In 1913, MacMahon found another important way of classifying permutations by defining the greater index of a permutation. For a permutation $a_1, a_2, \ldots a_n$ of $1, 2, 3, \ldots n$, MacMahon defined the greater index to be the sum $\sum_{i=1}^{n-1} \lambda(a_i)$, where $\lambda(a_i) = i$ if $a_i > a_{i+1}$, and $\lambda(a_i) = 0$ otherwise. Let $G(n,k)$ denote the number of permutations for which the greater index is equal to k. MacMahon proved that

$$\sum_{k=0}^{n(n-1)/2} G(n,k) q^k = \frac{(1-q)(1-q^2) \cdots (1-q^n)}{(1-q)^n}. \qquad (29.64)$$

This immediately gave him the result

$$G(n,k) = Z(n,k). \qquad (29.65)$$

In fact, MacMahon proved his theorems even more generally, for permutations of multisets. In a multiset, the elements need not be distinct. For example, $1^{m_1} 2^{m_2} \ldots r^{m_r}$ denotes a multiset with m_1 ones, m_2 twos, and so on. The concepts of inversion and greater index can be extended in an obvious way to multisets. So if $Z(m_1, m_2, \ldots, m_r; k)$ and $G(m_1, m_2, \ldots m_r; k)$ denote the number of permutations with k inversions and the number of permutations with greater index k, then MacMahon had

$$\sum Z(m_1, m_2, \ldots, m_r; k) q^k = \sum G(m_1, m_2, \ldots, m_r; k) q^k$$
$$= \frac{(1-q)(1-q^2) \cdots (1-q^{m_1+m_2+\cdots+m_r})}{(1-q) \cdots (1-q^{m_1})(1-q) \cdots (1-q^{m_2}) \cdots (1-q) \cdots (1-q^{m_r})}. \qquad (29.66)$$

Note that when $r = 2$, the expression on the right is the Gaussian polynomial $(m_1 + m_2, m_1)$, in Gauss's notation. Just as the Gaussian polynomial is the q-binomial coefficient, we can see that (29.66) is the q-multinomial coefficient.

MacMahon (1854–1929) studied at the military academy at Woolwich. He became a lieutenant in 1872, captain in 1881, and major in 1889. He returned to Woolwich as an instructor in 1882. This teaching post, along with his friendship with the mathematician George Greenhill, set the scene for MacMahon to exercise his mathematical talents. Starting in the early 1880s, he contributed numerous important papers to the subject of combinatorics and related topics, including symmetric functions and invariants. He was also a fast arithmetical calculator and constructed a table of partitions of integers up through 200. By studying this table, Ramanujan was able to discover the arithmetic properties of the partition function. MacMahon's calculations played a crucial role in Ramanujan's research, influential even today.

29.13 Exercises

1. Prove Bernoulli's formulas (29.17) and (29.18) for the probabilities to win.
2. Prove Euler's first identity (29.19).
3. For (m, μ) defined by (29.27), prove Gauss's formulas (29.28), (29.29), and (29.30).

4. Following F. H. Jackson, set $[\alpha] = \frac{1-q^\alpha}{1-q}$, $[n]! = [1][2]\cdots[n]$, and

$$(1-x)^{(-\alpha)} = \sum_{k=1}^{\infty}\left((1-q^k x)/(1-q^{k+\alpha}x)\right).$$

Show that

$$(1-x)^{(-\alpha)} = 1 + \frac{[\alpha]}{[1]!}x + \frac{[\alpha][\alpha+1]}{[2]!}qx^2 + \frac{[\alpha][\alpha+1][\alpha+2]}{[3]!}q^3 x^3 + \cdots.$$

See Jackson (1910).

5. Let $u'(x) \equiv \Delta u(x) = \frac{u(x)-u(qx)}{x-qx}$ and $\Delta^{-1}u(x) \equiv \int u(x)d_q x$. Show that

(a) $\int (x)u'(x)d_q x = u(x)v(x) - \int u(qx)v'(x)d_q x$.

(b) (i) $\Delta(1-x)^{(n+1)} = [n+1](1-qx)^{(n)}$.

 (ii) $\int x^m (1-qx)^{(n)}d_q x = -\frac{x^m}{[n+1]}(1-x)^{(n+1)} + \frac{[m]}{[n+1]}\int x^{m-1}(1-qx)^{(n+1)}d_q x$.

 (iii) $\int_0^1 x^m (1-qx)^{(n)}d_q x = \frac{[m]}{[n+1]}\int_0^1 x^{m-1}(1-qx)^{(n+1)}d_q x$

$$= \frac{\Gamma_q(m+1)(\Gamma_q(n+1)}{\Gamma_q(m+n+2)} \equiv B_q(m+1, n+1).$$

 (iv) $\int_0^1 t^{\beta-1}(1-qt)^{(\gamma-\beta-1)}(1-q^\alpha tx)^\alpha d_q t$

$$= B_q(\beta, \gamma - \beta)\left(1 + \frac{(1-q^\alpha)(1-q^\beta)}{(1-q)(1-q^\gamma)}\right)x$$

$$+ \frac{(1-q^\alpha)(1-q^{\alpha+1})(1-q^\beta)(1-q^{\beta+1})}{(1-q)(1-q^2)(1-q^\gamma)(1-q^{\gamma+1})}x^2 + \cdots.$$

(c)

$$1 + \frac{(1-q^\alpha)(1-q^\beta)}{(1-q)(1-q^\gamma)}q^{\gamma-\alpha-\beta}$$

$$+ \frac{(1-q^\alpha)(1-q^{\alpha+1})(1-q^\beta)(1-q^{\beta+1})}{(1-q)(1-q^2)(1-q^\gamma)(1-q^{\gamma+1})}q^{2(\gamma-\alpha-\beta)} + \cdots$$

$$= \frac{B_q(\beta, \gamma - \alpha - \beta)}{B_q(\beta, \gamma - \beta)}.$$

(d) $\int_0^\infty \frac{t^{m-1}}{(1+qy)^{(l+m)}}d_q t = \frac{\Gamma_q(m)\Gamma_q(l)}{\Gamma_q(m+l)}$, provided $l + m$ is an integer. See Jackson (1910).

6. Prove Cauchy's formula

$$\frac{(ax;q)_\infty}{(bx;q)_\infty} = \sum_{n=0}^{\infty} \frac{(b-a)(bq-a)\cdots(bq^{n-1}-a)q^{\binom{n}{2}} x^n}{(q;q)_n (bx;q)_n}.$$

See Cauchy (1882–1974), vol. 8, series 1, pp. 42–50.

7. Prove Ramanujan's quintuple product identity

$$H(x) \equiv \prod_{n=1}^{\infty}(1-q^n)(1-xq^n)(1-q^{n-1}/x)(1-x^2 q^{2n-1})(1-q^{2n-1}/x^2)$$

$$= \sum_{n=-\infty}^{\infty} (x^{3n} - x^{-3n-1}) q^{n(3n+1)/2}.$$

One method of proof is to assume $H(x) = \sum_{n=-\infty}^{\infty} c(n) x^n$. Then compute $H(qx)/H(x)$ and $H(1/x)/H(x)$ to determine $c(n)$ in terms of $c(0)$. To find $c(0)$, specialize x. This formula was discovered several times. It is possible that Weierstrass was aware of it, since it follows from a three-term relation for sigma functions, a part of elliptic functions theory, presented by Weierstrass in his lectures. This formula appears explicitly in a 1916 book on elliptic functions by R. Fricke. Again, Ramanujan found it around that same time and made extensive use of it. In this exercise, we name the formula after Ramanujan. For a detailed history of the formula and several proofs, see Cooper (2006). Also see the remarks in Berndt (1985–98), Part III, p. 83.

8. Prove the septuple product identity of Farkas and Kra:

$$(1+x)(1-x)^2 \prod_{n=1}^{\infty}(1-q^n)^2(1-q^n x)(1-q^n/x)(1-q^n x^2)(1-q^n/x^2)$$

$$= \sum_{-\infty}^{\infty}(-1)^n q^{(5n^2+n)/2}\left(\sum_{-\infty}^{\infty}(-1)^n q^{(5n^2+3n)/2} x^{5n+3} + \sum_{-\infty}^{\infty}(-1)^n q^{(5n^2-3n)/2} x^{5n}\right)$$

$$- \sum_{-\infty}^{\infty}(-1)^n q^{(5n^2+n)/2}\left(\sum_{-\infty}^{\infty}(-1)^n q^{(5n^2+n)/2} x^{5n+2} + \sum_{-\infty}^{\infty}(-1)^n q^{(5n^2-n)/2} x^{5n+1}\right).$$

This result generalizes the quintuple product identity. For a proof, see Farkas and Kra (2001), p. 271.

29.14 Notes on the Literature

For Jakob Bernoulli's series arising from probability theory, see Bernoulli (2006), translated with an extensive introduction by Sylla, pp. 176–180. Note that Bernoulli presented four problems leading to theta series in his *Ars Conjectandi*, but two of these had appeared earlier, in his 1685 paper. Hald (1990) gives a translation of the problems contained in this paper. Euler published his first paper on the partition of integers in 1741; it can be found in Eu. I-2, pp. 163–193. He gave a longer treatment of this work in chapter 16 of his *Introductio*; see Euler (1988). Euler conjectured the pentagonal number theorem at the end of his 1741 paper, but published a proof nine years later; see Eu. I-2, pp. 254–294. See Fuss (1968), vol. I, pp. 522–524 for Euler's June 1750 letter to Goldbach, containing his proof.

The charming English translation of the excerpt from Gauss's letter to Olbers was taken from Bühler (1981), p. 31, and the diary translation is from Dunnington (2004), p. 481. See Gauss (1863–1927), vol. 3, pp. 437–439 for his theorem concerning triangular and square numbers and pp. 461–464 for his proof of the triple product identity. Remmert (1998), p. 29, gives the translated quotation from Legendre's letter, on Gauss's claim that he knew the triple product identity. For Gauss's evaluation of the quadratic Gauss sum, see Gauss (1981), pp. 467–481. Also see Patterson (2007). For more papers on related topics, see Goldstein, Schappacher, and Schwermer (2007). Jacobi first published his proof of the triple product identity in his *Fundamenta Nova* of 1829, laying the foundations of a new theory of elliptic functions; see Jacobi (1969), vol. 1, pp. 232–234. For his version of the q-binomial theorem and the extension of Gauss's summation, see Jacobi (1969), vol. 6, pp. 163–171. Eisenstein (1975), vol. 1, pp. 117–121 is a reprint of his paper on the q-binomial theorem.

See Rodrigues (1839) for inversions in permutations, and see MacMahon (1978), vol. 1, pp. 508–563 for his results. Also see Altmann and Ortiz (2005) for more information about Rodrigues. For recent developments and proofs connected with the triple product identity, see Andrews (1986), pp. 63–64, Foata and Han (2001), and Wilf (2001). Askey (1975) discusses the discovery of Chu on pp. 59–60.

30

Partitions

30.1 Preliminary Remarks

We have seen how the theory of partitions originated with Philip Naudé's problems to Euler, who used generating functions to solve them. Euler employed the same idea to prove the following remarkable theorem:

> The number of different ways a given number can be expressed as the sum of different whole numbers is the same as the number of ways in which that same number can be expressed as the sum of odd numbers, whether the same or different.

For example, the number of ways six can be expressed as a sum of different whole numbers is four:

$$6, \quad 5+1, \quad 4+2, \quad 3+2+1.$$

And six can be expressed as a sum of odd numbers in the following four ways:

$$5+1, \quad 3+3, \quad 3+1+1+1, \quad 1+1+1+1+1+1.$$

To prove this in general, Euler gave the generating function for the number of partitions with distinct parts:

$$(1+q)(1+q^2)(1+q^3)(1+q^4)(1+q^5)(1+q^6)\cdots.$$

Observe that 4 is the coefficient of q^6 in the power series expansion of this product, for q^6 can be obtained as $q^6, q^5 q, q^4 q^2$, and $q^3 q^2 q$. On the other hand, Euler noted that the generating function for odd parts was

$$\frac{1}{(1-q)(1-q^3)(1-q^5)\cdots} = (1+q+q^{1+1}+\cdots)(1+q^3+q^{3+3}+\cdots)(1+q^5+\cdots).$$

To prove the theorem, Euler showed that the generating functions and therefore the coefficients of their series expansions were identical:

$$(1+q)(1+q^2)(1+q^3)(1+q^4)\cdots = \frac{1-q^2}{1-q} \cdot \frac{1-q^4}{1-q^2} \cdot \frac{1-q^6}{1-q^3} \cdot \frac{1-q^8}{1-q^4}\cdots$$

$$= \frac{1}{1-q} \cdot \frac{1}{1-q^3} \cdot \frac{1}{1-q^5} \cdots . \tag{30.1}$$

We noted earlier that in 1742 Euler conjectured and eight years later he proved the pentagonal number theorem

$$\prod_{n=1}^{\infty}(1-q^n) = \sum_{n=-\infty}^{\infty} (-1)^n q^{n(3n-1)/2}. \tag{30.2}$$

Though Euler did not give a combinatorial interpretation of this identity, A. M. Legendre found one and included it in the 1830 edition of his number theory book. To understand Legendre's interpretation, consider how q^6 would arise in the power series expansion of the infinite product

$$(-q)(-q^2)(-q^3) = -q^6, \ (-q)(-q^5) = +q^6, \ (-q^2)(-q^4) = +q^6, \ (-q^6) = -q^6.$$

When the partition of 6 contains an odd number of parts (e.g., $6 = 1 + 2 + 3$) then a corresponding -1 is contributed to the coefficient of q^6 in the series. When the number of parts is even, then $+1$ is contributed. Hence the coefficient of q^6 in the series is 0. Thus, if we denote by $p_e(n)$, $p_0(n)$ the number of partitions of n with an even/odd number of distinct parts, then Legendre's theorem states that

$$p_e(n) - p_0(n) = (-1)^m, \quad \text{when} \quad n = m(3m \pm 1)/2,$$
$$= 0, \quad \text{when} \quad n \neq m(3m \pm 1)/2.$$

Even before Euler, Leibniz once enquired of Johann Bernoulli whether he had considered the problem of finding the number of partitions of a given number. Leibniz thought the problem was important and mentioned its connection with the number of monomial symmetric functions of a given degree. For example, the partitions of three, $3, 2 + 1, 1 + 1 + 1$, correspond to the symmetric functions $\sum a^3, \sum a^2 b, \sum abc$. However, it seems that neither Leibniz nor Bernoulli pursued this topic any further.

After Euler, J. J. Sylvester (1814–1897) was the next mathematician to make major contributions to the theory of partitions. Sylvester entered St. John's College, Cambridge, in 1833 and came out as Second Wrangler in 1837. The great applied mathematician George Green was fourth. Sylvester, of Jewish heritage, was unwilling to sign the thirty-nine articles; consequently, he was unable to take a degree, to obtain a fellowship, or to compete for one of the Smith's prizes. It was only in 1855 that he received a professorship of mathematics at the Royal Military Academy at Woolwich. Unfortunately, in 1870 he was retired early from this position, when his mathematical creativity was at its peak. In 1875, when Johns Hopkins University was founded in Baltimore, Sylvester was elected the first professor of mathematics (1876–83). Sylvester enjoyed a happy and productive late career in Baltimore; he there founded the *American Journal of Mathematics* whose first volume appeared in 1878. Moreover, Sylvester very successfully trained a number of excellent mathematicians, inaugurating serious mathematical research in America. It is not surprising that many of these American mathematicians contributed to the theory of partitions, since research in that topic required abundant ingenuity but minimal background.

Sylvester's interest in partitions arose fairly early. In 1853, he published a paper on his friend Cayley's quick method for determining the degree of a symmetric function expressed as a polynomial in elementary symmetric functions. For that purpose, Cayley had employed a result Sylvester attributed to Euler: "to wit, that the number of ways of breaking up a number n into parts is the same, whether we impose the condition that the number of parts in any partitionment shall not exceed m, or that the magnitude of any one of the parts should not exceed m." To understand this last result, consider that the generating function for the number of partitions of an integer into at most m parts, with each part $\leq n$, can be inductively demonstrated to be equal to the Gaussian polynomial

$$\frac{(1-q^{m+n})(1-q^{m+n-1})\cdots(1-q^{m+1})}{(1-q^n)(1-q^{n-1})\cdots(1-q)}.$$

This polynomial remains unchanged when m and n are interchanged; hence follows the result used by Sylvester. The Gaussian polynomial also cropped up in the work of Cayley and Sylvester in invariant theory. As we shall see in chapter 34, they related the coefficients of the polynomial to the number of independent seminvariants. Cayley and Sylvester took an interest in partitions as a result of their researches on invariants. Though they both contributed to partition theory, Sylvester made the subject his own domain by establishing fundamental ideas and producing new researchers, in the form of his students.

A graphical proof of Euler's theorem would start out by representing a given partition as a graph. For example, write the partition $5 + 2 + 1$ of eight as

$$\begin{matrix} \bullet & \bullet & \bullet & \bullet & \bullet \\ \bullet & \bullet & & & \\ \bullet & & & & \end{matrix}$$

and then enumerate by columns. Thus, one obtains the conjugate partition $3 + 2 + 1 + 1 + 1$ of eight. It is immediately clear that if we have a partition of an integer N into n parts of which the largest is m, then its conjugate is a partition of N into m parts of which the largest is n. This at once gives us the theorem: The number of partitions of any integer N into exactly n parts with the largest part m and the number of partitions of N into at most n parts with the largest part at most m both remain the same when m and n are interchanged. The proof of this theorem using the generating function method is less illuminating, illustrating the power of the graphical method. Sylvester remarked that he learned the technique from its originator, N. M. Ferrers. In a footnote to his paper, Sylvester wrote, "I learn from Mr Ferrers that this theorem was brought under his cognizance through a Cambridge examination paper set by Mr Adams of Neptune notability." Here Sylvester was referring to the astronomer John Couch Adams, discoverer of Neptune.

It was within this very concrete graphical method that Sylvester and his American students, including Fabian Franklin, William Durfee, and Arthur Hathaway, made their original and important contributions to the theory of partitions. It is interesting to note that the other significant results obtained by American mathematicians at around the same time were in abstract algebra. At that time, this too was a topic requiring a minimal amount of background knowledge, unlike subjects such as the theory of abelian functions. Early American results in abstract algebra included Benjamin Peirce's (1809–1880) paper on linear associative algebras dating from 1869, published posthumously in 1881 by his son Charles Saunders Peirce (1839–1914) in Sylvester's new journal. B. Peirce introduced the important concepts of nilpotent and idempotent elements and the paper starts with his famous dictum "Mathematics is the science which draws necessary conclusions." C. S. Peirce added an appendix to the paper, proving a significant theorem of his own on finite dimensional algebras over the real numbers. In modern language, the theorem states: The only division algebras algebraic over the real numbers are the fields of real and complex numbers and the division ring of quarternions.

The German mathematician G. Frobenius (1849–1917) also discovered this theorem at about the same time as Peirce, though he published it in 1877. The Frobenius-Peirce theorem and Franklin's beautiful proof of Euler's pentagonal number theorem are the earliest major contributions by Americans to mathematics. We shall see details of Franklin's work later in this chapter; concerning C. S. Peirce, we simply note that he made outstanding contributions to mathematical logic and to some aspects of philosophy. The systematic philosopher Justus Buchler, who edited Peirce's philosophical writings, stated in the introduction, "Even to the most unsympathetic, Peirce's thought cannot fail to convey something of lasting value. It has a peculiar property, like that of the Lernean hydra: discover a weak point, and two strong ones spring up beside it. Despite the elaborate architectonic planning of its creator, it is everywhere uncompleted, often distressingly so. There are many who have small regard for things uncompleted, and no doubt what they value is much to be valued. In his quest for magnificent array, in his design for a mighty temple that should house his ideas, Peirce failed. He succeeded only in advancing philosophy."

After the researches of Sylvester and his young American students, P. A. MacMahon (1854–1929) dominated the topic of partitions. One of MacMahon's results was connected with Ramanujan's 1910 rediscovery of two identities, first found by Rogers in the 1890s during his work on q-series:

$$\sum_{m=0}^{\infty} \frac{q^{m^2}}{(q;q)_m} = \prod_{m=0}^{\infty} \left(1 - q^{5m+1}\right)^{-1} \left(1 - q^{5m+4}\right)^{-1}, \qquad (30.3)$$

$$\sum_{m=0}^{\infty} \frac{q^{m(m+1)}}{(q;q)_m} = \prod_{m=0}^{\infty} \left(1 - q^{5m+2}\right)^{-1} \left(1 - q^{5m+3}\right)^{-1}. \qquad (30.4)$$

In 1913, Srinivasa Ramanujan communicated these identities to G. H. Hardy, although by this time Rogers's work was forgotten. Ramanujan had no proof; Hardy unsuccessfully sought a proof, showing the identities to his colleagues. MacMahon was among those who saw the formulas. An expert in symmetric functions, invariant theory, partitions, and combinatorics, he had known Sylvester and his work. Thus, it was natural that MacMahon conceived of an interpretation of the identities in terms of partitions.

30.1 Preliminary Remarks

By expanding $(1-q^{5m+1})^{-1}$ and $(1-q^{5m+4})^{-1}$ as geometric series, the coefficient of q^n in the expression on the right-hand side of the first identity is clearly equivalent to the number of partitions of n into parts $\equiv 1$ or $4 \pmod 5$. For the left-hand side, observe that

$$m^2 = (2m-1) + (2m-3) + \cdots + 5 + 3 + 1,$$

or the sum of the first m odd parts. We can find a partition of n if $n - m^2$ is partitioned into at most m parts with the largest part added to $2m-1$, the next to $2m-3$ and so on. The parts in this partition of n differ by at least 2. Moreover, the partitions of n associated with a specific m are enumerated by

$$\frac{q^{m^2}}{(1-q)(1-q^2)\cdots(1-q^m)},$$

and the sum of these terms yields all the partitions of this form. We therefore have MacMahon's theorem, presented in his 1915 *Combinatory Analysis*: The number of partitions of n in which the difference between any two parts is at least 2, equals the number of partitions of n into parts $\equiv 1$ or $4 \pmod 5$. We note that in MacMahon's own statement of the theorem, instead of specifying that the parts differ by at least 2, he wrote that there were neither repetitions nor sequences. In a similar way, the second identity states: The number of partitions of n in which the least part is ≥ 2 and the difference between any two parts is at least 2, is equal to the number of partitions of n into parts $\equiv 2$ or $3 \pmod 5$. This arises out of the relation $m(m+1) = 2 + 4 + 6 + \cdots + 2m$.

Several proofs of the Rogers–Ramanujan identities have been given and they have been generalized both combinatorially and analytically. Issai Schur independently discovered the Rogers–Ramanujan identities and their partition theoretic interpretation; in 1917 he gave two proofs, one of which was combinatorial. However, as Hardy wrote in 1940, it is only natural to seek an argument that sets up a one-to-one correspondence between the two sets of partitions. No such bijective proof was known in Hardy's time, and it was not until 1981 that Adriano Garsia and Stephen Milne, working on the foundation established by Schur, published a proof of the MacMahon–Schur theorem, equivalent to the Rogers–Ramanujan identities. We note that Schur's combinatorial proof also motivated Basil Gordon's 1961 partition-theoretic generalization. See the exercises.

Issai Schur (1875–1941) was born in Russia but studied at the University of Berlin under Georg Frobenius who had a great influence on him. Schur made fundamental contributions to representation theory, to the related theory of symmetric functions, and also to topics in analysis such as the theory of commutative differential operators. A great teacher, he founded an outstanding school of algebra in Berlin. Dismissed from his chair by the Nazi government, he took a position in 1938 at the Hebrew University in Jerusalem.

Garsia and Milne's bijective proof of the Rogers–Ramanujan identities is based on their involution principle: Let $C = C^+ \cup C^-$, where $C^+ \cap C^- = \phi$, be the disjoint union of two finite components C^+ and C^-. Let α and β be two involutions on C, each of whose fixed points lie in C^+. Let F_α (resp F_β) denote the fixed-point set of α (resp β). Suppose $\alpha(C^+ - F_\alpha) \subset C^-$ and $\alpha(C^-) \subset C^+$ and similarly $\beta(C^+ - F_\beta) \subset C^-$ and

$\beta(C^-) \subset C^+$. Then a cycle of the permutation $\Delta = \alpha\beta$ contains either fixed points of neither α nor β, or exactly one element of F_α and one of F_β. This powerful involution principle has been successfully applied to several q-series identities. Garsia and Milne's proof of Rogers–Ramanujan was very long but soon afterward David Bressoud and Doron Zeilberger found a shorter proof.

Now observe that in Euler's theorem the parts are distinct and hence differ by at least one, whereas in MacMahon's theorem the parts differ by at least two. If we denote by $q_{d,m}(n)$ the number of partitions of n into parts differing by at least d, each part being greater than or equal to m, the Euler and MacMahon theorems take the form

$$q_{d,m}(n) = p_{d,m}(n),$$

where $p_{d,m}(n)$ is the number of partitions of n into parts taken from a fixed set $S_{d,m}$. H. L. Alder observed that for $d = 1$, m could be taken to be any positive integer. In fact, the number of partitions of n into distinct parts, with each part $\geq m$, was equal to the number of partitions of n into parts taken from the set $\{m, m+1, \ldots, 2m-1, 2m+1, 2m+3, \ldots\}$.

In 1946 D. H. Lehmer proved for $m = 1$, and in 1948 Alder proved for the general case: The number $q_{d,m}(n)$ is not equal to the number of partitions of n into parts taken from any set of integers whatsoever unless $d = 1$ or $d = 2$, $m = 1, 2$. Now the generating function for $q_{d,m}(n)$ is easily seen to be

$$\sum_{k=0}^{\infty} \frac{q^{mk+dk(k-1)/2}}{(1-q)(1-q^2)\cdots(1-q^k)}, \tag{30.5}$$

while the generating for partitions with parts from a fixed set $\{a_1, a_2, a_3, \ldots\}$ is $1/\prod_{k=1}^{\infty}(1-q^{a_k})$. Alder's proof consisted in showing that no matter how the a_k were chosen, the two generating functions could not be equal for the values of m and d excluded by the theorem.

When MacMahon interpreted the Rogers–Ramanujan identity in terms of partitions, Hardy and Ramanujan may have been spurred to examine the asymptotic behavior of $p(n)$, the number of partitions of n. MacMahon assisted them in this work by constructing a table of $p(n)$ for $n = 1, 2, \ldots, 200$. We later consider the impact of this on the work of Hardy and Ramanujan. For now, we note that this table was created by means of Euler's formula

$$p(n) = p(n-1) + p(n-2) - p(n-5) - p(n-7) + \cdots$$
$$+ (-1)^{m-1} p\left(n - \frac{1}{2}m(3m-1)\right) - (-1)^{m-1} p\left(n - \frac{1}{2}m(3m+1)\right)\cdots. \tag{30.6}$$

Note that $p(k) = 0$ for k negative. This formula is quite efficient for numerical work. Ramanujan enjoyed numerical computation and could do it with unusual rapidity and accuracy. It is therefore interesting that in his obituary notice of Ramanujan, Hardy wrote, "There is a table of partitions at the end of our paper.... This was, for the most part, calculated independently by Ramanujan and Major MacMahon; and Major MacMahon was, in general, slightly the quicker and more accurate of the two."

30.1 Preliminary Remarks

J. E. Littlewood once remarked that every positive integer was one of Ramanujan's personal friends. Thus, Ramanujan noticed in the tables something missed by others, the arithmetical properties of partitions. In his 1919 paper on partitions he wrote,

> On studying the numbers in this table I observed a number of curious congruence properties, apparently satisfied by $p(n)$. Thus
>
> (1) $p(4), \ p(9), \ p(14), \ p(19), \ \ldots \quad \equiv 0 \pmod{5}$,
>
> (2) $p(5), \ p(12), \ p(19), \ p(26), \ \ldots \quad \equiv 0 \pmod{7}$,
>
> (3) $p(6), \ p(17), \ p(28), \ p(39), \ \ldots \quad \equiv 0 \pmod{11}$,
>
> (4) $p(24), \ p(49), \ p(74), \ p(99), \ \ldots \quad \equiv 0 \pmod{25}$,
>
> (5) $p(19), \ p(54), \ p(89), \ p(124), \ \ldots \quad \equiv 0 \pmod{35}$,
>
> (6) $p(47), \ p(96), \ p(145), \ p(194), \ \ldots \quad \equiv 0 \pmod{49}$,
>
> (7) $p(39), \ p(94), \ p(149), \ \ldots \quad \equiv 0 \pmod{55}$,
>
> (8) $p(61), \ p(138), \ \ldots \quad \equiv 0 \pmod{77}$,
>
> (9) $p(116), \ \ldots \quad \equiv 0 \pmod{121}$,
>
> (10) $p(99), \ \ldots \quad \equiv 0 \pmod{125}$.
>
> From these data I conjectured the truth of the following theorem: If $\delta = 5^a 7^b 11^c$ and $24\lambda \equiv 1 \pmod{\delta}$ then
>
> $$p(\lambda), p(\lambda+\delta), p(\lambda+2\delta), \ldots \ \equiv 0 \pmod{\delta}.$$

Ramanujan gave very simple proofs of $p(5m+4) \equiv 0 \pmod{5}$ and $p(7m+5) \equiv 0 \pmod{7}$, using only Euler's pentagonal number theorem and Jacobi's formula for $\prod_{n=1}^{\infty}(1-q^n)^3$. Ramanujan's further efforts, to prove $p(25m+24) \equiv 0 \pmod{25}$ and $p(49m+47) \equiv 0 \pmod{49}$, led him deeper into the theory of modular functions. In particular, he found the following two remarkable identities:

$$p(4) + p(9)q + p(14)q^2 + \cdots = 5 \frac{\{(1-q^5)(1-q^{10})(1-q^{15})\cdots\}^5}{\{(1-q)(1-q^2)(1-q^3)\cdots\}^6} \tag{30.7}$$

$$p(5) + p(12)q + p(19)q^2 + \cdots = 7\frac{\{(1-q^7)(1-q^{14})(1-q^{21})\cdots\}^3}{\{(1-q)(1-q^2)(1-q^3)\cdots\}^4}$$

$$+ 49q \frac{\{(1-q^7)(1-q^{14})(1-q^{21})\cdots\}^7}{\{(1-q)(1-q^2)(1-q^3)\cdots\}^8}. \tag{30.8}$$

The rest of Ramanujan's conjecture concerning the divisibility of the partition function by $5^a 7^b 11^c$ is not completely correct. In 1934, on the basis of the extended tables for $p(n)$ constructed by Hansraj Gupta, Sarvadaman Chowla observed that $p(243)$ was not divisible by 7^3, though $24 \cdot 243 \equiv 1 \pmod{7^3}$. However, $p(243)$ is divisible by 7^2. The correct reformulation of Ramanujan's conjecture would state: Let $\delta = 5^a 7^b 11^c$, $\delta' = 5^a 7^{b'} 11^c$, where $b' = b$, if $b = 0, 1, 2$, and $b' = \lfloor (b+2)/2 \rfloor$, if $b > 2$. If $24\lambda \equiv 1 \pmod{\delta}$, then

$$p(\lambda + n\delta) \equiv 0 \pmod{\delta'}, \quad n = 0, 1, 2, \ldots. \tag{30.9}$$

In an unpublished manuscript, Ramanujan outlined a proof of his conjecture for arbitrary powers of 5. He may have had a proof for the powers of 7 as well, since he apparently began writing it down. George N. Watson's proof of Ramanujan's conjecture for powers of 5 is identical with the one contained in the unpublished manuscript. Watson also gave a proof of the corrected version for the powers of 7. In 1967, A. O. L. Atkin provided a proof for powers of 11, based on work of Joseph Lehner from the 1940s. Atkin and Lehner's proofs require the use of modular equations, a topic in which Ramanujan was a great expert. It is remarkable that he was able to conjecture an essentially correct result on so little numerical evidence, especially in higher powers.

The fact that $p(5n+4) \equiv 0 \pmod{5}$ suggests that partitions of $5n+4$ should be divisible into five classes with the same number of partitions in each class. Freeman Dyson got this idea around 1940 when he was in high school; as a second year student at Cambridge University, he found a way of making this division. For this purpose, he defined the concept of the rank of a partition: the largest part minus the number of parts. He checked this concept, applying it to the three cases $p(4)$, $p(9)$, and $p(14)$ and found it accurate; he also found that it worked for $p(5)$ and $p(12)$. He conjectured its truth for all $p(5n+4)$ and for $p(7n+5)$, but was unable to prove it. A decade later, Atkin and Peter Swinnerton–Dyer found a proof involving combinatorial arguments combined with ideas from modular function theory. Mock theta functions also made an appearance; Atkin and Swinnerton–Dyer rediscovered and used a number of identities for mock theta functions. Unbeknownst to them and the rest of the world, these identities were contained in Ramanujan's lost notebook, buried under a mountain of paper on the floor of Watson's study.

The rank of a partition can be defined graphically as the signed difference between the number of nodes in the first row and number of nodes in the first column. Consider the ranks of the partitions of 5:

Partition	Rank	
5	$5-1 \equiv 4$	$\pmod{7}$
4+1	$4-2 \equiv 2$	$\pmod{7}$
3+2	$3-2 \equiv 1$	$\pmod{7}$
3+1+1	$3-3 \equiv 0$	$\pmod{7}$
2+2+1	$2-3 \equiv 6$	$\pmod{7}$
2+1+1+1	$2-4 \equiv 5$	$\pmod{7}$
1+1+1+1+1	$1-5 \equiv 3$	$\pmod{7}$.

Dyson found that the concept of rank failed to classify the partitions of $11n+6$; he conjectured the existence of a crank for this purpose. Almost half a century later, a day after the 1987 Centenary Conference at the University of Illinois, celebrating the work of Ramanujan, Andrews and Frank G. Garvan discovered the crank: The crank of a partition is the largest part in the partition if it has no ones; otherwise, it is the number of parts greater than the number of ones, minus the number of ones. A nice property of the crank is that it works for 5, 7, and 11. Amazingly, Ramanujan discovered the generating functions for both the rank and the crank, and his results can again be found in his lost notebook, though he did not refer to these concepts.

Concerning the congruence properties of partitions, Ramanujan wrote, "It appears that there are no equally simple properties for any moduli involving primes other than these three." As we shall see, Ramanujan's intuition has been shown to be correct. However, in the late 1960s, Atkin found some more complicated congruences involving other primes. For example, he showed that

$$p(11^3.13n + 237) \equiv 0 \pmod{13},$$

$$p(23^3.17n + 2623) \equiv 0 \pmod{17}.$$

Atkin used computers to do the numerical work necessary for constructing these examples. In fact, Atkin was among the pioneers in the use of computers for number theory research. Concerning this aspect of his work, he wrote, "it is often more difficult to discover results in this subject than to prove them, and an informed search on the machine may enable one to find out precisely what happens." Atkin's aim was to understand partition identities, including Ramanujan's, from the more general viewpoint of modular function theory. His student Margaret Ashworth (1944–73) shared this perspective, although her researches were halted much too soon. Thus, Atkin and Ashworth did not succeed in fully developing their approach. Atkin himself made important contributions to the theory of modular forms and in 1970, Atkin and Lehner conceived the fundamental idea of new forms. These are eigenforms for Hecke operators, on the space of cusp forms for Hecke subgroups of the modular group. In fact, it was only recently that Ken Ono developed a theory of the kind Atkin may have been seeking. In 2000, Ono was able to prove that for any prime $l \geq 5$, there exist infinitely many congruences of the form $p(An + B) \equiv 0 \pmod{l}$. Soon after this, Scott Ahlgren extended the congruence to the case in which l is replaced by l^k. Subsequently, Ono and Ahlgren jointly extended these results and wrote a historical essay explaining that their work "provides a theoretical framework which explains every known partition function congruence." Ono and Ahlgren based their work on results in modular forms from the 1960s and 1970s due to Goro Shimura, Jean-Pierre Serre, and Pierre Deligne.

Confirming another conjecture of Ramanujan, in 2003 Ahlgren and Matthew Boylan proved that if l is prime and $0 \leq \beta \leq l$ is any integer for which

$$p(ln + \beta) \equiv 0 \pmod{l} \quad \text{for all } n \geq 0,$$

$$\text{then} \quad (l, \beta) \epsilon \{(5, 4), (7, 5), (11, 6)\}.$$

We note that all these cases of simple congruence were found by Ramanujan; his intuition that no other cases exist has been verified. In 2005, Karl Mahlburg succeeded in extending the partition congruences to the crank function. Let $M(m, N, n)$ be the number of partitions of n whose rank equals $m \pmod{N}$. Mahlburg's theorem states that for every prime $l \geq 5$ and integer $i \geq 1$, there are infinitely many nonnested arithmetical progressions $An + B$ such that simultaneously for every $0 \leq m \leq l^j - 1$

$$M(m, l^j, An + B) \equiv 0 \pmod{l^i}.$$

It is clear from the definition of M that

$$p(n) = M(0, N, n) + M(1, N, n) + \cdots + M(N - 1, N, n).$$

Therefore, Mahlburg's theorem implies the corresponding result for $p(n)$.

MacMahon and Hardy greatly admired Ramanujan's generating function for $p(5n+4)$. A number of proofs of this and the generating function for $p(7n+5)$ have subsequently been found. A recent proof by Hershel Farkas and Irwin Kra is based on the theory of Riemann surfaces and theta functions. In the final year of his life, Ramanujan introduced a new type of series, mock theta functions. These q-series, convergent in $|q| < 1$, also have connections with the theory of partitions, although Ramanujan's motivation was to study their asymptotic properties as q approached a root of unity. Ramanujan noted that the asymptotic behavior of theta series such as

$$\sum_{n=0}^{\infty} q^{n^2}/((1-q)(1-q^2)\cdots(1-q^n))^2$$

and $\sum_{n=0}^{\infty} q^{n^2}/((1-q)(1-q^2)\cdots(1-q^n))$

could be expressed in a neat and closed exponential form as q approached roots of unity. He conceived mock theta functions as those series with similar asymptotic properties, without being theta functions. He gave seventeen examples of mock theta functions, dividing them into four groups, named mock theta functions of orders 3, 5, 5, and 7. One of the third-order functions he mentioned was defined by

$$f(q) = 1 + \frac{q}{(1+q)^2} + \frac{q^4}{(1+q)^2(1+q^2)^2} + \cdots.$$

He noted that when $q = -e^{-t}$ and $t \to 0$

$$f(q) + \sqrt{\frac{\pi}{t}} \exp\left(\frac{\pi^2}{24t} - \frac{t}{24}\right) \to 4.$$

Ramanujan also stated a few identities connecting some of these functions with each other. For example, he mentioned the third-order function

$$\chi(q) = 1 + \frac{q}{1-q+q^2} + \frac{q^4}{(1-q+q^2)(1-q^2+q^4)} + \cdots$$

and the relation $\quad 4\chi(q) - f(q) = \dfrac{(1 - 2q^3 + 2q^{12} - \cdots)^2}{(1-q)(1-q^2)(1-q^3)\cdots}.$

After Ramanujan, G. N. Watson (1886–1965) was the first to study these functions. The title of his 1936 paper on this topic, "The Final Problem: An Account of the Mock Theta Functions," was borrowed from an Arthur Conan Doyle story. In this paper, Watson introduced three new third-order functions, and proved identities such as

$$f(q) \prod_{n=1}^{\infty} (1-q^n) = 1 + 4 \sum_{n=1}^{\infty} \frac{(-1)^n q^{n(3n+1)/2}}{1+q^n}.$$

Watson employed the identities to show that the third-order mock theta functions had the asymptotic properties asserted by Ramanujan and that they were not theta functions.

A year later, Watson proved that the fifth-order functions listed by Ramanujan had the asymptotic properties; he did not succeed in showing that they were not theta functions. Watson's proofs of some of the identities were long, and he wrote that he counted the number of steps in the longest to be twenty-four instead of the thirty-nine he had hoped for as a student of John Buchan.

Watson's papers motivated Atle Selberg (1917–2007) to prove asymptotic formulas for seventh-order functions. Selberg had been drawn to a study of Ramanujan's work by a 1934 article by Carl Störmer in a periodical of the Norwegian Mathematical Society. The next year, Selberg started reading Ramanujan's *Collected Papers*. In 1987, he described his impressions: "So I got a chance to browse through it for several weeks. It seemed quite like a revelation – a completely new world to me, quite different from any mathematics book I had ever seen – with much more appeal to the imagination, I must say. And frankly, it still seems very exciting to me and also retains that air of mystery which I felt at the time. It was really what gave the impetus which started my own mathematical work. I began on my own, experimenting with what is often referred to as q-series and identities and playing around with them."

In the 1960s, Andrews began his extensive work on mock theta functions. His work was further facilitated by his 1976 discovery of Ramanujan's Lost Notebook in the Trinity College library of Cambridge University. For example, among myriad formulas in this notebook, Ramanujan gave ten identities for the fifth-order functions. In 1987, Andrews and Garvan showed that these ten identities could be reduced to two conjectures on partitions. To state these conjectures, let $R_a(n)$ denote the number of partitions of n with rank congruent to a (mod 5). The first conjecture stated that for every positive integer n, $R_1(5n) - R_0(5n)$ was equal to the number of partitions of n with unique smallest part and all other parts less than or equal to the double of the smallest part. The second stated that $2R_2(5n+3) - R_1(5n+3) - R_0(5n+3) - 1$ was equal to the number of partitions of n with unique smallest part and all other parts less than or equal to one plus the double of the smallest part. A year later Dean Hickerson proved these conjectures.

We mention in passing that as a byproduct of his work on mock theta functions, Andrews discovered the identity

$$\left(\sum_{n=0}^{\infty} q^{\binom{n+1}{2}}\right)^3 = \sum_{n=0}^{\infty} \sum_{j=0}^{2n} \frac{q^{2n^2+2n-\binom{j+1}{2}}(1+q^{2n})}{(1-q^{2n+1})}.$$

An immediate consequence of this formula is that every positive integer can be expressed as a sum of at most three triangular numbers. This theorem was first stated by Fermat, who said he had a proof. The first published proof appeared in Gauss's *Disquisitiones*.

Though mock theta functions were shown to have connections with several areas of mathematics, it was not clear how they fit into any known general framework. The work of Sander Zwegers, Don Zagier, Ken Ono, and Kathrin Bringmann, 2002–2007, has shown that Ramanujan's twenty-two mock theta functions are examples of infinite families of weak Maass forms of weight 1/2. This understanding has led to further new results.

30.2 Sylvester on Partitions

In 1882, Sylvester collected together the investigations he and his students had done on partitions dating from 1877–1882 and published them in his newly founded journal as a long paper, "A Constructive Theory of Partitions, Arranged in Three Acts, an Interact and an Exodion." He presented Franklin's proof of Euler's pentagonal number theorem (30.2). Sylvester placed the smallest part at the top of his graphical representation. We present the proof in his own words, illustrating his habit of using periods very sparingly.

If a regular graph represent a partition with unequal elements, the lines of magnitude must continually increase or decrease. Let the annexed figures be such graphs written in ascending order from above downwards:

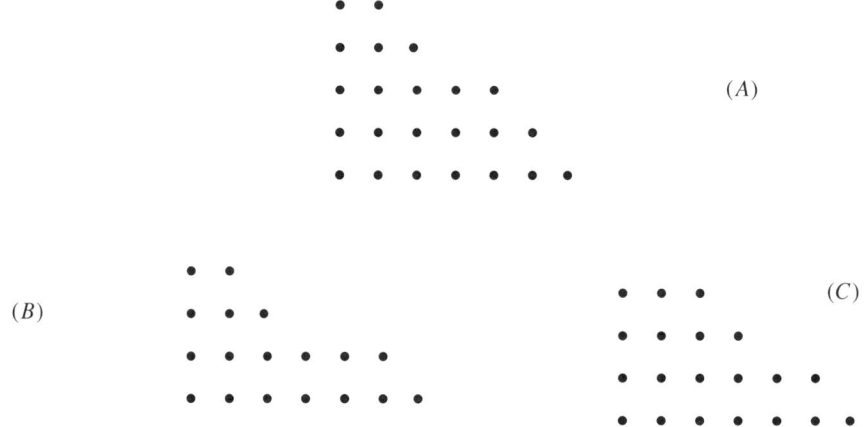

In (A) and (B) the graphs may be transformed without altering their content or regularity by removing the nodes at the summit and substituting for them a new slope line at the base. In C the new slope line at the base may be removed and made to form a new summit; the graphs so transformed will be as follows:

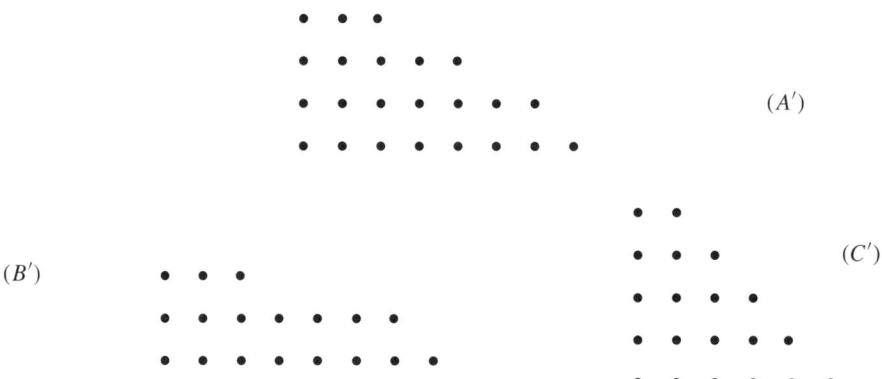

30.2 Sylvester on Partitions

A' and B' may be said to be derived from A, B by a process of contraction, and C' from C by one of protraction.

Contraction could not now be applied to A' and B', nor protraction to C' without destroying the regularity of the graph; but the inverse processes may of course be applied, namely, of protraction to A' and B' and contraction to C', so as to bring back the original graph A, B, C.

In general (but as will be seen not universally), it is obvious that when the number of nodes in the summit is inferior or equal to the number in the base-slope, contraction may be applied, and when superior to that number, protraction: each process alike will alter the number of parts from even to odd or from odd to even, so that barring the exceptional cases which remain to be considered where neither protraction nor contraction is feasible, there will be a one-to-one correspondence between the partitions of n into an odd number and the partitions of n into an even number of unrepeated parts; the exceptional cases are those shown below where the summit meets the base-slope line, and contains either the same number or one more than the number of nodes in that line; in which case neither protraction nor contraction will be possible, as seen in the annexed figures which are written in regular order of succession, but may be indefinitely continued:

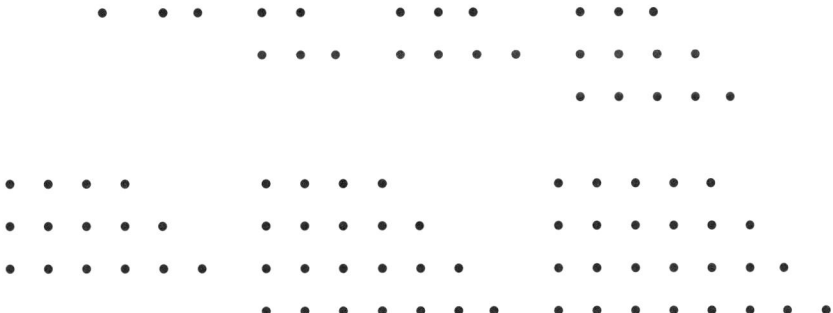

for the protraction process which *ought*, for example, according to the general rule, to be applicable to the last of the above graphs, cannot be applied to it, because on removing the nodes in the slope line and laying them on the summit, in the very act of so doing the summit undergoes the loss of a node and is thereby incapacitated to be surmounted by the nodes in the slope, which will have not now a less, but the same number of nodes as itself; and in like manner, in the last graph but one, the nodes in the summit cannot be removed and a slope line be added on containing the same number of nodes without the transformed graph ceasing to be regular, in fact it would take the form

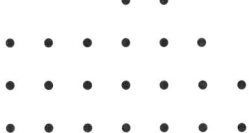

and so the last graph transformed according to rule [by protraction] would become:

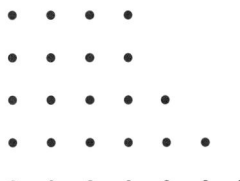

which, although regular, would cease to represent a partition into unlike numbers. The excepted cases then or unconjugate partitions are those where the number of parts being j, the successive parts form one or the other of the two arithmetical series

$$j, j+1, j+2, \ldots, 2j-1 \quad \text{or} \quad j+1, j+2, \ldots, 2j,$$

in which cases the contents are $\frac{3j^2-j}{2}$ and $\frac{3j^2+j}{2}$ respectively, and consequently since in the product of $1-x \cdot 1-x^2 \cdot 1-x^3 \cdots$ the coefficient of x^n is the number of ways of composing n with an even less the number of ways of composing it with an odd number of parts, the product will be completely represented by $\sum_{j=-\infty}^{\infty} (-1)^j x^{\frac{3j^2+j}{2}}$.

Sylvester's student Durfee introduced the important concept of the Durfee square for the purpose of studying self-conjugate graphs. These graphs remain unchanged when rows of nodes are changed to columns. Sylvester gave the partition of $27 = 7+7+4+3+2+2+2$ as an example; it has the self-conjugate graph as shown below.

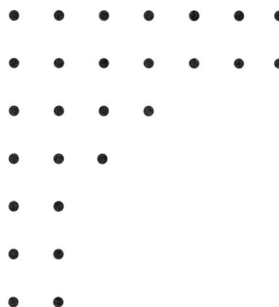

Note that the largest square in this graph is of size 3×3 in the upper left corner and the remaining nodes form two graphs with nine nodes each, partitioned into identical partitions, $3+2+2+2$, provided the nodes on the right-hand side of the square are read column-wise. The number of partitions of 9 in which the largest part is at most 3 is the coefficient of

$$x^9 \quad \text{in} \quad \frac{1}{(1-x)(1-x^2)(1-x^3)},$$

and this is the same as the coefficient of

$$x^{18} \quad \text{in} \quad \frac{1}{(1-x^2)(1-x^4)(1-x^6)}.$$

Sylvester applied this analysis to find the number of self-conjugate partitions of n. He considered all the partitions that could be dissected into a square of size m^2. The number of such partitions would be the coefficient of

$$x^{n-m^2} \quad \text{in} \quad \frac{1}{(1-x^2)(1-x^4)\cdots(1-x^{2m})}$$

or the coefficient of

$$x^n \quad \text{in} \quad \frac{x^{m^2}}{(1-x^2)(1-x^4)\cdots(1-x^{2m})}.$$

30.2 Sylvester on Partitions

Thus, the number of self-conjugate partitions of n was the coefficient of x^n in the series

$$1 + \frac{x}{1-x^2} + \frac{x^4}{(1-x^2)(1-x^4)} + \frac{x^9}{(1-x^2)(1-x^4)(1-x^6)} + \cdots.$$

There is yet another manner in which a self-conjugate partition can be dissected: by counting the number of nodes in the m angles or bends, as Sylvester called them. Thus, for the self-conjugate partitions of 27, there are three bends. The outermost right angle has thirteen nodes; the second has eleven; and the third three. It is easy to see that the number of nodes in each right angle of a self-conjugate partition will always be an odd number. Moreover, different right angles in the same partition will have different numbers of nodes. Thus, the number of self-conjugate partitions of n will be the coefficient of x^n in $(1+x)(1+x^3)(1+x^5)\cdots$. Therefore

$$\prod_{n=0}^{\infty}(1+x^{2n+1}) = \sum_{n=0}^{\infty} \frac{x^{n^2}}{(1-x^2)(1-x^4)\cdots(1-x^{2n})}.$$

Sylvester generalized this analysis of self-conjugate partitions by introducing an additional parameter a, whose exponent registered the number of parts in a partition, concluding that the coefficient of $x^n a^j$ in $(1+ax)(1+ax^3)\cdots(1+ax^{2j-1})\cdots$ was the same as in

$$\frac{x^{j^2} a^j}{(1-x^2)(1-x^4)\cdots(1-x^{2j})}.$$

Thus, he had Euler's formula

$$\prod_{n=0}^{\infty}(1+ax^{2n+1}) = \sum_{n=0}^{\infty} \frac{x^{n^2} a^n}{(1-x^2)(1-x^4)\cdots(1-x^{2n})},$$

but by a combinatorial argument. By means of a Durfee square analysis, Sylvester also obtained the identity needed by Jacobi to complete his proof of the triple product identity. Thus, in Jacobi's formula

$$\prod_{n=1}^{\infty} \frac{1}{(1-aq^n)} = \sum_{m=0}^{\infty} \frac{q^{m^2} a^m}{(1-q)(1-q^2)\cdots(1-q^m)} \cdot \frac{1}{(1-aq)(1-aq^2)\cdots(1-aq^m)},$$

the factor

$$\frac{q^{m^2} a^m}{(1-q)(1-q^2)\cdots(1-q^m)}$$

accounted for the square and the nodes to the right of it, while

$$\frac{1}{(1-aq)(1-aq^2)\cdots(1-aq^m)}$$

did the same for the nodes and the number of rows below the square. It is an interesting and instructive exercise to work out the details. Sylvester demonstrated that graphical analysis could also be used as a tool for the discovery of new identities. As an example, he presented

$$\sum_{n=0}^{\infty} \frac{(1+ax)(1+ax^2)\cdots(1+ax^{n-1})(1+ax^{2n})a^n x^{n(3n-1)/2}}{(1-x)(1-x^2)\cdots(1-x^n)} = \prod_{m=1}^{\infty}(1+ax^n). \quad (30.10)$$

Briefly, Sylvester considered all partitions with distinct parts to account for the product on the right-hand side. To obtain the series on the left-hand side, he considered a graph of an arbitrary partition of n with distinct parts. He supposed that the Durfee square (the largest square of nodes in the upper left corner) had θ^2 nodes. Again, there were two subgraphs, called by Sylvester appendages: one to the right of the square with either θ or $\theta-1$ rows and with unrepeated parts; and one below the square with $j-\theta$ rows and with unrepeated parts. Moreover, since the parts were distinct, Sylvester observed that the subgraph below the square had the largest part, at most θ or $\theta-1$, depending on whether the subgraph to the right had θ or $\theta-1$ rows. In the first case, because $1+2+\cdots+\theta = \theta(\theta+1)/2$, the number of distributions was the coefficient of $x^{n-\theta^2}a^{j-\theta}$ in

$$\frac{x^{(\theta^2+\theta)/2}}{(1-x)(1-x^2)\cdots(1-x^\theta)} \cdots (1+ax)(1+ax^2)\cdots(1+ax^\theta);$$

in the second case, it was the coefficient of $x^{n-\theta^2}a^{j-\theta}$ in

$$\frac{x^{(\theta^2-\theta)/2}}{(1-x)(1-x^2)\cdots(1-x^{\theta-1})} \cdot (1+ax)(1+ax^2)\cdots(1+ax^{\theta-1}).$$

By adding these two expressions, Sylvester obtained the θth term of his series and this proved the formula. Note that Euler's pentagonal number theorem follows from Sylvester's formula when one takes $a = -1$. Sylvester commented, "Such is one of the fruits among a multitude arising out of Mr. Durfee's ever-memorable example of the dissection of a graph (in the case of a symmetrical one) into a square, and two regular graph appendages."

30.3 Cayley: Sylvester's Formula

In 1882, Sylvester's mathematical correspondent and comrade, Cayley, responded to his friend's great paper on partitions by showing how the "very beautiful formula" (30.10) could be proved by an interesting analytic method. He expressed the series side of the formula as

$$\Omega = 1 + P + Q(1+ax) + R(1+ax)(1+ax^2)$$
$$+ S(1+ax)(1+ax^2)(1+ax^3) + \cdots \quad (30.11)$$

where
$$P = \frac{(1+ax^2)xa}{\mathbf{1}}, \quad Q = \frac{(1+ax^4)x^5a^2}{\mathbf{1\cdot 2}},$$
$$R = \frac{(1+ax^6)x^{12}a^3}{\mathbf{1\cdot 2\cdot 3}}, \quad S = \frac{(1+ax^8)x^{22}a^4}{\mathbf{1\cdot 2\cdot 3\cdot 4}}, \text{ etc.} \quad (30.12)$$

where the numbers in bold $\mathbf{1, 2, 3, 4, \ldots}$ denoted $1-x, 1-x^2, 1-x^3, 1-x^4, \ldots$. Cayley observed that the x exponents $1, 5, 12, 22, \ldots$ were the pentagonal numbers $(3n^2-n)/2$. Cayley then set

$$P' = \frac{ax^2}{\mathbf{1}}, \quad Q' = \frac{ax^3}{\mathbf{1}} + \frac{a^2x^7}{\mathbf{1\cdot 2}}, \quad R' = \frac{ax^4}{\mathbf{1}} + \frac{a^2x^9}{\mathbf{1\cdot 2}} + \frac{a^3x^{15}}{\mathbf{1\cdot 2\cdot 3}}, \text{ etc.}$$

where the x exponents were

$$2; \ 3, 3+4; \ 4, 4+5, 4+5+6; \quad \text{etc.}$$

He then noted that it was easily verified that

$$1 + P = (1+ax)(1+P'),$$
$$1 + P' + Q = (1+ax^2)(1+Q'),$$
$$1 + Q' + R = (1+ax^3)(1+R'),$$
$$1 + R' + S = (1+ax^4)(1+S'), \quad \text{etc.}$$

Cayley concluded from these relations that

$$\Omega \div (1+ax) = 1 + P' + Q + R(1+ax^2) + S(1+ax^2)(1+ax^3) + \cdots,$$
$$\Omega \div (1+ax)(1+ax^2) = 1 + Q' + R + S(1+ax^3) + T(1+ax^3)(1+ax^4) + \cdots,$$
$$\Omega \div (1+ax)(1+ax^2)(1+ax^3) = 1 + R' + S + T(1+ax^4) + \cdots,$$

and so on. When this was done infinitely often, the right-hand side would become 1, giving Cayley the complete proof of Sylvester's theorem.

Andrews pointed out that Cayley's method is more easily understood in terms of Euler's method of proving the pentagonal number theorem. In Sylvester's series, take the factor $1 + aq^{2n}$ in the $(n+1)$th term, counting 1 as the first term, and split it into two parts:
$$1 + ax^{2n} = 1 - x^n + x^n(1+ax^n). \quad (30.13)$$

Interestingly, this breaks the $(n+1)$th term into two parts, so that one can associate the second part of the term with the first part of the succeeding term. The result of this association is

$$\frac{(1+ax)(1+ax^2)\cdots(1+ax^{n-1})(1+ax^n)x^n a^n x^{n(3n-1)/2}}{(1-x)(1-x^2)\cdots(1-x^n)}$$
$$+ \frac{(1+ax)(1+ax^2)\cdots(1+ax^n)a^{n+1}x^{(n+1)(3n+2)/2}}{(1-x)(1-x^2)\cdots(1-x^n)}$$

$$= (1+ax) \cdot \frac{(1+ax^2)\cdots(1+ax^n)}{(1-x)\cdots(1-x^n)} \cdot a^n x^{n(3n+1)/2}(1+ax^{2n+1}). \tag{30.14}$$

Now observe that the factor multiplying $1 + ax$ is the $(n + 1)$th term of Sylvester's series except that a has been replaced by ax. So, if we denote Sylvester's series by $f(a)$, then by (30.14) we have

$$\begin{aligned} f(a) &= (1+ax)f(ax) \\ &= (1+ax)(1+ax^2)f(ax^2) \\ &= (1+ax)(1+ax^2)(1+ax^3)\cdots. \end{aligned}$$

Note that for convergence we would require $|x| < 1$, and therefore $x^n \to 0$ as $n \to \infty$. This implies

$$\lim_{n \to \infty} f(ax^n) = f(0) = 1.$$

We note that Gauss also used this method on various occasions and that it is possible that Cayley rediscovered Euler's method. We shall see in the next section how Ramanujan made brilliant use of this technique to prove the Rogers–Ramanujan identities.

30.4 Ramanujan: Rogers–Ramanujan Identities

Ramanujan discovered a new proof of the Rogers–Ramanujan identity after he saw Rogers's original proof, presented chapter 31. Ramanujan communicated the proof to Hardy in a letter of April 1919 and Hardy had it published in a 1919 paper. In this paper, Hardy also included another proof, sent by Rogers to MacMahon in October 1917. Ramanujan's proof started with a series very similar to Sylvester's series (30.10):

$$G(x) = 1 + \sum_{n=1}^{\infty} (-1)^n x^{2n} q^{n(5n-1)}(1-xq^{2n}) \frac{(1-xq)(1-xq^2)\cdots(1-xq^{n-1})}{(1-q)(1-q^2)\cdots(1-q^n)}. \tag{30.15}$$

He split this series into two parts, exactly as Cayley had done with Sylvester's series, by applying

$$1 - xq^{2n} = 1 - q^n + q^n(1 - xq^n)$$

to transform (30.15) into

$$G(x) = \sum_{n=1}^{\infty} x^{2n-2} q^{(5n^2-9n+4)/2}(1 - x^2 q^{2(2n-1)}) \frac{(1-xq)\cdots(1-xq^{n-1})}{(1-q)\cdots(1-q^{n-1})}, \tag{30.16}$$

where the empty product when $n = 1$ was set equal to 1. Ramanujan then set

$$H(x) = \frac{G(x)}{1 - xq} - G(xq)$$

30.4 Ramanujan: Rogers–Ramanujan Identities

and used the value of $G(x)$ from (30.16) and of $G(xq)$ from (30.15) to obtain

$$H(x) = xq - \frac{x^2 q^3}{1-q}\left((1-q) + xq^4(1-xq^2)\right)$$
$$+ \frac{x^4 q^{11}(1-xq^2)}{(1-q)(1-q^2)}\left((1-q^2) + xq^7(1-xq^3)\right)$$
$$- \frac{x^6 q^{24}(1-xq^2)(1-xq^3)}{(1-q)(1-q^2)(1-q^3)}\left((1-q^3) + xq^{10}(1-xq^4)\right) + \cdots.$$

Again, as in Cayley's argument, Ramanujan associated the second part of each term with the first part of the succeeding term to arrive at the relation

$$H(x) = xq(1-xq^2)G(xq^2) \quad \text{or}$$
$$G(x) = (1-xq)G(xq) + xq(1-xq)(1-xq^2)G(xq^2).$$

Setting

$$F(x) = \frac{G(x)}{(1-xq)(1-xq^2)(1-xq^3)\cdots},$$

Ramanujan obtained the relation

$$F(x) = F(xq) + xq F(xq^2). \tag{30.17}$$

He observed that it readily followed that

$$F(x) = 1 + \frac{xq}{1-q} + \frac{x^2 q^4}{(1-q)(1-q^2)} + \frac{x^3 q^9}{(1-q)(1-q^2)(1-q^3)} + \cdots. \tag{30.18}$$

Note that we can derive this last equation from

$$F(x) = 1 + A_1(q)x + A_2(q)x^2 + A_3(q)x^3 + \cdots.$$

Applying (30.17), he obtained

$$A_n(q) = \frac{q^{2n-1}}{1-q^n} A_{n-1}(q), \quad n = 1, 2, 3, \ldots,$$

where $A_0(q) = 1$. This implied (30.18). Ramanujan obtained the required identities by taking $x = 1$ and $x = q$ in (30.18). For $x = 1$, he got

$$1 + \frac{q}{1-q} + \frac{q^4}{(1-q)(1-q^2)} + \cdots = \frac{G(1)}{(1-q)(1-q^2)(1-q^3)\cdots}$$
$$= \frac{1 - q^2 - q^3 + q^9 + q^{11} - \cdots}{(1-q)(1-q^2)(1-q^3)\cdots}.$$

The series can be converted to a product by the triple product identity and the result follows.

30.5 Ramanujan's Congruence Properties of Partitions

Ramanujan was the first mathematician to study the divisibility properties of the partition function. In a paper published in 1919, he gave fairly simple proofs of the congruence relations $p(5m+4) \equiv 0 \pmod{5}$ and $p(7m+5) \equiv 0 \pmod{7}$. He started with Euler's generating function for $p(n)$,

$$\frac{q}{(1-q)(1-q^2)(1-q^3)\cdots} = \sum_{n=1}^{\infty} p(n-1)q^n, \tag{30.19}$$

and observed that the first congruence would follow if the coefficient of q^{5m} on the right-hand side were divisible by 5. Thus, it was sufficient to show that the same was true for the coefficient of q^{5m} in

$$\frac{q(1-q^5)(1-q^{10})(1-q^{15})\cdots}{(1-q)(1-q^2)(1-q^3)\cdots} = \frac{q(1-q^5)(1-q^{10})\cdots}{\{(1-q)(1-q^2)\cdots\}^5}\{(1-q)(1-q^2)\cdots\}^4.$$

Ramanujan then noted that $1-q^{5m} \equiv (1-q^m)^5 \pmod{5}$ and hence

$$\frac{(1-q^5)(1-q^{10})(1-q^{15})\cdots}{\{(1-q)(1-q^2)(1-q^3)\cdots\}^5} \equiv 1 \pmod{5}.$$

Thus, to prove that $p(5m+4) \equiv 0 \pmod{5}$, it was enough to show that the coefficient of q^{5m} in

$$q\{(1-q)(1-q^2)(1-q^3)\cdots\}^4$$
$$= q\{(1-q)(1-q^2)(1-q^3)\cdots\}^3(1-q)(1-q^2)(1-q^3)\cdots$$
$$= q\sum_{m=0}^{\infty}(2m+1)q^{m(m+1)/2}\sum_{n=-\infty}^{\infty}(-1)^n q^{n(3n+1)/2}$$

was divisible by 5. Observe that in the last step, Ramanujan used Jacobi's identity and Euler's pentagonal number theorem. Jacobi's identity can be derived from the triple product identity. He then noted that the exponent of q in the double sum was divisible by 5 when

$$1 + \frac{m(m+1)}{2} + \frac{n(3n+1)}{2} \equiv 0 \pmod{5} \quad \text{or}$$
$$8 + 4m(m+1) + 4n(3n+1) \equiv 0 \pmod{5} \quad \text{or}$$
$$(2m+1)^2 + 2(n+1)^2 \equiv 0 \pmod{5}.$$

Ramanujan noted that $(2m+1)^2 \equiv 0, 1, 4 \pmod{5}$ and $2(n+1)^2 \equiv 0, 2, 3 \pmod{5}$. So when the exponent of q was a multiple of 5, then $2m+1 \equiv 0 \pmod{5}$ and $n+1 \equiv 0 \pmod{5}$. Since the coefficient of this power of q was $2m+1$, a multiple of 5, Ramanujan's proof was complete. He gave a similar proof for the congruence modulo 7; see the exercises.

In the same paper, Ramanujan outlined a proof of (30.7). He intended to publish more details at a later date, but his premature death made this impossible. However, he wrote

30.5 Ramanujan's Congruence Properties of Partitions

notes giving these details, later found and published as part of in his lost notebook. In fact, this proof used only the pentagonal number theorem, Jacobi's identity

$$\prod_{n=1}^{\infty}(1-q^n)^3 = \sum_{m=0}^{\infty}(-1)^m(2m+1)q^{m(m+1)/2} \qquad (30.20)$$

and fifth roots of unity. By the pentagonal number theorem:

$$\prod_{n=1}^{\infty}(1-q^{n/5}) = \sum_{n=-\infty}^{\infty}(-1)^n q^{n(3n+1)/10}. \qquad (30.21)$$

Partition the series into five parts according to whether $n \equiv 0, \pm 1, \pm 2 \pmod 5$. For example, for $n \equiv 0 \pmod 5$, we have $n = 5m$ and the part of the series corresponding to these values of n would be given by

$$\sum_{m=-\infty}^{\infty}(-1)^m q^{m(15m+1)/2}.$$

Note that the subseries corresponding to $n = 5m - 1$ can once again be expressed a product:

$$-\sum_{m=-\infty}^{\infty}(-1)^m q^{(5m-1)(15m-2)/10} = -q^{1/5}\sum_{m=-\infty}^{\infty}(-1)^m q^{5m(3m-1)/2}$$

$$= -q^{1/5}\prod_{n=1}^{\infty}(1-q^{5n}).$$

Thus, (30.21) can be written as

$$\prod_{n=1}^{\infty}(1-q^{n/5}) = \sum_{m=-\infty}^{\infty}(-1)^m q^{m(15m+1)/2} + \sum_{m=-\infty}^{\infty}(-1)^m q^{(3m-1)(5m-2)/2}$$

$$+ q^{2/5}\left(\sum_{m=-\infty}^{\infty}(-1)^m q^{(3m+2)(5m+1)/2} - \sum_{m=-\infty}^{\infty}(-1)^m q^{m(15m+7)/2}\right)$$

$$- q^{1/5}\prod_{n=1}^{\infty}(1-q^{5n}).$$

Dividing by $\prod_{n=1}^{\infty}(1-q^{5n})$ gives

$$\prod_{n=1}^{\infty}\left(\frac{1-q^{n/5}}{1-q^{5n}}\right) = \xi_1 - q^{1/5} - \xi q^{2/5}, \qquad (30.22)$$

where ξ and ξ_1 are power series in q. Ramanujan applied Jacobi's identity to show that $\xi\xi_1 = 1$. So cube both sides of (30.22) and use Jacobi's identity (30.20) to get

$$\frac{\sum_{n=0}^{\infty}(-1)^n(2n+1)q^{n(n+1)/10}}{\prod_{n=1}^{\infty}(1-q^{5n})^3} = \left(\xi_1 - q^{1/5} - \xi q^{2/5}\right)^3. \qquad (30.23)$$

Since the exponent of q, given by $n(n+1)$, is either 0, 2, or 6 (mod 10), it follows that no power of q is of the form $2/5+$ an integer. This implies that the term

$$3q^{2/5}\xi_1 - 3\xi_1^2 \xi q^{2/5} = 3q^{2/5}\xi_1(1-\xi\xi_1)$$

on the right-hand side of (30.23) must be zero. This implies that $\xi_1 = \xi^{-1}$, and we can write

$$\prod_{n=1}^{\infty}\left(\frac{1-q^{5n}}{1-q^{n/5}}\right) = \frac{1}{\xi^{-1} - q^{1/5} - \xi q^{2/5}}. \tag{30.24}$$

Consider the expression $\lambda^{-1} - \lambda - 1$, where $\lambda = \xi q^{1/5} w$, and w is a fifth root of unity. Observe that if $\lambda^{-1} - \lambda = 1$, then by an elementary calculation $\lambda^{-5} - \lambda^5 = 11$. Thus,

$$\xi^{-5} - 11q - \xi^5 q^2 = \prod_{k=0}^{4}\left(\xi^{-1} - q^{1/5}w^k - \xi q^{2/5}w^{2k}\right).$$

It is now easy to check by long division that

$$\prod_{n=1}^{\infty}\left(\frac{1-q^{5n}}{1-q^{n/5}}\right) =$$

$$\frac{\xi^{-4} - 3q\xi + q^{1/5}(\xi^{-3} + 2q\xi^2) + q^{2/5}(2\xi^{-2} - q\xi) + q^{3/5}(3\xi^{-1} + q\xi^4) + 5q^{4/5}}{\xi^{-5} - 11q - q^2\xi^5}.$$

Now multiply across by $q^{1/5}$ and replace $q^{1/5}$ by $q^{1/5}e^{2\pi i k/5}$, $k=1,2,3,4$ to obtain five identities. Next, apply

$$q^{1/5}\prod_{n=1}^{\infty}(1-q^{n/5})^{-1} = \sum_{n=1}^{\infty} p(n-1)q^{n/5}$$

and add the five identities to get

$$\prod_{n=1}^{\infty}(1-q^{5n})\sum_{n=0}^{\infty} p(5n+4)q^n = \frac{5}{\xi^{-5} - 11q - q^2\xi^5}.$$

By replacing $q^{1/5}$ by $e^{2\pi i k/5}q^{1/5}$, $k=0,1,2,3,4$ and multiplying the five equations together, Ramanujan arrived at

$$\prod_{n=1}^{\infty}\left(\frac{1-q^{5n}}{1-q^n}\right)^6 = \frac{1}{\xi^{-5} - 11q - q^2\xi^5}.$$

Combining this with the previous equation gives the necessary result.

In his paper, Ramanujan also noted that

$$\xi^{-1} = \prod_{n=0}^{\infty} \frac{(1-q^{5n+2})(1-q^{5n+3})}{(1-q^{5n+1})(1-q^{5n+4})}. \tag{30.25}$$

This can be proved by using (30.24), the pentagonal number theorem and the quintuple product identity. Ramanujan observed in his paper that (30.7) implied that $p(25m+24)$ was divisible by 25. He argued that by (30.7),

$$\frac{p(4)x + p(9)x^2 + p(14)x^3 + \cdots}{5((1-x^5)(1-x^{10})(1-x^{15})\cdots)^4}$$

$$= \frac{x}{(1-x)(1-x^2)(1-x^3)\cdots} \frac{(1-x^5)(1-x^{10})(1-x^{15})\cdots}{((1-x)(1-x^2)((1-x^3)\cdots)^5},$$

and since the coefficient of x^{5n} on the right-hand side was a multiple of 5, it followed that $p(25m+24)$ was divisible by 25. It is interesting to see how Ramanujan used (30.25) to compute the Rogers–Ramanujan continued fraction as well as the generating function for $p(5m+4)$.

30.6 Exercises

1. Prove that the number of partitions of n into parts not divisible by d is equal to the number of partitions of n of the form $n = n_1 + n_2 + \cdots + n_s$, where $n_i \geq n_{i+1}$ and $n_i \geq n_{i+d-1} + 1$. See Glaisher (1883). James Whitehead Lee Glaisher (1848–1928) single-handedly edited two journals for over forty years: *Messenger of Mathematics* and *Quarterly Journal*. The *Messenger* carried the first published papers of many English mathematicians and physicists of the late nineteenth century, including H. F. Baker, E. W. Barnes, W. Burnside, G. H. Hardy, J. J. Thompson, and J. Jeans. Glaisher published almost four hundred papers, many of them in his own journals. G. H. Hardy wrote in his obituary notice, "He wrote a great deal of very uneven quality, and he was 'old fashioned' in a sense which is most unusual now; but the best of his work is really good." This best work included results in number theory and, in particular, the representation of numbers as sums of squares. See vol. 7 of Hardy (1966–1979).

2. Complete the following number theoretic proof, due to Glaisher, of Euler's theorem that the number of partitions of n into odd parts equals the number of partitions of n into distinct parts. Let

$$n = f_1 \cdot 1 + f_3 \cdot 3 + \cdots + f_{2m-1} \cdot (2m-1).$$

Here $f_1, f_3, \ldots$ represent the number of times $1, 3, \ldots$, respectively, occur in the partition of n into odd parts. Now write $f_1, f_3, \ldots$ in powers of two: $f_1 = 2^{a_1} + 2^{a_2} + \cdots + 2^{a_l}$, $f_3 = 2^{b_1} + 2^{b_2} + \cdots + 2^{b_l}$, …. Then

$$n = 2^{a_1} + 2^{a_2} + \cdots + 2^{a_l} + 2^{b_1} \cdot 3 + 2^{b_2} \cdot 3 + \cdots + 2^{b_l} \cdot 3 + \cdots + 2^{r_1} \cdot (2m-1)$$
$$+ \cdots + 2^{r_s} \cdot (2m-1)$$

gives a partition of n into distinct parts. See Glaisher (1883).

3. Show that the number of partitions of n into odd parts, where exactly k distinct parts appear, is equal to the number of partitions of n into distinct parts,

where exactly k sequences of consecutive integers appear. Show that this correspondence is one-to-one. This result was published by Sylvester in 1882. See Sylvester (1973), vol. 4, p. 45.

4. Let $p_{k,r}(n)$ denote the number of partitions of n into parts not congruent to 0, $\pm r \pmod{2k+1}$, where $1 \leq r \leq k$. Let $q_{k,r}(n)$ denote the number of partitions of n of the form $n = n_1 + n_2 + \cdots + n_s$ where $n_1 \geq n_{i+1}$, $n_i \geq n_{i+k-1} + 2$ and with 1 appearing as a part at most $r - 1$ times. Prove that then

$$p_{k,r}(n) = q_{k,r}(n).$$

See Gordon (1961).

5. Prove that if $p_m(n)$ denotes the number of partitions of n with rank m, then

$$\sum_{n=1}^{\infty} p_m(n) q^n = \frac{1}{(q,q)_{\infty}} \sum_{n=1}^{\infty} (-1)^{n-1} \left(q^{\frac{1}{2}n(3n-1)} - q^{\frac{1}{2}n(3n+1)} \right) q^{|m|n}.$$

See Atkin and Swinnerton–Dyer (1954).

6. Derive Szekeres's combinatorial interpretation of the Rogers–Ramanujan continued fraction. Let $B(n.k)$, $n \geq 1$, $k \geq 0$ represent the number of sequences of integers $b_1 \leq b_2 \leq \cdots \leq b_n = n$ with $b_i > i$ for $1 \leq i < n$ and $b_1 + b_2 + \cdots + b_{n-1} = \binom{n}{2} + k$. Observe that $B(n,k) = 0$ for $0 \leq k < n-1$ and for $k > \binom{n}{2}$. Also note that $B(n, n-1) = B(n, \binom{n}{2}) = 1$, $B(1,0) = 1$. Now show that

$$\frac{x}{1+} \frac{qx}{1+} \frac{q^2 x}{1+} \cdots = \sum_{0 \leq n-1 \leq k} (-1)^{n+1} B(n,k) x^n q^k.$$

See Szekeres (1968). George Szekeres (1911–2005) was trained as a chemical engineer in Hungary but his association with Paul Erdős, Esther Klein, and Paul Turán turned his interest to mathematics. In 1935, Erdős and Szekeres wrote a paper laying the foundation of Ramsey theory. This arose out of Szekeres's efforts to solve a problem (proposed by Klein who later became his wife): For all n there exists N such that for any N points in a plane there are n which form a convex n-gon. Szekeres and his family escaped to China from the Nazi government in Germany and moved to Australia after the war. His presence gave a boost to the development and teaching of mathematics in Australia where he was greatly admired.

7. Andrews gave an interpretation of the Rogers–Ramanujan continued fraction different from Szekeres's. Let

$$C(q) = 1 + \frac{q}{1+} \frac{q^2}{1+} \cdots = \sum_{m \geq 0} c_m q^m.$$

Also let $B_{k,a}(n)$ denote the number of partitions of n of the form $n = b_1 + b_2 + \cdots + b_s$, where $b_i \geq b_{i+1}$, $b_i - b_{i+k-1} \geq 2$ and at most $a - 1$ of the b_i are equal to one. Then prove that

$$c_{5m} = B_{37,37}(m) + B_{37,13}(m-4),$$
$$c_{5m+1} = B_{37,32}(m) + B_{37,7}(m-6),$$
$$c_{5m+2} = -\left(B_{37,23}(m-1) + B_{37,2}(m-8)\right),$$
$$c_{5m+3} = -\left(B_{37,28}(m) + B_{37,22}(m-1)\right),$$
$$c_{5m+4} = -\left(B_{37,17}(m-2) + B_{37,8}(m-5)\right).$$

Show that it follows that, in particular, $c_2 = c_4 = c_9 = 0$ and that the remaining c_n satisfy

$$c_{5m} > 0, \ c_{5m+1} > 0, \ c_{5m+2} < 0, \ c_{5m+3} < 0, \ c_{5m+4} < 0.$$

See Andrews (1981).

8. Prove Ramanujan's result that $p(7m+5) \equiv 0 \pmod{7}$ by the following method. Square Jacobi's identity to obtain

$$x^2 \prod_{n=1}^{\infty}(1-x^n)^6$$
$$= \sum_{\mu=-\infty}^{\infty} \sum_{\nu=-\infty}^{\infty} (-1)^{\mu+\nu}(2\mu+1)(2\nu+1)x^{2+\frac{1}{2}\mu(\mu+1)+\frac{1}{2}\nu(\nu+1)}.$$

Now show that the coefficient of x^{7n} in the sum is divisible by 49. Next observe that

$$(1-x^7)/(1-x)^7 \equiv 1 \pmod{7}$$

and deduce that the coefficient of x^{7n} in $x^2 / \prod_{n=1}^{\infty}(1-x^n)$ is a multiple of 7. See Ramanujan (2000), p. 212.

9. For a partition π of n, let $\lambda(\pi)$ denote the largest part of π; let $\mu(\pi)$ denote the number of ones in π; and let $\nu(\pi)$ denote the number of parts of π larger than $\mu(\pi)$. The crank $c(\pi)$ is defined as

$$c(\pi) = \lambda(\pi) \text{ if } \mu(\pi) = 0; \ c(\pi) = \nu(\pi) - \mu(\pi) \text{ if } \mu(\pi) > 0.$$

Let $M(m,n)$ denote the number of partitions of n with crank m. Then prove that

$$\sum_{m=-\infty}^{\infty} \sum_{n=0}^{\infty} M(m,n) a^m q^n = \frac{(q;q)_\infty}{(aq;q)_\infty (q/a;q)_\infty}.$$

See Andrews and Garvan (1988).

30.7 Notes on the Literature

Euler's theorem on the number of partitions with distinct parts is in Euler (1988), pp. 275–276. For Hardy's remark on MacMahon's calculating prowess, see Ramanujan

(2000), p. xxxv. For Ramanujan's papers, see Ramanujan (2000); his 1919 paper on the arithmetic properties of partitions can be seen on pp. 210–213. Berndt has added a seventy-page commentary at the end of this book, where the reader will find references to the work on the rank and crank of a partition. See Ramanujan (1988) for some of his recently discovered work. See p. 238 for his proof of (30.7). Sylvester (1973), vol. 4, pp. 1–83 is a reprint of his 1882 paper. See especially pp. 11–12. For his expressions of indebtedness to Ferrers, see vol. 1, p. 597 and vol. 3, p. 683. See Cayley (1889–1898), vol.12, pp. 217–219 for his paper on Sylvester's formula.

The Selberg quote is from Selberg (1989), vol. 1, p. 696. Alder (1969) gives a history of partition identities. See also Ahlgren and Ono (2001) for a history of the arithmetic properties of the partition function. The theory of modular forms has been extensively used in recent years to study the partition function; K. Ono and his students and collaborators have been leaders in this area. Modular forms are also important in the study of many arithmetical problems. For example, Fermat's last theorem is a consequence of the Shimura–Taniyama conjecture. See Shimura (2008) for an account of how he arrived at this conjecture. For Atkin's remarks on finding results mechanically, see Atkin (1968), p. 564.

31

q-Series and q-Orthogonal Polynomials

31.1 Preliminary Remarks

In the early nineteenth century, q-series proved their worth with their broad applicability to number theory, elliptic and modular functions, and combinatorics. Nevertheless, as late as 1840, no general framework for the study of q-series had been established; although q-series had been used to solve problems in other areas, it had not become a subject of its own. Finally, q-series came into its own when it was viewed as an extension of the hypergeometric series. In the 1840s Cauchy, Eisenstein, Jacobi, and Heine each presented the q-binomial theorem for general exponents. In 1846, Jacobi wrote a paper stating a q-extension of Gauss's $_2F_1$ summation formula. His interesting proof was based on Schweins's q-extension of the Chu-Vandermonde identity, a terminating form of Gauss's summation. The latter result gave the value of the series in terms of gamma functions. So Jacobi suggested a q-analog of $\Gamma(a)$:

$$\Omega(q,a) = \frac{(1-q)(1-q^2)(1-q^3)\cdots}{(1-q^{a+1})(1-q^{a+2})(1-q^{a+3})\cdots}. \tag{31.1}$$

It is interesting that at the end of his 1846 paper, Jacobi wrote a lengthy historical note mentioning the 1729 letter from Euler to Goldbach, containing Euler's description of his discovery of the gamma function by the use of infinite products. Jacobi had a keen interest in the history of mathematics, and some of his papers contain very helpful historical information. With Jacobi's work, the stage was set to obtain the q-extension of the hypergeometric series and this was soon accomplished by Heinrich Eduard Heine (1821–1881) who studied in Göttingen under Gauss and Stern and in Berlin under Dirichlet. Heine received his doctoral degree under the supervision of Dirksen and Ohm in 1842. He then spent a year in Königsberg with Jacobi and Franz Neumann. It is most likely that Jacobi encouraged Heine to work on hypergeometric series and its q-extension; Heine later edited a posthumous paper of Jacobi on this subject.

In his 1847 paper defining the q-hypergeometric series, Heine developed properties of the series $\phi(\alpha, \beta, \gamma, q, x)$ defined by

$$1 + \frac{(1-q^\alpha)(1-q^\beta)}{(1-q)(1-q^\gamma)}x + \frac{(1-q^\alpha)(1-q^{\alpha+1})(1-q^\beta)(1-q^{\beta+1})}{(1-q)(1-q^2)(1-q^\gamma)(1-q^{\gamma+1})}x^2 + \cdots. \quad (31.2)$$

Observe that as $q \to 1^-$, this series converges term by term to

$$1 + \frac{\alpha \cdot \beta}{1 \cdot \gamma}x + \frac{\alpha(\alpha+1) \cdot \beta(\beta+1)}{1 \cdot 2 \cdot \gamma(\gamma+1)}x^2 + \cdots,$$

the hypergeometric series $F(\alpha, \beta, \gamma, x)$. Heine took many results from Gauss's 1812 paper on hypergeometric series and extended them to the q-series ϕ. He listed the contiguous relations for ϕ, from which he derived continued fractions expansions for ratios of q-hypergeometric series, and he gave a very simple proof of the q-binomial theorem. The notation $\Omega(q, a)$ for the gamma function analog is also due to Heine, and as an analog of Gauss's $F(a, b, c, 1)$ sum, he presented

$$\phi(\alpha, \beta, \gamma, q, q^{\gamma-\alpha-\beta}) = \frac{\Omega(q, \gamma-1)\Omega(q, \gamma-\alpha-\beta-1)}{\Omega(q, \gamma-\alpha-1)\Omega(q, \gamma-\beta-1)}. \quad (31.3)$$

Heine applied the q-binomial theorem to obtain an important transformation now known as Heine's transformation:

$$\begin{aligned}
&\frac{(1-q^\gamma x)(1-q^{\gamma+1}x)\cdots}{(1-q^\beta x)(1-q^{\beta+1}x)\cdots}\left(1 + \frac{(1-q^\alpha)(1-q^\beta x)}{(1-q)(1-q^\gamma x)} \cdot z\right. \\
&\left. + \frac{(1-q^\alpha)(1-q^{\alpha+1})(1-q^\beta x)(1-q^{\beta+1}x)}{(1-q)(1-q^2)(1-q^\gamma x)(1-q^{\gamma+1}x)} \cdot z^2 + \cdots\right) \\
&= \frac{(1-q^\alpha z)(1-q^{\alpha+1}z)\cdots}{(1-z)(1-qz)\cdots}\left(1 + \frac{(1-q^{\gamma-\beta})(1-z)}{(1-q)(1-q^\alpha z)} \cdot q^\beta x\right. \\
&\left. + \frac{(1-q^{\gamma-\beta})(1-q^{\gamma-\beta+1})(1-z)(1-qz)}{(1-q)(1-q^2)(1-q^\alpha z)(1-q^{\alpha+1}z)} \cdot q^{2\beta}x^2 + \cdots\right).
\end{aligned} \quad (31.4)$$

He also defined a q-difference operator and found the second-order difference equation of which the hypergeometric equation was a limiting case. However, he did not define a q-integral.

The origin of the transformation (31.4) remained a puzzle; to which result in hypergeometric series did this correspond? C. Johannes Thomae (1840–1921) answered this question. Thomae studied at the University of Halle and was inspired by Heine to devote himself to function theory. Thomae moved to Göttingen in 1862 with the intention of working under Riemann who soon fell seriously ill. Thomae stayed on and in 1864, he earned his doctoral degree under Schering, one of the editors of Gauss's papers. Thomae then returned to teach at Halle for a few years before moving to Freiburg and then to Jena. Thomae wrote his paper on Heine's series in 1869 while at Halle. In this paper he defined the q-integral and showed that Heine's transformation was actually the q-extension of Euler's integral representation of the hypergeometric series. Thomae defined the q-integral by

$$\int_x^{xq^n} f(x) \frac{\Delta x}{x(q-1)} = \sum_{s=0}^{n-1} f(xq^s). \quad (31.5)$$

He explained that the integral was the inverse of the difference operator

$$\Delta f(x) = f(xq) - f(x), \tag{31.6}$$

also noting that it was better to define the q-gamma, or more correctly, the q-Π function by

$$\Pi(\alpha, q) = (1-q)^{-\alpha} \Omega(q, \alpha). \tag{31.7}$$

Indeed, with this definition,

$$\Pi(\alpha, q) = \frac{1-q^\alpha}{1-q} \cdot \Pi(\alpha - 1, q), \tag{31.8}$$

and $\Pi(\alpha, q) \to \Pi(\alpha) = \Gamma(\alpha + 1)$ as $q \to 1^-$. He also observed that the q-binomial theorem was equivalent to

$$\int_1^{q^\infty} s^\alpha \, p(\beta, sq) \, \Delta s = -\frac{\Pi(\alpha, q)\,\Pi(\beta, q)}{\Pi(\alpha + \beta + 1, q)}, \tag{31.9}$$

where $p(\beta, x) = \dfrac{(1-x)(1-xq)(1-xq^2)\cdots}{(1-xq^\beta)(1-xq^{\beta+1})(1-xq^{\beta+2})\cdots}.$ \hfill (31.10)

Moreover, he noted that as $q \to 1^-$, formula (31.9) reduced to Euler's beta integral formula

$$\int_0^1 s^\alpha (1-s)^\beta \, ds = \frac{\Pi(\alpha)\,\Pi(\beta)}{\Pi(\alpha + \beta + 1)}.$$

About forty years later, the able amateur mathematician Frank Hilton Jackson (1870–1960) redefined the q-integral and the q-gamma function. He took the q-integral to be the inverse of the q-derivative

$$\Delta_q \phi(x) = \frac{\phi(qx) - \phi(x)}{qx - x}, \tag{31.11}$$

so that the q-integral amounted to

$$\int_0^a f(x) \, d_q x = \sum_{n=0}^\infty f(aq^n)(aq^n - aq^{n+1}). \tag{31.12}$$

The reader may observe that the expression on the right is the Riemann sum for the division points $a, aq, aq^2, \ldots$ on $[0, a]$. Note that Fermat integrated $x^{m/n}$, where m and n were integers, by first evaluating its q-integral and then letting $q \to 1^-$. Jackson's notation for the q-gamma, $\Gamma_q(x)$, is still in use; he set

$$\Gamma_q(x) = \Pi(x - 1, q). \tag{31.13}$$

Jackson was a British naval chaplain; he apparently had a little difficulty in publishing some of his earlier works because their significance was not quite clear to referees. Jackson's lifelong program was to systematically develop the theory of q-hypergeometric (or basic hypergeometric) series by proving analogs of summation and transformation formulas for generalized hypergeometric series. Jackson conceived of q as the base of the series, analogous to the base of the logarithm. His terminology is now widely used. In recent years, Vyacheslav Spiridonov has worked out another important and productive generalization of hypergeometric functions, elliptic hypergeometric functions. A formal series, $\sum_{-\infty}^{\infty} c_n$, is called an elliptic hypergeometric series if $c_{n+1}/c_n = h(n)$, where $h(n)$ is some elliptic function of $n \in \mathbb{C}$.

Leonard James Rogers (1862–1933) gave a new direction to q-series theory through his researches in the early 1890s. Rogers studied at Oxford where his father was a professor of political economy. As a boy, Rogers was tutored in mathematics by A. Griffith, an Oxford mathematician with a strong interest in elliptic functions. Rogers's earliest work was in reciprocants, a topic in invariant theory. The second half of Sylvester's 1886 Oxford lectures on this subject were devoted to the work of Rogers. Around this same time, Rogers's interest turned to analysis and to the topic in which he did his most famous work, theta series and products and, more generally, q-series. In his Royal Society obituary notice of Rogers, A. L. Dixon recalled attending an 1887 course of lectures at Oxford in which Rogers manipulated q-series and products with great skill. In the period 1893–95, after a study of Heine's 1878 *Kugelfunctionen*, Rogers published four important papers on q-extensions of Hermite and ultraspherical polynomials. In his book, Heine included extra material, printed in smaller type, on basic hypergeometric series, clearly expecting that the q-extensions of some results in the book would be important and fruitful. Rogers showed that Heine was right. Rogers was initially struck by a lack of symmetry in Heine's transformation formula. In order to present Rogers's work succinctly and with transparency, we introduce some modern notation different from that of Rogers, though he also employed abbreviations. Let

$$(x)_n \equiv (x;q)_n = (1-x)(1-qx)(1-q^2x)\cdots(1-q^{n-1}x),$$
$$n = 1, 2, 3, \ldots, \infty. \tag{31.14}$$

We observe that Rogers wrote x_n for $1 - q^{n-1}x$; $x_n!$ for $(x)_n$; and (x) for $(x)_\infty$. We now replace $q^\alpha, q^\beta, q^\gamma$ in Heine's formula (31.4) by a, b, c and write it as

$$\phi(a, bx, cx, q, z) = \frac{(bx)_\infty (az)_\infty}{(cx)_\infty (z)_\infty} \phi(c/b, z, az, q, bx).$$

Here, $\phi(a,b,c,q,x) = \sum_{n=0}^{\infty} \frac{(a)_n (b)_n}{(q)_n (c)_n} x^n.$ \tag{31.15}

After reparametrization, we can write Heine's transformation as

$$\phi(a,b,c,q,x) = \frac{(ax)_\infty (b)_\infty}{(x)_\infty (c)_\infty} \phi(c/b, x, ax, q, b). \tag{31.16}$$

Perhaps Rogers's earlier work on invariant theory had made him sensitive to symmetry, so as a first step he wrote the transformation in symmetric form. He observed that, by the symmetry in a and b and also by a reapplication of Heine's transformation, he obtained two results from (31.16):

$$\phi(a,b,c,q,x) = \frac{(bx)_\infty (c/b)_\infty}{(x)_\infty (c)_\infty} \phi(b, abx/c, bx, q, c/b), \qquad (31.17)$$

$$\phi(a,b,c,q,x) = \frac{(abx/c)_\infty}{(x)_\infty} \phi(c/a, c/b, c, q, abx/c). \qquad (31.18)$$

Rogers set $a = \mu e^{-i\theta}$, $b = \gamma e^{-i\theta}$, $c = \mu\gamma$, and $x = \lambda e^{i\theta}$, observing that the last three formulas implied that

$$\psi(\lambda, \mu, \gamma, q, \theta) = (\lambda e^{i\theta})_\infty (\mu\gamma)_\infty \phi(\mu e^{-i\theta}, \gamma e^{-i\theta}, \mu\gamma, q, \lambda e^{i\theta})$$

was symmetric in λ, μ, γ and in θ and $-\theta$. He then went on to define a q-extension of the Hermite polynomials $A_n(\theta)$ by the relation

$$P(t) \equiv \frac{1}{\prod_{n=0}^{\infty}(1 - 2tq^n \cos\theta + t^2 q^{2n})} = \sum_{n=0}^{\infty} \frac{A_n(\theta)}{(q)_n} t^n. \qquad (31.19)$$

In a later paper, Rogers defined a q-extension of the ultraspherical or Gegenbauer polynomials and denoted it by $L_n(\theta)$, using the equation

$$\frac{P(t)}{P(\lambda t)} \equiv \prod_{n=0}^{\infty} \left(\frac{1 - 2\lambda t q^n \cos\theta + \lambda^2 t^2 q^{2n}}{1 - 2tq^n \cos\theta + t^2 q^{2n}} \right) = \sum_{n=0}^{\infty} \frac{L_n(\theta)}{(q)_n} t^n. \qquad (31.20)$$

In connection with the q-Hermite polynomials, Rogers raised the question: Suppose

$$a_0 + a_1 A_1(\theta) + a_2 A_2(\theta) + \cdots = b_0 + 2b_1 \cos\theta + 2b_2 \cos 2\theta + \cdots.$$

How are the coefficients a_0, a_1, a_2, ... and b_0, b_1, b_2, ... related to each other? He solved the problem and then applied the solution to the function

$$\prod_{n=1}^{\infty} \left(1 + 2q^{n-1/2} \cos\theta + q^{2n-1}\right).$$

From the triple product identity, the Fourier cosine expansion of this function was already known; Rogers found the expansion in terms of $A_r(\theta)$, and in particular, he got

$$a_0 = \sum_{n=0}^{\infty} \frac{q^{n^2}}{(q)_n}, \quad a_1 = \frac{q^{1/2}}{1-q} \sum_{n=0}^{\infty} \frac{q^{n(n+1)}}{(q)_n}.$$

When he expressed a_0, a_1 in terms of the bs, he obtained a series convertible to products by the triple product identity. The final results emerged as the Rogers–Ramanujan identities. Recall that we gave Ramanujan's derivation of these formulas in chapter 30. Although Rogers discovered these remarkable identities in 1894, they remained unnoticed until Ramanujan rediscovered them without proof. Then, in 1917, quite by chance, Ramanujan came across Rogers's paper while browsing through old journals. The ensuing correspondence between Ramanujan and Rogers led them both to new proofs of the identity now known by their names. Somewhat surprisingly, around 1980, the physicist R. J. Baxter rediscovered the Rogers–Ramanujan identities in the course of his work on the hard hexagon model.

Rogers left unanswered the question of the orthogonality of his q-extensions of the Hermite and ultraspherical polynomials. He had found three-term recurrence relations for these polynomials, but it was not until the 1930s that these relations were widely understood to imply that the polynomials were orthogonal with respect to some positive weights. Consider the statement of the spectral theorem for orthogonal polynomials: Suppose that a sequence of monic polynomials $\{P_n(x)\}$ with real coefficients satisfies a three-term recurrence relation

$$x P_n(x) = P_{n+1}(x) + \alpha_n P_n(x) + \beta_n P_{n-1}(x), \quad n \geq 1, \tag{31.21}$$

with $P_0(x) = 1$, $P_1(x) = x - \alpha_0$, α_{n-1} real and $\beta_n > 0$. Then there exists a distribution function μ, corresponding to a positive and finite Borel measure on the real line, such that

$$\int_{-\infty}^{\infty} P_m(x) P_n(x) d\mu(x) = \zeta_n \delta_{m,n}, \tag{31.22}$$

where $\zeta_n = \beta_1 \beta_2 \cdots \beta_n$. $\qquad(31.23)$

The converse is also true and straightforward to prove. This theorem is sometimes known as Favard's theorem since Jean Favard (1902–1965) published a proof in the *Comptes Rendus* (Paris) in 1935. However, the theorem appeared earlier in the works of O. Perron, M. Stone, and A. Wintner. In fact, J. Meixner applied this theorem to his work on orthogonal polynomials, with a reference to a 1934 result of Perron. Stieltjes's famous 1895 paper on continued fractions also contains a result yielding the spectral theorem, when combined with Riesz's representation theorem. It is unlikely that Rogers was aware of Stieltjes's work. We note that A. L. Dixon wrote of Rogers that he had only a vague notion of the work of other mathematicians. Probably an abstract result such as the spectral theorem for orthogonal polynomials would not have interested Rogers who loved special functions and formulas. He would have wanted to use the actual weight function, needed for computational purposes.

The Hungarian mathematician Gabor Szegő (1895–1985) took the first step toward finding an explicit weight function. Szegő studied in Hungary and Germany under Fejér, Frobenius, and Hilbert. His most famous book, *Problems and Theorems in Analysis*, was originally written in German in 1924, in collaboration with George Pólya. Szegő made significant contributions to orthogonal polynomials, on which he also wrote a very influential book first published in 1939. He founded the study of orthogonality on the unit circle; for a probability measure $\alpha(\theta)$, he defined this orthogonality by

$$\int_0^{2\pi} \phi_n(e^{i\theta}) \overline{\phi_m(e^{i\theta})} d\alpha(\theta) = 0, \quad m \neq n,$$

where $\phi_n(z) = \sum_{k=0}^{n} a_{k,n} z^k$.

31.1 Preliminary Remarks

Szegő was the first to appreciate the general program Rogers had in mind, as opposed to its specific though very important corollaries, such as the Rogers–Ramanujan identities. Rogers's papers inspired Szegő to discover the first nontrivial example of orthogonal polynomials on the circle, where

$$a_{k,n} = (-1)^k q^{-k/2} \frac{(1-q^n)(1-q^{n-1})\cdots(1-q^{n-k+1})}{(1-q)(1-q^2)\cdots(1-q^k)}.$$

Here note the connection with the Gaussian polynomial. Recall also that Gauss had expressed $\phi(-q)$ and $\phi(\sqrt{q})$ as finite products and evaluated the general quadratic Gauss sum from these expressions. Szegő found that the weight function $f(\theta)\,d\theta = d\alpha(\theta)$ in this case was

$$f(\theta) = \sum_{n=-\infty}^{\infty} q^{n^2/2} e^{in\theta},$$

and he applied the triple product identity to prove it. The weight function for Rogers's q-Hermite polynomial is also a theta function; the proof of orthogonality in that case also uses the triple product identity.

The q-ultraspherical polynomials were independently rediscovered by Feldheim and I. L. Lanzewizky in 1941. The Hungarian mathematician Ervin Feldheim (1912–1944) studied in Paris, since he was not admitted to the university in Budapest. His thesis was in probability theory but on his return to Hungary he contributed important results to the classical theory of orthogonal polynomials. One of these results was contained in a letter to Fejér written by Feldheim shortly before his tragic death at the hands of the Nazis. The letter was later found by Paul Turán, who described the incident, "Thus the letter had been resting among Fejér's letters for some 15 years. ... *On the next day* I received a letter from Szegő, in which he raised just a problem solved in Feldheim's letter! I sent this letter to Szegő and he published it with applications."

The origin of Feldheim and Lanzewizky's papers was the work of Fejér on a generalization of Legendre polynomials; this generalization also included the ultraspherical polynomials. Feldheim and Lanzewizky wished to determine those generalized Legendre polynomials that were also orthogonal. They used the spectral theorem and found conditions under which the generalized Legendre polynomials satisfied the appropriate three-term recurrence relation. At the end of his paper, Feldheim raised the problem of determining the weight or distribution function for orthogonality but was unable to resolve the question. Earlier works of Stieltjes and Markov could have helped him here, and from a remark in his paper, it seems that he may have been aware of their work. We should remember, however, that Feldheim was working under extremely difficult circumstances during the war. So, like Rogers himself, Feldheim and Lanzewizky did not give the relevant orthogonality relations and, in fact, they did not write the polynomials as q-extensions of the ultraspherical polynomials. Around 1980, Richard Askey and Mourad Ismail finally established the explicit orthogonality relation.

In his paper containing the Rogers–Ramanujan identities, Rogers also observed that the series

$$\chi(\lambda^2) = 1 + \frac{\lambda^2 q^2}{1-q} + \frac{\lambda^4 q^6}{(1-q)(1-q^2)} + \cdots \qquad (31.24)$$

satisfied the relation

$$\chi(\lambda^2) - \chi(\lambda^2 q) = \lambda^2 q^2 \chi(\lambda^2 q^2). \qquad (31.25)$$

This can be easily verified and implies the continued fraction expansion

$$\frac{\chi(\lambda^2)}{\chi(\lambda^2 q)} = \frac{1}{1+} \frac{\lambda^2 q^2}{1+} \frac{\lambda^2 q^3}{1+} \frac{\lambda^2 q^4}{1+} \cdots. \qquad (31.26)$$

This is now known as the Rogers–Ramanujan continued fraction because Ramanujan rediscovered and went much farther with it. In his first letter to Hardy, of January 16, 1913, Ramanujan stated without proof that

$$\frac{1}{1+} \frac{e^{-2\pi}}{1+} \frac{e^{-4\pi}}{1+} \frac{e^{-6\pi}}{1+} \cdots = \left(\sqrt{\frac{5+\sqrt{5}}{2}} - \frac{\sqrt{5}+1}{2}\right) e^{2\pi/5}, \qquad (31.27)$$

$$\frac{1}{1-} \frac{e^{-\pi}}{1+} \frac{e^{-2\pi}}{1-} \frac{e^{-3\pi}}{1+} \cdots = \left(\sqrt{\frac{5-\sqrt{5}}{2}} - \frac{\sqrt{5}-1}{2}\right) e^{\pi/5}, \qquad (31.28)$$

and that $\frac{1}{1+} \frac{e^{-\pi\sqrt{n}}}{1+} \frac{e^{-2\pi\sqrt{n}}}{1+} \frac{e^{-3\pi\sqrt{n}}}{1+} \cdots$ could be exactly determined if n were any positive rational quantity. In his next letter of February 27, 1913, Ramanujan wrote

$$\text{If } F(x) = \frac{1}{1+} \frac{x}{1+} \frac{x^2}{1+} \frac{x^3}{1+} \cdots, \text{ then}$$

$$\left\{\frac{\sqrt{5}+1}{2} + e^{-2\alpha/5} F(e^{-2\alpha})\right\} \left\{\frac{\sqrt{5}+1}{2} + e^{-2\beta/5} F(e^{-2\beta})\right\} = \frac{5+\sqrt{5}}{2}$$

with the condition $\alpha\beta = \pi^2, \ldots$. This theorem is a particular case of a theorem on the continued fraction

$$\frac{1}{1+} \frac{ax}{1+} \frac{ax^2}{1+} \frac{ax^3}{1+} \cdots,$$

which is a particular case of the continued fraction

$$\frac{1}{1+} \frac{ax}{1+bx+} \frac{ax^2}{1+bx^2+} \frac{ax^3}{1+bx^3+} \cdots,$$

which is a particular case of a general theorem on continued fractions.

Hardy was very impressed with Ramanujan's results on continued fractions. Concerning formulas (31.27) and (31.28) he wrote, "I had never seen anything in the least like them before. A single look at them is enough to show that they could only be written down by a mathematician of the highest class. They must be true because, if they were not true, no one would have had the imagination to invent them."

31.2 Heine's Transformation

Heine proved his transformation for the q-hypergeometric series by a judicious application of the q-binomial theorem. He proved the q-binomial theorem by the use of contiguous relations; as we mentioned in exercise 13 of chapter 4, Gauss may have employed this method to prove the binomial theorem. Heine required the contiguous relation

$$\phi(\alpha+1,\beta,\gamma,q,x) - \phi(\alpha,\beta,\gamma,q,x) = q^\alpha x \frac{1-q^\beta}{1-q^\gamma} \phi(\alpha+1,\beta,\gamma,q,x) \quad (31.29)$$

and the q-difference relation

$$\phi(\alpha,\beta,\gamma,q,x) - \phi(\alpha,\beta,\gamma,q,qx) = \frac{(1-q^\alpha)(1-q^\beta)}{1-q^\gamma} x\phi(\alpha+1,\beta+1,\gamma+1,q,x). \quad (31.30)$$

Note that the second relation is the analog of the derivative equation

$$\frac{d}{dx} F(a,b,c,x) = \frac{a \cdot b}{c} F(a+1,b+1,c+1,x).$$

Heine then supposed that $\beta = \gamma = 1$ so that he had the q-binomial series

$$\phi(\alpha,x) = \phi(\alpha,1,1,q,x) = 1 + \frac{1-q^\alpha}{1-q} x + \frac{(1-q^\alpha)(1-q^{\alpha+1})}{(1-q)(1-q^2)} x^2 + \cdots. \quad (31.31)$$

By (31.29),

$$\phi(\alpha+1,x) = \frac{1}{1-q^\alpha x} \phi(\alpha,x). \quad (31.32)$$

Combined with (31.30), this produced

$$\phi(\alpha,x) = \frac{1-q^\alpha x}{1-x} \phi(\alpha,qx) = \frac{(1-q^\alpha x)(1-q^{\alpha+1}x)\cdots(1-q^{\alpha+n}x)}{(1-x)(1-qx)\cdots(1-q^n x)} \phi(\alpha,q^{n+1}x). \quad (31.33)$$

The q-binomial theorem followed, since it was assumed that $|q| < 1$ and

$$\phi(\alpha,q^{n+1}x) \to \phi(\alpha,0) = 1 \quad \text{as} \quad n \to 0.$$

Then, to obtain the transformation formula (31.32), Heine started with the series

$$S = 1 + \frac{(1-q^\alpha)(1-q^\beta x)}{(1-q)(1-q^\gamma x)} z + \frac{(1-q^\alpha)(1-q^{\alpha+1})(1-q^\beta x)(1-q^{\beta+1}x)}{(1-q)(1-q^2)(1-q^\gamma x)(1-q^{\gamma+1}x)} z^2 + \cdots, \quad (31.34)$$

where he assumed $|q| < 1$, $|x| < 1$, and $|z| < 1$ for convergence. He multiplied both sides by $\phi(\gamma - \beta, q^\beta x)$ and used the product expression for this function, from the

q-binomial theorem, to get

$$\phi(\gamma - \beta, q^\beta x)S = \phi(\gamma - \beta, q^\beta x) + \frac{1 - q^\alpha}{1 - q} z \cdot \phi(\gamma - \beta, q^{\beta+1} x)$$
$$+ \frac{(1 - q^\alpha)(1 - q^{\alpha+1})}{(1 - q)(1 - q^2)} z^2 \cdot \phi(\gamma - \beta, q^{\beta+2} x) + \cdots. \tag{31.35}$$

In the next step, Heine expanded each of the ϕ on the right as series. We here employ the abbreviated notation given in (31.14).

$$1 + \frac{(q^{\gamma-\beta}; q)_1}{(q; q)_1} \cdot q^\beta x + \frac{(q^{\gamma-\beta}; q)_2}{(q; q)_2} \cdot q^{2\beta} x^2 + \cdots$$
$$+ z \left(\frac{(q^\alpha; q)_1}{(q; q)_1} + \frac{(q^{\gamma-\beta}; q)_1}{(q; q)_1} \cdot \frac{(q^\alpha; q)_1}{(q; q)_1} q^{\beta+1} x + \frac{(q^{\gamma-\beta}; q)_2}{(q; q)_2} \cdot \frac{(q^\alpha; q)_1}{(q; q)_1} q^{2\beta+2} x^2 + \cdots \right)$$
$$+ z^2 \left(\frac{(q^\alpha; q)_2}{(q^2; q)_2} + \frac{(q^{\gamma-\beta}; q)_1}{(q; q)_1} \cdot \frac{(q^\alpha; q)_2}{(q; q)_2} \cdot q^{\beta+2} x + \cdots \right) + \cdots.$$

He changed the order of summation and used the q-binomial theorem to obtain

$$\phi(\alpha, z) + \frac{(q^{\gamma-\beta}; q)_1}{(q; q)_1} \cdot q^\beta x \cdot \phi(\alpha, qz) + \frac{(q^{\gamma-\beta}; q)_2}{(q; q)_2} \cdot q^{2\beta} x^2 \phi(\alpha, q^2 z) + \cdots$$
$$= \phi(\alpha, z) \left(1 + \frac{(q^{\gamma-\beta}; q)_1 (z; q)_1}{(q; q)_1 (q^\alpha z; q)_1} \cdot q^\beta x + \frac{(q^{\gamma-\beta}; q)_2 (z; q)_2}{(q; q)_2 (q^\alpha z; q)_2} \cdot q^{2\beta} x^2 + \cdots \right).$$

Finally, Heine substituted this expression into the right-hand side of (31.35) and replaced $\phi(\alpha, z)$ and $\phi(\gamma - \beta, q^\beta x)$ by their product expressions, to arrive at the transformation (31.4). Heine found the q-extension of Gauss's summation of $F(\alpha, \beta, \gamma, 1)$ by taking $x = 1$, $z = q^{\gamma-\alpha-\beta}$ in his transformation:

$$\phi(\alpha, \beta, \gamma, q^{\gamma-\alpha-\beta})$$
$$= \frac{(q^\beta; q)_\infty (q^{\gamma-\beta}; q)_\infty}{(q^\gamma; q)_\infty (q^{\gamma-\alpha-\beta}; q)_\infty} \cdot \phi(\gamma - \alpha - \beta, 1, 1, q^\beta) \tag{31.36}$$
$$= \frac{(q^{\gamma-\alpha}; q)_\infty (q^{\gamma-\beta}; q)_\infty}{(q^\gamma; q)_\infty (q^{\gamma-\alpha-\beta}; q)_\infty}.$$

Note that this derivation is analogous to the method used to derive Gauss's formula from Euler's integral for the hypergeometric function. This suggests that Heine's transformation is the q-analog of Euler's integral formula; indeed, recall that Thomae proved this after defining the q-integral.

31.3 Rogers: Threefold Symmetry

Rogers applied his knowledge and experience of elliptic functions and invariant theory to develop the theory of q-Hermite and q-ultraspherical polynomials. From invariant

31.3 Rogers: Threefold Symmetry

theory, he brought a sense of symmetry and expertise in applying infinite series/products of operators. We may recall that in the first half of the nineteenth century, Arbogast, Français, Murphy, D. Gregory, and Boole made extensive use of operational calculus. Cayley and Sylvester appropriated these methods for invariant theory. Then, in the 1880s, J. Hammond and P. MacMahon applied these techniques to combinatorial and invariant theoretic problems. It is interesting to observe Rogers's use of algebra, combinatorics and analysis as he conceived of and solved new problems in analysis. In his second paper of 1893, Rogers converted Heine's transformation into an equation with threefold symmetry. He showed that the function

$$\psi(\lambda,\mu,\nu,q,\theta) = \phi(\mu e^{-i\theta}, \nu e^{-i\theta}, \mu\nu, q, \lambda e^{i\theta})(\lambda e^{i\theta})_\infty (\mu\nu)_\infty \qquad (31.37)$$

was symmetric in λ, μ, and ν, and also symmetric in θ and $-\theta$. He then set

$$\chi(\lambda,\mu,\nu,q,\theta) = \frac{\psi(\lambda,\mu,\nu,q,\theta)}{P(\lambda)P(\mu)P(\nu)}, \quad \text{where} \qquad (31.38)$$

$$P(\lambda) = \prod_{n=0}^{\infty}(1 - 2\lambda q^n \cos\theta + \lambda^2 q^{2n})^{-1} = \frac{1}{(\lambda e^{i\theta})_\infty (\lambda e^{-i\theta})_\infty}. \qquad (31.39)$$

In his 1847 paper, Heine discussed such products; for particular values of λ they are ubiquitous in elliptic function theory. Rogers defined a q-extension of the Hermite polynomials $A_r(\theta)$ as the coefficient of $\lambda^r/(q)_r$ in the series expansion of $P(\lambda)$:

$$P(\lambda) = \sum_{r=0}^{\infty} \frac{A_r(\theta)}{(q)_r} \lambda^r. \qquad (31.40)$$

From Euler's expansion, Rogers had

$$\frac{1}{(\lambda e^{i\theta})_\infty (\lambda e^{-i\theta})_\infty} = \left(1 + \frac{\lambda e^{i\theta}}{1-q} + \frac{\lambda^2 e^{2i\theta}}{(q)_2} + \cdots\right)\left(1 + \frac{\lambda e^{-i\theta}}{1-q} + \frac{\lambda^2 e^{-2i\theta}}{(q)_2} + \cdots\right). \qquad (31.41)$$

Hence Rogers obtained his expression for the q-Hermite polynomial:

$$A_r(\theta) = \sum_{n=0}^{r} \frac{(q)_r e^{i\theta(r-2n)}}{(q)_n (q)_{r-n}} = \sum_{n=0}^{r} \frac{(q)_r \cos(r-2n)\theta}{(q)_n (q)_{r-n}}. \qquad (31.42)$$

Note that the last step followed because $A_r(\theta)$ was an even function of θ. In his 1894 paper, Rogers noted the three-term recurrence relation for $A_r(\theta)$:

$$2\cos\theta\, A_{r-1}(\theta) = A_r(\theta) + (1 - q^{r-1})A_{r-2}(\theta). \qquad (31.43)$$

He obtained it as a consequence of the relation

$$P(\lambda q) = (1 - 2\lambda \cos\theta + \lambda^2) P(\lambda), \qquad (31.44)$$

or $$\sum_{r=0}^{\infty} \frac{A_r(\theta)\lambda^r q^r}{(q)_r} = (1 - 2\lambda \cos\theta + \lambda^2) \sum_{r=0}^{\infty} \frac{A_r(\theta)\lambda^r}{(q)_r}. \qquad (31.45)$$

In his 1893 paper, Rogers raised the problem of expanding $\chi(\lambda,\mu,\nu,q,\theta)$ in (31.38) as a series in q-Hermite polynomials. The series would take the form $A_0 + A_1 H_1 + A_2 H_2 + \cdots$ and the problem was to determine H_r. He showed that H_r, a homogeneous symmetric function of degree r in λ, μ, and ν, was the coefficient of k^r in the series expansion of

$$1/((k\lambda)_\infty (k\mu)_\infty (k\nu)_\infty). \tag{31.46}$$

Rogers gave an interesting proof of this expansion and we sketch it very briefly. He observed that for the function χ defined by (31.38),

$$\delta_\lambda \chi = \delta_\mu \chi = \delta_\nu \chi, \tag{31.47}$$

where δ was the difference operator defined by

$$\delta_q f(x) = (f(x) - f(qx))/x. \tag{31.48}$$

From (31.47), he was able to deduce that

$$\delta_\lambda H_r(\lambda,\mu,\nu) = \delta_\mu H_r(\lambda,\mu,\nu) = \delta_\mu H_r(\lambda,\mu,\nu). \tag{31.49}$$

He denoted the coefficient of $\lambda^\alpha \mu^\beta \nu^\gamma$ in H_r by $a_{\alpha,\beta,\gamma}$. Note that $\alpha + \beta + \gamma = r$. Rogers next showed that (31.49) implied the recurrence relations

$$(1-q^{\alpha+1})a_{\alpha+1,\beta,\gamma} = (1-q^{\beta+1})a_{\alpha,\beta+1,\gamma} = (1-q^{\gamma+1})a_{\alpha,\beta,\gamma+1}. \tag{31.50}$$

Combined with the initial condition $a_{r,0,0} = 1/(q)_r$, these relations uniquely defined the coefficients. At this point, Rogers remarked that because of uniqueness it was sufficient to produce a set of coefficients satisfying these conditions. He then quickly demonstrated that the coefficient of k^r in the series expansion of (31.46) was a homogeneous symmetric function of degree r in λ, μ, and ν whose coefficients satisfied the same initial condition and recurrence relations (31.50). This proved his result, though he gave no indication of how arrived at (31.46).

Rogers noted some interesting and important particular cases of his theorem. When $\lambda = 0$, he had

$$\frac{(\mu\nu)_\infty}{P(\mu)P(\nu)} = 1 + A_1(\theta)H_1(\mu,\nu) + A_2(\theta)H_2(\mu,\nu) + \cdots, \tag{31.51}$$

where $H_r(\mu,\nu)$ was the coefficient of k^r in

$$\frac{1}{(k\mu)_\infty (k\nu)_\infty} = \left(1 + \frac{k\mu}{1-q} + \frac{k^2\mu^2}{(q)_2} + \cdots\right)\left(1 + \frac{k\nu}{1-q} + \frac{k^2\nu^2}{(q)_2} + \cdots\right). \tag{31.52}$$

So $H_r(\mu,\nu) = \mu^r + \frac{(q)_r}{(q)_1(q)_{r-1}}\mu^{r-1}\nu + \frac{(q)_r}{(q)_2(q)_{r-2}}\mu^{r-2}\nu^2 + \cdots + \nu^r$. (31.53)

We note that when $\mu = x$ and $\nu = 1$,

$$H_r(x) = x^r + \frac{(q)_r}{(q)_1(q)_{r-1}}x^{r-1} + \frac{(q)_r}{(q)_2(q)_{r-2}}x^{r-2} + \cdots + 1. \tag{31.54}$$

These polynomials $H_r(x)$ are now called Rogers–Szegő polynomials, because Szegő proved the orthogonality of $H_r(-x/\sqrt{q})$ with respect to a suitable measure on the unit circle. When $\mu = xe^{i\theta}$ and $\nu = xe^{-i\theta}$ in (31.51), Rogers got

$$\prod_{n=0}^{\infty} \frac{(x^2)_\infty}{(1 - 2xq^n \cos(\theta + \phi) + x^2 q^{2n})(1 - 2xq^n \cos(\theta - \phi) + x^2 q^{2n})}$$
$$= \sum_{n=0}^{\infty} \frac{A_n(\theta) A_n(\phi)}{(q)_n} x^n. \qquad (31.55)$$

This result is now known as the q-Mehler formula. Rogers also applied (31.51) to prove the useful linearization formula for q-Hermite polynomials:

$$\frac{A_m(\theta) A_n(\theta)}{(q)_m (q)_n} = \sum_{k=0}^{\min(m,n)} \frac{A_{m+n-2k}(\theta)}{(q)_k (q)_{m-k} (q)_{n-k}}. \qquad (31.56)$$

Rogers also noted that, by definition, $A_r(\pi/2)$ was the coefficient of $k^r/(q)_r$ in $\prod(1 + k^2 q^{2n})^{-1}$, and hence

$$A_r(\pi/2) = (-1)^r (1-q)(1-q^3) \cdots (1-q^{r-1}) \qquad r \text{ even},$$
$$= 0 \qquad r \text{ odd}.$$

Recall that Gauss evaluated Gauss sums from this result; see the formula (29.30). Thus, though Rogers may not have known it, this result is due to Gauss.

31.4 Rogers: Rogers–Ramanujan Identities

In order to derive the Rogers–Ramanujan identities, Rogers expanded $P(\lambda)$ as a Fourier series and as a series in the q-Hermite polynomials. He then found a relation between the coefficients in these two series, yielding the famous identities. In his 1894 paper, Rogers raised and solved the problem: Suppose a function $f(\theta)$ is expanded as a Fourier cosine series and as a series in q-Hermite polynomials $A_n(\theta)$:

$$f(\theta) = a_0 + a_1 A_1(\theta) + a_2 A_2(\theta) + \cdots = b_0 + 2b_1 \cos\theta + 2b_2 \cos 2\theta + \cdots . \qquad (31.57)$$

Express the coefficient a_n in terms of a series of b_j and, conversely, b_n in terms of a series of a_j. Rogers found that

$$b_n = a_n + \frac{1-q^{n+2}}{1-q} a_{n+2} + \frac{(1-q^{n+4})(1-q^{n+3})}{(1-q)(1-q^2)} a_{n+4}$$
$$+ \frac{(1-q^{n+6})(1-q^{n+5})(1-q^{n+4})}{(1-q)(1-q^2)(1-q^3)} a_{n+6} + \cdots \qquad (31.58)$$

For the converse, he gave a_0, a_1 in terms of series of b, but for the general case he merely described the method by which the a_n could be obtained for higher values of n. He had

$$a_0 = b_0 - (1+q)b_2 + q(1+q^2)b_4 - \cdots + (-1)^r q^{r(r-1)/2}(1+q^r)b_{2r} + \ldots, \quad (31.59)$$

$$(1-q)a_1 = (1-q)b_1 - (1-q^3)b_3 + q(1-q^5)b_5 - \cdots$$
$$+ (-1)^r q^{r(r-1)/2}(1-q^{2r+1})b_{2r+1} + \cdots. \quad (31.60)$$

The derivation of (31.58) was simple. Rogers substituted the expression (31.42) for $A_n(\theta)$ in terms of $\cos k\theta$ on the left side of (31.57) and then equated the coefficients of $\cos n\theta$ on both sides. He noted the formula only for even n, but the method also yields the case for odd n. Rogers's method for finding (31.59) and (31.60) was quite elaborate and he did not give the general formula. In his third paper of 1893, on the expansion of infinite products, Rogers supposed

$$f(\theta) = C_0 + C_1 A_1(\theta) + C_2 A_2(\theta) + \cdots$$

to be given. He then asked how to find K_0, K_1, K_2, ... in the expansion

$$\frac{f(\theta)}{\prod_{n=0}^{\infty}(1 - 2\lambda q^n \cos\theta + \lambda^2 q^{2n})} = K_0 + K_1 A_1(\theta) + K_2 A_2(\theta) + \cdots. \quad (31.61)$$

Rogers expressed the result symbolically, in terms of the difference operator

$$\delta_\lambda \phi(\lambda) = \frac{\phi(\lambda) - \phi(\lambda q)}{\lambda}:$$

$$K_0 + K_1 x + K_2 x^2 + \cdots = \frac{1}{(x\lambda)_\infty (x\delta_\lambda)_\infty}(C_0 + C_1\lambda + C_2\lambda^2 + \cdots). \quad (31.62)$$

Note that he used an infinite product in the operator δ_λ.

In his 1894 paper containing the Rogers–Ramanujan identities, Rogers applied (31.62) to find the q-Hermite expansion of

$$P(\lambda) = \prod_{n=0}^{\infty}(1 - 2\lambda q^n \cos\theta + \lambda^2 q^{2n}). \quad (31.63)$$

He noted that for an analytic function ϕ

$$\frac{1}{(x\delta_\lambda)_\infty}\phi(\lambda) = \frac{1}{(\lambda\delta_x)_\infty}\phi(x). \quad (31.64)$$

31.4 Rogers: Rogers–Ramanujan Identities

He verified this by taking the special value $\phi(\lambda) = \lambda^m$. Since

$$\delta_\lambda \lambda^m = (1-q^m)\lambda^{m-1}, \tag{31.65}$$

$$\frac{1}{(x\delta_\lambda)_\infty}\lambda^m = \sum_{n=0}^\infty \frac{x^n \delta_\lambda^n}{(q)_n}\lambda^m = \sum_{n=0}^m \frac{x^n \lambda^{m-n}(1-q^m)\cdots(1-q^{m-n+1})}{(q)_n}$$

$$= \sum_{n=0}^m \frac{\lambda^n x^{m-n}(1-q^m)\cdots(1-q^{n+1})}{(q)_{m-n}}$$

$$= \frac{1}{(\lambda\delta_x)_\infty} x^m.$$

Therefore, (31.62) could be rewritten as

$$K_0 + K_1 x + K_2 x^2 + \cdots = \frac{1}{(\lambda x)_\infty} \cdot \frac{1}{(\lambda\delta_x)_\infty}(C_0 + C_1 x + C_2 x^2 + \cdots),$$

or $C_0 + C_1 x + C_2 x^2 + \cdots = (\lambda\delta_x)_\infty (\lambda x)_\infty (K_0 + K_1 x + \cdots). \tag{31.66}$

Rogers then argued that if $f(\theta) = P(\lambda) = C_0 + C_1 A_1(\theta) + \cdots$, then

$$K_0 + K_1 A_1(\theta) + \cdots = 1.$$

Substituting this value in (31.66) gave him

$$C_0 + C_1 x + C_2 x^2 + \cdots = (\lambda\delta_x)_\infty (\lambda x)_\infty$$

$$= \left(1 - \frac{\lambda\delta_x}{1-q} + \frac{q\lambda^2\delta_x^2}{(1-q)(1-q^2)} - \cdots\right)\left(1 - \frac{\lambda x}{1-q} + \frac{q\lambda^2 x^2}{(1-q)(1-q^2)} - \cdots\right).$$

The last step made use of Euler's special case of the q-binomial theorem. Rogers applied (31.65) with λ replaced by x in the last expression to conclude that the coefficient C_r of x^r was

$$\frac{(-1)^r q^{r(r-1)/2}\lambda^r}{(q)_r} \sum_{s=0}^\infty \frac{\lambda^{2s} q^{rs+s(s-1)}}{(q)_s}.$$

This gave the q-Hermite expansion of $P(\lambda)$. Changing λ to $-\lambda q$ yielded

$$(1 + 2\lambda q \cos\theta + \lambda^2 q^2)(1 + 2\lambda q^2 \cos\theta + \lambda^2 q^4)\cdots$$

$$= \chi(\lambda^2) + \frac{q\lambda}{1-q}\chi(\lambda^2 q)A_1(\theta) + \frac{q^3\lambda^2}{(q)_2}\chi(\lambda^2 q^2)A_2(\theta) + \frac{q^6\lambda^3}{(q)_3}\chi(\lambda^2 q^3)A_3(\theta) + \cdots, \tag{31.67}$$

where

$$\chi(\lambda^2) = 1 + \frac{\lambda^2 q^2}{1-q} + \frac{\lambda^4 q^6}{(1-q)(1-q^2)} + \frac{\lambda^6 q^{12}}{(1-q)(1-q^2)(1-q^3)} + \cdots. \tag{31.68}$$

From the triple product identity, Rogers knew the Fourier cosine expansion of the product in (31.67) when $\lambda = 1/\sqrt{q}$:

$$(1 + 2q^{1/2}\cos\theta + q)(1 + 2q^{3/2}\cos\theta + q^3)\cdots$$
$$= \frac{1}{(q)_\infty}(1 + 2q^{1/2}\cos\theta + 2q^2\cos 2\theta + 2q^{9/2}\cos 3\theta + \cdots).$$

He therefore set $\lambda = 1/\sqrt{q}$ in (31.67) and applied (31.59) and (31.60). Thus, the Rogers–Ramanujan identities were discovered. From (31.59) he obtained

$$a_0 = \chi(1/q) = 1 + \frac{q}{1-q} + \frac{q^4}{(1-q)(1-q^2)} + \frac{q^9}{(1-q)(1-q^2)(1-q^3)} + \cdots$$
$$= \frac{1}{(q)_\infty}\left(1 - (1+q)q^2 + (1+q^2)q^9 - (1+q^3)q^{21} + \cdots\right)$$
$$= \frac{1}{(q)_\infty}\left(1 + \sum_{m=1}^\infty (-1)^m q^{m(5m\pm 1)/2}\right).$$

(31.69)

The sum could be evaluated by the triple product identity

$$(x)_\infty (q/x)_\infty (q)_\infty = \sum_{n=-\infty}^\infty (-1)^n q^{n(n-1)/2} x^n.$$

Replacing q by q^5 and setting $x = q^2$ yielded

$$(q^2;q^5)_\infty (q^3;q^5)_\infty (q^5;q^5)_\infty = 1 + \sum_{n=1}^\infty (-1)^n q^{n(5n\pm 1)/2}.$$

So the right-hand side of the equation (31.69) was reduced to

$$\frac{(1-q^2)(1-q^7)\cdots(1-q^3)(1-q^8)\cdots(1-q^5)(1-q^{10})\cdots}{(1-q)(1-q^2)(1-q^3)\cdots}$$
$$= \frac{1}{(1-q)(1-q^4)(1-q^6)(1-q^9)\cdots} = \prod_{n=0}^\infty \frac{1}{(1-q^{5n+1})(1-q^{5n+4})}.$$

This completed the proof of the first identity. Similarly,

$$(1-q)a_1 = q^{1/2}\left(1 + \frac{q^2}{1-q} + \frac{q^6}{(1-q)(1-q^2)} + \frac{q^{12}}{(1-q)(1-q^2)(1-q^3)} + \cdots\right)$$
$$= \frac{1}{(q)_\infty}\left((1-q)q^{1/2} - (1-q^3)q^{9/2} + (q-q^6)q^{25/2} - \cdots\right).$$

31.4 Rogers: Rogers–Ramanujan Identities

In this way, he obtained the second identity

$$1 + \frac{q^2}{1-q} + \frac{q^6}{(1-q)(1-q^2)} + \cdots$$
$$= \frac{1}{(q)_\infty}(1 - q - q^4 + q^7 + q^{13} - q^{18} - \cdots) \quad (31.70)$$
$$= \prod_{n=0}^{\infty} \frac{1}{(1-q^{5n+2})(1-q^{5n+3})}.$$

To prove (31.62), Rogers observed that since

$$\frac{1}{\prod_{n=0}^{\infty}(1 - 2\lambda q^n \cos\theta + \lambda^2 q^{2n})} = 1 + \frac{\lambda}{1-q} A_1(\theta) + \frac{\lambda^2}{(1-q)(1-q^2)} A_2(\theta) + \cdots,$$

the linearization formula (31.56) for $A_n(\theta)$ implied that the left-hand side of (31.62) was a product of series in $A_n(\theta)$ and therefore was itself such a series. He also observed that the linearization of $A_m(\theta) A_n(\theta)$ contained a term independent of θ only when $m = n$, and in that case this term would be $(q)_n$. Hence, he had

$$K_0 = C_0 + C_1 \lambda + C_2 \lambda^2 + \cdots. \quad (31.71)$$

From the difference relation

$$\delta_\lambda \left(\frac{1}{P(\lambda)}\right) = \frac{1}{\lambda}\left(\frac{1}{P(\lambda)} - \frac{1}{P(\lambda q)}\right)$$
$$= \frac{1 - (1 - 2\lambda \cos\theta + \lambda^2)}{\lambda P(\lambda)} = \frac{2\cos\theta - \lambda}{P(\lambda)},$$

Rogers obtained

$$\delta_\lambda(K_0 + K_1 A_1(\theta) + K_2 A_2(\theta) + \cdots) = (C_0 + C_1 A_1 + \cdots)\frac{2\cos\theta - \lambda}{P(\lambda)}$$
$$= (2\cos\theta - \lambda)(K_0 + K_1 A_1(\theta) + \cdots).$$

He equated the coefficients of A_r on both sides and applied the recurrence relation

$$A_{r+1} + (1 - q^r) A_{r-1} = 2\cos\theta \, A_r \text{ to conclude}$$
$$(1 - q^{r+1}) K_{r+1} = \lambda K_r - K_{r-1} + \delta_\lambda K_r. \quad (31.72)$$

He showed inductively that (31.72) implied

$$K_r = H_r(\lambda, \delta_\lambda) K_0, \quad (31.73)$$

where H_r was defined as in (31.53). For $r = 0$, (31.72) gave him

$$(1 - q) K_1 = (\lambda + \delta_\lambda) K_0 = H_1(\lambda, \delta_\lambda) K_0.$$

He assumed the result true up to r so that (31.72) could be written as

$$(1-q^{r+1})K_{r+1} = (\lambda + \delta_\lambda) H_r(\lambda, \delta) K_0 - H_{r-1}(\lambda, \delta) K_0. \quad (31.74)$$

He noted that when $m+n = r$, the coefficient of $\lambda^m \delta_\lambda^n$ in $H_r(\lambda, \delta_\lambda)$ was $1/(q)_m (q)_{r-m}$. He defined the operator η by $\eta f(\lambda) = f(q\lambda)$, so that

$$\begin{aligned}\delta_\lambda \lambda^m \delta_\lambda^n f(\lambda) &= \lambda^{m-1} \delta_\lambda^n f(\lambda) - \lambda^{m-1} q^m \eta \delta_\lambda^n f(\lambda) \\ &= \lambda^{m-1} \delta_\lambda^n f(\lambda) - q^m (\lambda^{m-1} \delta_\lambda^n f(\lambda) - \lambda^m \delta_\lambda^{n+1} f(\lambda)) \\ &= (1-q^m) \lambda^{m-1} \delta^n f(\lambda) + q^m \lambda^m \delta_\lambda^{n+1} f(\lambda). \end{aligned}$$

The first term in the last expression cancelled with the term containing $\lambda^{m-1} \delta_\lambda^n$ in $H_{r-1}(\lambda, \delta_\lambda)$. So (31.73) implied that

$$\begin{aligned}(1-q^{r+1})K_{r+1} &= \sum \left(\frac{q^m}{(q)_m (q)_{r-m}} + \frac{1}{(q)_{m-1} (q)_{r-m+1}} \right) \lambda^m \delta_\lambda^{n+1} K_0 \\ &= (1-q^{r+1}) \sum \frac{\lambda^m \delta_\lambda^{r+1-m}}{(q)_m (q)_{r+1-m}} K_0 \\ &= (1-q^{r+1}) H_{r+1}(\lambda, \delta_\lambda) K_0.\end{aligned}$$

Thus (31.73) was proved. Moreover, by (31.71), Rogers had

$$\frac{C_0 + C_1 A_1 + C_2 A_2 + \cdots}{P(\lambda)} = (1 + H_1(\lambda, \delta_\lambda) A_1 + H_2(\lambda, \delta_\lambda) A_2 + \cdots) \\ \times (C_0 + C_1 \lambda + C_2 \lambda^2 + \cdots).$$

This meant, remarked Rogers, that $K_0 + K_1 x + K_2 x^2 + \cdots$ was equal to

$$\begin{aligned}&(1 + x H_1(\lambda, \delta_\lambda) + x^2 H_2(\lambda, \delta_\lambda) + \cdots)(C_0 + C_1 \lambda + C_2 \lambda^2 + \cdots) \\ &= \frac{1}{(x\lambda)_\infty (x\delta_\lambda)_\infty} (C_0 + C_1 \lambda + C_2 \lambda^2 + \cdots).\end{aligned} \quad (31.75)$$

This completed the proof of (31.62).

31.5 Rogers: Third Memoir

Rogers defined q-ultraspherical polynomials and derived some of their properties in his 1895 memoir. His definition used their generating function

$$\frac{P(\lambda x)}{P(x)} = \sum_{n=0}^\infty \frac{L_n(\theta)}{(q)_n} x^n. \quad (31.76)$$

He obtained the recurrence relation

$$L_r - 2\cos\theta \cdot L_{r-1}(1 - \lambda q^{r-1}) + L_{r-2}(1 - q^{r-1})(1 - \lambda^2 q^{r-2}) = 0$$

from the equation

$$\frac{P(\lambda x)}{P(x)}(1 - 2x\cos\theta + x^2) = \frac{P(\lambda q x)}{P(qx)}(1 - 2\lambda x \cos\theta + \lambda^2 x^2). \tag{31.77}$$

He observed that by the q-binomial theorem,

$$\frac{(\lambda x e^{i\theta})_\infty}{(x e^{i\theta})_\infty} = \sum_{n=0}^{\infty} \frac{(\lambda)_n}{(q)_n} x^n e^{in\theta};$$

an expression for $L_n(\theta)$ could be obtained by multiplying this series with the series for $(\lambda x e^{-i\theta})_\infty/(xe^{-i\theta})_\infty$. He also noted the following particular cases: when $\lambda = 0$, $L_r = A_r$;

when $\lambda = q$, $L_r = (q)_r \dfrac{\sin(r+1)\theta}{\sin\theta}$; when $\lambda \to 1$, $\dfrac{1-\lambda q^r}{1-\lambda} L_r \to (q)_r \cdot 2\cos r\theta$.

For yet another noteworthy result, Rogers supposed M_r to be the same function of μ as L_r of λ:

$P(\mu x)/P(x) = 1 + \sum M_r x^r/(q)_r$. Then

$$\frac{L_r}{(q)_r} = \sum_{0 \leq s \leq r/2} \frac{M_{r-2s}(1-q^{r-2s}\mu)}{(q)_{r-2s}} \frac{(\mu-\lambda)(\mu-q\lambda)\cdots(\mu-q^{s-1}\lambda)(\lambda)_{r-s}}{(q)_s (\mu)_{r-s+1}}. \tag{31.78}$$

Secondly, Rogers gave a formula now known as the linearization formula: For $s \leq r$

$$\frac{L_r L_s}{(q)_r (q)_s} = \sum_{t=0}^{s} \frac{(1-\lambda q^{r+s-2t}) L_{r+s-2t} (\lambda)_{s-t} (\lambda)_t (\lambda)_{r-t} (\lambda^2)_{r+s-t}}{(\lambda^2)_{r+s-2t} (q)_{s-t} (q)_t (q)_{r-t} (\lambda)_{r+s-t+1}}. \tag{31.79}$$

Note that the modern definition of the q-ultraspherical polynomial $C_n(\cos\theta; \lambda|q)$ is slightly different:

$$\frac{P(x)}{P(\lambda x)} = \sum_{n=0}^{\infty} C_n(\cos\theta; \lambda|q) x^n. \tag{31.80}$$

31.6 Rogers–Szegő Polynomials

Rogers found q-extensions for two systems of orthogonal polynomials, but he did not prove their orthogonality; Szegő was the first to take a significant step in that direction. He considered polynomials orthogonal on the unit circle. In a 1921 paper he showed how to associate polynomials orthogonal on $-1 \leq x \leq 1$ of weight function $w(x)$ with polynomials orthogonal on the unit circle of weight function $f(\theta) = w(\cos\theta)|\sin\theta|$. However, Szegő had been able to find only a few simple examples; in 1926, Rogers's work motivated him to discover that the polynomials

$$\phi_n(z) = a \sum_{k=0}^{n} \frac{(q)_n}{(q)_k (q)_{n-k}} (-q^{-1/2} z)^k,$$

$$\text{where} \quad a = \frac{(-1)^n q^{n/2}}{\sqrt{(q)_n}},$$

were orthogonal on the unit circle with respect to the weight function

$$f(\theta) = |D(e^{i\theta})|^2 \text{ and } D(z) = \sqrt{(q)_\infty} (-q^{1/2} z)_\infty.$$

Szegő proved this by first observing that by the triple product identity

$$f(\theta) = \sum_{n=-\infty}^{\infty} q^{n^2/2} e^{in\theta};$$

hence,
$$\frac{1}{2\pi} \int_0^{2\pi} f(\theta) e^{-in\theta} d\theta = q^{n^2/2}, \quad n = 0, \pm 1, \pm 2, \ldots. \tag{31.81}$$

He then took $\phi_n(z) = \sum_{k=0}^n a_k z^k$ to determine a_k, by requiring the relations

$$\frac{1}{2\pi} \int_0^{2\pi} f(\theta) \phi_n(z) \overline{z^k} d\theta = 0, \quad z = e^{i\theta}, \quad k = 0, 1, \ldots, n-1. \tag{31.82}$$

By (31.81), equation (31.82) gave $\sum_{s=0}^n a_s q^{(s-k)^2/2} = 0$ or

$$\sum_{s=0}^n a_s q^{-sk+s^2/2} = 0, \quad k = 0, 1, \ldots, n-1. \tag{31.83}$$

To solve this system of equations, Szegő recalled the Rothe–Gauss formula

$$(1+qx)(1+q^2 x) \cdots (1+q^n x) = \sum_{s=0}^n \frac{(q)_n}{(q)_s (q)_{n-s}} q^{s(s+1)/2} x^s.$$

He took $x = -q^{-k-1}$ to get

$$\sum_{s=0}^n \frac{(q)_n}{(q)_s (q)_{n-s}} (-1)^s q^{-s(k+1/2)+s^2/2} = 0, \quad k = 0, 1, \ldots, n-1. \tag{31.84}$$

By comparing (31.83) and (31.84), he concluded that

$$a_s = a(-1)^s \frac{(q)_n}{(q)_s (q)_{n-s}} q^{-s/2}, \quad s = 0, 1, \ldots, n.$$

The factor a was then chosen so that

$$\frac{1}{2\pi} \int_0^{2\pi} f(\theta) |\phi_n(z)|^2 d\theta = 1.$$

We note that the triple product identity can be applied to also obtain the orthogonality of the q-Hermite polynomials $A_n(\theta)$. The orthogonality relation here would be given by

$$\frac{1}{2\pi}\int_0^\pi A_m(\theta)\,A_n(\theta)\,|(e^{2i\theta})_\infty|^2\,d\theta = \frac{\delta_{mn}}{(q^{n+1})_\infty}. \tag{31.85}$$

31.7 Feldheim and Lanzewizky: Orthogonality of q-Ultraspherical Polynomials

The work of Feldheim and Lanzewizky arose out of the papers of Fejér and Szegő on some questions relating to generalized Legendre polynomials. To define Fejér's generalized Legendre polynomial, let

$$f(z) = a_0 + a_1 z + a_2 z^2 + \cdots \tag{31.86}$$

be analytic in a neighborhood of zero with real coefficients. Then

$$|f(re^{i\theta})|^2 = \sum_{n=0}^\infty a_n r^n e^{in\theta} \sum_{n=0}^\infty a_n r^n e^{-in\theta}$$

$$= \sum_{n=0}^\infty r^n \sum_{k=0}^n a_{n-k} a_k e^{i(n-2k)\theta}$$

$$= \sum_{n=0}^\infty r^n \sum_{k=0}^n a_{n-k} a_k \cos(n-2k)\theta.$$

The last step is valid because the left-hand side is real and the coefficients are real. The polynomials $p_n(x)$ defined by

$$p_n(\cos\theta) = \sum_{k=0}^n a_k a_{n-k} \cos(n-2k)\theta, \tag{31.87}$$

where $x = \cos\theta$, are the Fejér–Legendre polynomials; they have properties similar to those of Legendre polynomials. For example, Fejér and Szegő proved that under certain conditions on the coefficients, $p_n(x)$ had n zeros in the interval $(-1, 1)$ and that the zeros of $p_n(x)$ and $p_{n+1}(x)$ separated each other. Feldheim and Lanzewizky showed that these polynomials were orthogonal when $f(z)$ in (31.86) was the q-binomial series. Their result was stated in a different form, and they did not give the orthogonality relation. Feldheim used the theorem: For a sequence of polynomials $P_n(x)$ ($n = 0, 1, 2, \ldots$) to be orthogonal, it is necessary and sufficient that the recurrence relation

$$2b_n x P_n(x) = P_{n+1}(x) + \lambda_n P_{n-1}(x) \quad (n \geq 1, \lambda_n > 0) \tag{31.88}$$

hold true. Feldheim substituted (31.87) in (31.88) and applied the trigonometric identity $2x T_n(x) = T_{n+1}(x) + T_{n-1}(x)$ where $T_k(x) = \cos k\theta$, $x = \cos\theta$. He then wrote

(31.88) as

$$b_n \sum_{k=0}^{n} a_k a_{n-k} \Big(T_{n-2k+1}(x) + T_{n-2k-1}(x) \Big)$$
$$= \sum_{k=0}^{n+1} a_k a_{n-k+1} T_{n-2k+1}(x) + \lambda_n \sum_{k=0}^{n-1} a_k a_{n-k-1} T_{n-2k-1}(x),$$

or, with $a_{-1} = 0$,

$$b_n \sum_{k=0}^{n} (a_k a_{n-k} + a_{k-1} a_{n-k+1}) T_{n-2k+1}(x) = \sum_{k=0}^{n+1} (a_k a_{n-k+1} + \lambda_n a_{k-1} a_{n-k}) T_{n-2k+1}(x).$$

By equating coefficients,

$$b_n(a_k a_{n-k} + a_{k-1} a_{n-k+1}) = a_k a_{n-k+1} + \lambda_n a_{k-1} a_{n-k}.$$

Dividing by $a_{k-1} a_{n-k}$ and setting $b_n = a_{n+1}/a_n$, Feldheim obtained

$$\lambda_n = b_n(b_{k-1} + b_{n-k}) - b_{k-1} b_{n-k}, \quad k = 1, 2, \ldots, n. \tag{31.89}$$

He considered the three equations, obtained when $k = n, n-1$, and $n-2$:

$$\lambda_n = b_n(b_{n-1} + b_0) - b_{n-1} b_0 = b_n(b_{n-2} + b_1) - b_{n-2} b_1 = b_n(b_{n-3} + b_2) - b_{n-3} b_2.$$

He initially set $b_0 = 0$, since b_0 could be arbitrarily chosen, and solved for b_n to obtain

$$b_n = \frac{b_1 b_{n-2}}{b_1 + b_{n-2} - b_{n-1}} = \frac{b_2 b_{n-3}}{b_2 + b_{n-3} - b_{n-1}}. \tag{31.90}$$

With n replaced by $n - 1$, (31.90) gave

$$b_{n-1} = \frac{b_1 b_{n-3}}{b_1 + b_{n-3} - b_{n-2}}.$$

Solving for b_{n-3} produced

$$b_{n-3} = \frac{b_{n-1}(b_{n-2} - b_1)}{b_{n-1} - b_1} \quad \text{and therefore}$$

$$b_{n-3} - b_{n-1} = \frac{b_{n-1}(b_{n-2} - b_{n-1})}{b_{n-1} - b_1} \quad \text{and then}$$

$$b_n = \frac{b_2 b_{n-1}(b_{n-2} - b_1)}{b_2(b_{n-1} - b_1) + b_{n-1}(b_{n-2} - b_{n-1})} = \frac{b_2 b_{n-1}(b_{n-2} - b_1)}{b_{n-1}(b_2 + b_{n-2} - b_{n-1}) - b_1 b_2}$$
$$= \frac{b_1 b_{n-2}}{b_1 + b_{n-2} - b_{n-1}}.$$

The last equation was simplified to

$$(b_2 - b_1)(b_{n-2} - b_{n-1}) b_{n-1} b_{n-2} = b_1 b_2 (b_{n-2} - b_{n-1})(b_{n-1} - b_1), \quad n = 3, 4, 5, \ldots.$$

From these relations, Feldheim expressed the value of b_n in the simpler form

$$b_n = \frac{b_1^2 b_2}{b_1 b_2 - (b_2 - b_1) b_{n-1}}, \quad n = 1, 2, 3, \ldots. \tag{31.91}$$

Considering the general case where b_0 was not necessarily zero, he noted that since b_0 and b_1 could be arbitrarily chosen, all the b_k in (31.91) could then be replaced by $b_k - b_0$. Feldheim then rewrote this relation as

$$\beta_n = \frac{c}{c - \beta_{n-1}} = \frac{1}{1 - \frac{\beta_{n-1}}{c}}, \quad \beta_0 = 1, \beta_1 = 1, n = 1, 2, 3, \ldots \tag{31.92}$$

$$\text{where } \beta_n = \frac{b_n - b_0}{b_1 - b_0} \quad \text{and} \quad c = \frac{b_2 - b_0}{b_2 - b_1}.$$

To solve the Riccati difference equation (31.92), Feldheim expanded $\delta_n = 1 - \beta_n/c$ as the continued fraction

$$\delta_n = \frac{R_n}{S_n} = 1 - \frac{1/c}{1-} \frac{1/c}{1-} \cdots \frac{1/c}{1-},$$

so that R_n and S_n satisfied the recurrence relation

$$t_n = t_{n-1} - \frac{1}{c} t_{n-2} \tag{31.93}$$

with initial condition $R_0 = 1$, $S_0 = 1$; $R_1 = 1 - 1/c$, $S_1 = 1$. The linear equation (31.93) was solved by the quadratic

$$x^2 - x + 1/c = 0 \quad \text{to obtain} \quad x = \frac{\sqrt{c} \pm \sqrt{c-4}}{2\sqrt{c}}.$$

For real solutions, it was required that $c \geq 4$ so that Feldheim could set $c = 4\cosh^2 \xi$. Then

$$R_n = \frac{(1 + \tanh \xi)^{n+2} - (1 - \tanh \xi)^{n+2}}{2^{n+2} \tanh \xi}, \quad S_n = \frac{(1 + \tanh \xi)^{n+1} - (1 - \tanh \xi)^{n+1}}{2^{n+1} \tanh \xi}.$$

He substituted these values for δ_n and β_n to arrive at

$$b_n = b_1 + (b_1 - b_0) \frac{\sinh(n-1)\xi}{\sinh(n+1)\xi}, \quad n = 0, 1, 2, \ldots, \tag{31.94}$$

where $\xi \geq 0$, and b_0 and b_1 were arbitrary. Feldheim applied this to (31.89) to show that $\lambda_n > 0$ if $b_1 > b_0$. Thus, he found the orthogonal generalized Legendre polynomials. He also observed that he could obtain the ultraspherical polynomials as special cases.

Lanzewizky's paper was very brief and gave only statements of his results. In his first theorem, he noted that if $C_n = a_n/a_{n-1}$, then C_n satisfied the difference equation

$$C_{n+1} = \frac{C_1 C_n - C_2 C_{n-1}}{C_1 + C_n - C_2 - C_{n-1}}, \quad n = 3, 4, 5, \ldots. \tag{31.95}$$

He presented the solution to this difference equation as

$$C_{n+1} = C_2 + (C_2 - C_1)\frac{U_{n-2}(\xi)}{U_n(\xi)}, \quad \text{where} \tag{31.96}$$

$$\xi = \frac{1}{2}\sqrt{\frac{C_3 - C_1}{C_3 - C_2}} \quad \text{and} \quad U_n(\xi) = \frac{\sin(n+1)\arccos\xi}{\sin\arccos\xi}. \tag{31.97}$$

For orthogonality, he required either that $\xi \geq 1$ and $-\xi < C_1/2C_2 < 1$; or that $\xi = i\eta$ with $\eta > 0$ and $-\eta^2 < C_1/2C_2 < 1$; he observed that for $0 < \xi < 1$, orthogonality was not possible.

Askey has pointed out that it is more convenient to write the solution of the difference equation (31.95) as

$$C_n = \frac{\alpha(1 - \beta q^{n-1})}{1 - q^n}, \tag{31.98}$$

where α, β are real constants and $|q| \leq 1$. For $|q| < 1$, we get

$$\frac{a_n}{a_0} = C_1 \cdot C_2 \cdots C_n = \alpha^n \frac{(1-\beta)(1-\beta q)\cdots(1-\beta q^{n-1})}{(1-q)(1-q^2)\cdots(1-q^n)}, \tag{31.99}$$

and hence $f(re^{i\theta}) = a_0 \sum \frac{(\beta)_n}{(q)_n} \alpha^n r^n e^{in\theta}$,

where $|\alpha r| < 1$ for convergence. Orthogonality is obtained if

$$\frac{(1-q^{n+1})(1-\beta^2 q^n)}{(1-\beta q^n)(1-\beta q^{n+1})} > 0.$$

So one may take $\alpha = 1$, yielding

$$p_n(\cos\theta) = \sum_{k=0}^n \frac{(\beta)_k (\beta)_{n-k}}{(q)_k (q)_{n-k}} \cos(n-2k)\theta. \tag{31.100}$$

These are the q-ultraspherical polynomials denoted by $C_n(x;\beta|q)$ with $x = \cos\theta$.

In 1977, Richard Askey and James Wilson derived explicit orthogonality relations for some basic hypergeometric orthogonal polynomials. These relations included the orthogonality relation for the q-ultraspherical polynomials of Rogers. But it was later, upon reading Rogers's papers, that Askey recognized the full significance of the q-ultraspherical polynomials. Jointly with Mourad Ismail, he worked out the properties of these polynomials and discovered various methods for deriving their orthogonality relation:

$$\int_{-1}^1 C_n(x;\beta|q) C_m(x;\beta|q) w_\beta(x) \frac{dx}{\sqrt{1-x^2}}$$
$$= \frac{2\pi(1-\beta)}{1-\beta q^n} \cdot \frac{(\beta^2)_n}{(q)_n} \cdot \frac{(\beta)_\infty (\beta q)_\infty}{(\beta^2)_\infty (q)_\infty} \delta_{mn}, \quad 0 < q < 1, \tag{31.101}$$

where $w_\beta(\cos\theta) = \dfrac{(e^{2i\theta})_\infty (e^{-2i\theta})_\infty}{(\beta e^{2i\theta})_\infty (\beta e^{-2i\theta})_\infty}$, $-1 < \beta < 1$. (31.102)

Interestingly, one of these methods employed Ramanujan's summation formula, paralleling the use of the triple product identity in the derivation of the orthogonality relation for the q-Hermite polynomials.

31.8 Exercises

1. Show that
$$\frac{1}{1-x} + \frac{z}{1-qx} + \frac{z^2}{1-q^2x} + \cdots = \frac{1}{1-z} + \frac{x}{1-qz} + \frac{x^2}{1-q^2z} + \cdots.$$

    See Heine (1847).

2. Show that
$$\prod_{m=0}^{n-1} \Omega\left(q^n, a - \frac{m}{n}\right) = c\,\Omega(q, na),$$

    where $c = \dfrac{\left((1-q^n)(1-q^{2n})(1-q^{3n})\cdots\right)^n}{(1-q)(1-q^2)(1-q^3)\cdots}.$

    See Heine (1847).

3. Let
$$\phi(q) = \prod_{n=0}^{\infty}(1-q^{5n+1})^{-1}(1-q^{5n+4})^{-1}, \quad \psi(q) = \prod_{n=0}^{\infty}(1-q^{5n+2})^{-1}(1-q^{5n+3})^{-1}$$

    and then prove that

    (a) $\phi(q) = \displaystyle\prod_{n=1}^{\infty}(1+q^{2n}) \cdot \sum_{n=0}^{\infty} \dfrac{q^{n^2}}{(1-q^4)(1-q^8)\cdots(1-q^{4n})},$

    (b) $\phi(q^4) = \displaystyle\prod_{n=1}^{\infty}(1-q^{2n-1}) \cdot \sum_{n=0}^{\infty} \dfrac{q^{n^2}}{(q)_{2n}},$

    (c) $\psi(q) = \displaystyle\prod_{n=1}^{\infty}(1+q^{2n}) \cdot \sum_{n=0}^{\infty} \dfrac{q^{n(n+2)}}{(1-q^4)\cdots(1-q^{4n})}.$

    See Rogers, (1894) pp. 330–331.

4. Following Rogers, define $B_r(\theta)$ by
$$\prod_{n=1}^{\infty}(1 + 2xq^n\cos\theta + x^2q^{2n}) = \sum_{n=0}^{\infty} \frac{B_n(\theta)}{(q)_n} x^n.$$

Demonstrate that

(a)
$$B_{2n}(\theta) = q^{n(n+1)} \frac{(q)_{2n}}{(q)_n (q)_n} \left(1 + \frac{1-q^n}{1-q^{n+1}} \cdot 2q \cos 2\theta \right.$$
$$\left. + \frac{(1-q^n)(1-q^{n-1})}{(1-q^{n+1})(1-q^{n+2})} \cdot 2q^4 \cos 4\theta + \cdots \right)$$

(b)
$$B_{2n+1}(\theta) = q^{(n+1)^2} \frac{(q)_{2n+1}}{(q)_n (q)_{n+1}} \left(2\cos\theta + \frac{1-q^n}{1-q^{n+1}} \cdot 2q^2 \cos 3\theta \right.$$
$$\left. + \frac{(1-q^n)(1-q^{n-1})}{(1-q^{n+2})(1-q^{n+3})} \cdot 2q^6 \cos 5\theta + \cdots \right)$$

(c)
$$(q)_\infty \left(\sum_{n=0}^\infty \frac{q^{-n/2} B_n(\theta)}{(q)_n} \right) = 1 + 2 \sum_{n=1}^\infty q^{n/2} \cos n\theta.$$

See Rogers (1917), pp. 315–316.

5. Replace $2\cos 2k\theta$ by

$$(-1)^k (1+q^k) q^{k(k-1)/2}$$

in the expression for $B_{2n}(\theta)$ in exercise 4(a) and denote the result β_{2n}. Likewise, let β_{2n+1} denote the result of replacing $2\cos(2k+1)\theta$ by

$$(-1)^k (1-q^{2k+1}) q^{k(k-1)/2}$$

in 4(b). Show that

(a) $\beta_{2n+1} = q^{n+1}(1-q^{2n+1})\beta_{2n}$,

(b) $\beta_{2n+2} = q^{n+1} \dfrac{1-q^{2n+2}}{1-q^{n+1}} \beta_{2n+1}$.

In exercise 4(c), equate terms containing even multiples of θ, and replace these cosines as indicated earlier in this exercise. Show that this process leads to

$$\frac{1}{(q)_\infty} (1 - q^2 - q^3 + q^9 + q^{11} - \cdots) = 1 + \frac{\beta_2 q^{-1}}{(q)_2} + \frac{\beta_4 q^{-2}}{(q)_4} + \cdots$$
$$= 1 + \frac{q}{1-q} + \frac{q^4}{(q)_2} + \frac{q^9}{(q)_3} + \cdots.$$

Prove that the cosines of odd multiples of θ lead to the second Rogers–Ramanujan identity. See Rogers (1917), pp. 316–317. In the 1940s, W. N. Bailey elucidated the underlying structure of Rogers's method. In the 1980s, G. E. Andrews developed Bailey's idea into a powerful tool to handle q-series and mock theta functions. Andrews named this method Bailey chains and around the same time, P. Paule independently realized the significance of Bailey's method. See Andrews (1986).

6. Show that

$$\prod_{n=0}^{\infty}\left(\frac{1-q^{n+1}}{1-q^n x}\right) = \frac{1-q}{1-x} + q \cdot \frac{x-q}{(1-x)(1-qx)}(1-q^3)$$
$$+ q^4 \cdot \frac{(x-q)(x-q^2)}{(1-x)(1-qx)(1-q^2 x)}(1-q^5) + \cdots.$$

Consider the cases $x = 0$, $x = q^{1/2}$ and $x = -1$. Prove that

$$\sum_{n=1}^{\infty} \frac{c^n q^n}{1-q^n} = \sum_{n=1}^{\infty} \frac{c^n q^{n^2}}{1-q^n} + \sum_{n=1}^{\infty} \frac{c^{(n+1)} q^{n(n+1)}}{1-cq^n}.$$

See Rogers (1893a), p. 30.

31.9 Notes on the Literature

See Turán (1990), vol. 3, p. 2626, for the quotation concerning Feldheim; see Hardy (1978) for his quote on Ramanujan. Heine (1847) contains his work on the q-hypergeometric series. For the work of Rogers mentioned in the text, see our references to his seven papers, dating from 1893 to 1917, at the end of the book. Feldheim (1941) and Lanzewizky (1941) present their work on the orthogonality of the generalized Legendre polynomials of Fejér. For Szegő's 1926 paper, "Ein Beitrag zur Theorie der Thetafunktionen," see pp. 795–805 of Szegő (1982). Also see Askey's commentary in Szegő (1982), pp. 806–811. For Askey and Ismail on the orthogonality relations for the q-ultraspherical polynomials, see their paper in Cheney (1980), pp. 175–182. See Andrews (1986) for an interesting and detailed discussion, with good references, of the work of Heine, Thomae, Rogers, and Ramanujan.

32

Primes in Arithmetic Progressions

32.1 Preliminary Remarks

One of the great theorems of number theory states that any arithmetic progression $l, l+k, l+2k, \ldots$, where l and k are relatively prime, contains an infinite number of primes. Euler conjectured this result for the particular case $l = 1$, probably in the 1750s, though it appeared in print much later. It appears that the general form of this conjecture first appeared in Legendre's 1798 book on number theory. Then in 1837, Dirichlet proved the case with k prime and he published a demonstration of the general result three years later. Interestingly, the germ of the central idea in Dirichlet's proof came from Euler. Note that in a paper of 1737, Euler used the formula

$$1 + \frac{1}{2} + \frac{1}{3} + \frac{1}{4} + \cdots = \frac{1}{\left(1 - \frac{1}{2}\right)\left(1 - \frac{1}{3}\right)\left(1 - \frac{1}{5}\right)\cdots} \tag{32.1}$$

to prove that the series of the reciprocals of primes $\sum 1/p$ was divergent. Of course, this implied that the number of primes was infinite. It is obvious that the series and product in Euler's formula are divergent but, as discussed in chapter 26, this defect is easy to remedy. Euler studied numerous Dirichlet series and their infinite products, including

$$\frac{\pi}{4} = 1 - \frac{1}{3} + \frac{1}{5} - \frac{1}{7} + \cdots = \frac{1}{\left(1 - \frac{1}{3}\right)\left(1 + \frac{1}{5}\right)\left(1 - \frac{1}{7}\right)\left(1 - \frac{1}{11}\right)\left(1 + \frac{1}{13}\right)\cdots}, \tag{32.2}$$

where the primes of the form $4n+3$ appeared with a negative sign and those of the form $4n+1$ had a positive sign. This led him to the series

$$\frac{1}{3} - \frac{1}{5} + \frac{1}{7} + \frac{1}{11} - \frac{1}{13} - \frac{1}{17} + \frac{1}{19} + \cdots. \tag{32.3}$$

In a letter of October 28, 1752, to Goldbach, Euler wrote that he had found the sum of this series to be approximately 0.334980, implying that the series $\sum 1/p$ and $\sum 1/q$, where p and q were primes of the form $4n+1$ and $4n+3$, respectively, were both divergent. Euler's results were published in a posthumous paper of 1785, also containing the conjecture mentioned previously.

In the second edition of his 1808 book on number theory, Legendre gave an incorrect proof of Dirichlet's theorem. Legendre was not known for taking heed of the criticism of others and the flawed proof was again included in the 1830 third edition of his book. In his papers of 1837 and 1838 on this topic, Dirichlet pointed out that Legendre's proof depended upon an ingenious lemma and showed that Legendre's proof of the lemma was inadequate. Dirichlet unsuccessfully tried to prove the lemma, writing that he found the lemma at least as difficult to prove as the theorem deduced from it. In 1859, Athanase Dupré (1808–1869) published his proof that Legendre's lemma was false, for which he was awarded half the 1858 *Gran Prix* from the French Academy of Sciences.

In his 1837 paper presented to the Berlin Academy, Dirichlet wrote that he based his ideas on Euler's *Introductio in Analysin Infinitorum*. He expressed the sum of the reciprocals of the primes in the given arithmetic progression as an appropriate linear combination of the logarithms of the $p-1$ L-series arising from the $p-1$ characters modulo p. He then had to prove the divergence of this expression, based on the divergence of the series corresponding to the trivial character, $\ln L_0(1)$. Then, in order to maintain the singularity of $L_0(1)$, he had to show that the values $L_k(1)$ did not vanish. For L-series arising from complex characters, Dirichlet was easily able to do this. However, it was much more difficult to prove that the L-series produced by the real character, defined by the Legendre symbol, did not vanish. To tackle this problem, Dirichlet first reduced the infinite series to a finite sum and considered two cases of primes: those of the form $4m+3$, and then $4m+1$. The first case was relatively easy; for the second case, he used a result on Pell's equation, from the *Disquisitiones*. The appearance of Pell's equation may have alerted Dirichlet to the connection between L-functions for real characters and quadratic forms. In fact, this allowed him to prove in 1840 that the class number of the binary quadratic forms of a given determinant could be evaluated in terms of the value of the L-function at 1, implying that the function did not vanish.

With these papers, based on Euler's work on series, Dirichlet established analytic number theory as a distinct new branch of mathematics; he applied infinite series to the derivation of the class number formula, to the problem of primes in an arithmetic progression, and to the evaluation of Gauss sums, leading to a proof of the quadratic reciprocity law. Interestingly, Gauss wrote Dirichlet in 1838 that he had worked with similar ideas around 1801, but he regretted not finding the time to develop and publish them. Indeed, Gauss's unpublished papers included an incomplete manuscript partially outlining the theory of the class number formula.

We parenthetically note that the ancient Babylonians considered particular cases of Pell's equation $x^2 - ny^2 = 1$, where n is a non-square positive integer; in India, Brahmagupta in the 600s and Bhaskara in the 1100s gave procedures for solving it. William Brouncker can be credited with giving a general method for its solution in a 1657 letter to Wallis, in response to a challenge from Fermat; Lagrange finally gave a rigorous derivation in 1766.

Dirichlet's proof of the nonvanishing of the L-series was somewhat roundabout, but a more direct proof was published by the Belgian mathematician Charles de la Vallée Poussin (1866–1962) in his 1896 paper "Démonstration simplifiée du théorème de Dirichlet." Vallée Poussin took the L-functions to be functions of a complex variable and then employed analytic function theory to give his elegant proof. Interestingly, he made use of a construction also given by Dirichlet. Vallée Poussin, who made many contributions to various areas of analysis, studied at the university at Louvain under L.-P. Gilbert, whom he succeeded as professor of mathematics at the age of 26.

As early as 1861–62, Hermann Kinkelin of Basel studied L-functions of complex variables, proving their functional relation for characters modulo a prime power. And in 1889 Rudolf Lipschitz, using the Hurwitz zeta function, proved the latter result for general Dirichlet characters. In papers published between 1895 and 1899, Franz Mertens gave proofs of the nonvanishing of the L-series by elementary methods, that is, without the use of quadratic forms or functions of a complex variable. One of these proofs used a technique from an 1849 paper of Dirichlet on the average behavior of the divisor function. This result now has many elementary proofs, but the simplest may be due to Paul Monsky in 1994, based on the earlier elementary proof of A. Gelfond and Yuri Linnik, published in 1962.

32.2 Euler: Sum of Prime Reciprocals

In his 1737 paper "Variae Observationes circa Series Infinitas," Euler showed that the sum of the reciprocals of primes

$$\frac{1}{2}+\frac{1}{3}+\frac{1}{5}+\frac{1}{7}+\frac{1}{11}+\frac{1}{13}+ \text{etc.}$$

was of infinite magnitude and was, moreover, the logarithm of the harmonic series

$$1+\frac{1}{2}+\frac{1}{3}+\frac{1}{4}+\frac{1}{5}+ \text{etc.}$$

Taking the logarithm of (32.1), Euler got

$$\ln\left(1+\frac{1}{2}+\frac{1}{3}+\frac{1}{4}+\cdots\right) = -\ln\left(1-\frac{1}{2}\right)-\ln\left(1-\frac{1}{3}\right)-\ln\left(1-\frac{1}{5}\right)-\cdots$$

$$=\frac{1}{2}+\frac{1}{3}+\frac{1}{5}+\cdots+\frac{1}{2}\left(\frac{1}{2^2}+\frac{1}{3^2}+\frac{1}{5^2}+\cdots\right)$$

$$+\frac{1}{3}\left(\frac{1}{2^3}+\frac{1}{3^3}+\frac{1}{5^3}+\cdots\right)+\cdots$$

$$\equiv A+\frac{1}{2}B+\frac{1}{3}C+\cdots.$$

He could express this relation as

$$e^{A+\frac{1}{2}B+\frac{1}{3}C+\frac{1}{4}D+\text{etc.}} = 1+\frac{1}{2}+\frac{1}{3}+\frac{1}{4}+\frac{1}{5}+\frac{1}{6}+\frac{1}{7}+ \text{etc.}$$

He then observed that since the harmonic series diverged to ∞ and the series B, C, D, etc. were finite, the series

$$\frac{1}{2}B + \frac{1}{3}C + \frac{1}{4}D + \text{etc.}$$

was negligible and hence

$$e^A = 1 + \frac{1}{2} + \frac{1}{3} + \frac{1}{4} + \frac{1}{5} + \text{etc.}$$

By taking the logarithm of both sides, he obtained his result:

$$\frac{1}{2} + \frac{1}{3} + \frac{1}{5} + \frac{1}{7} + \frac{1}{11} + \frac{1}{13} + \frac{1}{17} + \text{etc.}$$

$$= \ln\left(1 + \frac{1}{2} + \frac{1}{3} + \frac{1}{4} + \frac{1}{5} + \text{etc.}\right)$$

Euler then noted that the harmonic series summed to $\ln \infty$ and hence the sum of the reciprocals of the primes was $\ln \ln \infty$. To understand this, recall that $\sum_{k=1}^{n} 1/k \sim \ln n$.

32.3 Dirichlet: Infinitude of Primes in an Arithmetic Progression

In 1758–59, Waring and Simpson, starting with $f(x) = \sum_{n=0}^{\infty} a_n x^n$, used roots of unity to obtain an expression for $\sum_{n=0}^{\infty} a_{mn+l} x^n$. In other words, they used characters of the additive group $\mathbb{Z}_m$ to extract from the power series the subsequence of terms in an arithmetic progression. Since the L-functions were multiplicative, Dirichlet had to define and use characters of the multiplicative group. An additional complication for Dirichlet was that he had to work with the logarithm of the L-functions and therefore had to prove their nonvanishing. In the case where $m = p$ was a prime, using some results of Gauss, he found an intricate proof of this fact, published in his 1837 paper "Beweiss des Satzes, dass jede unbegrenzte arithmetische Progression ...". Dirichlet supposed p to be a prime and set $\Omega^k = e^{2\pi i k/(p-1)}$, $k = 0, 1, \ldots, p-1$. He let L_k denote the L-function defined by the product

$$L_k(s) = \Pi \left(1 - \frac{w^{\gamma_q}}{q^s}\right)^{-1}, \quad \text{where } w = \Omega^k = e^{2\pi i k/(p-1)},$$

and the product was taken over all primes $q \neq p$. Recall that in the chapter on L-series, γ_q was defined by means of a generator of the multiplicative cyclic group of the integers modulo p. Then

$$\log L_k = \sum \frac{w^{\gamma_q}}{q^s} + \frac{1}{2} \sum \frac{w^{2\gamma_q}}{q^{2s}} + \frac{1}{3} \sum \frac{w^{3\gamma_q}}{q^{3s}} + \cdots.$$

To extract the primes in the arithmetic progression identical to 1 modulo p, Dirichlet first observed that for any integer h

$$1 + \Omega^{h\gamma} + \Omega^{2h\gamma} + \cdots + \Omega^{(p-2)h\gamma} = \begin{cases} p-1, & h\gamma \equiv 0 \pmod{p-1}, \\ 0 & h\gamma \not\equiv 0 \pmod{p-1}. \end{cases}$$

It followed that

$$\log(L_0 L_1 \cdots L_{p-2}) = (p-1)\left(\sum \frac{1}{q^s} + \frac{1}{2}\sum \frac{1}{q^{2s}} + \frac{1}{3}\sum \frac{1}{q^{3s}} + \cdots\right), \quad (32.4)$$

where the primes q in the first sum satisfied $q \equiv 1 \pmod{p}$; those in the second sum satisfied $q^2 \equiv 1 \pmod{p}$; those in the third $q^3 \equiv 1 \pmod{p}$; and so on. The second and later sums were convergent for $s \geq 1$; to show that $\sum 1/q$ was divergent, Dirichlet had to focus on the behavior of $L_0, L_1, \ldots, L_{p-2}$ as $s \to 1^+$. Dirichlet first expressed the series as an integral; he calculated that for any positive real number k,

$$S = \frac{1}{k^{1+\rho}} + \frac{1}{(k+1)^{1+\rho}} + \frac{1}{(k+2)^{1+\rho}} + \cdots$$

$$= \frac{1}{\Gamma(1+\rho)} \int_0^1 \log^\rho(1/x) \frac{x^{k-1}}{1-x} dx$$

$$= \frac{1}{\rho} + \frac{1}{\Gamma(1+\rho)} \int_0^1 \left(\frac{x^{k-1}}{1-x} - \frac{1}{\log(1/x)}\right) \log^\rho(1/x) dx.$$

He observed that the integral was convergent as $\rho \to 0+$. Note that since the series $L_0(s)$ is given by $\sum 1/m^s$ where the sum is over all integers m not divisible by p, Dirichlet could write

$$L_0(1+\rho) = \sum_{m=1}^{p-1} \sum_{l=0}^{\infty} \frac{1}{(m+lp)^{1+\rho}}.$$

Next, he applied the foregoing integral representation for the series to obtain

$$\sum_{l=0}^{\infty} \frac{1}{(m+lp)^{1+\rho}} = \frac{1}{p^{1+\rho}} \sum_{l=0}^{\infty} \frac{1}{(\frac{m}{p}+l)^{1+\rho}} = \frac{1}{p} \cdot \frac{1}{\rho} + \phi(\rho)$$

where $\phi(\rho)$ had a finite limit as $\rho \to 0^+$. So Dirichlet could conclude that

$$L_0(1+\rho) = \frac{p-1}{p} \cdot \frac{1}{\rho} + \phi(\rho),$$

where $\lim_{\rho \to 0^+} \phi(\rho)$ was finite. This implied that $\log L_0(1+\rho)$ behaved like $-\log \rho$ as $\rho \to 0^+$. Dirichlet also showed that the series $L_1(1), L_2(1), \ldots, L_{p-2}(1)$ were convergent. Thus, if $L_j(1) \neq 0$ for $j = 1, 2, \ldots, p-2$, then the product $L_0 L_1 \cdots L_{p-2}$ had to diverge as $\rho \to 0^+$, and the series $\sum 1/q$ for $q \equiv 1 \pmod{p}$ would also diverge. This proved that there existed an infinity of primes of the form $pl + 1$.

Dirichlet found a simple proof that for $j \neq \frac{p-1}{2}$, $L_j(1) \neq 0$. For such a j, $\Omega^{p-1-j} \neq \Omega^j$; Dirichlet therefore considered the product $L_j L_{p-1-j}$. Recall that

$$L_j(s) = \frac{1}{\Gamma(s)} \int_0^1 \frac{\frac{1}{x} f(x)}{1-x^p} \log^{s-1}(1/x) dx = \psi(s) + \chi(s)\sqrt{-1}.$$

Dirichlet noted that $L_j(s)$ was differentiable for $s > 0$, and hence by the mean value theorem he had

$$\psi(1+\rho) = \psi(1) + \rho \psi'(1+\delta\rho), \ \chi(1+\rho) = \chi(1) + \rho \chi'(1+\epsilon\rho),$$

32.3 Dirichlet: Infinitude of Primes in an Arithmetic Progression

where $0 < \delta < 1$ and $0 < \epsilon < 1$. Since $L_{p-1-j}(s)$ was the complex conjugate of $L_j(s)$, he got

$$L_{p-1-j}(s)L_j(s) = \psi^2(s) + \chi^2(s).$$

Next, if $L_j(1) = 0$, then $L_{p-1-j}(1) = 0$. This implied that $\psi(1) = 0$ and $\chi(1) = 0$. Thus,

$$\log L_j(1+\rho)L_{p-1-j}(1+\rho) = \log \rho^2 \left(\psi'^2(1+\delta\rho) + \chi'^2(1+\epsilon\rho) \right)$$

$$= -2\log\frac{1}{\rho} + \log(\psi'^2(1+\delta\rho) + \chi'^2(1+\epsilon\rho)).$$

These calculations implied that if $L_j(1) = 0$ for $j \neq \frac{p-1}{2}$, then

$$\log L_0 L_j L_{p-1-j} = -\log\frac{1}{\rho} + \phi(\rho),$$

and the term on the left-hand side tended to $-\infty$ as $\rho \to 0^+$. Clearly, $\log(L_0 L_1 \cdots L_{p-2})$ was positive from (32.4) so Dirichlet had come to a contradiction. This completed the proof for complex characters.

Dirichlet then dealt with the difficult case in which $j = \frac{p-1}{2}$. In this case, $\omega^{\frac{p-1}{2}} = e^{\pi i} = -1$ and hence the character and the series $L_{\frac{p-1}{2}}$ were real and given by

$$L_{\frac{p-1}{2}}(s) = \sum_{n=1}^{\infty} \left(\frac{n}{p}\right) \frac{1}{p^s}.$$

Recall Dirichlet's results from the L-series chapter: When $p \equiv 3 \pmod 4$,

$$\sum \left(\frac{n}{p}\right) \frac{1}{n} = \frac{\pi}{p\sqrt{p}} \left(\sum b - \sum a\right),$$

and when $p \equiv 1 \pmod 4$,

$$\sum \left(\frac{n}{p}\right) \frac{1}{n} = \frac{1}{\sqrt{p}} \log \frac{\prod \sin(b\pi/p)}{\prod \sin(a\pi/p)},$$

where a and b were quadratic residues and nonresidues, respectively, modulo p. For $p \equiv 3 \pmod 4$, Dirichlet noted that

$$\sum a + \sum b = \sum_{m=1}^{p-1} m = \frac{p(p-1)}{2} = \text{an odd integer.}$$

Hence for $p \equiv 3 \pmod 4$, $\sum b - \sum a$ could not be zero and $L_{\frac{p-1}{2}}(1) \neq 0$. For $p \equiv 1 \pmod 4$, Dirichlet used a result Gauss proved in section 357 of his *Disquisitiones*. This important result in cyclotomy stated that

$$2\prod_a \left(x - e^{2\pi i a/p}\right) = Y - Z\sqrt{p}, \quad 2\prod_b \left(x - e^{2\pi i b/p}\right) = Y + Z\sqrt{p}, \qquad (32.5)$$

where Y and Z were polynomials in x with integral coefficients; hence, Gauss had

$$Y^2 - pZ^2 = 4\prod_{k=1}^{p-1}\left(x - e^{2\pi ik/p}\right) = 4\frac{x^p - 1}{x - 1}.$$

Dirichlet set $g = Y(1)$, $h = Z(1)$ so that g and h were integers and $g^2 - ph^2 = 4p$; he could conclude that g was divisible by p. He could then set $g = pk$ to obtain $h^2 - pk^2 = -4$. Since p could not divide 4, he could write that $h \neq 0$. Next, when $x = 1$ in (32.5), he got

$$2\prod_a \left(1 - e^{2\pi ia/p}\right) = 2^{\frac{p+1}{2}}(-1)^{\frac{p-1}{4}} e^{\pi i \sum_a a} \prod_a \frac{e^{\pi ia/p} - e^{-\pi ia/p}}{2i}$$

$$= 2^{\frac{p+1}{2}}(-1)^{\frac{p-1}{4} + \sum_a a}\prod_a \sin(a\pi/p) = 2^{\frac{p+1}{2}}\prod_a \sin(a\pi/p).$$

Note that this last equation depends on the fact that when p is of the form $4n + 1$, a and $p - a$ are both quadratic residues, and the residues can be grouped in pairs. There are $\frac{p-1}{4}$ such pairs, and it follows that

$$\sum_a a = \sum_b b = \frac{1}{4}p(p-1).$$

Similarly, $\quad 2\prod_b \left(1 - e^{2\pi ib/p}\right) = 2^{\frac{p+1}{2}}\prod_b \sin(b\pi/p),$

and thus, because $h \neq 0$,

$$\frac{\prod_b \sin(b\pi/p)}{\prod_a \sin(a\pi/p)} = \frac{k\sqrt{p} + h}{k\sqrt{p} - h} \neq 1.$$

This proved that $L_{\frac{p-1}{2}}(s)$ did not vanish at $s = 1$ and also that the number of primes $\equiv 1$ (mod p) was infinite. To show that the number of primes $\equiv m$ (mod p) was infinite, Dirichlet gave a modified argument. He considered the sum

$$\log L_0 + \Omega^{-\gamma_m}\log L_1 + \Omega^{-2\gamma_m}\log L_2 + \cdots + \Omega^{-(p-2)\gamma_m}\log L_{p-2}$$

$$= (p-1)\left(\sum \frac{1}{q^{1+\rho}} + \frac{1}{2}\sum \frac{1}{q^{2+2\rho}} + \frac{1}{3}\sum \frac{1}{q^{3+3\rho}} + \cdots\right),$$

where the primes q in the first sum satisfied $q \equiv m$ (mod p) and those in the kth sum satisfied $q^k \equiv m$ (mod p). Since he had already proved that $\log L_1, \log L_2, \ldots, \log L_{p-2}$ were finite as $\rho \to 0^+$ and that $\log L_0$ behaved like $\log(1/\rho)$, Dirichlet could conclude that, when the sum was taken over primes $\equiv m \mod p$, $\sum 1/q$ diverged.

32.4 Class Number and $L_\chi(1)$

In a paper of 1838 published in *Crelle's Journal*, "Sur l'usage des séries infinies dans la théorie des nombres," Dirichlet worked out some particular cases of his class number

32.4 Class Number and $L_\chi(1)$

formula. In this formula, he expressed the class number in terms of $L_\chi(1)$, a necessarily nonvanishing quantity. In order to give a definition of class number, we first observe that for a, b, c integers, $b^2 - ac$ is called the determinant or discriminant of the quadratic form $ax^2 + 2bxy + cy^2$. Two quadratic forms with the same determinant are in the same class if a linear substitution $x = \alpha x' + \beta y'$ and $y = \gamma x' + \delta y'$ with $\alpha\gamma - \beta\delta = 1$ transforms one quadratic form into the other. This basic definition can be traced to Lagrange. In addition, Lagrange proved that there was a finite number of such classes (called the class number) for a given negative discriminant. Note that Lagrange worked with b instead of $2b$.

In his 1838 paper, Dirichlet considered quadratic forms of determinant $-q$ with q prime. He separated his proof into the two cases $q = 4\nu + 3$ and $q = 4\nu + 1$; we present Dirichlet's proof of the former case. He denoted by f the primes for which the discriminant $-q$ was a quadratic residue and by g the primes for which it was not:

$$\left(\frac{-q}{f}\right) = \left(\frac{f}{q}\right) = 1, \quad \left(\frac{-q}{g}\right) = \left(\frac{g}{q}\right) = -1.$$

He then considered the L-series relations

$$\prod \frac{1}{1 - 1/f^s} \prod \frac{1}{1 - 1/g^s} = \sum \frac{1}{n^s},$$

$$\prod \frac{1}{1 - 1/f^s} \prod \frac{1}{1 + 1/g^s} = \sum \left(\frac{n}{q}\right) \frac{1}{n^s},$$

$$\prod \frac{1}{1 - 1/f^{2s}} \prod \frac{1}{1 - 1/g^{2s}} = \sum \frac{1}{n^{2s}},$$

where the n were odd numbers not divisible by q. He deduced that

$$\frac{\sum \frac{1}{n^s} \cdot \sum \left(\frac{n}{q}\right) \frac{1}{n^s}}{\sum \frac{1}{n^{2s}}} = \prod \frac{1 + 1/f^s}{1 - 1/f^s} = \prod \left(1 + \frac{2}{f^s} + \frac{2}{f^{2s}} + \frac{2}{f^{3s}} + \cdots\right) = \sum \frac{2^\mu}{m^s}, \quad (32.6)$$

where the summation was over odd integers m divisible only by primes of the type f; μ was the number of distinct primes f by which m was divisible. Dirichlet denoted the inequivalent quadratic forms of determinant $-q$ by

$$ax^2 + 2bxy + cy^2, \; a'x^2 + 2b'xy + c'y^2, \ldots$$

and then observed that articles 180, 155, 156, and 105 of Gauss's *Disquisitiones* implied that

$$2 \sum \frac{2^\mu}{m^s} = \sum \frac{1}{(ax^2 + 2bxy + cy^2)^s} + \sum \frac{1}{(a'x^2 + 2b'xy + c'y^2)^s} + \cdots, \quad (32.7)$$

where the summation on the right was taken over positive as well as negative values of x and y relatively prime to one another. From this Dirichlet deduced that

$$2 \sum \frac{1}{n^s} \cdot \sum \left(\frac{n}{q}\right) \frac{1}{n^s} = \sum \frac{1}{n^{2s}} \cdot \sum \frac{1}{(ax^2 + 2bxy + cy^2)^s} + \cdots. \quad (32.8)$$

Without giving details, Dirichlet remarked in this paper that "by means of geometric considerations" it could be proved that

$$\sum \frac{1}{n^{2(1+\rho)}} \cdot \sum \frac{1}{(ax^2 + 2bsy + cy^2)^{1+\rho}} \sim \frac{q-1}{2q\sqrt{q}} \cdot \frac{\pi}{\rho}, \ \rho \to 0^+. \tag{32.9}$$

Thus, given h different inequivalent forms of determinant $-q$, that is, if h were the class number, then the right-hand side of (32.8) could be expressed as

$$\frac{h(q-1)}{2q\sqrt{q}} \cdot \frac{\pi}{\rho}, \ \rho \to 0^+. \tag{32.10}$$

On the other hand, since

$$\sum \frac{1}{n^s} = \left(1 - \frac{1}{2^s}\right)\left(1 - \frac{1}{q^s}\right) \zeta(s),$$

he had

$$\sum \frac{1}{n^{1+\rho}} \sim \frac{q-1}{2q} \cdot \frac{1}{\rho} \ \text{as } \rho \to 0^+. \tag{32.11}$$

He applied (32.10) and (32.11) to (32.8) and found a special case of his famous class number formula

$$h = \frac{2\sqrt{q}}{\pi} \sum \left(\frac{n}{q}\right) \frac{1}{n}. \tag{32.12}$$

This formula expressed the class number in terms of the value of an L-series at $s = 1$. Since the class number had to be at least one, the series had a nonzero value. In 1840, Dirichlet published a proof along similar lines of this result for arbitrary negative determinants. Recall that Dirichlet made liberal use of results from Gauss in his proofs; his contemporaries reported that his copy of the *Disquisitiones* was never kept on the shelf, but on his writing table, and that it always accompanied him on his travels. Through his lectures on number theory, published by Dedekind, Dirichlet made the work of Gauss accessible to all his students.

32.5 De la Vallée Poussin's Complex Analytic Proof of $L_\chi(1) \neq 0$

Before he published his famous work on the distribution of primes, de la Vallée Poussin published a paper presenting a simpler proof of Dirichlet's theorem on primes in arithmetic progressions, observing that his proof was more natural since it did not depend on the theory of quadratic forms. In this paper, presented to the Belgian Academy in 1896, de la Vallée Poussin defined $L_\chi(s)$, where χ was a character modulo an integer M, as a function of a complex variable s. By a simple argument, he showed that for the principal character χ_0, $L_{\chi_0}(s)$ was an analytic function for Re $s > 0$, except for a simple pole at $s = 1$. On the other hand, for any nonprincipal character, the corresponding L-function was analytic for Re $s > 0$ with no exception. De la Vallée Poussin's proof that $L_\chi(1) \neq 0$ employed a function similar to one constructed by Dirichlet in his discussion of quadratic forms of negative discriminant. He let χ be a real nonprincipal character; he let q_1 denote primes for which $\chi(q_1) = 1$ and let q_2 denote primes for which $\chi(q_2) = -1$. He set

$$\psi(s) = \frac{L_\chi(s) L_{\chi_0}(s)}{L_{\chi_0}(2s)}.$$

Then, for Re $s > 1$, he observed that

$$\psi(s) = \prod \frac{1+q_1^{-s}}{1-q_1^{-s}} = \prod (1 + 2q_1^{-s} + 2q_1^{-2s} + \cdots) = \sum_{n=1}^{\infty} \frac{a_n}{n^s},$$

where $a_n \geq 0$. In addition, since $L_{\chi_0}(2s)$ had a pole at $s = 1/2$, he deduced that $\psi(s) = 0$ at $s = 1/2$. Vallée Poussin also observed that there was at least one prime q_1. If not, then $\psi(s) = 1$ for Re $s > 1$ and by analytic continuation $\psi(1/2) = 1$. This contradicted $\psi(1/2) = 0$, and hence $a_n > 0$ for some n in $\sum a_n/n^s$.

In order to obtain a proof by contradiction, Vallée Poussin next assumed that $L_\chi(s) = 0$ at $s = 1$, so that this zero would cancel the pole of $L_{\chi_0}(s)$ at $s = 1$ and $L_\chi(s) L_{\chi_0}(s)$ would be analytic for Re $s > 0$. Again for Re $s > 1$, the derivatives of $\psi(s)$ were given by

$$\psi^{(m)}(s) = (-1)^m \sum_{n=1}^{\infty} \frac{a_n (\log n)^m}{n^s}, \quad m = 1, 2, 3, \ldots.$$

He let $a > 0$ so that $\psi(1 + a + t)$ had radius of convergence greater than $a + 1/2$ and

$$\psi(1+a+t) = \psi(1+a) + t\psi'(1+a) + \frac{t^2}{2!}\psi''(1+a) + \cdots.$$

Denoting $(-1)^m \psi(1+a)$ by A_m, it was clear that $A_m > 0$ and for $t = -(a+1/2)$

$$\psi(1/2) = \psi(1+a) + (a+1/2)A_1 + \frac{(a+1/2)^2}{2!} A_2 + \cdots.$$

Since $\psi(1/2) = 0$ and the all the terms on the right were positive, Vallée Poussin arrived at the necessary contradiction.

32.6 Gelfond and Linnik: Proof of $L_\chi(1) \neq 0$

Gelfond and Linnik's proof that $L_\chi(1) \neq 0$ for any real nonprincipal character χ modulo m was presented in their 1962 book. They made the observation that if ϕ denoted the Euler totient function and $T(n) = \sum_{k=1}^{n} \chi(k)$, then for any positive integer N

$$|T(N) - T(n-1)| < \phi(m) \quad \text{and} \quad \left| \sum_{k=n}^{N} \frac{\chi(k)}{k} \right| < \frac{2\phi(m)}{n}. \tag{32.13}$$

Note that the second inequality follows by partial summation. Gelfond and Linnik defined the function

$$U(x) = \sum_{n=1}^{\infty} \frac{\chi(n) x^n}{1 - x^n} = \sum_{n=1}^{\infty} \left(\sum_{d|n} \chi(d) \right) x^n \qquad (32.14)$$

and showed that

$$U(x) > \frac{1}{2\sqrt{1-x}}. \qquad (32.15)$$

They then proved that if $L_\chi(1) = 0$, then

$$U(x) = O\left(\ln \frac{1}{1-x} \right). \qquad (32.16)$$

Clearly, (32.15) and (32.16) were in contradiction to one another, indicating that the assumption $L_\chi(1) = 0$ had to be false. Next, to prove (32.15), Gelfond and Linnik made the important observation that

$$f_n \equiv \sum_{d|n} \chi(d) = \prod_{k=1}^{s} (1 + \chi(p_k) + \cdots + \chi^{\nu_k}(p_k)), \quad \left(n = p_1^{\nu_1} p_2^{\nu_2} \cdots p_s^{\nu_s} \right).$$

Since $\chi(p) = 1, -1,$ or 0, it followed that $f_n \geq 0$. Then, if n were a square, all the ν_k would be even and each of the s factors of f_n would be ≥ 1. Thus, $f_n \geq 1$ when n was a square. Hence, with $1 > x > 1/2, x > x_0$

$$U(x) > \sum_{n=1}^{\infty} x^{n^2} = \int_1^{\infty} x^{t^2} dt + O(1)$$

$$= \frac{1}{\sqrt{-\ln x}} \int_0^{\infty} e^{-t^2} dt + O(1) = \frac{\sqrt{\pi}}{2\sqrt{-\ln x}} + O(1)$$

$$= \frac{\sqrt{\pi}}{2} \frac{1}{(-\ln(1-(1-x)))^{1/2}} + O(1) > \frac{1}{2\sqrt{1-x}}.$$

We note that Gelfond and Linnik set $x > x_0$ for some x_0 such that the inequalities would hold. To prove (32.16), they set

$$S_n = \sum_{k=n}^{\infty} \frac{\chi(k)}{k}, \quad S_1 = L_\chi(1);$$

$$R_1(x) = U(x) - \frac{L_\chi(1)}{1-x}$$

$$= \sum_{n=1}^{\infty} \chi(n) \frac{x^n}{1-x^n} - \sum_{n=1}^{\infty} \frac{\chi(n)}{n} \frac{x^n}{1-x^n} + \sum_{n=1}^{\infty} (S_n - S_{n+1}) \frac{x^n}{1-x} - \frac{L_\chi(1)}{1-x}$$

$$= \sum_{n=1}^{\infty} \chi(n) \left(\frac{x^n}{1-x^n} - \frac{x^n}{n(1-x)} \right) - \sum_{n=0}^{\infty} S_{n+1} x^n. \qquad (32.17)$$

To see how they arrived at the last equation, observe that

$$\frac{1}{1-x}\left(\sum_{n=1}^{\infty}(S_n - S_{n+1})x^n - L_\chi(1)\right)$$
$$= \frac{1}{1-x}\left(S_1 x + \sum_{n=1}^{\infty} S_{n+1}x^n(1-x) - S_1\right) = -\sum_{n=0}^{\infty} S_{n+1}x^n.$$

Next, by (32.13), $S_n = O(1/n)$, and hence they could write the second sum in (32.17) as

$$\sum_{n=0}^{\infty} S_{n+1}x^n = O\left(\sum_{n=1}^{\infty} \frac{x^n}{n}\right) = O\left(\ln\frac{1}{1-x}\right).$$

By an application of Abel's summation by parts to the first sum in (32.17), they got

$$\left|\sum_{n=1}^{\infty} \chi(n)\left(\frac{x^n}{1-x^n} - \frac{x^n}{n(1-x)}\right)\right|$$
$$= \left|\sum_{n=1}^{\infty}(T(n) - T(n-1))\left(\frac{x^n}{1-x^n} - \frac{x^n}{n(1-x)}\right)\right|$$
$$= \left|\sum_{n=1}^{\infty} T(n)\left(\frac{x^n}{1-x^n} - \frac{x^n}{n(1-x)} - \frac{x^{n+1}}{1-x^{n+1}} + \frac{x^{n+1}}{(n+1)(1-x)}\right)\right|$$
$$< \frac{\phi(m)}{1-x}\sum_{n=1}^{\infty}\left|\frac{x^n}{1+x+\cdots+x^{n-1}} - \frac{x^{n+1}}{1+x+\cdots+x^n} - \frac{x^n}{n(n+1)} - \frac{(1-x)x^n}{n+1}\right|$$
$$< \frac{\phi(m)}{1-x}\sum_{n=1}^{\infty}\left(\frac{x^n}{1+x+\cdots+x^{n-1}} - \frac{x^{n+1}}{1+x+\cdots+x^n} - \frac{x^n}{n(n+1)}\right)$$
$$+ \phi(m)\sum_{n=1}^{\infty}\frac{x^n}{n+1}$$
$$= 2\phi(m)\sum_{n=1}^{\infty}\frac{x^n}{n+1} = O\left(\ln\frac{1}{1-x}\right).$$

Note that the final inequality was possible because the expression in the first sum was positive; then, since the series was telescoping, it would sum to $x^n(1-x)/(n+1)$. It then followed from (32.17) that if $L_\chi(1) = 0$, then $U(x) = O\left(\ln\frac{1}{1-x}\right)$; this completed the proof.

32.7 Monsky's Proof That $L_\chi(1) \neq 0$

In 1994, Paul Monsky showed that Gelfond's proof of $L_\chi(1) \neq 0$ could be considerably simplified. Use of the strong result (32.15) turned out to be avoidable. Observe that $\lim_{x \to 1-} U(x) = \infty$, because $U(x) > \sum_{n=1}^{\infty} x^{n^2}$. Monsky demonstrated that if $L_\chi(1)$ vanished, then $U(x)$ was bounded. This contradiction proved the result. His simplification took place in the first sum, $R_1(x)$, in (32.17), where

$$R_1(x) = -\sum_{n=1}^{\infty} \left(\frac{\chi(n)}{n(1-x)} - \frac{\chi(n)x^n}{1-x^n} \right) \equiv -\sum_{n=1}^{\infty} \frac{\chi(n)}{1-x} b_n.$$

Monsky first showed that $b_1 \geq b_2 \geq b_3 \geq \cdots$. He noted that

$$b_n - b_{n+1} = \frac{1}{n(n+1)} - \frac{x^n}{(1+x+\cdots+x^{n-1})(1+x+\cdots+x^n)}.$$

Then, by the inequality of the arithmetic and geometric means

$$1 + x + \cdots + x^{n-1} \geq nx^{(n-1)/2} \geq nx^{n/2} \text{ and}$$

$$1 + x + \cdots + x^n \geq (n+1)x^{n/2}.$$

Hence, $b_n \geq b_{n+1}$. Applying Abel's partial summation, Monsky wrote

$$\sum_{n=1}^{\infty} \frac{\chi(x)}{1-x} b_n \leq \frac{\phi(m)b_1}{1-x} = \phi(m). \tag{32.18}$$

He next assumed that $L_\chi(1) = 0$, and this implied $U(x) = R_1(x)$; but (32.18) in turn implied that $\lim_{x \to 1-} U(x)$ could not be infinite. Thus, he got a contradiction to prove the result.

32.8 Exercises

1. Investigate Chebyshev's assertion in an 1853 letter to Fuss that

$$\lim_{c \to 0} \left(e^{-3c} - e^{-5c} + e^{-7c} + e^{-11c} - e^{-13c} - e^{-17c} + e^{-19c} + e^{-23c} - \cdots \right)$$

diverges to $+\infty$. See Chebyshev (1899–1907), vol. 1, p. 697. See also Hardy (1966–1978), vol. 2, pp. 42–49, where Hardy and Littlewood derive it from the extended Riemann hypothesis for the series $1^{-s} - 3^{-s} + 5^{-s} - 7^{-s} + \cdots$. Note the editor's comment on this result on p. 98.

2. Let χ be a real nonprincipal character modulo m, and let $f(n) = \sum \chi(d)$, where the sum is over all divisors d of n. Show that if

$$G(x) = \sum_{n \leq x} f(n)/\sqrt{n}, \quad \text{then} \quad \lim_{x \to \infty} G(x) = \infty.$$

Show also that

$$G(x) = 2\sqrt{x}L(1,\chi) + O(1).$$

Conclude that if $Ł(1,\chi) \neq 0$, a contradiction ensues. See Mertens (1895).

3. Suppose χ is a primitive character mod d, that is, there does not exist a proper divisor m of d such that $\chi(a) = \chi(b)$ whenever $a \equiv b$ and ab is prime to d. Define the Gauss sum $G(\chi)$ by

$$G(\chi) = \sum_{a=1}^{d} \chi(a) e^{2\pi i a/d}.$$

Prove that

$$\overline{\chi}(n) G(\chi) = \sum_{a=1}^{d} \chi(a) e^{2\pi i n a/d}.$$

This result is due to Vallée Poussin (2000), vol. 1, pp. 358–362; he also defined the concept of a primitive character.

4. Define the Euler polynomials $E_n(t)$ by the relation

$$\frac{2e^{tx}}{1+e^x} = \sum_{n=0}^{\infty} \frac{E_n(t)}{n!} x^n.$$

Let χ be a primitive character (mod d) and let k be a positive integer such that $\chi(-1) = (-1)^k$. Prove that if q is the greatest integer in $(d-1)/2$, then

$$\frac{(k-1)!}{(2\pi i)^k} G(\overline{\chi}) L(k, \chi) = \frac{1}{2(2^k - \chi(2))} \sum_{a=1}^{q} \overline{\chi}(a) E_{k-1}(2a/d).$$

This formula is due to Shimura (2007), p. 35; in this book, Shimura observed that most books and papers give only one result on the values of the Dirichlet L-function; Shimura derived several new formulas for these values, including the foregoing example. For a discussion of Shimura's well-known conjecture related to Fermat's theorem, see Gouvêa (1994) and Shimura (2008).

5. Prove that if the set of positive integers is partitioned into a disjoint union of two nonempty subsets, then at least one of the subsets must contain arbitrarily long arithmetic progressions. This result was conjectured by I. Schur and proved in 1927 by van der Waerden, who studied under E. Noether. The reader may enjoy reading the proof in Khinchin (1998), a book originally written in 1945 as a letter to a soldier recovering from his wounds. In 1927, van der Waerden's theorem was a somewhat isolated result, but it has now become a part of Ramsey theory, an important area of combinatorics. See Graham, Rothschild, and Spencer (1990). Robert Ellis's algebraic methods in topological dynamics also have applications to this topic. See Ellis, Ellis, and Nerurkar (2000).

6. Prove that the primes contain arbitrarily long arithmetic progressions. For this result of Ben Green and Terrence Tao, see Green's article in Duke and Tschinkel (2007).

32.9 Notes on the Literature

See Eu. I-14, pp. 242–244 for Euler's theorem on the sum of the reciprocals of the primes. Dirichlet (1969), vol. I, pp. 313–342 is a reprinting of his 1837 paper in which he proved a particular case of his famous theorem on primes in arithmetic progressions. Pp. 357–74 reproduce a particular case of his theorem on the class number of quadratic forms; see pp. 360–364 for the material presented in the text. De la Vallée Poussin (2000), vol. 1, pp. 187–222, contains his proof of the theorem on primes in arithmetic progressions; see pp. 208–213 for his proof of the nonvanishing of $L(1, \chi)$. For Gelfond and Linnik's proof of this result, see their book originally published in 1962, an English translation of which appeared in 1966, pp. 46–47. Monsky (1994) provides the simplified version of this proof. A historical account of the topic of this chapter was given by Littlewood's student Davenport (1980). This book was very influential because of its treatment of the large sieve, a relatively new topic at the time of first publication in 1967. The extensive notes in each chapter refer to numerous papers and books on this and related topics.

33

Distribution of Primes: Early Results

33.1 Preliminary Remarks

Prime numbers appear to be distributed among the integers in a random way. Mathematicians have searched for a pattern or patterns in the sequence of primes, discovering many interesting features and properties of primes and sequences of primes, but many fundamental questions remain outstanding. In the area of prime number distribution, even apparently very elementary results can be enlightening. For example, in 1737, Euler proved that the series $\sum 1/p$, where p is prime, was divergent. He also knew that $\sum_{n=1}^{\infty} 1/n^2$ was convergent. By combining these results, one may see that the prime numbers are more numerous than the square numbers. Thus, for large enough x, we expect that $\pi(x)$, the number of primes less than or equal to x, satisfies $\pi(x) > \sqrt{x}$. In fact, extending this type of reasoning, we may expect that

$$\pi(x) > x^{1-\delta} \tag{33.1}$$

for any $\delta > 0$ and x correspondingly large enough. Recall that, in fact, Euler had a fairly definite idea of how the series of prime reciprocals diverged:

$$\sum 1/p = \ln(\ln \infty). \tag{33.2}$$

From this, it can easily be shown, by means of a nonrigorous, probabilistic argument, that the density of primes in the interval $(1, x)$ is approximately $1/\ln x$. In 1791 or 1792, when he was about 15 years old, Gauss conjectured just this result.

Gauss never published anything on the distribution of primes but in 1849 he wrote a letter to the astronomer J. F. Encke giving some insight into his thought in this area. Gauss recounted that he had started making a table of prime numbers from a very young age, noting the number of primes in each chiliad, or interval of a thousand. As a consequence of this work, around 1792 he wrote the following remark in the margin of his copy of J. C. Schulze's mathematical tables:

$$\text{Primzahlen unter } a \, (= \infty) \quad \frac{a}{la}.$$

We may understand this to mean that

$$\lim_{x \to \infty} \frac{\pi(x)}{x/\ln x} = 1. \qquad (33.3)$$

This is the prime number theorem. Let us see how Gauss may have come to this conclusion. Consider the following table:

x	$\pi(x)$	$\pi(x)/x$
10	4	0.4
100	25	0.25
1000	168	0.168
10000	1229	0.1229
100000	9592	0.09592
1000000	78498	0.078498

Look at the column for $\pi(x)/x$. If we divide 0.4, the number in the first row, by 2, 3, 4, 5, 6, then we get approximately the numbers in the second, third, fourth, fifth, and sixth rows. The result is even nicer if we change the 0.4 to 0.5 and then do the division. So if we write $\pi(x)/x$ as $1/f(x)$, then $f(x)$ has the property that $f(10^n) = nf(10)$ for $n = 2, 3, 4, 5, 6$. This calculation strongly suggests that $f(x)$ is the logarithmic function. Moreover, $1/f(10) = 0.4$ and $\ln 10 = 2.3$; this may have led Gauss to his conjecture that $f(x) = \ln x$.

In his letter to Encke, Gauss suggested the approximation $\pi(x) \approx \int_2^x \frac{dt}{\ln t}$. In fact, he gave the following table of values for $\pi(x)$ and the corresponding values of the integral:

x	$\pi(x)$	$\int_2^x \frac{dt}{\ln t}$	error
500000	41556	41606.4	+50.4
1000000	78501	79627.5	+126.5
1500000	114112	114263.1	+151.1
2000000	148883	149054.8	+171.8
2500000	183016	183245.0	+229.0
3000000	216745	216970.6	+225.6.

Observe that there are inaccuracies in this table. Gauss made mistakes in his extensive calculations of primes, but the number of his mistakes is surprisingly small. For example, his value of the number of primes less than a million was overestimated by three, while he underestimated those under three million by 72.

In an 1810 letter to the astronomer Olbers, F. W. Bessel (1784–1846) mentioned the logarithmic integral, now defined by

$$\text{li}(x) = \lim_{\epsilon \to 0} \left(\int_0^{1-\epsilon} + \int_{1+\epsilon}^x \right) \frac{dt}{\ln t} = \int_2^x \frac{dt}{\ln t} + 1.0451. \qquad (33.4)$$

Gauss had already computed this integral for several values of x and Bessel noted in his letter that he learned from Gauss that $\pi(4,000,000) = 33,859$, while the corresponding value of the logarithmic integral was 33,922.621995.

In his 1798 book on number theory, Legendre made a similar conjecture: that $x/(A \ln x + B)$ was a good approximation of $\pi(x)$ for suitable A and B. In the second edition of his book, published in 1808, he gave the values $A = 1$ and $B = -1.08366$. Gauss observed in his letter to Encke that as the value of x was made larger, the value of B must likewise increase. However, Gauss was unwilling to conjecture that $B \to -1$ as $x \to \infty$. It is interesting to note that while Gauss was writing these thoughts to Encke, the Russian mathematician Chebyshev was developing his ideas on prime numbers, showing that if B tended to a limit as $x \to \infty$, then the limit had to be -1.

In 1849, the Russian Academy of Sciences published a collection of Euler's papers on number theory. In 1847, the editor, Viktor Bunyakovski, solicited Chebyshev's participation in this project, thereby arousing his interest in number theory. Thus, in 1849 Chebyshev defended his doctoral thesis on theory of congruences, one of whose appendices discussed the number of primes not exceeding a given number. He there expressed doubt about the accuracy of Legendre's formula. He then went on to prove that if n was any fixed nonnegative integer and ρ was a positive real variable, then the sum

$$\sum_{x=2}^{x=\infty} \left(\pi(x+1) - \pi(x) - \frac{1}{\ln x} \right) \frac{\ln^n x}{x^{1+\rho}}, \tag{33.5}$$

considered as a function of ρ, approached a finite limit as $\rho \to 0$. We mention that Chebyshev wrote $\phi(x)$ for $\pi(x)$. From this theorem, he deduced that $\frac{x}{\pi(x)} - \ln x$ could not have a limit other than -1 as $x \to \infty$. He then observed that this result contradicted Legendre's formula, under which the limit was given as -1.08366.

Chebyshev wrote a second paper on prime numbers in 1850. This important work was apparently motivated by Joseph Bertrand's conjecture that for all integers $n > 3$, there was at least one prime between n and $2n - 2$. In 1845, Bertrand used this conjecture to prove a theorem on symmetric functions. In group theoretic terms, the theorem states that the index of a proper subgroup of the symmetric group S_n is either 2 or $\geq n$. Chebyshev proved Bertrand's conjecture using Stirling's approximation. He also showed that the series $\sum_p 1/(p \ln p)$ converged. The results he obtained implied the double inequality

$$0.92129 < \frac{\pi(x)}{x/\ln x} < 1.10555. \tag{33.6}$$

In a paper of 1881, Sylvester used Chebyshev's analysis to give improved bounds, obtaining 0.95695 for the lower bound and 1.04423 for the upper bound. Though Schur and others have succeeded in narrowing the gap between the bounds, it appears that Chebyshev's methods cannot be developed to give a proof of the prime number theorem. We note that, in order to prove Bertrand's conjecture, Chebyshev defined two arithmetical functions of interest even today:

$$\theta(x) = \sum_{p \leq x} \ln p \tag{33.7}$$

$$\psi(x) = \theta(x) + \theta(x^{1/2}) + \theta(x^{1/3}) + \theta(x^{1/4}) + \cdots. \tag{33.8}$$

In fact, Chebyshev proved the inequalities

$$Ax - \frac{5}{2}\ln x - 1 < \psi(x) < \frac{6}{5}Ax + \frac{5}{4\ln 6}\ln^2 x + \frac{5}{4}\ln x + 1, \qquad (33.9)$$

$$\text{where} \quad A = \ln\left(\frac{2^{1/2}3^{1/3}5^{1/5}}{30^{1/30}}\right) = 0.92129202\ldots.$$

He then used these inequalities to indicate a method for obtaining the result for $\pi(x)$, though he did not give the results explicitly. Chebyshev also proved that if $\lim_{x\to\infty} \psi(x)/x$ existed, its value was 1. Note that this implies that if $\lim_{x\to\infty} \frac{\pi(x)}{x/\ln x}$ exists, then this limit too must be 1.

At the end of his paper, Sylvester noted that for a proof of the prime number theorem, "we shall probably have to wait until some one is born into the world as far surpassing Tchebycheff in insight and penetration as Tchebycheff has proved himself superior in these qualities to the ordinary run of mankind." Chebyshev's elementary but powerful methods formed the basis of a new topic, elementary methods in analytic number theory, and also served as motivation for Alphonse de Polignac (1826–1863) and Franz Mertens (1840–1927) to firmly establish this new subject. In 1874, Mertens showed that Chebyshev's results could be used to obtain asymptotic formulas for the series

$$\sum_{p\leq x}\ln p/p \quad \text{and} \quad \sum_{p\leq x} 1/p.$$

Thus, he proved the following refinement of a result of de Polignac:

$$\sum_{p\leq x}\frac{\ln p}{p} = \ln x + O(1), \qquad (33.10)$$

where $O(1)$ denoted a quantity bounded as $x \to \infty$. Mertens also gave a more precise formulation of Euler's 1737 result (33.2):

$$\sum_{p\leq x}\frac{1}{p} = \ln\ln x + C + O\left(\frac{1}{\ln x}\right), \qquad (33.11)$$

$$\text{where} \quad C = \gamma - \sum_{k=2}^{\infty}\frac{1}{k}\sum_{p}\frac{1}{p^k} \qquad (33.12)$$

and γ denoted Euler's constant.

In his famous paper of 1859, Riemann introduced ideas through which the prime number theorem would eventually be proved. Riemann's interest in prime number theory was not surprising, surrounded as he was by great researchers in this field. Riemann began his paper by mentioning Gauss, Euler, and his good friend and teacher Dirichlet, writing that their attention to the subject would surely justify its further study. He did not mention Chebyshev, but he was familiar with the work of Chebyshev to whom he sent a copy of his paper. Also, we know that as a student Riemann studied Legendre's number theory book very carefully. Moreover, Dirichlet stated in a note of 1838 that his analytic methods for studying primes could provide a proof of Legendre's conjecture related to the prime number theorem; Dirichlet, however, did not publish any ideas in this direction.

Riemann based his investigation of $\pi(x)$ on Euler's product formula

$$\zeta(s) = \sum_{n=1}^{\infty} 1/n^s = \prod_p (1-p^{-s})^{-1}.$$

His innovation here was to take s to be a complex variable with $\operatorname{Re} s > 1$. He then defined $\zeta(s)$ as a contour integral, thereby extending its domain to the whole complex plane, except for the pole at $s = 1$. He used the Euler product to show that

$$\frac{\log \zeta(s)}{s} = \int_1^{\infty} f(x) x^{-s-1} dx, \quad \operatorname{Re} s > 1, \tag{33.13}$$

where $f(x) = F(x) + \frac{1}{2} F(x^{1/2}) + \frac{1}{3} F(x^{1/3}) + \cdots.$ \hfill (33.14)

We here write log because complex variables are involved. Riemann defined $F(x)$ as the number of primes less than x when x was not prime; but when x was a prime,

$$F(x) = \frac{F(x+0) + F(x-0)}{2}.$$

Thus, Riemann's $F(x)$ was essentially $\pi(x)$. He obtained the integral representation for $f(x)$ by a method we now call Mellin inversion. Actually, he applied the Fourier inversion to get

$$f(y) = \frac{1}{2\pi i} \int_{a-\infty i}^{a+\infty i} \frac{\log \zeta(s)}{s} y^s ds, \quad a > 1. \tag{33.15}$$

To evaluate this integral, Riemann defined the entire function

$$\xi(s) = (s-1) \Pi(s/2) \pi^{-s/2} \zeta(s), \tag{33.16}$$

where $\Pi(s) = s\Gamma(s)$. He then obtained an infinite product (or Hadamard product) for $\xi(s)$, given by

$$\xi(s) = \xi(0) \prod_{\rho} (1 - s/\rho). \tag{33.17}$$

To use this formula effectively in (33.15), one must first understand the distribution of the zeros ρ. It is easy to show that $0 \leq \operatorname{Re} \rho \leq 1$. It follows from the functional equation for $\zeta(s)$ that if ρ is a zero, then so is $1 - \rho$. Riemann then observed that the number of roots ρ whose imaginary parts lay between 0 and some value T was approximately

$$\frac{T}{2\pi} \log \frac{T}{2\pi} - \frac{T}{2\pi}, \tag{33.18}$$

where the relative error was of the order $1/T$. He sketched a one-sentence proof of this result and added that the estimate for the number of zeros with $\operatorname{Re} \rho = 1/2$ was about the same as in (33.18). He remarked that it was very likely, though his passing attempts to prove it had failed, that all the roots had Re $\rho = 1/2$. This is the famous Riemann hypothesis.

By combining (33.15) and (33.17), and assuming the truth of his hypothesis, Riemann derived the formula

$$f(x) = \mathrm{li}(x) - \sum_\alpha \left(\mathrm{li}(x^{\frac{1}{2}+\alpha i}) + \mathrm{li}(x^{\frac{1}{2}-\alpha i}) \right)$$
$$+ \int_x^\infty \frac{1}{t^2-1} \frac{dt}{t \log t} + \log \xi(0), \qquad (33.19)$$

where the sum $\sum_\alpha$ was taken over all positive α such that $\frac{1}{2} + i\alpha$ was a zero of $\xi(s)$. Note that $\xi(0) = 1/2$, though due to some confusion in Riemann's notation, he obtained a different value.

In the final remarks in his paper, Riemann noted first that $F(x)$ (or $\pi(x)$) could be obtained from $f(x)$ by the inversion

$$F(x) = \sum_{m=1}^\infty \frac{\mu(m)}{m} f(x^{1/m}),$$

where $\mu(m)$ was the Möbius function. We remark that Riemann did not use the brief notation $\mu(m)$. He also noted that the approximation $F(x) = \mathrm{li}(x)$ was correct only to an order of magnitude $x^{1/2}$, yielding a value somewhat too large, while better approximation was given by

$$\mathrm{li}(x) - \frac{1}{2}\mathrm{li}(x^{\frac{1}{2}}) - \frac{1}{3}\mathrm{li}(x^{\frac{1}{3}}) - \frac{1}{5}\mathrm{li}(x^{\frac{1}{5}}) + \frac{1}{6}\mathrm{li}(x^{\frac{1}{6}}) - \cdots. \qquad (33.20)$$

Apart from the Riemann hypothesis, the most difficult part of Riemann's paper was his factorization of $\xi(s)$. Indeed, Weierstrass had to develop his theory of product representations of entire functions before even the simpler aspects of $\xi(s)$ could be tackled. Then in 1893, Jacques Hadamard (1865–1963) worked out the theory of factorization of entire functions of a finite order and applied it to $\xi(s)$. That set the stage for his 1896 proof of the prime number theorem. Briefly, Hadamard first proved that $\zeta(s)$ had no zeros on the line $\mathrm{Re}\, s = 1$. Then, using earlier ideas of Cahen and Halphen, he applied Mellin inversion to an integral of a weighted average, say $A(x)$, of Chebyshev's arithmetical function $\theta(x)$. From this inversion, Hadamard derived the asymptotic behavior of $A(x)$ and this in turn yielded the asymptotic behavior of $\theta(x)$, that

$$\lim_{x \to \infty} \frac{\theta(x)}{x} = 1.$$

This proved the prime number theorem. It is interesting that in an 1885 letter to Hermite, Stieltjes claimed to have a proof of the Riemann hypothesis. Aware of this claim, Hadamard remarked that since Stieltjes had not published his proof, he himself would put forward a proof of the simpler result.

Also in 1896, C. J. de la Vallée-Poussin published his own proof of the prime number theorem (PNT), based on similar ideas. After these proofs appeared, research on the prime number theorem centered around efforts to simplify the proof and to understand its logical structure. E. Landau, G. H. Hardy, J. E. Littlewood, and N. Wiener were the main contributors to this endeavor. In 1903, Landau found a new proof of the prime number theorem, not dependent on Hadamard's theory of entire functions or on the functional relation for the zeta function. Landau required only that $\zeta(s)$ could be continued slightly to the left of $\operatorname{Re} s = 1$. This method could be extended to the Dedekind zeta function for number fields and Landau used it to state and prove the prime ideal theorem.

Hardy, Littlewood, and Wiener explicated the key role of Tauberian theorems in prime number theory. It became clear from their work that the prime number theorem was equivalent to the statement that $\zeta(1+it) \neq 0$ for real t. On the basis of this result, Hardy expected that the zeta function would play a crucial role in any proof of the PNT. But two years after Hardy's death, Atle Selberg and Paul Erdős found an elementary proof of the PNT, obviously without zeta function theory.

33.2 Chebyshev on Legendre's Formula

Recall that Euler proved the divergence of $\sum_p 1/p$, where p was prime, by comparing it with $\ln(\sum_{n=1}^\infty 1/n)$. In his 1849 paper, Chebyshev followed up on this work, proving in his first theorem the existence of the limit

$$\lim_{\rho \to 0} \left(\sum_p \frac{\ln p}{p^{1+\rho}} - \sum_{k=2}^\infty \frac{1}{k^{\rho+1}} \right) \tag{33.21}$$

and, more generally, of the limit after taking derivatives with respect to ρ,

$$\lim_{\rho \to 0} \left(\sum_p \frac{\ln^n p}{p^{1+\rho}} - \sum_{k=2}^\infty \frac{\ln^{n-1} k}{k^{\rho+1}} \right), \quad n = 1, 2, 3, \ldots. \tag{33.22}$$

In order to obtain information about $\pi(x)$, the number of primes less than x, he wrote the series in (33.22) as

$$\sum_{x=2}^\infty \left(\pi(x+1) - \pi(x) - \frac{1}{\ln x} \right) \frac{\ln^n x}{x^{1+\rho}}. \tag{33.23}$$

Then since

$$\frac{1}{\ln x} - \int_x^{x+1} \frac{dt}{\ln t} = O\left(\frac{1}{x}\right) \quad \text{as} \quad x \to \infty,$$

Chebyshev deduced the important corollary of the existence of the limit

$$\lim_{\rho \to 0} \sum_{x=2}^\infty \left(\pi(x+1) - \pi(x) - \int_x^{x+1} \frac{dt}{\ln t} \right) \frac{\ln^n x}{x^{1+\rho}}. \tag{33.24}$$

From this, Chebyshev proceeded to derive his second theorem: For any positive real number α, any positive integer n, and for infinitely many integer values of x,

$$\pi(x) > \int_2^x \frac{dt}{\ln t} - \frac{\alpha x}{\ln^n x} \tag{33.25}$$

and with the same conditions on α and n, for infinitely many integer values of x,

$$\pi(x) < \int_2^x \frac{dt}{\ln t} + \frac{\alpha x}{\ln^n x}. \tag{33.26}$$

Using (33.25) and (33.26), Chebyshev could state his remarkable result that if

$$\lim_{x \to \infty} \pi(x) / \int_2^x \frac{dt}{\ln t} \quad \left(\text{or} \quad \lim_{x \to \infty} \frac{\pi(x)}{x / \ln x} \right)$$

existed, then its value had to be 1. Of course, he was unable to show existence here, and that was the essence of the PNT. From (33.25) and (33.26), he also deduced that if

$$\lim_{x \to \infty} \left(\frac{x}{\pi(x)} - \ln x \right)$$

existed, then it had to be -1. He supposed the limit to be L, so that there would exist an N such that for $x > N$,

$$L - \epsilon < \frac{x}{\pi(x)} - \ln x < L + \epsilon.$$

But by (33.25), there would be an infinite number of integers $x > N$ such that

$$\frac{x}{\int_2^x \frac{dt}{\ln t} - \frac{\alpha x}{\ln^n x}} - \ln x > L - \epsilon,$$

$$\text{or} \quad L + 1 < \frac{x - (\ln x - 1)\left(\int_2^x \frac{dt}{\ln t} - \frac{\alpha x}{\ln^n x}\right)}{\int_2^x \frac{dt}{\ln t} - \frac{\alpha x}{\ln^n x}} + \epsilon. \tag{33.27}$$

Similarly, (33.26) implied an inequality in the other direction. At this point, Chebyshev remarked that by a principle of differential calculus (now called l'Hôpital's rule), the expression on the right-hand side of (33.27) could be made arbitrarily small as x became large, so that the result followed. He also remarked that this theorem determined that the limit of $\frac{x}{\pi(x)} - \ln x$ as x went to infinity, was -1, contradicting Legendre, who predicted the limit would be -1.08366.

Chebyshev's proof of (33.26) was similar to his argument for (33.25). He first supposed (33.26) to hold for only a finite number of positive integers x so that there would be an integer a larger than e^n and larger than the largest integer x for which (33.26) would hold. Then for $x > a$

$$\pi(x) - \int_2^x \frac{dt}{\ln t} \geq \frac{\alpha x}{\ln^n x}, \quad \frac{n}{\ln x} < 1. \tag{33.28}$$

33.2 Chebyshev on Legendre's Formula

Chebyshev showed that in this case the series in (33.24) would diverge, a contradiction proving the result. To demonstrate this divergence, he used Abel's summation by parts:

$$\sum_{x=a+1}^{s} u_x(v_{x+1} - v_x) = u_s v_{s+1} - u_a v_{a+1} - \sum_{x=a+1}^{s} v_x(u_x - u_{x-1}).$$

He took

$$v_x = \pi(x) - \int_2^x \frac{dt}{\ln t}, \quad u_x = \frac{\ln^n s}{x^{1+\rho}},$$

so that

$$\sum_{x=a+1}^{s} \left(\pi(x+1) - \pi(x) - \int_x^{x+1} \frac{dt}{\ln t} \right) \frac{\ln^n x}{x^{1+\rho}}$$

$$= \left(\pi(s+1) - \int_2^{s+1} \frac{dt}{\ln t} \right) \frac{\ln^n s}{s^{1+\rho}} - \left(\pi(a+1) - \int_2^{a+1} \frac{dt}{\ln t} \right) \frac{\ln^n a}{a^{1+\rho}}$$

$$- \sum_{x=a+1}^{s} \left(\pi(x) - \int_2^x \frac{dt}{\ln t} \right) \left(\frac{\ln^n x}{x^{1+\rho}} - \frac{\ln^n(x-1)}{(x-1)^{1+\rho}} \right).$$

By the mean value theorem,

$$\frac{\ln^n x}{x^{1+\rho}} - \frac{\ln^n(x-1)}{(x-1)^{1+\rho}} = \left(\frac{n}{\ln(x-\theta)} - (1+\rho) \right) \frac{\ln^n(x-\theta)}{(x-\theta)^{2+\rho}},$$

where $0 < \theta < 1$ and θ depended on x. The sum in the previous expression then took the form

$$\sum_{x=a+1}^{s} \left(\pi(x) - \int_2^x \frac{dt}{\ln t} \right) \left(1 + \rho - \frac{n}{\ln(x-\theta)} \right) \frac{\ln^n(x-\theta)}{(x-\theta)^{2+\rho}}. \tag{33.29}$$

Then for $x > a$,

$$1 + \rho - \frac{n}{\ln(x-\theta)} > 1 - \frac{n}{\ln a},$$

and by (33.28)

$$\pi(x) - \int_2^x \frac{dt}{\ln t} \geq \frac{\alpha x}{\ln^n x} \geq \frac{\alpha(x-\theta)}{\ln^n(x-\theta)}.$$

Observe that Chebyshev could derive the last inequality because $\frac{x}{\ln^n x}$ was an increasing function. Therefore, he could see that the sum (33.29) was greater than

$$\alpha \left(1 - \frac{n}{\ln a} \right) \sum_{x=a+1}^{s} \frac{1}{(x-\theta)^{1+\rho}}.$$

Chebyshev thus arrived at a contradiction: when $s \to \infty$, he had the infinite series (33.24) diverging as $\rho \to 0$.

Chebyshev's proof of the existence of the limit (33.21), his first theorem mentioned earlier, made use of the formula found in Euler and Abel:

$$\int_0^\infty \frac{e^{-x}}{e^x - 1} x^\rho \, dx = \int_0^\infty e^{-2x}(1 + e^{-x} + e^{-2x} + \cdots) x^\rho \, dx$$

$$= \sum_{m=2}^\infty \int_0^\infty e^{-mx} x^\rho \, dx = \sum_{m=2}^\infty \frac{1}{m^{1+\rho}} \int_0^\infty e^{-x} x^\rho \, dx. \quad (33.30)$$

To show that the limit (33.21) existed, Chebyshev rewrote the sums contained in it as

$$\frac{d}{d\rho}\left(\sum_p \ln\left(1 - \frac{1}{p^{1+\rho}}\right) + \sum_p \frac{1}{p^{1+\rho}}\right)$$

$$+ \frac{d}{d\rho}\left(\ln \rho - \sum_p \ln\left(1 - \frac{1}{p^{1+\rho}}\right)\right) + \left(\sum_{m=1}^\infty \frac{1}{m^{1+\rho}} - \frac{1}{\rho}\right), \quad (33.31)$$

and proved that each of the three expressions in parentheses was finite as $\rho \to 0$. He proved the more general result for (33.22), by showing that the derivatives of those expressions also had finite limits. Using (33.30), Chebyshev rewrote the third expression in (33.31) as a ratio of two integrals:

$$\sum_{m=2}^\infty \frac{1}{m^{1+\rho}} - \frac{1}{\rho} = \frac{\int_0^\infty \left(\frac{1}{e^x - 1} - \frac{1}{x}\right) e^{-x} x^\rho \, dx}{\int_0^\infty e^{-x} x^\rho \, dx}. \quad (33.32)$$

He noted that these integrals converged as $\rho \to 0$ and that the derivatives of (33.32) contained expressions of the form $\int_0^\infty \left(\frac{1}{e^x-1} - \frac{1}{x}\right) e^{-x} x^\rho (\ln x)^k \, dx$ or $\int_0^\infty e^{-x} x^\rho (\ln x)^k \, dx$; these integrals also had finite limits as $\rho \to 0$. To show that the middle expression in (33.31),

$$\ln \rho - \sum_p \ln\left(1 - \frac{1}{p^{1+\rho}}\right),$$

was finite as $\rho \to 0$, Chebyshev employed the Euler product

$$\sum_{m=1}^\infty \frac{1}{m^{1+\rho}} = \prod_p \left(1 - \frac{1}{p^{1+\rho}}\right)^{-1}.$$

After taking the logarithm of both sides and adding $\ln \rho$ to each side, he had

$$\ln \rho - \sum_p \ln\left(1 - \frac{1}{p^{1+\rho}}\right) = \ln\left(\left(1 + \sum_{m=2}^\infty \frac{1}{m^{1+\rho}}\right)\rho\right)$$

$$= \ln\left(1 + \rho + \left(\sum_{m=2}^\infty \frac{1}{m^{1+\rho}} - \frac{1}{\rho}\right)\rho\right). \quad (33.33)$$

Noting the expression on the right-hand side, Chebyshev thus proved the existence of the limit of the left-hand side as well as the limits of all its derivatives, as $\rho \to 0$. It is even simpler to show that the first expression in (33.31), and all its derivatives, have a finite limit as $\rho \to 0$. This proves Chebyshev's first theorem.

At the end of his 1849 paper, Chebyshev followed Legendre in assuming the prime number theorem to prove that

$$\frac{1}{2} + \frac{1}{3} + \frac{1}{5} + \cdots + \frac{1}{x} = \ln \ln x + c, \qquad (33.34)$$

where x was a very large prime and c was finite. Chebyshev corrected the corresponding formula in Legendre, who had $\ln(\ln x - 0.08366)$ on the right-hand side. Chebyshev also suggested a similar change in Legendre's formula for the product

$$\left(1 - \frac{1}{2}\right)\left(1 - \frac{1}{3}\right)\left(1 - \frac{1}{5}\right)\cdots\left(1 - \frac{1}{x}\right) = \begin{cases} c_0/\ln x & \text{in Chebyshev,} \\ c_0/(\ln x - 0.08366) & \text{in Legendre.} \end{cases} \qquad (33.35)$$

In 1874, Mertens proved, without assuming the then-unproved PNT, that $c_0 = e^{-\gamma}$, where γ was Euler's constant. Thus, Mertens's result implies

$$\prod_{p \leq x}\left(1 - \frac{1}{p}\right)^{-1} = e^{\gamma} \ln x + O(1). \qquad (33.36)$$

An approximate value of e^{γ} is 1.781, whereas, presumably on numerical evidence, Gauss gave the value of the constant to be 1.874.

33.3 Chebyshev's Proof of Bertrand's Conjecture

In his second memoir on prime numbers, Chebyshev proved Bertrand's conjecture, making effective and original use of Stirling's approximation. In the course of his discussion of the series for the logarithm of $n!$, he was led to define two related arithmetical functions $\theta(x)$ and $\psi(x)$:

$$\theta(x) = \sum_{p \leq x} \ln p, \qquad \psi(x) = \sum_{p^n \leq x} \ln p, \qquad (33.37)$$

where p was prime. Chebyshev immediately noted the clear relation between the two:

$$\psi(x) = \theta(x) + \theta(\sqrt{x}) + \theta(\sqrt[3]{x}) + \theta(\sqrt[4]{x}) + \cdots. \qquad (33.38)$$

Keeping in mind the preceding definitions and a result first noted by Legendre,

$$n! = \prod_{p \leq n} p^{\lfloor n/p \rfloor + \lfloor n/p^2 \rfloor + \lfloor n/p^3 \rfloor + \cdots}, \qquad (33.39)$$

Chebyshev observed that if $T(x) = \ln \lfloor x \rfloor!$, then

$$T(x) = \psi(x) + \psi(x/2) + \psi(x/3) + \cdots. \qquad (33.40)$$

Next, he set $a = \lfloor x \rfloor$ so that, by Stirling's formula,

$$T(x) = \ln a! < \frac{1}{2}\ln 2\pi + a \ln a - a + \frac{1}{2}\ln a + \frac{1}{12a},$$

$$T(x) = \ln(a+1)! - \ln(a+1) > \frac{1}{2}\ln 2\pi + (a+1)\ln(a+1) - (a+1) - \frac{1}{2}\ln(a+1).$$

Thus,

$$\frac{1}{2}\ln 2\pi + x \ln x - x - \frac{1}{2}\ln x < T(x) < \frac{1}{2}\ln 2\pi + x \ln x - x + \frac{1}{2}\ln x + \frac{1}{12x}. \tag{33.41}$$

From these bounds for $T(x)$, Chebyshev obtained bounds for $\psi(x)$. Of course, one might obtain an expression for $\psi(x)$ in terms of $T(x)$ from (33.40) by means of Möbius inversion. However, Chebyshev chose to work with the sum

$$T(x) - T(x/2) - T(x/3) - T(x/5) + T(x/30). \tag{33.42}$$

He showed that when the value of $T(x)$, taken from (33.40), was substituted in (33.42), the result was the alternating series

$$\psi(x) - \psi(x/6) + \psi(x/7) - \psi(x/10)$$

$$+ \psi(x/11) - \psi(x/12) + \psi(x/13) - \psi(x/15) + \cdots. \tag{33.43}$$

He observed that, in general, the coefficient of $\psi(x/n)$ would be

$$+1 \text{ if } n = 30m+k, \ k = 1, 7, 11, 13, 17, 19, 23, 29; \tag{33.44}$$
$$0 \text{ if } n = 30m+k, \ k = 2, 3, 4, 5, 8, 9, 14, 16, 21, 22, 25, 26, 27, 28; \tag{33.45}$$
$$-1 \text{ if } n = 30m+k, \ k = 6, 10, 12, 15, 18, 20, 24; \tag{33.46}$$
$$-1 \text{ if } n = 30m + 30. \tag{33.47}$$

Note here that the series (33.43) was alternating and that the absolute values of terms were nonincreasing, making the sum of the series less than the first term and greater than the sum of the first two terms. Thus, Chebyshev could conclude that

$$\psi(x) - \psi(x/6) \leq T(x) - T(x/2) - T(x/3) - T(x/5) + T(x/30) \leq \psi(x).$$

An application of the two inequalities (33.41) then yielded

$$Ax - \frac{5}{2}\ln x - 1 < T(x) - T(x/2) - T(x/3) - T(x/5) + T(x/30) < Ax + \frac{5}{2}\ln x,$$

$$\text{where} \quad A = \ln\left(2^{1/2}3^{1/3}5^{1/5}/30^{1/30}\right) = 0.92129202\ldots. \tag{33.48}$$

33.3 Chebyshev's Proof of Bertrand's Conjecture

In this way, Chebyshev obtained the two inequalities

$$\psi(x) > Ax - \frac{5}{2}\ln x - 1 \quad \text{and} \quad \psi(x) - \psi(x/6) < Ax + \frac{5}{2}\ln x. \tag{33.49}$$

The first inequality determined a lower bound for $\psi(x)$; Chebyshev obtained an upper bound from the second inequality by employing an interesting trick. He set

$$f(x) = \frac{6}{5}Ax + \frac{5}{4\ln 6}\ln^2 x + \frac{5}{4}\ln x$$

and by a simple calculation obtained

$$f(x) - f(x/6) = Ax + \frac{5}{2}\ln x.$$

Therefore, by the second inequality

$$\psi(x) - \psi(x/6) < f(x) - f(x/6) \quad \text{or} \quad \psi(x) - f(x) < \psi(x/6) - f(x/6).$$

Replacing x by $x/6, x/6^2, \ldots, x/6^m$ successively, he got

$$\psi(x) - f(x) < \psi(x/6) - f(x/6) < \psi(x/6^2) - f(x/6^2) < \cdots$$
$$< \psi(x/6^{m+1}) - f(x/6^{m+1}).$$

Taking m to be the largest integer for which $x/6^m \geq 1$, $x/6^{m+1}$ would have to lie between $1/6$ and 1. Therefore,

$$\psi(x/6^{m+1}) - f(x/6^{m+1}) < 1 \quad \text{and}$$

$$\psi(x) - f(x) < 1 \quad \text{or} \quad \psi(x) < f(x) + 1.$$

$$\text{Thus,} \quad \psi(x) < \frac{6}{5}Ax + \frac{5}{4\ln 6}\ln^2 x + \frac{5}{4}\ln x + 1. \tag{33.50}$$

Chebyshev obtained bounds for $\theta(x)$ from those of $\psi(x)$. He observed that (33.38) implied that

$$\psi(x) - \psi(\sqrt{x}) = \theta(x) + \theta(\sqrt[3]{x}) + \theta(\sqrt[5]{x}) + \cdots$$

$$\psi(x) - 2\psi(\sqrt{x}) < \theta(x) < \psi(x) - \psi(\sqrt{x}). \tag{33.51}$$

He concluded from the bounds for $\psi(x)$ in (33.49) and (33.51) that

$$Ax - \frac{12}{5}Ax^{1/2} - \frac{5}{8\ln 6}\ln^2 x - \frac{15}{4}\ln x - 3 < \theta(x)$$

$$< \frac{6}{5}Ax - Ax^{1/2} + \frac{5}{4\ln 6}\ln^2 x + \frac{5}{2}\ln x + 2. \tag{33.52}$$

With the help of these inequalities, Chebyshev was able to prove Bertrand's conjecture.

Chebyshev argued that if there were exactly m primes between the numbers l and L, then $\theta(L) - \theta(l)$ could be expressed as the sum of the logarithms of these primes and hence

$$m \ln l < \theta(L) - \theta(l) < m \ln L \quad \text{or} \tag{33.53}$$

$$\frac{\theta(L) - \theta(l)}{\ln L} < m < \frac{\theta(L) - \theta(l)}{\ln l}. \tag{33.54}$$

He denoted the upper and lower bounds of $\theta(x)$ in (33.52) by $\theta_I(x)$ and $\theta_{II}(x)$, respectively, and noted that by the last inequality, m was greater than $k = \theta_{II}(L) - \theta_I(l)$. Substituting the values of $\theta_{II}(L)$ and $\theta_I(l)$ and solving for l, Chebyshev obtained

$$l = \frac{5}{6}L - 2L^{1/2} - \frac{25 \ln^2 L}{16A \ln 6} - \frac{5}{6A}\left(\frac{25}{4} + k\right) \ln L - \frac{25}{6A} \tag{33.55}$$

and observed that between l and L there were more than k primes. He then took $k = 0$ and saw that there had to be at least one prime between

$$l = \frac{5}{6}L - 2L^{1/2} - \frac{25 \ln^2 L}{16A \ln 6} - \frac{125}{24A} \ln L - \frac{25}{6A} \quad \text{and} \quad L. \tag{33.56}$$

Finally, he remarked that for $L = 2a - 3$ and $a > 160$, the value of l in (33.56) was larger than a, and hence there was a prime between a and $2a - 3$. Since the conjecture could be confirmed to hold for values of $a \leq 160$, this completed the proof. In 1919, Ramanujan published a similar but very brief proof of Bertrand's conjecture. It may also be of interest to note that in 1932, when Paul Erdős was only 18 years of age, he found a proof quite similar to Ramanujan's.

Chebyshev closed his paper by giving bounds for $\pi(x)$, the number of primes less than x. He derived these bounds as a corollary to an interesting theorem on series: Supposing that for large enough x, $F(x)/\ln x$ was positive and decreasing, the series

$$F(2) + F(3) + F(5) + F(7) + F(11) + F(13) + \cdots$$

converged if and only if the series

$$\frac{F(2)}{\ln 2} + \frac{F(3)}{\ln 3} + \frac{F(4)}{\ln 4} + \frac{F(5)}{\ln 5} + \frac{F(6)}{\ln 6} + \cdots$$

converged. Clearly, this theorem implied the convergence of the series $\sum_p 1/(p \ln p)$. In fact, Chebyshev showed that the sum of the series lay between 1.53 and 1.73. To prove this theorem, Chebyshev took $\alpha, \beta, \gamma, \ldots, \rho$ to be prime numbers between the integers l and L. Then he defined U by

$$S = F(2) + F(3) + F(5) + \cdots + F(\alpha) + F(\beta) + F(\gamma) + \cdots F(\rho)$$
$$= S_0 + F(\alpha) + F(\beta) + F(\gamma) + \cdots + F(\rho) = S_0 + U.$$

33.3 Chebyshev's Proof of Bertrand's Conjecture

Since $\theta(x) - \theta(x-1) = \ln x$ for prime x, and $= 0$ for composite x, Chebyshev could conclude that

$$U = \frac{\theta(l) - \theta(l-1)}{\ln l} F(l) + \frac{\theta(l+1) - \theta(l)}{\ln(l+1)} F(l+1)$$

$$+ \frac{\theta(l+2) - \theta(l+1)}{\ln(l+2)} F(l+2) + \cdots + \frac{\theta(L) - \theta(L-1)}{\ln L} F(L).$$

He then applied summation by parts to obtain

$$U = -\theta(l-1)\frac{F(L)}{\ln l} + \left(\frac{F(l)}{\ln l} - \frac{F(l+1)}{\ln(l+1)}\right)\theta(l) + \left(\frac{F(l+1)}{\ln(l+1)} - \frac{F(l+2)}{\ln(l+2)}\right)\theta(l+1)$$

$$+ \cdots + \left(\frac{F(L)}{\ln L} - \frac{F(L+1)}{\ln(L+1)}\right)\theta(L) + \frac{F(L+1)}{\ln(L+1)}\theta(L). \tag{33.57}$$

He took l large enough that $F(x)\ln x$ was positive and increasing in $l-1 \leq x \leq L+1$ and obtained the inequalities

$$\theta_{II}(l-1)\frac{F(l)}{\ln l} - \theta_I(l-1)\frac{F(l)}{\ln l} + \sum_{x=l}^{L} F(x)\frac{\theta_{ii}(x) - \theta_{II}(x-1)}{\ln x} < U$$

$$< \theta_I(l-1)\frac{F(l)}{\ln l} - \theta_{II}(l-1)\frac{F(L)}{\ln l} + \sum_{x=l}^{L} F(x)\frac{\theta_I(x) - \theta_I(x-1)}{\ln x}. \tag{33.58}$$

Chebyshev noted that

$$\theta_{II}(x) - \theta_{II}(x-1)$$

$$= A - \frac{12}{5}A(\sqrt{x} - \sqrt{x-1}) - \frac{5}{8\ln 6}(\ln^2 x - \ln^2(x-1))$$

$$- \frac{15}{4}(\ln x - \ln(x-1))$$

and that this expression was bounded as $x \to \infty$. For example,

$$\sqrt{x} - \sqrt{x-1} = \sqrt{x} - \sqrt{x}\left(1 - \frac{1}{x}\right)^{1/2} \approx \frac{1}{2\sqrt{x}} \to 0,\ x \to \infty,\ \text{and}$$

$$\ln x - \ln(x-1) = \ln\frac{x-1}{x} = \ln\left(1 - \frac{1}{x}\right) \to 0 \text{ as } x \to \infty.$$

By a similar analysis with $\theta_I(x) - \theta_I(x-1)$, he noted that the two inequalities in (33.58) implied the theorem. Chebyshev obtained the bounds for $\pi(x)$ by taking $F(x) = 1$ and $l = 2$ in (33.58):

$$\frac{\theta_{II}(1)}{\ln 2} - \frac{\theta_I(1)}{\ln 2} + \sum_{x=2}^{L} \frac{\theta_{II}(x) - \theta_{II}(x-1)}{\ln x} < \pi(x)$$

$$< \frac{\theta_I(1)}{\ln 2} - \frac{\theta_{II}(1)}{\ln 2} + \sum_{x=2}^{L} \frac{\theta_I(x) - \theta_I(x-1)}{\ln x}.$$

33.4 De Polignac's Evaluation of $\sum_{p \leq x} \frac{\ln p}{p}$

Inspired by the work of Chebyshev, in the 1850s, Alphonse de Polignac published a number of papers in the *Comptes Rendus* and *Liouville's Journal*. Though his work was largely lacking in rigor, in 1857 de Polignac gave a fairly good proof of

$$\sum_{p \leq x} \frac{\ln p}{p} = \ln x + \epsilon,$$

where ϵ was a quantity small compared to $\ln x$. In fact, his proof implies that ϵ is bounded. Now by Chebyshev's work

$$\sum_{p^m \leq x} \ln p = \sum_{p \leq x} \left(\left\lfloor \frac{x}{p} \right\rfloor + \left\lfloor \frac{x}{p^2} \right\rfloor + \left\lfloor \frac{x}{p^3} \right\rfloor + \cdots \right) \ln p.$$

De Polignac denoted the left-hand side by $\ln F_o(x)$ and $\lfloor x/p^k \rfloor$ by $E\left(x/p^k\right)$. He let n be the largest integer such that $p^n \leq x$. Then

$$\sum_{k=1}^{n} \left\lfloor \frac{x}{p^k} \right\rfloor > \sum_{k=1}^{n} \frac{x}{p^k} - \sum_{k=1}^{n} 1, \text{ and}$$

$$\sum_{k=1}^{n} \left\lfloor \frac{x}{p^k} \right\rfloor < \sum_{k=1}^{n} \frac{x}{p^k} = \frac{x}{p-1} - \frac{x}{p^n(p-1)}.$$

Therefore,

$$x \sum_{p \leq x} \frac{\ln p}{p-1} - x \sum_{p \leq x} \frac{\ln p}{p^n(p-1)} > \sum_{p^m \leq x} \ln p$$

$$> x \sum_{p \leq x} \frac{\ln p}{p-1} - x \sum_{p \leq x} \frac{\ln p}{p^n(p-1)} - \ln x.$$

De Polignac then argued that $\sum_{p^m \leq x} \ln p = \sum_{n \leq x} \ln n = x \ln x +$ terms of smaller order, and $\sum \frac{\ln p}{p-1}$ was of the same order as $\sum \frac{\ln p}{p}$. Moreover, $\sum_{p \leq x} \frac{\ln p}{p^n(p-1)}$ was bounded, so that the required result followed from the two inequalities. In 1874, Franz Mertens, aware of de Polignac's work, but motivated by Chebyshev's second paper, proved that

$$\sum_{p \leq x} (\ln p)/p = \ln x + O(1).$$

33.5 Mertens's Evaluation of $\prod_{p \leq x} \left(1 - \frac{1}{p}\right)^{-1}$

According to Mertens, his interest in evaluating $\prod_{p \leq x} \left(1 - \frac{1}{p}\right)$ arose from the useful formulas he had seen in the third edition of Legendre's *Théorie des nombres*. Legendre's

33.5 Mertens's Evaluation of $\prod_{p \leq x}\left(1-\frac{1}{p}\right)^{-1}$

formulas stated without rigorous proof that for some constants A and C,

$$\sum_{p \leq G} \frac{1}{p} = \ln(\ln G - 0.08366) + C, \text{ and}$$

$$\prod_{p \leq G}\left(1-\frac{1}{p}\right) = \frac{A}{\ln G - 0.08366}.$$

Mertens proved the results

$$\sum_{p \leq G} \frac{1}{p} = \ln \ln G + \gamma - H + \delta, \tag{33.59}$$

where $H = \sum_{k=2}^{\infty} \frac{1}{k} \sum_{p} \frac{1}{p^k}$ and $\delta < \frac{4}{\ln(G+1)} + \frac{2}{G \ln G}$; and $\tag{33.60}$

$$\prod_{p \leq G}\left(1-\frac{1}{p}\right)^{-1} = e^{\gamma + \delta'} \ln G, \tag{33.61}$$

where $\delta' < \frac{4}{\ln(G+1)} + \frac{2}{G \ln G} + \frac{1}{2G}.$ $\tag{33.62}$

He began by observing that Dirichlet and Chebyshev had shown that for $\rho > 0$,

$$\zeta(1+\rho) = \frac{1+(\rho)}{\rho},$$

where (ρ) denoted a quantity tending to 0 as $\rho \to 0$. It followed from this and Euler's product for $\zeta(1+\rho)$ that

$$\ln \frac{1}{\rho} + (\rho) = \ln \zeta(1+\rho) = -\sum_p \ln(1-1/p^{1+\rho})$$

$$= \sum_p \frac{1}{p^{1+\rho}} + \frac{1}{2}\sum_p \frac{1}{p^{2+2\rho}} + \frac{1}{3}\sum_p \frac{1}{p^{3+3\rho}} + \cdots.$$

Mertens could then easily conclude that

$$\sum_p \frac{1}{p^{1+\rho}} = \ln \frac{1}{\rho} - H + (\rho). \tag{33.63}$$

He proceeded to complete the proof of (33.59) by showing that

$$\sum_{p > G} \frac{1}{p^{1+\rho}} = \ln \frac{1}{\rho} - \ln \ln G - \gamma + \delta + (\rho), \tag{33.64}$$

where $|\delta| < \frac{4}{\ln(G+1)} + \frac{2}{G \ln G}.$

He set $f(x) = \sum_{p \leq x} \frac{\ln p}{p}$. Then summation by parts gave him

$$\sum_{p > G} \frac{1}{p^{1+\rho}} = \sum_{n=G+1}^{\infty} \frac{f(n) - f(n-1)}{n^\rho \ln n}$$

$$= -\frac{f(G)}{(G+1)^\rho \ln(G+1)} + \sum_{n=G+1}^{\infty} f(n) \left(\frac{1}{n^\rho \ln n} - \frac{1}{(n+1)^\rho \ln(n+1)} \right).$$

Next set $f(n) = \ln n + D_n$. Recall that de Polignac had shown that D_n was small compared to $\ln n$, but Mertens required that D_n be bounded. This was easily achieved. Mertens showed that $D_n < 2$ by computing bounds explicitly, as Chebyshev had done in his work on primes. Now observe that

$$\ln n \left(\frac{1}{n^\rho \ln n} - \frac{1}{(n+1)^\rho \ln(n+1)} \right) = \frac{1}{n^\rho} - \frac{1}{(n+1)^\rho} - \frac{\ln\left(1 - \frac{1}{n+1}\right)}{(n+1)^\rho \ln(n+1)}$$

$$= \frac{1}{n^\rho} - \frac{1}{(n+1)^\rho} + \frac{1}{(n+1)^{1+\rho} \ln(n+1)}$$

$$+ \frac{\lambda}{2n(n+1)^{1+\rho} \ln(n+1)},$$

where $0 < \lambda < 1$. Applying this in the previous series and after cancellation of terms, he had

$$\sum_{p > G} \frac{1}{p^{1+\rho}} = \sum_{n=G+1}^{\infty} \frac{1}{n^{1+\rho} \ln n} + R, \tag{33.65}$$

where

$$R = \frac{\ln(G+1) - f(G)}{(G+1)^\rho \ln(G+1)} - \frac{1}{(G+1)^{1+\rho} \ln(G+1)} + \lambda \sum_{n=G+1}^{\infty} \frac{1}{2n(n+1)^{1+\rho} \ln(n+1)}$$

$$+ \sum_{n=G+1}^{\infty} D_n \left(\frac{1}{n^\rho \ln n} - \frac{1}{(n+1)^\rho \ln(n+1)} \right).$$

It is easy to show that $R = O\left(\frac{1}{\ln G}\right)$ and, in fact, Mertens proved that $|R| < \frac{4}{\ln(G+1)} + \frac{1}{G \ln G}$. To estimate the sum $\sum 1/(n^{1+\rho} \ln n)$, first note that

$$\sum_{n=G+1}^{\infty} \frac{1}{n^{1+t}} = \frac{G^{-t}}{t} - R', \tag{33.66}$$

where $R' = \frac{1+t}{2} \sum_{n=G+1}^{\infty} \frac{1}{n^{2+t}} + \frac{(1+t)(2+t)}{2 \cdot 3} \sum_{n=G+1}^{\infty} \frac{1}{n^{3+t}} + \cdots.$

To prove (33.66), observe that the binomial expansion of $(1 - 1/(n+1))^{-t}$ immediately implies

$$\frac{1}{tn^t} - \frac{1}{t(n+1)^t} = \frac{1}{(n+1)^{1+t}} + \frac{1+t}{2} \frac{1}{(n+1)^{2+t}} + \frac{(1+t)(2+t)}{2 \cdot 3} \frac{1}{(n+1)^{3+t}} + \cdots.$$

33.5 Mertens's Evaluation of $\prod_{p \leq x} \left(1 - \frac{1}{p}\right)^{-1}$

The required result followed when this formula was summed from $n = G$ to $n = \infty$. Now integrating (33.66) from ρ to 1, obtain

$$\sum_{n=G+1}^{\infty} \frac{1}{n^{1+\rho} \ln n} - \sum_{n=G+1}^{\infty} \frac{1}{n^2 \ln n} = \int_\rho^1 \frac{G^{-t}}{t} dt - \int_\rho^1 R' dt$$

$$= \int_{\rho \ln G}^\infty \frac{dx}{e^x - 1} - \int_{\rho \ln G}^\infty \left(\frac{1}{e^x - 1} - \frac{e^{-x}}{x}\right) dx$$

$$- \int_1^\infty \frac{G^{-x}}{x} dx - \int_\rho^1 R' dt.$$

The first integral in this expression could be written as

$$\int_{\rho \ln G}^\infty \frac{e^{-x}}{1 - e^{-x}} dx = \ln(1 - e^{-x})\Big|_{\rho \ln G}^\infty = -\ln(1 - G^{-\rho});$$

the second one, by Gauss's formula (23.72) for $\psi(1) = \Gamma'(1)/\Gamma(1) = -\gamma$, would be

$$\int_0^\infty - \int_0^{\rho \ln G} \left(\frac{1}{e^x - 1} - \frac{e^{-x}}{x}\right) dx = \gamma + (\rho).$$

Also

$$-\ln(1 - G^{-\rho}) = \ln \frac{1}{\rho} - \ln \ln G + (\rho),$$

so that

$$\sum_{n=G+1}^{\infty} \frac{1}{n^{1+\rho} \ln n} = \ln \frac{1}{\rho} - \ln \ln G - \gamma - \int_1^\infty G^{-x} \frac{dx}{x} + \sum_{n=G+1}^{\infty} \frac{1}{n^2 \ln n}$$

$$- \int_\rho^1 R' dt + (\rho).$$

Next,

$$\int_\rho^1 R' dt < \int_0^1 \left(\sum_{n=G+1}^{\infty} \frac{1}{n^{2+t}} + \sum_{n=G+1}^{\infty} \frac{1}{n^{3+t}} + \cdots\right) dt$$

$$< \sum_{n=G+1}^{\infty} \left(\frac{1}{n^2 \ln n} - \frac{1}{n^3 \ln n}\right) + \sum_{n=G+1}^{\infty} \left(\frac{1}{n^3 \ln n} - \frac{1}{n^4 \ln n}\right) + \cdots$$

$$< \sum_{n=G+1}^{\infty} \frac{1}{n^2 \ln n} < \sum_{n=G+1}^{\infty} \left(\frac{1}{(n-1)\ln(n-1)} - \frac{1}{n \ln n}\right) < \frac{1}{G \ln G}$$

and $\int_1^\infty \frac{G^{-x}}{x} dx < \frac{1}{G \ln G}.$

Hence,

$$\sum_{n=G+1}^{\infty} \frac{1}{n^{1+\rho} \ln n} = \ln \frac{1}{\rho} - \ln \ln G - \gamma + \frac{\lambda}{G \ln G} + (\rho).$$

When combined with (33.65), this gave (33.64). Mertens's version of Legendre's formula for $\sum_{p \leq G} \frac{1}{p}$ followed from this, (33.64), and (33.63). Mertens's formula for the product was an easy corollary.

In 1926, Hardy commented on Mertens's proof: "The proof is rather difficult to seize or to remember, since it depends on a combination of the method of Tchebycheff on the one hand and the theory of Dirichlet's series on the other, and it may be worth while to give an alternative proof." Hardy himself provided two proofs, the first published in 1927 using an integral analog of Littlewood's Tauberian theorem, and the second in 1935 using the analog of the simpler Tauber's theorem. Hardy's proofs are quite interesting, but we note that if Mertens's proof is recast in terms of the Stieltjes integral, a very simple proof of (33.64) emerges; note that the latter is the only complex argument in the proof. To begin this short proof, first observe that

$$\sum_{p \geq G+1} \frac{1}{p^{1+\rho}} = \int_{G+1}^{\infty} x^{\rho} \ln x \, df(x),$$

where $f(x) = \sum_{p \leq x} (\ln p)/p = \ln x + \epsilon$, $|\epsilon| \leq 3$. Then integration by parts and an easy calculation produce

$$\sum_{p \geq G+1} \frac{1}{p^{1+\rho}} = (1 + \epsilon \rho) \int_{\ln(G+1)}^{\infty} \frac{e^{-\rho u}}{u} du + O\left(\frac{1}{\ln G}\right).$$

Then by Gauss's formula for $\Gamma'(1)$, (23.72), this integral can be evaluated as

$$\int_{\rho \ln(G+1)}^{\infty} \frac{e^{-x}}{x} dx = -\int_{0}^{\infty} + \int_{0}^{\rho \ln(G+1)} \left(\frac{1}{e^x - 1} - \frac{e^{-x}}{x}\right) dx + \int_{\rho \ln(G+1)}^{\infty} \frac{1}{e^x - 1} dx$$

$$= \gamma + (\rho) + \ln(1 - (G+1)^{\rho})$$

$$= \gamma + (\rho) + \ln \frac{1}{\rho} + \ln \ln(G+1) + (\rho).$$

This proves (33.64) so that the proof of Mertens's formula can now be completed as before.

33.6 Riemann's Formula for $\pi(x)$

Riemann's eight-page paper of 1859, containing his formula for the number of primes less than a given number x, was actually an outline of a research program for the advancement of the theory of distribution of primes. He proved very few statements in this paper, but clearly set forth his conjectures and how some of them might be verified. It took fifty years of development in complex analysis to prove the first approximation of his formula, the prime number theorem. Almost a century after Riemann's paper appeared, Hardy's student and Oxford professor Edward Titchmarsh wrote, "The memoir in which Riemann first considered the zeta-function has become famous for the number of ideas it contains which have since proved fruitful, and it is by no means certain that these are even now exhausted."

33.6 Riemann's Formula for $\pi(x)$

Recall that Dirichlet and Chebyshev employed Mellin transforms to study prime numbers, but that they limited themselves to real variables. Riemann's great innovation was to employ complex variables. He expressed the Mellin transform of his arithmetic function $f(x)$ defined by (33.14) in terms of the zeta function; then by Mellin inversion, he expressed $f(x)$ as an integral in the complex plane. This made it possible to apply the powerful machinery of complex integration. Riemann observed that if

$$p^{-s} = s \int_p^\infty x^{-s-1} dx, \quad p^{-2s} = s \int_{p^2}^\infty x^{-s-1} dx, \ldots$$

were to be used in

$$\log \zeta(s) = -\sum_p \log(1 - p^{-s}) = \sum_p p^{-s} + \frac{1}{2} \sum_p p^{-2s} + \frac{1}{3} \sum_p p^{-3s} + \cdots,$$

he got

$$\frac{\log \zeta(s)}{s} = \int_1^\infty f(x) x^{-s-1} dx, \quad \operatorname{Re} s > 1.$$

He then applied the Fourier inversion formula to obtain an integral expression for $f(x)$:

$$f(y) = \frac{1}{2\pi i} \int_{a-\infty i}^{a+\infty i} \frac{\log \zeta(s)}{s} y^s ds, \quad a > 1.$$

Riemann set

$$\xi(s) = \Pi\left(\frac{s}{2}\right)(s-1)\pi^{-s/2} \zeta(s), \tag{33.67}$$

and by using (33.17), he obtained

$$\log \zeta = \frac{s}{2} \log \pi - \log(s-1) - \log \Pi(s/2) + \sum_\alpha \log(1 - s/\rho) + \log \xi(0).$$

Riemann noted, however, that when this expression was used in the integral for $f(y)$, the integral became divergent. So he applied integration by parts to get

$$f(s) = -\frac{1}{2\pi i} \frac{1}{\log x} \int_{a-\infty i}^{a+\infty i} \frac{d(\log \zeta(s)/s)}{ds} x^s ds.$$

Next he observed that

$$-\log \Pi(s/2) = \lim_{m \to \infty} \left(\sum_{n=1}^m \log(1 + s/2n) - \frac{s}{2} \log m \right) \quad \text{and therefore}$$

$$-\frac{d}{ds}(\log \Pi(s/2))/s = \sum_{n=1}^\infty \frac{d}{ds}(\log(1 + s/2n))/s.$$

Hence, every term in the expression for $f(s)$, except for the term

$$\frac{1}{2\pi i} \frac{1}{\log x} \int_{a-\infty i}^{a+\infty i} \frac{1}{ss} \log \xi(0) x^s ds = \log \xi(0),$$

took the form
$$\pm \frac{1}{2\pi i} \frac{1}{\log x} \int_{a-\infty i}^{a+\infty i} \frac{d}{ds} ((\log(1-s/\beta))/s) x^s ds.$$

To evaluate this integral, Riemann observed that
$$\frac{d}{d\beta} ((\log(1-s/\beta))/s) = \frac{1}{(\beta-s)\beta}.$$

Thus, for $\mathrm{Re}(s-\beta) > 0$, he had
$$-\frac{1}{2\pi i} \frac{d}{d\beta} \int_{a-\infty i}^{a+\infty i} \frac{\log(1-s/\beta)}{s} x^s ds = -\frac{1}{2\pi i} \int_{a-\infty i}^{a+\infty i} \frac{x^s ds}{(\beta-s)\beta} = \frac{x^\beta}{\beta}$$
$$= \begin{cases} \int_\infty^x t^{\beta-1} dt, & \text{when } \mathrm{Re}\,\beta < 0, \\ \int_0^x t^{\beta-1} dt, & \text{when } \mathrm{Re}\,\beta > 0. \end{cases}$$

Riemann could then conclude that
$$\frac{1}{2\pi i} \frac{1}{\log x} \int_{a-\infty i}^{a+\infty i} \frac{d}{ds} \left(\frac{\log(1-s/\beta)}{s} \right) x^s ds$$

$$= -\frac{1}{2\pi i} \int_{a-\infty i}^{a+\infty i} \frac{\log(1-s/\beta)}{s} x^s ds$$
$$= \begin{cases} \int_\infty^x \frac{t^{\beta-1}}{\log t} dt, & \text{when } \mathrm{Re}\,\beta < 0, \\ \int_0^x \frac{t^{\beta-1}}{\log t} dt, & \text{when } \mathrm{Re}\,\beta > 0. \end{cases}$$

Note that it was clear that
$$\int_0^x \frac{t^{\beta-1}}{\log t} dx = \int_0^{x^\beta} \frac{du}{\log u} = \mathrm{li}(x^\beta).$$

By using these results in the expression for $f(x)$, Riemann obtained his famous formula
$$f(x) = \mathrm{li}(x) - \sum_\beta (\mathrm{li}(x^\beta) + \mathrm{li}(x^{1-\beta})) + \int_x^\infty \frac{1}{t^2-1} \frac{dt}{t \log t} + \log \xi(0).$$

In writing this formula, Riemann assumed the truth of the Riemann hypothesis. He wrote $\beta = \frac{1}{2} + i\alpha$, and $1 - \beta = \frac{1}{2} - i\alpha$, so that the expression in the sum appeared as
$$\mathrm{li}\left(x^{\frac{1}{2}+i\alpha}\right) + \mathrm{li}\left(x^{\frac{1}{2}-i\alpha}\right).$$

One may verify that the integral evaluations as sketched by Riemann are indeed correct, and/or consult Harold Edwards's book, offering a detailed discussion of Riemann's paper.

33.7 Exercises

1. Using Chebyshev's notation, show that for $T(x) = \ln\lfloor x\rfloor! - 2\ln\lfloor x/2\rfloor!$

$$\psi(x) - \psi(x/2) \leq T(x) \leq \psi(x) - \psi(x/2) + \psi(x/3).$$

Apply Stirling's approximation to prove that $T(x) < 3x/4$ for $x > 0$ and $T(x) > 2x/3$ for $x > 300$. Use this to show that $\psi(x) < 3x/2$. Now show that $\psi(x) - 2\psi(\sqrt{x}) \leq \theta(x) \leq \psi(x)$ and, therefore,

$$\psi(x) - \psi(x/2) + \psi(x/3) \leq \theta(x) + 2\psi(\sqrt{x}) - \theta(x/2) + \psi(x/3) \quad (33.68)$$
$$< \theta(x) - \theta(x/2) + x/2 + 3\sqrt{x}. \quad (33.69)$$

Apply these results to show that

$$\theta(x) - \theta(x/2) > x/6 - 3\sqrt{x} \quad \text{for} \quad x > 300.$$

Show that this proves Bertrand's conjecture for $x \geq 162$. Finally, show that

$$\pi(x) - \pi(x/2) > (x/6 - 3\sqrt{3})/\ln x \quad \text{for} \quad x > 300.$$

See Ramanujan (2000), pp. 208–209.

2. Let $C(m,n)$ denote the binomial coefficient m choose n, that is, the number of ways of choosing n objects out of m distinct objects. Show that the exponent of a prime p in $C(2n,n)$ is given by

$$\sum_{k=1}^{\infty}(\lfloor 2n/p^k\rfloor - 2\lfloor n/p^k\rfloor).$$

Show that

$$d = \lfloor 2n/p^k\rfloor - 2\lfloor n/p^k\rfloor \leq 1,$$

that $d = 1$ for $\sqrt{2n} < p \leq 2n$, and that $d = 0$ for $p > 2n$ and for $2n/3 < p < n$. Use these results to conclude that

$$C(2n,n) \leq \prod_{p \leq \sqrt{2n}}(2n) \prod_{\sqrt{2n} < p \leq 2n/3} p \prod_{n < p \leq 2n} p.$$

Note that if Bertrand's conjecture is false for some n, then

$$C(2n,n) \leq (2n)^{\sqrt{2n}} \prod_{\sqrt{2n} < p \leq 2n/3} p. \quad (33.70)$$

3. Show that $2nC(2n,n) > 4^n$. Prove that for $a \geq 5$, $C(2a,a) < 4^{a-1}$ and that $\prod_{a<p<2a} \leq C(2a,a)$. Use the last two inequalities to show that $\prod_{10<p<n} p < 4^n$. Combine these results with (33.70) to show that if Bertrand's conjecture is false then we get a contradiction. See Erdős (1932). Paul Erdős (1913–1996) founded many aspects of combinatorics and popularized this area of mathematics by

continuously traveling all over the world and collaborating with hundreds of mathematicians.

4. Let $\phi(1) = 1$ and let $\phi(n)$, $n > 1$, be the number of numbers less than n and prime to n. Let $F(t) = \sum_{1 \le n \le t} \phi(t)$. Prove that $F(t) = 3t^2/\pi^2 + O(t \ln t)$. See Mertens (1874b), pp. 290–292.

5. Show that the integral $\Gamma(s) = \int_0^\infty x^{s-1} e^{-x} \, dx$ has the inversion

$$e^{-x} = \frac{1}{2\pi i} \int_{a-i\infty}^{a+i\infty} \Gamma(x) x^{-s} \, ds, \quad a > 0, \ \mathrm{Re}\, x > 0.$$

See Cahen (1894).

6. Prove Ramanujan's formula

$$\int_0^\infty \frac{\cos \pi x^2}{\sinh \pi x} \sin(2\pi t x) \, dx = \frac{\cosh \pi t - \cos \pi t^2}{2 \sinh \pi t}.$$

Use this formula to show that for $-3 < \mathrm{Re}\, s < 4$, we have

$$\frac{\Gamma(s)}{2^s}(1 - 2^s)(1 - 2^{1-s})\zeta(s)$$

$$= -\int_0^\infty \left[\left(\frac{\pi}{2}\right)^{s-1} \Gamma(s) \sin \frac{\pi s}{2} x^{-s} + x^{s-1} \right] \frac{\sin^2\left(\frac{x^2}{2\pi}\right)}{\sinh x} \, dx.$$

See Ramanujan (2000), p. 64, for his formula. For the other formula, see Mustafy (1966).

7. Show that for $\xi(s)$ defined by (33.67) and for $0 < \mathrm{Re}\, s < 1$, $\lambda > 0$,

$$\int_0^1 u^{-1/2} k(\lambda, u) \left(u^{s-1/2} + u^{1/2-s}\right) du = \frac{1}{2} B(s) \xi(s) \left(\lambda^{s/2 - 1/4} + \lambda^{1/4 - s/2}\right),$$

where $\displaystyle B(s) = \frac{\pi^{-1/2}}{4} \Gamma\left(-\frac{s}{2}\right) \Gamma\left(\frac{s-1}{2}\right),$

$$\phi(\lambda, u) = \int_0^\infty \left(\lambda^{1/4} e^{-\pi x^2 u \lambda} + \lambda^{-1/4} e^{-\pi x^2 u/\lambda}\right) \frac{x \, dx}{e^{2\pi x} - 1},$$

$$k(\lambda, u) = \frac{1}{4\pi} \left(\lambda^{1/4} + \lambda^{-1/4}\right) - u\phi(\lambda, u^2).$$

Prove also that $k(\lambda, u) > 0$ for $0 \le u \le 1$, and that a number $s_0 = \sigma_0 + it_0$ with $0 < \sigma_0 < 1$ is a zero of $\xi(s)$ if and only if for every $\lambda > 0$

$$\int_0^1 u^{-1/2} k(\lambda, u) \left(u^{s_0 - 1/2} + u^{1/2 - s_0}\right) du = 0.$$

See Mustafy (1972). Ashoke Kumar Mustafy had a thirty-year career in the Indian Administrative Service, including as Vice Chancellor of Lucknow University during 1973–75. In spite of his heavy administrative duties, he worked on mathematics six to seven hours per day and had time to discuss mathematics with a young boy like the author. Mustafy hoped that his result would be useful in proving the Riemann hypothesis; indeed, in this connection he communicated with André Weil who wrote that he found Mustafy's work promising.

33.8 Notes on the Literature

For a history of the prime number theorem, see Goldstein (1973) and Bateman and Diamond (1996). These two papers have been reprinted in Anderson, Katz, and Wilson (2009). Goldstein's paper also contains a translation of Gauss's letter to Encke. Bessel's letter to Olbers may be found on p. 238 of Erman (1852), vol. 1. Sylvester (1973), vol. 3, p. 545, contains the quotation concerning Chebyshev. For the latter's work on prime numbers, see Chebyshev (1899–1907), vol. 1, pp. 29–70. See Smith (1959), pp. 127–48, for an English translation of some parts of Chebyshev's two papers on primes. Delone (2005), an English translation by R. Burns of Delone's Russian original of 1947, gives a detailed commentary on Chebyshev's papers and a discussion of the major contributions to number theory of St. Petersburg mathematicians in the period 1847–1947. See de Polignac (1857) and Mertens (1874ba) for the results on $\sum (\ln p)/p$ and $\sum 1/p$. For Hardy's proofs of Mertens's theorems, see his two notes in Hardy (1966–1979), vol. 2, pp. 210–12 and 230–33. The quote concerning the proof of Mertens is on p. 210. Riemann (1990), pp. 177–185, has a reprint of his 1859 paper. Edwards (2001) presents a detailed and fascinating discussion of this paper and some of its consequences. Narkiewicz (2000) offers an excellent exposition of the development of the prime number theorem and provides a comprehensive list of references. Titchmarsh and Heath-Brown (1986), p. 254, has the quote on Riemann's paper.

34

Invariant Theory: Cayley and Sylvester

34.1 Preliminary Remarks

The invariant theory of forms, with forms defined as homogeneous polynomials in several variables, was developed extensively in the nineteenth century as an important branch of algebra but with very close connections to algebraic geometry. Several ideas and methods of invariant theory were influential in diverse areas of mathematics: topics as concrete as enumerative combinatorics and the theory of partitions and as general as twentieth-century abstract commutative algebra.

George Boole, the highly original British mathematician, may be taken as the founder of invariant theory, though early examples of the use of invariance can be found in the works of Lagrange, Laplace, and Gauss. Boole had almost no formal training in mathematics, but he carefully studied the work of great mathematicians, including Newton, Lagrange, and Laplace. In a paper on analytic geometry written in 1839, Boole took the first tentative steps toward the idea of invariance, but he gave a clearly formulated definition in his 1841 "Exposition of a General Theory of Linear Transformations." He wrote that he found his inspiration in Lagrange's researches on the rotation of rigid bodies, contained in the 1788 *Mécanique analytique*. Lagrange's result is most economically described in terms of matrices, a concept developed in the 1850s by Cayley. In modern terms, Lagrange's problem was to diagonalize a 3×3 symmetric matrix A; Lagrange expressed this in terms of binary quadratic forms. Given a quadratic form $x^t A x$, with x a three vector, the problem would be to find a matrix P such that $P P^t = I$, the identity matrix, and $P^t A P$ is a diagonal matrix. This means that if x_1, x_2, x_3 are the components of x, y_1, y_2, y_3 of $y = P^t x$, and $\lambda_1, \lambda_2, \lambda_3$ are the diagonal entries in the diagonal matrix, then

$$x^t A x = \lambda_1 y_1^2 + \lambda_2 y_2^2 + \lambda_3 y_3^2, \tag{34.1}$$

$$x_1^2 + x_2^2 + x_3^2 = y_1^2 + y_2^2 + y_3^2. \tag{34.2}$$

It is not surprising that this result of Lagrange also served as the starting point of the spectral theory of matrices. Cauchy, Weierstrass, and Frobenius were the primary developers of this aspect of matrix theory. But Boole took a different turn; he considered a homogeneous polynomial of degree n in m variables and applied a linear

34.1 Preliminary Remarks

transformation to the variables to obtain a new homogeneous polynomial of degree n in m variables. He wished to determine the relations between the coefficients of the two polynomials. Boole's method may perhaps be best understood by studying his simplest example. Let $Q = ax_1^2 + 2bx_1x_2 + cx_2^2$ be a binary quadratic form. Set its two partial derivatives equal to zero and then eliminate the variables x_1 and x_2. Thus,

$$2ax_1 + 2bx_2 = 0 \quad \text{and} \quad 2bx_1 + 2cx_2 = 0. \tag{34.3}$$

Elimination of the variables x_1, x_2 gives

$$\theta(Q) = b^2 - ac = 0. \tag{34.4}$$

Now apply the linear transformation

$$x_1 = py_1 + qy_2 \quad x_2 = ry_1 + sy_2, \tag{34.5}$$

where p, q, r, s are real numbers with $ps - qr \neq 0$, to get a new quadratic form $R = Ay_1^2 + 2By_1y_2 + Cy_2^2$. A calculation similar to the previous one gives $\theta(R) = B^2 - AC$. Boole pointed out that

$$\theta(R) = (ps - qr)^2 \theta(Q); \tag{34.6}$$

the quantity $ps - qr$ is the determinant of the linear transformation (34.5). In addition, the degrees of the homogeneous polynomials $\theta(Q)$ and $\theta(R)$ are defined as equal to the degree of each term, in this case 2.

More generally, Boole showed that, with Q_n a homogeneous polynomial of degree n in m variables, if R_n was the polynomial obtained after the application to Q_n of a linear transformation with determinant E, and if $\theta(Q_n)$ and $\theta(R_n)$ were obtained by the elimination process described earlier, then

$$\theta(R_n) = E^{\gamma n/m} \theta(Q_n). \tag{34.7}$$

Here γ represented the degree of $\theta(R_n)$ and $\theta(Q_n)$. In the 1841 paper, Boole stated but did not prove this theorem, though he gave a few examples to illustrate it. He indicated a proof in a paper appearing four years later. Note that the polynomial $\theta(Q_n)$ is termed an invariant because it satisfies the relation (34.7). Sylvester introduced the term invariant in a long paper on the subject published in 1853, and he coined many other terms used in invariant theory.

At the end of the second part of his 1841 paper, Boole wrote that mathematicians should find invariant theory a fertile area for research and discovery. Indeed, Boole's paper had an immediate impact on Cayley who, upon reading it in 1844, wrote to Boole of his enthusiasm for this new area of mathematics. Cayley was then a recent graduate of Cambridge University and had published an 1843 paper on determinants in which he introduced the concept of hyperdeterminants or multidimensional determinants. In a paper of 1844, "On the Theory of Linear Transformations," Cayley applied these hyperdeterminants to generate new invariants. Cayley's work arose out of his efforts to generalize some well-known results. For example, the invariant $ac - b^2$ for the binary quadratic was known to be the determinant

$$\begin{vmatrix} a & b \\ b & c \end{vmatrix},$$

while the invariant $abc + 2fgh - ah^2 - bg^2 - cf^2$ for the ternary quadratic $ax_1^2 + bx_2^2 + cx_3^2 + 2fx_1x_2 + 2gx_1x_3 + 2hx_2x_3$ was the determinant

$$\begin{vmatrix} a & f & g \\ f & b & g \\ g & h & c \end{vmatrix}. \tag{34.8}$$

The first fact was already contained in Boole; Cayley presented the second invariant in his paper. As an example of the role of hyperdeterminants, so named by Cayley in 1845, he considered the multilinear form

$$\sum \alpha_{ijkl} x_i y_j z_k w_l,$$

where the indices i, j, k, l assumed only the values 1 and 2. Each of the four pairs of variables $(x_1, x_2), (y_1, y_2), (z_1, z_2), (w_1, w_2)$ could then be linearly transformed by 2×2 matrices. So the multilinear form corresponded to a $2 \times 2 \times 2 \times 2$ matrix and Cayley used hyperdeterminants to compute an invariant for this form. He then specialized the multilinear form by setting $x_1 = y_1 = z_1 = w_1 = x$ and $x_2 = y_2 = z_2 = w_2 = y$; he then identified the coefficients to get the binary quartic

$$u = ax^4 + 4bx^3y + 6cx^2y^2 + 4dxy^3 + ey^4, \tag{34.9}$$

where $a = \alpha_{1111}$, $b = \alpha_{2111} = \alpha_{1211} = \alpha_{1121} = \alpha_{1112}$, and so on. By making a similar identification in the invariant for the multilinear form, he obtained the second-degree invariant for the binary quartic:

$$I_1 = ae - 4bd + 3c^2. \tag{34.10}$$

Cayley realized that his result was different from Boole's invariant $\theta(u)$; he communicated his result to Boole, who pointed out that there was also an invariant of the third degree:

$$I_2 = ace - b^2e - ad^2 - c^3 + 2bcd.$$

Cayley in turn showed that his invariant as well as Boole's third-order invariant could most easily be derived by a method Boole had explained in his very first paper, written in 1839. Boole then informed Cayley of yet another result, obtained by trial and error:

$$\theta(u) = I_1^3 - 27I_2^2. \tag{34.11}$$

This relation shows that the three invariants $\theta(u)$, I_1^3, I_2^2 were linearly dependent. Cayley was intrigued by this result and computed invariants with still greater fervor, though by means of new methods.

Boole soon abandoned invariant theory in favor of analysis and logic; because of the unwieldy computational difficulties of hyperdeterminants, Cayley also gave up using them to find invariants. Perhaps surprisingly, Gelfand, Kapranov, and Zelevinsky rediscovered and promoted the study of hyperdeterminants. In their 1994 *Discriminants, Resultants, and Multidimensional Determinants*, they wrote that although hyperdeterminants had been largely abandoned for 150 years, they found them to be important in their attempt to construct a general theory of hypergeometric functions in several variables.

However, the relation (34.11) suggested to Cayley and Sylvester an important problem, and they began work on it in the 1850s: Determine invariants $I_1, I_2, \ldots, I_s$ of a binary quantic such that all other invariants would be of the form $P(I_1, I_2, \ldots, I_s)$, for some polynomial P. The English mathematicians Arthur Cayley (1821–1895) and J. J. Sylvester were mathematical friends, reminding us of Euler and Goldbach before them and Hardy and Littlewood after them. Cayley and Sylvester met in 1847 as law students; they remained close friends for almost fifty years until Cayley's death, meeting as frequently as possible and exchanging hundreds of letters and hand-delivered notes. Both algebraists, they often worked simultaneously on the same topic. One may ask why they published no joint work. First, Cayley was a reserved and reticent person, while Sylvester was extremely ebullient and volatile. Moreover, Sylvester exhibited a strong need to maintain strict mathematical priority, both for himself and others. For example, in 1882 Sylvester wrote a paper on partitions, divided into a number of distinct sections, each with its own heading and authorship, indicating whether that portion of the argument should be credited to himself or to his student Franklin. In spite of the apparent separateness of their work, Cayley and Sylvester's mutual support and motivation surely led each of them to more progress than they might have achieved separately. E. T. Bell aptly labeled Cayley and Sylvester the invariant twins; we remark that they must have been fraternal twins.

In order to look at the work of Cayley and Sylvester after 1850, we give some definitions in slightly modernized form, largely following Hilbert's notation, presented in his 1897 lectures, published in 1993. Cayley and Sylvester worked primarily on invariants of binary quantics. These are polynomials in two variables, of the form

$$f(x_1, x_2) = a_0 x_1^n + \binom{n}{1} a_1 x_1^{n-1} x_2 + \binom{n}{2} a_2 x_1^{n-2} x_2^2 + \cdots + a_n x_2^n. \tag{34.12}$$

Suppose the linear transformation (34.5) with determinant $\delta = ps - qr \neq 0$ converts $f(x_1, x_2)$ into

$$A_0 y_1^n + \binom{n}{1} A_1 y_1^{n-1} y_2 + \binom{n}{2} A_2 y_1^{n-2} y_2^2 + \cdots + A_n y_2^n. \tag{34.13}$$

An invariant I of $f(x_1, x_2)$ is then a polynomial in the coefficients $a_0, a_1, \ldots, a_n$, denoted by $I(a_0, a_1, \ldots, a_n)$, such that for some integer p

$$I(A_0, A_1, \ldots, A_n) = \delta^p I(a_0, a_1, \ldots, a_n), \tag{34.14}$$

where $A_0, A_1, \ldots, A_n$ are given by (34.13). Next, a covariant of $f(x_1, x_2)$, denoted by $C(a_0, a_1, \ldots, a_n, x_1, x_2)$, is defined as a polynomial in $a_0, a_1, \ldots, a_n$ and in x_1, x_2, such that

$$C(A_0, A_1, \ldots, A_n, y_1, y_2) = \delta^p C(a_0, a_1, \ldots, a_n, x_1, x_2), \quad (34.15)$$

with $A_0, A_1, \ldots, A_n$ again defined by (34.13).

Within this notation, the invariant of the quadratic form is $a_1^2 - a_0 a_2$; for the quartic form, the invariants mentioned earlier would be

$$I_1 = a_0 a_4 - 4 a_1 a_3 + 3 a_2^2 \text{ and } I_2 = a_0 a_2 a_4 - a_1^2 a_4 - a_0 a_3^2 - a_2^3 + 2 a_1 a_2 a_3.$$

Two of these invariants are homogeneous polynomials of degree 2 and the third is of degree 3. If the coefficient a_k is assigned a weight k, then the weight of each term in $a_1^2 - a_0 a_2$ can be given the value 2 by adding the weights in each product. Thus, this invariant is said to be of weight 2. Similarly, the weights of the other two invariants are 4 and 6. Note also that the invariant $a_1^2 - a_0 a_2$ is the discriminant of the quadratic form while $I_1^3 - 27 I_2^2$ is the discriminant of the quartic form. In a similar way, the cubic form discriminant given by

$$a_0^2 a_3^2 - 3 a_1^2 a_2^2 + 4 a_1^3 a_3 + 4 a_0 a_2^3 - 6 a_0 a_1 a_2 a_3$$

is an invariant of that form, of degree 4 and weight 6.

In his 1854 paper, "An Introductory Memoir upon Quantics," Cayley showed that all invariants of (34.12) were homogeneous polynomials of a given degree, say θ, and weight p, identical to the integer in equation (34.14), and that the relation of these quantities was determined by

$$n\theta = 2p. \quad (34.16)$$

Indeed, this paper included a similar result for covariants. Cayley also found a computationally simpler way of generating invariants by means of differential operators. Interestingly, Cayley later noted that as early as the 1840s, he had observed that

$$\left(a \frac{\partial}{\partial b} + 2b \frac{\partial}{\partial c}\right)(b^2 - ac) = 0,$$

and this then led him to consider such operators even in connection with his researches on hyperdeterminants. So in 1854, Cayley defined the two operators

$$\Omega = a_0 \frac{\partial}{\partial a_1} + 2 a_1 \frac{\partial}{\partial a_2} + 3 a_2 \frac{\partial}{\partial a_3} + \cdots + n a_{n-1} \frac{\partial}{\partial a_n}, \quad (34.17)$$

$$O = n a_1 \frac{\partial}{\partial a_0} + (n-1) a_2 \frac{\partial}{\partial a_1} + (n-2) a_3 \frac{\partial}{\partial a_2} + \cdots + a_n \frac{\partial}{\partial a_{n-1}}. \quad (34.18)$$

He showed that an invariant I of (34.12) satisfied the equations

$$\Omega I = 0 \text{ and } O I = 0. \quad (34.19)$$

In fact, a seminvariant is defined as a homogeneous and isobaric (each term of the same weight) polynomial S satisfying $\Omega S = 0$.

Sylvester also conceived of the idea of the differential operator and published it before Cayley in an 1852 paper, "On the Principles of the Calculus of Forms." In this paper, Sylvester noted that he had discovered the differential operator before Cayley communicated it to him. He also remarked that the German mathematician Siegfried Aronhold, "as I collect from private information, was the first to think of the application of this method to the subject." It is probable, however, that Cayley has actual priority in this matter.

Cayley and Sylvester each proved that every seminvariant $I(a_0, a_1, \ldots, a_n)$ of degree θ and weight p would be an invariant under the condition $n\theta = 2p$. They generalized this to covariants and it became their favorite method of producing invariants and covariants, of various degrees and weights. Cayley's "Second Memoir upon Quantics" gave a combinatorial method for computing the number of invariants of degree θ and weight p by solving the equation

$$\Omega S(a_0, a_1, \ldots, a_n) = \sum \alpha_{k_0 k_1 \cdots k_n} \Omega a_0^{k_0} a_1^{k_1} \cdots a_n^{k_n} = 0. \tag{34.20}$$

Now the number of terms of the form $a_0^{k_0} a_1^{k_1} \cdots a_n^{k_n}$ that are homogeneous of degree θ and weight p is equal to the number of nonnegative integer solutions of the two equations:

$$k_0 + k_1 + \cdots + k_n = \theta, \tag{34.21}$$

$$k_1 + 2k_2 + \cdots + nk_n = p. \tag{34.22}$$

Let $\omega_n(\theta, p)$ denote this number. When the Ω operator is applied, we get terms of degree θ and weight $p-1$. So equation (34.20) consists of $\omega_n(\theta, p-1)$ equations in $\omega_n(\theta, p)$ variables. Cayley conjectured that the equations were independent and proceeded on this certainty. With this assumption, he was able to prove that the number of invariants of degree θ and weight p for a form of $f(x_1, x_2)$ of degree $n = 2p/\theta$ would be given by $\omega_n(\theta, p) - \omega_n(\theta, p-1)$.

Observe that the number of solutions of equation (34.22) is the number of partitions of p, where each part is at most n. This connection between the number of invariants and the number of partitions probably led Cayley and Sylvester in the mid-1850s to investigate partitions; Sylvester gave a course of lectures on the subject in 1857. Interestingly, in 1878 during his later career at Johns Hopkins, Sylvester was able to prove Cayley's conjecture of independence, while he was again working intensely on partitions with his students. This proof implied that $\omega_n(\theta, p)$ was the coefficient of x^p in the Gaussian polynomial

$$\frac{(1-x^{n+1})(1-x^{n+2})\cdots(1-x^{n+\theta})}{(1-x)(1-x^2)\cdots(1-x^\theta)}. \tag{34.23}$$

We observe parenthetically this in turn implies that the Gaussian polynomial is unimodal; recall that a polynomial $a_0 + a_1 x + a_2 x^2 + \cdots + a_n x^n$ is called unimodal if there exists an integer $m \leq n$ such that

$$a_0 \leq a_1 \leq a_2 \leq \cdots \leq a_m \geq a_{m+1} \geq a_{m+2} \geq \cdots \geq a_n . \tag{34.24}$$

In his second memoir, Cayley also considered the problem of determining the fundamental invariants of a binary quantic. He presented the list of such invariants for quadratic through the sextic forms, but due to an error in reasoning he believed that binary forms of order seven and more did not have a finite basis, that is, that there did not exist a finite number of invariants $I_1, I_2, \ldots, I_s$ of a form such that every invariant of that form could be written as a polynomial in these s invariants. This mistake was not corrected until the German mathematician Paul Gordan proved in 1868 that the covariants, and hence also the invariants, of any binary quantic had a finite basis. This was later extended by Hilbert to covariants of m-ary quantics in a very important paper published in 1890. In spite of repeated attempts, Sylvester and Cayley failed to prove Gordan's theorem by their own methods.

It is very interesting to recognize the origins of German invariant theory in number theory and algebraic geometry. Since Fermat, binary quadratic forms had been studied in number theory. In the 1770s, in order to study such forms, Lagrange applied linear transformations such as in (34.5), except that he took the values p, q, r, and s to be integers. In this context, Gauss mentioned equation (34.6) in article 158 of his 1801 *Disquisitiones*. He did not go beyond the observation that the determinant $\theta(R)$ of the form R divided by the determinant of the form $\theta(Q)$ was a square, $(ps - qr)^2$. Then in 1844, while studying number theoretic properties of the binary cubic, Eisenstein found the invariant

$$a_0^2 a_3^2 - 3a_1^2 a_2^2 + 4a_1^3 a_3 + 4a_0 a_2^3 - 6a_0 a_1 a_2 a_3. \tag{34.25}$$

In the same year, L. O. Hesse (1811–1874), a student of Jacobi, defined the important covariant

$$\frac{\partial^2 Q}{\partial x_1^2} \frac{\partial^2 Q}{\partial x_2^2} - \left(\frac{\partial^2 Q}{\partial x_1 x_2}\right)^2 \tag{34.26}$$

for any binary quantic. He introduced this covariant in order to study critical points of curves. More generally, for any homogeneous polynomial f of degree m in n variables $x_1, x_2, \ldots, x_n$, he defined the determinant

$$\left| \frac{\partial^2 f}{\partial x_i \partial x_j} \right|. \tag{34.27}$$

Also note that in 1841, Jacobi published an important paper on functional determinants, defining the Jacobian and drawing attention to this area of study. The determinant (34.27) is now called the Hessian, a name given by Sylvester; in 1949, Hesse's student, Siegfried Aronhold (1819–1894), who also studied with Jacobi and Dirichlet, initiated the symbolic algebraic approach for studying invariants and covariants that characterized German invariant theory until Hilbert took it in a different direction in the 1880s. Clebsch and Gordan made use of Aronhold's approach; Paul Gordan (1837–1912), who wrote his thesis in Berlin under Kummer and had as his only doctoral student the great Emmy Noether, mastered the symbolic method and thereby proved that the invariants and covariants of a binary quantic had a finite basis.

The problem of extending Gordan's result to forms in n variables was very difficult to tackle using the existing algorithmic methods. In 1890, Hilbert introduced new methods and solved the problem. Hilbert's proof depended on his lemma concerning solutions of a system of linear Diophantine equations. He proved the existence of a finite number of solutions of a special kind. This approach lent his theorem a nonconstructive character. Thus, Gordan is said to have commented, "That is not mathematics; that is theology!" Three years later, Hilbert gave a different proof, dependent on what is now known as the Hilbert basis theorem; it has now been reformulated in terms of ideals: If $I \subseteq K(x_1, x_2, \ldots, x_n)$ is any ideal in the ring of polynomials in n variables with coefficients in the field K, then there exists a finite number of polynomials $f_1, f_2, \ldots, f_m$ in I such that for all f in I

$$f = A_1 f_1 + A_2 f_2 + \cdots + A_m f_m \tag{34.28}$$

for some polynomials $A_1, A_2, \ldots, A_m$ in the ring. With the development of the machinery of Gröbner bases, Hilbert's second method of proof has become computationally quite significant.

Hilbert's work became the foundation for the development of commutative algebra in the twentieth century; it paved the way leading toward the abstract point of view in algebra and to the recent computational methods of ideal theory. In the area of commutative algebra, the chess champion Emanuel Lasker (1868–1941) in 1905 established the main facts behind the primary decomposition of ideals. Another important contributor to the theory of rings of polynomials was F. S. Macaulay (1862–1937), Littlewood's teacher of mathematics at St. Paul's School in London. Although he was an excellent mathematical researcher, Macaulay remained a secondary school teacher throughout his career. In a famous 1921 paper, "Idealtheorie in Ringbereichen," Emmy Noether pioneered the abstract approach to ring theory. It is interesting that while Gordan had an algorithmic and concrete approach to mathematics, his student became one of the founding stars of the abstract and conceptual approach to mathematics.

Although we do not go into detail on the topic, it may be worthwhile to comment briefly on the origins and development of elimination theory. Invariant theorists and mathematicians working with rings of polynomials in several variables found the method of elimination useful in various contexts. Boole made liberal application of elimination theory to produce invariants, and the topic led to the development of several aspects of algebra and algebraic geometry. Remarkably, in the twenty-first century, Eric Feron, an aerospace engineer, saw fit to translate Étienne Bézout's 1779 book on elimination theory as applied to polynomials in several variables, *Théorie général des équations algébriques*. In 2006 Feron wrote in the translator's foreword:

> Translating Bézout's research centerpiece became necessary to me after attending an illuminating presentation made by Pablo Parrilo at MIT sometime around 2002. His presentation was devoted to polynomially constrained polynomial optimization via sum-of-square arguments. It was illuminating because much of sum-of-square optimization methods rely on (i) using *polynomial multipliers*, and (ii) considering the various monomials appearing in the polynomial expressions as *independent*

variables, resulting in interesting algorithmic simplifications. Such was also Bézout's approach when dealing with systems of polynomial equations. I decided I needed to investigate the matter in more detail, by reading Bézout's work and writing the present translation.

Étienne Bézout (1730–1783) became interested in mathematics by studying Euler, and he made many practical mathematical applications, including a six-volume course for the French artillery. His most original investigations involved the analysis of polynomial equations in many variables. Bézout's theorem on the number of intersection points of two plane algebraic curves is a direct consequence of his researches.

Even before Bézout, elimination was used in the seventeenth and eighteenth centuries to derive the discriminant of a polynomial or the resultant of two polynomials. In fact, the word resultant was used to signify the result obtained after elimination. The resultant $R(f, g)$ of two polynomials is a polynomial in the coefficients of f and g, and assuming that the coefficients of the highest powers of f and g do not vanish, then $R(f, g) = 0$ if and only if f and g have a common root. In 1665–66, Newton calculated the resultant of two cubics by computing various symmetric functions of the roots, including sums of powers. He wrote out all the thirty-four terms of the resultant. In his published work on algebra, he eliminated the variable from the two equations by a different method.

It is an interesting coincidence that at about this same time, Seki Takakazu (1642–1708) was also thinking about the problem of elimination. Seki was a pioneer in the development of algebra in Japan; he joined with his student, Takebe Katahiro, to lay the foundation of early Japanese mathematics, or Wasan. Around 1670, Seki presented a method of obtaining the resultant by using determinants. Given two polynomial equations of degree n, he first converted them into n equations, each of degree $n-1$. He then applied a method that amounted to computing the n by n determinant so obtained. He explained the details of his method by taking small values of n, at least up to $n = 4$. And Zhu Shijie investigated resultants in the thirteenth century. As for determinants, Chinese mathematicians had earlier used them to solve simultaneous linear equations and the Japanese mathematicians of the seventeenth and eighteenth centuries were familiar with this aspect of Chinese algebra. Seki's method was rediscovered by Bézout. Consider their method for eliminating the x term from two cubics:

$$f = a_1 x^3 + b_1 x^2 + c_1 x + d_1, \quad g = a_2 x^3 + b_2 x^2 + c_2 x + d_2.$$

The three quadratic polynomials obtained from f and g would be $a_2 f - a_1 g$, $(a_2 x + b_2) f - (a_1 x + b_1) g$, and $(a_2 x^2 + b_2 x + c_2) f - (a_1 x^2 + b_1 x + c_1) g$. The 3×3 determinant formed by the coefficients of the three quadratics would then be the resultant.

Euler and Lagrange and others contributed to elimination theory in the eighteenth century; in the nineteenth century, Sylvester and Cayley were deeply interested in the topic, especially for its connection with invariant theory. The resultant of two binary quantics, for example, was their simultaneous invariant. In 1840, Sylvester published "A Method of Determining by Mere Inspection the Derivatives from Two Equations of Any Degree," giving the modern expression of the resultant of two polynomials of degrees m and n, respectively, as an $m + n$ by $m + n$ determinant. He explained the general rule and illustrated it by computing the 4×4 determinant obtained in the case of two quadratics. Since the computation of determinants is generally tedious, it is interesting to read the remark at the end of Sylvester's paper:

> Through the well-known ingenuity and kindly proferred help of a distinguished friend, I trust to be able to get a machine made for working Sturm's theorem, and indeed all problems of derivation, after the method here expounded; on which subject I have a great deal more to say, than can be inferred from this or my preceding papers.

The distinguished friend was surely Charles Babbage who at that time was developing his analytical engine to carry out repetitive numerical and algebraic calculations. Babbage was assisted in his endeavor by Ada Lovelace, the daughter of Lord Byron.

Cayley published several papers on elimination theory, reworking and simplifying the methods of earlier writers but also making very original contributions. The comments of Gelfand, Kapranov, and Zelevinsky in this connection are worth noting. In their book, they write that in a short paper of 1848, Cayley "outlined a general method of writing down the resultant of several polynomials in several variables. We were very surprised to find that Cayley introduced in this note several fundamental concepts of homological algebra: complexes, exactness, Koszul complexes, and even the invariant now sometimes called the Whiteside torsion or Reidemeister–Franz torsion of an exact complex. The latter invariant is a natural generalization of the determinant of a square matrix (which itself was a recent discovery back in 1848), so we prefer to call it the determinant of a complex. Using this terminology, Cayley's main result is that the resultant is the determinant of the Koszul complex."

Elimination theory suffered a decline as Emmy Noether's abstract approach came to the forefront. Algebraic algorithms had to be reworked into this new context. Thus, in his 1946 book on the foundations of algebraic geometry, André Weil constructed an abstract device intended to finally make elimination theory superfluous. However, algebraic equations in many variables are also studied by engineers, for whom the abstract approach is not ideal. Moreover, Shreeram Abhyankar, protesting Weil's attempt to eliminate elimination theory, pointed out that some useful mathematical information could be lost in a nonconstructive method. Weil might well agree, and this may be indicated by his exposition of Eisenstein and Kronecker's work on the constructive development of elliptic functions. And so elimination theory continues to flourish. A renewed interest in finding efficient algorithms has produced new methods such as Gröbner bases.

34.2 Boole's Derivation of an Invariant

In his two-part paper published in 1841, "Exposition of a General Theory of Linear Transformations," Boole argued that the concept of an invariant could be useful in algebra. He gave a method for the derivation of an invariant of a general form, of degree n and in m variables. Although the method was not of great use in the further development of invariant theory, it is interesting to observe Boole's originality in arriving at this important concept. We follow Boole closely; he supposed h_n and H_n were nth degree homogeneous functions of m variables $x_1, x_2, \ldots, x_m$ expressible linearly in terms of m variables $y_1, y_2, \ldots, y_m$. He also supposed that

$$h_n(x_1, x_2, \ldots, x_m) = h'_n(y_1, y_2, \ldots, y_m),$$
$$H_n(x_1, x_2, \ldots, x_m) = H'_n(y_1, y_2, \ldots, y_m),$$
(34.29)

where h'_n and H'_n were also homogeneous functions of degree n. In addition, Boole wrote these relations in the simple form

$$q = r \quad \text{and} \quad Q = R,$$
(34.30)

respectively. He differentiated both sides of the second equation with respect to y_1, $y_2, \ldots, y_m$, and by means of the chain rule he got

$$\frac{\partial Q}{\partial x_1}\frac{\partial x_1}{\partial y_1} + \frac{\partial Q}{\partial x_2}\frac{\partial x_2}{\partial y_1} + \cdots + \frac{\partial Q}{\partial x_m}\frac{\partial x_m}{\partial y_1} = \frac{\partial R}{\partial y_1},$$
$$\frac{\partial Q}{\partial x_1}\frac{\partial x_1}{\partial y_2} + \frac{\partial Q}{\partial x_2}\frac{\partial x_2}{\partial y_2} + \cdots + \frac{\partial Q}{\partial x_m}\frac{\partial x_m}{\partial y_2} = \frac{\partial R}{\partial y_2},$$
$$\vdots$$
$$\frac{\partial Q}{\partial x_1}\frac{\partial x_1}{\partial y_m} + \frac{\partial Q}{\partial x_2}\frac{\partial x_2}{\partial y_m} + \cdots + \frac{\partial Q}{\partial x_m}\frac{\partial x_m}{\partial y_m} = \frac{\partial R}{\partial y_m}.$$
(34.31)

We note that Boole did not use the modern partial derivative notation; he wrote $\frac{dQ}{dx_1}$ for $\frac{\partial Q}{\partial x_i}$ and similarly for the other derivatives. Boole then assumed the linear relationship

$$x_1 = \lambda_1 y_1 + \lambda_2 y_2 + \cdots + \lambda_m y_m$$
$$x_2 = \mu_1 y_1 + \mu_2 y_2 + \cdots + \mu_m y_m$$
$$\vdots$$
$$x_m = \rho_1 y_1 + \rho_2 y_2 + \cdots + \rho_m y_m,$$
(34.32)

so that he could replace $\partial x_1/\partial y_1, \partial x_2/\partial y_1, \ldots$ by $\lambda_1, \lambda_2, \ldots$. He argued that since the values $\lambda_1, \lambda_2, \ldots, \mu_1, \mu_2, \ldots$ were finite, the equations

$$\frac{\partial Q}{\partial x_1} = 0, \ \frac{\partial Q}{\partial x_2} = 0, \cdots, \ \frac{\partial Q}{\partial x_m} = 0$$
(34.33)

implied that

$$\frac{\partial R}{\partial y_1} = 0, \ \frac{\partial R}{\partial y_2} = 0, \cdots, \ \frac{\partial R}{\partial y_m} = 0.$$
(34.34)

He observed that, since the determinant of the linear transformation (34.32) could be zero, (34.34) did not imply (34.33).

Boole denoted by $\theta(Q)$ the expression obtained when the variables were eliminated from the polynomials

$$\frac{\partial Q}{\partial x_1}, \frac{\partial Q}{\partial x_2}, \ldots, \frac{\partial Q}{\partial x_m}.$$

34.2 Boole's Derivation of an Invariant

To eliminate x from two polynomials of degree n, he suggested the Euclidean algorithm. Initially, he had m polynomials in m variables. He could eliminate, for example, x_1 from the first two, and then from the second and third, and so on until there were $m-1$ polynomials in $m-1$ variables. A repetition of this method produced $m-2$ polynomials in $m-2$ variables. Ultimately, he had all the variables eliminated, obtaining an expression $\theta(Q)$ containing only the constants. Thus, if $\frac{\partial Q}{\partial x_i}=0$, $i=1,\ldots,n$, he had $\theta(Q)=0$. Moreover, since (34.33) implied (34.34), he also had $\theta(R)=0$; and a similar relation of mutual dependence also existed between $\theta(q)$ and $\theta(r)$.

More generally, Boole combined the two relations in (34.30) into one relation of the form $Q+hq=R+hr$. In this case, if h was such that

$$\theta(Q+hq)=0, \qquad (34.35)$$

then an analogous relation

$$\theta(R+hr)=0 \qquad (34.36)$$

would also be satisfied. Next, Boole let v be the number of terms in the homogeneous polynomials q, r, Q, R and denoted the coefficients in these polynomials by a_1 $a_2, \ldots, a_v$, $b_1, b_2, \ldots, b_v$, $A_1, A_2, \ldots, A_v$, $B_1, B_2, \ldots, B_v$, respectively. Then θ would be a polynomial ϕ in v unknowns, and he could write $\theta(Q)=\phi(A_1,A_2,\ldots,A_v)$. Now from (34.35) and (34.36), Boole reasoned that for any h for which

$$\phi(A_1+ha_1, A_2+ha_2, \ldots, A_v+ha_v)=0, \qquad (34.37)$$

he must also have

$$\phi(B_1+hb_1, B_2+hb_2, \ldots, B_v+hb_v)=0. \qquad (34.38)$$

The expression on the left-hand side of (34.37) was a polynomial in h where the term independent of h would be $\phi(A_1,A_2,\ldots,A_v)=\theta(Q)$, and the coefficient of the highest power of h would be $\phi(a_1,a_2,\ldots,a_n)=\theta(q)$. If the polynomial was divided across by this coefficient, the resulting monic polynomial would be identical with the monic polynomial obtained after the same procedure was applied to the left-hand side of (34.38). Since the coefficients of the polynomials could also be seen as Taylor coefficients, Boole could deduce that

$$\frac{\theta(Q)}{\theta(q)}=\frac{\theta(R)}{\theta(r)}, \text{ and} \qquad (34.39)$$

$$\frac{\left(a_1\frac{\partial}{\partial A_1}+a_2\frac{\partial}{\partial A_2}+\cdots+a_v\frac{\partial}{\partial A_v}\right)^\lambda \theta(Q)}{\theta(q)}$$

$$=\frac{\left(b_1\frac{\partial}{\partial B_1}+b_2\frac{\partial}{\partial B_2}+\cdots+b_v\frac{\partial}{\partial B_v}\right)^\lambda \theta(R)}{\theta(r)}. \qquad (34.40)$$

As an example, Boole noted the simple case

$$ax^2 + 2bxy + cy^2 = a'x'^2 + 2b'x'y' + c'y'^2,$$
$$Ax^2 + 2Bxy + Cy^2 = A'x'^2 + 2B'x'y' + C'y'^2.$$

The results corresponding to (34.39) and (34.40) were

$$\frac{AC - B^2}{ac - b^2} = \frac{A'C' - B'^2}{a'c' - b'^2}, \tag{34.41}$$

$$\frac{aC - 2bB + cA}{ac - b^2} = \frac{a'C' - 2b'B' + c'A'}{a'c' - b'^2}. \tag{34.42}$$

From (34.39) and at the conclusion of the first part of his paper, Boole arrived at the result that gave rise to algebraic invariant theory:

$$\frac{\theta(Q)}{\theta(R)} = \frac{\theta(q)}{\theta(r)} = E.$$

Boole maintained that E could not depend on the coefficients in Q and R (or q and r). Thus, it must depend only on the coefficients appearing in the linear transformation (34.32). Boole wrote that he had found E to be an appropriate power of the determinant of the linear transformation (34.32), illustrating this by means of the binary quadratic and the cubic. He then went on to state the theorem contained in equation (34.7); this in turn led to the definition of an invariant. In a paper of 1844, Boole gave details of a proof and gave the value of γ in (34.7) as $m(n-1)^{m-1}$.

Near the end of the second part of his 1841 paper, Boole wrote that "Linear transformations have hitherto been chiefly applied to the purpose of taking away from a proposed homogeneous function, those terms which involve the products of the variables. ... [T]he transformations, besides being linear, are understood to represent a geometrical change of axes." He went on to say that linear transformation could be applied to purely algebraic problems without geometric considerations. As an example he posed the problem: "To transform the function, $ax^3 + 3bx^2y + 3cxy^2 + dy^3$, to the form $a'x'^3 + d'y'^3$, a' and d' being given, and the transformation unrestricted by any other condition than that of linearity." After solving this problem by means of his method, he applied it to the solution of a cubic equation with the comment, "The doctrine of linear transformations may be elegantly applied to the solution of algebraic equations." In this connection, Cayley, well aware that a general quintic could not be solved in radicals, corresponded with Boole concerning the solution of a quintic and found that invariants could shed some light on their solution.

It is interesting that in 1930, as a schoolboy of 16, Mark Kac solved the cubic by an independently discovered method similar to Boole's solution. Kac wrote the cubic as a difference of two cubes, each of which was a linear function of the variable:

$$x^3 + px + q = A(x + m)^3 - B(x + n)^3.$$

By equating the coefficients of x, he found that $A = n/(n-m)$ and $B = m/(n-m)$ and that m and n were solutions of the quadratic equation

$$y^2 + \frac{3q}{p} y - \frac{1}{3} p = 0;$$

from this, he was able to derive Cardano's formula. Kac's paper was published in a Polish mathematics journal for students and because of this achievement he went on to become a mathematician. Luckily, the journal's editor had been unaware of Boole's work.

34.3 Differential Operators of Cayley and Sylvester

By the early 1850s, Cayley and Sylvester had discovered several elementary properties of invariants of binary quantics. They knew, for example, that these variants were homogeneous and isobaric polynomials satisfying certain partial differential equations. They found these equations independently, though Sylvester was the first to publish them in 1852, and they used them as important tools as their work in invariant theory progressed. We present Sylvester's derivation of the partial differential operators, with a slight change in notation, especially in our use of subscripts. Following Sylvester closely, suppose that

$$\phi = a_0 x_1^n + n a_1 x_1^{n-1} x_2 + \frac{1}{2} n(n-1) a_2 x_1^{n-2} x_2^2 + \cdots + a_n x_2^n \tag{34.43}$$

is a binary quantic and that $I(a_0, a_1, \ldots, a_n)$ is an invariant of ϕ. To derive the differential equation, use the special linear transformation

$$x_1 = y_1 + e y_2, \quad x_2 = y_2 \tag{34.44}$$

to obtain the quantic

$$a_0(y_1 + e y_2)^n + \binom{n}{1} a_1 (y_1 + e y_2)^{n-1} y_2 + \cdots + \binom{n}{k} a_k (y_1 + e y_2)^{n-k} y_2^k + \cdots + a_n y_2^n$$

$$= A_0 y_1^n + \binom{n}{1} A_1 y_1^{n-1} y_2 + \cdots + \binom{n}{k} A_k y_1^{n-k} y_2^k + \cdots + A_n y_2^n,$$

where $A_0 = a_0$, $A_1 = a_1 + e a_0$, $A_2 = a_2 + 2 e a_1 + e^2 a_0$. Note that in general,

$$A_k = a_k + \binom{k}{1} e a_{k-1} + \binom{k}{2} e^2 a_{k-2} + \cdots + e^k a_0. \tag{34.45}$$

Since the determinant of the linear transformation is 1, it follows from the definition (34.14) of an invariant that

$$I(A_0, A_1, \ldots, A_n) = I(a_0, a_1, \ldots, a_n). \tag{34.46}$$

Let $\Delta a_k = A_k - a_k$ and $\Delta I = I(A_0, A_1, \ldots, A_n) - I(a_0, a_1, \ldots, a_n)$. By Taylor's theorem in several variables

$$0 = \Delta I = \sum_{k \geq 1} \frac{1}{k!} \left(\Delta a_0 \frac{\partial}{\partial a_0} + \Delta a_1 \frac{\partial}{\partial a_1} + \Delta a_2 \frac{\partial}{\partial a_2} + \cdots \right)^k I$$

$$= \sum_{k \geq 1} \frac{1}{k!} \left(ea_0 \frac{\partial}{\partial a_1} + (2ea_1 + e^2 a_0) \frac{\partial}{\partial a_2} + \cdots \right)^k I.$$

Since this is true for every value of e, the coefficient of every power of e must be zero. In particular, the coefficient of the first power of e gives

$$\Omega I \equiv \left(a_0 \frac{\partial}{\partial a_1} + 2a_1 \frac{\partial}{\partial a_2} + 3a_2 \frac{\partial}{\partial a_3} + \cdots + na_{n-1} \frac{\partial}{\partial a_n} \right) I = 0. \tag{34.47}$$

As Sylvester pointed out, this differential equation could also be obtained by taking the derivative of (34.46) with respect to e and applying the chain rule:

$$0 = \frac{\partial I}{\partial A_0} \frac{\partial A_0}{\partial e} + \frac{\partial I}{\partial A_1} \frac{\partial A_1}{\partial e} + \cdots + \frac{\partial I}{\partial A_n} \frac{\partial A_n}{\partial e}$$

$$= A_0 \frac{\partial I}{\partial A_1} + 2A_1 \frac{\partial I}{\partial A_2} + \cdots + nA_{n-1} \frac{\partial I}{\partial A_n}.$$

Similarly, apply the transformation

$$x_1 = y_1, \quad x_2 = ey_1 + y_2 \tag{34.48}$$

to get the other differential equation

$$OI \equiv \left(na_1 \frac{\partial}{\partial a_0} + (n-1)a_2 \frac{\partial}{\partial a_1} + \cdots + a_n \frac{\partial}{\partial a_{n-1}} \right) = 0. \tag{34.49}$$

The operators Ω and O defined by (34.47) and (34.49) turned out to be quite important in invariant theory; Cayley and Sylvester made considerable use of them in their researches. The corresponding operators for covariants, defined by (34.15), are given by

$$\left(\Omega - x_2 \frac{\partial}{\partial x_1} \right) C = 0, \tag{34.50}$$

$$\left(O - x_1 \frac{\partial}{\partial x_2} \right) C = 0. \tag{34.51}$$

To prove that invariants are homogeneous and isobaric polynomials, take another special linear transformation

$$x_1 = e_1 y_1, \quad x_2 = e_2 y_2. \tag{34.52}$$

34.3 Differential Operators of Cayley and Sylvester

In this case, the quantic is transformed to

$$a_0 e_1^n y_1^n + \binom{n}{1} a_1 e_1^{n-1} e_2 y_1^{n-1} y_2 + \cdots + \binom{n}{k} a_k e_1^{n-k} e_2^k y_1^{n-k} y_2^k + \cdots + a_n e_2^n y_2^n$$

$$= A_0 y_1^n + \binom{n}{1} A_1 y_1^{n-1} y_2 + \cdots + \binom{n}{k} A_k y_1^{n-k} y_2^k + \cdots + A_n y_2^n,$$

where $A_0 = a_0 e_1^n, A_1 = a_1 e_1^{n-1} e_2, \ldots, A_k = a_k e_1^{n-k} e_2^k$.

Now suppose

$$I = (a_0, a_1, \ldots, a_n) = \sum \alpha_{s_0 s_1 \cdots s_n} a_0^{s_0} a_1^{s_1} \cdots a_n^{s_n}.$$

Since the determinant δ of the transformation (34.52) is $e_1 e_2$, we have by definition (34.14)

$$I(A_0, A_1, \ldots, A_n) = I(a_0 e_1^n, a_1 e_1^{n-1} e_2, \ldots, a_n e_2^n)$$

$$= \sum \alpha_{s_0 s_1 \cdots s_n} a_0^{s_0} e_1^{n s_0} a_1^{s_1} e_1^{(n-1)s_1} e_2^{s_2} \cdots a_k^{s_k} e_1^{(n-k)s_k} e_2^{k s_k} \cdots a_n^{s_n} e_n^{n s_n}$$

$$= \sum \alpha_{s_0 s_1 \cdots s_n} a_0^{s_0} a_1^{s_1} \cdots a_n^{s_n} e_1^{n s_0 + (n-1)s_1 + \cdots + (n-k)s_k + \cdots s_{n-1}} e_2^{s_1 + \cdots + k s_k + \cdots + n s_n}$$

$$= \delta^p I(a_0, a_1, \ldots, a_n) = e_1^p e_2^p \sum \alpha_{s_0 s_1 \cdots s_n} a_0^{s_0} a_1^{s_1} \cdots a_n^{s_n}.$$

Equating the coefficients of $a_0^{s_0} a_1^{s_1} \cdots a_n^{s}$ yields

$$n s_0 + (n-1) s_1 + \cdots + (n-k) s_k + \cdots + s_{n-1} = p, \qquad (34.53)$$

$$s_1 + 2 s_2 + \cdots + k s_k + \cdots + n s_n = p. \qquad (34.54)$$

By adding these equations, one obtains

$$n(s_0 + s_1 + \cdots + s_n) = 2p. \qquad (34.55)$$

This in turn implies that an invariant I is homogeneous of degree $\theta = s_0 + s_1 + \cdots + s_n$. If we define the weight of a_k to be k, then the weight of $a_0^{s_0} a_1^{s_1} \cdots a_n^{s_n}$ will be given by equation (34.54); this means that I is isobaric of weight p.

In his 1856 memoir, using the same method, Cayley proved a similar result for covariants. Suppose the covariant is given by

$$C(a_0, a_1, \ldots, a_n, x_1, x_2) = C_0 x_1^m + C_1 \binom{m}{1} x_1^{m-1} x_2 + \cdots + C_m x_2^m. \qquad (34.56)$$

Following the procedure we have presented, one may conclude that the coefficients $C_0, C_1, \ldots, C_m$ are homogeneous in $a_0, a_1, \ldots, a_n$ and are of the same degree; call the degree θ. Each coefficient is isobaric, and the weights of the coefficients are given by $C_i = p + i$, $i = 0, 1, \ldots, m$. The weight p of the coefficient C_0 is called the weight of the covariant and the integer m in (34.56) is called the order of the covariant. The argument yielding these results on covariants also shows that

$$m = n\theta - 2p. \qquad (34.57)$$

Cayley also used the differential equations (34.50) and (34.51) satisfied by the covariant to derive important relations among coefficients of a covariant. For example, from the equation

$$OC = x_1 \frac{\partial C}{\partial x_2},$$

where C is given by (34.56) and O is the differential operator (34.49), we have

$$OC_0 x_1^m + \binom{m}{1} OC_1 x_1^{m-1} x_2 + \cdots + \binom{m}{k} OC_k x_1^{m-k} x_2^k + \cdots + OC_m x_2^m$$

$$= \binom{m}{1} C_1 x_1^m + 2 \binom{m}{2} C_2 x_1^{m-1} x_2 + \cdots$$

$$+ (k+1) \binom{m}{k+1} C_{k+1} x_1^{m-k} x_2^k + \cdots + m C_m x_1 x_2^{m-1}.$$

Equating coefficients produces the relations

$$OC_k = (m-k) C_{k+1}, \quad k = 0, 1, \ldots, m. \tag{34.58}$$

The first m of these equations imply the relations

$$C_k = \frac{1}{m(m-1)\cdots(m-k+1)} O^k C_0, \quad k = 1, 2, \ldots, m. \tag{34.59}$$

Since all the coefficients $C_1, C_2, \ldots, C_m$ of the covariant C can be obtained from C_0, Sylvester called C_0 the source of the covariant.

By using the other differential equation for the covariant C, (34.50), Cayley derived the equations

$$\Omega C_k = k C_{k-1}, \quad k = 0, 1, \ldots, m. \tag{34.60}$$

In particular, the source satisfied the differential equation $\Omega C_0 = 0$. Cayley and Sylvester named any homogeneous and isobaric function P of $a_0, a_1, \ldots, a_n$ a semi-invariant, or seminvariant, if it was annihilated by Ω, that is, $\Omega P = 0$. Thus, the source of a covariant turned out to be a seminvariant. Clearly, not all seminvariants are invariants. But Cayley pointed out that if a seminvariant of degree θ and weight p satisfied equation (34.55), that is, $n\theta = 2p$, then the seminvariant would also be an invariant.

34.4 Cayley's Generating Function for the Number of Invariants

Cayley's ambition was to develop an algorithm capable of producing all the invariants of a given binary form. In this pursuit, it was important for him to determine the number of seminvariants of given degree and weight. By 1856, he had discovered a beautiful

34.4 Cayley's Generating Function for the Number of Invariants

connection between this problem and Gaussian polynomials. Recall that if I is any seminvariant of degree θ and weight p, and

$$I = \sum \alpha_{s_0 s_1 \cdots s_n} a_0^{s_0} a_1^{s_1} \cdots a_n^{s_n}, \text{ then}$$

$$s_0 + s_1 + \cdots + s_n = \theta; \quad s_1 + 2s_2 + \cdots + n s_n = p.$$

Next let $N(n, \theta, p)$ denote the number of seminvariants with given n, θ, and p, and let $\omega_n(\theta, p)$ denote the number of integer solutions of the two previous equations for s_k, with the constraint that $s_k \geq 0$. The differential operator Ω, when applied to I, keeps the degree of each term the same but reduces the weight by one. Note that

$$\Omega I = \sum \alpha_{s_0 s_1 \cdots s_n} \Omega a_0^{s_0} a_1^{s_1} \cdots a_n^{s_n} = 0. \tag{34.61}$$

The number of terms in (34.61) is $\omega_n(\theta, p-1)$, and the coefficient of each of these terms is zero. This implies that there are $\omega_n(\theta, p-1)$ homogeneous linear equations for $\omega_n(\theta, p)$ quantities. In his second memoir of 1856, Cayley correctly assumed that these equations were independent and concluded that

$$N(n, \theta, p) = \omega_n(\theta, p) - \omega_n(\theta, p-1). \tag{34.62}$$

Cayley was unable to prove his assumption but was so certain of its correctness that he based his invariant theory upon it. Sylvester provided a proof in 1878. Cayley argued that it was obvious that the number $\omega_n(\theta, p)$ would turn out to be the coefficient of $x^p z^\theta$ in the series expansion of

$$\frac{1}{(1-z)(1-xz)(1-x^2 z) \cdots (1-x^n z)}. \tag{34.63}$$

Indeed, this is not difficult to see if we expand by the geometric series:

$$(1 + z + z^2 + \cdots)(1 + xz + x^2 z^2 + \cdots) \cdots (1 + x^n z + x^{2n} z^2 + \cdots).$$

Clearly, the coefficient of $x^p z^\theta$ will be equal to the number of nonnegative integer solutions of the two equations involving $s_0, s_1, \ldots, s_n$.

Summarizing Cayley's work in obtaining the invariants for forms of low degree, we observe that in 1855 he expanded the generating function (34.63) for $\omega_n(\theta, p)$:

$$\frac{1}{(1-z) \cdots (1-x^n z)} = 1 + G_1(x) z + G_2(x) z^2 + G_3(x) z^3 + \cdots . \tag{34.64}$$

To see the connection with Gaussian polynomials, change z to xz to get

$$\frac{1}{(1-xz) \cdots (1-x^{n+1} z)} = 1 + G_1(x) xz + G_2(x) x^2 z^2 + G_3(x) x^3 z^3 + \cdots . \tag{34.65}$$

The two equations (34.64) and (34.65) imply

$$(1-z)(1 + G_1 z + G_2 z^2 + \cdots + G_m z^m + \cdots)$$
$$= (1 - x^{n+1} z)(1 + G_1 xz + \cdots + G_m x^m z^m + \cdots).$$

Now equate the coefficients of z^m on both sides to get

$$G_m - G_{m-1} = G_m x^m - G_{m-1} x^{m+n} \quad \text{or}$$

$$G_m(x) = \frac{(1-x^{m+n})}{1-x^m} G_{m-1}$$
$$= \frac{(1-x^{m+n})(1-x^{m+n-1})\cdots(1-x^{n+1})}{(1-x^m)(1-x^{m-1})\cdots(1-x)}. \tag{34.66}$$

Thus, $G_m(x)$ is a Gaussian polynomial and the coefficient of x^p in the polynomial $G_\theta(x)$ gives $\omega_n(\theta, p)$. Now Cayley realized that the number of seminvariants $N(n, \theta, p)$ could be expressed as the difference between the coefficients of x^p and x^{p-1} in the Gaussian polynomial $G_\theta(x)$, and this difference gave the number of invariants of degree θ and weight p provided that $n\theta = 2p$. Note that $N(n, \theta, p)$ is the coefficient of x^p in

$$\frac{(1-x^{\theta+1})(1-x^{\theta+2})\cdots(1-x^{\theta+n})}{(1-x^2)(1-x^3)\cdots(1-x^\theta)}. \tag{34.67}$$

Also observe that for invariants with weight $p = n\theta/2$, $\omega_n(\theta, n\theta/2)$ and $\omega_\theta(n, n\theta/2)$ are equal because they both turn out to be the coefficient of $x^{n\theta/2}$ in

$$\frac{(1-x)(1-x^2)\cdots(1-x^{\theta+n})}{(1-x)\cdots(1-x^\theta)(1-x)\cdots(1-x^n)}.$$

This immediately implies that

$$N(n, \theta, n\theta/2) = N(\theta, n, n\theta/2), \tag{34.68}$$

a result, known as Hermite's reciprocity theorem, established by Hermite in 1852 by a different method. Sylvester noted that this theorem was equivalent to stating that the number of partitions of any number p into at most m parts, with each part at most n, equaled the number of partitions of p into at most n parts, with each part at most m.

Cayley proceeded by applying these results to determine the full invariant systems for forms of degree $n = 2, 3, 4, 5, 6$. For example, by (34.67), when $n = 4$, the number of independent invariants of degree θ would be the coefficient of $x^{2\theta}$ in

$$\frac{(1-x^{\theta+1})(1-x^{\theta+2})(1-x^{\theta+3})(1-x^{\theta+4})}{(1-x^2)(1-x^3)(1-x^4)}.$$

Observe that in order to find the coefficient of $x^{2\theta}$, we must retain numerator terms of degree 2θ or less; this means that we should determine the coefficient of $x^{2\theta}$ in the power series expansion of

$$\frac{1-x^{\theta+1}(1+x+x^2+x^3)}{(1-x^2)(1-x^3)(1-x^4)} = \frac{1}{(1-x^2)(1-x^3)(1-x^4)} - \frac{x^\theta \cdot x}{(1-x)(1-x^2)(1-x^3)}.$$

This would be the same as the coefficient of $x^{2\theta}$ in

$$1/\bigl((1-x^2)(1-x^3)(1-x^4)\bigr)$$

34.4 Cayley's Generating Function for the Number of Invariants

minus the coefficient of x^θ in

$$x/\big((1-x)(1-x^2)(1-x^3)\big)$$

or minus the coefficient of $x^{2\theta}$ in

$$x^2/\big((1-x^2)(1-x^4)(1-x^6)\big).$$

Thus, we need the coefficient of $x^{2\theta}$ in

$$(1+x^3-x^2)/\big((1-x^2)(1-x^4)(1-x^6)\big).$$

We may drop the odd power term x^3; then, we need the coefficient of $x^{2\theta}$ in

$$1/\big((1-x^4)(1-x^6)\big)$$

or of x^θ in

$$\frac{1}{(1-x^2)(1-x^3)} = (1+x^2+x^4+\cdots)(1+x^3+x^6+\cdots). \tag{34.69}$$

This implies that the number of independent invariants of degree θ is equal to the number of integer solutions of $2m + 3n = \theta$. Clearly, in each case $\theta = 2$ or $\theta = 3$, there is exactly one invariant, called I_2 or I_3. For nonnegative integers m_1 and n_1, if $2m_1 + 2n_1 = \theta$, then $I_2^{m_1} I_3^{n_1}$ is an invariant of degree θ. It is easy to see that all linearly independent invariants of a given degree can be produced by this method. Hence, I_2 and I_3 generate the full invariant system of a binary form, or quantic, of order 4.

Cayley also showed how the differential operators could be used to determine the invariants I_2 and I_3. For instance, for I_2, since it is of degree 2, it must be of weight 4 by the relation $n\theta = 2p$. The binary form of degree 4 has coefficients a_0, a_1, a_2, a_3, a_4; therefore, the weight 4 and degree 2 monomials are a_0a_4, a_1a_3, and a_2^2. To find an invariant I of degree 2 and weight 4, Cayley could set

$$I = Aa_0a_4 + Ba_1a_3 + Ca_2^2$$

and then determine A, B, C by solving the differential equation $\Omega I = 0$, where Ω was defined by (34.47). One may easily check that

$$\Omega a_0a_4 = 4a_0a_3, \quad \Omega a_1a_3 = a_0a_3 + 3a_1a_2, \quad \Omega a_2^2 = 4a_1a_2.$$

Thus, from

$$\Omega I = (4A+B)a_0a_3 + (3B+4D)a_1a_2 = 0,$$

Cayley had $B = -4A$ and $C = 3A$. Hence, there was only one independent invariant in this case, given by

$$I_2 = a_0a_4 - 4a_1a_3 + 3a_2^2; \tag{34.70}$$

one may check that the equation $OI = 0$, from (34.49), is also satisfied. A similar calculation would determine the invariant of degree 3 and weight 6:

$$I_3 = a_0a_2a_4 - a_0a_3^2 - a_1^2a_4 + 2a_1a_2a_3 - a_2^3. \tag{34.71}$$

Cayley's result for $n = 5$ was that the number of independent invariants of degree θ was the coefficient of x^θ in

$$\frac{1 - x^6 + x^{12}}{(1-x^4)(1-x^6)(1-x^8)},$$

or the coefficient of x^θ in

$$\frac{1 - x^{36}}{(1-x^4)(1-x^8)(1-x^{12})(1-x^{18})}. \tag{34.72}$$

This allowed Cayley to conclude that there were no invariants of odd degree, but that there was one irreducible invariant of degree 4, one of degree 8, one of 12, and one of 18. However, these were connected by an equation of degree 36, that is, the square of the invariant of degree 18 was a polynomial function of the other three. Sylvester called such a relation a syzygy. Cayley attributed this result to Hermite. In fact, before studying Hermite's work, Cayley had thought that the degrees of the invariants of order 5 binary forms had to be divisible by 4.

In the case of $n = 7$, Cayley made a conceptual error. He stated that the number of independent invariants of degree θ was equal to the coefficient of x^θ in

$$\frac{1 - x^6 + 2x^8 - x^{10} + 5x^{12} + \cdots}{(1-x^4)(1-x^6)(1-x^8)(1-x^{12})},$$

where the numerator was equal to

$$(1-x^6)(1-x^8)^{-2}(1-x^{10})(1-x^{12})^{-5}(1-x^{14})^{-5}\cdots,$$

and where the series of factors did not terminate. Hence, he mistakenly concluded that the invariants did not have a finite basis. Gordan proved this to be incorrect.

34.5 Sylvester's Fundamental Theorem of Invariant Theory

The counting method for finding the fundamental invariants, and Cayley's conjecture in particular, were called into question when Cayley's mistake became evident. But in 1878, Sylvester succeeded in proving this basic result:

$$N(n, \theta, p) = \omega_n(\theta, p) - \omega_n(\theta, p-1).$$

In his spirited style, his paper began:

> I am about to demonstrate a theorem which has been waiting proof for the last quarter of a century and upwards. It is the more necessary that this should be done, because the theorem has been supposed to lead to false conclusions, and its correctness has consequently been impugned ... but the theorem itself is perfectly true, as I shall show by an argument so irrefragable that it must be considered for ever hereafter safe from all doubt or cavil. It lies at the basis of the investigations begun by Professor Cayley in his *Second Memoir upon Quantics*, which it has fallen to my lot, with no small labour and contention of mind, to lead to a happy issue, and thereby to advance the standards of the Science of Algebraical Forms to the most advanced point that has hitherto

34.5 Sylvester's Fundamental Theorem of Invariant Theory

been reached. The stone that was rejected by the builders has become the chief corner-stone of the building.

We follow Sylvester's reasoning very closely, but present it in slightly streamlined form. The proof depends on Sylvester's lemma that for a seminvariant $F(a_0, a_1, \ldots, a_n)$ of degree θ and weight p,

$$\eta = n\theta - 2p \geq 0. \tag{34.73}$$

To prove the lemma, begin with the observation that if U is any homogeneous, isobaric polynomial of degree θ and weight p, then

$$(\Omega O - O\Omega)U = (n\theta - 2p)U. \tag{34.74}$$

This was well-known when Sylvester wrote his paper, but he presented an argument:

$$\Omega O - O\Omega$$
$$= na_0 \frac{\partial}{\partial a_0} + 2(n-1)a_1 \frac{\partial}{\partial a_1} + 3(n-2)a_2 \frac{\partial}{\partial a_2} + \cdots + (n-1)a_{n-1} \frac{\partial}{\partial a_{n-1}}$$
$$- na_1 \frac{\partial}{\partial a_1} - 2(n-1)a_2 \frac{\partial}{\partial a_2} - \cdots - 2(n-1)a_{n-1} \frac{\partial}{\partial a_{n-1}} - na_n \frac{\partial}{\partial a_n}$$
$$= na_0 \frac{\partial}{\partial a_0} + (n-2)a_1 \frac{\partial}{\partial a_1} + (n-4)a_2 \frac{\partial}{\partial a_2} + \cdots$$
$$- (n-2)a_{n-1} \frac{\partial}{\partial a_{n-1}} - na_n \frac{\partial}{\partial a_n}. \tag{34.75}$$

If $\alpha_{s_0 s_1 \cdots s_n} a_0^{s_0} a_1^{s_1} \cdots a_n^{s_n}$ is any monomial in U, then (34.75) implies that

$$(\Omega O - O\Omega)U = n\left(a_0 \frac{\partial}{\partial a_0} + a_1 \frac{\partial}{\partial a_1} + \cdots + a_n \frac{\partial}{\partial a_n}\right)U$$
$$- 2\left(a_1 \frac{\partial}{\partial a_1} + 2a_2 \frac{\partial}{\partial a_2} + \cdots + na_n \frac{\partial}{\partial a_n}\right)U$$
$$= \sum \alpha_{s_0 s_1 \cdots s_n}(n\theta - 2p)a_0^{s_0} a_1^{s_1} \cdots a_n^{s_n}$$
$$= (n\theta - 2p)U = \eta U.$$

So $\Omega O - O\Omega \equiv \eta$. Moreover,

$$\Omega O^2 - O^2 \Omega = (\Omega O - O\Omega)O + O(\Omega O - O\Omega).$$

Since the differential operator O raises the weight by 1, we see that

$$(\Omega O - O\Omega)OU = (n\theta - 2(p+1))OU = (\eta - 2)OU. \tag{34.76}$$

Hence, $\Omega O^2 - O^2 \Omega = (\eta - 2)O + \eta O = 2(\eta - 1)O.$

By induction one can show that

$$\Omega O^r - O^r \Omega = r(\eta - r + 1)O^{r-1}. \tag{34.77}$$

For a seminvariant F, $\Omega F = 0$; and so

$$\Omega O^r F = r(\eta - r + 1) O^{r-1} F. \tag{34.78}$$

To conclude the proof of Sylvester's lemma, suppose that η is negative. In that case, $|r(\eta - r + 1)|$, for $r = 1, 2, 3, \ldots$ forms an increasing sequence of nonzero integers. Now $O^k F = 0$ for some $k \leq n\theta - p + 1$. To understand this statement, note that F, OF, $O^2 F$, ... have weights p, $p + 1$, $p + 2$, ..., but also have the same degree θ, and the greatest possible weight of any homogeneous polynomial of degree θ is $n\theta$, attained by a_n^θ. Thus, $p + k \leq n\theta + 1$. So let r be the value of k such that $O^r F = 0$. By (34.77), this implies $O^{r-1} F = 0$. It then follows that $\Omega O^{r-1} F = 0$, and hence $O^{r-2} F = 0$. A repeated application of this procedure gives $\eta F = 0$ or $F = 0$; hence, η cannot be negative, proving the lemma. Sylvester also showed by induction that

$$\begin{aligned}\Omega^q O^q F &= \eta(2\eta - 2)(3\eta - 6) \cdots (q\eta - (q^2 - q)) F \\ &= q!(\eta(\eta - 1)(\eta - 2) \cdots (\eta - q + 1)) F.\end{aligned} \tag{34.79}$$

Now we are equipped to prove Cayley's conjecture. Let $D_n(p, \theta)$ denote the number of linearly independent seminvariants of degree θ and weight p so that the conjecture can be formulated as

$$D_n(p, \theta) = \omega_n(p, \theta) - \omega_n(p - 1, \theta) \equiv \Delta_n(p, \theta). \tag{34.80}$$

Observe that this equality holds if the $\omega_n(\theta, p-1)$ equations satisfied by $\alpha_{s_0 s_1 \cdots s_n}$ in (34.61) are independent. In any case, we have $D_n(p, \theta) \geq \Delta_n(p, \theta)$. Note that $D_n(0, \theta) = \omega_n(0, \theta)$ since both sides equal 1. It is also clear that

$$\begin{aligned}&D_n(p, \theta) + D_n(p - 1, \theta) + \cdots + D_n(0, \theta) \\ &\geq \Delta_n(p, \theta) + \Delta_n(p - 1, \theta) + \cdots + \Delta_n(0, \theta) \\ &= \omega_n(p, \theta).\end{aligned} \tag{34.81}$$

If equality holds in this situation, then we see that $D_n(w, \theta) = \Delta_n(w, \theta)$ for all weights $w \leq p$. Since, for given n and θ, the weight w satisfies the inequality $n\theta - 2w \geq 0$, we have $w \leq n\theta/2$. So the largest value of the weight would be $n\theta/2$ when $n\theta$ is even, and it would be $(n\theta - 1)/2$ when $n\theta$ is odd. Let p stand for the maximum weight. Also let $[p]$, $[p-1]$, $[p-2]$, etc., denote semivariants of degree θ in variables a_0, a_1, ..., a_n and of weights p, $p-1$, $p-2$, etc., respectively. Then the number of linearly independent $[p]$s would be given by $D_n(p, \theta)$, and the number of linearly independent $[p-1]$s would be $D_n(p-1, \theta)$, and so on. So choose a set of $D_n(p, \theta)$ independent $[p]$s, $D_n(p-1, \theta-1)$ independent $[p-1]$s, etc. From this set construct a new set S in which all the forms have the same weight p. This can be done by applying the operator O^q to the $D_n(p-q, \theta)$ forms $[p-q]$, since the weights of the forms $O^q[p-q]$ are all p.

To prove that this set S of forms of weight p is linearly independent, we first show that any one set of $O^q[p-q]$ is independent; if not, then the members $O^q[p-q]$ of the set are connected by a linear equation. Apply the operator Ω^q to this equation. By

(34.79), $\Omega^q O^q [p-q]$ is a nonzero constant multiple of $[p-q]$. But this contradicts the independence of the $[p-q]$s. Thus, we have shown that the subset consisting of $O^q[p-q]$ is independent. Now suppose that a linear relation holds among/between any number of subsets of the form $O^q[p-q]$ for which m is the largest value of q. Operate on this linear equation by Ω^m. For $q < m$, this operation will introduce quantities of the form $\Omega^{m-q}[p-q]$, but these will in fact vanish because $[p-q]$ is a seminvariant and is hence annihilated by Ω. Thus, only forms of the type $[p-m]$ will remain after the application of Ω^m. This again gives us a contradiction because the seminvariants $[p-m]$ were chosen to be independent. We can therefore conclude that the set S is linearly independent. Therefore, the number of elements in S cannot exceed $\omega_n(p,\theta)$. By construction, the number of elements of S is given by

$$D_n(p,\theta) + D_n(p-1,\theta) + \cdots + D_n(0,\theta).$$

Hence, this sum is less than or equal to $\omega_n(p,\theta)$. Therefore, by (34.81), equality holds and we have proved Cayley's conjecture.

Sylvester's comments on his proof suggest that he may have been a keen student of Kant and valued mathematics as a creative endeavor. He wrote that his proof was accomplished "by aid of a construction drawn from the resources of the Imaginative reason, and founded on the reciprocal properties that have just been exhibited by the famous O and Ω." Later in the paper, he argued that proofs of this type showed that mathematics belonged among the liberal arts. "Whether we look to the advances made in modern geometry, in modern integral calculus, or in modern algebra, in each of these a free handling of the material employed is now possible, and an almost unlimited scope left to the regulated play of the fancy."

34.6 Hilbert's Finite Basis Theorem

David Hilbert (1862–1943) was one of the most influential mathematicians of his time. He is famous for advocating an abstract, structural approach to mathematical problems, though his work on invariant theory had its algorithmic aspect. Hilbert studied at Königsberg and attended lectures given by the outstanding teacher and number theorist Heinrich Weber (1842–1913). In 1882, Weber and Dedekind collaborated on an important paper in algebraic geometry, in which they presented Riemann surface theory from an algebraic perspective. It is clear that this work influenced Hilbert's later approach to invariant theory. We note parenthetically that Weber wrote a three-volume work on algebra, useful even today.

Hilbert proved his basis theorem for the general situation, beginning with any number of m-ary forms or quantics. In his 1890 paper, Hilbert employed a theorem of Max Noether, father of Emmy, to prove the basic lemma upon which he built his theory: If $F_1, F_2, F_3, \ldots$ is an infinite sequence of forms, that is, homogeneous polynomials in n variables $x_1, x_2, \ldots, x_n$ with coefficients in a field, then there exists an integer m such that every form in the sequence can be expressed as

$$F = A_1 F_1 + A_2 F_2 + \cdots + A_m F_m, \qquad (34.82)$$

where $A_1, A_2, \ldots, A_m$ are appropriate forms in the same n variables.

Using this lemma, Hilbert demonstrated that from an arbitrary collection of forms in n variables one can always choose a finite number such that every form in the collection is a linear combination of the chosen forms, as in (34.82). Hilbert proved this by contradiction, assuming the result false. Let $F_1 \neq 0$ be a form in the collection and let F_2 be a form in the collection, but not expressible as $A_1 F_1$. By our assumption, F_2 exists. Now let F_3 be a form not expressible as $A_1 F_1 + A_2 F_2$. Again, F_3 exists by supposition. In this way, we construct a sequence of forms $F_1, F_2, F_3, \ldots$ for which no number m exists to satisfy (34.82). This contradicts Hilbert's lemma. We remark that in more modern books, this theorem is formulated in terms of polynomial ideals.

Hilbert's basis theorem for invariants states that there exists a finite number of invariants $I_1, \ldots, I_m$ of a binary quantic or form Q such that any invariant of Q is some polynomial function of $I_1, \ldots, I_m$. To prove this using Hilbert's reasoning, let S denote the set of all invariants of Q. Though our treatment of this theorem is for only one form, note that Hilbert did not restrict himself to one form Q, but to a finite number of them. His conclusion on the finite basis for the simultaneous invariants is a generalization of the result for one form. Now these invariants are homogeneous and isobaric polynomials in the $n+1$ variables $a_0, a_1, \ldots, a_n$, the coefficients of the quantic. Hence there exist m invariants $I_1, I_2, \ldots, I_m$ such that every invariant I in S can be written as

$$I = Q_1 I_1 + \cdots + Q_m I_m. \qquad (34.83)$$

Now the forms $Q_1, Q_2, \ldots, Q_m$ can be chosen to be isobaric in $a_0, a_1, \ldots, a_n$, but they need not be invariants. To get invariants from $Q_1, Q_2, \ldots, Q_m$, Hilbert constructed an operator using O and Ω:

$$L = 1 - \frac{O}{1!} \frac{\Omega}{2!} + \frac{O^2}{2!} \frac{\Omega^2}{3!} - \frac{O^3}{3!} \frac{\Omega^3}{4!} + \cdots. \qquad (34.84)$$

This operator has the property that if F is any homogeneous and isobaric polynomial in $a_0, a_1, \ldots, a_n$, of degree θ_1 and weight p_1, such that $n\theta_1 - 2p_1 = 0$, then LF is either zero or an invariant. We shall present the proof of this property of the operator L after we have deduced Hilbert's theorem from it. For this purpose, apply L to (34.83) to get

$$LI = I = (LQ_1)I_1 + \cdots + (LQ_m)I_m. \qquad (34.85)$$

This follows from the easily proved facts that for any invariant I, $LI = I$ and $L(QI) = (LQ)I$. We must now show that LQ_i is either an invariant or zero. Since $I, I_1, \ldots, I_m$ in (34.85) are invariants, they satisfy the degree and weight condition $n\theta - 2p = 0$, though the $m+1$ invariants may have differing weights and degrees. Thus, the isobaric forms $Q_1, Q_2, \ldots, Q_m$ also satisfy the condition $n\theta - 2p = 0$. Hence $LQ_1, LQ_2, \ldots, LQ_m$ must each be either zero or an invariant. Clearly, all of them cannot be zero for then I would be zero. Thus, the nonzero $LQ_1, LQ_2, \ldots, LQ_m$ are members of the set S of invariants, and can once again be expressed in terms of $I_1, I_2, \ldots, I_m$. However, the LQ_i terms are of lower degree than I and the process will therefore terminate and every invariant I will be a polynomial in $I_1, I_2, \ldots, I_m$.

Hilbert did not bother to write down a proof of the required property of L. In his 1895 book on the algebra of quantics, Edwin Elliott, Sylvester's student at Oxford, gave a simple proof using the Cayley–Sylvester relation (34.74). Let G be a form in $a_0, a_1, \ldots, a_n$ with $\eta = n\theta - 2p \geq 0$. The weights of $\Omega G, \Omega^2 G, \Omega^3 G, \ldots$ are $p-1, p-2, p-3, \ldots$, respectively, and hence the quantities corresponding to η become $\eta+2, \eta+4, \eta+6, \ldots$, respectively. Thus, from (34.74) and (34.77), we have the relations

$$\Omega O G - O\Omega G = \eta G,$$
$$\Omega O^2 \Omega G - O^2 \Omega^2 G = 2(\eta+1)G,$$
$$\vdots$$
$$\Omega O^r \Omega^{r-1} G - O^r \Omega^r G = r(\eta + r - 1) O^{r-1} \Omega^{r-1} G.$$

Multiply the first equation by $1/\eta$, the second by $-1/(2\eta(\eta+1))$, ..., the rth by

$$(-1)^{r-1}/\bigl(r!\eta(\eta+1)\cdots(\eta+r-1)\bigr),$$

and so on. Add the resulting equations to obtain

$$\Omega O \left\{ \frac{1}{1 \cdot \eta} - \frac{1}{2!\eta(\eta+1)} O\Omega + \frac{1}{3!\eta(\eta+1)(\eta+2)} O^2 \Omega^2 - \cdots \right\} G = G. \quad (34.86)$$

This sum is finite since $\Omega^{p+1} G$ vanishes. Now replace G by ΩF where F is an isobaric form in $a_0, a_1, \ldots, a_n$ of weight $p+1$, and write (34.86) as

$$\Omega \left\{ 1 - \frac{1}{1 \cdot \eta} O\Omega + \frac{1}{1 \cdot 2 \cdot \eta(\eta+1)} O^2 \Omega^2 - \frac{1}{3!\eta(\eta+1)(\eta+2)} O^3 \Omega^3 + \cdots \right\} F = 0. \quad (34.87)$$

Now substitute p for $p+1$, enabling us to write $\eta \geq -2$ and $\eta + 2 \geq 0$. So if F is of weight p, we replace η by $\eta + 2$ in (34.87) to get

$$\Omega \left\{ 1 - \frac{1}{1 \cdot (\eta+2)} O\Omega + \frac{1}{2!(\eta+2)(\eta+3)} O^2 \Omega^2 - \cdots \right\} F = 0.$$

Therefore, when $\eta = n\theta - 2p = 0$, we have $\Omega(LF) = 0$, and this means that LF is either an invariant or is identically zero.

We note that in his doctoral thesis of 1885, Hilbert introduced the operator L, and other similar operators. He explained that L served as a generalization of transvection, an older method of producing covariants. The subject of Hilbert's dissertation was special binary forms determined by algebraic differential equations and he mainly applied them to spherical functions. He took up this topic at the suggestion of his advisor at Königsberg, Ferdinand Lindemann (1852–1939), who is known for proving the transcendence of π.

34.7 Hilbert's Nullstellensatz

Hilbert's aim in his 1893 paper on invariants was to subsume invariant theory under the general theory of algebraic function fields. This led him to a deeper proof of the basis theorem and to the creation of important new ideas fundamental to the development of twentieth-century commutative algebra and algebraic geometry. This proof of the basis theorem satisfied Gordan's requirement in that it be algorithmic. We briefly discuss one of Hilbert's important results, now known as the Nullstellensatz.

Hilbert proved that for any form or quantic, or system of forms, there existed a finite number of invariants $I_1, I_2, \ldots, I_k$ such that any other invariant I satisfied an algebraic equation

$$I^m + G_1 I^{m-1} + G_2 I^{m-2} + \cdots + G_m = 0, \tag{34.88}$$

where $G_1, G_2, \ldots, G_m$ were integral rational functions of $I_1, I_2, \ldots, I_k$. By homogeneity, the functions $G_1, G_2, \ldots, G_n$ could not have a constant term. With this result in hand, Hilbert considered forms whose coefficients had numerical values such that all the invariants $I_1, I_2, \ldots, I_k$ became zero, meaning that the value of all the invariants was zero, since by (34.88), $I^m = 0$, or $I = 0$. Hilbert called a form null if all its invariants were zero.

The converse of this theorem is of interest. Suppose $I_1, I_2, \ldots, I_k$ are invariants such that their vanishing implies the vanishing of all other invariants of that form or quantic. Hilbert showed that under these conditions, any invariant I of this quantic satisfied an equation of the type (34.88). Hilbert based his proof of this converse on the result now known as the Hilbert Nullstellensatz:

Suppose $f_1, f_2, \ldots, f_m$ are m homogeneous polynomials in $x_1, x_2, \ldots, x_n$, and suppose $F_1, F_2, F_3, \ldots$ are homogeneous polynomials in the same variables, such that they vanish for any values of the variables for which $f_1, \ldots, f_m$ all vanish. Then one can find an integer r such that every product $\Pi^{(r)}$ of r arbitrary functions from the sequence $F_1, F_2, F_3, \ldots$ can be represented in the form

$$\Pi^{(r)} = a_1 f_1 + a_2 f_2 + \cdots + a_m f_m,$$

where $a_1, a_2, \ldots, a_m$ are appropriately chosen homogeneous polynomials in $x_1, x_2, \ldots, x_n$.

34.8 Exercises

1. Suppose the binary cubic form is

$$q = ax^3 + 3bx^2y + 3cxy^2 + dy^3.$$

Show that
$$\theta(q) = (ad - bc)^2 - 4(b^2 - ac)(c^2 - bd).$$

See Boole (1841).

2. Suppose the ternary quadratic form is

$$q = ax^2 + by^2 + cz^2 + 2dyz + 2exz + 2fxy.$$

Show that

$$\theta(q) = abc + 3def - (ad^2 + be^2 + cf^2).$$

See Boole (1841).

3. Let $q = ax^4 + 4bx^3y + 6cx^2y^2 + 4dxy^3 + ey^4$. Show that

$$\begin{aligned}\theta(q) = {} & a^3e^3 - 6ab^2d^2e - 12a^2bde^2 - 18a^2c^2e^2 - 27a^2d^4 - 27b^4e^2 \\ & + 36b^2c^2d^2 + 54a^2cd^2e + 54ab^2ce^2 - 54ac^3d^2 - 54b^2c^3e - 64b^3d^3 \\ & + 81ac^4e + 108abcd^3 + 108b^3cde - 180abc^2de.\end{aligned}$$

See Boole (1844a).

4. Prove Sylvester's 1877 generalization of Taylor's theorem: Suppose f is a function of $a, b, c, \ldots$ and f_1 is the same function of

$$a_1 = a, \ b_1 = b + ah, \ c_1 = c + 2bh + ah^2, \ d_1 = d + 3ch + 3bh^2 + ah^3, \ \ldots,$$

and let Ω represent the operator

$$a\frac{\partial}{\partial b} + 2b\frac{\partial}{\partial c} + 3c\frac{\partial}{\partial d} + \cdots.$$

Then $f_1 = f + \Omega.fh + (\Omega.)^2 f \dfrac{h^2}{1 \cdot 2} + (\Omega.)^3 f \dfrac{h^3}{1 \cdot 2 \cdot 3} + \cdots.$

Thus, $f_1 = f$ if and only if $\Omega f = 0$. According to Sylvester, this last statement makes the theorem important in the calculus of invariants. See Sylvester (1973), vol. 3, pp. 88–92.

5. Find the independent invariants of degrees 4 and 8 for a binary form of order 5. See Cayley (1889–98), vol. 2, pp. 250–275.

6. Show that a binary quantic has exactly two linearly independent seminvariants of degree 5 and weight 5. See Elliott (1964), p. 132.

7. Show that a binary form of order $4n + 2$ has a covariant of the second order and third degree. See Elliott (1964), p. 157. Elliott attributes this result to Hermite.

34.9 Notes on the Literature

See Boole (1841) and (1844a) for his papers leading to the concept of an invariant. See Cayley (1889–98), vol. 2, for his early memoirs on invariants and for his researches on partitions. See especially his second memoir, pp. 265–267. For the quote on the connection of Cayley's work with homological algebra, see p. 4 of Gelfand, Kapranov, and Zelvinsky (1994). Sylvester's 1852 paper, "On the Principles of the Calculus of Forms," can be found in Sylvester (1973), vol. 1, pp. 328–363. He discussed the differential operators on pp. 352–362 and mentioned Aronhold on pp. 351–352. See vol. 3, pp. 117–126 for Sylvester's paper, "Proof of the Hitherto Undemonstrated Fundamental Theorem of Invariants." The quotations in the text are taken from pp. 117–118 and 123.

For Hilbert's results discussed in the text, see Hilbert (1970), vol. 2, pp. 199–257 and 287–344. English translations of these papers appear in Hilbert (1978), pp. 143–301. See Corry (2004) for the role of invariant theory in the development of the structural method in algebra. He also elaborates on the influence of Dedekind on Hilbert and E. Noether. For Kac's very early work on the cubic, see Kalman (2009).

K. Parshall's article in Rowe and McCleary (1989), vol. 1, pp. 157–206, gives a history of nineteenth-century invariant theory before Hilbert. Crilly's (2006) biography presents the development of Cayley's mathematical thought with interesting details, especially in connection with invariant theory. The reader may also enjoy Hilbert's (1993) lectures, given in 1897; the first sixty pages cover the work of Cayley and Sylvester. See also Elliott (1964); the first edition of 1895 presented a very readable exposition of nineteenth-century invariant theory in English, but it did not include the symbolic method of the German school. The 1903 book by Grace and Young (1965) filled this need. For recent works on invariant theory incorporating the classical methods of Cayley and Sylvester, see Olver (1999) and Sturmfels (2008).

35

Summability

35.1 Preliminary Remarks

The subject of summability theory encompasses the variety of methods for averaging sequences, series, and integrals; it also includes the relationships among the various methods. This topic originated in the attempts to assign a value to the sum of a divergent series. Guido Grandi (1671–1742) made one of the earliest attempts, giving the sum of the series $1 - 1 + 1 - 1 + \cdots$ to be $1/2$ by setting $x = 1$ in the formula

$$\frac{1}{1+x} = 1 - x + x^2 - x^3 + \cdots.$$

In a letter to Christian Wolf, published in 1713, Leibniz reasoned that since the sum of the first n terms of $1 - 1 + 1 - 1 + \cdots$ would be 0 or 1 depending on whether n was even or odd, the values 0 and 1 would occur with equal frequency, and hence $1/2$ was the most probable value of the sum. This method amounts to taking the limit of the averages of the partial sums assigned to the series $1 - 1 + 1 - 1 + \cdots$ as the number of terms gets larger and larger. Note also that $1 - x + x^2 - x^3 + \cdots$ may be seen as a type of weighted average of $1 - 1 + 1 - 1 + \cdots$. Newton also dealt with divergent series, although in unpublished work. A significant example is his transformation formula, now named after Euler,

$$\sum_{n=0}^{\infty} A_n x^{n+1} = \sum_{n=0}^{\infty} y^{n+1} \Delta^n A_0,$$

where $y = x/(1-x)$. Newton discovered this transformation in 1684, but it unfortunately remained unpublished for almost three centuries. He used it to evaluate the alternating series for $\ln(1+x)$ and for $\arctan x$, taking the absolute value of x to be greater than 1. Newton explained that this transformation could be applied to convert an alternating divergent series to a convergent one; then, the value of the divergent series would be given by the corresponding value of the convergent series. From this we can see that Newton's ideas on divergent series were groundbreaking.

Between 1720 and 1740, de Moivre, Stirling, Euler, and Maclaurin gained significant, though partial, insights into divergent asymptotic series. Their method, based on the Euler–Maclaurin summation formula, was to begin with a finite series and convert it to an infinite asymptotic series, yielding an excellent numerical approximation of the finite series. Interestingly, in the twentieth century, Ramanujan also used the Euler–Maclaurin formula in an attempt to construct a theory for summing divergent series.

Euler and Lagrange also made considerable use of divergent series in their work, though Euler's work was clearly more incisive. In 1749, Euler gave a brilliant application of summability by defining

$$1^n - 2^n + 3^n - 4^n + \cdots = \lim_{x \to 1^-} (1^n x - 2^n x^2 + 3^n x^3 - \cdots). \tag{35.1}$$

He used this to discover the functional relation for the zeta function. Recall that Euler's initial motivation may have been to study the series on the left, hoping it would illuminate the problem of summing the zeta value $\zeta(2n+1)$ where n was a positive integer. By generalizing (35.1), we may say that Euler defined the sum of the series $\sum_{n=0}^{\infty} a_n$ by the equation

$$\sum_{n=0}^{\infty} a_n = \lim_{x \to 1^-} \sum_{n=0}^{\infty} a_n x^n. \tag{35.2}$$

We have seen that in 1826 Abel proved that if $\sum a_n$ was convergent, then (35.2) would hold. For this reason we say that if the value of the limit in (35.2) is taken to be L, then the series $\sum a_n$ is Abel-summable or A-summable to L. Thus, although Euler defined this summability method, it is named after Abel. As a matter of fact, when n is a positive even integer, then the value of the series on the left-hand side of (35.1) sums to 0. Ironically, Abel called this situation "horrible" and quoted Horace: "Risum teneatis, amici." [Hold your laughter, friends.]

Interestingly, in the 1820s, Poisson applied Abel summability to the convergence of Fourier series. Recall that Fourier claimed in his famous 1807 memoir and other works that an arbitrary function could be expanded as a Fourier series; though he presented several ingenious arguments in favor of this proposition, he did not provide a real proof. In a paper published in 1820, Poisson attempted to demonstrate that the Fourier series of a continuous function converged to that function by showing that

$$\lim_{r \to 1^-} \left(\frac{1}{2} a_0 + \sum_{n=1}^{\infty} (a_n \cos n\theta + b_n \sin n\theta) r^n \right) = f(\theta), \tag{35.3}$$

where a_n and b_n were the Fourier coefficients of a continuous function $f(\theta)$. Poisson showed that the expression within parentheses in (35.3) could be expressed as

$$P(r,\theta) = \frac{1}{2\pi} \int_0^{2\pi} \frac{1-r^2}{1-2r\cos(\theta-\phi)+r^2} f(\phi) \, d\phi,$$

now called the Poisson integral. He then gave an argument that as r approached 1, the integral approached $f(\theta)$, but this argument was full of gaps. But even had Poisson's proof of (35.3) been complete, he undermined it from the beginning by falsely assuming the converse of Abel's theorem. Recall that Cauchy made a similar error at around the same time. Tauber and Littlewood later established that the converse of Abel's theorem required a growth condition on the coefficients. Cauchy, Abel, and others concluded that divergent series had no sum, effectively banishing this topic for nearly fifty years. It was only after the theory of convergent series was established on a sound footing, through the efforts of Gauss, Cauchy, Abel, Dirichlet, and Weierstrass, that mathematicians could confidently address the summability of divergent series.

The German mathematician Ferdinand Georg Frobenius (1849–1917) initiated the modern theory of summability by proving the first theorem establishing a relation between two different methods of summation. In a short paper of 1880 he showed that if $s_n = \sum_{k=0}^{n} a_k$ and

$$\frac{s_0 + s_1 + \cdots + s_n}{n+1} \to S \text{ as } n \to \infty, \tag{35.4}$$

$$\text{then } \lim_{x \to 1^-} \sum_{n=0}^{\infty} a_n x^n = S.$$

This theorem explained why Grandi and Leibniz obtained the same value for the sum of the series $1 - 1 + 1 - 1 + \cdots$. Frobenius was a student of Weierstrass, and he initially worked in differential equations and their series solutions. He branched out into number theory and algebra with particular emphasis on groups. In answering a question of Dedekind on group determinants, Frobenius created and developed the topic for which he is best known, group representation theory. Two years after Frobenius's important paper, Otto Hölder (1859–1937), who also studied with Weierstrass, extended that work. He defined

$$H_n^{(r+1)} = \frac{H_0^{(r)} + H_1^{(r)} + \cdots + H_n^{(r)}}{n+1}, \quad r = 0, 1, 2, \ldots, \tag{35.5}$$

where $H_k^{(0)} = s_k$. He pointed out that there were sequences $s_0, s_1, s_2, \ldots$ for which the limit (35.4) did not exist but such that there was an integer r for which the $\lim_{n \to \infty} H_n^{(r)}$ existed. Thus, such a series $\sum a_n$ is said to be (H, r) summable. Moreover, Hölder proved that if $\lim_{n \to \infty} H_n^{(r)} = S$, then $\lim_{x \to 1^-} \sum_{n=0}^{\infty} a_n x^n = S$.

The Italian mathematician Ernesto Cesàro (1859–1906), in spite of financial and other challenges, managed to learn mathematics from a number of good teachers, obtain positions in Italian universities, and publish prolifically in differential geometry and number theory. He studied under Eugène Catalan in Liège and spent a year in Paris attending lectures by Hermite and Gaston Darboux. He had wide interests, including mathematical physics. In 1890, Cesàro gave an important application of the summability method (35.4), shedding light on a classical question on products of infinite series: Suppose that $\sum a_n = A$ and $\sum b_n = B$, and let the Cauchy product of these two series be $\sum c_n$, where $c_n = a_0 b_n + a_1 b_{n-1} + \cdots + a_n b_0$. When does the Cauchy product converge? Cesàro proved that even when the product did not converge, the limit of the arithmetic means of the partial sums of the product would converge to AB. In other words, if

$C_n = c_0 + c_1 + \cdots + c_n$ then

$$\frac{C_0 + C_1 + \cdots + C_n}{n} \to AB \text{ as } n \to \infty. \tag{35.6}$$

Note that Cesàro's theorem generalized Abel's theorem that if $\sum a_n = A$, $\sum b_n = B$, and $\sum c_n = C$, then $AB = C$. In today's terminology, we would say that a series $\sum a_n$ is Cesàro summable or $(C, 1)$ summable to S if (35.4) holds true. We may also say that the Cauchy product of two series converging to A and B is Cesàro summable to AB. Cesàro next extended his result to not necessarily convergent series: If

$$A_n = \sum_{k=0}^{n} a_k, \ B_n = \sum_{k=0}^{n} b_k, \ C_n = \sum_{k=0}^{n} c_k \text{ and}$$

$$\frac{\sum_{k=0}^{n} A_k}{n+1} \to A, \quad \frac{\sum_{k=0}^{n} B_k}{n+1} \to B \quad \text{as } n \to \infty, \text{ then}$$

$$\frac{\sum_{k=0}^{n} C_k}{n+1} \to AB \quad \text{as } n \to \infty.$$

Cesàro also defined a more general form of convergence, starting with

$$\binom{k+n}{n} A_{n,k} = a_n + \binom{k+1}{1} a_{n-1} + \cdots + \binom{k+n}{n} a_0. \tag{35.7}$$

He defined a series $\sum a_n$ as summable (today we say (C, k) summable) to A if there was a k such that $\lim_{n \to \infty} A_{n,k} = A$. Note that it is not necessary for k to be a nonnegative integer. Since we can write

$$\binom{k+j}{j} = \frac{(k+1)(k+2) \cdots (k+j)}{j!},$$

we may take k to be a real number > -1.

In 1900, the Hungarian mathematician Lipót Fejér (1880–1959) delivered a big boost to the Cesàro summability method by proving that the Fourier series of any continuous function was $(C, 1)$ summable to the function. It is interesting that Fejér's result arose out of an earlier attempt to solve the Dirichlet problem for the unit circle: For a continuous function $f(\theta)$ on the unit circle, determine a harmonic function $\phi(x, y) = \Phi(r, \lambda)$ inside the unit disk such that $\Phi(r, \lambda)$ tends to $f(\theta)$ as $re^{i\lambda}$ approaches $e^{i\theta}$ from inside the unit disk. Note that $\phi(x, y)$ would be harmonic if it satisfied Laplace's equation

$$\frac{\partial^2 \phi}{\partial x^2} + \frac{\partial^2 \phi}{\partial y^2} = 0. \tag{35.8}$$

In 1870, Carl Neumann (1832–1925), son of Franz Neumann and one of the founders of the *Mathematische Annalen*, made an attempt at solving this problem by means of the harmonic function determined by the Poisson integral $P(r, \theta)$. He used Poisson's result, that $P(r,\theta)$ tended to $f(\theta)$ as $r \to 1^-$. Recall, however, that the proof given by Poisson was incomplete; this in turn undermined Neumann's proof. As a third year student at the Technical University in Hungary, Fejér spent 1899–1900 in Berlin, attending lectures by L. Fuchs, Schwarz, and Frobenius, all students of Weierstrass. Fejér learned of Neumann's attempt from Schwarz, who in 1871 had solved the Dirichlet problem by an alternative method. Examining the gap in Neumann's proof, Fejér proved the $(C, 1)$ summability of the Fourier series of a continuous function. This result, combined with Frobenius's theorem that $(C, 1)$ summability implied Abel summability, mended Poisson's proof. As a corollary, Fejér obtained the theorem that a continuous function could be uniformly approximated by trigonometric polynomials on a closed interval. Since the sine and cosine functions could be approximated by their Taylor polynomials, he further deduced Weierstrass's theorem on the uniform approximation of continuous functions by polynomials.

Similar to his first mathematical efforts, a number of Fejér's later papers presented elegant solutions to interesting but circumscribed problems, where both the problems and the solutions had significant implications in several areas. While a professor at Budapest, he had a broad influence on the development of mathematics in Hungary. His mathematical style, his outgoing personality, and his wide-ranging cultural interests attracted many good students, including Erdős, Pólya, Szegő, Turán, and von Neumann.

The Austrian mathematician Alfred Tauber (1866–1942) gave a new direction to summability theory with a result on a converse of Abel's theorem on series. He proved that if $\sum a_n$ was Abel-summable to A and $na_n \to 0$ (or $a_n = o(1/n)$) as $n \to \infty$, then $\sum a_n = A$. Tauber also proved an Abel summable series $\sum a_n$ to be convergent if and only if

$$\frac{a_1 + 2a_2 + \cdots + na_n}{n} \to 0 \text{ as } n \to \infty. \tag{35.9}$$

In a paper of 1907, the German analytic number theorist Edmund Landau (1877–1938) extended Tauber's theorem to series of the form $\sum_{n=1}^{\infty} a_n e^{-\lambda_n x}$, where $\lambda_1 < \lambda_2 < \cdots$ and $\lambda_n \to \infty$ as $n \to \infty$. Note that this covers power series as well as Dirichlet series. Landau also proved an integral analog of Tauber's theorem: If

$$J(x) = \int_1^{\infty} f(t) t^{-x} dt \to A \text{ as } x \to 0 \tag{35.10}$$

$$\text{and } f(t) = o\left(\frac{1}{t \ln t}\right) \text{ as } t \to \infty, \tag{35.11}$$

$$\text{then } J(0) = \int_1^{\infty} f(t) dt = A. \tag{35.12}$$

Recall that in 1749, Euler attempted to use Abel summability to prove the functional equation for the zeta function. In 1906, Landau vindicated Euler's efforts by proving that the Abel sum of the series $\sum_{n=1}^{\infty} (-1)^{n-1}/n^s$ yielded the value $(1 - 2^{1-s})\zeta(s)$, obtained by the analytic continuation of the zeta function. In this work, Landau employed an 1898 result of the Finnish mathematician Hjalmar Mellin. Landau, a student of Frobenius, also introduced the one-sided Tauberian condition on the coefficients of series, especially applicable in number theory. In 1903, Landau derived the prime number theorem without using Hadamard's theory of entire functions of finite order and a year later, he obtained an important generalization of Picard's theorem on entire functions. Concerning Landau's 1927 *Vorlesungen über Zahlentheorie*, Hardy wrote, "This remarkable work is complete in itself; he does not assume ... even a little knowledge of number-theory or algebra. It stretches from the very beginning to the limits of knowledge, in 1927, of the 'additive,' 'analytic,' and 'geometric' theories."

The preliminary summability results of Frobenius, Cesàro, Fejér, Tauber, and Landau laid the foundation for a cohesive theory of summability with wide applicability. The British mathematicians G. H. Hardy (1877–1947) and J. E. Littlewood (1885–1977) were the first to fully understand the potential and scope of this mathematical theory. Hardy's many mathematical contributions included the circle method, discovered jointly with Ramanujan in their work on the asymptotic theory of partitions; and the concept of maximal functions, developed in collaboration with Littlewood. His influence was felt as much through his teaching as in his research. He helped raise British standards of teaching in analysis by publishing his 1908 *A Course of Pure Mathematics*, still in print today. In his preface to the 1937 edition of this book, Hardy remarked that if he were to rewrite the book, "I should not write (to use Prof. Littlewood's simile) like 'a missionary talking to cannibals,' but with decent terseness and restraint." Hardy enjoyed mathematical collaboration, and his association with Littlewood was one of the most productive in the history of mathematics. They published over one hundred joint papers in analysis and analytic number theory. According to Harald Bohr, the Hardy–Littlewood collaboration was based on four rules:

1. When one wrote to the other, it was completely indifferent whether what they wrote was right or wrong.
2. When one received a letter from the other, he was under no obligation whatsoever to read it, let alone answer it.
3. Although it did not really matter if they both simultaneously thought about the same detail, still, it was preferable that they should not do so.
4. It was quite indifferent if one of them had not contributed the least bit to the contents of a paper under their common name.

Although both Hardy and Littlewood lived on the Trinity College grounds, within one or two hundred yards of one another, and ate their meals in the same dining hall, their rules suggest that most of their communications were via the written word. Littlewood, unlike Hardy, had an interest in applied mathematics. In collaboration with Mary Cartwright (1900–1998), he also made important contributions to nonlinear differential equations and topological dynamics. Concerning Littlewood, V. I. Arnold wrote: "In mathematics he was a direct successor of Newton and Poincaré, doing research even on artillery ballistics. I was surprised to discover his estimates of the time of preservation of an adiabatic invariant in a Hamiltonian system .... It is even more surprising that the 'theory of chaos' in dynamical systems, including 'Smale's horseshoe,' had been already developed and published by Littlewood."

In a 1909 paper, Hardy showed that if $\sum a_n$ was $(C, 1)$ summable to S and $a_n = O(1/n)$ then $\sum a_n$ converged to S. He noted that by combining this result with Fejér's

$(C, 1)$ summability of the Fourier series of a continuous function $f(x)$, one obtained Dirichlet's theorem on Fourier series. Take $f(x)$ to be monotonic, and apply the second mean value theorem

$$\int_0^{2\pi} f(x)\cos nx\, dx = f(0)\int_0^{\xi} \cos nx\, dx + f(2\pi)\int_{\xi}^{2\pi} \cos nx\, dx$$

to see that the Fourier coefficients are $O(1/n)$. Hardy was not successful in his attempt to prove the more general result that if $\sum a_n$ was Abel summable to S and $a_n = O(1/n)$, then $\sum a_n = S$. In fact, he thought the result could well be false. He suggested the problem to his former student Littlewood, who succeeded in solving it in the affirmative. In his "A Mathematical Education," Littlewood gave an account of his discovery of the proof. Surprisingly, as he grappled with the problem, he forgot that Hardy had already proved the Cesàro–Tauber theorem. In 1911, during his attempt to reprove this, he discovered the derivatives theorem. Note that in intuitive terms, the derivatives theorem states that the orders of magnitude of two derivatives of a function restrict the order of magnitude of the intermediate derivatives. Hardy and Littlewood made considerable use of this concept in their early work. But, as they mentioned in a paper of 1914, Hadamard had already proved the derivatives theorem and had published it in an 1897 paper on waves. Indeed, A. Kneser also independently obtained the theorem in the same year. Littlewood stated his Abel–Tauber theorem in the general form that included Dirichlet series:

If $0 < \lambda_1 < \lambda_2 < \cdots < \lambda_n \to \infty$ as $n \to \infty$,

$$\lim_{n\to\infty}\sum_{n=1}^{\infty} a_n e^{-x\lambda_n} = S, \tag{35.13}$$

and $\quad |a_n| < K(\lambda_n - \lambda_{n-1})/\lambda_n \quad$ for a constant K, $\tag{35.14}$

then $\quad \displaystyle\sum_{n=1}^{\infty} a_n = S.$

Observe that when $\lambda_n = n$, the sum in (35.13) reduces to a power series, whereas when $\lambda_n = \ln n$, one gets a Dirichlet series. Littlewood's condition $|a_n| = O(1/n)$ is quite natural, since is it easy to see that under this condition, if $\sum a_n x^n$ oscillates finitely as $x \to 1^-$, then so does the sequence $\sum_{k=1}^{n} a_k$ as $n \to \infty$. In fact, Littlewood pointed out that the condition $|a_n| = o(1/n)$ implied the much stronger result: that the limits of oscillation of $\sum a_n x^n$ as $x \to 1^-$ and of $\sum_{k=1}^{n} a_k$ as $n \to \infty$ were the same. These results must have suggested to him that in order for Abel summability to imply convergence, a weaker condition would suffice.

In an interesting 1910 paper, Landau proved Hardy's $(C, 1)$ summability theorem with a weaker one-sided Tauberian condition $na_n \geq -K$, where K was a constant. He mentioned that one-sided Tauberian arguments had been used by Hadamard and de la Vallée Poussin in their proofs of the prime number theorem (PNT). In a 1913 paper, Hardy and Littlewood proved a one-sided extension of Littlewood's theorem: If

$a_n \geq 0$, $\alpha > 0$, and

$$\lim_{x \to 1^-} (1-x)^\alpha \sum_{n=0}^{\infty} a_n x^n = A, \tag{35.15}$$

$$\text{then } \lim_{n \to \infty} \frac{\sum_{k=0}^{n} a_k}{n^\alpha} = \frac{A}{\Gamma(1+\alpha)}. \tag{35.16}$$

Hardy and Littlewood soon saw that this theorem had important implications in prime number theory. They showed that

$$\lim_{\xi \to 0^+} \xi \sum_{n=1}^{\infty} \Lambda(n) e^{-n\xi} = 1, \tag{35.17}$$

and since $\Lambda(n) \geq 0$, the hypothesis of their theorem, given by (35.15), was true with $\alpha = 1$, $x = e^{-\xi}$, $a_n = \Lambda(n)$, and $A = 1$. Recall that when n is a positive integer power of a prime p, $\Lambda(n) = \ln p$; otherwise, it is 0. Next, by (35.16)

$$\lim_{n \to \infty} \frac{1}{n} \sum_{k=1}^{n} \Lambda(k) = 1. \tag{35.18}$$

It was well-known that (35.18) was equivalent to the PNT, and to prove (35.17) Hardy and Littlewood needed the fact that with $s = \sigma + it$, $\zeta(s)$ had no zeros on $\sigma = 1$ and satisfied a very mild growth condition for large t and $1 \leq \sigma \leq 2$. This growth condition was so weak that they concluded that there should be a proof of the PNT requiring only $\zeta(1+it) \neq 0$ for real t. In looking for such a proof, they investigated the Lambert summability method. Note that a series $\sum a_n$ is Lambert summable to S if

$$\lim_{x \to 1^-} (1-x) \sum_{n=1}^{\infty} n a_n \frac{x^n}{1-x^n} = S. \tag{35.19}$$

In a paper written in 1919, Hardy and Littlewood proved that Lambert summability implied Abel summability. From this theorem they could easily derive a result equivalent to the PNT. Unfortunately, this did not give a new proof of the PNT because to prove their Lambert summability theorem, they had used the fact that, with μ the Möbius function,

$$g(n) = \sum_{m=1}^{n} \frac{\mu(m)}{m} = O\left(\frac{1}{(\ln n)^2}\right). \tag{35.20}$$

Thus, they relied on a result a little deeper than the PNT, since the PNT is equivalent to $g(n) = o(1)$ as $n \to \infty$. Though they failed to offer another proof of the PNT, their work set the stage for Wiener.

In 1928, Norbert Wiener (1894–1964) found a method to directly handle Lambert summability. Wiener received his doctoral degree from Harvard University at the age of 18 with a thesis in logic. He then spent a part of 1913 at Cambridge University to study under Bertrand Russell who advised him to study mathematics and physics, especially the papers of Einstein and Niels Bohr on relativity, Brownian motion and quantum theory. Wiener was greatly impressed and influenced by Hardy's course on real and complex variables and all of this bore fruit about a decade later. Wiener was a professor at M.I.T. from 1919 to his death in 1964. He interacted vigorously with his engineering colleagues. The electrical engineering department requested that he provide a rigorous basis for Heaviside's operational methods. This work led Wiener to a very fruitful study of a generalized harmonic analysis. He encountered a technical problem in his harmonic analysis research: Show that for a class of nonnegative functions $f(t)$

$$\lim_{T\to\infty} \frac{1}{T} \int_0^T f(t)\,dt = \lim_{\epsilon\to 0} \frac{2}{\pi\epsilon} \int_0^\infty f(t) \frac{\sin^2 \epsilon t}{t^2}\,dt. \qquad (35.21)$$

At this point in his researches, in 1926, Wiener was visiting Göttingen, as was his friend, the English mathematician A. E. Ingham. Wiener learned from Ingham that his problem was Tauberian in nature and that Hardy and Littlewood had worked on similar problems. Wiener corresponded with Hardy on this question but finally decided to follow his own approach, using Fourier transforms. In his autobiography, *I Am a Mathematician*, Wiener wrote that he also consulted Toeplitz's student R. Schmidt, who had published an important paper on Tauberian theory in 1925. Wiener had hoped to collaborate with Schmidt on this problem, for there was a connection in their approaches, but this collaboration did not work out. However, Schmidt suggested that, since his own method had failed for Lambert summability and the PNT, Wiener might test his own approach in those cases. Wiener was soon able to discover a comprehensive method, covering all known Tauberian results.

To get a sense of Wiener's work, begin by writing the Abel sum of $\sum a_n$ in the form

$$A = \lim_{r\to 1^-} (1-r) \sum_{n=0}^\infty s_n r^n = \lim_{x\to\infty} \frac{1}{x} \sum_{n=0}^\infty s_n e^{-n/x}. \qquad (35.22)$$

With this, we have another form of the Hardy–Littlewood theorem: If

$$\lim_{x\to\infty} \frac{1}{x} \sum_{n=0}^\infty s_n e^{-n/x} = A \quad \text{and} \quad s_n = O(1), \quad \text{then} \quad \lim_{x\to\infty} \frac{1}{x} \sum_{n\le x} s_n = A. \qquad (35.23)$$

Now we can write the integral analog: If $F(t)$ is bounded and

$$\lim_{x\to\infty} \frac{1}{x} \int_0^\infty e^{-t/x} F(t)\,dt = A, \quad \text{then} \quad \lim_{x\to\infty} \frac{1}{x} \int_0^x F(t)\,dt = A. \qquad (35.24)$$

Note that the first limit in (35.24) is a weighted average of the function $F(t)$ where the weight function is given by $e^{-t/x}$. More generally, let the weight function be expressed as $G(t/x)$ so that the integral takes the form

$$\int_0^\infty G(t/x) F(t)\,dt = \int_{-\infty}^\infty e^{u-y} G(e^{u-y}) F(e^y)\,dy = \int_{-\infty}^\infty K_1(u-y) f(u)\,du, \qquad (35.25)$$

after applying the change of variables $t = e^u$, $x = e^y$, $F(e^u) = f(u)$, $e^{u-y}G(u-y) = K_1(u-y)$. Wiener could then pose the very general question: Given a bounded function $f(u)$ and kernel K_1 integrable over $(-\infty, \infty)$, under what conditions does the equation

$$\lim_{y \to \infty} \int_{-\infty}^{\infty} K_1(u-y) f(u) \, du = A \int_{-\infty}^{\infty} K_1(u) \, du \tag{35.26}$$

imply

$$\lim_{y \to \infty} \int_{-\infty}^{\infty} K_2(u-y) f(u) \, du = A \int_{-\infty}^{\infty} K_2(u) \, du \tag{35.27}$$

for a different integrable kernel K_2? To determine a simple condition on K_1, Wiener assumed that K_2 was a convolution of K_1 with an integrable function R, that is

$$K_2(y) = \int_{-\infty}^{\infty} K_1(y-u) R(u) \, du. \tag{35.28}$$

Now note that the Fourier transform converts a convolution of two functions to the ordinary product of the transforms of the two functions. So, where $\widehat{K}$ denotes the Fourier transform of K,

$$\widehat{K}_2 = \widehat{K}_1 \cdot \widehat{R}. \tag{35.29}$$

The beauty of this relation is that it allows us to determine $\widehat{R}$ at all points if for all x

$$\widehat{K}_1(x) = \int_{-\infty}^{\infty} e^{-ixt} K_1(t) \, dt \neq 0. \tag{35.30}$$

This was Wiener's now-famous condition, that the existence of the first average would imply the existence of the second. In his 1932 paper "Tauberian Theorems," Wiener stated two forms of this theorem. Note that he wrote L_p for L^p. The first version of Wiener's Tauberian theorem: Let $f(x)$ be a bounded measurable function, defined over $(-\infty, \infty)$. Let $K_1(x)$ be a function in L_1, and let

$$\frac{1}{\sqrt{2\pi}} \int_{-\infty}^{\infty} K_1(x) e^{-iux} \, dx \neq 0 \tag{35.31}$$

for all real u. Let

$$\lim_{x \to \infty} \int_{-\infty}^{\infty} f(\xi) K_1(\xi - x) \, d\xi = A \int_{-\infty}^{\infty} K_1(\xi) \, d\xi. \tag{35.32}$$

Then if $K_2(x)$ is any function in L_1,

$$\lim_{x \to \infty} \int_{-\infty}^{\infty} f(\xi) K_2(\xi - x) \, d\xi = A \int_{-\infty}^{\infty} K_2(\xi) \, d\xi. \tag{35.33}$$

Conversely, let $K_1(\xi)$ be a function of L_1, and let $\int_{-\infty}^{\infty} K_1(\xi)\,d\xi \neq 0$. Let (35.32) imply (35.33) whenever $K_2(x)$ belongs to L_1 and $f(x)$ is bounded. Then (35.31) holds. In his initial 1928 form of the theorem, Wiener required a growth condition $O(1/\xi^2)$ at $\pm\infty$ for the kernels $K_1(\xi)$ and $K_2(\xi)$. In the 1932 version, he refined his theory by means of his well-known theorem on absolutely convergent Fourier series: If a nonvanishing function f has an absolutely convergent Fourier series, then $1/f$ has an absolutely convergent Fourier series. Although this was a difficult result, it emerged less than a decade later as a corollary of I. M. Gelfand's work on commutative Banach algebras. Wiener stated a second general theorem, directly applicable to infinite series, involving Stieltjes integrals; he derived a form of the PNT from this result. Thus, Wiener got his second Tauberian theorem: Let $f(x)$ be a function of limited total variation over every finite range, and let

$$\int_y^{y+1} |df(x)| \tag{35.34}$$

be bounded in y. Let $K_1(x)$ be a continuous function in L_1, and let

$$\sum_{k=-\infty}^{\infty} \max_{k \leq x \leq k+1} |K_1(x)|. \tag{35.35}$$

converge. Now assume

$$\frac{1}{\sqrt{2\pi}} \int_{-\infty}^{\infty} K_1(x) e^{iux}\,dx \neq 0 \quad (-\infty < u < \infty) \tag{35.36}$$

$$\text{and} \quad \lim_{x \to \infty} \int_{-\infty}^{\infty} K_1(\xi - x)\,df(\xi) = A \int_{-\infty}^{\infty} K_1(\xi)\,d\xi. \tag{35.37}$$

If $K_2(x)$ is a continuous function in L_1 satisfying the condition (35.35), then

$$\lim_{x \to \infty} \int_{-\infty}^{\infty} K_2(\xi - x)\,df(\xi) = A \int_{-\infty}^{\infty} K_2(\xi)\,d\xi. \tag{35.38}$$

Note that Wiener also stated a converse of this theorem. Then in 1938, H. R. Pitt (1914–2005) formulated a simple theorem containing both Wiener theorems as corollaries. Pitt took undergraduate courses from Hardy and Littlewood at Cambridge in the 1930s. After graduation in 1936, he studied under Wiener at M.I.T. In his 1938 paper "General Tauberian Theorems," Pitt proved: Suppose $K(x) \in L_1(-\infty, \infty)$ and its Fourier transform $\widehat{K}(t)$ does not vanish for any real t. If $f(x)$ is bounded, slowly oscillating, that is

$$f(y) - f(x) \to 0 \text{ when } y > x,\ x \to \infty,\ y - x \to 0, \tag{35.39}$$

$$\text{and} \int_{-\infty}^{\infty} K(x-t)f(t)\,dt \to A \int_{-\infty}^{\infty} K(x)\,dx,$$

$$\text{then } f(x) \to A \text{ as } x \to \infty.$$

The Serbian mathematician Jovan Karamata (1902–1967) also made an important contribution to Tauberian theory. In 1930, he published a two-page proof of the Hardy–Littlewood theorem, that Abel summability with a one-sided condition implied Cesàro summability. Karamata's proof used only the Weierstrass approximation theorem to prove his main result that if $a_n \geq 0$ and $\sum a_n$ was Abel summable to s, then for every Riemann integrable function $g(x)$,

$$\lim_{x \to 1^-} (1-x) \sum_{n=0}^{\infty} a_n x^n g(x^n) = s \int_0^1 g(t)\, dt. \qquad (35.40)$$

This elegant proof took researchers in Tauberian theory completely by surprise, since up to that time all the proofs of the Hardy–Littlewood theorem had required a fair amount of machinery. Karamata graduated from the University of Belgrade in 1925, where he came under the influence of Mihailo Petrović (1868–1943) who had studied at the École Normale in Paris under Hermite, Poincaré, and Picard. Petrović brought to Serbia the spirit of scientific research he learned in France. By the time he met Karamata, he had ceased to do mathematical research but he advised Karamata to study the latest mathematical discoveries. Karamata regarded himself as self-taught and would say that his teacher in classical analysis was Pólya and Szegő's *Aufgaben und Lehrsätze aus der Analysis*, published in 1925. In fact, the topic of Karamata's doctoral thesis was the development of Weyl's work on the uniform distribution of sequences $x_1, x_2, x_3, \ldots$ in the interval $(0,1)$. We observe that Weyl's theorems were given as a set of five problems in Pólya and Szegő's book. The first of these problems was to show that a sequence $x_1, x_2, x_3, \ldots$ in $(0,1)$ was uniformly distributed if and only if for every Riemann integrable function f

$$\lim_{n \to \infty} \frac{f(x_1) + f(x_2) + \cdots + f(x_n)}{n} = \int_0^1 f(x)\, dx. \qquad (35.41)$$

One may compare this with Karamata's theorem. Again, it is interesting to note that, following their section on uniform distribution, Pólya and Szegő's book posed a problem requiring the use of Weyl's formula as well as Frobenius's theorem on summability. Karamata also introduced the important concept of a regularly varying function.

35.2 Fejér: Summability of Fourier Series

In 1900, L. Fejér made an application of $(C,1)$ summability to Fourier series by proving that the Fourier series of f was $(C,1)$ summable to $(f(x+0) + f(x-0))/2$ at every point where $f(x \pm 0)$ existed. He assumed that f was bounded and integrable on $[0, 2\pi]$. Recall that the Fourier coefficients are given by

$$a_n = \frac{1}{\pi} \int_0^{2\pi} f(t) \cos nt\, dt, \quad b_n = \frac{1}{\pi} \int_0^{2\pi} f(t) \sin nt\, dt$$

and that the nth partial sum of a Fourier series is given by

$$s_n(x) = \frac{1}{2}a_0 + \sum_{k=1}^{n}(a_k \cos kx + b_k \sin kx)$$

$$= \frac{1}{2\pi}\int_0^{2\pi} f(t)\,dt + \sum_{k=1}^{n}\frac{1}{\pi}\int_0^{2\pi} f(t)\cos k(t-x)\,dt.$$

Fejér began his proof with the observation that

$$\sigma_{n-1} = \frac{1}{2} + \cos\theta + \cdots + \cos(n-1)\theta = \frac{1}{2}\frac{\cos(n-1)\theta - \cos n\theta}{1 - \cos\theta};$$

hence, $\quad\dfrac{\sigma_0 + \sigma_1 + \cdots + \sigma_{n-1}}{n} = \dfrac{1}{2n}\dfrac{1 - \cos n\theta}{1 - \cos\theta} = \dfrac{1}{2n}\left(\dfrac{\sin(n\theta/2)}{\sin(\theta/2)}\right)^2.$

Thus, for the arithmetic mean of the partial sums, he had

$$S_n(x) = \frac{s_0(x) + s_1(x) + s_2(x) + \cdots + s_{n-1}(x)}{n}$$

$$= \frac{1}{n\pi}\int_{-x/2}^{\pi - x/2} f(x + 2u)\left(\frac{\sin nu}{\sin u}\right)^2 du.$$

Fejér immediately perceived that this integral was simpler than the one found by Dirichlet for the partial sum $s_n(x)$ because the kernel $\sin^2 v / \sin^2 u$ was always nonnegative, unlike the corresponding kernel in Dirichlet's integral $\sin(2n-1)u / \sin u$. Fejér first considered the case where f was continuous at x. He let $\epsilon > 0$, so that there existed a $\delta > 0$ such that

$$|f(x+h) - f(x)| < \epsilon \quad\text{for } |h| \leq \delta.$$

We note that Fejér's notation interchanged ϵ and δ. He next wrote the integral for $S_n(x)$ in three parts:

$$S_n(x) = \frac{1}{2n\pi}\int_0^{x-\delta}\frac{1 - \cos n(t-x)}{1 - \cos(t-x)}f(t)\,dt$$

$$+ \frac{1}{2n\pi}\int_{x-\delta}^{x+\delta}\frac{1 - \cos n(t-x)}{1 - \cos(t-x)}f(t)\,dt$$

$$+ \frac{1}{2n\pi}\int_{x+\delta}^{2\pi}\frac{1 - \cos n(t-x)}{1 - \cos(t-x)}f(t)\,dt.$$

He assumed $|f(t)| \leq M$ in $[0, 2\pi]$. Then the absolute values of the first and third integrals were bounded by $2M/n(1 - \cos\delta)$. For the second integral, the positivity of the term multiplying $f(t)$ implied that

$$\int_{x-\delta}^{x+\delta}\frac{1 - \cos n(t-x)}{1 - \cos(t-x)}f(t)\,dt = (f(x) + \eta)\int_{x-\delta}^{x+\delta}\frac{1 - \cos n(t-x)}{1 - \cos(t-x)}\,dt,$$

where $|\eta| < \epsilon$. Fejér then noted that

$$\frac{1}{2n\pi}\int_0^{2\pi}\frac{1-\cos n(t-x)}{1-\cos(t-x)}dt = 1.$$

Hence, $\quad \dfrac{1}{2n\pi}\displaystyle\int_{x-\delta}^{x+\delta}\dfrac{1-\cos n(t-x)}{1-\cos(t-x)}dt$

$$= 1 - \left(\frac{1}{2n\pi}\int_0^{x-\delta}\frac{1-\cos n(t-x)}{1-\cos(t-x)}dt + \frac{1}{2n\pi}\int_{x+\delta}^{2\pi}\frac{1-\cos n(t-x)}{1-\cos(t-x)}dt\right).$$

He observed that each of the last two integrals was less that $2/n(1-\cos\delta)$. With all this information, he could conclude that for n large enough

$$|S_n(x) - f(x)| < 2\epsilon.$$

This proved Fejér's theorem for the case in which f was continuous at x. Assuming only the existence of the limits $f(x-0)$ and $f(x+0)$, Fejér broke the integral for $S_n(x)$ into two parts:

$$I_1(x) = \frac{1}{2n\pi}\int_0^x \frac{1-\cos n(t-x)}{1-\cos(t-x)}f(t)dt,$$

$$I_2(x) = \frac{1}{2n\pi}\int_x^{2\pi} \frac{1-\cos n(t-x)}{1-\cos(t-x)}f(t)dt.$$

Then by a similar argument

$$\lim_{n\to\infty} I_1(x) = \frac{1}{2}f(x-0),\ \lim_{n\to\infty} I_2(x) = \frac{1}{2}f(x+0).$$

Fejér went on to observe that if $f(x)$ was everywhere continuous, then $S_n(x)$ converged uniformly to $f(x)$. He also noted the following immediate corollaries of his theorem:

- If the Fourier series converges at a point of continuity of a function, then its sum is the value of the function at that point.
- A continuous function on a closed interval is a uniform limit of a sequence of polynomials. This is Weierstrass's approximation theorem.
- Poisson's integral yields a solution for Dirichlet's problem for the circle.

Hermann A. Schwarz was the first to prove the third result. He felt that a proof by Fourier series was probably not possible. As noted before, Fejér's motivation in the discovery of his theorem was to provide a proof using Fourier series.

Hardy recognized that his Tauberian theorem on $(C, 1)$ summability, combined with Fejér's theorem, immediately yielded a result on Fourier series: If the Fourier coefficients of a continuous function f are $a_n = O(1/n)$ and $b_n = O(1/n)$, the Fourier series of f at x converges to $f(x)$. Hardy then reasoned that since the Fourier coefficients of a periodic function f of bounded variation satisfied $a_n = O(1/n)$, $b_n = O(1/n)$, then the Fourier series of a such a function converged to $\frac{1}{2}(f(x+0)+f(x-0))$. In fact, this is the classical Dirichlet-Jordan theorem. Further, observe that since Cesàro summability implies Abel summability, it follows that for f as in Fejér's theorem, we have

$$\lim_{r \to 1^-} \left(\frac{1}{2}a_0 + (a_1 \cos x + b_1 \sin x)r + (a_2 \cos 2x + b_2 \sin 2x)r^2 + \cdots \right)$$
$$= \frac{1}{2}\big(f(x+0) + f(x-0)\big).$$

This equation simplifies to

$$\lim_{r \to 1^-} \left(\frac{1}{2\pi} \int_0^{2\pi} \frac{1-r^2}{1 - 2r\cos(x-t) + r^2} f(t)\, dt \right) = \frac{1}{2}\big(f(x+0) + f(x-0)\big).$$

When Hilbert saw Fejér's work, he requested Fejér to attempt a proof of a similar theorem for the Laplace series where a function $f(\theta, \phi)$ was expanded in terms of surface harmonics. Fejér was unsuccessful in this effort for some years. Finally, while looking at a book on Bessel functions, he saw F. G. Mehler's integral formula for Legendre polynomials:

$$P_n(\cos\theta) = \frac{2}{\pi} \int_\theta^\pi \frac{\sin(2n+1)(t/2)}{\sqrt{2(\cos\theta - \cos t)}}\, dt, \quad 0 < \theta < \pi.$$

With the help of this result, in 1908 Fejér was able to prove that the Laplace series of a bounded integrable function was $(C, 2)$ summable to the function at any point of continuity. In 1913, H. Gronwall proved that $(C, 2)$ could be replaced by $(C, 1)$.

35.3 Karamata's Proof of the Hardy–Littlewood Theorem

Karamata's short proof of Littlewood's theorem and the more general Hardy–Littlewood theorem relied on Weierstrass's approximation theorem. Karamata used it in the following form: For any Riemann integrable function $g(x)$ on $(0, 1)$ and every $\epsilon > 0$ there exist two polynomials $p(t)$ and $P(t)$ such that

$$p(t) \leq g(t) \leq P(t) \quad \text{for} \quad 0 \leq t \leq 1 \tag{35.42}$$

$$\int_0^1 \big(P(t) - p(t)\big)\, dt \leq \epsilon. \tag{35.43}$$

Karamata did not give the details of the proof of this result. It can be proved, however, by first taking $g(t)$ to be a continuous function. By Weierstrass's theorem, there are polynomials $p(t)$ and $P(t)$ differing by at most $\epsilon/4$ from $g(t) - \epsilon/4$ and $g(t) + \epsilon/4$, respectively, for all $t \in [0, 1]$. Clearly, the required result follows for $g(t)$ continuous. We next take $g(t)$ to be piecewise continuous, and the result follows because $g(t)$ can be approximated by continuous functions. Finally, for any Riemann integrable function $g(t)$, there are step functions $m(t)$ and $M(t)$ such that $m(t) \leq f(t) \leq M(t)$ and

$$\int_0^1 \big(M(t) - m(t)\big)\, dt < \epsilon/2.$$

Karamata's theorem: If $a_n \geq -K$, with $K \geq 0$ independent of n and

$$(1-x)\sum_{n=0}^{\infty} a_n x^n \to A \text{ as } x \to 1^-,$$

then $(1-x)\sum_{n=0}^{\infty} a_n g(x^n) x^n \to A \int_0^1 g(t)\,dt$

for every Riemann integrable function $g(t)$.

In Karamata's proof, it was obviously sufficient to take $K = 0$, for he could replace a_n by $a_n + K$. Karamata then supposed $g(x) = x^\alpha$, $\alpha \geq 0$. Then he had

$$(1-x)\sum_{n=0}^{\infty} a_n g(x^n) x^n = (1-x)\sum_{n=0}^{\infty} a_n x^{(\alpha+1)n}$$

$$= \frac{(1-x)}{1-x^{\alpha+1}}(1-x^{\alpha+1})\sum_{n=0}^{\infty} a_n x^{(\alpha+1)n} \to \frac{A}{\alpha+1} = A\int_0^1 t^\alpha\,dt,$$

as $x \to 1^-$. It followed by linearity that for every polynomial $P(x)$

$$(1-x)\sum_{n=0}^{\infty} a_n P(x^n) x^n \to A \int_0^1 P(t)\,dt.$$

He could next apply (35.42) and (35.43) because a_n was positive; Karamata's theorem followed.

To derive the Hardy–Littlewood theorem, Karamata set $x = e^{-1/n}$ and let $g(t)$ be the piecewise continuous function

$$g(t) = \begin{cases} 0 & 0 \leq t < 1/e, \\ 1/t & 1/e \leq t \leq 1. \end{cases}$$

He then arrived at $g(x^m) = 0$ for $m > n$, $g(x^m)x^m = 1$ for $m \leq n$, and $\int_0^1 g(t)\,dt = 1$, thereby reducing his theorem to the Hardy–Littlewood theorem. In other words, given the one-sided Tauberian condition $a_n \geq -K$, if the Abel sum of $\sum_{n=0}^{\infty} a_n x^n$ was A, then the Cesàro sum of $\sum a_n$ was also A.

35.4 Wiener's Proof of Littlewood's Theorem

Littlewood's Tauberian theorem of 1910 was the first difficult and deep Tauberian result to be proved. It is therefore interesting to see how Wiener derived this theorem from his general theorem. We restate Littlewood's result:

$$\text{If } \lim_{y \to 1^-} \sum_{n=0}^{\infty} a_n y^n = s \text{ and } n|a_n| < K, \text{ then } \sum_{n=0}^{\infty} a_n = s.$$

The first step in Wiener's proof was to express $\sum a_n y^n$ as an integral. For that purpose he showed that $s(x) = \sum_{n \leq x} a_n$ was bounded for $0 \leq x < \infty$. By hypothesis, $\sum_{n=0}^{\infty} a_n e^{-n/x}$

35.4 Wiener's Proof of Littlewood's Theorem

was bounded for $0 \leq x < \infty$ and (using $n|a_n| < K$)

$$|s(x) - \sum_{n=0}^{\infty} a_n e^{-n/x}| = |\sum_{n \leq x} a_n(1 - e^{-n/x}) - \sum_{n > x} a_n e^{-n/x}|$$

$$\leq \sum_{n \leq x} \frac{K}{n} \cdot \frac{n}{x} + \sum_{n > x} \frac{K}{n} e^{-n/x}$$

$$\leq 2K + K \int_x^{\infty} e^{-u/x} \frac{du}{u}$$

$$\leq 3K + K \int_1^{\infty} e^{-u} \frac{du}{u} = \text{constant}.$$

This showed that $s(x)$ was bounded so that he had

$$\sum_{n=0}^{\infty} a_n e^{-nx} = \int_{0^-}^{\infty} e^{-ux} ds(u) = \int_0^{\infty} xe^{-ux} s(u) du.$$

Hence, $s = \lim_{x \to 0^+} \int_0^{\infty} xe^{-ux} s(u) du = \lim_{\xi \to \infty} \int_{-\infty}^{\infty} e^{-\xi} e^{-e^{\eta-\xi}} s(e^{\eta}) e^{\eta} d\eta.$

So Wiener set $K_1(\xi) = e^{-\xi} e^{-e^{-\xi}}$ and observed that

$$\int_{-\infty}^{\infty} K_1(\xi) d\xi = \int_{-\infty}^{\infty} e^{-\xi} e^{-e^{-\xi}} d\xi = \int_0^{\infty} e^{-x} dx = 1.$$

Thus, $\lim_{\xi \to \infty} \int_{-\infty}^{\infty} K_1(\xi - \eta) s(e^{\eta}) d\eta = s \int_{-\infty}^{\infty} K_1(\xi) d\xi$ and

$$\frac{1}{\sqrt{2\pi}} \int_{-\infty}^{\infty} K_1(\xi) e^{-iu\xi} d\xi = \frac{1}{\sqrt{2\pi}} \int_0^{\infty} x^{iu} e^{-x} dx = \frac{1}{\sqrt{2\pi}} \Gamma(1+iu) \neq 0.$$

Therefore, $K_1(\xi)$ satisfied the hypotheses, (35.31) and (35.32), of his first Tauberian theorem. Wiener then chose $K_2(\xi)$ in such a manner that he obtained the $(C,1)$ summability of $\sum a_n$ to s. He set

$$K_2(\xi) = \begin{cases} 0 & \xi < 0, \\ e^{-\xi} & \xi > 0, \end{cases}$$

so that by his first Tauberian theorem,

$$s = s \int_0^{\infty} e^{-\xi} d\xi = s \int_{-\infty}^{\infty} K_2(\xi) d\xi = \lim_{\xi \to \infty} \int_{-\infty}^{\infty} K_2(\xi - \eta) s(e^{\eta}) d\eta$$

$$= \lim_{\xi \to \infty} \int_{-\infty}^{\xi} e^{\eta - \xi} s(e^{\eta}) d\eta = \lim_{x \to \infty} \frac{1}{x} \int_0^x s(y) dy.$$

Note that by applying Hardy's theorem that $(C,1)$ summability together with $a_n = O(1/n)$ implies convergence, the Hardy–Littlewood theorem follows. However,

Wiener included a simple argument to prove Hardy's theorem: For $\lambda > 0$,

$$s = \frac{(1+\lambda)s - s}{\lambda} = \lim_{x\to\infty} \frac{1}{\lambda x}\left(\int_0^{(1+\lambda)x} s(y)\,dy - \int_0^x s(y)\,dy\right)$$

$$= \lim_{x\to\infty} \frac{1}{\lambda x}\int_x^{(1+\lambda)x} s(y)\,dy = \lim_{x\to\infty}\left(s(x) + \frac{1}{\lambda x}\int_x^{(1+\lambda)x}(s(y) - s(x))\,dy\right).$$

The condition $a_n = O(1/n)$ then implied the necessary result:

$$\left|\frac{1}{\lambda x}\int_x^{(1+\lambda)x}(s(y) - s(x))\,dy\right| \leq \frac{1}{\lambda x}\int_x^{(1+\lambda)x}\sum_{x<n<y}\frac{K}{n}\,dy$$

$$\leq \sum_{\lfloor x\rfloor+1}^{\lfloor(1+\lambda)x\rfloor}\frac{K}{\lfloor x\rfloor} \leq \frac{\lfloor\lambda x\rfloor K}{\lfloor x\rfloor} < 2\lambda K,$$

for sufficiently large x. Hence $\overline{\lim}_{x\to\infty}|s(x) - s| < 2\lambda K$; or, because λ was an arbitrary positive number, $\lim_{x\to\infty}|s(x) - s| = 0$. This completed Wiener's proof of Littlewood's theorem.

35.5 Hardy and Littlewood: The Prime Number Theorem

In their 1921 paper, "On a Tauberian Theorem for Lambert Series ...," Hardy and Littlewood gave a very simple proof of the PNT based on the result that Lambert summability implied Abel summability. As we mentioned before, their proof of the Lambert summability theorem employed a result, due to Landau, stronger than the PNT. Thus, although they did not produce a new proof, their derivation of the PNT insightfully reveals its Tauberian character. In this derivation, Hardy and Littlewood employed a number-theoretic result describing the average behavior of the arithmetic function $d(n)$, the number of divisors of n. Dirichlet first proved this result by his ingenious hyperbola method in an 1849 paper on the average behavior of arithmetic functions.

Hardy and Littlewood first showed that the series

$$\sum_{n=1}^{\infty}\frac{\Lambda(n) - 1}{n}$$

was Lambert summable to -2γ, where γ was Euler's constant. Note here that the Lambert series could be written as

$$f(y) = y\sum_{n=1}^{\infty}\frac{(\Lambda(n) - 1)e^{-ny}}{1 - e^{-ny}} = y\sum_{n=1}^{\infty}(\Lambda(n) - 1)e^{-ny}(1 + e^{-ny} + e^{-2ny} + \cdots)$$

$$= y\sum_{n=1}^{\infty}c_n e^{-ny}, \text{ where } c_n = \sum_{d|n}(\Lambda(d) - 1) = \ln n - d(n).$$

Next, they observed that

$$\sum_{i=1}^{n} c_i = \ln n! - \sum_{i=1}^{n} d(i).$$

To estimate the logarithmic term, they applied Stirling's formula and to estimate the second term they used the Dirichlet divisor theorem:

$$\sum_{i=1}^{n} d(i) = n \ln n + (2\gamma - 1)n + O(\sqrt{n}).$$

These calculations gave them

$$\frac{1}{n} \sum_{i=1}^{n} c_i \sim -2\gamma \quad \text{as } n \to \infty.$$

By Frobenius's theorem, the last result implied that

$$\lim_{y \to \infty} f(y) = \lim_{y \to \infty} y \sum_{n=1}^{\infty} c_n e^{-ny} = -2\gamma.$$

This proved the Lambert summability of $\sum_{n=1}^{\infty} \frac{\Lambda(n)-1}{n}$. Hence, by their theorem that Lambert summability implies Abel summability, the series was Abel summable to -2γ. It was also clear that $(\Lambda(n) - 1)/n \geq -1$. Moreover, Hardy and Littlewood had earlier extended Littlewood's theorem and this extension showed that this one-sided Tauberian condition was sufficient to obtain the ordinary convergence of $\sum_{n=1}^{\infty} \frac{\Lambda(n)-1}{n}$. Recall that by (35.9), the convergence of $\sum a_n$ implied that

$$\frac{a_1 + 2a_2 + \cdots + na_n}{n} \to 0 \quad \text{as } n \to \infty.$$

For $a_k = \frac{\Lambda(k)-1}{k}$, the last condition translated to

$$\frac{\Lambda(1) - 1 + \Lambda(2) - 1 + \cdots + \Lambda(n) - 1}{n} \to 0 \quad \text{as } n \to \infty$$

or

$$\lim_{N \to \infty} \frac{1}{N} \sum_{n=1}^{N} \Lambda(n) = 1.$$

and this was equivalent to the prime number theorem.

In his 1971 paper, "The Quickest Proof of the Prime Number Theorem," Littlewood observed that in 1918 he and Hardy proved (35.17). He pointed out that though they had earlier proved the Tauberian theorem (with the one-sided condition mentioned above) necessary to deduce the quickest proof, they did not mention the PNT in their 1918 paper. We note that it was this Tauberian theorem for which Karamata gave his nice proof, described by Littlewood as "highly sophisticated."

35.6 Wiener's Proof of the PNT

In his work on the Tauberian theorem, one of Wiener's fundamental aims was to prove the prime number theorem by means of Lambert summability. Thus, he wished to determine the behavior of $\sum_{n \leq x} \Lambda(n)$ as $x \to \infty$ from the behavior of

$$\sum_{n=1}^{\infty} \Lambda(n) \frac{x^n}{1-x^n} \quad \text{as } x \to 1^-.$$

First, Wiener observed that

$$\sum_{n=1}^{\infty} \Lambda(n) \frac{x^n}{1-x^n} = \sum_{n=1}^{\infty} x^n \sum_{m|n} \Lambda(m) = \sum_{n=1}^{\infty} x^n \ln n$$

$$= \sum_{n=1}^{\infty} \ln n \frac{x^n - x^{n+1}}{1-x} = \sum_{n=1}^{\infty} \frac{x^{n+1}}{1-x}(\ln(n+1) - \ln n)$$

$$= \frac{x}{1-x} \sum_{n=1}^{\infty} \ln\left(1 + \frac{1}{n}\right) x^n = \frac{x}{1-x} \sum_{n=1}^{\infty} \left(\frac{1}{n} + O\left(\frac{1}{n^2}\right)\right) x^n$$

$$= \frac{x}{1-x} \left(\ln \frac{1}{1-x} + \sum_{n=1}^{\infty} O\left(\frac{1}{n^2}\right) x^n \right).$$

Note that the second line used summation by parts. Wiener next set $x = e^{-\xi}$ and multiplied by $-\xi$ to obtain

$$\sum_{n=1}^{\infty} \Lambda(n) \frac{\xi e^{-n\xi}}{e^{-n\xi} - 1} = \frac{\xi e^{-\xi}}{1 - e^{-\xi}} \left(\ln(1 - e^{-\xi}) - \sum_{n=1}^{\infty} O\left(\frac{1}{n^2}\right) e^{-n\xi} \right).$$

It followed from the right-hand side that as $\xi \to 0^+$, the series behaved like $\ln \xi$. Wiener therefore worked with the differentiated series. Upon differentiating the last equation, he arrived at

$$\sum_{n=1}^{\infty} \Lambda(n) \frac{d}{dn\xi} \frac{n\xi e^{-n\xi}}{e^{-n\xi} - 1} = \sum_{n=1}^{\infty} \Lambda(n) \frac{e^{-2n\xi} - e^{-n\xi} + n\xi e^{-n\xi}}{(e^{-n\xi} - 1)^2}$$

$$= \frac{e^{-\xi} - e^{-2\xi} - \xi e^{-\xi}}{(1 - e^{-\xi})^2} \left(\ln(1 - e^{-\xi}) + \sum_{n=1}^{\infty} O\left(\frac{1}{n^2}\right) e^{-n\xi} \right)$$

$$+ \frac{\xi e^{-\xi}}{1 - e^{-\xi}} \left(\frac{e^{-\xi}}{1 - e^{-\xi}} + \sum_{n=1}^{\infty} O\left(\frac{1}{n}\right) e^{-n\xi} \right)$$

$$= O(1)(O(\ln \xi) + O(1))$$

$$+ (1 + O(\xi)) \left(\frac{1}{\xi} + O(1) + O(\ln \xi) \right)$$

$$= 1/\xi + O(\ln \xi),$$

as $\xi \to 0^+$. Thus, he had

$$\lim_{\xi \to 0^+} \xi \sum_{n=1}^{\infty} \Lambda(n) \frac{e^{-2n\xi} - e^{-n\xi} + n\xi e^{-n\xi}}{(e^{-n\xi} - 1)^2} = 1.$$

Wiener wrote the sum as a Stieltjes integral so that he could apply his second Tauberian theorem. Toward that end, he set

$$g(y) = \sum_{n=1}^{\lfloor e^y \rfloor} \frac{\Lambda(n)}{n},$$

so that the previous equation containing the limit took the form

$$1 = \lim_{\xi \to 0^+} \int_0^\infty \eta\xi \frac{e^{-2\eta\xi} - e^{-\eta\xi} + \eta\xi e^{-\eta\xi}}{(e^{-\eta\xi} - 1)^2} \, dg(\ln \eta)$$

$$= \lim_{x \to \infty} \int_{-\infty}^\infty \frac{e^{y-x}(e^{-2e^{y-x}} - e^{-e^{y-x}} + e^{y-x}e^{-e^{y-x}})}{(e^{-e^{y-x}} - 1)^2} \, dg(y).$$

To understand the next step, compare the last expression with the corresponding expression in Wiener's theorem. On this basis, we can see how Wiener next wrote

$$K_1(x) = \frac{e^{-x}(e^{-2e^{-x}} - e^{-e^{-x}} + e^{-x}e^{-e^{-x}})}{(e^{-e^{-x}} - 1)^2}$$

and then

$$\int_{-\infty}^\infty K_1(x) \, dx = \int_0^\infty \frac{e^{-2\xi} - e^{-\xi} + \xi e^{-\xi}}{(e^{-\xi} - 1)^2} \, d\xi = \int_0^\infty \frac{d}{d\xi} \frac{\xi e^{-\xi}}{e^{-\xi} - 1} \, d\xi$$

$$= \lim_{\xi \to 0^+} \frac{\xi e^{-\xi}}{1 - e^{-\xi}} = 1.$$

Thus, he had $A = 1$ in the hypothesis of his theorem. One may check that $K_1(x)$ satisfies (35.35) and since $g(y)$ is monotomic,

$$\int_n^{n+1} |dg(x)| = \int_n^{n+1} dg(x).$$

Moreover, the latter expression is bounded for $-\infty < n < \infty$. Finally, Wiener had only to check that the Fourier transform of $K_1(x)$ did not vanish. So he computed

$$\frac{1}{\sqrt{2\pi}} \int_{-\infty}^\infty K_1(x) e^{-iux} \, dx = \frac{1}{\sqrt{2\pi}} \int_0^\infty \frac{d}{d\xi}\left(\frac{\xi e^{-\xi}}{e^{-\xi} - 1}\right) \xi^{iu} \, d\xi$$

$$= \lim_{\lambda \to 0^+} \frac{1}{\sqrt{2\pi}} \int_0^\infty \frac{d}{d\xi}\left(\frac{\xi e^{-\xi}}{e^{-\xi} - 1}\right) \xi^{iu+\lambda} \, d\xi.$$

Integration by parts converted the last expression to

$$\lim_{\lambda \to 0^+} \frac{iu+\lambda}{\sqrt{2\pi}} \int_0^\infty \frac{\xi^{iu+\lambda} e^{-\xi}}{1-e^{-\xi}} d\xi = \lim_{\lambda \to 0^+} \frac{iu+\lambda}{\sqrt{2\pi}} \int_0^\infty \xi^{iu+\lambda} \sum_{n=1}^\infty e^{-n\xi} d\xi$$

$$= \lim_{\lambda \to 0^+} \frac{\lambda+iu}{\sqrt{2\pi}} \sum_{n=1}^\infty \frac{\Gamma(\lambda+1+iu)}{n^{\lambda+1+iu}}$$

$$= \lim_{\lambda \to 0^+} \frac{\lambda+iu}{\sqrt{2\pi}} \zeta(\lambda+1+iu)\Gamma(\lambda+1+iu)$$

$$= iu\zeta(1+iu)\Gamma(1+iu).$$

The work of Hadamard and de la Vallée Poussin showed that $\zeta(1+iu)$ did not vanish for any real u, and hence the Fourier transform of $K_1(x)$ did not vanish.

Finally, Wiener had to choose $K_2(x)$ appropriately so that he got the PNT in the form $\lim_{N\to\infty} \left(\sum_{n=1}^N \Lambda(n) \right)/N = 1$. Note that in this application, $K_2(x)$ had to be continuous; it could not be the piecewise continuous function

$$K_2(x) = \begin{cases} 0, & x < 0, \\ e^{-x}, & x > 0, \end{cases}$$

although, if allowed, this would have yielded the result immediately. So Wiener defined two continuous functions

$$K_{21}(x) = \begin{cases} 0, & x < -\epsilon, \\ \dfrac{x+\epsilon}{\epsilon}, & -\epsilon \leq x < 0, \\ e^{-x}, & 0 \leq x, \end{cases}$$

$$K_{22}(x) = \begin{cases} 0, & x < 0, \\ \dfrac{x}{\epsilon} e^{-\epsilon}, & 0 \leq x < \epsilon, \\ e^{-x}, & \epsilon \leq x. \end{cases}$$

Here he verified that

$$\int_{-\infty}^\infty K_{21}(x)\,dx = 1+\epsilon/2 \quad \text{and} \quad \int_{-\infty}^\infty K_{22}(x)\,dx = e^{-\epsilon}(1+\epsilon/2).$$

Wiener's second Tauberian theorem then implied

$$1+\epsilon/2 = \lim_{x\to\infty} \int_{-\infty}^\infty K_{21}(x-y)\,dg(y)$$

$$= \lim_{x\to\infty} \left(\int_{-\infty}^\infty e^{y-x}\,dg(y) + \int_x^{x+\epsilon} \frac{\epsilon-y+x}{\epsilon}\,dg(y) \right)$$

$$\geq \overline{\lim_{x \to \infty}} \int_{-\infty}^{x} e^{y-x} \, dg(y) = \overline{\lim_{N \to \infty}} \frac{1}{N} \int_{0}^{N} \eta \, dg(\ln \eta)$$

$$= \overline{\lim_{N \to \infty}} \frac{1}{N} \sum_{n=1}^{N} \Lambda(n), \text{ and} \tag{35.44}$$

$$e^{-\epsilon}(1 + \epsilon/2) = \lim_{x \to \infty} \int_{-\infty}^{\infty} K_{22}(x-y) \, dg(y)$$

$$= \lim_{x \to \infty} \left(\int_{\infty}^{x-\epsilon} e^{y-x} \, dg(y) + \int_{x-\epsilon}^{x} \frac{x-y}{\epsilon} e^{-\epsilon} \, dg(y) \right)$$

$$= \lim_{x \to \infty} \left(\int_{-\infty}^{x} e^{y-x} \, dg(y) - \int_{x-\epsilon}^{x} \left(e^{y-x} - \frac{x-y}{\epsilon} e^{-\epsilon} \right) dg(y) \right)$$

$$\leq \overline{\lim_{x \to \infty}} \int_{\infty}^{x} e^{y-x} \, dg(y) = \overline{\lim_{N \to \infty}} \frac{1}{N} \sum_{n=1}^{N} \Lambda(n). \tag{35.45}$$

Note that in the above calculation, one may use the fact that

$$e^{y-x} - \frac{x-y}{\epsilon} e^{-\epsilon} \geq 0 \text{ for } x - \epsilon \leq y \leq x.$$

Wiener let $\epsilon \to 0$ in the inequalities (35.44) and (35.45) to get

$$1 \geq \overline{\lim_{N \to \infty}} \frac{1}{N} \sum_{n=1}^{N} \Lambda(n) \text{ and } 1 \leq \overline{\lim_{N \to \infty}} \frac{1}{N} \sum_{n=1}^{N} \Lambda(n).$$

These inequalities implied that $\lim_{N \to \infty} \frac{1}{N} \sum_{n=1}^{N} \Lambda(n)$ existed and was equal to 1. This proof of the PNT used only one property of the zeta function: that it did not vanish on the line consisting of points with real part equal to 1.

35.7 Kac's Proof of Wiener's Theorem

The basic principle behind Wiener's Tauberian theorem is simple but penetrating. Mark Kac illustrated this insight by producing a short proof of the 1928 form of Wiener's theorem. This proof uses only Fubini's theorem and the uniqueness of Fourier transforms; like Wiener's 1928 theorem, it is powerful enough to produce the PNT as a consequence.

Kac's theorem: Suppose

$$K_1(x) \in L^1(-\infty, \infty), \quad x^2 K_1(x) \in L^1(-\infty, \infty) \text{ and}$$

$$k_1(\xi) = \int_{-\infty}^{\infty} K_1(x) e^{i\xi x} \, dx \neq 0, \quad -\infty < \xi < \infty.$$

If $m(y)$ is a bounded measurable function such that for all x

$$\int_{-\infty}^{\infty} K_1(x-y) m(y) \, dy = 0,$$

then $m(y) = 0$ almost everywhere.

In proving this theorem, Kac realized that the condition $x^2 K_1(x) \in L^1$ implied that $k_1(\xi)$ was twice continuously differentiable. Thus, let Φ be the set of all twice continuously differentiable functions with compact support. Since $k_1(\xi) \neq 0$ for all ξ, it follows that every $\phi \in \Phi$ is of the form $k_1 \psi$ for some $\psi \in \Phi$. In short, $k_1 \Phi = \Phi$. Let $\phi \in \Phi$ and let F be the Fourier transform of ϕ; that is, let

$$F(x) = \int_{-\infty}^{\infty} \phi(\xi) e^{ix\xi} \, d\xi.$$

Because ϕ has compact support, $F(x)$ is defined for all $x \in \mathbb{C}$ and $F'(x)$ exists. Hence, F is an entire function, and ϕ can be chosen such that F is not identically zero. Thus, F has only a countable number of zeros. Since $F \in L^1(-\infty, \infty)$ and $|F(x)||k_1(x-y)||m(y)|$ is integrable as a function of two variables (x, y), we can apply Fubini's theorem and change the order of integration:

$$0 = \int_{-\infty}^{\infty} F(x) \left(\int_{-\infty}^{\infty} K_1(x-y) m(y) \, dy \right) dx$$

$$= \int_{-\infty}^{\infty} m(y) \left(\int_{-\infty}^{\infty} K_1(x-y) F(x) \, dx \right) dy$$

$$= \int_{-\infty}^{\infty} m(y) \left(\int_{-\infty}^{\infty} k_1(\xi) \phi(\xi) e^{i\xi y} \, d\xi \right) dy.$$

Because $k_1 \Phi = \Phi$, we can conclude that for all $\phi \in \Phi$, we have

$$0 = \int_{-\infty}^{\infty} m(y) \left(\int_{-\infty}^{\infty} \phi(\xi) e^{i\xi y} \, d\xi \right) dy.$$

Now Φ is closed under translation so that we can replace $\phi(\xi)$ by $\phi(\xi - \alpha)$ and change variables to arrive at

$$0 = \int_{-\infty}^{\infty} m(y) \left(\int_{-\infty}^{\infty} \phi(\xi) e^{i\xi y} \, d\xi \right) e^{i\alpha y} \, dy,$$

for all real α. By the definition of F, this gives

$$0 = \int_{-\infty}^{\infty} m(y) F(y) e^{i\alpha y} \, dy$$

for all real α; and the uniqueness of Fourier transforms implies $m(y)F(y) = 0$ for almost all y. Since F can be chosen to have countably many zeros, we conclude that $m(y) = 0$ almost everywhere.

35.8 Gelfand: Normed Rings

Wiener derived the final form of his Tauberian theorem by means of his famous theorem on nonvanishing and absolutely convergent Fourier series. About ten years later, in 1941, Izrail Gelfand provided a short and elegant derivation of this theorem, based on

35.8 Gelfand: Normed Rings

his theory of normed rings. In this effort, Gelfand utilized an abstract formulation of the Fourier transform, now known as the Gelfand transform. In a short note published in 1939, Gelfand developed the elements of the theory of commutative Banach algebras. These are algebras B over the complex numbers, containing a multiplicative identity e; such algebras are complete with respect to a norm $||\ ||$, such that $||e|| = 1$ and $||xy|| \leq ||x|| \cdot ||y||$ for x and y in B; thus, Gelfand named them normed rings. In two further short notes published in the same year, Gelfand gave applications of his theory to absolutely convergent Fourier series and integrals and to the ring of almost periodic functions. Gelfand used the work of Wiener and Pitt on absolutely convergent Fourier series and integrals as a springboard in his construction of Banach algebras. He succeeded in obtaining short proofs of the Wiener-Pitt results by revealing their essentially algebraic character.

In his two-page fundamental paper of 1939, Gelfand denoted a normed ring, or commutative Banach algebra, by R and observed as his first theorem that any maximal ideal M was closed in R. His third theorem, now called the Gelfand–Mazur theorem, stated that R/M was isomorphic to the field of complex numbers. This theorem originated with the 1918 result of Alexander Ostrowski (1893–1986), student of Landau and Klein, that a complete Archimedean field is isomorphic to either the field of real numbers or the field of complex numbers. In 1938 this was generalized by Stanislaw Mazur (1905–1981), student of Stefan Banach, who proved that a normed associative real division algebra was isomorphic to the field of real numbers, or to the field of complex numbers, or to the noncommutative field of quarternions. In his 1941 paper "Normierte Ringe," Gelfand gave a beautiful proof of the particular case of Mazur's theorem he needed. This proof employed Liouville's theorem that a bounded entire function is a constant.

Gelfand was then able to associate with each $x \in R$ a complex number $x(M)$, to obtain a complex valued function on the set of all maximal ideals of R. He defined a topology on the set of maximal ideals to make the set into a compact Hausdorff space and the functions $x(M)$ continuous. In his 1941 paper, Gelfand also noted the easily proved result that $x(M) \leq ||x||$. This depended on the lemma that for the multiplicative identity e and any $y \in R$, if $||e + y|| < 1$, then y was invertible. In fact, it is easily verified that

$$y^{-1} = -(e + (e+y) + (e+y)^2 + \cdots). \quad (35.46)$$

Now observe that if $x(M) = \lambda \in \mathbf{C}$, then $x = \lambda e + z$ where $z \in M$. Assume $\lambda \neq 0$, because if $\lambda = 0$, then the inequality $|x(M)| \leq ||x||$ is obvious. Then for $y = z/\lambda$, we have

$$\frac{|x(M)|}{||x||} = \frac{\lambda}{||\lambda e + z||} = \frac{1}{||e+y||} \leq 1,$$

because if $||e + y|| < 1$, then y would be invertible and not be in M.

We now have the result needed to understand Gelfand's simple proof of Wiener's theorem on nonvanishing absolutely convergent Fourier series. He let R be the set of all functions $f(t) = \sum_{n=-\infty}^{\infty} a_n e^{int}$ such that $\sum_{n=-\infty}^{\infty} |a_n| < \infty$; he then let

$||f|| = \sum_{n=-\infty}^{\infty} |a_n|$. This gave R the structure of a commutative Banach algebra. Gelfand argued that for any maximal ideal M and $e^{it} \in R$, $e^{it}(M)$ was some complex number a. He then obtained

$$|e^{it}(M)| = |a| \leq ||e^{it}|| = 1, \text{ and } 1/|a| \leq ||e^{-it}|| = 1,$$

and hence $a = e^{it_0}$ for some real number t_0. This meant that any trigonometric polynomial $\sum_{n=-N}^{N} a_n e^{int}$ corresponded to the number $\sum_{n=-N}^{N} a_n e^{int_0}$. And since the mapping $R \to R/M$ was continuous, to every function $f(t) \in R$, there corresponded a number $f(t_0)$. Therefore, the maximal ideal M consisted of all functions $f(t)$ such that $f(t_0) = 0$. It followed that if a function $f(t)$ did not vanish at any point, then $f(t)$ was not a member of any maximal ideal of R. Gelfand could then conclude that $f(t)$ had an inverse in R, proving the theorem. Gelfand's proof of the Gelfand–Mazur theorem, that $R' = R/M$ is isomorphic to $\mathbb{C}$, began by supposing that $x \in R'$ and $x \neq \lambda e$ for any complex number λ. Then $(x - \lambda e)^{-1}$ exists for every λ. Moreover,

$$\lim_{h \to 0} \frac{(x - (\lambda + h)e)^{-1} - (x - \lambda e)^{-1}}{h} = -(x - \lambda e)^{-2},$$

and $|\lambda^{-1}| \, ||(e - x/\lambda)^{-1}|| \to 0$ as $\lambda \to \infty$, (35.47)

since, for $|\lambda| > ||x||$, equation (35.46) implies that $||(e - x/\lambda)^{-1}|| \leq 1/(1 - ||x/\lambda||)$. Hence, for any multiplicative linear functional $\phi : R \to \mathbb{C}$, that is, $\phi(xy) = \phi(x)\phi(y)$, the function $\phi((x - \lambda e)^{-1})$ is a bounded entire function and therefore a constant. By (35.47), this constant must be zero. It follows that $(x - \lambda e)^{-1}$ is zero and the theorem is proved by the contradiction:

$$e = (x - \lambda e)^{-1}(x - \lambda e) = 0.$$

F. Riesz, S. Mazur, and others studied normed rings before Gelfand, but Gelfand's concept of the space of maximal ideals unified several isolated earlier results and opened up new avenues for further research. In fact, the space of maximal ideals became important in algebraic geometry also, though in that area Alexander Grothendieck showed that the space of prime ideals produced better results.

In 1930, Gelfand (1913–2009) moved to Moscow from Odessa without completing his secondary education. He had studied mathematics on his own from an early age; his lack of books spurred him to great creativity. At the age of 15, for example, he discovered the Euler–Maclaurin formula. He studied a textbook on differential calculus and Taylor series, but he had no book on the integral calculus. While investigating the problem of the area under $y = x^n$, he was led to consider the sums $1^n + 2^n + \cdots + m^n$. He soon found the Euler–Maclaurin formula and the generating function for the Bernoulli numbers by means of the Taylor series. In a similar way, he discovered Newton's formula for the sums of powers symmetric functions. In Moscow, he worked in odd jobs such as doorkeeper at the Lenin Library, while also teaching mathematics. The great Russian mathematician A. N. Kolmogorov (1903–1987) took an interest in Gelfand, who very soon found himself lecturing at the Moscow State University and studying with Kolmogorov who directed him to problems in functional analysis. This resulted in Gelfand's 1935 thesis, "Abstract Functions and Linear Operators." The theory of commutative normed rings was the subject of his 1938 doctoral thesis.

Gelfand made major contributions to several areas of mathematics such as representation theory, differential equations, computational mathematics, and biocybernetics. At the age of 80, in collaboration with M. Kapranov and A. Zelevinsky, he was starting to develop a theory of hypergeometric functions of many variables. Though he was unable to bring this work to perfection, he had seen the importance of such a theory when he was much younger. For example, in his 1956 lecture "On Some Problems of Functional Analysis," he gave his thoughts on the matter:

> It is known that almost all the special functions of one variable to be met with in mathematical physics may be obtained from the general hypergeometric function of Gauss by a suitable choice of parameters. These same functions appear as elements of representations of the simplest classical groups, namely the groups of rotations of the sphere and of the Lobacevskii plane. This connection lies in the nature of the matter, since the special functions make their appearance by way of considerations connected with this or that invariance of a problem under transformations of a space. Hence it is natural to construct the theory of hypergeometric functions of several variables, relying on results and methods of the theory of the representations of compact or locally compact Lie groups. It is thus necessary so to construct the theory of hypergeometric functions that it should contain the theory of general spherical functions, connected with the representations of semi-simple groups.

35.9 Exercises

1. Prove that if $A_1, A_2, A_3, \ldots, A_n, \ldots$ is a sequence such that the difference $A_{n+1} - A_n$ converges to a limit A as $n \to \infty$, then A_n/n converges to the same limit. Cauchy stated and proved this result in his *Analyse algébrique*; See Cauchy (1989) or Bradley and Sandifer (2009), pp. 35 and 42. Show that this result implies that if a series $\sum a_n$ converges to A, then it converges $(C, 1)$ to A.

2. Prove the theorem of Frobenius that Cesàro summability implies Abel summability. Observe that

$$\sum_{n=0}^{\infty} a_n x^n = (1-x)^2 \sum_{n=0}^{\infty} (A_0 + \cdots + A_n) x^n,$$

   where $A_n = \sum_{k=0}^{n} a_k$. See Frobenius (1880).

3. Prove Borel's theorem that ordinary convergence implies Borel summability, that is, if $a_n \to A$ as $n \to \infty$, then

$$e^{-x} \sum_{n=0}^{\infty} a_n x^n / n! \to A \text{ as } x \to \infty.$$

   See Hardy (1949), p. 80.

4. Show that the condition in the second theorem of Tauber given by (35.9) is implied by the condition $na_n \to 0$ as $n \to \infty$. See Tauber (1897).

5. Prove Cesàro's theorem that if $\sum a_n = A$ and $\sum b_n = B$ and $C_n = c_0 + c_1 + c_2 + \cdots + c_n$, where $c_n = a_0 b_n + a_1 b_{n-1} + \cdots + a_n b_0$, then
$$\frac{C_0 + C_1 + \cdots + C_n}{n+1} \to AB \text{ as } n \to \infty.$$
See Cesàro (1890).

6. Suppose $a(x)$ and $b(x)$ are continuous functions. Prove that if
$$\int_0^\infty a(x)\,dx = A, \quad \int_0^\infty b(x)\,dx = B,$$
then $\lim_{x \to \infty} \frac{1}{x} \int_0^x dt \int_0^t du \int_0^u a(w) b(u-w)\,dw = AB.$

Next, deduce that if $\int_0^\infty dx \int_0^x a(t) b(x-t)\,dt$ is convergent, then its value is AB.

7. If $\int_0^\infty a(x)\,dx = A$, $\int_0^\infty b(x)\,dx = B$, $|xa(x)| < K$, and $|xb(x)| < K$,
then $\int_0^\infty dx \int_0^x a(t) b(x-t)\,dt = AB.$

The theorems in this and the previous exercise are due to Hardy. See Hardy (1966–1979), vol. 6, pp. 210–212.

8. In 1971, at the age of 86 and in memory of his student Harold Davenport (1907–1969), Littlewood gave a short proof of the PNT depending on the following known results:

- The Hardy–Littlewood theorem of which Karamata gave a two-page proof;
- The functional equation of the zeta function;
- The Cahen–Mellin integral for e^{-y} in terms of the gamma function;
- The Dirichlet series for $-\zeta'(s)/\zeta(s)$ when Re $s > 1$;
- The complex zeros ρ of $\zeta(s)$ have a real part between 0 and 1;
- $\zeta'(s)/\zeta(s) = O(\log t) + s \sum_\rho 1/(\rho(s-\rho))$;
- For $s = -1 + it$, $\zeta'(s)/\zeta(s) = O(t^A)$, where A is a positive absolute constant; Littlewood remarked that A would not necessarily have the same value from one occurrence to the next;
- If $N(T)$ denotes the number of zeros of $\rho = \beta + i\gamma$ with $0 \le \gamma \le T$, then $N(T) = O(T^A)$.

Note that the last result is extremely weak and far more is and was known about $N(T)$, but Littlewood did not require a stronger result. First prove Littlewood's first lemma: Given a large positive T_0, there is a T, with $AT_0 < T < AT_0$, such that $\zeta'(s)/\zeta(s) = O(T^A)$ for $s = \sigma + iT$, $-1 \le \sigma \le 2$. The corresponding result for $s = \sigma - iT$ follows by symmetry. Next, demonstrate Littlewood's second lemma that for
$$y > 0, \quad -2\pi i \sum \Lambda(n) e^{-ny} = \int_{2-i\infty}^{2+i\infty} \Gamma(s) \frac{\zeta'(s)}{\zeta(s)} y^{-s}\,ds.$$

From this, Littlewood deduced the PNT in one page. Show how this can be done. See Littlewood (1982), vol. 2, pp. 951–955.

9. Show that $\zeta(s)$ has no zeros when the real part of s is one. The original proofs of Hadamard and de la Valée Poussin were slightly more complicated than later proofs. See Titchmarsh and Heath–Brown (1986), p. 48.

35.10   Notes on the Literature

See Ore (1974), p. 97 for the quote in Latin from Abel. Consult Fejér (1970), vol. 1 for his work on Cesàro summability. Arnold (2007), pp. 115–116, contains his remarks on Littlewood. Some readers may be interested in Arnold's insightful comments on a number of noted mathematicians, including his assessment of S. Kovalevskaya, that her originality has been much underestimated by the mathematical community. In this connection, also see Cooke (1984). See Littlewood (1982) for his papers on Tauberian theory and Hardy (1966–1979), vol. 6, for their joint work on this subject. Hardy (1937), mentioned in this chapter in connection with the missionary remark, is an excellent calculus text and it has been in print for over one hundred years. Wiener (1976–1985), vol. 2, contains his paper on Tauberian theory and its applications. Wiener (1958) consists of a course of lectures given at Cambridge University on Fourier transforms and their applications; it includes his proof of Littlewood's theorem and of the PNT, presented in our text. Hardy gave his remarks on Landau's book in Landau's obituary notice; see Hardy (1966–1979), vol. 7.

An interesting account of Wiener's contributions to mathematics can be found in the AMS *Bulletin*, vol. 72, 1, part II. See especially the article by Levinson explaining Wiener's progression from harmonic analysis to Tauberian theory. Masani (1990) is a comprehensive biography of Wiener, discussing his mathematical work, its myriad applications, the development of his thought, and good references. Karamata (1930) gives the proof presented in the text. In his original paper submitted for publication, Karamata introduced the concept of majorizability to obtain a new condition for the convergence of Abel summable series. This idea sheds light on Karamata's proof, but that portion of the paper was removed by E. Landau before he communicated it for publication. See Nikolić (2009). For a helpful history of summable series, from their origins through the 1920s, including a good bibliography, see Tucciarone (1973). Korevaar (2004) is an encyclopedic treatment of Tauberian theory, covering a century of developments, with numerous historical comments and references. See Gelfand (1987), vol. 1 for his early papers on normed rings; this volume also contains his lecture on functional analysis with the quote given in the text.

36

Elliptic Functions: Eighteenth Century

36.1 Preliminary Remarks

In 1847, Jacobi wrote to Fuss that Euler had been motivated to found elliptic function theory by reading Count Fagnano's *Produzioni Matematiche*. Indeed, in the work of the then unknown Fagnano, Euler discovered the key to the apparently intractable elliptic integral. Giulio Carlo Fagnano (1682–1766) studied theology and philosophy in Rome but avoided mathematics, though he was encouraged to study it. Many years later, after reading Malebranche's *Concerning the Search for Truth*, he taught himself mathematics with great devotion, and from 1714 to 1720, he published some interesting papers on integrals in little-known Italian journals. In 1718 he published his now-famous paper on dividing the lemniscate into several equal parts. His results were not noted at first, but were brought to light by an interesting chain of events. In the early 1740s, Fagnano was consulted concerning the possible instability of the dome of St. Peter's. In 1750, in compensation for his help, Fagnano's collected papers were published at the order of Pope Benedict XIV. Fagnano then applied for membership in the Berlin Academy. Euler was assigned the task of evaluating the quality of the mathematical portion of Fagnano's papers. Euler was intrigued by Fagnano's results on the lemniscatic integral $\int \frac{dx}{\sqrt{1-x^4}}$; these results inspired some of Euler's most brilliant work on integral calculus, laying the foundation for the theory of elliptic integrals and functions. It goes without saying that Fagnano was admitted to the Academy.

The early work of Jakob and Johann Bernoulli on the lemniscatic integral $\int \frac{dx}{\sqrt{1-x^4}}$ led Fagnano to investigate the topic. The equation of the lemniscate is given in cartesian coordinates by

$$(x^2 + y^2)^2 = a^2(x^2 - y^2) \qquad (36.1)$$

and in polar coordinates by

$$r^2 = a^2 \cos 2\theta, \qquad (36.2)$$

where a is a constant. The graph of the lemniscate resembles the symbol for infinity, the diameter of one side given by $r = a$ when $\theta = 0$. For convenience, take $a = 1$. The

36.1 Preliminary Remarks

cartesian coordinates x and y are then given by

$$2x^2 = r^2 + r^4, \qquad (36.3)$$
$$2y^2 = r^2 - r^4, \qquad (36.4)$$

where $0 \leq r \leq 1$. A simple calculation shows that if s denotes the arclength of the lemniscate, then

$$ds = \frac{dr}{\sqrt{1-r^4}},$$

or

$$s(r) = \int_0^r \frac{dt}{\sqrt{1-t^4}}. \qquad (36.5)$$

The lemniscate appeared in Jakob Bernoulli's solution to his own 1691 problem on the shape of an elastic band constrained by its own weight. Johann Bernoulli later encountered the same curve when he asked how to find a curve such that the time taken to traverse it was proportional to the distance from a fixed point. Jakob Bernoulli opined that the lemniscatic integral could not be evaluated in terms of the inverse trigonometric, logarithmic, or rational functions. However, he offered the series expansion

$$\int_0^1 \frac{dt}{\sqrt{1-t^4}} = \sum_{n=0}^{\infty} \frac{1 \cdot 3 \cdot 5 \cdots (2n-1)}{n! 2^n (4n+1)}, \qquad (36.6)$$

obtained by expanding the denominator by the binomial theorem and integrating term by term. The Bernoullis also investigated the problem of bisecting the arcs of curves such as the parabolic spiral.

Fagnano's most famous accomplishment was to bisect an arc of a lemniscate and trisect and quinsect the full arc from $r = 0$ to $r = 1$. His methods were such that these procedures could actually be accomplished using a straight edge and compass. His proofs were based on obtaining appropriate changes of variables and on transforming lemniscatic integrals into other lemniscatic integrals. For example, he found that if

$$t = \frac{1}{r}\sqrt{1 \pm \sqrt{1-r^4}}, \quad \text{then} \qquad (36.7)$$

$$\frac{dr}{\sqrt{1-r^4}} = \frac{\sqrt{2}\,dt}{\sqrt{1+t^4}}; \qquad (36.8)$$

and if

$$\frac{u\sqrt{2}}{\sqrt{1-u^4}} = \frac{1}{r}\sqrt{1 - \sqrt{1-r^4}}, \quad \text{then} \qquad (36.9)$$

$$\frac{dt}{\sqrt{1-r^4}} = \frac{2\,du}{\sqrt{1-u^4}}. \qquad (36.10)$$

In order to better understand (36.7), write it as

$$r^2 = \frac{2t^2}{1+t^4}. \tag{36.11}$$

Fagnano may have made this substitution on the basis of a similar transformation

$$r^2 = \frac{2t}{1+t^2}, \tag{36.12}$$

used to rationalize the integrand for arcsine: $\arcsin x = \int_0^x \frac{dr}{\sqrt{1-r^2}}$. Note that from the substitution given by (36.12), we have

$$\frac{dr}{\sqrt{1-r^2}} = \frac{2\,dt}{1+t^2}. \tag{36.13}$$

It was thus natural for Fagnano to consider (36.11), even though it did not rationalize the integrand, so that he instead obtained (36.8). To understand (36.9), compare it with (36.7) to get

$$t^2 = \frac{2u^2}{1-u^4}. \tag{36.14}$$

Making this substitution is only reasonable, since it produces

$$\frac{dt}{\sqrt{1+t^4}} = \frac{\sqrt{2}\,du}{\sqrt{1-u^4}}. \tag{36.15}$$

This means that if substitutions (36.11) and (36.14) are applied successively, the result is (36.10). Moreover, the relationship between r and u can be expressed by

$$r^2 = \frac{4u^2(1-u^4)}{(1+u^4)^2}. \tag{36.16}$$

Thus, any arc in the first quadrant of a lemniscate, with an endpoint at the origin, can be bisected using straight edge and compass. See this by observing that the arclength of the lemniscate is given by (36.5) and that the arclength corresponding to the radius vector r is double the arclength given by u, where r and u are related by (36.16). One may check that r and u can be obtained from one another by solving only quadratic equations. Also recall that one may use (36.3) and (36.4) to obtain the coordinates of the points from the radius vector. This shows that we have geometric constructibility.

Fagnano's use of $\sqrt{2}\,dt/\sqrt{1+t^4}$ in (36.8) may seem peculiar, since this expression does not appear to take the form of an arclength of a lemniscate. Watson and Siegel have both explained this very nicely in terms of later ideas due to Gauss and Abel. Set $t = e^{i\pi/4}v$ in (36.11) so that

$$r^2 = \frac{2iv^2}{1-v^4}$$

and

$$\frac{dr}{\sqrt{1-r^4}} = \frac{(1+i)\,dv}{\sqrt{1-v^4}}. \tag{36.17}$$

Moreover, by (36.14)

$$v^2 = \frac{-2iu^2}{1-u^4} \quad \text{and} \tag{36.18}$$

$$\frac{dv}{\sqrt{1-v^4}} = \frac{(1-i)\,du}{\sqrt{1-u^4}}. \tag{36.19}$$

Now note that these transformations produce points on the lemniscate, but they are imaginary points. Thus, Siegel points out that (36.17) and (36.19) are examples of "complex multiplication" of the lemniscatic integral and, when applied successively, produce the bisection. Indeed, Fagnano was familiar with the use of complex numbers in integrals; he discovered that

$$\int \frac{dt}{1+t^2} = \log\left(\frac{1+it}{1-it}\right)^{1/2i},$$

a result also published by Johann Bernoulli in 1702. Fagnano noted the amusing particular case

$$\pi = 2i \log\left(\frac{1-i}{1+i}\right).$$

Upon reading Fagnano, Euler perceived that the doubling of the arclength of the lemniscate corresponded to the double angle formula for the sine function. This in turn was a particular case of the addition formula for the sine function. Thus, he gradually understood that Fagnano's transformation formulas might be particular cases of an addition formula for elliptic integrals. Euler's earlier efforts to evaluate these integrals in terms of elementary functions having reached a dead end, he sensed in the work of Fagnano an innovative and productive direction for the theory of elliptic integrals.

Consider Euler's state of mind when he began reading Fagnano. As a student of Johann Bernoulli, he knew that the integral $\int dt/\sqrt{1-t^4}$ could probably not be evaluated in terms of logarithms or inverse trigonometric functions. Then in 1738, he reproved Fermat's theorem that the equation $z^2 = x^4 - y^4$ had no nontrivial integer solutions. Note that this was one result in number theory for which Fermat wrote down a proof! Now the substitution (36.12), rationalizing $dt/\sqrt{1-t^2}$, also provided the rational solutions of $z^2 = 1 - x^2$ or the integer solutions of $z^2 = y^2 - x^2$. Euler realized that $dt/\sqrt{1-t^4}$ could not be rationalized by substitution, since that would imply that Fermat's equation could have integer solutions, a contradiction.

Here note that Euler was well aware of the connection between Diophantine equations of the form $y^2 = ax^2 + bx + c$ and the integration of expressions of the form $\sqrt{ax^2 + bx + c}$. In a 1723 letter to Goldbach, Daniel Bernoulli made specific mention of this connection and so did Johann Bernoulli in his integral calculus lectures, published in the 1740s, long after he delivered them. See chapter 13, section 8 in this connection. Thus, it is safe to assume that Euler was aware that the elliptic integral could not be evaluated in terms of elementary functions. He was searching for a new path, and found it in Fagnano, upon whom he heaped praise. Within a few weeks of receiving Fagnano's *Produzioni Matematiche*, Euler gave a favorable report to the Berlin Academy, including some of his own reflections. He soon wrote a paper reworking and generalizing Fagnano's results and then went on to publish several more papers, which now fill two volumes of his *Opera Omnia*.

Euler's papers and letters to Goldbach indicate that he saw a close connection between $\int dt/\sqrt{1-t^2}$ and $\int dt/\sqrt{1-t^4}$. In fact, Euler's May 30, 1752 letter to Goldbach mentioned that

$$\frac{dx}{\sqrt{1-xx}} = \frac{dy}{\sqrt{1-yy}} \tag{36.20}$$

had the complete integral

$$yy + xx = cc + 2xy\sqrt{(1-cc)}, \text{ while} \tag{36.21}$$

$$\frac{dx}{\sqrt{1-x^4}} = \frac{dy}{\sqrt{1-y^4}} \tag{36.22}$$

had the complete integral

$$yy + xx = cc + 2xy\sqrt{(1-c^4)} - ccxxyy. \tag{36.23}$$

Now from (36.16) we see that

$$\int_0^r \frac{dt}{\sqrt{1-t^4}} = 2\int_0^u \frac{dt}{\sqrt{1-t^4}} \tag{36.24}$$

when

$$r = \frac{2u\sqrt{1-u^4}}{1+u^4}. \tag{36.25}$$

The corresponding result for the arcsine function is

$$\int_0^r \frac{dt}{\sqrt{1-t^2}} = 2\int_0^u \frac{dt}{\sqrt{1-t^2}} \tag{36.26}$$

when

$$r = 2u\sqrt{1-u^2}. \tag{36.27}$$

These two relations are equivalent to the double angle formula for $\sin x$, that is, $\sin 2x = 2\sin x \cos x$. And this is in turn a particular case of the addition formula

$$\sin(x+y) = \sin x \cos y + \cos x \sin y.$$

Next write this in terms of integrals as

$$\int_0^u \frac{dt}{\sqrt{1-t^2}} + \int_0^v \frac{dt}{\sqrt{1-t^2}} = \int_0^z \frac{dt}{\sqrt{1-t^2}} \quad \text{for } z = u\sqrt{1-v^2} + v\sqrt{1-u^2}. \quad (36.28)$$

Recall that Euler thought that he could view Fagnano's bisection of the lemniscatic arc as a particular case of a possible addition formula for the lemniscatic function. In 1752, Euler found the required addition formula:

$$s(u) + s(v) = s(r), \quad r = \frac{u\sqrt{1-v^4} + v\sqrt{1-u^4}}{1+u^2v^2}, \quad (36.29)$$

where $s(u)$ was the lemniscatic integral defined by (36.5). To understand the method by which Euler obtained (36.29), note first that upon integrating (36.20), one obtains

$$\arcsin x = \arcsin y \pm \arcsin c = \arcsin(y\sqrt{1-c^2} \pm c\sqrt{1-y^2}).$$

The last equation implies that the complete integral of (36.20) is

$$x = y\sqrt{1-c^2} \pm c\sqrt{1-y^2}.$$

This is actually the addition formula for the sine function, also given by (36.28), and it is equivalent to the complete integral (36.21) from Euler's letter. In a similar manner, Euler derived the addition formula for the lemniscatic function (36.29) by solving equation (36.23) for x or for y. None of his papers on this topic give an account of how he found the complete integral; they merely verify that the differential $dx/\sqrt{1-x^4}$ remained invariant under the transformation obtained by solving (36.23) for y in terms of x.

Euler also extended (36.29) to the more general quartic $P(x) = 1 + mx^2 + nx^4$. He proved that the complete integral of

$$\frac{dx}{\sqrt{P(x)}} = \frac{dy}{\sqrt{P(y)}}$$

turned out to be the equation

$$-nc^2x^2y^2 + x^2 + y^2 = c^2 + 2xy\sqrt{1+mc^2+nc^4},$$

where c was an arbitrary constant. Upon solving for y, one obtained

$$y = \frac{x\sqrt{P(c)} \pm c\sqrt{P(x)}}{1-nc^2x^2}.$$

Euler also obtained the addition formula for the case in which $P(x)$ was the general quartic $A + 2Bx + Cx^2 + 2Dx^3 + Ex^4$. By means of a fractional linear transformation, he reduced the general quartic to the particular case $1 + mx^2 + nx^4$. The slight drawback in Euler's technique was that it introduced complex coefficients, whereas he intended to use only real coefficients. This lacuna was filled by Legendre in a paper of 1793. Euler proved these results on the addition formula during the 1750s; during the next twenty years, he went on to prove similar results for elliptic integrals of the second and third kinds, to use terminology introduced by Legendre.

This body of Euler's results brought the theory of elliptic integrals to prominence, not only in the context of the integral calculus and Diophantine equations, but also in areas of applied mathematics such as elasticity and dynamics, where numerical evaluations were paramount. Since elliptic integrals could not be evaluated in terms of elementary functions, numerical methods were sought. The early work of Jakob Bernoulli showed this to be a tough problem. His 1694 paper discussed the elastic curve defined by

$$f(x) = \int_0^x \frac{t^2 \, dt}{\sqrt{1-t^4}} \tag{36.30}$$

with arclength $s(x)$ given by (36.5). Bernoulli determined the intervals within which the values of $f(1)$ and $s(1)$ would have to fall. In a 1704 paper, Bernoulli was able use series methods to specify these values within shorter intervals: $1.3088173 < s(1) < 1.3152635$ and $0.5983546 < f(1) < 0.6004034$. But Bernoulli's hypergeometric series (36.6) did not converge rapidly enough, so that these values were not too accurate. Then, in his *Methodus Differentialis*, James Stirling produced a vastly better evaluation, correct to fifteen decimal places.

Recall that Stirling's book contained several methods for transforming hypergeometric and other series into more rapidly convergent series. In proposition 11 of his book, he applied a specific method to Bernoulli's hypergeometric series, obtaining:

$$\int_0^1 \frac{dx}{\sqrt{1-x^4}} = 1.31102877714605987 \quad \text{and} \tag{36.31}$$

$$\int_0^1 \frac{x^2 \, dx}{\sqrt{1-x^4}} = 0.59907011736779611. \tag{36.32}$$

Incidentally, it was a source of pride to Euler when he found, through his 1737 work on the elastic curve, that the product of these two integrals was exactly $\pi/4$. To get this result today, one would evaluate the beta integrals in terms of the gamma function. And it is interesting that soon after 1737, Euler found this method of evaluating in terms of the gamma function. Euler also used the series method to obtain numerical approximations of some elliptic integrals. Again, these results were not accurate to many decimal places because the series did not converge rapidly enough. Using the transformation of integrals, Lagrange, Legendre, and Gauss found better methods, also of great theoretical significance. But these methods owed a debt to the work of John Landen.

In 1771, John Landen presented a fundamental transformation of elliptic integrals; he elaborated on this result in another paper published four years later. He stated his problem in geometric terms, expressing the length of the arc of any hyperbola in terms of two elliptic arcs. The Landen transformation can be stated as the theorem: If $\sin(2\phi - \theta) = k \sin\theta$, then

$$(1+k) \int_0^\theta (1 - k^2 \sin^2 u)^{-1/2} du = 2 \int_0^\phi \left(1 - \frac{4k}{(1+k)^2} \sin^2 u\right)^{-1/2} du. \tag{36.33}$$

36.1 Preliminary Remarks

We note the important particular case

$$(1+k)\int_0^\pi (1-k^2\sin^2\theta)^{-1/2}\,d\theta = 2\int_0^{\pi/2}\left(1-\frac{4k}{(1+k)^2}\sin^2\theta\right)^{-1/2}\,d\theta. \qquad (36.34)$$

In Legendre's notation

$$K(k) = \int_0^{\pi/2}(1-k^2\sin^2\theta)^{-1/2}\,d\theta \qquad (36.35)$$

so that (36.34) can be expressed as

$$K\left(\frac{2\sqrt{k}}{1+k}\right) = (1+k)K(k). \qquad (36.36)$$

This last result is also referred to as Landen's (quadratic) transformation. This and Euler's addition formula were the two pillars on which the early theory of elliptic integrals and functions was constructed. Mittag-Leffler wrote, "Landen does not seem, however, to have fully understood the value of his discovery."

But Lagrange quickly grasped the applicability of Landen's transformation to the numerical approximation of elliptic integrals, and in this connection, he discovered the concept of the arithmetic-geometric mean of two numbers. In 1784–1785, Lagrange presented these ideas in the Turin Academy journal under the title "Sur une nouvelle méthode de calcul intégral." Lagrange was then a member of the Berlin Academy, but was born in Turin and was a founder of its academy. Consequently, he was quite interested in the growth of the Turin Academy and published several papers in its journal. In his paper, Lagrange expressed Landen's transformation in the elegant form: If $p > q > 0$,

$$p' = p + \sqrt{p^2 - q^2}, \quad q' = p - \sqrt{p^2 - q^2} \qquad (36.37)$$

and

$$R(p, q, y) = \sqrt{(1 \pm p^2 y^2)(1 \pm q^2 y^2)}, \quad y' = yR/(1 \pm q^2 y^2), \qquad (36.38)$$

then

$$\frac{dy}{R} = \frac{dy'}{R'}, \qquad (36.39)$$

where $R' = R(p', q', y')$. He also observed that $p = (p'+q')/2$, the arithmetic mean of p' and q', and that $q = \sqrt{p'q'}$, the geometric mean of p' and q'. He used these relations to define two sequences. Let $p_0 = p$, $q_0 = q$ with $p > q$ and for any positive or negative integer n set

$$p_n = p_{n-1} + \sqrt{p_{n-1}^2 - q_{n-1}^2}, \quad q_n = p_{n-1} - \sqrt{p_{n-1}^2 - q_{n-1}^2}, \qquad (36.40)$$

or

$$p_n = \frac{p_{n+1} + q_{n+1}}{2}, \quad q_n = \sqrt{p_{n+1}q_{n+1}}. \qquad (36.41)$$

These relations define the bilateral sequence

$$\ldots p_{-1}, q_{-1}, p_0, q_0, p_1, q_1, \ldots.$$

In the positive direction, for increasing n, the q_n terms tend to zero and p_n terms tend to infinity. So for the purpose of approximate evaluation of the integral, whose form did not change by (36.39), Lagrange took $q_n = 0$ for large enough n. This reduced the elliptic integral to $\int dx/\sqrt{1 \pm p_n^2 x^2}$, an exactly computable integral. In the negative direction, Lagrange observed that the arithmetic means p_{-n} and the geometric means q_{-n} had the same limit, because $p_{-n} - q_{-n} \to 0$ as $n \to \infty$. Thus, for sufficiently large n, the elliptic integral could be approximately evaluated from the exactly computable integral $\int dx/(1 \pm p_{-n}^2 x^2)$.

Apparently Lagrange did not enjoy numerical calculation as much as Newton, Stirling, and Euler did, so he did not actually apply his method to find approximate values for elliptic integrals. This was left to Legendre, who effectively used iterated forms of (36.33) and (36.34) to construct numerical tables of such integrals.

We mention in passing that Euler used results obtainable from the addition formula to study Diophantine equations of the form $y^2 = p(x)$ where $p(x)$ was a polynomial of degree four with integer coefficients. Since Euler did not note the connection with elliptic integrals, we are not certain that he was aware of it. In 1834, Jacobi reviewed some of these papers of Euler and on that basis, he concluded that Euler knew of this relationship. In a similar way, Euler as well as Lagrange employed quadratic (or second-order) transformations (isogenies) to study the special Diophantine equations $z^2 = x^4 \pm y^4$ and $z^2 = 2x^4 - y^4$. This may be one reason that Lagrange did not refer to Landen in his 1784–85 paper on elliptic integrals. Another reason, of course, is that mathematicians in the eighteenth century were not in the habit of giving an exhaustive list of references. Thus, Mittag-Leffler assumed that Landen and Lagrange had independently discovered the Landen transform. In fact, on January 3, 1777, Lagrange wrote to his friend Condorcet that he had seen Landen's 1775 paper containing the theorem reducing the problem of the rectification of arcs of ellipses to a problem of hyperbolic arcs. Lagrange wrote that he found this a singular result and that he had not yet verified it. Apparently, he found the time to study Landen and went on to discover the arithmetic-geometric mean and its use in numerical evaluation of integrals. It is remarkable that he did not do more with it. Perhaps he was already beginning to lose interest in mathematical research. After 1785, he produced no further original mathematical results, though he did publish important and influential books, including *Mécanique analytique* of 1788 and *Fonctions analytiques* of 1797.

36.2 Fagnano Divides the Lemniscate

In 1691, Jakob Bernoulli observed that the arc length of the parabolic spiral $(a-r)^2 = 2ab\theta$ was given by

$$s = \int \sqrt{1 + \frac{r^2(a-r)^2}{a^2 b^2}}\, dr,$$

36.2 Fagnano Divides the Lemniscate

and that the integrand was an even function of $\frac{1}{2}a - r$. Therefore, he had

$$\int_{\frac{1}{2}a-c}^{\frac{1}{2}a} = \int_{\frac{1}{2}a}^{\frac{1}{2}a+c}.$$

He could then conclude that the length of the arc of the spiral joining the points corresponding to $r = \frac{1}{2}a - c$ and $r = \frac{1}{2}a$ equaled the length of the arc from $r = \frac{1}{2}a$ to $r = \frac{1}{2}a + c$, the two arcs were incongruent.

Fagnano extended this and other of Bernoulli's results to arcs of other curves, including the lemniscate, given by

$$(x^2 + y^2)^2 = x^2 - y^2 \quad \text{or} \quad r^2 = \cos 2\theta.$$

In 1718, Fagnano published his two-part work on the division of the lemniscate. In the first part he stated that if

$$u = \frac{\sqrt{1-z^2}}{\sqrt{1+z^2}}, \quad \text{then} \qquad (36.42)$$

$$\int \frac{dz}{\sqrt{1-z^4}} = \int -\frac{du}{\sqrt{1-u^4}}. \qquad (36.43)$$

Fagnano's statement of this result took an apparently more general form, but we can obtain it from this one by replacing u and z by u/a and z/a, where a is a constant. He observed that the result could be proved by differentiating (36.42) and substituting in (36.43). As an immediate consequence of this theorem, it was clear that

$$u^2z^2 + u^2 + z^2 - 1 = 0 \qquad (36.44)$$

was an integral of the equation

$$\frac{dz}{\sqrt{1-z^4}} = \pm\frac{du}{\sqrt{1-u^4}}. \qquad (36.45)$$

As mentioned earlier, Euler noticed this fact very quickly. We can write the theorem in modern form as

$$\int_0^z \frac{dt}{\sqrt{1-t^4}} = \int_u^1 \frac{dt}{\sqrt{1-t^4}}, \qquad (36.46)$$

when u and z are related by (36.42) or (36.44). This means that in Figure 36.1, if O, P, Q, and A denote points on the lemniscate corresponding to the values $0, z, u$ and 1 of the radius vector, then arc OP = arc QA in length.

In the last section of part one of his paper, Fagnano observed that the full lemniscatic arc OA would be bisected if the points P and Q coincided. This would happen when $z = u$ and then (36.44) would imply

$$z^4 + 2z^2 - 1 = 0, \quad \text{or} \quad z = u = \sqrt{\sqrt{2}-1}.$$

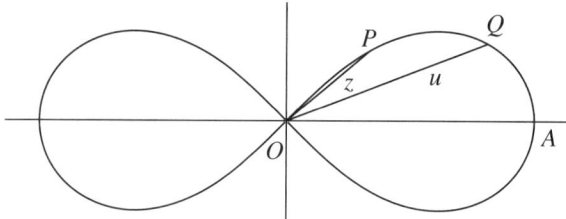

Figure 36.1. Fagnano's lemniscate.

Thus, this constructible number would express the distance between the point of bisection and the origin.

Fagnano started the second part of the paper on the division of the lemniscate with the theorem: If

$$x = \frac{\sqrt{1 \mp \sqrt{1-z^4}}}{z}, \text{ then} \tag{36.47}$$

$$\frac{\pm dz}{\sqrt{1-z^4}} = \frac{dx\sqrt{2}}{\sqrt{1+x^4}}. \tag{36.48}$$

His proof consisted of the observations that

$$dx = \frac{\pm dz\sqrt{1 \mp \sqrt{1-z^4}}}{z^2\sqrt{1-z^4}},$$

$$\frac{\sqrt{1+x^4}}{\sqrt{2}} = \frac{\sqrt{1 \mp \sqrt{1-z^4}}}{z^2}.$$

Note that in (36.48) the differential on the right-hand side is apparently not a lemniscatic differential. So Fagnano stated another theorem: If

$$x = \frac{u\sqrt{2}}{\sqrt{1-u^4}}, \text{ then} \tag{36.49}$$

$$\frac{du}{\sqrt{1-u^4}} = \frac{1}{\sqrt{2}} \times \frac{dx}{\sqrt{1+x^4}}. \tag{36.50}$$

Once again, his proof simply noted that differentiating (36.49) resulted in

$$\frac{dx}{\sqrt{2}} = \frac{du}{\sqrt{1-u^4}} \times \frac{1+u^4}{1-u^4},$$

and that (36.49) also implied

$$\sqrt{1+x^4} = \frac{1+u^4}{1-u^4}.$$

By combining these theorems, he obtained the result on the duplication of the lemniscatic arc starting at the origin: If

$$\frac{u\sqrt{2}}{\sqrt{1-u^4}} = \frac{1}{z}\sqrt{1-\sqrt{1-z^4}}, \quad \text{then} \tag{36.51}$$

$$\frac{dz}{\sqrt{1-z^4}} = \frac{2\,du}{\sqrt{1-u^4}}. \tag{36.52}$$

Note that if P corresponds to z and Q to u, and if z and u are related by (36.51), then (36.52) shows that arc $OP = 2$ arc OQ. This means that if the value of z is given, then u can be obtained by taking square roots and conversely. Hence, duplication and bisection can be done by straight edge and compass. Fagnano made the observation that (36.51) was equivalent to the relation

$$z = \frac{2u\sqrt{1-u^4}}{1+u^4}. \tag{36.53}$$

Recall that Euler saw this result as the extension of the double angle formula for arcsine, and it was perhaps this result that led him to the addition formula for the lemniscatic integral.

Fagnano trisected the full arc OA of the lemniscate by combining (36.42) and (36.43) with (36.51) and (36.52). To obtain the trisection in a simpler form, he presented another transformation: If

$$\frac{\sqrt{1-t^4}}{t\sqrt{2}} = \frac{1}{z}\sqrt{1-\sqrt{1-z^4}}, \quad \text{then} \tag{36.54}$$

$$\frac{dz}{\sqrt{1-z^4}} = -\frac{2\,dt}{\sqrt{1-t^4}}. \tag{36.55}$$

He then noted that he could obtain a point of trisection by setting $t = z$ and that the trisection point would be given by $t = \sqrt[4]{2\sqrt{3}-3}$. One may check that for $t = z$, (36.54) simplifies to the equation $t^8 + 6t^4 - 3 = 0$, and that $2\sqrt{3} - 3$ is a solution of $x^2 + 6x - 3 = 0$.

Fagnano went on to work out how the arc OA could be divided into five equal parts. He did not write down the details, but, based on his trisection method, the method would probably begin by taking points on the lemniscate for which the distances from the origin are $t, z, v,$ and u such that arc $Ot = 2$ arc Oz, arc $Oz = 2$ arc Ov, and arc $Ov = $ arc uA. Fagnano's formulas give the relations connecting t with z, z with v, and v with u. Finally, if we take $t = u$, then we get arc $Ot = 4/5$ arc uA and the equation for t reduces to

$$t^{24} + 50t^{20} - 125t^{16} + 300t^{12} - 105t^8 - 62t^4 + 5 = 0.$$

Although Fagnano did not publish this equation, it is likely that he obtained and solved it. Gauss derived the equation, and it is explicitly given in his collected works. The twenty-fourth degree polynomial has factors

$$t^8 - 2t^4 + 5 \quad \text{and} \quad t^8 + (26 \pm 12\sqrt{5})t^4 + 9 \pm 4\sqrt{5},$$

where we choose either both plus signs or both minus signs. Note that the first polynomial has only complex roots, but the real roots of the other two polynomials can be expressed in terms of square roots and are therefore constructible. The quinsection may be obtained by solving the polynomial with both negative signs. Fagnano stated the corollary that the quadrant of the leminscate could be divided algebraically into a number of equal parts if that number were of the form $2 \times 2^m, 3 \times 2^m, 5 \times 2^m$ for any positive integer m. He wrote that this was a "new and singular property" of his curve.

36.3   Euler: Addition Formula

Although Euler quickly perceived the importance of Fagnano's work on the lemniscatic integral, he could not at first locate any fundamental guiding principle among the large number of apparently ad hoc transformations applied somewhat randomly. It took him a little while to discover the required unifying ideas: the addition formula and the complete integral for the equation

$$\frac{m\,dx}{\sqrt{1-x^4}} = \frac{n\,dy}{\sqrt{1-y^4}}. \tag{36.56}$$

It appears that Euler spotted a hint: Fagnano's result that

$$(x^2+1)(y^2+1) = 2 \quad \text{or} \quad x^2 y^2 + x^2 + y^2 - 1 = 0$$

actually gave a special integral of the preceding differential equation, when $m = n = 1$. In his first paper on this topic, presented to the Academy in 1752, Euler gave some preliminary results on this hint. But soon after this, as his letter to Goldbach indicated, he discovered the general algebraic integral and published it in his 1756–57 paper. In the first theorem of this paper, Euler took $m = n = 1$ and stated that the differential equation

$$\frac{dx}{\sqrt{1-x^4}} = \frac{dy}{\sqrt{1-y^4}} \tag{36.57}$$

had the complete integral

$$xx + yy + ccxxyy = cc + 2xy\sqrt{(1-c^4)}. \tag{36.58}$$

Here note that by taking $c = 1$, one obtains Fagnano's result (36.44).

Euler argued that taking the differential of (36.58) gave

$$x\,dx + y\,dy + ccxy(x\,dy + y\,dx) = (x\,dy + y\,dx)\sqrt{(1-c^4)},$$

and hence

$$dx(x + ccxyy - y\sqrt{(1-c^4)}) + dy(y + ccxxy - x\sqrt{(1-c^4)}) = 0. \tag{36.59}$$

He solved (36.58) as a quadratic in y (and then in x), choosing the signs of the square roots so that $y = c$ when $x = 0$ to get

$$y = \frac{x\sqrt{(1-c^4)} + c\sqrt{(1-x^4)}}{1 + ccxx} \quad \text{and} \quad x = \frac{y\sqrt{(1-c^4)} - c\sqrt{(1-y^4)}}{1 + ccyy}.$$

These equations implied that

$$x + ccxyy = y\sqrt{(1-c^4)} - c\sqrt{(1-y^4)},$$

$$y + ccxxy - x\sqrt{(1-c^4)} = c\sqrt{(1-x^4)}.$$

Euler substituted these relations in (36.59) to obtain

$$-c\,dx\sqrt{(1-y^4)} + c\,dy\sqrt{(1-x^4)} = 0.$$

This was equivalent to (36.57), and the theorem was proved. Euler then noted that this theorem was equivalent to the formula

$$\int_0^u \frac{dt}{\sqrt{1-t^4}} + \int_0^c \frac{dt}{\sqrt{1-t^4}} = \int_0^x \frac{dt}{\sqrt{1-t^4}}$$

where

$$x = \frac{u\sqrt{(1-c^4)} + c\sqrt{(1-u^4)}}{1 + c^2 u^2}.$$

This was the famous addition formula for the lemniscatic integral and it generalized Fagnano's duplication formula obtained by taking $u = c$. Thus, Euler saw that the transformation that left the differential $dx/\sqrt{1-x^4}$ invariant also provided the addition formula.

Euler then considered the more general differential $dx/\sqrt{1+mx^2+nx^4}$ and proved in a similar way that it remained invariant under the transformation

$$cc - xx - yy + nccxxyy + 2xy\sqrt{(1 + mcc + nc^4)} = 0.$$

This in turn yielded an appropriate addition formula for this more general elliptic integral.

36.4 Cayley on Landen's Transformation

Landen's exposition of his transformation is not easy to read; in fact, G. N. Watson aptly described it as "clumsy." However, Cayley's 1876 text (reprinted in 1895) on elliptic functions, described in his preface as "founded upon Legendre's *Traité des fonctions elliptiques* and upon Jacobi's *Fundamenta Nova*, and Memoirs by him in *Crelle's Journal*," presents Landen's work in more felicitous notation and in such a manner as to outline its geometric underpinnings and make clear its essential and useful features.

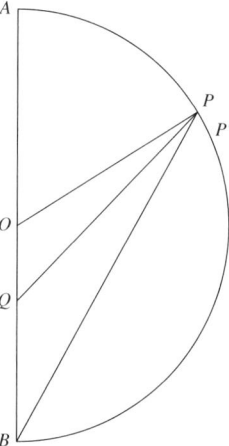

Figure 36.2. Cayley's diagram for Landen's transformation.

Summarizing Cayley, with reference to Figure 36.2, we begin by taking a point P on the circle with center O and another point Q, on the diameter AB. Set $QA = a$, $QB = b$, $A\hat{Q}P = \phi_1$, $A\hat{B}P = \phi$ so that $A\hat{O}P = 2\phi$. Now let $a_1 = (a+b)/2$, $b_1 = \sqrt{ab}$, $c_1 = (a-b)/2$. Then $OA = OB = OP = a_1$, $OQ = a_1 - b = (a-b)/2 = c_1$,

$$QP \sin\phi_1 = a_1 \sin 2\phi,$$
$$QP \cos\phi_1 = c_1 + a_1 \cos 2\phi,$$
$$QP^2 = c_1^2 + 2c_1 a_1 \cos 2\phi + a_1^2$$
$$= \frac{1}{2}(a^2 + b^2)(\cos^2\phi + \sin^2\phi) + \frac{1}{2}(a^2 - b^2)(\cos^2\phi - \sin^2\phi)$$
$$= a^2 \cos^2\phi + b^2 \sin^2\phi.$$

Therefore,

$$\sin\phi_1 = \frac{a_1 \sin 2\phi}{\sqrt{a^2 \cos^2\phi + b^2 \sin^2\phi}} \quad \text{and} \quad \cos\phi_1 = \frac{c_1 + a_1 \cos 2\phi}{\sqrt{a^2 \cos^2\phi + b^2 \sin^2\phi}}, \quad (36.60)$$

$$a_1^2 \cos^2\phi_1 + b_1^2 \sin^2\phi_1 = \frac{a_1^2 (a \cos^2\phi + b \sin^2\phi)^2}{a^2 \cos^2\phi + b^2 \sin^2\phi}.$$

A simple calculation produces

$$\sin(2\phi - \phi_1) = \frac{1}{2} \cdot \frac{(a-b) \sin 2\phi}{\sqrt{a^2 \cos^2\phi + b^2 \sin^2\phi}}, \quad (36.61)$$

$$\cos(2\phi - \phi_1) = \frac{1}{a_1} \cdot \sqrt{a_1^2 \cos^2\phi + b_1^2 \sin^2\phi}.$$

36.4 Cayley on Landen's Transformation

Take a point P' on the circle close enough to P so that we can regard PP' as tangent to the circle. An elementary geometric argument shows that

$$PQ d\phi_1 = PP' \sin P'\hat{P}Q = 2a_1 d\phi \cos(2\phi - \phi_1).$$

This is equivalent to

$$\frac{2 d\phi}{\sqrt{a^2 \cos^2 \phi + b^2 \sin^2 \phi}} = \frac{d\phi_1}{\sqrt{a_1^2 \cos^2 \phi_1 + b_1^2 \sin^2 \phi_1}}. \tag{36.62}$$

Thus, if ϕ and ϕ_1 are related by (36.61) and $0 \leq \phi \leq \pi/2$, then

$$\int_0^\phi \frac{dt}{\sqrt{a^2 \cos^2 t + b^2 \sin^2 t}} = \frac{1}{2} \int_0^{\phi_1} \frac{dt}{\sqrt{a_1^2 \cos^2 t + b_1^2 \sin^2 t}}. \tag{36.63}$$

In particular,

$$\int_0^{\pi/2} \frac{dt}{\sqrt{a^2 \cos^2 t + b^2 \sin^2 t}} = \frac{1}{2} \int_0^{\pi} \frac{dt}{\sqrt{a_1^2 \cos^2 t + b_1^2 \sin^2 t}}$$

$$= \int_0^{\pi/2} \frac{dt}{\sqrt{a_1^2 \cos^2 t + b_1^2 \sin^2 t}}. \tag{36.64}$$

Using the notation of Legendre, we set

$$k^2 = 1 - \frac{b^2}{a^2}, \quad k' = \frac{b}{a} \quad \text{and} \quad k_1^2 = 1 - \frac{b_1^2}{a_1^2}.$$

Then (36.63) can be written as

$$\int_0^\phi \frac{dt}{\sqrt{1 - k^2 \sin^2 t}} = \frac{1}{2} \frac{a}{a_1} \int_0^{\phi_1} \frac{dt}{\sqrt{1 - k_1^2 \sin^2 t}}$$

$$= \frac{1}{2}(1 + k_1) \int_0^{\phi_1} \frac{dt}{\sqrt{1 - k_1^2 \sin^2 t}}. \tag{36.65}$$

This is in fact Landen's transformation (36.33). To see this, compare the first equations in (36.60) and (36.61) to get

$$\sin(2\phi - \phi_1) = k_1 \sin \phi_1. \tag{36.66}$$

Now (36.33) follows from (36.65) by noting that $4k_1/(1 + k_1)^2 = k^2$.

Finally, we observe that if we write $y = \sin \phi_1$ and $x = \sin \phi$, then (36.66) can be expressed as the quadratic transformation

$$y = \frac{(1 + k')x \sqrt{1 - x^2}}{\sqrt{1 - k^2 x^2}}, \tag{36.67}$$

and (36.62) takes the form

$$\frac{(1+k_1)\,dy}{\sqrt{(1-y^2)(1-k_1^2y^2)}} = \frac{2\,dx}{\sqrt{(1-x^2)(1-k^2x^2)}}. \tag{36.68}$$

36.5 Lagrange, Gauss, Ivory on the agM

Lagrange was the first mathematician to observe the connection between the arithmetic geometric mean (agM) and elliptic integrals. His 1784–85 result (36.39) essentially expressed this connection. He made this discovery as he pursued a numerical method for evaluating elliptic integrals, and he did not further investigate the concept of the agM. In his 1818 paper on astronomy, Gauss gave a formula relating the agM of two positive real numbers with an elliptic integral and he worked out an extensive theory on this topic, but published very little of it. Gauss denoted the agM of two positive numbers a and b, with $a \geq b$, as $M(a,b)$. He considered the sequence

$$a_1 = \frac{a+b}{2}, \quad b_1 = \sqrt{ab}, \quad a_2 = \frac{a_1+b_1}{2}, \quad b_2 = \sqrt{a_1 b_1}, \quad \text{etc.}, \tag{36.69}$$

and noted that

$$b \leq b_1 \leq b_2 \leq \cdots \leq b_n \leq \cdots \leq a_n \leq \cdots \leq a_1 \leq a.$$

He observed that if $a = b$, then $a_n = b_n$ for all n. On the other hand, if $a > b$, then

$$\frac{a_n - b_n}{a_{n-1} - b_{n-1}} = \frac{a_{n-1} - b_{n-1}}{4(a_n + b_n)} = \frac{a_{n-1} - b_{n-1}}{2(a_{n-1} + b_{n-1}) + 4b_n}$$

and hence

$$a_n - b_n < \frac{a_{n-1} - b_{n-1}}{2} < \frac{a-b}{2^n}. \tag{36.70}$$

Consequently, the increasing sequence b_n and the decreasing sequence a_n converged to the same number denoted by $M(a,b)$.

Gauss also noted the simple properties of $M(a,b)$ given by the equations

$$M(a,b) = M(a_1,b_1) \quad \text{and} \quad M(na,nb) = nM(a,b) \quad \text{for any real } n > 0. \tag{36.71}$$

From these equations he deduced a number of relations; for instance, for $x = 2t/(1+t^2)$, he had

$$M(1+x, 1-x) = M\left(1, \frac{1-t^2}{1+t^2}\right) = \frac{1}{1+t^2} M(1-t^2, 1+t^2). \tag{36.72}$$

Gauss's theorem, published in 1818 but proved much earlier, stated that

$$\frac{1}{M(a,b)} = \frac{2}{\pi} \int_0^{\pi/2} \frac{d\theta}{\sqrt{a^2\cos^2\theta + b^2\sin^2\theta}}. \tag{36.73}$$

To see how Lagrange's transformation implies this theorem of Gauss, denote the integral in (36.73) by $I(a, b)$ and set $x = (\cot\theta)/b$ to get

$$I(a,b) = \frac{2}{\pi} \int_0^\infty \frac{dx}{\sqrt{(1+a^2x^2)(1+b^2x^2)}}. \tag{36.74}$$

Recall Lagrange's result (36.39), that if

$$x = y\sqrt{\frac{1+a_1^2 y^2}{1+b_1^2 y^2}}, \quad \text{then} \tag{36.75}$$

$$\frac{dx}{\sqrt{(1+a^2x^2)(1+b^2x^2)}} = \frac{dy}{\sqrt{(1+a_1^2 y^2)(1+b_1^2 y^2)}}. \tag{36.76}$$

When this result is applied to (36.74), we see that

$$I(a,b) = I(a_1, b_1). \tag{36.77}$$

Upon iteration, we conclude that if $c = M(a, b)$, then

$$I(a,b) = I(a_n, b_n) = I(c,c) = \frac{2}{\pi} \int_0^\infty \frac{dx}{1+c^2x^2} = \frac{1}{c}.$$

This proves Gauss's theorem. Note that (36.64) is identical to (36.77).

Gauss derived (36.73) by means of a different transformation. He set

$$\sin\theta = \frac{2a\sin\theta'}{(a+b)\cos^2\theta' + 2a\sin^2\theta'} = \frac{2a\sin\theta'}{a+b+(a-b)\sin^2\theta'}$$

and observed that (36.73) would follow. Jacobi provided more details on this transformation in section 38 of his *Fundamenta Nova*, published a decade after Gauss's paper. Since Jacobi was pursuing other threads, his presentation is not as direct as that in Cayley's 1876 treatise. Following Cayley, replace $\sin\theta$ and $\sin\theta'$ by y and x, respectively, to write Gauss's substitution as

$$y = \frac{(1+k)x}{1+kx^2}, \quad k = \frac{a-b}{a+b}. \tag{36.78}$$

This is the form in which Gauss's transformation is often presented, particularly in connection with the transformation theory of elliptic functions. One then perceives that proving the relation $I(a,b) = I(a_1, b_1)$ is equivalent to showing that

$$\frac{dy}{\sqrt{(1-y^2)(1-\lambda^2 y^2)}} = \frac{(1+k)dx}{\sqrt{(1-x^2)(1-k^2x^2)}}, \tag{36.79}$$

where $\lambda = 2\sqrt{k}/(1+k)$.

To prove (36.79), let $D = 1 + kx^2$, the denominator of y in (36.78). Then

$$1 - y = (1-x)(1-kx)/D,$$
$$1 + y = (1+x)(1+kx)/D,$$
$$1 - \lambda x = (1-\sqrt{k}x)/D,$$
$$1 + \lambda x = (1+\sqrt{k}x)/D.$$

Consequently,

$$\sqrt{(1-y^2)(1-\lambda^2 y^2)} = (1-kx^2)\sqrt{(1-x^2)(1-k^2x^2)}D^2$$
$$dy = (1+k)(1-kx^2)\,dx/D^2.$$

The last two formulas imply (36.79), and we have another proof of Gauss's theorem.

Gauss gave yet another proof, by means of power series, though he did not publish it. We reproduce Gauss's proof, but we use subscript and factorial symbols where Gauss did not. In this derivation, he assumed that $M(1+x, 1-x)$ had a series expansion so that he could write

$$\frac{1}{M(1+x, 1-x)} = A_0 + A_1 x^2 + A_2 x^4 + A_3 x^6 + \cdots, \quad A_0 = 1.$$

Using (36.72), he had

$$\frac{2t}{1+t^2} + A_1 \left(\frac{2t}{1+t^2}\right)^3 + A_2 \left(\frac{2t}{1+t^2}\right)^5 + \cdots = 2t(A_0 + A_1 t^4 + A_2 t^8 + \cdots).$$

He equated the coefficients of powers of t to obtain the relations

$$A_0 = 1,$$
$$0 = 1 - 4A_1,$$
$$A_1 = 1 - 12A_1 + 16A_2,$$
$$0 = 1 - 24A_1 + 80A_2 - 64A_3,$$
$$A_2 = 1 - 40A_1 + 240A_2 - 448A_3 + 256A_4, \quad \text{etc.}$$

Unlike earlier mathematicians, he also presented the general nth relation

$$M = 1 - 4A_1 \frac{n(n-1)}{2!} + 16A_2 \frac{(n+1)n(n-1)(n-2)}{4!}$$
$$- 64A_3 \frac{(n+2)(n+1)n(n-1)(n-2)(n-3)}{6!} + \cdots,$$

with the remark that $M = 0$ when n was even and $M = A_{(n-1)/2}$ when n was odd; in other words, M was the $(n+1)/2$th term of the series $A_0, A_1, A_2, \ldots$. Taking $0 = 0$ as the 0th equation, and abbreviating and labeling the equations as $[0], [1], [2], \ldots$, he wrote down the equations

$$1^2[2] - 0^2[0], \ 2^2[3] - 1^2[1], \ 3^2[4] - 2^2[2], \ 4^2[5] - 3^2[3], \ 5^2[6] - 4^2[4], \ldots.$$

In general, for the nth equation he had

$$n^2 N - (n-1)^2 L = (2n-1)\left(1 - 2^2 A_1 \frac{3n^2 - 3n + 2}{2!} + 2^4 A_2 \frac{n(n-1)(5n^2 - 5n + 6)}{4!}\right)$$
$$- (2n-1)\left(2^6 A_3 \frac{(n+1)n(n-1)(n-2)(7n^2 - 7n + 12)}{6!} + \cdots\right). \tag{36.80}$$

Gauss observed in a footnote that L and N were equal to $A_{(n-2)/2}$ and $A_{n/2}$, respectively, when n was even, and zero when n was odd. In another footnote, he wrote that the derivation connected with the forms $n^2 N - (n-1)^2 L$ was explained in article 162 of his *Disquisitiones Arithmeticae*. It may be of interest to note that in that article, Gauss discussed the problem of determining all transformations of the form $X = \alpha' x + \beta' y$, $Y = \gamma' x + \sigma' y$, given one known transformation $X = \alpha x + \beta y$, $Y = \gamma x + \sigma y$ of $AX^2 + 2BXY + CY^2$ to $ax^2 + 2bxy + cy^2$.

Continuing the proof, let $k = 2l - 1$. Gauss wrote the lth term of the expression in (36.80) (without the factor $2n - 1$) as

$$\pm 2^{k-1} A_{(k-1)/2} \frac{\left(n + \frac{k-5}{2}\right)\left(n + \frac{k-7}{2}\right) \cdots \left(n - \frac{k-3}{2}\right)(kn^2 - kn + (k^2 - 1)/4)}{(k-1)!}. \tag{36.81}$$

We note that the sign is plus when $l - 1$ is even and minus when $l - 1$ is odd. Gauss then divided each such term into two parts and added the second part of one term to the first part of the succeeding term. As a first step in this process, he observed that

$$kn^2 - kn + \frac{k^2 - 1}{4} = k\left(n - \frac{k-1}{2}\right)\left(n + \frac{k-3}{2}\right) + \frac{(k-1)^3}{4}.$$

Using this, he could express the term (36.81) as a sum of two terms

$$\pm \left(2^{k-1} k^2 A_{(k-1)/2} \frac{\left(n + \frac{k-3}{2}\right)\left(n + \frac{k-1}{2}\right) \cdots \left(n - \frac{k-1}{2}\right)}{k!}\right)$$
$$+ \left(2^{k-1} A_{(k-1)/2} \frac{n + \frac{k-5}{2} \cdots n - \frac{k-3}{2}}{(k-2)!} \cdot \left(\frac{k-1}{2}\right)^2\right).$$

When he added the second part of this expression to the first part of the succeeding expression, he obtained

$$2^{k-1} \frac{\left(n + \frac{k-3}{2}\right)\left(n + \frac{k-5}{2}\right) \cdots \left(n - \frac{k-1}{2}\right)}{k!} \left((k+1)^2 A_{(k+1)/2} - k^2 A_{(k-1)/2}\right).$$

Thus, he could express (36.80) as

$$n^2 N - (n-1)^2 L = (2n-1)\left((A_0 - 2^2 A_1) - 4\frac{n(n-1)}{2!3}(3^2 A_1 - 4^2 A_2)\right)$$
$$+ (2n-1)\left(4^2 \frac{(n+1)n(n-1)(n-2)}{4!5}(5^2 A_2 - 6^2 A_3)\right)$$
$$- (2n-1)\left(4^3 \frac{(n+2)(n+1)\cdots(n-2)(n-3)}{6!7}\right.$$
$$\left. \times (7^2 A_3 - 6^2 A_2) + \cdots\right).$$

To clarify the general result implied by this procedure, Gauss wrote the first few special cases:

$$0 = 1 - 4A_1$$
$$4A_1 - 1 = 3(1 - 4A_1) - 4(9A_1 - 16A_2)$$
$$0 = 5(1 - 4A_1) - 20(9A_1 - 16A_2) + 16(25A_2 - 36A_3)$$
$$16A_2 - 9A_1 = 7(1 - 4A_1) - 56(9A_1 - 16A_2) + 112(25A_2 - 36A_3)$$
$$- 65(49A_3 - 64A_4) \quad \text{etc.}$$

Thus, he found that

$$A_1 = \frac{1^2}{2^2}, \quad A_2 = \frac{3^2}{4^2} A_1 = \frac{1^2 \cdot 3^2}{2^2 \cdot 4^2}, \quad A_3 = \frac{1^2 \cdot 3^2 \cdot 5^2}{2^2 \cdot 4^2 \cdot 6^2}, \ldots,$$

and in general

$$A_n = \frac{1^2 \cdot 3^2 \cdots (2n-1)^2}{2^2 \cdot 4^2 \cdots (2n)^2}.$$

Gauss then related $1/M(1+x, 1-x)$ with the elliptic integral by evaluating the following integral as a series:

$$\frac{1}{\pi}\int_0^\pi (1 - x^2 \cos^2\theta)^{-1/2} d\theta = \frac{1}{\pi}\int_0^\pi \left(1 + \frac{1}{2}x^2\cos^2\theta + \frac{1}{2}\cdot\frac{3}{4}x^4\cos^4\theta + \cdots\right) d\theta$$
$$= 1 + \frac{1^2}{2^2}x^2 + \frac{1^2 \cdot 3^2}{2^2 \cdot 4^2}x^4 + \cdots. \qquad (36.82)$$

Since the series for the integral and the agM were the same, he concluded that

$$\frac{1}{M(1+x, 1-x)} = \frac{2}{\pi}\int_0^{\pi/2} \frac{d\theta}{\sqrt{1 - x^2 \sin^2\theta}}.$$

Finally, one may see that by taking $x = \sqrt{1 - b^2/a^2}$, we have Gauss's formula (36.73).

In a 1796 paper, James Ivory gave an interesting new method to prove the formula

$$K\left(\frac{2\sqrt{x}}{1+x}\right) = (1+x)K(x), \qquad (36.83)$$

where $K(x) = \frac{2}{\pi}\int_0^{\pi/2} d\theta/\sqrt{1-x^2\cos^2\theta}$. Legendre was the first to prove this result, for the purpose of numerically evaluating complete elliptic integrals; his proof used the Landen transformation. In his paper, Ivory did not mention the agM and, indeed, he may not have been aware of its significance even if he had noticed it in Lagrange's 1785 paper. In the cover letter accompanying his paper, Ivory explained that his aim was to present a simple method for obtaining the expansion

$$(a^2 + b^2 - 2ab\cos\phi)^n = A + B\cos\phi + C\cos 2\phi + \cdots. \tag{36.84}$$

Ivory started with the relation $\sin(\psi - \phi) = c\sin\psi$, took its fluxion (derivative), and simplified to get

$$\dot\phi = \frac{\sqrt{1-c^2\sin^2\psi} - c\cos\psi}{\sqrt{1-c^2\sin^2\psi}}\dot\psi.$$

He performed an elementary calculation to show that the numerator could be expressed as $\sqrt{1+c^2 - 2c\cos\phi}$. This led him to the equation

$$\frac{\dot\phi}{\sqrt{1+c^2 - 2c\cos\phi}} = \frac{\dot\psi}{\sqrt{1-c^2\sin^2\psi}}. \tag{36.85}$$

He then set

$$c' = \frac{1-\sqrt{1-c^2}}{1+\sqrt{1-c^2}} \tag{36.86}$$

to find that

$$\sqrt{1-c^2\sin^2\psi} = \frac{\sqrt{1+c'^2 + 2c'\cos 2\psi}}{1+c'}.$$

Thus, he could express (36.85) in the form

$$\frac{\dot\phi}{\sqrt{1+c^2 - 2c\cos\phi}} = \frac{(1+c')\dot\psi}{\sqrt{1+c'^2 + 2c'\cos 2\psi}}. \tag{36.87}$$

Ivory's contribution was a new method for evaluating the integrals

$$\int_0^\pi \frac{d\phi}{\sqrt{1+c^2 - 2c\cos\phi}} = \int_0^\pi \frac{(1+c')d\psi}{\sqrt{1+c'^2 + 2c'\cos 2\psi}}. \tag{36.88}$$

First, he observed that by the binomial theorem

$$(1+c^2 - 2c\cos\phi)^{-1/2} = (1-ce^{i\phi})^{-1/2}(1-ce^{-i\phi})^{-1/2}$$

$$= \left(1 + \frac{1}{2}ce^{i\phi} + \frac{1\cdot 3}{2\cdot 4}c^2 e^{2i\phi} + \cdots\right)$$

$$\times \left(1 + \frac{1}{2}ce^{-i\phi} + \frac{1\cdot 3}{2\cdot 4}c^2 e^{-2i\phi} + \cdots\right). \tag{36.89}$$

Ivory then noted that multiplying these two series gave him the cosine expansion when $n = -1/2$ in (36.84). This showed that the value of the integral on the left-hand side of (36.88) was simply the constant A. Since the constant term A was easy to evaluate from the product of the two series, he could express equation (36.88) as

$$\int_0^\pi \frac{d\phi}{\sqrt{1+c^2-2c\cos\phi}} = 1 + \frac{1^2}{2^2}c^2 + \frac{1^2\cdot 3^2}{2^2\cdot 4^2}c^4 + \cdots$$
$$= (1+c')\left(1 + \frac{1^2}{2^2}c'^2 + \frac{1^2\cdot 3^2}{2^2\cdot 4^2}c'^4 + \cdots\right). \quad (36.90)$$

Observe that this proves (36.83).

Ivory remarked that c' was smaller than c; as an example, he pointed out that when $c = 4/5$, then $c' = 1/4$. Thus, (36.90) was very useful for computational purposes. Ivory noted, as Legendre had done before him, that the formula's computational effectiveness was greatly improved by iteration. Now note that (36.90) can be expressed as a quadratic transformation of hypergeometric functions:

$$F(1/2, 1/2, 1, c^2) = (1+c')F(1/2, 1/2, 1, c'^2),$$

where c' is given by (36.86). Gauss may have been motivated to study quadratic transformations of hypergeometric functions because of this and similar results.

We have discussed two different quadratic transformations of elliptic integrals. We can rewrite Landen's transformation (36.33) in the form: If $\lambda = 2\sqrt{k}/(1+k)$, $\lambda^2 + \lambda'^2 = 1$, and $z = (1+\lambda')y\sqrt{(1-y^2)}/\sqrt{(1-\lambda^2 y^2)}$, then

$$\frac{(1+k)dz}{\sqrt{(1-z^2)(1-k^2z^2)}} = \frac{2dy}{\sqrt{(1-y^2)(1-k^2y^2)}}.$$

This is one quadratic transformation and the other is Gauss's transformation given by the equations (36.78) and (36.79). It is easy to check that if these transformations are applied one after the other, we get the duplication of the elliptic integral:

$$\frac{dz}{\sqrt{(1-z^2)(1-k^2z^2)}} = \frac{2dx}{\sqrt{(1-x^2)(1-k^2x^2)}},$$

$$z = 2x\sqrt{1-x^2}\sqrt{1-k^2x^2}/(1-k^2x^4).$$

36.6 Remarks on Gauss and Elliptic Functions

Gauss wrote in an 1816 letter to his friend Schumacher that he rediscovered the arithmetic-geometric mean in 1791 at the age of 14. From 1791 until 1800, Gauss made a series of discoveries advancing the theory of elliptic integrals and functions to new and extraordinary heights. Among his great achievements in this area were the inversion of the elliptic integral and the consequent discovery of double periodicity and the development of elliptic functions as power series and as double products, leading to series and product expansions in terms of trigonometric functions. This in turn brought him to the discovery of the theta functions and the triple product identity. The initial motivation behind Gauss's work on elliptic functions was the problem of the division of the lemniscate. Gauss solved the problem by means of complex multiplication of elliptic functions. Finally, in 1800, he extended his youthful work on the arithmetic-geometric mean by considering the agM of two complex numbers. He found that the agM in this case was countably many-valued, and his attempts to find a relation among the values led him deeper into the theory of theta and modular functions. In this connection, Gauss discovered an important transformation of theta functions:

$$\sum_{k=-\infty}^{\infty} e^{-\alpha(k+w)^2} = \sqrt{\frac{\pi}{\alpha}} \sum_{k=-\infty}^{\infty} e^{-\frac{n^2\pi^2}{\alpha}+n\omega\pi i}.$$

It is remarkable that Gauss published very little of these groundbreaking theories, although they surely rank among his greatest discoveries in pure mathematics. Perhaps he wished to first develop a coherent theory of functions of complex variables. Consider the fact that, though he initially discovered double periodicity through a formal use of complex numbers, his 1800 work defined an elliptic function by means of a ratio of two theta functions. His early definition of an elliptic function by means of the inversion of an elliptic integral would require the concept and careful use of analytic continuation, not then developed. We note that Gauss's 1811 letter to Bessel shows that he was making inroads into the mysteries of complex variables. The unpublished portion of Gauss's 1812 paper on hypergeometric series also gives some indication of his understanding of analytic continuation. However, it seems that after 1805–06, Gauss never found the time to completely develop his ideas in number theory, elliptic functions, or complex variables. From 1801 onward, he researched applied topics such as astronomy, geodesy, telegraphy, magnetism, crystallography, and optics. Of course, mathematical problems in these areas led him to interesting and important discoveries such as the method of least squares, trigonometric interpolation, numerical integration, the technique of fast Fourier transforms, and the theory of curved surfaces.

Thus, Gauss never wrote up a detailed account of his researches on elliptic functions, though he wrote a substantial amount on theta functions. In his fragmentary notes on elliptic functions, one may see some of the main results but usually there are no details of the methods he employed. However, in a letter to Schumacher, Gauss wrote that Abel's first paper on the theory of elliptic functions followed the same path he himself had trod in 1798 and that Abel's work relieved him of the burden of publishing that part of his work. He wrote a similar letter to Crelle and the entries in Gauss's diary from the period 1797–1800 bear out these assertions.

Gauss's investigations relating to elliptic functions, the agM, and theta functions took place during 1791–1800, the critical period when he was maturing into a most formidable mathematical mind. Gauss's early interest in mathematics was kindled by his association with Johann Bartels (1769–1836), a teacher's assistant at the school Gauss attended. Gauss was 11 years old when he and Bartels studied infinite series and the binomial theorem. Bartels later became professor of mathematics at the University of Kazan where he taught the great Russian mathematician N. I. Lobachevsky (1793–1856), one of the discoverers of non-Euclidean geometry. Having become known as a promising student, in 1791 Gauss met the Duke of Braunschweig and was presented with a table of logarithms by the minister of state. Greatly impressed with the genius of Gauss, the Duke provided financial support for Gauss to attend the Collegium Carolinum in Braunschweig (Brunswick). Entering the Collegium in 1792, Gauss became accomplished in languages and began studying the works of Newton, Euler, and Lagrange. He was impressed by Euler's pentagonal number theorem:

$$\prod_{n=1}^{\infty}(1-x^n) = \sum_{n=-\infty}^{\infty}(-1)^n x^{n(3n-1)/2}.$$

This result led him to investigate series whose exponents were square or triangular numbers. Gauss is reported to have said that in 1794 he knew the connection between such series and the agM. This is very likely because the series identities needed for this question could be easily proved by methods Gauss had seen in the works of Euler. To state the identities, set $A(x) = \sum x^{n^2}$ and $B(x) = \sum (-1)^n x^{n^2}$ where the sums are over all integers. Then

$$A(x) + B(x) = 2A(x^4), \qquad (36.91)$$

$$A^2(x) + B^2(x) = 2A^2(x^2), \qquad (36.92)$$

$$A(x)B(x) = B^2(x^2). \qquad (36.93)$$

The first identity is almost obvious; the second can be proved by first observing that the coefficient of x^n in $A^2(x)$ is the number of ways n can be expressed as a sum of two integer squares and then noting that this is the same as the number of ways $2n$ can be expressed as a sum of two integer squares. The third identity is a consequence of the first two because

$$A(x)B(x) = (A(x) + B(x))^2/2 - (A^2(x) + B^2(x))/2$$
$$= 2A^2(x^4) - A^2(x^2) = B^2(x^2).$$

It follows from (36.91) and (36.93) that the arithmetic mean and geometric mean of $A^2(x)$ and $B^2(x)$ are $A^2(x^2)$ and $B^2(x^2)$. This kind of reasoning must have been very familiar to Gauss in 1794, both from his investigations in collaboration with Bartels and from his study of Euler's *Introductio*. Gauss wrote up his results connecting series with the agM at a later date. In that manuscript, he derived the properties of the series by means of their product representations obtained from the triple product identity.

In 1794–95, Gauss does not seem to have been aware of the connection of these series or the agM with elliptic integrals. Then in October 1795, with the continued support of the Duke, Gauss registered as a student at the University of Göttingen where he had access to an excellent library. For example, in early 1796 he borrowed many volumes of the *Mémories de l'Académie de Berlin* from the library. The volumes contained several works of Lagrange on number theory, algebra, and other mathematical topics. Gauss first mentioned elliptic integrals in his mathematical diary on September 9, 1796. He gave the power series expansion of the inverse of the elliptic integral $\int (1-x^3)^{-1/2} dx$; he found it by Newton's reversion of series method. A few days later, he noted the series for the inverse of the more general integral $\int (1-x^n)^{-1/2}\,dx$. In January 1797, his interest in elliptic integrals became more serious. His notes indicate that he had already studied Stirling and Euler on this topic. He noted in his January 7 entry in the diary that

$$\int \sqrt{\sin x}\,dx = 2\int \frac{yy\,dy}{\sqrt{1-y^4}}, \quad \int \sqrt{\frac{1}{\sin x}}\,dx = 2\int \frac{dy}{\sqrt{1-y^4}}, \quad yy = \frac{\sin}{\cos}x,$$

and a day later he recorded that he had started investigating the elastic curve depending on $\int (1-x^4)^{-1/2}\,dx$. Later he crossed out the words "elastic curve" and replaced them with "lemniscate." In order to understand the reason for this change in point of view, we note that Gauss started his mathematical diary in March 1796 when he discovered the principles underlying the problem of dividing the circle into n equal parts. In particular, this problem required a study of the polynomials obtained when $\sin(nx)$ and $\cos(nx)$ were expressed in terms of $\sin x$ and $\cos x$. Note that the sine or cosine function can be defined in terms of the inverse of the integral $\int dx/\sqrt{1-x^2}$. Around March 1797, Gauss found that in order to divide the lemniscate into n equal parts, he had to study the properties of the lemniscatic function, defined as the inverse of the integral $\int dx/\sqrt{1-x^4}$.

On March 19, Gauss observed in his diary that the division of the lemniscate into n parts led to an algebraic equation of degree n^2. In fact, this follows from the addition formula for the lemniscatic integral. However, it appears from his September 1796 note that he was already thinking in terms of the inverse of the integral $\int dx/\sqrt{1-x^4}$. If we denote the inverse by sl x, then we see that he had discovered that sl(nx) could be expressed as a rational function of sl x and that the numerator was a polynomial of degree n^2 in sl x. Since only n solutions correspond to real division points, this discovery showed him that a majority of the solutions of the equation of degree n^2 had to be complex. It is possible that this led him to make an imaginary substitution in the integral $\int dx/\sqrt{1-x^4}$, and this in turn led him to the discovery of the double periodicity of the lemniscatic function. In his undated diary entry between March 19 and March 21, Gauss noted that the lemniscate was geometrically divisible into five equal parts. This is a remarkable statement; it shows that Gauss had not only found double periodicity but had also found an example of complex multiplication of elliptic functions. We note that in general, an elliptic function $\phi(x)$ has two fundamental periods whose ratio must be a complex number. Moreover, for any integer n, the addition formula for elliptic functions shows that $\phi(nx)$ is a rational function of $\phi(x)$. However, in the case where the ratio of the fundamental periods is a root of a quadratic with rational coefficients, there exists a complex number α such that $\phi(\alpha x)$ can be expressed as a rational function of $\phi(x)$. In this situation, we say that $\phi(x)$ permits complex multiplication by α. Apparently, in 1828 Abel was the first to study this phenomenon. It is not clear to what extent Gauss had investigated complex multiplication, but he certainly used it in connection with dividing the lemniscate into n equal parts, at least when $n = 5$. To be able to show that a fifth part of the lemniscatic curve could be obtained by geometric construction, Gauss had to solve two appropriate quadratic equations. The surviving fragments of Gauss's work on this problem do not contain these equations, but they can be found in Abel's first paper on elliptic functions.

Abel noted that $5 = (2+i)(2-i)$, and that for $y = \phi(x) \equiv \mathrm{sl}(x)$,

$$\phi((2+i)x) = yi \frac{1-2i-y^4}{1-(1-2i)y^4} \equiv z, \tag{36.94}$$

$$\phi(5x) = \phi((2-i)(2+i)x) = -zi \frac{1+2i-z^4}{1-(1+2i)z^4}. \tag{36.95}$$

We note that Abel here used complex multiplication of ϕ by $2 \pm i$. Next, using (36.95) to solve the equation $\phi(5x) = 0$, he had to first solve $z^4 = 1+2i$, and then solve

$$yi \frac{1-2i-y^4}{1-(1-2i)y^4} = (1+2i)^{1/4}. \tag{36.96}$$

Solve the latter by dividing the previous equation by the conjugate equation

$$-yi \frac{1+2i-y^4}{1-(1+2i)y^4} = (1-2i)^{1/4} \tag{36.97}$$

to obtain a quadratic in y^4. Note that all the equations can be solved by appropriate quadratic equations.

In his notes from this period, Gauss defined the lemniscatic sine and cosine functions by the equations

$$\mathrm{sinlemn}\left(\int_0^x dt/\sqrt{1-t^4}\right) = x, \quad \mathrm{coslemn}\left(\frac{1}{2}\omega - \int_0^x dt/\sqrt{1-t^4}\right) = x,$$

where $\omega = \int_0^1 (1-t^4)^{-1/2} dt$. Gauss sometimes abbreviated sine lemn and cos lemn as s and c. We use the more common sl and cl. By an application of Euler's addition formula for elliptic integrals, Gauss found the addition formulas for the elliptic functions sl and cl:

$$1 = ss + cc + sscc, \tag{36.98}$$

$$\mathrm{sl}(a \pm b) = \frac{sc' \mp ss'}{1 \mp scs'c'}, \tag{36.99}$$

$$\mathrm{cl}(a \pm b) = \frac{cc' \mp ss'}{1 \pm ss'cc'}, \tag{36.100}$$

where $s = \mathrm{sl}(a)$, $s' = \mathrm{sl}(b)$ and c, c' are similarly defined. He employed these formulas to express $\mathrm{sl}(n\phi)$ and $\mathrm{cl}(n\phi)$ in terms of $\mathrm{sl}(\phi)$ and $\mathrm{cl}(\phi)$. By a formal change of variables $t = iu$, Gauss obtained

$$i \int_0^x (1-t^4)^{-1/2} dt = \int_0^{ix} (1-u^4)^{-1/2} du,$$

or, in terms of the lemniscatic functions:

$$\mathrm{sl}(iy) = i\,\mathrm{sl}(y), \quad \mathrm{cl}(iy) = 1/\mathrm{cl}(y).$$

36.6 Remarks on Gauss and Elliptic Functions

Thus, with (36.99), he had the formula for complex arguments

$$\text{sl}(a+ib) = \frac{\text{sl}(a) + i\,\text{sl}(b)\text{cl}(a)\text{cl}(b)}{\text{cl}(b) - i\,\text{sl}(a)\text{sl}(b)\text{cl}(a)}. \tag{36.101}$$

Gauss used these formulas to determine that the periods of sl were 2ω and $2i\omega$. The ratio of the periods would then be $i = \sqrt{-1}$, a root of the quadratic equation $x^2 + 1 = 0$, so that complex multiplication by $\sqrt{-1}$ was possible. Gauss also found that the zeros and poles of $\text{sl}(\phi)$ were of the form $(m+in)\omega$ and $((m+1/2) + i(n+1/2))\omega$, where $m, n \in \mathbb{Z}$. These results allowed him to express the lemniscatic function as a quotient of two entire functions

$$\text{sl}(\phi) = \frac{M(\phi)}{N(\phi)}, \tag{36.102}$$

where M and N were double infinite products. Gauss's diary entry for March 29, 1797 confirms that by that time he was aware of all these results. In the same entry, he gave numerical evidence that $\log N(\omega)$ agreed with $\pi/2$ to six decimal places and noted that a proof would be an important advance in analysis. Gauss's fragmentary notes from this period also show that he had found the significant formula

$$M(\phi)^4 + N(\phi)^4 = N(2\phi). \tag{36.103}$$

He also wrote, perhaps based on numerical evidence, that

$$M(\omega/2) = N(\omega/2). \tag{36.104}$$

Subsequent entries in Gauss's diary suggest that he abandoned his intensive study of elliptic functions for a year. The ninety-second entry, written July 1798, concerned the lemniscatic function; he noted that he had "found out the most elegant things exceeding all expectations and that by methods which open to us a whole new field ahead." According to his notes, his result was

$$\text{sl}(\phi) = P(\phi)/Q(\phi), \quad \text{where} \tag{36.105}$$

$$P(\phi) = \frac{\omega}{\pi} s \left(1 + \frac{4ss}{(e^\pi - e^{-\pi})^2}\right)\left(1 + \frac{4ss}{(e^{2\pi} - e^{-2\pi})^2}\right)\left(1 + \frac{4ss}{(e^{3\pi} - e^{-3\pi})^2}\right)\cdots, \tag{36.106}$$

$$Q(\phi) = \left(1 - \frac{4ss}{(e^{\pi/2} - e^{-\pi/2})^2}\right)\left(1 - \frac{4ss}{(e^{3\pi/2} - e^{-3\pi/2})^2}\right)$$
$$\cdot \left(1 - \frac{4ss}{(e^{5\pi/2} - e^{-5\pi/2})^2}\right)\cdots. \tag{36.107}$$

From Abel, we may surmise that Gauss used the product formula for $\sin x$ to transform the products M and N into new products expressed in terms of the variable $s = \sin(\pi\phi/\omega)$. Gauss also gave the equations connecting M and N with P and Q:

$$M(\psi\omega) = e^{\pi\psi\psi/2} P(\psi\omega), \quad N(\psi\omega) = e^{\pi\psi\psi/2} Q(\psi\omega). \tag{36.108}$$

These are particular cases of Weierstrass's relations connecting his sigma function with the theta function. Note that when $\psi = 1, s = \sin(\pi\psi) = \sin\pi = 0$ and $Q(\omega) = 1$. Therefore, $N(\omega) = e^{\pi/2}$; thus, Gauss resolved the questions he raised the year before.

In the summer of 1798, Gauss discovered another important set of relations, the significance of which is better understood by observing that for $s = \sin\psi\pi$,

$$1 + 4s^2/(e^{n\pi} - e^{-n\pi})^2 = 1 + 4s^2 e^{-2n\pi}/(1 - e^{-2n\pi})^2$$
$$= (1 - 2e^{-2n\pi}\cos 2\psi\pi + e^{-4n\pi})/(1 - e^{-2n\pi})^2$$
$$= (1 - e^{-2n\pi}e^{2i\psi\pi})(1 - e^{-2n\pi}e^{-2i\psi\pi})/(1 - e^{-2n\pi})^2.$$

This equation converts the previous product for $P(\psi\omega)$ to

$$P(\psi\omega) = \frac{\omega}{\pi}\sin\psi\pi \prod_{n=1}^{\infty}(1 - e^{-2n\pi}e^{2i\psi\pi})(1 - e^{-2n\pi}e^{-2i\psi\pi})/(1 - e^{-2n\pi})^2. \quad (36.109)$$

Similarly, the product for Q can be rewritten as

$$Q(\psi\omega) = \prod_{n=1}^{\infty}(1 + e^{-(2n-1)\pi}e^{2i\psi\pi})(1 + e^{-(2n-1)\pi}e^{-2i\psi\pi})/(1 - e^{-(2n-1)\pi})^2. \quad (36.110)$$

From these results, Gauss found the Fourier series expansion of sl(ϕ):

$$\text{sl}(\psi\omega) = \frac{\pi}{\omega}\frac{4}{e^{\pi/2} + e^{-\pi/2}}\sin\psi\pi - \frac{\pi}{\omega}\frac{4}{e^{3\pi/2} + e^{-3\pi/2}}\sin 3\psi\pi + \cdots. \quad (36.111)$$

He also found Fourier series for $\log Q(\psi\omega)$, $\log P(\psi\omega)$, $\log \text{sl}(\psi\omega)$, etc. For example,

$$\log Q(\psi\omega) = -\frac{1}{2}\log 2 + \frac{\pi}{12} + \frac{2}{e^{\pi} - e^{-\pi}}\cos 2\psi\pi - \frac{1}{2}\frac{2}{e^{2\pi} - e^{-2\pi}}\cos 4\psi\pi + \cdots. \quad (36.112)$$

Note that products (36.109) and (36.110) are theta products revealing the form of the product representation for a general theta function.

It appears that Gauss had reached this point in his researches on the lemniscatic function by the end of summer 1798. And on September 28, he completed his studies at Göttingen and departed for Braunschweig. From a letter to his great friend Wolfgang Bolyai, we learn that Gauss was uncertain of his financial future. Note that Bolyai's son was the noted János Bolyai, discoverer of non-Euclidean geometry. Gauss's financial uncertainty remained until the end of the year when the Duke guaranteed him further support, suggesting that Gauss earn a doctoral degree in mathematics. Gauss accomplished this by submitting a thesis to the University of Helmstedt on the fundamental theorem of algebra, work he had completed a year earlier. He noted in his diary, "Proved by a valid method that equations have imaginary roots." In a later addendum to this diary entry, from October 1798, Gauss wrote that this method was published August 1799 as his dissertation; he received his degree in July 1799 on the recommendation of Johann Friedrich Pfaff (1765–1825). Gauss was then free to continue his mathematical researches with the Duke's financial assistance. This increased after Gauss turned down

36.6 Remarks on Gauss and Elliptic Functions

an offer from St. Petersburg in 1802 and continued until he was appointed director of the observatory at the University of Göttingen where he remained to the end of his life.

In spite of his financial insecurity during the fall of 1798, Gauss's creativity did not abate. In October 1798, Gauss noted, "New things in the field of analysis opened up to us, namely, investigation of a function etc." Gauss had earlier found the Fourier expansion of P/Q, but he was excited now to discover the Fourier expansions of the functions P and Q themselves:

$$P(\psi\omega) = 2^{3/4}\sqrt{\frac{\pi}{\omega}}(e^{-\pi/4}\sin\psi\pi - e^{-9\pi/4}\sin 3\psi\pi + e^{-25\pi/4}\sin 5\psi\pi - \cdots) \quad (36.113)$$

and

$$Q(\psi\omega) = 2^{-1/4}\sqrt{\frac{\pi}{\omega}}(1 + 2e^{-\pi}\cos 2\psi\pi + 2e^{-4\pi}\cos 4\psi\pi + \cdots). \quad (36.114)$$

As consequences, he noted

$$1 - 2e^{-\pi} + 2e^{-4\pi} - 2e^{-9\pi} + \cdots = \sqrt{\frac{\omega}{\pi}}, \quad (36.115)$$

$$e^{-\pi/4} + e^{-9\pi/4} + e^{-25\pi/4} + \cdots = \frac{1}{2}\sqrt{\frac{\omega}{\pi}}, \quad (36.116)$$

$$\sqrt{\frac{\omega}{\pi}} = 0.91357913815611682140724259.$$

He found this last value by computing $2e^{-\pi/4}$ to thirty-nine decimal places. Note that (36.115) in fact gives the period of the lemniscatic function as a theta series value.

By comparing the products for P and Q in (36.109) and (36.110) with the series for P and Q in (36.113) and (36.114), we see that by 1798 Gauss knew the triple product identity. In fact, to derive the series from the product, one requires not only the triple product identity but also an additional formula. Gauss could have derived this from what he already knew. First consider how the factor $e^{-\pi/4}$ arises in the series for P. The addition formula (36.101) implies Gauss's observation that

$$\mathrm{sl}\left(\omega\psi + \frac{\omega}{2} + i\frac{\omega}{2}\right) = \frac{-i}{\mathrm{sl}(\omega\psi)}. \quad (36.117)$$

Here we mention that Jacobi used a similar formula in his *Fundamenta Nova*. We also note that

$$e^{i\pi(\psi+1/2+i/2)} = ie^{-\pi/2}e^{i\psi\pi}.$$

When these two relations are applied to the formulas (36.105), (36.109), and (36.110), the result after a simple calculation for $q = e^{-\pi}$ is that

$$\frac{4\pi e^{-\pi/2}}{\omega} \cdot \prod_{n=1}^{\infty}\left(\frac{1-q^{2n}}{1+q^{2n-1}}\right)^2 = \frac{\omega}{\pi}\prod_{n=1}^{\infty}\left(\frac{1+q^{2n-1}}{1-q^{2n}}\right)^2.$$

This simplifies to

$$\frac{2\pi}{\omega} e^{-\pi/4} = \prod_{n=1}^{\infty} \left(\frac{1+q^{2n-1}}{1-q^{2n}} \right)^2.$$

When the value π/ω from this equation is substituted in (36.109), we arrive at

$$P(\psi\omega) = 2e^{-\pi/4} \sin\psi\pi \prod_{n=1}^{\infty} (1 - e^{-2n\pi} e^{2i\psi\pi})(1 - e^{-2n\pi} e^{-2i\psi\pi})/(1 + e^{-(2n-1)\pi})^2.$$

Here observe that the factor $e^{-\pi/4}$ is accounted for. After applying the triple product identity to this equation, we obtain the series

$$P(\psi\omega) = \frac{2e^{-\pi/4} \sum_{n=0}^{\infty} (-1)^n e^{-n(n+1)\pi} \sin(2n+1)\psi}{\prod_{n=1}^{\infty} (1 - e^{-2n\pi})(1 + e^{-(2n-1)\pi})}. \tag{36.118}$$

Since $n(n+1) + 1/4 = (2n+1)^2$, we see that this is Gauss's series except for the infinite product in the denominator. To eliminate this term, we apply Gauss's relations (36.103), (36.104), and (36.108) to get

$$P(\omega/2) = Q(\omega/2) = 2^{-1/4}.$$

Setting $\psi = 1/2$ and using the last relation in (36.109) and (36.110), we find

$$2^{-1/4} = \frac{\pi}{\omega} \prod_{n=1}^{\infty} (1 + e^{-2n\pi})^2 / (1 + e^{-(2n-1)\pi})^2,$$

$$2^{-1/4} = \prod_{n=1}^{\infty} (1 - e^{-(2n-1)\pi})^2 / (1 + e^{-(2n-1)\pi})^2.$$

From these two equations, a few lines of calculation yield

$$2^{1/4} \sqrt{\frac{\omega}{\pi}} = \prod_{n=1}^{\infty} \frac{1}{(1 - e^{-2n\pi})(1 + e^{-(2n-1)\pi})}.$$

Ultimately, we obtain Gauss's series (36.113) when we apply this equation to (36.118). Obtain series (36.114) similarly.

Although the triple product identity is difficult to prove ab initio, Gauss gave at least two documented proofs of this kind, thought to date from approximately 1808. However, it is possible that Gauss could have proved the triple product identity in 1798 by assuming Euler's pentagonal number theorem and in that case the proof would have been straightforward, the necessary technique having been established by Euler. Consider the product in the numerator of $Q(\psi\omega)$ in (36.110). For convenience, set $q = e^{-\pi}$ and $x = e^{2i\psi\pi}$, so that the product becomes

$$f(x) = \prod_{n=1}^{\infty} (1 + q^{2n-1} x)(1 + q^{2n-1}/x). \text{ Then}$$

36.6 Remarks on Gauss and Elliptic Functions

$$f(q^2 x) = \prod_{n=1}^{\infty}(1+q^{2n+1}x)(1+q^{2n-3}/x)$$

$$= \frac{1+1/qx}{1+qx}f(x) = \frac{1}{qx}f(x).$$

Now let $f(x) = \sum_{-\infty}^{\infty} a_n x^n$ so the previous equation becomes

$$\sum_{n=-\infty}^{\infty} a_n x^n = qx \sum_{n=-\infty}^{\infty} a_n q^{2n} x^n = \sum_{n=-\infty}^{\infty} a_n q^{2n+1} x^{n+1}.$$

By equating the coefficients of x^n, we see that

$$a_n(q) = a_{n-1}(q) q^{2n-1} = a_{n-2}(q) q^{2n-1+2n-3} = a_0(q) q^{n^2}.$$

Hence

$$\prod_{n=1}^{\infty}(1+q^{2n-1}x)(1+q^{2n-1}/x) = a_0(q) \sum_{n=-\infty}^{\infty} q^{n^2} x^n. \tag{36.119}$$

These simple calculations appear in Gauss's notes of 1799. Now to employ Euler's pentagonal number theorem

$$\prod_{n=1}^{\infty}(1-p^n) = \sum_{n=-\infty}^{\infty}(-1)^n p^{n(3n+1)/2}, \tag{36.120}$$

we set $q = p^{3/2}$ and $x = -p^{1/2}$ in (36.119) to get

$$\prod_{n=1}^{\infty}(1-p^{3n-1})(1-p^{3n-2}) = a_0(q) \sum_{n=-\infty}^{\infty}(-1)^n p^{n(3n+1)/2}. \tag{36.121}$$

Comparing (36.120) and (36.121), we arrive at

$$a_0(q) = 1/\prod_{n=1}^{\infty}(1-p^{3n}) = 1/\prod_{n=1}^{\infty}(1-q^{2n})$$

and this proves the triple product identity.

After his remarkable work of 1798, Gauss's journal entry of May 30, 1799, connected the agM with the lemniscatic integral: "We have proved that the arithmetic-geometric mean of 1 and $\sqrt{2}$ is π/ω to 11 places, which thing being proved a new field will certainly be opened up."

We derive the agM of $\sqrt{2}$ and 1 from some previously mentioned formulas of Gauss. By taking $\psi = 1$ in (36.114), we get

$$2^{1/2}\sqrt{\frac{\omega}{\pi}} = 1 + 2e^{-\pi} + 2e^{-4\pi} + 2e^{-9\pi} + \cdots.$$

In terms of the functions A and B in (36.91) and (36.93), the previous formula and (36.115) imply that

$$A^2(e^{-\pi}) = 2^{1/2}\omega/\pi, \quad \text{and} \quad B^2(e^{-\pi}) = \omega/\pi.$$

When this is combined with the fact that $A^2(x^2)$ and $B^2(x^2)$ are the arithmetic and geometric means, respectively, of $A^2(x)$ and $B^2(x)$, it follows that the agM of $\sqrt{2}$ and 1 is π/ω. This means that, if in 1795 Gauss knew the connection of the two series $\sum x^{n^2}$ and $\sum (-1)^n x^{n^2}$ with the agM, then in 1799 he had a proof of the result quoted from the diary. Since he enjoyed numerical computation, he also verified this result to eleven places. Felix Klein and Ludwig Schlesinger, the editors of Gauss's mathematical diary, have remarked that the May 30 entry could represent a conclusion or a conjecture. It is very likely that it was a conclusion and that when he spoke of a new field, Gauss had in mind a generalization to any two real numbers a and b instead of the pair 1, $\sqrt{2}$. As we have seen, Lagrange had already found this generalization in 1785. Gauss published his work in an astronomical paper of 1818, where he wrote that he discovered the result before he saw the paper of Lagrange.

It appears that up to 1798, Gauss did not investigate elliptic functions beyond the lemniscatic function, but with his discovery of the connection between the agM and the elliptic integral $\int_0^{\pi/2} d\theta/\sqrt{a^2\cos^2\theta + b^2\sin^2\theta}$, he began to explore the inversion of more general elliptic integrals. This culminated with his May 6, 1800 journal entry: "We have led the theory of transcendental quantities:

$$\int \frac{dx}{\sqrt{(1-\alpha x x)(1-\beta x x)}}$$

to the summit of universality." Gauss's notes show that he used the agM to define two theta functions whose ratio he demonstrated to be the elliptic function inverting the integral. Gauss's approach to elliptic functions as ratios of theta functions was the same as the point of view taken by Jacobi in his 1836 Königsberg lectures. Gauss started with the integral

$$\int \frac{du}{\sqrt{(1+\mu\mu\sin^2 u)}} = \phi = \psi\omega$$

and set

$$\mu = \tan v, \quad \frac{\pi}{M\sqrt{(1+\mu\mu)}} = \frac{\pi\cos v}{M\cos v} = \omega, \quad \frac{\pi}{\mu M\sqrt{(1+\frac{1}{\mu\mu})}} = \frac{\pi\cos v}{M\sin v} = \omega'.$$

Note that Gauss denoted the agM of $\sqrt{(1+\mu^2)}$ and 1 by

$$M(\sqrt{(1+\mu^2)}, 1) \equiv M\sqrt{(1+\mu\mu)}, \quad \text{so that}$$

$$M\cos v = M(1, \cos v) \quad \text{and} \quad M\sin v = M(1, \sin v).$$

Note also that the Lagrange-Gauss agM theorem implied that

$$\frac{\omega}{2} = \int_0^{\pi/2} \frac{du}{\sqrt{(1+\mu^2\sin^2 u)}} = \int_0^1 \frac{dx}{\sqrt{((1-x^2)(1+\mu^2 x^2))}},$$

$$\frac{\omega'}{2} = \frac{1}{\mu}\int_0^{\pi/2} \frac{du}{\sqrt{(1+\frac{1}{\mu^2}\sin^2 u)}} = \frac{1}{\mu}\int_0^1 \frac{dx}{\sqrt{(1-x^2)(1+\frac{1}{\mu^2}x^2)}}.$$

Gauss then wrote the elliptic function as

$$S(\psi\omega) = \frac{\pi}{\mu\omega}\left(\frac{4\sin\psi\pi}{e^{(\omega'/2\omega)\pi}+e^{-(\omega'/2\omega)\pi}} - \frac{4\sin 3\psi\pi}{e^{(3\omega'/2\omega)\pi}+e^{-(3\omega'/2\omega)\pi}} + \cdots\right)$$

$$= \frac{T(\psi\omega)}{W(\psi\omega)},$$

where the theta functions T and W were defined by the series

$$W(\psi\omega) = \sqrt{M}\cos v(1 + 2e^{-\omega'\pi/\omega}\cos 2\psi\pi + 2e^{-4\omega'\pi/\omega}\cos 4\psi\pi + \cdots),$$

$$T(\psi\omega) = \sqrt{\cot v}\sqrt{M}\cos v(2e^{-\omega'\pi/4\omega}\sin\psi\pi - 2e^{-9\omega'\pi/4\omega}\sin 3\psi\pi + \cdots).$$

To demonstrate that his elliptic function was actually the inversion of his original elliptic integral, he effectively showed that if

$$\int_0^u \frac{dt}{\sqrt{(1+\mu\mu\sin^2 t)}} = \phi,$$

then $s(\phi) = \sin u$. Without giving details, he next wrote down the zeros of W and T and extended to this elliptic function all the results he had obtained for the lemniscatic function.

36.7 Exercises

1. Show that if $t = \sqrt{1-u^2}/\sqrt{1+u^2}$, then

$$\int -dt\frac{\sqrt{1+t^2}}{\sqrt{1-t^2}} - \int du\frac{\sqrt{1+u^2}}{\sqrt{1-u^2}} = t^3\frac{\sqrt{1-t^4}}{1+t^4} + u^3\frac{\sqrt{1-u^4}}{1+u^4}.$$

   See Fagnano (1911), vol. 2, p. 453.

2. Show that the complete integral of $dx/\sqrt{(f+gx^3)} = dy/\sqrt{(f+gy^3)}$ is given by

$$f(x^2+y^2) + g^2c^2x^2y^2/(4f) - gcxy(x+y) - 2fxy - gc^2(x+y) - 2fc = 0.$$

   See Eu. I-20, p. 78.

3. Let $(x,y) = \prod_{n=1}^{\infty}(1+x^{2n-1}y)(1+x^{2n-1}/y)$ and $[x] = \prod_{n=1}^{\infty}(1-x^n)$. Show that

$$(x,\alpha y)\cdot(x,y/\alpha) = ((x^2,\alpha^2)\cdot(x^2,y^2) + x\alpha y(x^2,\alpha^2x^2)(x^2,x^2y^2))\cdot[x^4]^2/[x^2]^2.$$

   See Gauss, (1863–1927), vol. 3, p. 458.

4. Let

$$\prod_{n=1}^{\infty}(1+x^{2n-1}y)(1+x^{2n-1}/y) = Fx\sum_{m=-\infty}^{\infty}x^{m^2}(y^m+y^{-m}),$$

and let $[x]$ be defined as in the problem above. Show that

(a)
$$Fx = \frac{(1-x)^2(1-x^2)^2(1-x^3)^2 \cdots}{1-2x+2x^4-2x^9+\cdots} = \frac{[x]^2}{[x^2]^2} \cdot \frac{1}{1-2x+2x^4-2x^9+\cdots},$$

(b)
$$Fx = \frac{(1+x^2)(1+x^6)(1+x^{10})\cdots}{1-2x^4+2x^{16}-2x^{36}+\cdots} = \frac{[x^4]^2}{[x^2][x^8]} \cdot \frac{1}{1-2x^4+2x^{16}-\cdots},$$

(c)
$$[x^2]Fx = [x^8]Fx^4 = [x^{32}]Fx^{16} = [x^{128}]Fx^{64} = \text{etc.} = 1,$$

(d)
$$1-2x+2x^4-\cdots = \frac{[x]^2}{[x^2]} = \frac{1-x}{1+x} \cdot \frac{1-x^2}{1+x^2} \cdot \frac{1-x^3}{1+x^3}\cdots,$$

(e)
$$1+2x+2x^4+\cdots = \frac{[x^2]^5}{[x]^2[x^4]^2} = \frac{1+x}{1-x} \cdot \frac{1-x^2}{1+x^2} \cdot \frac{1+x^3}{1-x^3}\cdots.$$

See Gauss, (1863–1927), vol. 3, pp. 446–447. Observe that this was one of Gauss's proofs of the triple product identity.

5. Set
$$Px = 1 + 2\sum_{n=1}^{\infty} x^{n^2}, \quad Qx = 1 + 2\sum_{n=1}^{\infty} (-1)^n x^{n^2}, \quad Rx = 2\sum_{n=1}^{\infty} x^{(2n-1)^2/4}.$$

Note that we would write $Px = P(x)$, etc. Show that

(a) $Rx = 2x^{1/4}[x^4]^2/[x^2]$,
(b) $Px \cdot Qx = (Qxx)^2$; $\quad Px \cdot Rx = (R\sqrt{x})^2/2$,
(c) $Px + Qx = 2P(x^4)$; $\quad Px - Qx = 2R(x^4)$; $\quad (Px)^2 - (Qx)^2 = 2(Rxx)^2$,
(d) $Px + iQx = (1+i)Q(ix)$; $\quad Px - iQx = (1-i)P(ix)$,
(e) $(Px)^2 + (Qx)^2 = 2(Pxx)^2$; $\quad (Px)^4 - (Qx)^4 = (Rx)^4$,
(f) The arithmetic geometric mean of $(Px)^2$, $(Qx)^2$ is always 1,
(g)
$$\int_0^{2\pi} \frac{d\theta}{\sqrt{((Px)^4 \cos^2\theta + (Qx)^4 \sin^2\theta)}} = 2\pi.$$

We note that Gauss wrote $\cos\theta^2$ for $\cos^2\theta$, etc. See Gauss (1863–1927), vol. 3, pp. 465–467.

6. Show that

$$(1 + 2x + 2x^4 + 2x^9 + \cdots)^4 = 1 + \frac{8x}{1-x} + \frac{16xx}{1+xx} + \frac{24x^3}{1-x^3} + \frac{32x^4}{1+x^4} + \cdots$$

$$= 1 + \frac{8x}{(1-x)^2} + \frac{8xx}{(1+xx)^2} + \frac{8x^3}{(1-x^3)^2} + \frac{8x^4}{(1+x^4)^2} + \cdots.$$

See Gauss (1863–1927), vol. 3, p. 445.

7. Show that

$$1 + \left(\frac{1}{2}\right)^3 + \left(\frac{1\cdot 3}{2\cdot 4}\right)^3 + \left(\frac{1\cdot 3\cdot 5}{2\cdot 4\cdot 6}\right)^3 + c\cdots = 2\left(\frac{\tilde{\omega}}{\pi}\right)^2,$$

where

$$\frac{\tilde{\omega}}{2} = \int_0^1 \frac{dz}{\sqrt{(1-z^4)}}.$$

See Gauss (1863–1927), vol. 3, p. 425.

8. This exercise gives a proof of the transformation formula for a theta function, first published by Cauchy and Poisson. See chapter 37, section 11. Gauss worked out the details given here in a paper published only long after his death.

   (a) Expand $T = \sum_{k=-\infty}^{\infty} e^{-\alpha(k+\omega)^2}$ as a Fourier series

   $$T = A_0 + 2\sum_{n=1}^{\infty} A_n \cos n\omega P \quad,$$

   where $A_n = \int_0^1 T \cos n\omega P \, d\omega$ and $P = 2\pi$.

   (b) Show that

   $$A_n = \int_{-\infty}^{\infty} e^{-\alpha\omega\omega} \cos n\omega P \, d\omega = e^{-\frac{nn\pi i}{\alpha}} \sqrt{\frac{\pi}{\alpha}}.$$

   (c) Conclude that

   $$\sum_{k=-\infty}^{\infty} e^{-\alpha(k+\omega)^2} = \sqrt{\frac{\pi}{\alpha}} \cdot e^{-\alpha\omega\omega} \cdot \sum_{k=-\infty}^{\infty} e^{-\frac{\pi\pi}{\alpha}(k+\frac{\alpha\omega i}{\pi})^2}.$$

See Gauss (1863–1927), vol. 3, pp. 436–437.

36.8 Notes on the Literature

Fagnano (1911) is a reprint of his *Produzioni Matematiche*. See pp. 293–297, 304–313 of vol. 2 for material on the lemniscate. This volume contains several more articles by Fagnano on the integral calculus and on the lemniscatic calculus.

The proof of the addition formula for the elliptic integral given here is from Euler's 1756 paper "De Integratione Aequationis Differentialis $mdx/\sqrt{(1-x^4)} = mdy/\sqrt{(1-y^4)}$." See Eu. I-20, pp. 63–67. Volumes 20–21 contain the papers of Euler providing the foundation for the theory of algebraic functions and their integrals. Of course, Euler dealt only with the integrals arising from the algebraic equation $y^2 = p(x)$, where $p(x)$ was of degree 4.

See Landen (1771) and (1775) for his original contributions to elliptic integrals. Our exposition is based largely on Cayley (1895), pp. 327–330. Cayley's book contains a good account of Jacobi's *Fundamenta Nova*, elaborating the transformation theory of elliptic integrals of Landen, Legendre, and Gauss.

Lagrange's 1784–1785 paper on the elliptic integral and the agM is in Lagrange (1867–1892) vol. 2, pp. 251–312. See Ivory (1796) for his series evaluation of the complete elliptic integral of the first kind. Gauss's extensive work on the agM and his work on elliptic functions in general can be found in Gauss (1863–1927), vols. 3 and 10. Gauss's determination of the series for $1/M(1+x, 1-x)$ appears on pp. 367–369 of vol. 3. Pieper (1998) suggests that Gauss discovered the triple product identity between April and June 1800. He has also pointed out that this identity can be proved easily by applying Euler's pentagonal number theorem. Berggren, Borwein, and Borwein (1997) contains a number of interesting papers on the agM and its application to the computation of π.

There are several interesting historical accounts of the theory of elliptic functions and integrals. Weil (1983) deals with Euler's work on this topic and its relation to Diophantine equations. Varadarajan (2006) gives a brief analysis of Euler in terms of Riemann surfaces of genus one. Watson (1933) gives a very entertaining and detailed mathematical exposition of Fagnano, Landen, and Ivory's contributions to elliptic integrals. Watson also wrote without giving a reference that Jacobi called December 23, 1751 the birthday of elliptic functions. Later, André Weil observed, "According to Jacobi, the theory of elliptic functions was born between the twenty-third of December 1751, and the twenty-seventh of January 1752." See Weil (1983), p. 1. Ozhigova (2007), first published in 1988, (also reprinted in Bogolyubov, Mikhailov, and Yushkevich (2007)) refers to Jacobi's 1847 letter to Fuss, saying on p. 55 that Euler's study of Fagnano inaugurated the subject of elliptic functions. We observe that in his October 24, 1847 letter to Euler's great-grandson P. H. Fuss, Jacobi strongly recommended the publication of Euler's papers, arguing that they were very important to the advancement of science. As further support for his point, Jacobi mentioned that by reading of the minutes of the Berlin Academy, he discovered a critical date in the history of mathematics: when the Academy assigned Euler the task of refereeing Count Fagnano's mathematical work. Jacobi then stated that Euler's evaluation of these papers served to found the theory of elliptic functions. See Stäckel and Ahrens (1908), p. 23.

Cox (1984) contains a fascinating and enlightening resumé of Gauss's remarkable work on the agM of two complex numbers. He shows that Gauss may have had significant ideas on the modular group and some of its subgroups and their fundamental domains. The reader may wish to read this paper before reading Gauss's somewhat fragmentary original papers on the topic. Mittag-Leffler (1923) and Almkvist and Berndt (1988) are both interesting papers. The first is an insightful account of work on elliptic functions and integrals from 1718 to 1870; the second focuses on topics related to the quadratic transformation and the agM. The first chapter of Siegel (1969), vol. 1 contains perceptive remarks concerning Fagnano and Euler on the addition formula. Bühler (1981) and Dunnington (2004) are well-written biographies of Gauss. Bühler has more mathematical exposition, but the value of Dunnington is enhanced by the inclusion of an English translation with commentary by J. J. Gray of Gauss's diary; we have made use of this translation in the text.

37

Elliptic Functions: Nineteenth Century

37.1 Preliminary Remarks

The eighteenth century saw two major new results in elliptic functions: the addition formula of Euler and the second-order transformation of Landen and Lagrange. Gauss discovered yet another quadratic transformation, in connection with his proof that the agM of two positive numbers could be represented by an elliptic integral. These transformations changed the parameters in the elliptic integrals, without changing their basic form. In fact, Gauss went well beyond this elementary transformation theory, and before the end of the eighteenth century he had greatly refined elliptic function theory. He did not publish his work; it was rediscovered by Abel and Jacobi in the 1820s.

Adrien–Marie Legendre was the main contributor to elliptic integrals in the period between Lagrange and Abel. He reduced any elliptic integral $\int A(x)\,dx/\sqrt{R(x)}$, where $R(x)$ was a fourth-degree polynomial in x and $A(x)$ was a rational function in x and $R(x)$, to integrals of three kinds:

$$F(k, x) = \int_0^x \frac{dt}{\sqrt{(1-t^2)(1-k^2t^2)}}, \tag{37.1}$$

$$E(k, x) = \int_0^x \frac{\sqrt{1-k^2t^2}}{\sqrt{1-t^2}}\,dt, \text{ and} \tag{37.2}$$

$$\Pi(n, k, x) = \int_0^x \frac{dt}{(1+nx^2)\sqrt{(1-x^2)(1-k^2x^2)}}. \tag{37.3}$$

Legendre published two major works on elliptic integrals: *Exercices de calcul intégral* in 1811, and the three-volume *Traité des fonctions elliptiques* in 1825–1828. The first volume of the latter presented the received eighteenth-century theory of elliptic integrals with some improvements and additions; the second volume gave extensive and long-useful numerical tables, constructed by Legendre's own methods. At the age of seventy-five, upon learning of the more advanced results of Abel and Jacobi, Legendre did his best to give a flattering exposition of their work, and this was the topic of the third volume.

Legendre (1752–1833) studied at the College Mazarin in Paris, where he received an excellent education. Apparently, Legendre wished to be remembered for his works alone and not much is known of his personal life. In fact, it was only recently discovered that the portrait by which he had been identified for a century was actually that of an unrelated politician named Louis Legendre. Thus, the only portrait now available is a sketched caricature made by Julien–Léopold Boilly. Legendre's research on his two favorite subjects, number theory and elliptic functions, was immediately superseded after his books appeared. Nevertheless, Legendre's name became permanently associated with several mathematical objects, including the Legendre polynomials, the Legendre symbol, and the Legendre differential equation. Though he studied elliptic functions for almost forty years, Legendre apparently never considered inverting the integral. Abel was the first to publish this idea, inaugurating a great advance in this topic.

The mathematical career of Niels Henrik Abel began in 1821 with his attempt to solve the general quintic equation. His mathematics professors at the University of Christiania could find no errors in Abel's solution and communicated it to Ferdinand Degen in Copenhagen. Though Degen could not find the mistake, he made two suggestions: that Abel apply his method to specific examples, since that could reveal hidden errors; and that he abandon the sterile subject of algebraic equations to exercise his brilliance in the more fruitful subject of elliptic integrals. Degen's advice led Abel to find the mistake in his work and eventually to prove the impossibility of solving the quintic in radicals. He also began to work on elliptic integrals, and it is fairly certain that by 1823 he had inverted the elliptic integral to rediscover elliptic functions. We recall that Gauss had already done this without publishing it. Moreover, the problem of the division of elliptic functions carried Abel deeper into the theory of algebraic equations and ultimately to his famous theorem on solvable equations. In this manner, Abel found an extremely productive connection between elliptic functions and the theory of algebraic equations.

Abel's first work on elliptic functions, the first part of "Recherches sur les fonctions elliptiques," appeared in *Crelle's Journal* in September 1827. In this paper he defined the elliptic function $\phi\alpha = x$ when

$$\alpha = \int_0^x \frac{dt}{\sqrt{(1 - c^2 t^2)(1 + e^2 t^2)}}. \tag{37.4}$$

He showed that $\phi\alpha$ was a meromorphic function with two independent periods, $2w$ and $2i\tilde{w}$, given by

$$w = 2\int_0^{1/c} \frac{dx}{\sqrt{(1 - c^2 x^2)(1 + e^2 x^2)}}, \quad \tilde{w} = 2\int_0^{1/e} \frac{dx}{\sqrt{(1 - e^2 x^2)(1 + c^2 x^2)}}. \tag{37.5}$$

He then gave a new proof for the addition formula for elliptic functions and used it to express $\phi(n\alpha)$, where n was an integer, as a rational function of $\phi\alpha$, $f\alpha = \sqrt{(1 - c^2 \phi^2 \alpha)}$, and $F\alpha = \sqrt{(1 + e^2 \phi^2 \alpha)}$. This was analogous to expressing $\sin(nx)$ as a polynomial in $\sin x$ and $\cos x = \sqrt{(1 - \sin^2 x)}$. With n an odd integer, he noted that the rational function could be written as $xp(x)/q(x)$, where p and q were polynomials in $x = \phi\alpha$ and of degree $n^2 - 1$. Abel next showed that the solution of the equation $p(x) = 0$, whose roots were

$$\phi\left(\frac{aw+ib\tilde{w}}{n}\right),$$

for integers a and b, depended on an equation of degree $n+1$ which could be solved algebraically only in particular cases.

In the second part of his "Recherches," Abel applied his theory to the division of the lemniscate. Recall that Fagnano divided the full arc of the lemniscate in the first quadrant into two, three, and five equal parts. In his *Disquisitiones*, without reference to Fagnano, Gauss stated that the theory he had constructed for the division of a circle into n equal parts could be extended to the lemniscate, but he never gave details on this. In the case of the circle, Gauss was able to simplify the problem, so that he had only to prove that $\cos(2\pi/n)$ could be expressed in terms of square roots, when n was a prime of the form 2^k+1. He did this by showing that $\cos(2\pi/n)$ satisfied an appropriate algebraic equation of degree $(n-1)/2 = 2^{k-1}$.

Thus, in order to extend Gauss's theory to the lemniscate, Abel had to find the division point by working with $\text{sl}(w/n)$, where $2w$ was the period of the lemniscatic function. However, note that $\text{sl}(w/n)$ satisfies an equation of degree n^2-1, and this cannot be a power of 2, except when $n=3$. This drawback would apparently suggest that Gauss's theory for the circle could not be extended to the case of the lemniscate. But Abel found a resolution to this roadblock by discovering complex multiplication of elliptic functions.

The primes expressible as 2^k+1, except for 3, take the form $4m+1$ and can be written as sums of two squares, $4m+1 = a^2+b^2$, where $a+b$ is odd. Abel used this fact to solve the problem of dividing the lemniscate into $n = 4m+1$ parts. He showed that the complex number $\text{sl}(w/(a+ib))$ was the solution to an equation of degree $n-1 = 4m$ with coefficients of the form $c+id$, where c and d were rational numbers. To prove this, he used the addition formula for $\text{sl}\,\alpha$ to first prove that $\text{sl}((a+ib)\alpha)$ could be expressed as a rational function $xp(x)/q(x)$, $x = \text{sl}\,\alpha$, where $p(x)$ and $q(x)$ were polynomials of degree $n-1$. Next, he employed the Lagrange resolvent, just as Gauss had done for the cylotomic case, to show that $\text{sl}(w/(a+ib))$ could be evaluated by means of only square roots, providing n was of the form 2^k+1. Abel pointed out that the value of $\text{sl}(w/n)$ could then be found by means of square roots and so this value was constructible by straight edge and compass. The second part of the "Recherches" also dealt with the transformation of elliptic functions, but on this topic Jacobi had published earlier than Abel.

Carl Gustav Jacob Jacobi (1804–1851) studied at the University of Berlin, though he largely preferred to study on his own, especially Euler's works. His interest in elliptic integrals was aroused by the quadratic transformations in Legendre's *Exercices de calcul intégral*. In June 1827, Jacobi communicated a short note to the *Astronomische Nachrichten* giving two cubic transformations and two fifth-order transformations of elliptic integrals.

In fact, in 1825 Legendre had already discovered this cubic transformation, though it was published in *Traité des fonctions elliptiques* of 1827. Heinrich Schumacher, editor of the *Astronomiche Nachrichten*, noted that Jacobi did not refer to Legendre's book, though this was not surprising, since Jacobi had not seen Legendre's work at the time. In any case, Jacobi also had the new result on the fifth-order transformation. Then in

August 1827, Jacobi communicated to Schumacher a general odd-order transformation, allowing the division of an elliptic integral into an arbitrary odd number of parts. Unfortunately, Jacobi included no proof, and so Schumacher consulted his friend Gauss about the correctness of the results. Gauss replied that the results were correct but asked Schumacher not to communicate with him further on this topic. Gauss himself was planning to publish his twenty-five-year-old results on elliptic functions and wished to avoid priority disputes.

Schumacher published Jacobi's notes but urged him to supply the proofs as soon as possible. Legendre saw the paper and was eager to see the proofs. Jacobi told Legendre that he had only guessed the theorem for odd-order transformations; in November 1827 he was able to derive a proof by means of the inversion of elliptic integrals. Meanwhile, in September, the first part of Abel's "Recherches" had appeared. It is curious that Jacobi did not refer to Abel's paper and avoided the question of whether he had borrowed any idea from Abel. However, in letters to Legendre, Jacobi heaped praise on Abel's groundbreaking work. In 1828, the second part of the "Recherches" was published, in which Abel added an appendix explaining how his own results could prove Jacobi's theorem. Jacobi's proof was published after Abel had written the second part of his paper. Christoffer Hansteen reported that when Abel saw Jacobi's inversion of the elliptic integral without reference to him, he was visibly shocked. In fact, Abel wrote in a letter to Bernt Holmboe that he published his "Transformations des fonctions elliptiques" in order to supercede Jacobi; he called the paper his "knockout" of Jacobi. In 1828, Gauss wrote Schumacher that Abel's "Recherches" had relieved him of the duty of writing up a third of his investigations on elliptic functions. The other two thirds consisted of the arithmetic-geometric mean and the elliptic modular and theta functions.

Abel's early and tragic death in 1829 cut short the rivalry between Abel and Jacobi. In that same year, Jacobi published the results of two years' labor on elliptic functions, in his *Fundamenta Nova*. This work presented an extensive development of transformation theory and applied it to the derivation of series and product representations of elliptic functions, their moduli, and periods. The problem of converting products into series led Jacobi to the discovery of the triple product identity, though Gauss had anticipated him. In fact, these series and products were theta functions; thus, Jacobi had discovered that elliptic functions could be expressed as quotients of theta functions.

Jacobi earned his doctoral degree from Berlin with a thesis on partial fractions in 1825 and a year later he took a position at Königsberg. Because of his sharp wit and tongue, Jacobi might have faced obstacles to advancement. However, he gained quick recognition from French mathematicians and Legendre in particular, who had presented Jacobi's work to the French Academy in 1827. Finally, with the publication of the *Fundamenta Nova*, Jacobi became known as one of the most outstanding mathematicians in Europe. In a paper of 1835, Jacobi proved two important theorems on functions of one variable: First, he showed that such a function could not have two fundamental periods whose ratio was real; secondly, he showed that such a function could have only two fundamental periods whose ratio was complex. He argued that the functions would otherwise have arbitrarily small periods, a condition he assumed to be absurd. Jacobi had not yet conceived of an analytic function, but when Weierstrass and Cauchy later confirmed his assumption, the proof relied on the fact that the zeros of analytic functions were isolated.

Using a suggestion from Hermite that he use Fourier series, Joseph Liouville (1809–1882) in 1844 reproved Jacobi's first theorem. This work initiated Liouville's definitive theory of elliptic functions. According to Weierstrass, this work was very important, though Liouville published little of it; Weierstrass also criticized Briot and Bouquet for publishing Liouville's ideas without giving him sufficient credit. Liouville's innovation was to define elliptic functions as doubly-periodic functions, rather than as inverses of integrals. He showed that doubly-periodic functions could not be bounded and, in fact, had to have at least two simple poles. Except for two short notes, he did not publish these results, but in 1847 he began a series of lectures on this topic. These lectures were first published in 1880 by the longtime editor of *Crelle's Journal*, Carl Borchardt (1817–1880), who in 1847 had attended the lectures. Reportedly, Borchardt also showed the notes to Jacobi and informed Liouville that Jacobi was extremely impressed. A typeset manuscript of these lectures, said by Weierstrass to have been taken from the notes of Borchardt, was found among Dirichlet's papers after his death in 1859. Apparently, Liouville had intended to publish the notes in his own journal, but had perhaps asked his friend Dirichlet to review the proofs. Why did Liouville not see to it that the proofs were published? This sequence of events remains a mystery, even after Jesper Lützen's comprehensive and detailed book on Liouville, published in 1990. It is interesting to note that the book by Liouville's students, Briot and Bouquet, started with Liouville's approach, but proved the results by using the complex analytic methods of Cauchy and Laurent. Many standard textbooks of today make use of these methods.

At about the same time as Liouville, Gotthold Eisenstein (1823–1852) provided yet another important approach to elliptic functions. Eisenstein was dissatisfied with the inversion of the elliptic integral in Abel and Jacobi. He observed that, since integrals defined single-valued functions, the periodicity of their inverses must be problematic. Eisenstein was a number theorist of extraordinary vision. He viewed the theory of periodic functions as inseparable from number theory. In fact, in 1847 he published a 120-page treatise in *Crelle's Journal*, developing a new basis for elliptic function theory, using double series and double products; this approach was well suited for number theoretic applications. The Weierstrass elliptic function $\wp(z)$ first appeared in this work. Although this paper was soon republished in a collection of Eisenstein's papers, with a foreword by no less a personage than Gauss, it unfortunately did not receive recognition in the nineteenth century. In the preface to his 1975 book *Elliptic Functions According to Eisenstein and Kronecker*, André Weil brought this paper to the attention of the mathematical community. He wrote:

> It is not merely out of an antiquarian interest that the attempt will be made here to resurrect them [Eisenstein's ideas]. Not only do they provide the best introduction to the work of Hecke; but we hope to show that they can be applied quite profitably to some current problems, particularly if they are used in conjunction with Kronecker's late work which is their natural continuation.

Weil's treatment of Eisenstein is thorough and insightful as well as easily available. Thus, the reader may profitably consult Weil for Eisenstein's 1847 work.

Eisenstein's objection to the inversion of the elliptic integral was addressed by Cauchy in the 1840s and then by Riemann in the 1850s. Cauchy had been vigorously developing the theory of complex integration since 1814; this work provided him with the tools necessary to address this problem. Riemann was familiar with Cauchy's work, but he added his original idea of a Riemann surface to study Abelian, and in particular elliptic, functions.

37.2 Abel: Elliptic Functions

Abel's great paper of 1827, "Recherches sur les fonctions elliptiques," was published in two parts in volumes two and three of *Crelle's Journal*. In this paper, Abel defined an elliptic function as the inverse of the elliptic integral

$$\alpha = \int_0^x \frac{dx}{\sqrt{(1-c^2x^2)(1+e^2x^2)}}, \quad 0 \leq x \leq 1/c. \tag{37.6}$$

He expressed x as a function of α and set $x = \phi\alpha$. He noted that α was positive and increasing as x moved from 0 to $1/c$, and set

$$\frac{w}{2} = \int_0^{1/c} \frac{dx}{\sqrt{(1-c^2x^2)(1+e^2x^2)}}. \tag{37.7}$$

Thus, $\phi\alpha$ was positive and increasing in $0 \leq \alpha \leq w/2$ and

$$\phi(0) = 0, \quad \phi(w/2) = 1/c.$$

Moreover, since α changed sign when x was changed to $-x$, he had $\phi(-\alpha) = -\phi(\alpha)$. Abel then formally changed x to ix without a rigorous justification, just as Euler, Laplace, and Poisson had done earlier. Now in an 1814 paper published in 1827 and in papers published as early as 1825, Cauchy discussed functions of complex variables in a more systematic manner. Abel could have employed Cauchy's ideas to give a more rigorous foundation of his theory of elliptic functions. It is possible that Abel was not aware of this aspect of Cauchy's work. In any case, with the above change of variables, Abel set

$$xi = \phi(\beta i), \quad \text{where} \quad \beta = \int_0^x \frac{dx}{\sqrt{(1+c^2x^2)(1-e^2x^2)}}, \tag{37.8}$$

and observed that β was real and positive for $0 \leq x \leq 1/e$. He then set

$$\frac{\tilde{w}}{2} = \int_0^{1/e} \frac{dx}{\sqrt{(1-e^2x^2)(1+c^2x^2)}}, \tag{37.9}$$

so that $-i\phi(\beta i)$ was positive for $0 \leq \beta \leq \tilde{w}/2$; he also had

$$\phi(\tilde{w}i/2) = i/e. \tag{37.10}$$

Abel then defined two auxiliary functions

$$f\alpha = \sqrt{1 - c^2\phi^2\alpha}, \tag{37.11}$$

$$F\alpha = \sqrt{1 + e^2\phi^2\alpha}, \tag{37.12}$$

and noted that when c and e were interchanged, $f(\alpha i)$ and $F(\alpha i)$ were transformed into $F(\alpha)$ and $f(\alpha)$, respectively,

At this point, Abel observed that $\phi(\alpha)$ was already defined for $-w/2 \leq \alpha \leq w/2$, and $\phi(\beta i)$ for $-\tilde{w}/2 \leq \beta \leq \tilde{w}/2$; next he wished to define ϕ for all complex numbers. In order to achieve this, Abel employed the addition formula for ϕ:

$$\phi(\alpha + \beta) = \frac{\phi\alpha \cdot f\beta \cdot F\beta + \phi\beta \cdot f\alpha \cdot F\alpha}{1 + e^2 c^2 \phi^2 \alpha \cdot \phi^2 \beta}. \tag{37.13}$$

He also stated the addition formulas for the auxiliary functions $f\alpha$ and $F\alpha$, remarking that these formulas could be deduced from the results in Legendre's *Exercices* but he wanted to give an alternative derivation. He first deduced the easily proved formulas for the derivatives:

$$\phi'\alpha = f\alpha \cdot F\alpha, \quad f'\alpha = -c^2 \phi\alpha \cdot F\alpha \quad \text{and} \quad F'\alpha = e^2 \phi\alpha \cdot f\alpha.$$

Abel then let r designate the right-hand side of (37.13) and showed that

$$\frac{dr}{d\alpha}$$
$$= \frac{(1 - e^2 c^2 \phi^2 \alpha \phi^2 \beta)[(e^2 - c^2)\phi\alpha\phi\beta + f\alpha f\beta F\alpha F\beta] - 2e^2 c^2 \phi\alpha\phi\beta(\phi^2 \alpha + \phi^2 \beta)}{(1 + e^2 c^2 \phi^2 \alpha \phi^2 \beta)^2}.$$

By symmetry in α and β, Abel concluded that

$$\frac{dr}{d\alpha} = \frac{dr}{d\beta}.$$

He observed that this partial differential equation implied that $r = \psi(\alpha + \beta)$ for some function ψ. Moreover, since $\phi(0) = 0$, $f(0) = 1$, $F(0) = 1$, he set $\beta = 0$ in the expression for r on the right-hand side of (37.13) and found that $r = \phi\alpha$. But $r = \psi\alpha$ when $\beta = 0$. So he had $\phi\alpha = \psi\alpha$ or $\phi = \psi$. This proved the addition formula.

Abel deduced the periodicity of ϕ from the addition formula. He first set $\beta = \pm w/2$ and $\beta = \pm \tilde{w}i/2$ in (37.13). Observing that $f(\pm w/2) = 0$, and $F(\pm \tilde{w}i/2) = 0$, he then obtained the formulas

$$\phi(\alpha \pm w/2) = \pm \phi(w/2) f\alpha / F\alpha = \pm f\alpha/(cF\alpha),$$

$$\phi(\alpha \pm \tilde{w}i/2) = \pm \phi(\tilde{w}i/2) F\alpha / f\alpha = \pm i F\alpha/(ef\alpha).$$

These results implied that

$$\phi\left(\frac{w}{2} + \alpha\right) = \phi\left(\frac{w}{2} - \alpha\right), \tag{37.14}$$

$$\phi\left(\frac{\tilde{w}}{2}i + \alpha\right) = \phi\left(\frac{\tilde{w}}{2}i - \alpha\right), \tag{37.15}$$

$$\phi\left(\alpha \pm \frac{w}{2}\right)\phi\left(\alpha + \frac{\tilde{w}}{2}i\right) = \pm \frac{i}{ce}. \tag{37.16}$$

Replacing α by $\alpha + w/2$ in (37.14), and α by $\alpha + \tilde{w}i/2$ in (37.15), Abel found

$$\phi(\alpha + w) = \phi(-\alpha) = -\phi\alpha, \tag{37.17}$$

$$\phi(\alpha + \tilde{w}i) = -\phi\alpha. \tag{37.18}$$

By means of these formulas, he defined $\phi\alpha$ and $\phi(\alpha i)$ for all real α and then by the addition formula (37.13) he obtained $\phi(\alpha + \beta i)$ for any complex value $\alpha + \beta i$. Moreover, from (37.17) and (37.18), it followed that ϕ was doubly-periodic with periods $2w$ and $2\tilde{w}i$:

$$\phi(2w + \alpha) = -\phi(w + \alpha) = \phi\alpha,$$

$$\phi(2\tilde{w}i + \alpha) = -\phi(\tilde{w}i + \alpha) = \phi\alpha.$$

Abel also determined the zeros and poles of ϕ. For example, from (37.16) he obtained

$$\phi\left(\frac{w}{2} + \frac{\tilde{w}i}{2}\right) = \frac{1}{0}.$$

Then by (37.17) and (37.18),

$$\phi[(m + 1/2)w + (n + 1/2)\tilde{w}i] = \frac{1}{0},$$

when m and n were integers. Then with a little more work, Abel showed that $(m + 1/2)w + (n + 1/2)\tilde{w}i$ were all the poles of ϕ. Similarly, he showed that $mw + n\tilde{w}i$ were all the zeros of ϕ.

37.3 Abel: Infinite Products

Recall that one way of deriving the infinite product for $\sin x$ is to express $\sin(2n + 1)x$ by means of the addition theorem as a polynomial of degree $2n + 1$ in $\sin x$, factorize this polynomial, and then take the limit as n tends to infinity. Abel applied a similar procedure to obtain the product for ϕx. Abel deduced from the addition formula that for a positive integer n,

$$\phi(n + 1)\beta = -\phi(n - 1)\beta + \frac{2\phi(n\beta) f\beta \cdot F\beta}{1 + c^2 e^2 \phi^2(n\beta) \phi^2 \beta}.$$

After some further calculation, he proved by induction that

$$\phi(2n\beta) = \phi\beta . f\beta . F\beta . T, \qquad \phi(2n + 1)\beta = \phi\beta . T_1,$$

where T and T_1 were rational functions of $(\phi\beta)^2$. He then wrote

$$\phi(2n + 1)\beta = \frac{P_{2n+1}}{Q_{2n+1}}, \tag{37.19}$$

where P_{2n+1} and Q_{2n+1} were polynomials of degree $(2n+1)^2$ and $4n(n+1)$, respectively. He noted that the roots of $P_{2n+1} = 0$ were clearly given by

$$x = (-1)^{m+\mu} \phi \left(\beta + \frac{m}{2n+1} w + \frac{\mu}{2n+1} \tilde{w} i \right), \qquad (37.20)$$

for $-n \leq m, \mu \leq n$; by setting $\beta = \alpha/(2n+1)$, the roots were

$$x = (-1)^{m+\mu} \phi \left(\frac{\alpha}{2n+1} + \frac{mw + \mu \tilde{w} i}{2n+1} \right).$$

Abel next expressed $\phi(2n+1)\beta$ as a sum and as a product of terms of the form (37.20). His method was similar to Euler's derivation in the *Introductio in Analysin Infinitorum*, where Euler expressed $\sin(2n+1)x$ as a product of terms of the form $\sin\left(x + \frac{m\pi}{2n+1}\right)$. See chapter 16, section 4 in this connection. Abel wrote

$$P_{2n+1} = Ax^{(2n+1)^2} + \cdots + Bx,$$

$$Q_{2n+1} = Cx^{(2n+1)^2-1} + \cdots + D,$$

so that by (37.19), he had

$$(Ax^{(2n+1)^2} + \cdots + Bx) = \phi(2n+1)\beta \cdot (Cx^{(2n+1)^2-1} + \cdots + D).$$

He observed that the highest-power term had coefficient A, the second highest term had coefficient $-\phi(2n+1)\beta \cdot C$, and the last term was $-\phi(2n+1)\beta \cdot D$. Then, since the roots of the equation were given by (37.20), the sum of the roots could be obtained from the coefficient of the second highest term and the product of the roots from the last term. Thus, he had the equations

$$\phi(2n+1)\beta = \frac{A}{C} \sum_{m=-n}^{n} \sum_{\mu=-n}^{n} (-1)^{m+\mu} \phi \left(\beta + \frac{mw + \mu \tilde{w} i}{2n+1} \right) \qquad (37.21)$$

$$= \frac{A}{D} \prod_{m=-n}^{n} \prod_{\mu=-n}^{n} \phi \left(\beta + \frac{mw + \mu \tilde{w} i}{2n+1} \right). \qquad (37.22)$$

Abel set $\beta = \frac{w}{2} + \frac{\tilde{w}}{2} i + \alpha$, and let $\alpha \to 0$ to determine

$$\frac{A}{C} = \frac{1}{2n+1}. \qquad (37.23)$$

He then let $\beta \to 0$, to obtain

$$(2n+1) = \frac{A}{D} \prod_{m=1}^{n} \phi^2 \left(\frac{mw}{2n+1} \right) \prod_{\mu=1}^{n} \phi^2 \left(\frac{\mu \tilde{w} i}{2n+1} \right)$$

$$\times \prod_{m=1}^{n} \prod_{\mu=1}^{n} \phi^2 \left(\frac{mw + \mu \tilde{w} i}{2n+1} \right) \phi^2 \left(\frac{mw - \mu \tilde{w} i}{2n+1} \right). \qquad (37.24)$$

37.3 Abel: Infinite Products

This gave him an expression for A/D and he substituted it back in (37.22). To simplify the resulting product, he applied a consequence of the addition formula:

$$\frac{\phi(\beta+\alpha)\phi(\beta-\alpha)}{\phi^2\alpha} = -\frac{1-\frac{\phi^2\beta}{\phi^2\alpha}}{1-\frac{\phi^2\beta}{\phi^2\left(\alpha+\frac{w}{2}+\frac{\tilde{w}}{2}i\right)}}.$$

Thus, Abel obtained

$$\phi(2n+1)\beta = (2n+1)\phi\beta \prod_{m=1}^{n}\frac{N_{m,0}}{R_{m,0}} \prod_{\mu=1}^{n}\frac{N_{0,\mu}}{R_{0,\mu}} \prod_{m=1}^{n}\prod_{\mu=1}^{n}\frac{N_{m,\mu}}{R_{m,\mu}} \cdot \frac{\bar{N}_{m,\mu}}{\bar{R}_{m,\mu}}, \qquad (37.25)$$

where $\quad N_{m,\mu} = 1 - \dfrac{\phi^2\beta}{\phi^2\left(\frac{mw+\mu\tilde{w}i}{2n+1}\right)}, \quad \bar{N}_{m,\mu} = 1 - \dfrac{\phi^2\beta}{\phi^2\left(\frac{mw-\mu\tilde{w}i}{2n+1}\right)}, \qquad (37.26)$

$$R_{m,\mu} = 1 - \frac{\phi^2\beta}{\phi^2\left(\frac{w}{2}+\frac{\tilde{w}}{2}i+\frac{mw+\mu\tilde{w}i}{2n+1}\right)},$$

$$\bar{R}_{m,\mu} = 1 - \frac{\phi^2\beta}{\phi^2\left(\frac{w}{2}+\frac{\tilde{w}}{2}i+\frac{mw-\mu\tilde{w}e}{2n+1}\right)}.$$

He then set $\beta = \alpha/(2n+1)$, let $n \to \infty$, and used the formula

$$\lim_{n\to\infty}\frac{\phi^2(\alpha/(2n+1))}{\phi^2(\lambda/(2n+1))} = \frac{\alpha^2}{\lambda^2}$$

to obtain an infinite product for $\phi\alpha$. Abel carried out several pages of calculations to show that the limiting procedure was valid and that the product converged to $\phi\alpha$. It is not clear that Abel's justification was complete. Anyhow, Abel obtained the formula

$$\phi\alpha = \alpha \prod_{m=1}^{\infty}\left(1-\frac{\alpha^2}{(mw)^2}\right) \cdot \prod_{\mu=1}^{\infty}\left(1+\frac{\alpha^2}{(\mu\tilde{w})^2}\right)$$

$$\times \prod_{m=1}^{\infty}\left(\prod_{\mu=1}^{\infty}\left(\frac{1-\frac{\alpha^2}{(mw+\mu\tilde{w}i)^2}}{1-\frac{\alpha^2}{\left(\left(m-\frac{1}{2}\right)w+\left(\mu-\frac{1}{2}\right)\tilde{w}i\right)^2}}\right)\right.$$

$$\left.\times \prod_{\mu=1}^{\infty}\left(\frac{1-\frac{\alpha^2}{(mw-\mu\tilde{w}i)^2}}{1-\frac{\alpha^2}{\left(\left(m-\frac{1}{2}\right)w-\left(\mu-\frac{1}{2}\right)\tilde{w}i\right)^2}}\right)\right). \qquad (37.27)$$

Recall that in 1797 Gauss obtained a similar formula for the particular case of the lemniscatic function. Abel then expressed (37.27) in terms of sines, just as Gauss had done in 1798. This and other similarities in their work led Gauss to remark that Abel followed the same steps as he did in 1798. Abel next rewrote the double product in (37.27):

$$\prod_{m=1}^{\infty}\prod_{\mu=1}^{\infty} \frac{1+\frac{(\alpha+mw)^2}{\mu^2\tilde{w}^2}}{1+\frac{\left(\alpha+\left(m-\frac{1}{2}\right)w\right)^2}{\left(\mu-\frac{1}{2}\right)^2\tilde{w}^2}} \cdot \frac{1+\frac{(\alpha-mw)^2}{\mu^2\tilde{w}^2}}{1+\frac{\left(\alpha-\left(m-\frac{1}{2}\right)w\right)^2}{\left(\mu-\frac{1}{2}\right)^2\tilde{w}^2}} \cdot \left(\frac{1+\frac{\left(m-\frac{1}{2}\right)^2 w^2}{\left(\mu-\frac{1}{2}\right)^2\tilde{w}^2}}{1+\frac{m^2 w^2}{\mu^2\tilde{w}^2}}\right)^2.$$

Then by means of the product for $\sin x$ given by

$$\sin x = x \prod_{\mu=1}^{\infty}\left(1-\frac{x^2}{\mu^2 \pi^2}\right),$$

and using the addition formula for sine given by

$$\sin(a-b).\sin(a+b) = \sin^2 a - \sin^2 b,$$

he obtained

$$\phi\alpha = \frac{\tilde{w}}{\pi}\frac{s}{i}\prod_{m=1}^{\infty}\frac{1-s^2/A_m^2}{1-s^2/B_m^2}$$

where

$$s = \sin(\alpha\pi i/\tilde{w}), \quad A_m = \sin(mw\pi i/\tilde{w}), \quad B_m = \cos((m-1/2)w\pi i/\tilde{w}).$$

Finally, by the use of $\phi(i\alpha) = i\phi\alpha$, he obtained his product formula:

$$\phi\alpha = \frac{w}{\pi}\sin\frac{\alpha\pi}{w}\prod_{m=1}^{\infty}\frac{1+\frac{4\sin^2(\alpha\pi/w)}{(e^{m\tilde{w}\pi/w}-e^{-m\tilde{w}\pi/w})^2}}{1-\frac{4\sin^2(\alpha\pi/w)}{(e^{(2m-1)\tilde{w}\pi/(2w)}+e^{-(2m-1)\tilde{w}\pi/(2w)})^2}}. \qquad (37.28)$$

Abel also used the series (37.21) to obtain various other formulas, including

$$\phi\left(\frac{\alpha w}{2}\right) = \frac{4\pi}{w}\left(\frac{e^{\pi/2}}{1+e^{\pi}}\sin\frac{\alpha\pi}{2} - \frac{e^{3\pi/2}}{1+e^{3\pi}}\sin\frac{3\alpha\pi}{2} + \frac{e^{5\pi/2}}{1+e^{5\pi}}\sin\frac{5\alpha\pi}{2} - \cdots\right).$$

37.4 Abel: Division of Elliptic Functions and Algebraic Equations

In his 1827 paper, Abel considered Gauss's algebraic theory on the division of periodic functions and extended it to the division of doubly-periodic functions. To understand Abel's motivation, recall that from a study of Viète, Newton determined that for an odd number n, $\sin nx$ could be expressed as a polynomial of degree n in $\sin x$. Note that a similar result holds for $\cos nx$. Gauss proved that these polynomials could be solved algebraically; Euler and Vandermonde had earlier done this for values of n up to eleven. Abel determined from the addition theorem that for a positive integer n, $\phi(2n+1)\alpha$ was a rational function of $\phi\alpha$ such that the numerator took the form $xR(x^2)$, where R was a polynomial of degree $n(2n+2)$ and $x = \phi\alpha$. His problem was to find out whether R could be solved algebraically, that is, by radicals. He discovered that he could employ Lagrange resolvents, an idea due to Waring, Vandermonde, and Lagrange, as Gauss had also done. However, Abel's problem was more complicated than Gauss's and took him deeper into the theory of equations.

As a mathematical aside, we briefly discuss Abel's related contributions to the theory of equations. His work in elliptic function theory gave him glimpses into the nature of algebraically solvable equations. In particular, he sought to determine solvability in terms of the structure of the roots of the equation. In an 1826 letter to Crelle, Abel stated a result on the form of the roots of a solvable quintic. He later generalized this result to irreducible equations of prime degree, published posthumously in the first edition of his collected papers of 1837. This paper contained the remarkable theorem that an irreducible equation of prime degree was solvable by radicals if and only if all its roots were rational functions of any two of the roots. Galois rediscovered this theorem a few years later, but his work arose out of a study of those permutations of the roots preserving algebraic relations among the roots. Because the group theory of algebraic equations, developed by Galois, gained recognition before Abel's theory, based on structure of roots, Abel's theorems have now become recast and known in terms of groups. It might be fruitful to make a parallel study of the two approaches.

Recall that Abel proved that $\phi(2n+1)\beta$ was a rational function of $x = \phi\beta$ whose numerator took the form $xR(x^2)$ where R was a polynomial of degree $n(2n+2)$. Abel then proved the important theorem that the solutions of $R=0$ depended on the solutions of a certain equation of degree $2n+2$ with coefficients that were rational functions of c and e. He proceded to demonstrate that if the latter equation could be solved by radicals, then so could $R=0$. He went on to observe that, in general, this equation was not solvable by radicals but could be solved in particular cases, such as for $e=c$, $e=\sqrt{3}c$, $e=(2\pm\sqrt{3})c$, etc. The case $e=c$ corresponded to the lemniscatic function and had already been discussed in Gauss's unpublished work, at least in special cases.

Abel's proof of this theorem was lengthy. We present a brief summary, using his notation. First note that by (37.25) and the fact that the zeros of ϕ occur at $mw + in\tilde{w}$, it follows that the solutions of $R=0$ must be given by

$$r = \phi^2\left(\frac{mw \pm i\mu\tilde{w}i}{2n+1}\right).$$

By periodicity of ϕ, the number of different values of r can be reduced to the $n(2n+2)$ values given by

$$r_\nu = \phi^2\left(\frac{\nu w}{2n+1}\right), \quad r_{\nu,m} = \phi^2\left(\nu\frac{mw + i\tilde{w}}{2n+1}\right), \tag{37.29}$$

where $1 \le \nu \le n$, $0 \le m \le 2n$. Now let w' denote any quantity of the form $mw + i\mu\tilde{w}$ and define ψ by the equation

$$\psi\left(\phi^2\left(\frac{w'}{2n+1}\right)\right) = \theta\left(\phi^2\left(\frac{w'}{2n+1}\right), \phi^2\left(\frac{2w'}{2n+1}\right), \ldots, \phi^2\left(\frac{nw'}{2n+1}\right)\right), \tag{37.30}$$

where θ is a rational symmetric function of the n quantities. It is clear from the definition of ψ that

$$\psi\left(\phi^2\left(\frac{vw'}{2n+1}\right)\right) = \psi\left(\phi^2\left(\frac{w'}{2n+1}\right)\right), \quad 1 \le v \le n. \tag{37.31}$$

In particular,

$$\psi r_v = \psi r_1; \quad \psi r_{v,m} = \psi r_{1,m}, \quad 1 \le v \le n. \tag{37.32}$$

The aforementioned equation of degree $2n+2$ can be given by

$$(p - \psi r_1)(p - \psi r_{1,0})(p - \psi r_{1,1}) \cdots (p - \psi r_{1,2n})$$
$$= q_0 + q_1 p + q_2 p^2 + \cdots + q_{2n+1} p^{2n+1} + p^{2n+2}. \tag{37.33}$$

It is easy to see that $q_0, q_1, \ldots, q_{2n+1}$ are rational functions of c and e. Note that the sum of the kth powers of the roots of (37.33) are symmetric functions of the $n(2n+2)$ roots r_v and $r_{v,m}$ of $R = 0$, where r_v and $r_{v,m}$ are given by (37.29). To see this, observe that

$$(\psi r_1)^k = \frac{1}{n}[(\psi r_1)^k + (\psi r_2)^k + \cdots + (\psi r_n)^k],$$

$$(\psi r_{1,m})^k = \frac{1}{n}[(\psi r_{1,m})^k + (\psi r_{2,m})^k + \cdots + (\psi r_{n,m})^k], \quad 0 \le m \le 2n,$$

and

$$(\psi r_1)^k + (\psi r_{1,0})^k + (\psi r_{1,1})^k + \cdots + (\psi r_{1,2n})^k$$
$$= \frac{1}{n}\left[(\psi r_1)^k + (\psi r_2)^k + \cdots + (\psi r_n)^k\right]$$
$$+ \frac{1}{n}\left[(\psi r_{1,0})^k + (\psi r_{2,0})^k + \cdots + (\psi r_{n,0})^k\right]$$
$$\cdots \cdots \cdots$$
$$+ \frac{1}{n}\left[(\psi r_{1,2n})^k + (\psi r_{2,2n})^k + \cdots + (\psi r_{n,2n})^k\right].$$

Since the coefficients of the polynomial R are rational functions of c and e, we may now conclude that each sum of the kth powers of the roots of the polynomial (37.33) is a rational function of c and e. Since the power sum symmetric functions form a basis for the symmetric functions, it follows that $q_0, q_1, \ldots, q_{2n+1}$ are rational functions of c and e.

Next, we show that if $p = \psi r_1$ and $q = \theta r_1$ are rational symmetric functions of $r_1, r_2, \ldots, r_n$, then q can be determined in terms of p. Note that a similar result holds for $\psi r_{1,m}$ and $\theta r_{1,m}$. For $k = 0, 1, \ldots, 2n+1$, set

$$s_k = (\psi r_1)^k \theta r_1 + (\psi r_{1,0})^k \theta r_{1,0} + \cdots + (\psi r_{1,2n})^k \theta r_{1,2n}. \tag{37.34}$$

37.4 Abel: Division of Elliptic Functions and Algebraic Equations

We prove that s_k can be expressed as a rational function of c and e. Note that

$$(\psi r_1)^k \theta r_1 = (\psi r_v)^k \theta r_v = \frac{1}{n}[(\psi r_1)^k \theta r_1 + (\psi r_2)^k \theta r_2 + \cdots + (\psi r_n)^k \theta r_n];$$

$$(\psi r_{1,m})^k \theta r_{1,m} = (\psi r_{v,m})^k \theta r_{v,m} = \frac{1}{n}[(\psi r_{1,m})^k \theta r_{1,m} + \cdots + (\psi r_{n,m})^k \theta r_{n,m}].$$

When these values are substituted in (37.34), we observe that s_k is a symmetric rational function of the roots of $R = 0$; therefore, s_k, $k = 0, 1, \ldots, 2n+1$, are rational functions of c and e. We can apply Cramer's rule to solve these equations for $\theta r_1, \theta r_{1,0}, \ldots, \theta r_{1,2n}$ in terms of rational functions of $\psi r_1, \ldots, \psi r_{1,2n}$. This result in turn implies that the coefficients of the equation

$$(r - r_1)(r - r_2) \cdots (r - r_n) = r^n + p_{n-1}f r^{n-1} + p_{n-2} r^{n-2} + \cdots + p_1 r + p_0 \quad (37.35)$$

can be determined by the equation (37.33). There are $2n + 1$ additional equations of degree n with roots $r_{1,v}, \ldots, r_{n,v}$ for $0 \leq v \leq 2n$; the coefficients of these equations are also determined by (37.33).

In this way, Abel reduced the problem of solving the equation $R = 0$ of degree $n(2n \pm 2)$ to that of solving $2n + 2$ equations of the form (37.35). We demonstrate by means of the Lagrange resolvent (Gauss's method for solving the cyclotomic equation) that the solutions of these equations can be expressed in terms of the solutions to (37.33). Let

$$\phi^2\left(\frac{w'}{2n+1}\right), \quad \phi^2\left(\frac{2w'}{2n+1}\right), \quad \ldots, \quad \phi^2\left(\frac{nw'}{2n+1}\right)$$

denote the solutions of (37.35), where w' stands for w or $mw + i\tilde{w}$. By a theorem of Gauss, there exists a number α generating the numbers $1, 2, \ldots, 2n$ (modulo $2n + 1$). Then by the periodicity of ϕ, the set

$$\phi^2(\epsilon), \phi^2(\alpha\epsilon), \phi^2(\alpha^2\epsilon), \ldots, \phi^2(\alpha^{n-1}\epsilon),$$

where $\epsilon = w'/(2n+1)$, represents all the solutions of (37.35). We omit Abel's straightforward proof of this result.

Now let θ denote any imaginary root of $\theta^n - 1 = 0$, and define the Lagrange resolvent

$$\psi(\epsilon) = \phi^2(\epsilon) + \phi^2(\alpha\epsilon)\theta + \phi^2(\alpha^2\epsilon)\theta^2 + \cdots + \phi^2(\alpha^{n-1}\epsilon)\theta^{n-1}. \quad (37.36)$$

It is clear that $\psi(\epsilon)$ is a rational function of $\phi^2(\epsilon)$, expressible as $\psi(\epsilon) = \chi(\phi^2(\epsilon))$. By a simple calculation involving roots of unity, we can show that

$$\psi(\alpha^m \epsilon) = \theta^{-m} \psi(\epsilon) \quad \text{or} \quad \psi(\epsilon) = \theta^m \chi(\phi^2(\alpha^m \epsilon)).$$

This implies that $(\psi\epsilon)^n = [\chi(\phi^2(\alpha^m\epsilon))]^n$. Taking $m = 0, 1, \ldots, n-1$ and adding we arrive at

$$n(\psi\epsilon)^n = \left[\chi\left(\phi^2(\epsilon)\right)\right]^n + \left[\chi\left(\phi^2(\alpha\epsilon)\right)\right]^n + \cdots + \left[\chi\left(\phi^2(\alpha^{n-1}\epsilon)\right)\right]^n.$$

The expression on the right is a rational symmetric function of $\phi^2\epsilon, \phi^2(\alpha\epsilon), \ldots, \phi^2(\alpha^{n-1}\epsilon)$. That is, it is a rational symmetric function of the roots of (37.35). Therefore, $(\phi\epsilon)^n = v$ is a rational function of $p_0, p_1, \ldots, p_{n-1}$ and

$$\sqrt[n]{v} = \phi^2\epsilon + \theta\phi^2(\alpha\epsilon) + \theta^2\phi^2(\alpha^2\epsilon) + \cdots + \theta^{n-1}\phi^2(\alpha^{n-1}\epsilon). \tag{37.37}$$

Note also that v is a rational function of the roots of (37.33); so if (37.33) can be solved by radicals, then v can be expressed in terms of radicals. By changing θ to $\theta^2, \theta^3, \ldots, \theta^{n-1}$ and denoting the corresponding values of v by $v_2, v_3, \ldots, v_{n-1}$, we have

$$\sqrt[n]{v_k} = \phi^2(\epsilon) + \theta^k\phi^2(\alpha\epsilon) + \cdots + \theta^{k(n-1)}\phi^2(\alpha^{n-1}\epsilon), \quad k = 1, 2, \ldots, n-1. \tag{37.38}$$

When these $n-1$ equations are combined with the equation

$$-p_{n-1} = \phi^2(\epsilon) + \phi^2(\alpha\epsilon) + \cdots + \phi^2(\alpha^{n-1}\epsilon),$$

we can easily solve these n linear equations to get

$$\phi^2(\alpha^m\epsilon) = \frac{1}{n}(-p_{n-1} + \theta^{-m}\sqrt[n]{v_1} + \theta^{-2m}\sqrt[n]{v_2} + \cdots + \theta^{-(n-1)m}\sqrt[n]{v_{n-1}}), \tag{37.39}$$

for $m = 0, 1, \ldots, n-1$. It can also be shown that $s_k = \sqrt[n]{v_k}/(\sqrt[n]{v_1})^k$ is a rational function of $p_0, p_1, \ldots, p_{n-1}$. For this purpose, it is sufficient to check that s_k is unchanged by $\epsilon \to \alpha^m\epsilon$. This gives us Abel's final formula for $\phi^2(\alpha^m\epsilon)$:

$$\phi^2(\alpha^m\epsilon) = \frac{1}{n}\left(-p_{n-1} + \theta^{-m}v^{\frac{1}{n}} + s_2\theta^{-2m}v^{\frac{2}{n}} + \cdots + s_{n-1}\theta^{-(n-1)m}v^{\frac{n-1}{m}}\right), \tag{37.40}$$

for $m = 0, 1, \ldots, n-1$. This implies that if v can be expressed in terms of radicals, then $R = 0$ can be solved by radicals.

37.5 Abel: Division of the Lemniscate

Recall Abel's remark that in the case $e/c = 1$, the division points $\phi^2\left(\frac{kw'}{2n+1}\right)$ could be obtained by solving an algebraic equation by radicals. When $e = c = 1$, Abel's integral (37.6) is reduced to

$$\alpha = \int_0^x \frac{dx}{\sqrt{1-x^4}}, \quad \text{or} \quad x = \phi\alpha. \tag{37.41}$$

It is easy to check that

$$\phi(\alpha i) = i\phi\alpha \quad \text{and} \tag{37.42}$$

$$w/2 = \tilde{w}/2 = \int_0^1 \frac{dx}{\sqrt{1-x^4}}. \tag{37.43}$$

Abel applied the addition formula to show that for $m + \mu$ odd and $x = \phi\delta$,

$$\phi(m + \mu i)\delta = x\psi(x^2), \tag{37.44}$$

37.5 Abel: Division of the Lemniscate

for some rational function ψ. Then by changing δ to $i\delta$ and using (37.42), he obtained $\phi(m+\mu i)\delta = x\psi(-x^2)$, or $\psi(-x^2) = \psi(x^2)$. He therefore concluded that

$$\phi(m+\mu i)\delta = x \cdot \frac{T}{S}, \tag{37.45}$$

where T and S were polynomials in powers of x^4. This very significant result showed that the elliptic function $\phi\delta$ permitted complex multiplication, that is, $\phi(m+\mu i)\delta$ could be expressed as a rational function of $\phi\delta$. As an example, he noted that

$$\phi(2+i)\delta = ix \cdot \frac{1-2i-x^4}{1-(1-2i)x^4}, \tag{37.46}$$

a result proved by Gauss in an unpublished work, wherein he also divided the lemniscate into $5 = (2+i)(2-i)$ parts.

Abel showed how (37.45) could be applied to the problem of dividing the lemniscate into $4\nu+1$ parts. By Fermat's theorem on sums of two squares, Abel could write

$$\alpha^2 + \beta^2 = 4\nu + 1 = (\alpha + i\beta)(\alpha - i\beta),$$

where $\alpha + \beta$ was odd. With $m = \alpha$, $\mu = \beta$, and $\delta = w/(\alpha + i\beta)$, he could use (37.45) to obtain $x = \phi(\delta)$ as a root of $T = 0$. By using the periodicity of ϕ and the addition formula, Abel proved that

$$\pm \phi\left(\frac{w}{\alpha + i\beta}\right), \pm \phi\left(\frac{2w}{\alpha + i\beta}\right), \ldots, \pm \phi\left(\frac{\alpha^2+\beta^2-1}{2} \cdot \frac{w}{\alpha + \beta i}\right)$$

comprised all the roots of the polynomial T. By setting $T(x) = R(x^2)$, he obtained

$$\phi^2(\delta), \phi^2(2\delta), \phi^2(3\delta), \ldots, \phi^2(2\nu\delta) \tag{37.47}$$

as all the roots of $R = 0$. Next, Abel once again applied Gauss's method. He first showed that for a primitive root ϵ modulo $4\nu+1 = \alpha^2+\beta^2$, the set $\phi^2(\epsilon^m\delta)$, $m = 0, 1, \ldots, 2\nu-1$, was equal to the set given in (37.47). He then referred to the method of Lagrange resolvents to conclude that

$$\phi^2(\epsilon^m\delta) = \frac{1}{2\nu}\left(A + \theta^{-m} \cdot v^{\frac{1}{2\nu}} + s_2\theta^{-2m} \cdot v^{\frac{2}{2\nu}} + \cdots + s_{2\nu-1}\theta^{-(2\nu-1)m} \cdot v^{\frac{2\nu-1}{2\nu}}\right), \tag{37.48}$$

where θ was an imaginary root of $\theta^{2\nu} - 1 = 0$, and v, s_k were determined by the expressions

$$v = \left[\phi^2(\delta) + \theta \cdot \phi^2(\epsilon\delta) + \theta^2 \cdot \phi^2(\epsilon^2\delta) + \cdots + \theta^{2\nu-1} \cdot \phi^2(\epsilon^{2\nu-1}\delta)\right]^{2\nu}, \tag{37.49}$$

$$s_k = \frac{\phi^2(\delta) + \theta^k \cdot \phi^2(\epsilon\delta) + \cdots + \theta^{(2\nu-1)k} \cdot \phi^2(\epsilon^{2\nu-1}\delta)}{[\phi^2(\delta) + \theta \cdot \phi^2(\epsilon\delta) + \cdots + \theta^{2\nu-1} \cdot \phi^2(\epsilon^{2\nu-1}\delta)]^k}, \tag{37.50}$$

$$A = \phi^2(\delta) + \phi^2(\epsilon\delta) + \cdots + \phi^2(\epsilon^{2\nu-1}\delta). \tag{37.51}$$

Moreover, the expressions (37.49), (37.50), and (37.51) could be written as rational functions of the coefficients of $R = 0$. Recall that the coefficients of $R = 0$ took the form $a+bi$ with a,b rational. Thus, v, s_k and A were of the form $c+id$, with c and d rational.

Abel then noted that if $4v+1 = 1+2^n$, then $2v = 2^{n-1}$ and the values in (37.48) could be computed by repeatedly taking square roots. Thus, the values of $\phi(mw/(\alpha+i\beta))$ could be evaluated by taking square roots; hence, by applying the addition formula, the value of $\phi(w/(4v+1))$ could be so determined. This proved the result that the lemniscate could be geometrically divided into 2^n+1 parts, when this was a prime number. Recall that in his *Disquisitiones*, Gauss had stated that this was true.

37.6 Jacobi's Elliptic Functions

In his *Fundamenta Nova*, Jacobi presented a detailed account of his theory of elliptic functions. He inverted the elliptic integral

$$u = \int_0^\phi \frac{d\phi}{\sqrt{1-k^2\sin^2\phi}} = \int_0^x \frac{dx}{\sqrt{(1-x^2)(1-k^2x^2)}}$$

by defining the function $x = \sin\operatorname{am} u$, where $\phi = \operatorname{am} u$, calling ϕ the amplitude of u. He noted that, in general, any trigonometric function of ϕ, such as $\cos\phi = \cos\operatorname{am} u$, $\tan\phi = \tan\operatorname{am} u$, could be defined in this manner. Jacobi worked mainly with the functions $\sin\phi$, $\cos\phi$, and

$$\Delta\operatorname{am} u = \sqrt{1-k^2\sin^2\operatorname{am} u} = \frac{d\operatorname{am} u}{du}.$$

Following Gudermann, we employ modern notation for these functions: $\operatorname{sn} u$, $\operatorname{cn} u$, and $\operatorname{dn} u$. When we emphasize dependence on modulus k, we write $\operatorname{sn}(u,k)$, $\operatorname{cn}(u,k)$, and $\operatorname{dn}(u,k)$. The complementary modulus k', defined by $k^2 + k'^2 = 1$, is also important. Legendre denoted by K the complete elliptic integral obtained by taking $x = 1$ in the preceding integral; he denoted the corresponding complete integral for the modulus k' by K'.

Jacobi listed the addition theorems and related identities, results he obtained directly from those of Euler and Legendre:

$$\operatorname{sn}(u+v) = (\operatorname{sn} u \operatorname{cn} v \operatorname{dn} v + \operatorname{sn} v \operatorname{cn} u \operatorname{dn} u)/D,$$
$$\operatorname{cn}(u+v) = (\operatorname{cn} u \operatorname{cn} v - \operatorname{sn} u \operatorname{dn} u \operatorname{sn} v \operatorname{dn} v)/D,$$
$$\operatorname{dn}(u+v) = (\operatorname{dn} u \operatorname{dn} v - k^2 \operatorname{sn} u \operatorname{cn} u \operatorname{sn} v \operatorname{cn} v)/D,$$
$$\operatorname{sn}(u+v)\operatorname{sn}(u-v) = (\operatorname{sn}^2 u - \operatorname{sn}^2 v)/D,$$

where $D = 1 - k^2 \operatorname{sn}^2 u \operatorname{sn}^2 v$.

Jacobi then extended the domain of the elliptic functions by applying the transformation, later called Jacobi's imaginary transformation:

$$\sin\phi = i\tan\psi. \tag{37.52}$$

37.6 Jacobi's Elliptic Functions

This implied $\cos\phi = \sec\psi$ and $d\phi = id\psi/\cos\psi$ and

$$\frac{d\phi}{\sqrt{1-k^2\sin^2\phi}} = \frac{id\psi}{\sqrt{\cos^2\psi + k^2\sin^2\psi}} = \frac{id\psi}{\sqrt{1-k'k'\sin^2\psi}}.$$

Jacobi used this to write

$$\sin\text{am}(iu,k) = i\tan\text{am}(u,k'),$$
$$\cos\text{am}(iu,k) = \sec\text{am}(u,k'),$$
$$\tan\text{am}(iu,k) = i\sin\text{am}(u,k'),$$

and other similar formulas. From these results, Jacobi deduced that $\text{sn}(u,k)$ had periods $4K$ and $2iK'$; $\text{cn}(u,k)$ had periods $4K$ and $2K+2iK'$; and $\text{dn}(u,k)$ had periods $2K$ and $4iK'$. Moreover, in a period parallelogram, $\text{sn}\,u$ had zeros at $u=0$ and at $u=2K$ and had poles at iK' and $2K+iK'$. Jacobi had similar results for $\text{cn}\,u$ and $\text{dn}\,u$.

We note an application of Jacobi's imaginary transformation to the quadratic transformations discussed earlier. This will provide an introduction to the higher-order transformations appearing in the next two sections. Recall that Landen's quadratic transformation

$$y = \frac{(1+k')x\sqrt{1-x^2}}{\sqrt{1-k^2x^2}} \tag{37.53}$$

produces the differential relation

$$\frac{dy}{\sqrt{(1-y^2)(1-\lambda^2y^2)}} = \frac{(1+k')dx}{\sqrt{(1-x^2)(1-k^2x^2)}}, \tag{37.54}$$

where $\lambda = (1-k')/(1+k')$ or, in other words, $k = 2\sqrt{\lambda}/(1+\lambda)$. This algebraic relation between the moduli λ and k is called a modular equation. By means of this relation, we may write (37.54) as

$$\frac{(1+\lambda)dy}{\sqrt{(1-y^2)(1-\lambda^2y^2)}} = \frac{2dx}{\sqrt{(1-x^2)(1-k^2x^2)}}. \tag{37.55}$$

If we integrate the differential on the right-hand side from 0 to 1, we get $2K$. However, as x increases from 0 to $1/\sqrt{1+k'}$, y increases from 0 to 1; and as x continues to increase to 1, y decreases from 1 to 0. Thus, if Λ denotes the complete integral corresponding to the modulus λ, we get the equation

$$2(1+\lambda)\Lambda = 2K \quad \text{or} \quad K = (1+\lambda)\Lambda. \tag{37.56}$$

Now note that the second quadratic transformation of Gauss

$$z = \frac{(1+\lambda)y}{1+\lambda y^2} \tag{37.57}$$

produces the differential relation

$$\frac{dz}{\sqrt{(1-z^2)(1-\gamma^2z^2)}} = \frac{(1+\lambda)dy}{\sqrt{(1-y^2)(1-\lambda^2y^2)}}, \tag{37.58}$$

where $\gamma = 2\sqrt{\lambda}/(1+\lambda)$. We can therefore take $\gamma = k$ and apply (37.55) followed by (37.58) to obtain duplication:

$$\frac{dz}{\sqrt{(1-z^2)(1-k^2z^2)}} = \frac{2\,dx}{\sqrt{(1-x^2)(1-k^2x^2)}}.$$

One of Jacobi's earliest discoveries was that there were, similarly, two cubic transformations, and when these were applied consecutively, they produced triplication. He then extended this to general odd order transformations.

Jacobi's imaginary transformation (37.52) when written in terms of x and y amounts to setting

$$x = \frac{iX}{\sqrt{1-X^2}} \quad \text{and} \quad y = \frac{iY}{\sqrt{1-Y^2}}. \tag{37.59}$$

When these expressions for x and y are substituted in Landen's transformation (37.53), we obtain, after simplification, Gauss's form of the transformation:

$$\frac{Y}{\sqrt{1-Y^2}} = \frac{(1+k')X}{\sqrt{1-X^2)(1-k^2X^2)}}$$

$$\text{or} \quad Y = \frac{(1+k')X}{1+k'X^2}.$$

Moreover, the differential relation (37.55) converts to

$$\frac{(1+\lambda)\,dY}{\sqrt{(1-Y^2)(1-\lambda^2Y^2)}} = \frac{2\,dX}{\sqrt{(1-X^2)(1-k'^2X^2)}}.$$

Observe that, since X and Y increase simultaneously from 0 to 1, this relation can be written in terms of complete integrals:

$$2K' = (1+\lambda)\Lambda'.$$

Dividing this equation by (37.56) gives another form of the modular relation, also used by Legendre:

$$\frac{2K'}{K} = \frac{\Lambda'}{\Lambda}. \tag{37.60}$$

As one might expect, when Jacobi's imaginary transformation is applied to Gauss's transformation, one obtains Landen's transformation, except that k and λ are converted to their complements k' and λ'. These results also carry over to general transformations.

37.7 Jacobi: Cubic and Quintic Transformations

In a letter of June 13, 1827, Jacobi communicated to Schumacher, editor of the *Astronomische Nachrichten*, two cubic and two quintic transformations. Jacobi's first result stated: If we set

$$\sin\phi = \frac{\sin\psi \left(ac + \left(\frac{a-c}{2}\right)^2 \sin^2\psi\right)}{cc + \frac{a-c}{2} \cdot \frac{a+3c}{2} \sin^2\psi}, \tag{37.61}$$

37.7 Jacobi: Cubic and Quintic Transformations

we obtain

$$\frac{d\phi}{\sqrt{a^3c - \frac{a-c}{2}\left(\frac{a+3c}{2}\right)^3 \sin^2\phi}} = \frac{d\psi}{\sqrt{c^3a - \left(\frac{a-c}{2}\right)^3 \frac{a+3c}{2} \sin^2\psi}}. \tag{37.62}$$

If, in addition,

$$\sin\psi = \frac{\sin\theta\left(-3ac + \left(\frac{a+3c}{2}\right)^2 \sin^2\theta\right)}{aa - 3\frac{a-c}{2} \cdot \frac{a+3c}{2} \sin^2\theta} \quad \text{and} \tag{37.63}$$

$$\chi = \frac{a-c}{2c}\left(\frac{a+3c}{2a}\right)^3, \tag{37.64}$$

then we have

$$\frac{d\phi}{\sqrt{1 - \chi \sin^2\phi}} = \frac{3\,d\theta}{\sqrt{1 - \chi \sin^2\theta}}. \tag{37.65}$$

Note that (37.61) and (37.63) are Jacobi's two cubic transformations, and when applied in succession, they produce the triplication (37.65) for the modulus $k^2 = \chi$, given by (37.64).

Jacobi's second result stated: If we set $a^3 = 2b(1+a+b)$ and

$$\sin\phi = \frac{\sin\psi(1 + 2a + (aa + 2ab + 2b)\sin^2\psi + bb\sin^4\psi)}{1 + (aa + 2a + 2b)\sin^2\psi + b(b + 2a)\sin^4\psi}, \tag{37.66}$$

we get

$$\int \frac{d\phi}{\sqrt{(a-2b)(1+2a)^2 - (2-a)(b+2a)^2 \sin^2\phi}} = \int \frac{d\psi}{\sqrt{a - 2b - bb(2-a)\sin^2\psi}}.$$

Also, if

$$\alpha = \frac{2-a}{1+2a},$$

$$\beta = -\frac{b+2a}{1+2a} \cdot \frac{2-a}{a-2b},$$

$$\chi = \frac{2-a}{a-2b} \cdot \left(\frac{b+2a}{1+2a}\right)^2,$$

$$\sin\psi = \frac{\sin\theta(1 + 2\alpha + (\alpha\alpha + 2\alpha\beta + 2\beta)\sin^2\theta + \beta\beta\sin^4\theta)}{1 + (\alpha\alpha + 2\alpha + 2\beta)\sin^2\theta + \beta(\beta + 2\alpha)\sin^4\theta}, \tag{37.67}$$

then we have

$$\int \frac{d\phi}{\sqrt{1 - \chi \sin^2\phi}} = 5 \int \frac{d\theta}{\sqrt{1 - \chi \sin^2\theta}}. \tag{37.68}$$

Here (37.66) and (37.67) are Jacobi's quintic transformations, and they together produce the quinsection given by (37.68).

In his April 12, 1828 letter to Legendre, Jacobi wrote that he found (37.63) and (37.67) by trial and error. But he explained that he had found the cubic and quintic transformations (37.61) and (37.66) on the basis of the general algebraic theory of transformations he had developed in March 1827. For this theory, he considered the transformation $y = U/V$, where U and V were polynomials in x differing in degree by at most one, and such that

$$\frac{dy}{\sqrt{Y}} = \frac{1}{M}\frac{dx}{\sqrt{X}}, \tag{37.69}$$

where X and Y were quartics in x and y, respectively, and M was a constant depending on the constants in X and Y. In particular, he took $X = (1-x^2)(1-k^2x^2)$ and $Y = (1-y^2)(1-\lambda^2 y^2)$. By substituting $y = U/V$ in (37.69), he obtained the relation

$$\frac{dy}{\sqrt{(1-y^2)(1-\lambda^2 y^2)}} = \frac{\left(V\frac{dU}{dx} - U\frac{dV}{dx}\right)dx}{\sqrt{(V^2-U^2)(V^2-\lambda^2 U^2)}}.$$

He noted that if U and V were of degree p, the numerator of the expression on the right was a polynomial of degree $2p-2$, while the expression inside the radical was of degree $4p$. Moreover, since for any number α,

$$(V-\alpha U)\frac{dU}{dx} - \frac{d(V-\alpha U)}{dx}U = V\frac{dU}{dx} - U\frac{dV}{dx},$$

it followed that if any of the factors $V \pm U$, $V \pm \lambda U$ in the denominator had a square factor $(1-\beta x)^2$, then $1-\beta x$ was a factor of the numerator polynomial. Thus, if the denominator was of form $T^2 X$, where X was a quadratic and T was of degree $2p-2$, then

$$M = \frac{T}{V\frac{dU}{dx} - U\frac{dV}{dx}}$$

was a constant depending only on the constants in X and Y. Jacobi noted that the problem of finding $y = U/V$ was determinate because U/V had $2p+1$ constants of which $2p-2$ could be determined by requiring that $(V^2-U^2)(V^2-\lambda^2 U^2) = T^2 X$. This left three undetermined constants, and that number could not be reduced because x could be replaced by $(a+bx)/(1+dx)$, resulting in a similar relation. Thus, he looked for polynomials U and V such that $V+U = (1+x)AA$, $V-U = (1-x)BB$, $V+\lambda U = (1+kx)CC$, $V-\lambda U = (1-kx)DD$. He also noted that y was an odd function of x, and hence $U = xF(x^2)$ and $V = \phi(x^2)$. Moreover, the equation (37.69) remained invariant when y was replaced by $1/(\lambda y)$ and x by $1/(kx)$. This observation allowed him to determine explicit algebraic relations between k, λ, and the coefficients of U and V. In particular, it was possible to obtain for small values of p (the degree of U) the explicit algebraic relations satisfied by k and λ. These relations are called modular equations. So if either k or λ is given, the other can be found as one of the roots of this equation. The value of M can also be determined. It can be proved that if p is an odd prime, then the modular equation is irreducible and of order $p+1$. Thus, for a given k, there are $p+1$ different values of λ and each one leads to a distinct transformation of order p. We note that Legendre and Jacobi took k^2 to be between 0 and 1. The modular equation gave $p+1$ values of λ of which two were real, one greater than k and the other less than k. Jacobi denoted the smaller value by λ and the larger by λ_1; he called the transformation with the smaller λ the first transformation and the other the second transformation. He noted that when the two transformations were applied one after the other, the result was a multiplication by p of the differential. So if $y = U/V$ was the first transformation and $z = U_1/V_1$ the second, then

$$\frac{dz}{\sqrt{(1-z^2)(1-k^2z^2)}} = \frac{p\,dx}{\sqrt{(1-x^2)(1-k^2x^2)}}.$$

Jacobi worked out the algebraic theory of transformations only for the cubic and quintic cases; he needed the theory of elliptic functions to develop the higher-order transformations. He called this the transcendental theory of transformations. He may have obtained from Abel the idea of the elliptic function as the inverse of an elliptic integral, though he unfortunately never discussed this question. Jacobi did not go deeply into modular equations; the algebraic theory of modular equations was developed, starting in the 1850s, by Betti, Brioschi, Hermite, and Kronecker.

In the *Fundamenta Nova*, Jacobi gave details of how he found the first cubic transformation. First set $a/c = 2\alpha + 1$ in (37.61). Then the transformation would take the form: If

$$y = x(2\alpha + 1 + \alpha^2 x^2)/(1 + \alpha(\alpha + 2)x^2), \quad \text{then} \tag{37.70}$$

$$\frac{dy}{\sqrt{(1-y^2)(1-\lambda^2 y^2)}} = \frac{(2\alpha+1)\,dx}{\sqrt{(1-x^2)(1-k^2 x^2)}}, \tag{37.71}$$

where $k^2 = \alpha^3(2+\alpha)/(2\alpha+1)$ and $\lambda^2 = \alpha(2+\alpha)^3/(2\alpha+1)^3$.

Recall that since $U = xF(x^2)$ and $V = \phi(x^2)$, to derive this cubic transformation, Jacobi could take

$$V = 1 + bx^2 \quad \text{and} \quad U = x(a + a_1 x^2).$$

He then assumed that A was of the form $1 + \alpha x$ so that

$$V + U = (1+x)AA = 1 + (1+2\alpha)x + \alpha(2+\alpha)x^2 + \alpha\alpha x^3.$$

By equating the powers of x, he had

$$b = \alpha(2+\alpha), \quad a = 1 + 2\alpha, \quad \text{and} \quad a_1 = \alpha^2.$$

Note that this gives the preceding cubic transformation (37.70). To find the algebraic relation satisfied by k and λ, he changed x into $1/(kx)$ and y into $1/(\lambda y)$ in (37.70) to get

$$\frac{\lambda x\left((2\alpha+1)\alpha^2 + \alpha^4 x^2\right)}{\alpha^2 + \alpha^3(\alpha+2)x^2} = \frac{kx\left(\alpha(\alpha+2) + k^2 x^2\right)}{\alpha^2 + (2\alpha+1)k^2 x^2}.$$

By equating coefficients of various powers of x, Jacobi found

$$k^2 = \frac{\alpha^3(2+\alpha)}{2\alpha+1}, \quad \lambda^2 = \frac{k^6}{\alpha^8} = \alpha\left(\frac{2+\alpha}{2\alpha+1}\right)^3.$$

The complementary moduli were then given by

$$k'^2 = 1 - k^2 = \frac{(1-\alpha)(1+\alpha)^3}{2\alpha+1}, \quad \lambda'^2 = \frac{(1+\alpha)(1-\alpha)^3}{(2\alpha+1)^3}.$$

Observe that this immediately gives the modular equation $\sqrt{k\lambda} + \sqrt{k'\lambda'} = 1$. Moreover, he noted that with $D = 1 + \alpha(\alpha+2)x^2$,

$$1 - y = (1-x)(1-\alpha x)^2/D, \quad 1 + y = (1+x)(1+\alpha x)^2/D,$$
$$1 - \lambda y = (1-kx)(1-kx/\alpha)^2/D, \quad 1 + \lambda y = (1+kx)(1+kx/\alpha)^2/D,$$

and hence he arrived at the transformation (37.71):

$$\frac{dy}{\sqrt{(1-y^2)(1-\lambda^2 y^2)}} = \frac{(2\alpha+1)\,dx}{\sqrt{(1-x^2)(1-k^2 x^2)}}.$$

Jacobi wrote the modular equation in a slightly different form, by setting $k^{1/4} = u$ and $\lambda^{1/4} = v$, to get

$$u^4 - v^4 + 2uv(1 - u^2 v^2) = 0. \tag{37.72}$$

He showed how to obtain the second transformation from this modular equation. He first wrote (37.70) in terms of u and v by observing that $k^3/\lambda = \alpha^4$ or $\alpha = u^3/v$. Then (37.70) and (37.71) could be rewritten as

$$y = \frac{v(v+2u^3)x + u^6 x^3}{v^2 + v^3 u^2 (v+2u^3) x^2}; \tag{37.73}$$

$$\frac{dy}{\sqrt{(1-y^2)(1-v^8 y^2)}} = \frac{v+2u^3}{v} \frac{dx}{\sqrt{(1-x^2)(1-u^8 x^2)}}. \tag{37.74}$$

Jacobi then observed that the modular equation remained unchanged when u and v were changed to $-v$ and u, respectively. This gave him the second transformation

$$z = \frac{u(u-2v^3)y + v^6 y^3}{u^2 + u^3 v^2 (u-2v^3) y^2}; \tag{37.75}$$

$$\frac{dz}{\sqrt{(1-z^2)(1-u^8 z^2)}} = \frac{u-2v^3}{u} \frac{dy}{\sqrt{(1-y^2)(1-v^8 y^2)}}. \tag{37.76}$$

By the modular equation

$$\left(\frac{v+2u^3}{v}\right)\left(\frac{u-2v^3}{u}\right) = \frac{2(u^4-v^4) + uv(1-4u^2 v^2)}{uv} = -3,$$

he obtained triplication formula

$$\frac{dz}{\sqrt{(1-z^2)(1-u^8 z^2)}} = \frac{-3\,dx}{\sqrt{(1-x^2)(1-u^8 x^2)}}.$$

To get $+3$ instead of -3, it was sufficient to change z to $-z$.

In the case of the quintic transformation, Jacobi set

$$V = 1 + b_1 x^2 + b_2 x^4, \quad U = x(a_1 + a_2 x^2 + a_3 x^4), \quad A = 1 + \alpha x + \beta x^2.$$

From the equation $V + U = (1+x)AA$, he found

$$b_1 = 2\alpha + 2\beta + \alpha\alpha, \quad b_2 = \beta(2\alpha + \beta),$$

$$a_1 = 1 + 2\alpha, \quad a_2 = 2\beta + \alpha\alpha + 2\alpha\beta, \quad a_3 = \beta\beta.$$

He gave the details in section 15 of his *Fundamenta*. He presented the modular equation in the form

$$u^6 - v^6 + 5u^2v^2(u^2 - v^2) + 4uv(1 - u^4v^4) = 0.$$

In 1858, Hermite used this relation to solve a quintic equation, just as Viète solved a cubic by means of trigonometric functions.

37.8 Jacobi's Transcendental Theory of Transformations

Euler, Legendre, and others were aware of the fact that the addition formula for elliptic integrals solved the problem of the multiplication or division of an elliptic integral by an integer. In transformation theory, the multiplication was accomplished in two steps. The first step was to apply a transformation that gave a new elliptic integral with a modulus λ^2 smaller than the original modulus k^2. This was followed by a second transformation serving to increase the modulus. Jacobi discovered these facts about transformation theory by the summer of 1827, at least in the cases of the cubic and quintic transformations. To develop the theory in general, he had to invert the elliptic integral and work with elliptic functions. In his December 1827 paper, however, he gave only the first transformation because he did not define elliptic functions of a complex variable. It was after he introduced complex periods in the spring of 1828 that he was able to develop the complete transformation theory as presented in his *Fundamenta Nova*. He explained how the two transformations arose and also the manner in which they were related to the complementary transformations. To obtain a glimpse of the general theory, we consider the cubic transformation in some detail from the transcendental viewpoint. For the most part, we follow the exposition from Cayley's *Elliptic Functions*. It can be shown by means of the addition formula for the elliptic function $x = \operatorname{sn}(u, k)$ that if $z = \operatorname{sn}(3u, k)$, then

$$z = \frac{3x\left(1 - \frac{x^2}{a_1^2}\right)\left(1 - \frac{x^2}{a_2^2}\right)\left(1 - \frac{x^2}{a_3^2}\right)\left(1 - \frac{x^2}{a_4^2}\right)}{(1 - k^2 a_1^2 x^2)(1 - k^2 a_2^2 x^2)(1 - k^2 a_3^2 x^2)(1 - k^2 a_4^2 x^2)}, \quad (37.77)$$

where $a_1 = \operatorname{sn}\dfrac{4K}{3}$, $a_2 = \operatorname{sn}\dfrac{4iK'}{3}$, $a_3 = \operatorname{sn}\dfrac{4K + i4K'}{3}$, $a_4 = \operatorname{sn}\dfrac{-4K + i4K'}{3}$.

Also, it follows from a formula of Legendre that a_1, a_2, a_3, a_4 are the roots of

$$3 - 4(1 + k^2)x^2 + 6k^2x^4 - k^4x^8 = 0.$$

Note that Legendre knew that (37.77) was an integral of the differential equation

$$\frac{dz}{\sqrt{(1 - z^2)(1 - k^2 z^2)}} = \frac{3\,dx}{\sqrt{(1 - x^2)(1 - k^2 x^2)}}. \quad (37.78)$$

Now from Jacobi's algebraic theory presented in the previous section, it follows that the first transformation has the form

$$y = \frac{\frac{x}{M}\left(1 - \frac{x^2}{a_1^2}\right)}{1 - k^2 a_1^2 x^2}, \qquad (37.79)$$

where M is to be determined. Recall that Jacobi required the existence of a polynomial A such that $V - U = (1-x)A^2$ where $y = U/V$. This means that the value $x = 1$ can be required to correspond to $y = 1$. Taking these values for x and y in (37.79), we see that

$$M = -\frac{1 - a_1^2}{a_1^2(1 - k^2 a_1^2)} \quad \text{and}$$

$$1 - y = \left((1 - k^2 a_1^2 x^2) - \frac{x}{M}\left(1 - \frac{x^2}{a_1^2}\right)\right)\bigg/ D,$$

where $D = 1 - k^2 a_1^2 x^2$. We can rewrite the numerator of $1 - y$ as

$$(1-x)\left(1 - \left(\frac{1}{M} - 1\right)x - \frac{x^2}{Ma_1^2}\right).$$

Now let $A = 1 - x/f$ so that, for consistency, we require that

$$1 - \left(\frac{1}{M} - 1\right)x - \frac{x^2}{Ma_1^2} = 1 - \frac{2}{f}x + \frac{x^2}{f^2}.$$

Equating coefficients, we get

$$\frac{2}{f} = -\frac{1 - k^2 a_1^4}{1 - a_1^2} \quad \text{and} \quad \frac{1}{f^2} = -\frac{1}{Ma_1^2} = \frac{1 - k^2 a_1^2}{1 - a_1^2}. \qquad (37.80)$$

These relations are consistent because, by the addition formula and periodicity of $\operatorname{sn} u$,

$$\operatorname{sn}\frac{8K}{3} = -\operatorname{sn}\frac{4K}{3} = \frac{2\operatorname{sn}(4K/3)\operatorname{cn}(4K/3)\operatorname{dn}(4K/3)}{1 - k^2 \operatorname{sn}^4(4K/3)},$$

or $\quad 2\sqrt{1 - a_1^2}\sqrt{1 - k^2 a_1^2} = -(1 - k^2 a_1^4).$

Hence,

$$1 + y = (1+x)(1+x/f)^2 / D.$$

The next step is to determine λ by using the invariance of the transformation (37.79) under the change x to $1/(kx)$ and y to $1/(\lambda y)$. This gives

$$\lambda = M^2 k^3 a_1^4 = \frac{k^3(1 - a_1^2)^2}{(1 - k^2 a_1^2)^2}.$$

Note that since a_1 is real, we have $1 - a_1^2 < 1 - k^2 a_1^2$ and λ is smaller than k. It is also easy to check that

$$1 - \lambda y = (1-kx)(1-kfx)^2/D; \quad 1 + \lambda y = (1+kx)(1+kfx)^2/D. \qquad (37.81)$$

It follows that
$$\frac{M\,dy}{\sqrt{(1-y^2)(1-\lambda^2 y^2)}} = \frac{dx}{\sqrt{(1-x^2)(1-k^2 x^2)}}. \tag{37.82}$$

Then, by means of an algebraic calculation, obtain $\sqrt{\lambda k} + \sqrt{\lambda' k'} = 1$.

Now for the second transformation, we require that if it is applied after the first, we get triplication. Note that (37.79) implies (37.82). Therefore, we want a transformation

$$z = \frac{3My\left(1 - \frac{y^2}{\theta^2}\right)}{1 - \lambda^2 \theta^2 y^2} \tag{37.83}$$

such that

$$\frac{dz}{\sqrt{(1-z^2)(1-k^2 z^2)}} = \frac{3M\,dy}{\sqrt{(1-y^2)(1-\lambda^2 y^2)}}. \tag{37.84}$$

Thus, (37.78) must hold in this case. Next note that if the value of y given by (37.79) when substituted in (37.83) were to produce (37.77), then (37.78) would hold true. Moreover, it can be shown that if we take $\theta = -a_2 a_3 a_4/(M a_1^2)$, then

$$1 - \frac{y}{\theta} = \left(1 - \frac{x}{a_2}\right)\left(1 - \frac{x}{a_3}\right)\left(1 - \frac{x}{a_4}\right)\Big/D,$$

$$1 - \lambda\theta y = (1 - k a_2 x)(1 - k a_3 x)(1 - k a_4 x)\big/D,$$

and there are similar formulas for $1 + y/\theta$, $1 + \lambda\theta y$ where the sign of x is changed. So this value of θ in (37.83) indeed produces the desired result (37.84). Moreover θ is related to λ as a_1 to k, that is, θ is a solution of

$$3 - 4(1+\lambda^2)\theta^2 + 6\lambda^2\theta^4 - \lambda^4\theta^8 = 0.$$

In fact, it can be shown that θ may be taken to be the purely imaginary value

$$a_2 = \mathrm{sn}\,(4iK'/3).$$

This implies that θ^2 is real and negative and that (37.83) is a real transformation. Transformations similar to (37.79), wherein a_1 is replaced by a_3 or a_4, contain complex numbers.

In general, for an odd integer n, Jacobi gave the transformation formulas in the form

$$y = \frac{\frac{x}{M}\prod_{s=1}^{n}\left(1 - \frac{x^2}{\mathrm{sn}^2 2sw}\right)}{\prod_{s=1}^{n}(1 - k^2(\mathrm{sn}^2 2sw)x^2)},$$

where

$$w = \frac{m_1 K + m_2 i K'}{n}.$$

Denoting the denominator on the right-hand side by D, he showed that under the conditions

$$\lambda = k^n \prod_{s=1}^{n} \mathrm{sn}^4\,(K - 2sw), \tag{37.85}$$

$$\lambda' = k'^n / \prod_{s=1}^{n} \mathrm{dn}^4(2sw), \quad M = (-1)^{(n-1)/2} \prod_{s=1}^{n} ((\mathrm{sn}(K-2sw))/\mathrm{sn}(2sw))^2,$$

the following expressions for $1-y$, $1+y$, $1-\lambda y$, $1+\lambda y$ were consistent with each other and with the expression for y:

$$(1-y)D = (1-x) \prod_{s=1}^{n} \left(1 - \frac{x}{\mathrm{sn}(K-2sw)}\right)^2,$$

$$(1+y)D = (1+x) \prod_{s=1}^{n} \left(1 + \frac{x}{\mathrm{sn}(K-2sw)}\right)^2,$$

$$(1-\lambda y)D = (1-kx) \prod_{s=1}^{n} \left(1 - kx\,\mathrm{sn}(K-2sw)\right)^2,$$

$$(1+\lambda y)D = (1+kx) \prod_{s=1}^{n} \left(1 + kx\,\mathrm{sn}(K-2sw)\right)^2.$$

These equations implied the differential equation (37.84). He also rewrote the transformation formulas in the form

$$\mathrm{sn}\left(\frac{u}{M}, \lambda\right) = \frac{\mathrm{sn}\,u}{M} \left(\prod_{s=1}^{n} \left(1 - \frac{\mathrm{sn}^2 u}{\mathrm{sn}^2(2sw)}\right)\right) \div D, \qquad (37.86)$$

$$\mathrm{cn}\left(\frac{u}{M}, \lambda\right) = \mathrm{cn}\,u \left(\prod_{s=1}^{n} \left(1 - \frac{\mathrm{sn}^2 u}{\mathrm{sn}^2(K-2sw)}\right)\right) \div D,$$

$$\mathrm{dn}\left(\frac{u}{M}, \lambda\right) = \mathrm{sn}\,u \left(\prod_{s=1}^{n} (1 - k^2 \mathrm{sn}^2(K-2sw)\mathrm{sn}^2 u)\right) \div D,$$

where $\quad D = \prod_{s=1}^{n}(1 - k^2\mathrm{sn}^2(2sw)\mathrm{sn}^2 u), \quad$ and $\quad \mathrm{sn}\,u = \mathrm{sn}(u,k)$.

The real transformations corresponded to the cases $w = K/n$ and $w' = iK'/n$. Then, by applying the imaginary transformation, Jacobi obtained the transformations for the moduli w' and λ'. This meant that the transformation for K/n, that is the first transformation, was converted to the form of the second transformation, arising from iK'/n.

Jacobi obtained the relation between the complete integrals K and Λ by observing that for $w = K/n$ the least positive value for which $\mathrm{sn}(u/M, \lambda)$ vanished in (37.86) was given by $u/M = 2\Lambda$, while on the right-hand side it was given by $u = 2K/n$. Hence,

$$2M\Lambda = \frac{2K}{n} \quad \text{or} \quad \frac{K}{nM} = \Lambda. \qquad (37.87)$$

Note that this relation came from the first transformation, since w was taken to be K/n. Jacobi denoted the value of M by M_1 in the second transformation, where w was taken

37.8 Jacobi's Transcendental Theory of Transformations

to be iK'/n. Since $\operatorname{sn}^2(2sw)$ was negative in this case, Jacobi noted that the smallest value of u for which the right-hand side of (37.86) vanished was given by $u = 2K$. Hence, he obtained

$$2M_1 \Lambda_1 = 2K \quad \text{or} \quad \frac{K}{M_1} = \Lambda_1. \tag{37.88}$$

On the other hand, the transformations for the complementary moduli gave Jacobi the relations

$$\Lambda' = \frac{K'}{M} \quad \text{and} \quad \Lambda_1 = \frac{K'}{nM_1}. \tag{37.89}$$

The first relation combined with (37.87) produced the modular equation

$$\frac{\Lambda'}{\Lambda} = n\frac{K'}{K},$$

while the second together with (37.88) gave

$$\frac{K'}{K} = n\frac{\Lambda'_1}{\Lambda_1}.$$

Jacobi also found transformations easily derivable from the first and second transformations; he named these supplementary transformations and used them to obtain product expansions for elliptic functions. For example, he started with the second transformation

$$\operatorname{sn}\left(\frac{u}{M_1}, \lambda_1\right) = \frac{\operatorname{sn} u}{M_1} \left(\prod_{s=1}^{n} \left(1 - \frac{\operatorname{sn}^2 u}{\operatorname{sn}^2(2si K'/n)}\right) \right)$$

$$\div \prod_{s=1}^{n} \left(1 - \frac{\operatorname{sn}^2 u}{\operatorname{sn}^2((2s-1)i K'/n)}\right), \tag{37.90}$$

where $\operatorname{sn} u = \operatorname{sn}(u, k)$. He changed k into λ so that λ_1 then changed to k. Denoting the new value of M_1 by M', he had the relations

$$M_1 = \frac{K}{\Lambda_1}, \quad M' = \frac{\Lambda}{K} = \frac{1}{nM}, \quad \text{or} \quad n = \frac{1}{MM'}.$$

Then, replacing u by u/M in the transformation obtained after changing k to λ, he reached his first supplementary transformation

$$\operatorname{sn}(nu, k) = nM \operatorname{sn}\left(\frac{u}{M}, \lambda\right) N/D \tag{37.91}$$

$$\text{where} \quad N = \prod_{s=1}^{n} \left(1 - \frac{\operatorname{sn}^2(u/M, \lambda)}{\operatorname{sn}^2(2si \Lambda'/n, \lambda)}\right)$$

$$\text{and} \quad D = \prod_{s=1}^{n} \left(1 - \frac{\operatorname{sn}^2(u/M, \lambda)}{\operatorname{sn}^2((2s-1)i \Lambda'/n, \lambda)}\right).$$

Similarly, Jacobi had formulas for the functions sn and dn.

37.9 Jacobi: Infinite Products for Elliptic Functions

In 1828–29, Jacobi obtained his initial infinite products for the elliptic functions sn, cn, and dn. To do this, he took his order n supplementary transformations for these functions, such as (37.91) for sn, and let the integer n tend to infinity. He noted that since k^2 was less than 1, it followed that k^n tended to zero and hence by equation (37.85) $\lambda = 0$, $am(u, \lambda) = u$, and $\text{sn}(\theta, \lambda) = \sin\theta$. This then implied that the corresponding complete integral Λ was equal to $\pi/2$. Moreover, since by (37.87) and (37.89) $\Lambda = K/nM$ and $\Lambda' = K'/M$, it followed that

$$nM = \frac{2K}{\pi}, \quad \frac{\Lambda'}{n} = \frac{K'}{nM} = \frac{\pi K'}{2K}. \tag{37.92}$$

Jacobi also had

$$am(u/(nM), \lambda) = am(u/(nM), 0) = u/(nM) = \pi u/(2K)$$

and he set

$$\text{sn}(u/M, \lambda) = \sin(\pi u/(2K)) = y.$$

Replacing nu by u in (37.91), he let $n \to \infty$ to obtain the product formula:

$$\text{sn}\, u = \frac{2Ky}{\pi} \frac{\left(1 - \frac{y^2}{\sin^2 i\pi K'/K}\right)\left(1 - \frac{y^2}{\sin^2 2i\pi K'/K}\right)\left(1 - \frac{y^2}{\sin^2 3i\pi K'/K}\right)\cdots}{\left(1 - \frac{y^2}{\sin^2 i\pi K'/(2K)}\right)\left(1 - \frac{y^2}{\sin^2 3i\pi K'/(2K)}\right)\left(1 - \frac{y^2}{\sin^2 5i\pi K'/(2K)}\right)\cdots}.$$

In a similar way, he got

$$\text{cn}\, u = \sqrt{1-y^2}\, \frac{\left(1 - \frac{y^2}{\cos^2(\pi i K'/K)}\right)\left(1 - \frac{y^2}{\cos^2(2\pi i K'/K)}\right)\left(1 - \frac{y^2}{\cos^2(3\pi i K'/K)}\right)\cdots}{\left(1 - \frac{y^2}{\sin^2(\pi i K'/(2K))}\right)\left(1 - \frac{y^2}{\sin^2(3\pi i K'/(2K))}\right)\left(1 - \frac{y^2}{\sin^2(5\pi i K'/(2K))}\right)\cdots}, \tag{37.93}$$

and

$$\text{dn}\, u = \frac{\left(1 - \frac{y^2}{\cos^2(\pi i K'/(2K))}\right)\left(1 - \frac{y^2}{\cos^2(3\pi i K'/(2K))}\right)\left(1 - \frac{y^2}{\cos^2(5\pi i K'/(2K))}\right)\cdots}{\left(1 - \frac{y^2}{\sin^2(\pi i K'/(2K))}\right)\left(1 - \frac{y^2}{\sin^2(3\pi i K'/(2K))}\right)\left(1 - \frac{y^2}{\sin^2(5\pi i K'/(2K))}\right)\cdots}. \tag{37.94}$$

Recall that Abel obtained his similar product formula (37.28) for $\phi\alpha$ using a different method. Jacobi then set $e^{-\pi K'/K} = q$, $u = 2Kx/\pi$, and $y = \sin x$ to obtain

$$\sin\frac{m\pi i K'}{K} = \frac{q^m - q^{-m}}{2i} = \frac{i(1 - q^{2m})}{2q^m}, \tag{37.95}$$

$$\cos\frac{m\pi i K'}{K} = \frac{q^m + q^{-m}}{2} = \frac{1 + q^{2m}}{2q^m}, \tag{37.96}$$

$$1 - \frac{y^2}{\sin^2 \frac{m\pi i K'}{K}} = 1 + \frac{4q^{2m}\sin^2 x}{(1-q^{2m})^2} = \frac{1 - 2q^{2m}\cos 2x + q^{4m}}{(1-q^{2m})^2}, \tag{37.97}$$

$$1 - \frac{y^2}{\cos^2 \frac{m\pi i K'}{K}} = 1 - \frac{4q^{2m}\sin^2 x}{(1+q^{2m})^2} = \frac{1 + 2q^{2m}\cos 2x + q^{4m}}{(1+q^{2m})^2}. \tag{37.98}$$

37.9 Jacobi: Infinite Products for Elliptic Functions

He was then able to rewrite the products as

$$\operatorname{sn}\frac{2Kx}{\pi} = \frac{2AK}{\pi}\sin x$$
$$\cdot \frac{(1-2q^2\cos 2x+q^4)(1-2q^4\cos 2x+q^8)(1-2q^6\cos 2x+q^{12})\cdots}{(1-2q\cos 2x+q^2)(1-2q^3\cos 2x+q^6)(1-2q^5\cos 2x+q^{10})\cdots}$$

$$\operatorname{cn}\frac{2Kx}{\pi} = B\cos x$$
$$\cdot \frac{(1+2q^2\cos 2x+q^4)(1+2q^4\cos 2x+q^8)(1+2q^6\cos 2x+q^{12})\cdots}{(1-2q\cos 2x+q^2)(1-2q^3\cos 2x+q^6)(1-2q^5\cos 2x+q^{10})\cdots}$$

$$\operatorname{dn}\frac{2Kx}{\pi} = C$$
$$\cdot \frac{(1+2q\cos 2x+q^2)(1+2q^3\cos 2x+q^6)(1+2q^5\cos 2x+q^{10})\cdots}{(1-2q\cos 2x+q^2)(1-2q^3\cos 2x+q^6)(1-2q^5\cos 2x+q^{10})\cdots}.$$

Here

$$A = \left\{\frac{(1-q)(1-q^3)(1-q^5)\cdots}{(1-q^2)(1-q^4)(1-q^6)\cdots}\right\}^2,$$

$$B = \left\{\frac{(1-q)(1-q^3)(1-q^5)\cdots}{(1+q^2)(1+q^4)(1+q^6)\cdots}\right\}^2,$$

$$C = \left\{\frac{(1-q)(1-q^3)(1-q^5)\cdots}{(1+q)(1+q^3)(1+q^5)\cdots}\right\}^2.$$

Jacobi set $x = \frac{\pi}{2}$ and observed that since $\operatorname{sn} K = 1$ and $\operatorname{dn} K = \sqrt{1-k^2\operatorname{sn}^2 K} = \sqrt{1-k^2} = k'$,

$$k' = C \cdot C = C^2 \quad \text{or} \quad C = \sqrt{k'}.$$

To rewrite these formulas in a more useful form, he changed x to $x + \frac{i\pi K'}{2K}$ in the first equation, so that by the addition formula

$$\operatorname{sn}\left(\frac{2Kx}{\pi} + iK'\right) = \frac{1}{k\operatorname{sn}\frac{2Kx}{\pi}} \quad ;$$

$$\cos 2\left(x + i\frac{\pi K'}{2K}\right) = \frac{e^{2ix - \pi K'/K} + e^{-2ix + \pi K'/K}}{2} = \frac{1}{2}\left(qe^{2ix} + \frac{1}{q}e^{-2ix}\right);$$

$$e^{i(x + i\pi K'/(2K))} = \sqrt{q}\, e^{ix}.$$

Note that the first product formula could be written as

$$\operatorname{sn}\frac{2Kx}{\pi} = \frac{2AK}{\pi}\left(\frac{e^{ix} - e^{-ix}}{2i}\right). \tag{37.99}$$

$$\frac{(1-q^2 e^{2ix})(1-q^2 e^{-2ix})(1-q^4 e^{2ix})(1-q^4 e^{-2ix})\cdots}{(1-qe^{2ix})(1-qe^{-2ix})(1-q^3 e^{2ix})(1-q^3 e^{-2ix})\cdots}. \tag{37.100}$$

Observe that after applying $x \to x + i\frac{\pi K'}{2K}$, the formula would become

$$\frac{1}{k \operatorname{sn} \frac{2Kx}{\pi}} = \frac{2AK}{\pi} \left(\frac{\sqrt{q} e^{ix} - \frac{1}{\sqrt{q}} e^{-ix}}{2i} \right). \tag{37.101}$$

$$\frac{(1-q^3 e^{2ix})(1-q e^{-2ix})(1-q^5 e^{2ix})(1-q^3 e^{-2ix})\cdots}{(1-q^2 e^{2ix})(1-e^{-2ix})(1-q^4 e^{2ix})(1-q^2 e^{-2ix})\cdots}. \tag{37.102}$$

Multiply equations (37.100) and (37.102) to obtain Jacobi's result

$$\frac{1}{k} = \frac{1}{\sqrt{q}} \left(\frac{AK}{\pi} \right)^2 \quad \text{or} \quad A = \frac{\pi \sqrt[4]{q}}{\sqrt{k} K}.$$

Jacobi then set $x = \frac{\pi}{2}$ in (37.100) and applied $C = \sqrt{k'}$ to get

$$1 = \frac{2AK}{\pi} \left\{ \frac{(1+q^2)(1+q^4)(1+q^6)\cdots}{(1+q)(1+q^3)(1+q^5)\cdots} \right\}^2 = \frac{2\sqrt{k'} AK}{\pi B}.$$

Jacobi was then in a position to rewrite the products:

$$\operatorname{sn} \frac{2Kx}{\pi} = \frac{2q^{1/4}}{\sqrt{k}} \sin x \frac{(1-2q^2 \cos 2x + q^4)(1-2q^4 \cos 2x + q^8)\cdots}{(1-2q \cos 2x + q^2)(1-2q^3 \cos 2x + q^6)\cdots}, \tag{37.103}$$

$$\operatorname{cn} \frac{2Kx}{\pi} = \sqrt{\frac{k'}{k}} \cdot 2q^{1/4} \frac{\cos x (1+2q^2 \cos 2x + q^4)(1+2q^4 \cos 2x + q^8)\cdots}{(1-2q \cos 2x + q^2)(1-2q^3 \cos 2x + q^6)\cdots}, \tag{37.104}$$

$$\operatorname{dn} \frac{2Kx}{\pi} = \sqrt{k'} \cdot \frac{(1+2q \cos 2x + q^2)(1+2q^3 \cos 2x + q^6)\cdots}{(1-2q \cos 2x + q^2)(1-2q^3 \cos 2x + q^6)\cdots}. \tag{37.105}$$

Thus, from the products for A, B, and C, Jacobi had infinite products for $\frac{2K}{\pi}$, k' and k:

$$\frac{2K}{\pi} = \left\{ \frac{(1-q^2)(1-q^4)(1-q^6)\cdots}{(1-q)(1-q^3)(1-q^5)\cdots} \right\}^2 \cdot \left\{ \frac{(1+q)(1+q^3)(1+q^5)\cdots}{(1+q^2)(1+q^4)(1+q^6)\cdots} \right\}^2,$$

$$k' = \left\{ \frac{(1-q)(1-q^3)(1-q^5)\cdots}{(1+q)(1+q^3)(1+q^5)\cdots} \right\}^4,$$

$$k = 4\sqrt{q} \left\{ \frac{(1+q^2)(1+q^4)(1+q^6)\cdots}{(1+q)(1+q^3)(1+q^5)\cdots} \right\}^4.$$

After obtaining $\frac{2K}{\pi}$ as an infinite product, Jacobi applied the triple product identity to express this product as a theta series:

$$\sqrt{\frac{2K}{\pi}} = 1 + 2q + 2q^4 + 2q^9 + 2q^{16} + \cdots. \tag{37.106}$$

This formula laid the basis for Jacobi's results on the sums of squares, two of which reproved theorems of Fermat.

37.10 Jacobi: Sums of Squares

In 1750, Euler suggested that problems on sums of squares could most naturally be studied through the series whose powers were squares. In his 1828 work on elliptic functions, Jacobi followed Euler's suggestion, with great success. Though primarily an analyst, Jacobi had a strong interest in number theory, leading him to perceive that his famous formula (37.106) could be employed to obtain Fermat's theorems on sums of two and four squares. In fact, Jacobi also found analytic formulas implying results for sums of six and eight squares. He arrived at all these results, including (37.106), through his product expansions of the doubly-periodic elliptic functions. Recall that in an analogous manner, Euler evaluated the zeta values at the even integers by means of the infinite product expansions of the singly periodic trigonometric functions. Also note that the period K of the elliptic function was obtained as a value of a theta function, while, as in the Madhava–Leibniz formula, the period π of a trigonometric function was expressed as a value of an L-function.

To derive the formulas necessary to work with sums of squares, Jacobi first took the logarithmic derivatives of the product expansions for the elliptic functions sn, cn, and dn. First note

$$\log(1 - 2q^m \cos 2x + q^{2m}) = \log(1 - q^m e^{2ix}) + \log(1 - q^m e^{-2ix})$$

$$= -\sum_{l=1}^{\infty} \frac{q^{lm} \cos 2lx}{l}.$$

Combining this relation with the geometric series $1 - q^l + q^{2l} - \cdots = \frac{1}{1+q^l}$ gives us Jacobi's formulas; he simply wrote them down without details:

$$\log \operatorname{sn} \frac{2Kx}{\pi} = \log \left\{ \frac{2\sqrt[4]{q}}{\sqrt{k}} \sin x \right\} + \frac{2q \cos 2x}{1+q} + \frac{2q^2 \cos 4x}{2(1+q^2)} + \frac{2q^3 \cos 6x}{3(1+q^3)} + \cdots,$$

$$\log \operatorname{cn} \frac{2Kx}{\pi} = \log \left\{ 2\sqrt[4]{q} \sqrt{\frac{k'}{k}} \cos x \right\} + \frac{2q \cos 2x}{1-q} + \frac{2q^2 \cos 4x}{2(1+q^2)} + \frac{2q^3 \cos 6x}{3(1-q^3)} + \cdots,$$

$$\log \operatorname{dn} \frac{2Kx}{\pi} = \log \sqrt{k'} + \frac{4q \cos 2x}{1-q^2} + \frac{4q^3 \cos 6x}{3(1-q^6)} + \frac{4q^5 \cos 10x}{5(1-q^{10})} + \cdots.$$

To obtain the derivatives of these formulas, Jacobi observed that

$$\frac{d}{dx} \log \operatorname{sn}(2Kx/\pi) = \frac{2k'K}{\pi} \frac{\operatorname{cn}(2Kx/\pi)}{\operatorname{cn}(K - 2Kx/\pi)}, \qquad (37.107)$$

$$-\frac{d}{dx} \log \operatorname{cn}(2Kx/\pi) = \frac{2K}{\pi} \cdot \frac{\operatorname{sn}(2Kx/\pi)}{\operatorname{sn}(K - 2Kx/\pi)}, \qquad (37.108)$$

$$-\frac{d}{dx} \log \operatorname{dn}(2Kx\pi) = \frac{2k^2 K}{\pi} \operatorname{sn}(2Kx/\pi) \operatorname{sn}(K - 2Kx/\pi). \qquad (37.109)$$

Thus, he obtained

$$\frac{2k'K}{\pi} \cdot \frac{\operatorname{cn}\frac{2Kx}{\pi}}{\operatorname{cn}(K - \frac{2Kx}{\pi})} = \cot x - \frac{4q\sin 2x}{1+q} - \frac{4q^2\sin 4x}{1+q^2} - \frac{4q^3\sin 6x}{1+q^3} - \cdots,$$

$$\frac{2K}{\pi} \cdot \frac{\operatorname{sn}\frac{2Kx}{\pi}}{\operatorname{sn}(K - \frac{2Kx}{\pi})} = \tan x + \frac{4q\sin 2x}{1-q} + \frac{4q^2\sin 4x}{1+q^2} + \frac{4q^3\sin 6x}{1-q^3} + \cdots,$$

$$\frac{2k^2K}{\pi}\operatorname{sn}\frac{2Kx}{\pi}\operatorname{sn}(K - \frac{2Kx}{\pi}) = \frac{8q\sin 2x}{1-q^2} + \frac{8q^3\sin 6x}{1-q^6} + \frac{8q^5\sin 10x}{1-q^{10}} + \cdots.$$

Note that when $x = \pi/4$ in the second equation, we get the Lambert series for $2K/\pi$:

$$\frac{2K}{\pi} = 1 + \frac{4q}{1-q} - \frac{4q^3}{1-q^3} + \frac{4q^5}{1-q^5} - \cdots.$$

Also, the derivative of the second equation at $x = 0$ gives us the Lambert series for the square of $2K/\pi$:

$$\left(\frac{2K}{\pi}\right)^2 = 1 + \frac{8q}{1-q} + \frac{16q^2}{1+q^2} + \frac{24q^3}{1-q^3} + \cdots.$$

By further manipulation of the products, using differentiation and series expansions, Jacobi obtained formulas for the cubes and fourth powers, as given in sections 40–42 of his *Fundamenta Nova*:

$$\left(\frac{2K}{\pi}\right)^3 = 1 + 16\sum_{n=1}^{\infty} \frac{n^2 q^n}{1+q^{2n}} - 4\sum_{n=1}^{\infty}(-1)^{n-1}\frac{(2n-1)^2 q^{2n-1}}{1-q^{2n-1}}, \qquad (37.110)$$

$$\left(\frac{2K}{\pi}\right)^4 = 1 + 16\sum_{n=1}^{\infty} \frac{n^3 q^n}{1+(-1)^{n-1}q^n}. \qquad (37.111)$$

The reader may observe that by expressing the Lambert series in the last four equations as power series in q, we obtain the number of representations of an integer as the sum of two, four, six, and eight squares.

In the final paragraph of his *Fundamenta*, Jacobi gave a number theoretic interpretation of his analytic formula for the sums of four squares, but he did not write down interpretations for the other formulas. In 1865, Henry Smith gave these explicitly, in sections 95 and 127 of his report on number theory:

> The number of representations of any uneven (or unevenly even) number by the form $x^2 + y^2$ is the quadruple of the excess of the number of its divisors of the form $4n + 1$, above the number of its divisors of the form $4n + 3$.
>
> The number of representations of any number N as a sum of four squares is eight times the sum of its divisors if N is uneven, twenty-four times the sum of its uneven divisors if N is even.
>
> The number of representations of any number N as a sum of six squares is $4\sum(-1)^{(\delta-1)/2}(4\delta'^2 - \delta^2)$, δ denoting any uneven divisor of N, δ' its conjugate divisor. In particular if $N \equiv 1$, mod 4, the number of representations is $12\sum(-1)^{(\delta-1)/2}$; if $N \equiv -1$, mod 4, it is $-20\sum(-1)^{(\delta-1)/2}\delta^2$.
>
> The number of representations of any uneven number as a sum of eight squares is sixteen times the sum of the cubes of its divisors; for an even number it is sixteen times the excess of the cubes of the even divisors above the cubes of the uneven divisors.

In his July 1828 paper in *Crelle's Journal*, Jacobi gave a beautiful application of (37.106) to derive a very efficient proof of the transformation formula for a theta function:

$$\sqrt{\frac{1}{x}} = \frac{1 + 2e^{-\pi x} + 2e^{-4\pi x} + 2e^{-9\pi x} + 2e^{-16\pi x} + \cdots}{1 + 2e^{-\pi/x} + 2e^{-4\pi/x} + 2e^{-9\pi/x} + 2e^{-16\pi/x} + \cdots}. \tag{37.112}$$

Cauchy found this in 1817, and Poisson did so in 1823, though Jacobi referred only to Poisson. Jacobi observed that if the moduli k and k' were interchanged, then K and K' would also be interchanged. Thus, with $x = K'/K$, (37.106) implied

$$\sqrt{\frac{2K'}{\pi}} = 1 + 2e^{-\pi/x} + 2e^{-4\pi/x} + 2e^{-9\pi/x} + 2e^{-16\pi/x} + 2e^{-25\pi/x} + \cdots.$$

Dividing (37.106) by this equation gave him the required transformation. As we shall see in the next section, in 1836 Cauchy applied (37.112) to evaluate a Gauss sum, and in 1840 he provided a more succinct argument.

It is interesting to note that Euler foresaw, albeit vaguely, Jacobi's manner of proof for the four squares theorem and the importance of the transformation of the theta function. In a letter to Goldbach of August 17, 1750, Euler discussed the series

$$1 - x + x^4 - x^9 + x^{16} - x^{25} + \cdots.$$

He wrote that he had approximately evaluated to several decimal places this series for values of x close to 1, a remarkable calculation since the series is very slowly convergent. He commented that it would be very useful if a method could be found for efficiently summing the series for such values. And the transformation of theta functions accomplishes just this task. Moreover, in the same letter Euler mentioned Fermat's remarkable theorem that every number could be expressed as a sum of three triangular numbers, four squares, five pentagonal numbers, and so on. He remarked that the most natural way to prove this proposition might be to show that the coefficient of every power of x must be positive in the series:

$$(1 + x + x^3 + x^6 + \cdots)^3, \quad (1 + x + x^4 + x^9 + \cdots)^4, \quad \text{and so on}.$$

37.11 Cauchy: Theta Transformations and Gauss Sums

Cauchy's 1817 derivation of his transformation of the theta function depended upon the theorem now known as the Poisson summation formula. Cauchy was the first to discover this result, and he did so in the course of his work on the theory of waves. For the Poisson summation formula, consult chapter 25 on the Euler–Maclaurin series. Independent of Fourier's earlier work, Cauchy also discovered the reciprocity of the Fourier cosine transform, given by

$$f(x) = \sqrt{\frac{2}{\pi}} \int_0^\infty \phi(t) \cos tx \, dt; \qquad \phi(x) = \sqrt{\frac{2}{\pi}} \int_0^\infty f(t) \cos tx \, dt.$$

He gave his summation formula in the form of the relation

$$\sqrt{\alpha}\sum f(n\alpha) = \sqrt{\beta}\sum \phi(n\beta),$$

where $\alpha\beta = 2\pi$, and the summation was taken over all integers. Cauchy obtained his transformation formula by setting $f(x)$ equal to the function he called the reciprocal function, $e^{-x^2/2}$, and then setting $\phi(x) = e^{-x^2/2}$ in the summation formula. He then took $\alpha = \sqrt{2}a$ and $\beta = \sqrt{2}b$ and stated the transformation as

$$a^{1/2}\left(\frac{1}{2} + e^{-a^2} + e^{-4a^2} + e^{-9a^2} + \cdots\right) = b^{1/2}\left(\frac{1}{2} + e^{-b^2} + e^{-4b^2} + \cdots\right) \quad (37.113)$$

$$\text{when} \quad ab = \pi. \quad (37.114)$$

Note that (37.113) describes the transformation of the theta function (or theta constant)

$$\sum_{n=-\infty}^{\infty} e^{\pi i n^2 \tau}$$

under the mapping $\tau \to -1/\tau$.

Cauchy applied a very interesting idea to evaluate Gauss sums from (37.113). Taking n to be an integer, he set $\tau = \frac{2}{n} + i\frac{\alpha^2}{\pi}$ in the transformation formula, and let $\alpha \to 0$. The asymptotic behavior of the two sides of the formula then yielded the result. Note that the theta function is analytic in the upper half plane and every point of the real line is a singular point.

In 1840, Cauchy published his quadratic Gauss sum evaluation in *Liouville's Journal*. He noted that (37.113) could be rewritten as

$$a\left(\frac{1}{2} + e^{-a^2} + e^{-4a^2} + \cdots\right) = \sqrt{\pi}\left(\frac{1}{2} + e^{-\pi^2/a^2} + e^{-4\pi^2/a^2} + \cdots\right).$$

For $a = \alpha$, an infinitely small number, this reduced to

$$\alpha\left(\frac{1}{2} + e^{-\alpha^2} + e^{-4\alpha^2} + \cdots\right) = \sqrt{\pi}/2.$$

Cauchy remarked that this step could be verified by the fact that the limit as $\alpha \to 0$ of the product

$$\alpha\left(1 + e^{-\alpha^2} + e^{-4\alpha^2} + \cdots\right)$$

was the integral

$$\int_0^\infty e^{-x^2}\,dx = \sqrt{\pi}/2.$$

With n a positive integer and $a^2 = -2\pi\sqrt{-1}/n$, $b^2 = n\pi\sqrt{-1}/2$, Cauchy could obtain

$$e^{-(n+k)^2 a^2} = e^{-k^2 a^2}; \quad (37.115)$$

$$e^{-(2m)^2 b^2} + e^{-(2m+1)^2 b^2} = 1 + e^{-n\pi\sqrt{-1}/2}. \quad (37.116)$$

37.11 Cauchy: Theta Transformations and Gauss Sums

He then set $a^2 = \alpha^2 - 2\pi\sqrt{-1}/n$ and $b^2 = \beta^2 + n\pi\sqrt{-1}/2$ where α and β were infinitely small numbers and where $2\beta = n\alpha$. The last condition was needed to satisfy the requirement $ab = \pi$. After substituting these values of a^2 and b^2 in (37.113), he multiplied the equation by $n\alpha = 2\beta$ and remarked that the result was

$$a^{1/2}\Delta = b^{1/2}(1 + e^{-n\pi\sqrt{-1}/2}), \tag{37.117}$$

where Δ was the Gauss sum

$$\Delta = 1 + e^{\frac{2\pi\sqrt{-1}}{n}} + e^{4 \cdot \frac{2\pi\sqrt{-1}}{n}} + \cdots + e^{(n-1)^2 \cdot \frac{2\pi\sqrt{-1}}{n}}. \tag{37.118}$$

From (37.117) and (37.114) Cauchy completed his evaluation:

$$\begin{aligned}\Delta &= \frac{\pi^{1/2}}{a}\left(1 + e^{-n\pi\sqrt{-1}/2}\right) \\ &= \frac{n^{1/2}}{2}(1 + \sqrt{-1})(1 + e^{-n\pi\sqrt{-1}/2}).\end{aligned} \tag{37.119}$$

To see why (37.117) holds true, we note that by (37.115)

$$n\alpha(e^{-a^2} + e^{-4a^2} + e^{-9a^2} + \cdots) = n\alpha \sum_{k=1}^{n} e^{2\pi i k^2/n} \sum_{s=0}^{\infty} e^{-(k+sn)^2\alpha^2}. \tag{37.120}$$

Moreover, $\quad n\alpha \displaystyle\sum_{s=0}^{\infty} e^{-(k+sn)^2\alpha^2} = \int_0^{\infty} e^{-x^2}\, dx = \sqrt{\pi}/2,$

and hence the expression (37.120) equals $\Delta\sqrt{\pi}/2$. And, by using (37.116), we can show that

$$2\beta(1 + e^{-b^2} + e^{-4b^2} + e^{-9b^2} + \cdots) = (1 + e^{-n\pi\sqrt{-1}/2})\sqrt{\pi}/2.$$

Thus, Cauchy's equation (37.117) is verified.

In the first part of his 1859 *Report on the Theory of Numbers*, Smith noted that Cauchy's method could be applied to derive the more general reciprocity relation for Gauss sums. He set

$$\psi(k, n) = \sum_{s=0}^{n-1} e^{2\pi i k s^2/n}$$

and took

$$a^2 = \alpha^2 - \frac{2m\pi i}{n}, \quad b^2 = \beta^2 + \frac{ni\pi}{2m}, \quad i = \sqrt{-1}$$

in (37.113) to find the reciprocity relation

$$\psi(m, n) = \frac{1}{4}\sqrt{\frac{n}{m}}(1 + i)\psi(-n, 4m). \tag{37.121}$$

He also observed that from $\psi(-4\nu, 4m) = 4\psi(-\nu, m)$ and (37.121), it followed that

$$\psi(m, 4\nu) = 2\sqrt{\frac{\nu}{m}}(1 + i)\psi(-\nu, m),$$

so that the case with even n would depend upon the case with odd n.

Henry John Stephen Smith, son of an Irish lawyer, studied at Oxford, where mathematics was not then popular. He independently read in detail the number theoretic work of Gauss, Dirichlet, Eisenstein, Jacobi, Kummer, and others; he became the most outstanding British number theorist of the nineteenth century. An active member of the British Association for the Advancement of Science, he wrote his well-known report on number theory for the association. The report covered developments in number theory up to the 1850s. In spite of his important researches in number theory and elliptic functions, he worked alone without a following and was mostly ignored in his lifetime. Smith is most noted for his 1867 work was on the representation of numbers as sums of squares. Although it established Eisenstein's unproved theorems on sums of five and seven squares, this important paper remained unnoticed. In fact, as late as 1883, the Paris Academy offered a prize for the proof of Eisenstein's results. Fortunately, this brought Smith's work to the notice of mathematicians and also succeeded in gaining some prominence for the 18-year-old Hermann Minkowski (1864–1909), who offered his own highly original paper on the topic.

37.12 Eisenstein: Reciprocity Laws

Even before his great 1847 paper laying the foundations for a new theory of elliptic functions, Eisenstein used Abel's formulas to make some original applications of elliptic functions to number theory. In 1845, Eisenstein published "Application de l'algèbre à l'arithmétique transcendante," in which he used circular and elliptic functions to prove the quadratic and biquadratic reciprocity laws. We review some of the then-known number theoretic results results upon which Eisenstein based his work: Let p be an odd prime. Following Gauss, divide the residues modulo p, namely $1, 2, \ldots, p-1$, into two classes: $r_1, r_2, \ldots, r_{\frac{p-1}{2}}$ and $-r_1, -r_2, \ldots, -r_{\frac{p-1}{2}}$, so that every residue falls into exactly one class. Eisenstein took one class to be $1, 2, \ldots, \frac{p-1}{2}$. Note that then $-1, -2, \ldots, -\frac{p-1}{2}$ are identical (mod p) to $p-1, p-2, \ldots, \frac{p+1}{2}$. A number a, prime to p, is called a quadratic residue modulo p if the equation

$$x^2 \equiv a \,(\text{mod } p) \tag{37.122}$$

has a solution; otherwise, a is a quadratic nonresidue. Eisenstein used a result of Euler now known as Euler's criterion: A number a, prime to p, is a quadratic residue if and only if

$$a^{\frac{p-1}{2}} \equiv 1 \,(\text{mod } p). \tag{37.123}$$

From Fermat's theorem, $a^{p-1} \equiv 1 \,(\text{mod } p)$, and hence $a^{\frac{p-1}{2}} \equiv \pm 1 \,(\text{mod } p)$. With this, Euler's criterion can be proved: If a satisfies (37.122), then (37.123) follows by Fermat's theorem. From the fact that there are exactly $\frac{p-1}{2}$ quadratic residues (mod p), it follows that $x^{\frac{p-1}{2}} \equiv 1 \,(\text{mod } p)$ has at least $\frac{p-1}{2}$ solutions. However, the equation of degree $\frac{p-1}{2}$ has at most $\frac{p-1}{2}$ solutions. Hence, these comprise all the solutions.

37.12 Eisenstein: Reciprocity Laws

To state the law of quadratic reciprocity, we define the Legendre symbol $\left(\frac{a}{p}\right)$ by the equation

$$a^{\frac{p-1}{2}} \equiv \left(\frac{a}{p}\right) \pmod{p}. \tag{37.124}$$

Note that if a is a multiple of p, then we set $\left(\frac{a}{p}\right) = 0$. The law of quadratic reciprocity states that if p and q are odd primes, then

$$\left(\frac{p}{q}\right)\left(\frac{q}{p}\right) = (-1)^{\frac{p-1}{2} \cdot \frac{q-1}{2}}. \tag{37.125}$$

Note that (37.125) is equivalent to the statement that if q is a quadratic residue (mod p), then p is a quadratic residue (mod q) except when both p and q are of the form $4n + 3$. In the latter case, q is a quadratic residue (mod p), if and only if p is not a quadratic residue (mod q).

To begin his proof of the law of quadratic reciprocity, Eisenstein let r denote a number in $1, 2, \ldots, \frac{p-1}{2}$. Then

$$qr \equiv \pm r' \pmod{p}, \tag{37.126}$$

where r' was also contained in $1, 2, \ldots, \frac{p-1}{2}$. Eisenstein observed that since sine was an odd periodic function,

$$\sin\frac{2\pi qr}{p} = \pm \sin\frac{2\pi r'}{p}. \tag{37.127}$$

Therefore (37.126) could be rewritten as

$$qr \equiv r' \frac{\sin(2\pi qr/p)}{\sin(2\pi r'/p)} \pmod{p}. \tag{37.128}$$

Substituting the $(p-1)/2$ different values of r in this equation and multiplying, he obtained

$$q^{(p-1)/2}\Pi r \equiv \Pi r' \prod_{k=1}^{(p-1)/2} \frac{\sin(2\pi qk/p)}{\sin(2\pi k/p)} \pmod{p}. \tag{37.129}$$

Eisenstein saw that Πr and $\Pi r'$ were identical and concluded that

$$q^{(p-1)/2} \equiv \prod_{k=1}^{(p-1)/2} \frac{\sin(2\pi qk/p)}{\sin(2\pi k/p)} \pmod{p}. \tag{37.130}$$

Note that by Euler's criterion, Eisenstein had found a trigonometric expression for the Legendre symbol $\left(\frac{q}{p}\right)$. By reversing the roles of p and q, he obtained

$$p^{(q-1)/2} \equiv \prod_{l=1}^{(q-1)/2} \frac{\sin(2\pi pl/q)}{\sin(2\pi l/q)} \pmod{q}. \tag{37.131}$$

At this juncture, Eisenstein employed Euler's factorization

$$\frac{\sin px}{\sin x} = \frac{(-1)^{\frac{p-1}{2}}}{2^{p-1}} \prod_{k=1}^{(p-1)/2} \left(\sin^2 x - \sin^2 \frac{2\pi k}{p} \right),$$

to conclude that the product on the right side of (37.131) equaled

$$C \prod_{k=1}^{(p-1)/2} \prod_{l=1}^{(q-1)/2} \left(\sin^2 \frac{2\pi l}{q} - \sin^2 \frac{2\pi k}{p} \right), \qquad (37.132)$$

$$\text{where} \quad C = \frac{(-1)^{\frac{p-1}{2} \cdot \frac{q-1}{2}}}{2^{(p-1)(1-1)/2}}.$$

For Euler's factorization, see chapter 16, sections 4 and 5. Next, by symmetry, Eisenstein had a similar product for (37.130) with the same constant C, but with factors of the form

$$\sin^2 \frac{2\pi k}{p} - \sin^2 \frac{2\pi l}{q}.$$

Thus, each factor in (37.132) was the negative of the corresponding factor in the product for the expression in (37.130) and the number of such factors was $\frac{(p-1)}{2} \cdot \frac{q-1}{2}$. So Eisenstein could obtain the product in (37.131) by multiplying the product in (37.130) by $(-1)^{\frac{p-1}{2} \cdot \frac{q-1}{2}}$. Therefore, employing Euler's criterion, Eisenstein had the reciprocity law

$$\left(\frac{p}{q}\right) \cdot \left(\frac{q}{p}\right) = (-1)^{\frac{p-1}{2} \cdot \frac{q-1}{2}}.$$

Eisenstein gave a similar proof of the biquadratic (quartic) reciprocity law, but used the lemniscatic function instead of the sine function. Again, we consider the backdrop to his work. Even while he was working on the *Disquisitiones Arithmeticae*, Gauss started thinking about extending quadratic reciprocity to cubic and quartic residues. It appears that he very quickly realized that to state these reciprocity laws he had to extend the field of rational numbers by cube roots and fourth roots of unity. It is not clear when Gauss found the law of biquadratic or quartic reciprocity. On October 23, 1813, his mathematical diary noted, "The foundation of the general theory of biquadratic residues which we have sought for with utmost effort for almost seven years but always unsuccessfully at last happily discovered the same day on which our son is born." Strangely, in a letter of April 1807 to Sophie Germain (1776–1831), Gauss had made a similar claim, challenging her to determine the cubic and quartic residue character of 2. Perhaps he discovered the theorem in 1807 and proved it in 1813. In any case, Germain obtained some good results on this problem; she found the quartic character of -4. And Gauss wrote that, especially given the obstacles to women working in mathematics, he was very impressed with her accomplishments. However, Germain's main contribution to number theory was in connection with Fermat's last theorem; she discovered and applied the Germain primes p such that $2p+1$ was also prime.

Gauss published two papers on biquadratic reciprocity, in 1828 and 1832. The first paper contained a thorough treatment of the biquadratic character of 2 with respect

37.12 Eisenstein: Reciprocity Laws

to a prime $p = 4s + 1$. Note that by Euler's criterion, -1 is then a quadratic residue (mod p); further, p can be expressed as $a^2 + b^2$. Gauss denoted the two solutions of $x^2 \equiv -1 \pmod{p}$ by f and $-f$. He also took a to be odd and b to be even, and he took their signs such that $a \equiv 1 \pmod 4$ and $b \equiv af \pmod p$. His theorem stated that 2 satisfied $2^{\frac{p-1}{4}} \equiv 1, f - 1, -f \pmod p$ where $b/2$ was of the form 0, 1, 2, 3, (mod 4), respectively.

In his second paper, Gauss proved that where m and n were integers, the ring $\mathbb{Z}[\sqrt{-1}]$ consisting of $m + ni$, was a unique factorization domain. In 1859, Smith commented on this result that "By thus introducing the conception of imaginary quantity into arithmetic, its domain, as Gauss observes, is indefinitely extended; nor is this extension an arbitrary addition to the science, but is essential to the comprehension of many phenomena presented by real integral numbers themselves." It is clear from Gauss's second paper that since primes of the form $4s + 1$, where s is a positive integer, can be expressed as a sum of two squares, $a^2 + b^2 = (a + ib)(a - ib)$, they are not prime in the ring $\mathbb{Z}[i]$. However, primes of the form $4s + 3$ cannot be factored in $\mathbb{Z}[i]$. We see, therefore, that there are three classes of primes in $\mathbb{Z}[i]$: (a) primes of the form $i^k(4s + 3)$; (b) primes of the form $a + ib$ such that their norm, $N(a + ib) = a^2 + b^2$, is a prime of the form $4s + 1$ in $\mathbb{Z}$; (c) the primes $i^k(1 + i)$, whose norm is 2. Let $m = a + ib$ be a prime such that $a + b$ is an odd integer and $N(m) = p$. In this case, any number n, not a multiple of m in $\mathbb{Z}[i]$, leaves $p - 1$ possible residues when divided by m. For such m and n, the quartic symbol $(\frac{n}{m})_4$ takes the values ± 1 and $\pm i$ and is today defined by

$$\left(\frac{n}{m}\right)_4 = n^{\frac{p-1}{4}} \pmod{m}.$$

To prove quartic reciprocity, Eisenstein divided the $p - 1$ residues into four classes, with $(p-1)/4$ residues in each class, such that when r was in one class, $ir, -r, -ir$ each fell into a different one of the other classes. He noted that for any n in $\mathbb{Z}[i]$

$$nr \equiv r', ir', -r', -ir' \pmod{m}, \qquad (37.133)$$

where r' was in the same class as r. He set

$$w = 4 \int_0^1 \frac{dx}{\sqrt{(1 - x^4)}}.$$

Then by the periodicity of the lemniscatic function sl z, for m prime, Eisenstein had

$$\frac{\text{sl}(nrw/m)}{\text{sl}(r'w/m)} = 1, i, -1, \text{ or } -i,$$

corresponding to the four cases in (37.133). Hence in all cases,

$$nr \equiv r' \frac{\text{sl}(nrw/m)}{\text{sl}(r'w/m)} \pmod{m}.$$

From this he got the formula analogous to (37.130),

$$n^{(p-1)/4} \equiv \prod_r \frac{\text{sl}(nrw/m)}{\text{sl}(rw/m)} \pmod{m}. \qquad (37.134)$$

To obtain a formula with m and n interchanged, Eisenstein chose n to be another complex prime $c + id$ with $c + d$ odd and norm q. He divided the residues of nonmultiples of n into four classes represented by $\rho, i\rho, -\rho, -i\rho$ and concluded that

$$m^{(q-1)/4} \equiv \prod_\rho \frac{\mathrm{sl}(m\rho w/n)}{\mathrm{sl}(\rho w/n)} \pmod{n}. \tag{37.135}$$

Gauss defined the concept of a primary number so that he could express his results in unambiguous form. A number $c + id$, where $c + d$ was odd, was called primary if d was even and $c + d - 1$ was evenly even (that is, divisible by 4). This definition was adopted by Eisenstein. We remark by the way that Gauss also suggested a slightly different definition of a primary number, useful in some circumstances; this definition was employed by Dirichlet. It is easy to show and Gauss of course knew that $c + id$, with $c + d$ odd, was primary if and only if $c + id \equiv 1 \pmod{2 + 2i}$. It follows that the product of primary numbers is primary and that the conjugate of a primary number is primary. In his work on the division of the lemniscate, Abel showed that for a primary number m

$$\mathrm{sl}(mw/4) = 1.$$

He also proved that

$$\frac{\mathrm{sl}(mv)}{\mathrm{sl}\,v} = \frac{\phi(x^4)}{\psi(x^4)}, \tag{37.136}$$

where $x = \mathrm{sl}\,v$ and where $\phi(x)$ and $\psi(x)$ were polynomials of degree $(p-1)/4$. See equation (37.45). Eisenstein improved on this by proving that $\psi(x)$ in (37.136) satisfied

$$\psi(x) = i^\nu x^{\frac{p-1}{4}} \phi(1/x) \tag{37.137}$$

for some integer ν; he also showed that when m was primary, $\nu = 0$. To prove (37.137), he noted that $y = x\phi(x^4)/\psi(x^4)$ satisfied the differential equation

$$\frac{dy}{\sqrt{1-y^4}} = \frac{m\,dx}{\sqrt{1-x^4}}. \tag{37.138}$$

He set $y = 1/\eta$, $x = 1/(i^\mu \xi)$ where μ was an integer yet to be determined. This change of variables converted (37.138) to

$$\frac{i^\mu\,d\eta}{\sqrt{(\eta^4-1)}} = \frac{m\,d\xi}{\sqrt{(\xi^4-1)}}. \tag{37.139}$$

Eisenstein took μ such that (37.139) would be equivalent to

$$\frac{d\eta}{\sqrt{(1-\eta^4)}} = \frac{m\,d\xi}{\sqrt{(1-\xi^4)}}$$

and concluded

$$\eta = i^\mu \frac{\xi \psi(1/\xi^4)}{\phi(1/\xi^4)};$$

this immediately implied (37.137). Thus, by (37.136) and (37.137) he obtained

$$\operatorname{sl}(mv) = x \frac{\phi(x^4)}{i^v x^{p-1} \phi(1/x^4)}. \tag{37.140}$$

Eisenstein next set $v = w/4$, so that $\operatorname{sl} v = 1$ and for primary m, $\operatorname{sl}(mv) = 1$. Thus, for m primary in (37.137), he had $1 = i^{-v}$. He then assumed that n was also a primary prime so that

$$\frac{\operatorname{sl}(nv)}{\operatorname{sl} v} = \frac{f(x^4)}{x^{q-1} f(1/x^4)},$$

where $f(x)$ was a polynomial of degree $(q-1)/4$. He set

$$\alpha = \operatorname{sl}(\frac{rw}{m}), \quad \beta = \operatorname{sl}(\frac{\rho w}{n}),$$

so that the solutions of $\phi(x^4) = 0$ were of the form $\pm\alpha, \pm i\alpha$ and those of $f(x^4) = 0$ were of the form $\pm\beta, \pm i\beta$. Thus, he arrived at

$$\frac{\operatorname{sl}(mv)}{\operatorname{sl} v} = \frac{\Pi(x^4 - \alpha^4)}{\Pi(1 - \alpha^4 x^4)}, \quad \frac{\operatorname{sl}(nv)}{\operatorname{sl}(v)} = \frac{\Pi(x^4 - \beta^4)}{\Pi(1 - \beta^4 x^4)}.$$

When he combined these formulas with (37.134) and (37.135), he obtained

$$n^{(p-1)/4} \equiv \frac{\Pi(\alpha^4 - \beta^4)}{\Pi(1 - \beta^4 \alpha^4)} \pmod{m},$$

$$m^{(p-1)/4} \equiv \frac{\Pi(\beta^4 - \alpha^4)}{\Pi(1 - \alpha^4 \beta^4)} \pmod{n}.$$

Eisenstein observed that since there were $\frac{p-1}{4} \cdot \frac{q-1}{4}$ factors in the products, the fundamental theorem on biquadratic residues followed immediately.

Eisenstein studied the polynomial $\phi(x)$ in even greater detail later in his 1845 paper "Beiträge zur Theorie der elliptischen Functionen, I." For primary m, he proved that

$$\phi(x) = x^{p-1} + A_1 x^{p-5} + \cdots + m,$$

and showed that all the coefficients $A_1, A_2, \ldots, m$ were divisible by m. Then, in 1850, he published a paper using a generalization of what we now call Eisenstein's criterion to prove the irreducibility of $\phi(x)$. Suppose $f(x) = a_0 x^n + a_1 x^{n-1} + \cdots + a_n$, where $a_j \in \mathbb{Z}[i]$. Also suppose m is a prime in $\mathbb{Z}[i]$ such that m divides $a_1, \ldots, a_n$, but does not divide a_0, and m^2 does not divide a_n. Then $f(x)$ is irreducible over $\mathbb{Z}[i]$. Eisenstein included a statement and proof of this theorem in an 1847 letter to Gauss. But in 1846, Theodor Schönemann, a student of Jacobi and of the Swiss geometer Jakob Steiner, published this theorem for the case where $\mathbb{Z}[i]$ was replaced by $\mathbb{Z}$; this particular case is now known as Eisenstein's criterion. Eisenstein acknowledged this work in his 1850 paper.

37.13 Liouville's Theory of Elliptic Functions

Liouville's contributions to this topic are mainly contained in his lectures, published by Borchardt in 1880. However, as we mentioned earlier, Liouville began to grapple with elliptic functions as early as the 1840s. We briefly discuss his early thoughts, contained in his numerous notebooks. When Hermite remarked to Liouville that one could use Fourier series to prove Jacobi's theorem on the ratio of two independent periods of a function, Liouville was apparently motivated to prove this and wrote it up in his notebook on August 1, 1844. A few pages later he included a more direct proof, supposing that f had real periods α and α', independent over the rationals. Then, using the fact that α was a period, he noted that f had a Fourier expansion

$$f(x) = \sum A_j \cos\left(\frac{2j\pi x}{\alpha} + \epsilon_j\right).$$

Since α' was also a period, he had

$$A_j \cos\left(\frac{2j\pi x}{\alpha} + \epsilon_j\right) = A_j \cos\left(\frac{2j\pi x}{\alpha} + \epsilon_j + \frac{2j\pi \alpha'}{\alpha}\right).$$

Thus, Liouville concluded that either $A_j = 0$ or $2j\pi\alpha'/\alpha = 2m\pi$, where m was an integer. The last equation implied that α and α' were commeasurable, or dependent over the rationals.

These ideas soon led to the statement and proof of the theorem now famous as Liouville's theorem, that a bounded entire function is a constant. He first proved this for doubly-periodic functions, using Fourier series. He then extended it to functions bounded on the Riemann sphere. Assuming the result for periodic functions, he proved the extension by taking an analytic function $f(z)$ and assuming $|f(z)| \leq M$ for all z. Then the function $f(\operatorname{sn} z)$, f composed with the Jacobi elliptic function, would be a doubly-periodic bounded function and hence a constant. Liouville noted that an application of this theorem was that every algebraic equation had to have a root. He argued that if $p(x)$ was a polynomial and $1/p(x)$ did not become infinite for any complex x, then the same would be true for $1/p(\operatorname{sn} x)$, and this was a contradiction. It is interesting that though Liouville never published this application, it is usually the first one to be given in textbooks.

Liouville also proved that an elliptic function could not have only one simple pole. He noted that, on the other hand, if there were two simple poles, then the function would reduce to the usual elliptic function. In this connection, Liouville showed in his notebooks that if ϕ had two simple poles, α and β, then there would be a constant D such that

$$u = \bigl(\phi(\alpha + x) - D\bigr) + \bigl(\phi(\alpha - x) - D\bigr)$$

was a solution to

$$\left(\frac{du}{dx}\right)^2 = a + bu + cu^2 + du^3 + eu^4.$$

37.13 Liouville's Theory of Elliptic Functions

This meant that u was the inverse of the elliptic integral:

$$x = \int \frac{du}{\sqrt{a+bu+cu^2+du^3+eu^4}}.$$

The following theorems of Liouville on elliptic functions commonly appear in modern treatments of the topic:

- The number of poles equals the number of zeros, counting multiplicity.
- The sum of the zeros minus the sum of the poles (in a period parallelogram) is a period of the function.
- The sum of the residues is equal to zero. Liouville proved this for functions with only two poles.
- A doubly-periodic function with only one simple pole does not exist.

Liouville wrote that within his new approach, "integrals which have given rise to the elliptic functions and even moduli disappear in a way, leaving only the periods and the points for which the functions become zero or infinite." This important new principle, that a function may be largely defined by its singularities, was greatly extended by Riemann in his remarkable works on functions of a complex variable.

We present Liouville's proofs based on Borchardt's notes. Liouville considered a doubly-periodic function $\phi(z)$ with periods $2w$ and $2w'$ so that its values were completely defined by its values in the region

$$z = z_0 + uw + u'w', \quad -1 \le u \le 1, \quad -1 \le u' \le 1.$$

We would now refer to this region as the period parallelogram P_{z_0}. Liouville then assumed that $z = \alpha, z = \alpha_1, z = \alpha_2, \ldots, z = \alpha_{n-1}$ were the n roots of the equation $\phi(z) = \pm\infty$ in this region. Then there would exist constants $G, G_1, \ldots, G_{n-1}$ so that

$$\phi(z) - \left\{\frac{G}{z-\alpha} + \frac{G_1}{z-\alpha_1} + \frac{G_2}{z-\alpha_2} + \cdots + \frac{G_{n-1}}{z-\alpha_{n-1}}\right\} \quad (37.141)$$

was finite at $\alpha_1, \alpha_2, \ldots, \alpha_n$. In the case where there were multiple roots, so that the (say i) values $\alpha_p, \alpha_q, \ldots, \alpha_s$ coincided, the sum of simple fractions

$$\frac{G_p}{z-\alpha_p} + \frac{G_q}{z-\alpha_q} + \cdots + \frac{G_s}{z-\alpha_s}$$

had to be replaced by

$$\frac{G_p}{z-\alpha_p} + \frac{G_q}{(z-\alpha_p)^2} + \cdots + \frac{G_s}{(z-\alpha_p)^i}. \quad (37.142)$$

Liouville designated the sum of the fractions as the fractional part of $\phi(z)$ and denoted it by $[\phi(z)]$. He noted that this fractional part played an important role in the calculus of residues, and he showed that a doubly-periodic function without a fractional part was a constant. Liouville did not refer to poles, but we would now say that a doubly-periodic function must have poles. Note that in (37.141) all the poles are simple and in (37.142), α_p is a pole of order i.

Liouville next proved that there could be no doubly-periodic function with a fractional part, $[\phi(z)] = \frac{G}{z-\alpha}$. In other words, there did not exist a doubly-periodic function with just one simple pole in the period parallelogram. To prove this, Liouville set $z - \alpha = t$ so that the fractional part at α would be given by

$$[\phi(z)]^\alpha = [\phi(\alpha + t)]^0 = \frac{G}{t}.$$

Similarly, $[\phi(\alpha - t)]^0 = -\frac{G}{t},$

so that $[\phi(\alpha + t) + \phi(\alpha - t)]^0 = 0,$

and therefore $\phi(\alpha + t) + \phi(\alpha - t) = 2c$, where c was a constant. Liouville then set $f(t) = \phi(\alpha + t) - c$, to get $f(t) = -f(-t)$. Since $2w$ and $2w'$ were periods of f, he obtained

$$f(w) = -f(-w) = -f(-w + 2w) = -f(w) = 0, \tag{37.143}$$

$$f(w') = 0, \quad f(w + w') = 0. \tag{37.144}$$

He then defined a new function $F(t) = f(t) f(t + w)$, noting that this function had no singularities; the zeros canceled with the poles, based on (37.143) and (37.144). This implied that there were constants k, k', and k'' such that

$$f(t) f(t+w) = k, \quad f(t) f(t+w') = k', \quad f(t) f(t+w+w') = k''. \tag{37.145}$$

Liouville changed t to $t + w$ in the third equation, obtaining

$$f(t+w) f(t+w') = k''.$$

Finally, multiplying the first two equations and dividing by the fourth he arrived at

$$\left(f(t)\right)^2 = \frac{k k'}{k''}.$$

This implied that $\phi(z)$ was a constant and the result was proved.

Liouville then gave a simple construction of a doubly-periodic function with periods $2w$ and $2w'$ and poles at α and β. He set

$$\phi(z) = \sum_{i=-\infty}^{\infty} f(z + 2iw'), \quad \text{where}$$

$$f(z) = \frac{1}{\cos \frac{\pi}{w}(z-h) - \cos \frac{\pi}{w} h'} \quad \text{and} \quad \alpha = h + h', \beta = h - h'.$$

Liouville next analyzed the zeros of an elliptic function. He observed that a doubly-periodic function $\phi(z)$, with poles at α and β, could not have only one simple zero because its reciprocal would have one simple pole, an impossibility. He then showed that $\phi(z)$ could not have three zeros. Supposing a and b to be two of the zeros, he took another function $\psi(z)$ with periods $2w$ and $2w'$ and poles at a and b. He also set $\psi_1(z) = \psi(z) - \psi(\alpha)$. Clearly, $\psi_1(z)\phi(z)$ had only one pole at β, implying that $\psi_1(z)\phi(z) = $ constant. Now if ϕ had another zero at c, then ψ_1 would have a pole at c. This contradiction proved that $\phi(z)$ had zeros only at a and b. Liouville also proved that if two functions $\phi(z)$ and $\phi_1(z)$ had the same periods with simple poles at α and β, then there existed constants c, c', such that $\phi_1(z) = c\phi(z) + c'$. To prove this he set

$$[\phi(z)] = \frac{G}{z-\alpha} + \frac{H}{z-\beta}, \quad [\phi_1(z)] = \frac{G_1}{z-\alpha} + \frac{H_1}{z-\beta}, \qquad (37.146)$$

so that $\quad [G\phi_1(z) - G_1\phi(z)] = \dfrac{GH_1 - G_1 H}{z-\beta}.$

Hence, the result:
$$G\phi_1(z) - G_1\phi(z) = \text{constant}. \qquad (37.147)$$

Liouville proceeded to prove that for a doubly-periodic function ϕ, the sum of the zeros was equal to the sum of the poles, modulo some period of the function. He assumed that ϕ had poles at α and β, so that $\phi(\alpha + \beta - z)$ also had poles at α and β. Hence, by the previous result,

$$\phi(z) = c\phi(\alpha + \beta - z) + c'.$$

Replacing z by $\alpha + \beta - z$, he got the relation

$$\phi(\alpha + \beta - z) = c\phi(z) + c',$$

and by subtraction

$$(1+c)(\phi(z) - \phi(\alpha + \beta - z)) = 0.$$

Liouville noted that if $c = -1$, then

$$\phi(z) + \phi(\alpha + \beta - z) = c'.$$

To prove this impossible, he set $z = \frac{\alpha+\beta}{2} + t$, and $\phi\left(\frac{\alpha+\beta}{2} + t\right) - \frac{1}{2}c' = f(t)$, so that $f(t) = -f(-t)$. From this and seeing that $2w$ and $2w'$ were periods of $f(t)$, it followed that

$$f(0) = f(w) = f(w') = f(w+w') = 0.$$

Since f could not have four roots, he obtained the required contradiction and therefore

$$\phi(z) = \phi(\alpha + \beta - z).$$

By taking the reciprocal of ϕ, Liouville saw that $\phi(z) = \phi(a+b-z)$. He thus arrived at the relation

$$\phi(z) = \phi(\alpha + \beta - z) = \phi(\alpha + \beta - a - b + z)$$

and concluded that
$$\alpha + \beta = a + b + 2mw + 2m'w'.$$

Liouville went on to show that any doubly-periodic function could be written in terms of functions of the form ϕ. He presented the details for functions with simple poles. Suppose ψ is a function with periods $2w$ and $2w'$ and

$$[\psi(z)] = \frac{A}{z-\alpha} + \frac{A_1}{z-\alpha_1} + \frac{A_2}{z-\alpha_2} + \cdots.$$

Denote by $\phi(z; \alpha, \alpha_1)$ the function with the same periods as ψ and with simple poles at α and α_1 and let

$$[\phi(z; \alpha, \alpha_1)] = \frac{G_1}{z-\alpha} - \frac{G_1}{z-\alpha_1},$$

$$[\phi(z; \alpha_1, \alpha_2)] = \frac{G_2}{z-\alpha_1} - \frac{G_2}{z-\alpha_2},$$

$$[\phi(z; \alpha_2, \alpha_3)] = \frac{G_3}{z-\alpha_2} - \frac{G_3}{z-\alpha_3} \text{ and so on.}$$

Then

$$[\psi(z)] = B_1[\phi(z; \alpha, \alpha_1)] + B_2[\phi(z; \alpha_1, \alpha_2)] + B_3[\phi(z; \alpha_2, \alpha_1)] + \cdots$$

$$\text{with } B_1 = \frac{A}{G_1}, \quad B_2 = \frac{A+A_1}{G_2}, \quad B_3 = \frac{A+A_1+A_2}{G_3}, \text{ etc.}$$

Thus,

$$\psi(z) = B + B_1\phi(z; \alpha, \alpha_1) + B_2\phi(z; \alpha, \alpha_2) + B_3\phi(z; \alpha, \alpha_3) + \cdots.$$

This theorem was then employed to prove that any doubly-periodic function had exactly as many zeros as poles. Here let $\phi(z; \alpha, \beta; a, b)$ denote a function with periods $2w, 2w'$; poles at α, β; zeros at a, b, with $\alpha + \beta = a + b$. Suppose also that $\psi(z)$ is doubly-periodic with poles at $z = \alpha, \alpha_1, \ldots, \alpha_{n-1}$ and zeros at $z = a, a_1, a_2, \ldots, a_{i-1}$. If $i < n$, Liouville arbitrarily chose $n - i - 1$ numbers $a_i, a_{i+1}, \ldots, a_{n-2}$ and determined $b, b_1, \ldots, b_{n-2}$ by the system of equations

$$b = \alpha + \alpha_1 - a,$$
$$b_1 = \alpha_2 + b - a_1,$$
$$b_2 = \alpha_3 + b_1 - a_2,$$
$$\vdots$$
$$b_{n-2} = \alpha_{n-1} + b_{n-3} - a_{n-2}.$$

He next defined $w(z)$ as

$$\phi(z; \alpha, \alpha_1; a, b) \cdot \phi(z; \alpha_2, b; a_1, b_1) \cdot \phi(z; \alpha_3, b_1; a_2, b_2) \cdot \phi(z; \alpha_{n-1}, b_{n-3}; a_{n-2}, b_{n-2})$$

and noted that $w(z)$ had poles at

$$\alpha, \alpha_1, \alpha_2, \ldots, \alpha_{n-1}$$

and zeros at
$$a, a_1, a_2, \ldots, a_{n-2}, b_{n-2}.$$

If $i < n$, then the function $\frac{w(z)}{\psi(z)}$ had no poles but had zeros at $a_i, \ldots, a_{n-2}, b_{n-2}$, an impossibility. Similarly, for $i > n$, he took the function $\frac{\psi(z)}{w(z)}$ to get a similar contradiction. Thus, $i = n$, $\psi(z) = cw(z)$, and $\psi(z)$ had as many zeros as poles. Also, since ψ had zeros at $z = a, a_1, a_2, \ldots, a_{n-1}$ and $w(z)$ had zeros at $z = a, a_1, \ldots, a_{n-2}, b_{n-2}$, he could conclude that $b_{n-2} = a_{n-1}$. Liouville substituted these values in his system of equations to arrive at
$$\alpha + \alpha_1 + \alpha_2 + \cdots + \alpha_{n-1} = a + a_1 + a_2 + \cdots + a_{n-1}.$$

This implied that the sum of the zeros differed from the sum of the poles by $2mw + 2m'w'$, where m and m' were integers. In the applications, Liouville derived the differential equation and addition formula satisfied by a function ϕ with simple poles at α and β. He also explained how to obtain the Abel and Jacobi elliptic functions from his general results.

37.14 Exercises

1. If ϕ is the cosine transform of f, then
$$\sqrt{\alpha}\bigl(f(\alpha) - f(3\alpha) - f(5\alpha) + f(7\alpha) + f(9\alpha) - \cdots\bigr)$$
$$= \sqrt{\beta}\bigl(\phi(\beta) - \phi(3\beta) - \phi(5\beta) + \cdots\bigr),$$
where $\alpha\beta = \pi/4$; and for $\alpha\beta = \pi/6$
$$\sqrt{\alpha}\bigl(f(\alpha) - f(5\alpha) - f(7\alpha) + f(11\alpha) + f(13\alpha) - \cdots\bigr)$$
$$= \sqrt{\beta}\bigl(\phi(\alpha) - \phi(5\alpha) - \phi(7\alpha) + \cdots\bigr),$$
where the integers 1, 5, 7, 11, 13, ... are prime to 6. See Ramanujan (2000), p. 63.

2. Let $\omega/2$ denote the complete lemniscatic integral $\int_0^1 dx/\sqrt{1-x^4}$. Show that then
$$\frac{\omega}{4\pi} = \frac{e^{\pi/2}}{e^{\pi}-1} - \frac{e^{3\pi/2}}{e^{3\pi}-1} + \frac{e^{5\pi/2}}{e^{5\pi}-1} - \cdots,$$
$$\frac{\omega^2}{4\pi^2} = \frac{e^{\pi/2}}{e^{\pi}+1} - \frac{3e^{3\pi/2}}{e^{3\pi}+1} + \frac{5e^{5\pi/2}}{e^{5\pi}+1} - \cdots.$$
See Abel (1965), vol. 1, p. 351.

3. For $f(\alpha)$ and $F(\alpha)$ defined by (37.11) and (37.12), and ω by (37.7), show that
$$f\left(\frac{\alpha\omega}{2}\right) = \frac{4\pi}{\omega}\left(\frac{\cos(\alpha\pi/2)}{e^{\pi/2}-e^{-\pi/2}} - \frac{\cos(3\alpha\pi/2)}{e^{3\pi/2}-e^{-3\pi/2}} + \frac{\cos(5\alpha\pi/2)}{e^{5\pi/2}-e^{-5\pi/2}} - \cdots\right),$$
$$F\left(\frac{\alpha\omega}{2}\right) = \frac{4\pi}{\omega}\left(\frac{\cosh(\alpha\pi/2)}{e^{\pi/2}-e^{-\pi/2}} - \frac{\cosh(3\alpha\pi/2)}{e^{3\pi/2}-e^{-3\pi/2}} + \frac{\cosh(5\alpha\pi/2)}{e^{5\pi/2}-e^{-5\pi/2}} - \cdots\right).$$

4. Consider the equation

$$dy/\sqrt{(1-y^2)(1-e^2y^2)} = \sqrt{-n}\,dx/\sqrt{(1-x^2)(1-e^2x^2)}.$$

Show that if $n = 3$, then e satisfies the equation $e^2 - 2\sqrt{3}e = 1$; and if $n = 5$, then e satisfies $e^3 - 1 - (5 + 2\sqrt{5})e(e-1) = 0$. Kronecker called these values of e singular moduli. See Abel (1965), vol. 1, pp. 379–384.

5. Show that if the equation

$$dy/\sqrt{(1-c^2y^2)(1-e^2y^2)} = a\,dx/\sqrt{(1-c^2x^2)(1-e^2x^2)}$$

admits of an algebraic solution in x and y, then a is necessarily of the form $\mu' + \sqrt{-\mu}$ when μ and μ' are rational and μ is positive. For such values of a, the moduli e and c can be expressed in radicals. See Abel (1965), vol. 1, pp. 425–428. Kronecker made a deep study, related to algebraic number theory, of complex multiplication. For a historical discussion of this topic, see Vlăduţ (1991). Also see Takase (1994).

6. Show that the functions θ and θ_1 below satisfy $\partial^2 f/\partial x^2 = 4\partial f/\partial \omega$:

$$\theta(x) = 1 - 2e^{-\omega}\cos 2x + 2e^{-4\omega}\cos 4x - 2e^{-9\omega}\cos 6x + \cdots,$$

$$\theta_1(x) = 2e^{-\omega/4}\sin x - 2e^{-9\omega/4}\sin 3x + 2e^{-25\omega/4}\sin 5x - \cdots.$$

See Jacobi (1969), vol. 1, p. 259.

7. Jacobi defined the theta functions (with $q = e^{-\pi K'/K}$):

$$\Theta\left(\frac{2Kx}{\pi}\right) = 1 - 2q\cos 2x + 2q^4\cos 4x - 2q^9\cos 6x + \cdots,$$

$$H\left(\frac{2Kx}{\pi}\right) = 2q^{1/4}\sin x - 2q^{9/4}\sin 3x + 2q^{25/4}\sin 5x - \cdots.$$

Show that for $u = 2Kx/\pi$, we have

$$\Theta(u+2K) = \Theta(-u) = \Theta(u),\ H(u+2K) = H(-u) = -H(u);$$

$$\Theta(u+2iK') = -e^{\pi(K'-iu)/K}\Theta(u),\ H(u+2iK') = -e^{\pi(K'-iu)/K}H(u);$$

$$\operatorname{sn} u = H(u)/(\sqrt{k}\Theta(u)),\ \operatorname{cn} u = \sqrt{k'}H(u+K)/(\sqrt{k}\Theta(u)).$$

See Jacobi (1969), vol. 1, pp. 224–231.

8. Let $w = 2m\omega + 2n\omega'$, where m and n are integers and let $\tau = \omega'/\omega$, with $\operatorname{Im}\tau \neq 0$. Note that σ' denotes the derivative of σ. Define the Weierstrass sigma function by

$$\sigma(u) = u\prod_{m,n}{}'\left(1 - \frac{u}{w}\right)e^{u/w + u^2/(2w^2)},$$

where the product is taken over all m and n except $m = n = 0$. Show that

$$\sigma(u) = e^{2\eta\omega v^2}\frac{2\omega}{\pi}\sin v\pi \prod_n \left((1 - 2h^{2n}\cos 2v\pi + h^{4n})/(1 - h^{2n})^2\right),$$

where $v = u/(2\omega)$ $h = e^{\tau\pi i}$, $\eta = \pi^2/(12\omega) + \sum \csc^2 n\tau\pi$. Show also that

$$\sigma(u \pm 2\omega) = -e^{\pm 2\eta(u\pm\omega)}\sigma(u), \quad \eta = \sigma'(\omega)/\sigma(\omega),$$

$$\sigma(u \pm 2\omega') = -e^{\pm 2\eta'(u\pm\omega')}\sigma(u), \quad \eta' = \sigma'(\omega')/\sigma(\omega').$$

Prove Legendre's relation $\eta\omega' - \omega\eta' = \pi i/2$, when $\text{Im}\,\tau > 0$. See Schwarz (1893), pp. 5–9; note that these are Schwarz's notes of Weierstrass's lectures.

9. Set $\wp(u) = -d^2(\log \sigma(u))/du^2$. Show that $\wp'(\omega) = \wp'(\omega') = \wp'(\omega + \omega') = 0$. Set $\wp(\omega) = e_1$, $\wp(\omega + \omega') = e_2$, $\wp(\omega') = e_3$ and show that

$$(\wp'(u))^2 = 4(\wp(u) - e_1)(\wp(u) - e_2)(\wp(u) - e_3).$$

Also prove that

$$\wp(u) - \wp(v) = -\sigma(u+v)\sigma(u-v)/\left(\sigma^2(u)\sigma^2(v)\right),$$

$$\wp(u \pm v) = -\wp(u) - \wp(v) - \left((\wp'(u) \mp \wp'(v)/(\wp(u) - \wp(v))\right)^2.$$

The last result is the addition formula for Weierstrass's $\wp$-function. See Schwarz (1893), pp. 10–14.

10. Let $k^2 = (e_2 - e_3)/(e_1 - e_3)$. Set

$$\sigma_1(u) = e^{-\eta u}\sigma(\omega + u)/\sigma(\omega), \quad \sigma_3(u) = e^{-\eta' u}\sigma(\omega' + u)/\sigma(\omega').$$

Prove that

$$\sigma(u)/\sigma_3(u) = \text{sn}(\sqrt{e_1 - e_3} \cdot u, k)/\sqrt{e_1 - e_3} \text{ and}$$

$$\sigma_1(u)/\sigma_3(u) = \text{cn}(\sqrt{e_1 - e_3} \cdot u, k)/\sqrt{e_1 - e_3}.$$

See Schwarz (1893), pp. 30–35.

37.15 Notes on the Literature

Euler's letter to Goldbach may be found in Fuss (1968), pp. 530–532. Abel's elliptic functions papers from *Crelle's Journal* have been republished in vol. 1 of Abel (1965). See pp. 262–388 for his "Recherches." C. Houzel's article in Laudal and Piene (2004) summarizes Abel's mathematical work, mostly in elliptic functions, within 150 pages. See Jacobi (1969), vol. 1, for his work on elliptic functions. See Smith (1965b) for his number theory report; the quotation on Gauss's use of complex numbers in number theory may be found on p. 71. Eisenstein (1975), vol. 1, pp. 291–298 contains the paper on quadratic and quartic reciprocity. The history of the Schönemann–Eisenstein criterion is discussed in the interesting and historically informative books by Lemmermeyer (2000) and Cox (2004). The entertaining book by Dörrie (1965), on p. 118, attributes the criterion to Schönemann alone. Liouville (1880) presents his theory of elliptic functions. A fascinating discussion of the origin and development of Liouville's ideas on this topic is available in Lützen (1990). Consult Prasad (1933) for an interesting account of the work of Abel and Jacobi in elliptic functions. For Ramanujan's prolific work on modular equations, see Berndt's helpful summary in Andrews, Askey, Berndt, et al. (1988).

38

Irrational and Transcendental Numbers

38.1 Preliminary Remarks

The ancient Greek mathematicians were aware of the existence of irrational numbers; Eudoxus gave his theory of proportions to deal with that awkward situation. The Greeks also considered the problem of constructing a square with area equal to that of a circle. Later generations of mathematicians probably began to suspect that this was not possible; they were possibly almost certain that π was not rational. The sixteenth-century Indian mathematician and astronomer, Nilakantha, wrote in his *Aryabhatiyabhasya*, "If the diameter, measured using some unit of measure, were commensurable with that unit, then the circumference would not allow itself to be measured by means of the same unit so likewise in the case where the circumference is measurable by some unit, then the diameter cannot be measured using the same unit." He gave no indication of a proof in any of his works. It appears that the first proof of the irrationality of π was presented to the Berlin Academy by the Swiss mathematician J. H. Lambert (1728–1777) in a 1766 paper. He demonstrated that if $x \neq 0$ was a rational number, then $\tan x$ was irrational. He deduced this from the continued fraction expansion

$$\tan v = \cfrac{1}{v - \cfrac{1}{\cfrac{3}{v} + \cfrac{1}{\cfrac{5}{v} - \cdots}}}.$$

Then, since $\tan \pi/4 = 1$, it followed that π was irrational. Lambert's work was based on some results of Euler, who was a colleague of Lambert for about two years at the Berlin Academy. Later, in his 1794 book on geometry, Legendre gave a completely rigorous and concise presentation of Lambert's proof. In particular, he showed that the continued fraction

$$\cfrac{m}{n + \cfrac{m_1}{n_1 + \cfrac{m_2}{n_2 + \cfrac{m_3}{n_3 + \cdots}}}},$$

where m_i, n_i were nonzero integers, converged to an irrational number when $m_i/n_i < 1$ for all i beyond some i_0. Legendre went a little further than Lambert by observing that the continued fraction for $\tan x$ also implied that π^2 was irrational.

A century before Lambert, James Gregory tried to prove that π was transcendental. Since he was starting from scratch, it is not surprising that Gregory failed. C. Goldbach and D. Bernoulli carried on a correspondence in the 1720s, in which they mentioned that the series they had discovered could not represent rational numbers or even roots of rational numbers. Thus, a letter of Bernoulli dated April 28, 1729, commented, concerning the series

$$\log \frac{m+n}{n} = \frac{m}{n} - \frac{n^2}{2m^2} + \frac{n^3}{2m^3} - \frac{n^4}{4m^4} + \cdots, \quad (m, n \text{ positive integers, } m < n),$$

"[N]ot only can it not be expressed in rational numbers, but it cannot be expressed in radical or irrational numbers either." Unfortunately, he had to admit to Goldbach that he had no proof; a proof of the transcendence of this number follows from a theorem proved by Ferdinand Lindemann in 1882. Goldbach's reply to Bernoulli contained the remark that it was not known whether, in general, the number

$$\sum_{n=1}^{\infty} \frac{1}{n^2 + \frac{p}{q}n}$$

could be expressed as a root of a rational number. Note that, for example, when $p = 2$, $q = 1$, the sum is 3/2, while with $p = 1$, $q = 2$, the sum is $4\log(e/2)$. Goldbach was probably aware of the first result; though he may not have noticed it, he could have derived the second result from Brouncker's series for $\log 2$. Observe that the second result can also be derived from the Euler–Maclaurin summation formula or from Mengoli's inequalities, (25.40). In a later letter of October 20, Goldbach wrote, "Here follows a series of fractions, such as you requested whose sum is neither rational nor the root of any rational number:

$$\frac{1}{10} + \frac{1}{100} + \frac{1}{10000} + \frac{1}{100000000} + \text{etc}.$$

The general term is

$$\frac{1}{10^{2^{x-1}}}.\text{"}$$

Neither Goldbach nor Bernoulli could suggest any method for attacking these problems. In 1937, Kurt Mahler proved a more general theorem, as a consequence of which Goldbach's number was necessarily transcendental. In 1938, Rodion Kuzmin (1891–1949) also gave a proof of the transcendence of Goldbach's number

In his 1748 book, *Introductio in Analysin Infinitorum*, Euler made some insightful remarks on the values of the logarithm function, "Since the logarithms of numbers which are not powers of the base are neither rational nor irrational, it is with justice that they are called transcendental quantities." He did not clearly define his meaning of irrational, but from his examples we gather that he meant numbers expressible by radicals. A clear definition of a transcendental number was given by Legendre in his 1794 book: "It is probable that the number π is not even comprised among algebraic irrationals, that is, it cannot be the root of an algebraic equation of a finite number of terms whose coefficients are rational, but it seems very difficult to prove this proposition rigorously."

The first mathematician to rigorously prove the existence of transcendental numbers was Liouville. In 1840, he published two notes showing that e and e^2 could not be solutions of a quadratic equation. The 1843 publication by P. H. Fuss of the Euler, Goldbach, and Bernoulli correspondence further aroused Liouville's interest in transcendental numbers. He read a note on continued fractions to the French Academy in 1844. Given that a continued fraction was the root of an algebraic equation with integral coefficients (in modern terminology, an algebraic number), he gave the condition the terms of such a continued fraction had to satisfy. In a subsequent paper in the *Comptes Rendus*, he presented his famous criterion for a number to be algebraic of degree n: If x was such a number, then there existed an $A > 0$ such that for all rational $\frac{p}{q} \neq x$,

$$\left| x - \frac{p}{q} \right| > \frac{A}{q^n}.$$

He noted an almost immediate consequence of this:

$$\frac{1}{l} + \frac{1}{l^{(2!)}} + \frac{1}{l^{(3!)}} + \cdots + \frac{1}{l^{(n!)}} + \cdots$$

was a transcendental number, for $l > 1$ any integer.

It is clear that Liouville was attempting to prove the transcendence of e and it must have pleased him that his younger friend Hermite did so in 1873. Hermite used the basic identity

$$\int e^{-z} F(z) \, dz = -e^{-z} \gamma(z),$$

where $F(z)$ and $\gamma(z)$ were polynomials. Note that his can be proved by integration by parts and depends on the fact that $\frac{d}{dz} e^{-z} = -e^{-z}$. By means of this formula, Hermite defined certain polynomials with integer coefficients and employed them to obtain simultaneous rational approximations of e^x, for certain integer values of x. These in turn were sufficient to show that, except for the trivial cases, there could be no equation of the form

$$e^{z_0} N_0 + e^{z_1} N_1 + \cdots + e^{z_n} N_n = 0, \tag{38.1}$$

when $z_0, z_1, \ldots, z_n$ and $N_0, N_1, \ldots, N_n$ were all integers. We note that in the last portion of his paper, Hermite used his method to obtain the rational approximations

$$e = \frac{55291}{21344}, \qquad e^2 = \frac{158452}{12344}.$$

He left the problem of proving the transcendence of π to others. And soon afterwards, in 1883, Ferdinand Lindemann used Hermite's methods to prove this. Lindemann's theorem was a generalization of Hermite's: If $z_0, z_1, \ldots, z_n$ were distinct algebraic numbers and $N_0, N_1, \ldots, N_n$ were algebraic and not all zero, then equation (38.1) could not hold. The equation $1 + e^{i\pi} = 0$ implied the transcendence of π. Lindemann argued that if π were algebraic, then by the preceding theorem, $1 + e^{i\pi}$ could not equal zero. Lindemann's theorem also implied that when $x \neq 0$ was algebraic, then all the numbers e^x, $\arcsin x$, $\tan x$, $\sin^{-1} x$, and $\tan^{-1} x$ were transcendental. Moreover, if x was not equal to one, then $\log x$ was transcendental. Lindemann's proof was somewhat sketchy, but in 1885 Weierstrass gave a completely rigorous proof. In particular, he noted that Lindemann's theorem followed readily from the particular case in which $N_0, N_1, \ldots, N_n$ were integers. A number of mathematicians, including Hilbert, Hurwitz, Markov, Mertens, Sylvester, and Stieltjes improved and streamlined the proofs of Hermite and Lindemann without introducing any essentially new methods or results.

In his famous 1900 lecture at Paris, David Hilbert (1862–1943) gave a list of twenty-three problems for future mathematicians; the seventh of these was to prove the transcendence of certain numbers:

> I should like, therefore, to sketch a class of problems which, in my opinion, should be attacked as here next in order. That certain special transcendental functions, important in analysis, take algebraic values for certain algebraic arguments, seems to us particularly remarkable and worthy of thorough investigation. Indeed, we expect transcendental functions to assume, in general, transcendental values for even algebraic arguments; and, although it is well known that there exist integral transcendental functions, which even have rational values for all algebraic arguments, we shall still consider it highly probable that the exponential function $e^{i\pi z}$, for example, which evidently has algebraic values for all rational arguments z, will on the other hand always take transcendental values for irrational algebraic values of the argument z. We can also give this statement a geometrical form, as follows: *If, in an isosceles triangle, the ratio of the base angle to the angle at the vertex be algebraic but not rational, the ratio between base and side is always transcendental.* In spite of the simplicity of this statement and of its similarity to the problems solved by Hermite and Lindemann, I consider the proof of this theorem very difficult; as also the proof that *The expression α^β, for an algebraic base α and an irrational algebraic exponent β, e.g., the number $2^{\sqrt{2}}$, or $e^\pi = i^{-2i}$, always represents a transcendental or at least an irrational number.* It is certain that the solution of these and similar problems must lead us to entirely new methods and to a new insight into the nature of special irrational and transcendental numbers.

Hilbert's last comment has certainly turned out to be true. The resolution of Hilbert's seventh problem in the 1930s by the efforts of A. O. Gelfond and T. Schneider and the work of C. L. Siegel, the latter more directly inspired by that of Hermite and Lindemann, have initiated an era of tremendous growth and development in the theory of transcendental numbers. Hilbert himself was not very hopeful of a proof of his theorem within his lifetime, a theorem, as we have seen, also stated by Euler. Hilbert thought, in fact, that the Riemann hypothesis would be proved first.

The Russian mathematician Aleksandr O. Gelfond (1906–1968) took the first important step toward a proof of the Hilbert–Euler conjecture. He was a colleague at Moscow University of I. I. Privalov, whose influence in complex analysis is evident in Gelfond's work. Gelfond was a student of Aleksandr Khinchin who in 1922–23 studied the metrical properties of continued fractions, in which he obtained important results. Khinchin attracted several researchers to a whole range of problems in analytic number theory through his 1925–1926 seminar on this subject at Moscow. Gelfond's early work was influenced by a result in analytic functions due to Pólya: If an entire function assumes integral values for positive rational integral values of its argument and its growth is restricted by the inequality

$$|f(z)| < C 2^{\alpha|z|}, \quad \alpha < 1,$$

then it must be a polynomial. Roughly speaking, this means that a transcendental entire function taking integral values at integers must grow at least as fast as 2^z. Concerning the connection of this result with transcendental numbers, in his *Transcendental and Algebraic Numbers*, first published in Russian in 1952, Gelfond wrote,

> There is a very essential relationship between the growth of an entire analytic function and the arithmetic nature of its values for an argument which assumes values in a given algebraic field. If we assume in this connection that the values of the function also belong to some definite algebraic field, where all the conjugates of every value do not grow too rapidly in this field, then this at once places restriction on the growth of the function from below, in other words, it cannot be too small. This situation and its analogues for meromorphic functions can be used with success to solve transcendence problems. The first theorem concerning the relationship between the growth and the arithmetic value of a function was the Pólya theorem.

Hardy, Landau, and Okada successively managed to produce a streamlined proof of this result. We briefly sketch the argument, showing that very old ideas on interpolation have continued to play a role in function theory and transcendence theory. First, prove that if an entire function $f(z)$ satisfies

$$\varlimsup_{r \to \infty} \frac{\log M(r, f)}{r} < \log 2,$$

then
$$f(0) + z\Delta f(0) + \frac{z(z-1)}{2!}\Delta^2 f(0) + \frac{z(z-1)(z-2)}{3!}\Delta^3 f(0) + \cdots$$

converges uniformly to $f(z)$ in any finite region of the plane. Note that this is the Briggs–Harriot–Gregory–Newton interpolation series. Thus, if $f(z)$ is of exponential type less than $\log 2$, then $f(z)$ is represented by the interpolation series and can be evaluated at $z = -1$:

$$f(-1) = f(0) - \Delta f(0) + \Delta^2 f(0) - \cdots.$$

The convergence of the series implies that $|\Delta^n f(0)| < 1$ for $n > N$. Moreover, since $\Delta^n f(0)$ is an integer when $f(0), f(1), f(2), \ldots$ are all integers, we may conclude that $\Delta^n f(0) = 0$ for $n > N$. Thus, $f(z)$ is a polynomial and Pólya's theorem is proved.

In 1929, Gelfond took a step closer to solving Hilbert's problem when he used this type of interpolation series to obtain a key transcendence theorem: For $\alpha \neq 0, 1$ and algebraic, $\alpha^{\sqrt{-p}}$ is transcendental when p is a nonsquare positive integer. In particular, $2^{\sqrt{-2}}$ and $(-1)^{-i} = e^{\pi}$ are transcendental numbers. Gelfond gave details of only the particular case that e^{π} is transcendental. We present an outline of Gelfond's proof. First enumerate the Gaussian integers $m + in$ as a sequence $z_0, z_1, z_2, \ldots$, where one term precedes another if its absolute value is smaller; if the absolute values are the same, then the term with the smaller argument comes first. Then expand $e^{\pi z}$ as an interpolation series

$$\sum_{n=0}^{\infty} A_n (z - z_0) \cdots (z - z_{n-1}),$$

where, by Cauchy's theorem, A_n can be expressed as

$$\sum_{k=0}^{n} \frac{e^{\pi z_k}}{B_k}, \quad \text{where} \quad B_k = \prod_{\substack{j=0 \\ j \neq k}}^{n} (z_k - z_j).$$

This interpolation series converges to $e^{\pi z}$ because of the relatively slow growth of the function and the relatively high density of the interpolation points. Now if Ω_n is the least common multiple of $B_0, B_1, \ldots, B_n$, then, by the distribution of the primes of the form $4n + 1$ and $4n + 3$, it can be established that

$$|\Omega_n| = e^{(n \log n)/2 + O(n)} \quad \text{and} \quad |\Omega_n / B_k| = e^{O(n)}.$$

If one assumes that e^{π} is algebraic, then these estimates can be used to show that either $A_n = 0$ or that

$$|\Omega_n A_n| > e^{-O(n)}.$$

However, from the Cauchy integral for A_n, it follows that

$$|\Omega_n A_n| < e^{(-n \log n)/2 + O(n)}.$$

The two inequalities contradict one another for large enough n, unless $A_n = 0$ for all n larger than some value. Thus, one may argue that the interpolation series is finite and hence is a polynomial. This is a contradiction, so that Gelfond could conclude that his assumption that e^{π} was algebraic was false, proving his result.

In 1930, R. Kuzmin showed that, with some modifications in Gelfond's proof, one could prove the transcendence of $\alpha^{\sqrt{p}}$, with α and p as before. One implication of this was that $2^{\sqrt{2}}$ was transcendent, as Hilbert and Euler had conjectured. Since for general algebraic numbers β (in α^{β}), it was no longer possible to find useful upper bounds for Ω_n, a generalization along these lines was difficult. However, in 1933 K. Boehle was able to prove by this method that if $\alpha \neq 0, 1$ and β was an irrational algebraic number of degree $n \geq 2$, then at least one of the numbers $\alpha, \alpha^{\beta}, \alpha^{\beta^2}, \ldots, \alpha^{\beta^{n-1}}$ had to be transcendental. Carl Ludwig Siegel (1896–1981), who had been a student of Landau at Göttingen, also succeeded in proving Kuzmin's result after seeing Gelfond's proof of the transcendence of $\alpha^{\sqrt{-p}}$. But Siegel did not publish his proof in spite of Hilbert's suggestion that he do so. Siegel also made important and very original contributions to the theory of quadratic forms and to modular forms in several variables. His interest in

38.1 Preliminary Remarks

the history of mathematics led him to study Riemann's cryptic unpublished notes on the zeta function and to discover the Riemann–Siegel formula.

Though the method of Gelfond did not generalize to α^β, it suggested new lines of research. Gelfond himself applied it to a new proof of Lindemann's theorem; in 1943, his student A. V. Lototskii used it to show that certain infinite products represented irrational numbers; and in 1932 Siegel showed that if g_2 and g_3 were algebraic numbers, then at least one period of the Weierstrass $\wp$ function, satisfying the equation,

$$\wp'(z)^2 = 4\wp(z)^3 - g_2\wp(z) - g_3,$$

was transcendental. In particular, if $\wp(z)$ allowed complex multiplication, then both periods were transcendental. Siegel's student, Theodor Schneider (1911–1988), developed improved methods, allowing him to prove in 1935 that both periods were transcendental and even their ratio was transcendental, except when $\wp(z)$ permitted complex multiplication.

In 1934, Gelfond published a new method by which he obtained the complete proof of Hilbert's seventh problem. This proof made use of complex analysis, but some years later Gelfond and Linnik gave an interesting elementary proof of a special case, without recourse to analysis, except for Rolle's theorem. In his proof of the seventh problem, Gelfond assumed the result false. Thus, he posited that there existed algebraic numbers α, β, where $\alpha \neq 0, 1$ and β was not rational but $\alpha^\beta = \lambda$ was algebraic. On this assumption, there existed algebraic numbers α and β such that $\beta = \log\lambda/\log\alpha$ was an algebraic irrational number. He then constructed a function

$$f(z) = \sum_{k=0}^{N}\sum_{m=0}^{N} C_{k,m}\alpha^{kz}\lambda^{mz} = \sum_{k,m} C_{k,m} e^{(ak+bm)z} \quad (a = \log\alpha,\ b = \log\lambda),$$

where N was a suitably chosen large integer. Also, the $C_{k,m}$ were such that their absolute values and the absolute values of their conjugates were less than e^{2N^2}. Note that $f(z)$ could not be identically zero because b/a had to be irrational; moreover, the derivative of order s could be expressed as

$$f^{(s)}(z) = a^s \sum_{k=0}^{N}\sum_{m=0}^{N} C_{k,m}(k+\beta m)^s \alpha^{kz}\lambda^{mz}.$$

Gelfond proved that if α^β was an algebraic number, then it was possible to choose the $(N+1)^2$ nonzero algebraic numbers $C_{k,m}$ such that $f^{(s)}(z) = 0$ at $z = 0, 1, \ldots, r_2$ for $0 \leq s \leq r_1$, where r_1 was the greatest integer in $N^2/\log N$ and r_2 was the greatest integer in $\log\log N$. All this then implied that $f(z)$ had zeros of sufficiently high order at $0, 1, \ldots, r_2$. By an ingenious argument using Cauchy's integral formula, Gelfond then showed that $f(z)$ had a zero of even higher order at $z = 0$—in fact, of order at least $(N+1)^2 + 1$. Thus, he could conclude that the nonzero algebraic numbers $C_{k,m}$ satisfied the equations

$$a^{-s}f^{(s)}(0) = \sum_{k,m=0}^{N} C_{k,m}(k+\beta m)^s = 0, \quad 0 \leq s \leq (N+1)^2.$$

Taking the first $(N+1)^2$ equations, he obtained a system of equations with a Vandermonde determinant that had to be zero; this could happen if and only if there were integers

$$m_1, k_1;\quad m_2, k_2 \quad \text{such that} \quad \beta m_1 + k_1 = \beta m_2 + k_2.$$

This relation yielded the conclusion that β was rational, a contradiction to Gelfond's assumption, so that the theorem was proved.

In 1934, Schneider obtained an independent solution of Hilbert's seventh problem. His interest in transcendental numbers was aroused by a lecture of Siegel, who subsequently gave him a list from which to choose a dissertation topic. Schneider selected a problem on transcendental numbers; he reported, "After a few months, I gave him a work of six pages and then was told by Siegel that the work contained the solution of Hilbert's seventh problem." Schneider's proof was different in details from Gelfond's, but it too depended on the construction of an auxiliary function with a large number of zeros at specific points. In fact, both these mathematicians had adopted this technique from a previous work of Siegel on transcendence questions related to the values of Bessel functions. One may go further back and observe that in his proof of the transcendence of e, Hermite had also constructed a function of this kind!

Siegel's work of 1929 introduced another important method in the theory of transcendental numbers. Recall that Hermite's work depended on the fact that $\frac{d}{dx}e^x = e^x$. It was not until 1929, when Siegel published his paper on E-functions, that this idea was generalized to prove the transcendence of values of functions satisfying linear differential equations. E-functions are entire functions

$$\sum_{n=0}^{\infty} \frac{a_n}{n!} z^n,$$

where a_n are algebraic numbers satisfying certain arithmetic conditions. First, for any $\epsilon > 0$, a_n and all its conjugates are $O(n^{\epsilon n})$ as $n \to \infty$ and second, the least common denominator of $a_0, a_1, \ldots, a_n$ is also $O(n^{\epsilon n})$. Siegel considered a system of homogeneous linear differential equations of the first order

$$y'_k = \sum_{l=1}^{m} Q_{kl}(x) y_l, \quad \text{for} \quad (k=1,\ldots,m),$$

where the $Q_{kl}(x)$ were rational functions with coefficients in a number field K. To obtain transcendence results, Siegel required that some products of powers of the E-function $E_1, E_2, \ldots, E_m$, in fact, solutions of this system, satisfy a normality condition. Siegel formalized the concept of normality in his 1949 book; its meaning was only implicit in his 1929 paper. In spite of the fact that this condition was difficult to verify, thereby limiting the scope of its application, Siegel was able to employ it to rederive the classical theorem of Lindemann (and Weierstrass). He also proved a new theorem on the transcendence of a class of numbers related to the Bessel function: Observing that

$$K_\lambda(x) = \sum_{n=0}^{\infty} \frac{(-1)^n}{n!(\lambda+1)\cdots(\lambda+n)} \left(\frac{x}{2}\right)^n \quad (\lambda \neq -1, -2, \ldots),$$

one may verify that K_λ is an E-function and satisfies the differential equation

$$y'' + (2\lambda+1)/xy' + y = 0.$$

Siegel's theorem states that if λ is a rational number, $\lambda \neq \pm\frac{1}{2}, -1, \pm\frac{3}{2}, -2, \ldots$, and $\alpha \neq 0$ an algebraic number, then $K_\lambda(\alpha)$ and $K'_\lambda(\alpha)$ are algebraically independent. Note that complex numbers $\zeta_1, \zeta_2, \ldots, \zeta_n$ are called algebraically independent if for every nonzero polynomial $P(x_1, \ldots, x_n)$, in n variables with rational coefficients, we have $P(\zeta_1, \ldots, \zeta_n) \neq 0$. Otherwise, the ζ_j are algebraically dependent. Thus, if several numbers are algebraically independent, then each of them is transcendental. Therefore, $K_\lambda(\alpha)$ and $K'_\lambda(\alpha)$ are transcendental. Also, since the Bessel function $J_\lambda(x)$ may be expressed as

$$\frac{1}{\Gamma(\lambda+1)} \left(\frac{x}{2}\right)^\lambda K_\lambda(x),$$

it follows that except for $x = 0$, all the zeros of $J_\lambda(x)$ and $J'_\lambda(x)$ are transcendental numbers.

From his theorem, Siegel obtained the transcendence of certain continued fractions by noting that

$$i\sqrt{x}\,\frac{K_\lambda(2i\sqrt{x})}{K'_\lambda(2i\sqrt{x})} = \lambda+1+\cfrac{x}{\lambda+2+\cfrac{x}{\lambda+3+\ldots}}$$

Thus, Siegel's theorem implied that when 2λ was not an odd integer, the continued fraction was transcendental for every nonzero algebraic x. But when 2λ was an odd integer, Lindemann's theorem entailed the transcendence of the continued fraction. Siegel took the special case, when $\lambda = 0$ to obtain a nice result: the transcendence of

$$1 + \cfrac{1}{2+\cfrac{1}{3+\ldots}}$$

Siegel obtained the Lindemann-Weierstrass theorem using his method: Take algebraic numbers $a_1, \ldots, a_m$ linearly independent over the rational number field. The E-functions are $E_k(x) = e^{a_k x}$, $(k = 1, \ldots, m)$, and the μ power products take the form $e^{\rho_k x}$ ($k = 1, \ldots, \mu$) with μ different algebraic numbers ρ_k. The system of equations takes the form $y'_k = \rho_k y_k$ ($k = 1, \ldots, \mu$); verifying the normality condition in this case reduces to proving that any equation

$$P_1(x)e^{\rho_1 x} + \cdots + P_\mu(x)e^{\rho_\mu x} = 0,$$

where $P_i(x)$ are polynomials, implies that $P_1 = 0, \ldots, P_\mu = 0$. This is easy to show, and the Lindemann theorem follows from Siegel's theorem.

Siegel described the historical background to his work:

Lambert's work was generalized by Legendre who considered the power series

$$y = f_\alpha(x) = \sum_{n=0}^\infty \frac{x^n}{n!\alpha(\alpha+1)\cdots(\alpha+n-1)} \quad (\alpha \neq 0, -1, -2, \ldots)$$

satisfying the linear differential equation of second order $y'' + \alpha y' = y$. He obtained the continued fraction expansion

$$\frac{y}{y'} = \alpha + \cfrac{x}{\alpha + 1 + \cfrac{x}{\alpha + 2 + \cdots}}$$

and proved the irrationality of y/y' for all rational $x \neq 0$ and all rational $\alpha \neq 0, -1, -2, \ldots$. In the special case $\alpha = 1/2$ we have $y = \cosh(2\sqrt{x})$, $y' = \sinh(2\sqrt{x})/\sqrt{x}$, so that Legendre's theorem contains the irrationality of $(\tg a)/a$ for rational $a^2 \neq 0$. In more recent times, Stridsberg proved the irrationality of y and of y', separately, for rational $x \neq 0$ and rational $\alpha \neq 0, -1, \ldots$, and Maier showed that neither y nor y' is a quadratic irrationality. Maier's work suggested the idea of introducing more general approximation forms which enabled me to prove that the numbers y and y' are not connected by any algebraic equation with algebraic coefficients, for any algebraic $x \neq 0$ and any rational $\alpha \neq 0, \pm 1/2, -1, \pm 3/2, \ldots$. The excluded case of an integer $\alpha + \frac{1}{2}$ really is an exception, since then the function $f_\alpha(x)$ satisfies an algebraic differential equation of first order whose coefficients are polynomials in x with rational numerical coefficients; this follows from the explicit formulas

$$f_{k+\frac{1}{2}} = \frac{1}{2} \cdot \frac{3}{2} \cdots \left(k - \frac{1}{2}\right) D^k \cosh(2\sqrt{x}),$$

$$f_{\frac{1}{2}-k} = \frac{(-1)^k x^{k+\frac{1}{2}}}{\frac{1}{2} \cdot \frac{3}{2} \cdots (k - \frac{1}{2})} D^{k+1} \sinh(2\sqrt{x}) \quad (k = 0, 1, 2, \ldots).$$

For instance, in case $\alpha = \frac{1}{2}$, the differential equation is $y^2 - xy'^2 = 1$. In the excluded case, however, Lindemann's theorem shows that y and y' are both transcendental for any algebraic $x \neq 0$.

Due to the difficulty in verifying the normality condition, only these examples involving the exponential function and the Bessel function were obtained by this method between the publication of Siegel's paper in 1929 and his book in 1949. Finally, in 1988, F. Beukers, W. D. Brownwell, and G. Heckmann applied differential Galois theory to obtain a more tractable equivalent of the normality condition. They were able to verify the normality condition for a large class of hypergeometric functions. In their theory, the algebraic relations between the solutions of differential equations could be studied by means of the classification of linear algebraic groups. We note that the work of E. Vessiot, G. Fano, and E. Picard on linear differential equations during the late nineteenth century provided the foundation for differential Galois theory. Starting in 1948, E. Kolchin's work, itself based on the earlier 1932 book of J. F. Ritt, brought differential Galois theory to maturity.

In the period 1953–1959, Andrei Shidlovskii (1915–2007), student of Gelfond and teacher of V. A. Oleinikov, made major advances in the theory of E-functions. In 1954, he was able to replace Siegel's normality condition with a certain irreducibility condition, enabling him to work with some E-functions satisfying third- or fourth-order linear differential equations. A year later, he obtained stronger results; we give definitions before stating one of his theorems. Functions $f_1(z), f_2(z), \ldots, f_m(z)$ are homogeneously algebraically independent over $\mathbb{C}(z)$ if $P(f_1(z), \ldots, f_m(z)) \neq 0$ for every nonzero homogeneous polynomial in m variables with coefficients in $\mathbb{C}(z)$. Similarly, complex numbers $w_1, \ldots, w_m$ are said to be homogeneously algebraically independent over the field of algebraic numbers if $P(w_1, \ldots, w_n) \neq 0$ for every nonzero homogeneous polynomial P with algebraic numbers as coefficients. Now suppose

$$y'_k = \sum_{i=1}^{m} Q_{k,i} y_i \quad (k=1,\ldots,m), \quad Q_{k,i} \in \mathbb{C}(z), \qquad (38.2)$$

and suppose $T(z)$ is the least common denominator of all the m^2 rational functions $Q_{k,i}$. Shidlovskii's theorem may then be stated as: Let $f_1(z), f_2(z), \ldots, f_m(z)$ be a set of E-functions that satisfy the system of equations (38.2) and are homogeneously algebraically independent over $\mathbb{C}(z)$, and let ζ be an algebraic number such that $\zeta T(\zeta) \neq 0$. Then the numbers $f_1(\zeta), \ldots, f_m(\zeta)$ are homogeneously algebraically independent.

We may get an idea of the mathematical tradition within which Gelfond, Shidlovskii, and their students did their work by reading Mikhail Gromov's comments on his experience as a student in Russia: "There was a very strong romantic attitude toward science and mathematics: the idea that the subject is remarkable and that it is worth dedicating your life to. . . . that is an attitude that I and many other mathematicians coming from Russia have inherited." The accounts of the Gelfand seminars in Moscow, by Gromov, Landis and others, describe this attitude. The seminars extended to many hours of enthusiastic, colorful, and passionate discussion.

During the 1960s, Alan Baker, a student of Harold Davenport, effected another important and very productive development in transcendental number theory. He proved a substantial generalization of the Gelfond-Schneider theorem of 1934. We may state the latter in the form: If α and β are nonzero algebraic numbers and $\log \alpha$ and $\log \beta$ are independent over the rationals, then for any nonzero algebraic numbers α_1 and β_1,

$$\alpha_1 \log \alpha + \beta_1 \log \beta \neq 0.$$

In 1939, Gelfond obtained an explicit lower bound for $|\alpha_1 \log \alpha + \beta_1 \log \beta|$ in terms of the degrees and height of the four algebraic numbers. In a paper of 1948, Yuri Linnik and Gelfond pointed out that if a lower bound could be obtained for a similar three-term sum, then it would follow that the number of imaginary quadratic fields of class number one was finite; note that this result was one case within Gauss's class number problem. In 1966, Baker began to study this question by means of linear forms in logarithms. In that year, he established that if $\alpha_1, \alpha_2, \ldots, \alpha_n$ were nonzero algebraic numbers such that $\log \alpha_1, \log \alpha_2, \ldots, \log \alpha_n$ were independent over the rationals, then $1, \log \alpha_1, \ldots, \log \alpha_n$ would be independent over the field of algebraic numbers. As a corollary, Baker obtained the generalization of the Gelfond-Schneider theorem: If $\alpha_1, \alpha_2, \ldots, \alpha_n$ are algebraic but not 0 or 1; and if $\beta_1, \beta_2, \ldots, \beta_n$ are algebraic numbers such that $1, \beta_1, \beta_2, \ldots, \beta_n$ are linearly independent over the rationals, then $\alpha_1^{\beta_1} \alpha_2^{\beta_2} \cdots \alpha_n^{\beta_n}$ is transcendental. Baker also found an effectively computable lower bound for the absolute values of a nonvanishing linear form

$$|\beta_0 + \beta_1 \log \alpha_1 + \cdots + \beta_n \log \alpha_n|.$$

This result was applicable to a number of outstanding number theory problems, including Gauss's class number problem. But some of these number theoretic problems were also solved by other methods. In 1967, H. M. Stark solved the class number one problem by means of the theory of modular functions. Two years later, he published another paper, explaining that Kurt Heegner's 1952 solution of this problem was essentially sound. In constructing his proof, Heegner, a secondary school teacher, had made use of his deep understanding of Heinrich Weber's work on modular functions. Perhaps Heegner took it for granted that his readers would be equally familiar with Weber; this may have rendered his proof opaque. Indeed, Serre remarked that he found the paper very difficult to understand.

It is remarkable that mathematicians have managed to learn so much about transcendental numbers, for they have very strange properties. For example, transcendental numbers do not behave well under the usual algebraic operations. Moreover, even though it is true that if a number can be approximated sufficiently well by rational numbers, it must be transcendental, there nevertheless exist transcendental numbers that are not able to be approximated even as well as some quadratic irrational numbers. Indeed, Weil has stated that a preliminary version of Siegel's 1929 paper ended with the remark: "Ein Bourgeois, wer noch Algebra treibt! Es lebe die unbeschkränte Individualität der transzendenten Zahlen! [It's a bourgeois, who still does algebra! Long live the unrestricted individuality of transcendental numbers!]"

38.2 Liouville Numbers

In a paper of 1851, based on earlier work, Liouville constructed his transcendental numbers by proving that if x was the root of an irreducible polynomial of degree $n > 1$ with integer coefficients

$$f(x) = ax^n + bx^{n-1} + \cdots + gx + h,$$

then there existed a constant $A > 0$ such that $|x - p/q| > \frac{1}{Aq^n}$ for all rational numbers p/q. Although the absolute value sign was not in use at that time, Liouville made his meaning clear. To prove this theorem, Liouville supposed that $x, x_1, x_2, \ldots, x_{n-1}$ comprised all the roots of $f(x) = 0$ so that

$$f\left(\frac{p}{q}\right) = a\left(\frac{p}{q} - x\right)\left(\frac{p}{q} - x_1\right) \cdots \left(\frac{p}{q} - x_{n-1}\right).$$

He then set

$$f(p, q) = q^n f(p/q) = ap^n + bp^{n-1}q + \cdots + hq^n$$

so that he could write

$$\left|\frac{p}{q} - x\right| = \frac{|f(p, q)|}{q^n \left|a\left(\frac{p}{q} - x_1\right)\left(\frac{p}{q} - x_2\right) \cdots \left(\frac{p}{q} - x_n\right)\right|}.$$

Next, since $n > 1$, $f(p, q)$ was a nonzero integer, so that $|f(p, q)| \geq 1$. Moreover,

$$\left|a\left(x_1 - \frac{p}{q}\right)\left(x_2 - \frac{p}{q}\right) \cdots \left(x_n - \frac{p}{q}\right)\right|$$

38.2 Liouville Numbers

was bounded by a maximum value for values of p/q in a neighborhood of x (of, say, radius 1). Liouville denoted that maximum by A. It then became clear that

$$\left|\frac{p}{q} - x\right| > \frac{1}{Aq^n}$$

so that the proof was complete. Note that for points p/q outside the radius 1 around x, one has $|p/q - x| > 1$, so that the result holds. Liouville went on to show that the result was valid even when $n = 1$. In that case, $f(x) = ax + b = 0$ and so

$$\frac{p}{q} - x = \frac{ap + bq}{aq}.$$

If $x \neq p/q$, then $ap + bq \neq 0$, and

$$\left|\frac{p}{q} - x\right| \geq \frac{1}{aq} = \frac{1}{Aq}.$$

Liouville used his theorem to produce examples of transcendental numbers. He argued that a given number x could not be algebraic unless there was a constant A such that $|p/q - x| > \frac{1}{Aq^n}$. He took x to be defined by the series

$$x = \frac{1}{l} + \frac{1}{l^{2!}} + \frac{1}{l^{3!}} + \cdots + \frac{1}{l^{m!}} + \cdots,$$

where l was an integer ≥ 2. He let the partial sum up to the term whose denominator was $l^{m!}$ be p/q, so that $q = l^{m!}$. Liouville then observed that

$$x - \frac{p}{q} = \frac{1}{l^{(m+1)!}} + \frac{1}{l^{(m+2)!}} + \cdots \leq \frac{2}{l^{(m+1)!}} = \frac{2}{q^{m+1}}.$$

This inequality followed from the series

$$\frac{1}{l^{(m+1)!}}\left(1 + \frac{1}{l^{m+1}} + \frac{1}{l^{(m+1)(m+2)}} + \cdots\right) < \frac{1}{l^{(m+1)!}}\left(1 + \frac{1}{2} + \frac{1}{2^2} + \cdots\right) = \frac{2}{l^{(m+1)!}}.$$

By increasing m, he saw that for any fixed A and n he could not obtain $x - p/q > 1/(Aq^n)$, thus proving that x was transcendental. Liouville also noted the more general case; if he took

$$x = \frac{k_1}{l} + \frac{k_2}{l^{2!}} + \frac{k_3}{l^{3!}} + \cdots + \frac{k_m}{l^{m!}} + \cdots,$$

where $k_1, k_2, \ldots, k_m, \ldots$ were nonzero integers bounded by a constant, then x would be transcendental. He gave the example in which l could take the value 10 and the k_m could then take values between 1 and 9 inclusive. As an example of a slightly different kind, he considered

$$x = \frac{1}{l} + \frac{1}{l^4} + \frac{1}{l^9} + \cdots + \frac{1}{l^{m^2}} + \cdots.$$

For $q = l^{m^2}$,

$$x - \frac{p}{q} = \frac{1}{l^{(m+1)^2}} + \cdots < \frac{2}{l^{2m+1}q};$$

therefore, x could not be a root of a first-degree equation with rational coefficients and was hence irrational.

38.3 Hermite's Proof of the Transcendence of e

In his 1873 paper "Sur la fonction exponentielle," Hermite gave two proofs that e was transcendental. We give his second and more rigorous proof, following his notation for the most part, except that in some places we employ the matrix notation, not explicitly used by Hermite. About fifteen years before Hermite gave his proof, Cayley introduced the matrix notation and some of the elementary algebraic properties of matrices. It was some time, however, before the usefulness of matrices was generally recognized. We first sketch the structure of Hermite's argument. Take a relation of the form

$$e^{z_0} N_0 + e^{z_1} N_1 + \cdots + e^{z_n} N_n = 0, \tag{38.3}$$

where $z_0, z_1, \ldots, z_n$ are distinct nonnegative integers and $N_0, N_1, \ldots, N_n$ are any integers. It is clear that unless all the N are zero, e is an algebraic number. Hermite defined a set of $n(n+1)$ numbers η_j^i, $i = 0, 1, \ldots, n$ and $j = 1, 2, \ldots, n$, by the equation

$$\eta_j^i = \frac{1}{(m-1)!} \int_{z_0}^{z_j} \frac{e^{-z} f^m(z)}{z - z_i} \, dz, \tag{38.4}$$

where m was some positive integer and

$$f(z) = (z - z_0)(z - z_1)(z - z_2) \cdots (z - z_n). \tag{38.5}$$

He showed that the numbers η_j^i got arbitrarily small as m became large. To demonstrate this fact, Hermite's reasoning was that since e^{-z} was always positive, for any continuous functions $F(z)$, he had

$$\int_{z_0}^{Z} e^{-z} F(z) \, dz = F(\xi) \int_{z_0}^{Z} e^{-z} \, dz = F(\xi) \left(e^{-z_0} - e^{-Z} \right),$$

where ξ lay between z_0 and Z, the limits of integration. By choosing $Z = z_j$ and

$$F(z) = \frac{f^m(z)}{(m-1)!(z - z_i)},$$

he obtained

$$\eta_j^i = \frac{f^{m-1}(\xi)}{(m-1)!} \frac{f(\xi)}{\xi - z_i} \left(e^{-z_0} - e^{-z_j} \right).$$

This proved that $\eta_j^i \to 0$ as $m \to \infty$. Thus, the $(n+1) \times n$ matrix $\eta = \left(\eta_j^i \right)$, with η_j^i the element in the ith row and the jth column, depends on m. Denote this dependence by $\eta(m)$. Hermite determined a relation between $\eta(m)$ and $\eta(m-1)$ giving, by iteration, a relation between $\eta(m)$ and $\eta(1)$. Let us write the first relation as

$$\eta(m) = \Theta(m) \eta(m-1), \tag{38.6}$$

38.3 Hermite's Proof of the Transcendence of e

where $\Theta(m)$ is an $n+1 \times n+1$ matrix depending on $z_0, z_1, \ldots, z_n$. Thus,

$$\eta(m) = \Theta(m)\Theta(m-1)\cdots\Theta(2)\eta(1),$$

and we write (following Hermite) the element in the ith row and jth column of the matrix Θ as $\theta(j,i)$ where i and j run from 0 to n. Hermite showed that the θs were integers and that

$$\det \Theta(k) = \prod_{0 \le i < j \le n} (z_i - z_j)^2, \quad \text{for } k = 2, \ldots, m. \tag{38.7}$$

He then obtained an explicit expression for the elements of $\eta(1)$ in a suitable form: Let ζ denote any one of $z_0, z_1, \ldots, z_n$. He set

$$F(z) = \frac{f(z)}{z - \zeta} \quad \text{and}$$

$$\int e^{-z} F(z)\,dz = -e^{-z}\gamma(z), \tag{38.8}$$

where $\gamma(z) = F(z) + F'(z) + F''(z) + \cdots + F^{(n)}(z)$. Hermite noted that if

$$f(z) = z^{n+1} + p_1 z^n + p_2 z^{n-1} + \cdots + p_{n+1},$$

then $F(z) = z^n + (\zeta + p_1)z^{n-1} + (\zeta^2 + p_1\zeta + p_2)z^{n-2} + \cdots,$ \tag{38.9}

and the coefficients of the two polynomials were integers. From this he could conclude that

$$\gamma(z) = \Phi(z, \zeta) = z^n + \phi_1(\zeta)z^{n-1} + \phi_2(\zeta)z^{n-2} + \cdots + \phi_n(\zeta), \tag{38.10}$$

with $\phi_i(\zeta)$ a monic polynomial in ζ of degree i and with integer coefficients. We may let Φ denote the matrix with entry $\Phi(z_j, z_i)$ in the ith row and jth column where i and j run from 0 to n. Again, these entries must be integers and $\det \Phi = \det \Theta$. From (38.8) and (38.10), he then had

$$\int_{z_0}^{Z} \frac{e^{-z} f(z)}{z - \zeta}\,dz = e^{-z_0}\Phi(z_0, \zeta) - e^{-Z}\Phi(Z, \zeta). \tag{38.11}$$

This equation gave Hermite the required values of the entries of $\eta(1)$. For the final step, let $X = \Theta(m)\ldots\Theta(2)\Phi$ so that elements η_j^i of $\eta(m)$ are given by

$$\eta_j^i = e^{-z_0} X_{i0} - e^{-z_j} X_{ij}, \tag{38.12}$$

where the integers X_{ij} are the entries of X. Note that (38.12) gives rational approximations of $e^{z_j - z_0}$ for j running from 1 to n. Now, by (38.3) and (38.12),

$$e^{z_1}\eta_1^i N_1 + e^{z_2}\eta_2^i N_2 + \cdots + e^{z_n}\eta_n^i N_n$$
$$= e^{-z_0}(e^{z_1} N_1 + e^{z_2} N_2 + \cdots + e^{z_n} N_n) X_{i0}$$
$$\quad - (X_{i1} N_1 + X_{i2} N_2 + \cdots + X_{in} N_n)$$
$$= -(X_{i0} N_0 + X_{i1} N_1 + X_{i2} N_2 + \cdots + X_{in} N_n).$$

Hermite argued that, since the X_{ij} and the N_i were integers, the term on the right-hand side was an integer, but the term on the left-hand side could be made arbitrarily small because of the η_j^i. Therefore, he concluded that

$$X_{i0}N_0 + X_{i1}N_1 + \cdots + X_{in}N_n = 0, \quad i = 0, 1, \ldots, n.$$

We can write this system of equations as $XN = 0$, where the components of the vector N are $N_0, N_1, \ldots, N_n$. Since $\det X = (\det \Theta)^{m-1} \det \Phi = (\det \Theta)^m \neq 0$, we must have $N = 0$. This completes our outline of Hermite's proof.

Now let us see how Hermite obtained the basic formulas (38.6) and (38.7). To prove (38.6), Hermite showed that

$$\int_{z_0}^Z \frac{e^{-z} f^{m+1}(z)}{z - \zeta} dz = m\theta(z_0, \zeta) \int_{z_0}^Z \frac{e^{-z} f^m(z)}{z - z_0} dz + m\theta(z_1, \zeta) \int_{z_0}^Z \frac{e^{-z} f^m(z)}{z - z_1} dz$$

$$+ \cdots + m\theta(z_n, \zeta) \int_{z_0}^Z \frac{e^{-z} f^m(z)}{z - z_n} dz, \tag{38.13}$$

where $\theta(z, \zeta)$ was of the form

$$\theta(z, \zeta) = z^n + \alpha_1(\zeta) z^{n-1} + \alpha_2(\zeta) z^{n-2} + \cdots + \alpha_n(\zeta), \tag{38.14}$$

with $\alpha_1(\zeta), \alpha_2(\zeta), \ldots, \alpha_n(\zeta)$ monic polynomials in ζ, with integer coefficients, and where Z and ζ took values in $z_0, z_1, \ldots, z_n$. For this, he needed the auxiliary formula that there existed polynomials $\theta(z)$ and $\theta_1(z)$ of degree n, such that

$$\int \frac{e^{-z} G(z) f(z)}{z - \zeta} dz = \int \frac{e^{-z} G(z) \theta_1(z)}{f(z)} dz - e^{-z} G(z) \theta(z), \tag{38.15}$$

where $G(z) = (f(z))^m$. After taking the derivative of (38.15) and multiplying across by $f(z)/G(z)$, he had only to determine $\theta_1(z)$ and $\theta(z)$ so that

$$\frac{f(z)}{z - \zeta} f(z) = \theta_1(z) + \left[1 - \frac{G'(z)}{G(z)}\right] f(z) \theta(z) - f(z) \theta'(z). \tag{38.16}$$

He set $z = z_i$ in this equation, and got $0 = \theta_1(z_i) - mf'(z_i)\theta(z_i)$ or

$$\theta_1(z_i) = mf'(z_i)\theta(z_i), \quad i = 0, 1, \ldots, n. \tag{38.17}$$

Once the values $\theta(z_i)$ were found, the $n+1$ values determined the polynomials $\theta_1(z)$ and $\theta(z)$. To this end, he divided equation (38.16) by $f(z)$ to get

$$\frac{f(z)}{z - \zeta} = \frac{\theta_1(z)}{f(z)} + \left[1 - \frac{G'(z)}{G(z)}\right] \theta(z) - \theta'(z). \tag{38.18}$$

Next, by (38.17), the fractional part of $\left[1 - \frac{G'(z)}{G(z)}\right]\theta(z)$ canceled with $\frac{\theta_1(z)}{f(z)}$; and hence to determine $\theta(z)$, Hermite had to consider only the polynomial part of $\left[1 - \frac{G'(z)}{G(z)}\right]\theta(z)$. So he supposed

$$\theta(z) = \alpha_0 z^n + \alpha_1 z^{n-1} + \alpha_2 z^{n-2} + \cdots + \alpha_n.$$

38.3 Hermite's Proof of the Transcendence of e

By taking the logarithmic derivative of $G(z)$, he obtained

$$\frac{G'(z)}{G(z)} = \frac{m}{z-z_0} + \frac{m}{z-z_1} + \cdots + \frac{m}{z-z_n} = \frac{s_0}{z} + \frac{s_1}{z^2} + \frac{s_3}{z^3} + \cdots, \quad (38.19)$$

where $s_i = m(z_0^i + z_1^i + \cdots + z_n^i)$. Thus, comparing the coefficients of the polynomials on the two sides of (38.18) and using (38.9), he got the relations

$$1 = \alpha_0,$$
$$\zeta + p_1 = \alpha_1 - \alpha_0(s_0 + n),$$
$$\zeta^2 + p_1\zeta + p_2 = \alpha_2 + \alpha_1(s_0 + n - 1) - \alpha_0 s_1,$$
$$\vdots$$

These equations yielded the required coefficients of $\theta(z)$:

$$\alpha_0 = 1,$$
$$\alpha_1 = \zeta + p_1 + s_0 + n,$$
$$\alpha_2 = \zeta_2 + (s_0 + n - 1)\zeta_1 + (s_0 + n)(s_0 + n - 1) + s_1,$$
$$\vdots$$

where $\zeta_2 = \zeta^2 + p_1\zeta + p_2$, $\zeta_1 = \zeta + p_1$. Thus, α_i was shown to be a monic polynomial of degree i in ζ, and Hermite could write

$$\theta(z) = \theta(z,\zeta) = z^n + \alpha_1(\zeta)z^{n-1} + \alpha_2(\zeta)z^{n-2} + \cdots + \alpha_n(\zeta). \quad (38.20)$$

But in order to derive (38.13), Hermite set the limits of integration in (38.15) from z_0 to Z, where Z was one of the values $z_0, z_1, \ldots, z_n$; he arrived at

$$\int_{z_0}^{Z} \frac{e^{-z}G(z)f(z)}{z-\zeta} dz = \int_{z_0}^{Z} e^{-z}G(z)\frac{\theta_1(z)}{f(z)} dz. \quad (38.21)$$

By (38.17),

$$\frac{\theta_1(z)}{f(z)} = \frac{m\theta(z_0)}{z-z_0} + \frac{m\theta(z_1)}{z-z_1} + \cdots + \frac{m\theta(z_n)}{z-z_n}, \quad (38.22)$$

and then, in order to recognize the dependence on ζ, he wrote, as in (38.20), $\theta(z_i) = \theta(z_i, \zeta)$. And when (38.22) was substituted in (38.21), he got (38.13).

Now, in order to prove (38.7), observe that from (38.20), $\det \Theta$ can obtained by multiplying the determinants

$$\begin{vmatrix} z_0^n & z_0^{n-1} & \cdots & 1 \\ z_1^n & z_1^{n-1} & \cdots & 1 \\ \vdots & \vdots & \cdots & \vdots \\ z_n^n & z_n^{n-1} & \cdots & 1 \end{vmatrix} \times \begin{vmatrix} 1 & 1 & \cdots & 1 \\ \alpha_1(z_0) & \alpha_1(z_1) & \cdots & \alpha_1(z_n) \\ \vdots & \vdots & \cdots & \vdots \\ \alpha_n(z_0) & \alpha_n(z_1) & \cdots & \alpha_n(z_n) \end{vmatrix},$$

completing Hermite's proof of (38.7).

38.4 Hilbert's Proof of the Transcendence of e

In 1893, Felix Klein presented a series of lectures at the University of Chicago, including Hilbert's new and very efficient proof of the transcendence of e. Hilbert's elegant proof was based on ideas of Lindemann, Weierstrass, and Paul Gordan. To begin the proof, take ρ to be a positive integer, and set

$$I = z^\rho [(z-1)(z-2)\cdots(z-n)]^{\rho+1} e^{-z}.$$

Suppose e is not transcendental. Then we may set $a, a_1, a_2, \ldots, a_n$ to be integers such that

$$a + a_1 e + a_2 e^2 + \cdots + a_n e^n = 0.$$

Then

$$\frac{a}{\rho!}\int_0^\infty I\,dz + \frac{a_1 e}{\rho!}\int_1^\infty I\,dz + \frac{a_2 e^2}{\rho!}\int_2^\infty I\,dz + \cdots + \frac{a_n e^n}{\rho!}\int_n^\infty I\,dz$$
$$+ \left(\frac{a_1 e}{\rho!}\int_0^1 I\,dz + \frac{a_2 e^2}{\rho!}\int_0^2 I\,dz + \cdots + \frac{a_n e^n}{\rho!}\int_0^n I\,dz\right) = 0, \quad (38.23)$$

or $P_1 + P_2 = 0$, where P_2 is the sum inside the parentheses and P_1 is the part outside. In the term

$$\frac{a_k e^k}{\rho!}\int_k^\infty I\,dz,$$

with $k \geq 1$, contained in P_1, change z to $z+k$. We then have

$$\frac{a_k e^k}{\rho!}\int_0^\infty e^{-(z+k)} z^{\rho+1}(z+k)^\rho (z+k-1)^{\rho+1}\cdots(z+1)^{\rho+1}(z-1)^{\rho+1}\cdots(z+k-n)^{\rho+1}\,dz$$

$$= \frac{a_k}{\rho!}\int_0^\infty e^{-z} z^{\rho+1} \sum t_m z^m \,dz,$$

where $\sum t_m z^m$ is a polynomial in z with integer coefficients. Take one term in the sum, and evaluate as a gamma integral to get

$$\frac{a_k t_m}{\rho!}\int_0^\infty e^{-z} z^{\rho+m+1}\,dz = \frac{a_k t_m (\rho+m+1)!}{\rho!}.$$

Therefore,

$$\frac{a_k e^k}{\rho!}\int_k^\infty I\,dz$$

is an integer divisible by $\rho+1$. The first term $\frac{a}{\rho!}\int_0^\infty I\,dz$ in P_1 is easily seen to be

$$\pm a\,(n!)^{\rho+1} \pmod{\rho+1},$$

and hence

$$P_1 = \pm a\,(n!)^{\rho+1} \pmod{\rho+1}.$$

Take $\rho + 1$ to be a large prime so that $a(n!)^{\rho+1}$ is not divisible by $\rho + 1$. Notice that we can obviously choose $a \neq 0$, if e is algebraic. As for P_2, we can make it as small as we like. But $a(n!)^{\rho+1}$ is a nonzero integer, contradicting (38.23), and hence e cannot be algebraic. Note that Hermite's original proof still has the advantage that it obtains rational approximations of e raised to integer powers.

38.5 Exercises

1. Show that the sum of Goldbach's series is given exactly by

$$\sum_{1}^{\infty} \frac{1}{n^2 + \frac{p}{q}n} = \frac{q}{p}\left(\psi\left(\frac{p}{q}\right) + \gamma + \frac{q}{p}\right),$$

   where $\psi(x)$ is Gauss's digamma function. Observe that the value of $\psi(p/q)$ may be explicitly calculated for integer values of p and q, as Goldbach used them. See the results of Gauss, (23.69), (23.70), (23.71).

2. Suppose λ is a rational number not equal to a negative integer. Let

$$\phi_\lambda(z) = \sum_{n=0}^{\infty} \frac{z^n}{(\lambda+1)_n}$$

   and let ξ be a nonzero algebraic number. Show that $\phi_\lambda(\xi)$ is transcendental. Siegel stated this result without proof in his 1929 paper. The first published proof is due to Shidlovskii, dating from 1954. See Shidlovskii (1989), p. 185.

3. Suppose that the E-functions $f_1(z), \ldots, f_m(z)$ are algebraically independent over $\mathbb{C}(z)$ and form a solution of the system of linear differential equations

$$y'_k = Q_{k,0} + \sum_{i=0}^{m} Q_{k,i} y_i, \quad k = 1, \ldots, m; \quad Q_{k,i} \in \mathbb{C}(z).$$

   Let ξ be an algebraic number such that $\xi T(\xi) \neq 0$, with $T(\xi)$ as defined earlier. Show that under these conditions, the numbers $f_1(\xi), \ldots, f_m(\xi)$ are algebraically independent. See Shidlovskii (1989), p. 123.

4. Let

$$f(z) = \sum_{n=0}^{\infty} z^{2^n}.$$

   Show that if α is algebraic and $0 < |\alpha| < 1$, then $f(\alpha)$ is transcendental. This result is due to Kurt Mahler (1903–1985); though mostly self-taught, he regarded himself as a student of Siegel in his research. For this and other results, see the paper by J. H. Loxton and A. J. van der Poorten in Baker and Masser (1977), pp. 211–226.

5. Show that if

$$f_1(z) = \sum_{n=0}^{\infty} z^{2^n} \quad \text{and} \quad f_2(z) = \sum_{n=0}^{\infty} z^{3^n},$$

then for any two algebraic numbers α_1, α_2 in $0 < |z| < 1$, $f_1(\alpha_1)$ and $f_2(\alpha_2)$ are algebraically independent. See Kubota (1977).

6. Show that for an algebraic number $\beta \neq 0, 1$, the two numbers defined by the hypergeometric series $F(1/2, 1/2, 1, \beta)$ and $F(-1/2, 1/2, 1, \beta)$ are algebraically independent over the rationals. See Chudnovsky and Chudnovsky (1988).

7. Show that if α is algebraic and $0 < |\alpha| < 1$, the theta series $\sum_{n \geq 0} \alpha^{n^2}$ is transcendental. Recall that Liouville had shown this number to be irrational for $\alpha = 1/l$, where l was an integer > 1. See Nesterenko (2006) in Bolibruch, Osipov, and Sinai (2006).

38.6 Notes on the Literature

The quote from Nilakantha is a translation from Yushkevich (1964). Liouville (1851) contains Liouville's construction of transcendental numbers; see Lützen (1990), pp. 511–526 for a very interesting history of Liouville's work on these numbers. Hermite's proof of the transcendence of e was published in four parts in the *Comptes Rendus* (Paris) in 1873. It was reprinted in Hermite (1905–1917), vol. III, pp. 150–181. An English translation of a portion of Hermite's paper may be found in Smith (1959), vol. I, pp. 97–106. Hilbert (1970), vol. I, pp. 1–4, is a reprint of Hilbert's short proofs of the transcendence of e and π. These proofs were also presented by Felix Klein (1911), as lecture seven of his 1893 Evanston lectures. For the proof of the transcendence of e, see Klein (1911), pp. 53–55. Gelfond's quote concerning the relation between analytic functions and transcendental numbers may be found on p. 97 of Gelfond (1960); this work also contains some historical remarks on transcendental numbers. For earlier history, see Lützen (1990).

The English translations of the excerpts from the letters of Daniel Bernoulli and Goldbach were taken from Lützen (1990), pp. 513–514. See Yandell (2002), p. 404, for the quotation from Hilbert; see p. 199 for Schneider's remarks on how he realized that he had worked out Hilbert's seventh problem. Yandell gives an entertaining popular account of Hilbert's problems and those who made contributions to the solutions. See also Browder (1976) to read articles by experts on mathematical developments connected with Hilbert's problems (up to 1975). Siegel (1949), pp. 31–32 gives his historical remarks quoted in the text. The interview with Gromov by Raussen and Skau (2010) contains his comments on the Russian mathematical tradition; for the quote, see p. 392. For more on Russian mathematicians, see Zdravkovska and Duren (1993); and for an amusing description of Gelfand's seminar at Moscow by E. M. Landis, see pp. 68–69. Weil's quotation concerning Siegel and transcendental numbers can be found in Weil (1992), p. 53. For the theory of E-functions, related to material in this chapter, see Siegel (1949) and Shidlovskii (1989), nicely translated by Koblitz. See also articles in Baker (1988), one by Beukers and one by Beukers and Wolfart.

39

Value Distribution Theory

39.1 Preliminary Remarks

Value distribution theory addresses the problem of measuring the solution set of the equation $f(z) = b$, where f is some analytic function in some domain D and b is any complex number. For example, when f is a polynomial of degree n, the fundamental theorem of algebra, proved by Gauss and others, states that $f(z) = b$ has n solutions for a given b. The converse of this is an easier proposition. Algebraists since Descartes and Harriot recognized the important property of polynomials, that if $a_1, a_2, \ldots, a_n$ were a finite sequence of numbers, then there would be a polynomial of degree n, $(x - a_1)(x - a_2) \cdots (x - a_n)$, with zeros at exactly these numbers. After Euler found the infinite product factorization of the trigonometric and other functions, mathematicians could raise the more general question of the existence of a function with an infinite sequence $a_1, a_2, a_3, \ldots$ as its set of zeros. Of course, it was almost immediately understood that the product $\prod_{n=1}^{\infty}(x - a_n)$ might not converge; in special instances such as the gamma function, the proper modification was also determined, in order to ensure convergence. Gauss and Abel treated infinite products with some care in their work on the gamma and elliptic functions. But the answer to the general question had to wait for the development of the foundations of the theory of functions of a complex variable. In fact, Weierstrass, one of the founders of this theory, published an important 1876 paper dealing with the problem.

Karl Weierstrass (1815–1897) studied law at Bonn University, but after four years he failed to get a degree. With Christoph Gudermann as his mathematics teacher, Weierstrass became a Gymnasium teacher in 1841. Gudermann was a researcher in the area of power series representation of elliptic functions, and Weierstrass in turn made power series the basic technique in his work in complex analysis. His great accomplishment was the construction of a theory of Abelian functions; the 1854 publication of the first installment of his theory, secured him a professorship at Berlin. Weierstrass was a great teacher and had many great students, including H. A. Schwarz, G. Cantor, Leo Königsberger, and Sonya Kovalevskaya, whom he held in high regard. The Mathematics Genealogy Project counts his mathematical descendants as 16,585; the author would be included in that number. In order to lay a firm foundation for his theories, Weierstrass carefully developed the basic concepts of infinite series and infinite products.

Weierstrass took a sequence $\{a_n\}$ such that $\lim_{n=\infty} |a_n| = \infty$ was the only limit point. Note that if c is a finite limit point of the sequence, then a nonconstant function with zeros at $\{a_n\}$ cannot be analytic at c. And since a polynomial $f(x)$ is analytic at every finite x, it is reasonable to require that our function be analytic in the complex plane; Weierstrass called such a function an entire function and specified that $\lim_{n\to\infty} |a_n| = \infty$. He observed Cauchy's result that in the case of a finite number of zeros, an entire function f with zeros at $a_1, a_2, a_3, \ldots, a_n$ would take the form $f(x) = e^{g(x)} \prod_{k=1}^n (x - a_k)$, with g an entire function. Note that it is now standard practice to denote a complex variable by z or w but Weierstrass used x. Weierstrass noted that for infinite sequences, one might make the product conditionally convergent by arranging the factors in a particular order, but this was not possible in general. As an example, he gave the product always divergent for $x \neq 0$:

$$(1+x)\left(1+\frac{x}{2}\right)\left(1+\frac{x}{3}\right)\cdots.$$

Now, as we have seen in chapter 23, the reciprocal of the gamma function has zeros at the negative integers and by Euler's definition, attributed by Weierstrass to Gauss,

$$\frac{1}{\Gamma(x)} = \prod_{n=1}^\infty \left\{ \left(1+\frac{x}{n}\right)\left(\frac{n+1}{n}\right)^{-x} \right\} \quad \text{or}$$

$$\frac{1}{\Gamma(x)} = \prod_{n=1}^\infty \left\{ \left(1+\frac{x}{n}\right) e^{-x\log(\frac{n+1}{n})} \right\}.$$

In this context, instead of $\ln z$, we use the notation $\log z$, the logarithm of a complex number z; this is a multivalued function whose principal value is such that for $x > 0, \log x = \ln x$. Next, note that F. W. Newman had explicitly observed that convergence required the exponential factor $e^{-x/n}$. Although Weierstrass may not have been familiar with Newman's paper, he wrote that the product for $1/\Gamma(x)$ directed him toward a method for achieving convergence. He realized that with each factor $(1+x/a_n)$ it was necessary to include an exponential factor

$$e^{x/a_n + \frac{1}{2}(x/a_n)^2 + \cdots + \frac{1}{m_n}(x/a_n)^{m_n}},$$

where m_n was chosen in such a way that the product converged. For this purpose, Weierstrass defined the primary factors

$$E(x,0) = (1-x) \quad \text{and} \quad E(x,m) = (1-x)e^{\sum_{r=1}^m (x^r/r)}, \quad m = 1, 2, 3, \ldots.$$

Since

$$1 - x = e^{\log(1-x)} = e^{-\sum_{r=1}^\infty (x^r/r)} \quad \text{for } |x| < 1,$$

he had

$$E(x,m) = e^{-\sum_{r=1}^{\infty} x^{m+r}/(m+r)} \quad \text{for } |x| < 1.$$

Thus, m_n could be chosen so that for any fixed x, the series $\sum_{n=1}^{\infty} |x/a_n|^{m_n+1}$ converged. In fact, note that $m_n = n$ would work, since $|x/a_n|^n < (1/2)^n$ as long as $|a_n| > 2|x|$. Also, because $\lim_{n \to \infty} |a_n| = \infty$, this inequality would be valid for all but a finite number of a_n. Weierstrass proved that

$$\sum_{n=1}^{\infty} \log E\left(\frac{x}{a_n}, m_n\right)$$

converged absolutely and uniformly in any disk $|x| \leq R$, and so did the product

$$\prod_{n=1}^{\infty} E\left(\frac{x}{a_n}, m_n\right).$$

This product was an entire function with zeros at exactly $a_1, a_2, a_3, \ldots$.

The French mathematician Edmond Laguerre (1834–1886) used Weierstrass's product to classify transcendental entire functions according to their genus, just as polynomials may be classified by their degree. He defined a product to be of genus m if the integer m_n in each primary factor was a fixed integer m. Thus, a product of genus 0 is of the form $\prod_{n=1}^{\infty}(1 - x/a_n)$ while a product of genus 1 takes the form $\prod(1 - x/a_n)e^{-x/a_n}$. As we have seen in a different context, Laguerre was motivated by a desire to extend to transcendental functions the classical results on polynomials of Descartes, Newton, and others. Recall that in the course of discovering his extension of the Descartes rule of signs, Newton showed that if a polynomial with real coefficients $c_0 + c_1 x + c_2 x^2 + \cdots + c_n x^n$ had all roots real, then

$$(r+1)(n-r+1)c_{r+1}c_{r-1} \leq r(n-r)c_r^2, \quad r = 1, 2, \ldots, n-1;$$

the inequality would be strict when all roots were not equal. Note that by taking n infinite, we have the inequalities

$$(r+1)c_{r+1}c_{r-1} < rc_r^2, \quad r = 1, 2, 3, \ldots.$$

Laguerre raised the question: Given a transcendental entire function $f(x) = \sum_{n=0}^{\infty} c_n x^n$ with all real roots and real coefficients, will the coefficients satisfy these inequalities? In a paper of 1882, he showed that the result was true for $f(x)$ of genus 0 or 1. In the same year, he proved that, given a sequence of circles $|x| = r_n \to \infty$, such that $f'(x)/(x^n f(x))$ went to zero as $r_n \to \infty$, then $f(x)$ was of genus n. Laguerre also investigated the relationship between the zeros of a function of genus 0 or 1, with real zeros, and the zeros of its derivative.

In an 1883 paper on entire functions, Poincaré looked for a connection between the growth of a function and its genus p. He proved that for every $\epsilon > 0$,

$$\lim_{x \to \infty} |f(x)|/e^{\epsilon |x|^{p+1}} = 0,$$

and if $f(x) = \sum_{n=0}^{\infty} c_n x^n$, then

$$\lim_{n \to \infty} c_n \Gamma\left(\frac{n+\epsilon+1}{p+1}\right) = 0.$$

These results suggested that in order to measure growth of a function, one required a concept more refined than its genus. Consider the case of a monic polynomial $g(x)$ of degree n. For large x, $|g(x)|$ behaves like $|x|^n$. So if $M(r, g)$ is the maximum value of $|g(x)|$ on $|x| = r$, then

$$\lim_{r \to \infty} \frac{\log M(r, g)}{\log r} = n.$$

In 1896, Émile Borel defined the order ρ of a transcendental entire function f:

$$\rho = \overline{\lim_{r \to \infty}} \frac{\log \log M(r, f)}{\log r},$$

a concept implicitly contained in Hadamard's earlier work on the Riemann zeta function. Now Riemann had introduced the entire function

$$\xi(s) = \Gamma(1 + s/2)(s - 1)\pi^{-s/2}\zeta(s) \tag{39.1}$$

and by a brilliantly intuitive argument obtained its product formula. Hadamard's work on entire functions was motivated by the desire to provide justification for some of Riemann's results.

Jacques Hadamard (1865–1963) studied at the École Normale where his teachers included the outstanding mathematicians J. Tannery, Hermite, Picard, P. Appell, and G. Goursat. Hadamard wrote his doctoral thesis on the Taylor series of complex analytic functions, deriving results on the relation of the coefficients with the location of the singularities and with the radius of convergence. In his report on the thesis, Picard wrote that the abstract results appeared to lack practical value; Hermite suggested that Hadamard look for applications. Fifty years later, Hadamard recalled, "At that time, I had none [no applications] available. Now, between the time my manuscript was handed in and the day when the thesis was defended, I became aware of an important question which had been proposed by the Académie des Sciences as a prize subject; and precisely the results in my thesis gave the solution of that question. I had been led solely by my feeling of the interest of the problem and it led me the right way." The problem posed by the Académie was to prove Riemann's unproved assertions. Hadamard used his result on the relation between the coefficients and the growth of the function to prove that the exponent of convergence of the zeros of the function in (39.1) was at most 1. This effectively established Riemann's product formula for $\xi(s)$.

Hadamard's 1893 work implicitly contained the factorization theorem for functions of finite order: If $f(z)$ is of order ρ, then

$$f(z) = z^k P(z) e^{Q(z)},$$

where $Q(z)$ is a polynomial of degree $q \leq \rho$ and $P(z)$ is a product of genus $p \leq \rho$. Moreover, the order of $P(z)$ is equal to the exponent of convergence ρ_1 of the zeros z_n of P and $\rho_1 \leq \rho$. Note that the exponent of convergence is the infimum of the positive numbers α such that $\sum_{n=1}^{\infty} |z_n|^{-\alpha}$ converges. The work of Hadamard and Borel also implied a formula connecting the coefficients c_n of the Taylor series expansion of an entire function with the order of the function. In 1902, this was explicitly stated by the Finnish mathematician Ernst Lindelöf, son of Lorenz Lindelöf (1870–1946), a student of Mellin, as the relation

$$\rho = \varlimsup_{n \to \infty} \frac{-n \log n}{\log |c_n|}.$$

Lindelöf's interest in entire functions was aroused by his contact with Hadamard and others when he stayed in Paris in 1893–94 and then in 1898–99. When he returned to the University of Helsingfors in Finland, Lindelöf communicated this interest to his students, including Frithiof and Rolf Nevanlinna, who made fundamental contributions to the value distribution theory of meromorphic functions. Rolf Nevanlinna (1895–1980) founded value distribution theory as a quantitative generalization of Picard's theorem.

Charles-Émile Picard (1856–1941) proved that for an entire function $f(x)$, the equation $f(x) = a$ had a solution for every complex number a with at most one exception. The value of e^x is never 0, illustrating that exceptions might exist. Picard proved this in 1879 by an ingenious application of the multivalued inverse of the elliptic modular function $k^2(\tau)$, earlier studied by Abel, Jacobi, Hermite, Schwarz, and others. Speaking at his Jubilee celebration of 1936, Hadamard praised Picard's teaching as masterly. Referring to Picard's theorem, Hadamard addressed his teacher: "All mathematicians know, on the other hand, what a marvelous stimulus for research your mysterious and disconcerting theorem on entire functions was, and still is, because the subject has lost nothing of its topicality. I can say that I owe to it a great part of the inspiration of my first years of work." Indeed, Picard went on to extend his result to functions with an essential singularity. This theorem is a vast generalization of the Sokhotskii–Casorati–Weierstrass theorem that every complex number is a limit of the values assumed by a function in any neighborhood of an essential singularity.

Now in the case of a polynomial $f(x)$ of degree n, for every a, the equation $f(x) = a$ has n roots, counting multiplicity. Picard's theorem predicts that for a transcendental entire function $f(x)$, the equation $f(x) = a$ has an infinite number of solutions with at most one exceptional number a. It was then natural to seek a more precise measure of the number of solutions, or to inquire about their density. In 1896, Borel proved that, for entire functions of finite order, if f was of nonintegral order ρ, then the exponent of convergence of the zeros of $f - a$ equaled ρ for all complex numbers a. If f was of integral order ρ, then the same result would hold with at most one exceptional value of a, in which case the exponent of convergence was less than ρ.

With the development of complex function theory, attempts were made to prove Picard's theorem without using elliptic modular functions. Borel, Landau, Bloch and R. Nevanlinna found such proofs, opening up new paths in function theory and making the topic among the most popular in the mathematics of the early twentieth century. In working with meromorphic functions, Nevanlinna's difficulty in extending the results for entire functions to meromorphic functions was the lack of a concept corresponding to the maximum modulus of a function. Interestingly, in 1899, Jensen derived the basic formula for obtaining this expanded concept. He proved that if f was meromorphic in $|z| \leq r$ with zeros at a_j and poles at b_k inside $|z| < r$, then

$$\log|f(0)| = \frac{1}{2\pi} \int_0^{2\pi} \log|f(re^{i\phi})| d\phi - \sum \log \frac{r}{|a_j|} + \sum \log \frac{r}{|b_k|},$$

where sums were taken over all the zeros and all the poles respectively. Jensen thought that the formula might be important in studying the zeros of the Riemann zeta function and in particular in the proof of the Riemann hypothesis. In fact, Jensen's formula was useful in simplifying proofs of results in both prime number theory and entire functions.

In 1925, R. Nevanlinna defined the analog of the maximum modulus, the characteristic function of a meromorphic function f, as the sum of two functions: the mean proximity function, measuring the average closeness of f to a given complex number a; and the counting function, measuring the frequency with which f assumed the value a. In the same year, he went on to prove two fundamental theorems on the characteristic function, and his brother F. Nevanlinna recast them in a geometric context. The latter approach was further developed and extended in 1929 by T. Shimizu; by Lars Ahlfors in papers of the 1930s; and in 1960 by S. S. Chern. In the 1980s, Charles Osgood and Paul Vojta observed a close analogy between Nevanlinna theory and Diophantine approximation. The clarification and precise delineation of this analogy has had important consequences for both topics.

It is somewhat surprising that Jacobi had already found Jensen's result in 1827. Jacobi stated it only for polynomials, but as Landau pointed out, his argument can be extended to the general case. Jacobi used Fourier series to obtain his formula and was inspired by the work of Marc-Antoine Parseval (1755–1836) in Lagrange series. Parseval derived formulas for roots of equations in terms of definite integrals. Jacobi carried this program further by finding integral expressions for sums of powers of any number of roots of an equation, given in increasing order. Incidentally, Jacobi mentioned in his paper that as early as 1777, Euler had discovered the "Fourier coefficients"; Riemann was apparently not aware of this fact when he wrote his 1853 thesis on trigonometric series.

39.2 Jacobi on Jensen's Formula

Jacobi took a polynomial $f(x) = a + bx + cx^2 + \cdots + x^p$ with real coefficients and defined

$$\log(U^2 + V^2) = \phi\left(re^{+x\sqrt{-1}}\right) + \phi\left(re^{-x\sqrt{-1}}\right),$$

where $\phi(x) = \log f(x)$. He denoted the zeros of f by $\alpha', \alpha'', \alpha''', \ldots, \alpha^{(p)}$. These were taken in increasing order of absolute values, and Jacobi considered three separate cases: (i) r greater than all the roots, (ii) r less than all the roots, and (iii) r between $\alpha^{(k)}$ and $\alpha^{(k+1)}$. His formula is stated as

$$\frac{1}{2\pi}\int_{-\pi}^{\pi}\log(U^2+V^2)\,dx = \begin{cases} p\log r^2 \text{ in the first case,} \\ \log a^2 \text{ in the second case,} \\ k\log r^2 + \log(\alpha^{(k+1)})^2 + \log(\alpha^{(k+2)})^2 + \cdots + \log(\alpha^{(p)})^2 \\ \qquad \text{in the third case.} \end{cases}$$
(39.2)

Summarizing Jacobi's argument, suppose $\phi(x) = \log(a + bx + cx^2 + \cdots + x^p)$, where the coefficients $a, b, c, \ldots$ of the pth degree polynomial are real. Let $(x-\alpha')(x-\alpha'')(x-\alpha''')\cdots(x-\alpha^{(p)})$ represent the factorization of the polynomial. Since the polynomial has real coefficients, the complex roots appear in conjugate pairs. It is clear that

$$\phi(re^{+x\sqrt{-1}}) + \phi(re^{-x\sqrt{-1}})$$
$$= \log\left\{(a+br\cos x+cr^2\cos 2x+\cdots+r^p\cos px)^2 \right.$$
$$\left. + (br\sin x+cr^2\sin 2x+\cdots+r^p\sin px)^2\right\}.$$

Denote the expression inside the chain brackets by $U^2 + V^2$ so that

$$U^2+V^2 = \begin{cases} a^2+b^2r^2+c^2r^4+\cdots+r^{2p}, \\ +\ 2r\cos x(ab+bcr^2+cdr^4+\cdots), \\ +\ 2r^2\cos 2x(ac+bdr^2+cer^4+\cdots), \\ +\ 2r^2\cos 3x(ad+ber^2+cfr^4+\cdots), \\ +\ \cdots. \end{cases}$$
(39.3)

Now if $f(x) = a + bx + cx^2 + dx^3 + \cdots$, then the values of the Fourier integrals for $f(re^{ix})$ are given by

$$a^2+b^2r^2+c^2r^4+d^2r^6+\cdots = \frac{1}{2\pi}\int_{-\pi}^{\pi} f\left(re^{+x\sqrt{-1}}\right) f\left(re^{-x\sqrt{-1}}\right) dx,$$

$$ab+bcr^2+cdr^4+der^6+\cdots = \frac{1}{2\pi r}\int_{-\pi}^{\pi} f\left(re^{+x\sqrt{-1}}\right) f\left(re^{-x\sqrt{-1}}\right) \cos x\,dx,$$

$$ac+bdr^2+cer^4+dfr^6+\cdots = \frac{1}{2\pi r^2}\int_{-\pi}^{\pi} f\left(re^{+x\sqrt{-1}}\right) f\left(re^{-x\sqrt{-1}}\right) \cos 2x\,dx,$$

$$\cdots.$$
(39.4)

Note that $U^2+V^2 = f(re^{+x\sqrt{-1}})f(re^{-x\sqrt{-1}})$ and (39.3) gives the Fourier expansion of this function so that the Fourier coefficients can be computed by (39.4). We also have

$$\log(r e^{\sqrt{-1}x} - \alpha') = \log r e^{\sqrt{-1}x} + \log\left(1 - \frac{\alpha'}{r}e^{-\sqrt{-1}x}\right) \quad \text{when } r > |\alpha'|,$$

$$= \log(-\alpha') + \log\left(1 - \frac{r}{\alpha'}e^{\sqrt{-1}x}\right) \quad \text{when } |\alpha'| > r.$$

The logarithms on the right-hand sides can be expanded as a series by

$$-\log(1-t) = t + \frac{1}{2}t^2 + \frac{1}{3}t^3 + \frac{1}{4}t^4 + \cdots, \quad |t| < 1.$$

Jacobi used the above facts to give the series expansions for $\log(U^2 + V^2)$:

1.) With r greater than the absolute values of all the roots:

$$p \log r^2 - 2\sum_{1}^{p}\left(\frac{\alpha}{r}\cos x + \frac{\alpha^2}{2r^2}\cos 2x + \frac{\alpha^3}{3r^3}\cos 3x + \frac{\alpha^4}{4r^4}\cos 4x + \cdots\right).$$

2.) With r smaller than the absolute values of all the roots:

$$\log a^2 - 2\sum_{1}^{p}\left(\frac{r}{\alpha}\cos x + \frac{r^2}{2\alpha^2}\cos 2x + \frac{r^3}{3\alpha^3}\cos 3x + \frac{r^4}{4\alpha^4}\cos 4x + \cdots\right),$$

where a is the constant term in the polynomial $f(x)$.

3.) When $|\alpha^{(k)}| < r < |\alpha^{(k+1)}|$:

$$k \log r^2 - 2\sum_{1}^{k}\left(\frac{\alpha}{r}\cos x + \frac{\alpha^2}{2r^2}\cos 2x + \frac{\alpha^3}{3r^3}\cos 3x + \frac{\alpha^4}{4r^4}\cos 4x + \cdots\right)$$

$$+ 2\sum_{k+1}^{p}\left(\frac{1}{2}\log\alpha^2 - \frac{r}{\alpha}\cos x - \frac{r^2}{2\alpha^2}\cos 2x - \frac{r^3}{2\alpha^3}\cos 3x - \frac{r^4}{4\alpha^4}\cos 4x - \cdots\right),$$

where, as Jacobi explained, $\sum_{m}^{n}\psi(\alpha)$ denoted the sum of the quantities

$$\psi(\alpha^{(m)}), \psi(\alpha^{(m+1)}), \ldots, \psi(\alpha^{(n)}).$$

Jacobi integrated the series for $\log(U^2 + V^2)$ over the interval $(-\pi, \pi)$ to obtain

$$\int_{-\pi}^{\pi} \log(U^2 + V^2)\,dx$$

in the three cases. All the cosine terms vanished, and he obtained the formula (39.2). Jacobi also found formulas for $\sum_{1}^{k}\alpha^n$ and $\sum_{k+1}^{p}\frac{1}{\alpha^n}$ in terms of the nth Fourier coefficients of $\log(U^2 + V^2)$ and $\arctan V/U$.

39.3 Jensen's Proof

Jensen's 1899 rediscovery of Jacobi's formula succeeded in connecting the modulus of an analytic function with its zeros, and this occurred at just the right moment to fill a need in the theory of analytic functions. Jensen himself mentioned the possibility of applying his result to a proof of the Riemann hypothesis. Though he apparently did not pursue this topic further, and though his work has not as yet made a significant dent in the Riemann hypothesis, his formula is fundamental for the theory of entire functions. His proof, similar in some respects to Jacobi's, started with the formula

$$\log\left|1-\frac{z}{a}\right| = -\sum_{\nu=1}^{\infty}\frac{r^{\nu}}{2\nu}\left(\frac{e^{\nu\theta i}}{a^{\nu}}+\frac{e^{-\nu\theta i}}{\bar{a}^{\nu}}\right), \quad r=|z|<|a|,$$

$$= \log\frac{r}{|a|}-\sum_{\nu=1}^{\infty}\frac{1}{2\nu r^{\nu}}\left(a^{\nu}e^{-\nu\theta i}+\bar{a}^{\nu}e^{\nu\theta i}\right), r>|a|.$$

By integrating, he obtained

$$\frac{1}{2\pi}\int_{0}^{2\pi}\log\left|1-\frac{z}{a}\right|d\theta = \begin{cases} \log\frac{r}{|a|}, & \text{for } r>|a|, \\ 0, & \text{for } r<|a|. \end{cases} \quad (39.5)$$

Next he supposed $f(z)$ to be meromorphic in $|z| \leq r$ with zeros at $a_1, a_2, a_3, \ldots, a_n$ and poles at $b_1, b_2, \ldots, b_m$ in $|z| < r$ and with no singularities on $|z| = r$. Then he could express $f(z)$ in the form

$$f(z) = f(0)\frac{\prod_{k=1}^{n}\left(1-\frac{z}{a_k}\right)}{\prod_{k=1}^{m}\left(1-\frac{z}{b_k}\right)}e^{f_1(z)}, \quad (39.6)$$

where $$f_1(z) = \sum_{\nu=1}^{\infty}B_{\nu}z^{\nu} \quad \text{for} \quad |z|\leq r. \quad (39.7)$$

He took the real part of the logarithm of each side of (39.6), integrated over $(0, 2\pi)$, and applied (39.5) to get

$$\frac{1}{2\pi}\int_{0}^{2\pi}\log|f(re^{i\theta})|d\theta = \log|f(0)|+\log r^{n-m}\frac{|b_1||b_2|\cdots|b_m|}{|a_1||a_2|\cdots|a_n|}.$$

Note that the constant term in f_1 is zero, and hence there is no contribution from the integral of f_1.

39.4 Bäcklund Proof of Jensen's Formula

The Finnish mathematician R. J. Bäcklund, a student of Ernst Lindelöf, is credited with using a conformal mapping to prove Jensen's theorem in 1916 or 1918. This proof first assumed $g(z)$ to be analytic without zeros in $|z| \leq R$ and used Cauchy's integral formula to compute $\log g(0)$ as an integral. Then for a function f with zeros at $a_1, a_2, \ldots, a_n$ in $|z| \leq R$, consider a new function with no zeros in $|z| \leq R$:

$$g(z) = f(z)\frac{R^2-\bar{a}_1 z}{R(z-a_1)}\cdot\frac{R^2-\bar{a}_2 z}{R(z-a_2)}\cdots\frac{R^2-\bar{a}_n z}{R(z-a_n)}. \quad (39.8)$$

Since

$$\left|\frac{R^2 - \overline{a_k} z}{R(z - a_k)}\right| = 1 \quad \text{for} \quad |z| = R,$$

we have

$$|g(Re^{i\theta})| = |f(Re^{i\theta})|.$$

To fill out the details, start with Cauchy's integral formula

$$\log g(0) = \frac{1}{2\pi i} \int_{|w|=R} \log g(w) \frac{dw}{w} = \frac{1}{2\pi} \int_0^{2\pi} \log g(Re^{i\theta}) d\theta. \tag{39.9}$$

Now apply the expression for g in (39.8) and take the real part to find that

$$\log \frac{|f(0)| R^n}{|a_1||a_2|\cdots|a_n|} = \frac{1}{2\pi} \int_0^{2\pi} \log |f(Re^{i\theta})| d\theta.$$

This proof considered only analytic functions, but if one takes the case where $f(z)$ has poles at $b_1, b_2, \ldots, b_m$, one need merely multiply the right-hand side of (39.8) by

$$\frac{R(z-b_1)}{R^2 - \overline{b_1} z} \cdot \frac{R(z-b_2)}{R^2 - \overline{b_2} z} \cdots \frac{R(z-b_m)}{R^2 - \overline{b_m} z}$$

to obtain the result. Although it is difficult to ascertain exactly where Bäcklund gave this proof, we note that the use of the conformal mapping

$$\frac{R^2 - \overline{a} z}{R(z - a)}$$

is a beautiful and efficient innovation because it vanishes at $z = a$ and its value on $|z| = R$ is 1.

39.5 R. Nevanlinna's Proof of the Poisson–Jensen Formula

Rolf Nevanlinna gave an important extension of Jensen's formula; this result became the foundation of his theory of meromorphic functions. Suppose $f(x)$ is a meromorphic function in $|x| \leq \rho$ $(0 < \rho < \infty)$ with zeros and poles at a_h $(h = 1, \ldots, \mu)$ and b_k $(k = 1, \ldots, \nu)$, respectively. Let $x = re^{i\phi}$, $f(x) \neq 0, \infty$, and $r < \rho$. Then

$$\log |f(re^{i\phi})| = \frac{1}{2\pi} \int_0^{2\pi} \log |f(\rho e^{i\theta})| \frac{\rho^2 - r^2}{\rho^2 + r^2 - 2\rho r \cos(\theta - \phi)} d\theta$$
$$- \sum_{n=1}^{\mu} \log \left|\frac{\rho^2 - \overline{a_n} x}{\rho(x - a_n)}\right| + \sum_{k=1}^{\nu} \log \left|\frac{\rho^2 - \overline{b_k} x}{\rho(x - b_k)}\right|. \tag{39.10}$$

39.5 R. Nevanlinna's Proof of the Poisson–Jensen Formula

Nevanlinna called this the Poisson–Jensen formula, and his proof employed Green's formula

$$\int_\Gamma \left(U \frac{\partial V}{\partial n} - V \frac{\partial U}{\partial n} \right) ds = -\int_\Gamma (U \Delta V - V \Delta U) \, \partial \sigma. \tag{39.11}$$

Here U and V are twice continuously differentiable functions in a connected domain G with boundary Γ formed by a finite number of analytic arcs. The symbol Δ denotes the Laplacian and $\frac{\partial}{\partial n}$ represents the derivative in the direction normal to the boundary but pointing to the interior of G.

Take U to be a real-valued function $u(z)$ harmonic in $G \cup \Gamma$ except for logarithmic singularities at $z = c_1, c_2, \ldots, c_p$, so that

$$u(z) = \lambda_k \log |z - c_k| + u_k(z),$$

where λ_k is real, and u_k is continuous at $z = c_k$. Now let $V = g(z, x)$, where g denotes Green's function for the domain G with the singularity at an interior point $z = x$. This function is completely defined by the two conditions: The sum $g(z, x) + \log |z - x|$ is harmonic at all points interior to the domain G; also, $g(z, x)$ vanishes on the boundary Γ. Green's formula (39.11), discussed in our chapter 19, can be applied to U and V as chosen earlier if the points c_k and x are excluded by means of small circles around these points. Then, when the radii of these circles are allowed to tend to zero after the application of Green's formula, we get

$$u(x) = \frac{1}{2\pi} \int_\Gamma u(z) \frac{\partial g(z, x)}{\partial n} ds - \sum_{k=1}^{p} \lambda_k g(c_k, x). \tag{39.12}$$

Nevanlinna then took a meromorphic function f in the domain G with zeros at a_h ($h = 1, \ldots, \mu$) and poles at b_k ($k = 1, \ldots, \nu$). Then the function $u(z) = \log |f(z)|$ satisfied the conditions for (39.12) to hold so that he had

$$\log |f(x)| = \frac{1}{2\pi} \int_\Gamma \log |f(z)| \frac{\partial g(z, x)}{\partial n} ds - \sum_{h=1}^{\mu} g(a_n, x) + \sum_{k=1}^{\nu} g(b_k, x). \tag{39.13}$$

Nevanlinna noted that this important formula permitted him to compute the modulus, $|f|$, at any point inside G by using its values on the boundary of G and the location of the poles and zeros of f inside G. He took G to be a circle of radius ρ about the origin so that Green's function would be given by

$$g(z, x) = \log \left| \frac{\rho^2 - \bar{x} z}{\rho(x - z)} \right|.$$

By substituting this g, with $z = \rho e^{i\theta}$ and $x = r e^{i\theta}$, in (39.13), Nevanlinna obtained his Poisson-Jensen formula (39.10).

Further, to obtain Jensen's formula, he took $x = 0$ in (39.10), assuming that $f(0) \neq 0$ or ∞. Note that a slight modification was necessary in the cases $f(0) = 0$ or ∞. Thus, if

$$f(x) = c_\lambda x^\lambda + c_{\lambda+1} x^{\lambda+1} + \cdots,$$

and if $c_\lambda \neq 0$, then f had to be replaced by $x^{-\lambda} f$. He therefore had

$$\log |f(0)| = \frac{1}{2\pi} \int_0^{2\pi} \log |f(\rho e^{i\theta})| d\theta - \sum_{h=1}^{\mu} \frac{\rho}{|a_h|} + \sum_{k=1}^{\nu} \frac{\rho}{|b_k|}. \tag{39.14}$$

In his 1964 book on meromorphic functions, W. K. Hayman noted that the idea in the Bäcklund proof could be extended to yield a simple derivation of the Poisson–Jensen formula. Let g be an analytic function without zeros or poles in $|z| < R$. Now note that the mapping

$$w = \frac{R(\zeta - z)}{R^2 - \bar{z}\zeta}$$

maps the disk $|\zeta| \leq R$ conformally onto the unit disk and takes the point $\zeta = z$ to $w = 0$. This gave Hayman

$$\frac{dw}{w} = \frac{d\zeta}{\zeta - z} + \frac{\bar{z} d\zeta}{R^2 - \bar{z}\zeta} = \frac{(R^2 - |z|^2) d\zeta}{(R^2 - \bar{z}\zeta)(\zeta - z)}.$$

so that the result of Cauchy's theorem,

$$\log g(0) = \frac{1}{2\pi i} \int_{|w|=R} \log g(w) \frac{dw}{w},$$

could be replaced by

$$\log g(z) = \frac{1}{2\pi i} \int_{|\zeta|=R} \log g(\zeta) \frac{(R^2 - |z|^2) d\zeta}{(R^2 - \bar{z}\zeta)(\zeta - z)}.$$

Taking real parts of this formula and setting $z = r e^{i\theta}$ and $\zeta = R e^{i\phi}$,

$$\log |g(r e^{i\theta})| = \frac{1}{2\pi} \int_0^{2\pi} \log |g(R e^{i\phi})| \frac{(R^2 - r^2) d\phi}{R^2 - 2Rr \cos(\theta - \phi) + r^2}.$$

Following the Bäcklund approach, Hayman took

$$g(\zeta) = f(\zeta) \prod_{k=1}^{\mu} \left(\frac{R(\zeta - b_k)}{R^2 - \bar{b}_k \zeta} \right) \prod_{k=1}^{\nu} \left(\frac{R^2 - \bar{a}_k \zeta}{R(\zeta - a_k)} \right),$$

and the Poisson-Jensen formula followed.

39.6 Nevanlinna's First Fundamental Theorem

In his 1913 thesis, Georges Valiron (1884–1955), student of Émile Borel and teacher of Laurent Schwartz, expressed the sums on the right-hand side of (39.14) as integrals by means of the counting functions $n(r, 0)$ and $n(r, \infty)$. These functions denote the

39.6 Nevanlinna's First Fundamental Theorem

number of zeros and poles, counting multiplicity, of $f(x)$ in $|x| \leq r$. To efficiently implement Valiron's idea, Nevanlinna applied the Stieltjes integral to get

$$\sum \log \frac{\rho}{|a_h|} = \int_0^\rho \log \frac{\rho}{r} dn(r,0)$$

$$= \int_0^\rho \frac{n(r,0)}{r} dr \quad \text{and}$$

$$\sum \log \frac{\rho}{|b_k|} = \int_0^\rho \frac{n(r,\infty)}{r} dr.$$

He could then express (39.14) in the form

$$\log|f(0)| = \frac{1}{2\pi} \int_0^{2\pi} \log|f(\rho e^{i\theta})| d\theta - \int_0^\rho \frac{n(r,0)}{r} dr + \int_0^\rho \frac{n(r,\infty)}{r} dr. \quad (39.15)$$

Nevanlinna went on to write this in symmetric form, where he had the large values of the function on one side and the small values on the other. For that purpose, he set

$$\log^+ \alpha = \log \alpha, \quad \alpha \geq 1,$$
$$= 0, \quad 0 \leq \alpha < 1,$$

so that for $x > 0$, $\log x = \log^+ x - \log^+ 1/x$. He then defined the mean proximity function

$$m(\rho, f) = \frac{1}{2\pi} \int_0^{2\pi} \log^+ |f(\rho e^{i\theta})| d\theta$$

and the function

$$N(\rho, f) = \int_0^\rho \frac{n(r,\infty)}{r} dr.$$

Thus, he was able to rewrite (39.15) as

$$\log|f(0)| = m(\rho, f) - m(\rho, 1/f) + N(\rho, f) - N(\rho, 1/f),$$

$$\text{or} \quad T(\rho, f) = T(\rho, 1/f) + \log|f(0)|, \quad (39.16)$$

$$\text{where} \quad T(\rho, f) = m(\rho, f) + N(\rho, f).$$

Note that the term $m(\rho, f)$ is an average of $\log|f|$ on $|z| = \rho$ for large values of $|f|$, while the term $N(\rho, f)$ deals with the poles. So $T(\rho, f)$ acts as a measure of the large values of $|f|$ in $|z| \leq \rho$, while $T(\rho, 1/f)$ does the same for the small values of $|f|$. The function $T(\rho, f)$ has been named the Nevanlinna characteristic function, and it plays a fundamental role in the theory of meromorphic functions.

In the preceding formulation, we considered small and large values of f. More generally, Nevanlinna considered values of f close and/or equal to any fixed number

a by defining

$$N(r,a) = N\left(r, \frac{1}{f-a}\right) = \int_0^r \frac{n(t,a)}{t} dt, \qquad (39.17)$$

$$m(r,a) = m\left(r, \frac{1}{f-a}\right) = \frac{1}{2\pi} \int_0^{2\pi} \log^+ \left|\frac{1}{f(re^{i\theta})-a}\right| d\theta, \qquad (39.18)$$

$$T\left(r, \frac{1}{f-a}\right) = m(r,a) + N(r,a). \qquad (39.19)$$

By a simple argument, he showed that

$$|T(r,f) - T(r, f-a)| \leq \log^+ |a| + \log 2.$$

Combining this inequality with Jensen's formula (39.16), Nevanlinna arrived at his first fundamental theorem:

$$T\left(r, \frac{1}{f-a}\right) = T(r,f) - \log|f(0)-a| + \epsilon(r,a), \qquad (39.20)$$

$$|\epsilon(r,a)| \leq \log^+ |a| + \log 2.$$

Additionally, he proved that $N(r,a)$ and $T\left(r, \frac{1}{f-a}\right)$ were increasing convex functions of $\log r$. The result for $N(r,a)$ followed from (39.17), since

$$\frac{d N(r,a)}{d \log r} = n(r,a). \qquad (39.21)$$

Nevanlinna's proof for the convexity of $T(r, 1/(f-a))$ was rather lengthy, but in 1930 Henri Cartan obtained a simpler proof by first showing that

$$T(r,f) = \frac{1}{2\pi} \int_0^{2\pi} N(r, e^{i\theta}) d\theta + \log^+ |f(0)|. \qquad (39.22)$$

Since this immediately implied that

$$\frac{d T(r,f)}{d \log r} = \frac{1}{2\pi} \int_0^{2\pi} n(r, e^{i\theta}) d\theta,$$

the theorem was proved.

Note that the characteristic function $T(r,f)$ was Nevanlinna's analog of the maximum modulus $\log M(r,f)$, long sought after by complex function theorists. Recall that the logarithm of the maximum modulus, $\log M(r,f)$, was one of the essential objects in the study of entire functions; it was investigated by Hadamard, Borel, E. Lindelöf, and others. The efforts to extend the theory of entire functions to meromorphic functions required a suitable analog of $\log M(r,f)$ and Nevanlinna provided just that. Incidentally, in 1896 Hadamard proved that $\log M(r,f)$ was a convex function of $\log r$; this is usually known as the Hadamard three circles theorem.

39.7 Nevanlinna's Factorization of a Meromorphic Function

By use of the Poisson–Jensen formula, Nevanlinna was able to present a simple form of a canonical factorization of meromorphic functions of finite order. He gave this proof in the third chapter of his book, *Le théorème de Picard–Borel*, and we outline his argument. Note that a function meromorphic in the complex plane is said to be of order ρ if $\rho = \overline{\lim\limits_{r\to\infty}} (\log T(r, f)/\log r)$.

Nevanlinna stated the theorem: Suppose $f(x)$ is a meromorphic function of finite order with zeros and poles at $a_1, a_2, \ldots$ and $b_1, b_2, \ldots$, respectively. Let q be a integer such that

$$\lim_{r=\infty} \frac{T(r)}{r^{q+1}} = 0. \text{ Then}$$

$$f(x) = x^\alpha e^{\sum_0^q c_\nu x^\nu} \lim_{\rho=\infty} \frac{\prod_{|a_\nu|<\rho} \left(1 - \frac{x}{a_\nu}\right) e^{\frac{x}{a_\nu} + \cdots + \frac{1}{q}\left(\frac{x}{a_\nu}\right)^q}}{\prod_{|b_\nu|<\rho} \left(1 - \frac{x}{b_\nu}\right) e^{\frac{x}{b_\nu} + \cdots + \frac{1}{q}\left(\frac{x}{b_\nu}\right)^q}},$$

where α is an integer.

To prove this, assuming $f(0) \neq 0$, Nevanlinna differentiated the Poisson–Jensen formula $q+1$ times to get

$$D^{q+1} \log f(x) = \sum_{|a_\mu|<\rho} \frac{(-1)^q q!}{(x-a_\mu)^{q+1}} - \sum_{|b_\nu|<\rho} \frac{(-1)^q q!}{(x-b_\nu)^{q+1}} + S_\rho(x) + T_\rho(x),$$

where

$$S_\rho(x) = q! \sum_{|a_\mu|<\rho} \left(\frac{\overline{a_\mu}}{\rho^2 - a_\mu x}\right)^{q+1} - q! \sum_{|b_\nu|<\rho} \left(\frac{\overline{b_\nu}}{\rho^2 - b_\nu x}\right)^{q+1},$$

$$I_\rho(x) = \frac{(q+1)!}{2\pi} \int_0^{2\pi} \log\left|f(\rho e^{i\theta})\right| \frac{2\rho e^{i\theta} d\theta}{(\rho e^{i\theta} - x)^{q+1}}.$$

He then showed that $S_\rho(x)$ and $I_\rho(x)$ uniformly converged to zero for $|x| \leq r$ as $\rho \to \infty$. Taking this for granted, we have

$$D^{q+1} \log f(x) = (-1)^{q+1} q! \lim_{\rho \to \infty} \left\{ \sum_{|b_\nu|<\rho} \left(\frac{1}{x-b_\nu}\right)^{q+1} - \sum_{|a_\mu|<\rho} \left(\frac{1}{x-a_\mu}\right)^{q+1} \right\}.$$

Because of the uniform convergence, he could integrate $q+1$ times to get

$$\log f(x) = \sum_0^q c_\nu x^\nu + \lim_{\rho=\infty} \sum_{|a_\mu|<\rho} \left[\log\left(1 - \frac{x}{a_\mu}\right) + \frac{x}{a_\mu} + \cdots + \frac{1}{q}\left(\frac{x}{a_\mu}\right)^q\right]$$

$$- \sum_{|b_\nu|<\rho} \left[\log\left(1 - \frac{x}{b_\nu}\right) + \frac{x}{b_\nu} + \cdots + \frac{1}{q}\left(\frac{x}{b_\nu}\right)^q\right].$$

The result follows after exponentiation, since in case $f(0) = 0$, $f(0)$ can be replaced by $f(x)/x^\alpha$ for a suitable positive integer α.

39.8 Picard's Theorem

In his 1953 *A Mathematician's Miscellany*, later published with additional material as *Littlewood's Miscellany*, J. E. Littlewood raised and answered the question: "whether a dissertation of 2 lines could deserve and get a Fellowship." He answered in the affirmative, giving examples, including Picard's theorem for which there was a one-line statement and a one-line proof:

> (Theorem.) An integral [entire] function $f(z)$ never 0 or 1 is a constant. (Proof.) $\exp\{i\Omega(f(z))\}$ is a bounded integral function.

Littlewood explained that $\tau = \Omega(w)$ was the inverse of the modular function $w = k^2(\tau)$, arising in the theory of elliptic functions. The function $k^2(\tau)$ gave an analytic map from the half-plane $\{\tau \in \mathbb{C} : \text{Im}\,\tau > 0\}$ onto $\mathbb{C}\setminus\{0, 1\}$. Although the inverse Ω was many-valued, for any branch of it, $\Omega(f(z))$ extended analytically to give an entire function from $\mathbb{C}$ into $\{\tau \in \mathbb{C} : \text{Im}\,\tau > 0\}$. Note further that this argument implies that $\exp\{i\Omega(f(z))\}$ is a bounded analytic function; hence, by Liouville's theorem, it is a constant. Therefore, f is a constant. This was Picard's proof, but recall that in 1879, the study of the inverse of the modular function Ω was not well-established. So Picard used some care to prove that it was possible to define a single-valued branch of $\Omega(f(z))$ on the complex plane. Littlewood imagined what a referee's report could have been:

> Exceedingly striking and a most original idea. But, brilliant as it undoubtedly is, it seems more odd than important; an isolated result, unrelated to anything else, and not likely to lead anywhere.

It was clearly difficult to foresee the large number of interesting developments of complex function theory that would arise from Picard's theorem.

39.9 Borel's Theorem

Recall the Hadamard–Borel factorization theorem: An entire function of finite order $f(z)$ can be written in the form $z^k P(z) e^{Q(z)}$, where $Q(z)$ is a polynomial and $P(z)$ is the canonical product constructed from the zeros of f. Note here that the order of the entire function $P(z)$ is equal to the exponent of convergence of the zeros of f. We may deduce from this theorem, since $e^{Q(z)}$ is of integral order, that if f is of nonintegral order ρ, then the order of $P(z)$ must be ρ. This in turn implies that for an entire function $f(z)$ of nonintegral order ρ and any complex number x, the exponent of convergence of the zeros of $f(z) - x$ is also ρ. In keeping with the notation of Weierstrass, Valiron, and Nevanlinna, we sometimes employ x to represent a complex number or variable. In 1900, Borel showed that for entire functions of integral order ρ, the exponent of convergence of the zeros of $f(z) - x$ was equal to ρ except for at most one value of x. These exceptions became known as the Borel exceptional values.

Outlining Borel's proof, first suppose a and b are two exceptional values of x. Then by Hadamard's thereom

$$f(z) - a = z^{\alpha_1} P_1(z) e^{Q_1(z)} \quad \text{and} \quad f(z) - b = z^{\alpha_2} P_2(z) e^{Q_2(z)}, \qquad (39.23)$$

where Q_1 and Q_2 are polynomials of degree ρ and P_1 and P_2 are canonical products of order less than ρ. By subtracting the equations and multiplying by e^{-Q_2}, we have

$$z^{\alpha_1} P_1 e^{Q_1-Q_2} = z^{\alpha_2} P_2 + (b-a)e^{-Q_2}.$$

The term on the right-hand side has order equal to ρ and hence the polynomial $Q_1 - Q_2$ must be of degree ρ. Now differentiate the equation

$$z^{\alpha_1} P_1 e^{Q_1} - z^{\alpha_2} P_2 e^{Q_2} = b - a$$

to get

$$\left(z^{\alpha_1} P_1 Q_1' + (z^{\alpha_1} P_1)'\right) e^{Q_1} - \left(z^{\alpha_2} P_2 Q_2' + (z^{\alpha_2} P_2)'\right) e^{Q_2} = 0.$$

The coefficients of e^{Q_1} and e^{Q_2} are entire functions of order less than ρ, since the order of the derivative does not exceed the order of the function. So we can factorize these coefficients by the Hadamard–Borel theorem to obtain

$$z^{\alpha_3} P_3 e^{Q_3} e^{Q_1} - z^{\alpha_4} P_4 e^{Q_4} e^{Q_2} = 0,$$

where Q_3 and Q_4 are polynomials of degree at most $\rho - 1$, with P_3 and P_4 canonical products of orders less than ρ. Now rewrite the last equation as

$$e^{Q_1-Q_2+Q_3-Q_4} \equiv z^{\alpha_4-\alpha_3} P_4/P_3.$$

The degree of the polynomial $Q_1 - Q_2 + Q_3 - Q_4$ is ρ, and hence the left-hand side is an entire function of order ρ. On the other hand, the order of the function on the right-hand side is less than ρ. This contradiction proves the theorem.

39.10 Nevanlinna's Second Fundamental Theorem

In a paper of 1925, in what is now called his second fundamental theorem, R. Nevanlinna gave a far-reaching generalization of Picard's theorem. Nevanlinna's result showed that the term $N(r,a)$ was the dominant part of the characteristic function and that most of the roots of the equation $f(z) = a$ were simple. In his influential 1929 book on the Picard–Borel theorem, he discussed his theorem. He supposed $f(x)$ to be a meromorphic function and $z_1, z_2, \ldots, z_q$ ($q \geq 3$) distinct complex numbers, finite or not. Then

$$(q-2)T(r,f) < \sum_{\nu=1}^{q} N(r,z_\nu) - N_1(r) + S(r), \tag{39.24}$$

where $N_1(r) = N(r, 1/f') + (2N(r,f) - N(r,f'))$

and where the expression S satisfied:
1. For any positive number λ,

$$\int_{r_0}^{r} \frac{S(t)}{t^{\lambda+1}} dt = O\left(\int_{r_0}^{r} \frac{\log T(t,f)}{t^{\lambda+1}} dt\right). \tag{39.25}$$

2. Moreover,

$$S(r) < O\left(\log T(r, f) + \log r\right) \tag{39.26}$$

except for a set of finite linear measure. And if $f(x)$ was of finite order, that is,

$$\varlimsup_{r \to \infty} \frac{\log T(r, f)}{\log r} < \infty, \quad \text{then}$$

$$S(r) = O(\log r) \tag{39.27}$$

without restriction.

The proof of this theorem is lengthy and requires the computation of several estimates, the most important of which shows that $m(r, f'/f)$ is in general negligible in comparison with $T(r, f)$. For this quantity Nevanlinna proved that

$$m(r, f'/f) = O\left(\log\left(r\,T(r, f)\right)\right),$$

except on a set of finite linear measure when f is of infinite order, and

$$m(r, f'/f) = O(\log r),$$

without restriction, when f is of finite order.

To derive Picard's theorem, suppose f is an entire function that does not assume the values a and b. Take $q = 3$ and $z_1 = a$, $z_2 = b$, and $z_3 = \infty$ in (39.24) and since $N_1(r)$ is positive, we have $T(r, f) < S(r)$, contradicting (39.26). Thus, Picard's theorem is proved.

Nevanlinna combined the two fundamental theorems to derive an elegant extension of Picard's theorem. By the first fundamental theorem

$$\lim_{r=\infty} \frac{m(r, a) + N(r, a)}{T(r, f)} = 1.$$

He set

$$\delta(a) = \varliminf_{r=\infty} \frac{m(r, a)}{T(r, f)} = 1 - \varlimsup_{r=\infty} \frac{N(r, a)}{T(r, f)},$$

and by the second fundamental theorem

$$\sum_{\nu=1}^{q} \delta(a_\nu) \leq 2.$$

Observe that from this, Borel's theorem can be deduced. If a is a Borel exceptional value of an entire function, the reader may easily verify that $\delta(a) = 1$ and that $\delta(\infty) = 0$. By the preceding inequality, we know that there cannot be more than one exceptional value, completing the derivation.

39.11 Exercises

1. Suppose that all the roots of
$$f(x) = a_0 + a_1 x + a_2 x^2 + \cdots + a_n x^n$$
are real and that $\theta(x)$ is an entire function of genus 0 or 1. Suppose also that $\theta(x)$ is real for real x and all its zeros are real and negative. Prove that all the roots of
$$g(x) = a_0 \theta(0) + a_1 \theta(1) x + \cdots + a_n \theta(n) x^n$$
are real; in fact, that $f(x)$ and $g(x)$ have the same number of positive zeros and the same number of negative zeros. See Laguerre (1972), vol. 1, p. 201.

2. Show that if $f(z) = \sum a_n z^n$ is an entire function of finite order ρ, then
$$\rho = \limsup_{n \to \infty} \frac{n \log n}{\log(1/|a_n|)}.$$
See Lindelöf (1902).

3. Let $f(z) = \sum a_n z^n$ be an entire function of finite order and let
$$m(r) = \max(|a_n| r^n), \quad n = 0, 1, 2, \ldots.$$
Prove that
$$\lim_{r \to \infty} \frac{\log M(r, f)}{\log m(r, f)} = 1.$$
See Valiron (1949), p. 32.

4. Let $f(z)$ be of finite order ρ and of finite type $\tau = \lim_{r \to \infty} \frac{\log M(r)}{r^\rho}$. If
$$L = \limsup_{r \to \infty} r^{-\rho} n(r, f) \quad \text{and} \quad l = \liminf_{r \to \infty} r^{-\rho} n(r, f),$$
then $L e^{l/L} \leq e\rho\tau$. See Shah (1948) and Boas (1954), p. 16. Swarupchand Mohanlal Shah (1905–1996) received his appointment at Aligarh Moslem University (India) from André Weil, who served there as department head from 1931 to 1933. In 1942, Shah received his Ph.D. from the University of London under Hardy's student Titchmarsh. Returning to Aligarh, Shah served as head of the department from 1953 to 1958 when he reached the mandatory retirement age in India; he then took up a second mathematics career in the United States. He taught for more than twenty years in the United States, at Kansas and at Kentucky. Shah published hundreds of papers in complex analysis and gave a boost to a number of young mathematicians by encouraging them and collaborating with them.

5. Show that if $a \neq 0$ and $f(x) = a + a_1 x + \cdots$ is analytic at the origin, then there is a number L, depending only on a and a_1, such that if $f(x)$ is analytic in the disk $|x| < L$, then $f(x)$ must take the value 0 and/or 1 somewhere in the disk. See Landau (1904). In 1905, Constantin Carathéodory found an expression for L in terms of the fundamental branch of the inverse of the elliptic modular function.

Carathéodory used what is now called Schwarz's lemma, a result he extracted from Schwarz's work; he showed its importance, thereby elevating it to the status of an important lemma. Georg Pick then generalized this lemma.

6. Suppose $f(z)$ is meromorphic and has only a finite number of poles, and that $f(z), f^{(l)}(z)$ have only a finite number of zeros for some $l \geq 2$. Show that then

$$f(z) = \frac{P_1(z)e^{P_2(z)}}{P_3(z)},$$

with P_1, P_2, P_3 polynomials. Further, if $f(z)$ and $f^{(l)}(z)$ have no zeros, then either $f(z) = e^{Az+B}$ or else $f(z) = (Az+B)^{-n}$. This result is due to J. Clunie. See Hayman (1964), p. 67.

7. Let $f(z)$ be a meromorphic function of order ρ, where $0 \leq \rho \leq 1/2$; $\delta(a, f) > 0$ when $\rho = 0$ and $\delta(a, f) \geq 1 - \cos \pi \rho$ when $\rho > 0$. Show that then a is the only deficient value of $f(z)$; in particular, a meromorphic function of order zero can have at most one deficient value. This result is due to the German mathematician Oswald Teichmüller (1913–1943) for functions with positive poles and negative zeros; to the Russian mathematician A. A. Goldberg for the general case. See Hayman (1964), p. 114.

39.12 Notes on the Literature

Weierstrass gave a rigorous treatment of infinite products in his 1876 work *Zur Theorie der eindeutigen analytischen Functionen.* See Weierstrass (1894–1927), vol. 2, pp. 76–101. For proofs of Jensen's formula, see Jacobi (1969), vol. 6, pp. 12–20, and Jensen (1899); also see Bäcklund (1918) in this connection. Picard (1879) contains his theorem on the number of exceptional values of entire functions. For Borel's theorem, see Valiron (1949), pp. 72–73; also consult Borel (1900). Material on the Nevanlinna characteristic function and his two fundamental theorems may be found in his 1929 book, republished as Nevanlinna (1974). Neuenschwander (1978a) gives a history of the Casorati-Weierstrass theorem. Cartan's proof and Hayman's proof are both presented in Hayman (1964). See Cherry and Ye (2001), M. Ru (2001), and Bombieri and Gubler (2006) for treatments of the remarkable analogy between the Diophantine equations and value distribution or Nevanlinna theory. This parallel has been worked out in some detail and has led to significant advances in both areas. See the charming and witty book of Littlewood (1986), p. 40, for his amusing comments on Picard's theorem. Littlewood (1986) was edited with a foreword by his friend Béla Bollobás, and it contains a reprint of Littlewood's 1953 *A Mathematician's Miscellany*, along with photographs and some additional material. Hadamard's account of finding applications of his thesis results can be found on p. 56 of Maz'ya and Shaposhnikova (1998), translated by P. Basarab–Horwath; the Hadamard quote on Picard's theorem is on p. 36.

40

Univalent Functions

40.1 Preliminary Remarks

Weierstrass constructed a theory of functions using power series as the basic object, in contrast with Riemann, who studied analytic functions as mappings, specifically conformal mappings. The Bieberbach conjecture was rooted in these dual aspects of analytic function theory; it simultaneously viewed a function as a mapping and as a series. Thus, Bieberbach considered conformal mappings, such as those studied by Riemann, and then speculated on the magnitude of the coefficients, assuming the first two to be zero and one, respectively. A function f analytic in a domain D, an open and connected subset of the complex plane, is called univalent in D if it does not assume any value more than once. A univalent function f maps D conformally onto its image domain $f(D)$. Riemann was the first to study conformal mappings in the context of complex function theory. In his 1851 doctoral dissertation, he stated his famous theorem, now called the Riemann mapping theorem, that any simply connected proper subdomain D of the complex plane could be conformally mapped onto the unit disk $|z| < 1$. Note here that the mapping must be one-to-one and analytic. This mapping f is unique if we require that for a given point z_0 in the domain D, $f(z_0) = 0$ and $f'(z_0) > 0$. Observe that since the inverse of a univalent function is also univalent, it is of interest to consider functions univalent on the unit disk. We denote by S the set of normalized univalent functions on the unit disk, that is, univalent functions for which $f(0) = 0$ and $f'(0) = 1$. The Taylor expansion of f would take the form

$$f(z) = z + a_2 z^2 + a_3 z^3 + \cdots + a_n z^n + \cdots. \tag{40.1}$$

In a paper of 1916, Ludwig Bieberbach (1886–1982) proved that $|a_2| \leq 2$ and then, in a footnote, conjectured that $|a_n| \leq n$. Attempts to prove this conjecture led to valuable developments in the theory of analytic functions of one variable, lending it additional significance. The functional analyst Louis de Branges's 1984 proof of this conjecture concluded an era in the theory of functions, comparable, albeit on a smaller scale, to the 350-year era in number theory brought to an end by Andrew Wiles's 1994 resolution of Fermat's problem.

Riemann gave a sketch of a proof of his mapping theorem, but in 1871 his student, F. Emil Prym, found a flaw in his line of reasoning, apart from Riemann's use of an unproved variational principle rigorously established by Hilbert only a half century later. Note that George Green had made use of this principle in his famous work of 1828 on electricity and magnetism. In spite of its shaky foundations, however, the significance of Riemann's mapping theorem was immediately recognized. In 1867, Dirichlet's student Elwin B. Christoffel (1829–1900) showed that the upper half plane could be conformally mapped onto polygonal regions by means of functions defined by integrals. Note that the upper half plane may be mapped onto the unit disk by a fractional linear transformation. About two years later, Christoffel's result was independently rediscovered by H. A. Schwarz. In the 1870s, Carl Neumann and Schwarz used potential theoretic methods to prove the mapping theorem for regions bounded by analytic arcs. In the years around 1900, Hilbert brought renewed attention to the Riemann mapping problem and its generalization, the uniformization theorem, with the statement of his twenty-second problem and with his proofs of the Dirichlet principle.

In 1907, Paul Koebe (1882–1945) and Henri Poincaré proved the uniformization theorem that every simply connected Riemann surface was conformal to one of the three: the unit disk, the complex plane, or the extended complex plane. Poincaré's work was a continuation of methods and ideas he had developed in the early 1880s, when he established the theory of Fuchsian and Kleinian groups and the related theory of automorphic functions. Felix Klein played an equally important role in this development. In fact, Klein and Poincaré corresponded regularly in 1881–82 while creating these theories by differing approaches and techniques. In one of his proofs of the uniformization theorem, Koebe showed that the set S of normalized univalent functions was a normal family. Now a family F of analytic functions defined on a domain D is called normal if every sequence of functions f_n in F has a subsequence converging uniformly on each compact subset of D. The concept of a normal family is due to Paul Montel (1876–1975), a student of Borel and Lebesgue. In a June 1935 letter to Zermelo, Carathéodory discussed the history of this concept:

> The word and the notion "normal family" comes from Montel, who had shaped it around 1904. This notion has emerged from a further development of the Weierstrass double-series theorem stemming from Stieltjes (around 1895). If one notes that for all *analytic* functions $f(z)$, which are regular for $|z| < 1$ and satisfy the condition $|f(z)| < 1$ there, all coefficients of the power series $a_0 + a_1 z + \cdots = f(z)$ are uniformly limited, it follows that from every set $\{f(z)\}$ of such functions one can choose a uniformly convergent sequence on every circle $|z| \leq r < 1$. This led Montel to give the name "normal families" to all sets of functions which possess an analogous property. So, one was able to show that all functions which are regular in a domain G and are $\neq 0, 1$ constitute a normal family; the Picard theorem follows on from here easily. The notion of the limiting oscillation which allows us to speak of families that are normal in a *point* comes from me.

Constantin Carathéodory (1873–1950) was a German mathematician of Greek descent. He initially studied engineering at the Military School of Belgium and was involved with the construction of the Assiut dam in Egypt. Abandoning his engineering career due to an increasing attraction to mathematics, Carathéodory attended H. A. Schwarz's Berlin colloquia; he received his doctoral degree from Göttingen in 1904 under Minkowski for a thesis on the calculus of variations; his peripatetic career was then spent at a number of institutions in Germany, Greece, and the United States. Hans Rademacher was his 1916 student at Göttingen. Carathéodory's interest in function theory was aroused when Pierre Boutroux, nephew of H. Poincaré, visited Göttingen in 1905. Boutroux was then trying to simplify E. Borel's recent proof of Picard's theorem on entire functions and discussed this problem with Carathéodory. In his autobiographical notes, Carathéodory recalled this encounter:

> Boutroux had noticed that his proof was successful only because in the case of conformal mappings there was a remarkable rigidity, which, by the way, he was not able to put into formulae. Boutroux's discovery did not let me rest and six weeks later I was able to prove Landau's sharpening of the Picard theorem in a few lines by using the theorem which is today called the lemma of Schwarz. I produced this theorem with the help of Poisson's integral; only through Erhard Schmidt, whom I had informed of my findings, did I learn not only that the theorem already exists in the work of Schwarz, but also that it can be gained by absolutely elementary means. Indeed, the proof, which Schmidt informed me about, cannot be improved. Thus, I gained a further field of activity apart from the calculus of variations.

Schmidt's proof of Schwarz's lemma, a form of which was used by Schwarz in 1869 for his proof of the Riemann mapping theorem, is the one usually found in complex analysis textbooks. It was Carathéodory, however, who revealed the importance of the lemma by giving several significant applications of it. It is due to his efforts that Schwarz's lemma and its generalizations became so useful in complex function theory.

In his important 1912 paper, Carathéodory applied Schwarz's lemma to prove a result on kernel convergence, a key concept within geometric function theory. Suppose $G_1, G_2, \ldots, G_n, \ldots$ is an infinite sequence of simply connected domains in the complex plane, containing the origin but not coinciding with the whole complex plane. Suppose also that $f_n(z)$ is a conformal mapping of the unit disk onto the domain G_n with $f_n(0) = 0$ and $f_n'(0) > 0$. Carathéodory's theorem related the geometric behavior of the domains G_n with the analytic behavior of the functions f_n; this result was later employed by Löwner (Loewner) to develop his parametric method for the study of univalent functions. Carathéodory applied it to determine the boundary behavior of conformal mappings. As he wrote in his letter to Hilbert in connection with this theorem, "A first application of this theorem is, for instance, the proof of continuity of the conformal mapping as a function of its boundary, even if the boundary is a non-analytic curve and the Cauchy theorem cannot be applied." To state Carathéodory's convergence theorem, first suppose the origin is an interior point of $\bigcap G_n$; then the kernel of the sequence $\{G_n\}$ is defined as the largest domain G containing the origin such that each compact subset of G is contained in every G_n, with the possible exception of a finite number of G_n. Note that it is easy prove that G exists. Next, if the origin is not an interior point of $\bigcap G_n$, then the kernel is defined by $G = \{0\}$. The sequence $\{G_n\}$ is said to converge to the kernel G if every subsequence of $\{G_n\}$ has G as kernel. When convergence occurs, either $G = \{0\}$ or G is simply connected. Also, let $\{f_n\}$ be a sequence of univalent functions on the unit disk with $f_n(0) = 0$ and $f_n' > 0$; moreover, let f_n map the unit disk to G_n. On this basis, the theorem states that a sequence of functions $\{f_n\}$ converges uniformly on compact subsets of the unit disk to a function f if and only if $\{G_n\}$ converges to the kernel $G \neq \mathbb{C}$. If convergence occurs, then either $G = \{0\}$, in which case $f = 0$, or $G \neq \{0\}$, in which case f is a conformal mapping from the unit disk to G.

To prove this theorem, Carathéodory used Schwarz's lemma combined with Koebe's one-quarter theorem. The latter theorem was not fully proved until Bieberbach did so in 1916; for Carathéodory's convergence theorem, Koebe's weaker result, for some positive constant not necessarily 1/4, was sufficient.

Carathéodory's results were used a decade later by Löwner to construct his parametric theory of univalent functions. The Czech mathematician Karel Löwner (1893–1968) was a student of Georg Pick, and his name was later spelled Karl Löwner and then, after emigration to America, Charles Loewner. He studied in the German section of the University of Prague, writing his thesis in 1917 on convex conformal mappings under the direction of Pick, who himself did notable work in complex analysis. Pick's invariant form of the Schwarz lemma appears in several books on geometric function theory. Note that Pick was a student of Weierstrass's student Königsberger. Löwner's thesis contained interesting results on the growth of convex univalent functions and their derivatives. He also proved that the Bieberbach conjecture would hold for the subclass of convex univalent functions, and in fact, $|a_n| \leq 1$. In an important paper of 1923, Löwner developed a powerful method for dealing with the class of univalent functions. Bieberbach was very impressed by this method and inserted "I" (i.e., part I) in the title of Löwner's paper, implying that Löwner should work further in this area; unfortunately, Löwner did not return to the coefficient problem. In his paper, he defined a subset S_1 of S consisting of single slit mappings, univalent functions mapping the unit disk onto the complex plane minus one analytic Jordan arc extending to infinity. Using Carathéodory's theorem, he showed that S_1 was a dense subset of S in the topology defined by uniform convergence on compact subsets. Next, he proved that any function in S_1 could be obtained from the identity mapping by a series of successive infinitesimal transformations. He gave a fairly simple differential equation to effect this transformation; in fact, he gave two forms of this equation, one of which he himself used to prove that $|a_i| \leq i$ for $i = 2, 3$; the other form was used by de Branges to derive the complete result.

An alternative approach to the coefficient problem for univalent functions, using area inequalities, was initiated in a 1914 paper by the Swedish-American mathematician Thomas Hakon Gronwall (1877–1932); Bieberbach's work was an independent discovery of the same idea. This method would play a part in the development of univalent functions and in the proof of the Bieberbach conjecture. Gronwall received his doctoral degree under Mittag–Leffler, but also learned from mathematicians such as H. von Koch, I. Fredholm, and E. Phragmén. He received an engineering degree from Berlin in 1902 and then worked at various steel works in the United States. In 1912 he returned to his first love and in the next two years published almost two dozen papers ranging over the topics of Fourier series, analytic functions, conformal mappings, and special functions. Consequently, he was invited to Princeton as an instructor in 1913, and was promptly promoted. Gronwall soon left Princeton to take up a number of other pursuits, but not before J. W. Alexander (1888–1971), the famous topologist, had completed his thesis on univalent functions under him. In fact, Alexander had been a protégé of O. Veblen (1880–1960) and had already published couple of papers in topology when Veblen suggested that he do his thesis in analysis under Gronwall. Apparently, Veblen feared that topology might be a passing fad!

40.1 Preliminary Remarks

In 1916, Bieberbach rediscovered one of Gronwall's area inequalities and employed it to prove his theorem on the second coefficient in the Taylor expansion of a normalized univalent function. In this paper, he also obtained a result on the growth of a univalent function; he later used this to prove that $|a_n| = O(n^2)$. Then in 1923, Littlewood improved on this, showing that the order of the nth coefficient had to be n. Seven years later, Littlewood made another significant contribution to this topic in collaboration with his pupil Paley. R. E. A. C. Paley (1907–1933), graduated from Trinity College in 1929. He wrote his dissertation under Littlewood on nondifferentiable functions and was elected to a Trinity Fellowship in 1930. He was quickly blossoming into one of the leading British mathematicians of his generation when his life was cut short by a skiing accident in the Rocky Mountains. In his very brief career, he published almost thirty papers in several aspects of analysis and collaborated with such outstanding mathematicians as Littlewood, N. Wiener, and A. Zygmund. Littlewood and Paley proved that the coefficients of any odd univalent function in S are bounded by a constant independent of the function. More precisely, for all $F \in S$, and

$$F(z) = z + c_3 z^3 + c_5 z^5 + \cdots, \tag{40.2}$$

there exists an absololute constant A independent of F such that $|c_{2n+1}| \leq A$. In a footnote they observed, "No doubt the true bound is by $A = 1$." This conjecture makes sense in light of an earlier result of I. I. Privalov, that $A = 1$ for odd starlike functions. A set $E \subset \mathbb{C}$ is called starlike with respect to a point $w_0 \in E$ if the line segment joining w_0 to every point $w \in E$ lies entirely in E. A starlike function is a conformal mapping of the unit disk onto a domain starlike with respect to the origin. This conjecture implied the Bieberbach conjecture, but it was proved false in 1933 by the Hungarian mathematicians M. Fekete and G. Szegő. They used Löwner's theory to establish that

$$|c_5| \leq \frac{1}{2} + e^{-2/3} = 1.013\ldots, \tag{40.3}$$

and that the inequality was sharp. A modification of the Paley–Littlewood conjecture was suggested by a result of the French mathematician Jean Dieudonné (1906–1992). Dieudonné, a founding member of the Bourbaki group, proved in 1931 that if an odd univalent function is real on the real axis, then

$$|c_{2n-1}| + |c_{2n+1}| \leq 2, \quad \text{and} \quad |c_3| \leq 1. \tag{40.4}$$

Then in 1936, M. S. Robertson applied the method of Fekete to prove that

$$|c_3| + |c_5| \leq 2, \tag{40.5}$$

even when $F(z)$ was not real on the real axis. Combining this with Dieudonné's result, Robertson conjectured that the Littlewood–Paley conjecture was true on the average for an odd univalent function:

$$\sum_{k=1}^{n} |c_{2k-1}|^2 \leq n. \tag{40.6}$$

Observe that this implies the Bieberbach conjecture: If $f \in S$ is given by (40.1), and the odd function $F(z) = (f(z^2))^{1/2}$ by (40.2), then the relation between the coefficients of these two functions is given by

$$a_n = c_1 c_{2n-1} + c_3 c_{2n-3} + \cdots + c_{2n-1} c_1, \quad n \geq 1. \tag{40.7}$$

De Branges's proof of the Bieberbach conjecture in actuality demonstrated a more general result, Milin's conjecture; this concerned the logarithmic coefficients of the univalent function. Thus, de Branges's method did not directly yield Bieberbach's conjecture. We note that logarithmic coefficients in connection with univalent functions were first considered by Helmut Grunsky (1904–1986). Grunsky was an excellent analyst with a long and varied career; Ahlfors remarked that his thesis on extremal problems in conformal mappings was "a truly remarkable piece of work." In 1939, while at Berlin, Grunsky showed that an analytic function

$$g(z) = z + b_0 + \frac{b_1}{z} + \cdots$$

in a neighborhood of ∞ would extend to an injective and analytic function in the disk $|z| > 1$ (i.e. $g \in \Sigma$), if and only if its Grunsky coefficients, defined by

$$\log \frac{g(z) - g(\zeta)}{z - \zeta} = -\sum_{k=1}^{\infty} \sum_{l=1}^{\infty} b_{kl} z^{-k} \zeta^{-l}, \tag{40.8}$$

satisfied the Grunsky inequalities

$$\left| \sum_{k=1}^{\infty} \sum_{l=1}^{\infty} b_{kl} x_k x_l \right| \leq \sum_{k=1}^{\infty} \frac{|x_k|^2}{k}, \tag{40.9}$$

where $\{x_k\}$ was a sequence of complex numbers. Clearly, the Grunsky coefficients provided a characterization of the property of univalence. Grunsky's proof of the theorem employed contour integration and was not difficult, although the expressions of the Grunsky coefficients b_{kl} in terms of the coefficients b_k of g were very complicated. Perhaps this is one reason that the effectiveness of Grunsky's inequality was not noticed until around 1960 when it was used by Z. Charzynski and M. Schiffer to reprove the result that $|a_4| \leq 4$. In 1955, Schiffer and P. R. Garabedian had already proved $|a_4| \leq 4$ by means of a powerful variational technique developed by Schiffer in the 1930s. Soon after the work of Charzynski and Schiffer, a generalization of Gronwall's area theorem in terms of the Grunsky coefficients was noted by a number of mathematicians, including J. A. Jenkins, Milin, and C. Pommerenke. Schiffer had already made this observation in 1948. For Pommerenke's formulation, let $g \in \Sigma$, and let $x_1, x_2, \ldots, x_m$ be complex numbers not all zero. Then

$$\sum_{k=1}^{\infty} \left| \sum_{l=1}^{m} b_{kl} x_l \right|^2 \leq \sum_{k=1}^{m} \frac{1}{k} |x_k|^2, \tag{40.10}$$

where equality holds if and only if the area of $\mathbb{C} \setminus g(|z| > 1)$, that is, the complement of the image of $|z| > 1$, is zero. Note that when $m = \infty$, (40.10) and (40.9) are equivalent.

In a 1964 paper, I. M. Milin applied the area method to study the properties of $\{A_n(\zeta)\}$, defined by

$$\log \frac{z-\zeta}{F(z)-F(\zeta)} = \sum_{n=1}^{\infty} A_n(\zeta) z^{-n}, \tag{40.11}$$

where $F \in \Sigma$. Soon after this, I. E. Bazilevich worked directly with $\log(f(z)/z)$ and proved an interesting inequality about its coefficients. In his account of the motivations behind his conjecture, Milin wrote, "In this way I developed the conviction that the property of univalence reveals itself rather simply through area theorems or other methods in the form of restrictions on the coefficients of the logarithmic function (40.11) and $\log \frac{f(z)}{z} = 2\sum_{n=1}^{\infty} \gamma_n z^n$, and that it is necessary to construct an 'apparatus of exponentiation' to transfer the restrictions from logarithmic coefficients to coefficients of the functions themselves."

It was with this in mind that in 1966, N. A. Lebedev and Milin worked out the exponential inequality: If $\sum_{k=1}^{\infty} A_k z^k$ is an arbitrary power series with positive radius of convergence and

$$\exp\left(\sum_{k=1}^{\infty} A_k z^k\right) = \sum_{k=0}^{\infty} D_k z^k,$$

then $$\sum_{k=0}^{n-1} |D_k|^2 \leq n \exp\left\{\frac{1}{n} \sum_{v=1}^{n-1} \sum_{k=1}^{v} \left(k|A_k|^2 - \frac{1}{k}\right)\right\}.$$

Now note that if we write $(f(z)/z)^{1/2} = \sum c_{2n+1} z^{2n+1}$, then

$$\left(\frac{f(z)}{z}\right)^{1/2} = \exp\left(\frac{1}{2} \log \frac{f(z)}{z}\right)$$

implies that for γ_n, as defined in Milin's quotation,

$$\sum_{n=0}^{\infty} c_{2n+1} z^n = \exp\left\{\sum_{n=1}^{\infty} \gamma_n z^n\right\}. \tag{40.12}$$

Applying the Lebedev-Milin inequality, we obtain

$$\sum_{k=0}^{n-1} |c_{2k+1}|^2 \leq n \exp\left\{\frac{1}{n} \sum_{v=1}^{n} \sum_{k=1}^{v} \left(k|\gamma_k|^2 - \frac{1}{k}\right)\right\}. \tag{40.13}$$

Milin observed this inequality in 1970 in the course of writing his book on univalent functions. He perceived that if the inequalities

$$\sum_{v=1}^{n} \sum_{k=1}^{v} \left(k|\gamma_k|^2 - \frac{1}{k}\right) \leq 0, \ n = 1, 2, 3, \ldots, \tag{40.14}$$

were true, then Robertson's conjecture (and hence Bieberbach's conjecture) followed from (40.13). For Koebe's function $f_\theta(z)$, given by (40.24), $|\gamma_k| = 1/k$ so that equality holds in (40.14). Milin did not state (40.14) as a conjecture in his book, although he had evidence to support it. For instance, inspired by a result of Pommerenke, Milin obtained the equality

$$\sum_{k=1}^{n} |\gamma_k|^2 \leq \sum_{k=1}^{n} \frac{1}{k} + \delta, \quad \delta < 0.312. \tag{40.15}$$

It was in a 1971 paper that A. Z. Grinshpan, with the approval of Lebedev and Milin, referred to inequality (40.14) as Milin's conjecture.

Lebedev, Milin, E. G. Emelyanov, and Grinshpan were members of the Leningrad or St. Petersburg school of geometric function theory. In 1984, these mathematicians and other members of the Leningrad Seminar joined together and exerted considerable effort to reformulate in classical form de Branges's proof of the Milin conjecture, making it more accessible to the community of geometric function theorists. The Leningrad school was founded by G. M. Goluzin (1906–1952). Goluzin entered Leningrad University in 1924 and remained there in various capacities until his death. He was appointed professor of mathematics in 1938, and from then on he led the seminar and built up a school of function theorists. Goluzin made major contributions to the theory of univalent functions and developed a variation on Schiffer's technique of interior variations, applying it to several problems and deriving a number of deep results. In an early paper, he applied Löwner's parametric method to obtain a sharp bound on $|\arg f'(z)|$ for $f \in S$. Another easily stated theorem of Goluzin is that $|a_n| < \frac{3}{4}en$, an improvement on Littlewood; of course, he derived the Goluzin inequality: If $g \in \Sigma$, z_ν lie in the set $|z| > 1$ and $\gamma_\nu \in \mathbb{C}$, $\nu = 1, 2, \ldots, n$, then

$$\left| \sum_{\mu=1}^{n} \sum_{\nu=1}^{n} \gamma_\mu \gamma_\nu \log \frac{g(z_\mu) - g(z_\nu)}{z_\mu - z_\nu} \right| \leq \sum_{\mu=1}^{n} \sum_{\nu=1}^{n} \gamma_\mu \overline{\gamma_\nu} \log \frac{1}{1 - (z_\mu \overline{z_\nu})^{-1}}. \tag{40.16}$$

We observe that this inequality can be derived from Grunsky's; conversely, this implies Grunsky. In 1972, the American mathematician C. H. FitzGerald exponentiated Goluzin's inequality to obtain what is now called FitzGerald's inequality, from which he derived several coefficient inequalities. For example, he showed that $|a_n| < \sqrt{\frac{7}{6}} n < 1.081 n$, and in 1978, D. Horowitz, using the same method, made an improvement, obtaining

$$|a_n| < \left(\frac{1,659,164,137}{681,080,400} \right)^{1/14} n \approx 1.0657 n.$$

Until de Branges, this was the best result on the Bieberbach conjecture for all n.

40.2 Gronwall: Area Inequalities

In his 1914 paper, Gronwall derived results on the growth of a univalent function and its derivative. These depended on the measure of the area of the image of a disk under the conformal transformation given by the univalent function. Gronwall gave two main

40.2 Gronwall: Area Inequalities

applications of this idea. In the first, he assumed $f(x) = \sum_{n=1}^{\infty} a_n x^n$ to be univalent and the area of the image of the unit disk under f to be at most A. Then for $|x| \leq r < 1$, he showed that

$$|f(x)| \leq \sqrt{\frac{A}{\pi}} \sqrt{\log \frac{1}{1-r^2}}.$$

He used the change of variables formula for an integral to conclude that the area $A(r)$ of the image of the disk $|x| \leq r < 1$ was

$$A(r) = \int_0^r d\rho \int_0^{2\pi} |f'(\rho e^{i\theta})|^2 \rho \, d\theta = \pi \sum_{n=1}^{\infty} n|a_n|^2 r^{2n}. \tag{40.17}$$

Note that

$$f'(\rho e^{i\theta}) = \sum_{n=1}^{\infty} n a_n \rho^{n-1}$$

and term by term integration is possible because of absolute convergence. From (40.17) Gronwall concluded, after letting r tend to one, that

$$\pi \sum_{n=1}^{\infty} n|a_n|^2 \leq A. \tag{40.18}$$

Using (40.18),

$$|f(x)| \leq \sum_{n=1}^{\infty} |a_n| r^n,$$

and the Cauchy–Schwarz inequality, he found the required result:

$$|f(x)|^2 \leq \sum_{n=1}^{\infty} n|a_n|^2 \sum_{n=1}^{\infty} \frac{r^{2n}}{n} = \frac{A}{\pi} \log \frac{1}{1-r^2}.$$

For the second application, Gronwall considered a function

$$f(x) = 1/x + \sum_{n=1}^{\infty} a_n x^n,$$

where, without the term $1/x$, the series converged for $|x| < 1$. Such series had been discussed earlier, but Gronwall derived an important inequality for them, called the area theorem. For this purpose, it is convenient to let $z = 1/x$ and consider the class Σ of functions

$$g(z) = z + b_0 + \frac{b_1}{z} + \frac{b_2}{z^2} + \cdots, \tag{40.19}$$

analytic and one-to-one in $|z| > 1$ except for a simple pole at ∞ with residue 1. For these functions, Gronwall proved that

$$\sum_{n=1}^{\infty} n|b_n|^2 \leq 1. \tag{40.20}$$

To verify this, he again applied the area method. He did not give all the details, but one may apply Green's theorem to see that if the closed curve C_r is the image of $|z| = r > 1$ under $g(z)$, then it encloses a positive area

$$0 < \frac{1}{2}\int_0^{2\pi} \overline{g(re^{i\theta})} g'(re^{i\theta}) re^{i\theta}\, d\theta$$
$$= \pi\left\{r^2 - \sum_{n=1}^{\infty} n|b_n|^2 r^{-2n}\right\}. \tag{40.21}$$

The necessary inequality follows by letting $r \to 1^+$.

40.3 Bieberbach's Conjecture

In 1916, apparently unaware of Gronwall's earlier work, Bieberbach reproved the area theorem and deduced his inequality for the second coefficient of functions in the set S of normalized univalent functions. To achieve this, he used an idea he called Faber's trick: Supposing f is a function in S, then $F(z) = (f(z))^{1/2}$ is an odd univalent function. To prove this, observe that $f(z)$ vanishes only at $z = 0$ and hence a single valued branch of the square root can be chosen in

$$F(z) = z(1 + a_2 z^2 + a_3 z^4 + \cdots)^{1/2}.$$

Clearly, $F(z)$ is odd. It is univalent because if $F(z_1) = F(z_2)$, then $f(z_1^2) = f(z_2^2)$; moreover, the univalence of $f(z)$ implies $z_1 = \pm z_2$. If $z_1 = -z_2$, then $F(z_1) = F(z_2) = -F(z_1)$. This implies $F(z_1) = 0$ or $z_1 = 0$, proving the result. To apply the area theorem, Bieberbach noted that

$$F(z) = z + \frac{1}{2}a_2 z^3 + \cdots,$$

and used $F(z)$ to construct a function $g(z)$ in the class Σ:

$$g(z) = \frac{1}{F(1/z)} = z - \frac{1}{2}a_2\frac{1}{z} + \cdots = z + \sum_{n=1}^{\infty} b_n z^{-n}.$$

Hence, by (40.20), he obtained $|b_1| \leq 1$ or $|a_2| \leq 2$. In a footnote, Bieberbach went on to conjecture that $|a_n| \leq n$ for the coefficients of a normalized univalent function on the unit disk. He was able to verify Koebe's conjecture that the image of the open unit disk under any $f \in S$ would always contain a circle of radius $1/4$ with the origin as center.

More precisely, he showed that if $f(z) \neq w$ in $|z| < 1$, then $|w| \geq 1/4$ (an improvement on Koebe). This was an immediate corollary of the bound for a_2. For if $f(z) \neq w$, then

$$\frac{w f(z)}{w - f(z)} = z + \left(a_2 + \frac{1}{w}\right) z^2 + \cdots \in S$$

and hence

$$\left| a_2 + \frac{1}{w} \right| \leq 2, \quad \left| \frac{1}{w} \right| \leq 2 + |a_2| \leq 4 \text{ or } |w| \geq 1/4. \tag{40.22}$$

The example (often called Koebe's function)

$$f(z) = \frac{z}{(1-z)^2} = z + 2z^2 + 3z^3 + \cdots + nz^n + \cdots, \tag{40.23}$$

or more generally

$$f_\theta(z) = \frac{z}{(1 - e^{i\theta} z)^2} = z + 2e^{i\theta} z^2 + 3e^{2i\theta} z^3 + \cdots + n e^{(n-1)\theta} z^n + \cdots, \tag{40.24}$$

shows that $|a_2| = 2$ actually occurs for functions in S. We can also write (40.23) as

$$w = f(z) = \frac{1}{4} \left(\frac{1+z}{1-z} \right)^2 - \frac{1}{4}.$$

From this representation, it is easy to see that $f(z)$ maps $|z| < 1$ conformally onto the w-plane cut from $-\frac{1}{4}$ to $-\infty$ along the negative real axis. Note that $f(z) \neq -\frac{1}{4}$ in $|z| < 1$. Moreover, because $f_\theta(z) = e^{-i\theta} f(ze^{i\theta})$, this function maps $|z| < 1$ conformally onto the w-plane cut radially from $-\frac{1}{4} e^{-i\theta}$ to $-\infty e^{-i\theta}$.

In this paper, Bieberbach obtained another important result on the growth of a normalized univalent function $f(z)$:

$$\frac{r}{(1+r)^2} \leq |f(z)| \leq \frac{r}{(1-r)^2}, \quad |z| = r, \ (0 < r < 1). \tag{40.25}$$

40.4 Littlewood: $|a_n| \leq en$

In 1923, Littlewood proved that Bieberbach's conjecture was correct up to the order of magnitude. His paper with the result given in the title of this section appeared in 1925. Littlewood derived his inequality for the coefficients a_n from the inequality,

$$\frac{1}{2\pi} \int_0^{2\pi} |f(re^{i\theta})| \, d\theta < \frac{r}{1-r}, \quad 0 < r < 1, \tag{40.26}$$

where $f \in S$. He considered the univalent function

$$\phi(z) = (f(z^2))^{1/2} = z + b_3 z^3 + \cdots$$

and by (40.25) concluded that

$$|\phi(te^{i\theta})| \leq t/(1-t^2).$$

This result in turn implied that ϕ transformed $|z| \leq t < 1$ to a region whose area $A(t)$ was less that $\pi t^2/(1-t^2)^2$. He combined this with the equation

$$\pi \sum_{n=1}^{\infty} n|b_n|^2 t^{2n} = \int_0^t r\,dr \int_0^{2\pi} |\phi'(re^{i\theta})|^2\,d\theta = A(t)$$

to derive the inequality

$$\sum_{n=1}^{\infty} n|b_n|^2 t^{2n-1} \leq t/(1-t^2)^2.$$

Integrating from 0 to $r < 1$, he obtained

$$\sum_{n=1}^{\infty} |b_n|^2 r^{2n} \leq r^2/(1-r^2).$$

He next observed that the series on the left-hand side of this inequality was given by

$$\frac{1}{2\pi}\int_0^{2\pi} |\phi(re^{i\theta})|\,d\theta = \frac{1}{2\pi}\int_0^{2\pi} |f(r^2 e^{2i\theta})|\,d\theta = \frac{1}{2\pi}\int_0^{2\pi} |f(r^2 e^{i\psi})|\,d\psi.$$

At this point, to derive the necessary result for a_n, Littlewood could apply Cauchy's formula, with $r = 1 - \frac{1}{n}$, to obtain

$$|a_n| = \frac{1}{2\pi}\left|\int_{|z|=r} \frac{f(z)}{z^{n+1}}\,dz\right| \leq \frac{1}{2\pi r^n}\int_0^{2\pi} |f(re^{i\theta})|\,d\theta \leq \frac{1}{r^{n-1}(1-r)},$$

$$\leq \left(1 + \frac{1}{n-1}\right)^{n-1} n < en. \tag{40.27}$$

In the same paper, Littlewood also showed that if $M(r, f)$ denoted the maximum of $|f(z)|$ on the circle $|z| = r$, and f was univalent, then for $\lambda > 1/2$

$$\frac{1}{2\pi}\int_{-\pi}^{\pi} |f(re^{i\theta})|^\lambda\,d\theta \leq A_\lambda \rho^\lambda (1-\rho)^{-2\lambda+1}. \tag{40.28}$$

40.5 Littlewood and Paley on Odd Univalent Functions

Littlewood and Paley stated their main theorem of 1932: If

$$f(z) = z + a_3 z^3 + a_5 z^5 + \cdots$$

40.5 Littlewood and Paley on Odd Univalent Functions

is an odd univalent function, then there is an absolute constant A, such that $|a_n| \leq A$. To prove this result, they used Bieberbach's growth theorem for univalent functions, an inequality from Littlewood's 1925 paper, and a new inequality given by

$$\frac{1}{2\pi}\int_{-\pi}^{\pi} |\sigma'(\rho e^{i\theta})|^2 d\theta \leq C\rho^{-1}(1-\rho)^{-1} M^2(\rho^{1/2}, \sigma), \tag{40.29}$$

where C denoted a constant and σ was in S, the set of normalized univalent functions. In their proof of this, they assumed that $\sigma(z) = z + c_2 z^2 + c_3 z^3 + \cdots$ and applied Gronwall's formula to conclude that the area of the image of $|z| < \rho$ under σ was given by

$$\pi \sum n|c_n|^2 \rho^{2n} \leq \pi M^2(\rho, \sigma).$$

Using this they arrived at the required result:

$$2\pi \sum n^2 |c_n|^2 \rho^{2n} \leq 2\pi \text{Max}(n\rho^n) \sum n|c_n|^2 \rho^n$$
$$\leq \frac{A\rho}{1-\rho} A M^2(\rho^{1/2}, \sigma),$$

for some absolute constant A. We note that in Littlewood and Paley's paper, every absolute constant was denoted by the same symbol, A. We shall follow their convention. They constructed two other univalent functions related to $f(z)$, defined by the relations

$$\phi(z) = \left(f(\sqrt{z})\right)^2 = z + 2a_3 z^2 + \cdots,$$
$$\psi(z) = \left(f(z^3)\right)^{1/3} = z + \frac{1}{3} a_3 z^7 + \cdots.$$

They proved the univalence of $\phi(z)$ by noting that $\phi(z) = w$ implied $f(\sqrt{z}) = \pm\sqrt{w}$. They reasoned that since f was odd, only a pair of equal and opposite values were possible for $\sqrt{z}$, and hence only one value was possible for z. They also used a simple argument to demonstrate $\psi(z)$ to be univalent.

To prove their theorem, Littlewood and Paley applied Cauchy's theorem to the coefficients of $f'(z)$:

$$|na_n| \leq \frac{\rho^{-n+1}}{2\pi} \int_{-\pi}^{\pi} |f'(\rho e^{i\theta})| d\theta.$$

Thus, it was sufficient for them to show that

$$\int_{-\pi}^{\pi} |f'(\rho e^{i\theta})| d\theta < A/(1-\rho). \tag{40.30}$$

They noted that by combining (40.30) with the inequality for na_n, $\rho = 1 - 1/n$, they obtained the required inequality for a_n. To prove (40.30), they observed that since $f(z) = \psi^3(z^{1/3})$, it followed that for $z = \rho e^{i\theta}$,

$$\frac{1}{2\pi} \int_{-\pi}^{\pi} |f'(z)| d\theta = \frac{1}{6\pi} \int_{-3\pi}^{3\pi} |f'(z)| d\theta = \frac{\rho^{-2/3}}{6\pi} \int_{-3\pi}^{3\pi} |\psi^2(z^{1/3}) \psi'(z^{1/3})| d\theta.$$

On applying the Cauchy–Schwarz inequality to the last integral, they found

$$\frac{1}{2\pi}\int_{-\pi}^{\pi}|f'(z)|d\theta \le \rho^{-2/3}\left(\frac{1}{6\pi}\int_{-3\pi}^{3\pi}|\psi(z^{1/3})|^4 d\theta\right)^{1/2}\left(\frac{1}{6\pi}\int_{-3\pi}^{3\pi}|\psi'(z^{1/3})|^2 d\theta\right)^{1/2}. \tag{40.31}$$

Denoting the two integrals on the right by P and Q, respectively, and applying $|\psi(z)|^4 = |\phi(z^2)|^{2/3}$ combined with the change of variables $t = 2\theta$, they estimated

$$P = \frac{1}{12\pi}\int_{-6\pi}^{6\pi}|\phi(\rho^2 e^{it})|^{2/3} dt = \frac{1}{2\pi}\int_{-\pi}^{\pi} < A\rho^{4/3}(1-\rho^2)^{-1/3}. \tag{40.32}$$

The last inequality followed from Littlewood's inequality (40.28). To estimate Q, they first used (40.29) to arrive at

$$Q = \frac{1}{2\pi}\int_{-\pi}^{\pi}|\psi'(\rho^{1/3}e^{it})|^2 dt \le A\rho^{-1/3}(1-\rho^{1/3})^{-1}M^2(\rho^{1/3},\psi). \tag{40.33}$$

An application of the growth estimate to $M^2(\rho^{1/3},\psi)$ would not produce the necessary result, so they used $\psi^3(z^{1/3}) = \phi^{1/2}(z^2)$ to get

$$M^2(\rho^{1/3},\psi) = M^2(\rho^2,\phi^{1/6}) < \rho^{2/3}(1-\rho^2)^{-2/3}.$$

Combining this with (40.33), they obtained

$$Q < A\rho^{1/3}(1-\rho^{1/3})^{-1}(1-\rho^2)^{-2/3} < A(1-\rho)^{-5/3}.$$

Taking this inequality with (40.32) and (40.31) gave the required result:

$$\frac{1}{2\pi}\int_{-\pi}^{\pi}|f'(\rho e^{i\theta})|d\theta < A\rho^{-2/3}\rho^{2/3}(1-\rho^2)^{-1/6}(1-\rho)^{-5/6} < A(1-\rho)^{-1}.$$

This completed their ingenious proof that the coefficients of odd univalent functions were bounded.

40.6 Karl Löwner and the Parametric Method

Carathéodory's theorem was used a decade later by Löwner to construct his parametric theory of univalent functions. To describe Löwner's method, we must first define slit mappings. A single-slit mapping is a function mapping a domain conformally onto the complex plane minus a single Jordan arc. Löwner showed that such mappings were dense in S, the set of all conformal mappings of the unit disk with $f(0) = 0$ and $f'(0) = 1$. More exactly, for each $f \in S$, there exists a sequence of single-slit mappings $f_n \in S$ such that $f_n \to f$ uniformly on compact subsets of the unit disk.

We follow Duren's presentation to summarize the argument from Löwner's 1923 paper. It is sufficient to consider functions f mapping the unit disk onto a domain bounded by a (closed) analytic Jordan curve because, for any $f \in S$, the function

$f(rz)$, $0 < r < 1$, is also univalent with the required image and $f_r(z) = \frac{1}{r}f(rz) \in S$. By letting $f \to 1^-$, we get functions $f_r \in S$ such that $f_r \to f$ uniformly on compact subsets of the unit disk. So assume that $f \in S$ maps the unit disk onto a domain G bounded by an analytic Jordan curve C. Choose a point $w_0 \in C$ and let Γ be any Jordan curve from ∞ to w_0. Denote by Γ_n the Jordan curve consisting of Γ followed by a part of C joining w_0 to a point $w_n \in C$. Let G_n represent the complement of Γ_n in the complex plane, and let g_n map the unit disk onto G_n with $g_n(0) = 0$ and $g'_n(0) > 1$. We note that such a function g_n exists, by the Riemann mapping theorem. Now choose a sequence of points $w_n \in C$ such that $w_n \to w_0$ and $\Gamma_n \subset \Gamma_{n+1}$. Then G is the kernel of the sequence $\{G_n\}$. By Carathéodory's kernel convergence theorem, we must have $g_n \to f$ uniformly on compact subsets of the unit disk and, by Cauchy's theorem, $g'_n(0) \to f'(0) = 1$. Hence, $f_n = g_n/g'_n(0)$ is a sequence of single-slit mappings converging to f uniformly on compact subsets of the unit disk. Thus, we conclude with Löwner that the single-slit mappings are dense in S.

Now suppose $f \in S$ is a single-slit mapping taking the unit disk onto a domain G, the complement of a Jordan arc Γ extending from a point w_0 in the complex plane to ∞. Also suppose that $w = \psi(t)$, $0 \le t < T$, is a continuous, one-to-one parametrization of Γ with $\psi(0) = w_0$. Let Γ_t denote that part of Γ from $\psi(t)$ to ∞, and let G_t represent the complement of Γ_t. Let $g_t(z) = g(z,t)$ be the conformal mapping of the unit disc onto G_t, with $g(0,t) = 0$ and $g'(0,t) = \gamma(t) > 0$, so that $g(z,t)$ has the series expansion

$$g_t(z) = g(z,t) = \gamma(t)\left\{z + c_1(t)z^2 + c_2(t)z^3 + \cdots\right\}, \tag{40.34}$$

where $g(z,0) = f(z)$. By an application of the Schwarz lemma, $\gamma(t)$ may be seen to be a monotonically increasing function of t. Thus, by reparametrization, we can take $\gamma(t) = e^t$. Moreover, T will then be ∞. So we can write

$$g_t(z) = g(z,t) = e^t\left\{z + \sum_{n=2}^{\infty} b_n(t)z^n\right\}, \quad 0 \le t < \infty, \tag{40.35}$$

in what is called the standard parametrization. Löwner then considered the family of mappings

$$f_t(z) = g_t^{-1}(f(z)) = e^{-t}\left\{z + \sum_{n=2}^{\infty} a_n(t)z^n\right\}, \quad 0 \le t < \infty. \tag{40.36}$$

It is easy to see that the functions f_t map the unit disk onto the unit disk minus an arc extending inward from the boundary, and that $e^t f_t \in S$. By using the growth estimates of Bieberbach and Gronwall, he was able to conclude that

$$\lim_{t \to \infty} e^t f_t(z) = f(z). \tag{40.37}$$

And it is obvious that $f_0(z) = z$, the identity function. So the function $e^t f_t(z)$ starts at the identity function $f(z) = z$ and ends at $f(z) \in S$ as $t \to \infty$. Löwner determined the differential equation satisfied by this one parameter family of functions,

$$\frac{\partial f_t}{\partial t} = -f_t \frac{1+\chi(t)f_t}{1-\chi(t)f_t}, \tag{40.38}$$

where $\chi(t)$ was a continuous complex valued function with $|\chi(t)| = 1$, $0 \leq t < \infty$. He also gave the equation satisfied by the family of functions $g_t(z)$. By (40.36), $g_t(f_t(z)) = f(z)$. Setting $\zeta = f_t(z)$, we have $g_t(\zeta) = f(z)$; take the derivative with respect to t to get

$$\frac{\partial g_t}{\partial \zeta} \frac{\partial \zeta}{\partial t} + \frac{\partial g_t}{\partial t} = 0.$$

When this is substituted in (40.38), the result is the differential equation for $g_t(z)$:

$$\frac{\partial g_t}{\partial t} = \frac{\partial g_t}{\partial z} z \frac{1+\chi(t)z}{1-\chi(t)z}, \quad 0 \leq t < \infty, \tag{40.39}$$

where $g_0(z) = f(z)$ and $\lim_{t\to\infty} g_t(z) = z$.

Löwner applied his parametric method to the Bieberbach conjecture. In his paper he deduced only that $|a_2| \leq 2$ and $|a_3| \leq 3$. Bieberbach suggested that Löwner call his paper part I of a work in progress, since it was clear that he had a general method applicable to the coefficient problem. To understand Löwner's derivation of the inequalities for the second and third coefficients, note that since the class of univalent functions S is invariant under rotation, it is sufficient to prove that $\text{Re}(a_3) \leq 3$. Now substitute the series (40.36) for f_t into the differential equation (40.38) and equate the coefficients of z^2 and z^3 on both sides to get the two relations

$$a_2'(t) = -2e^{-t}\chi(t) \quad \text{and} \tag{40.40}$$

$$a_3'(t) = -2e^{-2t}[\chi(t)]^2 - 4e^{-t}\chi(t)a_2(t). \tag{40.41}$$

Since $a_2(0) = 0$, and $\lim_{t\to\infty} a_n(t) = a_n$, where a_n is the nth Taylor coefficient of the univalent function $f \in S$, we may integrate equation (40.40) to get

$$a_2 = \int_0^\infty a_2'(t)\,dt = -2\int_0^\infty e^{-t}\chi(t)\,dt; \tag{40.42}$$

and hence

$$|a_2| \leq 2\int_0^\infty e^{-t}\,dt = 2 \text{ because } |\chi(t)| = 1.$$

Substituting (40.40) into (40.41) then produces

$$a_3'(t) = 2a_2(t)a_2'(t) - 2e^{-2t}[\chi(t)]^2; \tag{40.43}$$

integrate to obtain

$$a_3 = 4\left(\int_0^\infty \chi(t)e^{-t}\,dt\right)^2 - 2\int_0^\infty \chi^2(t)e^{-2t}\,dt.$$

We next set $\chi(t) = e^{i\theta(t)}$ to get

$$\operatorname{Re}(a_3) = 4\left\{\left(\int_0^\infty \cos\theta(t)\,e^{-t}\,dt\right)^2 - \left(\int_0^\infty \sin\theta(t)\,e^{-t}\,dt\right)^2 \right.$$
$$\left. - \int_0^\infty \cos^2\theta(t)\,e^{-2t}\,dt\right\} + 1.$$

By the Cauchy–Schwarz inequality

$$\left(\int_0^\infty \cos\theta(t)\,e^{-t}\,dt\right)^2 \leq \int_0^\infty \cos^2\theta(t)\,e^{-t}\,dt \int_0^\infty e^{-t}\,dt \leq \int_0^\infty \cos^2\theta(t)\,e^{-t}\,dt,$$

so that

$$\operatorname{Re}(a_3) < 4\int_0^\infty \cos^2\theta(t)\left(e^{-t} - e^{-2t}\right)dt + 1$$
$$< 4\int_0^\infty \left(e^{-t} - e^{-2t}\right)dt + 1 = 3.$$

Löwner also wrote down expressions for $a_n'(t)$ and then, after integration, the expressions for $a_n(t)$. However, these are generally too complex to be conveniently utilized.

40.7 De Branges: Proof of Bieberbach's Conjecture

In proving the Bieberbach conjecture, by first proving the Milin conjecture, de Branges applied Löwner's theory of 1923. Though it may appear that Löwner's theory could have been applied at any time, it was not until Milin's contribution that there was a route connecting Löwner to Bieberbach; recall that Milin stated his conjecture for logarithmic coefficients only in 1971. De Branges's great insight was to use special functions to prove Milin's conjecture. And it took the boldness of an independent thinker such as de Branges to make such an attempt. As we have mentioned, de Branges was a functional analyst. He had developed the theory of square summable power series within that context and wished to apply it to various problems, including the Bieberbach conjecture. His extensive functional analytic machinery, so useful to his insights and manner of thought, proved to be a roadblock for others attempting to understand his proof. In the spring of 1984, de Branges presented his proof to the members of the Leningrad (St. Petersburg) geometric function theory seminar. The members of the seminar generously expended a good deal of effort to help him simplify it and express it in classical form. This version of the proof was soon written up by Milin and published as a preprint by the Steklov Institute in Leningrad. FitzGerald and Pommerenke used this preprint to obtain further technical simplifications, also independently found by de Branges. It is the simplified form of the proof that we shall discuss here.

Consider the logarithmic coefficients of the function $g_t(z)$ defined by equation (40.35). Thus,

$$\log\left(\frac{g(z,t)}{e^t z}\right) = \sum_{k=1}^{\infty} c_k(t) z^k, \quad |z| < 1, \tag{40.44}$$

with $0 \leq t < \infty$, and $c_k(0) = 2\gamma_k$, where $2\gamma_k$ are the logarithmic coefficients of the function $f \in S$. Here recall that $g(z,t) = g_t(z)$. If equation (40.44) is differentiated with respect to t and then z and the results are substituted in (40.39) and simplified, we get

$$1 + \sum_{n=1}^{\infty} c'_n(t) z^n = \frac{1 + \chi(t)z}{1 - \chi(t)z}\left(1 + \sum_{n=1}^{\infty} n c_n(t) z^n\right)$$
$$= \left(1 + 2\chi(t)z + 2\chi(t)^2 z^2 + \cdots\right)\left(1 + \sum_{n=1}^{\infty} n c_n(t) z^n\right). \tag{40.45}$$

Equate the coefficients of z^n on both sides to get the differential equations satisfied by the coefficients $c_n(t)$:

$$c'_n(t) = 2\chi(t)^n + n c_n(t) + 2\sum_{m=1}^{n-1} \chi(t)^{n-m} m c_m(t), \quad n = 1, 2, \ldots. \tag{40.46}$$

Note that this is a differential equation for logarithmic coefficients; recall that Löwner had obtained similar differential equations for the coefficients of $g_t(z)$, although they quickly became unwieldy when one tried to solve for them inductively. To prove Milin's conjecture, de Branges made effective use of these differential equations, by introducing some special functions. Recall the Milin conjecture:

$$\sum_{m=1}^{n}\left(m|c_m(0)|^2 - \frac{4}{m}\right)(n - m + 1) \leq 0, \quad n = 1, 2, 3, \ldots.$$

De Branges defined a function

$$\phi(t) = \sum_{m=1}^{n}\left(m|c_m(t)|^2 - \frac{4}{m}\right)\tau_{n,m}(t), \tag{40.47}$$

where certain properties were required of $\tau_{n,m}(t)$, including

$$\tau_{n,m}(0) = n - m + 1, \quad m = 1, 2, \ldots, n. \tag{40.48}$$

Now, if we can choose $\tau_{n,m}(t)$, such that $\phi'(t) \geq 0$ and $\phi(\infty) = 0$, then we would automatically have $\phi(0) \leq 0$, Milin's conjecture. To compute $\phi'(t)$, first set

$$b_0(t) = 0, \quad b_m(t) = \sum_{m=1}^{n} m c_m(t) \chi(t)^{-m}, \quad m = 1, 2, \ldots \tag{40.49}$$

40.7 De Branges: Proof of Bieberbach's Conjecture

and set

$$\tau_{n,n+1}(t) \equiv 0, \quad \text{for } 0 \leq t < \infty, \tag{40.50}$$

so that, by a straightforward calculation,

$$\phi'(t) = \sum_{m=1}^{n} \left(|b_m - b_{m-1}|^2 - 4 \right) \frac{\tau'_{n,m}}{m} + \sum_{m=1}^{n} \left(2|b_m|^2 + 4\text{Re}(b_m) \right) \left(\tau_{n,m} - \tau_{n,m+1} \right). \tag{40.51}$$

This expression for $\phi'(t)$ takes a very simple form if the functions $\tau_{n,m}(t)$ satisfy the difference differential equation

$$\tau_{n,m} - \tau_{n,m+1} = -\frac{\tau'_{n,m}}{m} - \frac{\tau'_{n,m+1}}{m+1}. \tag{40.52}$$

In that case,

$$\phi'(t) = -\sum_{m=1}^{n} |b_m + b_{m-1} + 2|^2 \frac{\tau'_{n,m}}{m}, \tag{40.53}$$

and Milin's conjecture is proved, provided that de Branges's system of functions $\tau_{n,m}$ satisfying (40.48), (40.50), and (40.52) also satisfy

$$\tau'_{n,m}(t) \leq 0, \quad 0 \leq t < \infty, \quad m = 1, 2, \ldots, n. \tag{40.54}$$

Thus, from these equations we must determine the form of the functions $\tau_{n,m}(t)$. We may solve successively for $\tau_{n,n}$, $\tau_{n,n-1}$, and so on. So by (40.48), (40.50), and (40.52), we may write

$$\tau_{n,n} = -\frac{\tau'_{n,n}}{n} \quad \text{or}$$

$$\tau_{n,n}(t) = Ae^{-nt} = e^{-nt},$$

because $\tau_{n,n}(0) = 1$. Next we solve

$$\frac{\tau'_{n,n-1}}{n-1} + \tau_{n,n-1} = 2e^{-nt}$$

to get

$$\tau_{n,n-1}(t) = -2(n-1)e^{-nt} + 2ne^{-(n-1)t}.$$

Note that in general we obtain

$$\tau_{n,m}(t) = m \sum_{k=0}^{n-m} \frac{(-1)^k (2m+k+1)_k (2m+2k+2)_{n-m-k}}{(m+k)k!(n-m-k)!} e^{-(m+k)t}, \tag{40.55}$$

when $m = 1, 2, \ldots, n$, since $\tau_{n,n+1} \equiv 0$.

Now recall that the truth of the nth Milin inequality implies the truth of the Bieberbach conjecture for the $(n+1)$th coefficient. Thus, to show that $|a_2| \leq 1$, it is enough to check that $\tau'_{1,1}(t) = -e^{-t}$ is negative, and this fact is obvious. For the third coefficient, we need to check the derivatives of two polynomials in e^{-t}:

$$\tau_{2,1}(t) = -2e^{-2t} + 4e^{-t} \quad \text{and} \quad \tau_{2,2}(t) = e^{-2t}.$$

Observe that their derivatives are

$$4e^{-t}(e^{-t} - 1) \leq 0 \quad \text{and} \quad -2e^{-2t} < 0 \quad (\text{for } 0 \leq t < \infty),$$

respectively. Hence, $|a_3| \leq 3$. In this manner, de Branges verified the Bieberbach conjecture up to the sixth coefficient, although the computations for the last two cases were complicated. At this point in early February 1984, the stage was set for de Branges to request his colleague at Purdue, Walter Gautschi, a numerical analyst with an interest in special functions, to check the calculations by computer. Gautschi was swamped with work at the time, but he was unable to resist the challenge; he attended de Branges's seminar and reported, "I was immediately struck by the clarity, freshness, and elegance of Louis' talk and began to appreciate how those inequalities came about. To my delight, they could be written in terms of orthogonal polynomials – currently a subject very much on my mind." Gautschi developed the necessary algorithms and managed to verify the Milin conjecture up to $n = 30$. Wondering if the inequalities could be proved analytically, he consulted Richard Askey. It turned out Askey and George Gasper had proved a slightly more general inequality less than a decade earlier.

As it turned out, $\tau'_{n,m}(t)$ could be expressed as a sum of Jacobi polynomials:

$$\tau'_{n,m}(t) = -me^{-mt} \sum_{k=0}^{n-m} P_k^{(2m,0)}(1 - 2e^{-t}). \tag{40.56}$$

In a 1976 paper, Askey and Gasper had proved that for any real $\alpha > -1$,

$$\sum_{k=0}^{n} P_k^{(\alpha,0)}(x) \geq 0, \quad -1 \leq x \leq 1. \tag{40.57}$$

This immediately implied de Branges's inequalities: $\tau'_{n,m}(t) \leq 0$ for $0 \leq t < \infty$. Askey and Gasper's investigation of these sums of Jacobi polynomials arose out of their study of several classical inequalities for trigonometric functions. Their insight was that the correct generalization for these classical inequalities was in the context of Jacobi polynomials, within which the powerful machinery of hypergeometric functions could be applied.

One proof of the Askey–Gasper inequality employed a theorem of Clausen on the square of a $_2F_1$ hypergeometric function and also a connection coefficient result of Gegenbauer. Note that we have stated these results in the exercises of chapters 27 and 28. This second result is also known as the Gegenbauer-Hua formula. Askey has pointed out that Gegenbauer's formula having been forgotten, Hua rediscovered it in the course of his work in harmonic analysis, carried out in the 1940s and 1950s. Note that Askey also rediscovered this formula in the 1960s. Hua Loo–Keng (1910–1985) taught himself mathematics, and by the age of 19, he was writing papers; these came to the notice of a professor at Qinghua University in Beijing. Hua was consequently appointed to a position at that university, and his career was launched. In 1936, he traveled to Cambridge to work with Hardy, Littlewood, and Davenport on problems in additive number theory. Later, he did research in several complex variables, automorphic functions, and group theory. His wide interests helped him lead the development of modern mathematics in China; in fact, in the 1960s, he turned his attention to mathematical problems with immediate practical applicability. Hua's student, Chen Jing–Run (1933–1996) made important contributions to the Goldbach conjecture.

40.8 Exercises

1. Show that if $0 \leq t \leq 1$ and $\alpha > -2$, then

$$_3F_2\left(\begin{matrix}-n, n+\alpha+2, (\alpha+1)/2 \\ (\alpha+3)/2, \alpha+1\end{matrix}; t\right) > 0.$$

   See the article by Askey and Gasper in Baernstein (1986).

2. Show that if

$$|a_1| \geq \sum_{i=2}^{\infty} |a_i|, \text{ then the function } \sum_{n=1}^{\infty} a_n z^n / n$$

   maps the interior of the unit circle upon a star-shaped region with center at the origin.

3. Show that, with the same condition on the coefficients as in exercise 2, the function $\sum_{n=1}^{\infty} a_n z^n / n^2$ maps the interior of the unit circle upon a convex region. See Alexander (1915) for this and for exercise 2.

4. Prove that if f is an analytic mapping of the unit disk into itself and if z, z_1, z_2 are in the unit disk, then

$$\frac{|f(z_1) - f(z_2)|}{|1 - \overline{f(z_1)}f(z_2)|} \leq \frac{|z_1 - z_2|}{|1 - \overline{z_1}z_2|};$$

$$\frac{|f'(z)|}{1 - |f(z)|^2} \leq \frac{1}{1 - |z|^2}.$$

   Since the Poincaré metric $ds = 2|dz|/(1-|z|^2)$ defines a noneuclidean length element (infinitesimal) in the unit disk, these inequalities may be interpreted to mean that the analytic mapping f decreases the noneuclidean distance between two points and the noneuclidean length of an arc. This invariant form of Schwarz's lemma is due to Georg Pick; see Pick (1915).

5. This exercise and the next mention results in geometric function theory, taking a direction different from that discussed in the text. If Γ is a finitely generated Kleinian group with region of discontinuity Ω, then Ω/Γ is of finite type. Read

the proof of this theorem in Ahlfors (1982), vol. 2, pp. 273–290. Note that this proof has a gap, later filled by Lipman Bers and Ahlfors himself. The theory of Kleinian groups was initiated by Poincaré. A nice history of this topic is given by Gray (1986). Poincaré (1985), translated into English by Stillwell, offers many important papers of Poincaré in this area and also provides a useful introduction by putting the papers into historical perspective and relating their results to some modern work. For Poincaré's pioneering work on topology, the reader may enjoy the article by Karanbir S. Sarkaria in James (1999), pp. 123–167.

6. If a Kleinian group is generated by N elements, then Area $(\Omega/\Gamma) \leq 4\pi(N-1)$ and Ω/Γ has at most $84(N-1)$ components. Bers proved this theorem in 1967; read a proof in Bers (1998), vol. 1, pp. 459–477.

40.9 Notes on the Literature

The quotations concerning Carathéodory are in Georgiadou (2004); see p. 63 for the Schwarz lemma, p. 75 for the letter to Hilbert, and p. 82 for the remarks on the history of the idea of normal families. For the quotation from Milin, see his article in Baernstein (1986). See Gronwall (1914) for his proof of the area theorem and Bieberbach (1916) for the conjecture and the proof of the Koebe one-quarter theorem. Loewner (1988) (also spelled Löwner), pp. 45–64, contains a reprint of his paper on differential equations and his proof of $|a_3| \leq 3$. The reader may also enjoy reading his 1917 paper on univalent functions mapping the unit disk to a convex region; see pp. 1–18. Littlewood (1982), vol. II, pp. 963–1004, especially pp. 980–981, gives his result that the Bieberbach conjecture holds true up to the order of magnitude; his paper with Paley can be found on pp. 1046–1048. De Branges (1985) contains the proof of the Bieberbach conjecture. Several papers deal with historical aspects of the conjecture and proof. See the relevant articles in Baernstein (1986), especially the riveting account by Gautschi, Askey's amusing and informative note, and de Branges's own report, recounting his experience in Russia. Also see Fomenko and Kuzmina (1986). Pommerenke (1985) and FitzGerald (1985) contain the reactions to the proof by two experts in univalent function theory. The books of Hayman (1994) and Gong (1999) present proofs of the conjecture. Duren (1983) is an extremely readable book with clear exposition and explanation of results on univalent functions; it was written just before the conjecture was proved. To read Hua's proof of the Hua–Gegenbauer formula, consult the elegant book by Hua (1981), pp. 38–39. For Askey's mention of Hua's and his own rediscovery of the Gegenbauer formula, and an account of his work with Gasper, see Schoenberg (1988), vol. 1, p. 192. For a history of the Riemann mapping theorem, see Gray (1994). The comments of Ahlfors on Grunsky's thesis may be found in Ahlfors (1982), vol. 1, pp. 493–501.

41

Finite Fields

41.1 Preliminary Remarks

Finite fields are of fundamental importance in pure as well as applied mathematics. Applications to coding, combinatorial design, and switching circuits have been made since the mid-1900s. Gauss himself first conceived of the theory of finite fields between 1796 and 1800, although he published very little of his work in this area, so that it did not exert the influence it might have. Gauss's work arose in the context of divisibility problems in number theory. The origins of these questions may, in turn, be traced to the work of Fermat who pursued the topic in the course of tackling a problem on perfect numbers posed to him by the amateur mathematician Frénicle de Bessy, through Mersenne. This question boiled down to showing that $2^{37} - 1$ was not prime. In a letter to Frénicle dated October 18, 1640, Fermat wrote that a prime p would divide $a^n - 1$ for some n dividing $p - 1$; this is now called Fermat's little theorem. Moreover, if $N = nm$, where n was the smallest such number, then p would also divide $a^N - 1$. The second part of the result is easy to understand by observing that

$$a^N - 1 = a^{mn} - 1 = (a^n - 1)\left(a^{(m-1)n} + a^{(m-2)n} + \cdots + a^n + 1\right).$$

Fermat intended to write a treatise on his number theoretic work, but never did so.

Because Fermat failed to publish his proofs, Euler had to rediscover them; this effort, like many of his projects, stretched over decades. He investigated the structure of the set of integers modulo a prime p, among many other questions. He conjectured but did not completely prove that there existed an integer a such that $a, a^2, \ldots, a^{p-1}$ modulo p produced the integers $1, 2, \ldots, p-1$, though not in that order. In modern terminology, letting $\mathbb{Z}_p$ denote the integers modulo p, and setting $\mathbb{Z}_p^\times = \mathbb{Z}_p - \{0\}$, then Euler's conjecture was that $\mathbb{Z}_p^\times$ would be cyclic. This was proved in full in the 1790s by Gauss and published in his *Disquisitiones Arithmeticae*. Euler worked with $\mathbb{Z}[x]$, the ring of polynomials with coefficients in $\mathbb{Z}$, as did Lagrange. In 1768, Lagrange proved the basic theorem that any such polynomial of degree m would have at most m roots modulo p.

Gauss was able to delve deeply into the theory of the ring of polynomials over finite fields when he perceived that the number theoretic properties of this ring were analogous to those of the ring of integers; the irreducible polynomials here played the role of the prime numbers. Gauss proved the fundamental theorem that every irreducible polynomial $P(x) \neq x$ of degree m in $\mathbb{Z}_p[z]$ must divide $X^{p^m-1} - 1$ in $\mathbb{Z}_p[x]$. Gauss also gave a formula for the number of irreducible polynomials of degree m. In his derivation, he applied Möbius inversion without an explicit statement of the general formula. In 1832, August Möbius (1790–1868), a student of Pfaff and Gauss, published this formula, although it was not much noted. In fact, in 1857, Dedekind and Liouville published proofs of the inversion formula without reference to Möbius.

Also during the period 1796–1800, Gauss studied the Galois theory of cyclotomic extensions of the field $\mathbb{Z}_p$ by explicitly constructing the subfields of the splitting field of the polynomial $x^\nu - 1$ over $\mathbb{Z}_p$, where ν was a positive integer not divisible by p. He saw this theory as analogous to his cyclotomic theory over the field of rational numbers. Gauss applied his results to obtain a new proof of the law of quadratic reciprocity. He intended to include his work on the extensions of finite fields as the eighth section of his 1801 *Disquisitiones*, but omitted it to make room for his theory of binary quadratic forms, completed after 1798. In fact, binary quadratic forms occupied more than half of the published book, so that Gauss could include in the text only references to his unpublished work on finite fields.

In 1830, Galois published his theory of algebraic extensions of finite fields by using numbers analogous to complex numbers, known in the nineteenth century as Galois imaginaries. These numbers were required in order to extend the field $\mathbb{Z}_p$. For example, if a polynomial $F(x)$ of degree ν was irreducible over $\mathbb{Z}_p$, then Galois assumed i to be the imaginary solution of $F(x) = 0$; he then showed that the set consisting of the p^ν expressions $a_0 + a_1 i + \cdots + a_{\nu-1} i^{\nu-1}$ could be given the structure of a field and that $F(x)$ could be completely factored in this field. It is interesting to note that Gauss preferred to avoid imaginary roots. For example, in the unpublished eighth section of the *Disquisitiones*, he wrote:

> It is clear that the congruence $\xi \equiv 0$ does not have real roots if ξ has no factors of dimension one; but nothing prevents us from decomposing ξ, nevertheless, into factors of two, three or more dimensions, whereupon, in some sense, *imaginary* roots could be attributed to them. Indeed, we could have shortened incomparably all our following investigations, had we wanted to introduce such imaginary quantities by taking the same liberty some more recent mathematicians have taken; but nevertheless, we have preferred to deduce everything from [first] principles. Perhaps, we shall explain our view on this matter in more detail on another occasion.

In 1845, T. Schönemann, unaware of Galois's work, published a paper on algebraic extensions of $\mathbb{Z}_p$. By application of his theory, he partially recovered Kummer's result on the factorization of a prime $q \neq p$ in the cyclotomic field generated by a pth root of unity. Schönemann also applied his theory to prove the irreducibility of the cyclotomic polynomial $x^{q-1} + x^{q-2} + \cdots + x + 1$. In 1857, Dedekind began to develop a theory of finite fields, in order to generalize Kummer's theory of ideals in cyclotomic fields and to place it on a firm logical foundation. In 1871, Dedekind published his first version of this generalization as his theory of algebraic numbers. In this work, given a polynomial irreducible over the rational numbers, he delineated the relation between the factorization of this polynomial modulo p and the prime ideal factorization of the

41.1 Preliminary Remarks

ideal generated by p in the number field arising out of the polynomial. We remark that Dedekind was familiar with the work of Schönemann and of Galois. Note that Galois's 1830 paper was republished by Liouville in the 1840s and that J. A. Serret's 1854 algebra book discussed the work of Galois in detail.

Richard Dedekind (1831–1916) was the Ph.D. student of Gauss and mathematical friend of Dirichlet. Dedekind performed the valuable service of editing the works of Riemann and Gauss and the lectures of Dirichlet. In his 1857 paper, Dedekind carefully showed that results in elementary number theory, including Fermat's theorem of Euler's generalization that $a^{\phi(m)} \equiv 1 \bmod m$, with a and m relatively prime, could be carried over to the ring of polynomials over finite fields. Dedekind also stated and proved the corresponding law of quadratic reciprocity. In 1902, Hermann Kühne proved the general reciprocity theorem in $F_q[x]$, a finite field with $q = p^m$ elements. This theorem was rediscovered in 1925 by Friedrich K. Schmidt, and then again in 1932 by Leonard Carlitz. The reciprocity laws are more easily proved for polynomials in $F_q[x]$ than for integers. In 1914, Heinrich Kornblum (1890–1914), student of Landau, further developed this analogy by defining L-functions for $F_p[x]$ and proving an analog of Dirichlet's theorem on primes in arithmetic progressions. Unfortunately, Kornblum was killed in World War I, but Landau published the work in 1919. It may also be of interest to note that Gauss used the zeta function for $\mathbb{Z}_p[x]$ to find a formula for the number of irreducible polynomials of degree n, although he did not express it in such terms. Gauss's formula implies that if (n) denotes the number of irreducible polynomials of degree n, then

$$(n) - \frac{p^n}{n} = O\left(\frac{p^{n/2}}{n}\right).$$

If we set $x = p^n$, then $n = \log_p x$, and we may express the last relation as

$$(n) = \frac{x}{\log_p x} + O\left(\frac{\sqrt{x}}{\log_p x}\right).$$

Note the similarity in appearance between this equation and the conjectured form of the number of primes less than x, following from the unproven Riemann hypothesis on the nontrivial zeros of the Riemann zeta function. Note that in 1973, Pierre Deligne established the Riemann hypothesis for the zeta function of smooth projective varieties over finite fields. He based his proof on the novel framework for algebraic geometry created by Alexander Grothendieck and his collaborators, including Deligne. Weil conjectured this theorem in 1949, so that it was also known as Weil's conjecture. Even earlier, special cases of the Riemann hypothesis over finite fields had been proved; Emil Artin (1898–1962) presented the earliest example in 1921, when he defined and studied the zeta function of a quadratic extension of the field $\mathbb{Z}_p(x)$ and proved the Riemann hypothesis for that case. Artin's advisor, Gustav Herglotz (1881–1953), proposed the problem after reading Kornblum's posthumously published thesis.

41.2 Euler's Proof of Fermat's Little Theorem

In 1736, in his second paper on number theory, Euler presented an inductive proof of Fermat's theorem. He stated the result that if $a^p - a$ is divisible by p, then $(a+1)^p - (a+1)$ is also divisible by p. To see this, consider that by the binomial theorem,

$$(a+1)^p = a^p + \frac{p}{1}a^{p-1} + \frac{p(p-1)}{1 \cdot 2}a^{p-2} + \cdots + \frac{p}{1}a + 1 \text{ or}$$

$$(a+1)^p - a^p - 1 = \frac{p}{1}a^{p-1} + \frac{p(p-1)}{1 \cdot 2}a^{p-2} + \cdots + \frac{p}{1}a.$$

The right-hand side has p as a factor in each term, and therefore p divides the left-hand side

$$(a+1)^p - a^p - 1 = (a+1)^p - (a+1) - a^p + a.$$

This implies the required result, that if p divides $a^p - a$, it must also divide $(a+1)^p - (a+1)$. Since the result is true for $a = 1$, it is true for all positive integers a. Moreover, if p does not divide a, then since p divides $a(a^{p-1} - 1)$, we obtain the result that p divides $a^{p-1} - 1$.

For Euler's multiplicative proof of Fermat's theorem, we follow the concise presentation in Gauss's *Disquisitiones*. Suppose a prime p does not divide a positive integer a. Then there are at most $p-1$ different remainders when $1, a, a^2, \ldots$ are divided by p. So let a^m and a^n have the same remainder with $m > n$. Then $a^{m-n} - 1$ is divisible by p. Let t be the least integer such that p divides $a^t - 1$. If $t = p-1$, our proof is complete. If $t \neq p-1$, then $1, a, a^2, \ldots, a^{t-1}$ have t distinct remainders when divided by p. Thus, we can choose an integer b, not divisible by p and not among $1, a, a^2, \ldots, a^{t-1}$ modulo p, and consider the numbers $b, ab, a^2b, \ldots, a^{t-1}b$. Each of these numbers also leaves a different remainder after division by p and each of these is different from the previous set of remainders. Hence, we have $2t \leq p-1$ remainders. If $2t = p-1$, then our proof is complete. If not, then we can continue the process until some multiple of t is $p-1$. This completes the proof of the theorem.

41.3 Gauss's Proof that $\mathbb{Z}_p^\times$ Is Cyclic

In one of his two proofs that $\mathbb{Z}_p^\times$ is cyclic, Gauss used a theorem of Lagrange: Assuming $A \not\equiv 0$ modulo p, the congruence

$$Ax^m + Bx^{m-1} + Cx^{m-2} + \cdots + Mx + N \equiv 0 \pmod{p}$$

has at most m noncongruent solutions. Gauss presented a proof of this result, similar to that of Lagrange but more succinct. It is easy to see that a congruence of degree 1 has at most one solution. Assume, with Gauss, that m is the lowest degree for which the result is false; then $m \geq 2$. Suppose that the preceding congruence has at least $m+1$ roots, $\alpha, \beta, \gamma, \ldots$, where $0 \leq \alpha < \beta < \gamma < \cdots \leq p-1$. Now set $y = x + \alpha$ so that the congruence takes the form

$$A't^m + B'y^{m-1} + C'y^{m-2} + \cdots + M'y + N' \equiv 0 \pmod{p}.$$

Note that this congruence has at least $m+1$ solutions $0, \beta - \alpha, \gamma - \alpha, \ldots$. Since $y \equiv 0$ is a solution, we must have $N' \equiv 0$. Thus,

$$y(A'y^{m-1} + B'y^{m-2} + C'y^{m-3} + \cdots + M') \equiv 0 \pmod{p}.$$

If y is replaced by any of the m values $\beta - \alpha, \gamma - \alpha, \ldots$, then the identity is satisfied but y is not zero. This means that the m values $\beta - \alpha, \gamma - \alpha, \ldots$ are solutions of the $m-1$ degree congruence

$$A'y^{m-1} + B'y^{m-2} + C'y^{m-3} + \cdots + M' \equiv 0, \quad (A' \equiv A \neq 0).$$

This contradicts the statement that m is the least integer for which the result is false, proving the theorem. Gauss thought that this theorem was significant; in his *Disquisitiones* he discussed its history, pointing out that Euler had found special cases, Legendre had given it in his dissertation, and Lagrange had been the first to state and prove it.

Gauss used this result to prove the proposition that there always exist primitive $(p-1)$th roots of unity modulo p. Suppose that $p - 1 = a^\alpha b^\beta c^\gamma \cdots$, where $a, b, c, \ldots$ are distinct primes. The first step is to show the existence of integers $A, B, C, \ldots$ of orders $a^\alpha, b^\beta, c^\gamma, \ldots$ respectively. Note that by Lagrange's theorem given above, the congruence

$$x^{(p-1)/a} \equiv 1 \pmod{p}$$

has at most $(p-1)/a$ solutions. Hence there is an integer g, $1 \leq g \leq p-1$, not a solution of the congruence. Now let h be an integer, $1 \leq h \leq p-1$, congruent to $g^{(p-1)/a^\alpha}$. It is clear that $h^{a^\alpha} \equiv 1 \pmod{p}$, but that no power d less that a^α will give $h^d \equiv 1$. This is because d must take the form a^j with $j < k$, so that $h^d = g^{(p-1)/a^{\alpha-j}} \not\equiv 1$ by the definition of g. We may take A to be h and similarly find $B, C, \ldots$. We can now show that the order of $y = ABC \cdots$ is $p - 1$. To see this, suppose, without loss of generality, that the order of y divides $(p-1)/a$. Since $b^\beta, c^\gamma, \ldots$ also divide $(p-1)/a$, it follows that

$$1 \equiv y^{(p-1)/a} \equiv A^{(p-1)/a} B^{(p-1)/a} C^{(p-1)/a} \cdots \equiv A^{(p-1)/a} \pmod{p}.$$

This implies that a^α divides $(p-1)/a$, an impossibility. Thus, the theorem is proved. This argument of Gauss can be extended to arbitrary finite fields. By a different argument, Gauss showed that the number of primitive roots of unity modulo p was equal to $\phi(p-1)$.

41.4 Gauss on Irreducible Polynomials Modulo a Prime

To count the number of irreducible polynomials of a given degree modulo a prime p, Gauss started with the observation that the number of monic polynomials (mod p)

$$x^n + Ax^{n-1} + Bx^{n-2} + Cx^{n-3} + \cdots$$

was p^n, because each of the n coefficients $A, B, C, \ldots$ would take exactly p values

$$0, 1, 2, \ldots, p-1.$$

Thus, there were p polynomials of degree 1, all irreducible. Gauss then remarked that it followed from the theory of combinations that the number of (monic) reducible degree-two polynomials was $p(p+1)/2$. So the number of irreducible ones would be given by

$$p^2 - p(p+1)/2 = (p^2 - p)/2.$$

To determine the irreducible polynomials of higher degree, Gauss devised a notation and method. He let (a) denote the number of irreducible polynomials of degree a and (a^α) the number of polynomials of degree αa factorizable into α irreducible polynomials of degree a. Gauss represented the number of polynomials of degree $\alpha + 2\beta + 3\gamma + \cdots$ with α factors of degree 1, β factors of degree 2, γ factors of degree 3, etc. by $(1^\alpha 2^\beta 3^\gamma 4^\delta \cdots)$. It followed that

$$(1^\alpha 2^\beta 3^\gamma 4^\delta \cdots) = (1^\alpha)(2^\beta)(3^\gamma)(4^\delta) \cdots.$$

Again, Gauss remarked that the theory of combinations implied that

$$(a^\alpha) = \frac{(a)}{1} \cdot \frac{[(a)+1]}{2} \cdot \frac{[(a)+2]}{3} \cdots \frac{[(a)+\alpha-1]}{\alpha}.$$

Though Gauss did not bother, it is not difficult to prove this: Let $p_1(x), p_2(x), \ldots, p_{(a)}(x)$ be the irreducible polynomials of degree a. In a factorization of a polynomial of degree αa, let y_i factors be $p_i(x)$, $i = 1, 2, \ldots, (a)$. Then (a^α) will be the number of nonnegative solutions of the equation

$$y_1 + y_2 + \cdots + y_{(a)} = \alpha.$$

Solving this equation yields the same value given by Gauss. He then noted that

$$p = (1),$$
$$p^2 = (1^2) + (2),$$
$$p^3 = (1^3) + (1 \cdot 2) + (3),$$
$$p^4 = (1^4) + (1^2 \cdot 2) + (1 \cdot 3) + (2^2) + (4),$$

and so on. Using the formula for (a^α), he found the following eight values:

$(1) = p$,
$(2) = (p^2 - p)/2$,
$(3) = (p^3 - p)/3$,
$(4) = (p^4 - p^2)/4$,
$(5) = (p^5 - p)/5$,
$(6) = (p^6 - p^3 - p^2 + p)/6$,
$(7) = (p^7 - p)/7$,
$(8) = (p^8 - p^4)/8$.

41.4 Gauss on Irreducible Polynomials Modulo a Prime

Solving these equations led him to:

$$p = (1), \qquad p^5 = 5(5) + (1),$$
$$p^2 = 2(2) + (1), \qquad p^6 = 6(6) + 3(3) + 2(2) + (1),$$
$$p^3 = 3(3) + (1), \qquad p^7 = 7(7) + (1),$$
$$p^4 = 4(4) + 2(2) + (1), \qquad p^8 = 8(8) + 4(4) + 2(2) + (1).$$

These results then suggested to Gauss that

$$p^n = \alpha(\alpha) + \beta(\beta) + \gamma(\gamma) + \delta(\delta) + \cdots,$$

where $\alpha, \beta, \gamma, \delta, \ldots$ were all the divisors of n. He sketched a proof, using generating functions. Although there are a few missing lines in Gauss's manuscript, it is not difficult to fill in the details. He wrote that the product

$$\left(\frac{1}{1-x}\right)^{(1)} \left(\frac{1}{1-x^2}\right)^{(2)} \left(\frac{1}{1-x^3}\right)^{(3)} \cdots \qquad (41.1)$$

could be developed into the series

$$1 + Ax + Bx^2 + \cdots = P, \qquad (41.2)$$

where $A = p$, $B = p^2$, $C = p^3$, ... Then by taking the logarithmic derivative of (41.1), he got

$$\frac{x\,dP}{P\,dx} = \frac{(1)x}{1-x} + \frac{2(2)x^2}{1-x^2} + \frac{3(3)x^3}{1-x^3} + \cdots. \qquad (41.3)$$

The required result followed after expanding the terms as an infinite series and equating coefficients. Gauss gave no details, but to prove (41.2), let f denote a monic irreducible polynomial. Since p^n is the number of monic polynomials of degree n, we see that by unique factorization of polynomials

$$\frac{1}{1-px} = 1 + px + p^2 x^2 + \cdots + p^n x^n + \cdots \qquad (41.4)$$

$$= \sum_{n=0}^{\infty} (\text{number of monic polynomials of degree } n)\, x^n$$

$$= \prod_f \left(1 + x^{\deg f} + x^{\deg f^2} + \cdots\right)$$

$$= \prod_f \left(1 - x^{\deg f}\right)^{-1}$$

$$= \prod_{d=1}^{\infty} \left(1 - x^d\right)^{-(d)},$$

where the notation $\prod_f$ stands for the product over all irreducible polynomials. This proves Gauss's assertion that product (41.1) equals the series (41.2). Now by (41.3)

$$\frac{px}{1-px} = \sum_{d=1}^{n} \frac{d(d)x^d}{1-x^d} = \sum_{n=1}^{\infty} \left(\sum_{d|n} d(d)\right) x^n.$$

Gauss then equated the coefficients of x^n on each side to get

$$p^n = \sum_{d|n} d(d). \tag{41.5}$$

He then inverted this formula to get (n) in terms of p^n, stating that if $n = a^\alpha b^\beta c^\gamma \cdots$ where $a, b, c, \ldots$ were distinct primes, then

$$n(n) = p^n - \sum p^{n/a} + \sum p^{n/ab} - \sum p^{n/abc} + \cdots. \tag{41.6}$$

Gauss wrote that, for example, when $n = 36$, he had

$$36(36) = p^{36} - p^{18} + p^{12} + p^6.$$

As a corollary of (41.6), Gauss observed that if $n = a^\alpha$, with a prime, then

$$p^n \equiv p^{n/a} \pmod{n}.$$

And for $\alpha = 1$ and a prime to p, the result was

$$p^{a-1} \equiv 1 \pmod{a}.$$

Note that it is also easy to see from (41.6) that $n(n) > 0$. Thus, there are irreducible polynomials of every degree n. Gauss gave no proof of the inversion (41.6) of (41.5). This means that Gauss knew the Möbius inversion formula before 1800 when he wrote up his researches on polynomials over the integers modulo p; Möbius's paper appeared in 1832.

41.5 Galois on Finite Fields

Although the French mathematician Évariste Galois's (1811–1832) research career lasted less than four years, his accomplishments have had lasting value and importance. Galois's premature death came about as a result of a tragic duel, the cause of which is not fully understood, but was possibly related to Galois's political activities. In his number theory report of 1859–66, Smith wrote on Galois:

> His mathematical works are collected in Liouville's Journal, vol. xl. p. 381. Obscure and fragmentary as some of these papers are, they nevertheless evince an extraordinary genius, unparalleled, perhaps, for its early maturity, except by that of Pascal. It is impossible to read without emotion the letter in which, on the day before his death and in anticipation of it, Galois endeavours to rescue from oblivion the unfinished researches which have given him a place for ever in the history of mathematical science.

41.5 Galois on Finite Fields

Galois published his first paper in 1828 on purely periodic continued fractions, a topic studied by Euler and Lagrange in the 1760s. Euler had shown that a periodic continued fraction would satisfy a quadratic equation with integer coefficients. Lagrange proved the more difficult converse, that a quadratic irrational number, a number satisfying a quadratic equation with integer coefficients, could be expressed as a periodic continued fraction. Galois explicitly proved a theorem implicitly in Lagrange: For integers $a_0, a_1, \ldots, a_n$, if the continued fraction

$$a_0 + \cfrac{1}{a_1+} \cfrac{1}{a_2+} \cdots \cfrac{1}{a_n+} \cfrac{1}{a_0+} \cdots \cfrac{1}{a_n+} \cfrac{1}{a_0+} \cdots$$

is a solution of a polynomial equation with integer coefficients, then the continued fraction

$$-\cfrac{1}{a_n+} \cfrac{1}{a_{n-1}+} \cdots \cfrac{1}{a_0+} \cfrac{1}{a_n+} \cdots \cfrac{1}{a_0+} \cfrac{1}{a_n+} \cdots$$

is also a solution of the same polynomial equation.

From a very early age, Galois had plans to develop the theory of algebraic extensions of fields. In this context, in 1830 Galois wrote his paper "Sur la théorie des nombres," rediscovering Gauss's unpublished results in this area and creating the theory of finite fields. Galois started with an equation $F(x) = 0$ modulo p, with $F(x)$ having integer coefficients and irreducible modulo p. Note that by this he meant that there could not exist polynomials $\phi(x)$, $\psi(x)$, and $\chi(x)$ with integer coefficients such that

$$\phi(x) \cdot \psi(x) = F(x) + p\chi(x).$$

After the initial portion of the paper, Galois omitted the modulo p, writing simply $F(x) = 0$. This means that he was assuming that the coefficients of the polynomials were taken from the finite field, integers modulo p. Galois argued that since $F(x)$ was irreducible, the equation $F(x) = 0$ had no solutions in integers (more precisely and in modern terms, no solutions in the finite field). He supposed F to be of degree ν and denoted by i an imaginary solution of $F(x) = 0$. Galois explained this imaginary solution by drawing an analogy with complex numbers. He let α denote any one of the $p^\nu - 1$ expressions

$$a + a_1 i + a_2 i^2 + \cdots + a_{\nu-1} i^{\nu-1}, \tag{41.7}$$

where $a, a_1, a_2, \ldots, a_{\nu-1}$ took values in the finite field, that is, $a_i = 0, 1, \ldots, p-1$, though all the as could not be zero. Then $\alpha, \alpha^2, \alpha^3, \ldots$ would all be expressions of the form (41.7), since if the degree of i were ν or higher, then $F(x) = 0$ could be used to express i^ν in the form (41.7). Next, since there were only $p^\nu - 1$ different such expressions, Galois had $\alpha^k = \alpha^l$ for two different integers k and l, or $\alpha^l(\alpha^{k-l} - 1) = 0$ (for $k > l$). From the irreducibility of F, Galois arrived at $\alpha^{k-l} = 1$. Letting n be the least positive integer such that $\alpha^n = 1$, Galois noted that $1, \alpha, \alpha^2, \ldots, \alpha^{n-1}$ were distinct; moreover, if β were another expression of the form (41.7), then $\beta, \beta\alpha, \ldots, \beta\alpha^{n-1}$ would be n additional elements distinct from each other and from the α^j. Moreover, if $2n < p^\nu - 1$, then there would be yet another element in (41.7) distinct from the known $2n$ elements. Because this process could be continued, Galois concluded that n divided $p^\nu - 1$, meaning that $\alpha^{p^\nu-1} = 1$ for every α of the form (41.7). This is Galois's generalization of Fermat's theorem. Here Galois also observed that by the known methods of number theory (in fact, by Gauss's published argument outlined earlier), there was a primitive root α for which $n = p^\nu - 1$. Moreover, any primitive root had to satisfy a congruence of degree ν irreducible modulo p.

Note that Galois's generalization of Fermat's theorem implied that all members of (41.7), including 0, were roots of the polynomial $x^{p^\nu} - x$. And every irreducible $F(x)$ would divide $x^{p^\nu} - x$ modulo p. We now continue to follow Galois. Because

$$(F(x))^{p^n} = F\left(x^{p^n}\right) \pmod{p},$$

the roots of $F(x) = 0$ had to be $i, i^p, i^{p^2}, \ldots, i^{p^{\nu-1}}$. Thus, he saw that all the roots of $x^{p^\nu} = x$ were polynomials in any root α of an irreducible polynomial of degree ν. To find all the irreducible factors of $x^{p^\nu} - x$, he factored out all polynomials dividing $x^{p^\mu} - x$ for $\mu < \nu$. The remaining product of polynomials was then a product of irreducible polynomials of degree ν. Galois pointed out that, since each of their roots was expressible in terms of a single root, these were obtainable through Gauss's method. Recall that Galois did not write modulo p repeatedly because he saw the coefficients of F to be elements of the finite field $\mathbb{Z}_p$.

Galois then gave an example in which $p = 7$ and $\nu = 3$. He here showed how to find the generator α of the multiplicative group of this field, as well as the irreducible polynomial equation satisfied by α. He noted that $x^3 - 2$ was an irreducible polynomial of degree 3 modulo 7 and hence the roots of $x^{7^3} - x$ would be $a + a_1 i + a_2 i^2$ where a, a_1, a_2 took values $0, 1, \ldots, 6$ and $i^3 = 2$. Galois denoted i by $\sqrt[3]{2}$ and then wrote the roots as

$$a + a_1 \sqrt[3]{2} + a_2 \sqrt[3]{4}.$$

To find a primitive root of $x^{7^3} - x$, Galois noted that $7^3 - 1 = 2 \cdot 3^2 \cdot 19$, so that he needed primitive roots of the three equations:

$$x^2 = 1, \quad x^{3^2} = 1, \quad x^{19} = 1.$$

He observed that $x = -1$ was a primitive root of the first equation. He cleverly noted that

$$x^{3^2} - 1 = (x^3 - 1)(x^3 - 2)(x^3 - 4) \pmod{7},$$

so that where $i^3 = 2$, i was a primitive root of the second equation. For the third equation, Galois took $x = a + a_1 i$ so that

$$(a + a_1 i)^{19} = 1.$$

Expanding by the binomial theorem, referred to by Galois as Newton's formula, and employing

$$a^{m(p-1)} = 1, \quad a_1^{m(p-1)} = 1, \quad i^3 = 2, \quad p = 7,$$

he reduced the expression modulo 7 to

$$3\left[a - a^4 a_1^3 + (a^5 a_1^2 + a^2 a_1^5)i^2\right] = 1,$$
$$\text{so that} \quad 3a - 3a^4 a_1^3 = 1, \quad a^5 a_1^2 + a^2 a_1^5 = 0.$$

Galois saw that these two equations were satisfied (modulo 7) by $a = -1$ and $a_1 = 1$. He therefore concluded that $-1 + i$ was a primitive root of $x^{19} = 1$. Thus, the product $i - i^2$ of the three primitive roots, $-1, i$, and $-1 + i$, was a primitive root of $x^{7^3-1} = 1$. By eliminating i from

$$i^3 = 2 \quad \text{and} \quad \alpha = i - i^2,$$

he obtained the irreducible equation for the primitive root α,

$$\alpha^3 - \alpha + 2 = 0.$$

Thus, α would generate all the nonzero elements of a finite field of 7^3 members. Galois ended his paper with the observation that an arbitrary polynomial $F(x)$ of degree n has n real or imaginary roots. The real roots, assuming no multiple roots, could be found from the greatest common divisor of $F(x)$ and $x^{p-1} - 1$. Note that this can be obtained by means of the Euclidean algorithm. And the imaginary roots of degree 2 could be obtained from the greatest common divisor of $F(x)$ and $x^{p^2-1} - 1$; this process could clearly be continued.

41.6 Dedekind's Formula

With his characteristic systematic approach, in his paper of 1857, Dedekind explained how to develop the theory of polynomials over finite fields such that the analogy with the ring of integers was completely clear. We present just one formula from his paper, deriving an elegant expression for the product of all irreducible polynomials of degree d. The arguments given by Galois showed that if $F_d(x)$ was the product of all irreducible polynomials of degree d, then

$$x^{p^n} - x = \prod_{d \mid n} F_d(x).$$

By the multiplicative form of the Möbius inversion formula, a proof of which Dedekind provided, since it was not generally known in 1857, he obtained

$$F_n(x) = \frac{(x^{p^n} - x) \prod \left(x^{p^{n/ab}} - x\right) \cdots}{\prod \left(x^{p^{n/a}} - x\right) \prod \left(x^{p^{n/abc}} - x\right) \cdots}.$$

Here $a, b, c, \ldots$ denoted the distinct prime divisors of n. Thus,

$$\prod \left(x^{p^{n/a}} - x\right) = \left(x^{p^{n/a}} - x\right)\left(x^{p^{n/b}} - x\right)\left(x^{p^{n/c}} - x\right)\cdots.$$

With the use of the Möbius function μ, Dedekind's formula can now be written as

$$F_n(x) = \prod_{d|n}\left(x^{p^{n/d}} - x\right)^{\mu(d)}.$$

Note that Möbius stated his inversion formula for a sum; Dedekind extended it to cover a product. We observe that the symbol μ for the Möbius function was introduced by Mertens in 1874.

41.7 Exercises

1. Let P be a monic irreducible polynomial in $R = \mathbb{Z}_p[x]$, and let $|P|$ denote the number of elements in R/P. Set

    $$\zeta_R(s) = \prod (1 - |P|^{-s})^{-1},$$

    where the product is taken over all monic irreducible polynomials in R. Determine $\zeta_R(s)$ and compare your result with equation (41.4).

2. The last entry in Gauss's diary, dated July 9, 1814, reads (in translation):

    > I have made by induction the most important observation that connects the theory of biquadratic residues most elegantly with the lemniscatic functions. Suppose $a + bi$ is a prime number, $a - 1 + bi$ divisible by $2 + 2i$, the number of all solutions to the congruence
    >
    > $$1 \equiv xx + yy + xxyy \pmod{a + bi},$$
    >
    > including $x = \infty$, $y = \pm i$, $x = \pm 1$, $y = \infty$ is $= (a-1)^2 + bb$.

    Prove Gauss's theorem. Note that the diary was discovered in 1897 and published in 1903. See Ireland and Rosen (1982), pp. 166–168, where a proof using Gauss and Jacobi sums is given. In 1921, Herglotz gave the first proof of Gauss's last entry by using complex multiplication of elliptic functions. Chapter 10 of Lemmermeyer (2000) gives an excellent discussion of this topic, including useful historical notes. See also Weil (1979), vol. 3, p. 298, for some perceptive historical remarks, pointing out the connection between Gauss's diary entry and the lemniscatic function.

3. Let p be an odd prime, and let Q, R be irreducible polynomials of degrees π and ρ in $\mathbb{Z}_p[x]$. With f any polynomial in this ring, let (f/Q) denote the unique element of $\mathbb{Z}_p^\times$ such that

    $$f^{(|Q|-1)/2} \equiv (f/Q) \pmod{Q}.$$

    Show that

    $$\left(\frac{R}{Q}\right)\left(\frac{Q}{R}\right) = \left(\frac{-1}{p}\right)^{\pi\rho}.$$

    See Dedekind (1930), vol. I, pp. 56–59, for a proof of this analog of the law of quadratic reciprocity.

4. Generalize the Euler totient function ϕ to the ring $\mathbb{Z}_p[x]$; state and prove a formula analogous to $\phi(m) = m(1 - 1/p_1) \cdots (1 - 1/p_k)$, where $p_1, \ldots, p_k$ comprise all the distinct prime factors of the positive integer m. See Dedekind (1930), vol. I, pp. 50–51.
5. State a generalization of Dirichlet's theorem on primes in an arithmetic progression to the ring $F_q[x]$, where F_q is a finite field with $q = p^n$ elements, p a prime. Rosen (2002) offers a statement and a proof of this theorem and a reference to Kornblum's paper. Compare Rosen's proof with that of Kornblum.
6. Let $q = p^n$ and $a_0, a_1, \ldots, a_r$ be nonzero elements of F_q. Let $n_0, n_1, \ldots, n_r$ be positive integers, and let d_i denote the greatest common divisor of $q - 1$ and n_i. Let N_1 represent the number of solutions in F_q of the equation

$$a_0 x_0^{n_0} + a_1 x_1^{n_1} + \cdots + a_r x_r^{n_r} + 1 = 0.$$

Prove that

$$|N_1 - q^r| \leq (d_0 - 1) \cdots (d_r - 1) q^{r/2}.$$

See Weil (1979), vol. I, pp. 399–410. On the basis of some of his earlier theorems and this result, Weil made four conjectures for zeta functions of smooth projective varieties over a finite base field; see pp. 409–410 for a statement of these conjectures, one of which was the Riemann hypothesis.
7. Define the Ramanujan $\tau(n)$ function by the formula

$$q \prod_{n=1}^{\infty} (1 - q^n)^{24} = \sum_{n=1}^{\infty} \tau(n) q^n.$$

Assuming the convergence of the series and the product, show that

$$\sum_{n=1}^{\infty} \tau(n) n^{-s} = \prod_p \left(1 - \tau(p) p^{-s} + p^{11-2s}\right)^{-1},$$

where the product is over all the primes. This result was conjectured by Ramanujan and proved in 1917 by Louis J. Mordell (1888–1972). See Hardy (1978), pp. 161–165. Ramanujan also conjectured that $|\tau(p)| \leq 2p^{11/2}$. This was deduced by Pierre Deligne from his 1974 proof of the characteristic p Riemann hypothesis.

41.8 Notes on the Literature

Euler's additive proof of Fermat's theorem using the binomial theorem may be found in Eu. I-2, pp. 33–37. See Eu. I-2, pp. 493–518, for the paper containing Euler's multiplicative proof. The proof in the text may be found in articles 49 and 50 of Gauss's *Disquisitiones*, reprinted in Gauss (1863–1927), vol. 1, pp. 40–42. Gauss (1966) is an English translation of the *Disquisitiones*. For Lagrange's theorem, see section 42 of the *Disquisitiones*, pp. 34–35 of Gauss (1863–1927), and for the proof that the multiplicative group of integers modulo p is cyclic, see section 55, pp. 44–46. Dedekind's formula on the product of irreducible polynomials of degree n appears in Dedekind (1930), vol. 1, pp. 65–66. For Galois's paper on finite fields, see Galois (1897), pp. 15–23. The quote from Smith about Galois may be seen in Smith (1965b), p. 149. The quotation from Gauss may be found in Gauss (1863–1927), vol. 2, p. 217; the translation in the text is by Günther Frei (2007), p. 180; Frei gives an excellent treatment of Gauss's researches in finite fields. See Katz (1976) for a brief sketch of Deligne's proof of Weil's conjecture. Roquette (2002) and (2004) give the early history of the characteristic p Riemann hypothesis.

References

Abel, N. H. 1826. Untersuchungen über die Reihe $1 + (m/1)x + (m(m-1)/2)x^2 + \cdots$. *J. Reine Angew. Math.*, **1**, 311–339.

Abel, N. H. 1965. *Oeuvres complètes*. New York: Johnson Reprint.

Acosta, D. J. 2003. Newton's rule of signs for imaginary roots. *Amer. Math. Monthly*, **110**, 694–706.

Ahlfors, L. V. 1982. *Collected Papers*. Boston: Birkhäuser.

Ahlgren, S., and Ono, K. 2001. Addition and counting: The arithmetic of partitions. *Notices Amer. Math. Soc.*, **48**, 978–984.

Alder, H. L. 1969. Partition identities – from Euler to the present. *Amer. Math. Monthly*, **76**, 733–746.

Alexander, J. W. 1915. Functions which map the interior of the unit circle upon simple regions. *Ann. Math.*, **17**, 12–22.

Allaire, P., and Bradley, R. E. 2004. Symbolical algebra as a foundation for calculus: D. F. Gregory's contribution. *Historia Math.*, **29**, 395–426.

Almkvist, G., and Berndt, B. 1988. Gauss, Landen, Ramanujan, the arithmetic-geometric mean, ellipses, π, and the Ladies Diary. *Amer. Math. Monthly*, **95**, 585–607.

Altmann, S., and Ortiz, E. L. 2005. *Olinde Rodrigues and His Times*. Providence, R.I.: Amer. Math. Soc.

Anderson, M., Katz, V., and Wilson, R. (eds) 2004. *Sherlock Holmes in Babylon*. Washington, D.C.: Math. Assoc. Amer.

Anderson, M., Katz, V., and Wilson, R. (eds) 2009. *Who Gave You the Epsilon?* Washington, D.C.: Math. Assoc. Amer.

Andrews, G. E. 1981. Ramanujan's "lost" notebook. III. The Rogers-Ramanujan continued fraction. *Adv. Math.*, **41**, 186–208.

Andrews, G. E. 1986. *q-Series: Their Development and Application in Analysis, Number Theory, Combinatorics, Physics, and Computer Algebra*. Providence: Amer. Math. Soc.

Andrews, G. E., and Garvan, F. G. 1988. Dyson's crank of a partition. *Bull. Amer. Math. Soc.*, **18**, 167–171.

Andrews, G. E., Askey, R. A., Berndt, B. C., Ramanathan, K. G., and Rankin, R. A. (eds). 1988. *Ramanujan Revisited*. Boston: Academic Press.

Andrews, G. E., Askey, R., and Roy, R. 1999. *Special Functions*. New York: Cambridge Univ. Press.

Arnold, V. I. 1990. *Huygens and Barrow, Newton and Hooke*. Boston: Birkhäuser. Translated by E. J. F. Primrose.

Arnold, V. I. 2007. *Yesterday and Long Ago*. New York: Springer.

Artin, E. 1964. *The Gamma Function*. New York: Holt, Reinhart and Winston. Translated by Michael Butler.

Ash, J. M. 1976. *Studies in Harmonic Analysis*. Washington, D.C.: Math. Assoc. Amer.

Askey, R. 1975. *Orthogonal polynomials and special functions*. Philadelphia: Society for Industrial and Applied Mathematics.

Atkin, A. O. L. 1968. Multiplicative congruence properties. *Proc. London Math. Soc.*, **18**, 563–576.

Atkin, A. O. L., and Swinnerton–Dyer, P. 1954. Some properties of partititons. *Proc. London Math. Soc.*, **4**, 84–106.

Babbage, C., and Herschel, J. 1813. *Memoirs of the Analytical Society*. Cambridge: Cambridge Univ. Press.

Bäcklund, R. 1918. Über die Nullstellen der Riemannschen Zetafunktion. *Acta Math.*, **41**, 345–375.

Baernstein, A. 1986. *The Bieberbach Conjecture*. Providence, R.I.: Amer. Math. Soc.

Bag, A. K. 1966. Trigonometrical series in the Karanapaddhati and the probable date of the text. *Indian J. Hist. of Sci.*, **1**, 98–106.

Baillaud, B., and Bourget, H. (eds). 1905. *Correspondance d'Hermite et de Stieltjes*. Paris: Gauthier-Villars.

Baker, A. 1988. *New Advances in Transcendence Theory*. New York: Cambridge University Press.

Baker, A., and Masser, D. W. 1977. *Transcendence Theory*. New York: Academic Press.

Barnes, E. W. 1908. A new development of the theory of the hypergeometric function. *Proc. London Math. Soc.*, **6**(2), 141–177.

Baron, M. E. 1987. *The Origins of the Infinitesimal Calculus*. New York: Dover.

Barrow, I. 1735. *Geometrical Lectures*. London: Austen. Translated by E. Stone.

Bateman, H. 1907. The correspondence of Brook Taylor. *Bibliotheca Math.*, **7**, 367–371.

Bateman, P. T., and Diamond, H. G. 1996. A hundred years of prime numbers. *Amer. Math. Monthly*, **103**(9), 729–741. Reprinted in Anderson, Katz, and Wilson (2009), pp. 328–336.

Becher, H. W. 1980. Woodhouse, Babbage, Peacock, and modern algebra. *Historia Math.*, **7**, 389–400.

Beery, J., and Stedall, J. 2009. *Thomas Harriot's Doctrine of Triangular Numbers: The 'Magisteria Magna'*. Zürich: European Mathematical Society.

Berggren, L., Borwein, J., and Borwein, P. 1997. *Pi: A Source Book*. New York: Springer-Verlag.

Berndt, B. C. 1985–1998. *Ramanujan's Notebooks*. New York: Springer-Verlag.

Bernoulli, D. 1982–1996. *Die Werke von Daniel Bernoulli*. Basel: Birkhäuser.

Bernoulli, J. 1993–1999. *Die Werke von Jakob Bernoulli*. Basel: Birkhäuser.

Bernoulli, J. 2006. *The Art of Conjecturing, Translation of Ars Conjectandi*. Baltimore: Johns Hopkins Univ. Press. Translated by E. D. Sylla.

Bernoulli, Joh. 1968. *Opera omnia*. Hildesheim, Germany: G. Olms Verlag.

Bernoulli, N. 1738. Inquisitio in summam seriei $1 + \frac{1}{4} + \frac{1}{9} + \frac{1}{16} + \frac{1}{25} + \frac{1}{36} +$ etc. *Comment. Petropolitanae*, **10**, 19–21.

Bers, L. 1998. *Selected Works of Lipman Bers*. Providence: Amer. Math. Soc.

Bézout, É. 2006. *General Theory of Algebraic Equations*. Princeton: Princeton Univ. Press. Translated by E. Feron.

Bieberbach, L. 1916. Über die Koeffizienten derjenigen Potenzreihen. *S.-B. Preuss Akad. Wiss.*, **138**, 940–955.

Binet, J. 1839. Mémoire sur les intégrales définies Eulériennes. *Journal de l' École Polytéchnique*, **16**, 123–340.

Bissell, C. C. 1989. Cartesian geometry: The Dutch contribution. *Mathematical Intelligencer*, **9**(4), 38–44.

Boas, R. P. 1954. *Entire Functions*. New York: Academic Press.

Bogolyubov, N. N., Mikhaĭlov, G. K., and Yushkevich, A. P. (eds). 2007. *Euler and Modern Science*. Washington, D.C.: Math. Assoc. Amer.

Bohr, H., and Mollerup, J. 1922. *Laerebog i Matematisk Analyse*. Copenhagen: Jul. Gjellerups Forlag.

Bolibruch, A. A., Osipov, Yu. S., and Sinai, Ya. G. (eds). 2006. *Mathematical Events of the Twentieth Century*. New York: Springer.

Bombieri, E., and Gubler, W. 2006. *Heights in Diophantine Geometry*. New York: Cambridge Univ. Press.

Boole, G. 1841. Exposition of a general theory of linear transformations, Parts I and II. *Cambridge Math. J.*, **3**, 1–20, 106–111.

Boole, G. 1844a. Notes on linear transformations. *Cambridge Math. J.*, **4**, 167–71.
Boole, G. 1844b. On a general method in analysis. *Phil. Trans. Roy. Soc. London*, **124**, 225–282.
Boole, G. 1847. *The Mathematical Analysis of Logic*. London: George Bell.
Boole, G. 1877. *A Treatise on Differential Equations*. London: Macmillan.
Borel, É. 1900. *Leçons sur les fonctions entières*. Paris: Gauthier-Villars.
Bornstein, M. 1997. *Symbolic Integration*. New York: Springer-Verlag.
Boros, G., and Moll, V. 2004. *Irresistible Integrals*. New York: Cambridge Univ. Press.
Borwein, J., Bailey, D., and Girgensohn, R. 2004. *Experimentation in Mathematics: Computational Paths to Discovery*. Natick, Mass.: A K Peters.
Bos, H. J. M. 1974. Differentials, higher-order differentials and the derivative in the Leibnizian calculus. *Arch. Hist. Exact Sci.*, **14**, 1–90.
Bottazzini, U. 1986. *The Higher Calculus*. New York: Springer-Verlag.
Bourbaki, N. 1994. *Elements of the History of Mathematics*. New York: Springer-Verlag. Translated by J. Meldrum.
Boyer, C. B. 1943. Pascal's formula for the sums of the powers of integers. *Scripta Math.*, **9**, 237–244.
Boyer, C. B., and Merzbach, U. C. 1991. *A History of Mathematics*. New York: Wiley.
Bradley, R. E., and Sandifer, C. E. (eds). 2007. *Leonhard Euler: Life, Work and Legacy*. Amsterdam: Elsevier.
Bradley, R. E., and Sandifer, C. E. 2009. *Cauchy's* Cours d'analyse. New York: Springer.
Bressoud, D. 2002. Was calculus invented in India? *College Math. J.*, **33**(1), 2–13. Reprinted in Anderson, Katz, and Wilson (2004), 131–137.
Bressoud, D. 2007. *A Radical Approach to Real Analysis*. 2nd edn. Washington, D.C.: Math. Assoc. Amer.
Bressoud, D. 2008. *A Radical Approach to Lesbesgue's Theory of Integration*. New York: Cambridge Univ. Press.
Brezinski, C. 1991. *History of Continued Fractions and Padé Approximations*. New York: Springer.
Bronstein, M. 1997. *Symbolic Integration*. Heidelberg: Springer-Verlag.
Browder, F. E. (ed). 1976. *Mathematical Developments Arising from Hilbert Problems*. Providence: Amer. Math. Soc.
Budan de Boislaurent, Ferdinand-François-Désiré. 1822. *Nouvelle méthode pour la résolution des équations numérique....* Paris: Dondey-Dupré.
Bühler, W. K. 1981. *Gauss: A Biographical Study*. New York: Springer-Verlag.
Burn, R. P. 2001. Alphose Antonio de Sarasa and logarithms. *Historia Math.*, **28**, 1–17.
Burnside, W. S., and Panton, A. W. 1960. *The Theory of Equations*. New York: Dover.
Butzer, P. L., and Sz.-Nagy, B. 1974. *Linear Operators and Approximation II*. Basel: Birkhäuser.
Cahen, E. 1894. Sur la fonction $\zeta(s)$ de Riemann et sur des fonctions analogues. *Ann. Sci. École Norm. Sup.*, **11**, 75–164.
Campbell, G. 1728. A method of determining the number of impossible roots in affected aequations. *Phil. Trans. Roy. Soc.*, **35**, 515–531.
Campbell, P. J. 1978. The origin of "Zorn's Lemma". *Historia Math.*, **5**, 77–89.
Cannon, J. T., and Dostrovsky, S. 1981. *The Evolution of Dynamics : Vibration Theory from 1687 to 1742*. New York: Springer-Verlag.
Cantor, G. 1932. *Gesammelte Abhandlungen*. Berlin: Springer.
Cardano, G. 1993. *Ars Magna or the Rules of Algebra*. New York: Dover. Translated by T. R. Witmer.
Carleson, L. 1966. On convergence and growth of partial sums of Fourier series. *Acta. Math.*, **116**, 135–157.
Cauchy, A.-L. 1823. *Résumé des leçons donnés à l'École Royale Polytechnique sur le calcul infinitésimal*. Paris: De Bure.
Cauchy, A.-L. 1829. *Calcul différentiel*. Paris: De Bure.
Cauchy, A.-L. 1843. Sur l'emploi légitime des séries divergentes. *Compt. Rend.*, **17**, 370–376.
Cauchy, A.-L. 1882–1974. *Oeuvres complétes*. Paris: Gauthier-Villars.
Cauchy, A.-L. 1989. *Analyse algébrique*. Paris: Jacques Gabay.
Cayley, A. 1889–1898. *Collected Mathematical Papers*. Cambridge: Cambridge Univ. Press.

Cayley, A. 1895. *Elliptic Functions*. London: Bell.
Cesàro, E. 1890. Sur la multiplication des séries. *Bull. Sci. Math.*, **14**, 114–20.
Chabert, J.-L. 1999. *A History of Algorithms: From the Pebble to the Microchip*. New York: Springer. Translated by C. Weeks.
Chebyshev, P. L. 1899–1907. *Oeuvres de P. L. Tchebychef*. St. Petersburg: Académie Impériale des Sciences.
Cheney, E. (ed). 1980. *Approximation Theory III*. New York: Academic Press.
Cherry, W., and Ye, Z. 2001. *Nevanlinna Theory of Value Distribution*. New York: Springer.
Child, J. M. 1916. *Geometrical Lectures of Isaac Barrow*. Chicago: Open Court.
Child, J. M. 1920. *The Early Mathematical Manuscripts of Leibniz*. Chicago: Open Court.
Chudnovsky, D. V., and Chudnovsky, G. V. 1988. Approximations and complex multiplication according to Ramanujan. Pages 375–472 of: Andrews, G. E., Askey, R. A., Berndt, B. C., Ramanathan, K. G., and Rankin, R. A. (eds), *Ramanujan Revisited: Proceedings of the Ramanujan Centenary Conference held at the University of Illinois, Urbana-Champaign, Illinois, June 1Ð5, 1987*. Boston: Academic Press.
Clairaut, A.-C. 1739. Recherches générales sur le calcul intégral. *Mémoires de l'Académie Royale des Sciences*, **1**, 425–436.
Clairaut, A.-C. 1740. Sur l'intégration ou la construction des équations différentielles du premier ordre. *Mémoires de l'Académie Royale des Sciences*, **2**, 293–323.
Clarke, F. M. 1929. *Thomas Simpson and his Times*. Baltimore: Waverly Press.
Clausen, T. 1828. Ueber die Fälle, wenn die Reihe von der Form $y = 1 + \cdots$ etc. ein Quadrat von der Form $z = 1 + \cdots$ etc. hat. *J. Reine Angew. Math.*, **3**, 89–91.
Cohen, H. 2007. *Number Theory, Volume II: Analytic and Modern Tools*. New York: Springer.
Cooke, R. 1984. *The Mathematics of Sonya Kovalevskaya*. New York: Springer-Verlag.
Cooke, R. 1993. Uniqueness of trigonometric series and descriptive set theory 1870–1985. *Arch. Hist. Exact Sci.*, **45**, 281–334.
Cooper, S. 2006. The quintuple product identity. *Int. J. Number Theory*, **2**, 115–161.
Corry, L. 2004. *Modern Algebra and the Rise of Mathematical Structures*. Basel: Birkhäuser.
Cotes, R. 1722. *Harmonia Mensurarum*. Cambridge: Cambridge Univ. Press.
Cox, D. A. 1984. The arithmetic-geometric mean of Gauss. *L'enseignement mathématique*, **30**, 275–330.
Cox, D. A. 2004. *Galois Theory*. Hoboken: Wiley.
Craik, A. D. D. 2000. James Ivory, F.R.S., mathematician: "The most unlucky person that ever existed". *Notes and Records Roy. Soc. London*, **54**, 223–247.
Craik, A. D. D. 2005. Prehistory of Faà di Bruno's formula. *Amer. Math. Monthly*, **112**, 119–130.
Crilly, T. 2006. *Arthur Cayley*. Baltimore: Johns Hopkins Univ. Press.
D'Alembert, J. 1761–1780. *Opuscules mathématiques*. Paris: David.
Dauben, J. 1979. *Georg Cantor*. Princeton: Princeton Univ. Press.
Davenport, H. 1980. *Multiplicative Number Theory*. New York: Springer-Verlag.
Davis, P. J. 1959. Leonhard Euler's Integral: A historical profile of the gamma function. *Amer. Math. Monthly*, **66**, 849–869.
de Beaune, F., Girard, A., and Viète, F. 1986. *The Early Theory of Equations: On Their Nature and Constitution*. Annapolis, Md.: Golden Hind Press. Translated by Robert Smith.
De Branges, L. 1985. A proof of the Bieberbach conjecture. *Acta Math.*, **157**, 137–162.
de Moivre, A. 1730a. *Miscellanea analytica de seriebus et quadraturis*. London: Tonson and Watts.
de Moivre, A. 1730b. *Miscellaneis analyticis supplementum*. London: Tonson and Watts.
de Moivre, A. 1967. *The Doctrine of Chances*. New York: Chelsea.
Dedekind, R. 1930. *Gesammelte Mathematische Werke*. Braunschweig: F. Vieweg.
Dedekind, R. 1963. *Essays on the Theory of Numbers*. New York: Dover.
Delone, B. N. 2005. *The St. Petersburg School of Number Theory*. Providence: Amer. Math. Soc.
Descartes, R. 1954. *La Géométrie*. New York: Dover. Translated by D. E. Smith and M. L. Latham.
Dieudonné, J. 1981. *History of Functional Analysis*. Amsterdam: Elsevier.

Dirichlet, P. G. L. 1862. Démonstration d'un théorème d'Abel. *J. Math. Pures App.*, **7**(2), 253–255. Also in Werke I, 305–306.

Dirichlet, P. G. L. 1969. *Mathematische Werke*. New York: Chelsea.

Dirichlet, P. G. L., and Dedekind, R. 1999. *Lectures on Number Theory*. Providence: Amer. Math. Soc. Translated by John Stillwell.

Dörrie, H. 1965. *100 Great Problems of Elementary Mathematics*. New York: Dover. Translated by D. Antin.

Duke, W., and Tschinkel, Y. 2005. *Analytic Number Theory: A Tribute to Gauss and Dirichlet*. Providence: Amer. Math. Soc.

Dunham, W. 1990. *Journey Through Genius*. New York: Wiley.

Dunnington, G. 2004. *Gauss: Titan of Science*. Washington, D.C.: Math. Assoc. Amer.

Duren, P. L. 1983. *Univalent Functions*. New York: Springer-Verlag.

Dutka, J. 1984. The early history of the hypergeometric series. *Arch. Hist. Exact Sci.*, **31**, 15–34.

Dutka, J. 1991. The early history of the factorial function. *Arch. Hist. Exact Sci.*, **43**, 225–249.

Edwards, A. W. F. 2002. *Pascal's Arithmetical Triangle*. Baltimore: Johns Hopkins Univ. Press.

Edwards, H. M. 1977. *Fermat's Last Theorem*. Berlin: Springer-Verlag.

Edwards, H. M. 1984. *Galois Theory*. New York: Springer-Verlag.

Edwards, H. M. 2001. *Riemann's Zeta Function*. New York: Dover.

Edwards, J. 1954. *An Elementary Treatise on the Differential Calculus*. London: Macmillan.

Edwards, J. 1954b. *Treatise on Integral Calculus*. New York: Chelsea.

Eie, M. 2009. *Topics In Number Theory*. Singapore: World Scientific.

Eisenstein, F. G. 1975. *Mathematische Werke*. New York: Chelsea.

Elliott, E. B. 1964. *An Introduction to the Algebra of Quantics*. New York: Chelsea.

Ellis, D. B., Ellis, R., and Nerurkar, M. 2000. The topological dynamics of semigroup actions. *Trans. Amer. Math. Soc.*, **353**, 1279–1320.

Ellis, R. L. 1845. Memoir to D. F. Gregory. *Cambridge and Dublin Math. J.*, **4**, 145–152.

Engelsman, S. B. 1984. *Families of Curves and the Origins of Partial Differentiation*. Amsterdam: North-Holland.

Enros, P. 1983. The analytical society (1812–1813). *Historia Math.*, **10**, 24–47.

Erdős, P. 1932. Beweis eines Satz von Tschebyschef. *Acta. Sci. Math.*, **5**, 194–198.

Erman, A. 1852. *Briefwechsel zwischen Olbers und Bessel*. Leipzig: Avenarius and Mendelssohn.

Euler, L. 1911–2000. *Leonhardi Euleri Opera omnia*. Berlin: Teubner.

Euler, L. 1985. An essay on continued fractions. *Math. Syst. Theory*, **18**, 295–328.

Euler, L. 1988. *Introduction to Analysis of the Infinite*. New York: Springer-Verlag. Translated by J. D. Blanton.

Euler, L. 2000. *Foundations of Differential Calculus. English Translation of First Part of Euler's* Institutiones calculi differentialis. New York: Springer-Verlag. Translated by J. D. Blanton.

Fagnano, G. C. 1911. *Opere matematiche*. Rome: Albrighi.

Farkas, H. M., and Kra, I. 2001. *Theta Constants, Riemann Surfaces and the Modular Group*. Providence, R.I.: Amer. Math. Soc.

Fatou, P. 1906. Séries trigonométriques et séries de Taylor. *Acta Math.*, **30**, 335–400.

Feigenbaum, L. 1981. *Brook Taylor's "Methodus incrementorum": A Translation with Mathematical and Historical Commentary*. Ph.D. thesis, Yale University.

Feigenbaum, L. 1985. Taylor and the method of increments. *Arch. Hist. Exact Sci.*, **34**, 1–140.

Feingold, M. 1990. *Before Newton*. New York: Cambridge Univ. Press.

Feingold, M. 1993. Newton, Leibniz and Barrow too. *Isis*, **84**, 310–338.

Fejér. 1970. *Gesammelte Arbeiten*. Basel: Birkhäuser.

Feldheim, E. 1941. Sur les polynomes généralisés de Legendre. *Bull. Acad. Sci. URSS. Sér. Math. [Izvestia Adad. Nauk SSSR]*, **5**, 241–254.

Ferraro, G. 2004. Differentials and differential coefficients in the Eulerian foundations of the calculus. *Historia Math.*, **31**, 34–61.

Ferreirós, J. 1993. On the relations between Georg Cantor and Richard Dedekind. *Historia Math.*, **20**, 343–63.

FitzGerald, C. H. 1985. The Bieberbach conjecture: Retrospective. *Notices Amer. Math. Soc.*, **32**, 2–6.

Foata, D., and Han, G.-N. 2001. The triple, qunituple and sextuple product identities revisited. Pages 323–334 of: Foata, D., and Han, G.-N. (eds), *The Andrews Festschrift: Seventeen Papers on Classical Number Theory and Combinatorics*. New York: Springer.

Fomenko, O. M., and Kuzmina, G. V. 1986. The last 100 days of the Bieberbach conjecture. *Mathematical Intelligencer*, **8**, 40–47.

Forrester, P. J., and Warnaar, S. O. 2008. The Importance of the Selberg Integral. *Bull. Amer. Math. Soc.*, **45**, 498–534.

Fourier, J. 1955. *The Analytical Theory of Heat*. New York: Dover. Translated by A. Freeman.

Français, J. F. 1812–1813. Analise transcendante. Memoire tendant à démontrer la légitimité de la séparation des échelles de différentiation et d'intégration des fonctions qu'elles affectent. *Annales de Gergonne*, **3**, 244–272.

Frei, G. 2007. The unpublished section eight: On the way to function fields over a finite field. Pages 159–198 of: Goldstein, C., Schappacher, N., and Schwermer, J. (eds), *The Shaping of Arithmetic after C. F. Gauss's* Disquisitiones Arithmeticae. New York: Springer.

Friedelmeyer, J. P. 1994. *Le calcul des dérivations d'Arbogast dans le projet d'algébrisation de l'analyse à fin du xviiie siècle*. Nantes, France: Université de Nantes.

Frobenius, G. 1880. Ueber die Leibnizsche Reihe. *J. Reine Ang. Math.*, **89**, 262–264.

Fuss, P. H. 1968. *Correspondance mathématique et physique*. New York: Johnson Reprint.

Galois, E. 1897. *Oeuvres mathématiques*. Paris: Gauthier-Villars.

Gårding, L. 1994. *Mathematics in Sweden before 1950*. Providence: Amer. Math. Soc.

Gauss, C. F. 1863–1927. *Werke*. Leipzig: Teubner.

Gauss, C. F. 1966. *Disquisitiones Arithmeticae (An English Translation)*. New Haven, Conn.: Yale Univ. Press. Translated by A. A. Clarke.

Gauss, C. F. 1981. *Arithmetische Untersuchungen*. New York: Chelsea.

Gegenbauer, L. 1884. Zur Theorie der Functionen $C_n^\nu(x)$. *Denkschriften der Akademie der Wissenschaften in Wien, Math. Naturwiss. Klasse*, **48**, 293–316.

Gelfand, I. M. 1988. *Collected Papers*. New York: Springer-Verlag.

Gelfand, I. M., Kapranov, M. M., and Zelevinsky, A. V. 1994. *Discriminants, Resultants, and Multidimensional Determinants*. Boston: Birkhäuser.

Gelfond, A. O. 1960. *Transcendental and Algebraic Numbers*. New York: Dover. Translated by Leo F. Boron.

Gelfond, A. O., and Linnik, Yu. V. 1966. *Elementary Methods in the Analytic Theory of Numbers*. Cambridge: MIT Press. Translated by D. E. Brown.

Georgiadou, M. 2004. *Constantine Carathéodory*. New York: Springer-Verlag.

Glaisher, J. W. L. 1878. Series and products for π and powers of π. *Messenger of Math.*, **7**, 75–80.

Glaisher, J. W. L. 1883. A theorem in partitions. *Messenger of Math.*, **12**, 158–170.

Goldstein, C., Schappacher, N., and Schwermer, J. (eds). 2007. *The Shaping of Arithmetic after C. F. Gauss's* Disquisitiones Arithmeticae. New York: Springer.

Goldstein, L. J. 1973. A history of the prime number theorem. *Amer. Math. Monthly*, **80**, 599–615. Correction, 1115. Reprinted in Anderson, Katz, and Wilson (2009), pp. 318–327.

Goldstine, H. H. 1977. *A History of Numerical Analysis*. New York: Springer-Verlag.

Gong, S. 1999. *The Bieberbach Conjecture*. Providence: Amer. Math. Soc.

Gordon, B. 1961. A combinatorial generalization of the Rogers-Ramanujan identities. *Amer. J. Math.*, **83**, 393–399.

Gouvêa, F. Q. 1994. A marvelous proof. *Amer. Math. Monthly*, **101**, 203–222.

Gowing, R. 1983. *Roger Cotes*. New York: Cambridge Univ. Press.

Grabiner, J. V. 1981. *The Origins of Cauchy's Rigorous Calculus*. Cambridge, Mass.: MIT Press.

Grabiner, J. V. 1990. *The Calculus as Algebra*. New York: Garland Publishing.

Grace, J. H., and Young, A. 1965. *The Algebra of Invariants*. New York: Chelsea.

Graham, G., Rothschild, B. L., and Spencer, J. H. 1990. *Ramsey Theory*. New York: Wiley.

Grattan–Guinness, I. 1972. *Joseph Fourier 1768–1830*. New York: MIT Press.

Grattan–Guinness, I. 2005. *Landmark Writings in Western Mathematics*. Amsterdam: Elsevier.
Graves, R. P. 1885. *Life of Sir William Rowan Hamilton*. London: Longmans.
Gray, J. 1986. *Linear Differential Equations and Group Theory from Riemann to Poincaré*. Boston: Birkhäuser.
Gray, J. 1994. On the history of the Riemann mapping theorem. *Rendiconti del Circolo Matematico di Palermo*, **34**, 47–94. Series II, Supplemento 34.
Gray, J., and Parshall, K. H. (eds). 2007. *Episodes in the History of Modern Algebra (1800–1950)*. Providence: Amer. Math. Soc.
Green, G. 1970. *Mathematical Papers*. New York: Chelsea.
Greenberg, J. L. 1995. *The Problem of the Earth's Shape from Newton to Clairaut*. New York: Cambridge Univ. Press.
Gregory, D. F. 1865. *The Mathematical Writings of Duncan Farquharson Gregory*. Cambridge: Deighton, Bell, and Co. Edited by W. Walton.
Gronwall, T. H. 1914. Some remarks on conformal representation. *Ann. Math.*, **16**, 72–76.
Grootendorst, A. W., and van Maanen, J. A. 1982. Van Heuraet's letter (1659) on the rectification of curves. *Nieuw Archief Wiskunde*, **30**(3), 95–113.
Gucciardini, N. 1989. *The Development of Newtonian Calculus in Britain 1700–1800*. New York: Cambridge Univ. Press.
Gupta, R. C. 1977. Paramesvara's rule for the circumradius of a cyclic quadrilateral. *Historia Math.*, **4**, 67–74.
Hadamard, J. 1898. Théorème sur séries entières. *Acta Math.*, **22**, 55–64.
Haimo, D. T. 1968. *Orthogonal Expansions and Their Continuous Analogues*. Carbondale: Southern Illinois Univ. Press.
Hald, A. 1990. *A History of Probability and Statistics and Their Applications Before 1750*. New York: Wiley.
Hamel, G. 1905. Eine basis aller Zahlen und die unstetiggen Lösungen der Funktionalgleichung: $f(x+y) = f(x) + f(y)$. *Math. Ann.*, **60**, 459–462.
Hamilton, W. R. 1835. *Theory of Conjugate Functions or Algebraic Couples*. Dublin: Philip Dixon Hardy.
Hamilton, W. R. 1945. Quaternions. *Proc. Roy. Irish Acad.*, **50**, 89–92.
Hardy, G. H. 1905. *The Integration of Functions of a Single Variable*. Cambridge: Cambridge Univ. Press.
Hardy, G. H. 1937. *A Course in Pure Mathematics*. Cambridge: Cambridge Univ. Press.
Hardy, G. H. 1949. *Divergent Series*. Oxford: Clarendon Press.
Hardy, G. H. 1966–79. *Collected Papers*. Oxford: Clarendon.
Hardy, G. H. 1978. *Ramanujan*. New York: Chelsea.
Hardy, G. H., Littlewood, J., and Pólya, G. 1967. *Inequalities*. New York: Cambridge Univ. Press.
Harkness, J., and Morley, F. 1898. *Introduction to the Theory of Analytic Functions*. London: Macmillan.
Hawking, S. 2005. *God Created the Integers*. Philadelphia: Running Press.
Hawkins, T. 1975. *Lebesgue Theory*. New York: Chelsea.
Hayman, W. K. 1964. *Meromorphic Functions*. Oxford: Clarendon Press.
Hayman, W. K. 1994. *Multivalent Functions*. New York: Cambridge Univ. Press.
Heine, E. 1847. Untersuchungen über die Reihe …. *J. Reine Angew. Math.*, **34**, 285–328.
Hermite, C. 1873. *Cours d'analyse*. Paris: Gauthier-Villars.
Hermite, C. 1891. *Cours de M. Hermite (rédigé en 1882 par M. Andoyer)*. Cornell Univ. Lib. Reprint.
Hermite, C. 1905–1917. *Oeuvres*. Paris: Gauthier-Villars.
Herschel, J. F. W. 1820. *A Collection of Examples of the Applications of the Calculus of Finite Differences*. Cambridge: Cambridge Univ. Press.
Hewitt, E., and Hewitt, R. E. 1980. The Gibbs-Wilbraham phenomenon. *Arch. Hist. Exact Sci.*, **21**, 129–160.
Hilbert, D. 1970. *Gesammelte Abhandlungen*. Berlin: Springer.
Hilbert, D. 1978. *Hilbert's Invariant Theory Papers*. Brookline, Mass.: Math. Sci. Press.

Hilbert, D. 1993. *Theory of Algebraic Invariants*. New York: Cambridge Univ. Press. Translated by Reinhard C. Laubenbacher.

Hobson, E. W. 1957a. *The Theory of Functions of a Real Variable*. New York: Dover.

Hobson, E. W. 1957b. *A Treatise on Plane and Advanced Trigonometry*. New York: Dover.

Hoe, J. 2007. *A Study of the Jade Mirror of the Four Unknowns*. Christchurch, N.Z.: Mingming Bookroom.

Hofmann, J. E. 1974. *Leibniz in Paris*. New York: Cambridge Univ. Press.

Hofmann, J. E. 1990. *Ausgewählte Schriften*. Zürich: Georg Olms Verlag.

Horiuchi, A. T. 1994. The *Tetsujutsu sankei* (1722), an 18th century treatise on the methods of investigation in mathematics. Pages 149–164 of: Sasaki, C., Sugiura, M., and Dauben, J. W. (eds), *The Intersection of History and Mathematics*. Basel: Birkhäuser.

Hua, L. K. 1981. *Starting with the Unit Circle*. New York: Springer-Verlag.

Hutton, C. 1812. *Tracts on mathematical and philosophical subjects*. London: Rivington and Company.

Ireland, K., and Rosen, M. 1982. *A Classical Introduction to Modern Number Theory*. New York: Springer-Verlag.

Ivory, J. 1796. A new series for the rectification of the ellipses. *Trans. Roy. Soc. Edinburgh*, **4**, 177–190.

Ivory, J. 1812. On the attractions of an extensive class of spheroids. *Phil. Trans. Roy. Soc. London*, **102**, 46–82.

Ivory, J. 1824. On the figure requisite to maintain the equilibrium of a homogeneous fluid mass that revolves upon an axis. *Phil. Trans. Roy. Soc. London*, **114**, 85–150.

Ivory, J., and Jacobi, C. G. J. 1837. Sur le développement de $(1 - 2xz + z^2)^{-1/2}$. *J. Math. Pures App.*, **2**, 105–106.

Jackson, F. H. 1910. On q-definite integrals. *Quart. J. Pure App. Math.*, **41**, 193–203.

Jacobi, C. G. J. 1969. *Mathematische Werke*. New York: Chelsea.

Jahnke, H. N. 1993. Algebraic analysis in Germany, 1780–1849: Some mathematical and philosophical issues. *Historia Math.*, **20**, 265–284.

James, I. M. 1999. *History of Topology*. Amsterdam: Elsevier.

Jensen, J. L. W. V. 1899. Sur un nouvel et important théorèm de la théorie des fonctions. *Acta Math.*, **22**, 359–364.

Jensen, J. L. W. V. 1906. Sur les fonctions convexes et les inégalités entre les valeurs moyennes. *Acta Math.*, **30**, 175–193.

Johnson, W. P. 2002. The curious history of Faà di Bruno's formula. *Amer. Math. Monthly*, **109**, 217–234.

Johnson, W. P. 2007. The Pfaff/Cauchy derivative and Hurwitz type extensions. *Ramanujan J. Math.*, **13**, 167–201.

Kac, M. 1979. *Selected Papers*. Cambridge, Mass.: MIT Press.

Kalman, D. 2009. *Polynomia and Related Realms*. Washington, D.C.: Math. Assoc. Amer.

Karamata, J. 1930. Über die Hardy-Littlewoodschen Umkehrungen des Abelschen Stetigkeitssatzes. *Math. Z.*, **32**, 219–320.

Katz, N. M. 1976. An overview of Deligne's proof of the Riemann hypothesis for varieties over finite fields. Pages 275–305 of: Browder, F. E. (ed), *Mathematical Developments Arising from Hilbert Problems*. Providence: Amer. Math. Soc.

Katz, V. J. 1979. The history of Stokes' theorem. *Math. Mag.*, **52**, 146–156.

Katz, V. J. 1982. Change of variables in multiple integrals: Euler to Cartan. *Math. Mag.*, **55**, 3–11.

Katz, V. J. 1985. Differential forms – Cartan to de Rham. *Arch. Hist. Exact Sci.*, **33**, 307–319.

Katz, V. J. 1987. The calculus of the trigonometric functions. *Historia Math.*, **14**, 311–324.

Katz, V. J. 1995. Ideas of calculus in Islam and India. *Math. Mag.*, **3**(3), 163–174. Reprinted in Anderson, Katz, and Wilson (2004), pp. 122–130.

Katz, V. J. 1998. *A History of Mathematics: An Introduction*. Reading, Mass.: Addison-Wesley.

Khinchin, A.Y. 1998. *Three Pearls of Number Theory*. New York: Dover.

Khrushchev, S. 2008. *Orthogonal Polynomials and Continued Fractions*. New York: Cambridge Univ. Press.

Kichenassamy, S. 2010. Brahmagupta's derivation of the area of a cyclic quadrilateral. *Historia Math.*, **37**(1), 28–61.
Klein, F. 1911. *Lectures on Mathematics*. New York: Macmillan.
Klein, F. 1979. *Development of Mathematics in the 19th Century*. Brookline, Mass.: Math. Sci. Press. Translated by M. Ackerman.
Knoebel, A., Laubenbacher, R., Lodder, J., and Pengelley, D. 2007. *Mathematical Masterpieces*. New York: Springer.
Knopp, K. 1990. *Theory and Application of Infinite Series*. New York: Dover.
Knuth, D. 2003. *Selected Papers*. Stanford, Calif.: Center for the Study of Language and Information (CSLI).
Kolmogorov, A. N. 1923. Une série de Fourier-Lebesgue divergente presque partout. *Fund. Math.*, **4**, 324–328.
Kolmogorov, A. N., and Yushkevich, A. P. (eds). 1998. *Mathematics of the 19th Century: Vol. III: Function Theory According to Chebyshev; Ordinary Differential Equations; Calculus of Variations; Theory of Finite Differences*. Basel: Birkhäuser.
Koppelman, E. 1971. The calculus of operations and the rise of abstract algebra. *Arch. Hist. Exact Sci.*, **8**, 155–242.
Korevaar, J. 2004. *Tauberian Theory*. New York: Springer.
Kronecker, L. 1968. *Mathematische Werke*. New York: Chelsea.
Kubota, K. K. 1977. Linear functional equations and algebraic independence. Pages 227–229 of: Baker, A., and Masser, D. W. (eds), *Transcendence Theory*. New York: Academic Press.
Kummer, E. E. 1840. Über die Transcendenten, welche aus wiederholten Integrationen rationaler Formeln entstehen. *J. Reine Angew. Math.*, **21**, 74–90, 193–225, 328–371.
Kummer, E. E. 1975. *Collected Papers*. Berlin: Springer-Verlag.
Kung, J. P. S. 1995. *Gian-Carlo Rota on Combinatorics*. Boston, Mass.: Birkhäuser.
Lacroix, S. F. 1819. *Traité du calcul différentiel et du calcul intégral*. Vol. 3. Paris: Courcier.
Lagrange, J. L. 1867–1892. *Oeuvres*. Paris: Gauthier-Villars.
Laguerre, E. 1972. *Oeuvres*. New York: Chelsea.
Landau, E. 1904. Über eine Verallgemeinerung des Picardschen Satzes. *S.-B. Preuss Akad. Wiss.*, **38**, 1118–1133.
Landau, E. 1907. Über einen Konvergenzsatz. *Göttinger Nachrichten*, **8**, 25–27.
Landen, J. 1758. *A Discourse Concerning The Residual Analysis*. London: Nourse.
Landen, J. 1760. A new method of computing the sums of certain series. *Phil. Trans. Roy. Soc. London*, **51**, 553–565.
Lanzewizky, I. L. 1941. Über die orthogonalität der Fejér-Szegöschen polynome. *D. R. Dokl. Acad. Sci. URSS*, **31**, 199–200.
Laplace, P. S. 1812. *Théorie analytique des probabilités*. Paris: Courcier.
Lascoux, A. 2003. *Symmetric Functions and Combinatorial Operators on Polynomials*. Providence, R.I.: Amer. Math. Soc.
Laudal, O.A., and Piene, R. 2002. *The Legacy of Niels Henrik Abel*. Berlin: Springer.
Laugwitz, D. 1999. *Bernhard Riemann 1826–1866*. Boston: Birkhäuser. Translated by A. Shenitzer.
Lebesgue, H. 1906. *Leçons sur les séries trigonométriques*. Paris: Gauthier-Villars.
Legendre, A.M. 1811–1817. *Exercices de calcul intégral*. Paris: Courcier.
Leibniz, G. W. 1971. *Mathematische Schriften*. Hildesheim, Germany: Georg Olms Verlag.
Lemmermeyer, F. 2000. *Reciprocity Laws*. New York: Springer-Verlag.
Lewin, L. 1981. *Polylogarithms and Associated Functions*. Amsterdam: Elsevier.
Lindelöf, E. 1902. Mémoire sur la théorie des fonctions entières de genre fini. *Acta Soc. Sci. Fennicae*, **31 (1)**, 1–79.
Liouville, J. 1837a. Note sur le développement de $(1-2xz+z^2)^{-1/2}$. *J. Math. Pures App.*, **2**, 135–139.
Liouville, J. 1837b. Sur la sommation d'une série. *J. Math. Pures Appl.*, **2**, 107–108.
Liouville, J. 1851. Sur de classes très étendus de quantités dont la valeur n'est ni algébrique, ni même réductible à des irrationelles algébriques. *J. Math. Pures Appl.*, **16**, 133–142.

Liouville, J. 1880. Leçons sur les fonctions doublement périodiques. *J. Reine Angew. Math.*, **88**, 277–310.

Littlewood, J. E. 1982. *Collected Papers*. Oxford: Clarendon Press.

Littlewood, J. E. 1986. *Littlewood's Miscellany*. New York: Cambridge Univ. Press.

Loewner, C. 1988. *Collected Papers*. Boston: Birkhäuser.

Lusin, N. 1913. Sur la convergence des séries trigonométriques de Fourier. *Compt. Rend.*, **156**, 1655–1658.

Lützen, J. 1990. *Joseph Liouville*. New York: Springer-Verlag.

Maclaurin, C. 1729. A second letter to Martin Folkes, Esq.: Concerning the roots of equations, with the demonstration of other rules in algebra. *Phil. Trans. Roy. Soc.*, **36**, 59–96.

Maclaurin, C. 1742. *A Treatise of Fluxions*. Edinburgh: Ruddimans.

Maclaurin, C. 1748. *A Treatise of Algebra*. London: Millar and Nourse.

MacMahon, P. A. 1978. *Collected Papers*. New York: MIT Press.

Mahoney, M. S. 1994. *The Mathematical Career of Pierre de Fermat (1601–1665)*. Princeton: Princeton Univ. Press.

Malet, A. 1993. James Gregorie on tangents and the Taylor rule. *Arch. Hist. Exact Sci*, **46**, 97–138.

Malmsten, C. J. 1849. De integralibus quibusdam definitis. *J. Reine Angew. Math.*, **38**, 1–38.

Manders, K. 2006. Algebra in Roth, Faulhaber and Descartes. *Historia Math.*, **33**, 184–209.

Manning, K. R. 1975. The emergence of the Weierstrassian approach to complex analysis. *Arch. Hist. Exact Sci.*, **14**, 297–383.

Maor, E. 1998. *Tirgonometric Delights*. Princeton: Princeton Univ. Press.

Martzloff, J. C. 1997. *A History of Chinese Mathematics*. New York: Springer.

Masani, P.R. 1990. *Norbert Wiener*. Basel: Birkhäuser.

Maxwell, J. C. 1873. *A Treatise on Electricity and Magnetism*. Oxford: Clarendon Press.

Maz'ya, V., and Shaposhnikova, T. 1998. *Jacques Hadamard*. Providence: Amer. Math. Soc. Translated by Peter Basarab-Horwath.

McClintock, E. 1881. On the remainder of Laplace's series. *Amer. J. Math.*, **4**, 96–97.

Mertens, F. 1874a. Ein beitrag zur analytischen Zahlentheorie. *J. Reine Angew. Math.*, **78**, 46–62.

Mertens, F. 1874b. Ueber einige asymptotische Gesetze der Zahlentheorie. *J. Reine Angew. Math.*, **77**, 289–338.

Mertens, F. 1875. Über die Multiplicationsregel für zwei unendliche Reihen. *J. Reine Angew. Math.*, **79**, 182–184.

Mertens, F. 1895. Über das nichtverschwinden Dirichlet-Reihen mit reellen Gliedern. *S.-B. Kais. Akad. Wissensch. Wien*, **104**(Abt. 2a), 1158–1166.

Meschkowski, H. 1964. *Ways of Thought of Great Mathematicians*. San Francisco: Holden-Day.

Mikami, Y. 1974. *The Development of Mathematics in China and Japan*. New York: Chelsea.

Mittag-Leffler, G. 1923. An introduction to the theory of elliptic functions. *Ann. Math.*, **24**, 271–351.

Miyake, K. 1994. The establishment of the Takagi-Artin class field theory. Pages 109–128 of: Sasaki, C., Sugiura, M., and Dauben, J. W. (eds), *The Intersection of History and Mathematics*. Basel: Birkhäuser.

Moll, V. 2002. The evaluation of integrals: A personal story. *Notices Amer. Math. Soc.*, 311–317.

Monsky, P. 1994. Simplifying the proof of Dirichlet's theorem. *Amer. Math. Monthly*, **100**, 861–862.

Montmort, P. R. de. 1717. De seriebus infinitis tractatus. *Phil. Trans. Roy. Soc.*, **30**, 633–675.

Moore, G. H. 1982. *Zermelo's Axiom of Choice*. New York: Springer-Verlag.

Morrison, P., and Morrison, E. 1961. *Charles Babbage and His Calculating Engines: Selected Writings by Charles Babbage*. New York: Dover.

Muir, T. 1960. *The Theory of Determinants in the Historical Order of Development*. New York: Dover.

Mukhopadhyaya, A. 1998. *A Diary of Asutosh Mookerjee*. Calcutta: Mitra and Ghosh Publ.

Murphy, R. 1833. On the inverse method of definite integrals, with physical applications. *Trans. Cambridge Phil. Soc.*, **4**, 353–408.

Murphy, R. 1835. Second memoir on the inverse method of definite integrals. *Trans. Cambridge Phil. Soc.*, **5**, 113–148.

Murphy, R. 1837. First memoir on the theory of analytic operations. *Phil. Trans. Roy. Soc. London*, **127**, 179–210.

Murphy, R. 1839. *A Treatise on the Theory of Algebraical Equations*. London: Society for Diffusion of Useful Knowledge.

Mustafy, A. K. 1966. A new representation of Riemann's zeta function and some of its consequences. *Norske Vid. Selsk. Forh. (Trondheim)*, **39**, 96–100.

Mustafy, A. K. 1972. On a criterion for a point to be a zero of the Riemann zeta function. *J. London Math. Soc. (2)*, **5**, 285–288.

Narasimhan, R. 1991. The coming of age of mathematics in India. Pages 235–258 of: Hilton, P., Hirzebruch, F., and Remmert, R. (eds), *Miscellanea Mathematica*. New York: Springer.

Narkiewicz, W. 2000. *The Development of Prime Number Theory*. New York: Springer.

Nesterenko, Yu. V. 2006. Hilbert's seventh problem. Pages 269–282 of: Bolibruch, A. A., Osipov, Yu. S., and Sinai, Ya. G. (eds), *Mathematical Events of the Twentieth Century*. Berlin: Springer. Translated by L. P. Kotova.

Neuenschwander, E. 1978a. The Casorati-Weierstrass theorem. *Historia Math.*, **5**, 139–166.

Neuenschwander, E. 1978b. Der Nachlass von Casorati (1835–1890) in Pavia. *Archive Hist. Exact Sci.*, **19**, 1–89.

Neumann, O. 2007a. Cyclotomy: From Euler through Vandermonde to Gauss. Pages 323–362 of: Bradley, R. E., and Sandifer, C. E. (eds), *Leonhard Euler: Life, Work and Legacy*. Amsterdam: Elsevier.

Neumann, O. 2007b. The *Disquisitiones Arithmeticae* and the theory of equations. Pages 107–128 of: Goldstein, C., Schappacher, N., and Schwermer, J. (eds), *The Shaping of Arithmetic after C. F. Gauss's* Disquisitiones Arithmeticae. New York: Springer.

Nevai, P. 1990. *Orthogonal Polynomials: Theory and Practice*. New York: Kluwer.

Nevanlinna, R. 1974. *Le théorème de Picard–Borel et la théorie des fonctions méromorphes*. New York: Chelsea.

Newman, F. W. 1848. On $\Gamma(a)$, Especially When a Is Negative. *Cambridge and Dublin Math. J.*, **3**, 57–63.

Newton, I. 1959–1960. *The Correspondence of Isaac Newton*. Cambridge: Cambridge Univ. Press. Edited by H. W. Turnbull.

Newton, I. 1964–1967. *The Mathematical Works of Isaac Newton, Introduction By D. T. Whiteside*. New York: Johnson Reprint.

Newton, I. 1967–1981. *The Mathematical Papers of Isaac Newton*. Cambridge: Cambridge Univ. Press. Edited by D. T. Whiteside.

Nicole, F. 1717. Traité du calcul des différences finies. *Histoire de l'Academie Royale des Sciences*, 7–21.

Nikolić, A. 2009. The story of majorizability as Karamata's condition of convergence for Abel summable series. *Historia Math.*, **36**, 405–419.

Olver, P. J. 1999. *Classical Invariant Theory*. Cambridge: Cambridge Univ. Press.

Ore, Ø. 1974. *Niels Henrik Abel: Mathematician Extraordinary*. Providence: Amer. Math. Soc.

Ozhigova, E. P. 2007. The part played by the Petersburg Academy of Sciences (the Academy of Sciences of the USSR) in the publication of Euler's collected works. Pages 53–74 of: Bogolyubov, N. N., Mikhaĭlov, G. K., and Yushkevich, A. P. (eds), *Euler and Modern Science*. Washington, D.C.: Mathematical Association of America.

Parameswaran, S. 1983. Madhava of Sangamagramma. *J. Kerala Studies*, **10**, 185–217.

Patterson, S. J. 2007. Gauss sums. Pages 505–528 of: Goldstein, C., Schappacher, N., and Schwermer, J. (eds), *The Shaping of Arithmetic after C. F. Gauss's* Disquisitiones Arithmeticae. New York: Springer.

Peano, G. 1973. *Selected Works*. London: George Allen and Unwin. Edited by H. C. Kennedy.

Petrovski, I. G. 1966. *Ordinary Differential Equations*. Englewood Cliffs, N.J.: Prentice-Hall. Translated by R. A. Silverman.

Picard, É. 1879. Sur une propriété des fonctions entières. *Comptes Rendus*, **88**, 1024–1027.

Pick, G. 1915. Über eine Eigenschaft der konformen Abbildung kreisförmiger Bereiche. *Math. Ann.*, **77**, 1–6.

Pieper, H. 1998. *Korrespondenz zwischen Legendre und Jacobi*. Leipzig: Teubner.

Pieper, H. 2007. A network of scientific philanthropy: Humboldt's relations with number theorists. Pages 201–234 of: Goldstein, C., Schappacher, N., and Schwermer, J. (eds), *The Shaping of Arithmetic after C. F. Gauss's Disquisitiones Arithmeticae*. New York: Springer.

Pierpoint, W. S. 1997. Edward Stone (1702–1768) and Edmund Stone (1700–1768): Confused Identities Resolved. *Notes and Records Roy. Soc. London*, **51**, 211–217.

Pierpont, J. 2000. The history of mathematics in the nineteenth century. *Bull. Amer. Math. Soc.*, **37**, 9–24.

Pietsch, A. 2007. *History of Banach Spaces and Linear Operators*. Boston: Birkhäuser.

Pingree, D. 1970–1994. Census of the exact sciences in sanskrit. *Amer. Phil. Soc.*, **81, 86, 111, 146, 213**.

Plofker, K. 2009. *Mathematics in India*. Princeton, N.J.: Princeton Univ. Press.

Poincaré, H. 1886. Sur les intégrales irrégulières des équations linéaires. *Acta Math.*, 295–344.

Poincaré, H. 1985. *Papers on Fuchsian Functions*. New York: Springer-Verlag. Translated by John Stillwell.

Poisson, S. D. 1823. Suite du mémoire sur les intégrales définies et sur la sommation des séries. *J. de l'École Polytechnique*, **12**, 404–509.

Poisson, S. D. 1826. Sur le calcul numérique des intégrales définies. *Mémoire l'Académie des Sciences*, **6**, 571–602.

Polignac, A. de. 1857. Recherches sur les nombres premiers. *Compt. Rend.*, **45**, 575–580.

Pommerenke, C. 1985. The Bieberbach conjecture. *Math. Intelligencer*, **7**(2), 23–25; 32.

Prasad, G. 1931. *Six Lectures on the Mean-Value Theorem of the Differential Calculus*. Calcutta: Calcutta Univ. Press.

Prasad, G. 1933. *Some Great Mathematicians of the Nineteenth Century*. Benares, India: Benares Mathematical Society.

Pringsheim, A. 1900. Zur Geschichte des Taylorschen Lehrsatzes. *Bibliotheca Mathematica*, **3**, 433–479.

Rajagopal, C. T. 1949. A neglected chapter of Hindu mathematics. *Scripta Math.*, **15**, 201–209.

Rajagopal, C. T., and Rangachari, M. S. 1977. On the untapped source of medieval Keralese mathematics. *Arch. Hist. Exact Sci.*, **18**, 89–102.

Rajagopal, C. T., and Rangachari, M. S. 1986. On medieval Keralese mathematics. *Arch. Hist. Exact Sci.*, **35**, 91–99.

Rajagopal, C. T., and Vedamurtha Aiyar, T. V. 1951. On the Hindu proof of Gregory's series. *Scripta Math.*, **17**, 65–74.

Rajagopal, C. T., and Venkataraman, A. 1949. The sine and cosine power series in Hindu mathematics. *J. Roy. Asiatic Soc. Bengal, Sci.*, **15**, 1–13.

Ramanujan, S. 1988. *The Lost Notebook and Other Unpublished Papers*. Delhi: Narosa Publishing House.

Ramanujan, S. 2000. *Collected Papers*. Providence: AMS Chelsea.

Raussen, M., and Skau, C. 2010. Interview with Mikhail Gromov. *Notices Amer. Math. Soc.*, **57**, 391–403.

Remmert, R. 1991. *Theory of Complex Functions*. New York: Springer-Verlag. Translated by Robert Burckel.

Remmert, R. 1996. Wielandt's theorem about the Γ-function. *Amer. Math. Monthly*, **103**, 214–220.

Remmert, R. 1998. *Classical Topics in Complex Function Theory*. New York: Springer-Verlag. Translated by Leslie Kay.

Riemann, B. 1990. *Gesammelte Mathematische Werke*. New York: Springer-Verlag. Edited by R. Dedekind, H. Weber, R. Narasimham, and E. Neuenschwander.

Riesz, F. 1913. *Les systèmes d'équations linéaires a une infinité d'inconnues*. Paris: Gauthier-Villars.

Riesz, F. 1960. *Oeuvres complètes*. Budapest: Académie des Sciences de Hongrie.

Riesz, M. 1928. Sur les fonctions conjugées. *Math. Z.*, **27**, 218–44.

Rigaud, S. P. 1841. *Correspondence of Scientific Men of the Seventeenth Century*. Oxford: Oxford Univ. Press.

Rodrigues, O. 1816. Mémoire sur l'attraction des sphéroids. *Correspondance sur l'École Polytechnique*, **3**, 361–385.

Rodrigues, O. 1839. Note sur les inversions, ou dérangements produits dans les permutations. *J. Math. Pures Appl.*, **4**, 236–240.

Rogers, L. J. 1893a. Note on the transformation of an Heinean series. *Messenger of Math.*, **23**, 28–31.

Rogers, L. J. 1893b. On a three-fold symmetry in the element's of Heine's series. *Proc. London Math. Soc.*, **24**, 171–179.

Rogers, L. J. 1893c. On the expansion of some infinite products. *Proc. London Math. Soc.*, **24**, 337–352.

Rogers, L. J. 1894. Second memoir on the expansion of certain infinite products. *Proc. London Math. Soc.*, **25**, 318–343.

Rogers, L. J. 1895. Third memoir on the expansion of certain infinite products. *Proc. London Math. Soc.*, **26**, 15–32.

Rogers, L. J. 1907. On function sum theorems connected with the series $\sum_{n=1}^{\infty} \frac{x^n}{n^2}$. *Proc. London Math. Soc.*, **4**, 169–189.

Rogers, L. J. 1917. On two theorems of combinatory analysis and some allied identities. *Proc. London Math. Soc.*, **16**, 321–327.

Roquette, P. 2002. The Riemann hypothesis in characteristic p, its origin and development. Part I. The formation of the zeta-functions of Artin and of F. K. Schmidt. *Mitt. Math. Ges. Hamburg*, **21**, 79–157.

Roquette, P. 2004. The Riemann hypothesis in characteristic p, its origin and development. Part II. The first steps by Davenport and Hasse. *Mitt. Math. Ges. Hamburg*, **23**, 5–74.

Rosen, M. 2002. *Number Theory in Function Fields*. New York: Springer.

Rothe, H. A. 1811. *Systematisches Lehrbuch der Arithmetik*. Erlangen: Barth.

Rowe, D. E., and McCleary, J. 1989. *The History of Modern Mathematics*. Boston: Academic Press.

Roy, R. 1990. The discovery of the series formula for π by Leibniz, Gregory and Nilakantha. *Math. Mag.*, **63**(5), 291–306. Reprinted in Anderson, Katz, and Wilson (2004), pp. 111–121.

Ru, M. 2001. *Nevanlinna Theory and its Relation to Diophantine Approximation*. Singapore: World Scientific.

Sandifer, C. E. 2007. *The Early Mathematics of Leonhard Euler*. Washington, D. C.: Math. Assoc. Amer.

Sarma, K. V. 1972. *A History Of The Kerala School Of Hindu Astronomy*. Hoshiarpur, India: Punjab Univ.

Sarma, K. V. 1977. *Tantrasangraha of Nilakantha Somayaji*. Hoshiarpur, India: Panjab Univ.

Sarma, K. V. 2008. *Ganita-Yukti-Bhasa of Jyesthadeva*. Delhi: Hindustan Book Agency.

Sarma, K. V., and Hariharan, S. 1991. Yuktibhasa of Jyesthadeva. *Indian J. Hist. Sci.*, **26**(2), 185–207.

Sasaki, C. 1994. The adoption of Western mathematics in Meiji Japan, 1853–1903. Pages 165–186 of: Sasaki, C., Sugiura, M., and Dauben, J. W. (eds), *The Intersection of History and Mathematics*. Basel: Birkhäuser.

Sasaki, C., Sugiura, M., and Dauben, J. W. (eds). 1994. *The Intersection of History and Mathematics*. Basel: Birkhäuser.

Schellbach, K. 1854. Die einfachsten periodischen Functionen. *J. Reine Angew. Math.*, **48**, 207–236.

Schlömilch, O. 1843. Eineges über die Eulerischen Integrale der zweiten Art. *Archiv Math. Phys.*, **4**, 167–174.

Schlömilch, O. 1847. *Handbuch der Differenzial- und Integralrechnung*. Greifswald, Germany: Otte.

Schneider, I. 1968. Der Mathematiker Abraham de Moivre (1667–1754). *Arch. Hist. Exact Sci*, **5**, 177–317.

Schneider, I. 1983. Potenzsummenformeln im 17. Jahrhundert. *Historia Math.*, **10**, 286–296.

Schoenberg, I. J. 1988. *Selected Papers*. Boston: Birkhäuser.

Schwarz, H. A. 1893. *Formeln und Lehrsätze zum Gebrauche der elliptischen Funktionen*. Berlin: Springer.

Schwarz, H. A. 1972. *Abhandlungen*. New York: Chelsea.
Scriba, C. J. 1964. The inverse method of tangents. *Arch. Hist. Exact Sci.*, **2**, 113–137.
Segal, S. L. 1978. Riemann's example of a continuous "nondifferentiable" function. *Math. Intelligencer*, **1**, 81–82.
Selberg, A. 1989. *Collected Papers*. New York: Springer-Verlag.
Sen Gupta, D. P. 2000. Sir Asutosh Mookerjee – educationist, leader and insitution-builder. *Current Sci.*, **78**, 1566–1573.
Shah, S. M. 1948. A note on uniqueness sets for entire functions. *Proc. Indian Acad. Sci., Sect. A*, **28**, 519–526.
Shidlovskii, A. B. 1989. *Transcendental Numbers*. Berlin: Walter de Gruyter. Translated by N. Koblitz.
Shimura, G. 2007. *Elementary Dirichlet Series and Modular Forms*. New York: Springer.
Shimura, G. 2008. *The Map of My Life*. New York: Springer.
Siegel, C. L. 1949. *Transcendental Numbers*. Princeton, N.J.: Princeton Univ. Press.
Siegel, C. L. 1969. *Topics In Complex Function Theory*. New York: Wiley.
Simmons, G. F. 1992. *Calculus Gems*. New York: McGraw-Hill.
Simon, B. 2005. OPUC on One Foot. *Bull. Amer. Math. Soc.*, **42**, 431–460.
Simpson, T. 1759. The invention of a general method for determining the sum of every second, third, fourth, or fifth, etc. term of a series, taken in order; the sum of the whole being known. *Phil. Trans. Roy. Soc.*, **50**, 757–769.
Simpson, T. 1800. *A Treatise of Algebra*. London: Wingrave.
Smith, D. E. 1959. *A Source Book in Mathematics*. New York: Dover.
Smith, D. E., and Mikami, Y. 1914. *A History of Japanese Mathematics*. Chicago: Open Court.
Smith, H. J. S. 1965a. *Collected Mathematical Papers*. New York: Chelsea. Edited by J. W. L. Glaisher.
Smith, H. J. S. 1965b. *Report on the Theory of Numbers*. New York: Chelsea.
Spence, W. 1819. *Mathematical Essays*. London: Whittaker. Edited by J. F. W. Herschel.
Srinivasiengar, C. N. 1967. *The History of Ancient Indian Mathematics*. Calcutta: World Press.
Stäckel, P., and Ahrens, W. 1908. *Briefwechsel zwischen C. G. J. Jacobi und P. H. Fuss*. Leipzig: Teubner.
Stedall, J. A. 2000. Catching proteus: The collaborations of Wallis and Brouncker, I and II. *Notes and Records Roy. Soc. London*, **54**, 293–331.
Stedall, J. A. 2003. *The Greate Invention of Algebra: Thomas Harriot's* Treatise on Equations. Oxford: Oxford Univ. Press.
Steele, J. M. 2004. *The Cauchy-Schwarz Master Class*. Cambridge: Cambridge Univ. Press.
Steffens, K.-G. 2006. *The History of Approximation Theory*. Boston: Birkhäuser.
Stieltjes, T. J. 1886. Recherches sur quelques séries semi-convergentes. *Ann. Sci. Éc. Norm.*, **3**, 201–258.
Stieltjes, T. J. 1993. *Collected Papers*. New York: Springer-Verlag.
Stirling, J. 1730. *Methodus differentialis*. London: Strahan.
Stirling, J. 2003. *Methodus Differentialis*. London: Springer. Translated by I. Tweddle.
Stone, E. 1730. *The Method of Fluxions both Direct and Inverse*. London: W. Innys.
Strichartz, R. S. 1995. *The Way of Analysis*. London: Jones and Bartlett.
Struik, D. J. 1969. *A Source Book in Mathematics*. Cambridge, Mass.: Harvard Univ. Press.
Stubhaug, A. 2000. *Niels Henrik Abel and His Times*. New York: Springer.
Sturm, C. 1829. Analyse d'un mémoire sur la résolution des équations numériques. *Bulletin des Sciences de Férussac*, **11**, 419.
Sturmfels, B. 2008. *Algorithms on Invariant Theory*. Wien: Springer.
Sylvester, J. J. 1973. *Mathematical Papers*. New York: Chelsea.
Szegő, G. 1982. *The Collected Papers of Gabor Szegő*. Boston: Birkhäuser. Edited by R. Askey.
Szekeres, G. 1968. A combinatorial interpretation of Ramanujan's continued fraction. *Canadian Math. Bull.*, **11**, 405–408.
Takagi, T. 1990. *Collected Papers*. Tokyo: Springer-Verlag.

Takase, M. 1994. Three aspects of the theory of complex multiplication. Pages 91–108 of: Sasaki, C., Sugiura, M., and Dauben, J. W. (eds), *The Intersection of History and Mathematics*. Basel: Birkhäuser.

Thomson, W., and Tait, P. G. 1890. *Treatise on Natural Philosophy*. Cambridge: Cambridge Univ. Press.

Tignol, J.-P. 1988. *Galois' Theory of Algebraic Equations*. New York: Wiley.

Titchmarsh, E. C., and Heath-Brown, D. R. 1986. *The Theory of the Riemann-Zeta Function*. Oxford: Oxford Univ. Press.

Truesdell, C. 1960. *The Rational Mechanics of Flexible or Elastic Bodies, 1638–1788. Introduction to Vols. 10 and 11, Second Series of Euler's* Opera omnia. Zurich: Orell Füssli Turici.

Truesdell, C. 1984. *An Idiot's Fugitive Essays on Science*. New York: Springer-Verlag.

Tucciarone, J. 1973. The development of the theory of summable divergent series from 1880 to 1925. *Arch. Hist. Exact Sci.*, **10**, 1–40.

Turán, P. 1990. *Collected Papers*. Budapest: Akadémiai Kiadó.

Turnbull, H. W. 1933. James Gregory: A study in the early history of interpolation. *Proc. Edinburgh Math. Soc.*, **3**, 151–172.

Turnbull, H. W. 1939. *James Gregory Tercentenary Memorial Volume*. London: Bell.

Tweddle, I. 1984. Approximating $n!$ Historical origins and error analysis. *Amer. J. Phys.*, **52**, 487–488.

Tweddle, I. 1988. *James Stirling: "This About Series And Such Things"*. Edinburgh: Scottish Academic Press.

Tweddle, I. 2003. *James Stirling's* Methodus differentialis, *An Annotated Translation of Stirling's Text*. London: Springer.

Tweedie, C. 1917–1918. Nicole's contributions to the foundations of the calculus of finite differences. *Proc. Edinburgh Math. Soc.*, **36**, 22–39.

Tweedie, C. 1922. *James Stirling: A Sketch of His Life and Works along with His Scientific Correspondence*. Oxford: Oxford Univ. Press.

Valiron, G. 1949. *Lectures on the General Theory of Integral Functions*. New York: Chelsea.

Van Brummelen, G. 2009. *The Mathematics of the Heavens and the Earth*. Princeton: Princeton Univ. Press.

Van Brummelen, G., and Kinyon, M. 2005. *Mathematics and the Historian's Craft*. New York: Springer.

Van Maanen, J. A. 1984. Hendrick van Heuraet (1634–1660?): His life and work. *Centaurus*, **27**, 218–279.

Varadarajan, V. S. 2006. *Euler Through Time*. Providence: Amer. Math. Soc.

Viète, F. 1983. *The Analytic Art*. Kent, Ohio: Kent State Univ. Press.

Vlăduţ, S. G. 1991. *Kronecker's Jugendtraum and Modular Equations*. Basel: Birkhäuser.

Wali, K. C. 1991. *Chandra: A Biography of S. Chandrasekhar*. Chicago: Univ. Chicago Press.

Wallis, J. 2004. *The Arithmetic of Infinitesimals*. New York: Springer. Translation of *Arithmetica Infinitorum* by J. A. Stedall.

Waring, E. 1779. Problems concerning interpolations. *Phil. Trans. Roy. Soc. London*, **69**, 59–67.

Waring, E. 1991. *Meditationes Algebraicae*. Providence: Amer. Math. Soc. Translated by D. Weeks.

Watson, G. N. 1933. The marquis and the land-agent. *Math. Gazette*, **17**, 5–17.

Weber, H. 1895. *Lehrbuch der Algebra*. Braunschweig: Vieweg.

Weierstrass, K. 1894–1927. *Mathematische Werke*. Berlin: Mayer and Müller.

Weil, A. 1979. *Collected Papers*. New York: Springer-Verlag.

Weil, A. 1984. *Number Theory: An Approach Through History from Hammurapi to Legendre*. Boston: Birkhäuser.

Weil, A. 1989a. On Eisenstein's copy of the *Disquisitiones*. *Adv. Studies Pure Math.*, **17**, 463–469.

Weil, A. 1989b. Prehistory of the zeta function. Pages 1–9 of: Aubert, K. E., Bombieri, E., and Goldfeld, D. (eds), *Number Theory, Trace Formulas, and Discrete Groups*. Boston: Academic Press.

Weil, A. 1992. *The Apprenticeship of a Mathematician*. Boston: Birkhäuser.

Whiteside, D. T. 1961. Patterns of mathematical thought in the later seventeenth century. *Arch. Hist. Exact Sci.*, **1**, 179–388.

Whittaker, E. T., and Watson, G. N. 1927. *A Course of Modern Analysis*. Cambridge: Cambridge Univ. Press.

Wiener, N. 1958. *The Fourier Integral and Certain of Its Applications*. New York: Dover.

Wiener, N. 1979. *Collected Works*. Cambridge: MIT Press.

Wilbraham, H. 1848. On a certain periodic function. *Cambridge and Dublin Math. J.*, **3**, 198–201.

Wilf, H. S. 2001. The number-theoretic content of the Jacobi triple product identity. Pages 227–230 of: Foata, D., and Han, G.-N. (eds), *The Andrews Festschrift: Seventeen Papers on Classical Number Theory and Combinatorics*. New York: Springer.

Woodhouse, R. 1803. *The Principles of Analytical Calculation*. Cambridge: Cambridge Univ. Press.

Yandell, B. H. 2002. *The Honors Class*. Natick, Mass.: A K Peters.

Young, G. C., and Young, W. H. 1909. On derivatives and the theorem of the mean. *Quart. J. Pure Appl. Math.*, **40**, 1–26.

Young, G. C., and Young, W. H. 2000. *Selected Papers*. Lausanne, Switzerland: Presses Polytechniques. Edited by S. D. Chatterji and H. Wefelscheid.

Yushkevich, A . P. 1964. *Geschichte der Mathematik in Mittelalter*. Leipzig: Teubner.

Yushkevich, A. P. 1971. The concept of function up to the middle of the 19th century. *Arch. Hist. Exact Sci.*, **16**, 37–85.

Zdravkovska, S., and Duren, P. 1993. *Golden Years of Moscow Mathematics*. Providence: Amer. Math. Soc.

Zolotarev, E. 1876. Sur la série de Lagrange. *Nouvelles Annales*, **15**, 422–423.

Index

Abel, Niels Henrik, 57, 58, 60, 61, 63–65, 67, 69, 70, 91, 182, 195, 291, 292, 302, 305–307, 310, 316, 317, 325, 371, 417, 447, 461, 499, 503, 508, 509, 514, 519, 602, 691, 692, 703, 704, 750–753, 755, 777, 780, 801, 803–805, 816–827, 829–832, 837, 844, 852, 856, 863–866, 887, 891
Abhyankar, Shreeram, 146, 729
absolutely convergent Fourier series, 759, 772, 773
Academia Algebrae, 19, 26
Acosta, D. J., 96
Acta Eruditorum, 122, 126, 197, 219, 240, 243, 271, 605
Adams, John Couch, 629
addition formula for elliptic functions, 803, 804, 817, 839, 865
addition formula for elliptic integrals, 781, 814
Aepinus, Franz, 59
Ahlfors, Lars, 553, 892, 912, 928
Ahlgren, Scott, 635, 652
Aiyar, T. V. V., 14
Akhiezer, N. I., 590, 597
Alder, H. L., 632, 652
d'Alembert, Jean, 56, 203, 212, 214, 215, 221, 246, 254, 267, 313, 348, 354, 355, 365, 369, 370, 401, 402, 410, 426
Alexander, J. W., 910, 927
algebraic analysis, 69, 205, 290, 367, 373
algebraically independent numbers, 876, 877, 885
Almagest, 13, 158
Almkvist, Gert, 814
Altmann, Simon, 597, 626
Analyse algébrique, 56, 67–69, 154, 166, 175, 221, 295, 308, 775
analytic continuation, 753, 801
analytical engine, 729
Analytical Society, 291, 370, 398
Anderson, Alexander, 151
Anderson, Marlow, 15, 719

Andrews, George, 345, 475, 608, 626, 634, 637, 643, 650, 651, 678, 679
Andrews, George, 548
Angeli, Stephano degli, 99, 264
Apollonius, 71
Appell, Paul, 405, 890
approximate quadrature, 11, 161, 168, 175, 225, 578, 582
Arbogast, Louis, 368–370, 374–376, 379, 385, 391, 398, 663
Archimedean spiral, 78
Archimedes, 6, 16, 25, 99
Archimedes's formula, 6, 13, 16
area inequalities, 910, 911, 914
arithmetic-geometric mean, 794, 798, 799, 801, 809, 810, 814–816
Arithmetica Infinitorum, 28, 29, 50, 52, 120, 148, 446
Arithmetica Logarithmica, 159
Arithmetica Universalis, 83, 94, 96, 328
Arithmeticorum Libri Duo, 17
Arnold, Vladimir, 119, 754, 777
Aronhold, Siegfried, 725, 726, 747
Ars Conjectandi, 18, 19, 26, 347, 605
Artin, Emil, 450, 462–464, 471, 475, 931
Artis Analyticae Praxis, 82
Artis Magnae, 311
Artmann, B., 535
Aryabhatyabhasya, 2, 3, 867
Ash, J. Marshall, 443
Ashworth, Margaret, 635
Askey, Richard, 390, 473, 475, 548, 597, 619, 659, 676, 679, 926–928
Askey–Gasper inequality, 926
Atkin, A. O. L., 634, 635, 650, 652
Aubrey, John, 31

Babbage, Charles, 368, 370, 371, 380, 398, 399, 729
Bäcklund, R.J., 895, 896, 898, 906

Baernstein, Albert, 927, 928
Bag, Amulya Kumar, 12, 15
Bailey, D., 546
Bailey, W. N., 678, 679
Baillaud, B., 49
Baker, Alan, 877, 885, 886
Baker, H. F., 649
Banach, Stefan, 773
Banach algebras, 759, 773, 774
Barnes, E. W., 473, 553, 554, 649
Barnes's integral, 473, 554
Baron, Margaret, 119
Barrow, Isaac, 97–100, 104, 106–115, 118, 119, 121, 141, 147, 156, 238, 260, 261
Bartels, Johann, 801, 802
Basarab–Horwath, P., 906
Bateman, Harry, 199, 220
Bateman, P. T., 719
Baxter, R. J., 658
Bazilevich, I. E., 913
de Beaune, Florimond, 127, 138, 202, 210, 260, 344
Beery, Janet, 175
Bell, Eric Temple, 723
Berggren, Lennart, 814
Berkeley, George, 80
Berndt, Bruce, 571, 625, 652, 814, 866
Bernoulli, Daniel, 18, 180, 232, 246, 252–254, 259, 262, 263, 265, 266, 285, 286, 288, 290, 401–403, 426, 445, 474, 494, 499, 507, 517, 577, 781, 868, 869, 886
Bernoulli, Jakob, 18–26, 97, 122, 178, 190, 193, 197, 210, 227, 262, 264, 265, 271, 278, 287, 330, 477, 481, 495, 497, 505, 515, 516, 601, 602, 605, 623, 625, 778, 779, 784, 786, 787
Bernoulli, Johann, 18, 19, 38, 122, 123, 131–133, 136–139, 152, 157, 178, 202, 210–212, 219–222, 224, 226, 227, 238, 240, 243, 261–264, 271–273, 287, 289, 291, 304, 330, 346, 347, 367, 368, 373, 385, 403, 445, 447, 453, 515, 523, 525–527, 570, 628, 778, 779, 781
Bernoulli, Johann II, 18
Bernoulli, Niklaus, 19
Bernoulli, Niklaus I, 18, 19, 179, 186, 245, 246, 253, 265, 290, 298–300, 310, 330, 346, 347, 366, 477, 517, 527, 528, 545
Bernoulli, Niklaus II, 18, 265, 445
Bernoulli numbers, 18, 22–24, 26, 304, 337, 479, 485, 494, 495, 499, 505, 509, 511, 513, 518, 519, 522, 523, 544, 774
Bernoulli polynomials, 20, 23, 24, 403, 494, 498, 499, 505, 507, 508, 517
Bers, Lipman, 928
Bertrand, Joseph, 348, 697, 705, 708, 717
Bertrand's conjecture, 697, 705, 707, 717
Bessel, F. Wilhelm, 160, 286, 361, 450, 452, 453, 480, 550, 573, 696, 719, 801, 874–876

beta integral, 41, 42, 101, 119, 222, 233, 234, 237, 324, 349, 453, 460, 461, 473, 475, 551, 589, 600, 655, 784
  multidimensional, 473, 475
Betti, Enrico, 837
Beukers, Frits, 876, 886
Bézout, Étienne, 313, 374, 727, 728
Bhaskara, 4, 9, 10, 13, 36, 681
Bieberbach, Ludwig, 907, 910, 911, 916, 917, 919, 921–923, 928
Bieberbach conjecture, 907, 910–913, 916, 917, 922, 923, 926, 928
Binet, Jacques, 480, 486–491, 493
Binet's integral formulas, 480, 486, 490
binomial coefficients, 21, 53, 159, 556, 611
  additive rule, 51
  multiplicative rule, 51
binomial theorem, 12, 51, 52, 54–58, 60–62, 65, 66, 68–70, 101, 140, 142, 147, 148, 154, 159, 176, 184, 203, 210, 248, 249, 289, 292, 293, 372, 374, 384, 386, 604, 618, 661, 779, 799, 801, 932, 938, 941
  non-commutative, 383
biquadratic reciprocity, 854
al-Biruni, 158
Bissell, Christopher, 79
Blanton, J. D., 199
Bloch, André, 892
Boas, Ralph, 905
Bodenhausen, Rudolf von, 123, 129, 139
Boehle, K., 872
Bohr, Harald, 33, 37, 81, 444, 450, 462–464, 475, 754
Bohr, Niels, 756
Boilly, Julien, 817
du Bois-Reymond, Paul, 440
Bolibruch, A. A., 886
Bollobás, B., 906
Bolyai, János, 806
Bolyai, Wolfgang, 806
Bolza, Oskar, 366
Bolzano, Bernard, 56, 204–206, 218, 219, 348, 440
Bombelli, Rafael, 33, 34
Bombieri, Enrico, 906
Bond, Henry, 104
Bonnet, Pierre Ossian, 205
Book on Polygonal Numbers, 17
Boole, George, 283, 284, 287, 369, 371–373, 387–390, 398, 399, 512, 663, 720–723, 727, 729–733, 746, 747
Boole summation formula, 512
Borchardt, Carl, 820, 858, 859
Borel, Émile, 362, 658, 775, 890–892, 898, 900, 902–904, 908, 909
Boros, George, 243, 244
Borwein, Jonathan, 546, 814
Borwein, Peter, 814
Bos, Henk, 139
Bose, S. N., 284

Bottazzini, Umberto, 259, 288, 426
Bouquet, Jean Claude, 268, 820
Bourbaki, Nicolas, 33, 50, 139, 350, 351, 365, 911
Bourget, H., 49
Boutroux, Pierre, 909
Boyer, Carl, 13, 25
Boylan, Matthew, 635
brachistochrone, 132, 133, 138, 139, 347, 351
Bradley, Robert E., 69, 398, 775
Brahmagupta, 13, 158, 681
de Branges, Louis, 907, 910, 912, 914, 923–926, 928
Brashman, N. D., 590
Bressoud, David, 15, 27, 221, 443, 573, 632
Briggs, Henry, 18, 51, 53, 159, 160
Bringmann, Kathrin, 637
Brinkley, John, 378, 379, 399
Brioschi, Francesco, 837
Briot, C., 268, 820
Brisson, Barnabé, 368
Bromhead, Edward, 350, 370
Bronstein, Manuel, 228, 244
Bronwin, Brice, 162, 171, 172
Brouncker, William, 14, 31–34, 38, 40, 43, 50, 141, 681, 868
Brownwell, W. D., 876
Brun, Viggo, 69
Buchan, John, 637
Buchler, Justus, 630
Budan, François, 93, 94
Bullialdus, Ismael, 22
Bunyakovski, Viktor, 86, 697
Burckel, Robert, 573
Burn, R. P., 288
Burns, Robert, 719
Burnside, William, 649
Burnside, William Snow, 94
Byron, Lord George Gordon, 729

Cahen, Eugène, 554, 700, 718, 776
Cambridge and Dublin Mathematical Journal, 369, 372
Cambridge Mathematical Journal, 372
Campbell, George, 84, 92, 96
Campbell, Paul J., 287
Cannell, D. Mary, 350
Cannon, John T., 286
Cantor, Georg, 68, 206, 428, 429, 431, 432, 436–439, 443, 887
Cantor set, 431, 432
Cantor's uniqueness theorem, 428, 437, 439, 474
Carathéodory, Constantin, 905, 906, 908–910, 920, 921, 928
Cardano, Girolamo, 51, 311–313, 315, 325, 326, 733
Carleson, Lennart, 442, 443
Carlitz, Leonard, 931
Carlson, Fritz, 571
Carson, J. R., 373

Cartan, Élie, 350
Cartan, Henri, 350, 351, 900, 906
Cartier, Pierre, 392
Cartwright, Mary L., 754
Casorati, Felice, 14, 891
Casorati–Weierstrass theorem, 891, 906
Castillione, Johan, 156
Catalan, Eugène, 751
Cataldi, Pietro Antonio, 33, 34
catenary, 122, 123, 129–131, 139, 269
Cauchy, Augustin–Louis, 56, 57, 60–64, 67–69, 85, 86, 89, 96, 152, 154, 161, 163, 166, 167, 174, 175, 203–206, 213, 216–221, 268, 290, 295–298, 308, 329, 335–338, 345, 346, 348–350, 352–354, 361, 365, 369, 371, 417, 418, 427, 430, 441, 449, 461, 462, 467, 479, 480, 488, 489, 491, 498, 501, 509, 519, 538, 551, 553, 586, 604, 618, 621, 625, 653, 720, 751, 775, 813, 819–821, 849–851, 872, 888, 898, 909, 918, 919, 921
Cauchy product, 61, 63, 67, 751, 752
Cauchy's integral formula, 361, 873, 895, 896
Cavalieri, Bonaventura, 29, 99, 168
Cayley, Arthur, 284, 371, 373, 395–397, 577, 606, 608, 617, 629, 642–645, 652, 663, 720–726, 728, 729, 732–740, 742, 743, 745, 747, 748, 791, 792, 795, 814, 880
Cesàro, Ernesto, 751, 752, 754, 776
Chabert, Jean–Luc, 175
Chandrasekhar, S., 99, 119
Charzynski, Z., 912
Chebyshev, Pafnuty, 50, 162, 163, 172–175, 578, 590–598, 692, 697, 698, 700–712, 714, 715, 717, 719
Chen Jing–Run, 927
Cheney, E. W., 679
Chern, S. S., 892
Cherry, William, 906
Child, J. M., 79, 114–119
Chowla, Sarvadaman, 633
Christoffel, Elwin B., 908
Chu, Shih–Chieh (also Zhu Shijie), 184, 199, 601, 619, 728
Chu–Vandermonde identity, 154, 620
  q-extension, 620, 653
Chudnovsky, D. V., 886
Chudnovsky, G. V., 886
Clairaut, Alexis–Claude, 246, 267, 268, 346, 351, 352, 354–356, 365, 366, 369, 403
Clarke, Frances M., 259
Clausen, Thomas, 520, 552, 570, 572, 926
Clavis Mathematicae, 28
Clavius, Christoph, 71
Clebsch, R. Alfred, 726
Clunie, J., 906
Cohen, Henri, 514
Collins, John, 31, 54, 55, 69, 98, 100, 102, 104, 119, 141, 142, 156, 159–161, 175–177, 201, 202, 207, 208, 243, 328, 377, 496, 498, 509

Colson, John, 79, 156
Commercium Epistolicum, 156, 208, 243
complete induction, 17, 51
complex multiplication of elliptic functions, 801, 803–805, 818, 831, 864, 873
Comptes Rendus, 488, 658, 710, 869, 886
condensation test, 67
Condorcet, Marie-Jean, 24, 786
continued fraction
  periodic, 937
  Rogers–Ramanujan, 650, 660
continued fractions
  approximants, 10, 15
  convergents, 3, 12, 32, 35, 38
  divergent series, 186
  hypergeometric series, 560, 571
  infinite, 31, 33, 34
  infinite products, 41
  integrals, 43
  Riccati's equation, 32, 46
  series, 38, 602
continuous nowhere differentiable function, 205, 440
Cooke, Roger, 443, 777
Cooper, Shaun, 625
Corry, Leo, 748
Cotes, Roger, 32, 46, 160–162, 170, 224–226, 230–234, 238, 240, 243, 290, 293–295, 298, 300, 369, 377, 384, 448, 575, 576, 579, 582
covariant, 724–727, 734–736, 745, 747
Cox, David, 814, 865
Craig, John, 128
Craik, Alex D., 397, 597
Cramer, Gabriel, 340, 374, 405, 413, 829
Crelle, August Leopold, 57, 69, 292, 350, 373, 421, 801, 827, 865
Crelle's Journal, 57, 69, 93, 292, 417, 604, 618, 686, 791, 817, 820, 821, 849
Crilly, Tony, 748
cubic transformations of elliptic integrals, 818, 834, 837, 839
Cuming, Alexander, 477, 478

Damodara, 1
Dangerfield, Rodney, 574
Darboux, Gaston, 440, 592, 751
Dary, Michael, 54
Dauben, Joseph W., 443
Davenport, Harold, 694, 776, 877, 927
Davis, P. J., 475
De Analysi per Aequationes Infinitas, 121, 140–142, 144, 148, 156
De Computo Serierum, 142, 177, 199, 526
De Methodis Serierum et Fluxionum, 121, 123, 141, 142, 146, 156
De Numeris Triangularibus, 158
De Numerosa Potestatum Resolutione, 87

De Quadratura Curvarum, 142, 201, 203, 261, 262, 270
Dedekind, Richard, 68, 206, 219, 221, 298, 425, 426, 428, 429, 439, 443, 450, 461, 462, 471, 475, 688, 701, 743, 748, 751, 930, 931, 939–942
Degen, Ferdinand, 817
Deligne, Pierre, 635, 931, 941, 942
Delone, B. N., 719
Descartes, René, 29, 71–76, 79, 81–83, 87, 88, 92–94, 97, 98, 101, 120, 127, 135, 138, 260, 313, 318, 319, 325, 327, 328, 338, 887, 889
Descartes's rule of signs, 82, 92, 318, 325, 889
Despeaux, S. E., 399
Dhyanagrahopadesadhyaya, 158
Diamond, H. G., 719
Dieudonné, Jean, 86, 96, 911
difference equations, 245, 250, 252, 256, 259, 285, 380, 381, 387, 398, 403, 425, 483, 484, 675, 676
  linear, 245, 246
  linear with variable coefficients, 246
  non-homogeneous, 246
  operational method, 368
differential equations, 260–262, 283, 287, 290, 379, 388, 523, 548, 584, 736, 817, 924, 928
  adjoint, 266, 276, 287
  algebraic, 745
  algebraic coefficients, 553
  asymptotic series, 480
  definite integral, 286, 565
  hypergeometric, 569, 587, 588
  infinite order, 368, 406
  linear, 262, 263, 272, 278, 569, 874, 876
  non-homogeneous, 254, 256, 257, 266, 274, 277
  variable coefficients, 389
digamma function, 551, 885
Dini, Ulisse, 205
Diophantine approximation, 892
Diophantine equations, 4, 238, 727, 781, 784, 786, 814, 906
Diophantus, 17, 71
Dirac, Paul, 373
Dirichlet, Peter Lejeune, 57, 60, 65, 68, 371, 373, 404, 409, 410, 417, 418, 420–427, 449, 450, 460, 461, 466, 469, 470, 474, 475, 521, 525, 532–536, 540–542, 545, 546, 552, 554, 618, 619, 653, 680–689, 694, 698, 711, 714, 715, 726, 751, 753, 755, 761, 762, 766, 820, 852, 856, 908, 931, 941
Dirichlet's divisor theorem, 767
Dirichlet–Jordan theorem, 762
Dirksen, Enno, 653
Disquisitiones Arithmeticae, 320, 321, 323–325, 520, 535, 540, 541, 602, 681, 685, 687, 688, 726, 797, 818, 832, 854, 929, 930, 932, 933, 941, 942
Divergent Series, 371

divergent series, 58, 152, 177, 178, 186, 296, 371, 402, 457, 468, 469, 479, 480, 483, 488, 499, 507, 509, 513, 515, 518, 519, 531, 749–751
  asymptotic, 497, 749
Dixon, A. L., 656, 658
Doctrine of Chances, 246, 247, 249, 477, 485
Doetsch, Gustav, 373
Domingues, J. C., 27
Dong Youcheng, 557
Donkin, William F., 369
Dörrie, Heinrich, 865
Dostrovsky, Sigalia, 286
Doubilet, Peter, 331
double gamma function, 554
double zeta values, 528, 532, 545
Doyle, Arthur Conan, 636
Dunham, William, 325
Dunnington, Guy Waldo, 325, 626, 815
Duoji Bilei, 557
Dupré, Athanase, 681
Duren, Peter, 886, 920, 928
Durfee, William, 629, 640–642
Dutka, J., 475, 573
Dyson, Freeman J., 474, 634

Edwards, A. W. F., 21, 26, 27
Edwards, Harold, 198, 325, 546, 716, 719
Edwards, Joseph, 154, 284, 287
Egorov, Dimitri, 442
Einstein, Albert, 284, 756
Eisenstein, F. Gotthold, 309, 324, 520, 535, 536, 543, 546, 604, 618, 619, 626, 653, 726, 729, 820, 852–857, 865
Eisenstein's criterion, 857, 865, 866
elimination theory, 329, 727–729
Elliott, Edwin B., 745, 747, 748
elliptic functions, 284, 602, 625, 626, 662, 778, 791, 800, 801, 804, 815–817, 819–821, 826, 832, 839, 844, 847, 852, 858, 859, 865, 866, 940
elliptic hypergeometric functions, 656
elliptic integral, complete, 814, 832
Ellis, David, 693
Ellis, R. Leslie, 371, 372, 399
Ellis, Robert, 693
Elstrodt, J., 426
Emelyanov, E.G., 914
Encke, J. F., 695–697, 719
Encyclopedia Britannica, 341, 345
Engelsman, Steven, 365
Enros, P. C., 398
entire functions, 450, 473, 554, 700, 701, 805, 890–892, 895, 900, 906
  finite order, 700, 754, 891, 903
  infinite product, 889
  Picard's theorem, 754
  transcendental numbers, 874
  zeta function, 890

Erdős, Paul, 650, 701, 708, 717, 753
Erman, Adolph, 719
Euclid, 81, 88, 120, 163, 321, 557
Eudoxus, 204, 867
Euler, Leonhard, 23, 24, 28, 32, 38, 40–51, 54, 56, 58–61, 69, 86, 98, 100, 104, 139, 145, 151, 152, 157, 180, 181, 184–187, 192, 195, 198, 199, 204, 205, 211, 212, 221, 222, 226, 227, 232–235, 237, 243, 246, 248, 249, 251, 254, 259, 262–267, 272–279, 281, 282, 285–302, 304, 306, 307, 310, 313, 319, 320, 324, 325, 329–332, 340, 344, 346–349, 351–354, 356–358, 365, 366, 368, 369, 373, 379, 384, 385, 390, 392, 396, 400–404, 406, 407, 409–412, 422, 424–426, 444–450, 453–462, 465–475, 477–480, 486–489, 491, 494–496, 499–501, 504, 507–533, 535–540, 542–545, 547–549, 551, 552, 554, 555, 561–565, 569–571, 573, 577, 594, 599–601, 604–608, 616–618, 623, 626–630, 632, 633, 638, 641–644, 646, 649, 651, 653–655, 662, 663, 667, 680–683, 694, 695, 697–699, 701, 704, 705, 711, 723, 728, 749, 750, 753, 766, 778, 781–787, 789–791, 802–804, 808, 809, 814–816, 818, 821, 824, 826, 832, 839, 847, 849, 852–855, 865, 867, 869–872, 887, 888, 892, 929, 931–933, 937, 941
Euler numbers, 544, 557
Euler polynomials, 693
Euler totient function, 521, 689, 941
Euler's transformation, 556, 563, 564
Euler–Maclaurin series, 24, 498, 518, 522, 523, 849
Euler–Maclaurin summation formula, 18, 23, 24, 221, 368, 375, 376, 403, 421, 472, 494–505, 509, 511–514, 517–520, 522, 523, 750, 774, 868
exact differential, 354
Exercices de calcul intégral, 575, 816
Exercices de mathématiques, 335
Exercitatio Geometrica de Dimensione Figurarum, 142, 177
Exercitationes Geometricae, 99, 159, 168
Exercitationes of D. Bernoulli, 246

Faà di Bruno, 286, 343, 344, 374, 397
Faber, G., 520, 916
Fagnano, Giulio Carlo, 778–783, 786–791, 811, 813–815, 818
Farkas, Hershel M., 625, 636
Fatio de Duillier, Jean Christophe, 178
Fatio de Duillier, Nicolas, 138, 261, 262, 271, 287
Fatou, Pierre, 440, 441
Faulhaber, Johann, 16, 18–21, 23, 26, 71, 83, 344
Faulhaber polynomials, 26
Favard, J., 658
Favard's theorem, 658
Feigenbaum, Lenore, 219, 221, 287
Feingold, Mordechai, 119
Fejér, Lipót, 658, 659, 673, 679, 752–754, 760–763
Fekete, M., 911

Feldheim, Ervin, 659, 673–675, 679
Fermat, Pierre, 16–18, 25, 27, 29, 31, 71–73, 79, 120, 126, 128, 346, 532, 637, 652, 655, 681, 693, 726, 781, 831, 846, 847, 849, 852, 854, 907, 929, 931, 932, 938, 941
Fermat's little theorem, 198, 929, 932
Feron, Eric, 727
Ferrari, Lodovico, 311, 312, 314
Ferraro, Giovanni, 288
Ferreirós, José, 439
Ferrers, N. M., 629, 652
del Ferro, Scipione, 311, 312
Fibonacci sequence, 245, 248, 253
figurate numbers, 16–18, 21, 22, 30, 53, 151, 159, 197
Fior, Antonio Maria, 311, 312
Fischer, Ernst, 86
FitzGerald, C. H., 914, 923, 928
FitzGerald's inequality, 914
Foata, Dominique, 626
Fomenko, O. M., 928
Fonctions analytiques, 157, 166, 204, 213, 786
Fontaine, Alexis, 346, 351–353, 369
Forrest, Challand, 350
Forrester, P. J., 474
Fourier, Joseph, 92–96, 284, 350, 404–406, 409, 410, 412–418, 420, 425, 426, 449, 498, 499, 503, 602, 750, 849, 892
Fourier coefficients, 403, 406, 410, 426, 429, 436, 750, 755, 760, 762, 893, 894
Fourier inversion formula, 715
Fourier series, 762
Fourier transforms, 369, 535, 536, 757–759, 769, 770, 772, 773, 777
   fast, 801
Français, François, 379
Français, Jacques F., 369, 379, 380, 387, 398, 663
Franklin, Fabian, 344, 629, 630, 638, 723
Fréchet, Maurice, 68, 86
Fredholm, Ivar, 406, 910
Frei, Günther, 942
Frénicle de Bessy, 929
Fricke, R., 625
Friedelmeyer, Jean-Pierre, 398
Frobenius, Georg, 397, 440, 630, 631, 658, 720, 751, 753, 754, 760, 767, 775
Fuchs, Lazarus, 266, 753
Fundamenta Nova, 538, 604, 626, 791, 795, 807, 814, 819, 832, 837, 839, 848
Fuss, P. H., 50, 198, 474, 532, 543–545, 626, 692, 778, 814, 865, 869

Galileo, 120–122
Galois, Évariste, 316, 317, 577, 827, 876, 930, 931, 936–939, 942
gamma function, 33, 37, 151, 165, 324, 444–450, 457, 475, 478–480, 497, 519, 559, 565, 589, 621, 653, 776, 784, 888
   duplication formula, 449
   multiplication formula, 449
   reflection formula, 227, 233, 520
Ganita Kaumudi, 17
Garabedian, P. R., 912
Garsia, Adriano, 631, 632
Garvan, Frank, 634, 637, 651
Gasper, George, 926–928
Gauss, Carl Friedrich, 11, 50, 56, 68, 87, 93, 139, 151, 162, 166, 167, 170, 173, 181, 187, 189, 222, 224, 231, 290, 316, 320–325, 329, 334, 335, 344, 346, 349, 350, 361, 366, 371–374, 389, 393, 417, 424–426, 446, 448, 449, 457–461, 465, 466, 469, 470, 472, 473, 475, 478, 479, 489, 490, 492, 497, 499, 535, 537, 540, 541, 547–553, 557–564, 569–573, 575, 576, 579–582, 587, 590, 592, 595, 597, 601–605, 608–621, 623, 626, 637, 644, 653, 654, 659, 661, 662, 665, 681, 683, 685–688, 695–698, 705, 713, 714, 719, 720, 726, 751, 775, 780, 784, 789, 794–798, 800–820, 825–827, 829, 831–833, 852, 854–857, 865, 877, 885, 887, 888, 929–938, 940–942
Gauss sums, 320, 323, 324, 421–424, 426, 449, 535, 602, 603, 611, 612, 614, 626, 659, 665, 681, 693, 849–851, 940
Gauss's quadratic transformation, 565
Gauss's summation formula, 486, 604, 619, 626
Gauss's transformation of elliptic integrals, 795, 834
Gaussian polynomials, 374, 611, 617, 737
Gautschi, Walter, 926, 928
Gegenbauer, Leopold, 589, 596, 926, 928
Gegenbauer polynomials, 589, 596, 656–659, 675
Gegenbauer–Hua formula, 596, 926, 928
Gelfand, Izrail M., 723, 729, 759, 772–775, 777, 877, 886
Gelfond, Aleksandr, 473, 682, 689–691, 694, 870–874, 876, 877, 886
Gelfond–Schneider theorem, 874, 877
Gellibrand, Henry, 159
Genocchi, Angelo, 286
Geometriae Pars Universalis, 73, 99, 102
La Géométrie, 29, 71, 72, 74, 82, 120, 313, 318
Georgiadou, Maria, 928
Gerling, Christian, 321
Germain, Sophie, 854
German combinatorial school, 373, 399
Gerver, J., 440
Gilbert, L. P., 682
Girard, Albert, 326–328, 344
Girard–Newton formulas, 516, 517, 522
Girgensohn, R., 546
Glaisher, J. W. L., 153, 373, 649
Goldbach, Christian, 32, 46, 50, 180, 186, 254, 445–447, 474, 477, 495, 517, 528, 529, 532, 543, 545, 601, 606, 626, 653, 680, 723, 781, 782, 790, 849, 865, 868, 869, 885, 886, 927
Goldberg, A. A., 906
Goldstein, Catherine, 626

Goldstein, Larry J., 719
Goldstine, Herman H., 597
Goluzin, G. M., 914
Goluzin's inequality, 914
Gong, Sheng, 928
Gordan, Paul, 726, 727, 740, 746, 884
Gordon, Basil, 631, 650
Gosper, Bill, 559
Goursat, G., 890
Gouvêa, Fernando, 693
Gowing, Ronald, 243
Grabiner, Judith, 221
Grace, J. H., 748
Graham, Ronald, 693
Grandi, Guido, 749, 751
Grattan-Guinness, Ivor, 26, 27, 50, 96, 597
Graves, Charles, 399
Graves, John T., 393, 395, 396, 399
Graves, Robert, 399
Gray, J. J., 399, 815, 928
Green, B., 693
Green, George, 346, 349, 350, 354, 359–362, 365, 366, 397, 399, 628, 897, 908
Green's theorem, 350, 361, 916
Greenberg, John L., 366
Greenhill, Alfred G., 623
Gregory, David, 73, 133, 134, 138, 142, 177
Gregory, Duncan, 369, 371, 372, 384–387, 391, 392, 398, 399, 663
Gregory, James, 3, 7, 31, 51, 52, 54, 55, 69, 73, 97, 99–105, 115, 119, 133, 141, 151, 159–161, 165, 168, 176, 177, 181, 198, 199, 201–203, 206–208, 211, 220, 223, 243, 260, 264, 289, 372, 377, 384, 498, 505, 868
Gregory–Newton interpolation formula, 160, 165, 176, 180, 198, 202, 377, 384, 619
Griffith, A., 656
Grinshpan, A. Z., 914
Gröbner basis, 727, 729
Gromov, Mikhail, 877, 886
Gronwall, Thomas H., 763, 910–912, 914–916, 919, 921, 928
Grootendorst, A. W., 79
Grothendieck, Alexander, 96, 774, 931
Grundriss der Allgemeinen Arithmetik, 374
Grunsky, Helmut, 912
Grunsky's inequality, 912, 914
de Gua de Malves, J., 92, 94
Gubler, Walter, 906
Gucciardini, Niccoló, 80
Gudermann, Christoph, 57, 70, 373, 399, 563, 564, 832, 887
Gupta, Hansraj, 633
Gupta, R. C., 13, 15

Hadamard, Jacques, 86, 247, 250, 259, 700, 701, 754, 755, 770, 777, 890, 891, 900, 902

Hadamard product, 699
Hadamard three circles theorem, 900
Hadamard–Borel factorization theorem, 902
Haimo, Deborah T., 443
Hald, Anders, 259, 493, 626
Halley, Edmond, 178, 200, 290
Halphen, George–Henri, 700
Hamel, Georg, 68
Hamel basis, 68
Hamilton, William Rowan, 133, 372, 378, 379, 393–397, 399
Hamilton's theory of couples, 372, 393, 394, 396
Hammond, J., 663
Han, G.-N., 626
Hankel, Hermann, 362, 432, 443, 450
Hansteen, Christoffer, 819
Hardy, G. H., 63, 67, 81, 92, 96, 240, 244, 371, 440, 443, 473, 514, 571, 621, 630–632, 636, 644, 649, 651, 660, 679, 692, 701, 714, 719, 723, 754–757, 759, 762, 765–767, 775–777, 871, 905, 927, 941
Hardy–Littlewood maximal function, 442, 754
Hardy–Littlewood Tauberian theorem, 757, 760, 763, 764
Hariharan, S., 15
Harkness, J., 66, 70
Harmonia Mensurarum, 225, 243, 448
Harriot, Thomas, 18, 29, 72, 82, 87, 88, 96, 101, 104, 158–160, 175, 181, 260, 313, 317, 327, 887
Harriot–Briggs interpolation formula, 160, 176, 198, 202
Hasegawa Ko, 155
Hathaway, Arthur, 629
Hawking, Stephen, 443
Hawkins, Thomas, 443
Hawlitschek, Kurt, 26
Hayashi, Takao, 15
Hayman, Walter, 898, 906, 928
al-Haytham, 6, 13, 14, 16, 194
Heath–Brown, Roger, 719, 777
Heaviside, Oliver, 367, 369, 373, 757
Hecke, Erich, 635, 820
Heckmann, G., 876
Heegner, Kurt, 878
Heine, Eduard, 206, 428, 573, 587, 604, 618, 621, 653, 654, 656, 661–663, 677, 679
Heine's q-hypergeometric transformation, 654, 656, 661–663
Hellinger, Ernst, 86
Herglotz, Gustav, 931, 940
Hérigone, P., 120
Hermann, Jakob, 57, 178, 227, 263
Hermite, Charles, 36, 49, 162, 170–175, 222, 228, 237, 238, 240, 244, 317, 490, 577, 656, 657, 662, 663, 700, 738, 740, 747, 751, 760, 820, 837, 839, 858, 869, 870, 874, 880–883, 885, 886, 890, 891
Hermite polynomials, 658, 663
Herschel, John W., 291, 370, 371, 380–382, 398
Hesse, Otto, 726

van Heuraet, Hendrik, 71–73, 75, 76, 79, 99, 120, 328
Hewitt, E., 439
Hewitt, R., 439
Hickerson, Dean, 637
Hilbert, David, 68, 86, 91, 96, 205, 406, 440, 553, 658, 723, 726, 727, 743–746, 748, 763, 870–874, 884, 886, 908, 909, 928
Hilbert basis theorem, 727, 743, 744, 746
Hill, G. W., 405, 406
Hindenburg, C. F., 373–375, 548, 549
Hipparchus, 13, 158
Historia et Origo Calculi Differentialis, 197
Hobson, E. W., 156, 220, 348
Hofmann, Joseph E., 25, 79, 138
Hölder, Otto, 85, 86, 464, 751
Holmboe, Bernt, 58, 63, 69, 70, 509, 519, 819
homogeneous functions, 347, 351–353, 729, 730, 732
homogeneous polynomials, 721
l'Hôpital, G., 79, 80, 97, 123, 131, 136, 262, 346, 354, 367
l'Hôpital's Rule, 216
Horace, 750
Horiuchi, A. T., 546
Horowitz, D., 914
Horsley, Samuel, 156
Hoskin, Michael, 157
Houzel, Christian, 865
Hua Loo-Keng, 926–928
Hudde, Jan, 18, 72–76, 79, 125, 140, 222, 346
Hudde's rule, 72, 75, 76, 79, 125
Huygens, Christiaan, 287
L'Huillier, S., 13
Hunt, Richard, 442, 443
Hurwitz, Adolf, 397, 682, 870
Hutton, Charles, 70
Huxley, Thomas, 50
Huygens, Christiaan, 18, 19, 32, 50, 72, 122, 123, 126, 129, 131, 139, 245, 261, 269, 283, 287, 354
hypergeometric series, 154, 181, 189, 390, 448, 475, 498, 547, 549, 552, 555, 559, 563–566, 573–575, 604, 618, 654, 656, 784, 801, 886
  contiguous relations, 181, 549
  convergence, 458
  q-extension, 653, 661, 679

inequality
  arithmetic and geometric means, 81, 83–85, 87
  Cauchy–Schwarz, 86, 463
  Hölder, 85, 86
  Jensen's, 89
  Maclaurin's, 92
  Minkowski's, 86, 90
infinite determinant, 405
Ingham, A. E., 757
Institutiones Calculi Differentialis, 24, 58, 181, 199, 251, 385, 514, 518
integral test, 56, 501

integrating factor, 261, 262, 271, 272, 277, 287, 355, 356
intermediate value theorem, 56, 204, 215, 217–219, 221
Introductio in Analysin Infinitorum, 232, 248, 254, 289, 290, 295, 300, 373, 449, 530, 540, 599, 601, 606, 626, 681, 802, 824, 869
invariant theory, 371, 617, 630, 656, 662, 663, 720, 721, 723, 726, 728, 729, 732–734, 737, 743, 746, 748
Invention nouvelle en l'algèbre, 326
Ireland, Kenneth, 940
Ismail, Mourad, 659, 676, 679
isobaric polynomial, 725
Ivory, James, 577, 583–585, 597, 794, 798–800, 814

Jackson, Frank Hilton, 624, 655, 656
Jacobi, C. G. J., 18, 20, 23, 24, 26, 161, 166, 167, 175, 187, 224, 324, 346, 349, 390, 391, 398, 403, 460, 461, 475, 480, 498, 499, 505, 506, 514, 538, 548, 551, 552, 557, 564–566, 573, 576, 577, 582, 583, 585, 587–590, 594, 596, 597, 602–604, 606, 617–621, 626, 633, 641, 646, 647, 653, 726, 778, 786, 791, 795, 807, 810, 814, 816, 818–820, 832–849, 852, 857, 858, 863–866, 891–895, 906, 926
Jacobi polynomials, 587, 589, 595, 596, 926
  discrete, 594, 596
Jacobi sums, 324, 940
Jacobi's imaginary transformation, 832–834
Jahnke, H. N., 399
James, I. M., 928
Jartoux, Pierre, 157
Jeans, James, 649
Jenkins, J. A., 912
Jensen, Johan L., 85, 86, 89, 96, 463, 892, 894–898, 906
Jensen's formula, 892, 895–897, 900, 906
Jiuzhang Suanshu, 557
Johnson, Warren, 154, 157, 397
Jones, Sir William, 14
Jones, William, 156, 226, 230, 243
Jordan, Charles, 193
Journal littéraire d'Allemagne, 525
Jyesthadeva, 1, 2, 5–10, 14

Kac, Mark, 68, 732, 733, 748, 771, 772
Kalman, Dan, 748
Kant, Immanuel, 743
Kapranov, M. M., 723, 729, 775
Karamata, Jovan, 760, 763, 764, 767, 776, 777
Karanapaddhati, 12, 15
al-Karji, 51
al-Kashi, 51, 140
Katz, Nicholas M., 942
Katz, Victor J., 15, 287, 365, 719
Khayyam, Omar, 51

Khinchin, A. Y., 693, 871
Kinckhuysen, Gerard, 96
Kinkelin, Hermann, 682
Kinyon, Michael, 221
Klein, Esther, 650
Klein, Felix, 70, 397, 399, 553, 569, 773, 810, 884, 886, 908
Kneser, Adolf, 755
Knoebel, Arthur, 139
Knopp, Konrad, 474, 573
Knuth, Donald, 20, 26, 399
Ko-yuan Mi-lu Chieh-fa, 157
Koblitz, Neal, 886
von Koch, Niels Helge, 406, 910
Koebe, Paul, 908, 910, 913, 916, 917, 928
Kolchin, E., 876
Kolmogorov, A. N., 442, 597, 774
Koppelman, E., 399
Korevaar, Jacob, 777
Kornblum, Heinrich, 931, 941
Kovalevskaya, S., 268, 777, 887
Kra, Irwin, 625, 636
Kramp, Christian, 373–375, 399
Kriyakramakari, 2–4, 13, 15
Kronecker, Leopold, 324, 325, 329, 332, 362, 428, 429, 438–440, 499, 589, 729, 820, 837, 864
Kronecker's Jugendtraum, 440
Kubota, K. K., 886
Kühne, Hermann, 931
Kummer, Ernst, 181, 189, 190, 291, 307, 428, 465–467, 469, 475, 535, 551, 552, 563–565, 567, 569, 571, 573, 587, 727, 852, 930
Kummer's hypergeometric transformation, 571
Kung, Joseph, 344
Kuzmin, Rodion, 869, 872
Kuzmina, Galina V., 928
Kyuseki Tsuko, 155

L-series, 57, 450, 515, 520, 524, 525, 533, 542, 545, 681–683, 687, 693, 931
  class number, 688
  complex variable, 682, 688
  functional equation, 520, 536
  infinite product, 540
Lacroix, S. F., 24, 25, 27, 195, 196, 199, 204, 212, 213, 221, 368, 370, 380, 404, 469
Lagrange, J. L., 50, 86, 88, 93, 152, 154, 157, 161, 165–167, 174, 175, 181, 194, 195, 198, 199, 203–205, 213–218, 221, 246, 254–257, 259, 266–268, 276, 277, 287, 290, 291, 313–316, 346, 348, 349, 353, 358, 359, 365, 367–370, 374–378, 398, 401, 403, 404, 407–410, 425, 426, 478, 497–499, 505, 532, 563, 574, 575, 577, 579, 582, 585, 590, 681, 687, 720, 726, 728, 750, 784–786, 794, 795, 799, 802, 810, 814, 816, 818, 827, 829, 831, 892, 929, 932, 933, 937, 941

Lagrange inversion formula, 152, 157, 374, 587, 588
  remainder, 152
Lagrange resolvents, 320, 323
Lagrange's transformation of elliptic integrals, 795
Lagrange–Waring interpolation formula, 161
Laguerre, Edmond, 94, 95, 285, 336–338, 345, 577, 597, 889, 905
Laguerre polynomials, 254, 285, 577
Lambert, J. H., 48, 848, 867, 868, 875
Lampe, E., 552, 573
Landau, Edmund, 91, 701, 753–755, 766, 773, 777, 871, 872, 892, 905, 909, 931
Landen, John, 55–58, 70, 291, 302–304, 307, 310, 329, 340, 369, 517, 518, 784–786, 791, 799, 814, 816, 833
Landen's transformation of elliptic integrals, 791, 793, 800, 834
Landis, E. M., 877, 886
Lanzewizky, I. L., 659, 673, 675, 679
Laplace, Pierre-Simon, 86, 203, 204, 215, 216, 218, 221, 246, 247, 255, 257, 259, 267, 350, 368, 369, 377, 378, 381, 398, 404, 449, 467, 478, 498, 499, 575–577, 720, 752, 821
Laplace series, 763
Lascoux, Alain, 345
Lasker, Emanuel, 727
Latham, Marcia L., 79, 325
Laubenbacher, Reinhard, 139
Laudal, Olav, 865
Laurent, Pierre, 820
Lebedev, N. A., 913, 914
Lebedev-Milin inequality, 913
Lebesgue, Henri, 86, 362, 428, 429, 432, 440, 443, 908
Lectiones Geometricae, 118, 119, 147
Lectiones Mathematicae, 136
Legendre, Adrien-Marie, 93, 291, 425, 446, 449, 467, 474, 480, 492, 498, 541, 554, 574–578, 584, 585, 595, 597, 604, 628, 659, 675, 680, 681, 696–698, 701, 702, 705, 710, 714, 783–786, 791, 793, 799, 800, 814, 816–819, 822, 832, 834–836, 839, 853, 865, 867–869, 875, 876, 933
Legendre polynomials, 557, 574–578, 581, 583, 586, 587, 595, 596, 622, 659, 673, 763, 817
  discrete, 594, 597
Lehmer, D. H., 632
Lehner, Joseph, 634, 635
Leibniz, G. W., 3–5, 7–9, 18, 19, 38, 52, 56, 69, 77–79, 98, 103, 119–123, 126–131, 138, 139, 142, 147–149, 151, 156, 157, 160, 163, 176–178, 181, 189, 190, 195, 197, 202, 203, 210–213, 220, 223, 224, 226–228, 230, 239, 243, 246, 260–262, 264, 268–271, 278, 283, 287, 289, 306, 330, 345–348, 366–370, 373, 375, 380, 384, 387, 389, 392, 447, 497, 515, 517, 586, 628, 749, 751
Lemmermeyer, Franz, 865, 940

lemniscate, 440, 778–781, 787, 789, 803, 813, 818, 940
  division of, 779, 786, 788, 789, 801, 803, 818, 830, 831
lemniscatic integral, 778, 779, 783, 790, 803, 809, 863
  addition formula, 789, 791
  complex multiplication, 781
Lerch, Mathias, 490
Levinson, Norman, 777
Lewin, Leonard, 307, 310
Li Shanlan, 556, 557, 573
Lie, Sophus, 69, 599, 775
Lilavati, 4, 10, 13, 36
Lindelöf, Ernst, 553, 891, 895, 900, 905
Lindelöf, Lorenz, 348, 891
Lindemann, Ferdinand, 745, 868, 870, 873–876, 884
line integrals, 354, 362, 364
Linnik, Yuri V., 682, 689, 690, 694, 873, 877
Liouville, Joseph, 65, 94, 266, 585–587, 597, 773, 820, 858–863, 866, 869, 878, 879, 886, 902, 930, 931
Liouville numbers, 878
Liouville's Journal, 597, 710, 850, 936
Lipschitz, Rudolf, 682
Littlewood, John E., 63, 66, 92, 96, 443, 633, 692, 694, 701, 714, 723, 727, 751, 754–757, 759, 763–767, 776, 777, 902, 906, 911, 914, 917–920, 927, 928
Littlewood's Tauberian theorem, 764
Lobachevsky, Nikolai I., 775, 801
Löwner, Karl, 909–911, 914, 920–924, 928
Logarithmotechnia, 99, 141
Lototskii, A. V., 873
Lovelace, Ada, 729
Loxton, J. H., 885
Lusin, Nikolai, 441, 442

van Maanen, J. A., 79
Macaulay, F. S., 727
Machin, John, 200, 220
Maclaurin, Colin, 56, 61, 84, 88, 96, 102, 201, 203, 208, 221, 329, 330, 332, 333, 344, 369, 478, 479, 494, 496, 497, 501, 502, 514, 749
Maclaurin series, 201, 208, 219–221, 416, 501, 595
MacMahon, Percy A., 330, 340, 341, 344, 345, 617, 622, 623, 626, 630–632, 636, 644, 651, 663
Madhava, 1, 3, 4, 7, 8, 10, 13–16, 141, 517
Madhava–Leibniz series, 40, 46, 527, 847
Madhava-Gregory series, 8
Mahisamangalam, Narayana, 2
Mahlburg, Karl, 635
Mahler, Kurt, 868, 885
Mahoney, Michael, 27
Maier, W., 876
Malebranche, Nicolas, 778
Malmsten, Carl J., 520, 543
Manders, Kenneth, 344
Manfredi, Gabriele, 264

Manning, K. R., 399
Maor, Eli, 13
Markov, A. A., 578, 659, 870
Martzloff, Jean-Claude, 573
Masani, Pesi, 777
Mascheroni, Lorenzo, 472
Masser, David, 885
The Mathematical Analysis of Logic, 371
Mathematical Dissertations, 168
Matheseos Universalis Specimina, 142, 177, 181, 189, 199
Matsunaga Ryohitsu, 155
Maurolico, F., 17, 18
Maxwell, James Clerk, 350, 363–365
Mazur, Stanislaw, 773, 774
McCleary, John, 748
McClintock, Emory, 154
Mécanique analytique, 720, 786
Mécanique céleste, 350
Meditationes Algebraicae, 194, 313, 314, 329, 330
Mehler, F. G., 665, 763
Mehler's integral for Legendre polynomials, 763
Mehta, Madan Lal, 474
Meixner, J., 658
Mellin, Hjalmar, 536, 553, 554, 699, 700, 715, 753, 776, 891
Mellin inversion formula, 553
Mencke, Otto, 243
Mengoli, Pietro, 102, 119, 190, 197, 515, 868
Menshov, Dmitrii, 442
Méray, Charles, 206
Mercator (Kaufmann), Nicholas, 73, 96, 99, 104, 108, 140, 141, 159, 222, 268, 269, 289, 477
Mercator, Gerhard, 99, 104
Mercator's Great World Map, 104
meromorphic functions, 85, 817, 871, 892, 895–901, 903, 906
  finite order, 901
  value distribution, 891
Mersenne, Marin, 72, 119, 929
Mertens, Franz, 67, 682, 692, 698, 705, 710–712, 714, 718, 719, 870, 940
Merzbach, Uta, 13
Meschkowski, Herbert, 443
Messenger of Mathematics, 284, 649
method of undetermined coefficients, 144, 149, 203
Methodus Differentialis of Newton, 160, 163, 168
Methodus Differentialis of Stirling, 151, 160, 165, 189, 191, 199, 451, 474, 476, 483, 491, 494, 784
Methodus Incrementorum, 178, 198, 221, 279, 287
Mikami, Yoshio, 155, 157
Milin, I. M., 912–914, 923, 924, 926, 928
Milin's conjecture, 923–925
Milin's inequality, 913, 926
Milne, Stephen, 631, 632
Ming An-tu, 157
Minkowski, Hermann, 86, 90, 91, 852, 908

Miscellanea Analytica, 231, 234, 246, 476, 479, 485, 491
Mittag–Leffler, M. Gösta, 553, 785, 786, 814, 910
Mittal, A. S., 79
Miyake, K., 440
Möbius inversion, 331, 700, 706, 930, 936, 939, 940
Möbius, A. F., 362, 930, 936, 940
mock theta functions, 634, 636, 637, 679
modular equations, 634, 836, 837
Mohr, Georg, 52
de Moivre's factorization, 231, 234
de Moivre, Abraham, 23, 120, 149–151, 202, 203, 211, 219, 222, 226, 231–234, 243, 245–250, 254, 258, 259, 293, 330, 350, 369, 373, 374, 476–479, 481–483, 485, 488, 489, 491–493, 496, 497, 499, 749
de Moivre's series, 476, 477, 481, 488
Moll, Victor, 243, 244
Mollerup, Johannes, 33, 37, 444, 450, 462–464, 475
Monge, Gaspard, 24, 283, 404
Monge's equation, 283, 284, 287
Monsky, Paul, 682, 691, 692, 694
Montel, Paul, 908
Montmort, Pierre R. de, 178–184, 186, 190, 192, 198, 199, 245, 246, 251, 253, 330, 477, 601
Montucla, J. E., 313
Moore, Gregory H., 287
Mordell, Louis J., 941
Morley, F., 66, 70
Morrison, Emily, 398
Morrison, Philip, 398
Muir, Thomas, 374, 399
Mukhopadhyay, Asutosh, 283, 284, 287
Mukhopadhyay, Shyamadas, 284
Murphy, Robert, 94, 369, 371, 372, 382–384, 387, 393, 398, 399, 577, 583, 584, 597, 663
Mustafy, A. K., 79, 718, 719

Sz-Nagy, Béla, 443
Napier, MacVey, 577
Napoleon Bonaparte, 404
Narasimhan, Raghavan, 287, 443
Narayana Pandita, 2, 17, 20, 25, 27
Narkiewicz, W., 719
Naudé, Phillipe, 599, 605, 627
Nave, Annibale della, 312
Neile, William, 31, 71, 73
Nerurkar, Mahesh, 693
Nesterenko, Yu. V., 886
Netto, Eugen, 535
Neuenschwander, Erwin, 906
Neumann, Carl, 752, 753, 908
Neumann, Franz, 653, 752
Neumann, Olaf, 325
Nevai, Paul, 574, 598
Nevanlinna, Frithiof, 553, 891, 892
Nevanlinna, Rolf, 85, 553, 891, 892, 896–904, 906

Nevanlinna's characteristic function, 892, 899, 903, 906
Newman, Francis W., 450, 456, 457, 475, 888
Newman–Schlömilch product, 457
Newton, Isaac, 1, 3, 4, 12, 14, 28, 31, 33, 50–56, 67, 69, 73, 76–84, 87, 92, 94–100, 104, 118–121, 123–126, 133–157, 160–165, 168–170, 174–182, 184, 189, 190, 195, 199–204, 208–210, 213, 216, 219, 221–224, 226, 228–231, 239, 243, 245, 251, 260–264, 268, 270, 271, 284, 287, 289, 295, 306, 314, 328–334, 340, 344, 346, 354, 367, 369, 370, 374, 389, 444, 450, 452, 453, 458, 459, 476–478, 482, 496, 509, 510, 517, 525, 526, 542, 575, 576, 579, 582, 583, 720, 728, 749, 754, 774, 786, 802, 826, 889, 938
Newton's divided difference formula, 161, 164
Newton's polygon, 141, 145
Newton's recurrence rule, 328
Newton's rule for complex roots, 83, 84
Newton's transformation, 142, 177, 178, 181, 182, 184, 187, 245, 549
Newton–Bessel interpolation formula, 160, 164, 199 450
Newton–Cotes quadrature formulas, 162
Newton–Gauss interpolation formula, 176, 198
Newton–Stirling interpolation formula, 158, 160, 164
Nicole, François, 179–181, 190–192, 198, 199
Nikolić, A., 777
Nilakantha, 1–3, 5, 14, 867, 886
Noether, Emmy, 464, 693, 727, 729, 748
Noether, Max, 743
normal family, 908
normed rings, 772–775
Norwood, Richard, 104
nowhere-dense sets, 432
Nullstellensatz of Hilbert, 746
numerical integration, 161–163, 170, 172, 574, 579, 582, 583, 590, 592, 594, 597, 801
Nuñez, Pedro, 104

Okada, Y., 871
Olbers, Wilhelm, 602, 626, 696, 719
Oldenburg, Henry, 4, 52, 69, 98, 144, 148, 156, 160, 223, 239
Oleinikov, V. A., 876
Olver, P. J., 748
Ono, Ken, 635, 637, 652
Opera Geometrica, 29
operational calculus, 367, 371, 373, 384, 387, 398, 426, 663
Opus Geometricum, 56, 122
Ore, Øystein, 777
Oresme, N., 67
orthogonal matrices, 594
orthogonal polynomials, definition, 574
Ortiz, E. L., 399, 597, 626

Osgood, Charles, 892
Osipov, Y, 886
Osipovsky, T, 349
Ostrogradski, Mikhail V., 228, 346, 349, 460
Ostrowski, Alexander, 773
Oughtred, William, 28, 29, 31, 82, 120, 140, 328
Ozhigova, E. P., 814

Paccioli, Luca, 311
Paley, R. E. A. C., 911, 918, 919, 928
Panton, Arthur W., 94
Papperitz, Erwin, 569
Pappus, 71
Paramesvara, 1, 13, 15
Paramesvara's radius formula, 1, 13
Parameswaran, S., 13, 15
Pardies, Ignace-Gaston, 122, 129, 131, 139
Parrilo, Pablo, 727
Parseval, Marc-Antoine, 892
Parshall, Karen H., 399, 748
partial differential calculus, 351, 352
partially ordered set, 331
partitions, 330, 342, 343, 374, 600, 623, 627–632,
    634, 636, 639, 649, 650, 652, 725, 738, 747, 754
  congruence properties, 633, 634, 646
  crank, 634, 651
  distinct parts, 632, 642, 649, 651
  Durfee square, 640, 642
  graphical methods, 638
  rank, 634, 635, 637
  self-conjugate, 641
Pascal, Blaise, 9, 17, 21, 25, 51, 53, 77, 98, 100, 101,
    119, 122, 223, 245, 936
Patterson, S. J., 626
Paule, Peter, 679
Peacock, George, 370, 371, 399
Peano, Giuseppe, 220, 286, 287, 348
Peirce, Benjamin, 630
Peirce, C. S., 397, 630
Pengelley, David, 139
pentagonal number theorem, 606, 626, 628, 630, 638,
    642, 643, 646, 647, 808, 809, 814
permutations
  greater index, 623
  inversions, 622, 626
Perron, Oskar, 658
Petrović, Mihailo, 760
Petrovski, I. G., 287
Pfaff, Johann Friedrich, 26, 181, 187, 373, 374, 458,
    548, 549, 556, 557, 562–564, 569, 806, 930
Pfaff's hypergeometric transformation, 390–392, 571
Philosophical Transactions, 31, 84, 133, 149, 162,
    195, 201, 204, 232, 290, 310, 378
Phragmén, E., 910
Picard, Émile, 760, 876, 890–892, 902–904, 906, 909
Picard's theorem, 754, 891, 892, 902, 904, 906,
    908, 909

Pick, Georg, 906, 910, 927
Piene, Ragni, 865
Pieper, Herbert, 573, 814
Pierpoint, W. S., 78, 80
Pierpont, James, 569, 573
Pietsch, A., 96
Pingree, David, 14
Pitt, H. R., 759, 773
Plana, Giovanni, A., 465, 490, 499
Plana–Abel formula, 499, 508, 509, 514
Plofker, Kim, 15
Poincaré metric, 927
Poincaré, Henri, 338, 350, 405, 406, 480, 754, 760,
    890, 908, 909, 928
Poisson, S. D., 286, 349, 350, 377, 403, 417, 421, 460,
    461, 475, 480, 487, 498, 499, 503–505, 507, 514,
    535, 538, 604, 750, 752, 753, 762, 813, 821, 849,
    896–898, 909
Poisson integral, 441, 443, 750, 752
Poisson summation formula, 421, 498, 499, 503,
    535, 849
Poisson–Jensen formula, 897, 898, 901
Polignac, Alphonse de, 698, 710, 712, 719
Pólya, George, 92, 96, 337–340, 473, 658, 753,
    760, 871
Pommerenke, Christian, 912, 914, 923, 928
van der Poorten, Alfred, 885
Popoff, A., 152
Prag, Adolf, 157
Prasad, Ganesh, 220, 866
prime number theorem, 696, 698, 700–702, 705, 719,
    754–757, 759, 766–768, 770, 771, 776, 777
primes in arithmetic progressions, 449, 532, 540, 680,
    688, 694
Principia, 97–99, 119, 121, 133, 163, 178, 201, 225,
    261, 263, 370
The Principles of Analytical Calculation, 369
Pringsheim, Alfred, 221
Privalov, I. I., 442, 871, 911
Produzioni Matematiche, 778, 782, 813
de Prony, Gaspard Riche, 203, 204
Prym, F. Emil, 907
Ptolemy, 13, 158
Ptolemy's formula, 13

q-binomial theorem, 374, 603, 604, 615, 618, 619,
    621, 626, 653, 654, 661, 662, 671
q-difference operator, 654
q-Hermite polynomials, 657, 659, 664, 665, 673, 677
q-integral, 654, 655, 662
q-ultraspherical polynomials, 659, 662, 670, 673,
    676, 679
quadratic forms, 604, 681, 682, 688, 720, 726,
    872, 930
  class number, 421, 681, 687, 694
  negative determinant, 687, 689

quadratic reciprocity, 426, 604, 854
  law, 424, 425, 614, 681, 853, 931, 940
Quarterly Journal of Pure and Applied Mathematics, 373, 649
quaternions, 372, 393–397, 399

Raabe, Joseph, 23, 403, 498
Rademacher, Hans, 908
radius of curvature, 120, 126, 135–139, 283, 284
Rajagopal, C. T., 14
Raman, C. V., 284
Ramanujan, S., 571, 608, 621–623, 625, 630, 632–637, 644–649, 651, 652, 657, 660, 668, 677, 679, 708, 717, 718, 750, 754, 863, 866, 941
Ramanujan's integral formula, 571
Ramanujan's summation formula, 621
Ramanujan's tau function, 941
Ramasubramanian, K., 14
Rangachari, M. S., 14
ratio test, 56, 61, 67
Raubel, Claude, 74
Raussen, Martin, 886
recurrent series, 245–250, 258, 259, 577
Remmert, Reinhold, 474, 573
resultant, 228, 329, 723, 728, 729
Riccati, Jacopo, 264, 265, 278, 279
Riccati difference equation, 675
Riccati equation, 32, 38, 46, 48, 50, 261, 264–266, 278, 287
  generalized, 265
Riemann, G. Bernhard, 87, 139, 350, 361, 362, 365, 373, 410, 427–436, 438–440, 443, 450, 519, 520, 536–539, 546, 552–554, 567–569, 573, 654, 698–700, 714–716, 719, 760, 763, 820, 821, 859, 873, 890, 892, 907, 908, 931
Riemann hypothesis, 692, 700, 716, 719, 870, 892, 895, 931
  characteristic p, 931, 941, 942
Riemann integral, 429
Riemann mapping theorem, 361, 907–909, 921, 928
Riemann P function, 553, 568
Riemann surface, 636, 743, 814, 821, 908
Riemann zeta function, 462, 519, 520, 539, 890, 892, 931
  functional equation, 546, 753, 776
Riemann–Lebesgue lemma, 436
Riemann–Schwarz derivative, 427, 428, 431, 435
Riesz, Frigyes, 86, 90, 91, 96, 404–406, 658, 774
Riesz, Marcel, 442
Ritt, Joseph, 876
Robertson, M. S., 911, 913
Robertson's conjecture, 913
Robins, Benjamin, 204
Robinson, G., 163
Rodrigues, Olinde, 577, 584, 585, 597, 617, 622, 626
Rodrigues formula, 577, 583–585
Rodrigues-type formula, 587, 588, 594, 595

Rogers, L. J., 85, 86, 308, 606, 608, 630, 632, 644, 649, 656–660, 662–671, 676–679
Rogers–Ramanujan identities, 608, 631, 632, 644, 657, 658, 660, 665, 666, 678
Rogers–Szegő polynomials, 665, 671
Rolle, Michel, 84, 205
Rolle's theorem, 84, 205, 873
Roquette, Peter, 942
Rosen, Michael, 940, 941
Roth, Peter, 71, 83, 344
Rothe, H. A., 373, 374, 399, 603, 618
Rothe-Gauss theorem, 603, 604, 618, 672
Rothschild, Bruce L., 693
Rowe, David, 748
Roy, Ranjan, 15, 475
Ruffini, Paolo, 205, 316, 317, 369
Russell, Bertrand, 287, 756

Saalschütz, L., 548
Sadratnamala, 13
Saha, M. N., 284
Salmon, George, 283
Sampo Yenri Hyo, 155
Sandifer, C. Edward, 69, 325, 775
Sarkaria, Karanbir S., 928
Sarma, K. V., 14, 15
Sasaki Chikara, 440
Savile, Henry, 29
Schappacher, Norbert, 626
Schellbach, Karl, 153, 309
Schentz, Georg, 371
Schering, Ernst Julius, 654
Scherk, Heinrich Ferdinand, 373
Schiffer, Max, 912, 914
Schlesinger, Ludwig, 810
Schlömilch, Oscar, 220, 348, 450, 457, 470, 475, 520
Schmidt, Erhard, 86, 406, 909
Schmidt, F. K., 931
Schmidt, Robert, 757
Schneider, Ivo, 26, 493
Schneider, Theodor, 870, 873, 874
Schoenberg, Isaac J., 928
Schoenemann, Theodor, 324
Schönemann-Eisenstein criterion, 857, 865, 866
van Schooten, Frans, 29, 72, 73, 75, 76, 98, 120, 135, 328
van Schooten, Frans the Elder, 72
Schulze, J. C., 695
Schumacher, Heinrich, 800, 801, 818, 819, 834
Schur, Issai, 91, 631, 693, 697
Schwartz, Laurent, 898
Schwarz, H. A., 57, 86, 348, 428, 436–438, 440, 443, 569, 753, 762, 865, 887, 891, 906, 908–910, 927
Schwarz's lemma, 909
Schwarzian derivative, 291

Schweins, Franz F., 373, 603, 620, 653
Schwermer, Joachim, 626
Scriba, C. J., 79, 287
De Sectionibus Conicis, 29
Segal, Sanford L., 440
von Segner, J. A., 92
Seidel, Philipp Ludwig, 57
Seki Takakazu, 728
Selberg, Atle, 461, 473, 475, 637, 652, 701
Selberg's integral formula, 474
seminvariant, 725, 736–738, 741–743, 747
Sen Gupta, D. P., 287
Serre, Jean-Pierre, 635, 878
Serret, Joseph, 205, 931
Servois, François, 369
Shah, S. M., 905
Shaw, Gideon, 202, 207, 208
Shidlovskii, Andrei B., 876, 877, 885, 886
Shimizu, Tatsujiro, 892
Shimura, Goro, 635, 652, 693
Siegel, Carl Ludwig, 780, 781, 815, 870, 872–876, 878, 885, 886
Simmons, George F., 139
Simon, Barry, 574, 598
Simpson's rule, 168–170
Simpson, Thomas, 79, 168, 247, 249, 258, 259, 525, 683
Sinai, Y, 886
Singer, M. F., 228
singular solutions of differential equations, 267, 268, 279–281, 284, 286, 287
Siyuan Yujian, 184
Skau, Christian, 886
Smale, Stephen, 754
Smale's horseshoe, 754
Smith, David E., 79, 157, 243, 325, 719, 886
Smith, H. J. S., 424, 426, 430–432, 443, 848, 851, 852, 855, 865, 936, 942
Smith, John, 160
Smith, Robert, 225, 226, 243
Sokhotskii, I. V., 891
Spence, William, 291, 304, 305, 307, 308, 310
Spencer, Joel H., 693
Spiridonov, V., 656
Sridhara, 312
Srinivasa, M. D., 14
Srinivasienger, C. N., 12
Sriram, M. S., 14
de Stainville, J., 152
Stark, H. M., 877
Stedall, Jacqueline, 50, 82, 96, 175
Steele, J. Michael, 91
Steffens, Karl-Georg, 598
Stieltjes, Thomas Joannes, 33, 36–38, 48–50, 174, 175, 480, 490, 493, 596, 658, 659, 700, 870, 908
Stieltjes integral, 714, 759, 769, 899
Stillwell, John, 928

Stirling, James, 56, 150, 151, 157, 160–163, 165, 170, 175, 180, 181, 187–194, 198, 199, 201, 203, 250–252, 259, 289, 369, 377, 378, 444, 450–453, 458, 474, 476–480, 483–486, 489, 491–494, 496, 497, 499, 515, 518, 547, 549, 565, 705, 706, 749, 784, 786, 803
Stirling numbers, 557
Stirling's approximation, 476–478, 487, 489, 697, 717
Stirling's hypergeometric transformations, 187, 188
Stirling's method of ultimate relations, 250
Stirling's series, 476, 483
Stokes, George, 57, 349, 350, 354, 362–364, 373, 480
Stokes's theorem, 350, 362–365
Stoltz, Otto, 348
Stone, Edmund, 78–80, 118, 138, 369
Stone, Marshall, 658
Strichartz, Robert S., 440
Stridsberg, E., 876
Struik, Dirk Jan, 119, 139
Stubhaug, Arild, 70
Sturm, Charles-François, 94, 95, 729
Sturmfels, Bernd, 748
summability
    Abel, 750, 753, 755, 756, 760, 763, 766, 767, 775
    Borel, 775
    Cesàro, 752, 753, 755, 760, 762, 763, 765, 775
    Lambert, 756, 757, 766–768
summation by parts, 64, 179, 195, 199, 434, 691, 709, 712, 768
Swinnerton–Dyer, Peter, 634, 650
Sylla, Edith D., 26, 625
Sylow, Peter Ludwig, 69
Sylvester, James Joseph, 45, 46, 50, 83, 84, 95, 96, 198, 314, 325, 371, 373, 460, 608, 617, 628–630, 638, 640–644, 650, 652, 656, 663, 697, 698, 719–721, 723, 725, 726, 728, 729, 733, 734, 736–738, 740–743, 745, 747, 748, 870
Systematisches Lehrbuch der Arithmetik, 603
Les systèmes d'équations linéaires, 404
Szegő, Gabor, 603, 658, 659, 665, 671–673, 679, 753, 760, 911
Szekeres, George, 650

Tait, Peter Guthrie, 350, 363, 365
Takagi, Teiji, 440
Takase, Masahito, 864
Takebe Katahiro, 152, 157, 178, 527, 542, 546, 569, 728
Taniyama, Yutaka, 652
Tannery, Paul, 890
Tantrasangraha, 2, 14
Tao, Terrence, 693
Tartaglia, Niccoló, 311, 312
Tauber, Alfred, 63, 714, 751, 753–755, 775
tautochrone, 351

Taylor, Brook, 61, 88, 102, 134, 142, 156, 178–182, 190, 195, 196, 198–205, 211, 212, 215, 216, 218, 220, 221, 227, 240, 246, 267, 279, 280, 287, 289, 353, 354, 365, 367–369, 381, 383, 398, 401, 402, 497, 506, 731, 734, 890, 891, 922
Taylor polynomials, 753
Taylor series, 176, 180, 200–206, 210, 211, 221, 352, 354, 375, 497–499, 505, 518, 774, 907
 remainder, 204, 212, 213, 215, 216
Tetsujutsu Sankei, 527, 546
Théorie analytique des probabilités, 215
Théorie analytique de la chaleur, 412
Théorie des nombres, 710
theta functions, 602, 603, 636, 800, 801, 810, 811, 819, 864
Thibaut, Bernhard, 374
Thomae, Karl Johannes, 654, 662, 679
Thompson, J. J., 649
Thomson, William (Lord Kelvin), 350, 362–365, 373, 475
Tignol, John-Pierre, 325
Titchmarsh, Edward C., 714, 719, 777, 905
Toeplitz, Otto, 86, 757
Tonelli, Leonida, 90
Tonini, Zuanne de, 312
Toplis, John, 350
Torricelli, Evangelista, 25, 29, 72, 98, 197, 346
Tractatus de Cycloide, 73
Traité des fonctions elliptiques, 816, 818
Treatise of Algebra, 84, 329
Treatise on Electricity and Magnetism, 350
Treatise on Natural Philosophy, 350
Treatise on Plane and Spherical Trigonometry, 370
Trigonometria Britannica, 159
triple product identity, 603, 604, 609, 615–618, 621, 626, 641, 645, 646, 657, 659, 668, 672, 673, 677, 800, 802, 807–809, 812, 814, 819, 846
Truesdell, Clifford A., 139, 263, 286, 288, 402, 426, 445, 474
Tschinkel, Yuri, 426, 693
Tucciarone, John, 777
Turán, Paul, 557, 573, 650, 659, 679, 753
Turnbull, Herbert W., 69, 102, 105, 115, 119, 175, 199, 207, 208, 220, 243
al-Tusi, 51
Tweddle, Ian, 150, 157, 175, 199, 259, 474, 489, 491–493, 514
Tweedie, Charles, 199, 493

univalent functions, 907, 909, 910, 912–914, 917–920, 922, 928

Valiron, Georges, 898, 899, 902, 905
Vallée-Poussin, Charles de la, 681, 682, 688, 689, 693, 694, 700, 755, 770, 777
value distribution theory, 891
Van Assche, Walter, 596

Van Brummelen, Glen, 15, 221
van der Pol, B., 373
van der Waerden, Bartel, 693
Vandermonde, Alexander T., 313, 316, 320, 325, 369, 405, 603, 619, 620, 826, 827
Vandermonde determinant, 166, 335, 336, 413, 874
Varadarajan, V. S., 546, 814
Variyar, Shankar, 1–5, 15
Veblen, Oswald, 910
Verman, Sankara, 13
Vessiot, Ernest, 876
Viète, François, 28, 29, 79, 82, 87, 88, 96, 140, 151, 154, 295, 316–318, 325, 326, 328, 344, 826, 839
St. Vincent, Grégoire, 56, 108, 122, 223
Vojta, Paul, 892
von Neumann, John, 753
Vorsselman de Heer, P. C. O., 551, 563

Wali, Kameshwar C., 119
Wallis, John, 10, 14, 28–38, 40–42, 45, 46, 50, 52, 53, 55, 56, 69, 73, 93, 98, 102, 119, 120, 140, 147–149, 156, 229, 308, 405, 444, 446, 447, 453–455, 458, 479, 483, 484, 486, 491, 600, 681
Wang Lai, 557
Waring, Edward, 161, 165, 166, 174, 175, 194, 247, 249, 313, 314, 319, 325, 329–331, 334, 340, 341, 343, 345, 369, 575, 579, 683, 827
Waring-Lagrange interpolation formula, 165–167, 582
Warnaar, S. O., 474
Watson, G. N., 199, 634, 636, 637, 780, 791, 814
Weber, Heinrich, 335, 336, 344, 743, 878
Weierstrass, Karl, 14, 57, 58, 62, 66, 70, 85, 205, 206, 268, 348, 362, 373, 399, 428, 437, 438, 440, 443, 450, 457, 535, 553, 589, 625, 700, 720, 751, 753, 763, 806, 819, 820, 864, 865, 870, 873–875, 884, 887–889, 902, 906–908, 910
Weierstrass approximation, 587, 760, 762, 763
Weil, André, 27, 197, 198, 350, 351, 365, 435, 443, 520, 535, 543, 546, 719, 729, 814, 820, 878, 886, 905, 931, 941, 942
Weil, Simone, 435
Weil's conjecture, 931, 942
Weyl, Hermann, 91, 760
Whish, Charles, 14
Whiteside, Derek T., 14, 50, 79, 96, 98, 119, 121, 133, 134, 138, 155–157, 168, 174, 199, 201, 203, 219, 230, 231, 287, 329
Whittaker, E. T., 163, 199
Wielandt, Helmut, 450, 474
Wiener, Norbert, 369, 373, 425, 701, 756–759, 764–766, 768–774, 777, 911
Wiener's Tauberian theorem, 758, 759, 765, 768–772
Wilbraham, Henry, 439
Wiles, Andrew, 907
Wilf, Herbert, 626
Wilson, James, 676
Wilson, John, 181, 194, 195, 198, 199

Wilson, Robin, 15, 719
Wintner, Aurel, 658
Witmer, T. Richard, 325
de Witt, Jan, 19, 72
Wolf, Christian, 749
Wolfart, J., 886
Woodhouse, Robert, 369–372, 399
Wren, Christopher, 31, 73, 98
Wright, Edward, 104

Yandell, Benjamin H., 886
Yenri Hakki, 155
Yenri Tetsujutsu, 152
Young, Alfred, 748
Young, Grace Chisholm, 205, 220, 441

Young, W. H., 205, 206, 220, 348, 362, 429, 440, 441
Yuktibhasa, 2, 5, 7, 8, 13–15
Yuktidipika, 2, 3, 5, 15
Yushkevich, Adolf P., 598, 886

Zagier, Don B., 637
Zdravkovska, Smilka, 886
Zeilberger, Doron, 632
Zelevinsky, A. V., 723, 729, 775
Zermelo, Ernst, 68, 908
Zolotarev, Egor I., 152, 157, 578
Zorn's lemma, 287
Zwegers, Sander P., 637
Zygmund, Antoni, 443, 911